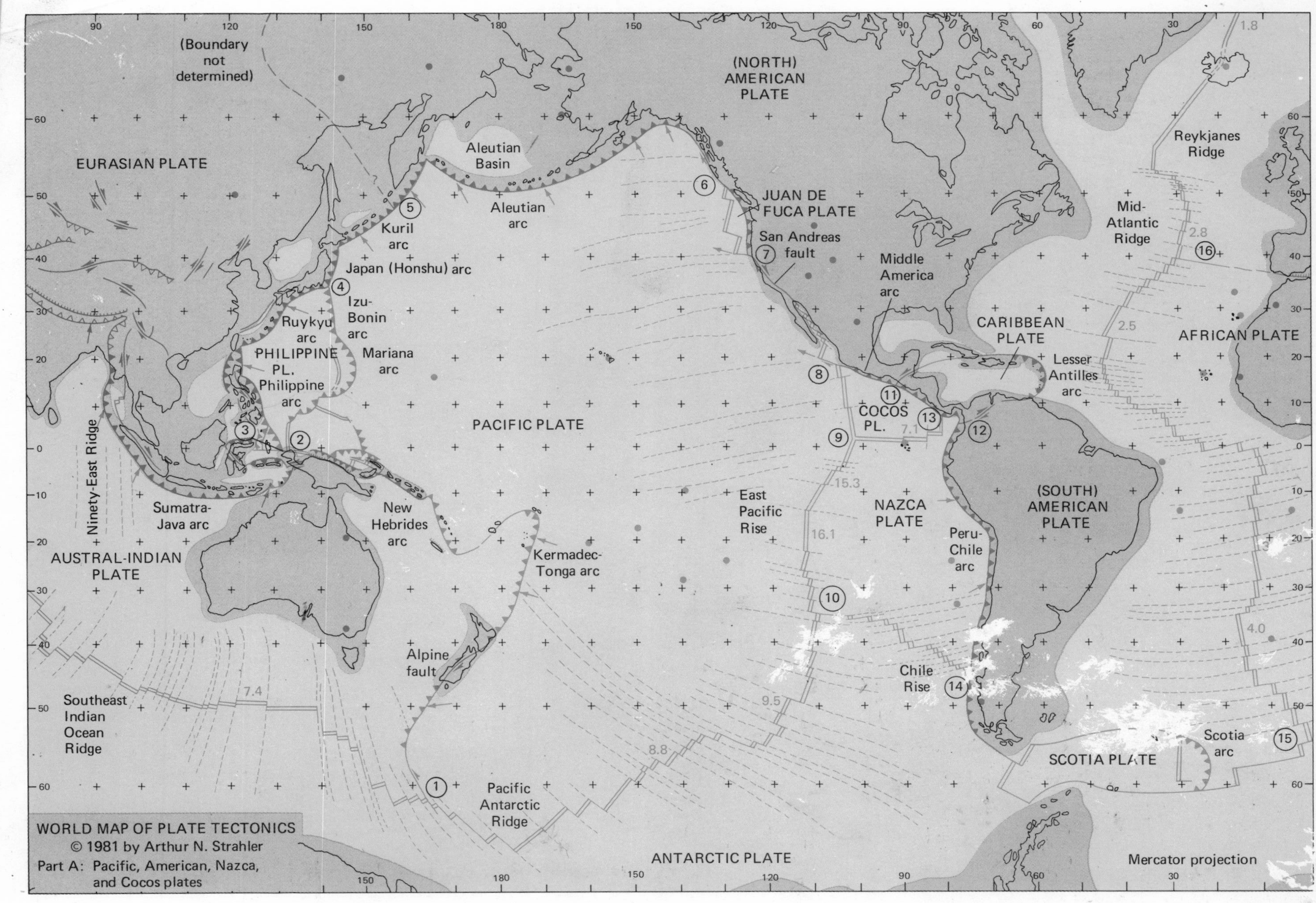
EURASIAN PLATE
(Boundary not determined)
(NORTH) AMERICAN PLATE
Aleutian Basin
Aleutian arc
Kuril arc
Japan (Honshu) arc
Izu-Bonin arc
Ruykyu arc
PHILIPPINE PL.
Philippine arc
Mariana arc
PACIFIC PLATE
Ninety-East Ridge
Sumatra-Java arc
AUSTRAL-INDIAN PLATE
New Hebrides arc
Kermadec-Tonga arc
Alpine fault
Southeast Indian Ocean Ridge
Pacific Antarctic Ridge
JUAN DE FUCA PLATE
San Andreas fault
Middle America arc
CARIBBEAN PLATE
Lesser Antilles arc
COCOS PL.
East Pacific Rise
NAZCA PLATE
Peru-Chile arc
(SOUTH) AMERICAN PLATE
Chile Rise
SCOTIA PLATE
Scotia arc
ANTARCTIC PLATE
AFRICAN PLATE
Mid-Atlantic Ridge
Reykjanes Ridge
Mercator projection
WORLD MAP OF PLATE TECTONICS
© 1981 by Arthur N. Strahler
Part A: Pacific, American, Nazca, and Cocos plates

Physical Geology

Physical Geology

Arthur N. Strahler

1817

HARPER & ROW, PUBLISHERS, New York
Cambridge, Hagerstown, Philadelphia, San Francisco,
London, Mexico City, São Paulo, Sydney

Cover photo: The Himalayan continental suture seen in false-color imagery obtained by NASA Landsat. The Ganges Plain lies at the bottom of the image, the main Himalayan Range with Mount Everest at the top. (See Color Plate C.2 for explanation and source.)

Sponsoring Editor: *Malvina Wasserman*

Special Projects Editor: *Marylou Mosher*

Project Editors: *Carol Pritchard-Martinez/Jon Dash*

Designer: *Dare Porter*

Senior Production Manager: *Kewal K. Sharma*

Compositor: *Allservice Phototypesetting Co. of Arizona*

Printer and Binder: *The Murray Printing Company*

Art Studio: *J & R Services*

Physical Geology

Library of Congress Cataloging in Publication Data

Strahler, Arthur Newell, 1918–
Physical geology.

Bibliography: p.
Includes index.
1. Physical geology. I. Title.
QE28.2.S67 551 80-28222
ISBN: 0-06-046462-3

Contents in Brief

*Optional chapters

Contents

CHAPTER ONE
Geology—An Overview

CHAPTER TWO
Matter and Energy—A Review

CHAPTER THREE
Introduction to Minerals

CHAPTER FOUR
Igneous Rocks and Igneous Activity

• Optional Topics

CHAPTER FIVE
Sediments and Sedimentary Rocks

• Optional Topics

CHAPTER NINE
The Ocean Basins and Their Sediments

• Optional Topics

CHAPTER TEN
Tectonics of the Oceanic Lithosphere

• Optional Topics

CHAPTER FOURTEEN

Landforms of Weathering and Mass Wasting

• Optional Topics

CHAPTER FIFTEEN

Ground Water and its Geologic Activity

CHAPTER SIXTEEN

The Geologic Work of Running Water

• Optional Topics

CHAPTER TWENTY
Astrogeology—The Geology of Outer Space

CHAPTER TWENTY-ONE
Geologic Resources of Materials and Energy

• Optional Topics

Preface

A physical geology course attuned to the 1980s will span the full range of both traditional and modern geology. Traditional geology, at a beginning level, is drawn largely from knowledge accumulated prior to about 1950. Traditional physical geology deals with mineralogy, petrology, structural geology, stratigraphy, and geomorphology. We can also include basic principles of seismology, gravity, geochemistry, and radiochemistry in the traditional mode, for they too were well established before 1950. There were, to be sure, clear precursors of plate tectonics in the form of strong geological arguments favoring Wegener's continental drift scenario, but the growth of modern geology required geophysical investigations of the ocean floors and oceanic crust that were not to begin in earnest until after World War II.

Prior to that time, geology based on observation of the continental crust had piled up a long list of unanswered questions. How, for example, could we explain such odd features as a mélange, an aulacogen, a taphrogen, or an ophiolitic suite? Indeed, we then had no satisfying explanation of such commonplace features as volcanic arcs, batholiths, and orogens. The answers to a host of fundamental geologic questions lay hidden in the earth's crust and mantle beneath the ocean basins and continental margins.

Modern geology, deriving much of its substance from studies of marine geology and plate tectonics, is a product of the past three decades and has been widely recognized as a major scientific revolution. The awesome power of plate tectonics to explain the elements of global crustal geology under a unified and coherent theory is a theme I have endeavored to develop throughout this book. Plate tectonics is given a general overview in the first chapter, along with such topics as earth structure, classes of rocks, origin of the solar system, and geologic time.

The second chapter, a review of some elementary principles of physics and chemistry, is for optional use by students with little background in physical science; it may be omitted entirely by others.

The six chapters that follow cover fundamentals of traditional geology, including mineralogy, petrology, stratigraphy, structural geology, radiochemistry, geochronology, and seismology. With these essential foundations of crustal geology carefully established, we turn to modern geology.

Chapter 9, on the ocean basins, their floors and sediments, sets the stage for four chapters on plate tectonics. I give special attention to geophysical evidence that lithospheric plates exist and move, that they undergo accretion along spreading boundaries and consumption along subduction boundaries. The traditional concept of the geosyncline is recast in a modern mold in the context of sedimentary, accretionary, tectonic, and magmatic processes that can be observed in action today along stable or active continental margins. Here, the classical principle of uniformitarianism is revitalized as we see—through the penetrating eye of the seismic reflection profiler—the growth of an accretionary prism off the Java coast and equate it to a Franciscan Complex that now forms part of the American continental crust.

It has become increasingly difficult to make a sharp or meaningful distinction between physical and historical geology. The Wilson cycle of opening and closing of ocean basins is a case in point, involving as it does a succession of historical events ranging in age from early Paleozoic time to the present. We have no hesitation about including a review of the history of the closing of the Iapetus Ocean followed by rifting of Pangaea and the excursions of its fragments. The only part of traditional historical geology that can be omitted without damage to the study of plate tectonics is organic evolution and the succession of changing life forms through time. On the other hand, organic activity that has contributed to the production of marine sediments is an integral part of the total geologic process related to plate tectonics.

Exogene processes powered by solar energy are the subject of the next six chapters. For the most part, this material is traditional descriptive geomorphology, but with some added insights into the dynamics of the fluid processes. Modern geology appears in depth in the chapter on glaciers and the Pleistocene glaciations, where the evidence of numerous cycles of paleoglaciation are revealed through the changing oxygen-isotope ratios within deep sea cores.

The final two chapters can be regarded as optional material, but this is not to understate their importance in physical geology. One of these chapters deals with astrogeology, the geology of planetary space. It has been a field of enormous scientific discovery over the past two decades, and many geologists have been deeply involved in that research. The final chapter reviews the geology of both mineral and energy resources, bringing plate tectonics into the explanation of ore deposits and petroleum accumulations.

I have made a special effort to explain geology in clear and simple language, avoiding mathematics and using elementary chemistry sparingly. I use metric units throughout, in many cases with SI equivalents. Scientific terms are defined as they first appear, while a full alphabetical glossary at the end of the book is available for quick reference. At the end of each chapter, key facts and concepts are recapitulated in statements that reinforce the text and show the framework of the chapter. Diagrams and maps, of which there are over 475, are an integral part of the presentation of physical geology. Block diagrams and structure sections, in particular, will be an enormous aid in clarifying many complex geological relationships. A special effort has been made to provide diagrams that show careful attention to correct scaling and accuracy of detail. Photographs, which number more than 250, have been carefully chosen for their geologic content.

For many introductory geology courses the book offers more text material than will be covered in one semester or one quarter. Because of the breadth of scope of the text within each chapter, the instructor has an exceptional opportunity to select those topics best suited to the needs of the course. Judgment as to which topics are to be covered rests with the individual instructor or department and will depend on several factors, including length of the course and its goals, learning capacity of the students, and the strong points of the local geological environs. In the detailed table of contents, topics that may be considered for omission are indicated by a bullet. Some of these are topics containing slightly more advanced or detailed material; others are case studies or environmental topics. An *Instructor's Manual* (available from Harper & Row) gives helpful suggestions on the selection of topics for emphasis as well as background information useful to the instructor.

I am greatly indebted to a number of technical reviewers who read individual chapters or groups of chapters. Specialists in the subjects named, they made many suggestions for improving the manuscript, and I carried these out in detail to eliminate errors and misconceptions. I am, however, solely responsible for the text as it stands. In addition, several anonymous reviewers who advised the publisher on the merits of the manuscript made valuable suggestions as to the content and organization of the chapters. Their advice has led to important improvements in the usability of the book for an introductory physical geology course.

Arthur N. Strahler

About the Author

Arthur N. Strahler holds the Ph.D. degree in geology from Columbia University. He was appointed to the Columbia University faculty in 1941, serving as Professor of Geomorphology from 1958 to 1967 and as Chairman of the Department of Geology from 1958 to 1962. A Fellow of the Geological Society of America, he is the author of several textbooks, including *The Earth Sciences*, Second Edition, 1971, *Principles of Earth Science*, 1976, and *Principles of Physical Geology*, 1977.

Reviewers of Selected Chapters

Bruce A. Bolt, Professor of Seismology, Seismographic Station, Department of Geology and Geophysics, University of California, Berkeley. (Earthquakes and seismology.)

William R. Dickinson, Professor of Geology, Department of Geosciences, University of Arizona, Tucson. (Plate tectonics.)

Fred A. Donath, Professor of Geology, Department of Geology, University of Illinois, Urbana, assisted by Polly Lee Knowlton. (Structural geology.)

William R. Farrand, Professor of Geology, Department of Geology and Mineralogy, University of Michigan, Ann Arbor. (Glacial and Pleistocene geology.)

Billy P. Glass, Professor of Geology, Department of Geology, University of Delaware, Newark. (Astrogeology.)

Kurt E. Lowe, Emeritus Professor and former Chairman, Department of Geology, The City College, City University of New York. (Mineralogy and petrology.)

Walter C. Pitman III, Professor of Geology, Department of Geological Sciences and the Lamont–Doherty Geological Observatory, Columbia University, New York. (Marine magnetics and plate tectonics.)

Stephen S. Streeter, Research Associate, Lamont–Doherty Geological Observatory of Columbia University, New York. (Micropaleontology and marine geology.)

Lee J. Suttner, Professor of Geology, Department of Geology, Indiana University, Bloomington. (Sedimentology and sedimentary petrology.)

Physical Geology

1

CHAPTER ONE

Geology-An Overview

Geology is the science of the planet Earth. Most of you are probably approaching your first course in geology with some reasonably accurate ideas on the subject. Rocks and minerals, along with the ways in which they are put together and changed through time, are, of course, basic ingredients of a geology course. You are probably also aware that the time dimension is extremely important in geology, just as it is in astronomy. Geology must deal with time in spans of millions and even billions of years to relate the major events in the history of our planet.

Physical and Historical Geology

Many of you know that geology includes the study of fossils—the remains of ancient animals and plants preserved in rocks—and that life of the geologic past shows an orderly progression of forms from primitive to advanced types. The branch of geology dealing with ancient life and the progress of organic evolution is *historical geology*. This book deals mostly with another branch, *physical geology*, which is concerned with the physical and chemical properties of the earth and the physical and chemical changes that occur in rocks. Because these changes and events occurred throughout the entire span of geologic time as recorded in rocks—about 3.8 billion years—the study of physical geology requires frequent reference to the time units and events of historical geology. We will need to study the table of geologic time, with its long sequence of time units, in order to chart the growth and breakup of continents and ocean basins, and the repeated building and destruction of mountain ranges.

Topics of Physical Geology

Here are some of the major topics included in our introduction to physical geology:

- Materials that make up the solid earth. Rocks and minerals; their classification in terms of chemical composition and physical properties; the ways in which they originate and change as time passes.
- The organization of rocks into structural units, such as layers or massive bodies, in a wide range of scales. On the largest scale is the organization of rocks into the major shells which form the entire earth from its inner core to its crust. On a smaller scale is the organization of the crust into continents and ocean basins. On a still smaller scale is the organization of rock units within the continents and ocean basins.
- The bending and breaking of rock masses by the action of enormous internal forces. Here, too, the range of scale involved in these processes is great and may be seen in the movement of an entire continent or in the mere crumpling of a rock layer a few centimeters in thickness.
- The accumulation of heat at various places far below the earth's surface, leading to melting of rock.

As this molten rock moves upward toward the surface, it causes major changes in the nature of the surrounding rock and solidifies into new solid rock bodies, or it builds great chains of volcanoes.

- Changes in rock exposed to the atmosphere and subjected to erosion by forces of running water (streams), wind, glacial ice, and waves. Landscape features of many sizes and shapes are interpreted in terms of these processes, both physical and chemical, which derive their energy from the sun's rays.

Tools of Geologic Investigation

We cannot investigate the topics listed above without a number of scientific tools and methods using laws of physics and chemistry. For example, we must understand the nature of radioactivity, a process in which atoms of a particular element undergo spontaneous internal changes to become atoms of another element, and emit energy in the process. Radioactivity will prove to be the key to establishing the ages in years of the events of geologic history throughout its entire span. Radioactivity will also provide us with an explanation for the accumulation of heat within the earth in quantities great enough to melt rock and to power the motion of rock masses the size of continents.

A second example of an essential scientific tool is knowledge of the phenomenon of earthquake waves. Sent out by sudden breaking of rock at some point in the outer zone of the earth, earthquake waves travel thousands of kilometers over the earth's surface and through its interior. When records are made of these waves, their travel paths can be interpreted to show the physical properties and layering of rock at depths impenetrable by any other means.

Geology and Earth Resources

Geology might seem like a welcome escape from the problems of today's technological society—shortages of energy supplies, mass killings of humans by nuclear weapons, and the destruction of our natural environment by urbanization and pollution. Yet geology is heavily involved in all these problems. Where do nearly all of our important sources of energy come from? Answer: petroleum, natural gas, coal, uranium ore—all from rocks in the earth's outer crust. Where do all the metals of our manufactured products originally come from? Answer: from ores formed by geologic processes within rocks of the earth's crust.

The search for new oil fields and for new ore deposits has been a motivating force in the development of the research tools needed to advance geology as a science. A case in point is geophysical exploration. As you probably know, one important means of exploration for oil is the study of shock waves sent into the earth by human-made explosions. These artificial earthquake waves not only lead to the discovery of oil-bearing rock structures, but also give information of purely scientific interest about those rock layers. Underground nuclear test explosions also generate earthquake-type waves; these are studied by geologists for clues on the deeper structure of the earth's rock layers.

Geology and Natural Hazards

Today, as cities expand and suburbs spread into the countryside, geology is being called upon to help guide urban planning. Many hazards can be avoided by applying what we know about the physical nature of rock and soil close to the surface. Unfortunately, however, human interaction with the earth is not always rewarding. Throughout this book you will find brief accounts of ways in which human activities can induce unwanted and harmful geologic effects. Although this subject is not developed in depth, you will gain some insight into the environmental role that geology plays.

On the other hand, geologic processes themselves often impose severe hazards upon people and their structures. A couple of well-known hazards are volcanic eruptions and earthquakes, events that have taken enormous tolls in lives and property through the ages. We will give accounts of such hazards, with particular emphasis upon earthquakes. Two of our great cities—Los Angeles and San Francisco—face the threat of a great earthquake that may very well take a frightful toll in human lives and cause destruction of property beyond anything yet experienced in history. Most of us would rather not think about such an eventuality, but geologists are striving to develop skill in predicting when and where destructive earthquakes will occur. It is interesting to learn that both the Soviet Union and the People's Republic of China are also mounting an intensive scientific effort toward improving methods of earthquake prediction. Information is being exchanged internationally in this race against time—a race in which no nation stands to lose and all of humanity stands to gain.

A Strategy for the Study of Geology

Some people might think it logical to begin the study of geology by taking the largest features first, then working down to smaller and smaller details. Following that program, we would consider first the gross structure of the entire planet, then the nature of continents and ocean basins, then such landscape features as volcanoes and mountain ranges, and finally individual varieties of rocks and minerals. The very last item would be the atomic structure of minerals. This text, however, will only adhere to this program in first sketching some broad outlines of the geology of our planet and its changes through time. Beginning with Chapter 2, we reverse the sequence with a detailed investigation of the atomic structure of minerals, and in subsequent chapters work up to rocks, their groupings into larger bodies, and finally to the nature of continents and ocean basins.

There are good reasons why a serious study of geology should begin with minerals and rocks. We cannot analyze the larger rock units and groupings meaningfully without understanding the chemical composition and physical properties of the forms of mineral matter that make up those larger features. Attention to small details that can be observed directly and proven by laboratory study has been essential to the advance of modern geology as a science. By observing detailed differences in the chemical composition and physical structure of minerals and rocks from place to place, geologists have been led directly to inferences regarding the processes and changes that have taken place on a much larger scale. Principles of physics and chemistry, proven by laboratory observations, are applied by geologists to large masses of rock and their processes of movement and change.

What Is Rock?

Most persons have seen many kinds of rock and have a good general grasp of its physical properties. Rock consists of mineral matter, as distinct from the organic matter produced by plants and animals. In other words, rock is composed of inorganic matter. Rock consists of minerals in a solid state. Most kinds of rock consist of several different minerals in combination. Commonly, the minerals occur as individual small grains, so the rock is a physical mixture of mineral grains.

FIGURE 1.1 The earth's core, mantle, and crust. The crust is much too thin to be shown to true scale. (Redrawn, by permission of the publisher, from A. N. Strahler, *Physical Geography*, 4th ed., Figure 23.1. Copyright © 1975 John Wiley & Sons, Inc.)

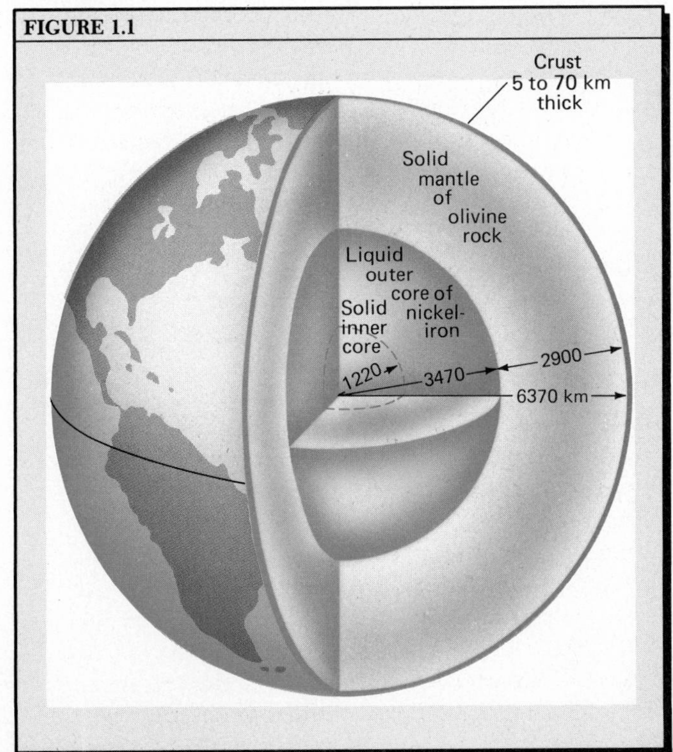

What, then, is a mineral? The answer to this question, with a well-devised and complete scientific definition of the term, will appear in Chapter 3. An intervening chapter reviews the nature of matter and energy, so that you can apply basic principles of physics and chemistry to the study of minerals.

The word *rock* is applied by geologists to all solid mineral matter, whether exposed or lying beneath a thin cover of soil and vegetation. Although most rock is commonly so hard that it cannot easily be broken except by using a hammer or pick, some kinds of rock are soft enough to pulverize easily with the fingers. Usually, rock is extremely old in comparison with living things and human-made objects. Nearly all solid rock is at least a million years old, and much of it is hundreds or thousands of millions of years old. Here, too, we encounter a striking exception: Molten lava being erupted at this very moment from one volcano or another solidifies into an extremely hard, dense rock almost before our eyes.

The Earth's Rock Shells

Like the other three inner planets of our solar system—Mercury, Venus and Mars—Planet Earth is a spherical ball of rock consisting largely of compounds of the elements oxygen, silicon, aluminum, and iron. The four inner planets differ greatly from four great outer planets—Jupiter, Saturn, Uranus, and Neptune—which are giants by comparison and are composed largely of hydrogen and helium.

The average radius of the earth sphere is about 6400 km (more exactly, 6370 km). Nearly all of this radius is taken up with two zones or regions: a core and a mantle (Figure 1.1). The spherical inner *core*, with a radius of 3470 km, consists of metallic iron mixed with a small proportion of nickel. The outer two-thirds of the core is in a liquid state, intensely hot and under enormous confining pressure. There is evidence that the inner portion of the core is solid. Surrounding the core is the *mantle*, an enormous rock shell nearly 2900 km thick. Thus, the mantle occupies nearly half the earth's radius. Dense mantle rock consists mostly of the elements silicon, oxygen, iron, and magnesium in a solid state, although it has a very high temperature and is under great confining pressure.

Extremely thin in comparison with the mantle is the outermost earth shell, the *crust*, ranging in thickness from about 5 to 70 km. If spread evenly over the entire earth, its average thickness would be about 17 km.

Such a thin layer cannot be shown to true scale on our cutaway drawing of the earth, Figure 1.1. Rock of the crust differs from that of the mantle in containing substantial proportions of the lighter metals: aluminum, sodium, calcium, and potassium, along with abundant silicon and oxygen. For this reason, crustal rock is less dense than mantle rock, just as a light metal such as aluminum is less dense than a heavy metal such as iron. Mantle rock is, in turn, less dense than the iron-nickel mixture of the core. Thus, our first generalization about the planetary structure is that it consists of shells that decrease in density outward from the center. (The concept of density of matter is explained in Chapter 2.)

The three earth shells we have described—core, mantle, and crust—are different in chemical composition, the primary basis on which each is defined. Superimposed on the outer shells (the outer mantle and crust) are layers defined according to the physical state of the rock—whether the rock is hard and brittle

FIGURE 1.2

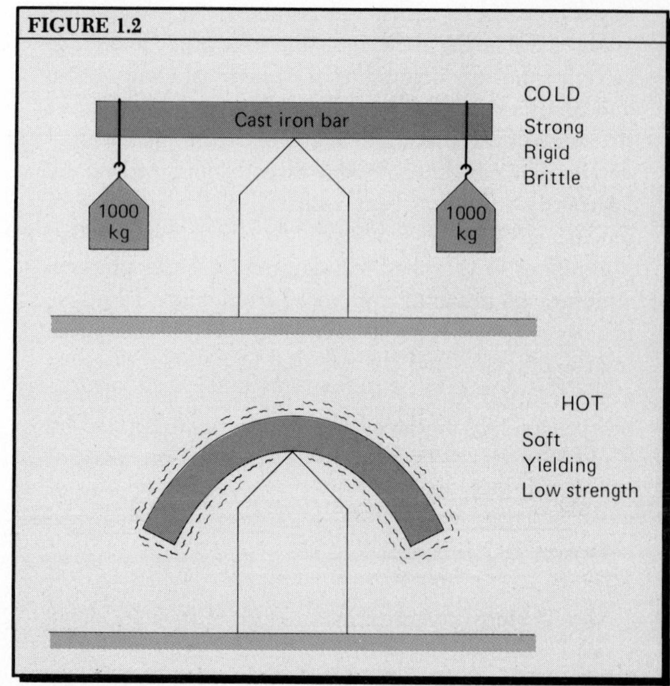

FIGURE 1.2 A cold bar of cast iron can support heavy weights, though it will snap if overloaded. The same bar heated to redness becomes soft and sags under its own weight.

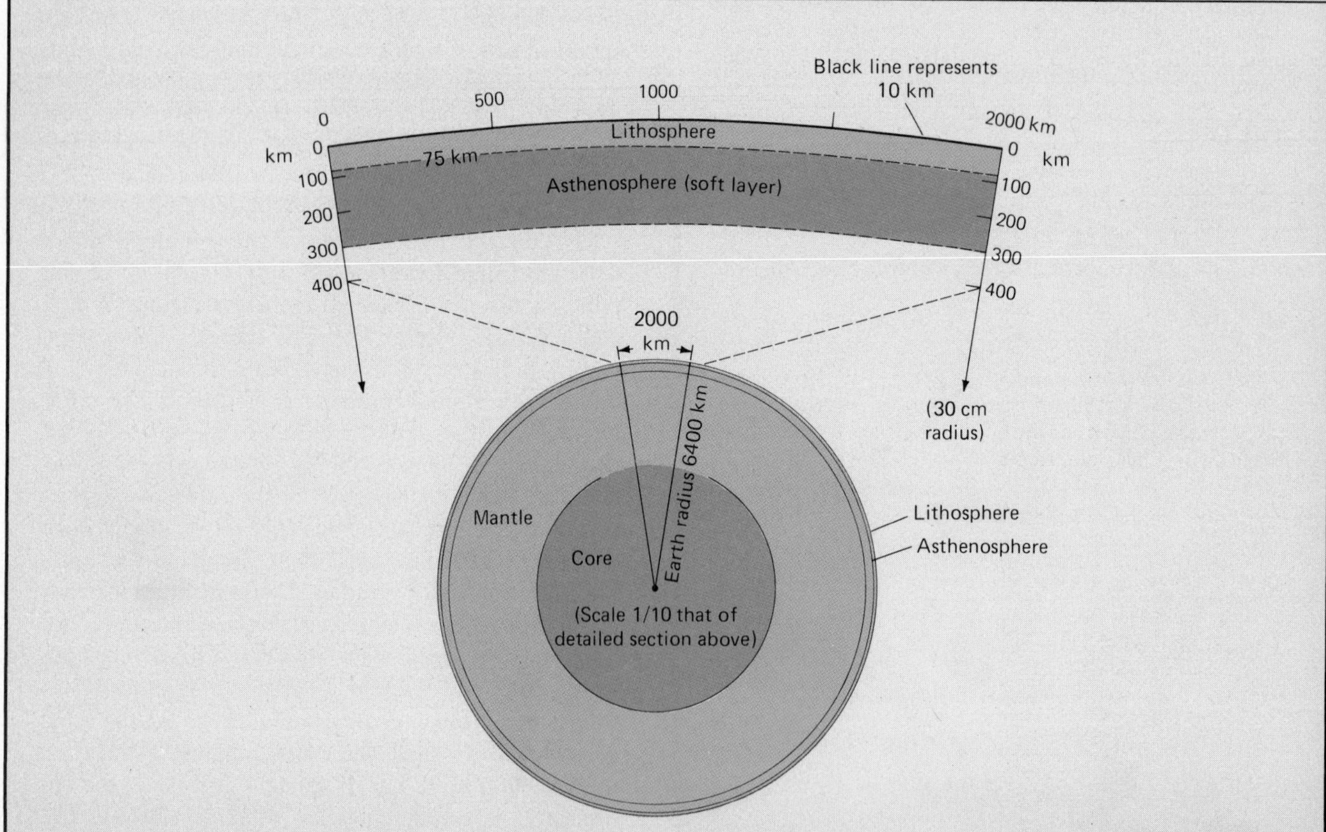

FIGURE 1.3 The lithosphere and asthenosphere drawn to true scale. The curvature of the upper diagram fits a circle 30 cm in radius. The black line at the top is scaled to represent a thickness of 10 km; it will contain about 98% of the earth's surface features, including the ocean floors and high mountains and plateaus. The complete circle below is drawn on a scale 1/10 as great as the upper diagram. Seen in true scale, the lithosphere is a very thin shell compared with the mantle and core. (Redrawn, by permission of the publisher, from A. N. Strahler, *Elements of Physical Geography,* 2nd ed., Figure 2.10. Copyright © 1979 by John Wiley & Sons, Inc.)

FIGURE 1.3

or soft and plastic. Depending largely on temperature, the same rock can be in either of these conditions, just as a bar of cast iron is strong, hard, and brittle when cold, but becomes soft and plastic when heated to a high temperature (Figure 1.2).

Lithosphere and Asthenosphere

The earth has an outer layer of hard, brittle rock known as *lithosphere*. This layer includes the entire crust and a portion of the upper mantle. Thickness of the lithosphere ranges from under 50 to over 125 km, with a rough average of perhaps 75 km, as shown in Figure 1.3. Below the lithosphere lies the *asthenosphere,* a soft layer in the upper mantle. The word is derived from the Greek root *asthenēs,* meaning "weak." The asthenosphere is in a soft condition because its temperature is high—about 1400 °C—and close to its melting point. The rock behaves much like an ingot of white-hot iron that will hold its shape when resting on a flat surface, but is easily formed into bars or sheets when squeezed between rollers.

Temperature increases steadily inward from the earth's surface, so that the change from lithosphere to asthenosphere is gradual, rather than abrupt. We know from the behavior of earthquake waves as they travel through the outer mantle that the asthenosphere extends to a depth of about 300 km, below which the strength of the mantle rock again begins to increase. The weakest portion of the asthenosphere lies at a depth of roughly 200 km. We present the evidence in detail in Chapter 8.

The important concept you should derive from the stated facts is this: The rigid, brittle lithosphere forms a hard shell capable of moving bodily over the soft, plastic asthenosphere. This motion is exceedingly slow and is distributed through a thickness of many tens of kilometers. A simple model of the motion is a deck of playing cards resting on a tabletop, its top ten cards glued together to form a solid block, representing the lithosphere (Figure 1.4). A horizontal force against the edge of the deck will move the upper block horizontally, while the motion becomes a slipping between the free cards beneath it. The card at the bottom of the deck will remain fixed to the table. Slippage on a great number of very thin parallel layers is described as *shearing.* Layers involved in shearing are no thicker than atoms or molecules, so that we could not actually see one layer gliding over another. The entire asthenosphere seems to move by the kind of flow we observe in a tacky liquid—like thick syrup—but the asthenosphere is not a true liquid. It is difficult for us to understand how rock behaves under such conditions as exist in the asthenosphere, where temperatures are very high and the rock is under enormous confining pressure from the weight of the overlying rock. We discuss this subject in greater detail in Chapter 8.

FIGURE 1.4 The shearing motion of soft rock in the asthenosphere resembles the slip of cards in a deck. We imagine the cards in the upper part of the deck to be glued together so as to move as a solid plate, representing the lithosphere.

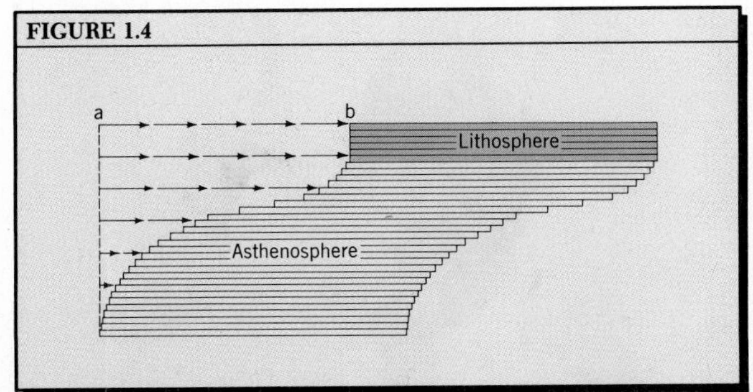

Lithospheric Plates

If the earth consisted of a perfectly spherical lithosphere with no flaws or fractures, it is conceivable that the entire lithosphere could move as a whole with respect to the rest of the earth (middle and lower mantle, and core). If that kind of motion actually occurred, it would move as a unit all of the earth's continents and ocean basins. We can visualize a situation in which western Europe might be moving toward the earth's north pole, so that there would come a time when the north pole would be situated in Ireland. At the same time, New Zealand would be nearing the south pole, while the city of Cape Town, South Africa, would be approaching the equator.

It is more realistic to suppose that the lithosphere has tended to break into large sections, the size of a whole continent or a whole ocean basin. There is strong evidence to support the statement that the lithosphere is quite thin in some areas—notably beneath the deep ocean floors—and quite thick in other areas—notably under the continents. Thin lithosphere might be expected to fracture quite easily, while thick lithosphere would tend to hold together much better.

In fact, breakup of the lithosphere has formed a number of *lithospheric plates*, each of which has some freedom to move independently of the plates around it. Like great slabs of floating ice on the polar sea, lithospheric plates can be seen to be pulling apart in some places and colliding in others. In the case of floating ice, two plates pulling apart leave a widening gap of exposed water (Figure 1.5). This gap can be filled with new ice as the top of the exposed water freezes. When two ice plates collide, they often come together in crushing impacts that raise great welts, called "pressure ridges." These welts remind us of mountain chains found along the margins of continents. When collision occurs, a thin ice plate might be expected to be forced down beneath a thicker plate.

The downdiving plate would melt and finally merge with the surrounding water.

Because ice is less dense than water, an ice plate floats easily and resists being pushed down below the water surface. Thus, it would not be wise to carry the floating-ice-plate model any further in search of a true model of lithospheric plates. Unlike ice, a given kind of solid rock is denser than the same rock in the molten condition, when the two are compared at the same pressure. Thus, lithosphere carried down into the asthenosphere will be denser than the surrounding rock, which is heated close to melting and may, in fact, be partly molten. A lithospheric plate, then, shows a tendency to sink. Given a suitable opportunity, the free edge of a plate may plunge into the asthenosphere, nosediving at a steep angle—but here we are getting ahead of our story line.

FIGURE 1.5 Sea ice of the Arctic Ocean illustrates some aspects of the behavior of lithospheric plates. The water on which the ice floats can be imagined to represent the asthenosphere. The open water at the left is formed by the pulling apart of two ice plates. The rugged ice zone extending across the photo from lower left to upper right (a pressure ridge) represents a miniature mountain belt produced by collision between two plates. (Official U.S. Coast Guard photo.)

Continental Lithosphere and Oceanic Lithosphere

We know that continents rise high in elevation above sea level—if they did not, they would not be continents. We know that the floors of the ocean basins lie at an average depth of some 4 km below sea level. In this case, our zero reference is the surface of the ocean, which tends to form a nearly perfect sphere. Figure 1.6 shows the continental surfaces rising above sea level and the ocean basin floors below that level. Some good scientific evidence has now established that the lithosphere under the continents is thicker than the lithosphere under the ocean basins. The reasons for this difference in thickness will become clear in later chapters. In Figure 1.6, we show continents to be underlain by a thick layer of *continental lithosphere* and the ocean basins by a thin layer of *oceanic lithosphere*.

Besides being thicker, continental lithosphere is also different in chemical composition from oceanic lithosphere. Continental lithosphere includes a special upper crustal layer of rock having lower-than-average density (Figure 1.6). Continental lithosphere is, therefore, more buoyant than oceanic lithosphere, and this property contributes to the higher upper surface of a continent. To explain this principle in simple terms,

FIGURE 1.5

consider what happens if we take two wooden blocks of exactly the same size and shape, one block of oak (a dense wood), the other of balsa (a wood of very low density). Placed in water, the balsa block floats much higher than the oak block. Our schematic diagram of continental lithosphere (Figure 1.6) includes an upper crustal layer of lower rock density. Notice that this crustal layer is missing from oceanic lithosphere. After we have completed a study of minerals and rocks, it will be a simple matter to assign rock names to the layers that make up the lithosphere.

Plate Tectonics

Our next step is to visualize the lithosphere broken into lithospheric plates that move with respect to one another. Figure 1.7*A* shows some of the major features of plate interactions. In Figure 1.7*B* we can see two plates, *X*, and *Y*, both made up of oceanic lithosphere, pulling apart along a common plate boundary. This activity tends to create a gaping crack in the crust, but molten rock from the mantle below rises continually to fill the crack. The rising molten rock, called ***magma***, solidifies in the crack and is added to the two edges of the spreading lithospheric plates. In this way, new solid lithosphere is continually formed. At the distant boundary of oceanic plate *Y*, the oceanic lithosphere is shown to be pushing against a thick mass of continental lithosphere, plate *Z*. Because of its greater crustal buoyancy, the continental plate remains in place, while the thinner, denser oceanic plate bends down and plunges into the asthenosphere. The process of downsinking of one plate beneath the edge of another is called ***subduction***.

The leading edge of the descending plate is cooler than the surrounding asthenosphere—cooler enough, in fact, that this descending slab of brittle rock is denser than its surrounding asthenosphere. Consequently, once subduction has begun, the slab can be said to "sink under its own weight." Gradually, however, the slab is heated by the surrounding asthenosphere and thus it eventually softens. The underportion, which is mantle rock in composition, simply reverts to asthenosphere as it softens. The thin upper

FIGURE 1.6 Continental lithosphere is thicker than oceanic lithosphere. The crust of the continents is also thicker than the crust of the oceans. For purposes of illustration, the continental lithosphere is shown as being twice as thick as the oceanic lithosphere.

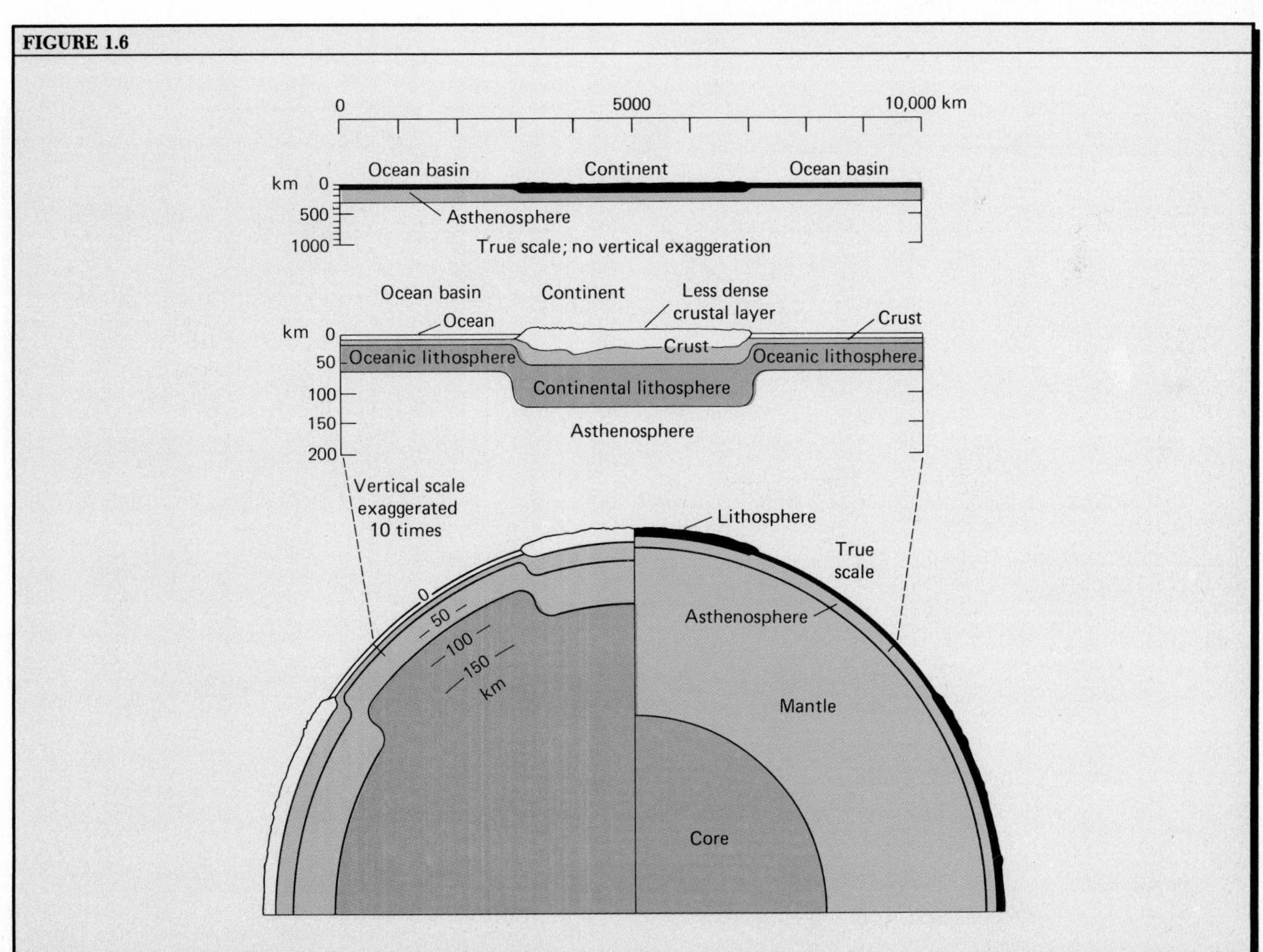

crust, formed of less dense mineral matter, may actually melt and become magma, which tends to rise because it is less dense than the surrounding material. Figure 1.7*A* shows some magma pockets formed from the upper edge of the slab. They are pictured as rising like hot-air balloons through the overlying continental lithosphere. Reaching the earth's surface, quantities of this magma build volcanoes, which tend to form a chain parallel with the deep *oceanic trench* that marks the line of descent of the oceanic plate.

Viewed as a unit, plate *Y* (Figure 1.7*B*), appears as a single lithospheric plate simultaneously undergoing *accretion* (growth by addition) and *consumption* (by softening and melting), so that the plate might conceivably maintain its size, without necessarily expanding or diminishing. Actually, our model can also call for a plate of oceanic lithosphere either to grow or to diminish, and we also have models that allow for the creation of new plates of oceanic lithosphere where none previously existed. In this respect, the theory is quite flexible.

The general theory of lithospheric plates, their relative motions, and their boundary interactions is called *plate tectonics*. *Tectonics* is a noun meaning "the study of tectonic activity." *Tectonic activity,* in turn, refers to all forms of breaking and bending of rock of the lithosphere.

FIGURE 1.7 Schematic cross sections showing some of the important elements of plate tectonics. Diagram *A* is greatly exaggerated in vertical scale so as to emphasize surface and crustal features. Only the uppermost 30 km is shown. Diagram *B* is drawn to true scale and shows conditions to a depth of 250 km. Here the actual relationships between lithospheric plates can be examined, but surface features can scarcely be shown. Diagram *C* is a pictorial rendition of plates on a spherical earth and is not to scale. (Drawn by A. N. Strahler.)

FIGURE 1.8 A transform fault involves the horizontal motion of two adjacent lithospheric plates, one sliding past the other.

FIGURE 1.9 A schematic diagram of a single rectangular lithospheric plate with four boundaries.

FIGURE 1.7 A, B

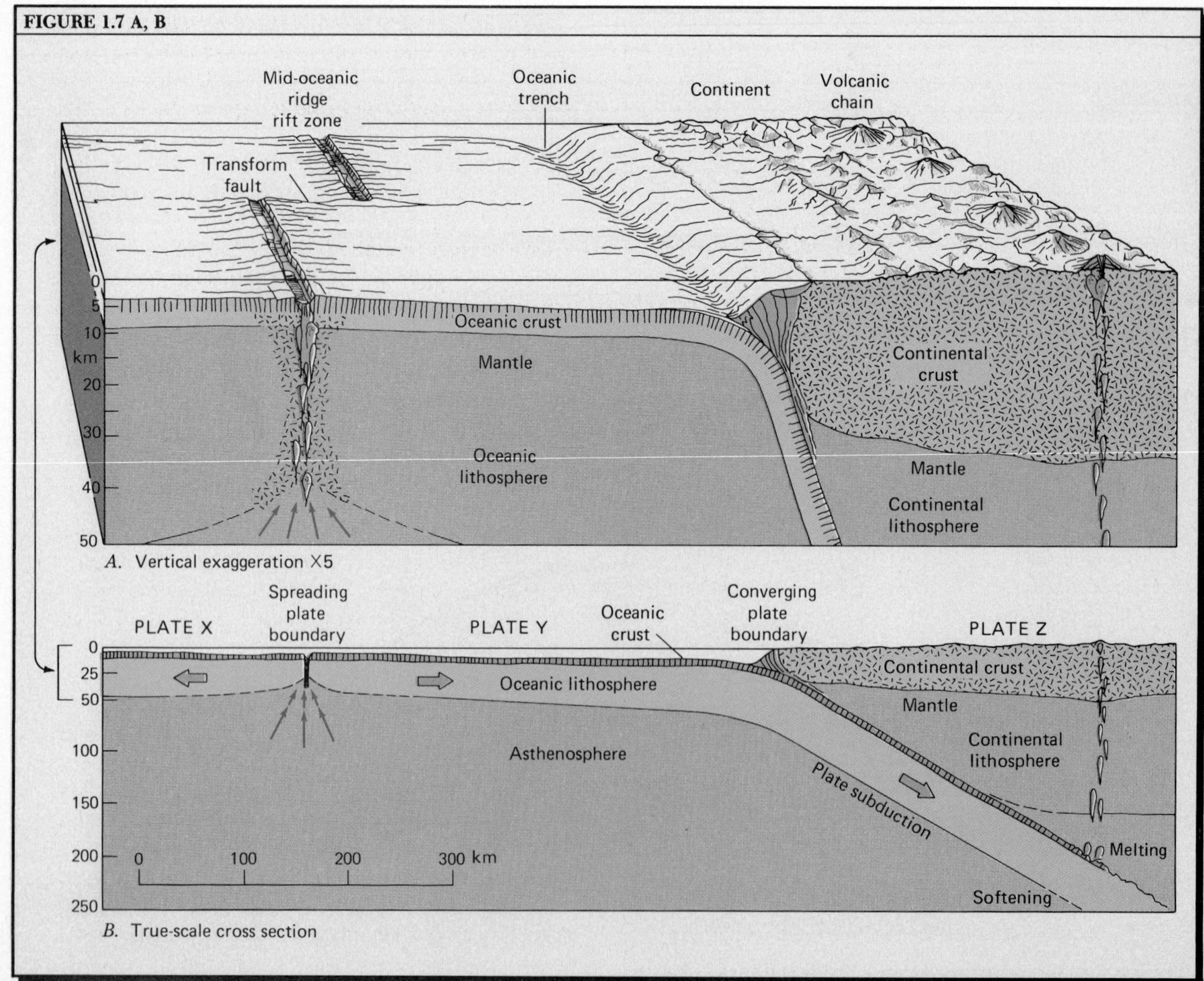

FIGURE 1.7C

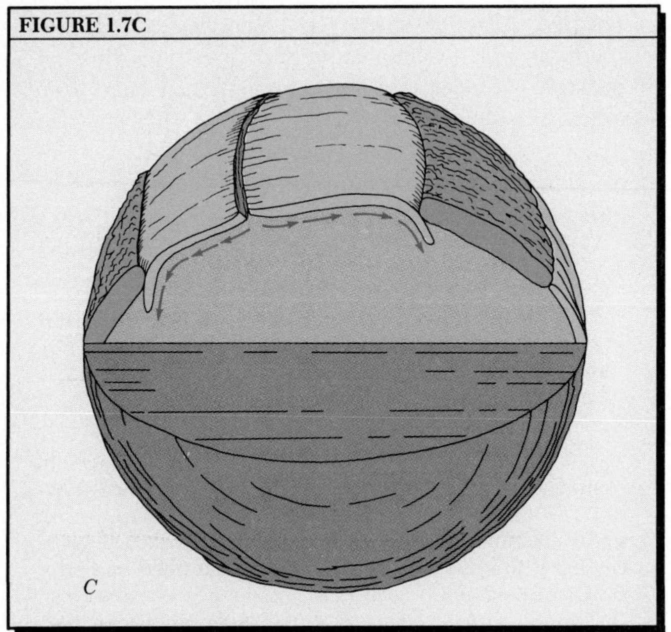

FIGURE 1.8

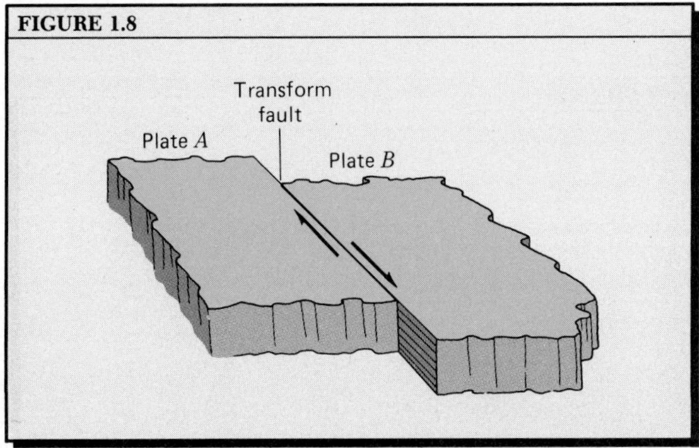

FIGURE 1.9

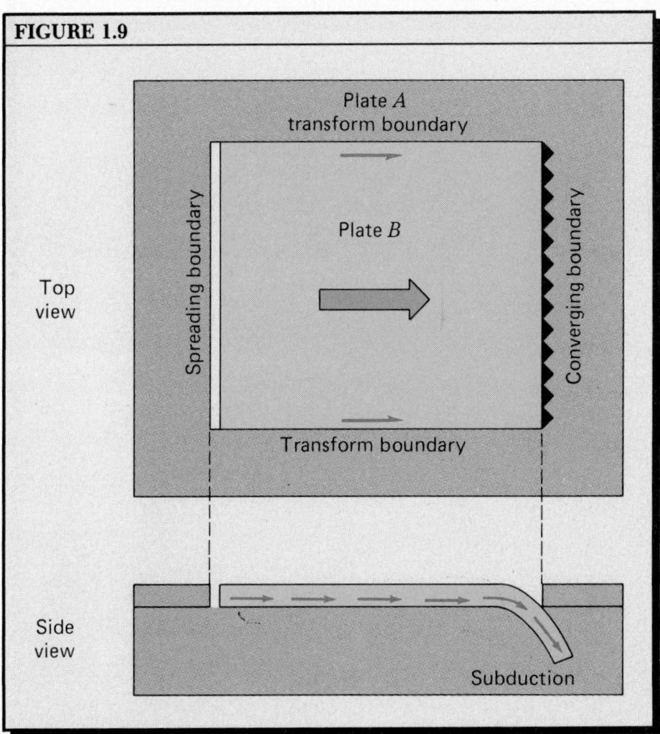

We have yet to consider a third type of lithospheric plate boundary. Two lithospheric plates may be in contact along a common boundary on which one plate merely slides past the other with no motion that would cause the plates to separate or to converge (Figure 1.8). The plane along which motion occurs is a nearly vertical fracture extending down through the entire lithosphere; it is called a *transform fault*. The word *fault* is used by geologists for any plane of fracture (a crack) along which there is motion of the rock mass on one side with respect to that on the other.

Thus, in summary, there are three major kinds of active plate boundaries:

Spreading boundaries. New lithosphere is being formed by accretion.

Converging boundaries. Subduction is in progress; lithosphere is being consumed.

Transform boundaries. Plates are gliding past one another.

Let us put these three boundaries into a pattern to include an entire lithospheric plate. As shown in Figure 1.9, we have visualized a moving rectangular plate like a window, set in the middle of a surrounding stationary plate. The moving plate is bounded by transform faults on two parallel sides. Spreading and converging boundaries form the other two parallel sides. Some familiar mechanical devices come to mind in visualizing this model. One is the sunroof top of an automobile; it has a window that opens by sliding backward along parallel sidetracks to disappear under the fixed roof. Another is the old-fashioned rolltop desk. In Chapter 10, we will investigate the various different arrangements of plate boundaries, which can be curved as well as straight, while individual plates can pivot as they move. Thus, there are many geometric variations to consider.

Continental Rupture and New Ocean Basins

Plate tectonics provides for a most remarkable geologic event—a *continental rupture*, the splitting apart of a plate of continental lithosphere. When this occurs, an entire continent is split into two parts, which begin to separate as shown in Figure 1.10. At first the crust is both lifted and stretched apart. Then a long narrow valley, called a *rift valley*, appears (Block *A*). The widening crack in its center is continually filled in with magma rising from the mantle below. The magma solidifies to form new crust in the floor of the rift valley. Crustal blocks slip down along a succession of steep faults, creating a mountainous landscape. As separation continues, a narrow ocean appears; down its center runs a spreading plate boundary (Block *B*). Plate accretion then takes place to produce new oceanic crust and lithosphere. The Red Sea is an example of a narrow ocean formed by continental rupture. Its straight, steep coasts are features we would expect after such deformation. The widening of the ocean basin can continue until a large ocean has formed and the continents are widely separated (Block *C*).

The Global System of Lithospheric Plates

It is now known that there are six major lithospheric plates and several minor ones (Figure 1.11). The American plate includes not only the American continents, but also the oceanic lithosphere of the western Atlantic Ocean. The Pacific plate is unique in being the only major plate consisting almost entirely of oceanic lithosphere. In general, the western boundary of the Pacific plate is one of subduction; the eastern boundary is one of spreading and plate accretion. The African plate, which includes a large border of oceanic lithosphere, is beginning to be split apart by continental rifting in East Africa and along the Red Sea.

Many details of the global system of plates will be touched on in later chapters. At this point, you should simply get a general impression of the arrangement of the major plates and take note of the three kinds of plate boundaries as shown by special map symbols.

Continental Drift— The Breakup of Pangaea

Although modern plate tectonics became scientifically acceptable within only the past two decades, the concept of the breakup of an early supercontinent into fragments that drifted apart and made the modern continents is many decades old. Almost as soon as good navigational charts showed the continental outlines, educated persons became intrigued with the close correspondence in outline between the eastern coastline of South America and the western coastline of Africa. In 1666, a French moralist, François Placet, interpreted the matching coastlines as proof that the two continents became separated during the Noachian Flood. In 1858, Antonio Snider-Pelligrini produced a map to show the American continents nested closely against Africa and Europe. He went beyond the purely geometric fitting to suggest that the reconstructed single continent explains the close similarity of fossil plant types in coal-bearing rocks in both Europe and North America.

In the early twentieth century, two Americans, Frank B. Taylor and Howard B. Baker, published articles presenting evidence for the hypothesis that the New World and Old World continents had drifted apart. Nevertheless, credit for a full-scale hypothesis of breakup of a single supercontinent and the drifting apart of individual continents belongs to a German scientist, Alfred Wegener, a meteorologist and geophysicist who became interested in various lines of geologic evidence that the continents had once been united. He first presented his ideas in 1912 and his major work on the subject appeared in 1922. A storm of controversy followed, and many American geologists denounced the hypothesis.

Wegener had reconstructed a supercontinent named *Pangaea,* which existed intact about 300 million years ago in a geologic time unit called the Carboniferous Period. Figure 1.12 shows Wegener's original 1922

FIGURE 1.10 Schematic block diagrams showing stages in continental rupture and the opening up of a new ocean basin. The vertical scale is greatly exaggerated to emphasize surface features. *A*. The crust is uplifted and stretched apart, causing it to break into blocks that become tilted on faults. *B*. A narrow ocean is formed, floored by new oceanic crust. *C*. The ocean basin widens, while the stable continental margins subside and receive sediments from the continents. (Drawn by A. N. Strahler.)

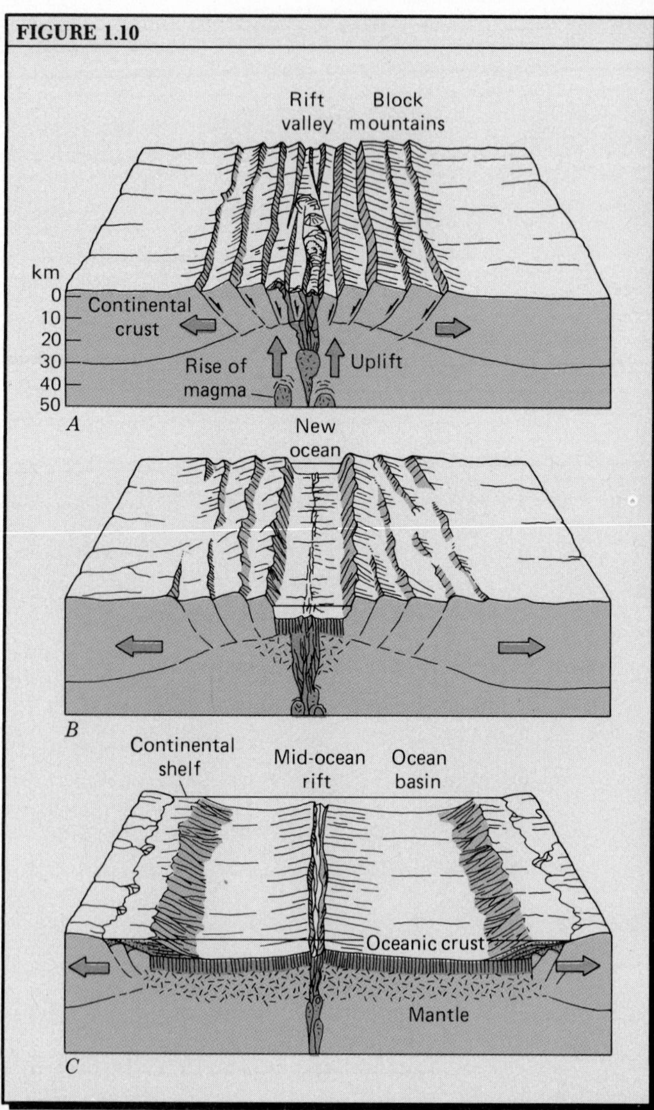

FIGURE 1.11 World map of lithospheric plates.

FIGURE 1.11

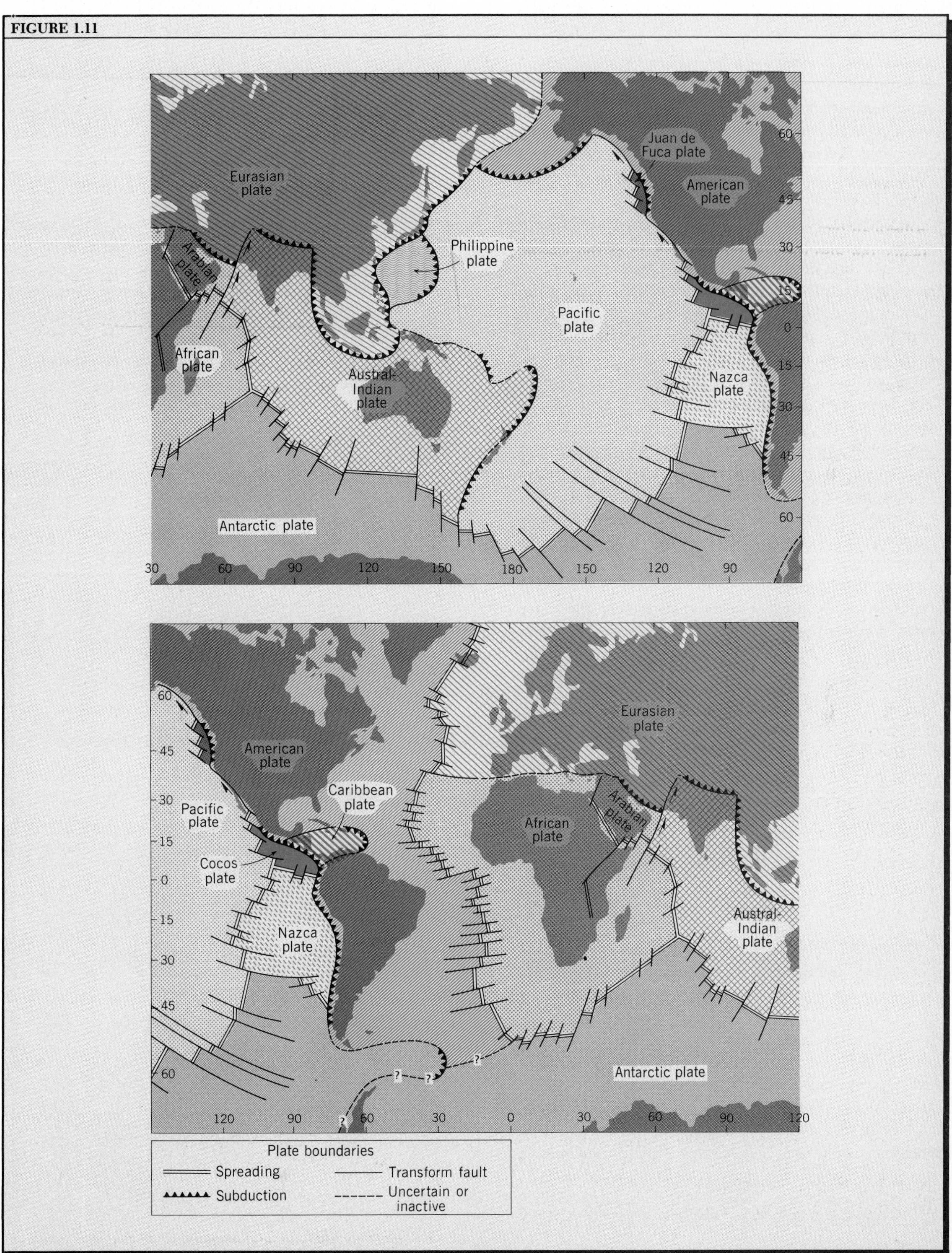

maps depicting the breakup of Pangaea. Wegener visualized the Americas as fitted closely against Africa and Europe, while the continents of Antarctica and Australia, together with the subcontinents of peninsular India and Madagascar were grouped closely around the southern tip of Africa. Starting about 200 million years ago, continental rifting began as the Americas pulled away from the rest of Pangaea, leaving a great rift that became the Atlantic Ocean. Later, the other fragments rifted apart, pulling away from Africa and from each other and causing the opening up of the ancestral Indian Ocean.

In Chapter 13, we will review several lines of hard geologic evidence for the existence of Pangaea. The evidence for a single supercontinent seemed quite convincing to many geologists, even during the great debate of the 1920s and 1930s, but the separation of the continents—a process then known as *continental drift*—was strongly opposed on physical grounds. Wegener had proposed that the continental layer of less dense rock had moved like a great floating raft through a "sea" of denser oceanic crustal rock. Geologists could show by use of laws of physics that this mechanism was physically impossible, because rigid crustal rock could not behave in such a fashion.

Wegener's scenario of continental drift took on new meaning in the 1960s and 1970s, when the theory of plate tectonics first emerged. The modern interpretation is that continental drift involves entire lithospheric plates, much thicker than merely the outer crust of either the continents or the ocean basins. Plate motions over a soft, plastic asthenosphere have allowed the continents to be carried along according to the general timetable postulated by Wegener. Some changes have been made in the timetable of events. There have also been various improvements in the fitting together of the original pieces of the supercontinent Pangaea. But what seems so remarkable is not that Wegener's original maps, shown in Figure 1.12, were somewhat in error, but that they are so much like those of the best modern interpretations.

Three Classes of Rock

Geologists recognize three major classes of rocks: igneous rock, sedimentary rock, and metamorphic rock. These classes can be understood and interpreted in terms of plate tectonics. *Igneous rocks* are those which have solidified from a molten state, which we have already referred to as magma, and which can originate in large masses along two types of plate boundaries. Along spreading plate boundaries, upwelling rock of the upper mantle reaches its melting point as pressure is relieved. The magma thus formed moves upward to occupy the space created by plate spreading. Along converging plate boundaries, large masses of magma form near the upper boundary of the descending plate. This magma rises to penetrate the overlying continental lithosphere, where it may solidify into igneous rock, or reaches the earth's surface, where it builds a chain of volcanoes.

Sedimentary rocks, composed of mineral matter derived from the chemical decay and physical breakup of any previously existing variety of rock, come in many varieties. Their finely divided mineral matter, consisting of solid mineral grains, is known as *sediment*. Some types of sediment are produced by biological and chemical synthesis rather than by rock decay. Sediment is transported by rivers, waves and ocean currents, winds, or glacial ice, and eventually is deposited in layers on the ocean floor, on the beds of lakes, or on low-lying land surfaces. Sediment can accumulate in numerous layers until a total thickness of many tens, hundreds, or thousands of meters is present. Compressed by the weight of overlying layers, the sediment is compacted and hardened into dense rock.

FIGURE 1.12 Alfred Wegener's maps showing the form of Pangaea and its breakup into the modern continents. (From A. Wegener, 1924, *The Origin of the Continents and Ocean Basins*, Methuen and Company, London, p. 6, Figure 1.)

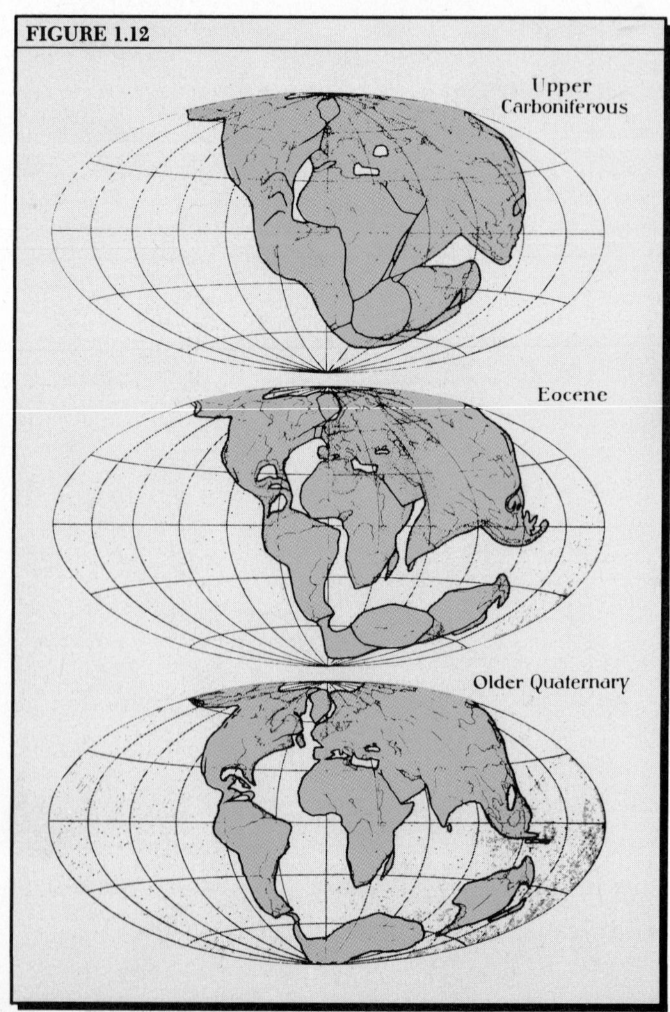

Chemical changes also may contribute to the total hardening process, which is termed *lithification*.

In terms of plate tectonics, we can point out one likely place for the accumulation of great thicknesses of sediment. Just offshore of the continent along converging plate boundaries, where oceanic lithosphere is plunging beneath continental lithosphere, there is usually a deep oceanic trench (Figure 1.13). This long, narrow trough may receive a large supply of sediment

FIGURE 1.13 Some typical features of an active subduction zone. *A.* A great vertical exaggeration is used to show surface and crustal details. Sediments scraped off the moving plate form tilted wedges that accumulate in a rising tectonic mass. Between the tectonic crest and the mainland is a shallow trough in which sediment brought from the land is accumulating. Metamorphic rock is forming above the descending plate. Magma rising from the upper layer of the descending plate reaches the surface to build a chain of volcanoes. *B.* A true-scale cross section shows the entire thickness of the lithospheric plates. (Drawn by A. N. Strahler.)

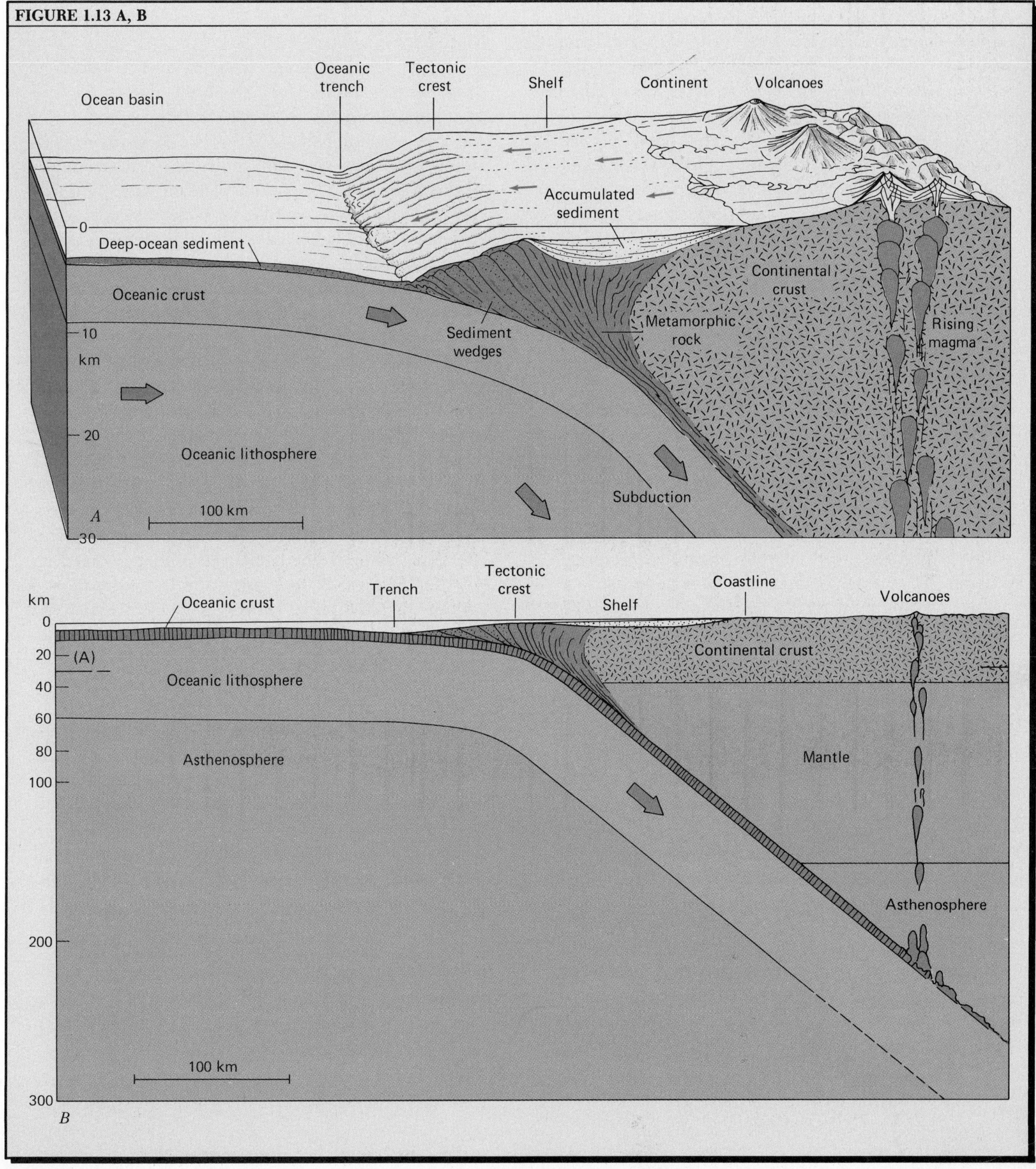

from the continent, as rivers bring the sediment to the shore and ocean currents transport it into deeper water. But sediment layers are also accumulating on the ocean floor of the moving lithospheric plate. As the plate undergoes subduction, its overlying sediment layer is carried toward the trench, as if on a continually moving conveyor belt. As plate motion continues, great masses of sediment are caught in the geologic trap and badly mauled. As Figure 1.13 shows, sediment arriving at the floor of the trench from both oceanic and continental sources is scraped off the descending plate and accumulates in a series of inclined, wedge-shaped bodies. The stack of wedges forms an enormous mass of deformed sediment that rises higher as time passes. The inner, older wedges are tilted to steeper angles as the pile grows. The crest of the rising pile of wedges, labeled the "tectonic crest" on the drawing, now forms a barrier to much of the sediment that is swept off the continent. This trapped sediment accumulates in an offshore trough, which may in time fill up to become a shallow continental shelf. Not all the sediment remains on the shelf, for some spills over low places in the tectonic crest and slides down the steep outer slope of the mass of deformed sediments.

Some of the sediment beneath the stack of wedges is dragged down to greater depths by the descending plate. Here rocks of the third major class are formed—the *metamorphic rocks*. Rocks of this general class are created by physical or chemical changes in preexisting types of rock, which may be either sedimentary or igneous. During the process of change, the rock remains in a solid state, but undergoes profound alterations in mineral composition, rock texture (sizes of grains or particles), and internal structure (crumpling of layers, fracturing). Sediment accumulating above a descending lithospheric plate is an important source of metamorphic rock. Here enormous confining pressures exist and there occurs a kneading action as material is caught between the descending oceanic plate and the continental plate. Here the deformed sediment is sheared between the opposing rigid masses. As water is forced out of the sediment, it becomes denser, and its atoms and molecules are reformed into new minerals. This change from sedimentary to metamorphic rock is called *metamorphism*.

Great masses of metamorphic rock are also created in other kinds of intense contact between two lithospheric plates, illustrated, for example, in Figure 1.14. Diagram *A* shows some details of the accumulated sediments along the stable margin of a continent. The continental margin was formed much earlier, following the opening up of an ocean basin. (You will find this type of stable continental margin on both sides of the widening ocean basin shown in Diagram *C* of Figure 1.10.) In the advanced stage shown in Diagram *A* of Figure 1.14, thick wedges of sediment have accumulated over the contact of continental and oceanic lithosphere. The wedge that lies upon the continental lithosphere consists of sands and muds brought to the shoreline by streams and spread over the shallow floor of a continental shelf. The continental margin slowly subsides, and the sediment wedge thickens. In deep water over the adjacent oceanic lithosphere, another sediment wedge accumulates; it is formed of sands and muds carried down the steep continental slope by submarine currents. As this deep-water sediment accumulates, it, too, forms a thick wedge, and the oceanic lithosphere subsides until it is downbent toward the continent.

At this time, a great tectonic event, called an *orogeny*, may occur. The downbent edge of the oceanic lithosphere can break free and begin to sink into the soft asthenosphere, as shown in Diagram *B*. Plate subduction is now in action, and a new plate boundary has been formed. The oceanic plate on the right is now moving toward the continental plate. Powerful compressional forces, which severely crumple and squeeze the two sediment wedges, are exerted upon the margin of the thick continental lithosphere; some of the deeper sediment is melted to form new igneous rock. The combination of heat and pressure causes the crumpled sediments to be changed into a belt of metamorphic rock that rises in elevation to form a great alpine mountain range. This mass of deformed rocks is called an *orogen*.

A second form of orogeny involving the production of metamorphic and igneous rock is called *continental collision* and involves two lithospheric plates (Figure 1.15). Plate *X* at the left of the figure is a relatively stationary plate of continental lithosphere capped by continental crust. Plate *Y* consists of both oceanic and continental lithosphere moving as a single unit. As plate *X* moves toward plate *Y*, the intervening ocean basin is gradually narrowed. Sediment carried on the moving plate is scraped off at the subduction zone and mixed with sediment brought from the continents, forming a rising stack of tectonic wedges. Finally, the two continental lithospheric masses collide, crushing and crumpling the sedimentary materials between them. Under this enormous kneading pressure, a mass of metamorphic rock is formed and welded between the joined continental plates. This new, permanent rock mass is called a *continental suture,* and is a distinctive kind of orogen.

The Rock Transformation Cycle

In order to relate the three major rock classes to plate tectonics, we have greatly oversimplified the origin of these kinds of rocks. Many important details remain to be filled out in later chapters. Moreover, we have said nothing about the chemical composition of the minerals in these rocks. We have, however, given enough information to derive the basic concept of a *rock transformation cycle* in which one class of rock is changed into a second, which in turn is changed into a third. Figure 1.16 shows this cycle in its simplest form.

Any one of the three rock classes can, by melting under high temperatures, form magma and lead to a new generation of igneous rock. Plate tectonics supplies us with a mechanism that has operated through all of recorded geologic time to produce large new patches of igneous rock from place to place.

FIGURE 1.14 These block diagrams show how metamorphic rock might be formed from sediment wedges of a stable continental margin. *A.* Sediment wedges have been deposited in a long period of stability during which the crust has undergone sinking to a lower level. *B.* A new subduction zone is formed, thrusting the sediment wedges against the continental lithosphere and forming a great mass of highly contorted sediments and metamorphic rocks. *C.* True-scale cross section of Diagram *B.* (Drawn by A. N. Strahler.)

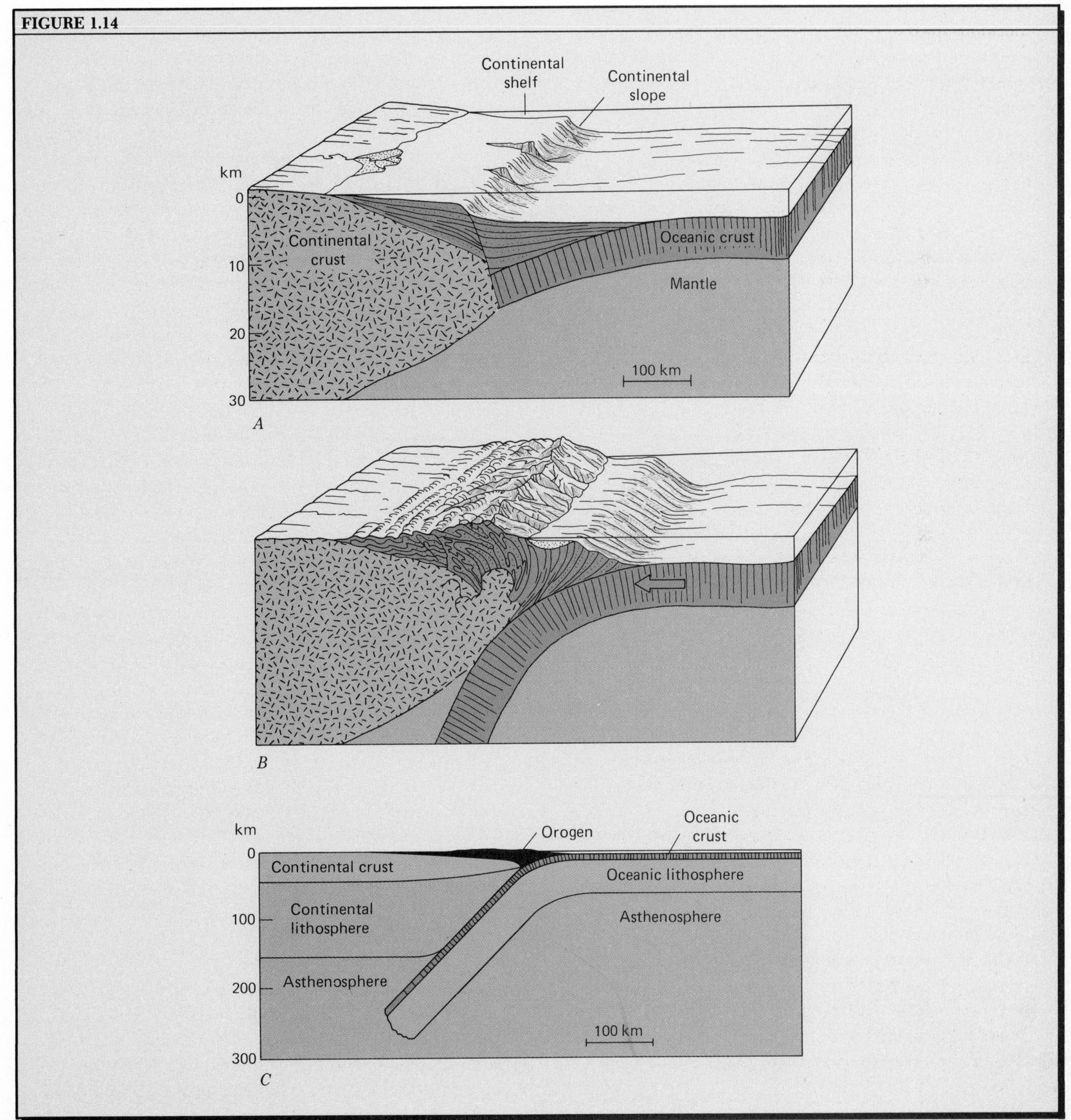

Rocks of all three classes can reach the earth's surface, where they are broken up into sediment, leading to the production of sedimentary rock. Sedimentary rock, in turn, can be transformed into metamorphic rock or melted to form magma and new igneous rock. Sedimentary rock can also be the direct source of more sediment and more sedimentary rock. Igneous rock, if caught up in an environment of high pressure and powerful deforming stresses, such as has happened

FIGURE 1.15 Four stages in the collision of two lithospheric plates, starting with closing of an ocean basin and ending in an orogeny that represents a suture between the plates. The vertical scale of these diagrams is greatly exaggerated to emphasize surface features and the difference in thickness between continental crust and oceanic crust. (Drawn by A. N. Strahler.)

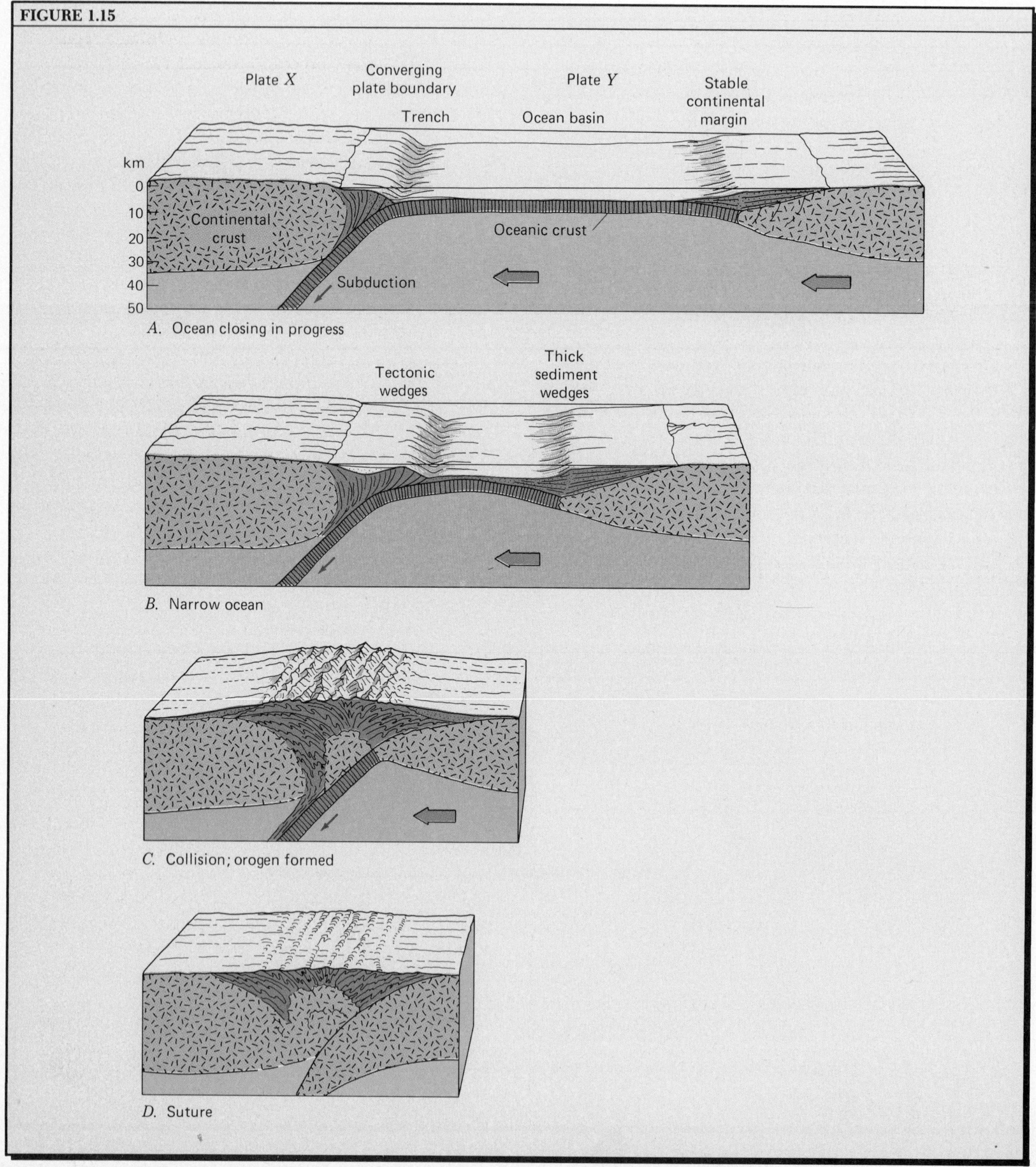

during continental collisions, is turned into metamorphic rock.

The rock transformation cycle is an example of a material flow system, of which there are many examples in natural science. Energy is required to drive material flow systems.

Relative Ages of Continents and Ocean Basins

Our final reference to plate tectonics in this brief introductory overview concerns a deeply significant concept in the interpretation of the history of our planet. We have seen that new oceanic lithosphere is being created by accretion at spreading plate boundaries and that oceanic lithosphere is simultaneously being consumed in converging plate boundaries. If so, oceanic lithosphere cannot be very old, measured in terms of the duration of geologic history. In contrast, the patches of continental lithosphere are remarkably enduring, because they are thick, buoyant masses that do not easily undergo subduction. Once formed, continental lithosphere tends to remain intact, and much of that lithosphere must be very old indeed.

As one continental collision after another formed new sutures of metamorphic and igneous rock, the continental lithosphere must have grown in extent. This would mean, of course, that the continents themselves have increased in area, at least through the earlier portions of geologic time. We will find some convincing evidence that the continental crust, which is the outermost layer of the continental lithosphere, is of great geologic age, for the most part—much of it is from one to three billion years old. In contrast, rock of the ocean floor, formed in spreading boundaries, has proved to be quite young geologically—none is older than about 200 million years.

FIGURE 1.16 A simplified schematic diagram of the cycle of rock transformation.

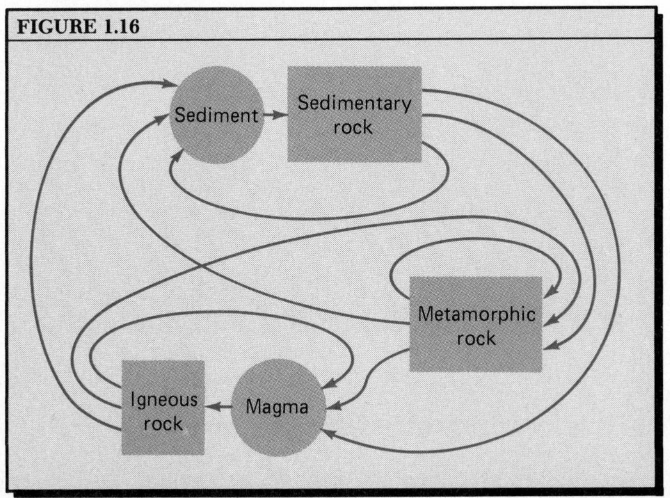

The Origin of Our Earth and Solar System

No overview of geology would be complete without a broad-brush sketch of the origin of our planet as a member of the solar system. The details will be filled out in Chapter 20. Our solar system consists of our sun—a true star—and numerous solid objects that orbit the sun in more-or-less circular paths. The nine major planets are among these orbiting objects.

Our sun, a typical medium-sized star, is a large sphere of extremely hot, densely compressed gas. Within the sun's interior, energy is continually generated by a complex process of nuclear fusion in which hydrogen (90% of the sun's material) is transformed into helium (8%). The sun's chemical composition is actually not so different from that of the great outer planets, which also consist largely of hydrogen and helium, but it is in marked contrast to that of the innermost planets, which consist largely of oxygen, silicon, aluminum, and iron. It is now generally agreed by scientists that the sun and planets evolved through the same process at the same time, and that the formation of all these bodies occurred about 4.5 billion years ago. Here is a brief scenario of the formative events.

In the vast open space among stars of our galaxy was a very diffuse body of hot interstellar dust and gas. This interstellar mass, called a *nebula*, is thought to be the starting point of a solar system. The solar nebula began to contract under the mutual gravitational attraction of all its particles to one another. Gradually the nebula began to cool and assume a wheellike shape—very thin and somewhat larger in diameter than our present solar system (Figure 1.17). A dense, luminous mass of gas that was to become the sun formed an enlarged central hub.

Then, with further cooling of the nebular disk, gases began to condense rapidly. Minute particles of iron and compounds of silicon and oxygen solidified first. Then later, as the remaining gases cooled, the highly volatile compounds—water, methane, and ammonia—condensed into icy grains, which collided and began to stick together, a process called *accretion*.

The resulting aggregations of matter, called *planetesimals*, grew rapidly, but many were unstable and disrupted into countless small solid fragments; these orbited the sun along with the remaining large planetesimals. By a series of collisions, not violent enough to heat the growing planets to the point of complete melting, the larger planetesimals swept up most of the smaller fragments and quickly grew to their final planetary dimensions.

A timetable for this phase of earth history has been suggested as follows: The solar nebula was in process of rapid contraction about five billion years ago (−5.0 b.y.); accretion of the planets was largely completed at about −4.7 to −4.5 b.y.

Earth and the other three inner planets were probably not in a completely molten state at any time during accretion. However, there must have been some local melting where exceptionally large objects impacted the surface. The earth at this stage was a mixture of the minerals that we now find segregated into the great earth shells—core, mantle, and crust. That segregation was to follow as heat produced by radioactivity accumulated and caused melting deep within the earth. In any event, the earth had acquired its layered structure and was internally stable by about −3.8 b.y., when the first crustal rocks of which we have any record were formed.

This brief overview of planetary origin may raise in your mind more questions than it answers. For example, what happened to the volatile substances such as methane and ammonia, abundantly present in early stages of accretion? These may have been driven off the growing planetesimals closest to the sun through heating and the pressure of a fast-moving wind of charged particles that continually sweeps outward from the sun's surface. Some water remained, however, locked up in the rocky minerals of the growing planet Earth. This water, with other gases, was later to emerge slowly from volcanoes, ultimately to form the world ocean and the atmosphere.

Geologic Time

The vastness of geologic time, which we can say starts with the accretion of the planets, is almost beyond human comprehension. This same vastness of time is one of the most fascinating aspects of geology, setting off both geology and astronomy from other physical and natural sciences. Figure 1.18 is a table of geologic time devised in such a way as to emphasize the enormity of the five-billion-year span of our planet's history. For convenience, we begin five billion years ago (−5 b.y.) and recognize five *aeons* of time, each one enduring a billion years.

The formation of the planets of the solar system was in progress as the First Aeon began. The earth's growth to form a solid spherical body was complete about −4.7 b.y. There is no record of this event in earth rocks, but a moon rock called the Genesis Rock has been found to contain mineral particles of this age. Throughout the First Aeon, the earth's history is completely unknown, and it is not until the Second Aeon that the oldest earth rock (so far discovered) appears, with an age of −3.8 b.y. It is speculated that the origin of living matter, an event called *biogenesis*, took place in shallow ocean water in the time span of −3.7 to 3.5 b.y., but there are no remains or evidence of such life until near the end of that aeon. The moon became geologically "dead" (inactive in terms of crustal movement or volcanic activity) at the end of the aeon.

The Third Aeon (−3.0 to −2.0 b.y.) saw the development of primitive one-celled organisms—forms of algae—and these are preserved in rocks of that aeon. The nuclei (oldest central parts) of the continents were growing in size. The ocean waters, which had been accumulating from gases emitted by the solid earth, were nearly at their present volume. The Fourth Aeon saw a continuation of much the same activity as in the previous aeon, with a continued growth of the continents and the continental lithosphere that underlies them.

Lithospheric plates were continually in motion in the Second, Third, and Fourth aeons, but we know very little about them. The metamorphic rocks that formed from repeated continental collisions are found in great abundance in the continental interiors today.

The Fifth Aeon, covering the last billion years, witnessed the very rapid evolution of many complex life forms; much of that organic change was concentrated into a short period around −0.5 b.y. The second half of the aeon saw most of the evolution of advanced life forms as we know them today.

Figure 1.18 shows the Fifth Aeon subdivided progressively into smaller time units, each being the youngest one-tenth part of the previous one. Human evolution takes place largely within the final ten mil-

FIGURE 1.17 (*Above*) An artist's conception of a disklike solar nebula surrounding a central solar mass (NASA). (*Below*) An edge-on view of the nebula.

lion years; human civilizations do not appear until the final 5000-year time block. Names of the geologic time units shown in Figure 1.18 and many of the events noted by labels will be mentioned in later chapters. At this point it is not necessary to memorize this material. Rather, it will be enough to grasp the vastness of geologic time and the sparseness of the geologic record of events throughout the first four aeons.

A Need for Skepticism

In presenting this brief overview of modern physical geology, we have made many statements about the earth as if they are facts to be accepted by you as proven beyond doubt. The scientific evidence is lacking for many of the assertions we have made. You, as a newcomer to the science of geology, have every right to be skeptical about such assertions.

Marshalling the scientific data to prove the correctness of our statements is a major task of later chapters of this book—particularly important in supporting the theory of plate tectonics, which is relatively new in geology. You should be aware that a few geologists do not accept the theory of plate tectonics; they continue to bring forward arguments and evidence that they feel support alternative explanations. These sincere dissenters continue to publish their arguments in respected scientific journals.

On the other hand, the number of these dissenters is now relatively small. The weight of evidence favoring the basic principles of plate tectonics is strong—so strong, in fact, as to be judged overwhelming by the majority of active research geologists in the English-speaking world. Nevertheless, you must weigh the evidence for yourself and draw your own conclusions.

Key Facts and Concepts

Physical and historical geology *Geology,* the science of the earth, has two sides or phases. *Physical geology* is concerned with physical and chemical properties of the earth and the physical and chemical changes that occur in rocks. *Historical geology* is concerned with ancient life and the progress of organic evolution.

Scope of physical geology Physical geology deals with the composition and origin of rocks and their organization into structural units on many scales of magnitude. Physical geology also deals with rising *magma* and igneous activity resulting from the accumulation of heat within the earth and with *tectonic activity*—crustal bending and breaking—powered by great internal forces. Physical geology also studies changes in exposed rock surfaces caused by erosion processes, leading to the production of varied landscape features.

The earth's rock shells *Rock,* consisting of minerals in the solid state, ranges in age from recent to 3.8 billion years. The central region of the earth sphere is occupied by a largely molten inner *core,* 3475 km in radius, consisting mostly of iron. Surrounding the core is the *mantle,* a shell of dense solid rock 2900 km thick, consisting of silicon, oxygen, iron, and magnesium. The *crust* is a comparatively thin outer rock layer (5 to 70 km thick); it consists of rock less dense than the mantle and composed largely of silicon, oxygen, and the metals aluminum, sodium, calcium, and potassium.

Lithosphere and asthenosphere A brittle outer layer, the *lithosphere,* includes the crust and upper mantle. Beneath lies the *asthenosphere,* a zone within the mantle in which rock is close to its melting point and is in a soft, weak physical condition. Large segments of the brittle lithosphere, behaving as *lithospheric plates* can move freely but slowly over the soft asthenosphere, which yields by flowing in a plastic manner.

Continental lithosphere and oceanic lithosphere The lithosphere beneath the ocean basins is comparatively thin (average about 60 km) and has a thin crust (5 km); lithosphere beneath the continents is thick (average 150 km) and has a thick crust (30–50 km). Plates consisting of *oceanic lithosphere* tend to move more freely than plates consisting of *continental lithosphere,* which extend more deeply into the asthenosphere.

Plate tectonics *Plate tectonics* is the general theory of lithospheric plates, their relative motions, and their boundary interactions. Lithospheric plates move apart along *spreading boundaries,* where magma is rising and the plates are undergoing *accretion.* Plates come together along *converging boundaries,* where plate *subduction* is in progress and lithosphere is being consumed. Two plates may glide past one another along a *transform boundary.* The global system consists of six major and several minor plates.

Continental rupture A plate consisting of continental lithosphere may split apart, forming a new ocean basin, which widens and becomes floored by new oceanic lithosphere. The modern continents came into existence largely through the rupturing of a major plate bearing a single continent, *Pangaea.* The basins of the Atlantic and Indian Oceans formed during the past 200 million years by rifting apart of the plate fragments of Pangaea. Thus, the oceanic lithosphere is much younger than the continental lithosphere.

Continental collision Continued subduction can lead to an *orogeny* by the collision of two masses of continental lithosphere, welding the plates together in a *suture* and deforming the rocks into an *orogen.*

Classes of rock The three major classes of rock are

FIGURE 1.18

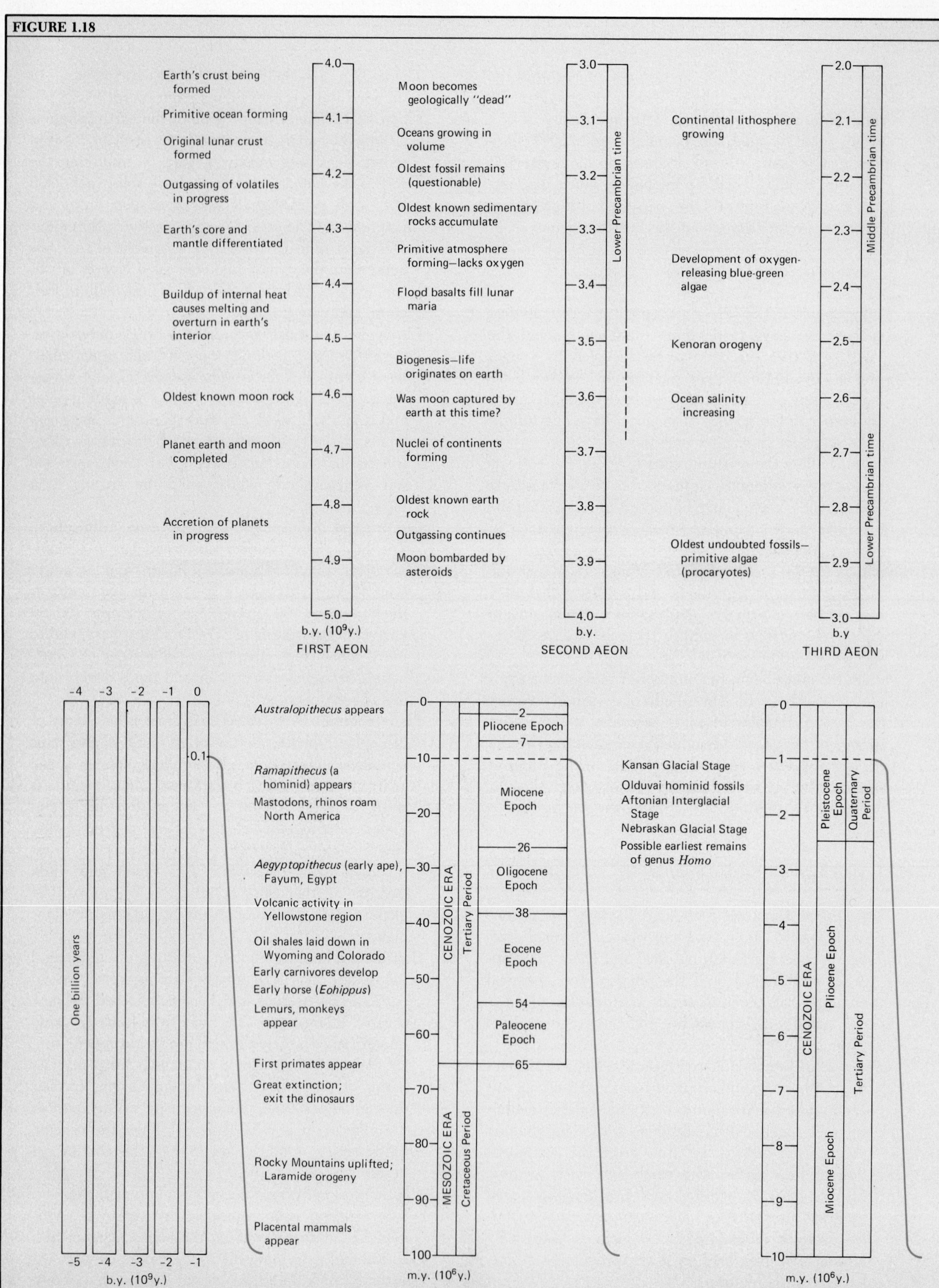

FIGURE 1.18 An annotated table of geologic time.

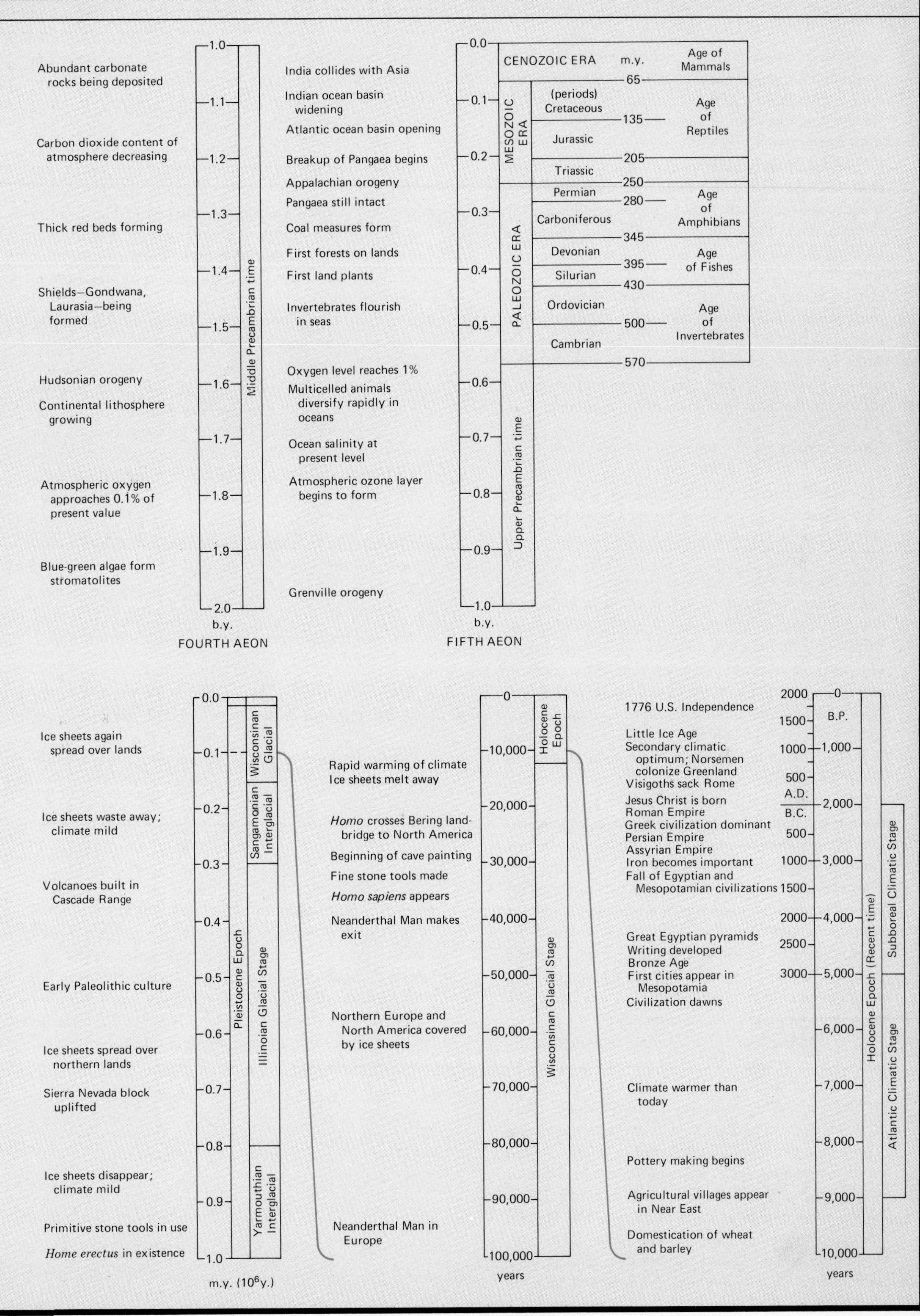

Abundant carbonate rocks being deposited
Carbon dioxide content of atmosphere decreasing
Thick red beds forming
Shields—Gondwana, Laurasia—being formed
Hudsonian orogeny
Continental lithosphere growing
Atmospheric oxygen approaches 0.1% of present value
Blue-green algae form stromatolites
Middle Precambrian time
1.0
1.1
1.2
1.3
1.4
1.5
1.6
1.7
1.8
1.9
2.0
b.y.
FOURTH AEON
India collides with Asia
Indian ocean basin widening
Atlantic ocean basin opening
Breakup of Pangaea begins
Appalachian orogeny
Pangaea still intact
Coal measures form
First forests on lands
First land plants
Invertebrates flourish in seas
Oxygen level reaches 1%
Multicelled animals diversify rapidly in oceans
Ocean salinity at present level
Atmospheric ozone layer begins to form
Grenville orogeny
0.0
0.1
0.2
0.3
0.4
0.5
0.6
0.7
0.8
0.9
1.0
b.y.
FIFTH AEON
CENOZOIC ERA
m.y.
Age of Mammals
65
MESOZOIC ERA
(periods)
Cretaceous
135
Age of Reptiles
Jurassic
205
Triassic
250
PALEOZOIC ERA
Permian
280
Age of Amphibians
Carboniferous
345
Devonian
395
Age of Fishes
Silurian
430
Ordovician
Age of Invertebrates
500
Cambrian
570
Upper Precambrian time
Ice sheets again spread over lands
Ice sheets waste away; climate mild
Volcanoes built in Cascade Range
Early Paleolithic culture
Ice sheets spread over northern lands
Sierra Nevada block uplifted
Ice sheets disappear; climate mild
Primitive stone tools in use
Home erectus in existence
0.0
0.1
0.2
0.3
0.4
0.5
0.6
0.7
0.8
0.9
1.0
m.y. (10^6 y.)
Pleistocene Epoch
Wisconsinan Glacial
Sangamonian Interglacial
Illinoian Glacial Stage
Yarmouthian Interglacial
Rapid warming of climate
Ice sheets melt away
Homo crosses Bering land-bridge to North America
Beginning of cave painting
Fine stone tools made
Homo sapiens appears
Neanderthal Man makes exit
Northern Europe and North America covered by ice sheets
Neanderthal Man in Europe
0
10,000
20,000
30,000
40,000
50,000
60,000
70,000
80,000
90,000
100,000
years
Holocene Epoch
Wisconsinan Glacial Stage
1776 U.S. Independence
Little Ice Age
Secondary climatic optimum; Norsemen colonize Greenland
Visigoths sack Rome
Jesus Christ is born
Roman Empire
Greek civilization dominant
Persian Empire
Assyrian Empire
Iron becomes important
Fall of Egyptian and Mesopotamian civilizations
Great Egyptian pyramids
Writing developed
Bronze Age
First cities appear in Mesopotamia
Civilization dawns
Climate warmer than today
Pottery making begins
Agricultural villages appear in Near East
Domestication of wheat and barley
2000
1500
1000
500
A.D.
B.C.
500
1000
1500
2000
2500
3000
0
B.P.
1,000
2,000
3,000
4,000
5,000
6,000
7,000
8,000
9,000
10,000
years
Holocene Epoch (Recent time)
Subboreal Climatic Stage
Atlantic Climatic Stage

igneous, sedimentary, and *metamorphic.* All are interrelated in origin through a *rock transformation cycle,* governed by plate tectonics and powered by internal energy sources and solar energy.

Origin of the earth and solar system Our solar system originated from a diffuse body of hot interstellar dust and gas called a *nebula.* As the disk-shaped nebula contracted, the inner portion condensed into the sun, while the outer parts of the disk condensed to form the planets.

Geologic time The immensity of geologic time is a central theme of geology. Five *aeons* of time can be recognized, each one billion years in duration. Following its formation as a planet in the first aeon, the earth became internally geologically stable. Lithospheric plates were then able to form and to collide, leading to the growth of primitive continents.

Questions and Problems

1. Describe the major ideas and conceptions you now hold of geology as a science, just as you embark on a first course in physical geology. For example, how do you visualize the condition of earth's interior? How do you view the total duration of geologic time, since the earth was formed as a planet, as compared with the length of time that advanced forms of life have existed (marine invertebrates, dinosaurs, or humans, for example)? What ideas do you have about the origin of volcanoes and mountains? Do you know what causes a major earthquake? Where and how did you develop an interest in geology? Summarize these answers in an essay of about 500 words.
2. Recall from news events of the past several years some examples of the ways in which geologic activities have had a severe impact on human beings. Have geologic events affected your life directly? For example, have you experienced an earthquake? Have you seen the eruption of a volcano? Have you seen any houses or roadways destroyed by landslides or mudslides, or by the undermining action of storm waves? Do you know of cases in which property damage could have been avoided by applying geologic knowledge prior to construction?
3. Do you find the basic ideas of plate tectonics totally foreign to your way of thinking about the earth? To feel more at home with the notion of huge plates jostling each other with enormous violence, think for a while about earthquakes. A great earthquake is a very real and terrifying personal experience, even for those persons lucky enough to escape without bodily harm or property damage. In Figure 1.7, you see the three principal kinds of active plate boundaries: spreading boundary on the ocean floor, converging boundary along an oceanic trench, and transform fault. Think about the kinds of motions that occur on each of these boundaries. Use your hands to demonstrate the three kinds of plate motions. If these boundaries are the major lines along which geologic action is taking place, it follows that the bulk of powerful earthquakes must occur along the boundaries. As we shall find in Chapter 8, this conclusion is correct. Along which of the three kinds of plate boundaries would you expect the greatest earthquakes to occur, as a general rule? Name some countries where this risk is high. Then write a short essay on the origin of earthquakes in relation to plate tectonics.
4. It is hard to appreciate the enormous duration of geologic time, just as one has difficulty in conceiving of the enormous distance to a star other than our sun or to another galaxy. Using the information in Figure 1.18, try an exercise that will focus upon particular events in the table of geologic history. Let the starting point in time be the origin of our planet 4.7 billion years ago. Make a list of several interesting points in geologic history, for example:

Start of the breakup of Pangaea	200 million years ago
Extinction of the dinosaurs	70 million years ago
Earliest appearance of man (genus *Homo*)	3 million years ago
First agriculture	10,000 years ago
Birth of Jesus Christ	2,000 years ago

Make a fraction in which the denominator is the age of the earth in years, thus:

$$\frac{x}{4{,}700{,}000{,}000}$$

For x substitute each of the events on the list, reducing the numerator to unity and making a simple fraction. The reduced fraction tells you the proportional part of geologic time since the event occurred. Another way to express these events is to let 4.7 billion years be represented by one calendar year of 365 days. Convert each event to time in days, hours, minutes, and seconds.

2

CHAPTER TWO

Matter and Energy—A Review

The real world with which geology is concerned consists of two components: matter and energy. Physics and chemistry are the basic sciences that deal with the nature of matter and energy and the formulation of laws that govern their behavior. Modern physical geology depends so heavily upon the use of physics and chemistry that two major areas of geology—geophysics and geochemistry—overlap between geology and physics and between geology and chemistry, respectively (Figure 2.1). In this crucial overlap zone have taken place many of the great recent advances in geology in the realm of plate tectonics. One could make a similar statement about the role that physics and chemistry have played in modern biology.

FIGURE 2.1 Physical geology as a capstone supported by the physical sciences—physics and chemistry—which are the foundation blocks. In the intermediate position are geophysics and geochemistry, which apply principles of physics and chemistry to the solving of problems in physical geology.

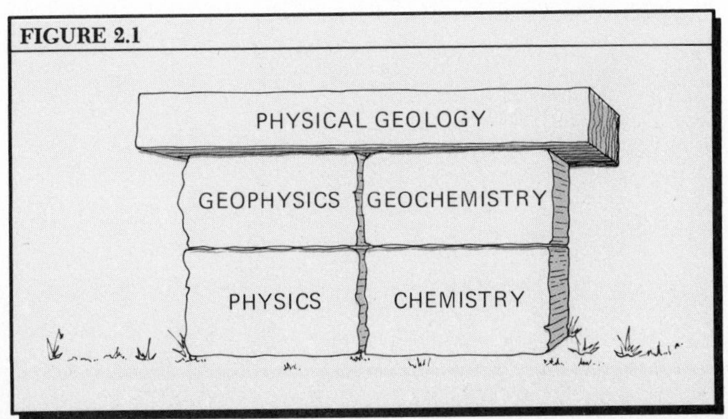

In this chapter we review some important principles of physics and chemistry essential to an understanding of modern geology. If you are already familiar with this material, pass over it lightly, making sure, however, that you understand the topics. For those of you who have not studied physical science, we hope this presentation will be helpful, even though it is brief.

Matter and Energy

To define the terms *matter* and *energy* is not an easy task because they represent concepts that include everything of which we have any comprehension in the real world. To start, we can use the word "substance" for the word "matter," but this only postpones the problem. A substance occupies space; it is "tangible" because it can be seen, felt, tasted, measured, weighed, stored, and so forth. Matter has the mysterious property of *gravitation*, the mutual attraction between any two aggregates (lumps, groups, pieces) of matter. Rather than attempt a simple definition of matter, we shall proceed a little later to investigate the nature of matter in some depth.

Energy, often defined in terms of its effects, is "the ability to do work." Energy is somehow always involved in the motion of matter, but energy can be stored in matter that does not appear outwardly to have motion. A brick poised on the brink of a high parapet contains a store of energy even though the brick is at rest. That stored energy will become very obvious if the brick is nudged slightly and swiftly falls to the sidewalk below. Whatever energy is, then, it can move or flow from place to place and it can also be held in storage in various ways. We frequently refer to energy as something "expendable." For example, someone may say, "I used up a lot of energy playing those two sets of tennis." Actually, energy cannot be destroyed—that is, be removed from existence—by being "used"; it can only change from one form to another and move from one place to another. The same statement applies to matter, which cannot be destroyed, but only changed or moved from place to place.

Laws of Conservation of Matter and Energy

Two basic laws that emerged early in the history of science are the *law of conservation of matter* and the *law of conservation of energy*. The first of these laws simply states that within the universe as a whole, matter cannot be created or destroyed. The second law says the same thing with respect to energy. Not long

FIGURE 2.2 The forms of matter—a classification of the kinds of substances. (From John R. Holum, 1975, *Elements of General and Biological Chemistry*, 4th ed. John Wiley & Sons, Inc., New York, Figure 2.1. Reproduced by permission.)

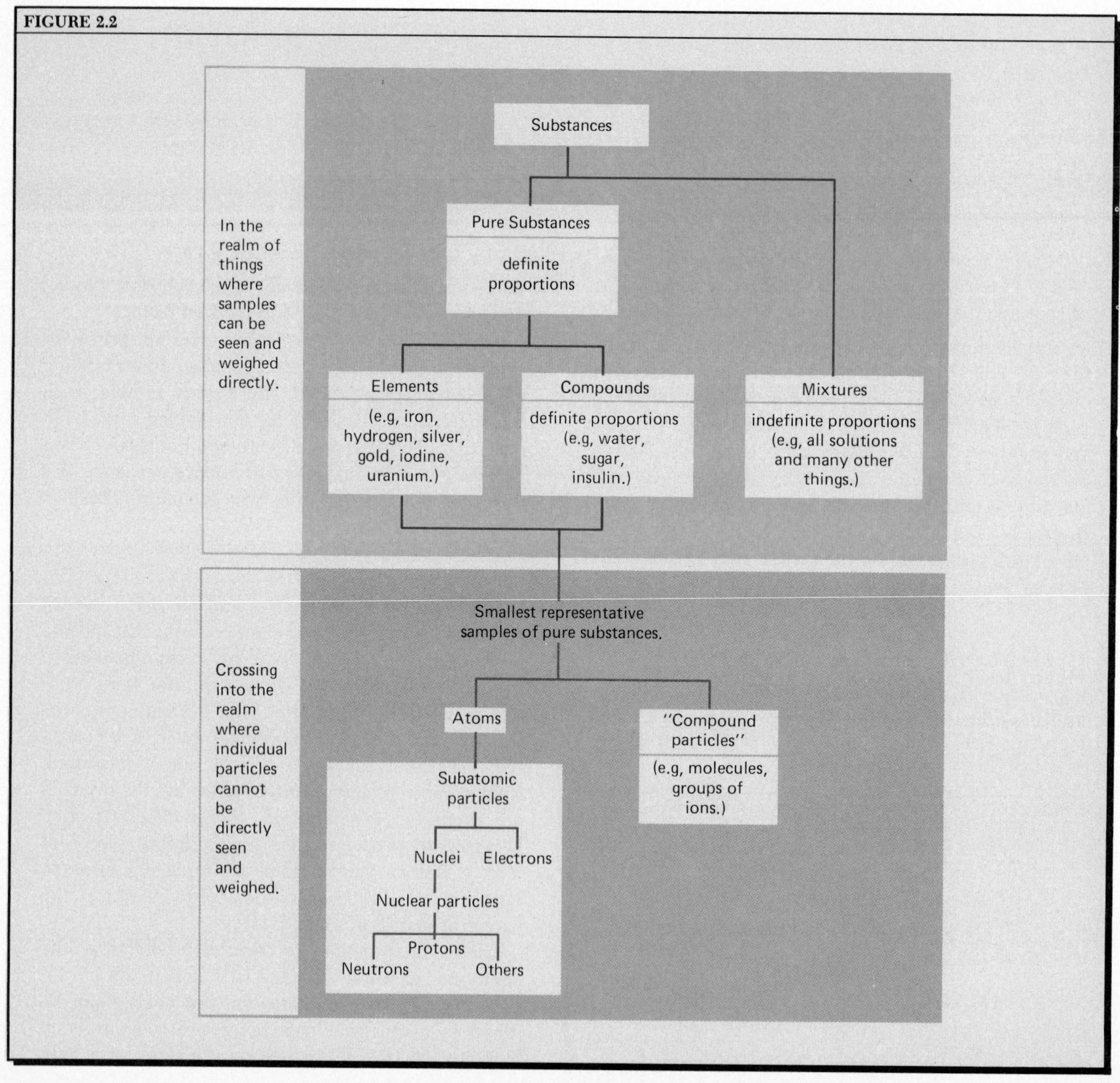

after Marie and Pierre Curie in 1898 isolated two radioactive elements, polonium and radium, scientists were forced to conclude that matter can be transformed into energy, and thus it became necessary to combine the two laws into a single law of conservation of matter and energy: The sum total of all matter and energy in the universe must remain constant.

The conversion of matter to energy through radioactivity within mineral substances of the solid earth has been of enormous importance to geologic processes over the entire 4-billion-year history of the planet. At the appropriate point we shall investigate this subject in some detail.

Classes of Matter

Let us now investigate the various kinds or classes of matter in the universe. A system (illustrated in Figure 2.2) devised by Professor John R. Holum, a chemist, collectively describes all matter by the word "substances." Next, *pure substances* are distinguished from *mixtures*, which consist of two or more pure substances mixed together in indefinite proportions. Pure substances are either *elements* or *compounds*.

Examples of elements are several metals in the pure state: for example, iron, gold, copper, or silver. Although never found in a condition of absolute purity, nuggets (lumps) or crystals of each of these *native elements* occur naturally in rock. Take native copper: In 1857 a single mass of native copper weighing 380,000 kg (about 420 U.S. tons) was discovered on the Keweenaw Peninsula in northern Michigan. Long before Europeans came to North America, Indians were using pieces of native copper as a malleable metal out of which to make various tools and ornaments. In addition, we are all familiar with flakes and nuggets of gold prospectors seek in the beds of streams. A large gold nugget is shown in Figure 2.3.

FIGURE 2.3 This great gold nugget was found near Greenville, California, during the gold rush of 1848. It is about 20 cm long and weighs 2.3 kg. (Smithsonian Institution Photo No. 2254.)

FIGURE 2.3

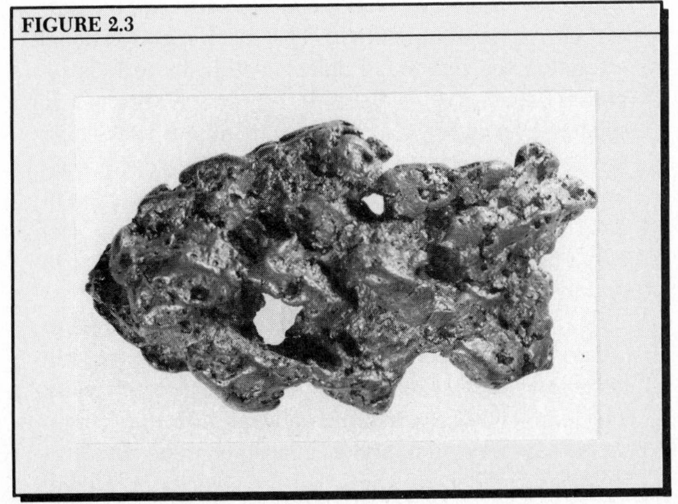

Compounds are also familiar: pure water, carbon dioxide gas, or crystals of amethyst quartz, for example. As we have described them, both elements and compounds exist "in the realm of things where samples can be seen and weighed directly," to use Professor Holum's words. The lower box in Figure 2.2 takes us into "the realm where individual particles cannot be directly seen and weighed." The smallest representative samples of pure substances are of two classes: *atoms* and *compound particles*. The latter consist of groups of atoms, occurring either as molecules or as ions. We shall describe atoms, molecules, and ions in detail later in this chapter.

Atoms and Compounds

The notion that atoms represent the smallest units of matter was expressed as early as 500 B.C. by Greek philosophers. (The Greek word, *atomos*, means "indivisible," or "not cuttable.") The idea of the atom was revived by John Dalton (1766–1844), an English chemist and physicist, who proposed the theory that all matter consists of indestructible atoms. A given element, he reasoned, consists of only one variety of atom, having a given weight and other distinctive properties. Atoms of another element must have a different weight and a different set of properties. Atoms of two or more different elements can combine to form *compounds*. An important point in Dalton's theory is that the atoms of the component elements of a given compound must always combine in a definite ratio by weight, and must always involve whole numbers of atoms.

Mixtures of substances, referred to in the upper part of Figure 2.2, usually consist of different compounds mixed together physically in varying proportions. For example, most rocks exposed at the earth's surface are mixtures of several compounds; seawater is a mixture of water and several different dissolved salts, each of which is a distinctive type of compound.

Much later than Dalton's time came the discovery that atoms themselves are made up of various kinds of parts, which we can refer to as *subatomic particles* (see Figure 2.2). If a single atom were to be enormously expanded, it would be found to consist of a dense central nucleus orbited at high speed by a large number of extremely minute particles called electrons.

Even the nucleus of the atom consists, in turn, of different kinds of parts, called *nuclear particles*. The only two classes of nuclear particles we will refer to are neutrons and protons, but you should be aware that there are many others.

Mass and Density

This is a good point at which to think in a very precise way about two of the most important terms relating to matter—mass and density. The measure of quantity of matter is called *mass*. The mass of a particular isolated solid object, such as a pebble or a billiard ball, is the quantity of matter contained within that object. A practical method of determining the mass of an object at the earth's surface is to weigh it. However, when we place the object on an ordinary scale and read the weight in grams or kilograms, we are actually measuring the force with which the earth's gravitational force pulls down upon the object. Performing this same operation on the moon, where the pulling force is much less, would give a much different weight on the same scale, but the mass of the object would be exactly the same as on earth, or as anywhere else in the solar system.

A practical solution to the problem of establishing a unit of mass is to designate some particular quantity of matter as the standard mass (unit mass). As the French Revolution drew to a close, the French National Assembly organized the metric system and defined its units. As originally planned, the unit of mass was to be the *gram* (g), a mass represented by one cubic centimeter (1 cc) of perfectly pure water at a temperature of 4 °C. Later, for practical reasons, the standard of mass was established as the quantity of matter in a cylindrical metal block of platinum-iridium alloy equivalent in weight to 1000 g of pure water. This standard unit of mass is the *kilogram* (kg); it rests in a vault at Sèvres, France. A *metric ton* (MT) is 1000 kg.

A second observation we can all make is that the quantity of matter (mass) in an object of a given volume varies a great deal, depending upon the substance of which the object is composed. Given two cubes of exactly the same dimensions, one made of rock crystal (quartz) and the other of iron, you need only lift each cube in turn to realize that the mass of the iron cube is much greater than that of the quartz cube. The term *density* describes the quantity of matter (mass) in a unit volume of space. Using the gram as the unit of mass, the density of a substance is defined as the mass in grams per cubic centimeter (g/cc). Because the meter (m) and kilogram (kg) are now designated as the official units of length and mass in the SI (System Internationale), it is considered proper to define density in terms of 1000 kg (10^3 kg) per cubic meter (m^3). The numbers come out the same in any case. Using grams and centimeters as units, the density of the mineral quartz is approximately 2.6 g/cc, that of iron is about 8.0 g/cc. Figure 2.4 shows how four similar cubes, each of a different substance, cause a coil spring to be extended to greater lengths in proportion to increased density.

The concept of density, applied to minerals and rocks, is of great importance in understanding the classification of rocks and the structure of the earth in terms of a system of layers, or shells.

Gravitation and Matter

We mentioned earlier in this chapter that the nature of matter is closely tied in with the phenomenon of gravitation, the attraction that every particle of matter in the universe exerts upon every other particle. We can, in fact, define mass as the property of being susceptible to this attraction. Probably all of you have memorized at one time or another the *law of gravitation* formulated by Sir Isaac Newton. It goes as follows: Any two bodies attract one another with a force that is directly proportional to the product of their masses and inversely proportional to the square of the distance separating them.

Written formally as an equation, the law of gravitation reads:

$$F_g = \frac{G M_1 M_2}{D^2}$$

where

M_1 and M_2	are the two masses
D	is the distance separating the masses

and

F_g	is the gravitational force between them

(The letter G is a number known as the universal constant of gravitation; its numerical value is not important here.)

So far as geologic processes at or near the earth's solid surface are concerned, what counts is the gravitational attraction of the earth—a truly enormous mass—exerted upon very tiny masses, such as molecules of gas or liquid, and particles of soil, rock, or organic matter. (The infinitesimally small force with which the particle attracts the entire earth can be written off as inconsequential.) The earth's gravitational attraction for very small objects within its surface region is called *gravity*. Because the earth is very nearly spherical, gravity has an almost constant value over the entire surface. In other words, the earth's attraction for a given quantity of matter (such as a mass of 1 kg) will be the same over the entire globe, subject only to minor corrections that are not particularly important in understanding most geologic processes.

If a particle of matter is placed in a true vacuum (absence of substance) at the earth's surface, it will fall faster and faster toward the earth's center of mass. This uniform increase in the velocity with time represents an *acceleration*, and is a constant value. By careful measurement, we know that the velocity of fall will increase by 9.8 m per second for each second of fall. The *acceleration of gravity* is, then, 9.8 m/sec^2. The acceleration of gravity multiplied by the quantity of mass upon which it acts constitutes the *force of gravity*. Stated in a simple equation,

$$F_g = M \times g$$

where

F_g is the force of gravity
M is the quantity of mass

and

g is the acceleration of gravity

Because the acceleration of gravity is nearly constant everywhere at sea level, the force of gravity varies according to the quantity of mass. We can therefore measure the quantity of mass with reasonable accuracy by the force registered upon a scale of the type that measures the amount of stretching of a coil spring. The standard unit of force is the *newton* (N). For one kilogram mass, the force of gravity is equal to 9.8 N.

States of Matter

The physical condition, or state, in which we find matter at a given place and time is a subject of great importance in geology.

The three common states of matter—solid, liquid, and gaseous—apply both to pure substances (elements, compounds) and to mixtures. Using only the simplest concepts of atoms and compounds, as Dalton visualized them, we can describe the three states of matter in terms of observable behavior. For this purpose, the atoms or molecules that comprise matter can be visualized as uniform spheres, all physically alike.

FIGURE 2.4 The concept of density of matter is illustrated by cubes of four different substances suspended from a coil spring. The distance through which the spring is stretched is proportional to the density of the substance in g/cc.

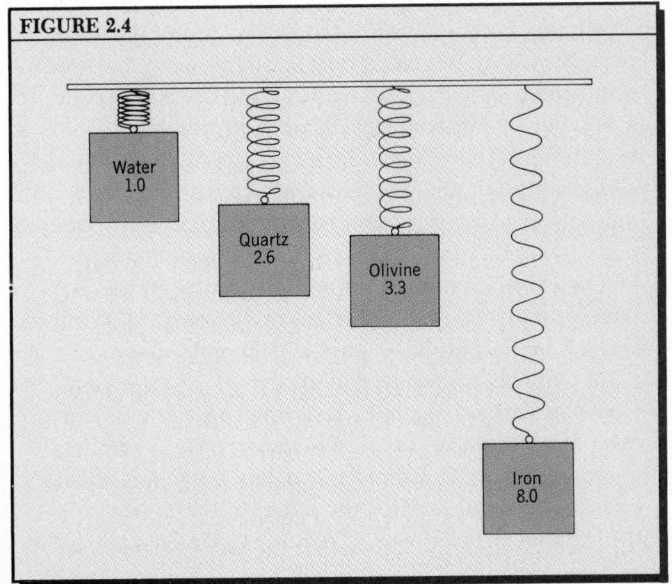

Certain states of matter are well known to everyone. Every day we drink water in its *liquid state*. We may also have close by a supply of ice cubes—water in the *solid state*. The *gaseous state* of water, called water vapor, cannot be seen, but is easily sensed by the human skin in summer when the relative humidity of the air is high.

To define these states further, a *gas* is a substance that expands easily and rapidly to fill any small, empty container. Atoms or molecules of the gas are in high-speed motion (Figure 2.5), and empty space between them is vast in comparison with their dimensions. The particles move in random directions and frequently collide. They rebound like perfect spheres at each impact, changing direction abruptly. They also strike and rebound off the walls of the container. A gas is usually very much less dense than a liquid or solid consisting of the same chemical substance. For example, the gaseous water vapor in warm, moist air has a density only about 1/100,000 that of liquid water.

A *liquid* is a substance that flows freely in response to unbalanced forces but maintains a free upper surface so long as it does not fill the container or cavity in which it is held. The molecules of a liquid compound—water, for example—move more or less freely past one another as individuals or small groups. For many practical purposes, liquids can be considered to be incompressible (not capable of being compressed). Under rather strong confining pressures (such as would exist at the bottom of the deep ocean), liquids are compressed only slightly into a smaller volume.

Both gases and liquids are classed as *fluids* because both substances freely flow. Put simply, both these substances flow downhill under the force of gravity wherever possible. Fluids of different density tend to come to rest in layers with the fluid of greatest density at the bottom and that of least density at the top.

A *solid* is a substance that resists changes of shape and volume. Solids are typically capable of withstanding large unbalanced forces (i.e., strong stresses) without yielding permanently, although they undergo a small amount of elastic bending. When yielding does occur, it is usually by sudden breakage, or rupture. Because most rock of the outermost layer of the earth is in the solid state, these principles have important applications in many areas of geology, including the behavior of lithospheric plates and the occurrence of earthquakes.

Changes of state are accompanied by either an energy input into or an energy output from the substance undergoing change. We are all familiar with this principle in the preparation of food. To boil water (change liquid to a vapor state), a great deal of heat must be applied; conversely, to freeze water, a great deal of heat must be removed. Our next step will be to examine the nature and forms of energy that can be easily observed to cause changes in pure substances and mixtures behaving as gases, liquids, and solids.

Kinds of Energy

We stated earlier that energy is commonly defined as the ability to do work. In mathematical terms of physics, energy is the product of force and distance:

$$E = F \times d$$

where

E is energy
F is the force applied

and

d is the distance through which the force acts

Thus, energy is the ability to move an object (exert a force) for a certain distance. Energy is stored and transported in a variety of ways. Some of the recognized forms of energy are: mechanical energy, heat energy, energy transmitted by radiation through space (electromagnetic energy), chemical energy, electrical energy, and nuclear energy.

Mechanical Energy

Mechanical energy—energy associated with the motion of matter—has two forms: kinetic energy and potential energy. *Kinetic energy* is the ability of a mass in motion to do work. Thus, an automobile traveling down a highway possesses kinetic energy because it is a mass in motion. Should this mass strike a telephone pole, its ability to do work upon its own body and upon the telephone pole will be quite obvious. The energy it will release in collision will increase with the weight (mass) of the car, and it will also increase with the square of the auto's speed. Kinetic energy, then, is proportional to the quantity of mass in motion multiplied by the square of its velocity. Stated as a simple equation,

$$E_k = M \times v^2$$

where

E_k is kinetic energy
M is quantity of mass in motion

and

v is velocity in a straight line

Kinetic energy is obvious in many forms of geologic processes acting at the earth's surface—a rolling boulder, a flowing stream, the pounding surf. Lithospheric plates in motion display kinetic energy. Although they seem to barely creep along, their mass is so enormous that the total kinetic energy in a single moving plate is very large indeed.

Potential energy, or energy of position, is equal to the kinetic energy an object would attain if it were allowed to fall under the influence of gravity. Imagine a brick balanced on the edge of a tabletop, then falling. The kinetic energy the brick possesses at the moment it hits the floor, as we have seen, is proportional to the mass of the brick multiplied by the square of its velocity. Because the brick is being accelerated by gravity at a constant 9.8 m/sec^2, we could find its velocity at impact by measuring the distance it will fall and using the acceleration formula from physics. If the brick is again lifted to the tabletop, the work done in lifting gives the block a renewed quantity of potential energy. This energy will be released when the brick is again allowed to fall.

It should be obvious at this point that the floor is merely a convenient stopping place for both the brick and our discussion; if we sawed a hole in the floor and allowed the brick to fall further, it would possess even more kinetic energy at its impact on the floor below. Therefore, with respect to the lower floor level, the brick would have a greater potential energy when we return it to the tabletop. Thus, the three factors which determine the magnitude of potential energy are the mass of the object, the vertical distance it will fall, and the acceleration of gravity.

Looking around us outdoors, we can spot many instances of the existence of potential energy in a landscape, for example, a boulder poised at the top of a steep mountain face. In fact, the entire mountain represents a large reservoir of potential energy judged in reference to the level of the floor of an adjacent valley. Going even further, an entire continent possesses potential energy with reference to sea level or to any lower level we might choose to use as a reference. The potential for any mass of rock, no matter how small or how large, to move toward the earth's center is always present. Geologists sometimes call this potential energy source "gravitational energy" and use it to explain a number of important earth phenomena. We should be careful at this point to note that gravitational force and gravity are not, in themselves, forms of energy.

Mechanical energy can be transmitted from one place to another in the form of *wave motion,* in which kinetic energy is carried through matter by an impulse passed along from one particle to the next. A sound wave is one example—a push on air molecules at one point will be transmitted outward in all directions. Geologists are particularly interested in the phenomenon of earthquake waves, which carry large amounts of energy for great distances, not only over the ground surface, but also in paths deep within the solid earth; the familiar Richter scale of earthquake intensity measures the quantity of energy released by an earthquake. In all mechanical forms of wave motion matter is displaced (up and down, sidewise, or forward and backward) in a rhythmic manner, but as frictional resistance within the moving substance withdraws energy, the waves gradually die out as they travel farther from the source.

Heat Energy

Sensible heat is another form of energy of paramount importance in geologic processes. Kinetic energy can readily be converted into sensible heat energy through the mechanism of friction. Braking a moving automobile is a familiar example. As the automobile slows to a stop (losing kinetic energy), the brake drums become intensely hot. What is probably not apparent, however, is just how much heat is related to how much motion. Laws of physics tell us that "a little goes a long way." The energy required to heat one cubic centimeter of water one degree Celsius is about the same as that contained in a mass of nearly 100 metric tons moving at a speed of one centimeter per second.

FIGURE 2.5 Some properties of gases, liquids, and solids.

Sensible heat represents kinetic energy, but it is of an internal form, rather than the external form we see in moving masses. Thus, a cupful of water resting completely motionless on the table has internal energy because of constant motion of the water molecules on a scale too small to be visible. This internal motion is the sensible heat of the substance, and its level of intensity is measured by the thermometer. For gases, the internal motion is in the form of high-speed travel of free molecules in space but with frequent collisions with other molecules. The energy level in the gaseous state of water (water vapor) is thus higher than within the liquid water. Ice, on the other hand, represents a lower energy level than liquid water, for here the molecules are locked into place in a fixed geometrical arrangement (see Figure 2.5). For these molecules in the solid state, the motion is one of vibration without relative motion.

When ice melts, work must be done to overcome the crystalline bonds between molecules. This work requires an input of energy, but it does not raise the temperature of the substance. Instead, the energy seems mysteriously to disappear. Since energy cannot actually be lost, it is placed in storage in a form known as *latent heat*. Should the water freeze and again become ice, the latent heat will be released as sensible

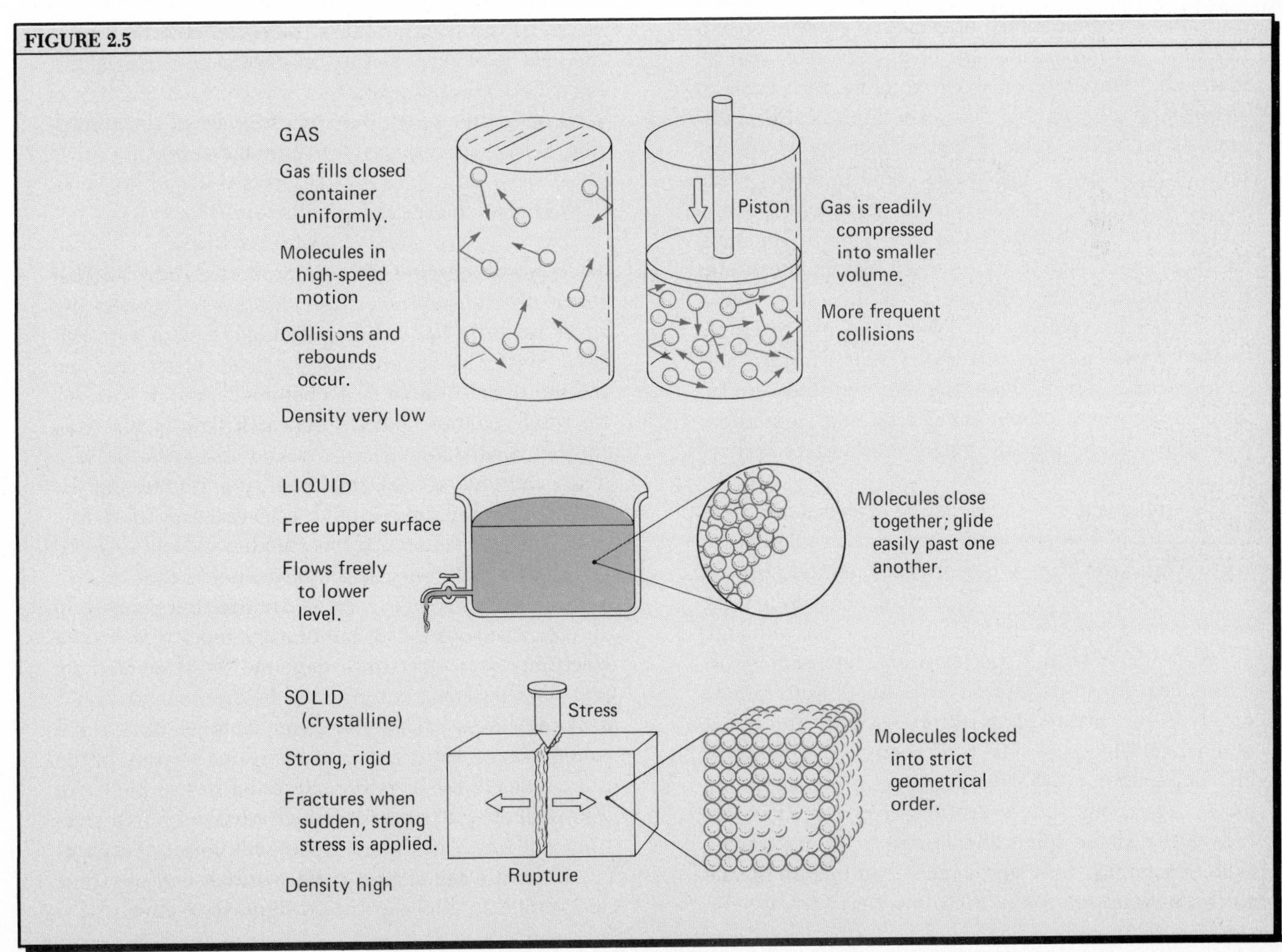

heat. A similar transformation from sensible to latent heat takes place during *evaporation*, when a liquid becomes a gas, because work must be done to overcome the bonds between molecules of the liquid. When water vapor returns to the liquid state, a process of *condensation*, latent heat is released as sensible heat.

Both sensible heat and latent heat are forms of stored energy. (Potential energy is also a form of stored energy.) Sensible heat in rock is important to geologists because the physical properties of rock change as temperature increases. Generally speaking, as rock is heated, it expands and becomes less dense. Heated to a critical temperature—its melting point—rock changes state from a solid to a liquid. Heat energy is largely responsible for the motions of lithospheric plates and the building of mountain ranges associated with those plate motions. One crucial question we must investigate is the source of the great reservoir of heat in the solid earth. Is it heat left over from some early event in the history of our planet, or is it heat being continually generated in some mysterious way?

Electromagnetic Energy

Sensible heat may be lost directly to the surroundings through conduction, but even in a vacuum objects lose heat. A fundamental law of physics states that all matter at temperatures above absolute zero radiates *electromagnetic energy*. We can think of this radiation as taking the form of waves traveling in straight lines through space. The waves come in a very wide range of lengths, but all travel at the same speed—300,000 km/sec—regardless of their length. Together, the total assemblage of waves of all lengths constitutes the *electromagnetic spectrum*. It includes *visible light*, with its rainbow of colors, and also invisible shorter waves such as *ultraviolet rays*, *X rays*, and *gamma rays*. Besides these, the spectrum includes invisible long waves known as *infrared rays* (sometimes called heat rays), and still longer microwaves and radio waves.

The temperature of the radiating objects determines which part of the spectrum carries the radiated energy. Thus, the sun, whose surface temperature is about three times greater than that of molten steel, radiates most strongly in the visible portion, although all other parts of the spectrum are represented. At normal earth temperatures, however, objects radiate mostly in the infrared part of the spectrum and not at all in the visible part. The total energy of radiation is also dependent on the temperature of the radiating object. Thus, one square centimeter of the sun's surface emits about 1.5 million times as much energy each second as does one square centimeter of the earth's surface.

Electromagnetic energy received from the sun powers a group of important geologic processes constantly at work on the earth's surface. Electromagnetic energy, upon arriving at the earth's surface, is continually converted into mechanical energy and sensible heat, which in turn are consumed in the breakup of exposed rock, in causing its chemical change, and in the transport of the resulting rock particles to new places of rest.

Chemical Energy

Yet another form of energy is *chemical energy*, absorbed or released by matter when chemical reactions take place. These reactions involve the coming together of atoms to form molecules, the recombining of molecules into new compounds, and the reversion to simpler forms of matter.

A familiar example that relates indirectly to certain geologic processes can be felt in the setting of wet cement or plaster. When water is added to these substances, they spontaneously harden into a rocklike mass. At the same time, you will notice that the mixture becomes quite warm to the touch. A chemical reaction is occurring, involving the liberation of heat stored as energy in the reacting chemicals. In other types of reactions, the material mixture becomes colder as the reaction takes place, showing that sensible heat is being taken up in storage as chemical energy. Geologic processes that involve such changes in mineral matter provide many examples of the absorption, release, and storage of chemical energy.

Electrical Energy

Chemical energy can be converted into another form of energy—*electrical energy*—a process we make use of in the ordinary flashlight cell or auto battery. Within the auto battery, lead plates are immersed in sulfuric acid. A chemical reaction between the acid and the plates produces a flow of electrons through a wire connected to the two battery terminals. Thus, the flow of electrical energy is represented by the flow of electrons through a conducting substance, usually a metal. Later in this chapter, we will describe the electron as a very small particle of matter.

Electrical energy can also be stored temporarily in an *electrical charge*. A familiar example is the static electricity you sometimes generate on your clothing and bodies as you walk across a carpeted floor on a cold, dry day. When you touch another object, the stored electricity is released in a small spark, often accompanied by a rather unpleasant sensation. An electrical charge consists of vast numbers of free electrons collected on the surface of an object. Accumulated charges are of two kinds, positive and negative, designated by plus and minus signs, respectively. It is important to note that objects having charges of the opposite sign (+ −) tend to be drawn together by an attracting force, whereas objects carrying charges of the same sign (+ + or − −) tend to be pushed apart by a repelling force. These concepts will be useful to recall as we investigate the nature of atoms.

Nuclear Energy

Although this list of forms of energy is not complete, we close it with *nuclear energy,* an important subject in geology. To explain the nature of nuclear energy requires a knowledge of the internal structure of the atom and the working of its component particles. Nuclear energy is a subject we will develop quite thoroughly on later pages, because it is generally believed that most of the earth's internal heat originates as nuclear energy. We can only point out here that the internal atomic structure of certain elements present in rock is unstable. These unstable elements spontaneously change into other elements, at the same time releasing energy to surrounding substances. This process of atomic change is called radioactivity, and it involves the steady conversion of matter into energy.

Flow Systems of Energy and Matter

Understanding the varied processes that form rocks and that shape those rocks into many different configurations requires geologists to think in terms of *flow systems* of both matter and energy. A flow system is simply a series of connected pathways through which either energy or matter moves more or less continuously. An *energy system* traces the flow of energy from a point of entry to a point of emergence. As energy flows through such a system, it may change form many times and may be detained temporarily in storage from place to place. In this process, the energy flow makes use of matter as the medium of motion and of storage.

For example, planet Earth can be regarded as an energy system receiving its energy input in the form of electromagnetic radiation from the sun (Figure 2.6). As solar energy falls on the earth's surface, some is reflected directly back to space, but the remainder is absorbed by the earth's atmosphere, oceans, and land surfaces. This absorbed energy is now in the form of sensible heat, which, in turn, may be transformed into kinetic energy of moving air, water, and rock particles. As friction acts on such motions, the kinetic energy is transformed back into sensible heat, and finally back to electromagnetic radiation, which streams outward into space and is lost forever. We refer to this total earth system as an *open system* because energy continually enters and leaves it.

Energy flow systems set in motion and sustain *material flow systems* (flow systems of matter). The matter involved in such systems is not only transported from place to place along certain pathways, but it can also undergo both changes of state and chemical changes. The matter traveling through the system can also be held temporarily in storage at certain points. A glacier is a simple example of a material flow system. Resting on the floor of a deep mountain canyon (Figure 2.7), the glacier receives solid nourishment from the atmosphere in the form of snow, which compacts into ice. Gradually the great ice mass flows downvalley to lower levels, where the ice disappears by melting or evaporation (change of state). The glacier is an open

FIGURE 2.6 A schematic diagram of the sun–earth–space energy system. Energy is produced within the sun by hydrogen fusion and is radiated into space from the sun's surface as shortwave electromagnetic radiation. Upon being intercepted by the earth, some energy is reflected directly back into space; the remainder is absorbed and stored as sensible heat by the earth's atmosphere, oceans, and lands. The earth loses energy to space as long wave radiation, completing the system. (From A. N. Strahler, *Jour. of Geography,* vol. 69, no. 2, p. 72. © 1970 by *The Journal of Geography;* reproduced by permission.)

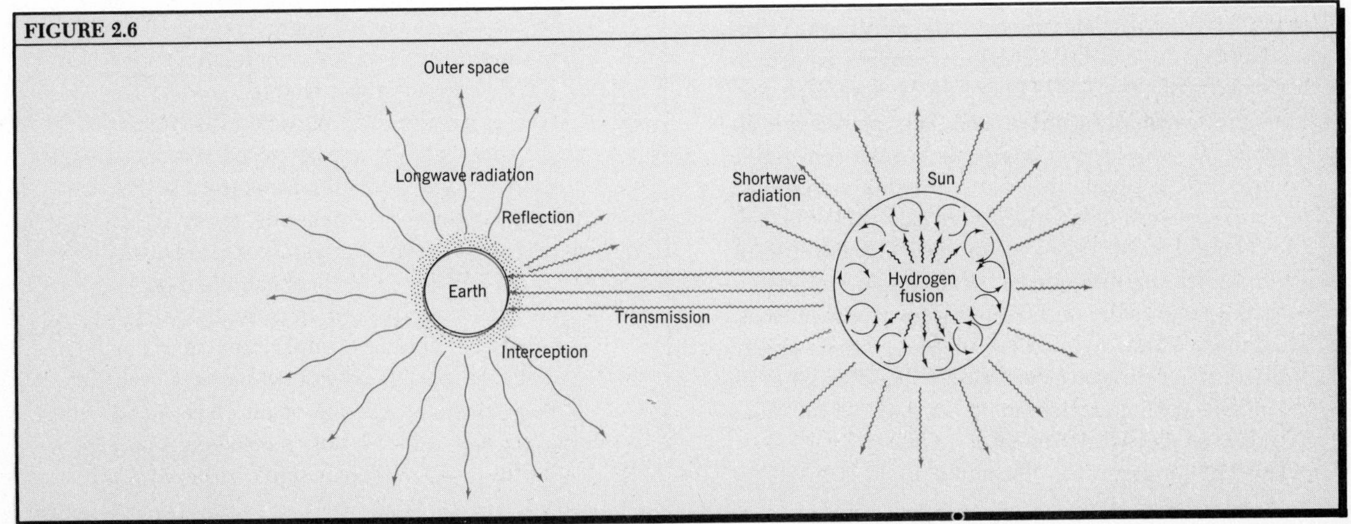

flow system because matter both enters and leaves. Some flow systems are said to be *closed systems*, because all, or nearly all, of the matter remains within the system and is simply recirculated (recycled).

The concept of flow systems of energy and matter will appear a number of times in later chapters of this book. As each system is explained, your understanding of natural systems will deepen and your appreciation of our earth as a dynamic planet will heighten.

The Atomic Structure of Matter

In the remainder of this chapter, we move to a different perspective, or vantage point, to take a much closer look at the nature of matter. As Professor Holum explains it (Figure 2.2), we are "crossing into the realm where individual particles cannot be directly seen and weighed." Even today, with all the sophisticated equipment of the scientific laboratory, we cannot take a photograph of a single individual atom.

How small is an atom? The diameter of a typical atom is on the order of one ten-billionth of a meter (10^{-10} m), which can surely have no meaning to a human being. All present knowledge about the internal structure of an atom and the arrangement of atoms into orderly groupings has had to be obtained by indirect methods using vast numbers of atoms at one time. For example, X rays fired into a mineral crystal are deflected by massed atoms in such a way that the arrangement of those atoms can be calculated from a photographic image of the scattered rays.

Structure of the Atom

The atom consists of a dense but very small central *nucleus* and one or more *electrons* moving at high speed in complex paths enveloping the nucleus. Nearly all the atom's mass is in its nucleus, which is extremely dense. Because the nucleus occupies only about one-trillionth of the volume of the atom, most of the atom is just empty space.

The nucleus of any atom (except hydrogen) consists of two kinds of particles: *protons* and *neutrons*. A proton has a positive electric charge, whereas a neutron is not electrically charged. Otherwise, protons and neutrons are essentially similar, and their masses are almost equal. An electron is a particle with a negative electric charge—opposite, that is, to the charge of a proton. The mass of an electron is extremely small—only about 1/2000 that of a proton. The number of electrons orbiting the nucleus of a normal atom is always the same as the number of protons in its nucleus. Therefore, within a stable atom there are as many positive as negative electric charges. The two kinds of charges balance or cancel each other out, and the atom is said to be electrically neutral.

The last paragraph is illustrated in Figure 2.8 by

FIGURE 2.7 Schematic diagram of a valley glacier as an open material flow system. The system input is in the form of snow in the upper part of the glacier. Compacted snow layers are transformed into glacial ice, which flows slowly downhill. The output of the system is in the lower part of the glacier, where the ice disappears by melting and surface evaporation. The system can achieve a balance, or steady state, in which the dimensions of the glacier hold nearly constant over many years.

FIGURE 2.8 A simplified model of the atom, showing the atomic structure of the first four elements. Called the Bohr model, after Neils Bohr, a Danish physicist, who proposed this concept in 1913, the model is useful in making simple calculations about the physical properties and behavior of atoms. Bohr's great contribution was to show that the energy level increases outward in quantum steps from one electron shell to the next.

FIGURE 2.7

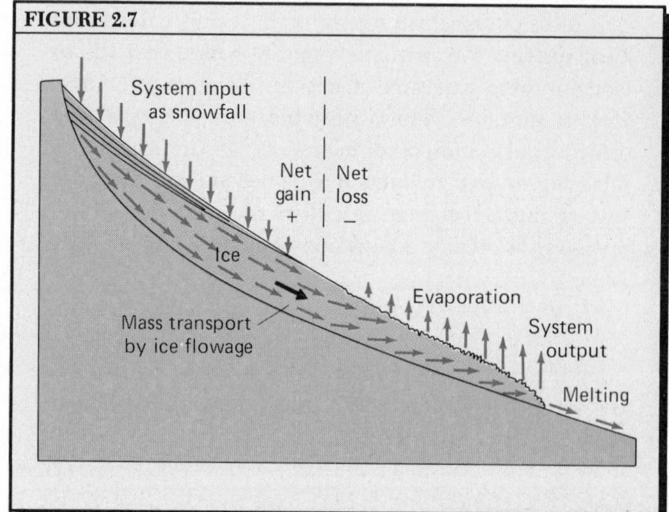

FIGURE 2.8

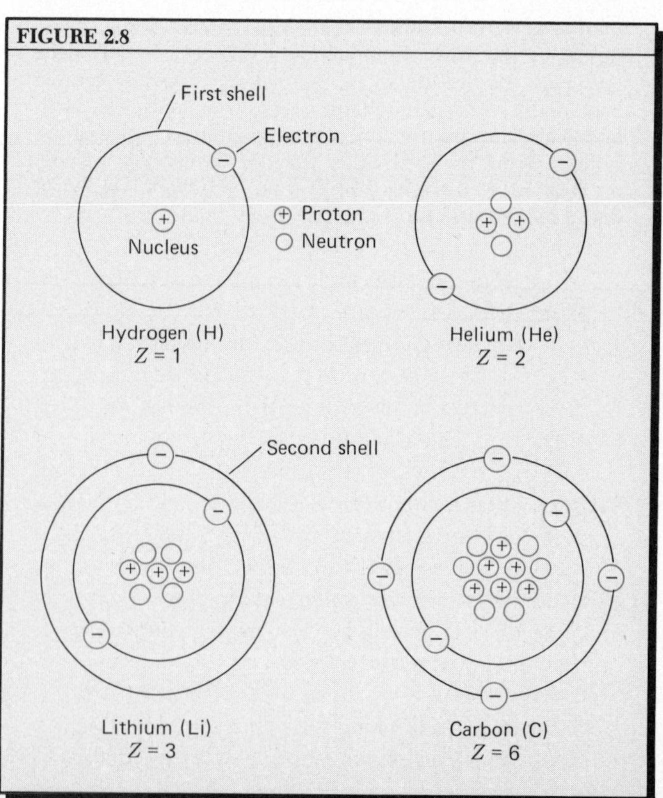

schematic diagrams of atoms of four different elements: hydrogen, helium, lithium, and carbon. Each electron has its own path configuration, called an *orbital*, with a certain specified distance, or radius, from the nucleus. We have shown the electrons to be orbiting the nucleus in more or less circular paths (only in one plane) much as the planets orbit our sun, or the moon orbits the earth. In reality, electrons deviate greatly from such simple paths, so that the actual collection of paths of a single electron describes a space occupying three dimensions.

Hydrogen (symbol H) is the simplest of the elements in terms of the numbers of particles; it has a nucleus consisting of only one proton (one positive electric charge) which is orbited by only one electron (one negative charge). To visualize the actual path followed by the single electron we might imagine a small, fast-flying insect incessantly circling a central object (the nucleus). Imagine that the insect carries an intense light source the size of a pinpoint, and that in the total darkness of a closed room we photograph its trail of light by leaving a camera shutter open for a long time. As the pinpoint light source moves around the nucleus, it leaves an image in the form of a line. The travel path varies in distance from the nucleus, being sometimes very close and sometimes far out.

Our photograph would show the innumerable flight circuits fused into a disk of light grading in tone from brightest at the center to darkest at the outer edge, where it would merge with darkness. In three dimensions we can picture a spherical ball of light grading inward to its brightest intensity closest to the nucleus. The probability of the electron being at a particular location in the light ball would be proportional to the light intensity. The imaginary ball of light is called an "electron cloud," but there is only one electron. In Figure 2.8, the orbital of the hydrogen atom is represented by a single circle. Keep in mind that this is only a graphic device to simplify the business of showing that atoms can be bonded together through their electrons.

Helium (He) has two protons, two neutrons, and two electrons. Notice in Figure 2.8 that both hydrogen and helium have a single orbital, whereas lithium (Li) and carbon (C), have two orbitals, one inside the other. Multiple orbitals of electrons are organized into concentric zones, called *shells*, and these are numbered consecutively outward (from one to seven). Within each shell, the number of electrons that can be present is limited. The innermost shell (shell one) can hold only two electrons; shell two can hold eight electrons. The concept of electron shells, which we have greatly oversimplified here, will be most useful in understanding how atoms of different elements combine into stable groups.

The atom of each element is uniquely described by its *atomic number*, which is the number of protons in its nucleus. Thus, all the 103 known elements can be arrayed in order of consecutive atomic numbers from one through 103. Atomic numbers of the four elements shown in Figure 2.8, designated by the capital letter Z, are as follows:

hydrogen	(H)	1
helium	(He)	2
lithium	(Li)	3
carbon	(C)	6

The total number of protons and neutrons within an atom is called the *mass number*, symbol *A*. In three of the four examples shown in Figure 2.8, the mass number is exactly twice the atomic number, because each of these atoms has the same number of neutrons as protons.

Isotopes

Our next step is to take note of the fact that for a given element the number of neutrons may be either fewer or greater than the number of protons. Because neutrons have no electric charge, such differences do not affect the balance of charges between protons and electrons, but only the mass of the atom; its mass number is changed accordingly. Figure 2.9 illustrates three forms of the element hydrogen, each with a different nucleus. While the number of protons (atomic number, Z) remains 1, the number of neutrons is shown to be 0, 1, and 2, from left to right, and the mass numbers are $A = 1$, $A = 2$, $A = 3$. Also shown are three forms of the element carbon. The carbon atom always contains six protons ($Z = 6$), but its number of neutrons can vary. We have shown three cases: $A = 12$, $A = 13$, and $A = 14$. The varied forms of atoms shown for each element are called *isotopes*. Each isotope of a given element has a different number of neutrons, and hence a different mass number, from other isotopes of the same element.

The number of isotopes that exist for a given element varies greatly. There are, however, only three isotopes of hydrogen. Common hydrogen ($A = 1$) makes up over 99.98% of all hydrogen atoms occurring in nature, while the two other forms, deuterium and tritium, are very rare. Tritium, which is produced artificially, is extremely unstable and rapidly disintegrates. Carbon has six isotopes, but of these, Carbon-12, with its equal numbers of protons and neutrons, makes up 98.89% of all naturally occurring carbon atoms.

Why, then, should we bother to explain what an isotope is? Geologists make use of isotopes to determine the age in years of certain mineral or organic substances. The isotope Carbon-14, although extremely rare, allows us to determine the age in years of

Table 2.1 Energy Levels in Atoms

Energy level, or quantum number	Name of shell	Number of electrons: Needed to fill the outermost shell	Number of electrons: Held in all shells including the outermost, when all spaces are full
1	*K*	2	
2	*L*	8	10
3	*M*	8	18
4	*N*	8	36
5	*O*	8	54

many substances ranging from a few centuries old to about 40,000 years old. Isotopes of the elements uranium, thorium, and lead allow us to determine the ages of the oldest rocks on earth—between 3 and 4 billion years old. In Chapter 7 we will explain how isotopes are used for this purpose.

Energy Levels in the Atom

Let us return to the arrangement of shells of electrons within an atom. The numbers one through seven we used refer to energy levels. An electron possesses kinetic energy because of its high speed of travel. Measured outward from one shell to the next, the energy possessed by an electron steps upward to higher values. The increase in energy from one level to the next is referred to as a *quantum* increase; thus, the integer numbers one through seven are called *quantum numbers.* If an electron were to be moved from one shell to the next adjacent shell it would need to gain or lose a certain quantity (or quantum) of energy. In that case, the atom would need to import or export a certain quantity of energy.

While the numbers one through seven designate the energy levels of the electron shells, each shell is named with a capital letter, *K, L, M, N, O,* etc. Table 2.1 shows the numbers of electrons needed to fill a shell when it is the outermost shell. The final column gives the total number of electrons held in all shells, when the outermost is filled. (The reason the final number is not equal to the sum of the numbers needed to fill the outermost shell is that, once full, some additional electrons can be introduced into orbitals of the earlier-formed shells.)

Activity of Elements

Let us now assemble our information about energy levels, shells, and electrons into a two-by-two table, as shown in Figure 2.10. The table is limited to energy levels one through four (shells *K, L, M,* and *N*), but these include the eight most abundant elements in the earth's crust as well as the important elements making up the atmosphere, the ocean and its dissolved salt, and pure water (a compound of hydrogen and oxygen). A few rather rare elements (gallium, germa-

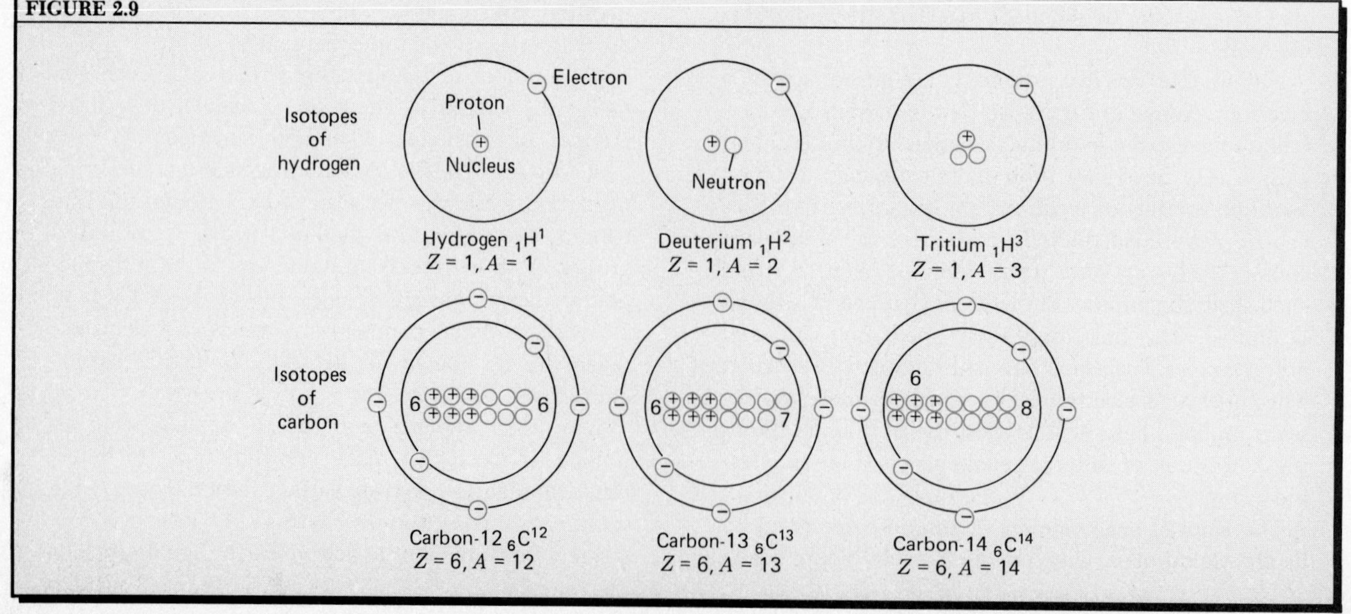

FIGURE 2.9 The concept of isotopes illustrated by three forms of the hydrogen atom and three forms of the carbon atom.

nium) are included to fill the blanks in the table. Our goal is to explain how different elements can combine into compounds, and why this event occurs readily with some pairs of elements, but not with others.

First, we note that the last column (extreme right) contains three elements—neon, argon, and krypton—which have a full complement of eight electrons (an *octet*) in the outer shell. When the octet is complete, the atom is extremely stable. The same can be said of helium, with two electrons in the *K*-shell. All four elements occur as gases; they belong to a group called the *noble gases*. Because none of the four is known to form a compound with any other element, they are described as being chemically *inert*.

Next, we look at the column of elements at the extreme left. Hydrogen, lithium, sodium, and potassium have only one electron in the outer shell. They are highly reactive elements because they readily lose their outer-shell electron to another element in order to achieve a stable condition. In the case of lithium, sodium, and potassium, loss of an electron leaves the next lower shell complete. For example, with potassium the *N* shell disappears, leaving the *M* shell with its octet complete. Loss of an electron from the *L* shell of lithium dispenses with that shell and leaves only the *K* shell of two electrons, which is stable. The case of hydrogen is somewhat different; loss of its electron leaves only the nucleus of the atom, which is a single proton.

When an atom from the first column (hydrogen, lithium, sodium, and potassium) loses an electron, it is left with a single positive electrical charge. Bear this fact in mind as we go further.

Look now at the elements in the next-to-last column: fluorine, chlorine, bromine. They have seven electrons in the outer shell, one less than a full octet. They, too, are highly reactive elements because they can easily gain a single electron to achieve a full octet. If this gain of one electron occurs, the atom acquires a single negative electric charge. If you have not already guessed, an atom of the first column, upon meeting an atom of the next-to-last column, should immediately pass an electron to it.

Ions

Actually, atoms that have lost or gained an electron of the outer shell are no longer atoms; they are *ions*. (An atom must have a neutral charge; thus, the number of electrons must equal the number of protons.) An ion with positive electrical charge is called a *cation* and is designated by use of a positive sign superscript next to the element symbol, thus:

Sodium atom	**Sodium cation**
Na	Na^+

FIGURE 2.10 This simple table of elements is useful in predicting how elements combine into compounds. The first 20 elements are shown in unbroken succession, followed by several elements needed to fill the fourth energy level. The eight most abundant elements in the earth's crust are shown with a color tint. Elements that are metals are shown with a gray overprint.

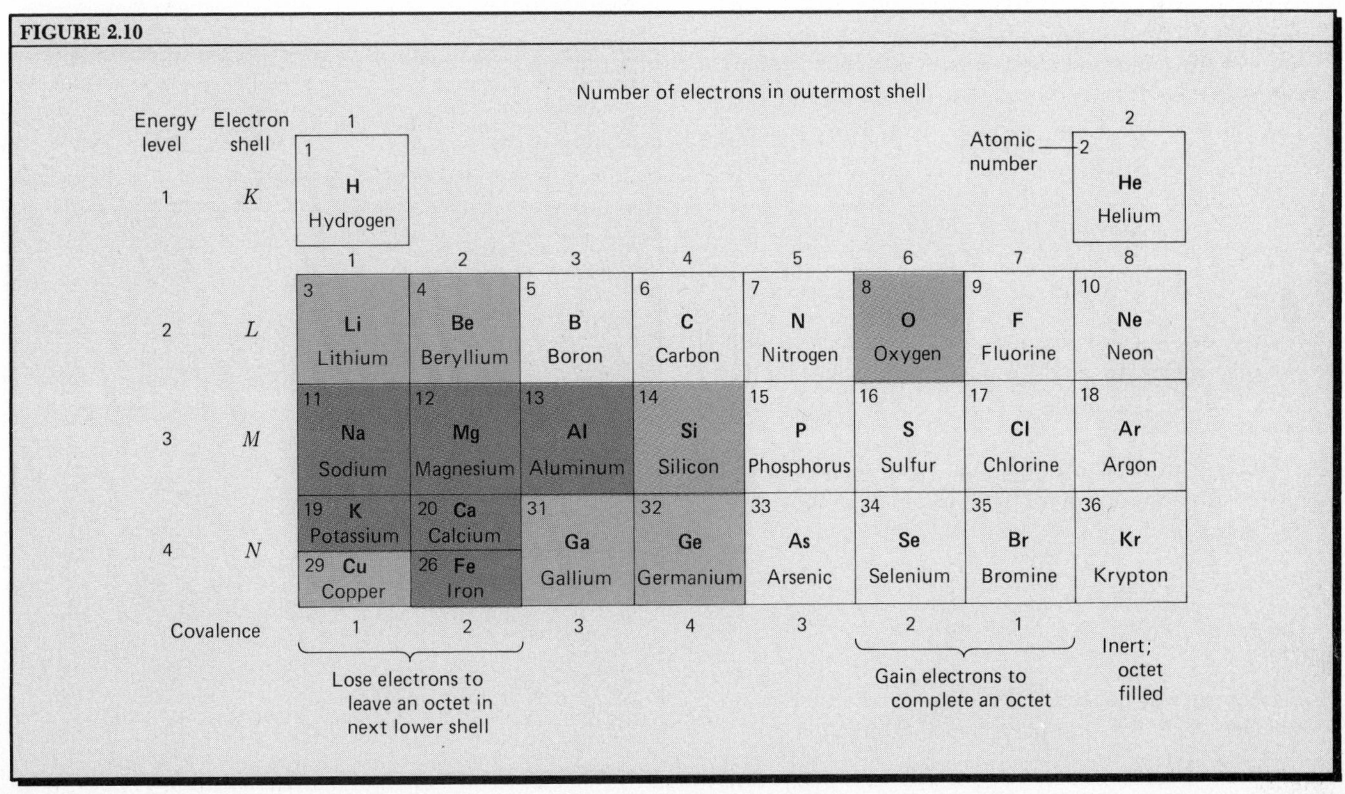

An ion with negative electric charge is called an *anion*, and is designated by a negative sign. In naming the anions, the name of the element is changed to end in "ide." Thus, the anion of chlorine is a chloride ion:

Chlorine atom	**Anion of chlorine (chloride ion)**
Cl	Cl^-

Ionic Compounds

This brings us to our first type of compound. As shown in Figure 2.11, the transfer of one electron from an atom of sodium to an atom of chlorine leaves two ions of stable configuration, each with a filled octet in the outer shell. Recall that opposite electrical charges attract one another. Therefore the transfer of an electron, which leaves ions of opposite charge in close contact, causes a strong *chemical bond* between the ions. This particular type of bond is called an *ionic bond.*

The substance produced by electron transfer is called an *ionic compound.* In the example we have used, the compound is sodium chloride, with the formula NaCl. This compound is ordinary table salt; its mineral name is halite. In Chapter 3 we will investigate the arrangement of ions in a crystal of halite. Because the sodium and chloride ions each have a single electrical charge, the compound NaCl will always be formed from equal numbers of atoms of each element.

Returning to our table (Figure 2.10), consider the elements of the second column, those with two electrons in the outermost shell. These elements can lose two electrons to leave the next lower shell complete. Therefore, they form cations with two positive charges, for example:

Magnesium ion	Mg^{2+}
Calcium ion	Ca^{2+}

If magnesium cations are to form ionic bonds with chloride anions, to produce magnesium chloride, the ratio must be two of the latter to one of the former: $MgCl_2$.

Atoms with six electrons in the outer shell (oxygen, sulfur) can also form ions; these would be anions with two negative charges:

Oxide ion	O^{2-}
Sulfide ion	S^{2-}

Cations can also be formed by loss of three electrons in the outer shell. An example is the aluminum ion, Al^{3+}.

FIGURE 2.11 Atoms of sodium and chlorine become ions with the loss and gain, respectively, of one electron from the outer shell. Their combination forms an ionic compound held by an ionic bond.

FIGURE 2.12 Electron sharing, found in the covalent bond, results in the formation of molecules of a compound. Electrons are shared in pairs, completing the octet of electrons in the outer shells of all atoms in the molecule.

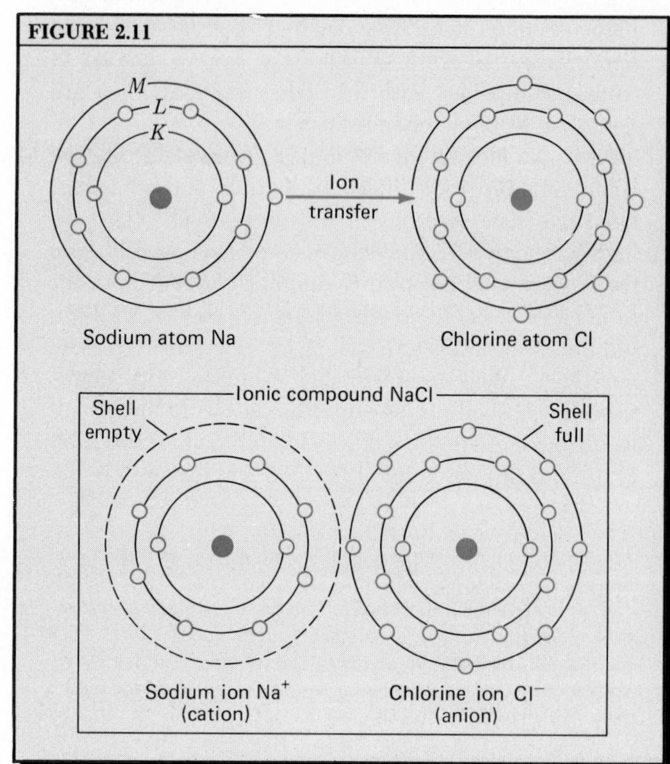

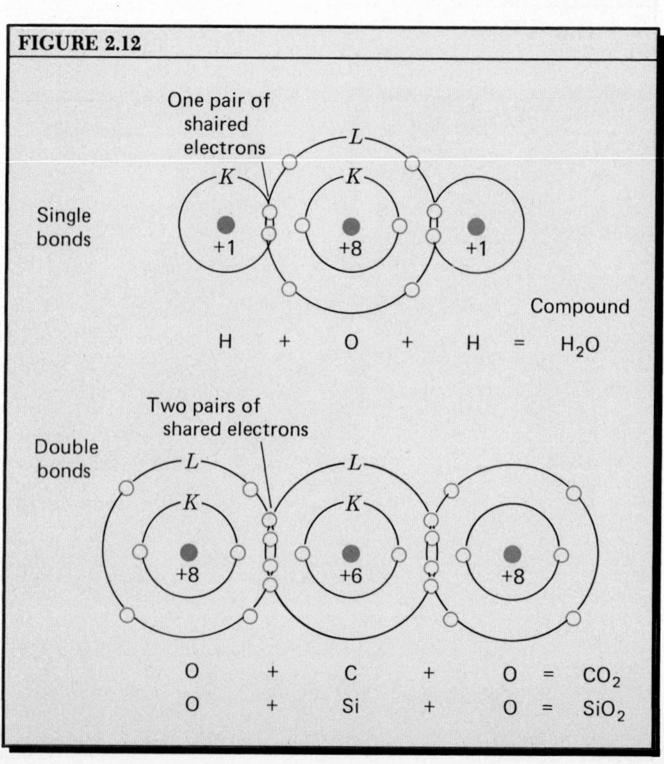

Electron Sharing

Atoms with four and five electrons in the outer shell do not readily form ions. Important examples are carbon (C) and silicon (Si). They form compounds by sharing electrons with other atoms, rather than by ion transfer; the process is called *electron sharing.*

Figure 2.12 shows how two single bonds join an atom of oxygen (O) with two atoms of hydrogen (H) to form the water molecule, H_2O. Each hydrogen atom shares two electrons with the oxygen atom, giving hydrogen a complete pair in its *K* shell.

The bonds that hold atoms together when electrons are shared are called *covalent bonds* to distinguish them from ionic and other kinds of bonds. The three atoms bonded in this way form a molecule (H_2O) which, as a whole, is electrically neutral. (You should count the electrons in the molecule and make sure that their number equals the sum of the positive charges in the three atomic nuclei.)

Figure 2.12 also shows how the common compound carbon dioxide, CO_2, can be formed when each of two oxygen atoms shares four of its electrons in the *L* shell with one carbon atom (C), which needs four more electrons in its *L* shell to complete an octet. We can also use the same diagram for a molecule of the compound silicon dioxide, which is familiar as the common mineral quartz.

FIGURE 2.13 Two ions of the same element are joined by a covalent bond to form a molecule. The three molecules shown are normally in the gaseous state.

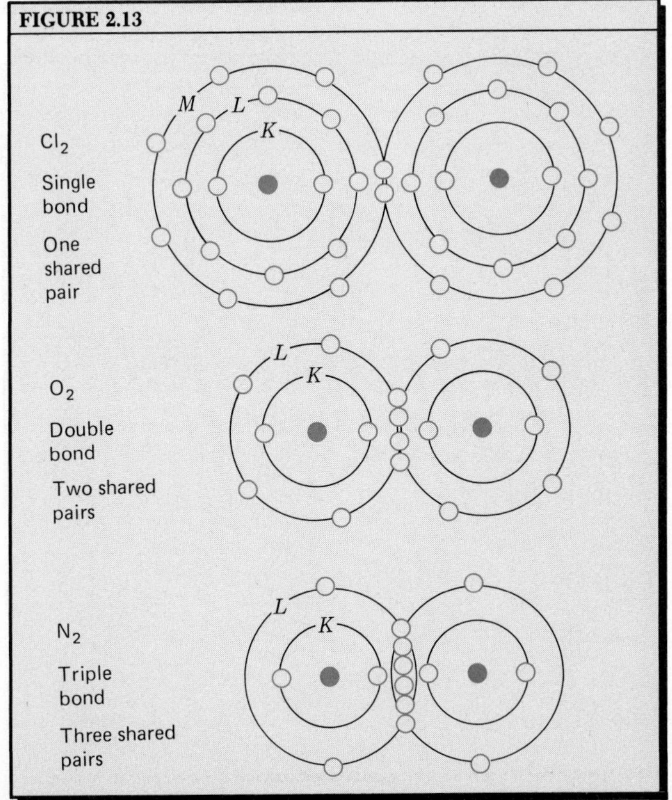

Several elements listed in Figure 2.10 normally exist in the gaseous state at temperatures usually found near the earth's surface. Two of these, oxygen (O) and nitrogen (N), form 99% of the atmosphere. The three elements with seven electrons in the outer shell—fluorine (F), chlorine (Cl), and bromine (Br)—are also normally in the gaseous state. These elements occur as molecules because they are unstable as individual atoms.

When two atoms of one of these elements are paired into a single molecule, the atoms share as many electrons as necessary to complete the octets of both (Figure 2.13). For example, chlorine needs one missing electron, and thus two chlorine atoms joined in a molecule share an electron, completing both octets, and creating one covalent bond. The chlorine atom is given by the formula Cl_2. Molecules of oxygen and nitrogen have the formulas O_2 and N_2, respectively requiring two and three covalent bonds.

Chemical Reactions

The *chemical formula* for a compound expresses by means of the element symbols the combination of the two or more elements it contains. For example, the formula for the compound iron sulfide is FeS. A knowledge of the number of electrons in the outermost electron shell will allow you to judge the proportions in which the component elements will combine. Those electrons that can be gained, lost, or shared are called *valence electrons*. Consulting Figure 2.10, we find that iron (Fe) has two valence electrons in its outer shell. This number of electrons is called the *covalence* of the atom. With a covalence of two, the Fe atom can form two covalent bonds. Sulfur (S), with six electrons in its outer shell, can gain two electrons, so its covalence is also two. In reacting to form a compound, Fe and S will combine in a one-to-one relationship, sharing two covalent bonds. Therefore, the *chemical equation* describing this *chemical reaction* is written

$$Fe + S \rightarrow FeS$$

The arrow indicates that the reaction takes place in the direction from left to right.

A second example is the reaction of aluminum (Al) with sulfur (S) to form the compound aluminum sulfide, Al_2S_3. From Figure 2.10, we find that the aluminum atom has three valence electrons, and these can be lost; its covalence is three. We already know that the covalence of sulfur is two. It is obvious that three atoms of sulfur are needed to form a compound with two atoms of aluminum, and a total of six covalent bonds is needed. The equation is written

$$2Al + 3S \rightarrow Al_2S_3$$

Following Dalton's principles, we must always state the numbers of atoms as whole numbers.

Polyatomic Ions

The ions we have described so far consist of only a single element. Another important class of ions exists, the *polyatomic ions,* in which atoms of two or three different elements are linked by covalent bonds. Unlike a molecule, which has a neutral charge, the polyatomic ion is charged positively or negatively, as is an ion of a single element.

Figure 2.14 shows two polyatomic ions important in geology, the sulfate ion and the carbonate ion. In the sulfate ion, SO_4^{2-}, each of four oxygen atoms has a covalent bond with the one sulfur atom. The entire ion contains 50 electrons (negative charges) and 48 protons (positive charges), so that the ion bears a double negative charge. The carbonate ion, CO_3^{2-}, consists of three oxygen atoms linked by covalent bonds to a single carbon atom. Two of the three bonds are single; the third is a double bond.

Figure 2.14 also shows how a polyatomic ion can join with another ion of opposite charge to form an ionic compound. The metallic ion of calcium, Ca^{2+}, has an empty *M* shell and a full *L* shell. All outer shells of the polyatomic ion are filled. Because both ions have double electric charges, they will combine in a one-to-one ratio to form calcium sulfate, $CaSO_4$, and calcium carbonate, $CaCO_3$. Calcium sulfate occurs in nature as the mineral anhydrite; calcium carbonate, as the mineral calcite. Both are very common minerals.

Notice that the reactions shown in Figure 2.14 use two arrows, pointing in opposite directions. This device means that we are dealing with a ***reversible reac-***

FIGURE 2.14 Two common polyatomic ions are shown here: the sulfate ion and the carbonate ion. They readily form ionic bonds with the calcium ion to make up ionic compounds abundant in one major class of rocks.

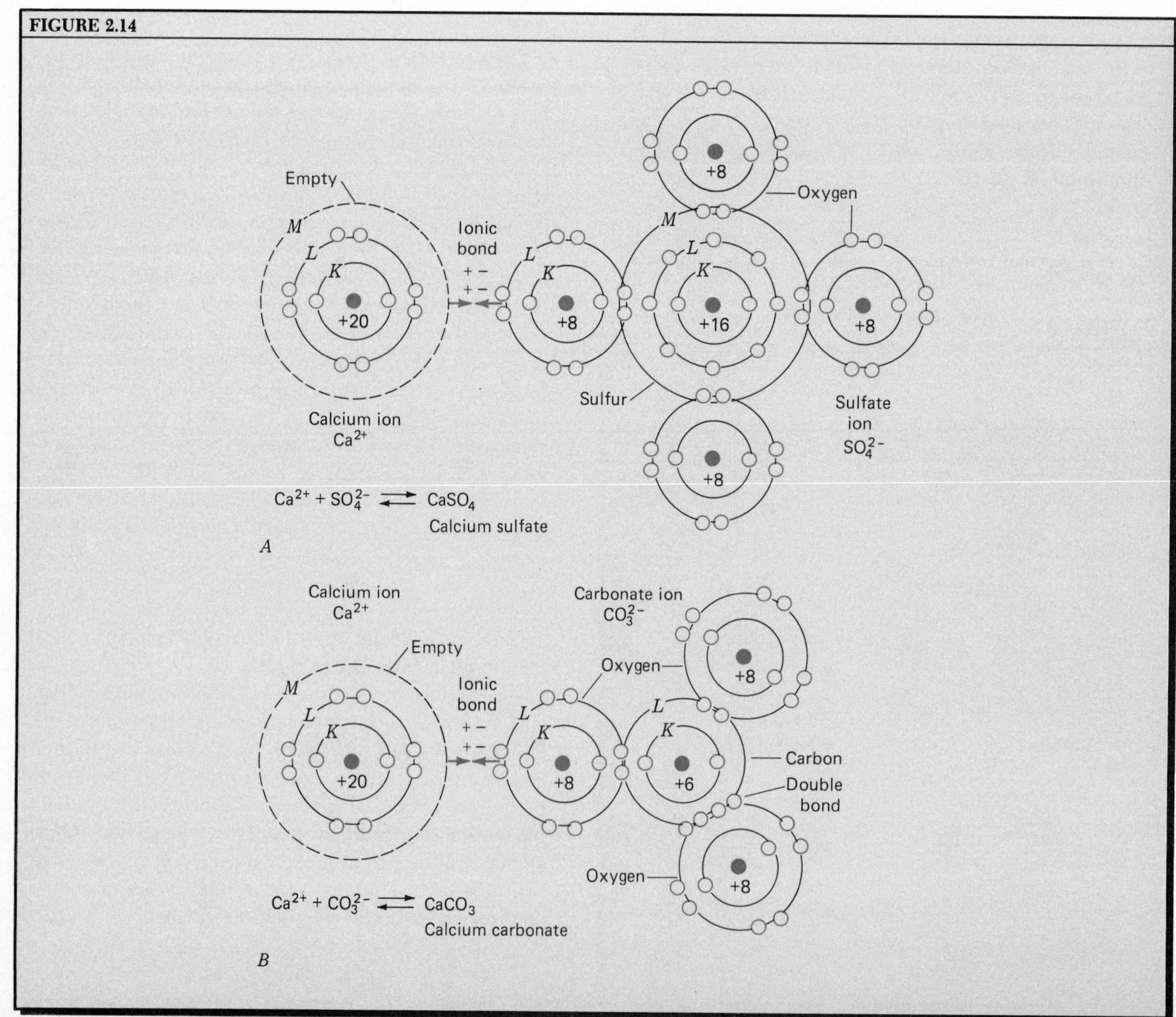

tion, and that the compound on the right can readily separate into the component ions. This separation occurs when the mineral is dissolved in water.

Polar Molecules

Although the molecules we have described are electrically neutral, with equal numbers of positive and negative charges inside, those two kinds of charges may tend to concentrate at opposite ends of the molecule. Thus, at one place on the outer surface of the molecule is a positive charge and at another place on the surface, a negative charge. This type of molecule is described as a ***polar molecule***. It behaves much like a small bar magnet, in that the negative pole of one molecule will tend to attract the positive pole of another molecule. The charges on the molecule surface are much weaker than the unit charge of an ion and can be referred to as ***partial charges***.

The water molecule is strongly polar. One representation of the water molecule shows it as a large spherical object (the oxygen atom) with two small hemispheres attached (the two hydrogen atoms), as shown in Figure 2.15. The hydrogen "bumps" are positively charged; the far side of the larger sphere bears a negative partial charge. As a result, these molecules tend to attract one another, and to attract ions of other elements.

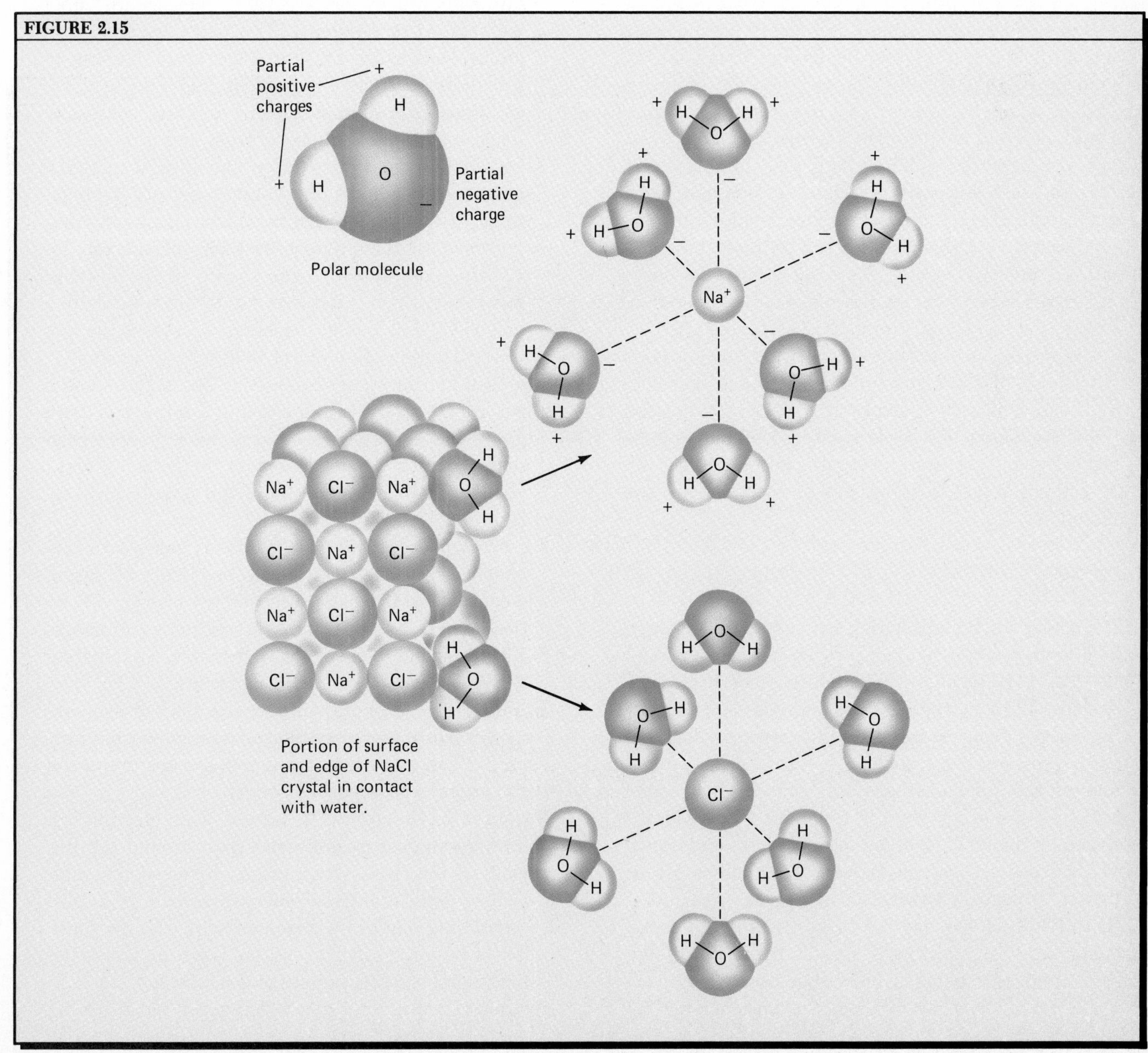

FIGURE 2.15 A schematic representation of the water molecule, showing its polar nature. Partial charges form on different parts of the surface of the molecule. When they surround ions of sodium and chlorine in a solution, the water molecules orient themselves to present an opposite charge toward the ion. (From John R. Holum, 1975, *Elements of General and Biological Chemistry*, 4th ed. John Wiley & Sons, Inc., New York, Figure 5.6. Reproduced by permission.)

The polar nature of the water molecule explains why water is such a good solvent for many ionic compounds. For example, we know that common table salt, sodium chloride (NaCl), rapidly dissolves in water. What happens is that the water molecules bombard the surfaces of the salt crystals, dislodging individual ions of sodium and chlorine (Figure 2.15). A free ion is quickly surrounded by several water molecules, which are turned so that their charged surface areas face the ion and attract it with a weak bond.

Metallic Bonds

Earlier in this chapter we referred to the occurrence of certain metals in rock as native elements and we use iron, gold, copper, and silver as examples. Most of the metals have only one, two, or three electrons in the outermost shell, and these are quite easily shared with other atoms. Since their outer electron shells can readily overlap with those of nearby electrons at several places at once, metal atoms are closely packed, with loosely held electrons moving freely throughout the entire mass. The free-moving electrons are often described as an *electron sea*, and the "communal" sharing of electrons characterizes the *metallic bond*.

These metals have the property of being malleable, meaning that they can be shaped by hammer blows and can be drawn out into thin sheets and wires without fracturing. Although the atoms of the metal maintain an orderly spacing, layers of atoms can easily glide past each other as the metal is hammered or drawn; the bonds remain intact at all times.

We also know that some metals like this are good conductors of electricity, since our transmission lines and household wiring consist of copper or aluminum wire. The conduction of electricity is facilitated by the freely moving electrons in a metal, because electricity is nothing more than a flow of electrons in one direction.

Matter and Energy in Geology

Without an understanding of matter and energy, and their relation to one another, no geologist can hope to carry on useful research in modern physical geology. Without applying basic physics and chemistry, today's geologist could do little more to interpret earth processes than could a geologist who worked around the time of the Civil War. In fact, some branches of geology were in a state of scientific stagnation by the late 1800s, because physicists and chemists had not yet learned enough about the nature of matter to provide answers to many geological questions. Radioactivity had not been discovered; nuclear energy was yet unknown. Ages of rocks and of the earth itself were being erroneously estimated and varied widely. The source of heat for volcano-building was still largely a mystery. Scientific tools were entirely lacking for probing deep into the earth's crust.

Only in the early 1900s, when physicists began to record and interpret deep earthquake waves, did we begin to form a picture of the states of mineral matter within the earth.

Key Facts and Concepts

Matter and energy *Matter* is tangible substance that occupies space and has the property of exerting gravitational attraction. *Energy,* the ability to do work, is involved in the motion of matter and can be stored in various forms. The *law of conservation of matter and energy* states that the sum total of all matter and energy in the universe must remain constant.

Classes of matter Pure substances can be distinguished from mixtures. Pure substances are either *elements* or *compounds*. The latter are groups of atoms—*molecules* or *ions*. The *atom* consists of subatomic particles—a *nucleus* and one or more *electrons*. The atomic nucleus, in turn, consists of nuclear particles, such as neutrons and protons.

Mass, density, and gravitation Quantity of matter is *mass;* mass per unit of volume is *density.* Gravitational attraction between two bodies is directly proportional to the product of their masses and inversely proportional to the square of the distance separating them. *Acceleration of gravity* with a value of 9.8 m/sec^2 at the earth's surface is the rate of acceleration of an object falling in a vacuum.

States of matter Matter can exist in one of three states: gas, liquid, solid. Gases are readily compressed. Both gases and liquids flow readily under unbalanced stresses and both are classed as *fluids*. Solids resist changes of shape and volume and usually yield by rupture.

Kinds of energy Common forms of energy are mechanical, sensible heat, electromagnetic, chemical, electrical, and nuclear. Mechanical energy can be transmitted by *wave motion* within a substance. Electromagnetic energy is transmitted by radiation within the *electromagnetic spectrum*.

Flow systems of energy and matter Both energy and matter can move in systems of interconnected flow paths. Energy systems are open systems, requiring the import and export of energy.

Structure of the atom The *orbital* of an electron can be diagrammed as a circular path around the nucleus of the atom. Orbitals are organized into concentric shells numbered outward from 1 to 7. The *atomic number* of an element is equal to the number of protons in its nucleus; the *mass number* is the total number of its protons and neutrons.

Isotopes *Isotopes* of a given element differ in having

different numbers of neutrons and thus differ in mass number, but not in atomic number.

Energy levels in the atom Energy possessed by an electron shows a quantum increase outward from one shell to the next, designated by *quantum numbers* 1 through 7. Electron shells are designated by letters *K, L, M, N, O,* etc. Shell *K* requires 2 electrons to be full; shells *L, M, N,* and *O* each requires 8 electrons to be full. Elements containing filled shells are extremely stable; those with only one electron in the outer shell are highly reactive.

Ions Atoms that have lost or gained an electron of the outer shell are *ions;* they may be positively charged (*cations*) or negatively charged (*anions*). Ions may also be formed by linkage of the atoms of two or three different elements.

Compounds formed by bonding An *ionic compound* is formed by electron transfer, creating a *chemical bond,* called an *ionic bond,* between the ions. Compounds can also be formed by *electron sharing,* and for this, *covalent bonds* are formed. The *chemical formula* of a compound expresses its element composition by element symbols. Proportions of elements are determined by the *valence electrons;* their number determines the *covalence* of the atom. A *chemical equation* describes the *chemical reaction* by which bonds are formed.

Polar molecules A *polar molecule,* of which water is an example, bears a positive charge at one place on its outer surface and a negative charge at another place. These are *partial charges* by means of which polar molecules attract one another or attract ions.

Metallic bonds Atoms of metals easily share electrons, which are loosely held and move freely throughout the entire mass as an *electron sea.* This form of sharing characterizes the *metallic bond.*

Questions and Problems

1. You just happen to be near when a great mass of solid rock breaks away from the edge of a sheer cliff and falls several hundred meters through the air, striking the ground at the base of the cliff and breaking into many fragments. Very soon you hear a great roaring sound and feel the ground shake. You also feel a strong gust of wind coming from the place where the rock mass came to rest. Describe in sequence the forms of energy that have been produced and transformed during this entire event. In what form was the energy prior to the breaking of the rock mass? Into what forms was this initial quantity of energy changed? Suppose that the rock mass fell through the air a vertical distance of 500 m before first impacting the ground. Neglecting the effect of friction with the air, what would be the approximate speed of fall of the mass at the point of impact? Would the temperature of the rock be different after impact than it was when at rest at the top of the cliff? Explain.

2. Using the schematic diagram of a glacier (Figure 2.7) for guidance, devise a schematic flow diagram to show the pathways of matter in the form of water as it enters, moves through, and leaves the glacier system. Then write a simple equation to show the balance between input and output, taking into account also the amount of matter in storage in the system.

3. Figure 2.11 shows how an ionic compound can be formed by electron sharing, using sodium and chlorine atoms to form sodium chloride. Using the information shown in Figure 2.10, name and give the formulas of other ionic compounds of chlorine that are possible. What other elements might be expected to form ionic compounds with sodium? Name and give the formula for those compounds. Metals having two electrons in the outer shell can also form ionic compounds. For example, write the formulas for magnesium bromide and calcium oxide, making sure that the ions are combined in the correct ratio in each case.

4. Using diagrams such as those in Figure 2.12, show how electron sharing takes place to produce covalent bonds in a compound composed of aluminum (Al) and oxygen (O). In what ratio do these elements combine? Give the formula of this compound, which is important in the study of minerals and rocks.

5. Write the chemical equation that describes the chemical reactions between the following pairs of elements or ions:

   A. copper (Cu) and the sulfate ion (SO_4^{2-})
   B. sodium (Na) and the carbonate ion (CO_3^{2-})
   C. magnesium (Mg) and chlorine (Cl)

6. Given a cube of rock salt (NaCl) measuring 2 cm on a side and a liter of water in a pyrex beaker, how would you go about dissolving the salt completely in the water in the shortest possible time?

3

CHAPTER THREE

Introduction to Minerals

Mineralogy, the science of minerals, is based on chemistry and physics, which developed in close relationship through the eighteenth and nineteenth centuries. The advance of knowledge about minerals depended on a chain of discoveries about the atom and the way in which atoms are arranged in orderly geometrical structures within crystals. Typically, minerals occur in pure form as crystals with distinctive outward shapes. One of the first persons to study crystals in a scientific way was Nicholas Steno (1638–1686), a Dane whose fields of accomplishment included both anatomy and geology—a remarkable combination! His important discovery in mineralogy was that the angles between corresponding faces of quartz crystals are constant, no matter what the size of the crystal or its place of origin (Figure 3.1).

Over a century later, mineralogy took a great step forward when a French scientist, Rene J. Haüy (1743–1823), proposed the theory that mineral crystals are constructed of building blocks of molecular size. Because the blocks are identical for a given mineral, its outward form as a crystal must always show the same angular relationships. (Recall from Chapter 2 that John Dalton's theory of atoms and compounds was developed at about this time.) Following Haüy's work, the study of crystal form and structure, called *crystallography,* developed rapidly into an exact science.

It was not until X rays were discovered (1895) that the internal arrangement of atoms within Haüy's "building blocks" could be verified and studied. In 1912, German physicist Max von Laue and his coworkers demonstrated that X rays traveling through a mineral crystal were bent (diffracted) in such a way as to produce a distinctive geometric pattern of dots on a photographic plate (see Figure 3.7). Then, in 1914, an English father-and-son team, W. H. Bragg and W. L. Bragg, used these dot patterns to develop a

FIGURE 3.1 These mineral crystals are of the variety of quartz known to collectors as *rock crystal;* they were found near Hot Springs, Arkansas. (Courtesy of Ward's Natural Science Establishment, Inc., Rochester, N.Y.)

FIGURE 3.1

method of measuring exactly the spacing of atoms within a crystal and of reconstructing their arrangement in three-dimensional space. For this work the Braggs received the 1915 Nobel Prize for Physics.

Mineralogy is also based on chemistry, because it is essential to determine the elements present in a given mineral compound and the chemical formula. Thus, in mineralogy, chemical analysis is of equal importance to the study of atomic structure. As each new mineral was identified and named, its chemical formula and its physical properties were determined. North Americans can take pride in the pioneering work of James D. Dana, a graduate of Yale University, who published the first comprehensive treatise on mineralogy in 1837. Titled *A System of Mineralogy,* the work has gone through six editions as the world standard of reference on mineralogy; more recently, it has been further supplemented and revised by others.

Table 3.1 Composition of the Earth

Rank	Element	Symbol	Earth average (% by mass)
1	Iron	Fe	34.6 } 92.0
2	Oxygen	O	29.5 }
3	Silicon	Si	15.2 }
4	Magnesium	Mg	12.7 }
5	Nickel	Ni	2.4
6	Sulfur	S	1.9
7	Calcium	Ca	1.1
8	Aluminum	Al	1.1
9	Sodium	Na	0.57
10	Chromium	Cr	0.26
11	Manganese	Mn	0.22
12	Cobalt	Co	0.13
13	Phosphorus	P	0.10
14	Potassium	K	0.07
15	Titanium	Ti	0.05

Source: Data from Brian Mason (1966), *Principles of Geochemistry,* 3d ed., New York, John Wiley, Table 3.7.

Element Composition of the Earth and Its Crust

Because all matter consists of elements, it is helpful to preface a study of minerals with a look at the list of the most abundant elements in the solid earth and its crust. Table 3.1 lists the fifteen most abundant elements of the entire earth in order of their percentage of the earth's total mass. Iron is first (more than one-third of the earth's mass) because the earth's large central core is composed mostly of metallic iron, and iron is also an important element in the mantle. Oxygen, about 30% of the total earth mass, is found in all important minerals of the mantle and crust. Silicon, linked with oxygen in the common minerals forming the igneous rocks, comes third; magnesium is fourth. Altogether, these four elements make up about 92% of the earth's total mass. Of the remaining elements, nickel, most of which is alloyed with iron in the core, is in fifth place. The percentages given in the table are calculated indirectly from certain assumptions as to the composition of the mantle and core. Although the figures are only expert guesses, they are considered sound within reasonable limits of error.

Next, look at the list of the eight most abundant elements of the earth's crust, given in Table 3.2. Recall from Chapter 1 that the crust is the outermost rock layer of the earth, averages about 17 km in thickness, and makes up only about 4/10 of 1% of the total earth mass. Crustal rock is available for direct study by geologists, so that an average value of its element composition can be calculated from a large number of samples from both continents and ocean basins. About 95% of the crust consists of igneous rocks and of metamorphic rocks derived directly from igneous rock. Sedimentary rock, together with metamorphic rock derived directly from sedimentary rock, makes up the remaining 5% of the crust. Thus, the figures in the list can be taken to represent the average element composition of igneous rock. The order of listing is according to percent by mass. The table also gives percent by volume and percent of atoms present.

Several points are of interest in Table 3.2. Notice,

Table 3.2 The Eight Most Abundant Elements in the Earth's Crust

Rank	Element	Symbol	Percent by mass	Percent by volume	Percent of atoms present
1	Oxygen	O	46.6	93.8	62.6
2	Silicon	Si	27.7	0.9	21.2
3	Aluminum	Al	8.1	0.5	6.5
4	Iron	Fe	5.0	0.4	1.9
5	Calcium	Ca	3.6	1.0	1.9
6	Sodium	Na	2.8	1.3	2.6
7	Potassium	K	2.6	1.8	1.4
8	Magnesium	Mg	2.1	0.3	1.8

Source: Data from Brian Mason (1966), *Principles of Geochemistry,* 3d ed., New York, John Wiley, Table 3.4, p. 48.
Note: Figures have been rounded to nearest one-tenth.

first, that the eight elements make up between 98% and 99% of the crust by mass and that almost half this mass is oxygen. Measured in other ways, the importance of oxygen is even greater; in numbers of atoms it makes up over 60% of the total and, being an atom of comparatively large radius, it represents almost 94% by volume. Notice that silicon is in second place with about 28% by mass, or roughly half the value for oxygen. Aluminum and iron occupy intermediate positions, while the last four elements—calcium, sodium, potassium, and magnesium—are subequal in the range of 2% to 4% by mass.

To extend the table, we add that the ninth most abundant element is titanium, followed in order by hydrogen, phosphorus, barium, and strontium. It is interesting to note that the metals copper, lead, zinc, nickel, and tin, which play such an important role in our modern technology, are present only in very small proportions and are indeed scarce. Fortunately, these and other rare but important elements have been concentrated locally into ores from which they can be extracted in useful quantities.

The list of element abundances in the crust should now be compared with that for the earth as a whole (Table 3.1). Iron is much less abundant in the crust than in the whole earth, because most iron is in the core. Oxygen and silicon are of major rank in both tables. Magnesium is less abundant in the crust than in the whole earth while aluminum is much more abundant in the crust—it moves up from eighth position to fourth. Potassium, too, is much more abundant in the crust. We can make the inference that in the early aeons of earth history, as the crust came into existence, a sorting-out process took place by which crustal rock became enriched in potassium and aluminum, but depleted of iron and magnesium. An important problem in earth history is to explain how the various shells of our planet acquired unique combinations of abundances of elements. This problem is dealt with in Chapter 7, in connection with the origin and early history of the earth.

What Is a Mineral?

Each mineral variety, or *mineral species*, bears a name; most end in the syllable *ite*. Examples mentioned in Chapter 2 are halite and calcite. Thus, the word *mineral* refers collectively to the more than 2000 mineral species whose chemical and physical properties have been determined. In this sense, a mineral can be defined as a naturally occurring homogeneous solid that is an inorganic substance, having either an orderly atomic structure and a definite chemical composition or one that is variable between stated limits. Let us look more closely into the meaning of individual terms in this rather long and ponderous definition.

"Naturally occurring" means, of course, that synthetic mineral substances are excluded. Corundum, an exceedingly hard mineral, consists of aluminum oxide (Al_2O_3). Because of its hardness, corundum was mined for use as an abrasive powder. Today, the same compound manufactured in large quantities as a synthetic material has largely taken the place of the naturally occurring mineral. Corundum also occurs in rare forms as the gemstones ruby (red) and sapphire (blue). Both rubies and sapphires of excellent quality are now produced synthetically by fusing powdered aluminum oxide in an intensely hot flame.

By the word "homogeneous," we mean simply that a mineral consists of a single chemical unit, which may be a single element (in the case of minerals that are native metals) or a single compound. (Chapter 2 explains the terms used in the previous sentence.) In other words, mixtures of compounds are excluded. Minerals must be in the solid state. For example, water in the form of ice found in snow crystals or in glaciers is a true mineral, but the liquid and gaseous forms of water are not minerals.

An "inorganic substance" is one that is not composed of organic matter, which according to the chemist consists of compounds of carbon and hydrogen. Because many organic compounds are synthesized and do not exist in nature, we are particularly interested in distinguishing a mineral from a naturally occurring organic compound produced by a living organism. Take the case of ordinary coal, a black, rocklike substance composed largely of carbon, hydrogen, and oxygen. It originated as an accumulation of the remains of woody plants and is therefore of organic origin. Geologists consider coal to be a variety of rock, but it is not a mineral. Next, consider the common mineral calcite, composed of calcium carbonate ($CaCO_3$). Mineral calcite may be secreted by certain organisms to form such hard parts as external shells or internal skeletons. Although produced by living organisms solid calcium carbonate in shells is a true mineral and is no different from calcite formed by natural processes which do not involve organisms.

What is meant by the statement that a mineral has "a definite chemical composition"? The proportions of elements in many minerals are always exactly the same. For example, take calcite, $CaCO_3$. Recall from Chapter 2 that this substance is an ionic compound in which one calcium ion (Ca^{2+}) unites with one carbonate ion (CO_3^{2-}) by transfer of two electrons (see Figure 2.14*B*). For other minerals, the proportions of two or more elements may be variable within stated limits in a single compound. Take, for example, the mineral olivine, with the formula $(Mg,Fe)_2SiO_4$. The element symbols for magnesium (Mg) and iron (Fe) are separated by a comma and placed together within parentheses; this notation means that the ratio or proportion of the two metallic ions can vary through a great range—from almost entirely magnesium to almost entirely iron. (This range in proportions is said to be a *solid solution.*) Thus, an almost infinite number of compositional differences are included in the formula for olivine. The composition is, however, definite in limiting the range of element proportions and restricting the list of elements it contains.

The "orderly atomic structure" of a mineral refers to the arrangement in space of the atoms or ions within the crystalline solid. Atomic structure is fully explained later in this chapter. As in the case of the chemical formula, the atomic structure within a single mineral must accommodate the varying proportions of the atoms or ions of two elements, one of which can replace the other. For example, in olivine an ion of either magnesium or iron always occupies precisely the same position in space with respect to the surrounding atoms of oxygen and silicon. A particularly striking example of the importance of atomic structure as a mineral property is seen in the minerals graphite and diamond. Both have the identical chemical composition—pure carbon—but they are at opposite ends of the scale of mineral hardness, and they have other greatly differing physical properties, as we will explain later in this chapter.

There exist in nature some kinds of mineral substances that fulfill all the qualifications of a mineral except that they do not show an orderly arrangement of atoms. These noncrystalline mineral substances may be of a type described as *amorphous minerals*. One class of amorphous minerals has the same structure as ordinary glass, which solidifies so quickly from a molten state that the atoms do not have a chance to arrange themselves in an orderly three-dimensional pattern. Another class of amorphous minerals are the *gel minerals*, also called *mineraloids*. Gels are mixtures of water with other compounds, and are illustrated by such familiar food substances as gelatine and jellies. A gel mineral is believed to form from a liquid gel trapped in a cavity in rock. As the gel mass dries out, the remaining mineral compound becomes a solid in which the molecules have no orderly arrangement. A familiar gel mineral is the opal, a semiprecious gemstone with soft iridescent colors.

Though it may seem as if we have labored too long and too hard to produce a definition of the word "mineral," we have in doing so presented a great deal of information about the fundamental properties of minerals. Fortunately, not all definitions in geology prove as difficult and elusive as this one. Science depends on a consensus as to the exact meaning of any term that is used repeatedly.

Minerals and Rocks

Minerals are the chemical substances of which rocks are composed. A particular mineral species can usually be recognized by a distinctive set of physical properties, such as color, hardness, density, and manner of fracturing. Our problem here is that hundreds of mineral species—each bearing a different name and displaying a different set of characteristics—have been identified, whereas this brief introduction to geology must limit emphasis to only a few. Fortunately, this process is greatly simplified if we consider only those minerals making up the bulk of common rocks.

Although rocks are composed of particles of one or more kinds of minerals, as it happens only a few mineral combinations commonly occur in nature. In this chapter, six important minerals or mineral groups are singled out for emphasis. Together they make up most of the bulk of the igneous rocks, and only a few additional minerals of secondary importance to these rocks will still need mention. Several other mineral species, named and described in Chapters 5 and 7, are important constituents of the sedimentary and metamorphic rocks.

Chemical Grouping of Minerals

Mineralogists use a system of twelve major chemical classes that brings within each class minerals with physical similarities and which often occur in the same geologic environment. A full description of the major classes is not necessary in an introduction to geology, but a selection of several representative classes, shown in Table 3.3, will be meaningful in understanding a number of topics in geology. The table gives examples of minerals in each group, together with their chemical formulas, which the principles of basic chemistry, given in Chapter 2, should be adequate to make meaningful.

First, *native elements* are a chemical class distinguished from compounds. Examples of native elements are native copper and gold, both of which are metals, and native sulfur and diamond, which are nonmetals.

Compounds make up the remaining chemical groups shown in the table and include oxides, sulfides, carbonates, halides, sulfates, silicates, hydrous silicates, and hydroxides. An example of the *oxides* is the common mineral hematite, sesquioxide of iron. *Sulfides* are a class including most of the common ore minerals (minerals that are a source of industrial metals). *Carbonates*, *halides*, and *sulfates* are classes of major importance in sediments and sedimentary rocks; they are formed readily in surface environments, whether on land or on the floors of lakes or the ocean. The *silicates*, minerals that include the silicate ion, make up the bulk of all igneous rocks. The *hydrous silicates* and *hydroxides* are minerals characterized by the inclusion of the hydroxide ion (OH^-) or the water molecule (H_2O) in their chemical formulas. Minerals of these classes are formed at or near the earth's surface, where

Table 3.3 Some Chemical Classes of Common Minerals

Chemical group or class		Representative mineral		Symbol
Native Elements	Metals	Native copper		Cu
		Gold		Au
	Nonmetals	Native sulfur		S
		Diamond		C
	Description	**Representative mineral**	**Composition**	**Formula**
Compounds				
Oxides	Elements in combination with oxygen	Quartz	Silicon dioxide	SiO_2
		Hematite	Sesquioxide of iron	Fe_2O_3
		Ice	Solid state of water	H_2O
Sulfides	Elements in combination with sulfur	Galena	Lead sulfide	PbS
		Sphalerite	Zinc sulfide	ZnS
Carbonates	Elements in combination with carbonate ion (CO_3)	Calcite	Calcium carbonate	$CaCO_3$
		Dolomite	Carbonate of calcium and magnesium	$CaMg(CO_3)_2$
Halides	Compounds of the halogen elements: chlorine, bromine, iodine, fluorine	Halite (rock salt)	Sodium chloride	NaCl
		Fluorite	Calcium fluoride	CaF_2
Sulfates	Elements in combination with sulfate ion (SO_4)	Anhydrite	Calcium sulfate	$CaSO_4$
		Gypsum	Hydrous calcium sulfate	$CaSO_4 \cdot 2H_2O$
Silicates	Elements in combination with silicate ion (SiO_4)	Orthoclase feldspar	Aluminosilicate of potassium and sodium	$(K,Na)AlSi_3O_8$
		Olivine	Silicate of magnesium and iron	$(Mg,Fe)_2\ SiO_4$
Hydrous silicates	(Clay minerals) Compounds derived by union of water with silicate minerals	Kaolinite	Hydrous alumino-silicate derived from feldspars	
		Illite	Complex hydrous aluminosilicate derived from micas	
Hydroxides	Compounds derived by union of water with oxides of the metals iron, aluminum, manganese	Limonite	Hydrous sesquioxide of iron	$2Fe_2O_3 \cdot 3H_2O$
		Bauxite	Hydrous sesquioxide of aluminum	$Al_2O_3 \cdot 2H_2O$

abundant water is present; they are typically associated with the decay of rock and the formation of soil.

Physical Properties of Minerals

As indicated above, each mineral species has a distinctive set of physical properties, and these can be used to identify an unknown mineral specimen. Some physical properties can be determined by the unaided eye, or with the help of a strong magnifying lens, but others require the use of laboratory equipment.

Crystal Symmetry and External Crystal Form

Our definition of a mineral requires that it occur in a solid state in which the atoms or ions have an orderly internal arrangement. In other words, minerals are *crystalline solids*. The crystalline condition expresses itself in two ways: internal atomic structure, which will be discussed in later paragraphs, and external form. The ideal external form of a crystalline solid is a *crystal*, a regular geometrical solid bounded by *crystal faces* that are plane surfaces intersecting along straight lines to give straight edges. The edges, in turn, meet in sharp corners.

Crystals develop their external form during *crystallization*, the process of addition of atoms or ions one at a time, each fitting into its proper place in a structural plan or design known as a *crystal lattice*. If the surface of the growing mineral fragment is surrounded by a fluid (a liquid or a gas), crystal faces will be created, and the external form will be visible. Crystals can be formed from a water solution in which an ionic compound has been dissolved. Various compounds that are soluble salts will produce growing crystals in a saturated solution as the water evaporates. Crystals also form when a pure compound in the liquid state passes into the solid state as it cools below the melting point. Take, for example, the freezing of water, in which thin, needlelike ice crystals can be seen to grow. Silicate minerals (olivine, for example) form crystals when magma (molten rock) cools and solidifies. If the crystallizing mineral occupies a small opening in solid rock, the outer boundary of the crystal will conform to the irregular surface of the cavity and thus the true external crystal form will not be seen. Even so, the internal structure will exactly fulfill the correct arrangement of atoms in the space lattice.

FIGURE 3.2 Fluorite crystals of cubical form, partly coated by minute calcite crystals. (Ward's Natural Science Establishment, Inc., Rochester, N.Y.)

FIGURE 3.3 External crystal forms and their symmetry. *A*. Octahedron. *B*. Hexagonal dipyramid.

FIGURE 3.2

FIGURE 3.3

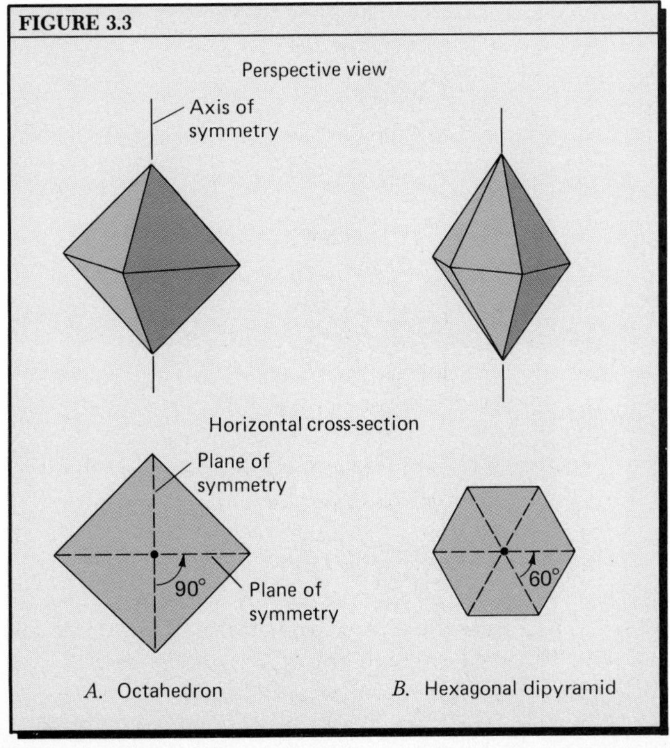

Mineral crystals of nearly perfect form are often obtained from the inside surface of a spherical rock cavity called a *geode* (see Chapter 15) or from an irregular cavity known to miners as a *vug*. For example, beautiful crystals of amethyst quartz (SiO_2) and fluorite (calcium fluoride, CaF_2) are found in such cavities and are valued as ornamental specimens (Figure 3.2). The cavity was filled with water at the time the crystals grew. These crystals are not complete in form, attached as they are to the wall of the cavity.

Mineral crystals possess *crystal symmetry*, which simply means that the surface of the solid figure shows similar or balanced forms on either side. A crystal also possesses an *axis of symmetry*, which is an imaginary line passing through the crystal in such a position that the crystal may be turned (rotated) on the axis and will repeat the same appearance a certain number of times with each rotation of 360°. For example, the octahedron shown in Figure 3.3 has its axis of symmetry passing through the upper and lower points of the crystal. The octahedron presents the same configuration to an observer each time that the crystal is rotated 90°. The hexagonal dipyramid shown next to the octahedron will repeat with each rotation of 60° and repeats six times with each complete rotation of 360°. A crystal may also possess a *plane of symmetry*, which is an imaginary plane that will divide the crystal into halves, each of which is a perfect mirror image of the other.

Mineral crystals fall into six *crystal systems* (Figure 3.4), each of which is defined in terms of its *crystallographic axes*—the axes that govern the position of crystal faces. For each system except the hexagonal, there are three crystallographic axes; for the hexagonal system, there are four. The simplest system is the *isometric system*, in which the three axes are perpendicular to one another and are of equal length. Common crystal forms in this system are the cube, as in halite, and the octahedron, as in diamond. Some further details of the other systems are given in the caption of Figure 3.4.

In many cases, a geologist can recognize a mineral by its external crystal form, because a given mineral may repeatedly show a particular *crystal habit*—a preference for a particular form or combination of forms. Crystal habit also refers to distinctive ways in which certain minerals grow together in groups (called aggregates). For example, one mineral may form an aggregate of many parallel flattened blades, while another mineral usually forms rounded masses of thin, needlelike, radiating crystals.

FIGURE 3.4 Axes and representative external crystal forms of the six crystal systems. I. *Isometric system*. Three axes of equal length lie at right angles to each other. Because the axes are identical, any one of the three can be placed in the vertical position; all three are designated by the letter a (a_1, a_2, a_3). II. *Tetragonal system*. Three axes make right angles with each other. The two horizontal axes, a and b, are of equal length, but the third, or c axis, is of different length. The example shown, zircon, consists of a prism of four rectangular vertical sides and a dipyramid (double pyramid) forming the top and bottom of the crystal. III. *Hexagonal system*. Four axes are present. Three of the axes are horizontal and of equal length, intersecting in angles of 60° and 120°. They are designated a_1, a_2, and a_3. The third, or c axis, is perpendicular to the a axes and is of different length. IV. *Orthorhombic system*. Three axes intersect at right angles to each other, but they are of unequal lengths. The horizontal axes are a_1 and a_2, the perpendicular axis is the c axis. The crystal of staurolite shown is tabular in form, with the a_1 axis much shorter than the a_2. V. *Monoclinic system*. The three axes are of unequal length. Whereas the c and b axes intersect at a right angle, the a axis intersects the c axis obliquely forming an acute angle (less than 90°). The "forward tilt" of the a axis gives the gypsum crystal a distinctive tabular form resembling a parallelogram when viewed from the side. VI. *Triclinic system*. The three axes are of unequal length and intersect at oblique angles. No single plane of symmetry can cut the crystal into two halves that are mirror images.

Cleavage and Fracture

Many minerals show *cleavage*, a pronounced tendency to split along smooth planar (flat) surfaces of weakness. These surfaces, known as *cleavage planes*, bear a close relationship to both internal atomic structure and the external crystal form. One set of cleavage planes may be very strongly developed, as in the familiar case of mica, which can easily be split into thin sheets (see Figure 3.12). In other minerals, there are three intersecting sets of parallel cleavage planes, enabling the mineral to be broken into similar prisms or rhombs of many sizes (Figure 3.5). The key to an understanding of mineral cleavage lies in the atomic structure of crystals and is explained later.

Minerals lacking cleavage break along various characteristic forms of *fracture surfaces*. For example, the curved fracture surfaces of a glass constitute *conchoidal fracture*, seen in quartz (see Figure 3.16). Other fracture types are described as even, uneven, splintery, or hackly.

Specific Gravity

For a given chemical composition, each mineral species has a certain *specific gravity*, which is the ratio of its density to the density of water at 4 °C (see Chapter 2 for an explanation of density). Most of the abundant minerals in igneous rocks have specific gravities in the

FIGURE 3.4

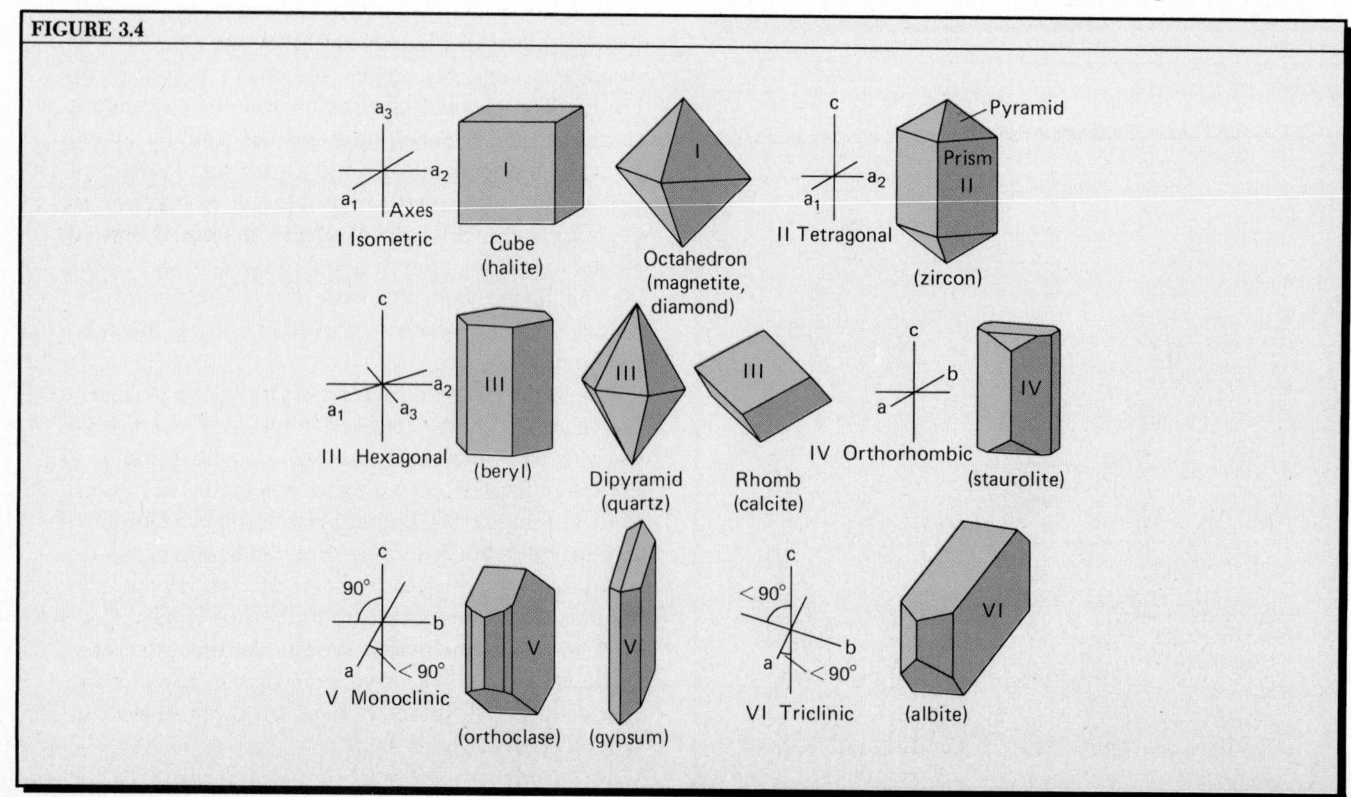

range of 2.7 (quartz) to 3.4 (olivine). Other minerals have much higher specific gravities—for example, hematite is about 5.0; galena, 7.5; and native copper, 8.8. Mineral specific gravity is a property of great importance because it determines the density of a given rock, which is a mixture of minerals; rock density in turn determines the gross layered structure of the earth.

Hardness

The degree to which a mineral surface resists being scratched is known as its *mineral hardness*. Hardness is geologically important because it determines how easily a mineral is worn away by the abrasive action of streams, waves, wind, and glaciers in the processes of erosion and transportation. Minerals themselves make tools of differing hardness by means of which other minerals can be tested. Ten standard minerals constitute the *Mohs scale* of hardness, ranging from the softest, talc (no. 1) to the hardest, diamond (no. 10). Each mineral on the scale will scratch all those of lower number, but will be scratched by those of higher number. The complete scale is as follows:

1. talc (softest)
2. gypsum (2½, fingernail)
3. calcite (3, copper coin)
4. fluorite
5. apatite (5½ to 6, knife blade, plate glass)
6. orthoclase (6½ to 7, steel file)
7. quartz
8. topaz
9. corundum
10. diamond (hardest)

FIGURE 3.5 These cleavage rhombohedrons of clear calcite are of a variety known as Iceland spar. About half natural size. (Ward's Natural Science Establishment, Inc., Rochester, N.Y.)

FIGURE 3.5

Luster

The appearance of a mineral surface under reflected light is referred to as its *mineral luster*, described by several descriptive adjectives, such as metallic, adamantine (diamondlike), vitreous, resinous, pearly, or silky.

Color

Certain minerals possess a distinctive *mineral color* that facilitates recognition, but many mineral species have varieties that differ conspicuously in color. Among the rock-forming minerals, the range is from colorless (quartz, calcite, gypsum) and white (orthoclase feldspar), through olive green (olivine) to black (biotite, hornblende). The reddish-brown and yellow-brown typical of iron oxides such as hematite and limonite lend earth-red coloration to many sedimentary rocks (red sandstones, red shales).

Streak

When a mineral specimen is rubbed across the unglazed surface of a white porcelain plate, it may leave a *mineral streak* of powder of distinctive color, such as the red-brown streak of hematite. This streak is consistently useful in identifying the mineral, but most common rock-forming minerals leave a white streak of no particular help in identification.

Optical Properties

A completely different group of mineral properties, called *optical properties*, relates to the effect of a transparent mineral upon light rays which pass through it. These optical properties are of great value in mineral identification and are evaluated by means of a microscope using polarized light rays. A particularly important diagnostic property is the degree to which the mineral bends, or refracts, light rays passing through it.

Atomic Structure of Minerals

Atomic structure refers to the orderly internal arrangement of atoms within a crystalline solid. Discovery of the atomic structure of minerals came long after the science of crystallography had reached a high level of perfection in describing and classifying the outward form of mineral crystals. We mentioned early in this chapter that the Braggs in 1914 had been successful in determining the atomic structure of minerals. They used the *X-ray diffraction method*, which had been suggested in 1912 by Max von Laue of the University of Munich and experimentally developed by his graduate students.

A narrow X-ray beam is passed through a small cleavage fragment of a mineral, and the rays are allowed to fall on a photographic plate (Figure 3.6). The developed plate, called a *Laue photograph,* shows many small spots arranged in a geometrical pattern around a central bright spot marking the direct X-ray beam (Figure 3.7). The researchers reasoned that the pattern of spots was produced by X rays that had been bent off course (diffracted) by atoms or ions within the crystal. They found it possible to calculate the arrangement of atoms in three-dimensional space, and to measure the actual distances separating layers of atoms.

By 1914, the Braggs had made use of the Laue method to determine the atomic structure of halite (NaCl) and went on to work out the atomic structure of many minerals. In later years, the X-ray method was extended to powdered mineral matter, making it unnecessary to find rare, well-formed crystals. A greatly improved modern type of X-ray diffraction apparatus uses a powdered mineral to derive a distinctive graphlike "signature" from the mineral. This signature can be used to identify a mineral much as an individual human being might be identified from a set of fingerprints.

The atomic structure of halite makes a good example to begin with. The space relationships between ions of sodium (Na^+) and chlorine (chloride ion, Cl^-) are shown by a "ball-and-stick" model in which ions of both elements are represented by small spheres and their bonds are represented by thin rods (Figure 3.8A). The distance separating the ions is greatly exaggerated by this method. Recall from Chapter 2 that the chemical bonds in halite are ionic bonds and that the mineral is an ionic compound. Because each ion bears a single electric charge, each chloride ion must be surrounded by six sodium ions spaced at equal distances from the chloride ion.

In Figure 3.8A, six sodium ions that surround a single chloride ion have been isolated by a network of dotted lines forming an octahedron. One sodium ion lies at each of the six points of the octahedron, while the chloride ion occupies the center. Bonds connecting the chloride ion with its sodium ions lie in three equidistant straight lines intersecting at right angles. You can also see that the centers of all ions lie simultaneously in three sets of parallel planes at right angles to one another. This arrangement is called a *space lattice,* and it consists of *unit cells,* which are all alike. In the case of halite, the unit cell is a cube with chloride ions at the corners (or a cube with sodium ions at the corners). For this reason, the space lattice of halite is called a *cubic space lattice*. Crystals of halite show the cubic form; they belong to the isometric system in which all three axes of symmetry are of equal length and at right angles (see Figure 3.4).

Halite has perfect *cubic cleavage*. If it is crushed

FIGURE 3.6 Schematic diagram of X-ray diffraction by a fragment of a mineral crystal.

FIGURE 3.7 A Laue photograph of the mineral beryl, a silicate of beryllium and aluminum, which crystallizes in the hexagonal system. The sixfold symmetry is evident in the pattern of dots. (Courtesy of C. S. Hurlbut, Jr., Geological Museum, Harvard University.)

FIGURE 3.6

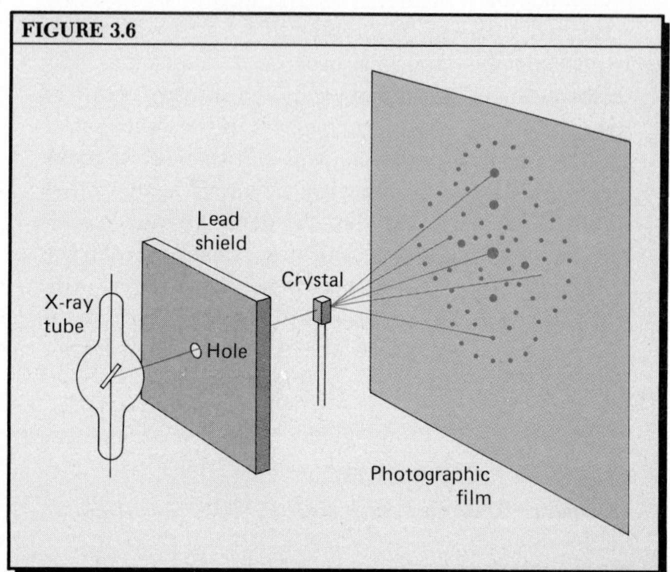

FIGURE 3.7

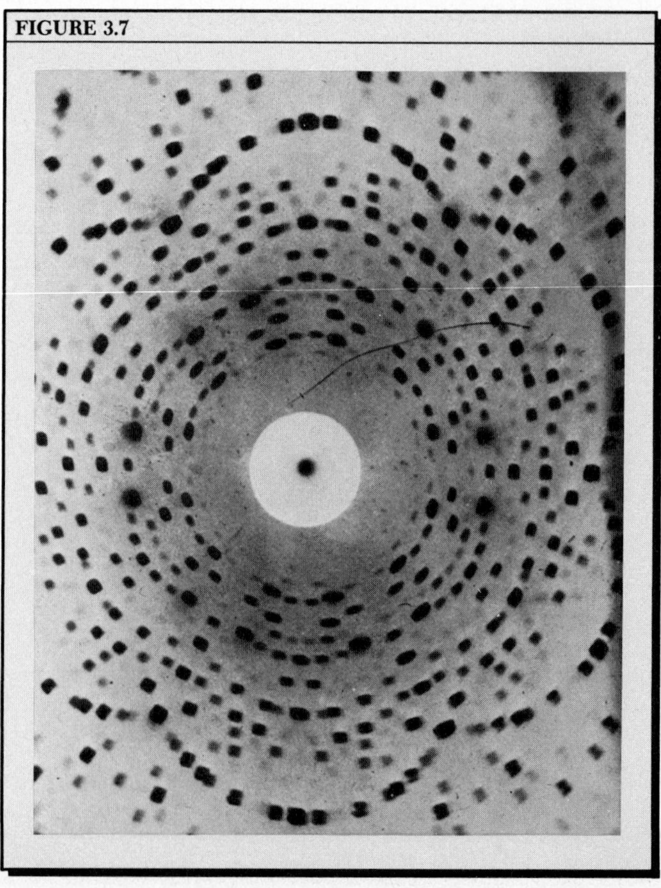

and the fragments examined with a magnifying lens or binocular microscope, they will appear to be cubes (Figure 3.8*B*). Each large cube always breaks up into smaller cubes. The cubic cleavage of halite is a direct consequence of the geometry of the space lattice. A cleavage plane represents the surface of least resistance to breakage within a given unit cell. Cleavage planes that cut the bonds at right angles encounter the smallest number of bonds. Imaginary cleavage planes, if made to pass diagonally through the unit cell, would encounter more bonds in the same distance.

*Ionic Radii and Ionic Bonds

All ions of a given element are of exactly the same size, which is to say that they have the same *ionic radius*. The ionic radius is given in units of *angstroms;* one angstrom (symbol Å) is equal to 10^{-8} cm (0.000,000,01 cm). The ionic radius of one element may be quite different from that of another. Generally, cations are small, while anions are large.

Figure 3.8*C* shows a model of closely packed ions of halite. The ionic radius of the sodium cation is 0.97 Å; of the chloride anion, 1.81 Å. The large chloride ions are packed together as closely as possible, so that the six chloride ions of the unit cell are nearly touching. The central sodium ion very nicely fits the central open space and touches all of the six anions surrounding it. When a cation exactly fits into the open space among its surrounding anions, the compound is in its most stable configuration. Where the cation is smaller than the opening, the compound is also stable. If, however, the cation is larger than the minimum packing space, the compound is unstable because it does not allow the surrounding anions to touch. The electrical forces of attraction between the centers of ions of opposite charges are then weaker than when the anions are in contact and closer to the central cation.

*Covalent Bonding in Minerals

Covalent bonding, explained in Chapter 2, involves the sharing of electrons in the outer shells of atoms. Examples given in Chapter 2 are the covalent compounds water (H_2O) and carbon dioxide (CO_2) (see Figure 2.12).

An excellent example of a mineral showing covalent bonding is *diamond*, composed entirely of carbon

FIGURE 3.8 Atomic structure of halite, NaCl. *A*. Expanded cubic space lattice with octahedral ionic bonding and unit cell outlined. *B*. Cubical cleavage of halite. *C*. Close-packed arrangement of ions of sodium and chlorine.

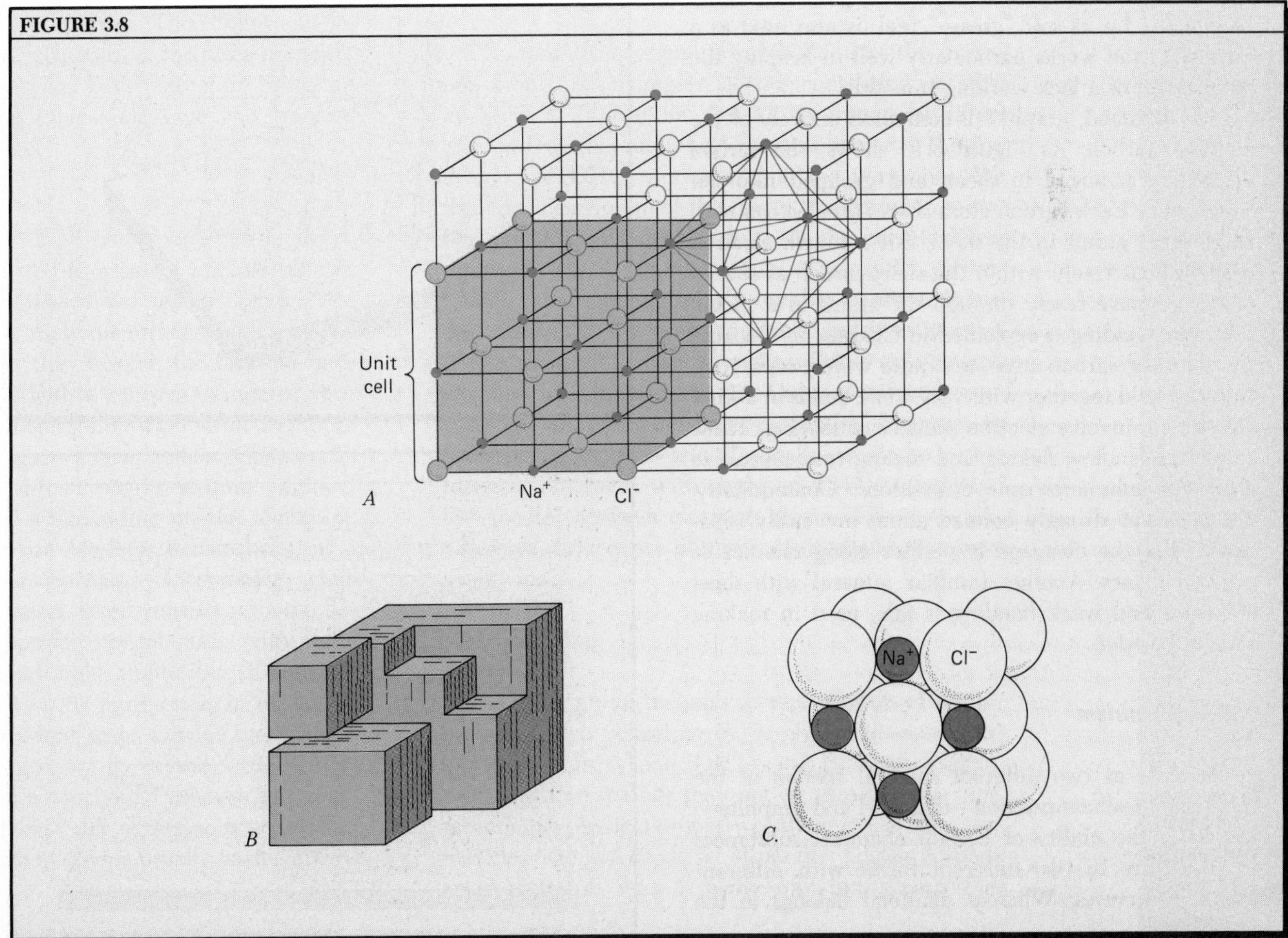

Table 3.4 Silicate Minerals Abundant in Igneous Rocks

Mineral or mineral group	Composition, formula	Specific gravity	Hardness, Mohs scale
Olivine	Silicate of magnesium and iron $(Mg,Fe)_2SiO_4$	3.3–4.4	6½–7
Pyroxene group (Augite)	Silicate of calcium, magnesium, iron, and aluminum (complex formulas)	3.2–3.3	5½
Amphibole group (Hornblende)	Silicate of calcium, magnesium, iron, and aluminum, with hydroxyl ion (complex formulas)	3.2	5½
Mica group (Muscovite, biotite)	Aluminosilicate of potassium, magnesium, and iron with water (complex formulas)	2.8–3.2	2½–3
Plagioclase feldspar group	Aluminosilicate of sodium and calcium Albite, $NaAlSi_3O_8$ Anorthite, $CaAl_2Si_2O_8$	2.62–2.76	6
Potash feldspar group	Aluminosilicate of potassium, also with sodium Orthoclase, microcline, $(K,Na)AlSi_3O_8$	2.5–2.6	6
Feldspathoid group (Nepheline)	Aluminosilicate of sodium and potassium Nepheline, $(Na,K)AlSiO_4$	2.6	5½–6
Quartz	Silicon dioxide, SiO_2	2.65	7

sheets that show some elastic flexibility. In examining a rock specimen containing grains of mica, the point of a sharp knife or needle can be used to pry up cleavage sheets of mica crystals. Hardness of the micas is quite low, in the range from 2½ to 3 on the Mohs scale.

Two species within the mica group deserve special mention. *Muscovite,* the light-colored mica, appears colorless and transparent in thin cleavage sheets (Figure 3.12) and has long been used, under the common name "isinglass," for small windows in the doors of furnaces, ovens, and stoves. It is also an excellent electrical insulator. The formula for muscovite is $KAl_2(AlSi_3O_{10})(OH)_2$. *Biotite,* the dark mica, is deep green or brown in thin sheets, but black in thick masses. Its formula is $K(Mg,Fe)_3(AlSi_3O_{10})(OH)_2$, and indicates that the presence of magnesium and iron give biotite a somewhat higher specific gravity than muscovite—2.8–2.9 for muscovite versus 2.8–3.2 for biotite.

Feldspar Group

The *feldspar group* consists of aluminum silicate with one or two of the ions potassium (K), sodium (Na), or calcium (Ca). Depending on which of these three metals are present, the feldspars can be subdivided into two major subgroups. Feldspars with Na

FIGURE 3.11 Stylized drawings of cleavage pieces of augite and hornblende, emphasizing the prismatic cleavage and its characteristic angles. In each case the top surface and left side of the specimen represent cleavage surfaces, which parallel the C-axis of the crystal structure.

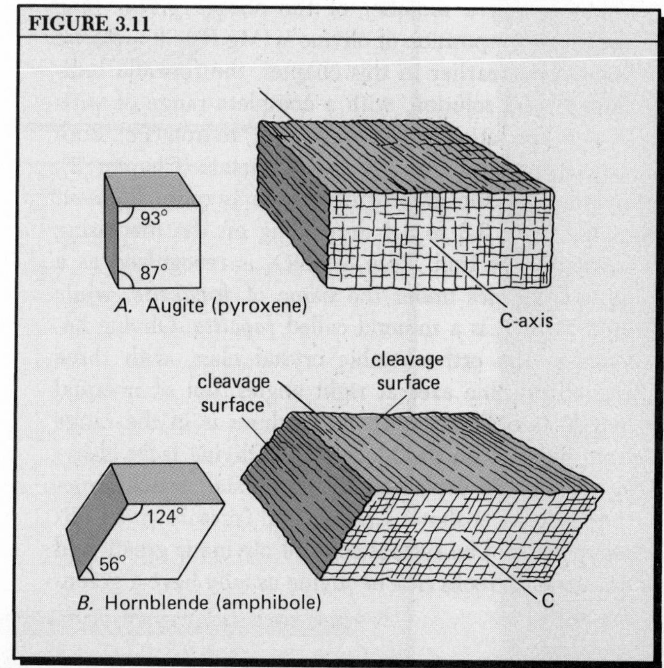

Table 3.4 Silicate Minerals Abundant in Igneous Rocks *(Continued)*

Crystal system	Cleavage or fracture	Luster	Color
Orthorhombic	No cleavage; conchoidal fracture	Vitreous	Yellowish-green to bottle green
Monoclinic	Imperfect prismatic cleavage at 87° and 93° angles	Vitreous	Dark green to black
Monoclinic	Imperfect prismatic cleavage at 56° and 124° angles	Vitreous or silky	Dark green to black
Monoclinic	Perfect cleavage in one plane only	Vitreous to pearly	Black or dark brown (biotite), clear (muscovite)
Triclinic	Good prismatic cleavage at 86° angle	Vitreous	White or gray, also iridescent
Monoclinic (Orthoclase) Triclinic (Microcline)	Good prismatic cleavage at about right angles	Vitreous	White, pale yellow, gray, or pink
Hexagonal	Poor cleavage in one plane	Vitreous or greasy	Colorless, white, yellowish
Hexagonal	No cleavage; conchoidal fracture	Vitreous or greasy	Colorless and various colors

and/or K make up the *alkali feldspars;* those with Na and/or Ca, the *plagioclase feldspars*. Thus, K and Ca do not occur together. We can represent these relationships by a simple triangular diagram with K, Na, and Ca at the corners (Figure 3.13).

Both the alkali and plagioclase feldspars contain solid solutions. As a subgroup, the plagioclase feldspars include two species that are end members of a solid solution series. *Albite*, $NaAlSi_3O_8$, represents the *alkalic* end of the series, while *anorthite*, $CaAl_2Si_2O_8$, represents the *calcic* end (Figure 3.13). Table 3.5 gives details of the composition of the plagioclase feldspars. Four additional mineral species in the series are recognized by name in the table, along with the percentages of the albite and anorthite components. The mineral names are not important here, but the figures in this table deserve close attention because the changes in chemical composition from one end to the other are of fundamental importance in understanding the classification of igneous rocks. As the series progresses from the alkalic end (albite) toward the calcic end (anorthite), not only do the proportions of Na and Ca reverse, but there are important changes in the percentage content of aluminum (calculated as *alumina*, Al_2O_3) and silicon (calculated as *silica*, SiO_2). The percentage of alumina nearly doubles from alkalic to calcic extremes, while at the same time silica decreases by about one-third. We shall find in Chapter 4 that plagioclase of the alkalic type is associated with quite different igneous rocks from those in which the calcic type is dominant.

The plagioclase feldspars crystallize in the triclinic system, in which the three crystallographic axes are of unequal length and intersect in oblique angles (see Figure 3.4). These feldspars have perfect cleavage in one set of planes and good cleavage in another, intersecting at an angle of 86° (Figure 3.14). Thus, fractured plagioclase often appears to consist of a set of rough, platelike layers. Specific gravity ranges from 2.6 to nearly 2.8; hardness is 6. Plagioclase varieties show a rather wide range of colors: Albite is typically an opaque white, resembling porcelain; other types are gray; the intermediate types (andesite and labradorite) often exhibit a beautiful iridescent play of many different colors.

The alkali feldspars are represented by a solid solution series between Na and K. They are usually rich in potassium and are often called *potash feldspars*. *Orthoclase*, a potash feldspar with the formula $(K,Na)Al_3SiO_8$, crystallizes in the monoclinic system. It has perfect cleavage in one set of planes, and good cleavage in another set at right angles. Orthoclase has a vitreous luster and ranges from colorless to white, gray, or flesh-colored. Specific gravity is 2.5 to 2.6; hardness is 6.

Another potash feldspar is *microcline*. It has the

Table 3.5 The Plagioclase Feldspar Group

	Name	Percent albite	Percent anorthite	Percent sodium as Na_2O	Percent calcium as CaO	Percent alumina Al_2O_3	Percent silica SiO_2
Alkalic	Albite	100–90	0–10	11	0.0–0.8	20	67
Alkalic	Oligoclase	90–70	10–30	10	3	23	64
Intermediate	Andesine	70–50	30–50	6	8	26	58
Intermediate	Labradorite	50–30	50–70	4	12	30	53
Calcic	Bytownite	30–10	70–90	3	15	32	49
Calcic	Anorthite	10–0	90–100	0.2–0.8	19	35	44

Source: Based on data of W. A. Deer, R. A. Howie, and J. Zussman (1966), *An Introduction to the Rock-Forming Minerals,* New York, John Wiley, pp. 324–325, Table 31.

same chemical formula as orthoclase, but it crystallizes in the triclinic system. Microcline has perfect cleavage in one set of planes and good cleavage in another set at nearly right angles (Figure 3.15). Color usually ranges from white to pale yellow, but is sometimes red or green. Specific gravity and hardness are about the same as for orthoclase.

*Feldspathoid Group

Closely related to the alkali feldspars is a group of silicate minerals called the *feldspathoids*. Like the alkali feldspars, they are silicates of sodium and potassium, but they contain only about two-thirds as much silica. An important member of this group is the mineral *nepheline,* with the formula $(Na,K)AlSiO_4$. The lower content of silica is obvious when compared with orthoclase, $(K,Na)Si_3O_8$. Nepheline crystallizes in the hexagonal system. Unlike the feldspars, it has poor cleavage. It is colorless, white, or yellowish, with a specific gravity of 2.6; its hardness is 5½ to 6.

Quartz

Quartz consists of silicon dioxide, SiO_2. It is one of the commonest minerals among rocks of all three major classes. Quartz crystallizes in the hexagonal system (see Figure 3.4). Quartz crystals typically show the form of a hexagonal (six-sided) prism topped by a hexagonal pyramid (Figure 3.1). It is a hard mineral (7 on the Mohs scale) and it lacks cleavage. It breaks with a conchoidal fracture, so that broken pieces of crystalline quartz resemble massive chunks of broken glass (Figure 3.16). Colorless quartz is called *rock crystal; amethyst* is a violet variety; *rose quartz* is rose-red or pink; quartz may also be smoky or milky in appearance. The specific gravity of quartz is 2.65.

FIGURE 3.12 A. Biotite, or black mica, has highly lustrous cleavage surfaces. (Ward's Natural Science Establishment, Inc., Rochester, N.Y.) *B.* Muscovite is a pale variety of mica and appears colorless in thin cleavage sheets. Surfaces at the top and sides of this specimen are crystal surfaces. In both examples the vertical surfaces facing the observer are cleavage planes. (American Museum of Natural History.)

FIGURE 3.12

Quartz also occurs in a number of *microcrystalline* varieties, with individual crystals too small to be distinguished by the unaided eye. Several of these forms are mentioned in Chapter 5 as occurring abundantly in sedimentary rocks. Among the microcrystalline types are a number of familiar semiprecious gemstones—carnelian, agate, onyx, and jasper.

FIGURE 3.13 Triangular diagram showing the range in composition of alkali feldspars and plagioclase feldspars.

FIGURE 3.14 A cleavage fragment of albite, the alkalic end member of the plagioclase feldspars. Cleavage is perfect in planes running vertically through the mineral parallel to the plane of the page. Imperfect planes at nearly right angles parallel the base of the specimen. (American Museum of Natural History.)

FIGURE 3.13

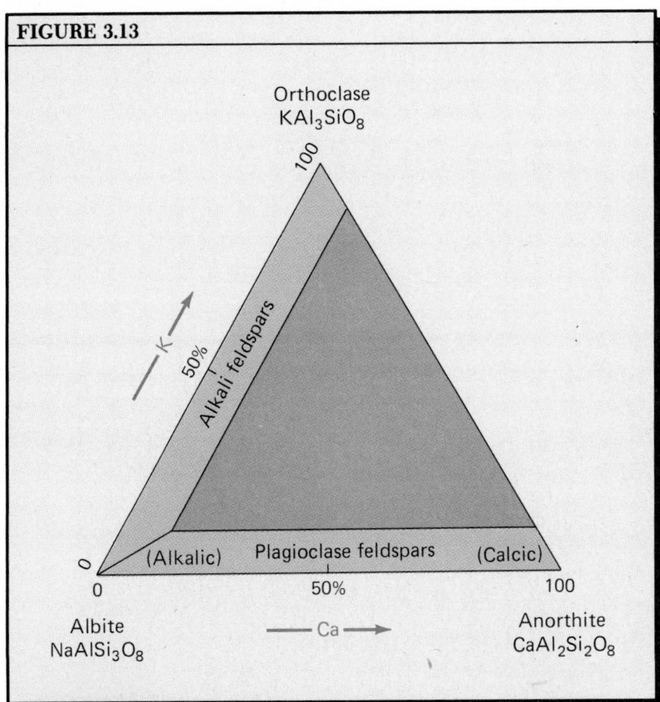

FIGURE 3.14

*Crystal Lattice Structure of Silicate Minerals

The building block of the crystal lattice structure of the silicate minerals is the *silicon-oxygen tetrahedron*, pictured in unexpanded (close-packing) form in Figure 3.17A. A single cation of silicon (Si^{4+}) is surrounded by four oxygen anions (O^{2-}). The silicon ion is small in radius (0.39Å) compared with the oxygen ion (1.40Å). The ratio between these two radii (the

FIGURE 3.15 A cleavage fragment of microcline, one of the alkali feldspars. Two sets of cleavage planes intersect at an angle of about 87.5°, which is very close to a right angle. (Ward's Natural Science Establishment, Inc., Rochester, N.Y.)

FIGURE 3.16 This broken fragment of a large crystal of clear quartz shows conchoidal fracture. The thin, needlelike objects within the quartz are enclosed crystals of another mineral, tourmaline. (Ward's Natural Science Establishment, Inc., Rochester, N.Y.)

FIGURE 3.15

FIGURE 3.16

FIGURE 3.17

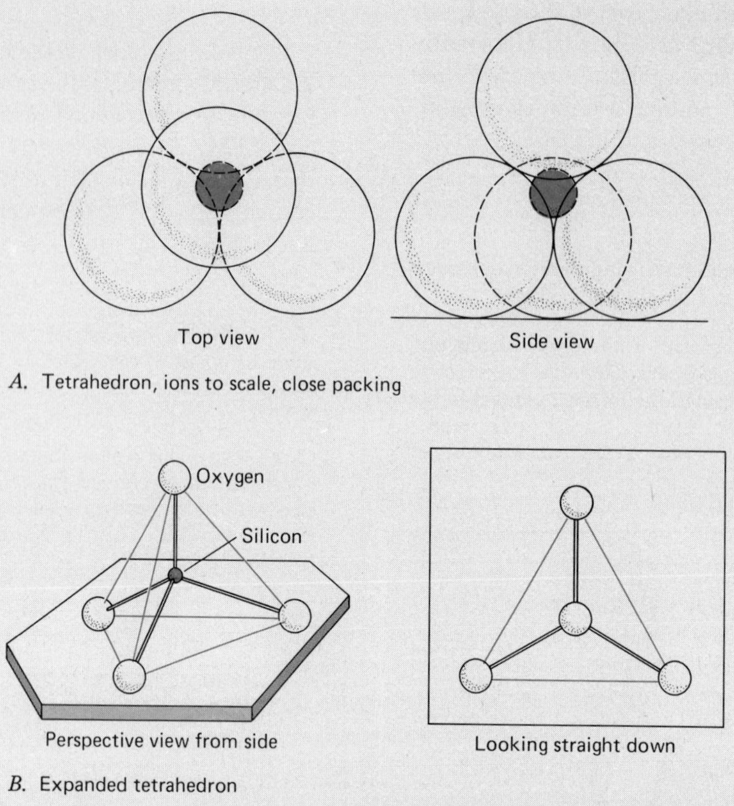

A. Tetrahedron, ions to scale, close packing

B. Expanded tetrahedron

FIGURE 3.18

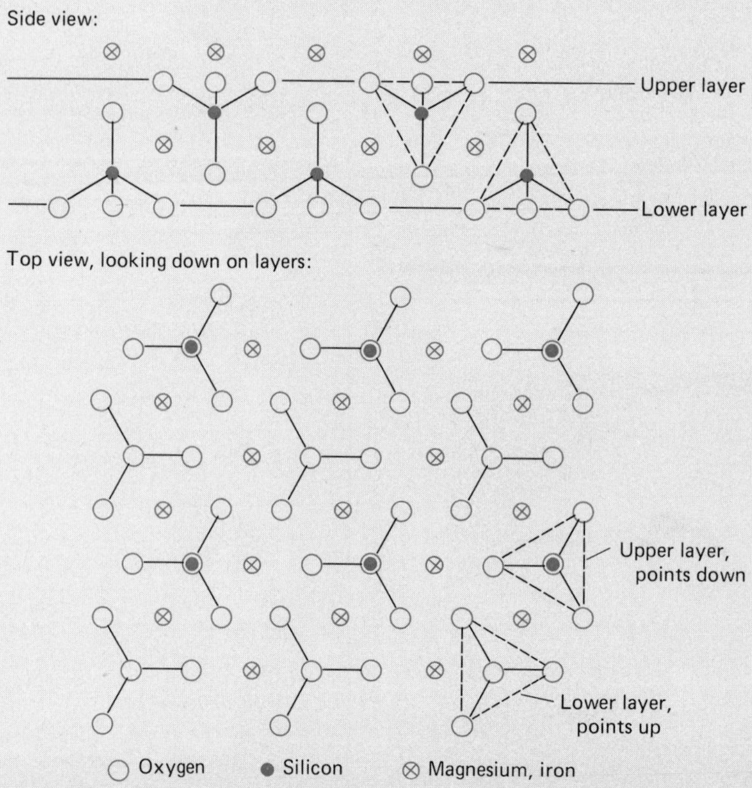

radius ratio) is thus about 0.28, so that the silicon ion is neatly accommodated in the space between the four oxygen ions, which are in close contact, and thus the arrangement is very stable and the bonds are very strong. (The bonding is about half ionic and about half covalent.)

The oxygen ions nevertheless have the capacity to form additional bonds with other ions and with other tetrahedra. The reason such bonding is possible is that each of the four oxygen ions of an isolated tetrahedron requires one additional electron to complete its octet (see Figure 2.12). Thus, each oxygen ion can belong in common to two joined tetrahedra, and the resulting electron-sharing fills the octets of all oxygen ions.

Tetrahedra can form a variety of stable linkages in such forms as rings, chains, sheets, and three-dimensional frameworks. Figure 3.17*B* shows an expanded silicon–oxygen tetrahedron using the stick-and-ball representation.

Mineralogists recognize seven classes of silicates based on the various ways in which silicon–oxygen tetrahedra can be arranged. A complete description of these classes, including their specialized names, is beyond the scope of an introduction to minerals. We will, however, briefly describe the structure associated with each of the six silicate minerals or mineral groups already described, following the same sequence of presentation.

The structure of olivine illustrates a system of *isolated tetrahedra* (also known as ***independent tetrahedra***), as shown in Figure 3.18. The silicon–oxygen tetrahedra are arranged in an orderly pattern in pairs of layers, alternately inverted (upside down) so that the points of the upper layer fit between points of the layer beneath. In the spaces between the tetrahedra are ions of iron and magnesium. Bonding of tetrahedra is carried out by means of the intervening Fe and Mg ions, each of which has two electrons in its outer shell.

FIGURE 3.17 The silicon–oxygen tetrahedron shown in close-packed arrangement (*A*) and expanded in a ball-and-stick model (*B*).

FIGURE 3.18 Atomic structure of olivine. Layers of independent silicon–oxygen tetrahedra are alternately inverted, while ions of magnesium and iron occupy positions between tetrahedra.

FIGURE 3.19 A single chain of silicon–oxygen tetrahedra forms the basic structure of the pyroxene group. The c-axis is in a horizontal position, running from left to right. *A*. Close-packed ions in a chain. *B*. Expanded tetrahedra seen in plan view from above and in perspective from the side. (Modified from GENERAL CRYSTALLOGRAPHY: A BRIEF COMPENDIUM by W. F. de Jong. W. H. Freeman and Company. Copyright © 1959.)

FIGURE 3.19

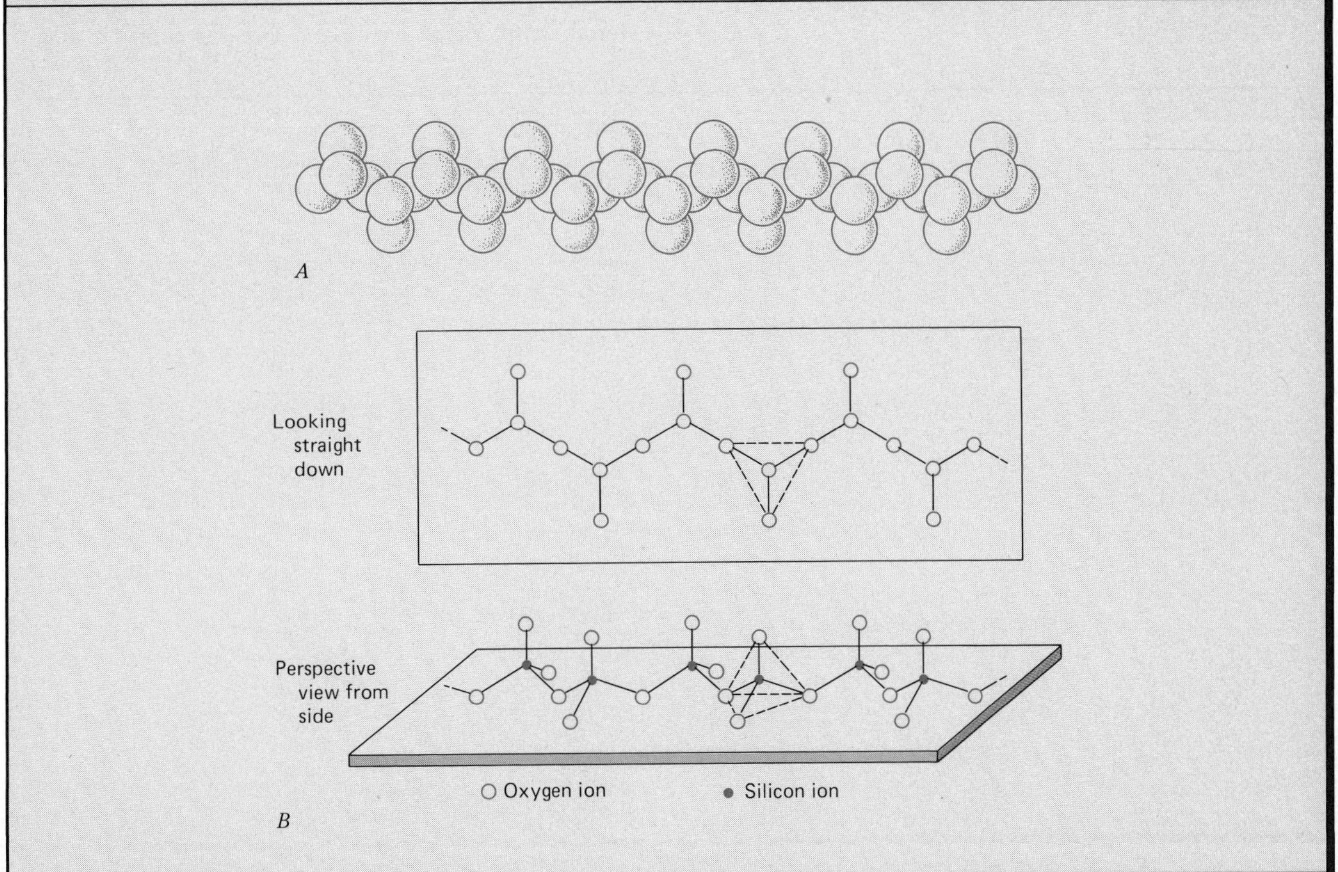

These electrons are shared with surrounding oxygen ions of the tetrahedra. Because the lattice structure of olivine allows for no planes of weakness in the bonding, olivine has no cleavage.

Structure of the pyroxene group consists of *single chains* of tetrahedra in which an oxygen ion is shared between each two adjacent tetrahedra in the chain (Figure 3.19). The chains run parallel with the vertical axis of the crystal form (*c-axis* in Figure 3.4). Spaces between chains are occupied by ions of Mg, Fe, or Ca and by Al ions (see Figure 3.21). Bonding between chains is made through these intervening cations, which share electrons with the oxygen ions of the tetrahedra. Bonds between chains are weaker than those within the chains, so two sets of planes of weakness exist within the lattice structure at about right angles to one another, giving rise to the 87°–93° cleavage angles of the pyroxenes. The three-dimensional geometrical relationships among tetrahedrons and cations are extremely complex, but we have greatly simplified them in Figure 3.21.

The amphiboles have a structure of *double chains*, with sharing of oxygen ions between adjacent single chains (Figure 3.20). In this way, six-sided rings (hexagons) of tetrahedra are formed. Between the double chains are located the cations of Fe, Mg, Na, and Al.

FIGURE 3.20 Double chain of silicon–oxygen tetrahedra seen in plan view from above and in perspective. (Modified from GENERAL CRYSTALLOGRAPHY: A BRIEF COMPENDIUM by W. F. de Jong. W. H. Freeman and Company. Copyright © 1959.)

The arrangement is such that the weaker bonds between intervening cations and oxygen ions of the chains form two sets of intersecting planes of weakness through the crystal (Figure 3.21). The angles between the two sets of planes are 124° and 56°, explaining the distinctive cleavage structure of the amphiboles. As in the case of the pyroxenes, the three-dimensional geometry of the complete lattice structure of the amphiboles is greatly simplified in Figure 3.21.

The structure of the mica group is of *sheets* of tetrahedra. Each sheet consists of two layers of tetrahedra linked into a hexagonal net, as shown in Figure 3.22. Alternate layers have the tetrahedra points downward (inverted). Within the sheet, magnesium ions occupy a layer between layers of tetrahedra whose points are opposed, forming strong bonds with adjacent oxygen ions. Hydroxyl ions are fitted into the spaces between the oxygen ions in the tetrahedra points. The potassium ions lie in a plane between the sheets of tetrahedra; here the tetrahedral bases face each other. Bonds between the potassium ion layer and the adjacent tetrahedral layers are of the weak type illustrated earlier in this chapter by the structure of graphite. As a result, cleavage is excellent in this plane, whereas bonds are strong in other directions within each sheet.

The structure of quartz, the feldspars, and the feldspathoids belongs to a class called *frameworks*. A framework is a continuous network of tetrahedra linked at the corners so that all oxygen ions are shared. Several such frameworks are possible. In the feldspars, the aluminum ion may occupy the center of an oxygen tetrahedron, replacing silicon, and the metallic ions

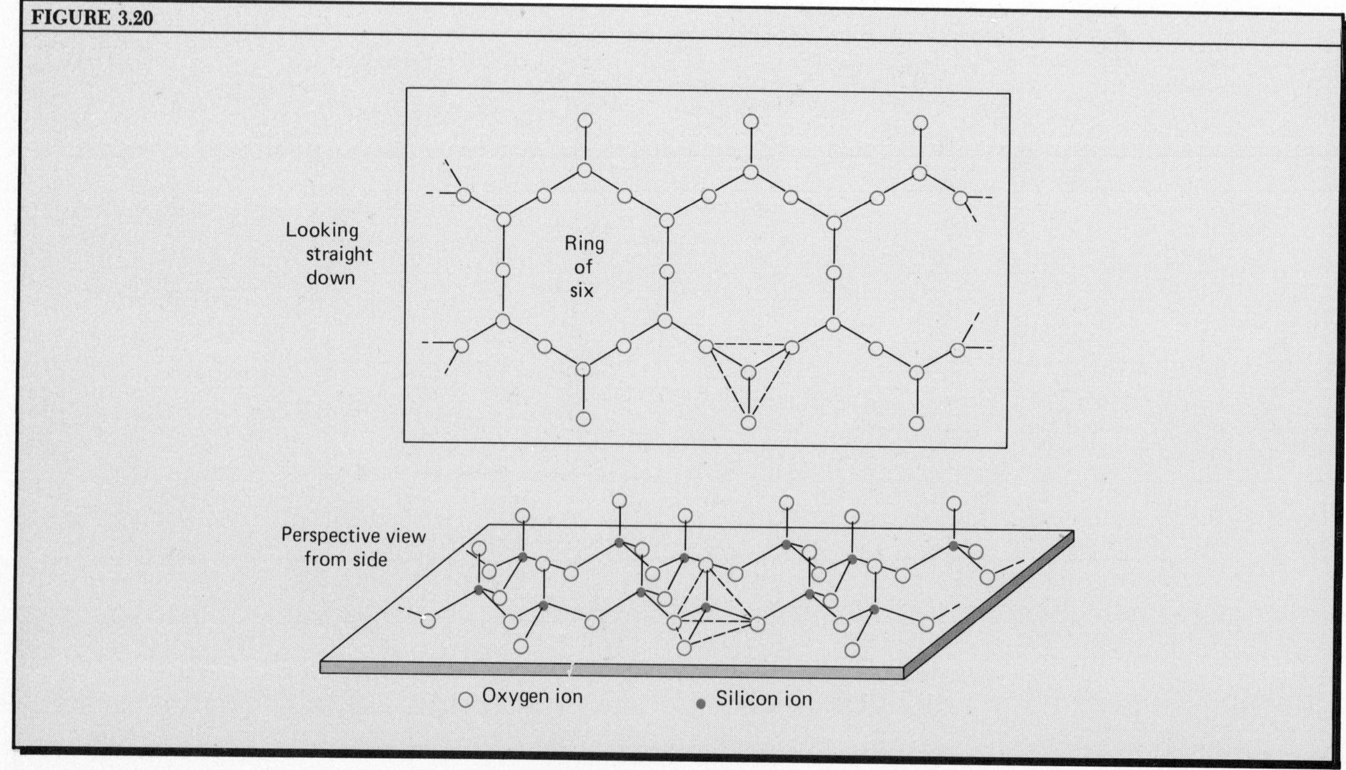

Na, K, and Ca, occupy voids between tetrahedra. Tetrahedra containing a central Al ion form bonds with ions of Na, K, and Ca between the tetrahedra. Certain sets of bonds with these intervening ions are weaker than others, with the result that cleavage is developed along two intersecting planes. In quartz, the entire framework consists of silicon–oxygen tetrahedra and the bonding is equally strong in all directions (Figure 3.23). As a result quartz shows no cleavage.

Mafic and Felsic Mineral Groups

We have presented the silicate minerals and mineral groups in a purposeful sequence in which specific gravity decreases in order from olivine to quartz. Notice also that in a general way the list runs from dark to light-colored minerals and that as the color lightens, there is a corresponding decrease in the proportion of the heavy metals magnesium and iron. In this sequence, the proportion of aluminum ranges in reverse from zero in olivine to high in potash feldspar.

Geologists have found it meaningful to divide the silicate minerals forming the bulk of igneous rocks into two classes: *mafic minerals* and *felsic minerals*. The list is as follows:

Mafic group	**Specific gravity**
Olivine	3.3–4.4
Pyroxene group	3.2–3.3
Amphibole group	3.2
Biotite mica	2.8–3.2
Felsic group	
Muscovite mica	2.8–2.9
Feldspar group	
Plagioclase	2.6–2.8
Potash feldspar	2.5–2.6
Quartz	2.65

The word "mafic" is coined from "magnesium" and "ferric"; the word "felsic" from the words "feldspar"

FIGURE 3.21 Schematic diagrams of the structure of *A.* pyroxene and *B.* amphibole showing the relationship of cleavage to atomic structure. Triangles are silicon–oxygen tetrahedra. The chains are shown in cross section. A pencil held perpendicular to the page would represent the long dimension of the chains, which is also the c-axis of the crystal structure. Cleavage fractures follow a series of zigzag steps between chains.

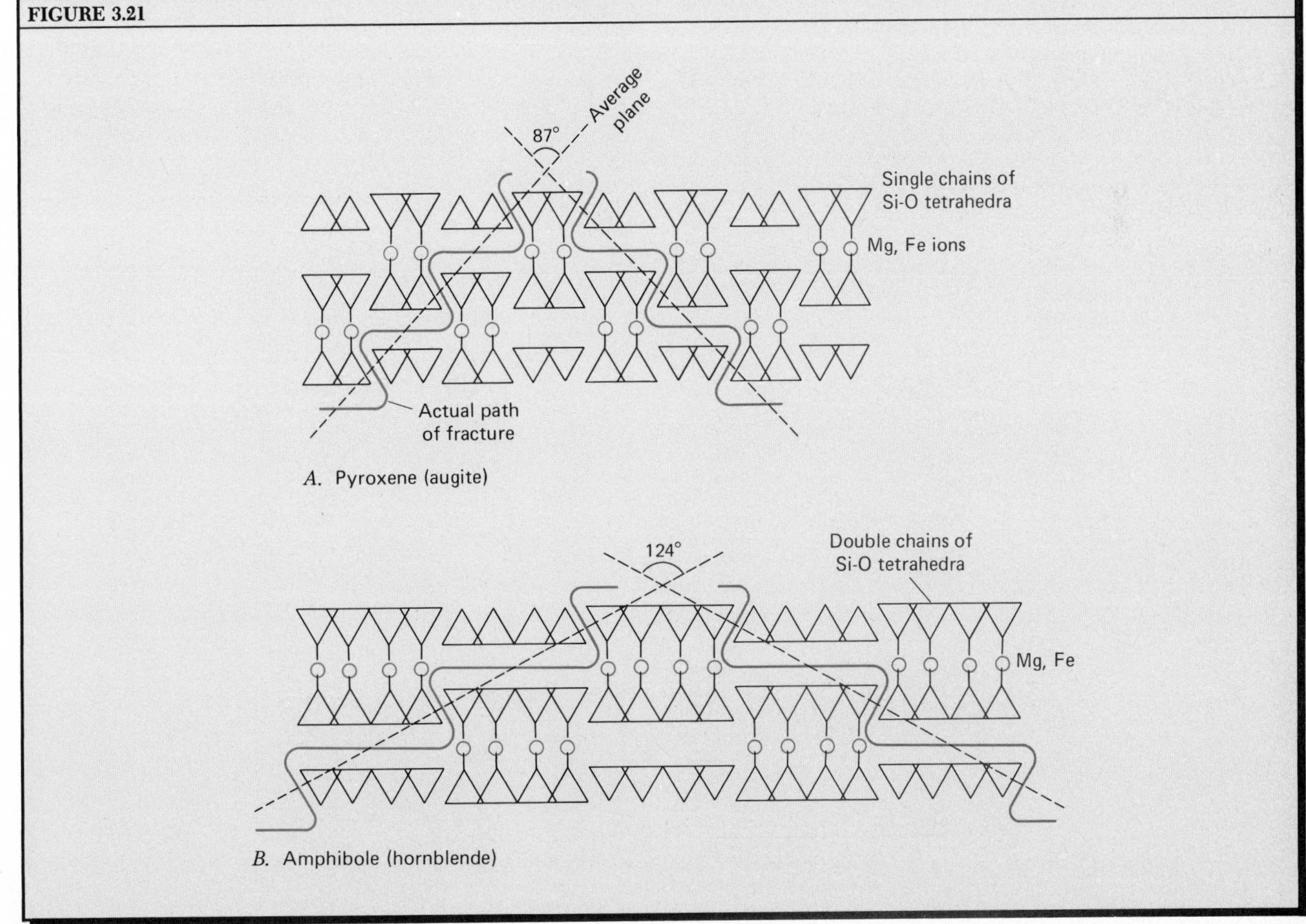

and "silica." In the study of igneous rocks, we will find that a predominance of mafic minerals produces rocks of a mafic rock type, whereas a predominance of felsic minerals produces rocks of a felsic rock type. Mafic rock forms the igneous crust beneath the ocean basins, whereas felsic rock forms the uppermost crustal layer of the continents. Mafic rocks are denser than felsic rocks, a fact of great importance in explaining the physical relationships between continental and oceanic crust. The mantle beneath the crust consists almost entirely of the two mafic minerals of high specific grav-

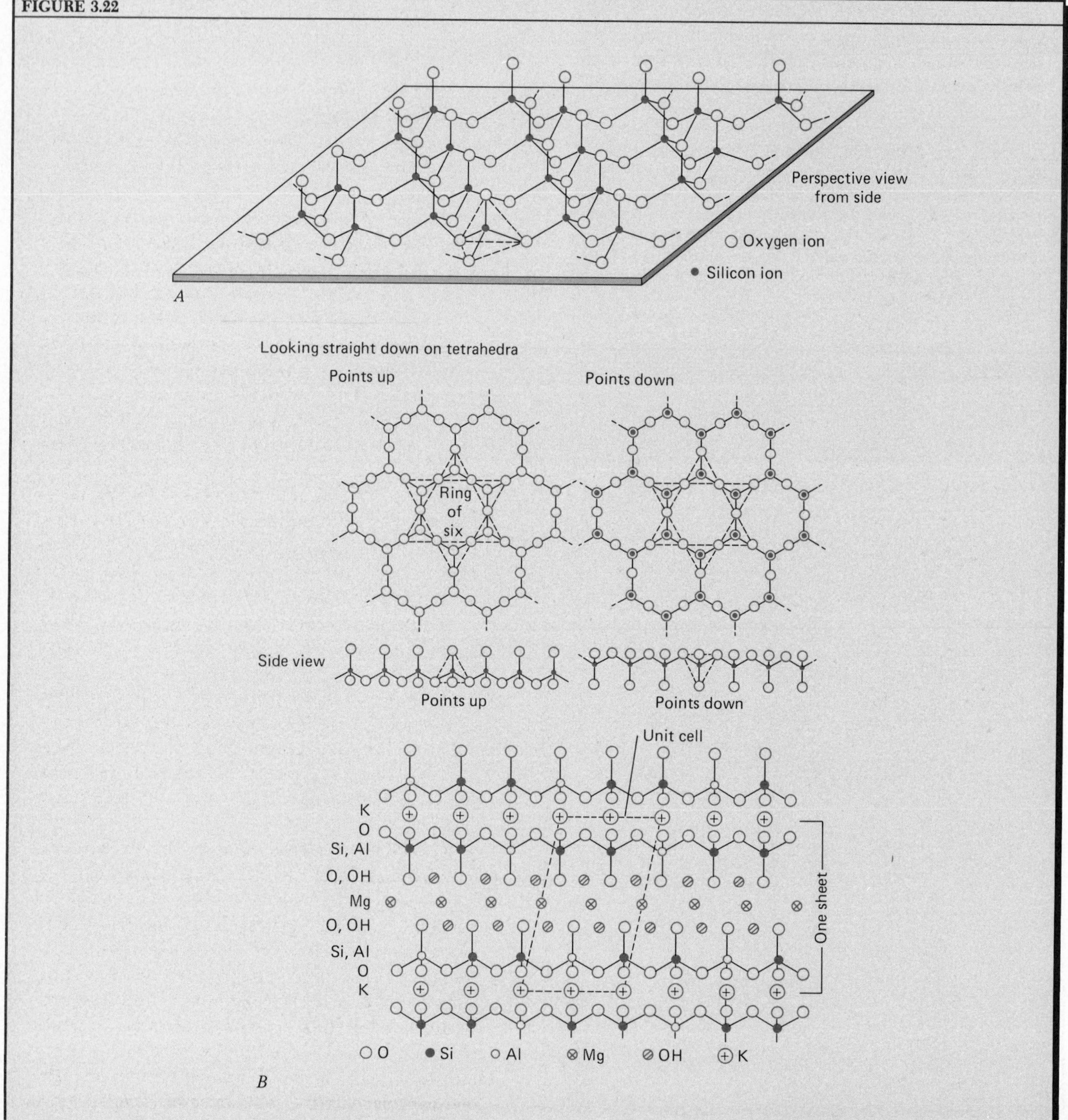

FIGURE 3.22 *A*. Perspective view of a single layer of silicon–oxygen tetrahedra in the atomic structure of mica. (Modified from GENERAL CRYSTALLOGRAPHY: A BRIEF COMPENDIUM by W. F. de Jong. W. H. Freeman and Company. Copyright © 1959.) *B*. Structure of mica sheets, with a detail of the arrangement of ions and tetrahedra within a single sheet. (Modified from W. A. Deer, R. A. Howie, and J. Zussman, 1966, *An Introduction to the Rock-Forming Minerals*, Longman Group Ltd., London, Figures 69 and 71, pp. 194–195.)

ity—olivine and pyroxene. Because of this, mantle rock is designated as yet a third rock type—ultramafic rock (Chapter 4).

With this description of the principal silicate minerals and mineral groups, we are ready to tackle the subject of the igneous rocks, their origin, and their classification. In terms of chemical composition and atomic structure, the silicates are extremely complex. We have only been able to skim lightly over this difficult area of physical geology.

FIGURE 3.23 Atomic structure of tridymite, a mineral closely related to quartz and of the same composition. *A.* Perspective view of the framework of silicon–oxygen tetrahedra. (Modified from W. L. Bragg, 1937, *Atomic Structure of Minerals*, Cornell University Press, Ithaca, N.Y.) *B.* and *C.* Layers of tetrahedra seen in plan and in cross section.

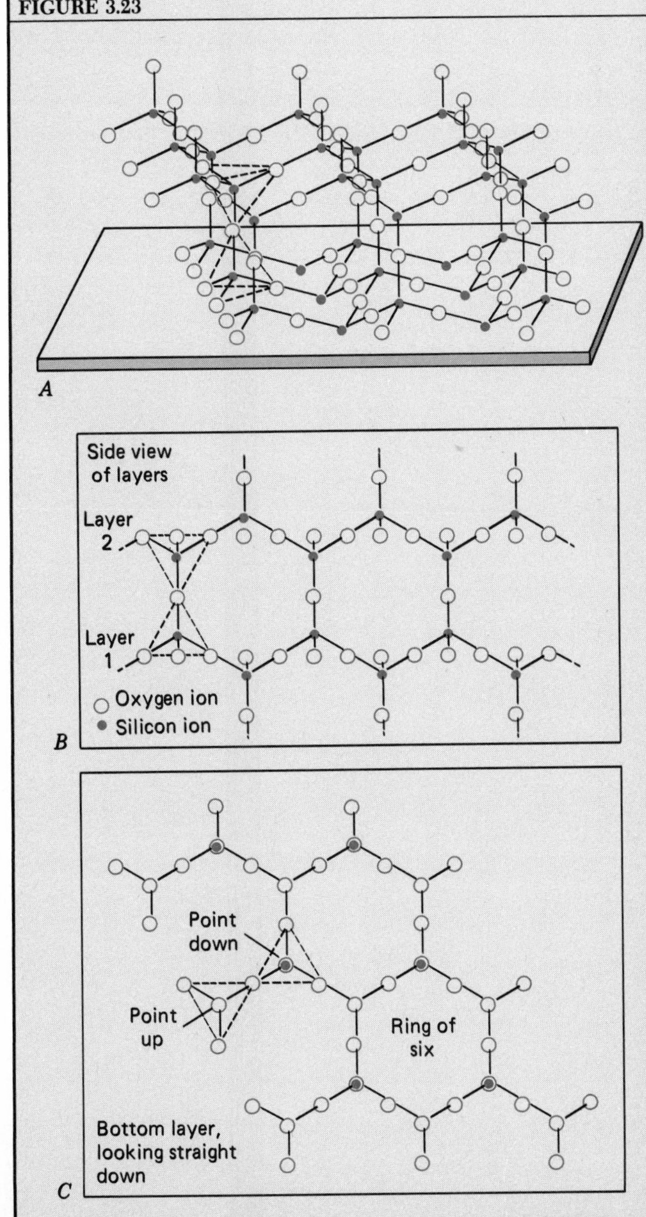

Key Facts and Concepts

Mineralogy *Mineralogy* deals with the chemical composition, physical properties, and atomic structure of minerals. A *mineral* is a naturally occurring solid inorganic substance, having either an orderly atomic structure and a definite chemical composition or one that is variable between stated limits.

Composition of the earth's crust The earth's crust is formed principally of the chemical elements oxygen (47%), silicon (28%), aluminum (8%), and iron (5%), with lesser amounts of calcium, sodium, potassium, and magnesium. Both iron and magnesium are much less abundant in the crust than in the earth as a whole.

Chemical classes of minerals Minerals can be grouped into major classes according to chemical composition: native elements, oxides, sulfides, carbonates, halides, sulfates, silicates, hydrous silicates, and hydroxides.

Minerals as crystalline solids The crystalline structure of a mineral expresses itself in both external crystal form and internal atomic structure. Mineral crystals possess *crystal symmetry* and fall into six *crystal systems*. External crystal form is an expression of the *crystal lattice* structure of atoms or ions of which the mineral is composed.

Physical properties of minerals Besides crystal structure, physical properties of minerals include *cleavage*, fracture, *specific gravity, hardness*, luster, color, streak, and various optical properties.

Atomic structure of minerals The orderly internal arrangement of atoms within a mineral can be studied by the *X-ray diffraction method*. Systematic arrangement of ions and atoms forms a *space lattice* consisting of *unit cells*. Strong bonds between atoms may be covalent, involving electron sharing. Weak bonding between atom layers results in excellent cleavage.

Silicate minerals *Silicate minerals*, making up most of the bulk of all igneous rocks, are *olivine*, four aluminosilicate groups (*pyroxene, amphibole, mica*, and *feldspar*), and *quartz*. The feldspar group includes the *alkali feldspars* and the *plagioclase feldspars*.

Crystal lattice structure of the silicate minerals The building block of the silicate minerals is the *silicon–oxygen tetrahedron*. *Isolated tetrahedra* bonded to intervening ions of iron and magnesium characterize olivine. The pyroxenes have *single chains* of tetrahedra; the amphiboles, *double chains*. *Sheets* of tetrahedra make up the mica group. Quartz and the feldspars show *framework* structures of tetrahedra with shared oxygen ions.

Felsic and mafic mineral groups Quartz and the feldspars make up the *felsic minerals*, specific gravity 2.6–2.8. Biotite mica, amphiboles, pyroxenes, and olivine comprise the *mafic minerals*, specific gravity 2.9–3.3.

Questions and Problems

1. Examine some crystals of quartz, preferably specimens from several different localities. Compare these closely with some crystals of clear calcite, using a hand lens. Look next at the surfaces of some broken fragments of quartz and calcite. Put yourself in the time (circa 1800–1820) and place (Paris) of the French scientist Abbé Haüy and speculate about the internal structure of crystalline minerals. List all facts you can observe from the specimens; use this list to support your theory of the internal structure of crystals, how this structure relates to their external form, and how the crystal breaks into smaller fragments.
2. Make a careful comparison of the element list in Table 3.1 with that in Table 3.2. In the space in front of the element name in Table 3.1 write its rank number as shown in Table 3.2. Make separate lists of those crustal elements that have moved up or down in rank, and compare these lists with the chemical compositions of the silicate minerals shown in Table 3.4. Are the elements in your "down" list found in the felsic group or in the mafic group? Are any of the elements on your "up" list found in the composition of olivine? Make a summary statement in which you speculate about the composition of the earth's crust, mantle, and core in terms of the elements and minerals that dominate their makeup. (Consult Chapter 1 for information on the crust, mantle, and core.)
3. Examine the list of ten minerals that the Mohs scale of hardness comprises. Insert the mineral name *graphite* (hardness between 1 and 2) between talc and gypsum. Place *halite* (hardness 2½) on the line beside gypsum. Insert *mica group* (hardness 2½–3) between gypsum and calcite; *pyroxene* and *amphibole* (5½) between apatite and orthoclase; *plagioclase* (6) beside orthoclase; and olivine (6½–7) between orthoclase and quartz. Based upon what you have learned of the atomic structure of diamond, graphite, and the silicate minerals, draw some conclusions as to the general relationships between mineral hardness and the kinds of atomic structures and bonds found in these minerals. (Note: Ionic bonding in halite is explained in Chapter 2.) The atomic structures of quartz and diamond are quite similar, so why is diamond so much harder than quartz? Diamond dust is used to shape and polish a cut diamond. How can abrasion take place between particles of the same mineral substance?

Igneous Rocks and Igneous Activity

Igneous rocks make up about 80% of the mass of the earth's crust. It is not surprising, then, that the subject of igneous rocks holds a very prominent place in physical geology.

In Chapter 3, we stated that the eight most abundant crustal elements provide nearly all the essential chemical components of eight silicate minerals or mineral groups. Hydrogen, in the form of the hydroxyl ion (OH), should be added to complete this element list. The igneous rocks are composed almost entirely of the eight silicate minerals or mineral groups (the silicate minerals make up about 99% of the bulk of all igneous rocks). Also present in igneous rocks are a number of other common minerals, but their importance is minor.

Magma

Igneous rock is formed from *magma,* a high-temperature mixture of the chemical ingredients of silicate minerals. Magma normally includes substances in liquid, solid, and gaseous states. A large proportion of most magmas consists of a hot liquid, or *melt,* present because the temperature of the magma is above the melting points of certain magma ingredients. In a melt, the metallic ions move about more or less freely, without being organized into crystal lattice structures. However, the silicon–oxygen tetrahedra probably remain largely intact and may be linked to one another in complex but irregular chains and networks. Suspended in the melt of most magmas are crystals of minerals formed during early stages of magma cooling. If the proportion of suspended crystals to liquid is high, it gives the magma some of the physical properties of a solid. It may help to visualize this mixture as resembling slush, produced by rapid melting of snow, in which ice crystals are mixed with liquid water. Slush has enough strength to hold its form on a flat pavement but contains a large proportion of free water. In addition to both liquids and solids, magma also contains various gases, which are dissolved in the melt.

Magma that appears at the earth's surface, coming into direct contact with air or water, is called *lava,* and its emergence from the solid earth provides us with sure proof that magma exists. Because lava is rapidly cooling and losing its dissolved gases, however, it cannot provide us with an accurate picture of the magma that lies far beneath the earth's surface. But it is important to note that magma formed deep within the crust undergoes various chemical and physical changes as it moves upward through the crust. Some of the chemical components originally present may be separated out, while new chemical components may be added from surrounding rock with which the magma comes in contact. The study of magma is thus extremely difficult and relies on many inferences based on indirect scientific evidence. The study of plate tectonics has shed a great deal of light on major sources of magmas and why the composition of emerging magma often varies from place to place.

There is general agreement among geologists that magma can be formed by the melting of solid rock in certain depth zones in the asthenosphere. You will recall from Chapter 1 that the asthenosphere is a soft layer of the upper mantle within which rock temperature is close to its melting point. Evidence from earthquake waves, explained in Chapter 8, indicates that the mantle is closest to its melting point in the depth of 100 to 200 km (see Figure 4.1*A*). It seems most probable that only a small fraction of the mantle rock at this depth range actually melts, while the larger proportion of the mass remains crystalline. This phenomenon is called *partial melting*. The melted fraction, which is a liquid less dense than the crystalline fraction, tends to rise through the spongelike crystalline mass and to collect in magma pockets at shallower depths. It is supposed that the enormous volume of mafic magma continually rising along the spreading plate boundaries collects in magma chambers near the base of the oceanic crust at a depth of 4 to 6 km below the ocean floor. Magma that rises from a location above a descending lithospheric plate at a subduction (converging) boundary is produced under quite different conditions and depths, as explained later in this chapter.

Temperatures and Pressures in Magma

Temperature increases rapidly with depth into the earth. This increase in rock temperature, as measured directly with thermometers in deep mines, averages about 1 °C per 30 m. Downward through the crust and into the upper mantle temperatures can only be estimated. Figure 4.1*A* is a graph showing the estimated increase in temperature with depth. Two curves are shown, one typical of conditions under the continents, the other for conditions under the ocean basins. Both curves show that while the temperature increases, the rate at which it does so falls off rapidly after about 300 km. Temperatures within the asthenosphere are largely in the range from 1000 °C to 1500 °C. For comparison, the melting point of iron at the earth's surface is about 1500 °C, while molten iron in a blast furnace reaches a temperature of 2000 °C. It might seem like a simple procedure to measure the melting point of a specimen of igneous rock and, using the temperature–depth curves in Figure 4.1*A*, find the depth at which melting is to be expected and magma to be formed.

Unfortunately, the actual relationships are not that simple. The temperature at which silicate minerals will melt depends upon several factors. Each mineral has its own melting point for a given environment. For example, in the surface environment, under the atmospheric pressure that prevails, pure olivine melts in the temperature range of about 1600 °C to 1800 °C, whereas plagioclase feldspar melts in the range from 1200 °C to 1400 °C.

Downward into the earth, the confining pressure to which all rock is subjected increases steadily under the load of the overlying rock layer. The unit of pressure used by geologists is the ***kilobar***, which is equal to one thousand bars, a ***bar*** being approximately equal to the pressure of the earth's atmosphere at sea level—about 1 kg per sq cm. Figure 4.1*B* is a graph to show the approximate increase of confining pressure with depth. An increase in pressure causes the temperature of the melting point of a given mineral or rock to increase. For example, the melting point of iron at a depth of about 100 km, under a confining pressure of 40 kb, is about 1650 °C, which is 150 °C higher than at the surface.

This relationship between melting point and pressure applies only to conditions in which no water is present. In presence of water, the melting-point tem-

FIGURE 4.1 *A*. Estimated increase in temperature with depth under continents and under ocean basins. The dashed line shows the melting point of dry peridotite. (Based on data of W. G. Ernst, 1969, ***Earth Materials***, Prentice-Hall, Englewood Cliffs, N.J., p. 107, Figure 5–9.) *B*. Estimated increase of confining pressure with depth.

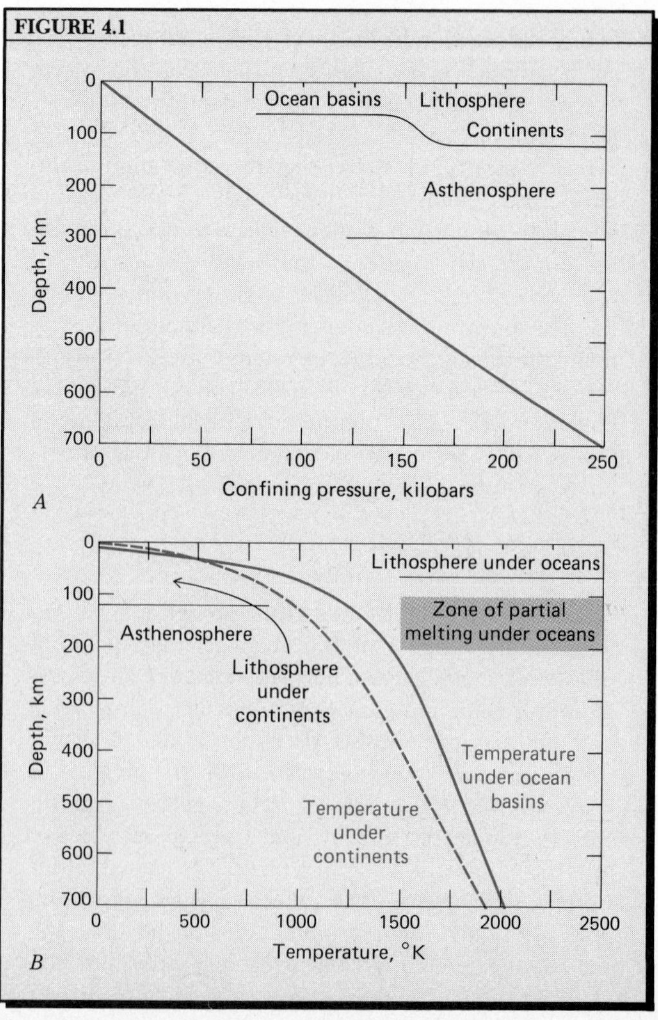

Table 4.1 Composition of Gases from Internal Earth Sources Compared with Total Earth Volatiles

	Volatiles of Earth's hydrosphere and atmosphere	Gases in hot springs, fumaroles, and geysers	Volcanic gases from basaltic lava of Mauna Loa and Kilauea
Water, H_2O	92.8	99.4	57.8
Total carbon, as CO_2	5.1	0.33	23.5
Sulfur, S_2	0.13	0.03	12.6
Nitrogen, N_2	0.24	0.05	5.7
Argon, A	trace	trace	0.3
Chlorine, Cl_2	1.7	0.12	0.1
Fluorine, F_2	trace	0.03	—
Hydrogen, H_2	0.07	0.05	0.04

Source: Data from W. W. Rubey (1952), *Geol. Soc. Amer. Bull.*, vol. 62, p. 1137, Table 6. Figures in table represent percentages by weight.

perature is considerably lowered. For example, using a confining pressure equal to a depth of 20 km, a dry rock sample consisting of about 60% pyroxene and 40% calcic plagioclase feldspar has a melting point of about 1300 °C to 1400 °C. The same rock sample containing a substantial proportion of water melts at only 700 °C to 1000 °C. Thus, the importance of water in magma is very great. Even a small amount of water in a magma can greatly lower the temperature at which the silicates will remain molten. A rising magma, experiencing a progressive lowering of temperature as it moves upward into cooler zones of surrounding rock, can remain molten at considerably lower temperatures with water present than if the magma consisted of the same silicates without water. Because of this effect, magma with water can reach closer to the earth's surface before solidifying, and it can pour out upon the surface as lava in greater volumes than if water were absent.

Volatiles in Magma

Water in magma occurs as a dissolved gas, and other dissolved gases are also present. Termed *volatiles*, these are chemical substances which remain in the liquid or gaseous state at a much lower temperature than that of the mineral-forming silicates. Thus, the volatiles become separated from the magma as temperatures drop and the silicate minerals crystallize. We can understand from this why, when a specimen of an igneous rock is melted in the laboratory, the melt it produces cannot be a true copy of the magma from which the rock originated.

Although we do not know a great deal about the content of volatiles in magmas at depth in the crust and upper mantle, much has been learned by sampling gases emitted from volcanoes along with emerging lava (Table 4.1).

The emission of volatiles with magma at the earth's surface is a geological process called *outgassing;* it has acted throughout all geologic time and continues today. As water is the dominant constituent of the volatiles contained in lava, we infer that it is by volume about 90% of the gas contained in an average magma. Estimates of the proportion of water in magmas show a range from 0.5 to 8%, while fresh igneous rocks commonly contain about 1% water, entrapped in small cavities within the minerals during their crystallization. Besides water, the list of volcanic volatiles includes carbon (as carbon dioxide gas), sulfur, nitrogen, argon, chlorine, fluorine, and hydrogen (Table 4.1).

Outgassing is responsible for the formation of the earth's atmosphere and hydrosphere. The *atmosphere* is the gaseous envelope that surrounds the earth, while the *hydrosphere* is the total free water of the globe, including not only the oceans but all fresh water on the land and beneath the land surfaces. Nitrogen, oxygen, argon, and carbon dioxide together make up over 99% of the atmosphere, and it is clear from Table 4.1 that these ingredients reached the earth's surface through outgassing. Molecular oxygen (O_2) was formed from water and released to the atmosphere through the process of photosynthesis in green plants.

Outgassing of water was also responsible for the growth of the early world ocean. The process was probably very rapid in the first aeon of geologic time. Salts accumulated in seawater through the decomposition of silicate minerals rich in ions of sodium, magnesium, calcium, and potassium. It is thought that as early as 3 billion years ago the salinity (salt concentration) of seawater had reached a value comparable to what it is today. (Salts in seawater are described in detail in Chapter 5.) Chlorine, which today makes up about 55% of the total weight of all matter dissolved in seawater, and sulfur, found in much smaller proportion, were contributed as volatiles through outgassing.

Crystallization of a Silicate Magma

Crystallization, the process of change from liquid to solid, begins to take place in a silicate magma at a certain critical combination of temperature and pressure. However, all minerals do not begin to crystallize at the same time, nor does a mineral, once formed, necessarily remain intact and unchanged from that point on. Instead, as temperature falls, the early-formed minerals may later be changed in composition, or they may be dissolved and their ions reformed into new minerals. This process of change is referred to as *reaction*, and the orderly series of such changes is referred to as a *reaction series*.

Reaction can take one of two forms. First is *continuous reaction*, in which the earlier formed mineral is gradually changed in composition by the substitution of ions of one element in the magma for another in the mineral. An example is the plagioclase feldspar series (see Table 3.5). The first feldspar to crystallize is of calcic composition near the anorthite end of the series. As the temperature of the magma continues to fall, sodium ions in the magma are substituted for calcium ions in the crystallized feldspar, causing a gradual change in the plagioclase toward the alkalic end of the series.

A second kind of reaction is *discontinuous reaction*, which takes place at a certain temperature during magma cooling. It is quite different from reaction that occurs over a range of temperatures during the crystallization of plagioclase feldspars. In discontinuous reaction an early-crystallized mineral—olivine, for example—reacts with the remaining liquid, whose composition is constantly changing during cooling to form a mineral of different composition—in this case, pyroxene. Pyroxene is stable at the lower temperature. Two minerals related in this way are known as a *reaction pair*.

The Bowen Reaction Series

Let us assume we are dealing with a silicate magma of average crustal composition, for which it is possible to recognize a complete reaction series that will take place during crystallization. The series is called the *Bowen reaction series*—after N. L. Bowen, an American scientist who developed the concept. The total series consists of two converging branches, one of continuous, the other of discontinuous, reactions (Figure 4.2).

The mafic minerals follow the discontinuous reaction series, starting with olivine and ending in biotite mica. The feldspars follow a continuous series, in which calcic plagioclase forms about the same time as olivine, while alkalic plagioclase is formed at about the same time as the amphiboles and biotite mica.

An interesting feature of the series is the progression of silicon–oxygen tetrahedra arrangements in the discontinuous reaction series of the mafic minerals (Figure 4.2). Olivine, with its independent tetrahedra, has the simplest arrangement, followed by the single-chain structure of the pyroxenes, the double-chain structure of the amphiboles, and finally by the highly complex sheet structure of the micas. This arrangement suggests that stability of the simpler, independent structures is greater at high temperatures. It seems that more complex structures, in order of their complexity, are stable only at lower temperatures.

The high-silica felsic minerals, potash feldspar, muscovite mica, and quartz generally crystallize last and at the lowest temperatures.

Fractionation

So far, we have assumed that the crystallizing magma retains all crystallized minerals with the remaining liquid melt, but this may not be the case. It is thought that in many cases the crystallized minerals become separated from the remaining silicate melt in a process called *fractionation*. For example, early-formed crystals may be left behind as the magma migrates upward, or the crystals may simply settle out to form a layer at the base of the magma body. Removal of crystallized minerals changes the average chemical composition of the remaining liquid magma and consequently modifies the series of reactions that can follow.

Let us consider how the Bowen reaction series is related to the formation of igneous rocks of various mineral compositions. Suppose that fractionation takes place after olivine and pyroxene have been formed. If these crystalline minerals remain behind while the melt moves to a shallower zone, they form a dark rock called peridotite. If the reaction series is allowed to progress to the stage in which pyroxene (augite) and intermediate plagioclase feldspar are formed, the separated crystals form a rock known as gabbro.

The remaining components of the liquid magma are now comparatively richer in silica, aluminum, and potassium because most of the calcium, iron, and magnesium have been used up. This remaining magma may crystallize after migrating to a different physical location from that of the earlier-formed minerals. The result may be an igneous rock, predominantly composed of quartz and potash feldspar, such as granite. The final residual matter of the magma, a watery solution rich in silica, remains fluid at comparatively lower temperatures. From this solution are deposited rock veins of a type known as *granite pegmatite*, which consists of large crystals of quartz, potash feldspar, and muscovite or biotite mica. This stage in magma crystallization is called the *hydrothermal stage*. ("Hydrothermal" literally translates as "hot water.") Ultimately, the water and other minor volatile constituents reach the earth's surface, to escape in fumaroles (vents emitting hot gases) and hot springs.

We have described *magmatic differentiation*, a general process in which the original magma, with its full range of component elements, is separated into rocks of quite different mineral composition. The process can be responsible for an arrangement of igneous rocks into a series ranging from *mafic igneous rock* rich in iron, magnesium, and calcium to *felsic igneous rock* rich in aluminum, sodium, and potassium, and with a high silica content. Our classification of igneous rocks is thus based upon principles of the Bowen reaction series and processes of mineral fractionation and magmatic differentiation.

From the description of magmatic differentiation, you might be led to think that there exists at great depth a single kind of source magma, a *primary magma*, containing all the ingredients necessary to produce the full range of rocks from mafic through felsic. Actually, for many decades, geologists favored such a concept. They were unaware, of course, of the basic principles of plate tectonics, which require that new crust of mafic composition be continually formed by rising magma at spreading plate boundaries and that subduction carry the edges of lithospheric plates down into the mantle where the upper surfaces of the plates are melted and transformed into new magma.

In earlier paragraphs we suggested a modern viewpoint in which partial melting within the asthenosphere produces the mafic magma that eventually rises to form new crust in spreading plate boundaries. Partial melting provides a type of fractionation, but it works in reverse from the fractionation occurring in a cooling magma.

Plate tectonics provides a remarkable means of *recycling* rock of any one of the three rock classes—igneous, sedimentary, or metamorphic—by converting existing rock into magma. Of course, magma produced by melting preexisting rock will have a distinctive chemical makeup—felsic or mafic—depending upon the composition of the preexisting rock. Thus, while there is some excellent geological evidence that fractionation of magma and magmatic differentiation have occurred at certain places in the earth's crust, with the origin of several different types of magma, the concept of a vast primary magma does not follow as a valid consequence.

Textures of Igneous Rocks

Igneous rocks are classified according to two different classes of information, or categories. The first is according to chemical composition, as apparent from our study of magma and its differentiation into mafic and felsic types. Second is on the basis of the texture of the igneous rock. *Texture*, as the word applies to rocks, relates to the sizes and patterns of the mineral crystals present in the rock. Texture also distinguishes between the crystalline and the noncrystalline (glassy) state of mineral matter. (This topic was discussed in Chapter 3.)

FIGURE 4.2 The Bowen reaction series.

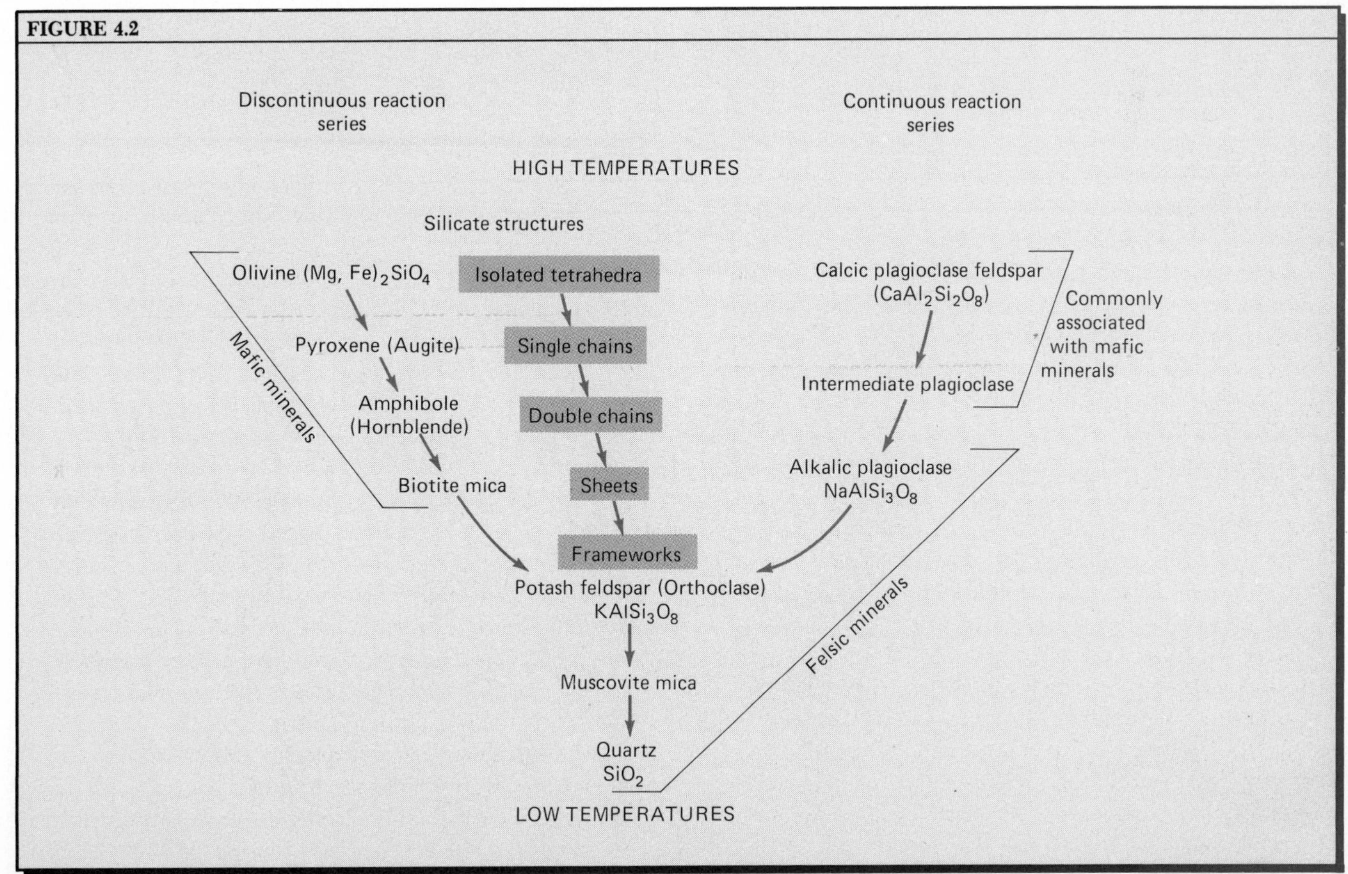

Geologists use the term texture with somewhat the same meaning as it has in the classification of fabrics. We say that burlap or canvas has a "coarse texture," in contrast to the "fine texture" of chiffon. Cloth woven on a loom to produce threads that cross at right angles (warp and woof) contains a texture quite different from that of a knitted fabric. A garment can also be made from a sheet of film plastic that shows no structure. Perhaps this last material can be considered the equivalent of glass in the classification of rock texture.

The size of mineral crystals in an igneous rock depends largely upon the rate of cooling of the magma through the stages of crystallization. As a general rule, rapid cooling results in small crystals, slow cooling in large crystals. Extremely sudden cooling will result in formation of a natural glass, which is noncrystalline.

The rate of cooling depends on where the magma is located as it cools. Large bodies of magma trapped deep beneath the surface cool very slowly because the surrounding rock conducts the heat very slowly. Bodies of rock formed this way belong to the class of *intrusive igneous rocks*, whose mineral crystals are usually large enough to be identified with the unaided eye. Rapid cooling occurs in lava, which belongs to the class of *extrusive igneous rocks*. Lava forms thin layers that lose heat rapidly to the atmosphere or to overlying ocean water. Thus, lava usually has mineral crystals too small to identify with the unaided eye, or it may solidify as a glass.

Among the crystalline igneous rocks, texture falls into two classes: (1) *Phaneritic texture* consists of crystals large enough to be seen with the unaided eye or with the help of a small hand lens (Figure 4.3). (2) *Aphanitic texture* consists of crystals too small to

FIGURE 4.3 Closeup view of the broken surface of fresh granite. The rock is coarse-grained, a typical phaneritic texture. The light-colored grains are quartz and feldspar; the dark grains are largely biotite mica. (A. N. Strahler.)

FIGURE 4.4 A sketch of the appearance of mineral grains of igneous rock, as seen in the microscope, enlarged about five times. An extremely thin slice of the rock is viewed by means of light passing through the slice from below. *Left*, granite; *right*, olivine gabbro. Minerals are identified by letters as follows: Q—quartz, K—potash feldspar, F—plagioclase feldspar, B—biotite, H—hornblende, P—pyroxene, O—olivine. Compare the cleavage patterns of hornblende and pyroxene. The banded appearance of plagioclase feldspar is caused by a crystal phenomenon known as "twinning."

FIGURE 4.3

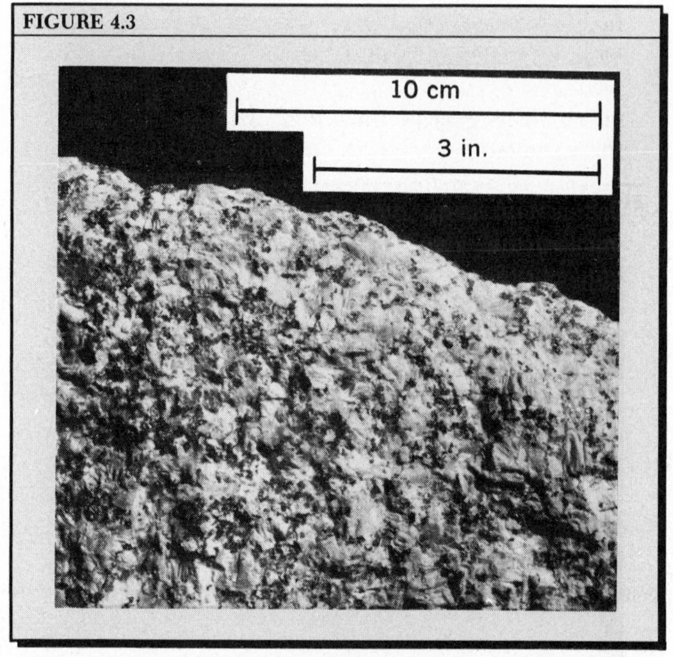

FIGURE 4.4

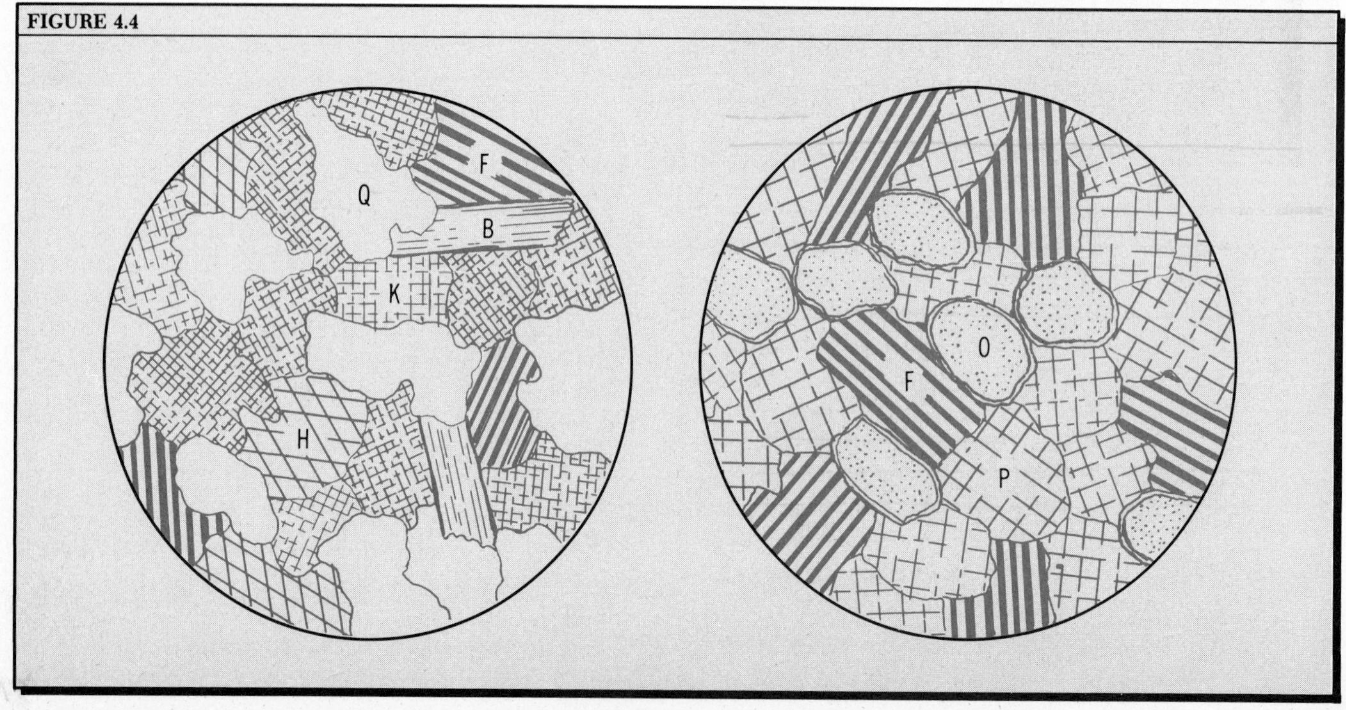

be distinguished as individual particles without the aid of a microscope.

The phaneritic igneous rocks have crystal grains ranging in diameter from 0.05 mm to over 10 mm. An accepted grade scale is as follows:

Fine-grained	0.05–1 mm
Medium-grained	1–5 mm
Coarse-grained	5–10 mm
Pegmatite texture	over 10 mm

Particles under 0.05 mm fall into the aphanitic texture class. Pegmatite texture consists of crystals that are often several centimeters long and in rare cases may be as long as a few meters.

FIGURE 4.5 This specimen of andesite porphyry has large feldspar phenocrysts scattered through an aphanitic groundmass. The specimen is about 8 cm wide. (A. N. Strahler.)

FIGURE 4.6 Specimens of scoria (*left*) and black volcanic glass (obsidian) (*right*). The glass shows conchoidal fracture. (A. N. Strahler.)

FIGURE 4.5

FIGURE 4.6

Though it can be seen with the naked eye, the crystalline fabric of the phaneritic rock is best studied under a specialized microscope in which very thin slices of the rock are examined under polarized light (Figure 4.4), which makes possible mineral identification based upon distinctive optical properties. For the most part, external crystal forms are lacking or only poorly developed. On the other hand, evidences of cleavage show up well as linear markings within the grains.

Where crystals in the rock are all within the same size range, the texture is described as *equigranular.* Where a few large crystals, called *phenocrysts,* are embedded in a matrix, or *groundmass,* of smaller crystals, we have *porphyritic texture;* the rock is designated as a *porphyry* in addition to its proper name (Figure 4.5). The groundmass may consist of crystals ranging in size from fine-grained to coarse-grained, or it may consist of a glass.

Extrusive igneous rocks have distinctive textures resulting from the rapid expansion of volatile gases as confining pressure is reduced and cooling occurs. The specimen of volcanic *scoria* shown in Figure 4.6 is full of cavities formed by gas bubbles, and is said to have *scoriaceous texture.* Because the cavities go by the name of *vesicles,* an alternative adjective for this texture is *vesicular texture.* An extreme case of scoriaceous texture is seen in *pumice,* formed of magma frothed by expanding gases into a glassy rock of such low density that it easily floats on water. Pieces of pumice are used as abrasive blocks, while powdered pumice serves as an abrasive cleaning powder. In contrast, the volcanic glass also pictured in Figure 4.6 is dense and free of such cavities. Notice its conchoidal fracture.

The Principal Igneous Rocks

Figure 4.7 organizes the commoner varieties of igneous rocks into an orderly classification based upon mineral composition. Typical percentages by volume of the component minerals or mineral groups are suggested by a set of numbers for each rock variety. The numbers given here are to serve only as approximate guide values, because, as the continuous curved boundary lines on the field of the graph suggest, one rock variety can grade into the next. Consequently the assigned rock name can apply to transitional zones between it and the adjacent varieties.

Two sets of rock names are given, one applying to large intrusive rock bodies, usually of medium to coarse crystal grain texture (phaneritic), the second set applying to the extrusive rocks, largely formed by the solidification of lava. Lavas are typically aphanitic or glassy and are often scoriaceous in texture.

In terms of bulk composition, most igneous rock of the earth's crust belongs to a *granite–gabbro series*, which follows closely upon the mineral order of the Bowen reaction series. Since these rocks are customarily presented in sequence from the felsic end toward the mafic end, we shall follow this practice, although it is the reverse of the order suggested by the reaction series and magmatic differentiation.

Granite is a felsic rock dominated in composition by the feldspars and quartz. Potash feldspar of the orthoclase variety is the most important mineral, while alkalic plagioclase may be present in moderate amounts or absent. Quartz, which accounts for perhaps a quarter of the rock, reaches its most abundant proportions in granite. Biotite and hornblende are common accessory minerals. (Magnetite, not shown on the chart, is commonly present.)

FIGURE 4.7 Mineral composition of common igneous rocks. Numbers give approximate mineral percentages by volume. (Based on a diagram by B. Mason, 1966, *Principles of Geochemistry*, 3d ed., John Wiley & Sons, Inc., New York, p. 101, Figure 5.3.)

Granite is a grayish to pinkish igneous rock depending upon the variety of potash feldspar present. Its specific gravity, about 2.7, is comparatively low among the igneous rocks. Most granites are sufficiently coarse in texture for the component minerals to be identified with the unaided eye. (Granite is pictured in Figures 4.3 and 4.4.) The grayish cast of the quartz grains, with their glassy luster, sets them apart from the milky white or pink feldspars. Black grains of biotite or hornblende contrast with the light minerals. The extrusive equivalent of granite is *rhyolite*, a light gray to pink form of lava.

Granite grades into the relatively less important rocks *granodiorite* and *tonalite*, with the extrusive equivalents of *quartz latite* and *dacite*, respectively. These rock names are not important, but you should notice that potash feldspar and quartz decrease in proportion, while plagioclase feldspar increases and moves from the alkalic end toward the intermediate varieties.

Diorite is an important intrusive rock; its extrusive equivalent, *andesite*, occurs very widely in lavas associated with volcanoes. Diorite is dominated by plagio-

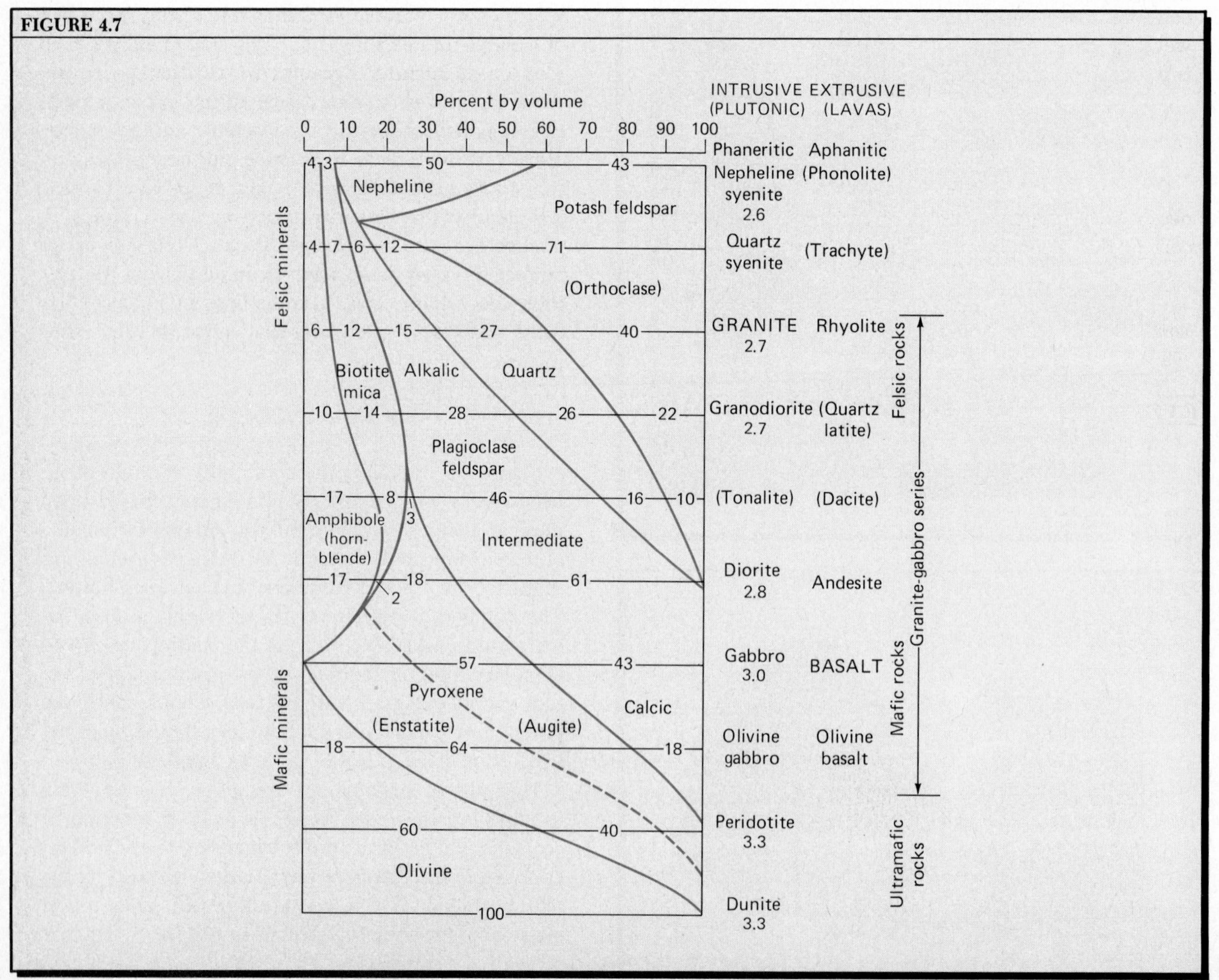

clase feldspar of intermediate composition, while quartz is a very minor constituent. At this point in the granite–gabbro series, pyroxene makes its appearance and is of the augite variety. Amphibole, largely hornblende, is also important, and some biotite is present.

Gabbro is an important though not abundant intrusive rock. It is greatly overshadowed in importance by its extrusive equivalent, ***basalt***, which makes up huge areas of lava flows and is the predominant igneous rock underlying the floors of the ocean basins. Gabbro and basalt are composed almost entirely of pyroxene and intermediate to calcic plagioclase feldspar with or without minor amounts of olivine. As olivine increases, at the expense of reduced plagioclase and pyroxene, *olivine gabbro* and *olivine basalt* appear in the series. Gabbro and basalt are dark rocks—dark-gray, dark-green, to almost black—and of relatively high specific gravity, 3.0. They are designated as mafic rocks in contrast with the felsic rocks related to granite.

An intrusive igneous rock often found in close association with gabbro is *anorthosite*, composed almost entirely of intermediate or calcic plagioclase feldspar. A small percentage of pyroxene and olivine may be present. There are also some rocks of an intermediate composition between anorthosite and gabbro. One well-known occurrence of anorthosite on a large scale is in the Adirondack Mountains of New York State.

Composition of the Earth's Crust

Our knowledge of the granite–gabbro series of igneous rocks can now be put to use to describe the composition of the earth's crust. In Chapter 1, we presented the crustal layers in a very general way, emphasizing rock density. Figure 1.6 showed the crust of the continental lithosphere as being thicker than the crust of the oceanic lithosphere. Moreover, the continental crust was shown to have a less dense upper layer, not present in the oceanic crust.

Figure 4.8 shows further details of the crust and its composition. The *oceanic crust*, an average of about 6 km thick, is composed of mafic igneous rock of the composition of basalt and gabbro. The upper part is basalt, which solidified from lava reaching the seafloor in the spreading plate boundary between two oceanic lithospheric plates. The same magma, solidifying in the lower part of the crust, below the basalt, is thought to consist of gabbro, the intrusive equivalent of basalt.

The *continental crust* averages some 30 to 35 km in thickness generally, but extends down to depths of 65 km or more beneath the high mountain belts. An upper, felsic rock layer can be distinguished from a basal, mafic layer. Except in a few places, no sharp boundary separates these layers. The upper, felsic layer, often described by the term *granitic rock*, has a chemical composition about that of granite or granodiorite, with a density of about 2.7 near the surface. Much of it is intrusive rock of granitic composition or metamorphic rock derived from granite. (Sedimentary rock, which covers the granitic rock in many places, makes up only a very small percentage of the crust.) Although little is known with certainty about the composition of the lower, mafic layer of the continental crust, it is usually considered to have a chemical composition similar to basalt and gabbro. The mafic layer should not, however, be visualized as a continuation of the oceanic crust, or thought to have had a similar origin. The mafic layer of the continental crust is probably vastly older than the mafic oceanic crust, just

FIGURE 4.8 Schematic cross section of the crust beneath the continents and ocean basins. The vertical scale of the drawing is greatly exaggerated.

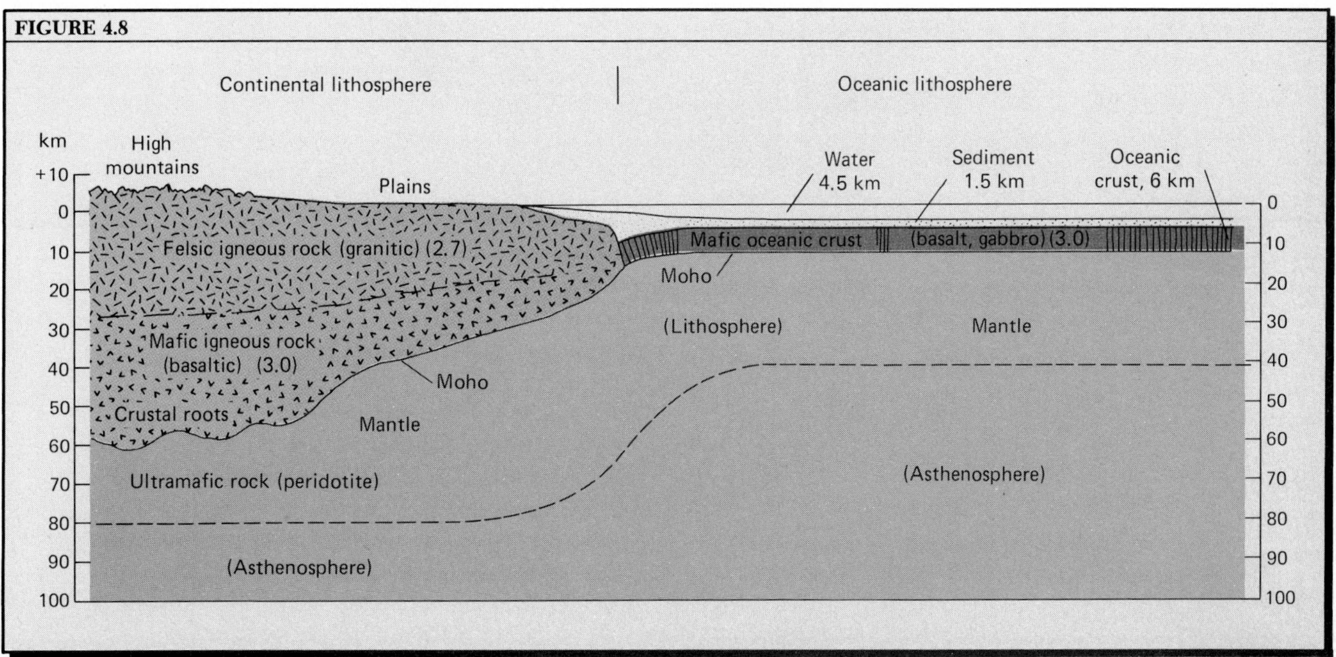

as the upper, felsic layer is also undoubtedly vastly older than the oceanic crust.

The contact between the crust and mantle is a sharply defined boundary, called the *Moho* after the seismologist who discovered it through the analysis of earthquake waves; it is explained in detail in Chapter 8. Its position is shown in Figure 4.8; below it lies the mantle.

The Ultramafic Rocks

Continuing down the igneous rock series shown in Figure 4.7, we arrive at *peridotite*, a rock composed almost entirely of two mineral constituents, olivine and pyroxene. Although widespread in occurrence, peridotite occurs in relatively small intrusive bodies. It is a dark rock of high specific gravity, 3.3, and belongs to a group designated as *ultramafic rocks*. Peridotite is believed to be the major rock type making up the earth's mantle (Figure 4.8).

At the bottom of the graph in Figure 4.7 is *dunite*, a rock consisting almost entirely of olivine. A rare rock at the earth's surface, it may possibly be abundant in the mantle.

**The Syenite Group*

Just as the ultramafic rocks extend beyond gabbro, Figure 4.7 shows the igneous rock series extended beyond granite in the direction of the felsic minerals. In this upper part of the graph, we find the *syenite group*—igneous rocks of considerable interest scientifi-

FIGURE 4.9 Block diagram showing the important types of plutons, along with a volcano and lava flow. The dikes and volcanic pipe depicted on the land surface have been revealed by long erosion of overlying rock, whereas the volcano and lava flow are comparatively recent features. (Drawn by A. N. Strahler.)

FIGURE 4.10 A granite batholith exposed by erosional removal of the overlying country rock. Portions of the rock that projected down into the batholith remain as roof pendants, along with inclusions of country rock surrounded by granite. (Drawn by A. N. Strahler.)

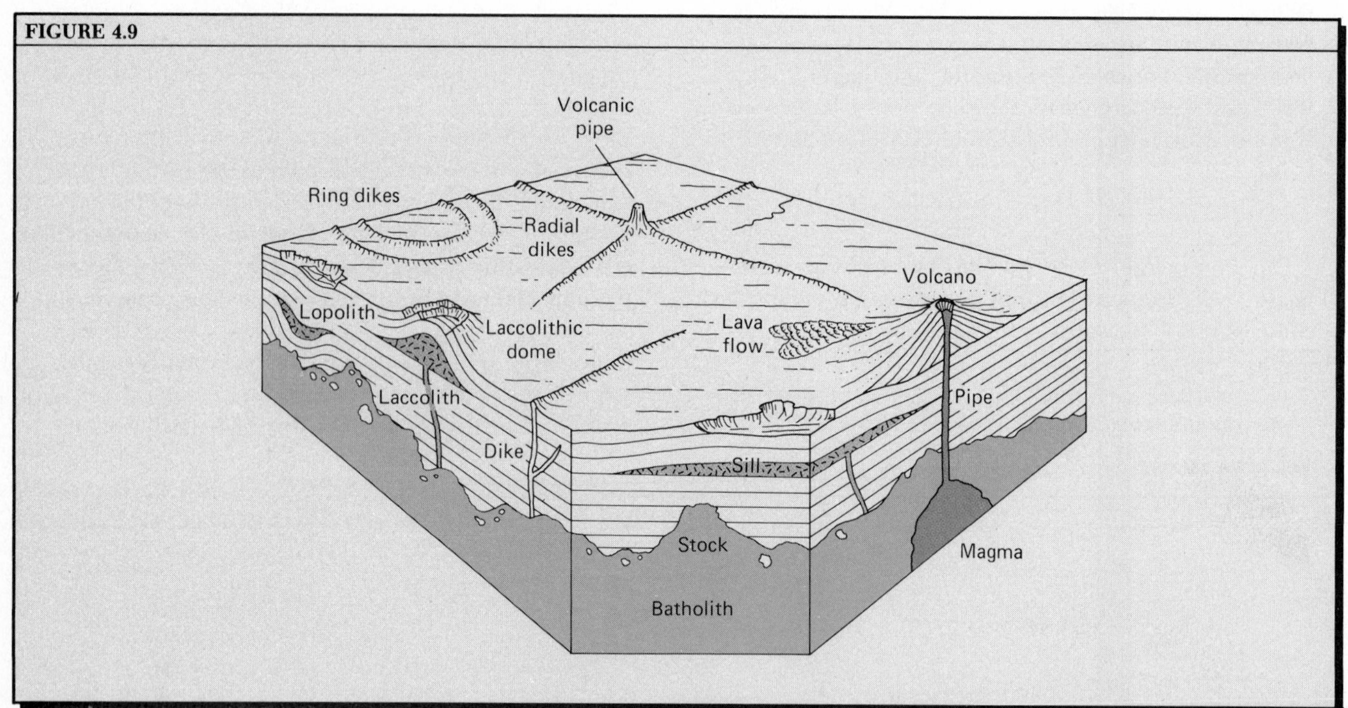

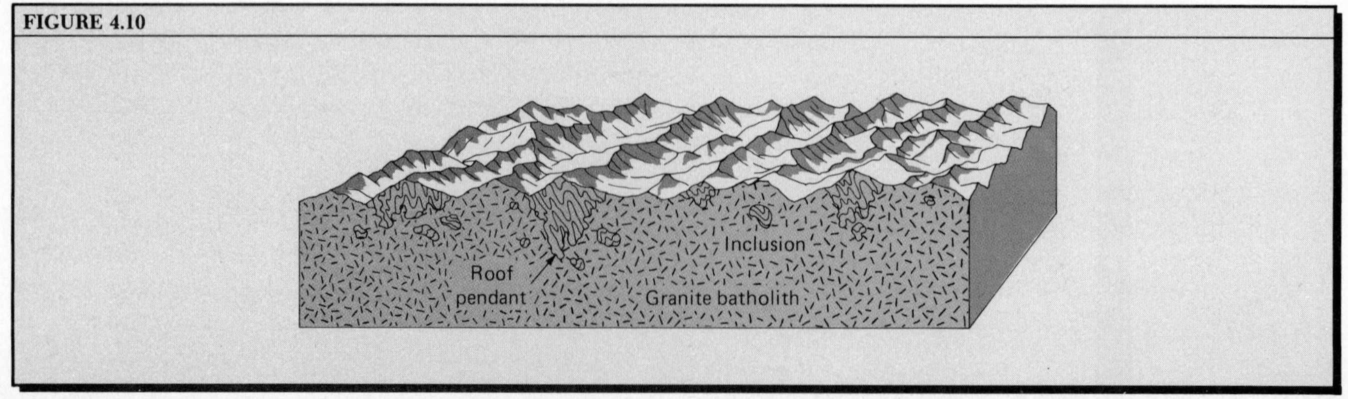

cally, despite their relatively rare occurrence. The name syenite comes from the Egyptian city of Syene, on the Nile River, where the rock was quarried in ancient times. *Quartz syenite* and its extrusive equivalent, *trachyte,* are dominated by potash feldspar and contain relatively minor amounts of quartz, plagioclase feldspar, biotite, and hornblende. As the syenite series is extended further, quartz disappears entirely, and its place is taken by *nepheline* or other feldspathoid minerals, a result of silica deficiency in the magma. Quartz and nepheline cannot be found together in the same igneous rock, and this fact is shown in the arrangement of boundary curves in Figure 4.7. The extreme end of the syenite series is represented by *nepheline syenite* and its extrusive equivalent, *phonolite*. The syenite rocks are light-colored and of relatively low specific gravity, around 2.6.

Forms of Intrusive Rock Bodies

Intrusive rock bodies are called *plutons;* they come in a wide range of sizes and shapes. The larger plutons consist of phaneritic rock, usually of medium-grained and coarse-grained crystal sizes. Small plutons or plutons in the form of thin, platelike masses are usually fine-grained or aphanitic in texture because of relatively rapid cooling in contact with surrounding rock. (An exception is pegmatite, described later.)

FIGURE 4.11 Xenoliths, which are fragments of older country rock, various kinds of igneous and metamorphic rock, incorporated in a granite batholith. (A. N. Strahler.)

FIGURE 4.11

It is helpful to divide the plutons into two major groups. *Discordant plutons* are intrusive bodies that cut across the structures of older rock; they are often massive and show irregular, ragged contacts with the surrounding rock. Other discordant types fill cracks cutting across structures in the surrounding rock. *Concordant plutons* have boundaries parallel with natural layers and planes in the enclosing rock. They represent magma bodies that forced their way through older rock, spreading apart the rock along natural planes of parting. Typically, the natural parting consists of the separation planes within sedimentary rock layers (see Chapter 5). Because of the manner in which the concordant plutons are controlled by layered rocks, they are typically platelike bodies with flat, parallel sides. (Such rock bodies are described as *tabular intrusions*.)

Largest of the discordant plutonic bodies is the *batholith* (Figure 4.9). Typically, the batholith consists of granite or a related rock variety such as granodiorite, but a few batholiths consist of anorthosite. A single batholith may have an area of several thousand square kilometers. An example is the Idaho batholith, of which 40,000 sq km are exposed, equal to the combined areas of New Hampshire and Vermont. Another example is the Sierra Nevada batholith of California, underlying much of the Sierra Nevada range.

Because batholiths are formed at considerable depth within the crust they can be seen only after erosion over millions of years has removed the overlying crustal rocks, exposing the mountain roots. As a batholith is uncovered, there appear first smaller bodies of the plutonic rock known as *stocks,* which occupy areas of less than 100 sq km. Remnants of the *country rock* (older rock into which the igneous rock was intruded) will be found extending down into the batholith as *roof pendants,* but with continued erosion of the land surface a vast expanse of uninterrupted plutonic rock will appear (Figure 4.10).

Close examination of rock exposures of a batholith will often reveal the presence of irregularly shaped inclusions of country rock, which may be of almost any variety (Figure 4.11). These inclusions, termed *xenoliths,* suggest that the entry of magma to form a batholith is in part at least a mechanical process, known as *stoping* (rhymes with "roping"). In this process, masses of the brittle country rock are wrenched loose and become incorporated into the magma. Batholiths are thick enough to prevent the bottom being exposed to observation.

Figure 4.9 shows other forms of plutons. The country rock is depicted as having a layered structure, and when magma intruding these layers extends relatively far as a horizontal sheet, the concordant pluton is named a *sill* (Figure 4.12). Where magma pressure lifts the overlying layers into a dome, the result is a *laccolith,* which is also a concordant pluton. In thick sills and large laccoliths, the rock texture is phaneritic.

A third type of concordant pluton is the *lopolith;* it is a saucer-shaped layer, conforming to the curvature of layers above and below (Figure 4.9).

Thin, platelike plutons that are discordant, cutting across the natural planes and other structures of the country rock, are called *dikes* (Figure 4.9). Dikes may also cut through massive plutons, such as batholiths (Figure 4.13). In such cases, the massive pluton had been formed at an earlier date and makes up part of the country rock. Dikes come in a wide range of thicknesses from a few centimeters to hundreds of meters. Their length can range to many kilometers, so that when exposed at the surface by erosion, they can often be traced for several kilometers across the landscape as wall-like features (Figure 4.14). In some instances, dikes occur in large numbers oriented in about the same compass direction; these groups are called *dike swarms.* Quite different in geometrical pattern are *ring dikes,* which form a circular pattern when exposed at the surface (Figure 4.9). Ring dikes appear to form above an invading magma body, as magma fills fractures caused by upward pressure of the domelike igneous mass below.

Shrinkage in volume, occurring during cooling of thin sills and dikes, results in a system of fractures that produces long rock columns of prismatic form with four, five, or six sides. This structure is termed *columnar jointing* (see Figure 4.20). The columns are oriented with the long dimension at right angles to the enclosing country-rock surfaces. Consequently, where a sill lies in a horizontal attitude, the columns are vertical; where a dike is vertical, the columns are horizontal. (We shall refer again to columnar jointing as a common feature in lava sheets.)

After a magma has largely crystallized, watery, silica-rich solutions remain. These hydrothermal solutions are forced to penetrate fractures in either the newly formed igneous body or the adjacent country rocks. Minerals deposited from hydrothermal solutions often take the form of irregular dikes with the texture of pegmatite (mentioned earlier), and they are called *pegmatite dikes* (Figures 4.15 and 4.16). Pegmatite also occurs in irregular masses within the igneous body, as shown in Figure 4.15. The rate of growth of unusually large pegmatite crystals, such as those

FIGURE 4.12 The horizontal dark band in this cliff face is a sill of diabase (a rock similar to gabbro). Near the center of the view, the intruding magma has broken diagonally across the bedding to penetrate between strata higher in the series, giving the sill a stepped form. The diagonal segment is thus a discordant intrusion—a thick dike. Banks Island, Northwest Territories, Canada. (Geological Survey of Canada, Photo No. 131185.)

FIGURE 4.12

shown in Figure 4.16, is thought to be rapid in the watery medium.

Mineral deposits filling narrow, irregular fracture openings in any rock mass are called *mineral veins,* and they may be hydrothermal in origin, but sometimes they are deposited by surface waters moving down through open fractures in shallow rock masses.

Last on our list of discordant intrusive bodies is a tall, vertical, tubelike igneous body. Commonly called a *volcanic pipe,* it is usually interpreted as igneous rock congealed in the deeper portions of the tube through which a volcano received its magma (Figures 4.9 and 4.17). One interesting variety is the *diamond pipe* (also called a *diatreme*), for it consists of ultra-

FIGURE 4.13 A dike of basalt cutting through granite, Cohasset, Massachusetts. (John A. Shimer.)

FIGURE 4.14 A vertical dike of igneous rock exposed by erosion to produce a wall-like landform. West Spanish Peak (*background*) is an intrusive igneous body exposed by erosion of surrounding sedimentary strata. The locality is south central Colorado. (G. W. Stose, U.S. Geological Survey.)

FIGURE 4.13

FIGURE 4.14

FIGURE 4.15 Occurrence of pegmatite bodies in relation to an intrusive igneous body and the country rock.

FIGURE 4.16 Enormous spodumene crystals from the Etta pegmatite deposit, Pennington County, South Dakota. The hammer rests upon a single large crystal. (J. J. Norton, U.S. Geological Survey.)

FIGURE 4.15

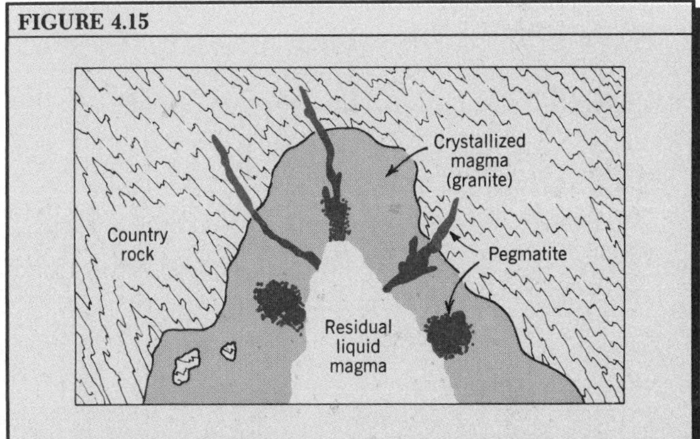

FIGURE 4.16

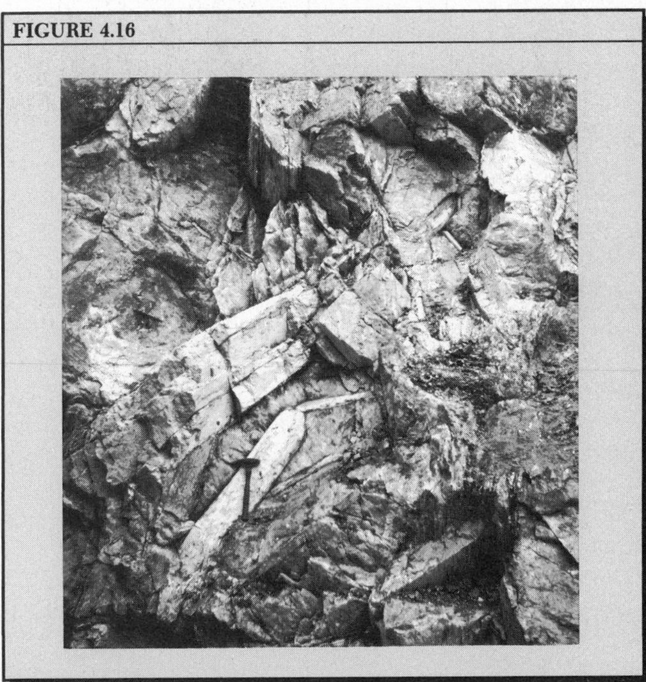

mafic rock believed to have been forced up from the upper mantle. (This feature is discussed in Chapter 21.) Igneous rock of volcanic pipes is usually mixed with broken fragments of country rock through which the pipe was forcibly opened under the enormous pressures of gas-rich rising magma.

Lava Types and Their Properties

Magma reaching the earth's surface as molten lava gives rise to a wide variety of igneous phenomena with external forms as many and varied as the internal structures and textures. To a large extent this great variety reflects the chemical and physical properties of the liquid lava. (The word "lava" is used for both the liquid magma and the solid rock formed when the magma cools.)

We have already named and described the kinds of extrusive igneous rocks. Generally speaking, each extrusive rock is similar in composition to a particular intrusive type, and these range from felsic to mafic, with extreme types—the syenite group and the ultramafic group—forming the ends of a continuous series (Figure 4.7). Lavas are classified in a similar manner, but with minor differences in the groups recognized. A threefold classification of lava types is as follows:

Parent magma type	Class of lava	Kind of extrusive rock
Felsic	Acidic (Silicic)	Rhyolite
Intermediate	Intermediate	Andesite
Mafic	Basic	Basalt

According to this classification, *acidic lavas* (*silicic lavas*) contain 70% or more of silica (SiO_2); *intermediate lavas,* 50% to 70%; *basic lavas,* less than 50%.

Properties associated with lavas are temperature, viscosity, and explosiveness. *Viscosity,* a property of all fluids (including both liquids and gases), describes the degree to which the fluid resists flowage when subjected to unbalanced forces. Everyday words that describe different states of viscosity are "runny" and "tacky." For example, in making caramel, we begin with a mixture of water and dissolved sugar. At low temperature, this liquid is runny—i.e., it has low viscosity. After the mixture is boiled for several minutes and then cooled it thickens and becomes tacky—i.e., it develops high viscosity. Fluids of high viscosity flow sluggishly as compared to fluids of low viscosity in the same situation.

The basic lavas, most commonly of the composition of basalt magma, have high temperatures and low viscosity as they emerge from the solid earth. Although direct measurements are few, temperatures of basic (basaltic) magma have been observed in the range of 1100 °C to 1200 °C. Lava of this type moves rapidly and travels far over comparatively gentle land slopes, often spreading into thin sheets before solidifying.

As a general rule, basic magma is not explosive in behavior, and even in large quantities it tends to erupt quietly. Explosiveness is related to the presence of gases under high pressure in the magma. Basic magma that is low in water content (a "dry" magma) tends not to be explosive because the proportion of volatiles is small. Besides, the proportion of silicon–oxygen tetrahedra in the magma is smaller than in the felsic magmas. As we noted earlier, a linkage of these tetrahedra into groups exists in magma and may be one cause of its greater viscosity in the felsic types.

The acidic lavas, of felsic composition, have temperatures in the range from 800 °C to 1000 °C—somewhat lower than for the basic lavas. The viscosity of acidic lava is generally high, so that this lava flows more slowly and congeals closer to its place of emergence than the basic type. Explosiveness of the acidic lavas is often very great, and it produces some extremely violent types of volcanic eruptions. Lavas of the intermediate class have properties between the basic and acidic types, but they tend to be more closely akin to the acidic types in physical behavior.

Lava Flows

A tongue or sheet of lava formed at a particular time and place is called a *lava flow.* Lava may emerge from the localized central vent of a volcano, in which case it travels away from the vent as a narrow tongue following any one of a number of radial paths. Lava (usually of basaltic composition) may also emerge from a long

FIGURE 4.17 Agathla Peak, a volcanic pipe in northern Arizona. Behind the peak are horizontal sedimentary strata which formerly enclosed the igneous rock. (Barnum Brown, American Museum of Natural History.)

FIGURE 4.17

FIGURE 4.18 *A*. Vertical aerial photograph of the King's Bowl Rift, Snake River Plain, Idaho. Long dimension of the photograph spans about 600 m; north is toward the top. Dark areas are basaltic lava flows. The fissure is 2 to 3 m wide and over 240 m deep. An explosion crater lies near the center (light-colored slope). (U.S. Dept. of Agriculture.) *B*. View south along the rift, with a spatter cone in the distance. (Martin Prinz.)

FIGURE 4.18 A

FIGURE 4.18 B

crack, or *fissure*, in the brittle surface rock, and then it flows out as a succession of thin sheets, known as *fissure flows*, covering considerable surface area (Figure 4.18). In highly fluid basalt flows, liquid lava may be seen exposed in the central zone of the flow and close to the vent, but usually, because of rapid cooling, the surface and margins of a lava flow are encrusted with solid lava.

Lava flows move downhill under the force of gravity to occupy any available valley or topographic depression. Where the solidified crust of a flow forms a strong structural arch, fluid lava beneath may move out to lower levels, leaving a hollow lava tube or tunnel within the flow.

Flow surfaces may be extremely rough (*aa texture*), where the lava is highly charged with gases and produces a scoriaceous texture. Other basalt flows have smooth, glassy outer surfaces convoluted into billowy and ropy configurations (*pahoehoe texture*) (Figure 4.19).

Upon solidification, the interiors of many lava flows exhibit the same type of columnar jointing previously mentioned as characteristic of thin sills and dikes (Figure 4.20). Lavas may flow into the ocean (Figure 4.21) or may erupt from vents or fissures beneath the ocean. Upon cooling, these water-quenched lavas develop a characteristic *pillow structure* and are known as *pillow lavas* (Figure 4.22).

FIGURE 4.19 A fast-moving basaltic lava has flowed down a paved highway and rapidly solidified as pahoehoe (ropy lava) with a smooth, glassy surface. Mauna Ulu pit crater, island of Hawaii. (Dr. Willard H. Parsons.)

FIGURE 4.19

Pyroclastic Materials

Quite different from lava flows as a class of extrusive rock are *pyroclastic materials,* consisting of *pyroclasts*—rock and mineral fragments blown out from a volcanic vent under pressure of rapidly expanding gases present in the magma. When projected into the air the pyroclastic fragments are either solid or in a plastic state that occurs immediately prior to solidification. Fragments of the country rock are commonly included in pyroclastic masses.

Pyroclasts are classified in terms of the sizes of the fragments. The largest are huge solid blocks, and the smallest range down in size to the finest dusts. Spindle-shaped or spherical masses a few centimeters to 30 cm or more in diameter, known as *volcanic bombs,* result from the congealing of blebs of fluid lava thrown high into the air (Figure 4.23). Blebs of plastic lava falling close to a small vent may build a small *spatter cone* (Figure 4.24). Smaller particles of scoriaceous lava, ranging in diameter from 4 to 25 mm, are called *lapilli.* Particles under about 4 mm constitute *volcanic ash* (Figure 4.25). Ash particles range downward in size to the fine *volcanic dust* (smoke particles) which can be transported for thousands of kilometers in the upper atmosphere. Upon microscopic examination, particles of fine ash will be found to take the form of minute *shards* (angular fragments) of volcanic glass.

Tephra is a convenient term widely used by geologists to describe collectively all the varieties and sizes of pyroclastic materials. Thus, all extrusive igneous material (except volatile gases) consists of lava, of tephra, or of a combination of the two.

FIGURE 4.20 Columnar structure in basaltic lava, Palisades of the Columbia River, Washington. The columns are about 5 m long. (G. K. Gilbert, U.S. Geological Survey.)

FIGURE 4.21 During the great 1914 eruption of the Japanese volcano, Sakurajima, andesite lavas reached the sea, generating dense plumes of condensed steam. (T. Nasaka.)

FIGURE 4.20

FIGURE 4.21

Pyroclastic Deposits and Rocks

Accumulations of tephra—*pyroclastic deposits*—form a class intermediate between the igneous and the sedimentary rocks. Pyroclasts are transported by or descend through air or water, but this occurs immediately at the time of volcanic eruption, and thus the original impelling force is volcanic.

Volcanic breccia, a crude mixture of large and small

FIGURE 4.22 Resembling a great heap of solidified spaghetti, this mass of basaltic pillow lava lies exposed in the wall of a desert canyon, etched into relief by weathering processes. Near Suhaylah, Sultanate of Oman, Arabian Peninsula. (E. H. Bailey, U.S. Geological Survey.)

FIGURE 4.22

pyroclasts that have fallen close to a vent, is a form of rock more closely allied to the igneous than to the sedimentary class. Volcanic breccia is also found in volcanic pipes and may include fragments of country rock. Quite different is *tuff*, volcanic ash that has been transported by winds or water and deposited in layers. It may become compacted into a stratified rock sometimes classed as sedimentary (Figure 4.26).

A unique form of volcanic material is that carried in *ash flows*, consisting of a highly heated mixture of gases and frothed lava. This phenomenon is called a *glowing avalanche* (in French, *nuée ardente*) (see Figure 4.34). Moving as a dense, cloudlike tongue down the slopes of a volcano, the glowing avalanche leaves a fine-textured rock layer resembling tuff but fused into hard layers by the high temperature. This rock is called a *welded tuff*.

Volcanism—Volcano Building

When igneous extrusion takes place repeatedly from an opening called a *volcanic vent*—or from closely grouped vents—the igneous matter accumulates to form a mound or peak. All such centralized accumula-

FIGURE 4.23 Volcanic bombs. *A*. During the 1943 eruption of Parícutin Volcano, Mexico, these bombs bounded down the steep slope, leaving tracks in the freshly fallen ash. (From F. M. Bullard, *Volcanoes of the Earth*, Univ. of Texas Press, Austin. Copyright © 1976, by Fred M. Bullard.) *B*. Two breadcrust bombs from the slopes of Mount Rainier, Washington. The bomb at the right has been broken open to show the scoriaceous texture of the interior. (D. R. Crandell, U.S. Geological Survey.)

FIGURE 4.23 A

FIGURE 4.23 B

FIGURE 4.24 A spatter cone built up of blebs of plastic lava, Sunset Crater, northern Arizona. (A. N. Strahler.)

FIGURE 4.25 During a large eruption in 1914, the Japanese volcano Sakurajima discharged great quantities of volcanic ash and pumice. Far from the crater, the ash layer partly buried this village. Torrential rains then formed deep rillmarks in the ash, washing some of it off the roofs. (T. Nasaka.)

FIGURE 4.24

FIGURE 4.25

tions come under the broad classification of *volcanoes*. There is a wide range in the sizes and shapes of volcanoes, as well as a wide range in the forms of extrusive rock of which they are constructed. The term *volcanism* covers all forms of volcano building as well as the emission of lava flows and tephra, called *volcanic eruption*.

The extremely high gas pressures causing a volcano to emit great quantities of lava and tephra are thought to develop in a *magma reservoir* situated perhaps 3 to 5 km below the volcano. Here the magma is in the process of slow cooling and partial crystallization. As crystallization proceeds, the water contained in the melt increases in proportion in the remaining solution, causing an enormous rise in pressure. This pressure increase is believed to be of an order of magnitude easily capable of causing volcanic explosion. Some insight into this process comes from laboratory experimentation with cooling of silicate melts.

Because basic lavas have low viscosity, gas can escape more easily than it can from acidic lavas with high viscosity. Hence, acidic lavas retain gases and are the more explosive of the two classes. The more viscous acidic lavas also solidify rapidly and thus build up large masses with steep slopes close to the vent. Moreover, the acidic magmas tend to plug the vent upon cooling, so that a subsequent eruption builds up a great deal more pressure, is extremely violent, and may greatly alter or even destroy the volcano. The coarser tephra fragments blown from a vent fall nearby, building up steep sides of loose consistency that attain a uniform angle of slope. This contributes to the smoothness and perfection of the conical form of the volcano, whose vent, or *volcanic crater*, takes the form of a bowl-shaped depression at the volcano summit. Volcanoes range in a series from those formed almost entirely of highly fluid lavas to those formed entirely of tephra.

In general, the form of a volcano depends upon the viscosity of the lava and the degree to which gases are present under confining pressure in the emerging magma.

Basaltic Shield Volcanoes

The continued outpouring of great quantities of highly fluid basaltic lava from a radiating series of fissures produces the *basaltic shield volcano*. Unquestionably the greatest assemblage of shield volcanoes is the Hawaiian Islands, each of which is formed of one or more such volcanoes (Figure 4.27). The Hawaiian domes were built upward from the floor of the Pacific Ocean basin, which averages about 4900 m below sea level in this region. The highest volcano, Mauna Loa, rises to an elevation over 4000 m above sea level. Measured upward from their bases on the ocean floor, these volcanoes are on the order of 8 km high, which is vastly greater than the height of most other forms of volcanoes. Side slopes of the Hawaiian volcanoes are usually gentle—not more than 4° to 5° from the horizontal in the freshly built condition. Lava flows emerge from fissures on the flanks of the dome and travel long distances before solidifying. There is comparatively little explosive activity and little tephra.

A characteristic feature of the Hawaiian form of shield volcano is the broad steep-walled *central depression* 3 km or more wide and about 100 m deep (Figure 4.28). The central depression is produced by a

FIGURE 4.26 Called The Pinnacles, these spires have been carved by rainbeat from layers of soft volcanic ash and pumice, deposited as a glowing avalanche (nuée ardente) during the eruption that resulted in collapse of Mount Mazama to produce Crater Lake (see Figure 4.37). Crater Lake National Park, Oregon. (Oregon Dept. of Transportation, Photo 8505.)

FIGURE 4.27 (*Above*) Schematic block diagram of a large basaltic shield volcano such as those making up the Hawaiian Islands. The vertical scale of the diagram is greatly exaggerated. (*Below*) A cross section drawn to true scale, showing the relationship between the volcano and the underlying crust and upper mantle. The crust has sagged down under the load of the volcano. (Drawn by A. N. Strahler.)

FIGURE 4.28 The summit of Mauna Loa, seen from the air, consists of a chain of pit craters leading to the central depression in the distance. The summit elevation is 4170 m. On the distant horizon can be seen the snow-capped summit of Mauna Kea, an extinct shield volcano. (U.S. Army Air Corps.)

FIGURE 4.26

FIGURE 4.27

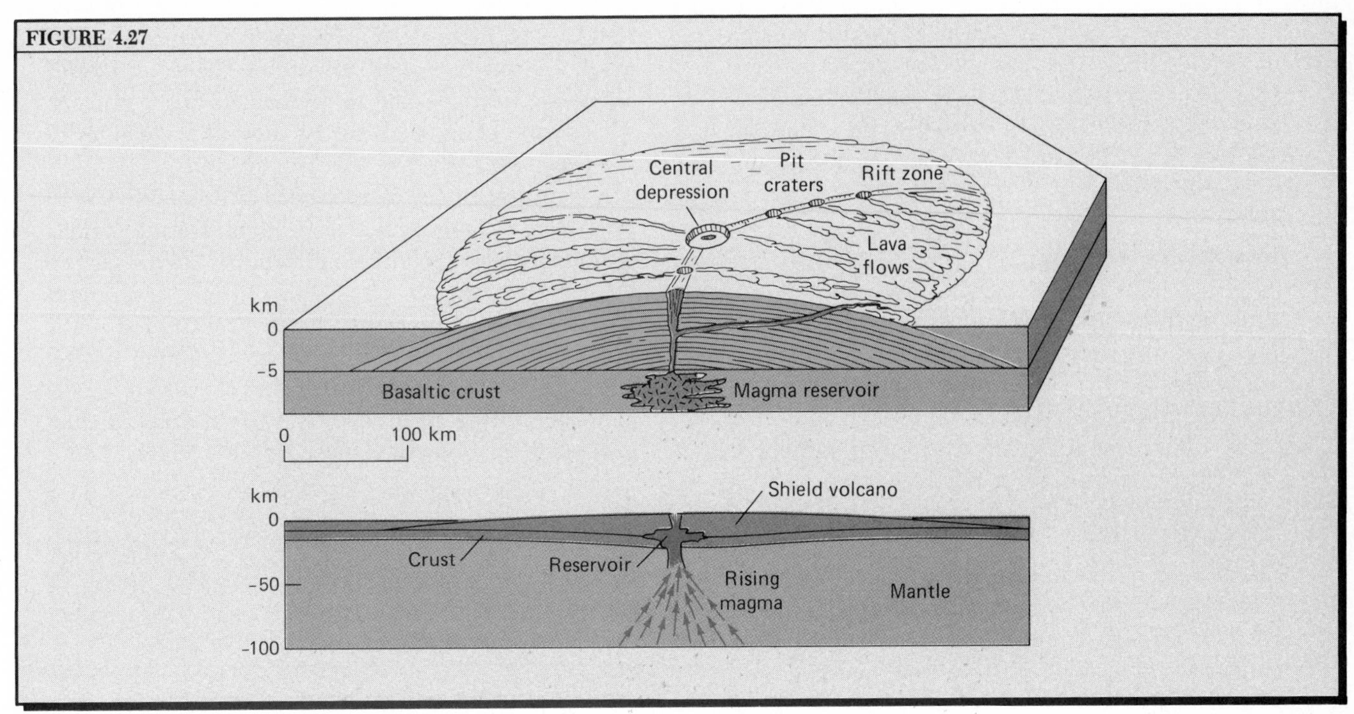

FIGURE 4.28

subsidence that follows withdrawal of basaltic magma from below. Upon the floor of the depression are smaller *pit craters*, 0.5 km across or less. Molten basalt is often exposed in the floor of an active pit crater and may give a spectacular display of power in a fire fountain (Figure 4.29).

Basaltic Cinder Cones

While on the subject of volcanoes formed of basaltic magma, we turn from the largest volcanoes to the smallest—the *basaltic cinder cone* (Figure 4.30). This small cone, usually only 100 to 300 m high and often less than 1 km in basal diameter, is formed entirely of tephra. Cinder cones have a relatively large central depression or crater, but scoriaceous basalt is ejected from a small pipe.

Solidification of the magma as it is blown out results in pyroclasts of many sizes, from large angular blocks and bombs through lapilli to ash. The larger tephra fragments accumulate close to the vent, building up a cone, whereas the finer particles of ash are carried in the wind to fall in a surrounding apron. An ash layer up to several centimeters deep may be found within a radius of a few kilometers. In some cases a basaltic lava flow emerges from the same vent, spreading in a tonguelike stream away from the cone and continuing for several kilometers down the nearest stream valley.

Cinder cones commonly occur in groups of as many as several dozen. One of the best examples of a cinder-cone field is that surrounding the San Francisco Peaks in northern Arizona. Other fine groups of cinder cones lie north and east of Mount Lassen in northern California and in Craters of the Moon National Monument, Idaho.

Composite Volcanoes

Most of the world's great steep-sided volcanic cones are composed of both lava and tephra layers, producing what is termed a *composite volcano* (or *stratovolcano*) (Figure 4.31). The lava is commonly of acidic (felsic) or intermediate composition—rhyolite or andesite. These types are of relatively high viscosity, which means that they do not flow readily but congeal close to the vent. The lava may emerge from a set of radial fissures extending down the flanks of the cone. Pyroclasts blown from the central crater rain down upon the surrounding slopes of the cone, which are built up to angles of 20° to 30° (Figure 4.32). An internal structure consisting of alternating layers of lava and tephra results. A composite cone may grow to a height of a few thousand meters and have a basal diameter of several kilometers. Typically the cone steepens toward the summit, giving the beautiful proportions so greatly admired in such large cones as Fujiyama in Japan, Mayon in the Philippines, and Shishaldin in the Aleutian Islands (Figure 4.33). The high elevation of many great composite volcanoes causes them to extend above the snowline and to accumulate small glaciers. The mantle of snow about the summit contributes to the beauty of the peak.

A characteristic of the behavior of composite volcanoes is their highly explosive eruption. During periods of dormancy lava solidifies in the upper central part of the volcano, beneath the crater, forming a *volcanic plug* that strongly resists renewed extrusion until great pressures have been built up. Eruption throws out huge blocks of broken lava and showers of lapilli and ash. During such activity, a great cloud of dust and condensed steam hangs over the crater. Some

FIGURE 4.29 Hawaii Volcanoes National Park. *A.* Halemaumau, a pit crater on Kilauea seen in 1952. *B.* A fire fountain on the floor of Halemaumau during the eruption of July 1961. (National Park Service, U.S. Dept. of the Interior.)

FIGURE 4.29 A

FIGURE 4.29 B

FIGURE 4.30

FIGURE 4.31

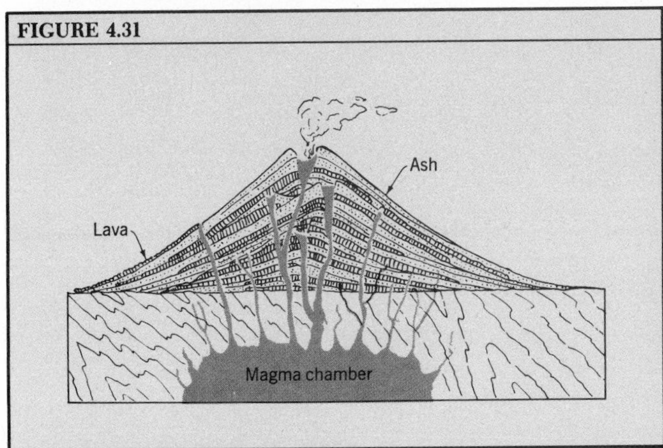

active composite volcanoes that erupt highly viscous lavas repeatedly produce incandescent ash flows (Figure 4.34).

Viscous Lava Domes

A distinctive type of volcanic eruption is associated with highly viscous lavas of the acidic type. The rhyolite lava emerges from a vent in the shape of a bulbous mass (not unlike the start of a bubble being blown from bubble gum). As more lava emerges, the mass sags and tends to collapse, but the newer lava keeps forcing the mass to expand outward. The result is a *viscous lava dome* (sometimes called a *cumulo-dome*) (Figure 4.35). As the dome expands, its upper surface solidifies into a crust, which cracks repeatedly to form a rough blocky exterior. The Mono Craters of California are a good example of a group of lava domes.

Calderas

On occasion, the explosive eruption of a volcano blows out an enormous mass of magma and previously solidified lava from a considerable depth. This event may be accompanied by a collapse or subsidence of

FIGURE 4.30 A fresh cinder cone, partly surrounded by a recent basaltic lava flow, Dixie State Park, near St. George, Utah. (Frank Jensen.)

FIGURE 4.31 This idealized cross section of a composite volcanic cone shows feeders rising from a magma reservoir beneath.

FIGURE 4.32 Summit and crater of an active composite volcano in central Java. (Luchtvaart-Afdeeling, Ned. Ind. Leger., Bandoeng.)

FIGURE 4.32

the central part of the cone to produce a deep, steep-sided crater. Thus, we may distinguish between a crater produced by the gradual construction of a pyroclastic rim (as in the cinder cone) and an *explosion crater* produced by destruction in a violent eruption. Explosion craters of a large stratovolcano are commonly less than 1 km in diameter and represent only a small proportion of the diameter of the cone at its base.

A much larger explosion depression, the *caldera*, may be from 5 to 15 km or more in diameter. It represents a large proportion of the total cone diameter (Figure 4.36), and its formation is one of the most violent of natural catastrophes. Perhaps the best-known event of this kind was the explosive destruction in 1883 of the Indonesian volcano, Krakatoa. Some 80 cu km of rock are estimated to have disappeared from the volcano, demolishing the cone and leaving a caldera about 6 km across. Much of this lost material is believed to have disappeared by subsidence into a cavity left by the loss of magma, but enormous quantities of volcanic dust and pumice spread outward. The explosion produced a great seismic sea wave (a tsunami) that caused the death of many thousands of coastal inhabitants of the islands of Java and Sumatra. (Seismic sea waves are described in Chapter 8.)

In 1912 another such explosion demolished the volcano Katmai, on the Alaskan Peninsula, producing a caldera 5 km wide and 500 to 1000 m deep. As far away as Kodiak, 160 km distant, the ash fall from this explosion totaled 25 cm, and the sound of the explosion was heard at Juneau, 120 km away.

Of the calderas produced in prehistoric time, perhaps the best-known is the basin of Crater Lake, Oregon (Figure 4.37). The caldera is about 9 km in diameter and surrounded by steep cliffs, rising to heights of 150 to 600 m above the lake. The lake is up to 600 m deep and covers 50 sq km. The original volcano, given the name Mount Mazama, probably rose 1200 m higher than the present caldera rim and was an imposing composite volcano resembling Mount Hood and other volcanoes of the Cascade Range. An interesting feature of the Crater Lake caldera is that the rim is notched with the cross sections of valleys and glacial troughs truncated by destruction of the cone. More recently a small cinder cone, Wizard Island, and its associated lava flow were built up in the floor of the caldera.

Igneous Activity and Plate Tectonics

The theory of plate tectonics has made possible new interpretations of the places where igneous activity occurs over the globe and the kinds of magmas that have reached shallow depths in the crust or have emerged as extrusives.

Igneous activity at spreading plate boundaries sepa-

FIGURE 4.33 Mount Shishaldin, an active composite volcano on Unimak Island in the Aleutian chain. The summit rises to an elevation just over 2800 m. A plume of condensed steam marks the summit crater. (U.S. Navy Department, from the National Archives.)

FIGURE 4.34 A glowing avalanche (nuée ardente) descending the slopes of Mount Pelée on the island of Martinique. The top of the cloud rises to a height of about 4 km above sea level. The photograph was taken in 1902, only a few months after a similar cloud had destroyed the city of St. Pierre, not far away. (A. Lacroix.)

FIGURE 4.33

FIGURE 4.34

FIGURE 4.35 Volcanic features built by highly viscous acidic lavas. In the foreground is a viscous lava dome similar to the type found in the Mono Craters, California; it is about 0.5 km in diameter. In the background is a dome-shaped mass of viscous lava—a cumulo-dome—that has emerged from the floor of an old crater. (Drawn by A. N. Strahler.)

FIGURE 4.36 Schematic, simplified diagrams of the formation of a caldera by collapse of a composite volcano. *A*. As the final eruption begins, magma fills the reservoir and stands high in the volcanic pipe. *B*. Violent eruption of gas and pumice occurs as the magma is draining downward, partly emptying the reservoir. *C*. Collapse of the cone fills the vacated space with a mass of broken blocks. Magma solidifies below and around the blocks. An ash layer covers the caldera floor. (Suggested by drawings from Howel Williams, 1942, Carnegie Institution of Washington, Publ. 540, p. 104, Figure 29.)

FIGURE 4.35

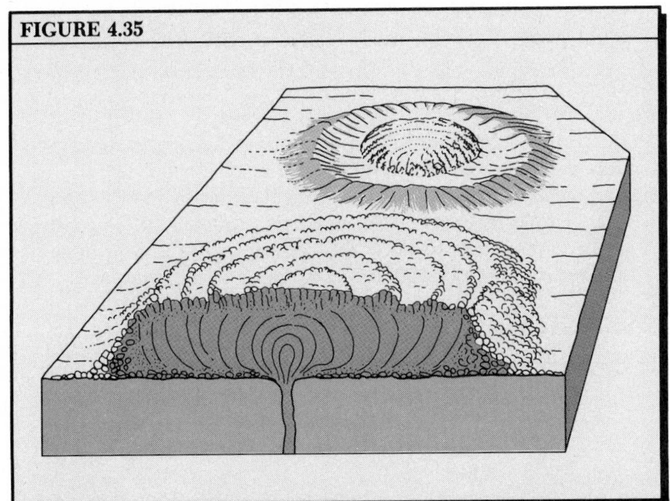

FIGURE 4.36

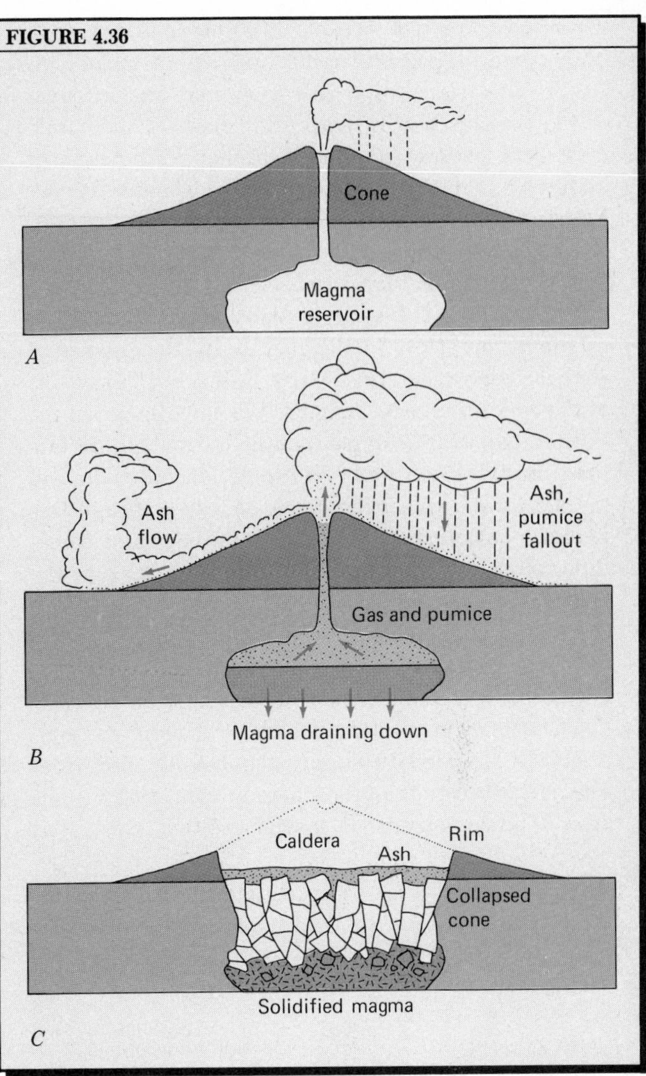

FIGURE 4.37 Crater Lake occupies a caldera that remained following the explosive demolition of Mount Mazama, about 6600 years ago. The lake surface elevation is about 1180 m. The upper part of Wizard Island, a cinder cone, can be seen above the nearside of the crater rim, just to right of center. The peak rising beyond the farside of the rim is Mount Scott, a parasitic volcanic cone on the flank of Mount Mazama. Crater Lake National Park, Oregon. (Oregon Dept. of Transportation, Photo 6711.)

FIGURE 4.37

rating two areas of oceanic lithosphere is predominately one of rise of basaltic magma to make new basalt. This basalt forms the uppermost igneous layer of the oceanic crust (Figure 4.38). Beneath the basalt, a layer of gabbro is formed simultaneously, as large bodies of mafic magma solidify. In Chapter 10, we investigate in detail the igneous structure of the crust along spreading boundaries.

Basaltic magma that rises beneath spreading boundaries of oceanic lithospheric plates is now thought to originate far below in the mantle by the process of partial melting, described early in this chapter as one of the major sources of magma. We have described the composition of the upper mantle as that of peridotite—mostly olivine and pyroxene. Included in this peridotite are minor amounts of calcic plagioclase feldspar, which, together with pyroxene (and sometimes olivine), makes up basalt. Partial melting of the peridotite would free the necessary pyroxene and calcic plagioclase feldspar, which would move slowly upward, accumulating in bodies of basalt magma beneath the spreading centers. It is important to note that enormous quantities of basaltic magma are continuously required to supply the basaltic rock that must be added to the lithosphere at spreading boundaries. The hypothesis of partial melting meets this requirement, because the bulk of the mantle is so enormous.

At a few places along spreading plate boundaries of a mid-oceanic ridge, volcanic islands have been formed (Figure 4.39). Here, basalt lavas have accumulated faster than they could be moved away from the spreading boundary by plate motion. A particularly important example is Iceland, which lies on the spreading boundary between the North American plate and the Eurasian plate (see Figure 1.11). Iceland has several active volcanoes, the larger ones basaltic shield volcanoes, the smaller ones, basaltic cinder cones. Except for Iceland, however, only a relatively few isolated volcanic cones rise above sea level on the spreading plate boundaries, and many of these volcanoes are inactive.

A second class of volcanic islands rising from the oceanic lithosphere form *volcanic island chains* or isolated groups of volcanoes that are within the oceanic

FIGURE 4.38 Schematic cross sections showing the relationship to plate tectonics of various forms of igneous activity. *A*. Cross section with vertical exaggeration to show crustal and surface features of the uppermost 50 km. *B*. Natural-scale cross section to show lithospheric plates and asthenosphere in the upper 250 km. (Drawn by A. N. Strahler.)

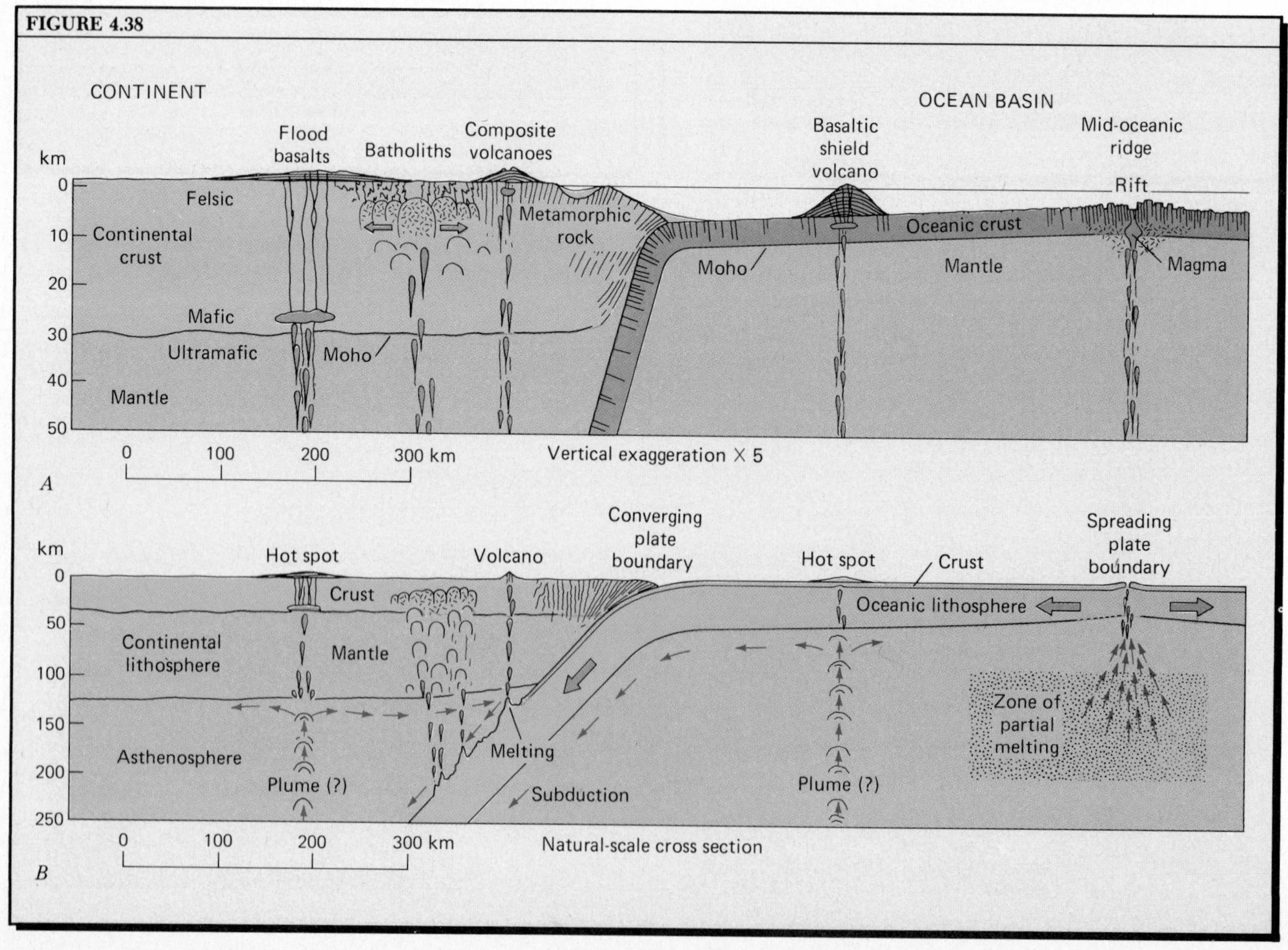

lithospheric plates, rather than on plate boundaries. The Hawaiian Islands are an outstanding example (Figure 4.39). Volcanic chains and groups of this class are thought to have been formed above a magma source called a *hot spot*. Some geologists think the hot spot is located over a slowly rising column of heated mantle rock known as a *mantle plume* (Figure 4.38). Whereas the mantle plume and its associated hot spot remain approximately fixed in position relative to the asthenosphere for long periods of time, the overlying oceanic lithospheric plate is drifting past it in continual motion. As a pulse of volcanic activity occurs above the hot spot, a volcanic island, or island group, comes into being, but is then carried away with the moving plate. A short time later, a new pulse of volcanic activity occurs above the hot spot, forming a new volcanic island. In time, a chain of islands results. We investigate this phenomenon in Chapter 10, for it provides direct evidence of the direction and the speed of plate motion.

Also related to hot spots is a form of isolated volcanic activity occurring within continental lithospheric plates, far from presently active boundaries. From a hot spot beneath the continental lithospheric plate, enormous volumes of basaltic magma may rise to the surface, emerging through fissures, and pouring out upon the landscape as thick basalt flows (Figure 4.40). Called *flood basalts*, these outpourings continue until a total thickness of some thousands of meters of basalt layers have accumulated (Figure 4.41). Often cited as examples are two great expanses of flood basalts (Figure 4.42): the Columbia Plateau of Oregon, Washington, and Idaho, where basalts covering about 130,000 sq km have a total thickness ranging from 600 to 1200 m, and a volume of at least 250,000 cu km; and the Deccan Plateau of peninsular India.

Besides flood basalts, some small, isolated volcanic areas are interpreted as igneous activity located over a hot spot. One example is the Yellowstone National Park volcanic area where many hot springs represent hydrothermal activity over a small body of hot magma. Lava flows formerly produced in this area are rhyolites formed from acidic magma.

Volcanic eruptions of felsic and intermediate lavas are largely concentrated directly above active subduction zones. *Volcanic arcs* of great andesitic cones,

FIGURE 4.39 Volcanic activity of the earth. Dots show the locations of volcanoes known or believed to have erupted within the past 12,000 years, as determined by radiocarbon dating, geological evidence, and other techniques. Each dot represents a single volcano or a cluster of volcanoes. (Data originally compiled by the Smithsonian Institution, Washington, D.C. Based on a map published in 1979 by World Data Center A for Solid Earth Geophysics at the Geophysical and Solar-Terrestrial Data Center, National Oceanic and Atmospheric Administration, Boulder, Colorado.)

FIGURE 4.39

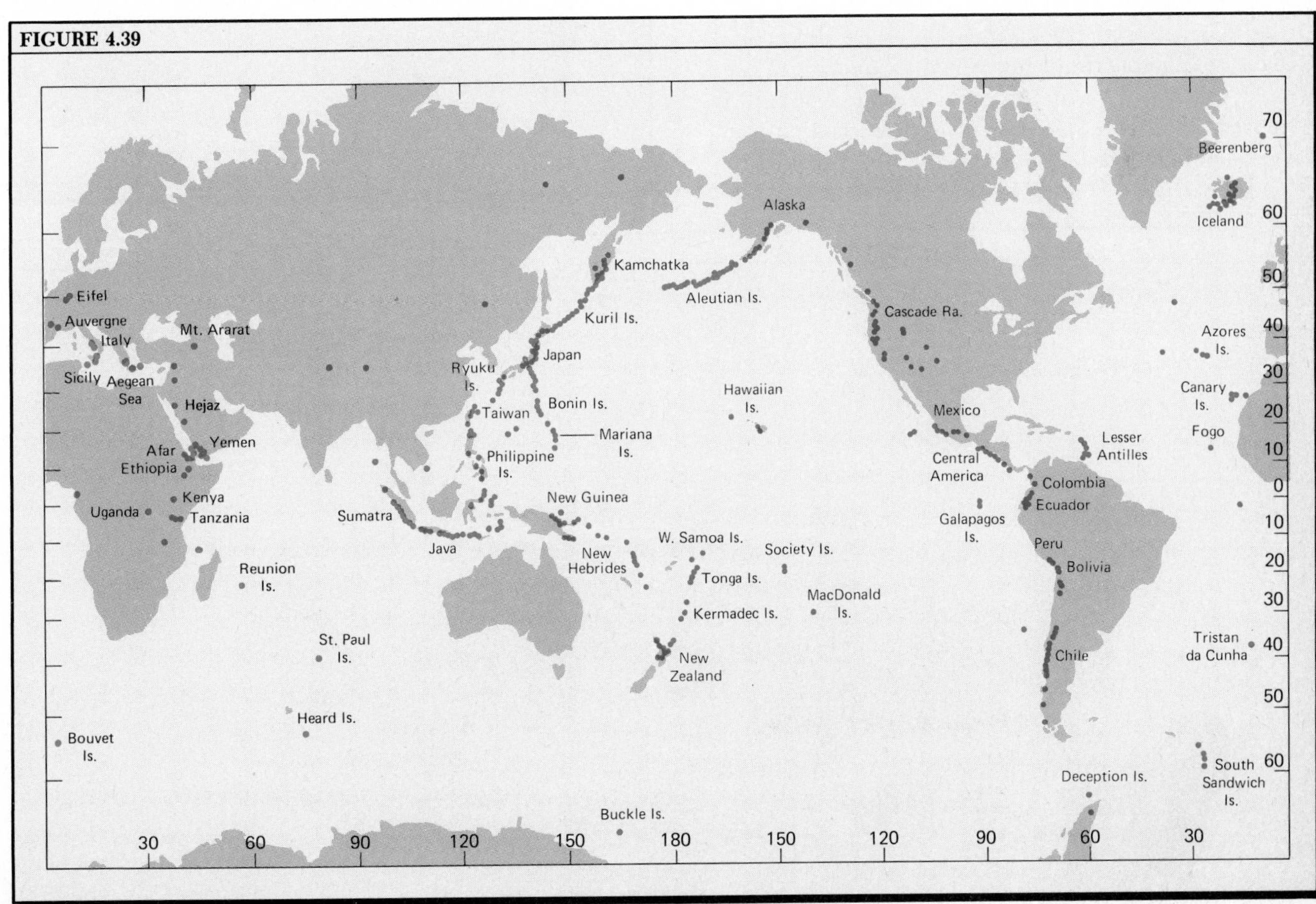

many of them active today, occupy positions above these zones. The subduction zones bounding the Pacific plate form a great "ring of fire" that borders the Pacific Ocean (Figure 4.39). Note in Figure 1.11 that the Pacific, Nazca, and Cocos plates consist almost entirely of oceanic lithosphere. In North and South America, oceanic lithosphere of the Pacific, Nazca, and Cocos plates is being subducted beneath continental lithosphere of the American plate. Arcs of andesitic volcanoes mark these subduction zones in the Aleutian Islands, the Cascade Range, Central America, and the Andes Range. (The word "andesite" is derived from its abundance as a rock type in the Andes Range.) Other arcs of andesitic volcanoes follow the western boundary of the Pacific plate, making several island arcs extending south from the Japanese Islands through the Philippines, and into the southwestern Pacific. Another similar volcanic arc forms the Indonesian crescent, including the islands of Java and Sumatra. Here, the Austral–Indian plate is being subducted beneath the margin of the Eurasian plate. Rhyolite lavas are also present in volcanic eruptions of the "ring of fire," but in smaller volumes than the andesite lavas.

Igneous activity forming volcanic arcs along the margins of the continental lithosphere largely involves magmas which also give rise to plutons of granite, granodiorite, and diorite. Reaching the surface, these magmas emerge as lavas of acidic (felsic) and intermediate types, with lava flows and pyroclastics of rhyolite and andesite. We have shown that, for the most part, the magmas arise from a position above descending slabs of lithospheric plates, where subduction of oceanic lithosphere is in progress (Figure 4.38). It is thought that the upper surface of a descending plate becomes highly heated in contact with hot rock of the asthenosphere. The descending layer of basaltic oceanic crust, forming a thin cap on the plate, is rich in water because of its prior contact with seawater. It melts readily, releasing magma of basaltic composition. The depth at which this process begins is estimated to be about 120 km, but melting then extends to depths of 200 km or more on the upper surface of the plate.

FIGURE 4.40 This schematic block diagram suggests the inferred relationship between flood basalts and a reservoir of basalt magma located near the base of the continental crust or in the uppermost mantle. (Drawn by A. N. Strahler.)

FIGURE 4.41 Flood basalts of the Columbia Plateau region. The basalt layers have been eroded to produce steep cliffs rimming broad, flat-topped mesas, Dry Falls, Grand Coulee, central Washington. (John S. Shelton.)

FIGURE 4.40

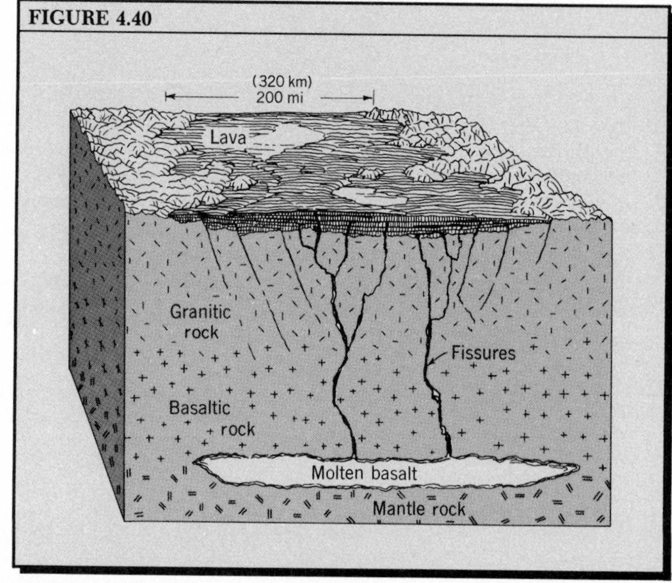

FIGURE 4.41

A difficult question is how this process can result in extrusion of enormous volumes of andesite and rhyolite at the continental surface. One hypothesis states that the rising basaltic magma undergoes magmatic differentiation, as described earlier in this chapter. The more felsic mineral ingredients of the melt are less dense than the more mafic ingredients and rise more rapidly. In this way, magmas of felsic and intermediate composition become separated from mafic magmas. The mafic magmas solidify near the base of the continental crust whereas the felsic magmas solidify as plutons in the upper layer of the crust or are extruded to form volcanoes. The hypothesis of differentiation has major weaknesses and does not seem adequate to explain the enormous volume of felsic and intermediate igneous rock in the continental crust.

A hypothesis that appears more successful is based on the recycling principle described in Chapter 1. It says that the felsic and intermediate magmas come largely from the melting of preexisting rock of the continental crust. The role of the rising basaltic magma from the descending plate is that of a carrier of the heat necessary to melt the crustal rock. At the continental margin, where subduction has been in progress for many tens of millions of years, crustal rock consists largely of deformed, water-rich sediments that have been plastered against the continental plate above the descending plate (see Figure 1.13). For reasons that will become clear in the next chapter, sedimentary rock as a whole has the necessary chemical components to produce felsic magma. For the same reason, metamorphic rock formed from sedimentary rocks would, by melting, also yield felsic magmas. Finally, the continental crust contains large masses of felsic plutonic igneous rocks produced in earlier tectonic events. These plutons are available for remelting.

FIGURE 4.42 Approximate present surface extent of the Columbia Plateau basalts (*upper map*) and the Deccan Plateau basalts of India (*lower map*).

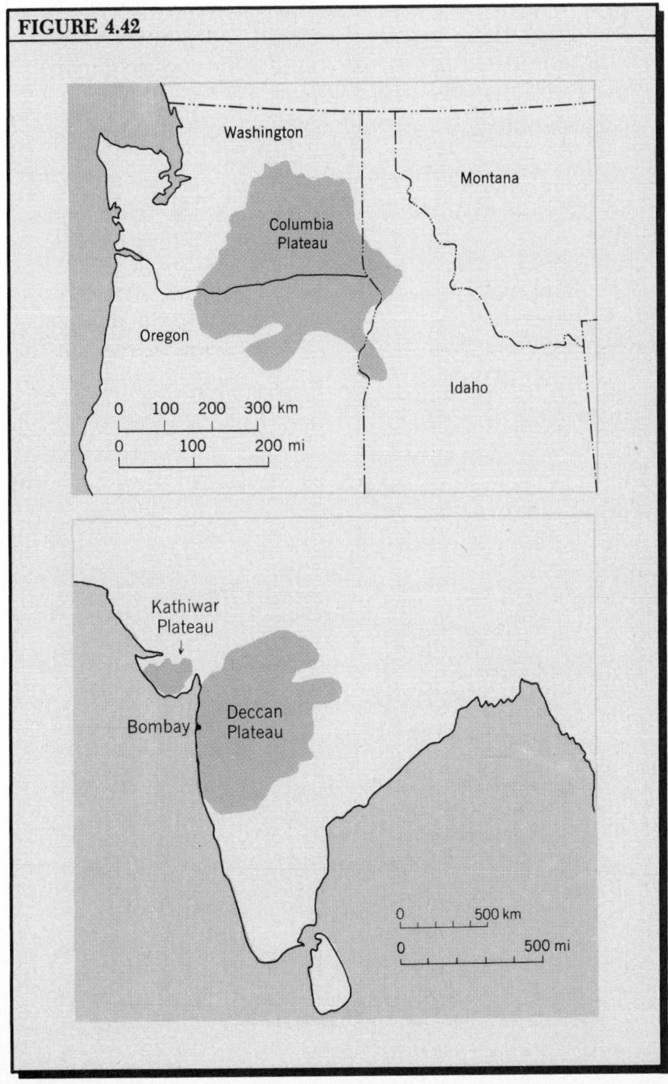

One of the major problems in determining the origin of batholiths of granitic composition is to explain what happened to the country rock that was displaced by the invading magma—assuming that the magma was produced from other, deeper sources than the country rock. If, instead, we envision the magma as produced directly by melting of the older country rock, the problem of finding room for the batholith largely disappears. Geologists have an alternative explanation for large plutons that relates to plate tectonics. The crust above a descending lithospheric plate is stretched horizontally and tends to become thinner. Rising magma takes advantage of this mass motion, pushing the older country rock aside, rather than upward. Under this hypothesis, there is no limit to the volume of new plutonic rock that can be accommodated in the crust.

In this brief summary of the relationships between igneous activity and plate tectonics, we have not described all types of magma intrusions or all regions of volcanic activity. What we have covered is, however, enough to give the most important relationships on a global scale and to reinforce the bare outlines of plate tectonics sketched in Chapter 1.

°*Volcanic Eruption as an Environmental Hazard*

Among the great natural environmental hazards to human life and property are several kinds of geologic phenomena. They include volcanic eruptions, earthquakes, seismic sea waves (tsunami), landslides, and floods. Through recorded history volcanic eruptions have been the cause of many disasters in which the toll in human lives has been great and the damage to property high. Volcanoes erupting close to cities have been a particularly serious threat. We offer three examples.

Mount Vesuvius is an active composite volcano situated on the Bay of Naples in southern Italy. Although the city of Naples is not close enough to be directly threatened by lava flows and falls of tephra, explosive activity of Mount Vesuvius can be watched by Neapolitans across the bay. Not so fortunate are a number of towns and villages situated on the densely populated lower slopes of the volcanic cone. One of those towns, Torre Annunziata, had been destroyed by lava flows three times in 900 years before succumbing again in the 1906 eruption of Vesuvius.

Undoubtedly the best publicized of all Vesuvius' eruptions occurred in A.D. 79, following a long period of inactivity. Severe earthquakes began to occur in A.D. 63 and continued up to the time of the major eruption, which destroyed about half the previous cone and left a large explosion crater. The rim of that former crater persists today, partly encircling the new cone, and is known as Monte Somma. Two small cities were destroyed in the eruption—Pompeii and Herculaneum—but by quite different forms of disaster.

As described in letters written by Pliny the Younger, the eruption was first observed as an enormous treelike cloud of dust and gases rising from the volcano summit. There followed a rain of ash, lapilli, and bombs, as well as large fragments of pumice. The fall of tephra continued for more than two days, during which darkness prevailed through the daytime hours. Pompeii was gradually buried under a layer of ash and pumice that totaled 5 to 8 m in thickness. Although roofs collapsed under the weight of the ash, walls and contents of the buildings survived and are remarkably well preserved to this day. Herculaneum, on the other hand, suffered destruction by a great tongue of mud (a mudflow) consisting of volcanic ash mixed with water from torrential rains that fell upon the upper slopes of the cone throughout the eruption. Little of the city remained intact under the impact and burial by the flowing mud. Since that time, Vesuvius has undergone many other violent eruptions, the latest in 1944.

Our second example of explosive violence comes from the Island of Martinique in the West Indies. Here, Mount Pelée, a composite volcano built of acidic lava, lies only a short distance from the port city of St. Pierre. The cone rises to a summit elevation of about 1200 m. On May 8, 1902, following several days of muffled explosions that sent up ash from the crater, four deafening blasts took place. These were followed by the rise of an enormous black cloud from the crater; another black cloud shot horizontally outward, then moved down the mountain slopes with great speed, headed directly for the city. A surviving eyewitness aboard a ship in the harbor described the arrival of the glowing cloud of ash and gases that in an instant reduced all buildings to burning rubble (Figure 4.43). Of the nearly 40,000 persons believed to be present in the city at the time, only two survived. One was a prisoner in an underground dungeon. A similar glowing avalanche from Mount Pelée, shown in Figure 4.34, was photographed about seven months after the destruction of St. Pierre.

Our third example is from Heimaey, a small island off the coast of Iceland. The island consists largely of a basaltic volcano, apparently dormant for the past several thousand years, and its lava flows. The fishing village of Vestmannaeyjar, with its fine harbor, had a population of over 5000 persons. On January 23, 1973, a fissure located about 1 km from the center of the town began to erupt tephra. Quickly a cinder cone grew up, reaching a height of 100 m within two days. Basaltic lava then began to emerge from the fissure and formed a flow which advanced toward the town. Strong winds carried the particles of lapilli and ash over the town, where it began to accumulate in a thick layer on roofs and streets. Gradually, houses nearest the center of eruption became buried in tephra, although efforts were made to remove the layer as it accumulated (Figure 4.44).

Within six hours of the start of the eruption most of the inhabitants had been evacuated from the island, along with the entire fishing fleet. There remained only a work force of 500 to fight against the advancing lava flow threatening to destroy the town and fill the harbor. With the advice of Icelandic geologists, a plan was formulated to try and stop the forward motion of the lava flow. Powerful pumps were brought in, and cold seawater was sprayed directly on the front of the

FIGURE 4.43 The city of St. Pierre, island of Martinique, West Indies, as it appeared a few weeks following total destruction by a glowing avalanche (nuée ardente) in 1902. In the distance you can see the slopes of Mount Pelée, from which the cloud of incandescent dust and gas originated. (A. Lacroix.)

FIGURE 4.43

lava. At the peak of the operation 47 pumps were in action. Altogether, over 5 million metric tons of water were sprayed on the flow.

Apparently, the chilling of the lava margin was effective in creating a barrier to the movement of the more fluid lava behind it, for the advance was finally halted, though lava continued to flow from the volcano to the sea in a direction away from the town.

Five months and five days after it started, the eruption was officially declared over. The addition of new flows to the harbor shoreline improved the degree of protection from the open sea. The area of the island was increased by about 20% by new flows. Damaged buildings were dug out, and most were repaired and reoccupied. Some of the tephra was used to extend the runways of the island's airport.

Predicting Volcanic Eruptions

Prediction of the occurrence of a devastating volcanic eruption has been under study for many years by geologists and geophysicists. Vulcanologists—scientists who study volcanic activity—have maintained an observatory on the rim of Kilauea volcano in Hawaii and have attempted to predict activity in both Kilauea and Mauna Loa, the two active volcanoes on the island. Phenomena that are monitored include minor earthquakes and changes in ground level. Hazards to humans are not great on Hawaii, aside from occasional property destruction in the path of a fast-moving basaltic lava flow.

Mount Baker, a composite volcano in the Cascade Range of Washington which last erupted in the 1840s, is being closely observed because it has been recently showing signs of renewed eruptive activity. As one of the highly explosive acidic type, the volcano is potentially dangerous because it is capable of emitting a great quantity of tephra, including large volumes of fine ash. At the time of its last explosion, the fall of volcanic ash killed numbers of fish in the Baker River and started a large forest fire. Since 1975, a crater on Mount Baker has been emitting hot gases from several small vents, called *fumaroles*. So far, hazards to humans have been limited to snow avalanches and flows of mud caused by increased volcanic heating of surface materials. As yet, none of the phenomena being monitored suggests that a large eruption is pending.

During the past 500 years, another Cascade volcano, Mount St. Helens, has erupted about once per century. It came to life in 1980 after a 120-year period of inactivity with emissions of steam and ash and the formation of a new crater. The renewed activity followed only a few days after swarms of minor earthquakes began to be recorded in the immediate area. Scientists issued warnings of a possible serious eruption and the area was closed to visitation by the public.

No one, however, was prepared for the catastrophic eruption of Mount St. Helens that occurred on May 18 when an outward bulge that had developed on the upper flank of the volcano broke away in a great landslide, triggered by an earthquake of moderate inten-

FIGURE 4.44 Homes partly buried in basaltic tephra during the 1973 eruption of a small volcano on the island of Heimaey, Iceland. Condensing water vapor issuing from the ground forms wisps of fog around the damaged structures. (Klaus D. Francke/Peter Arnold.)

FIGURE 4.44

sity. Evidently, the landslide relieved the confining pressure on a mass of dacite magma, heavily charged with gas, that had previously risen within the volcano. The magma exploded horizontally to become a huge cloud of incandescent gas and ash. The glowing avalanche, traveling at a speed of about 50 m/sec, instantly killed all animals and humans in its path, while trees were blown down over a forested area of 500 sq km. The landslide debris, mixed with new ash, became a tongue of mud, which flowed 20 km down a river valley. Following the initial explosion, a vertical cloud of fine ash rose to an elevation of 13 km. Carried by prevailing winds, ash settled upon the cities of Yakima and Spokane. In the weeks that followed several lesser explosions occurred and the crater became greatly enlarged and deepened. Then, in mid-June, a dome of viscous lava began to rise in the crater, and this was interpreted as possibly signaling the onset of the final phase of the eruptive sequence.

The present state of the art of predicting volcanic eruptions is not yet very far advanced, despite a few notable successes. Soviet scientists claim that a vulcanologist predicted an eruption in July 1975 of a new volcano in the Kamchatka Peninsula. One scientist specializing in this field has commented that "the only key to forecasting is better understanding of volcanic processes."

Key Facts and Concepts

Magma *Magma,* a high-temperature mixture of the chemical ingredients of the silicate minerals, includes substances in the liquid, solid, and gaseous states. Magma appears at the earth's surface as *lava.* Magma rising along spreading plate boundaries may originate by *partial melting* of mantle rock, rising to collect in pockets at shallower depth. Increased confining pressure raises the melting point of a silicate magma.

Volatiles in magma *Volatiles* held in rising magma reach the earth's surface during *outgassing,* contributing to the atmosphere and hydrosphere. The most important constituent of outgassed volatiles is water. Presence of water lowers the melting-point temperature of silicate magma.

Crystallization of silicate magma As temperature falls within a magma and crystallization occurs, reactions take place in a *reaction series.* Reaction may be *continuous,* as in the plagioclase feldspars, or *discontinuous,* as in the case of the *reaction pair* olivine–pyroxene. The *Bowen reaction series* describes the reaction series of a typical silicate magma; it begins with mafic minerals and ends with felsic minerals. *Fractionation* may lead to separation of early-crystallized minerals, resulting in *magmatic differentiation.*

Textures of igneous rocks *Texture* is typically *phaneritic* in the intrusive igneous rocks; *aphanitic* in the *extrusive igneous rocks.* Texture may be *equigranular* or *porphyritic.* Extrusive rocks may show *scoriaceous texture,* or may form as *volcanic glass.*

The granite–gabbro series The abundant igneous rocks are contained in a *granite–gabbro series,* in which *granite* (*rhyolite*) lies at the felsic end, *diorite* (*andesite*) in an intermediate position, and *gabbro* (*basalt*) toward the mafic end. *Ultramafic rock,* such as *peridotite* and *dunite,* lie at the extreme end of the series and consist dominantly of olivine. At the extreme felsic end of the sequence is the *syenite group* and rocks deficient in silica.

Composition of the crust *Oceanic crust* is of mafic composition, largely basalt and gabbro, with peridotite forming the mantle beneath. *Continental crust* has an upper felsic (*granitic*) layer, which grades downward into a lower mafic layer.

Intrusive rock bodies *Plutons,* which are intrusive rock bodies, may be *discordant* or *concordant* in terms of their contact relationships. Discordant bodies include the *batholith, stock, dike,* and *volcanic pipe.* Concordant bodies include the *sill, laccolith,* and *lopolith.* Thin tabular bodies show *columnar jointing,* as do lava flows. *Pegmatite dikes* form from hydrothermal solutions.

Lavas Lavas are classified as *acidic* (*silicic*), *intermediate,* and *basic* in composition, giving rise to rhyolite, andesite, and basalt, respectively. Dry basaltic lavas typically have low viscosity and high temperature, erupting quietly; acidic lavas, of high viscosity, give highly explosive eruptions. Lava takes the form of a *lava flow,* which may emerge from a *fissure.* Lava eruption in ocean water develops *pillow structure,* forming *pillow lava.*

Pyroclastic materials Fragments blown from a volcanic vent are *pyroclasts;* they take the form of *volcanic bombs, lapilli, ash,* or *dust,* collectively called *tephra. Pyroclastic deposits* include *volcanic breccia, tuff,* and *ash flows* formed of *welded tuff.*

Volcanoes Extrusion from a *magma reservoir* through a *volcanic vent* leads to building of a *volcano* in the process of *volcanism.* Highly fluid basaltic lava forms the *basaltic shield volcano.* Scoriaceous basaltic tephra may form small *cinder cones.* Lavas of felsic (rhyolitic) and intermediate (andesitic) composition form the *composite volcano* with conical form and steep sides. Highly viscous acidic lava forms the *viscous lava dome.* Composite volcanoes may be partially demolished by explosion, producing an *explosion crater,* or largely destroyed by explosion and subsidence to produce a *caldera.*

Igneous activity and plate tectonics Igneous activity is largely concentrated along or near active plate boundaries. Basalt flows and dikes form simultaneously with gabbro at spreading boundaries between oceanic lithospheric plates. Other basalt volcanic activity occurs in isolated *hot spots*, possibly associated with *mantle plumes*. *Flood basalts* occur within continental lithosphere. Volcanism involving acidic and intermediate lavas occurs largely above active subduction zones, where the original magma source is probably located on the upper surface of a descending oceanic lithospheric plate.

Volcanic eruption as a hazard Many volcanic eruptions have been disastrous to humans and their artifacts. Prediction of eruptions has proved difficult and uncertain.

Questions and Problems

1. Considering the possible sources of magma in relation to plate tectonics, along which type of plate boundary would you expect the greatest quantity of magma to be produced on a worldwide basis? Is it necessary that magma be produced at the same rate above a subducting oceanic plate as at the spreading margin of the same plate? Explain.
2. It is thought that basaltic magma rising from mantle rock beneath the oceanic spreading plate boundaries contains some water. Speculate on the possible source of this water. Is it likely that ocean water soaks through the crust and penetrates the mantle? Could water be carried down into the mantle along with a subducting lithospheric plate? If so, in what form might this water exist?
3. If we accept the hypothesis that volatiles have been continuously brought to the earth's surface by igneous activity throughout the earth's history as a planet, does it follow that the oceans and atmosphere have been continuously increasing in total mass? Does it also follow that the degree of salinity (saltiness) of the oceans has been increasing through geologic time?
4. Returning to Chapter 3, review the crystal lattice structure of plagioclase feldspar. How does continuous reaction within the plagioclase series take place in terms of ions within the silicate framework? In the case of discontinuous reaction of the olivine–pyroxene pair, can the lattice structure of the oxygen–silicon tetrahedra remain intact during the reaction process? Explain.
5. Do you think that the process of fractionation of a mafic magma is adequate by itself to explain the formation of the felsic (granitic) crust of the continents over the long span of geologic time? How else might felsic igneous rock be formed, i.e., by a process other than magmatic differentiation?
6. Columnar jointing in platelike igneous bodies—dikes, sills, lava flows—always occurs in such a way that the columns are perpendicular to the surface of contact between the igneous body and the older country rock (or in the case of a lava flow, the upper surface of contact with the atmosphere). What causes the joint fractures to form and why do they show the orientation we have described?
7. Suppose you were to examine an exposure of ancient pillow lavas among rocks of the continental mainland. Presumably the lava was extruded on the ocean floor and was subsequently elevated to its present position on the continent. Would it be possible to determine whether the lava extrusion occurred in shallow water off a coastline or in deep water (4 to 5 km) of the ocean floor along a spreading plate boundary? Explain.
8. Imagine yourself as a field geologist studying numerous exposures of tephra produced by an ancient caldera explosion. Assume that the caldera can no longer be easily identified. How might you attempt to locate the probable position of the caldera through your field studies?

5

CHAPTER FIVE

Sediments and Sedimentary Rocks

The interface between lithosphere and atmosphere represents a special environment for minerals and rocks. This surface environment is one of relatively low temperature and low confining pressure in contrast to the high-temperature and high-pressure environment in which plutonic igneous rocks are formed deep within the earth's crust.

The surface environment is one of instability for the silicate minerals, for they succumb readily to the presence of free oxygen, carbon dioxide, and water. We may like to think that no substance is more enduring than the granite we use in monuments to signify an everlasting tribute. The truth is that igneous rock is one of the most decay-susceptible of mineral assemblages to face the "elements." Much the same statement can be made of most varieties of metamorphic rocks, for they too were produced in an environment of high pressure and temperature and are not well adapted to endure lengthy exposure to atmospheric conditions.

The chemical change of silicate minerals in the presence of water, oxygen, and various natural acids is greatly aided by a group of physical forces that break the hard well-knit minerals of igneous and metamorphic rocks into fragments. Fragmentation increases the mineral surface area exposed to chemical reagents. Were it not for physical breakup, chemical alteration of rock would have proceeded very slowly throughout the geological past, and the course of earth history would have been quite different.

A Second Look at the Rock Cycle

In Chapter 1, we briefly described a cycle of rock transformation involving rocks of all three major classes—igneous, sedimentary, and metamorphic (Figure 1.15). We related the various steps in that rock cycle to plate tectonics, which acts as a great natural material flow system. It is now helpful to take a new look at the rock cycle to examine the sources of energy that furnish it with power. We are particularly interested in the way that igneous rock undergoes physical breakup and chemical change to give rise to sedimentary rock.

Figure 5.1 helps us relate the rock cycle to the contrasting physical–chemical environments found deep within the earth's crust or at the earth's surface. The diagram represents a schematic vertical cross section of the continental crust, to a depth of about 30 to 40 km. Throughout the rock cycle, large masses must be moved from the deep environment of high temperatures and pressures to the surface environment of low temperatures and pressures. We see that this change of environment can be accomplished by two processes: (1) Rise of magma brings igneous rock to various intermediate positions, where it solidifies into intrusive igneous rock bodies. By extrusion, magma may reach the surface to form volcanic igneous rocks. (2) Rock formed at depth can appear at the surface of the earth by uncovering as a result of the combined processes of crustal uplift and *denudation*, a general term describing the removal of rock at the earth's surface by rock breakup and erosion. (These processes are treated in detail in Chapters 14 through 19.) In this way, large bodies of igneous or metamorphic rock formed at depths of 10 to 40 km can migrate upward from the deep environment to the surface.

Transition from the surface to that deep environment can happen by burial and downsinking of the earth's crust. Both sedimentary layers and extrusive volcanic rocks can eventually reach the deep environment—10 to 20 km or more—where either metamorphism or remelting can take place.

The cycle of rock transformation can operate only because sources of energy exist. The energy required both to melt solid rock and to force enormous masses of magma upward through the crust arises internally, probably largely as heat produced from the process of spontaneous decay of radioactive substances (see Chapter 7). The internal energy supply comes from the time the earth originated as a planet.

The energy required to break surface rocks into sediment and to transport that sediment to sites of deposition is supplied by solar radiation. Solar energy is thus the external energy source, as contrasted with radioactivity which is the internal energy source. Energy of the sun's rays activates atmospheric motion and powers the hydrologic cycle in which water is carried up into the atmosphere as water vapor, then condensed into clouds, and precipitated in turn producing rain and snow. Falling on the land, the rain and snow feed streams and glaciers and cause erosion of surface forms. The sun's rays provide heat used in chemical reactions that affect surface rock, and solar energy is geologically active in many other ways.

A study of the rock cycle leads to the conclusion that any given form of rock is the product of its physical–chemical environment at the time of its formation. When it is transported to a different set of environmental conditions, the rock undergoes physical or chemical changes to bring the minerals it is composed of into an equilibrium with those new conditions.

Rock Weathering

Geologists use the term *rock weathering* to describe all external processes, acting at or near the earth's surface, by means of which rock undergoes chemical decomposition and physical *disintegration*—the process by which solid rock masses are broken into small fragments. This ongoing fragmentation is a purely physical change and is therefore also called *mechanical weathering*. On the other hand, *chemical weathering* of rock is a process of *decomposition* in which existing rock-forming minerals are changed in chemical composition. In decomposition, the preexisting minerals are transformed into minerals of different compositions and physical properties. Keep in mind, however, that disintegration by mechanical weathering complements decomposition. Fine mineral and rock particles

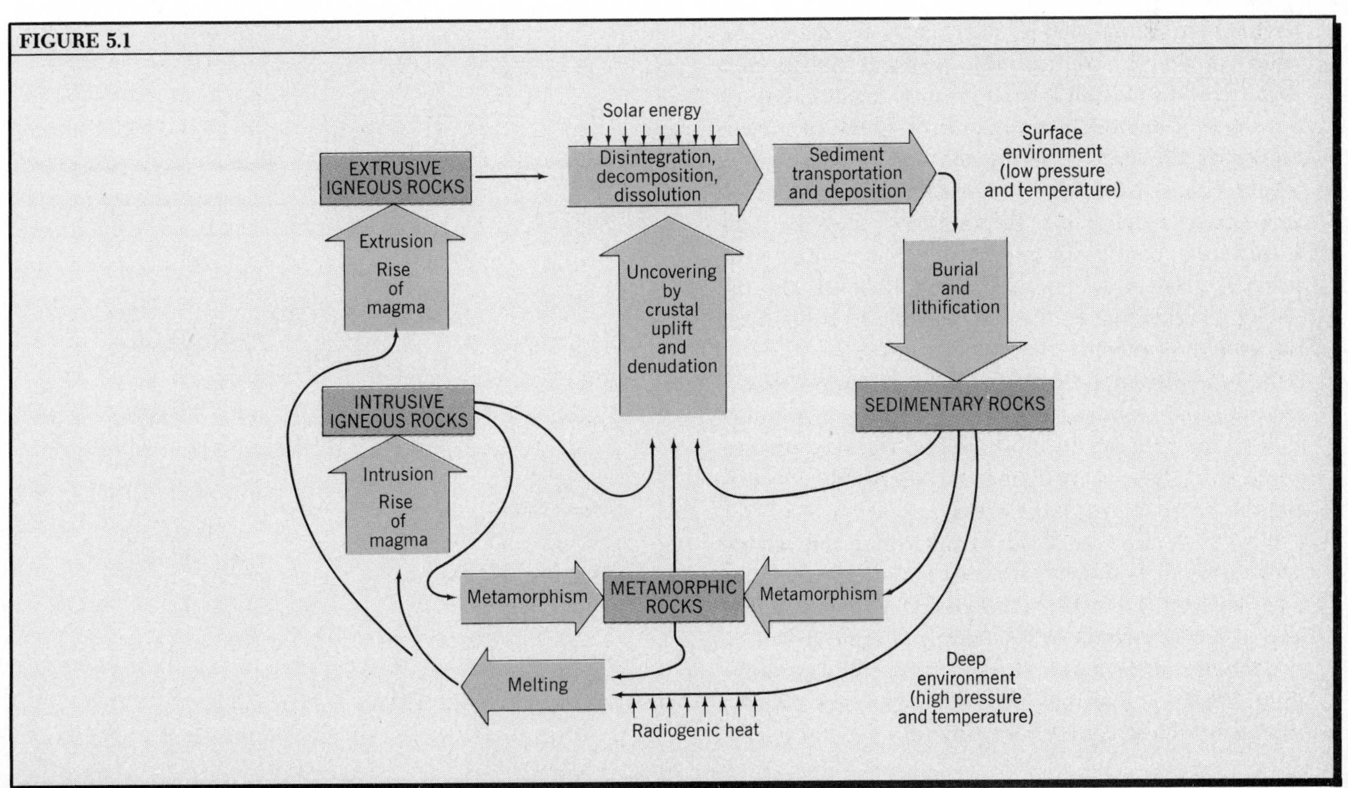

FIGURE 5.1 This schematic diagram of the rock transformation cycle shows how the three rock classes are related to the surface environment of low pressure and low temperature and the deep environment of high pressure and high temperature.

produced by mechanical weathering are much more susceptible to chemical change than are mineral grains tightly held within large masses of solid rock.

Bedrock and Regolith

Before beginning our discussion of weathering, we want to examine the earth materials commonly found in the upper few meters where weathering takes place. Examination of a freshly cut cliff, such as that in a new highway excavation or a quarry wall, will often reveal a succession of layers of different earth materials. Starting at the base of the exposure, there may be solid rock, called *bedrock;* that is still in its original place and relatively unchanged (Figure 5.2). Bedrock usually shows innumerable cracks or *joints*, the result of stresses mostly associated with a previous history of crustal bending and breaking (tectonic processes). Where intersecting sets of joints are present, the bedrock is easily disintegrated into blocks (joint blocks).

Lying above the bedrock may be a layer of soft mineral matter, the *regolith*. (The prefex *rego* comes from the Greek word for "blanket.") Regolith may be formed in place by decomposition and disintegration of the bedrock that lies directly beneath it; this type is called *residual regolith*. It can be seen to grade downward into the unaltered bedrock. Within the regolith, individual mineral grains or small groups of mineral particles are easily separated from one another.

Regolith is a term that can be used broadly to refer to any layer of relatively loose or soft mineral particles lying on the bedrock. Gravels, sands, floodplain silts laid down by streams, or rubble left by a disappearing glacier are all forms of regolith, but they are unique in having been transported by such agents as streams, ice, wind, or waves. We call this material *transported regolith* to distinguish it from residual regolith. Figure 5.3 shows a deposit of transported regolith of a type known as alluvium, covering the bottom of a valley, where it has been left by a stream. Many kinds of transported regolith can be identified, and we shall learn more about them in later chapters dealing with streams, glaciers, waves, and winds. Regolith also includes a surface layer, the soil, capable of supporting the growth of plants.

In some places, the regolith has been stripped away, exposing the bedrock, which then appears at the surface as an *outcrop*. In other places, the regolith can reach a thickness of many meters, completely concealing the bedrock over large areas.

Geologists use a set of terms, describing the various ways in which bedrock is reduced to individual masses and particles, to refer to the geometrical forms that may occur as a result of weathering (Figure 5.4).

Granular disintegration consists of the grain-by-grain breakdown of rock masses composed of discrete mineral crystals. The term usually refers to the weathering of coarse-grained igneous and metamorphic rocks—such as granite and gabbro—and of some coarse-textured sedimentary rocks. Individual mineral grains simply separate from one another along their natural contacts and produce a coarse sand or gravel in which each particle has much the same shape and size as it did in the original rock.

FIGURE 5.2 Bedrock, regolith, and soil. (Drawn by A. N. Strahler.)

FIGURE 5.3 Residual regolith on the hillside is distinguished from transported regolith in the valley bottom. Rock outcrops project through the regolith. (Drawn by A. N. Strahler.)

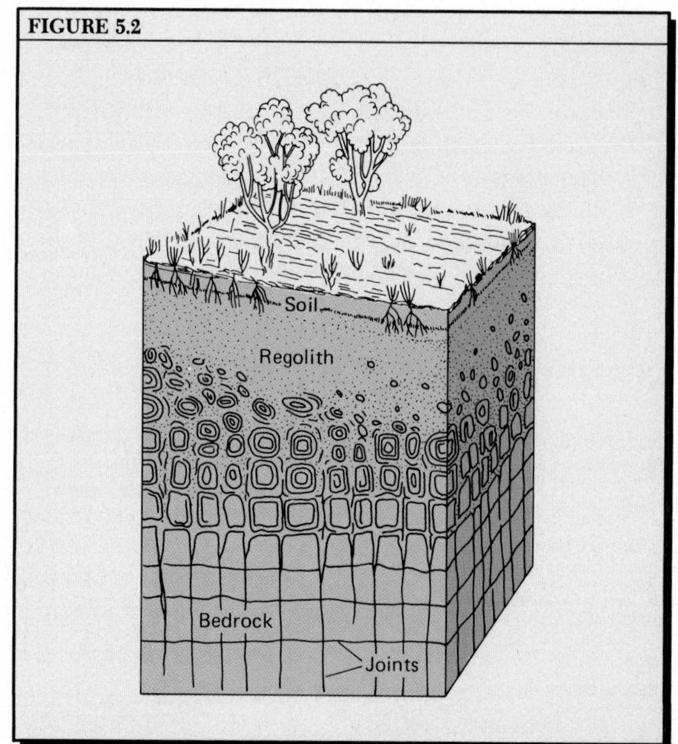

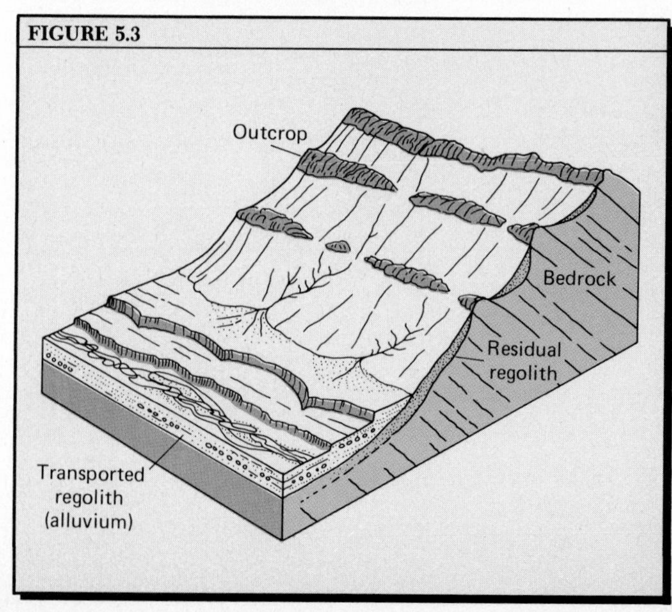

In *exfoliation,* solid rock bodies come apart in shells and plates, or in scales, conforming roughly with the configuration of the outer surface. The planes of fracture are thus new fractures rather than previously existing natural surfaces of breakage. Exfoliation can occur on a very large scale, with individual plates of rock several meters thick, or on a very small scale, with layers a millimeter or less in thickness. Exfoliation affects many varieties of rocks and can be caused by both physical and chemical processes.

Most bedrock is so fractured by joints that it is rare to find flawless bodies of rock (or monoliths) more than a meter or two across. Most joints occur in parallel sets, and there are often two or more sets intersecting at large angles. Consequently most bedrock is already broken into blocks from a few centimeters to several meters across. When stresses are exerted upon jointed rock, the rock comes apart rather readily along these planes by a type of disintegration called *joint-block separation.* Of course, single joint blocks are susceptible to granular disintegration or exfoliation, or both.

FIGURE 5.4 Four geometrical patterns of rock breakup. (Redrawn, by permission of the publisher, from A. N. Strahler and A. H. Strahler, *Modern Physical Geography,* Figure 17.2. Copyright © 1978 by John Wiley & Sons, Inc.)

FIGURE 5.4

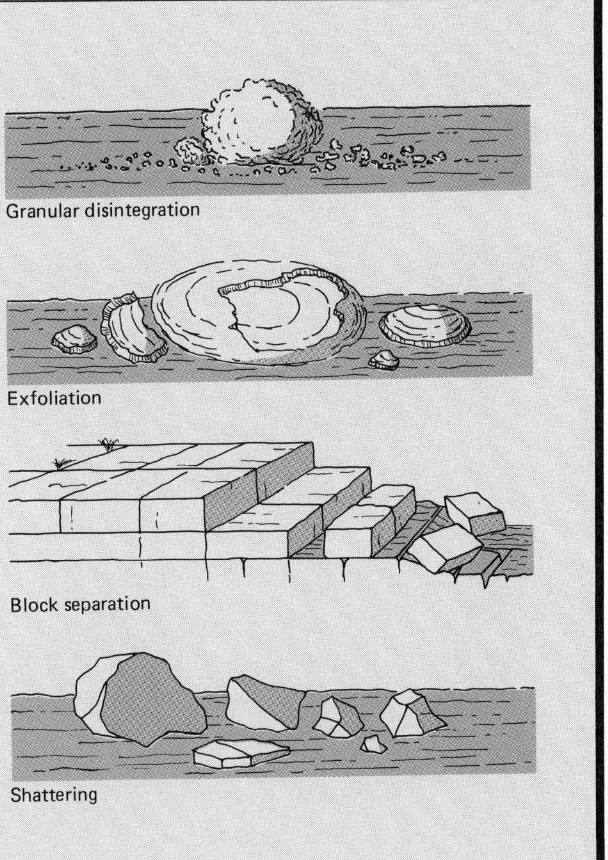

Granular disintegration

Exfoliation

Block separation

Shattering

Where a very hard rock is subjected to severe stresses, it may rupture into highly irregular angular blocks. In this form of breakup, described as *shattering,* the fractures may cut across mineral grains and other structures in the rock. A good illustration of shattering is the effect of blasting bedrock with explosives, or of a hammer blow on a boulder of hard rock such as granite.

Mechanical Weathering

We will first look more closely at the mechanical weathering processes, which involve physical stresses that act upon rock and cause disintegration. The mechanical processes constitute the initial, or primary, breakdown of bedrock into fragments whose surfaces are in turn exposed to chemical weathering.

In middle- and high-latitude climates, and at high altitudes, alternate freezing and melting of water, called *frost action,* provides a powerful mechanism for rock breakup. Water that has penetrated joint planes and other natural openings in the rock is transformed into ice crystals with needlelike form. The pressures of growing masses of such crystals cause joint blocks to be heaved up and pried free of the parent mass. Freezing of water that has soaked into the pore spaces of a rock can cause shattering into angular fragments.

In the dry climates of low and middle latitudes, an important agent of rock disintegration is *salt-crystal growth,* a process quite similar physically to ice-crystal growth. The dry climates have long droughts in which evaporation can occur continuously, causing water deep in the rock to be drawn surfaceward by capillary force. Near the exposed rock surface, this moisture steadily evaporates, permitting dissolved salts to be deposited in openings in the rock. Although minute in size and appearing fragile, the growing salt crystals are capable of exerting powerful stresses. Even the hardest rocks (also concrete, mortar, and brick) can be reduced to a sand by continued action of the process of crystal growth.

Rock disintegration has been attributed to temperature changes alone, because most crystalline solids expand when heated and contract when cooled. From this fact, we might reason that the heating of rock will cause expansion of the minerals and may break the rock. It has been directly observed that sudden and intense heating by forest and brush fires causes severe flaking and scaling of exposed rocks. It is also known that primitive mining methods include the building of fires upon a quarry floor to cause slabs to break free. But it cannot be proved that the daily temperature cycle of solar heating and nightly cooling produces sufficiently great stresses to cause fresh, hard rock to break apart. Laboratory tests have shown that rocks can stand the equivalents of centuries of daily heating and cooling without showing signs of disintegration. But even if daily temperature changes are not capable of causing rock disintegration, the repeated expansion and contraction may assist in breaking up rocks already affected by other stresses and by chemical decay.

Another kind of mechanical weathering is the action of the roots of growing plants, exerting pressure upon the confining walls of regolith or rock. This process is especially important in the breakup of rock already weakened by other physical and chemical means.

Closely related to mechanical weathering is the rupturing of otherwise solid bedrock as a result of the spontaneous expansion rock undergoes when it is relieved of the confining pressure of overlying and surrounding rock.

Mechanical weathering produces many interesting, varied surface features both on the exposed rock of outcrops and on the surface of the regolith. These are landscape features, and will be described and explained in Chapter 14, the first of our chapters on landscape evolution. Here our main theme is the way in which solid rock becomes sediment.

Of course, rock disintegration is also caused by many kinds of mechanical geologic processes other than those we have just mentioned. As rock fragments are carried by streams, glaciers, and winds, they are subjected to impacts and surface abrasion that reduce their size and alter their shapes. The impact of breaking waves on rocky cliffs, and the tumbling action that the particles experience in the surf zone produce other forms of sediment. These diverse mechanical processes of sediment production and change are explained at greater length in Chapters 16 through 19.

Chemical Weathering

Chemical weathering consists of several important chemical reactions, all of which may occur more or less simultaneously. Consider first that all surface water—whether in the form of raindrops, soil water, ground water, or water in streams, lakes, and the oceans—contains in solution the gases of the atmosphere. Disregarding nitrogen, the major atmospheric component but a comparatively inactive element, the principal gases important in chemical weathering are oxygen and carbon dioxide. Oxygen in water is readily available for the process of *oxidation*, in which oxygen combines with such metallic ions as may be available. Carbon dioxide in solution forms a weak acid capable of reacting with certain susceptible minerals. Other acids of organic origin are also active in the water found in regolith and rock. In addition to the chemical processes caused by the gases it contains, water itself is capable of dissolving minerals directly, a process we can see daily in the solution of table salt (halite).

FIGURE 5.5 Two tombstones in the graveyard of Christ Church, Cambridge, Massachusetts, provide a lesson in resistance of rocks to chemical weathering. The marble monument, dated 1852, is strongly etched by acids in rainwater (*A*), whereas the older slate monument, dated 1743, remains sharp in every detail (*B*). (M. E. Strahler.)

FIGURE 5.5 A

FIGURE 5.5 B

All the chemical processes mentioned above require water, with which the earth's surface is abundantly endowed. Even the most nearly rainless deserts are easily supplied with water vapor diffused through the atmosphere, and while water is not obviously present much of the time in these areas, occasional rainfall and condensation of dew allow mineral decomposition to be active. If there is any surface environment in which rocks can escape the chemical processes of decay, it is in the perpetually frozen layer found below the surface in arctic and antarctic lands.

The rates at which chemical weathering processes act in various climates can often be estimated by a study of the surfaces of stone buildings and monuments. Inscriptions and carvings show varying degrees of erasure, depending upon the kind of rock, the amount of time involved, and the intensity of the weathering. Ancient Egyptian monuments, such as statues and obelisks of granite or syenite are often seen in remarkably unblemished condition where they have been subjected only to the warm, dry climate of the desert region bordering the Nile River. Here, chemical processes act extremely slowly. When one of these monuments is taken from Egypt and set up in the urban environment of a European or American city, such as London or New York, severe surface damage occurs in only decades.

A classic example of this is the obelisk of Thothmes III, which originally stood in an Egyptian temple in Heliopolis (an ancient city near present-day Cairo). In 1879, this obelisk was transported to New York City and set up in Central Park, where it has been known as Cleopatra's Needle. Weathering quickly began to subdue the sharp corners, edges, and inscriptions of the monument and within half a century it was a sorry reminder of the art treasure that had endured thirty centuries without visible deterioration. Sulfuric acid, produced in heavily polluted air in the urban environment, is a potent agent of chemical weathering not present in the natural environment. Its effects have been particularly severe on all forms of masonry in industrialized Europe and North America.

Tombstones are another form of monument that tell us a lot about the susceptibility of various rock types to chemical weathering. A good example comes from the grounds of Christ Church in Cambridge, Massachusetts (Figure 5.5). Whereas a marble headstone carved in 1852 is now severely corroded, its inscription nearly illegible, a headstone of slate dating back to 1743 is scarcely changed from its original condition. The silicate minerals making up the slate are little affected by carbonic and sulfuric acid, whereas the mineral calcite that makes up most of the marble headstone reacts easily with those acids.

Chemical Weathering and Particle Size

Chemical weathering is greatly facilitated by the mechanical disintegration of rock into very fine mineral particles. To appreciate this principle, consider the way in which the surface area of mineral particles is increased by their breakdown into smaller and smaller sizes. Table 5.1 illustrates this increase by assuming that we start with a single mineral cube with a height of 1 cm (pebble size); it has a volume of 1 cc and a surface area of 6 sq cm. Next, we slice the cube into cubes 0.1 cm in height (size of coarse sand), yielding 1000 cubes with a total surface area of 60 sq cm. Subdividing the original cube into 10^{12} cubes, each with a height of 0.0001 cm (fine clay size), gives a total surface area of 60,000 sq cm. We can easily see that the surface exposed to chemical reaction is now greatly increased. Finally, when the original cube is subdivided into 10^{15} cubes, each with a height of 0.000,01 cm (0.1 micron), the total surface area has increased to 600,000 sq cm; if spread out into a single continuous surface, this would yield an area 6 m by 10 m. Particles of this final dimension, smaller than about 1 micron (0.001 mm), are of a size class called *colloids*. Such particles are so small that they remain indefinitely suspended in a mixture with water, and they respond to the impacts of fast-moving water molecules. The observed motion of the colloidal particles under these impacts is called brownian movement.

Mineral particles of colloidal dimensions are chemically active in ways impossible for larger particles. As we shall see, individual colloidal particles are electrically charged and can hold onto ions in a surrounding water solution. In other words, a form of ionic bonding is possible between colloids and free ions of many kinds.

Table 5.1 Surface Area Resulting from the Subdivision of a One-Centimeter Cube

Cube dimension; height or width of side, cm	Particle description	Number of particles	Total surface area, sq cm
1	Pebble	1	6
0.1	Coarse sand	10^3	60
0.0001	Fine clay	10^{12}	60,000
0.000,01 (0.1 micron)	Colloidal clay	10^{15}	600,000

The Clay Minerals

The breakup of large mineral particles into smaller ones initially occurs through forces of mechanical weathering. As particles become smaller, however, chemical weathering acts to cause further breakup. Mineral colloids are produced by the chemical processes that decompose a mineral, transforming it into another mineral of different chemical composition. The result is to produce a family of minerals called the *clay minerals*.

Clay minerals have the following general properties or characteristics. They are originally derived by chemical weathering from the common silicate minerals making up igneous rocks. They take the form of extremely minute flakes or scales, often of colloidal dimensions (see Figures 5.6 and 5.7). The particles of clay minerals are in a crystalline state, and their ions are arranged in crystal lattices. For the clay minerals, the lattice structure takes the form of flat, parallel layers of extreme thinness; for this reason the clay minerals are layer silicates. (Micas are also layer silicates.) The chemical bonds that hold together the ions within each layer are strong, whereas the bonds between layers are weak. Because of this structure, water molecules and various free ions can easily penetrate between the layers of the clay mineral, leading to its further chemical alteration and to its physical disruption. Absorption of water molecules between layers can lead to a large volume increase, or swelling, in one major group of clay minerals; these are therefore classed as *expanding clays*. However, clay minerals of other groups are not subject to expansion by taking up water, and thus these are classed as *nonexpanding clays*. Individual particles of clay minerals, when immersed in water, are capable of holding positively charged ions (cations). These cations form a surrounding layer, held to the clay particle by electrical charges. Water molecules are also attracted and held in a similar manner.

Transformation of silicate minerals into clay minerals involves the process of *hydrolysis*, in which water unites chemically with the mineral. (We explain this reaction in detail later.) The important point for this introductory discussion is that hydrolysis, like oxidation, is a response to the surface environment and produces minerals that are highly stable in the presence of water, oxygen, and weak acids.

There are many clay minerals, and the chemical compositions of most are complex. They commonly occur in mixtures along with other minerals in clay-rich regolith and clay sediments. Because they are both common and complex, we should consider the properties of three important clay mineral groups.

The *kaolin group* is familiar to us as the china clay (also called kaolin directly) used in the manufacture of chinaware and pottery, and in large quantities as a filler in the manufacture of paper. The word "kaolin" comes from a Chinese word, *Kauling*, meaning "high hill," and is the name of a mountain from which ceramic clay was once mined for export. Clays of the kaolin group are of the nonexpanding type. A common mineral of this group is *kaolinite* with the composition $Al_2Si_2O_5(OH)_4$. One way in which kaolinite can form is by the hydrolysis of orthoclase ($KAlSi_3O_8$), which will also be explained in detail later.

The *illite group* consists largely of the clay mineral *illite*, a potassium aluminum silicate, also of the nonexpanding type. Its name honors the state of Illinois, where local state geologists first identified and described the mineral.

The *montmorillonite group* gets its name from Montmorillon, a town in France near which mineralogists collected and identified the mineral *montmorillonite*. (More recently, many mineralogists have preferred to use the name *smectite* in place of montmorillonite.) The most interesting feature of this clay mineral group is that the clays are of the expanding type. Thus, when a mass of the dry clay becomes soaked with water it undergoes a large volume expansion, accompanied by loss of strength. It makes a poor material for foundations of buildings and is responsible for the disastrous downhill movement of large masses of saturated regolith (Chapter 14). When it dries, the clay contracts greatly. Regolith rich in montmorillonite typically shows deep, wide soil cracks in dry seasons. Clays of the montmorillonite group are often found in residual regolith formed over mafic and ultramafic rocks—for example, basalt.

*Crystal Lattice Structure of the Clay Minerals

Two basic types of layers, or sheets, make up any given clay mineral. One type of layer consists of linked silicon–oxygen tetrahedra; it is the same type of sheet structure found in mica (Figure 3.22). The second type of layer is composed of octahedrons, each consisting of six oxygen ions (O^{2-}) surrounding a central metallic cation, commonly referred to as a *base cation*, such as Mg^{2+}, Al^{3+}, or Fe^{2+}. Figure 5.6*A* shows an expanded octahedron. The outer points of the octahedron may also be occupied by hydroxide ions (OH^-). The octahedrons are arranged in sheets in which the oxygen ions or hydroxide ions are shared to form a strongly bonded layer.

The clay minerals are characterized by alternation of tetrahedral sheets with octahedral sheets, bonded by sharing of the oxygen ions making up the free points of the silicon–oxygen tetrahedrons. Two or three such sheets, bonded together, form the unit layer of the clay mineral, and we shall be concerned with only two such combinations of sheets. One is a unit layer formed by one tetrahedral sheet bonded to one octahedral sheet.

Table 5.2 **Summary of Properties of the Principal Clay Mineral Groups**

Clay mineral group	Sheet structure		Contents of interlayer space	Behavior when wetted
Kaolin (kaolinite)	1:1	One octahedral One tetrahedral	Empty	Nonexpanding
Illite	2:1	Two tetrahedral One octahedral	K ions (No H_2O or OH)	Nonexpanding
Montmorillonite	2:1	Two tetrahedral One octahedral	H_2O molecules bonded with Ca, Na ions	Expanding

This arrangement is designated by the symbol 1:1. The second consists of one octahedral sheet between two tetrahedral sheets. This arrangement is designated by the symbol 2:1. The tetrahedral sheets are opposed in their symmetry, so that the oxygens of the tetrahedron points face inward in contact with the octahedral sheet. In Figure 5.6, parts *B* and *C* are simplified schematic drawings showing both 1:1 and 2:1 arrangements.

FIGURE 5.6 *A*. An expanded tetrahedron of oxygen ions. The metallic cation in the center is equidistant from all six of the oxygen ions. *B*., *C*. Schematic side views of lattice layer structure of the clay minerals kaolinite and illite.

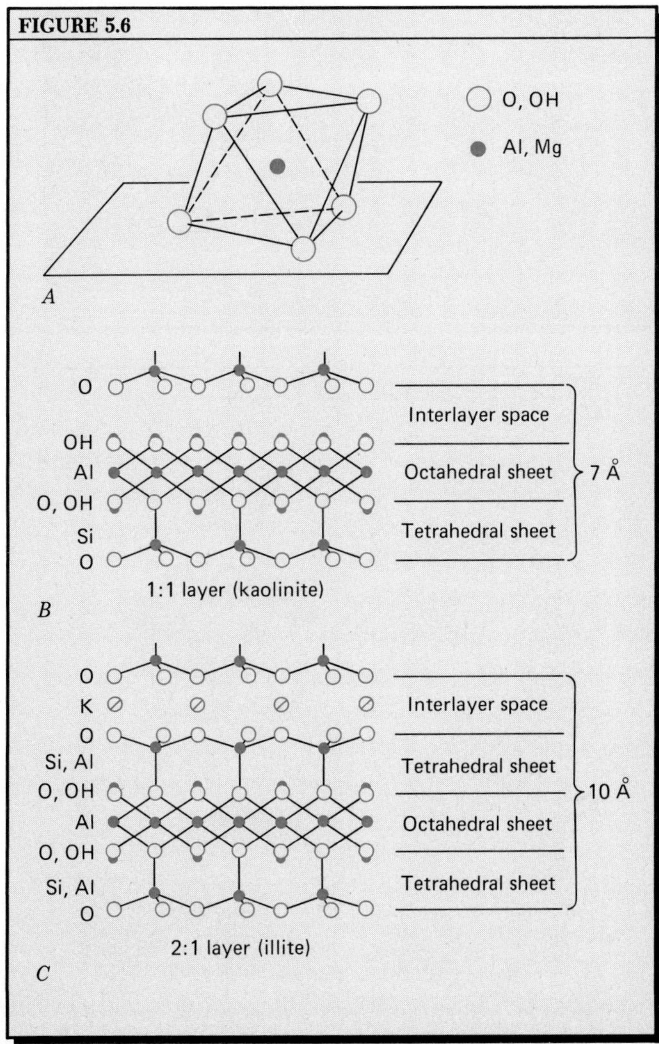

Stacked unit layers, one above the next, form the total mineral structure. Between the lattice layer groups is what we can call the *interlayer space*. This space may be unoccupied, or it may be occupied by layers of bonded base cations, by hydroxide ions, or by water molecules. However, the bonds are weak between components of the interlayer space and the lattice sheets above and below, so that unit layers are easily penetrated by migrating ions and water molecules, and the clay mineral is susceptible to physical disruption as well as to chemical change.

With this background information, we can examine the crystal lattice structure of the three principal groups of clay minerals. Table 5.2 lists some important distinguishing features of each group, which includes a number of individual mineral species.

Kaolin Group. The kaolin group has 1:1 lattice layer structure and is nonexpanding. The most important feature setting apart the kaolin group from the other two groups is the dominant presence of aluminum ions, which occupy central positions in the octahedral sheet. Indeed, in several important mineral species within the group, aluminum is the only base cation present. For example, the common mineral kaolinite, for which the group is named, has the chemical formula $Al_2Si_2O_5(OH)_4$. The interlayer space of kaolin group minerals is typically unoccupied. Figure 5.7A shows tabular crystals of kaolinite greatly magnified.

Illite Group. The illite group consists largely of the clay mineral of the same name; it has the 2:1 lattice layer structure and is nonexpanding. The key feature distinguishing the illite group from the other groups is the dominance of potassium as the base cation, and the presence of potassium ions in the interlayer space. The chemical formula for the mineral illite is $KAl_2(AlSi_3)O_{10}(OH)_2$. As with kaolinite, aluminum ions can occupy the central positions in the octahedral sheet, but there can also be present small proportions of magnesium and iron ions, partly replacing the aluminum ions.

Montmorillonite Group. Lattice layer structure of the montmorillonite group is of the 2:1 configuration, and it is of the expanding type. Although the lattice structure is very similar to that of the illite group, a distinguishing feature of the montmorillonite group is that the interlayer space is occupied by layers of water molecules, weakly bonded to calcium and sodium ions. Expansion of the clay mass results when more water penetrates the interlayer space and forms additional layers. When the clay dries, the loss of water molecules from the interlayer space causes shrinkage of the mass.

In terms of chemical composition, the montmorillonite group lacks the potassium ion (dominant in the illite group), and is rich in ions of magnesium and iron, which occupy positions in the center of the octahedral sheet, along with aluminum ions. Aluminum ions also can replace some of the silicon ions in the tetrahedral sheet. Figure 5.7*B* shows microscopic fragments of illite and montmorillonite. The thin, platelike form of the particles is typical, and it clearly reflects the layer lattice structure of these clays.

*Acid Reactions Affecting Minerals

Now that we know the physical–chemical nature of the lattice layer clay minerals, we can consider the nature of the weathering processes that produce them from the aluminosilicate minerals. Pure water alone has little effect upon the aluminosilicate minerals of igneous rock. An acid must be present, serving as an active agent to make a successful attack upon the bonded ions of the crystal lattice. To fully understand this chemical weathering process, we may need to review a few fundamental facts of chemistry.

The water molecule (H_2O), consisting of two hydrogen atoms bonded to one oxygen atom, is internally strong and is not easily broken apart under the typical environmental conditions prevailing in the soil and surface bodies of water. However, on infrequent occasions, two water molecules collide with sufficient force to cause the hydrogen atom of one of the water molecules to be broken off and adhere to the other water molecule. The two unlike bodies are now ions, one negatively, the other positively, charged. The following equation tells what has happened as a result of the collision:

$$\underset{\text{water molecule}}{H_2O} + \underset{\text{water molecule}}{H_2O} \rightleftarrows \underset{\text{hydroxide ion}}{HO^-} + \underset{\text{hydronium ion}}{H_3O^+}$$

The negatively charged ion (the anion), with only one atom of hydrogen linked to one oxygen atom, is known as a *hydroxide ion*. The positively charged ion (the cation), with three hydrogen atoms, is a *hydronium ion*. The hydronium ion is usually represented in shorthand form by the symbol H^+; we will use that short form in later paragraphs. It is also common practice to refer to the hydronium ion as a *hydrogen ion*,

FIGURE 5.7 *A*. Electron microscope photograph of kaolinite crystals, magnified about 20,000 times. (Paul F. Kerr.) *B*. Fragments of the clay minerals illite (*sharp outlines*) and montmorillonite (*fuzzy outlines*), which have settled from suspension in tidal waters of San Francisco Bay. Enlargement about 20,000 times. (Harry Gold. Courtesy of R. B. Krone, San Francisco District Corps of Engineers, U.S. Army.)

FIGURE 5.7 A

FIGURE 5.7 B

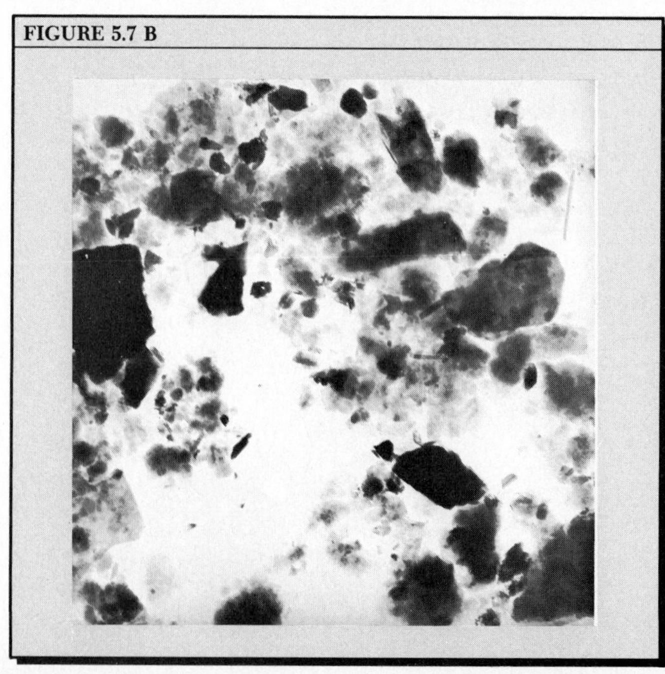

and we will do so even though it is impossible for the nucleus of a hydrogen atom to exist in the free state in a solution, separately from a water molecule.

The ionization of water by molecule collisions is very easily reversed, restoring the transferred hydrogen atom to its owner and yielding two whole water molecules. This form of ionization affects only a minute proportion of the molecules in pure water at any one time. There will at all times be exactly the same numbers of each kind of ion when they are created by molecular collision, as they are always formed in pairs.

An *acid* is a chemical compound of a special class capable of producing large numbers of hydrogen (hydronium) ions in pure water. As the concentration of hydronium ions, unmatched by hydroxide ions, increases, the *acidity* of a solution is said to increase. Another group of compounds, a *base*, when added to water releases large numbers of hydroxide ions. As the concentration of hydroxide ions, unmatched by hydronium ions, increases, the *alkalinity* of the solution is said to increase.

The acid most commonly involved in chemical reactions with minerals is *carbonic acid* (H_2CO_3). This acid forms readily in rainwater and is present in all surface water such as that in streams and lakes, as well as in water held in the regolith and bedrock. Carbon dioxide gas (CO_2), always present in the atmosphere, dissolves in water to form carbonic acid, thus:

$$\underset{\text{water}}{2H_2O} + \underset{\text{carbon dioxide}}{CO_2} \rightarrow \underbrace{\underset{\text{hydronium ion}}{H_3O^+} + \underset{\text{bicarbonate ion}}{HCO_3^-}}_{\text{carbonic acid}}$$

This acid is familiar as the active agent in carbonated beverages. Under high pressure, carbon dioxide gas is forced to dissolve in water, but when a bottle cap is removed, releasing the pressure, the reaction reverses and CO_2 gas forms bubbles that escape into the atmosphere. Under natural conditions, carbonic acid is present in a very weak concentration because the pressure exerted on atmospheric carbon dioxide is very low.

The compound sodium hydroxide (NaOH) is an example of a base. It consists of one sodium atom (Na) and one atom each of oxygen and hydrogen. When placed in water, sodium hydroxide simply "comes apart" or dissociates into one sodium cation (Na^+) and one hydroxide anion (OH^-).

The degree of acidity or alkalinity of a solution of ions in water (an aqueous solution) is stated in terms of a number called the *pH*, which indicates the degree of concentration of the hydronium ions in the solution. A pH number of 7.0 represents a neutral solution, with no excess of either hydronium or hydroxide ions. Numbers lower than 7.0 indicate acidity, which increases as the numbers become lower. This reverse relationship seems confusing to most persons; to explain it requires some knowledge of logarithms.* Numbers higher than pH 7.0 indicate alkalinity, which increases as the number becomes larger. The natural range of acidity and alkalinity found within soil water solutions is from about as low as a pH 4 for strongly acid soil conditions, to about as high as pH 11 in excessively alkaline soil conditions.

Consider next a simple reaction of natural carbonic acid in the soil solution upon a particular mineral. We chose the mineral calcite, consisting of calcium carbonate, with the formula $CaCO_3$. In the crystal lattice of this mineral, ions of calcium (Ca^{++}) lie in alternate layers with carbonate ions (CO_3^{--}). A small particle of calcite is now immersed in an aqueous solution of carbonic acid, with the following reaction:

$$\underset{\text{calcium ion}}{Ca^{2+}} + \underset{\text{carbonate ion}}{CO_3^{2-}} + \underbrace{H_3O^+ + HCO_3^-}_{\text{carbonic acid}} \rightarrow$$

$$\underset{\text{bicarbonate ion}}{2HCO_3^-} + \underset{\text{water}}{H_2O} + \underset{\text{calcium ion}}{Ca^{2+}}$$

We see that the extra hydrogen atom of the hydronium ion has moved to join the carbonate ion, forming a new bicarbonate ion. The positive hydrogen charge has canceled one negative charge of the carbonate ion. There are now two bicarbonate ions instead of one; an ordinary water molecule remains. The calcium ion has been freed from its crystal lattice and can wander about in the water solution. Since calcium is one of the important nutrients for plant growth, the free calcium ion can be drawn up into the root of a plant and then be used in the formation of some large, complex organic molecule in a living cell. On the other hand, the calcium ion may be taken out of the regolith and carried by a river to the ocean.

*Hydrolysis of the Aluminosilicate Minerals

Alteration of an aluminosilicate mineral to form a layer lattice clay mineral takes place through hydrolysis. The prefix "hydro" can be associated with the word "hydroxide," because hydrolysis involves the addition of hydroxide ions to the parent silicate compound.

Hydrolysis requires an acid to supply hydrogen in the form of hydronium ions to begin the reaction. (From this point on, we shall refer to the hydronium

*The pH number is the logarithm to the base 10 of the reciprocal of the weight in grams of hydronium ions per liter of water.

ion as a hydrogen ion and show it in equations as H^+.) Carbonic acid is normally present in water held in the regolith and in openings in the bedrock. Also present may be organic acids, generated in the decay of plant matter. Hydrolysis of orthoclase feldspar to produce kaolinite is a fairly simple case to use as an illustration. The reaction is as follows:

$$\underset{\text{orthoclase}}{2KAlSi_3O_8} + \underset{\text{hydrogen ions}}{2H^+} + \underset{\text{water}}{9H_2O} \rightarrow$$

$$\underset{\text{kaolinite}}{Al_2Si_2O_5(OH)_4} + \underset{\text{silicic acid}}{4H_4SiO_4} + \underset{\text{potassium ions}}{2K^+}$$

From this reaction we see that the change in formula from orthoclase to kaolinite has involved

A. loss of all potassium to the environment
B. gain of four hydrogens (as OH ions)
C. gain of one oxygen
D. loss of one silicon
E. retention of aluminum

Loss of silicon is typical of hydrolysis, as is the freeing of base cations. The lost silicon has gone into silicic acid, which can escape from the place of weathering, a process sometimes called *desilication* and particularly active in warm, moist climates. On the other hand, the silicic acid can easily break up, and the silica may be deposited as silicon dioxide (SiO_2), one of the forms of the mineral quartz. Retention of the original aluminum is typical of hydrolysis.

A second fairly simple example of hydrolysis is that in which orthoclase feldspar is transformed into illite:

$$\underset{\text{orthoclase}}{3KAlSi_3O_8} + \underset{\text{hydrogen ions}}{2H^+} + \underset{\text{water}}{12H_2O} \rightarrow$$

$$\underset{\text{illite}}{KAl_3Si_3O_{10}(OH)_2} + \underset{\text{silicic acid}}{6H_4SiO_4} + \underset{\text{potassium ions}}{2K^+}$$

In this case, one potassium ion has been retained in the clay mineral. Refer to the lattice diagram of illite (Figure 5.6) and notice that potassium ions are held in the interlayer space.

Lattice layer clay minerals are produced not only by hydrolysis of parent aluminosilicate minerals, but in some instances by the alteration of one clay mineral to form another. The alteration takes place within the regolith and may involve losses and/or gains of base cations, and the growth or removal of one of the lattice layers. Such changes are extremely complex, and little is known about them.

Mineral Oxides and Hydroxides in the Regolith

In addition to forming the lattice layer clay minerals we have studied, the chemical decomposition of silicate minerals of igneous rocks produces important groups of minerals classed as *oxides* and *hydroxides*. Most mineral oxides resulting from chemical decomposition are compounds formed by chemical union of one of the common metallic ions (Fe, Al, Mg, Ca) with oxygen. When water is combined chemically with one of these oxides, the compound becomes a mineral hydroxide.

The iron oxide compounds are produced by chemical weathering in regolith and in a shallow zone in the bedrock where atmospheric oxygen can penetrate open pore spaces. One of the commonest oxides is *hematite*, with the composition Fe_2O_3. In this ratio of two atoms of iron to three of oxygen, the compound is called *sesquioxide of iron*. (Another mineral form of iron oxide, common in igneous rocks, is magnetite, with the composition Fe_3O_4; it is described later.) Hematite is reddish brown to black and may occur either in crystalline form or in an earthy form known as red ocher. Sesquioxide of iron readily combines with water to form a hydrous iron sesquioxide called *goethite*, with the composition $Fe_2O_3 \cdot H_2O$. Addition of more water to this hydroxide of iron yields *limonite*, a soft mineral with the composition $2Fe_2O_3 \cdot 3H_2O$. Fine particles of limonite give regolith and many weathered rocks a yellow–brown color.

Aluminum ions are also released in large amounts in the chemical decomposition of silicate minerals. Oxidation produces sesquioxides of aluminum (Al_2O_3) in several forms. In combination with water, sesquioxide of aluminum is represented by the mineral *diaspore*, $Al_2O_3 \cdot H_2O$. Addition of more water results in the formation of *gibbsite*, $Al_2O_3 \cdot H_2O$. Both minerals are important constituents of residual regolith that forms over bedrock rich in aluminosilicate minerals.

Manganese ions released in the chemical alteration of mafic silicate minerals combine with oxygen to produce manganese sesquioxide, and with the addition of water becomes the hydrous mineral *manganite*, $Mn_2O_3 \cdot H_2O$. The hydrous sesquioxides of iron, aluminum, and manganese are commonly mixed together in regolith. They are typically associated with moist warm climates and are exceptionally stable compounds.

Silicon is released from the silicate minerals as silica, SiO_2. This is, of course, the same chemical composition as quartz formed from magma in the felsic igneous rocks. Grains of quartz in regolith and bedrock are subject to being dissolved and redeposited as silica in finely crystalline forms. Chalcedony, described later in this chapter, is one such form of secondary silica.

*Oxidation and Reduction as Chemical Processes

Oxidation, in the chemical sense, is defined as the loss of an electron (or electrons) from an atom or ion and the transfer of that electron to another atom or ion which serves as an *ion receptor. Reduction* is the corresponding gain of the transferred electron by the atom or ion of another species. Thus, both oxidation and reduction occur simultaneously in the same chemical reaction, for if one atom loses electrons (is oxidized) the other atom must gain them (be reduced). A simple example is the reaction of one atom of oxygen with two atoms of hydrogen, thus:

$$\underset{\text{oxygen}}{O} + \underset{\text{hydrogen}}{2H} \rightarrow \underset{\text{water}}{H_2O}$$

(Electron sharing in this reaction is illustrated in Figure 2.12.) Hydrogen and oxygen differ in their ability to hold electrons: Hydrogen holds its single electron quite loosely, whereas oxygen holds its eight electrons tightly. Oxygen has a very strong power to attract electrons; in fact, the attractive power of oxygen is higher than for any other element except fluorine. The word "oxidation" is well chosen because oxygen is the leading abundant element in causing the electron transfer process. Free oxygen in the form of molecular oxygen (O_2) makes up about 21% of the volume of the atmosphere. Molecular oxygen is dissolved in surface water and is readily available in regolith to engage in oxidation of ions released by chemical weathering of rock-forming minerals.

Acid reactions and hydrolysis of the mafic aluminosilicate minerals release ions of iron, manganese, calcium, sodium, and aluminum. Many of these ions are retained in the crystal lattices of the clay minerals, as we have seen. However, free ions of these bases are available and susceptible to oxidation.

Consider the case of iron. The iron cation as it exists in the crystal lattice of the silicate minerals has two positive charges; its symbol is written as Fe^{2+}. Iron in this form is known as *ferrous iron*. In the oxidation process, ferrous iron is released from its place in the parent mineral and is oxidized, transferring an electron to an oxygen atom with which it joins. The loss of one electron raises the positive charge of the iron ion to three, transforming it into *ferric iron* (Fe^{3+}). The ferric iron must become linked to oxygen atoms in the ratio of two atoms of iron to three atoms of oxygen. The formula for the compound that results is Fe_2O_3, which can be read as "ferric iron oxide." This sesquioxide represents the ultimate extent to which oxidation can proceed.

FIGURE 5.8 Felsic dike (angular blocks at center) in deeply altered mafic rock, Sangre de Cristo Mountains, New Mexico. (A. N. Strahler.)

FIGURE 5.8

Mineral Susceptibility to Chemical Decay

The Bowen reaction series (Figure 4.2) can be applied to the relative susceptibility of the silicate minerals to chemical decay by hydrolysis and accompanying oxidation. Geologists are well aware that in humid climates the mafic minerals decompose more rapidly than the felsic minerals. This fact is strikingly illustrated by a rock exposure in which a dike of felsic rock is surrounded by a pluton of mafic rock (Figure 5.8). The dike rock in this illustration is fresh in appearance and the exposed joint blocks have sharp corners and edges. By contrast, the surrounding mafic rock has been changed into a soft mass of partly altered crystals and clay minerals. The centers of a few joint blocks remain intact as spheroidal bodies.

Susceptibility to hydrolysis and oxidation follows the same sequence as the order of crystallization in the Bowen reaction series. Olivine and calcic feldspar are the most easily affected, followed by the pyroxenes, amphiboles, biotite, and sodic plagioclase feldspar. Potash feldspars are generally less susceptible than the more mafic types. Muscovite mica is comparatively resistant to alteration. Quartz is in a class by itself and almost immune to chemical change beyond very slow direct solution in water. An explanation of the relationship between mineral susceptibility to decay and the order of crystallization is found in the environment of mineral crystallization. Olivine and calcic plagioclase were crystallized at the highest temperatures and pressures and consequently the environment of their formation is farthest removed with respect to atmospheric conditions. Muscovite and quartz, crystallized at the lowest temperature and pressure, stand the least removed from atmospheric conditions.

The clay minerals, oxides, and hydroxides formed by decay of the silicate minerals make up much of the bulk of residual regolith overlying igneous rock bodies. From this environment of origin on the continents, the alteration products are transported by running water, wind, and glaciers to distant sites of sediment deposition. The alteration products of chemically weathered rocks, together with unaltered minerals (principally quartz, muscovite, and feldspar) and small unaltered rock fragments, become the sediment from which a major group of sedimentary rocks is formed.

Sediments and Sedimentary Rocks

We have used the term sediment a number of times under the simplified definition given in Chapter 1. To be more specific, *sediment* is fragmented mineral and organic matter derived directly or indirectly from preexisting rock and from life processes, transported by and deposited from air, water, or ice. It is obvious that many different mineral and organic substances qualify as sediment and that the rocks derived from accumulations of sediment are equally diverse in character. Figure 5.9 organizes both sediment types and sedimentary rocks according to processes of origin, physical properties, and chemical composition.

The first level of classification of sedimentary rocks is into clastic and nonclastic divisions. The adjective "clastic" comes from the Greek word *klastos*, meaning "broken," and describes *clastic sediment*, consisting of particles removed individually from a parent rock source. The *clastic sedimentary rocks* are therefore made up of mineral and rock fragments, and are further subdivided according to whether their sediment is pyroclastic or detrital. *Pyroclastic sediments*, covered collectively by the term tephra, were described in Chapter 4. They have been transported through air and in some cases also through water. The *detrital sediments* are mineral fragments derived by the weathering of preexisting rocks of any type. Detrital fragments may consist of individual or broken mineral crystals or groups of mineral particles as rock fragments. Detrital sediment also includes transported

FIGURE 5.9 A classification of sediments and sedimentary rocks.

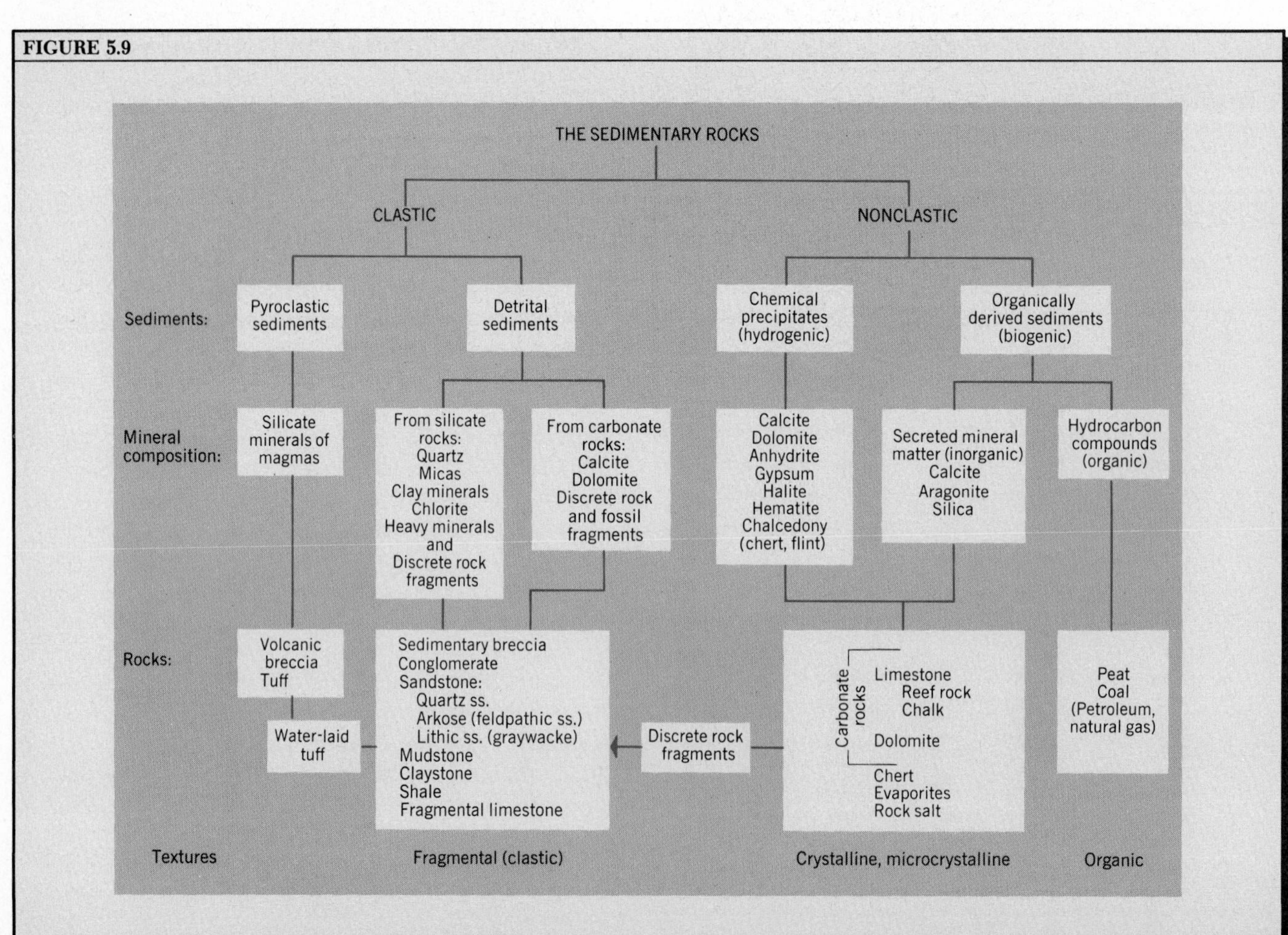

products of chemical weathering, such as the clay minerals, the oxides, and the hydroxides.

The *nonclastic sedimentary rocks* are made of sediments of two basic types: chemical precipitates and organically derived sediments. *Chemical precipitates* (also known as *hydrogenic sediments*) are inorganic compounds representing solid mineral matter precipitated from an aqueous solution in which the component ions have been transported. The *organically derived sediments* (also known as *biogenic sediments*) consist of the remains of plants or animals as well as mineral matter produced by the activities of plants and animals. For example, the shell matter secreted by animals is a crystalline inorganic substance and therefore a true mineral. Separated from the organic matter of these animals, shells constitute an inorganic sediment. On the other hand, accumulating plant remains, consisting of hydrocarbon compounds, form a truly organic sediment. As used here, the adjectives "organic" and "inorganic" agree in meaning with the chemist's classification of compounds as well as with the mineralogist's definition of a mineral. We shall need to be careful to distinguish between organically derived mineral matter and organic sediment (hydrocarbon compounds).

Once the sediments are classified and understood, the essential component minerals can be listed, as shown in the second row of boxes in Figure 5.9.

The third row of boxes in Figure 5.9 gives the names of the common sedimentary rocks. Texture, as well as mineral composition, is important in the naming of a sedimentary rock, and the names of textures are given on the bottom line in Figure 5.9. Textures of the clastic rocks are described as *fragmental textures* and are defined according to the sizes of the grains. Textures of the inorganic nonclastic rocks are either crystalline or microcrystalline. Organic textures, found in peat and coal, depend upon the way in which the original plant matter is deposited and compacted.

Grade Scale of Mineral Particles

Because the naming of clastic rocks depends in part on the sizes of component mineral grains, it is important to establish a system of *size grades*. For this purpose, geologists widely accept a system known as the *Wentworth scale* (Table 5.3). The units of length are millimeters or, for the finer grades, microns. In scanning down the list of numbers forming the limits of the successive classes, it is immediately evident that each number is half the value of that which precedes it and twice the value of that which follows it. The Wentworth scale is therefore a constant-ratio, or logarithmic, scale.

Table 5.3 The Wentworth Scale of Size Grades

	Grade name		mm		Microns
			4096		
Gravel	Boulders	Very large			
			2048		
		Large			
			1024		
		Medium			
			512		
		Small			
			256		
	Cobbles	Large			
			128		
		Small			
			64		
	Pebbles	Very coarse			
			32		
		Coarse			
			16		
		Medium			
			8		
		Fine			
			4		
		Very fine			
			2		
	Sand	Very coarse			
			1		1000
		Coarse			
			0.5	1/2	500
		Medium			
			0.25	1/4	250
		Fine			
			0.125	1/8	125
		Very fine			
			0.0625	1/16	62
	Silt	Coarse			
			0.0312	1/32	31
		Medium			
			0.016	1/64	16
		Fine			
			0.008	1/128	8
		Very fine			
			0.004	1/256	4
	Clay	Coarse			
			0.002		2
		Medium			
			0.001		1
		Fine			
			0.0005		0.5
		Very fine			
			0.00024		0.24
		(Colloids down to 0.001 microns)			

In the study of sediments and sedimentary rocks, the size grades of the component particles coarser than silt, but not so large as coarse pebbles, are determined by the use of sieves whose openings are spaced according to the Wentworth scale. Particles that pass through a given sieve opening but are caught upon the sieve of the next smaller mesh are referred to the named size grade given in Table 5.3. The sizes of coarse pebbles and larger grades of particles are generally determined by direct measurement. The dimensions of silt and clay particles are determined indirectly by the rate at which they settle in a still column of water.

The Wentworth scale is used in naming the detrital rocks. For example, sandstone consists of mineral grains of the size grade of *sand:* less than 2 mm but greater than 0.0625 mm (1/16 mm). Siltstone consists of particles in the grade range of *silt:* less than 0.0625 mm (62 microns) but greater than 0.004 mm (4 microns). Claystone consists largely of particles of *clay* grade: smaller than 0.004 mm (4 microns). (As used here, "clay" refers to particle size, not to mineral composition.) Conglomerate consists largely of particles of the size grade of *pebbles* (between 64 and 256 mm), or of larger size grades (*cobbles, boulders*). Note that the term *gravel*, shown in Table 5.3, is used to include pebbles, cobbles, and boulders.

Sorting of Size Grades

Texture of the detrital sediments relates not only to the dominant size grade, but also to the mixture of size grades that may be present in a given sample of sediment. *Sorting* of sediment refers to the range in size grades present (Figure 5.10). Perfect sorting—an ideal condition not found in nature—would refer to a sediment consisting entirely of particles of a single diameter. Excellent sorting does exist in some natural sediments, for example the sand shown in Figure 5.11. Nearly all of the quartz grains in this sample fall in the range from ¼ to 1 mm (medium sand and coarse sand). Poor sorting exists when the range in size grades is very large, and a substantial proportion of the sample belongs in both the very small and the very large grades. A poorly sorted sediment, for example, might include representative amounts of clay, silt, sand, and pebble grades.

Texture of detrital sediments also relates to the shapes of the grains, whether well rounded (as in Figure 5.11) or highly angular. Generally speaking, the degree of rounding of grains becomes better as sorting improves, because the process of transportation by water or wind tends to grind the particles as well as to sort them out by sizes.

Lithification of Sediment Layers

Throughout the geologic past, layers of sediment have accumulated to great total thicknesses in certain favorable areas on the continents and on the ocean floor. Such accumulations, which have amounted to many thousands of meters, have necessarily produced a series of physical changes in the sediment. The uppermost, freshly deposited layers consist of soft muds, clays, and silts, or of incoherent sands saturated with water. As new layers are added, the underlying layers experience progressively deeper burial. Under increasing pressures imposed by the overlying load, water is excluded from the sediment, which becomes denser and more strongly coherent. Ultimately, hard rock layers are produced, a process termed *lithification*. In some instances, chemical change has also occurred in the sediment following deposition. A more general term for the total of all processes of physical and chemical change affecting sediment during its conversion into solid rock is *diagenesis*.

Sedimentary rocks are usually recognizable through the presence of distinct layers resulting from changes in particle size and composition during the period of deposition. These layers are termed *strata*, or simply *beds*. The rock is described as being *stratified*, or *bedded* (Figure 5.12), and the planes of separation

FIGURE 5.10 Sorting of sediment relates to the range of size grades present.

FIGURE 5.11 Well-rounded quartz grains from the St. Peter sandstone of Ordovician age. The grains shown here average about 1 mm in diameter. (A. McIntyre, Columbia University.)

FIGURE 5.10

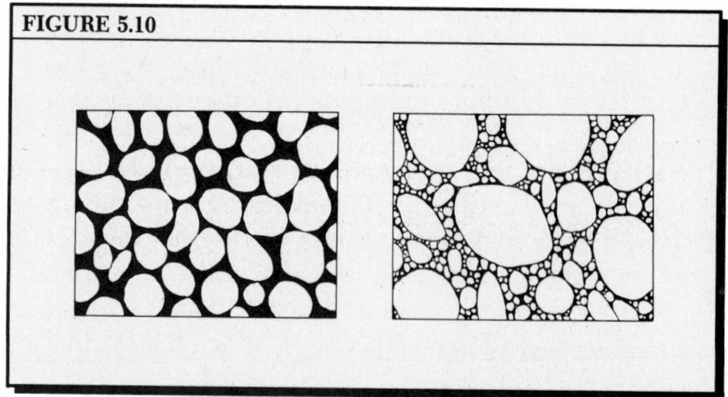

FIGURE 5.11

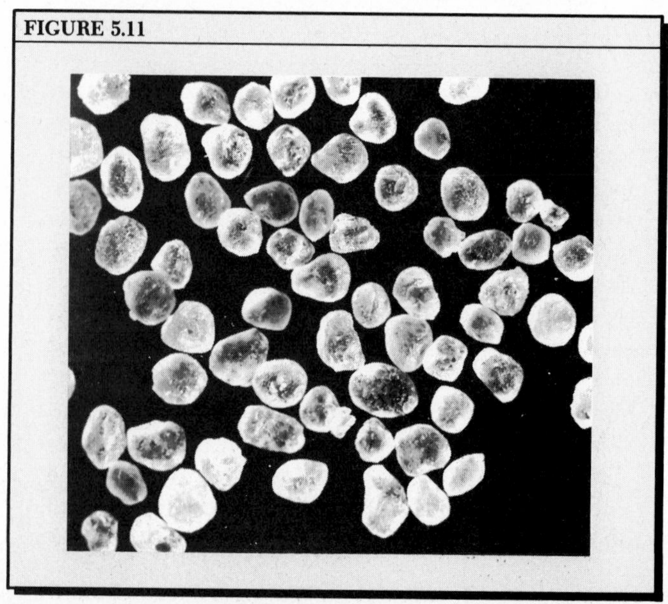

Table 5.4 Heavy Detrital Minerals Compared to Quartz

Mineral	Composition	Specific gravity	Hardness
Quartz	SiO_2	2.65	7
Magnetite	$Fe(FeO_2)_2$, also Fe_3O_4	4.9–5.2	5½–6½
Ilmenite	$FeTiO_3$	4.3–5.5	5–6
Zircon	$ZrSiO_4$	4.4–4.8	7½
Garnet group	Aluminosilicate of Ca,Mg, Mn,Fe	3.4–4.3	6½–7½

between layers are ***stratification planes***, or ***bedding planes***. Bedding planes in their original condition are commonly horizontal, but they may have become steeply tilted (see Figure 5.16) or distorted into wavelike folds by subsequent movements of the earth's crust.

The Clastic Sediments

The pyroclastic varieties of the clastic sediments have been discussed in connection with igneous rocks and the structure of volcanic cones. We shall therefore turn directly to the detrital sediments and rocks derived from them.

The most abundant particles of detrital sedimentary rocks are of quartz, rock fragments, or feldspar. Fragments of unaltered fine-grained parent rocks can easily be identified in coarse sandstone by microscopic examination. Rock fragments are typically second in abundance to quartz grains in sandstone but may be the chief component in the coarser grades of detritus. Mica and other minerals generally make up less than 3% of an average sandstone. Clay minerals, principally kaolinite, montmorillonite, and illite, dominate the colloidal fraction of the finer-grained detrital sediments.

FIGURE 5.12 The Badlands of South Dakota have been eroded in horizontal sedimentary strata that are mostly weak mud and clay layers. (Douglas Johnson.)

FIGURE 5.12

**The Heavy Detrital Minerals*

Quartz and the aluminosilicate minerals make up the bulk of all igneous and of many metamorphic rocks, but both types of rocks contain a number of other important minerals in minor proportions. Some of these latter constituents are highly resistant both to physical abrasion (wearing action) and to chemical decomposition. These strongly resistant, durable minerals remain intact during transportation. Because of their relatively greater specific gravity, as compared with that of quartz and other felsic minerals, these detrital minerals are referred to as the ***heavy minerals***. In Table 5.4, four heavy minerals are compared with quartz in terms of specific gravity and hardness.

Magnetite, an oxide of iron, is a dense black mineral with a submetallic luster. It crystallizes in the isometric system and typically displays the octahedral crystal form (Figure 5.13*A*). Magnetite is strongly attracted to a magnet, which will usually emerge with a coating of magnetite grains when dragged through dark-colored sands of beach or stream bed (Figure 5.13*B*).

Ilmenite, an oxide of iron and titanium, resembles magnetite in outward appearance but is only slightly magnetic. It is a common associate of magnetite in black beach sands. *Zircon*, a silicate of zirconium, is even harder than quartz and has a density comparable to that of magnetite. Unlike the dark metallic minerals magnetite and ilmenite, zircon can be transparent and pale in color. A highly durable mineral, zircon is important in providing a means of radiometric age determination, explained in Chapter 7. Minerals of the *garnet group*—aluminosilicates of calcium, magnesium, manganese, or iron—are abundant in the metamorphic rocks but also occur in igneous rocks. Although not as dense as the three detrital heavy minerals named above, garnet is a durable mineral and will be found in company with those minerals in the dark sands of beaches and stream beds.

Because of their greater relative density, the heavy detrital minerals are easily separated from the less dense quartz, feldspar, and mica by processes of water transportation or by winds. Consequently these minerals form local concentrations as dark layers in many types of sand accumulations.

The Detrital Sedimentary Rocks

Coarsest of the *detrital sedimentary rocks* are the *sedimentary breccias,* consisting of angular fragments in a matrix of finer particles (Figure 5.14). Many such breccias result from rock breakage during the collapse of strata into cavities opened out by the dissolving away of limestone that lies beneath. In some cases, breccias represent ancient submarine or terrestrial landslides, and as such, they are comparatively rare. *Volcanic breccias* are equivalent rocks in the pyroclastic group—that is, they consist of large angular blocks of volcanic rock in a matrix of fine tephra.

As mentioned earlier, *conglomerate* consists of pebbles, cobbles, or boulders, usually quite well rounded in shape, embedded in a fine-grained matrix of sand or silt (Figure 5.15). The main distinction between a conglomerate and a breccia is that the large fragments in the breccia are angular. Rounding of the conglomerate particles may be a result of abrasion during transportation in stream beds or along beaches. For example, one variety of conglomerate represents lithified stream gravel bars and gravel beaches. The rounding of cobbles and boulders found in coarse conglomerates may have occurred during weathering where bedrock was exposed at the surface. (This process is described in Chapter 14.)

The *sandstones* are composed of grains within the sand grade size and are designated as coarse-, medium-, or fine-grained in accordance with the grade sizes on the Wentworth scale. Sandstones can be differentiated into at least four types on the basis of composition. Perhaps the most abundant and familiar form is *quartz sandstone,* in which quartz is, of course, the predominant constituent. Pictured in Figure 5.11 are some beautifully rounded quartz grains taken from a sample of sandstone; they were shaped by the action of waves on a beach. In some sandstones, rounding of grains was perfected by wind transport in ancient dunes. Quartz sandstones can contain up to 5 or 10% of heavy detrital minerals, small flakes of muscovite mica, grains of feldspar, and rock fragments. Quartz sandstones are commonly lithified sediment deposits of the shallow oceans bordering a continent or of shallow inland seas. The quartz grains have traveled a long distance, while the finer and softer particles have been sorted out and removed during transportation. As we implied above, certain quartz sandstones were formed from large continental deposits of desert dunes. Also, some quartz sandstones are formed largely of the quartz derived from preexisting sandstones, from which only the quartz grains have been recycled.

Lithification of quartz sands into hard sandstones requires *cementation* by mineral deposition in the interstices between grains. Cementation is accomplished by the import of the cementing matter as ions in solution by slowly moving ground water. (Ground water and its paths of motion are explained in Chapter 15.)

FIGURE 5.13 *A.* Octahedral (eight-sided) crystals of magnetite from the Magnet Cove locality, Arkansas. (American Museum of Natural History.) *B.* This variety of magnetite, known as lodestone, attracts iron filings much as does a bar magnet. (Ward's Natural Science Establishment, Inc., Rochester, N.Y.)

FIGURE 5.13 A

FIGURE 5.13 B

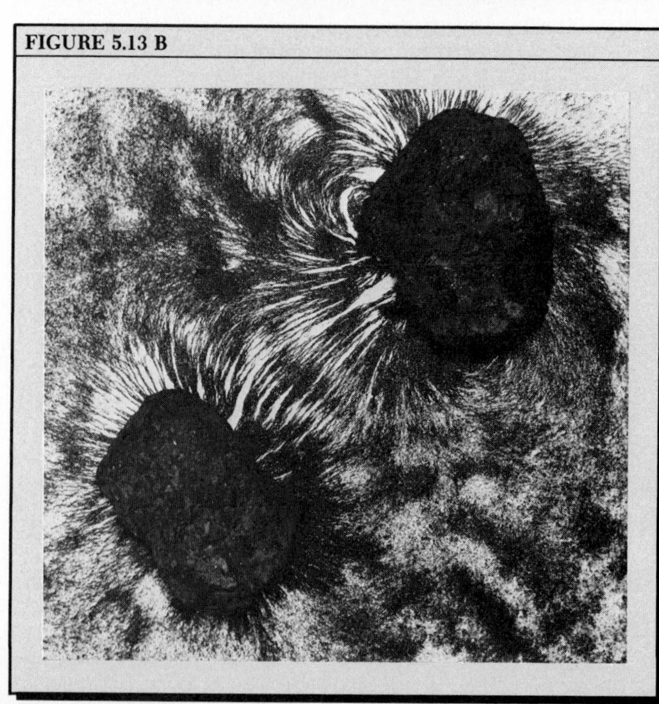

The cementing mineral may be silica (SiO_2), calcium carbonate ($CaCO_3$), iron oxide minerals, or clay minerals.

A second type of sandstone is called *arkose*, or *feldspathic sandstone*. It is characterized by having 30% or more feldspar grains, derived from partial chemical weathering of an igneous or metamorphic rock body. Arkose signifies the rapid erosion of an upland region of igneous or metamorphic rock and the relatively short transportation distance of the weathering products.

FIGURE 5.14 This coarse sedimentary breccia of Miocene age contains angular fragments of Precambrian granite. Riverside County, California. (Warren Hamilton, U.S. Geological Survey.)

FIGURE 5.15 Freshly broken conglomerate, consisting of well-rounded quartzite pebbles in a matrix of fine sand and silt. The specimen is about 10 cm wide; it comes from Hertfordshire, England, where the rock is known locally as "puddingstone." (Ward's Natural Science Establishment, Inc., Rochester, N.Y.)

FIGURE 5.14

FIGURE 5.15

A third type of sandstone may be termed *lithic sandstone*. It contains 15% or more of fine-grained rock particles. This kind of sandstone is generally dark, or speckled because of the presence of variously colored rock particles. Some lithic sandstones that are particularly dense and hard contain an admixture of clay minerals. The name *graywacke* has long been used by field geologists for this type of "dirty" sandstone, for it is gray to almost black. Rock particles of graywacke are often of metamorphic types, while the quartz grains present in it are often angular.

Lithic sandstones are not restricted to any particular environment of deposition but are derived from areas of complex bedrock geology that have exposures of varied igneous, metamorphic, or older sedimentary rocks. The composition and texture of graywacke suggests that the sediment of which it is formed was poorly sorted, partly unweathered, and transported in a highly turbulent flow of water.

A mixture of water with particles of silt and clay sizes is termed a *mud*. The sedimentary rock hardened from such a mixture is a *mudstone* (see Figure 5.12). The compaction and consolidation of clay layers, then, leads to the formation of *claystone*.

Many sedimentary rocks of mud composition are laminated in such a way that they break up easily into small flakes and plates. A rock that breaks apart in this way is described as *fissile* and is generally called a *shale* to distinguish it from nonfissile mudstone and claystone, which break up into blocks. Shale is fissile because clay particles lie in parallel orientation with the bedding and thus form natural surfaces of parting.

Most particles of clay size in claystones consist of clay minerals derived from the alteration of the silicate minerals, the most common of which are kaolinite and illite. Shales and mudstones usually contain substantial proportions of quartz and feldspar, mainly in silt-size grains, in addition to the clay minerals. Some sand-size quartz grains may also be present. Compaction of clay sediments into rock is largely a process of exclusion of water under the pressure of the overlying sediments. Because the clay minerals consist of minute flakes and scales, the proportion of water held in the initial sediment is very large. Once thoroughly compacted, claystone and clay shale do not soften appreciably when exposed to water, but under impact they break apart easily.

Shales make up the largest proportion of all sedimentary rocks in terms of volume. They can be subdivided by color, which may be red, gray, black, or green. Red shales, owing their color to finely disseminated hematite (oxide of iron), often are associated with red siltstones and red sandstones. Collectively known as *red beds*, these strata are interpreted as having been deposited in an environment of abundantly available oxygen, such as might be found on river floodplains in arid climates. Gray and black shales are interpreted as having been deposited in deep water in which oxygen was deficient. The dark color is due to disseminated carbon compounds of organic nature, possibly produced by anaerobic bacteria, which live in an environment deficient in oxygen. Dead organisms produced in the shallow surface water layer above rain down to the bottom, furnishing organic matter that is not readily decomposed and is added to the sediment.

A rather unique form of fine-textured clastic sediment is *bentonite,* formed by the alteration of beds of volcanic ash. The glassy shards of silicate mineral composition making up volcanic ash tend to be altered to montmorillonite. Bentonite layers are found interbedded with claystones and mudstones, but can be associated with almost any kind of sedimentary rock.

Volcanic ash is in some instances a true volcanic tuff settled out from the atmosphere, or it may be ash that has settled out in standing water or has been transported by streams and redeposited in standing water as *water-laid tuff.* This sedimentary rock has been given an intermediate position between pyroclastic and detrital rocks in Figure 5.9.

Minerals of the Nonclastic Sediments and Rocks

The nonclastic rocks are formed from sediment that may be either directly precipitated from solution (hence, inorganic in origin) or secreted by the activity of plants and animals (hence, organically derived). As stated earlier, inorganic precipitates have been designated as *hydrogenic sediments* to distinguish them from the second type, the organically derived sediments, which can be designated as *biogenic sediments* (Figure 5.9).

Perhaps the most important class of minerals of the nonclastic sediments are the *carbonate minerals (carbonates*), compounds of the calcium or magnesium ion, or both, with the carbonate ion. Calcium carbonate ($CaCO_3$) is the composition of one of the most abundant and widespread of minerals, *calcite* (Table 5.5). A soft mineral easily scratched with the point of a knife, calcite is most easily recognized in large crystals by the excellent cleavage in three directions, forming rhombohedrons (Figure 3.5). Many common forms of calcite, especially in the sedimentary rocks, show no identifiable crystalline structure to the unaided eye. A standard test for the presence of calcite is application of a drop of dilute hydrochloric acid to the mineral surface, as this results in strong effervescence (frothing).

Of the same composition as calcite but much less abundant is *aragonite.* Found in certain invertebrates, it makes up the pearly material of the shell. Aragonite is a harder mineral than calcite and lacks its good cleavage.

Dolomite, a close relative of calcite, is a carbonate of both calcium and magnesium. Denser and harder than calcite, dolomite effervesces with dilute hydrochloric acid only when in the powdered form. Like calcite, dolomite has excellent rhombohedral cleavage, but in the compact form found in rocks, particles with crystal faces are rare.

Sulfate compounds are also important as minerals of the hydrogenic sediments. *Anhydrite,* calcium sulfate, is a fairly soft mineral, commonly found in a granular state that gives a sugary appearance to broken surfaces. Derived directly from anhydrite by union with water (hydrolysis) or precipitated directly from solution is the mineral *gypsum,* a hydrous calcium sulfate. Gypsum is one of the softest of common minerals, defines hardness number 2 on the Mohs scale, and can be

Table 5.5 Important Hydrogenic and Biogenic Minerals

	Mineral name	Composition	Specific gravity	Hardness, Mohs scale	Crystal system
Carbonates	Calcite	Calcium carbonate $CaCO_3$	2.72	3	Hexagonal
	Aragonite	Calcium carbonate $CaCO_3$	2.9–3	3½–4	Orthorhombic
	Dolomite	Calcium-magnesium carbonate $CaMg(CO_3)_2$	2.9	2½–4	Hexagonal
Evaporites	Anhydrite	Calcium sulfate $CaSO_4$	2.7–3	3–3½	Orthorhombic
	Gypsum	Hydrous calcium sulfate $CaSO_4 \cdot 2H_2O$	2	2–2½	Monoclinic
	Halite	Sodium chloride NaCl	2.1–2.3	2–2½	Cubic
	Hematite	Sesquioxide of iron (ferric) Fe_2O_3	4.9–5.3	5½–6½	Hexagonal
	Chalcedony (chert, flint)	Silica SiO_2	2.6	7	Hexagonal

scratched with the fingernail. It has a low specific gravity, 2.2 to 2.4, and is typically fibrous in appearance, often with a silky luster.

Halite, or rock salt, was described in Chapter 3 (see Figure 3.8). We are all familiar with halite, which in the refined pure form is common table salt. Ease of solubility in water is its most obvious property. Halite belongs to a class of minerals known as *evaporites* (Table 5.5). Along with certain other highly soluble salts, halite is deposited from ocean water and the water of salt lakes when evaporation is sustained in an arid climate. Halite has been widely produced commercially by evaporation of sea water, but by far the greatest deposits are in rock strata.

We turn next to *hematite,* a sesquioxide of iron (Fe_2O_3), described earlier in this chapter. Hematite occurs in many forms and is particularly important as a sedimentary form of iron ore (Chapter 21). It is particularly widespread in certain sequences of sedimentary strata (Table 5.5). Notice that hematite represents the ferric form of iron oxide, as distinct from the ferrous oxide (FeO) found in magnetite. (The composition of magnetite, given in Table 5.4 as Fe_3O_4, can also be written as $FeO \cdot Fe_2O_3$, a combination of ferrous and ferric oxides.) In terms of sediments and sedimentary rocks, this distinction is important because hematite contains the greater proportion of oxygen, a consequence of exposure to free atmospheric oxygen. We have already noted that the mineral limonite is a hydrous form of sesquioxide of iron and can be derived from hematite by hydrolysis.

Finally, to this list of common sedimentary minerals we add the form of quartz known as *chalcedony.* It is described as *microcrystalline* (or *cryptocrystalline*) in structure, because although it is in the crystalline state individual crystals cannot be recognized, and the mineral appears as a hard, compact substance. Otherwise the mineral properties of quartz apply to chalcedony (Table 5.5). Most persons are familiar with a banded variety of chalcedony known as *agate,* which is ornamental when highly polished.

In sedimentary rocks like limestones, a hard, hornlike form of chalcedony, *chert,* is widespread and of great importance in forming nodules. It may also occur in layers of relatively pure *bedded chert.* (*Flint* is the popular name for a black variety of chert whose dark color is due to inclusions of organic matter.) Chert nodules form through replacement of carbonate sediment by silica in seawater trapped in the sediment. This replacement probably occurs before deep burial takes place. However, some chert beds were formed by the cementation of large quantities of siliceous skeletons of *diatoms* (microscopic one-celled plants) and *radiolaria* (one-celled animals), or of sponge spicules (small body-supporting parts of certain sponges). The organic origin of silica in sedimentary rocks is indicated in Figure 5.9 by its inclusion under organically derived mineral matter, as well as under the class of chemical precipitates.

*Salts in Seawater

To gain an appreciation of the way in which vast quantities of nonclastic sediments have accumulated on the ocean floors throughout the geologic past, we must investigate the salt content of seawater. The carbonate, sulfate, and evaporite minerals have been precipitated from dissolved substances in seawater. Organisms have obtained from seawater the calcareous and siliceous substances they have secreted as solid mineral matter.

At present the salinity of the oceans is sustained at a fairly constant value by a chemical system in which the amount of matter entering the ocean in each unit

Table 5.5 Important Hydrogenic and Biogenic Minerals *(Continued)*

Cleavage or fracture	Luster	Color
Perfect cleavage in 3 directions at 75° angle. Forms rhombs.	Vitreous to earthy	Colorless, white, or yellowish
Cleavage poor. Conchoidal fracture.	Vitreous on crystal surfaces, greasy on fracture surfaces	Colorless, white, or yellow
Perfect cleavage in 3 directions at 74° angle. Forms rhombs.	Vitreous to pearly	White; also yellow or brown
Cleavage poor in 3 directions. Conchoidal fracture.	Vitreous to pearly	Colorless, white, or grayish
Cleavage good in 3 directions. Forms thin sheets.	Pearly to silky	White, gray, or yellow
Perfect cleavage in 3 directions. Forms cubes, prisms.	Vitreous	Colorless to white
No cleavage.	Metallic or dull	Steel gray, reddish brown, or iron black. Streak cherry red or reddish brown
No cleavage.	Waxy to dull	Colorless, white, or any color

Table 5.6 Major Salts in Seawater

		Concentration, g/kg (parts per thousand)	Percent of total salt
Anions			
Chlorine	Cl^-	19.0	55.0
Sulfate radical	SO_4^{2-}	2.5	7.7
Carbonic acid radical	HCO_3^-	0.14 (varies)	0.4
Bromine	Br^-	0.065	0.2
Cations (bases)			
Sodium	Na^+	10.5	30.6
Magnesium	Mg^{2+}	1.3	3.7
Calcium	Ca^{2+}	0.40	1.2
Potassium	K^+	0.38	1.1
Neutral			
Boric acid	H_3BO_3	0.024	0.07

Source: Data from D. W. Hood (1966), in R. W. Fairbridge, ed., *Encyclopedia of Oceanography*, New York, Reinhold, Table 1, p. 793.

of time is balanced by the removal of an equal amount of matter as sedimentary deposits on the ocean floors. This sytem is thought to be in a steady state of action, holding constant the proportions of elements and the total salinity in seawater. Salinity of seawater averages about 35 parts per thousand (or 3.5%) by weight (see Chapter 9). Table 5.6 gives the average composition of salts in seawater in terms of the principal ions in solution.

The chlorine, sulfate, carbonic acid, bromine, and boric acid listed in Table 5.6 are among the volatile substances that are thought to have entered the atmosphere and hydrosphere by outgassing, a process explained in Chapter 4. The base cations—sodium, magnesium, calcium, and potassium—have been derived from the alteration of silicate minerals in igneous rocks exposed on the continents. The base cations thus released, together with silicon, aluminum, and many other elements, are carried to the oceans by streams which drain the weathered rock surfaces. Table 5.7 gives representative percentages of the four principal bases dissolved in stream waters.

Notice that the percentage of calcium in stream waters is high, and especially so from basalt rock areas, yet calcium constitutes only about 1% of the concentration in seawater. Sodium has a higher percentage concentration (31%) in seawater than in stream waters (13% to 27%). Thus, it appears that the concentrations of salt constituents in seawater depend not only on how rapidly the constituents enter the ocean, but also on how rapidly they are removed by precipitation.

The average time that an element remains dissolved in the ocean before removal is known as the *residence time*. Table 5.8 gives the residence times of the four principal bases and silicon, together with their concentrations in seawater. Notice that the concentrations of these constituents are in very nearly the same proportions as their residence times. Sodium, which has a very low rate of chemical reaction in the marine environment, remains much longer in the oceans than the others, and this is reflected in its predominant concentration. Calcium enters the oceans in the largest percentage of the four bases, but is relatively easily re-

Table 5.7

	Rock of the drainage basin: Igneous, all varieties	Rock of the drainage basin: Basalts only
Calcium	52%	73%
Magnesium	11	12
Sodium	27	13
Potassium	10	2
	100%	100%

Source: Data from W. W. Rubey (1951), *Geol. Soc. Am. Bull.*, vol. 62, pp. 1121, 1123.

Table 5.8

Element	Residence time, millions of years	Concentration in seawater
Sodium	260	31 %
Magnesium	45	3.7
Calcium	8	1.2
Potassium	11	1.1
Silicon	0.01 (10,000 years)	0.003%

Source: Data from E. D. Goldberg (1961), in Mary Sears, ed., *Oceanography*, Washington, D.C., Am. Assoc. Adv. Sci., p. 586.

moved as calcium carbonate to become a sedimentary deposit. Silicon, released in large amounts in the weathering of igneous rocks, has an extremely short residence time, which explains why silicon is present in very small amounts in seawater. The same can be said of aluminum (not shown in table), which is also released in the chemical weathering of aluminosilicate minerals in the igneous and metamorphic rocks.

We can guess from the dominance of calcium among the bases which enter seawater, together with its comparatively short residence time, that sediment of calcium carbonate composition will prove to be the dominant nonclastic contribution to sedimentary rocks of the geologic record.

The Carbonate Rocks

The *carbonate rocks* are formed largely from the carbonate minerals. *Limestone,* as broadly defined, is a carbonate rock in which calcite is the predominant mineral. Because either clay minerals or silica (as quartz grains, chalcedony, or chert) can be present in considerable proportions, limestones show a wide variation in chemical and physical properties. No single description is particularly helpful in identifying a specimen of limestone, although the test with dilute hydrochloric acid gives assurance of the presence of calcite. Limestones range in color from white through gray to black, in texture from obviously granular to very fine-grained, and in density from light and porous to very dense. Description of a few common kinds of limestone will serve to accentuate these variations.

FIGURE 5.16 Contorted beds of cherty limestone, near Atoka, Oklahoma. (Photo by Lofman; EXXON Corporation.)

FIGURE 5.16

The most abundant limestones are of marine origin and are formed by inorganic precipitation, or as byproducts of respiration and photosynthesis of organisms, or by the release of clay-size particles of aragonite from decaying green algae. Usually dense and fine-grained, with colors ranging from gray to almost black, these marine limestones show well-developed bedding and may contain abundant fossils. Dark color may be due to finely divided carbon. Many limestones have abundant nodules and inclusions of chert and are described as *cherty limestones* (Figure 5.16).

An interesting example of a form of hydrogenic limestone can be studied on the shallow banks surrounding the Bahama Islands, east of Florida. Here a white sediment of sand-size particles is accumulating in the tidal zone. Each sand grain proves, on microscopic examination, to be made up of concentric layers of aragonite precipitated from seawater. The layers surround a minute core particle that may be a shell fragment or a mineral grain. Because the particles resemble the roe (eggs) of fish, they are called *oolites.* When cemented into solid rock, layers of oolites form *oolitic limestone.*

Another interesting variety of limestone is *chalk,* a soft, pure white rock of low density (Figure 5.17). When examined under a microscope, chalk proves to be made up of extremely minute platelike particles, called *coccoliths,* derived from the disintegration of a form of algae. Thus, chalk is a limestone of biogenic origin.

Some important accumulations of biogenic limestone consist of the densely compacted skeletons of corals and the secretions of associated algae—they can be seen forming today as coral reefs along the coasts of warm oceans (Chapter 9). Rocks formed of these deposits are referred to as *reef limestones.* These limestones are in part fragmental, since the action of waves breaks the coral formations into small fragments that accumulate among the coral masses or nearby. Limestones composed of broken carbonate particles are also produced in other kinds of coastal environments. As a group, they are all mentioned in Figure 5.9 by the general name of *fragmental limestone.* Although the mineral matter is of biogenic origin, the rock as a whole can be classed as a detrital rock.

In the diagenesis that may accompany the burial of limestone strata under great loads of overlying sediment, the calcite and aragonite may recrystallize into close-fitting grains to produce limestones of granular texture; these are *crystalline limestones.*

Dolomite (sometimes called *dolostone* to avoid confusion with the mineral) is a rock composed largely of the mineral of the same name, but it poses a problem of origin, because the mineral is not secreted by organisms as shell material. Direct precipitation from solution in seawater is not considered adequate to explain the great thicknesses of dolomite rock that are found in the geologic record. The most widely held explanation for the formation of dolomite rock is that it has resulted from the alteration of limestone by a replacement process in which magnesium ions in seawater are substituted for a part of the calcium ions. It is not known whether the replacement, referred to as *dolomitization*, occurs immediately after deposit or over long periods by the slow movement of salt water through the rock. As seen in outcrops, dolomite rock is not easily distinguished from limestone, even though under the hydrochloric acid test it is only faintly effervescent. Dolomite is slightly harder and denser than limestone.

Strata Formed of Evaporites

Although many kinds of evaporites are produced from seawater, the great bulk of these deposits consists of the sulfates of calcium as gypsum and anhydrite, and of sodium chloride as halite.

The evaporites occur in association with one another in sedimentary strata, usually with marine sandstones and shales, but in some instances with chemically precipitated limestones and dolostones. Although details differ, most hypotheses of origin which try to explain the thick sequences of evaporites in the geologic record require a special set of environmental conditions. First, we assume an arid climate in which evaporation on the average exceeds precipitation. Such climates are widespread in tropical latitudes today and can be presumed to have been present in the past. Second, a shallow evaporating basin is required—perhaps a large, shallow bay or lagoon cut off from the open sea by a barrier bar (Figure 5.18). A narrow inlet, through which ocean water could enter to replace water lost by evaporation, is necessary to account for great thickness of evaporite beds. A slow subsidence of the area of deposition also accommodates the accumulating beds.

An entirely different environment of evaporite accumulation is found today in arid intermountain basins within the continents. The basin floors, known as playas, receive the dissolved products of streams flowing in from the surrounding mountains and dispose of the water by evaporation (see Chapter 17).

When seawater is evaporated under laboratory conditions, there is a definite order of precipitation of salts, in reverse order of solubility. First, calcium carbonate ($CaCO_3$) is precipitated, followed in order by calcium sulfate ($CaSO_4$), sodium chloride (NaCl), complex salts of magnesium, chlorine, and sulfate, and finally by the compounds of potassium. It is interesting to note that chemically precipitated limestone is the first member of the series, hence we might choose to classify this rock as an evaporite. Calcium sulfate is precipitated largely as anhydrite. Most gypsum found in sedimentary rocks is thought to be derived from anhydrite by later union with water. Halite follows anhydrite and is usually the last member of the series to be deposited in any large quantity. The more soluble salts are found in the geologic record very rarely, but they are common in playas.

Enormous accumulations of halite have formed at several points in geologic time. A good example is the accumulation of salt beds and associated layers of red shales and sandstones, gypsum, and anhydrite of the Permian basin in Kansas and Oklahoma. In the Permian Period, dozens of halite beds were deposited here, many of great thickness. A few individual salt beds 90 to 120 m in thickness are known (Figure 5.19). Thick anhydrite beds also occur in the series—one bed is over 400 m thick. Important salt beds of older geologic ages occur beneath Michigan, Ohio, and western New York. (This occurrence is described in Chapter 12.)

An entirely new concept of the origin of thick salt beds has been supplied by the theory of plate tectonics. Recall from Chapter 1 that in the early stages of continental rifting, a narrow ocean appears (see Figure 1.10*B*). It seems likely that this new ocean would, for long periods of time, be connected with a major open ocean by narrow straits. These connecting straits would restrict interchange of water with the open ocean and might at times be closed, and thus isolate the narrow ocean. If the narrow ocean happened to lie in a dry tropical zone with a desert climate, the trapped ocean water would evaporate rapidly, leading to the precipitation of thick salt beds. This is a scenario we will pick up in Chapter 21, for it is an important part of the history of major petroleum accumulations.

FIGURE 5.17 The chalk cliffs of Normandy, France, on the English Channel coast. Waves have carved a sea arch in the strong limestone. (Photographer not known.)

FIGURE 5.17

Peat, Coal, and Petroleum

In a swamp or bog environment, where water saturation persists, the remains of dead plants accumulate faster than they can be destroyed by bacteria. Bacterial activity breaks down the carbohydrate molecules into carbon dioxide and water, an oxidation process. Only partial decomposition occurs because oxygen is deficient in the stagnant water and the organic acids released by the decay process inhibit further bacterial activity. The product of this environment is *peat*, a soft fibrous material ranging in color from brown to black. Freshwater bogs containing peat deposits occur in vast numbers in those parts of North America and Europe which were subjected to glaciation (see Chapter 18). Peat from these bogs has been widely used as a low-grade fuel. A second environment of peat formation is salt marshes in tidal waters (see Chapter 19). Because of the slowly rising sea level following disappearance of glacial ice, tidal peat has accumulated in successive layers and in places reaches a thickness of 10 m or more.

In the geologic past, under the load of accumulating

FIGURE 5.18 A schematic diagram of an evaporating basin in which evaporites are accumulating. (Drawn by A. N. Strahler.)

FIGURE 5.19 The main corridor of the Carey Salt Company mine in Lyons, Kansas. Lyons residents, shown touring the mine in 1970, were concerned that it might be used for the disposal of radioactive wastes. (United Press International Photo.)

FIGURE 5.18

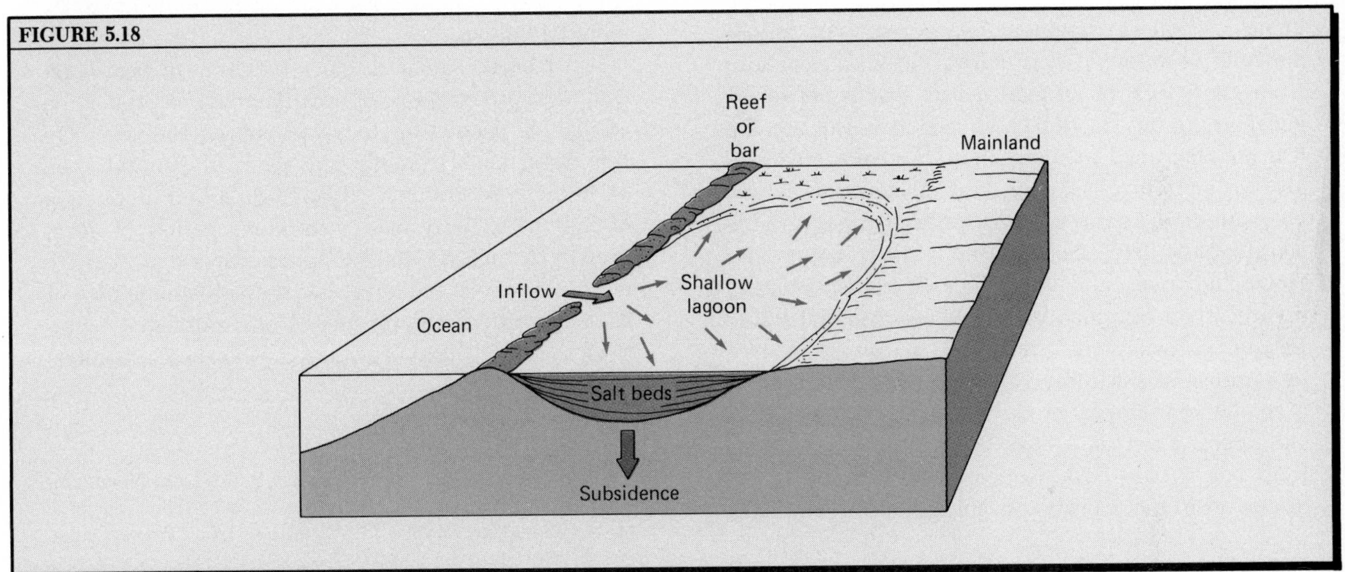

FIGURE 5.19

sediments, layers of peat have become compacted into *lignite* or "brown coal," a low-grade fuel intermediate in stage between peat and coal. Lignite, brown and with woody texture, tends to break up spontaneously into small fragments when exposed to the air (Figure 5.20).

Upon further compaction, lignite is transformed into *coal*. The first-formed and commonest type of coal is *bituminous coal*, or "soft coal." The bituminous form has in a few places been transformed into *anthracite*, or "hard coal," where the enclosing strata were subjected to intense pressures of folding during orogeny. Bituminous coal is a dense black substance which typically breaks into cubical or prismatic blocks. Anthracite is a hard, jet-black substance with a high luster and conchoidal fracture.

Lignite and coal are composed largely of the elements carbon, hydrogen, and oxygen, with minor amounts of nitrogen and sulfur. Coal also contains from 4% to 12% of mineral matter that is not of organic origin; this fraction is referred to as *ash*, because it is not consumed in combustion. For convenience in analysis and emphasis upon heat efficiency as a fuel, composition of coal is given in terms of three variable constituents—fixed carbon (the carbon atoms not bound into molecules with other elements), volatiles (hydrocarbon compounds), and water. Ash and sulfur remain fairly constant. Figure 5.21 shows the relative proportions of the three variable constituents for representative examples of lignite, bituminous coal, and anthracite. There is a complete series of gradations from one to the next. Notice that water is largely driven off in the transition from lignite to bituminous coal. Although the proportion of volatiles undergoes a relative rise from lignite to bituminous coal, the volatiles are almost entirely driven off when anthracite is formed.

Coal occurs in layers, known as *coal seams*, interbedded with sedimentary strata, which are usually thinly bedded shales, sandstones, and limestones. Collectively, such accumulations are known as *coal measures*. Individual coal seams range in thickness from a few centimeters to tens of meters (see Figure 21.13). Although coals are found in a wide range of ages in the geologic record, those of the Carboniferous Period between 280 and 345 million years ago are particularly outstanding. Vast expanses of the continents were then featureless plains close to sea level and were covered alternately by swamps bearing forest vegetation and by shallow seas in which sedimentary strata were laid down.

It has been estimated that some 30 m of peat is required to produce 1 m of coal. The rate of production of coal has been estimated at 10 cm per 100 years. On this basis the Mammoth coal seam of Pennsylvania, 15 m thick, would have required an initial production of 450 m of peat in a continuous period of plant growth lasting over 450,000 years. These figures are to a large degree speculative, but they give some idea of the extraordinary uniformity of environmental conditions that must have prevailed during the Carboniferous Period.

FIGURE 5.20 Specimens of coals. *A*. Lignite from North Dakota. (M. E. Strahler.) *B*. Bituminous coal from Virginia. (U.S. Geological Survey.) *C*. Anthracite from Pennsylvania. (M. E. Strahler.)

Although petroleum is neither a mineral nor a rock, it is closely associated with the clastic sedimentary rocks. Liquid *petroleum*, often called *crude oil*, is a mixture of several liquid hydrocarbon compounds of organic origin found locally concentrated in certain kinds of sedimentary strata. The origin and occurrence of petroleum are described in Chapter 21.

Sedimentary Rocks and the Rock Cycle

In this chapter we have traveled over another third of a full circuit in the cycle of rock transformation. With a knowledge of both igneous and sedimentary rocks and the common minerals of which they are composed we are ready to continue into the final third of the rock cycle—the making of metamorphic rocks. One of the most important concepts of geology has been made clear in this chapter: the interaction of two great energy sources—internal and external. The earth's internal energy source produces igneous rock under an environment of high temperatures and high pressures. The external energy source—solar energy—produces sediment under an environment of low temperature and low pressure. Investigation of the metamorphic rocks will take us again into the realm of high temperatures and high confining pressures, and to a study of the kneading action that deforms rock as if it were a mass of soft bread dough.

FIGURE 5.21 Composition of representative examples of

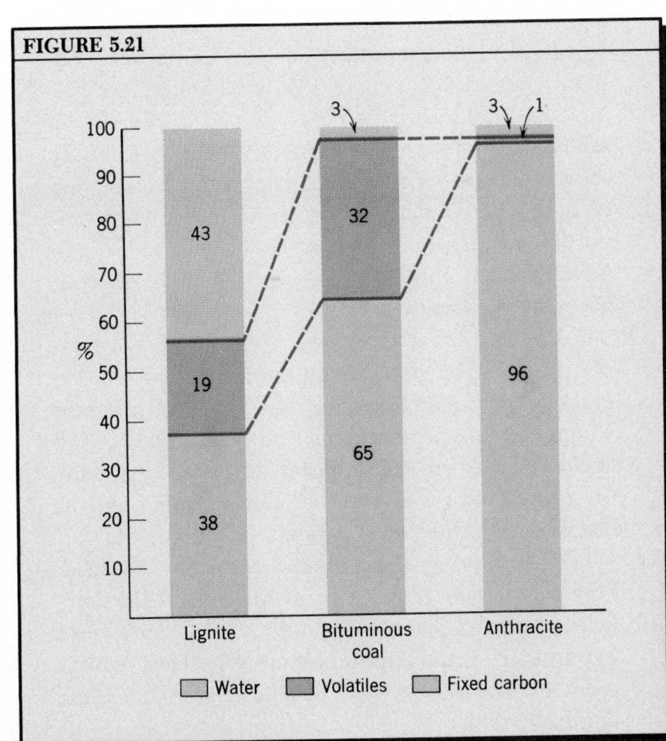

In this chapter, we have seen that the biosphere—the world of living organisms—plays an important role in geology by contributing enormous volumes of organically derived (biogenic) sediment.

Production of sediment and sedimentary rock has gone on throughout at least the past 3.8 billion years. It is thought that much of the enormous bulk of sedimentary rock produced during this time has been transformed into metamorphic rock, which in turn has been remelted to form new igneous rock. Much of that new igneous rock is felsic in composition and has steadily accumulated to increase the volume of the continental crust.

Key Facts and Concepts

Environments of the rock cycle Through the effects of plate motions, rock formed at depth under conditions of high pressure and temperature reaches the surface environment of low pressure and temperature in which *denudation*, powered by solar energy, produces sediment.

Rock weathering Rock weathering takes place by *disintegration* through *mechanical weathering* and by *decomposition* through *chemical weathering*. By weathering, *bedrock* is altered to *regolith*, which becomes sediment when transported. Mechanical weathering processes include *frost action*, *salt-crystal growth*, temperature change, plant-root wedging, and spontaneous rock expansion. Chemical weathering involves *oxidation*, carbonic acid action, and *hydrolysis*.

Clay minerals *Clay minerals*, hydrous silicates derived by chemical alteration of silicate minerals, occur as platelike colloidal particles and have layer lattice structure. Three major groups are *kaolin*, *illite*, and *montmorillonite* (smectite).

Mineral oxides and hydroxides Mineral oxides, formed during decomposition of silicate minerals, include *hematite* (*sesquioxide of iron*), *limonite* (hydrous form), *diaspore* (sesquioxide of aluminum), and *manganite*. These oxides accumulate in residual regolith in warm, moist climates.

Mineral susceptibility Susceptibility of silicate minerals to chemical decomposition follows the Bowen reaction series in reverse of the sequence of crystallization, the mafic minerals being most susceptible.

Sediment *Sediment* is fragmented mineral matter and organic matter derived directly or indirectly from preexisting rock and from life processes, transported by and deposited from air, water, or ice. *Clastic sediments* may be *pyroclastic* or *detrital; nonclastic sediments* may be *chemical* (*hydrogenic*) *precipitates* or *organically derived* (*biogenic*) *sediments*.

Grade scale of sediments Sediment particles fall into

size grades, described by the *Wentworth scale*. Grades include *boulders, cobbles, pebbles, sand, silt*, and *clay*. *Sorting* describes the range in size grades present. Sediment particles are also described in terms of texture, based on degree of angularity and rounding.

Lithification and diagenesis *Lithification*, the transformation of sediment into rock, may involve compaction, extrusion of water, cementation, and changes in chemical composition. The total process of change affecting sediment during lithification is *diagenesis*.

Strata Sedimentary rocks occur as *strata* (*beds*), separated by *stratification planes* (*bedding planes*) and forming *stratified* (*bedded*) rocks.

Clastic sediments The most abundant detrital sediments are particles consisting of quartz, rock fragments, feldspar, and clay minerals. *Heavy detrital minerals* include magnetite, ilmenite, zircon, and the garnet group.

Detrital sedimentary rocks *Detrital sedimentary rocks* include *sedimentary breccia, volcanic breccia, conglomerate, sandstone, mudstone, claystone*, and *shale*. Sandstone types include *quartz sandstone, arkose* (*feldspathic sandstone*), and *lithic sandstone* (*graywacke*).

Hydrogenic and biogenic minerals Most important of the nonclastic sediments are those formed of *carbonate minerals* (*carbonates*) represented by *calcite, aragonite*, and *dolomite*. Also important are sulfate minerals (*anhydrite, gypsum*), *evaporites* (*halite*), and oxides (*hematite, silica*). Silica commonly takes the form of *chalcedony* and *chert*.

Salts in seawater Salts precipitated from seawater consist of ionic compounds of sodium, magnesium, calcium, and potassium cations (bases) principally in combination with chloride and sulfate anions. Average time that an element remains in the ocean is its *residence time*, which is long for sodium and magnesium but short for calcium, potassium, and silicon.

Carbonate rocks *Carbonate rocks* include *limestone* and *dolomite* (*dolostone*). Limestone comes in many varieties, including cherty, oölitic, reef, fragmental, and crystalline forms. Dolomite may result from replacement of calcium ions by magnesium ions in seawater in the process of *dolomitization*.

Evaporite strata Evaporite strata, formed principally by evaporation of seawater in shallow subsiding basins, often occur as thick salt beds.

Hydrocarbon sediments Altered remains of plant matter form beds of *peat, lignite*, and *bituminous coal*. *Anthracite* is formed by alteration of bituminous coal under high pressure during tectonic activity. *Coal seams* accumulate as *coal measures*. *Liquid petroleum* (crude oil) is derived from organic sediments and accumulates in certain kinds of sedimentary rocks.

Questions and Problems

1. Do you think that the total volume of sedimentary rock present in the earth's crust has been increasing continuously through geologic time, or has it remained about constant? Take into account the ways in which sediments and sedimentary rocks can be transformed in the rock cycle.
2. What principal minerals might you expect to find in residual regolith that has formed over a long period of time in a warm, moist climate? Assume that the underlying rock is a mafic igneous type. Would you expect to find carbonate minerals in this regolith? Explain. Would silicon be present in the same ratio as in the underlying rock? Explain.
3. Using direct visual examination aided by a magnifying lens, how could you tell whether the rock fragments present at an outcrop were formed by frost action or by salt-crystal growth?
4. Rounding of mineral grains—large grains of quartz, for example—can take place in several ways: tumbling in the bed of a stream, by abrasion in the surf on a beach, or by abrasion in movement by wind over dry sand dunes. How might you distinguish among these causes of rounding, based on observation of the grains under a hand lens or a binocular microscope?
5. If you wished to select a tombstone that would hold a clear, sharp inscription for several centuries, what kind of rock would you choose? List several rock varieties in order of resistance to weathering. Which common rock would be a poor choice? If you had a choice of a burial ground anywhere in North America, what location would you choose as least subject to chemical decomposition of the tombstone? Explain your choice.
6. You are considering the purchase of a hillside lot on which to build a new house. A geological study shows that the bedrock consists of a thick formation of dark-colored shale in which there is a high proportion of montmorillonite. Describe and explain the risks that might be involved in placing your house at this location.
7. You are examining sandstone strata formed of beds of arkose. If you could study the individual mineral grains under magnification, what kinds of minerals would prove to be abundant, and what would be the typical shapes of the mineral grains? Make some inferences as to the source of this sediment and its distance from the present location of the strata.
8. When examining sedimentary rocks in the field, how might you be able to distinguish chalcedony from calcite? Calcite from dolomite? Calcite from gypsum? Gypsum from halite? What test equipment or materials would you need to make these determinations?

CHAPTER SIX

Strata and Their Deformation

The unique property that all sedimentary rocks have in common is that they are deposited in layers, one layer following the next to form a succession of strata. A particular layer, or stratum, supplies the geologist with two quite different categories of scientific information. First, the stratum records a singular event in geologic time. Of the total succession of all strata that were ever deposited, those strata which have survived to the present make up the record of geologic time. Second, because sediments tend to accumulate as horizontal layers, strata provide the geologist with geometric planes of reference. Deformation—the bending or breaking of rock—is usually very easy to detect in sedimentary rocks.

If I were to hand you a polished cube of granite, you would be at a complete loss to orient the cube correctly in the position it occupied in the quarry face from which it was taken. A natural outcrop of massive granite usually lacks any visible features by which a geologist can tell whether that entire mass might have been steeply tilted since it solidified from a magma. In contrast, almost any large exposure of sedimentary rock has conspicuous planes of stratification. Using the proper precautions to make sure that these planes are of a type that would have originally been horizontal, the geologist can measure any subsequent deformation the rock mass has experienced.

One type of deformation we commonly find is a contortion of the strata into folds. In the early days of geology as a science, a Scottish geologist, James Hall, observed highly contorted strata exposed on the Berwickshire coast of his native land. His field sketch of these folds, published in 1812, is reproduced here as Figure 6.1. Hall realized that these strata had once lain horizontally. He inferred that they were originally in a soft condition, like clay or loose beach sand, at the time the folding took place because the hard, brittle rocks they have since become would have fractured into many fragments if forced to yield by powerful compressive stresses.

Our chapter falls naturally into two parts. The first part is based on the concept of strata as making up the

FIGURE 6.1 Contorted strata on the coast of Berwickshire, Scotland, sketched by Sir James Hall, Bart., 1812.

FIGURE 6.1

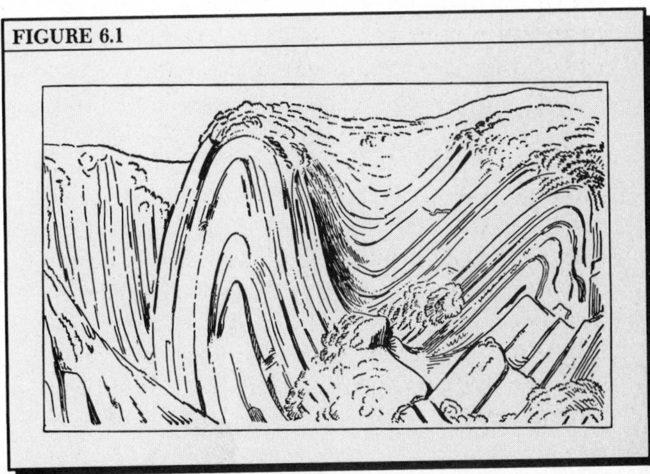

record of geologic time; the second part on the concept of strata as indicators of rock deformation. What you learn about deformation of strata will be put to use in the next chapter to help explain the metamorphic rocks, and in the following chapter to help explain how earthquakes originate.

Stratigraphy

A branch of geology important since the early 1800s is *stratigraphy*—the study of strata in terms of their relative ages and their distribution over the continents of the entire earth. Stratigraphy includes an interpre-

FIGURE 6.2

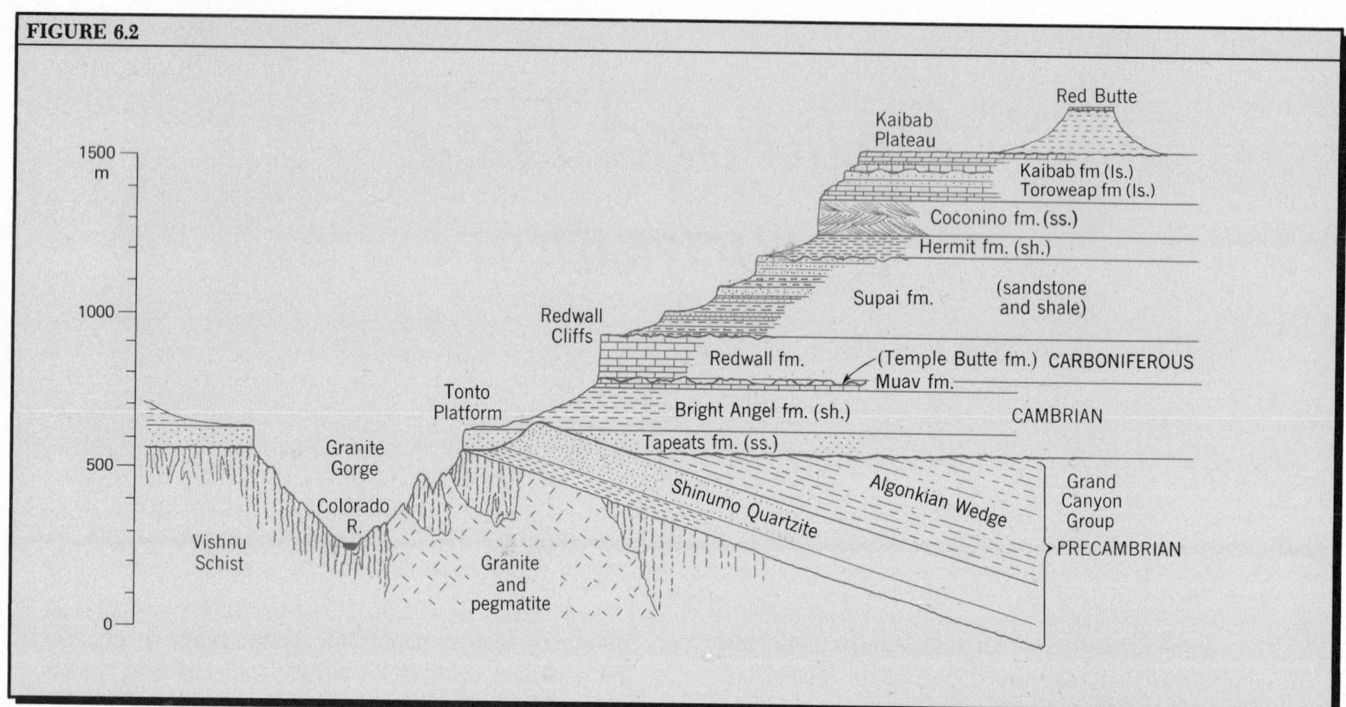

FIGURE 6.2A

tation of the environments under which strata were deposited and the geographical relationships of land and ocean areas that formerly prevailed. Stratigraphy is best considered as a major branch of historical geology, but the basic principles of stratigraphy we cover here are essential to an appreciation of physical geology.

Although there are many good exposures of horizontal strata throughout the plateaus and plains of the interior of North America, the most spectacular exposures are found in the Colorado Plateau region of Arizona, Utah, Colorado, and New Mexico. Here, in a dry climate, the edges of the strata are laid bare in high cliffs and in the walls of deep canyons. The Grand Canyon of the Colorado River in northern Arizona displays about 900 m of varied strata in a nearly horizontal attitude (Figure 6.2). We shall use this familiar natural wonder to illustrate several principles of stratigraphy.

FIGURE 6.2 Stratigraphy of the walls of Grand Canyon. *A.* Paleozoic strata in horizontal layers extend from the Tonto Platform to the canyon rim. (A. N. Strahler.) *B.* Details of formations in the upper cliffs. (Douglas Johnson.) *C.* The Inner Gorge and the Colorado River. (Douglas Johnson.)

FIGURE 6.2 B

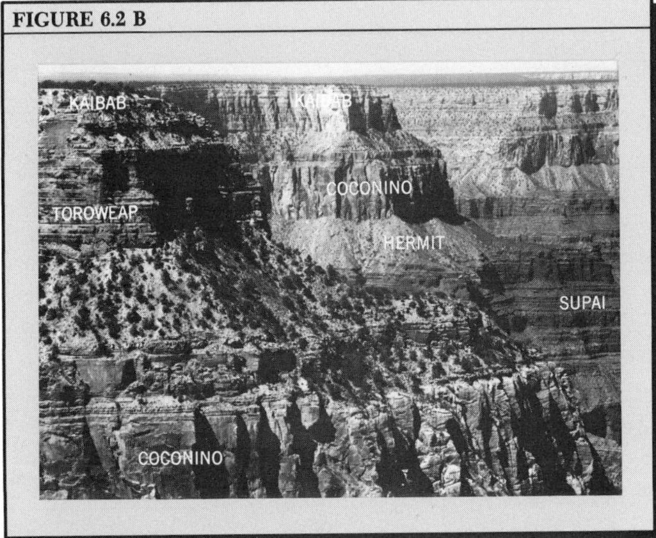

FIGURE 6.2 C

The Principle of Superposition

A geologic principle so simple as to seem self-evident is that within a series of sedimentary strata whose attitude is approximately horizontal, each bed is younger than the bed beneath, but older than the bed above it. This age relationship could not be otherwise in the case of sediment layers deposited from suspension in water or air. Thus, the first inference we can make concerning the strata exposed in the walls of the Grand Canyon is that they are arranged in order of decreasing age from bottom to top.

Despite the simplicity of the *principle of superposition*, as this age-layering principle is termed, there is a possible cause for uncertainty. It might just happen that the strata have been bodily overturned during tectonic activity, as often happens in severe deformation of strata, and that the uppermost beds are therefore actually the oldest. The geologist routinely checks against this possibility of error by examining closely certain details of the sedimentary rock. Features such as ripple marking, curvature of fine layers (crossbedding) in certain sandstones, and orientation of fossil shells give evidence of whether the strata are overturned or in their original attitude.

Looking at the upper walls of the Grand Canyon, your eye spans about 900 m of strata in almost perfectly horizontal, parallel arrangement (Figure 6.2*A*). The entire sequence of strata consists of several major layers, each with a distinctive appearance and composition. Each of these layers is referred to as a geologic *formation* and has been given a name. At the base, forming the edge of the Tonto Platform, is the Tapeats formation, a sandstone layer about 60 m thick. Above this is a soft, gray, sandy shale layer, about 150 m thick, named the Bright Angel formation; it forms smooth, gentle slopes. Above this, forming a great sheer wall 150 m high, are three formations of limestone: the Muav, Temple Butte, and Redwall formations. Still higher are layers of red sandstone and shale, totaling about 300 m in thickness, making up the Supai and Hermit formations. These are overlain by a pure creamy-white sandstone layer, the Coconino formation, whose sheer 90-m cliff is easily seen in the upper canyon walls. Forming the canyon rim are the Toroweap and Kaibab formations of limestone and dolomite, together about 150 m thick.

Stratigraphic Correlation

For nearly two centuries, stratigraphers have devoted their research efforts largely to relating the sequence of strata in one region with that in another.

Matching strata in terms of their relative geologic ages is called *stratigraphic correlation*. Physical attributes of strata such as lithology (rock type), texture, thickness, and color are often of little help in correlation because these properties can change greatly with horizontal distance within a unit of strata. Keep in mind, too, that modern methods of determining the absolute age of a rock sample were not available until well into the twentieth century.

A rather crude method you can use to match the age of strata in one locality with strata in another is to travel the ground between those localities, closely observing the exposed strata along the line of march. For example, you may be successful in following a particular quartz sandstone layer throughout the entire distance, seeing it continuously exposed in a cliff many kilometers long (Figure 6.3). Even so, it would be risky to conclude that its age is the same throughout the entire distance. An experienced geologist would be aware of the possibility that the sandstone accumulated as a rather narrow sand beach of an ocean shoreline and that the sand became a broad, continuous layer because that shoreline constantly shifted landward over a long period of time.

This principle might be illustrated by the building of a railroad line across a continent. The Union Pacific line was built westward from Omaha, Nebraska, starting in 1865, but progress was slow and it was not until May of 1869 that a connection was made at Ogden, Utah, with a line built eastward from Sacramento. After its completion, the Omaha–Ogden track was continuous for a distance of about 1500 km, but it was four years older at Omaha than at Ogden.

Correlation by Fossils

Of all sources of information as to the relative ages of strata, perhaps the most helpful to the stratigrapher are *fossils*, those ancient plant and animal remains or impressions preserved by burial in sedimentary strata. The scientific study of the fossil remains of animals and plants is *paleontology*. About the year 1800, an English civil engineer and geologist, William Smith, studied the kinds of fossils in strata exposed in canal excavations. He found that the fossil species were identical in all parts of a formation that he could prove to be continuous over the entire area. Fossil species in higher or lower strata were found to be distinctively different, but they occurred consistently in the same order in widely separate localities. Once the fossil order was established by direct observation, the fossils themselves became the evidence for relative ages of strata elsewhere in the world. For example, the fossils in certain strata in Wales were studied early in the nineteenth century, and these rocks became established as the original standard for the Cambrian Period of geologic time (*Cambria* is the Latin name for Wales). One distinctive fossil animal, the trilobite, was abundant in the Cambrian seas, and consequently some of its various species serve as guide fossils for the Cambrian Period throughout the world.

Relative age is merely a determination that strata in one location are older or younger than strata in another location or that they are of the same age. The actual age of those strata in years-before-present is their *absolute age*. As an analogy, we might observe from a distance a herd of African elephants. By carefully noting such features as height, length of tusks, condition of skin, and behavior toward one another, we could probably arrange the individuals in order of relative age. Without some specialized knowledge of the species, however, we might be completely in

FIGURE 6.3 In the Grand Canyon region a single formation of sandstone or limestone exposed in a cliff can often be followed for many tens of kilometers. (Drawn by A. N. Strahler.)

FIGURE 6.4 This fossil trilobite from the Bright Angel shale of Grand Canyon establishes the formation as being of Cambrian age. The head is to the left; about 1.5 times natural size. (Dept. of the Interior, Grand Canyon National Park.)

FIGURE 6.3

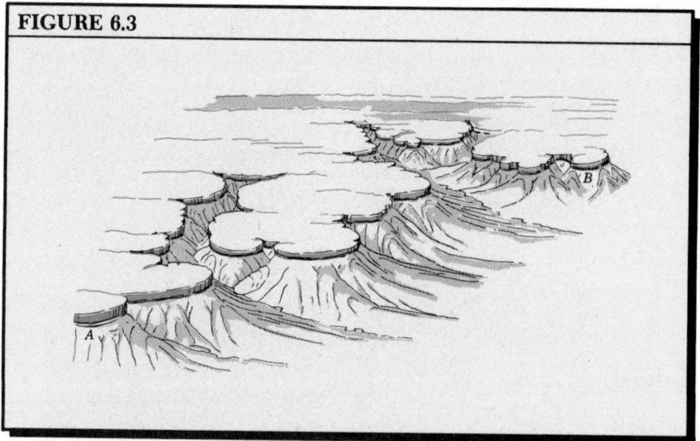

FIGURE 6.4

the dark as to the absolute ages of the individuals. Stratigraphers of the nineteenth century were, indeed, very much in the dark about the absolute ages of the strata they recorded, even though the relative ages had been worked out correctly in considerable detail from their fossil content.

The value of fossils in telling us the relative ages of rock strata arises from the fact that all forms of plant and animal life have continually and systematically undergone change, termed *organic evolution*, with passage of time. Suppose we have before us a complete, or nearly complete, description of past life forms as determined from fossils. If we know the relative geologic age to which each fossil form belongs, it is often a simple matter to give the relative age of any sedimentary layer merely by extracting a few fossils from the rock and comparing them with the reference forms. A fossil species particularly well suited to determination of age of strata is known as an *index fossil.*

Although this practice works well in many cases, there are some difficulties. For one thing, many strata contain no fossils, usually because the conditions under which the sediment was deposited were unsuitable for maintenance or preservation of plant or animal life. A second problem is that some organisms showed such slow evolutionary changes that almost identical fossil forms survived over a long span of geologic time.

The determination of geologic age is much more convincing when an entire natural assemblage of animal forms, or *fauna,* is studied, because a distinctive combination of animal types is far less likely to be duplicated than is a single type. Originally, a working sequence of ages of index fossils and faunas had to be set up from observations of the positions of those fossils in the stratigraphic column. The principle of *succession of faunas* is simply that each formation has a different fauna (or flora) from that in the formations above and below it.

In the Bright Angel formation of the Grand Canyon, many fossil trilobites have been found (Figure 6.4). Because these trilobites belong to particular varieties resembling those found in the original Cambrian strata of Wales, the geologist can state that the two sets of strata are of approximately the same age. Thus, a gap of thousands of kilometers is bridged through the use of particular index fossils, even though there is no direct connection between the two exposures of strata.

Worldwide studies by stratigraphers and paleontologists over the last 150 years have yielded an extremely detailed and nearly complete reference table of the divisions and subdivisions of geologic time, together with index fossils for all ages since complex life forms became abundant on earth. Radiometric age determination, explained in Chapter 7, finally made it possible to assign absolute ages in millions of years to the divisions of geologic time.

The Table of Geologic Time

In Chapter 1, we presented an overview of the five aeons of earth history in order to give you a feeling for the enormity of geologic time. Continuous geologic time for which we have fragmentary rock records goes back to about 3.8 billion years before the present (−3.8 b.y.). This span has been subdivided into blocks of time on the basis of long episodes of sedimentary deposition during which large areas of the continental crust were generally quiet and stable. These periods of deposition are interspersed with brief episodes of tectonic activity. Events of plate tectonics, including rapid subduction and continental collision, caused thick sequences of strata to undergo faulting, folding, intrusion, and alteration to metamorphic rock. For many decades, geologists have used the word *orogeny* to describe an episode of intense deformation of thick sequences of strata. "Orogeny" comes from the Greek words *oros*, relating to mountains, and *genos,* referring to birth or origin.

In terms of life forms, the largest time bracket is the *eon*. *Precambrian time*, which is all time before the start of the Cambrian Period, constitutes the *Cryptozoic Eon*. Cryptozoic, derived from the Greek words *kryptos*, "hidden," and *zoo*, "life," signifies the obscurity and simplicity of life of Precambrian time. The time span of abundant life which followed constitutes the *Phanerozoic Eon*, derived from the Greek word *phaneros*, "visible."

Phanerozoic time is made up of three *geologic eras:* the *Paleozoic Era*, the *Mesozoic Era*, and the *Cenozoic Era*, listed in order from earliest to latest. Translating from the Greek roots, they can be paraphrased as the eras of ancient (*paleos*), middle (*mesos*), and recent (*kainos*) life, respectively.

The inference of these names is clear. While geologic processes operated in repetitive cycles, each one about the same as the next, life was changing through time in a one-way, irreversible stream. Thus, each era is distinct from the next in terms of organic composition. The geologist cannot assign a given stratum of limestone or shale to a given era on the basis of physical rock properties alone, for similar rocks were formed in all eras. Instead, it is the distinctiveness of the remains of life forms enclosed in that stratum that permits it to be assigned its place in geologic time.

The *table of geologic time* gives ages and durations of the three eras, together with their subdivisions into *periods* (Table 6.1). Notice that the Paleozoic Era, with six periods, had a duration of 320 million years (m.y.); the Mesozoic Era, with three periods, lasted only 185 m.y.

The Cenozoic Era endured only 65 m.y., or about the average duration of a single long period in the preceding two eras. Table 6.1 lists two periods within

Table 6.1 The Table of Geologic Time

Eon	Era	Period		Duration, m.y.	Age, m.y.	Orogenies
Phanerozoic Eon	Cenozoic	Quaternary		2	2	Cascadian
		Tertiary		63	65	
	Mesozoic	Cretaceous		70	135	Laramian
		Jurassic		70	205	Nevadian
		Triassic		45	250	
	Paleozoic	Permian		30	280	Appalachian (Hercynian)
		Carboniferous	Pennsylvanian	45	325	
			Mississippian	20	345	Acadian
		Devonian		50	395	(Caledonian)
		Silurian		35	430	Taconian
		Ordovician		70	500	
		Cambrian		70	570	
				Duration, b.y.	**Age, b.y.**	
Cryptozoic Eon	Precambrian Time	Upper Precambrian (Algonkian)		0.3–0.4	0.9–1.0	Grenville
		Middle Precambrian		0.6–0.8	1.6–1.7	Hudsonian
				0.7–0.9	2.4–2.5	Kenoran
		Lower Precambrian (Archean)		0.9–1.0		
		Oldest dated rocks			3.6–3.8	
		Earth accretion completed			4.6–4.7	
		Age of universe			17–18?	

Sources: D. Eicher (1976), *Geologic Time*, Second Edition, Englewood Cliffs, N.J., Prentice-Hall, end paper; M. Kay and E. H. Colbert (1965), *Stratigraphy and Earth History*, New York, Wiley, p. 74.

the Cenozoic Era: Tertiary Period and Quaternary Period. Although these terms are still widely used in geological writing, current publications on plate tectonics use, instead, a system of seven *epochs* within the Cenozoic Era. These epochs are listed in Table 6.2, along with their durations and ages. All seven names end in the syllable "cene," taken from the first syllable of "Cenozoic." The last of the Cenozoic epochs is the Holocene, sometimes referred to as "recent time." The Holocene Epoch dates from the disappearance of glacial ice of the Pleistocene Epoch, an event occurring only about 11,000 years ago.

In later chapters, particularly those on plate tectonics, we shall refer repeatedly to the periods within the Paleozoic and Mesozoic eras and to the epochs within the Cenozoic Era. We suggest that you mark with an index tab the pages on which Tables 6.1 and 6.2 are found, so that you can quickly refer to them for the names, ages, durations, and sequence of the periods and epochs.

Precambrian time, enduring five to six times as long as the Paleozoic, Mesozoic, and Cenozoic eras combined, has proved extremely difficult to subdivide on any basis that has global continuity. Table 6.1 subdivides Precambrian time into three parts: upper, middle, and lower. In addition, we include two particular time units—Algonkian and Archean—because they figure in the geologic history of the Grand Canyon discussed in this chapter. Almost all the Precambrian rocks are igneous or metamorphic. Great thicknesses of sedimentary and volcanic rock were intensely deformed, then intruded by granites, and these complexes in turn were changed into metamorphic rock. Erosion has removed enormous volumes of Precambrian rock, but it still makes up most of the upper part of the continental crust.

Disconformities in the Walls of Grand Canyon

The horizontal strata of the Grand Canyon, extending upward from the lowest formation—the Tapeats sandstone—to the canyon rim, are of Paleozoic age. A question not answered by the principle of super-

Table 6.2 Epochs of the Cenozoic Era

Epoch	Duration (m.y.)	Age (m.y.)
		0
Holocene (recent)		
		(11,000 yr)
Pleistocene	2	
		2
Pliocene	11	
		13
Miocene	12	
		25
Oligocene	11	
		36
Eocene	22	
		58
Paleocene	7	
		65
(Cretaceous Period)		

position is this: Were all the sandstone, shale, and limestone strata of the Grand Canyon walls deposited in quick succession, so that in terms of available geologic time within the Paleozoic Era (320 m.y.) we may consider them all as being of approximately the same age? Or do they represent different periods within the Paleozoic Era? In that case the lowest formation, the Tapeats sandstone, would be much older than the rim formation, the Kaibab limestone, and there might conceivably be a difference in age of as much as 300 m.y. in the two formations. Assuming such a wide age difference exists, we are faced with the possibility that the entire sequence of rocks, 900 m thick, represents slow, continuous deposition of sediment without interruption of any consequence throughout the entire span of Paleozoic time.

A quite different possibility is that each formation was deposited in a short time span, but that the records of episodes of deposition are themselves separated by long intervals of time when no deposition took place. It is a reasonable guess that for long spans of time, each tens of millions of years long, there was no deposition of sediment.

We must therefore complicate still further the interpretation of Grand Canyon strata, as illustrated in Figure 6.5. Perhaps the bottom three formations, Tapeats sandstone, Bright Angel shale, and Muav limestone, were deposited in rather rapid succession in a shallow sea, as shown in Part *A*. Additional formations of which we have no record were possibly added to these three (Part *B*). All these formations are said to be *conformable strata* because they form an unbroken succession of strata, deposited in a more or less continuous sequence under the same general conditions. Only minor interruptions in deposition have affected a conformable sequence. For example, deposition may have been interrupted annually during a season in which no sediment was brought in from a distant source.

Following the deposition of the conformable strata, a uniform upward movement of the crust brought these formations above sea level, allowing erosion to remove great quantities of rock (Parts *C* and *D*). With only the Tapeats, Bright Angel, and Muav formations remaining, a downsinking of the crust occurred, depressing them below sea level and producing a shallow sea in which a new period of deposition began (Part *E*). This submergence allowed deposition of the Redwall limestone formation to take place directly upon the older Muav formation.

Thus, the thin line that we now see between the Muav and Redwall formations is the sole indicator of a vast lost period. That is to say, it records a time for which no rock has been retained. A surface of separation between two formations, representing a long gap of time, is termed a *disconformity*. The history of

FIGURE 6.5 Sequence of events leading to the development of a disconformity in the walls of the Grand Canyon. (Drawn by A. N. Strahler.)

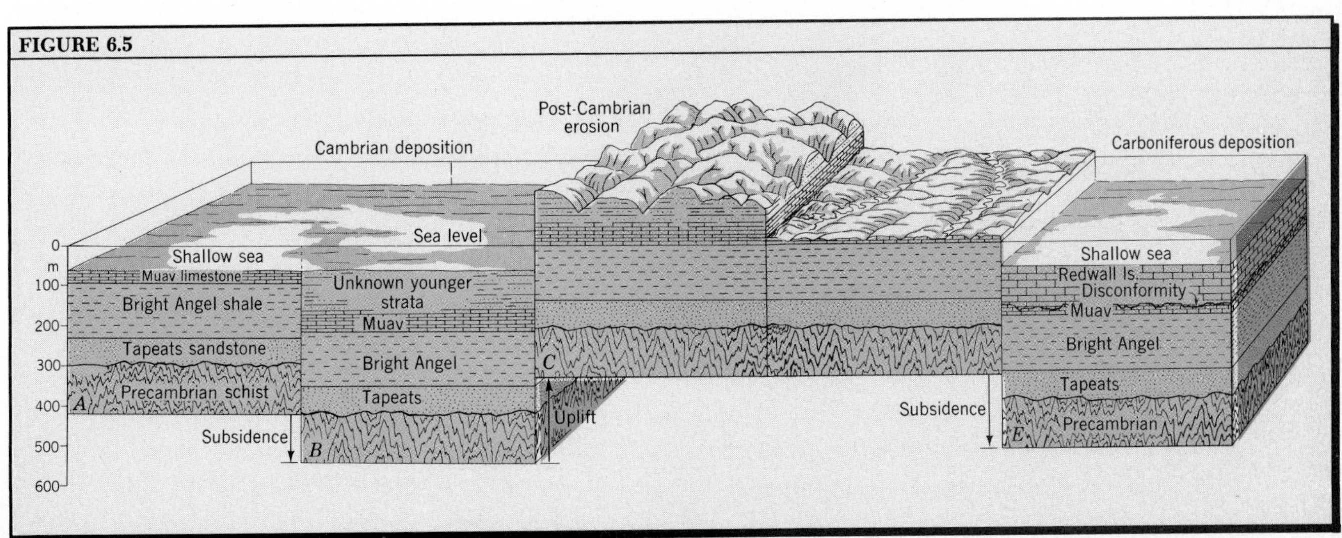

events we have described here for the Grand Canyon formations of lower Paleozoic ages is the actual history as worked out by stratigraphers.

Rising and sinking motions of the earth's crust over large areas and without appreciable deformation of the surface rocks are called *epeirogenic movements*. The process is referred to as *epeirogeny* to distinguish it from the orogeny that results in deformation of strata. Epeirogenic movements have been of great importance in the history of the stable interior portions of the continental lithosphere. Here, a negative epeirogenic movement (downsinking) of only a few hundred meters allowed shallow marine waters to spread over a large proportion of the continent, whereas a positive epeirogenic movement of the same magnitude caused the shallow sea to retreat to the continental margins. This process was repeated innumerable times, producing many disconformities in the geologic record.

Unconformities and Orogenies

The Inner Gorge of Grand Canyon provides the geologist with a now-classic example of Precambrian rocks far from those seen widely exposed in Canada. Because of their unusual structure and arrangement, these rocks allow us to develop further concepts in stratigraphic interpretation.

Precambrian rocks in Grand Canyon lie beneath the Cambrian Tapeats sandstone, the rim rock of the Tonto Platform which forms the brink of the Inner Gorge (Figure 6.2*C*). Looking down into the narrow Inner Gorge, you notice that the walls are here completely lacking in horizontal bedding planes. Instead, they have an extremely rough surface with sets of nearly vertical partings giving a grooved appearance to the rock walls. This rock, the Vishnu schist, is a metamorphic rock rich in quartz, mica, and hornblende, and is dated as early Precambrian in age, older than 2.4 b.y. (Schist is explained in Chapter 7.) Here and there, bands of coarse-grained diorite and granite cut through the schist, but no fossils or indications of life have been found in it, which is not surprising for a

FIGURE 6.6 This set of block diagrams shows the manner in which the great wedges of Algonkian strata came into existence in the lower Grand Canyon. (Based on data of C. O. Dunbar, 1960, *Historical Geology*, John Wiley & Sons, Inc., New York, p. 96, Figure 53.)

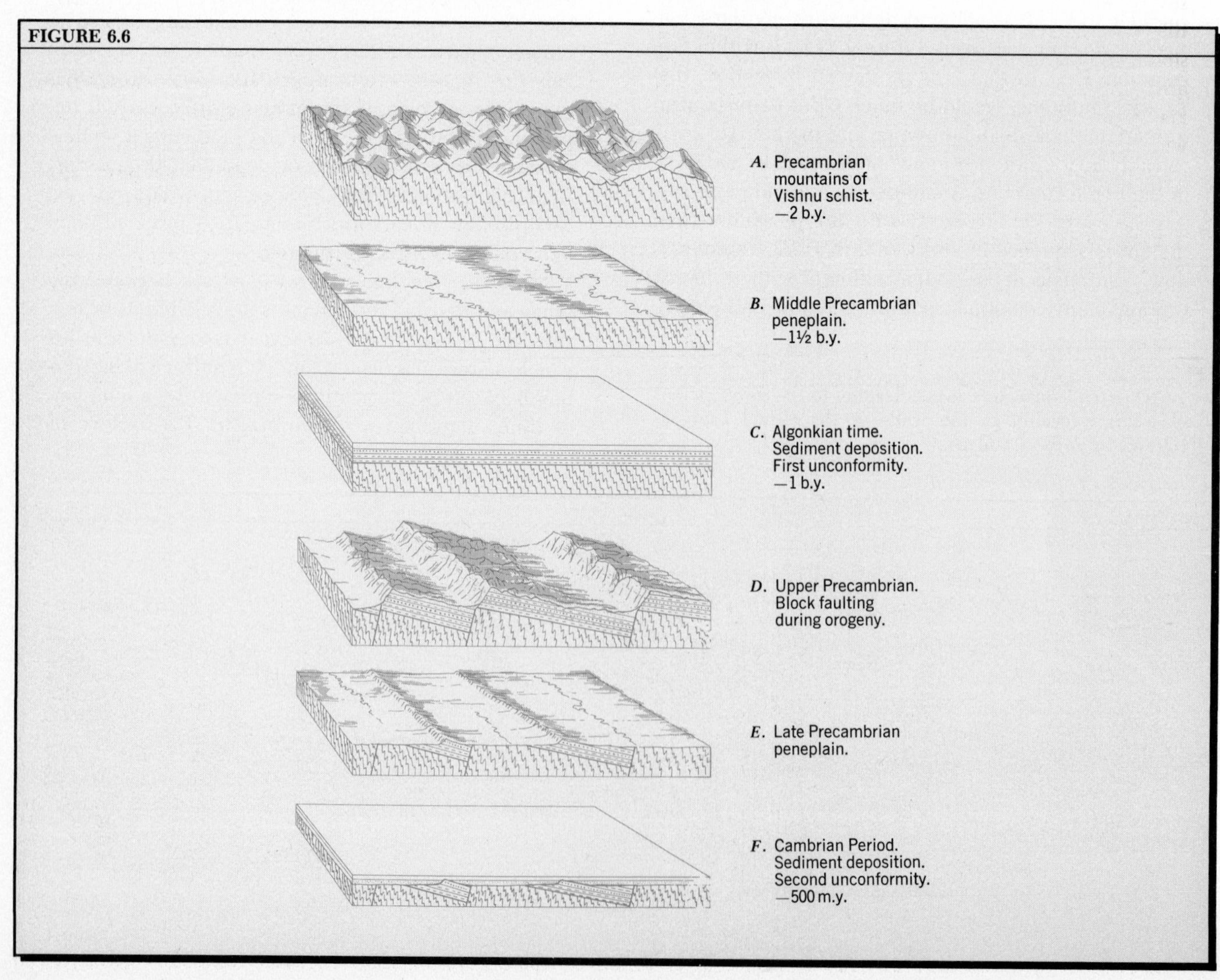

metamorphic rock. In the Grand Canyon region, the lower Precambrian division in which this rock was formed prior to undergoing metamorphism is named Archean time. The orogeny in which the Vishnu schist was formed and intruded by igneous rock is perhaps equivalent to the Kenoran Orogeny (see Table 6.1).

The surface of separation between the Cambrian strata and the underlying Vishnu schist is called a *nonconformity.* It represents a long period of erosion in which a great but undetermined thickness of the older metamorphic and igneous rock was removed from the area. Erosion finally reduced the land surface to a low plainlike surface, called a *peneplain*. (Development of a peneplain is explained in Chapter 17.) The peneplain was then submerged beneath the ocean by epeirogenic sinking of the crust. In this shallow ocean, the conformable sequence of Cambrian and younger strata was deposited on the Archean basement.

As we follow along the rim of the Tonto Platform, continuing our study of the walls of the Inner Gorge below us, a new geologic feature enters the picture (Figure 6.2). A sloping wedge of tilted sedimentary strata appears between the Tapeats sandstone and the Vishnu schist. The wedge continues to thicken until several thousand meters of strata are exposed. This tilted sedimentary series consists of shales and sandstones and belongs to the Grand Canyon Group. In this area, the Grand Canyon Group is assigned to Algonkian time and is placed in the late Precambrian; its age would perhaps be −1.7 b.y. or somewhat younger.

FIGURE 6.7 A wedge of Algonkian strata (*A*) lies beneath the Tapeats sandstone (*T*) of the Tonto Platform, but rests upon the Vishnu schist (*V*) of Archean age. (Sketched by A. N. Strahler from a photograph.)

From the principle of superposition, it is evident that the Algonkian sedimentary strata are younger than the Archean Vishnu schist, upon which they rest, but they are older than the Cambrian Tapeats formation, beneath which they lie. Thus, even without knowing their exact position in geologic time, it is fairly certain that the Algonkian rocks belong to the late or middle Precambrian.

An explanation of the Algonkian rock wedge in Grand Canyon is given in the series of block diagrams of Figure 6.6. Strata of the early Precambrian were crumpled and altered into the Vishnu schist by one of the earliest orogenies (Kenoran) of which we have any record (Diagram *A*), possibly caused by an early continental collision. Next, these mountains were reduced by erosion to a peneplain (Diagram *B*). After this great episode of denudation was complete, about −1.3 b.y., the crust began to subside, and the region was submerged beneath the water of a shallow sea. A nonconformity came into existence as clastic sediments accumulated rapidly on the ocean floor, reaching a total thickness of several thousand meters (Diagram *C*). After the sediment deposition was completed, a second orogeny took place. In this area, tectonic activity took the form of crustal fracturing that involved the rising, sinking, and tilting of large crustal blocks (Diagram *D*). Prolonged erosion removed the block mountains, creating a second peneplain above which a few of the harder sandstone masses projected as ridges (Diagram *E*). This topography existed at the close of the Precambrian time. Epeirogenic downwarping now took place, causing the region to become a shallow sea, which received the Cambrian sediments (Diagram *F*). As a result, the Cambrian layers in places rest directly upon the Vishnu schist (a nonconformity), but in other

places they rest upon a thick wedge of Algonkian strata (Figure 6.7).

The line of separation seen between the Cambrian strata and the Algonkian wedge strata in the canyon wall is referred to as an *angular unconformity*. "Angular" means that the strata of one rock group are set at an angle to the strata of the other group. The angular unconformity is evidence not only of a vast erosion period that intervened between the formation of the two groups of strata, but also of an orogeny that followed the deposition of the older strata and caused them to be tilted. This angular unconformity is shown in detail in Figure 6.2. Notice that a very hard formation, the Shinumo quartzite, stood as a residual mass above the general level of the late Precambrian peneplain. It protrudes through the Tapeats sandstone, which was deposited around the high mass but did not cover it. Figure 6.8 summarizes the three relationships that can exist between two groups of rocks separated by a large time gap in which a major episode of erosion occurred. All three relationships go under the general name of *unconformity*. The disconformity, which we explained in connection with the line of separation between the Muav limestone and the Redwall limestone, represents an erosion interval produced merely by simple vertical rising and sinking of the crust (epeirogenic movement) with no intervening episode of tilting or folding. Thus, in a disconformity, the strata above and below the line are horizontal and parallel. In both the nonconformity and the angular unconformity, overlying younger strata are discordant in attitude with structures that lie below the separation line.

The interpretation we have given for the angular unconformity seen in the Grand Canyon was soundly established by American geologists over a century ago. Those who made that interpretation could only guess at the actual ages of the rocks involved, for it was not until the early 1900s that the discovery of radioactivity made possible the dating of rocks in terms of millions of years. Neither did those early geologists have any clear notion of how crustal rocks could be deformed during an orogeny. In later chapters, we will show how plate tectonics provides a single, unified theory to cover both the deposition of thick layers of sediments and their later deformation.

Interpretation of Crosscutting Relationships

Relative ages of groups of rocks that are in contact with one another can often be determined by using the principle of *crosscutting relationships*. Any two of the three major rock groups—igneous, sedimentary, and metamorphic—may be involved in a crosscutting relationship. A rock group is younger than any adjacent group whose structure it cuts across. Examples of this principle were given in Chapter 4, in connection with the relationships between different kinds of igneous bodies. Intrusive bodies such as batholiths, stocks, pipes, dikes, and sills are obviously younger than the country rocks they intrude. In the case of the discordant intrusions, the outer surface of the pluton cuts across structures in the country rock. In the case of concordant intrusions—a sill or laccolith, for example—the older rock layers that are forced apart by the intruding magma usually show some crosscutting

FIGURE 6.8 Three kinds of unconformities.

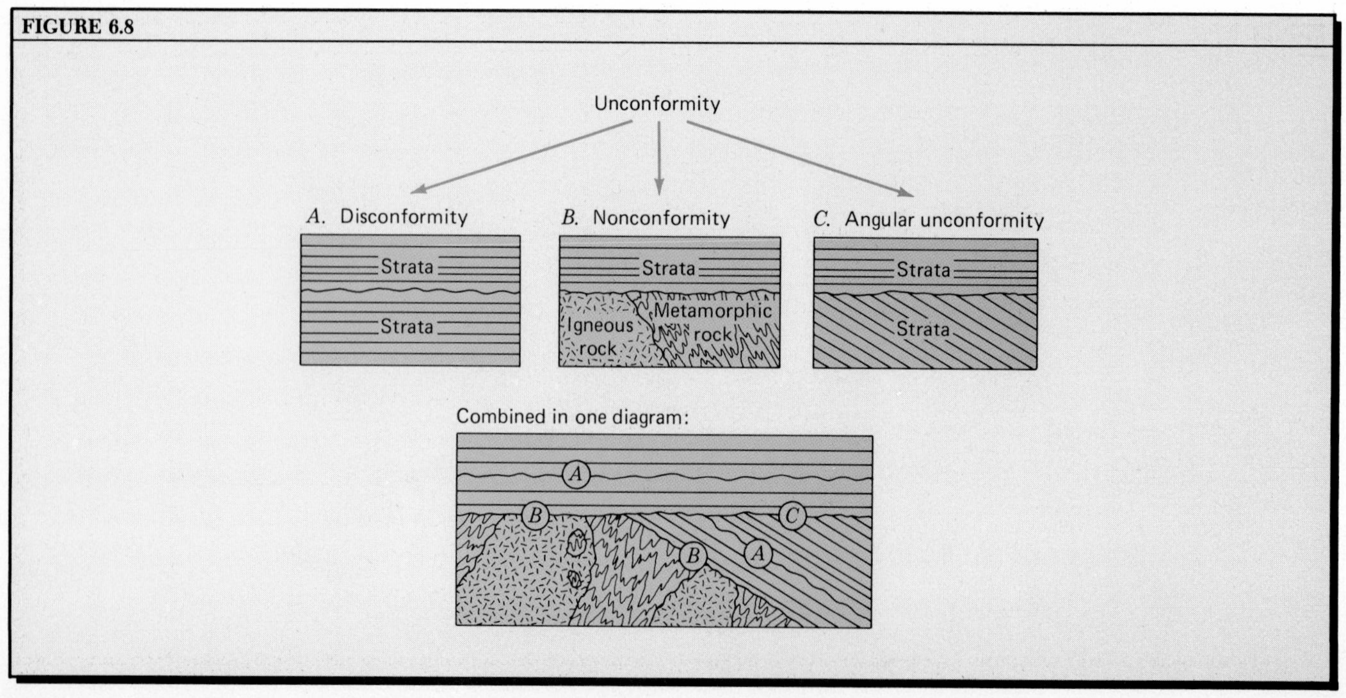

breaks that interrupt the continuity of the layers.

Figure 6.9 is a composite block diagram showing several kinds of crosscutting relationships as well as the three kinds of unconformities. The rock bodies are designated by letters arranged in alphabetical order from youngest to oldest. We must assume that these relationships are visible through exposures in canyon walls or in mine shafts. The principle of superposition is also called into use in determining relative ages of rock units shown in the diagram. For example, the fresh lava flow, *A*, has poured out over the surface of the alluvial fan, *B*. Caldera *C* is older than both the lava flow and fan, as judged by the deep erosion affecting its flanks. A field investigation might reveal that an ash fall from the caldera explosion has been overlain by both the fan alluvium and the lava flow. The caldera, in turn, rests on strata of the sedimentary sequence *D*. Igneous dike *E* is an interesting feature, because it intruded sedimentary sequence *F* but was leveled by erosion along the disconformity over which sequence *D* was deposited. This relationship suggests that erosion must have removed a thick mass of younger strata.

Sill *H* is older than stratigraphic unit *I*, which it has forcibly separated into a lower and upper sequence. A dike feeding the sill cuts across the lower sequence of strata. Batholith *J* is younger than the metamorphic rocks, *K*, which it has intruded, so that the metamorphic unit is the oldest rock shown. The intrusive igneous body labeled with a question mark is younger than groups *H*, *I*, *J*, and *K*, but it could also be younger than *D*, *E*, *F*, and *G*. There exists the possibility that the pluton is of the same age as the extrusive activity that produced the caldera and its feeder pipe. To place this pluton more accurately in the sequence, we would need some absolute rock ages based on the radiometric method of age determination.

Environments of Sediment Deposition

A stratigrapher continually faces the problem of interpreting sequences of strata in terms of the environment in which the sediments were deposited. There are many questions to be answered. Were the sediments deposited on land, exposed to the atmosphere, or were they deposited on the ocean floor? Each of these alternatives raises further questions. If the sediments are terrestrial (deposited on land), were they transported and deposited by streams or by wind? If marine, were they once part of a delta forming in shallow water just off the coast, or were they laid down in the floor of a deep oceanic trench? The answers must come from the physical and chemical nature of the rock itself and from the fossils (if any) that are contained in the rock.

Geologists recognize three major groups of environments of sediment deposition: marine, terrestrial, and littoral. The *marine environment* includes all ocean water and ocean bottom lying below the level of low tide. The *terrestrial environment* includes all land surfaces lying above high-tide level. Inland basins below sea level, but cut off from ocean waters, belong in the terrestrial environment, as do freshwater lakes. The *littoral environment* lies in the intertidal zone, which is between the limits of low tide and high tide and is alternately exposed to the atmosphere and covered by seawater.

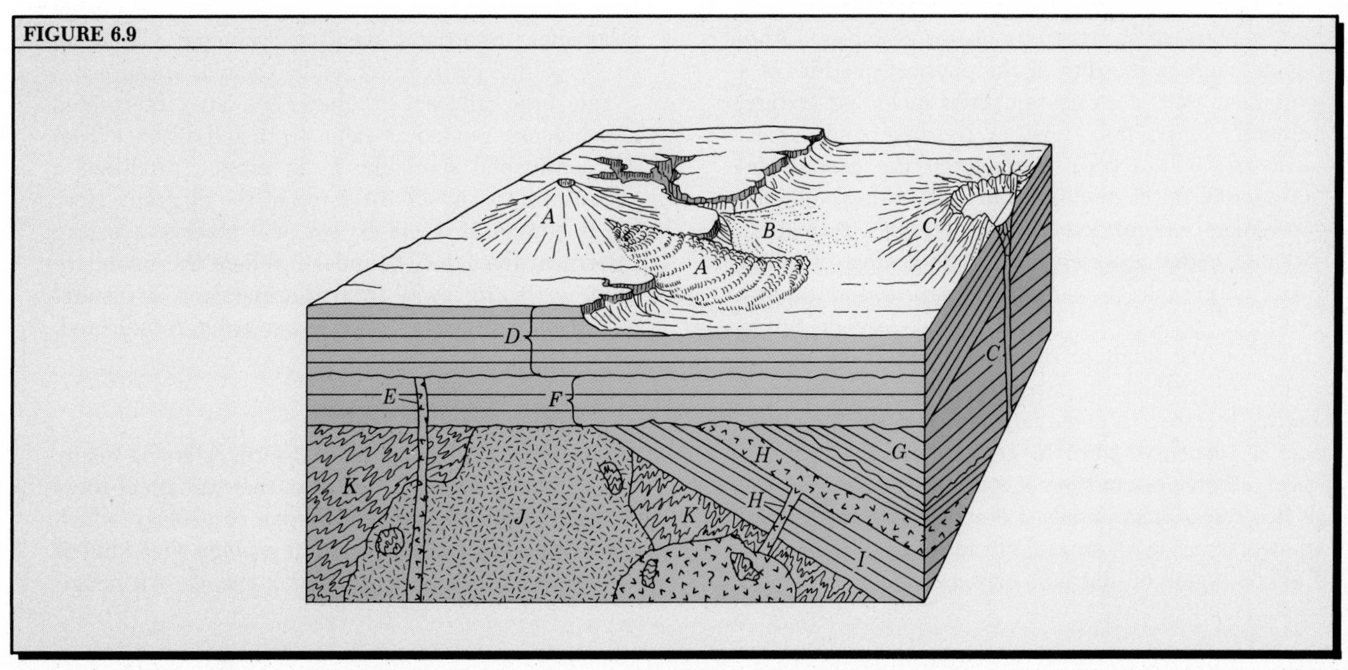

FIGURE 6.9 An imaginary set of igneous and sedimentary rock units to illustrate the principle of crosscutting relationships. (Drawn by A. N. Strahler.)

Sediments formed in the marine environment span a great range in terms of their kind and arrangement into strata. Sediments of the shallow continental shelves commonly take the form of well-sorted quartz sands and thin layers of clays and lime muds, and are often rich in marine fossils. Sediments deposited in very deep water adjacent to a continent are sands and muds that accumulate in total thicknesses of many kilometers, but they may be almost devoid of visible fossil remains. On deep ocean floors far from the continents, sediments are deposited very slowly and consist of very fine particles, which may include the hard parts of minute organisms that lived in the ocean water directly above, near the surface. In later chapters, we shall describe several of the marine environments and their sediments.

Terrestrial environments and their sediments also show a wide variety. First, there are sediments that have accumulated in freshwater lakes, marshes, and swamps. Stream-laid sediments form another class, and these take the form of deposits on river floodplains, desert alluvial fans, or the higher parts of river deltas. Different still are desert environments in which sand and silt are deposited by the action of wind. Glaciers, too, are carriers of sediment, which they deposit beneath moving ice. All of these varied terrestrial environments and their sediment deposits are described in detail in later chapters. Here, we will concentrate upon the interpretation of strata of the marine environment deposited in water of shallow or moderate depth on continental shelves with gently sloping floors.

*Lithofacies of Marine Deposition

Stratigraphers have a very special word for the general appearance, composition, and structure of a rock unit—*facies;* it comes from the Latin word for "face" or "form." The letter *c* has an *sh* sound (fay-shees). Sedimentary rocks have two classes of facies. *Lithofacies* refers to the sum of the physical qualities of a particular unit of strata—qualities including texture, mineral composition, type of bedding, and typical structures within beds. *Biofacies* is the general description of the character of the fossil fauna found in a particular level, or *faunal zone,* within a formation. Because certain species serve as indicators of water conditions, such as water temperature and salinity (saltiness) or whether the water is clear or turbid (muddy), the kinds of organisms that make up the fauna are particularly important in the description of marine biofacies. A given fauna is considered to represent a particular point in geologic time, whereas a given lithofacies can show a considerable range in age as it is traced horizontally. Let us see how this principle works, using an example from strata of the Grand Canyon region of northern Arizona.

Figure 6.10 summarizes the interpretation of strata of Cambrian age exposed at several study localities where geologists took detailed measurements of the thickness of the strata and recorded data on their lithofacies and biofacies. The results are displayed in a *lithofacies–time diagram.* The three formations involved are the Tapeats, the Bright Angel, and the Muav, described briefly earlier in this chapter. Lithofacies are indicated by patterns, boundaries between formations are shown by a heavy line, and time lines, called *isochrones,* run horizontally across the diagram.

To interpret this diagram, you should cover it with a sheet of paper. Now move the paper gradually up, keeping its bottom edge horizontal as you uncover the diagram. As you do this, you see emerging the sequence of deposition. Where the wavy line of an unconformity is above the isochrone, it can be interpreted as dry land. The first Cambrian formation to appear is the Tapeats, a sandstone, deposited as a sandy nearshore lithofacies.

The ancient shoreline evidently moved from west to east across this area, as shown on the maps below. These are *paleogeographic maps;* they reconstruct the environment at a given point in time. Encroachment of an ocean upon the land, accompanied by sediment deposition, is called a *transgression.* Stopping at the isochrone for Time *A,* we see that the nearshore lithofacies is being followed by an offshore facies consisting of shales, siltstones, and fine sandstones, making up the Bright Angel formation (Map *A*). Within the Bright Angel formation, we come to a faunal zone, establishing an isochrone at that level. At Time *B,* a carbonate lithofacies has appeared, represented by limestone beds of the Muav formation. In the meantime, the shoreline and the nearshore facies have disappeared off the diagram toward the east. At Time *C,* most of the region lies in the zone of carbonate lithofacies. Shortly thereafter, younger Cambrian strata were deposited and finally the seas withdrew.

The three different lithofacies we have interpreted are obviously related to water depth and distance from the continental shoreline. Later erosion produced a disconformity, shown at the top of the diagram. Near the isochrone of Time *B,* you will observe a zigzag pattern in the facies boundary. Where the boundary shifts westward, away from the mainland, a general withdrawal of the sea, called a *regression,* is indicated.

Rock Deformation

Rock deformation is a blanket term referring to any and all changes in shape or form that can affect rocks of the lithosphere, and the branch of geology which investigates it is called *structural geology.* One kind of deformation involves only volume change: An imaginary cube of rock surrounded by a large body of rock might be reduced or increased in volume while still remaining a cube (Figure 6.11*A*). Another kind of deformation results in a change in shape: Our imaginary rock cube might be deformed in such a way that it is changed into a rhomb with acute and obtuse angles (Figure 6.11*B*). Yet another kind of deformation is by fracture: The imaginary rock cube could also be deformed by fracturing, when a crack develops through the cube, allowing the part of the cube on one side of the crack to slip past the other part (Figure 6.11*C*). We will give scientific names to these different kinds of deformation as we investigate each in more detail.

FIGURE 6.10 A generalized lithofacies–time diagram of three Cambrian formations exposed in the walls of Grand Canyon. (Greatly simplified from data of E. D. McKee, 1945, *Cambrian History of the Grand Canyon Region*, Carnegie Inst. of Washington, Washington, D.C., Publ. 563, Part I, p. 14, Figure 1.)

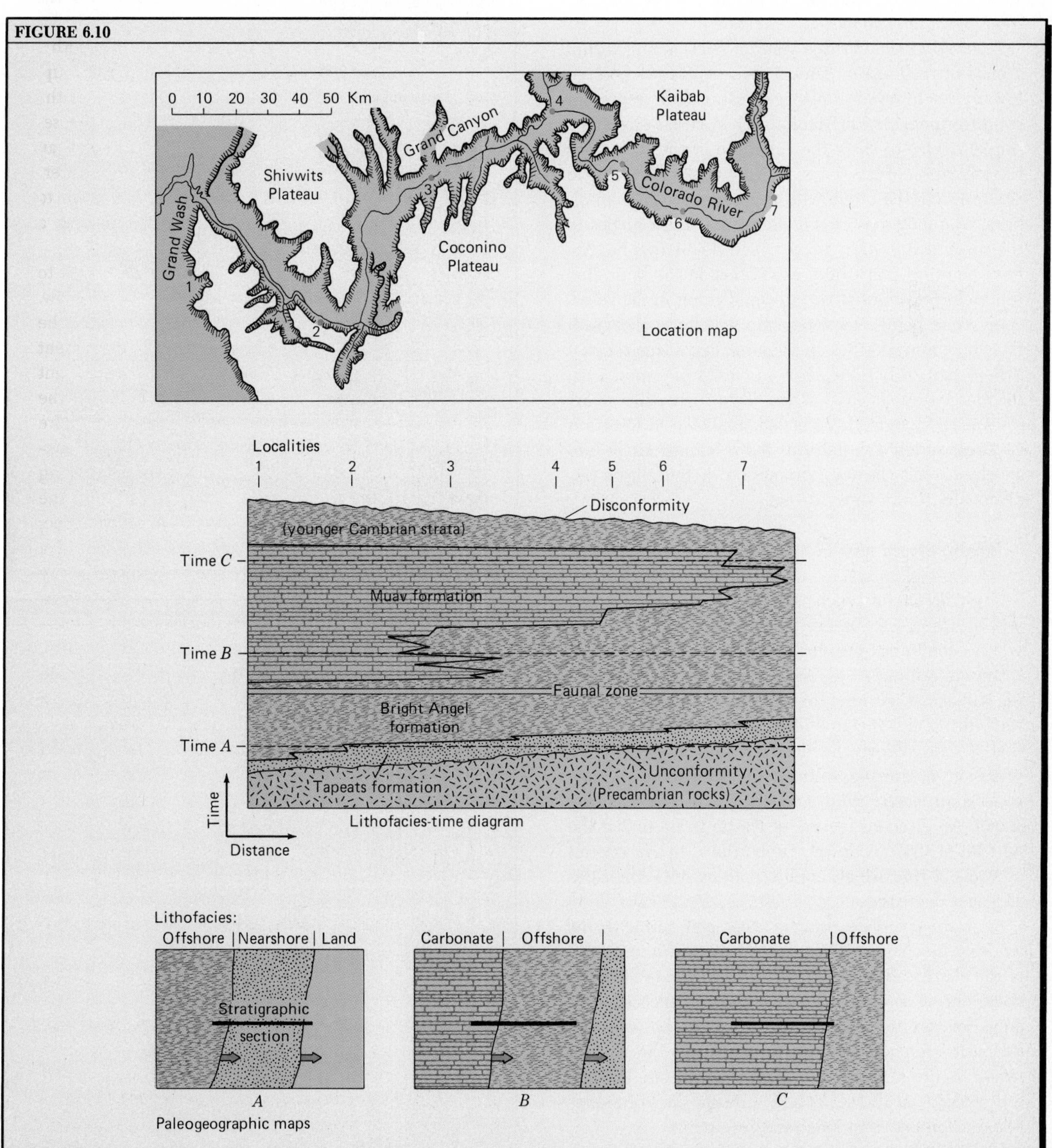

Kinds of Stress

Deformation of rock is described as *strain*, the yielding of a solid to applied stress, whether by a change in volume or shape, or by fracturing. *Stress* is the force acting across a unit surface area; it has the same general meaning as the term *pressure*. As we explained in Chapter 4, geologists state confining pressure within the earth in terms of the kilobar, equal to 1000 kg/cm sq. Stresses applied in laboratory experiments to deform rock are stated in *bars*, one bar being one-thousandth of a kilobar.

Figure 6.12 shows three kinds of stress acting within a mass of rock, using pairs of arrows. Arrows pointed toward each other along the same line represent *compressional stress* (Diagram A). We use a circle as a simple device to show the condition of the rock prior to application of the stress, for when the rock yields to compression, the circle will be compressed into an ellipse. The long axis of the ellipse is at right angles to the *compression axis*, along which the arrows lie. Arrows pointing away from each other in the same line represent *tensional stress;* it would result in the reference circle being elongated into an ellipse (Diagram B), whose long axis lies on the same line as the *tension axis*. A pair of parallel arrows pointed in opposite directions indicates *shear stress*. Two such pairs of arrows at right angles are needed, as shown in Diagram C. Their action also deforms the reference circle into an ellipse, but the long dimension of the ellipse lies diagonally to the stress arrows.

Elastic Strain and Plastic Flow

Crystalline solids, such as most rocks, are both strong and brittle as we observe them at the earth's surface, where confining pressures are very small. Under strong deforming stress, crystalline rocks such as granite, sandstone, or limestone deform elastically, like a steel spring, but the amount of deformation is so small that it can scarcely be detected without the use of precision measuring instruments. We refer to such rock as an *elastic solid*, and say that it shows *elastic strain*. An essential feature of elastic strain is that the amount of strain is exactly proportional to the amount of stress, a rule which applies both as stress increases and as it decreases.

An ordinary scale using a steel coil spring as its working mechanism illustrates a practical application of the elastic solid. As you add kilogram weights to the scale, one at a time, the dial registers a succession of one-kilogram increases in total weight because the spring is stretched in direct proportion to the applied stress. As the same weights are removed one by one, the reading on the scale falls through the same one-kilogram intervals to reach the zero point.

Elastic strain has its limit, as we know from every-

FIGURE 6.11 Types of rock deformation.

FIGURE 6.12 Three kinds of stress that affect rock.

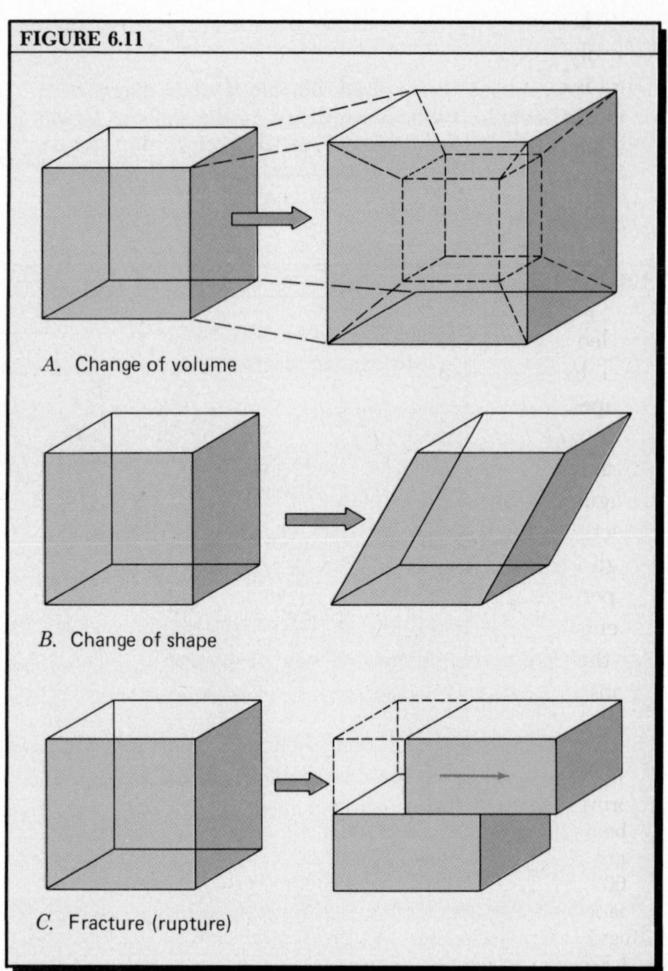

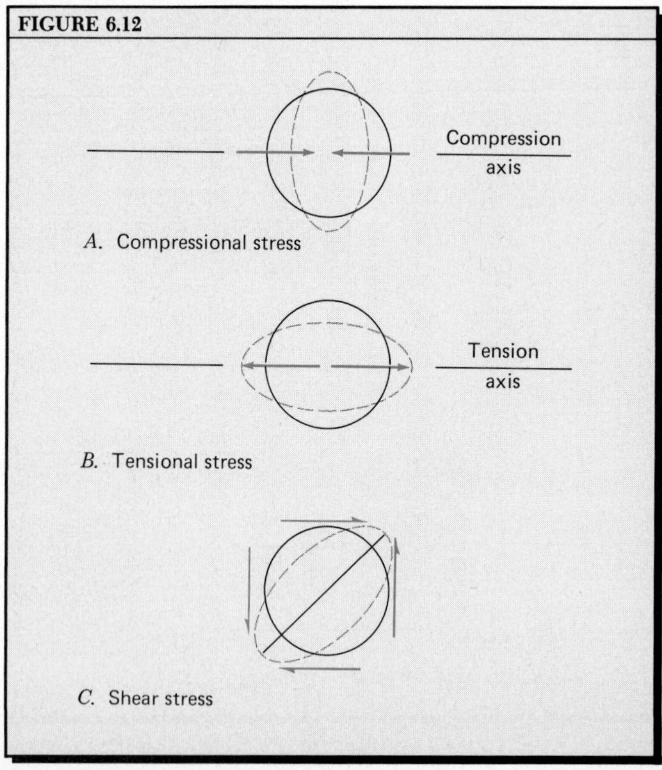

day experience. One of two things happens when this strength limit is exceeded. A *brittle solid* subjected to stress exceeding its *elastic limit* fractures with a sharp break. A steel spring, if it happens to be brittle, may simply snap in two. This behavior is called *rupture* and is characteristic of most rocks near the earth's surface. A second possible mode of response of a steel spring is to acquire a permanent deformation when stressed beyond its elastic limit. What has happened in this instance is that the molecules of the crystalline metal have been dislocated into a new arrangement, which is permanent. In metals, this property of acquiring a permanent change in shape without rupture is called *ductility. Ductile solids* (metals, for example) can be drawn out or hammered into a variety of shapes without rupturing.

In rock, the permanent change that results from deforming stress is often described as *plastic flow* to distinguish it from elastic strain. Rocks begin to show ductility when the confining pressure is increased to high values, as deep in the crust, and this can be done experimentally in the laboratory. Laboratory experiments with rock samples make use of small cylinders of the rock carefully cut and polished to exact dimensions. A rock cylinder is inserted into a metal or plastic jacket and placed in an enclosing fluid-filled chamber in which confining pressure can be increased to high values. At any desired value of confining pressure, the rock cylinder is subjected to compressional stress at its ends. After the rock has responded to this stress, the enclosing jacket is cut away and the rock core examined.

FIGURE 6.13 Four small cylinders of limestone, each deformed under confining pressure so as to be shortened by about 15 percent. *A*. Shear fracture (faulting) resulted at 200 bars. *B*. Combined plastic (ductile) flowage and faulting at 700 bars. *C*. Largely plastic flowage at 900 bars. *D*. Deformation almost entirely by plastic flowage at 1800 bars. Photographs are approximately 1½ times actual scale. (Courtesy of Fred A. Donath.)

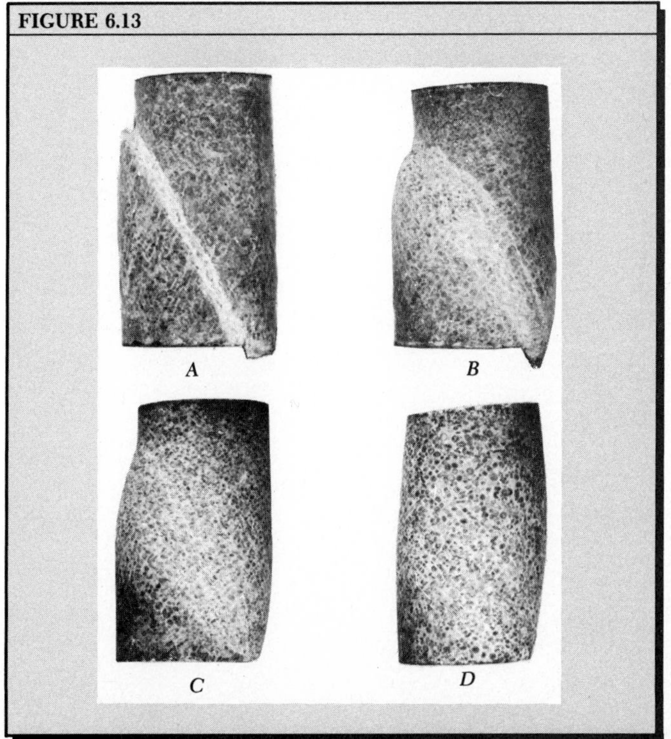

As Figure 6.13 shows, the rock behaves quite differently according to the different levels of confining pressure that were used. Under low pressure, the cylinder ruptures with a set of sharp diagonal fractures. Under higher pressure, the fractures are very numerous and very small. Under still higher pressure, there are no visible fractures; instead, the rock has deformed by plastic flow.

Dilatation of Rock

Dilatation is a form of strain involving only a change in volume, but not in shape (Figure 6.11*A*). It takes place in response to an increase or decrease in confining pressure within the earth. Dilatation of rock may be only elastic strain, or it may involve a permanent volume change. As we explained in Chapter 4, the confining pressure to which all rock is subjected increases steadily with depth into the earth. (Figure 4.1*B* showed the pressure increase with depth.)

Because confining pressure is equal in intensity in all directions at a given point within the rock, it acts to reduce the volume of the rock by compressing the atoms, ions, or molecules into a smaller space. An important effect of decreased volume is that rock density is increased, because the same amount of mass occupies a smaller volume. In reverse fashion, as a mass of crustal rock rises and the overlying rock is removed by denudation, the confining pressure decreases, the rock expands, and its density is reduced. These density changes have important applications in several areas of geology, as we shall find in later chapters.

The increase of confining pressure with increased depth is extremely important in the change (diagenesis) of soft, water-rich sediments into hard (lithified) sedimentary rock. As the confining pressure increases, water is driven out of the sediment, and the mineral particles become more closely packed. This form of sediment diagenesis is not pure dilatation, because some matter has been excluded from the rock. It is, however, a kind of dilatation caused by pressure increase acting over a long period of time. Shale, in particular, undergoes a density increase as depth of burial increases both because of water loss and because the flakelike grains of clay minerals are rotated and rearranged to fit more closely. For example, one well drilled for petroleum exploration passed down through 1800 m of a shale formation. Samples taken from the well showed a density increase from about 2.1 g/cc near the surface to over 2.4 g/cc at a depth of 1800 m.

Fracturing in Rock

One of the forms of deformation observed in laboratory experiments with rock cylinders is *shear fracture,* resulting from unbalanced stresses. By this we mean that stress acting from one direction is greater than stress acting from another direction at the same point in the mass. When shear fracture occurs, one mass or layer of rock moves past another along a surface that separates the two masses. Ideally, this plane of motion, the *shear plane,* is a true plane surface.

In brittle rock at the earth's surface, shear planes are formed by rupturing that occurs when the shear stress exceeds the elastic limit of the rock. Sliding then occurs along a single shear plane, relieving the stress. An isolated rock fracture of this kind is called a *fault* (Figure 6.14, Diagram *A*), and its motion occurs along a *fault plane* and is called *faulting.* Shear planes may also occur closely spaced in the rock, with only thin parallel layers of rock separating them (Diagram *B*). One can imagine multiple rock layers, each as thin as one centimeter or one millimeter. In that case, the shear planes would pass between individual mineral grains or groups of grains. Finally, imagine that the shear planes pass between single layers of molecules, atoms, or ions. As shown in Diagram *C,* shearing is, for all practical purposes, distributed uniformly through the substance along an almost infinite number of planes. This phenomenon is true plastic strain and can also be called plastic flow. The same phenomenon, when seen in fluids such as air (which is a gas) and water (which is a liquid), is called *fluid flow.* Substances that deform by plastic flow and fluid flow possess the physical property of viscosity, which we defined in Chapter 4 in connection with lavas.

An important difference between plastic flow and fluid flow is that for plastic flow to occur a substantial shearing stress must build up before the shearing starts. In fluid flow, the shearing motion begins with the slightest shearing stress, no matter how small, and the rate of shear increases in direct proportion to the increase in shear stress. The point is that shearing requires the expenditure of energy to overcome the bonds that hold the particles together, be they atoms, ions, molecules, or larger aggregates of matter. The stronger these bonding forces, the higher the viscosity of a substance. Thus, the viscosity of magma is vastly greater than the viscosity of water—actually by a factor of about 10^{13}. Viscosity of solid rock of the asthenosphere is even greater than that of magma at the same depth. This means, of course, that rock flowage by plastic shear is exceedingly slow by any standards we can imagine. We can get some idea of the slowness of rock flowage by noting that lithospheric plates moving away from a spreading boundary in the mid-oceanic ridge separate at a speed on the order of from 2 to 10 cm/yr. The plates are carried along by

FIGURE 6.14 Forms of shearing. *A.* Shear fracture illustrated by a single shear plane, or fault. *B.* Multiple shear planes. *C.* An infinite number of shear planes characterizing plastic flow.

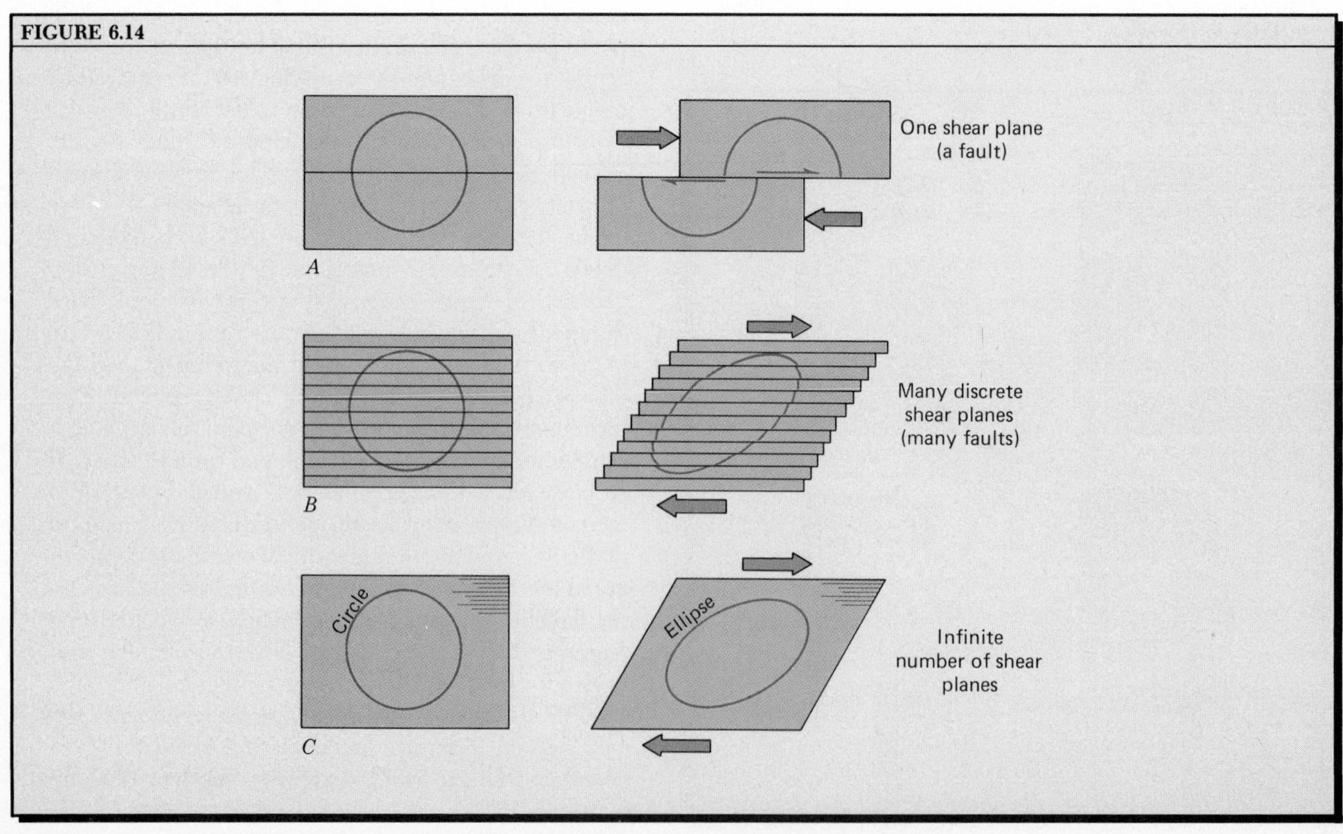

motion distributed through a plastic asthenosphere many tens of kilometers thick. On the other hand, geologic time spans are vastly longer than any we can comprehend in our present world, so the seeming slowness of rock flow is only relative.

Rocks deep in the crust under extremely high confining pressures and high temperatures behave very differently from the same rocks near the earth's surface. When shear stresses are applied steadily to deep-seated rock over long periods of time, the rock exhibits plastic strain and flows very much like a highly viscous fluid. Even though this rock flowage involves the continual rearrangement of atoms and ions in the space lattices of the minerals making up the rock, the rock is, strictly speaking, in the solid state at all times. We know this because earthquake waves pass through this rock, a phenomenon that would be impossible in a true fluid (Chapter 8).

Many decades ago, geologists recognized this dual behavior of deep rock—acting simultaneously as both a solid and a fluid—by the adjective *elastico-viscous*. The prefix *elastico*, from the word "elastic," describes the behavior of a solid; *viscous* describes the behavior of a fluid. In looking about for some common substance that can illustrate elastico-viscous behavior, geologists often cite the example of asphalt (roofing tar). At moderate atmospheric temperatures (10 °C–20 °C), asphalt is a black, glassy solid and will break along conchoidal fractures when struck a hammer blow. The same mass of tar, left standing at the same temperature on a pavement, will gradually collapse and form a pancakelike sheet, exhibiting flowage quite suggestive of a basaltic lava flow.

Besides deformation by shear fracture, brittle rock can be deformed by *extension fracture*, in which a crack appears and the masses on either side of the crack are forced apart. In this case, relative motion of the masses is one of separation and is at right angles to the plane of the fracture. Extension fractures are commonly seen in human-made construction materials that tend to shrink with time. A good example is an asphalt–concrete (macadam) pavement. As the more volatile substances in the asphalt evaporate, the asphalt–concrete layer contracts in volume. Irregular cracks, extension fractures usually without perceptible shearing movement, form in the pavement and gradually widen.

Another example is the kind of thin crack we often see in a ceramic plate or cup that has been struck a sharp blow at the rim. Under the impact, the crack is opened slightly but there is no slippage in the plane of the crack.

In the two examples we have used, fracture caused by shrinkage of the material is associated with tensional stress, whereas that caused by impact is associated with compressional stress. In rocks, extensional fractures are commonly called *joints*, and the process by which they are formed is called *jointing*, to distinguish it from faulting. We turn next to a study of these two forms of expression of fracturing in rocks.

FIGURE 6.15 Strike and dip of a fault plane.

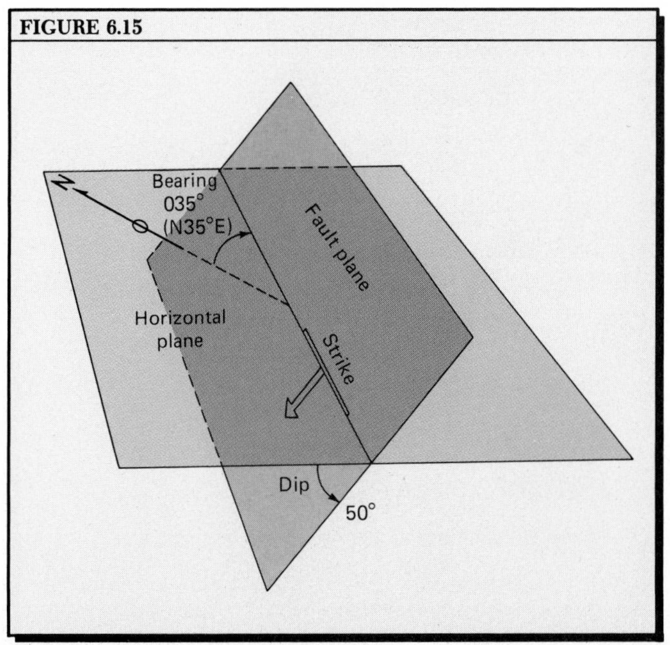

Kinds of Faults

For the purpose of description, geologists idealize a fault plane as a perfect geometrical plane. An ideal plane may have any conceivable orientation in space, ranging from the vertical to the horizontal. To describe the orientation in space of such a plane, geologists use a combination of two terms: dip and strike. The *dip* of the plane is the acute angle which the plane makes with respect to an imaginary horizontal surface (Figure 6.15). The angle of dip can thus range from zero—for a horizontal plane—to a maximum of 90 °—for a vertical plane. The *strike* of the plane is the compass direction taken by the horizontal line of intersection of the inclined plane with the imaginary horizontal plane (Figure 6.15). Strike is measured in degrees of compass bearing with respect to geographic North. Dip and strike are used to describe the orientation of fracture surfaces and bedding surfaces in rock.

To demonstrate the principal varieties of faults, we can use simple block models. Motion on a fault is easily illustrated by a single pair of blocks in which the fault plane cuts obliquely through the rectangular planes of the reference figure (Figure 6.16, Diagram A). Holding the right-hand block fixed, we can slide the left-hand block on the fault plane in different ways to illustrate various kinds of faults. First, however, let us take the two blocks that make up our fault model and separate them by a few centimeters, exposing the slanting surfaces that are in contact on the fault plane. As shown in Diagram *B* of Figure 6.16, one of these surfaces overhangs the vertical and is called the *hanging wall.* The slanting surface of the other block is called the *foot wall.* These are venerable terms used by miners who encountered faults in mine excavations. Using the two terms, we are able to describe the relative motion of the two blocks by statements such as these: The hanging wall moved up relative to the foot wall; the hanging wall moved down relative to the foot wall.

Table 6.3 is an orderly classification of faults using the various terms we have thus far defined. Three classes of faults are recognized: dip-slip, strike-slip, and oblique-slip faults. In a pure *dip-slip fault,* motion is entirely in a direction parallel with the dip of the fault plane, as shown in Diagrams C and *D* of Figure 6.16. In a pure *strike-slip fault* motion is entirely in the direction of the strike of the fault plane and is horizontal (Diagram *E*). In an *oblique-slip fault* the relative motion combines that of dip-slip and strike-slip (Diagram *F*). Consider next, the common types of faults within the first two classes.

In the *normal fault* (Diagram *C*) the block with the foot wall moves upward, while the block with the hanging wall moves downward. The up-moving block is commonly referred to as the *upthrown block,* the down-moving block, as the *downthrown block.* The fault plane dips toward the downthrown block. Obviously, this motion requires that the horizontal distance, *h,* separating reference points on opposite sides of the fault must increase as the vertical distance, *v,* increases. (The vertical distance of displacement is called the *throw.*) Thus, the two blocks have moved apart horizontally during faulting but have also retained contact on the fault plane.

In the *reverse fault,* relative motion is the opposite

FIGURE 6.16 Kinds of faults.

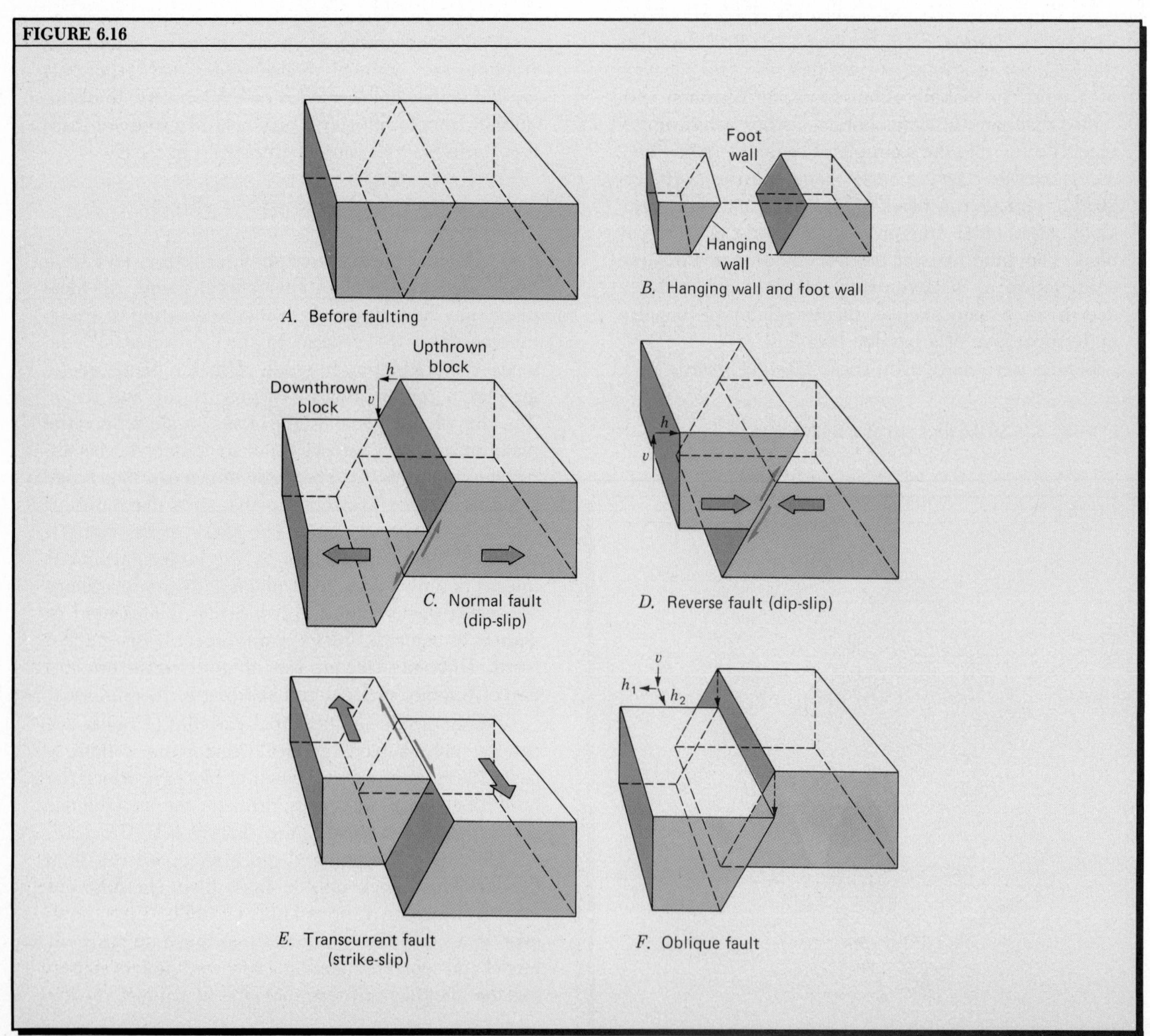

Table 6.3 Classification of Faults

Fault class	Fault types	Relative motion of fault blocks
Dip-slip		All motion is in direction of dip of fault plane.
	Normal	Hanging wall moves down.
	Reverse	Hanging wall moves up.
Strike-slip	Transcurrent	All motion is in direction of strike of fault plane.
	Right-lateral	Opposite block moves right.
	Left-lateral	Opposite block moves left.
Oblique-slip		Block motion combines vertical and horizontal components of dip-slip and strike-slip classes.

of that in the normal fault (Diagram *D*). Here, the fault plane dips beneath the upthrown block. Like the normal fault, the reverse fault is a dip-slip fault, but the hanging wall block has moved up and the foot wall block has moved down. The horizontal distance between reference points on the upper surface of the two fault blocks has been decreased.

FIGURE 6.17 Three forms of faults and their surface expression. (Redrawn, by permission of the publisher, from A. N. Strahler, *Physical Geography,* 4th ed., Figure 29.6. Copyright © 1975 by John Wiley & Sons, Inc.)

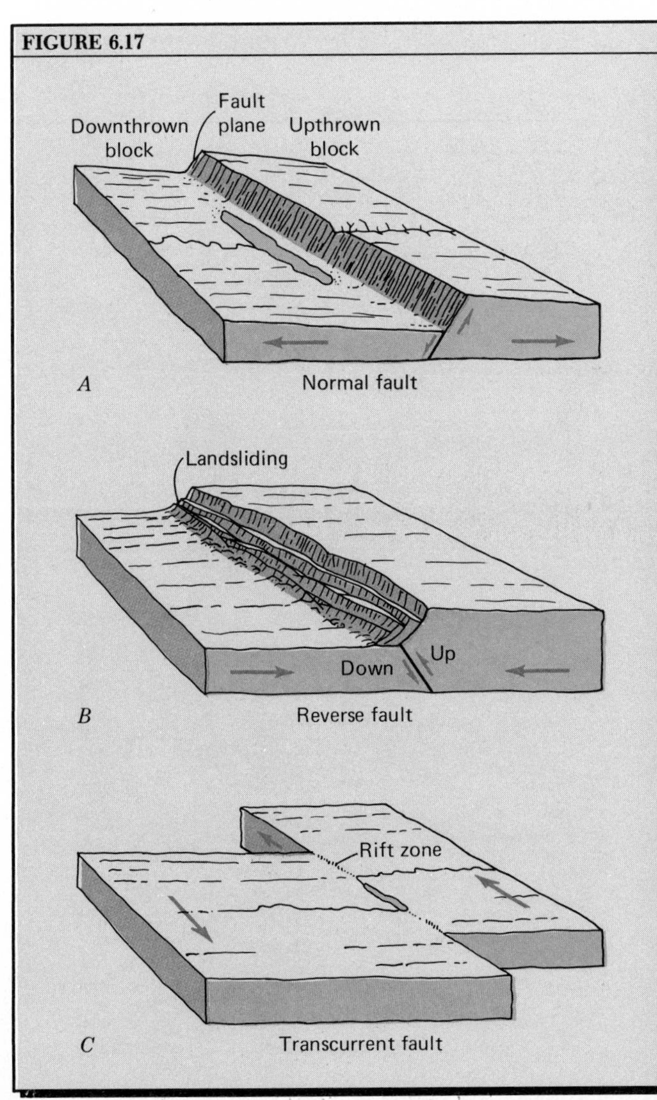

The strike-slip fault, shown in Diagram *E*, is often called a *transcurrent fault*. (We use the latter term throughout this book.) In most transcurrent faults, the fault plane is steeply inclined and may be nearly vertical. Relative motion on a transcurrent fault can be in either of two directions. Imagine that you are standing on one block, facing the fault. If the opposite block has moved toward your right, the fault is a *right-lateral transcurrent fault*. The fault shown in Diagram *E* is of this type. If the block opposite you has moved toward your left, the fault is a *left-lateral transcurrent fault*.

The oblique-slip fault, shown in Diagram *F*, has motion simultaneously in the direction of both strike and dip of the fault plane. Either the hanging wall or the foot wall may move upward. In the illustration, the hanging wall has moved downward, as in the normal fault.

Surface Expression of Faults

Faults that break the earth's surface often appear as distinctive landscape features formed by displacement occurring in almost instantaneous movements. A single displacement may amount to a few millimeters or as much as 10 m or more. Movements of this type are repeated many times at intervals of several decades or longer, so that throw accumulates, and the surface features change accordingly. Figure 6.17 illustrates these surface forms. The normal fault is expressed as a steep wall called a *fault scarp* (Figure 6.18).

Normal faults often occur in pairs, striking parallel with one another, but with opposite sides downthrown or upthrown. The result is a downdropped fault block, called a *graben*, lying between the faults, or an uplifted block, called a *horst* (Figure 6.19). Grabens can form broad, flat-floored valleys many kilometers wide and tens of kilometers long, bounded by steep fault scarps. Horsts can form elongate plateaus or tablelands.

The reverse fault (Diagram *B* of Figure 6.17) also forms a steep fault scarp, but because the rock on the upthrown side tends to overhang, it breaks off and slumps to lower levels under the force of gravity, burying the fault under rock rubble. The transcurrent fault (Diagram *C*) may give little surface expression where it crosses a flat plain, but there is usually present a trenchlike feature called a *rift*. Surface features of transcurrent faults are described in more detail in Chapter 8.

Not all normal faults are simple, clean breaks. The fault may be split in such a manner that the end of one fault is overlapped by the end of another (Figure 6.20*A*). Between the overlapping ends is a sloping ramp, or *fault splinter*. Strangely enough, you would be able to walk on the splinter from the downthrown block to the upthrown block without crossing any fault line. The total displacement between downthrown and upthrown blocks is often taken up in a number of parallel, closely spaced fault slices (Figure 6.20*B*). At the surface, these slices form a series of steplike levels, some of which may be fault splinters. The term *step faulting* is sometimes applied to this multiple-fault arrangement.

Keep in mind that the faults we have described can cut through any kind of rock, whether sedimentary, igneous, or metamorphic. A single fault can simultaneously cut across all three rock types as well as across older, inactive faults.

FIGURE 6.18

FIGURE 6.18 This fault scarp near Hebgen Lake, Montana, was produced during the earthquake of August 17, 1959. The light area in the center is exposed rock of the fault plane. (Irving J. Whitkind, U.S. Geological Survey.)

FIGURE 6.19

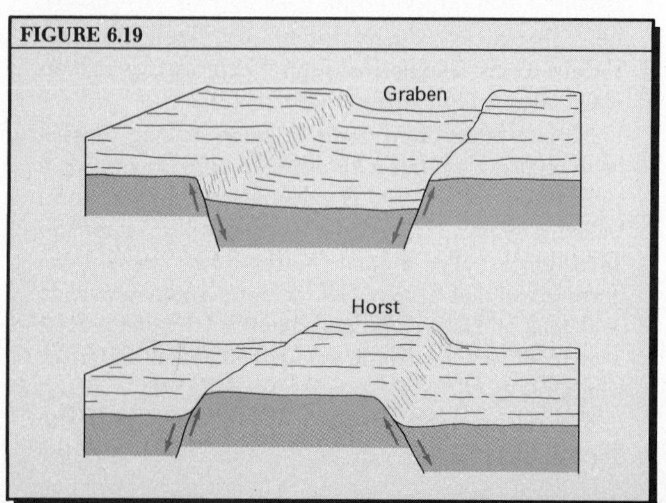

FIGURE 6.19 Graben and horst. (Redrawn, by permission of the publisher, from A. N. Strahler, *Physical Geography*, 4th ed., Figure 29.8. Copyright © 1975 by John Wiley & Sons, Inc.)

FIGURE 6.20

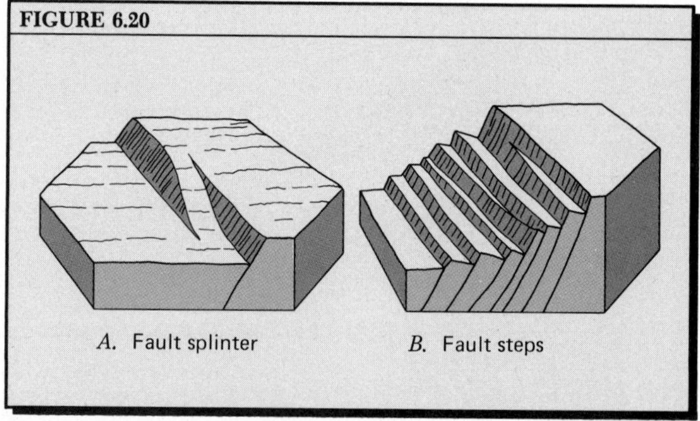

FIGURE 6.20 Normal faulting may be complicated by the presence of fault splinters and fault steps. (Redrawn, by permission of the publisher, from A. N. Strahler, *Physical Geography*, 3rd ed., Figure 33.10. Copyright © 1969 by John Wiley & Sons, Inc.)

FIGURE 6.21

FIGURE 6.21 These mudcracks on the dried bed of a river average about 30 cm across. (G. K. Gilbert, U.S. Geological Survey.)

Jointing in Rock

Joints, the extension fractures produced in brittle rock, are present in nearly all masses of rock we see exposed at the earth's surface. Here and there we do find some large rock masses, called *monoliths*, that consist of rock completely free of joint fractures, but these are the exception rather than the rule. The ancient Egyptians were fortunate in finding near the Nile River some monoliths of syenite out of which to hew their great statues and obelisks.

FIGURE 6.22 Sheeting structure in the Rock of Ages Granite Quarry, Barre, Vermont. (Rock of Ages Corporation, a Nortek Company, Barre, Vermont.)

FIGURE 6.23 Joint blocks of massive sandstone in the Trinidad formation, Cottonwood Canyon, Colfax County, New Mexico. (J. R. Stacy, U.S. Geological Survey.)

FIGURE 6.22

FIGURE 6.23

Jointing can occur from the action of a number of different patterns of tensional and compressional stresses. One cause of joints is rock dilatation, which we have described as a change in rock volume without an accompanying change in shape. One important example of joints caused by shrinkage (negative dilatation) is *columnar jointing*, seen in lava flows, dikes, and sills (Chapter 4; see Figure 4.20). Another phenomenon of mass shrinkage is seen in sediments in the form of *mudcracks* on the upper surfaces of mud deposited in shallow water (Figure 6.21). As the water evaporates and the mud dries, the uppermost layer shrinks uniformly in its horizontal dimensions. When the mud reaches a brittle stage, it breaks up into plates of rectangular, pentagonal, and hexagonal outline separated by vertical cracks. Although the cracked layer is often no more than a few centimeters thick, deeper cracks, called *desiccation cracks*, can affect clay-rich sediment layers as much as one meter thick and produce a columnar jointing of the layer that persists after the rock has become lithified.

The volume expansion type of rock dilatation—or positive dilatation—produces an entirely different type of rock jointing. Called *sheeting structure*, the phenomenon consists of thick shells of rock that seem to "peel" away from a monolith exposed at the surface. Sheeting structure is most commonly seen in massive igneous rock, such as granite of a batholith or stock, but it also occurs in some massive metamorphic rocks and even in thick masses of clastic sedimentary rocks, such as sandstone or conglomerate. In the case of rocks that were deeply buried during their formation, the rock was originally under a condition of slightly reduced volume because of the high confining pressure. In the process of *unloading*, denudation uncovers the rock until it arrives at the surface. During unloading, volume expansion is spontaneous and causes one layer after the next to break free from the rock beneath. Rock shells from 5 to 10 m thick are typical and can be seen in the walls of granite and marble quarries (Figure 6.22). As commercial quarrying takes place, the rock rifts loose in great slabs, often with explosive violence. If a vertical saw cut is made in the rock, the cut immediately narrows and rock expansion continues slowly for many days thereafter. In deep rock tunnels, spontaneous rock expansion causes rock shells to break free from the sides and ceiling of the tunnel. Called "popping rock" by miners, the phenomenon is a hazard to humans and has caused many serious injuries. In Chapter 14, we describe some landscape features resulting from spontaneous rock expansion accompanying denudation.

By far the most widespread forms of jointing in rocks are extension fractures caused by tectonic activity. These are well displayed in massive layers of horizontal sedimentary strata, particularly the brittle clastic rocks such as sandstones and siltstones which typically yield by fracturing (Figure 6.23). The fracture planes are commonly almost perpendicular in attitude and form in sets of parallel planes. The total joint system comprises two such sets of intersecting vertical joints. A third system of planes of weakness is provided by the horizontal bedding planes. As a result, the rock comes apart in rectangular *joint blocks*. Tectonic activity that has produced such joint sets in sedimentary strata is believed to be a broad warping of the crust, exerting a buckling action on the platelike sedimentary layers. This same tectonic action affects underlying igneous or metamorphic rocks on which the strata rest, but in those rock types bedding planes are lacking and the blocklike nature of the joints is not so obvious.

FIGURE 6.24 Synclines and anticlines.

FIGURE 6.25 Folds in strata of Cambrian age, near Sullivan River, Rocky Mountains of British Columbia, Canada. (Geological Survey of Canada, Photo No. 154571.)

FIGURE 6.26 Plunging folds.

FIGURE 6.27 Block diagram of plunging anticlines of the Jura Mountains of Switzerland and France. (Redrawn, by permission of the publisher, from A. N. Strahler, *Physical Geography*, 3rd ed., Figure 33.7. Copyright © 1969 by John Wiley & Sons, Inc.)

FIGURE 6.24

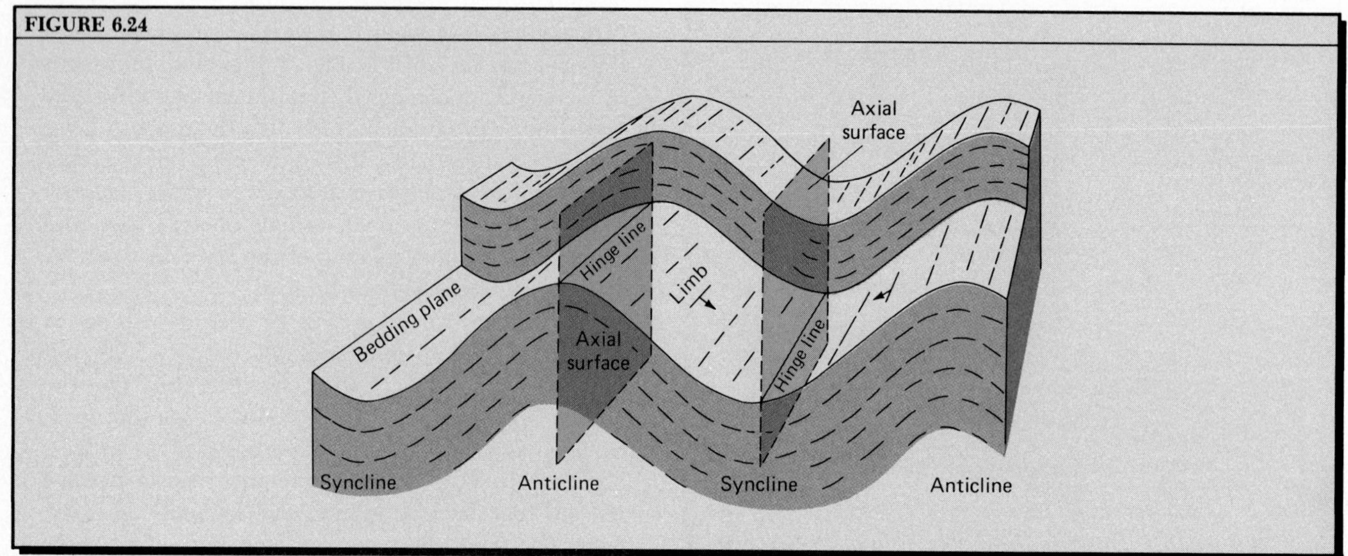

FIGURE 6.25

Folds in Strata

Horizontal compression acting upon flat-lying sedimentary strata produces wavelike undulations called *folds,* consisting of alternate synclines and anticlines. A *syncline* is a downfold (Figure 6.24). The root *cline* comes from the Greek word meaning "to lean" or "incline"; the prefix *syn* means "together" or "toward." Thus, the term syncline describes the manner in which the strata on opposite sides of the downfold dip inward toward the center line. An *anticline* is an archlike upfold in which the strata dip away from the center line, or fold crest (Figure 6.25). The prefix *anti* means "away from."

For either an anticline or a syncline, the *hinge line* is the line that connects points of maximum curvature on a single bedding plane along the fold crest or trough. An imagined surface that passes through the hinge lines of all bedding planes in a given anticline or syncline is called the *axial surface* (Figure 6.24); it divides a fold into two parts, called the *limbs.* It is obvious from Figure 6.24 that an anticline shares each of its limbs with the synclines next to it and that each syncline shares its two limbs with the adjacent anticlines. You might say that every anticline consists of two half-synclines, while every syncline consists of two half-anticlines.

In the simple case shown in Figure 6.24, the axial surfaces are vertical planes and divide each fold into exactly equal halves, one of which is the mirror image of the other. These idealized folds are, then, perfectly symmetrical.

FIGURE 6.26

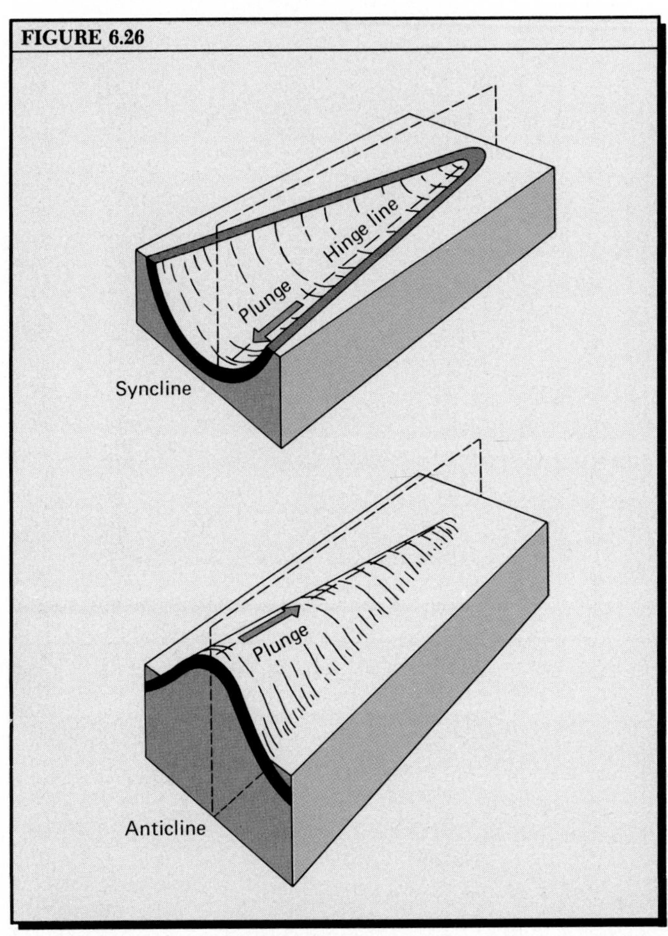

Plunging Folds

The hinge lines shown in Figure 6.24 are perfectly horizontal, but in nature hinge lines are rarely horizontal over long distances. Instead, the hinge line of an anticline or a syncline descends or rises in elevation as it is traced along its length. The descent of a hinge line is described as *plunge* and the fold itself is called a *plunging fold.* Figure 6.26 illustrates the geometry of plunging folds. A region that displays many plunging anticlines of limestone strata is the Jura Mountains of France and Switzerland (Figure 6.27).

FIGURE 6.27

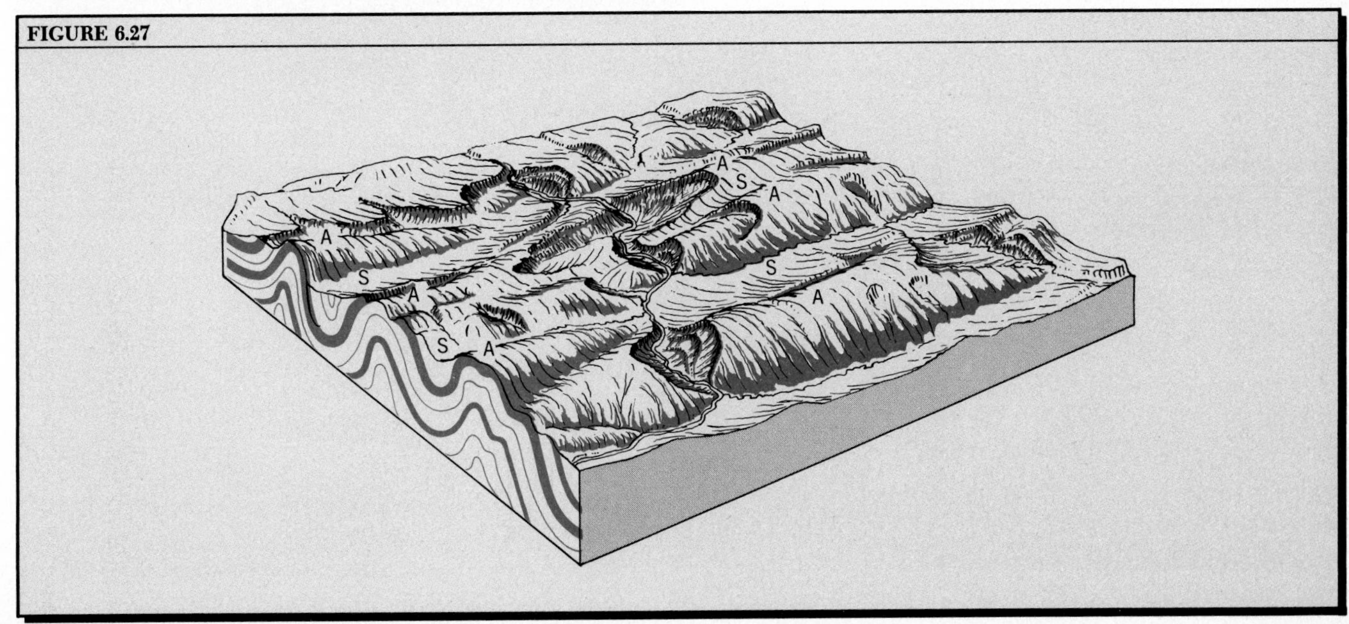

Figure 6.28 shows how individual beds or formations are exposed at the surface by erosion in a region of plunging folds. The block diagram above shows a plunging syncline and a plunging anticline. A vertical cross section of the folds appears on the front of the block; geologists call this a *structure section*. The surface exposure of the folded strata is shown in perspective on the upper surface of the block diagram. The land surface is idealized as a horizontal plane; it truncates the folds. We imagine that erosion has entirely removed the missing upper parts of the folds.

Below the block diagram in Figure 6.28 is a *geologic map;* lines on the map show the bedding planes that form the contacts between individual beds or formations in the sedimentary sequence. Each bed or formation can be assigned a different pattern or color. The beds are numbered according to age, No. 1 being the youngest and No. 10 the oldest. Dashed lines mark the

FIGURE 6.28

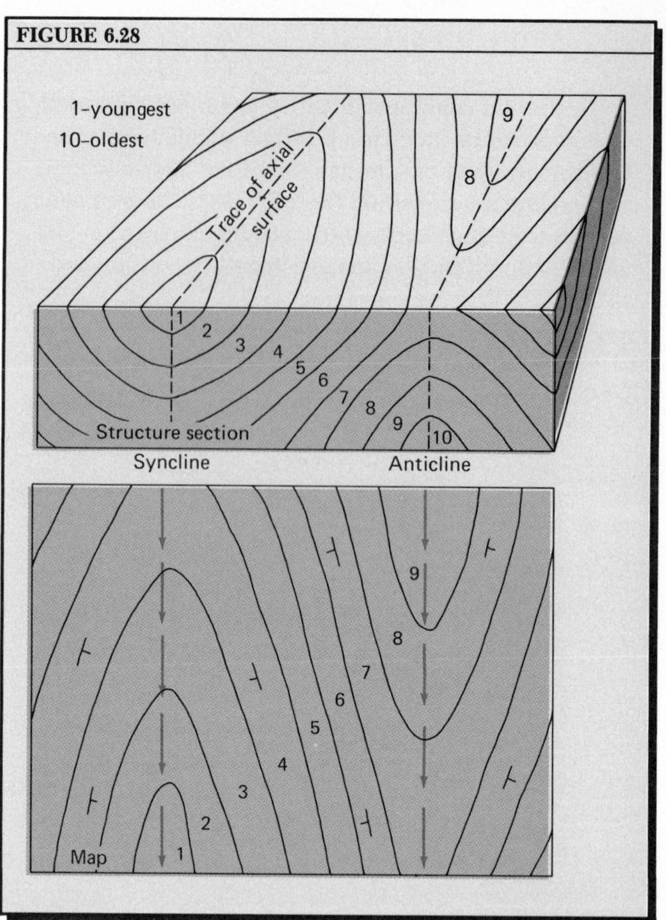

FIGURE 6.28 Idealized block diagram and map of plunging folds truncated by erosion.

FIGURE 6.29 Sheep Mountain in Wyoming is a plunging anticline, strongly eroded under a semiarid climate. The fold is asymmetrical, with strata of the nearer limb being much steeper in dip than those of the farther limb (*left*). The Bighorn River cuts across the fold axis, exposing arched strata in the walls of a narrow watergap. (John S. Shelton.)

FIGURE 6.29

traces of the axial surfaces. Arrows on the map show the direction of plunge, which is the same for both folds. The T-shaped marks are ***strike-and-dip symbols*** (see Figure 6.15); they are placed at various points on the limbs of the folds. As to scale of the map, the area shown might be 1 km or 10 km across, or it might represent an area only 1 m across. Folds come in all sizes, from minute crenulations a millimeter or less in amplitude to great folds several kilometers in amplitude.

An important principle illustrated by our block diagram and map is this: In a surface exposure of truncated folds, the youngest strata will be found along the trace of the axial surface of a syncline, while the oldest strata will be found along the trace of the axial surface of an anticline. This statement applies whether the folds are plunging or not. For example, in the case of the plunging folds shown in Figure 6.28, the age of exposed strata changes along the trace of the axial surface in the direction of plunge. For the syncline, progressively younger strata are encountered as we follow the trace down the direction of plunge. For the anticline, progressively older strata are encountered as we follow the trace opposite to the direction of plunge. Figure 6.29 shows a surface exposure of a plunging anticline.

Inclined and Overturned Folds

The axial surface of a fold is rarely a true vertical plane, so that perfect *vertical folds*, shown in Figures 6.28 and 6.30*A*, are rarely found in nature. In *inclined folds*, the axial surfaces are dipping (Diagram *B* of Figure 6.30). Where the axial surface assumes a low angle of dip, the steeper limb of the fold becomes inclined beyond a vertical attitude (Diagram *C*). In this case the limb is described as overturned, and we are dealing with *overturned folds* (Figure 6.31). Where folds have been overturned to the point that the axial surfaces are in a horizontal attitude, we have *recumbent folds* (Diagram *D* and Figure 6.32).

Folds vary a great deal with respect to the degree to which the limbs are compressed. Folds such as those

FIGURE 6.30 Kinds of folds, seen in cross section.

FIGURE 6.31 Overturned folds, Glacier National Park, Montana. A narrow ravine carrying meltwater marks the trace of the inclined axial surface of a syncline. (Douglas Johnson.)

FIGURE 6.32 Recumbent folds in slate of Ordovician age, Bay d'Espoir, Newfoundland. (Geological Survey of Canada, Photo No. 154575.)

FIGURE 6.30

A. Vertical folds

B. Inclined folds

C. Overturned folds

D. Recumbent folds

(Vertical) (Recumbent)

E. Isoclinal folds

F. Monocline

FIGURE 6.31

FIGURE 6.32

shown in Figures 6.24, 6.26, and 6.28 are described as *open folds* because the opposed limbs form a large angle. With increased compression, the limbs are brought together into a more nearly parallel arrangement described as *tight folding*. Where the limbs have been brought into a true parallel relationship, *isoclinal folds* result (Diagram *E* of Figure 6.30).

To add to our assortment of fold types, we recognize a *monocline*, which is a single fold limb, or flexure in a series of otherwise horizontal beds (Diagram *F* of Figure 6.30). Monoclines are often associated with vertical movements of crustal blocks, rather than with horizontal compression.

*Deformation Within Folds

Suppose we attempt to draw an idealized cross section through some imaginary folds. First, we draw a single wavy line to represent a single bedding plane. To show several beds, we need more wavy lines—some above the original line and some below. Now, one of two graphic forms will emerge. As shown in Diagram *A* of Figure 6.33, successive lines representing bedding planes will maintain a constant distance of separation, but the wave form will be forced to change both upward and downward from the initial line. Folds of this kind are called *concentric folds*. As shown in Diagram *B*, the initial wave form is exactly duplicated as each line is drawn. The finished drawing gives the appearance that the beds are thinner on the limbs and thicker on the hinge regions. These folds are known as *similar folds*. Figure 6.34 is a detailed drawing of similar folds. When the thickness of beds is measured along any vertical line, *a–b*, the thickness is the same on the limbs as on the hinges. Measured perpendicular to the bedding lines, however, the thickness has been reduced on the limbs and increased at the hinges. Thus, line *c–d* is much shorter than line *a–b*. From this observation, we conclude that material has been moved by plastic flow from the region of the limbs into the region of the hinges, as suggested by the small arrows.

The mechanical processes involved in the folding of strata are actually rather complex in detail. Looking at an exposure of open similar folds from a distance of several hundred meters, you might suppose that the folding of the strata was accomplished entirely by plastic flow. Under a great load of overlying rock one would expect plastic flow to have occurred, and this would be particularly true of limestone beds. Calcite, the common mineral in limestone, is capable of deforming in a plastic manner by the internal gliding of layers of ions along cleavage planes within the mineral crystals. Shale commonly appears to have undergone plastic flowage in which the platelike clay minerals have glided readily past one another. Shale layers usually show marked thickening or thinning where they lie between brittle rock layers, as shown in Figure 6.35. A quartz sandstone bed between shale beds would probably show numerous extension fractures (joints) but no observable thickening or thinning. Opening of these joints with slight shear movements between joint blocks accommodates the total deformation of the brittle bed. Beds that deform easily by plastic flow are referred to as *incompetent beds*, whereas brittle beds that fracture into joint blocks and fault blocks are *competent beds*.

FIGURE 6.33 Concentric folds and similar folds.

FIGURE 6.34 A symmetrical fold showing thinning of strata on the fold limbs and thickening on the fold hinges. The arrows show the movement of rock by plastic flow.

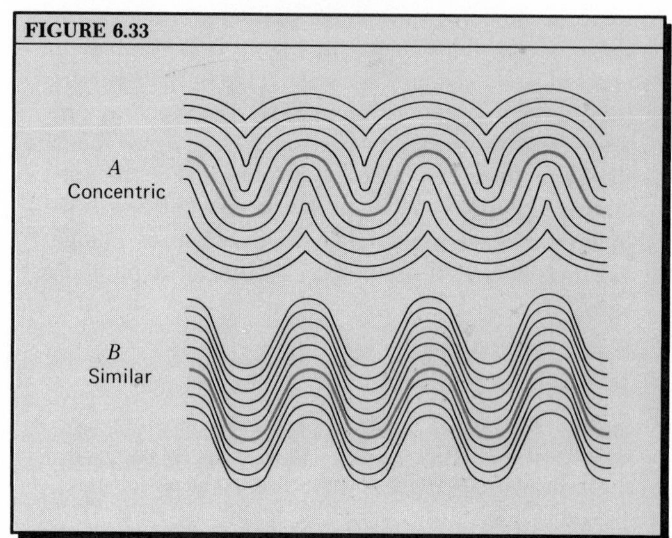

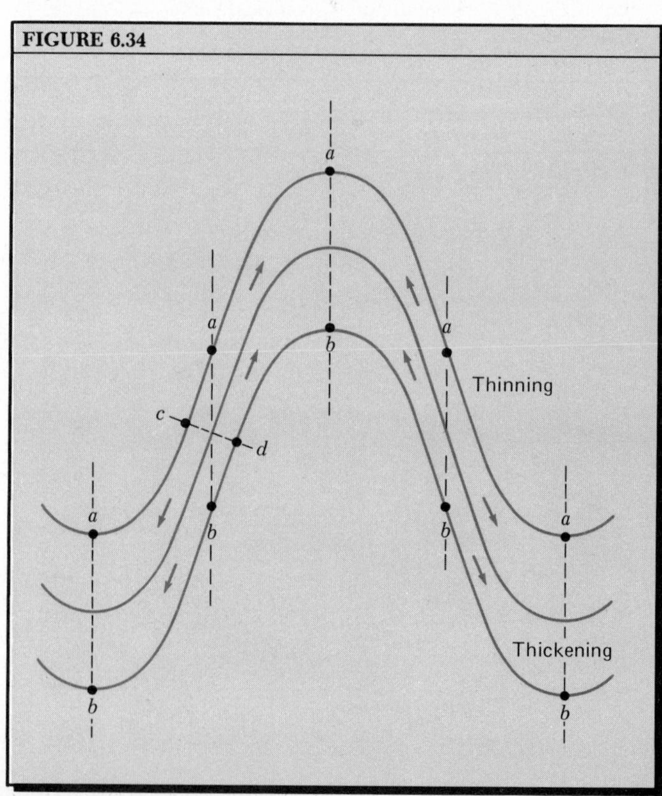

Where a shale bed is folded between competent beds of sandstone, one of two kinds of secondary structure usually develops (Figure 6.35). The shale on the limbs may be crumpled into a large number of very small *parasitic folds*. The axial surfaces of the parasitic folds are more or less parallel with the axial surface of the large fold. In other cases, the incompetent bed develops *axial planar cleavage*, consisting of discrete shear fractures that run diagonally across the bed. The cleavage planes are more or less parallel with the axial surface of the large fold. Because the orientation of these features helps in reconstructing the larger fold outlines, both parasitic folds and axial planar cleavage are useful to the geologist, who can see only a few small outcrops at the surface.

FIGURE 6.35 Parasitic folds and axial planar cleavage in incompetent beds on the limbs of a fold.

FIGURE 6.36 A circular stratigraphic dome with a nearly concentric system of ridges and valleys. Called the Richat Dome, it is located in Mauritania, West Africa. (Gemini IV photograph, NASA.)

FIGURE 6.35

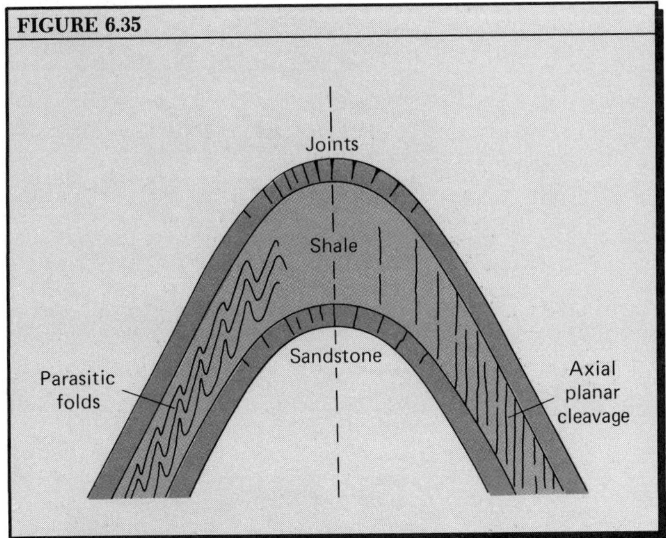

FIGURE 6.36

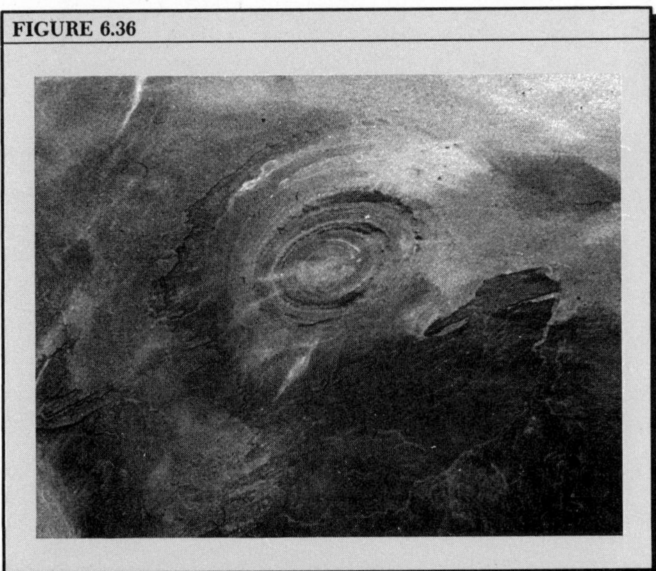

Stratigraphic Domes

Yet another type of fold deformation seen in sedimentary strata is the *stratigraphic dome*, a circular or elliptical uparching of strata (Figure 6.36). In many cases, doming is a form of tectonic activity that also affects underlying igneous or metamorphic rock on which the strata were deposited (Figure 6.37*A*). This type of dome is not caused by igneous intrusion beneath the sediments (Chapter 4). Evidently, the doming of the crust originates through forces acting deep in the lithosphere. We will discuss the possible origin of these tectonic domes in Chapter 12. Figure 6.37*B* shows the cross section of a stratigraphic dome over a laccolith; it is called a *laccolithic dome* to distinguish it from the stratigraphic dome that is of tectonic origin.

A special type of stratigraphic dome is associated with a column of rock salt (halite) that forces its way upward through sedimentary strata. This feature, called a salt dome, is described in detail in Chapter 21, in connection with important accumulations of petroleum trapped in the flanks of the dome.

Overthrusting and Alpine Structure

Finally, in our description of the ways in which sedimentary strata are deformed by tectonic processes, we come to complex combinations of folds and overthrust

FIGURE 6.37 Two types of stratigraphic domes. *A*. Tectonic doming affects the entire basement beneath the strata. *B*. Laccolithic doming is caused by igneous intrusion between strata.

FIGURE 6.37

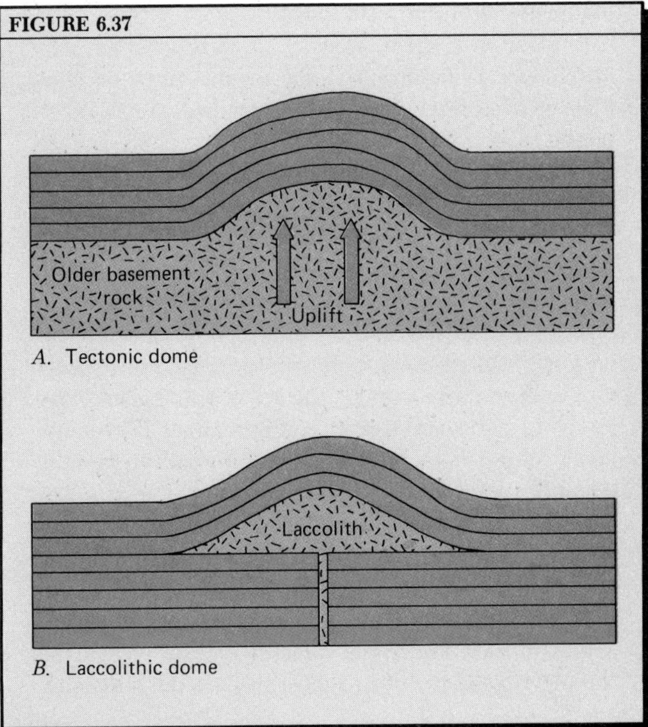

faults. These represent a severe form of tectonic activity that occurs in long, narrow belts of active mountain-making (Chapter 12). Because this activity has occurred in comparatively recent geologic time in the European Alps, we refer to it as *alpine structure*. European geologists have studied the Alps intensively for many decades. Gradually, they have unraveled some remarkably complex contortions of strata.

Under strong compression, strata undergo folding in a form–sequence seen in Figure 6.38. First, an anticline is formed. The axial surface of the fold is then rotated forward and results in an overturned anticline, which then flattens down upon itself to become a recumbent fold. Notice that the strata of the underside of the recumbent fold are upside down as compared with the way they were deposited. At this point, an *overthrust fault* may develop by rupture at the base of the recumbent fold, after which the upper portion of the fold is carried along as a contorted *thrust sheet* (Figure 6.39). A thrust sheet may have traveled forward for several tens of kilometers (Figure 6.40).

In the European Alps, highly contorted overthrust sheets are known as *nappes* (from the French word meaning "cover sheet" or "tablecloth"). Figure 6.41 is a cross section of complex nappe structure found in the Alps. One nappe lies above another, and all have been subjected to folding as a group. The first nappe to form lies at the base of the pile. Successive nappes formed one above the next. Today, these structures are exposed in the steep sides of alpine peaks and deep glacial troughs. Much of the upper part of the total structure has been removed by erosion. To a great degree, strata that have been subjected to this form of intense folding and overthrusting have also been altered to metamorphic rocks. (Metamorphism is the main subject of our next chapter.)

Looking at a great series of nappes, whether they are viewed in nature in alpine mountains or on true-scale cross sections, it is hard to visualize an individual nappe as a strong rock layer capable of being pushed uphill or even horizontally by compressive stresses acting in the crust. Folding within a nappe suggests that the rock was weak and behaved as a plastic solid. We might liken a nappe to a layer of soft pastry dough rolled out on a pastry board. Even if the dough is separated from the table by a layer of flour, so that it does not stick, it is impossible to force the entire sheet of dough to move over the board as a unit merely by applying horizontal pressure at one edge. The dough layer simply does not have the internal strength to transmit the horizontal stress.

Reasoning on these lines, geologists suggested an alternative hypothesis to explain nappes. Suppose, in the case of the sheet of dough, we tilt the pastry board. When the tilt becomes sufficiently steep, the dough begins to slide under the force of gravity. When the leading edge of the dough sheet reaches the horizontal

FIGURE 6.38 Development of overturned and recumbent folds and an overthrust sheet or nappe. (Suggested by drawings of A. Heim, 1922, *Geologie der Schweiz*, vol. II-1, Tauschnitz, Leipzig.)

FIGURE 6.39 Idealized diagram of a thrust sheet that has moved over a low-angle overthrust fault. The leading edge of the sheet has advanced over alluvial materials eroded from the surface of the sheet. (Drawn by A. N. Strahler.)

FIGURE 6.38

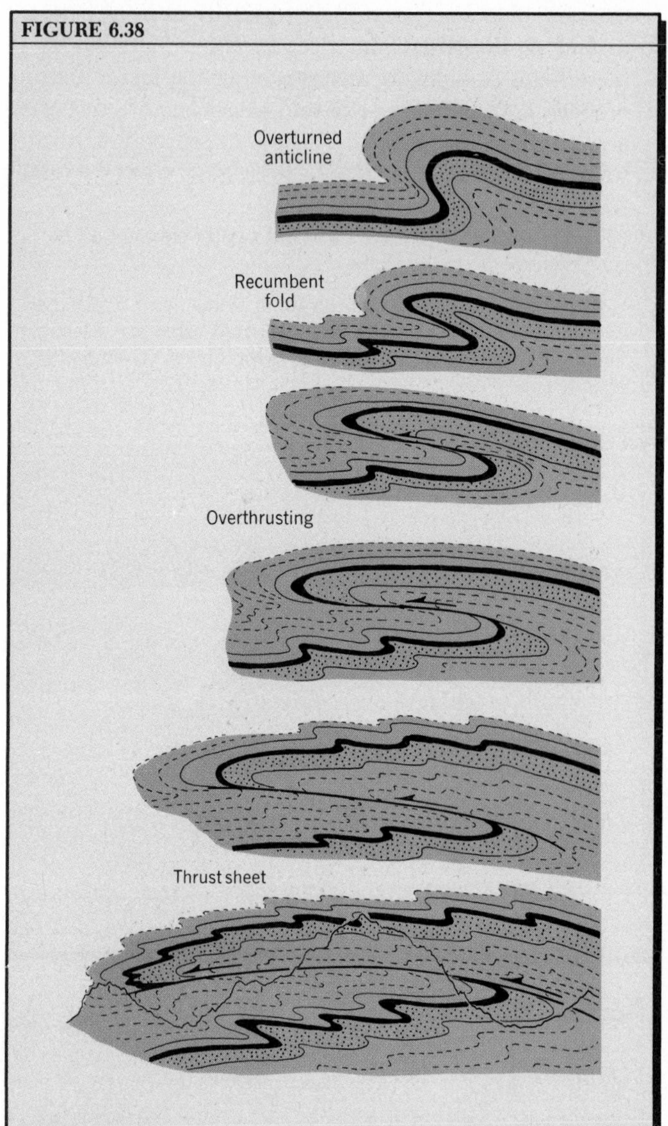

FIGURE 6.39

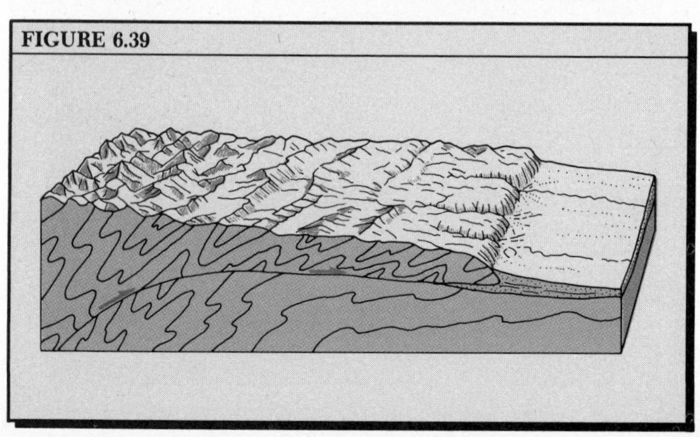

table top, it begins to buckle into folds, which pile up behind the leading edge. Our model illustrates the hypothesis of *gravity gliding* of overthrust sheets and nappes, proposed in the 1880s by E. Reyer, a geologist familiar with the structure of the Alps. It postulates sufficient crustal uplift occurring along a mountain axis to generate the required gravitational force parallel with the base of a series of strata.

Gravity gliding may explain not only nappes, but also certain other types of folding and overthrust faulting. Figure 6.42 suggests these types. Diagram *A* shows sedimentary strata becoming deformed into open folds during gliding on a basal thrust surface, or *sole*. This mechanism has been proposed to explain the Jura anticlines pictured in Figure 6.27. Diagram *B* shows thrusting affecting strata lying above a major thrust surface. Thrust slices pile up one behind the other in a pattern described as imbricate, an overlapping arrangement of plates, as in the shingles on a roof. *Imbricate thrusts,* such as those shown in Diagram *B*, are illustrated in Figure 12.16. Diagram *C* shows one interpretation of a succession of nappes. They are produced by the collision of two lithospheric plates which squeeze a great thickness of strata between them. The deformed sediments rise in a great arch, then glide under gravity down the continental grade to lower levels. The process is repeated until several nappes have formed, each one gliding downhill over the earlier ones. In Chapter 12, we shall relate these models to plate tectonics.

The Principle of Uniformity in Geology

In building a strong scientific base for geology through mineralogy, petrology, stratigraphy, and structural geology—all basic topics we have covered in these early chapters—geologists have been guided by a sound principle. The *principle of uniformity* maintains that geologic processes in action today—volcanic eruptions, faulting and earthquakes, stream floods, breaking surf, and the like—can account for nearly all forms of rock bodies and rock structures that make up the entire geologic record. The principle is a modern statement of an early doctrine known as *uniformi-*

FIGURE 6.40 The Lewis overthrust fault is marked by the light-colored slanting line about midway up the mountainside (arrow). The rock mass above the thrust moved from left to right for a distance of 24 km. Northern Rocky Mountains, Montana. (Douglas Johnson.)

FIGURE 6.41 Structure section through a portion of the Helvetian Alps, Switzerland, showing nappes. Horizontal and vertical scales are the same. Nappes are labeled *A, B, C,* and *D*. (Simplified from A. Heim, 1922, *Geologie der Schweiz,* vol. II-1, Tauschnitz, Leipzig.)

FIGURE 6.40

FIGURE 6.41

tarianism, set forth in the late 1700s and early 1800s. Two natives of Scotland—James Hall, mentioned earlier in this chapter, and his close associate, James Hutton—were largely responsible for introducing uniformitarianism to geology. Together, these two men investigated the folded strata of the Scottish coast and discovered a major unconformity.

During the late 1700s, geologic thought was strongly influenced by other scientists who explained rock deformation and fossils of extinct animals as the result of one or more great catastrophes that had affected the earth's crust. In one popular version of early history, the supposed catastrophic events were viewed as part of the six-day creation of the earth, believed to have occurred in the year 4004 B.C. But James Hutton's field observations convinced him that it was impossible to explain accumulations of fossil-bearing strata, unconformities, and intrusions of granite in a single catastrophic event lasting only a few days. Vast spans of time were needed for such features to be formed, he reasoned.

Hutton first published the results of his field studies and their interpretation in 1795 in a work entitled *Theory of the Earth.* The doctrine of uniformitarianism defended by Hutton, Hall, and some of their close associates encountered great opposition from leading geologists, who visualized catastrophic upheavals of the earth through the action of supernatural forces. Fortunately, support for the principle of uniformity came from another Scottish geologist, Sir Charles Lyell, a successful textbook writer whose three-volume work, *Principles of Geology,* went through twelve editions. Lyell's work came to dominate English geology in the mid-1800s, and stratigraphy became firmly established along lines we have explained in this chapter.

Looked upon today in the context of modern science, the principle of uniformity means simply that all phenomena of geology must be explained through the laws of science we accept as valid. The fundamental laws of physics, chemistry, and biology must have applied uniformly from early Precambrian time to the present. No explanation of an event that occurred in the Cambrian Period or the Triassic Period can be considered acceptable if it violates a scientific law we apply today. Supernatural forces are ruled out of the physical explanations of deposition and deformation of strata.

The principle of uniformity does not, however, require that the intensity of each geologic process shall have been uniform throughout all time. To the contrary, we accept as a valid conclusion that deformation of strata and intrusion of plutons may have been much more intense at one point in time than another. This concept is applied in the interpretation of orogenies as spasmodic events of fairly short duration when compared with longer spans of sediment deposition under stable crustal conditions. There are some good reasons to suppose that the overall total intensity of igneous and tectonic activity has been declining since early Precambrian time. The total release of heat energy from radioactive elements within crustal rocks has almost certainly been steadily decreasing (as we explain in Chapter 7). Nevertheless, we believe that the physical laws of radioactivity have not changed since earliest geologic time, for to think otherwise would violate the principle of uniformity.

Stratigraphy and Structural Geology

This chapter has given a detailed picture of the deposition of strata through geologic time and the ways in which strata are deformed by tectonic activity. Here is a point to ponder: Nearly all the facts and principles covered in this chapter were established before 1900, and most (with the exception of the laboratory study of rock strain under confining pressure) were known well over a century ago. Even the rather daring proposal of gravity gliding dates back almost a full century.

But it would be a serious mistake to judge this material as obsolete. That it has stood the test of time and remains part of the basic fabric of geology is a tribute to the pioneer geologists who made their field observations accurately and in detail.

On the other hand, the detailed observations on

FIGURE 6.42 Tectonic features produced by gravity gliding. *A.* Folds, such as those of the Jura Mountains, *B.* Imbricate thrust sheets. *C.* Nappe structure formed during continental collision.

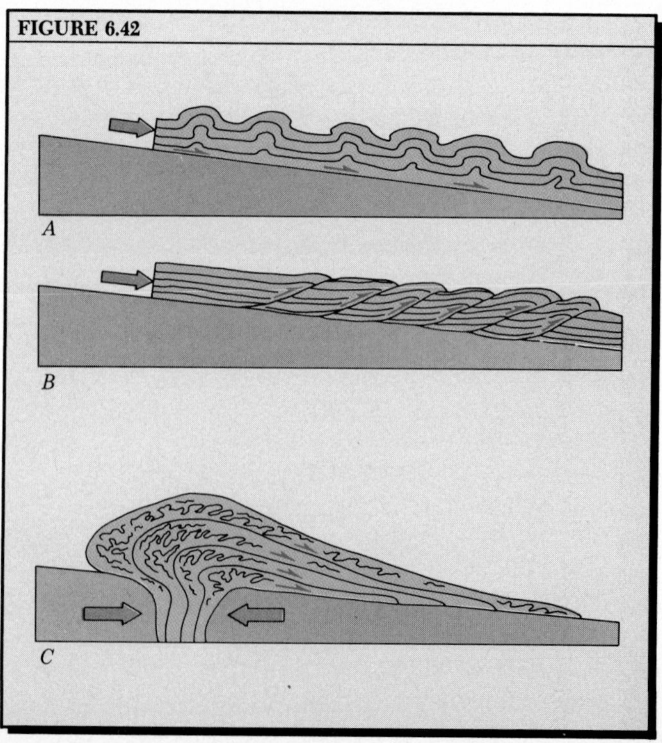

stratigraphy and structural geology contained in this chapter remained largely unexplained until long after 1900. The nature of the forces that cause epeirogeny and orogeny did not become clear until the theory of plate tectonics was assembled in working form in the 1960s. The reason for this long delay in development of a unified theory of geology is obvious. The study of stratigraphy and structural geology was carried out entirely on the continents, whereas the clues to what forces caused crustal evolution lay concealed in the floors of the oceans. Only when physics became a tool of geology, through geophysics, could the ocean basins be explored.

We have laid a good foundation in this and our earlier chapter on igneous rocks for an investigation of the third great class of rocks—the metamorphic types. Deformation during orogeny is one of the mechanisms needed to produce enormous masses of metamorphic rocks. Both igneous and sedimentary rocks are involved in metamorphism, which is one of the major processes of change within the rock cycle. Thus, the next chapter will complete our study of the rock cycle.

Key Facts and Concepts

Stratigraphy *Stratigraphy* deals with strata in terms of their relative ages, distribution, and their interpretation in terms of environments of deposition. Groups of strata of similar lithologic types are designated as *formations*.

Stratigraphic correlation *Stratigraphic correlation*, the matching of strata according to similar *relative* age, makes use of the *principle of superposition*, younger strata being always deposited above older strata. Correlation relies on *fossils* through the principle that *organic evolution* has produced changes in life forms over time. *Index fossils* are useful in correlation, while *faunas* are particularly reliable through the principle of *faunal succession*.

Geologic time A *table of geologic time*, established by stratigraphic correlation, consists of the record of long periods of sedimentary deposition punctuated by brief episodes of *orogeny*. Largest of the time units is the *eon*, of which there are two: *Cryptozoic* and *Phanerozoic*. Phanerozoic time consists of three *geologic eras: Paleozoic, Mesozoic, Cenozoic*. Within each era are *geologic periods*, subdivided into *epochs*.

Unconformities *Unconformities*, representing important time gaps in the deposition of strata, take three forms. *Disconformity* is a time gap between two sets of *conformable strata;* it represents an important erosion period involving only *epeirogenic movements (epeirogeny)*. *Nonconformity* separates strata from eroded masses of older igneous or metamorphic rock. *Angular unconformity* separates eroded strata deformed by orogeny from overlying younger strata. Together with other kinds of *crosscutting relationships*, unconformities are used in determining relative ages of rock units.

Environments of sediment deposition Major environments of sediment deposition are *marine, terrestrial*, and *littoral*. Sediments of each environment are characterized by a distinctive *facies*. Facies may involve physical and lithologic properties (*lithofacies*) or faunal properties (*biofacies*). *Faunal zones* represent time lines (*isochrones*), whereas lithofacies often cut across time lines, indicating *transgression* or *regression*.

Rock deformation *Rock deformation* includes changes in shape or form that affect rocks. Deformation consists of *strain* occurring in response to applied *stress*.

Stress Stress acting within rock may be *compressional stress*, directed along a *compression axis*, or *tensional stress*, acting along a *tension axis*. *Shear stress* consists of two parallel stresses acting in opposite directions.

Strain in rock Rock that is a *brittle solid* behaves as an *elastic solid*, deforming by *elastic strain* up to an *elastic limit*, above which *rupture* occurs. Rock behaving as a *ductile solid* under high confining pressure exhibits *ductility* and responds to deforming stress by *plastic flow*. *Dilatation* is a form of strain involving only an increase or decrease in volume.

Fracturing in rock Yielding of rock to shear stresses may result in *shear fracture* with motion on a *shear plane*. Shear fracture may take the form of a *fault* with motion (*faulting*) on a *fault plane*. Shearing on innumerable minute planes constitutes plastic flow, a process active in the asthenosphere. *Elastico-viscous behavior* characterizes rock of the asthenosphere. Deformation under tensional stress takes the form of *extension fracture*, with pulling apart on a widening crack. *Joints* in rock are an expression of extension fracture in the process of *jointing*.

Kinds of faults A *fault plane* is described in terms of its *dip* and *strike* and the relationships between *hanging wall* and *foot wall*. Three classes of faults are *dip-slip, strike-slip*, and *oblique-slip*. The fault classes are expressed as *normal, reverse, transcurrent*, and *oblique-slip* faults. Faults are expressed at the surface by the *fault scarp*, the *horst*, and the *graben*.

Jointing Dilatation leads to the formation of columnar jointing, sediment cracks (*desiccation cracks, mud cracks*), and *sheeting structure* caused by *unloading*. Joints form in sets and systems, producing *joint blocks*.

Folds in strata *Folds*, wavelike undulations in strata, take the form of *synclines* and *anticlines*. A fold is described by its *hinge line, axial surface*, and *limbs*.

Plunging folds result from the descent or ascent of the hinge line. Folds may be *vertical, inclined, overturned, recumbent, open, tight,* or *monoclinal* depending on the attitude of the axial surface and degree of compression of the limbs. *Concentric folds* are distinguished from *similar folds.* Minor folds and cleavage occur between competent beds.

Stratigraphic domes The *stratigraphic dome,* a circular uplift, includes the *laccolithic dome* caused by igneous intrusion.

Alpine structure Intense folding and overthrusting produces *alpine structure,* involving *overthrust faults, thrust sheets, nappes,* and *imbricate thrusts,* with possible *gravity gliding.*

Uniformitarianism The *principle of uniformity* in geology, based on the early doctrine of *uniformitarianism,* requires that all phenomena of geology be explained through laws of science accepted as valid today.

Questions and Problems

1. Suppose that you have arrived at a large outcrop of sedimentary strata that are all perfectly parallel and nearly vertical. The strata consist of rather thin beds of sandstone, shale, and limestone. Some of the beds contain numerous fossils of marine shellfish, resembling modern clams or oysters. How would you go about deciding which of the two surfaces of a particular bed was originally the upper surface? Think of various kinds of features, markings, and structures that might give clues to this question. Use your knowledge of features you have observed at the beach at low tide.
2. Stratigraphic correlation has been built upon the principle of organic evolution, essentially in the same form it was explained by Charles Darwin over a century ago. Although organic evolution is now accepted by nearly all paleontologists and biologists, some persons have been taught to believe otherwise. What is your personal view on this issue? What alternative theory could explain the differences in faunas observed from one formation to the next, as for example, in the sequence of strata seen exposed in the Grand Canyon?
3. Imagine that you have found in the Inner Gorge of Grand Canyon a good exposure of the nonconformity between the Precambrian rocks (schist, diorite, granite) and the overlying Cambrian Tapeats sandstone. If the Tapeats sandstone were deposited on an erosion surface, what evidence might you expect to find in proof of that interpretation? (Refer to the subject of rock weathering, Chapter 5.)
4. How might you be able to distinguish in a sedimentary formation the lithofacies of a large river delta from the lithofacies of a thick deposit of dune sand that formed in an interior desert region? List some specific features of the sediment that is deposited on a delta and make a similar list for the features of numerous sand dunes, one built upon the next.
5. Rock deformation by plastic flowage takes place in a deep environment of high pressure. How might you demonstrate plastic flowage by using some commonplace materials found in the household or in industry? Will a single substance provide a successful model of both folding (compressional stress) and jointing (extensional stress)?
6. Gravity gliding to produce overturned folds and nappes is sometimes vividly demonstrated by the deformation affecting a layer of snow that has fallen on a sloping roof or upon the windshield or rear window of an automobile. This spontaneous folding occurs only under a set of favorable conditions. Analyze the physical conditions that must be present to cause the snow layer to form folds. Can you reproduce the same kind of folding by pushing a shovel into a snow layer on a flat pavement? What physical conditions are different in the latter case?

7

CHAPTER SEVEN

Metamorphic Rocks and the Continental Crust

Rock metamorphism consists of physical and chemical changes in rock in an environment of high temperature, high confining pressure, or intense shearing action, or some combination of two or three of those factors, but without melting. The definition usually excludes chemical changes caused by hydrothermal solutions penetrating rock at shallow depths where pressures are very low. Generally speaking, rock metamorphism occurs under confining pressures of at least 2 kilobars (kb).

Rock metamorphism can result in the formation of new minerals, new rock textures, new rock structures, or a combination of such changes. The formation of new minerals by recrystallization of preexisting minerals is usually the most distinctive and important aspect of metamorphism that affects large masses of rock over wide areas. Crystal lattices are broken down and re-created, using different combinations of the same ions that were present in the earlier minerals. Another aspect of rock metamorphism is the importation of ions and atoms of rock-forming minerals from an outside source, or the export of various substances to an outside region (usually the overlying rock). Geologists refer to this import–export process as *metasomatism*. When it occurs, the rock undergoes corresponding changes in chemical composition. One important example of metasomatism during rock metamorphism is the loss of volatiles—particularly water and carbon dioxide.

Classification of the Metamorphic Rocks

The metamorphic rocks can be broadly grouped into two major classes: cataclastic rocks; recrystallized rocks (Table 7.1). The *cataclastic rocks* have experienced mechanical disruption (breaking, crushing) of the original minerals without appreciable chemical change. This process of change can be described as *dynamic metamorphism*. The *recrystallized rocks* have, as the name indicates, undergone a recrystallization of the original minerals. Recrystallization is considered to be a chemical change, because it usually produces minerals of chemical formulas and crystal lattice structures different from the parent minerals.

Within the class of recrystallized rocks we must distinguish two subclasses: contact metamorphic rocks; regional metamorphic rocks. The *contact metamorphic rocks* are formed by recrystallization under high temperature in country rock immediately adjacent to an intruding magma that solidifies into a pluton. The rock is not subjected to tectonic forces (bending and breaking) during the process of change, but new mineral substances emanating from the magma can be added to the country rock (i.e., metasomatism can take place). The *regional metamorphic rocks* undergo recrystallization during the process of being deformed by shearing, often under conditions of high pressure or high temperature, or both, and often to the accompa-

Table 7.1 The Metamorphic Rocks

Class	Process	Rock types
Cataclastic Rocks	Dynamic metamorphism: crushing and shearing at low temperatures.	Friction breccia Mylonite
Recrystallized Rocks		
Contact metamorphic rocks	Recrystallization and metasomatism at high temperature in aureoles adjacent to plutons.	Hornfels
Regional metamorphic rocks	Dynamothermal metamorphism: recrystallization, shearing, metasomatism.	Greenschist Amphibolite Pyroxene granulite Blueschist Eclogite
Metamorphic-Igneous Rocks (transitional)	Layer-by-layer igneous replacement: granitization, metasomatism.	Banded gneiss Injection gneiss Migmatites

niment of loss or gain of mineral components by metasomatism. The adjective "regional" refers to the occurrence of this subclass of rocks over large areas and in great crustal thicknesses. The total process is called *dynamothermal metamorphism*.

A third class of rocks, intermediate between metamorphic rocks and igneous rocks, is shown in Table 7.1. These are rocks characterized by a banded appearance on an exposed surface, caused by layering. The bands represent alternate layers of igneous rock and metamorphic rock. It is thought that the igneous material penetrated or replaced layers of a preexisting rock of sedimentary origin. Various degrees of transition from metamorphic to igneous rock can be included in this class.

Metamorphic Minerals

Recrystallization under high temperatures produces a distinctive group of metamorphic minerals. Most of these are different from the minerals we have thus far encountered in our study of igneous and sedimentary rocks. On the other hand, some of the same minerals found in the igneous rocks persist or reappear during

Table 7.2 Representative Metamorphic Minerals

Mineral name	Composition	Specific gravity
Kyanite	Aluminum silicate Al_2SiO_5	3.6
Andalusite	Aluminum silicate Al_2SiO_5	3.2
Sillimanite	Aluminum silicate Al_2SiO_5	3.2
Almandite (var. of garnet)	Aluminosilicate of iron $Fe_3Al_2Si_3O_{12}$	4.2
Wollastonite	Calcium silicate $CaSiO_3$	2.8–2.9
Staurolite	Hydrous aluminosilicate with iron $Fe_2Al_9O_6(SiO_4)_4(O,OH)_2$	3.7
Chlorite	Hydrous aluminosilicate with iron and magnesium $(Mg,Fe)_3(Si,Al)_4O_{10}(OH)_2\ (Mg,Fe)_3(OH)_6$	2.6–3.3
Epidote	Hydrous aluminosilicate with calcium and iron $Ca_2(Al,Fe)Al_2O(SiO_4)(Si_2O_7)(OH)$	3.2–3.5
Talc	Hydrous silicate of magnesium $Mg_3Si_4O_{10}(OH)_2$	2.7–2.8
Serpentine group (Chrysotile asbestos)	Hydrous silicate of magnesium $Mg_3Si_2O_5(OH)_4$	3–5
Graphite	Carbon C	3.1

recrystallization, such as quartz, biotite mica, pyroxene, amphibole, and feldspar. Calcite and dolomite also persist in recrystallized carbonate sedimentary rocks. Metamorphic minerals not encountered in earlier chapters are described in Table 7.2. The eleven minerals on this list have been selected because of their importance in the classification and naming of metamorphic rocks and to show a wide range of chemical diversity.

The first three minerals, *kyanite, andalusite,* and *sillimanite,* are of identical composition. All are aluminosilicates with the formula Al_2SiO_5, but each has a different space lattice structure. They are polymorphs of Al_2SiO_5. (Polymorphs are explained in Chapter 3.) They form by recrystallization of rock with abundant felsic components, such as quartz and feldspar. Figure 7.1 shows that each of these minerals forms under a different combination of pressure and temperature. Andalusite forms under conditions of comparatively low pressure and low temperature; kyanite at high pressure and low temperature; sillimanite at high temperature and moderate pressure (Figure 7.2). For this reason, the polymorphs can serve as indicators of the environment in which recrystallization took place.

Almandite is one of the *garnet group*—aluminosilicates of magnesium, iron, calcium, or manganese that crystallize in the isometric system. Almandite, the iron garnet, is a red mineral familiar as a semiprecious gemstone. Almandite crystals often grow to diameters of several centimeters in metamorphic rock (Figure 7.3).

Wollastonite, a silicate of calcium, is usually associated with the metamorphism of limestone. Under high temperatures the following reaction occurs:

$$CaCO_3 + SiO_2 \rightarrow CaSiO_3 + CO_2$$

The carbon dioxide is driven off as a volatile gas—an example of one of the forms of metasomatism.

Staurolite, a hydrous aluminosilicate of iron, is formed in a middle range of pressure and temperature. It is a particularly interesting mineral because of its habit of forming as twin crystals penetrating each other at right angles to make a natural cross (Figure 7.4).

Chlorite, a soft hydrous silicate of iron and magnesium, is similar in some ways to the micas, with sheet structure that forms thin cleavage flakes. Chlorite is associated with metamorphic conditions of comparatively low temperature and is one of the earliest-formed of the metamorphic minerals.

Epidote, a hydrous aluminosilicate of calcium and iron, is a green mineral that typically forms in elongate prisms. It is associated with the metamorphism of rocks rich in mafic minerals, such as pyroxene and

Table 7.2 Representative Metamorphic Minerals *Continued*

Hardness, Mohs scale	Crystal system and habit	Color	Occurrence in rocks
7	Triclinic system Bladed crystals	Blue, bluish gray, green, white	Blueschist
7½	Orthorhombic system Prismatic crystals, square cross sections	Gray, whitish	Greenschist Amphibolite
6–7	Orthorhombic system Long, slender crystals	Brown, green, white	Amphibolite Granulite
7	Isometric system Dodecahedrons, trapezohedrons	Deep red, brownish red	Amphibolite
5–5½	Triclinic system Fibrous or columnar	White or gray	Crystalline limestone (marble)
7–7½	Orthorhombic system Prisms, esp. crosslike twins	Brown	Amphibolite
2–2½	Monoclinic system Foliated, scaly	Green	Greenschist
6–7	Monoclinic system Elongate prisms	Green to black	Epidote amphibolite
1	Monoclinic system Foliated	Green, gray, white	Talc schist
2½	Monoclinic system	Green	Hydrothermal alteration
1–2	Hexagonal system	Dark gray, black, Metallic luster Gray streak, Sectile	Schist, gneiss, marble

amphibole, and it is formed in a middle range of temperatures.

Talc is a very soft, scaly mineral, a hydrous silicate of magnesium, and is classed as a clay mineral. Talc occurs in metamorphic rocks that are derived from mafic minerals such as olivine, pyroxene, and amphibole. In massive form as a rock, talc is known as *soapstone*. Because it is immune to action of acids, soapstone was once widely used for table tops in chemical laboratories. In powder form, talc has many industrial uses and is perhaps most familiar as talcum powder.

The *serpentine group* consists of two minerals that are hydrous magnesium silicates—antigorite and chrysotile. These minerals are derived from mafic minerals, particularly olivine, pyroxene, and amphibole. Chrysotile is well known as *asbestos*, formed of minute flexible crystals that resemble textile fibers (Figure 7.5), widely used in fire-resistant industrial products and as an insulator. It is now established that the minute fibers of asbestos cause irreversible lung diseases (asbestosis and lung cancer) in humans.

Graphite, comparatively minor among the metamorphic minerals, illustrates a substance probably derived from hydrocarbon compounds of organic origin that were present in sedimentary rock prior to its metamorphism. In some instances, seams of coal have been converted partly into graphite during metamorphism. (The structure of graphite is described in Chapter 3.)

Cataclastic Rocks

The cataclastic metamorphic rocks result from mechanical deformation without appreciable chemical change and recrystallization. They form under intense shearing stresses which cause grain fragmentation. Grains that consist of individual mineral crystals or groups of crystals are crushed and pulverized as they are rotated (turned over and over).

Cataclastic rocks are most commonly produced during the process of faulting. They form a thin layer along the fault plane where strong crushing and grinding action takes place. One product of this action is *friction breccia*, a rock consisting of rather large angular fragments in a matrix of small fragments (Figure 7.6). The largest fragments may measure a meter or

FIGURE 7.1 Kyanite, andalusite, and sillimanite form in three different ranges of pressure and temperature.

FIGURE 7.2 Bladed kyanite crystals (clear) and prismatic staurolite crystals (dark) in mica schist, St. Gothard, Switzerland. (Courtesy of C. S. Hurlbut, Jr., Geological Museum, Harvard University.)

FIGURE 7.1

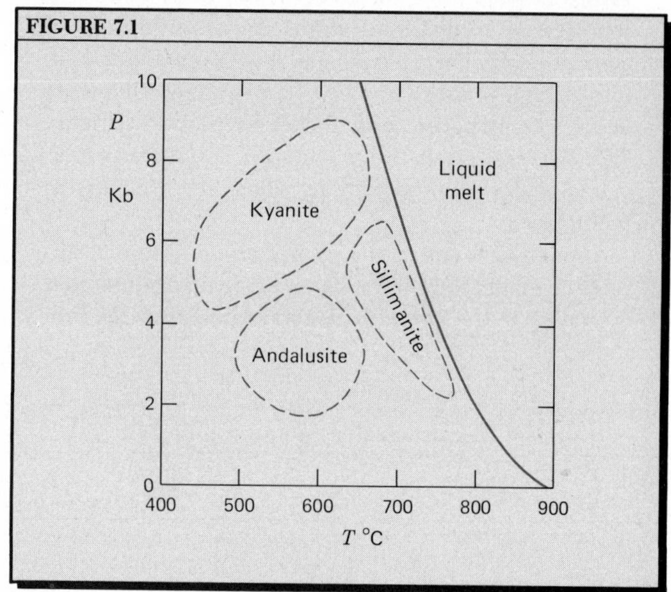

FIGURE 7.2

more across, while the smallest are a millimeter or smaller in diameter. Where the pulverized rock along a fault plane consists of very small particles—0.01 to 0.1 mm—the rock is called *mylonite*. It is a dense, fine-grained rock often with a streaked or banded appearance, and may outwardly resemble chert. (Chert is a sedimentary form of silica.) Some recrystallization may occur in mylonite, enabling the grains to adapt their shapes so as to fill the entire rock volume.

Contact Metamorphic Rocks

Contact metamorphic rocks form in an environment of high temperature in comparatively shallow crustal locations where confining pressure is not great. The high temperature is provided by intrusive magma, which has entered the country rock to solidify as a pluton, and thus a strong temperature contrast exists between the magma and the country rock. Shearing stress is practically absent under such conditions. The country rock close to the magma body is literally baked, like fired brick or tile, made of mud or clay. The resulting rock is *hornfels*, a word of German origin. The prefix *horn* refers to the hornlike appearance of the rock; the word *fels* means "rock." A direct English translation would be close to "hornstone." Hornfels has very fine-grained texture because cooling followed

FIGURE 7.3

FIGURE 7.3 Garnet crystals in schist. The larger crystal is about 2 cm in width. (A. N. Strahler.)

FIGURE 7.4

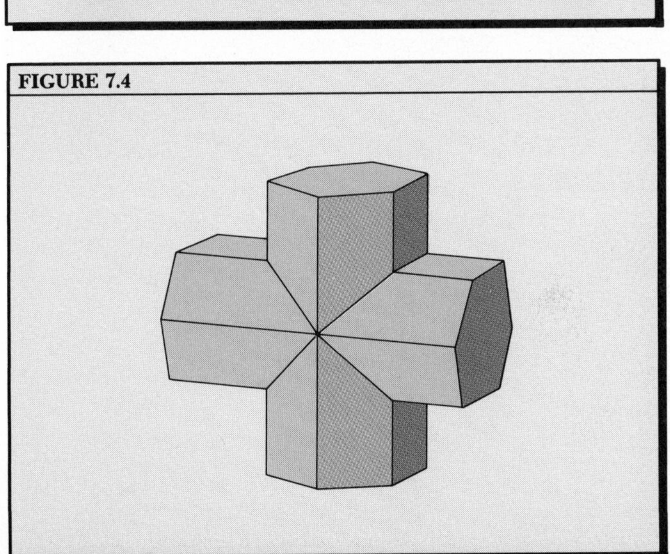

FIGURE 7.4 A crystal model of penetration twins of staurolite. The mineral name comes from the Greek word *stauros*, meaning "cross." A typical twin crystal would measure 2 to 3 cm in width. Called "fairy stones" in North Carolina and Georgia, the reddish-brown crosses are popular as amulets and charms. In Brittany, France, a local legend holds that the crosses were dropped from heaven.

FIGURE 7.5

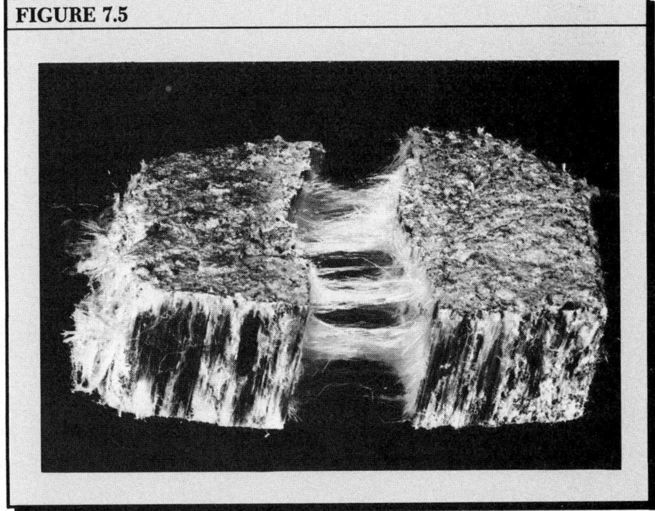

FIGURE 7.5 This specimen of asbestos is chrysotile, a fibrous form of the mineral serpentine. (Ward's Natural Science Establishment, Inc., Rochester, N.Y.)

FIGURE 7.6

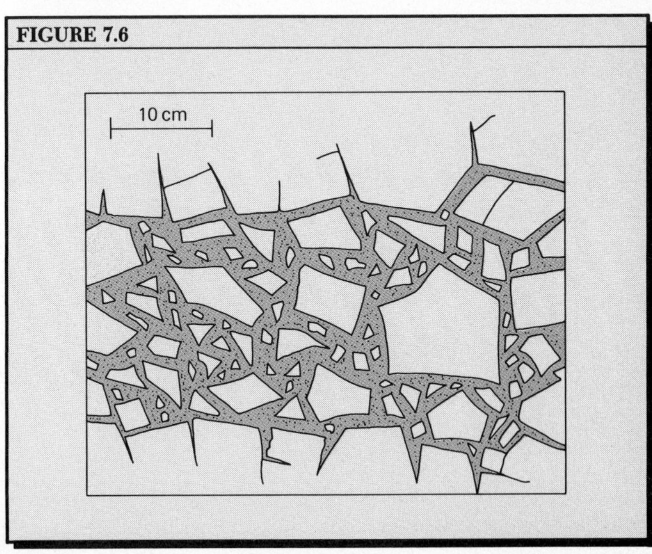

FIGURE 7.6 Schematic cross section of a friction breccia, such as one might find along a fault. The angular fragments are of the same composition as the rock on either side of the fault zone.

rapidly after recrystallization of the parent minerals.

Hornfels forms an *aureole*, which is a metamorphic layer surrounding the magma body (Figure 7.7). The aureole conforms with the outline of the pluton and may be subdivided into two or more zones of somewhat different mineral content. Where the magma is of granite composition, the aureole contains abundant quartz and potash feldspar. The inner zone may contain wollastonite and andalusite, while the outer zone may have amphibole and mica, which are hydrous minerals.

Texture of Regional Metamorphic Rocks

As regional metamorphism takes place, intense shearing action accompanied by recrystallization brings about new textures and structures. These features give the metamorphic rocks their distinctive appearances.

Consider first the structural changes that take place in a thick mass of black shale, a fine-grained sedimentary rock rich in clay minerals such as kaolinite and illite. Shale shows stratification (bedding) resulting from fluctuations in the particle size grades deposited from one layer to the next. Typically, regional metamorphism of shale results in crumpling of the beds

FIGURE 7.7 Schematic block diagram of contact metamorphic aureoles adjacent to plutons. At the upper left is a sill of gabbro intruded between shale layers, forming a narrow zone of hornfels. At the lower right is an aureole adjacent to a granite batholith intruding a mass of older igneous rock. (Drawn by A. N. Strahler.)

FIGURE 7.8 The development of slaty cleavage is associated with shearing within a mass of shale. *A*. The original bedding is horizontal. *B*. Compression causes close folding of the shale strata. The reference circle in diagram A is now distorted into an ellipse. *C*. Slaty cleavage forms in planes at right angles to the axis of compression and cuts across the original bedding.

FIGURE 7.7

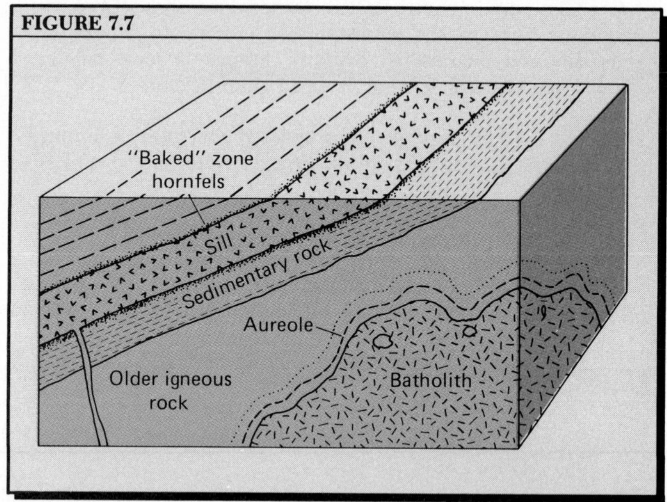

FIGURE 7.8

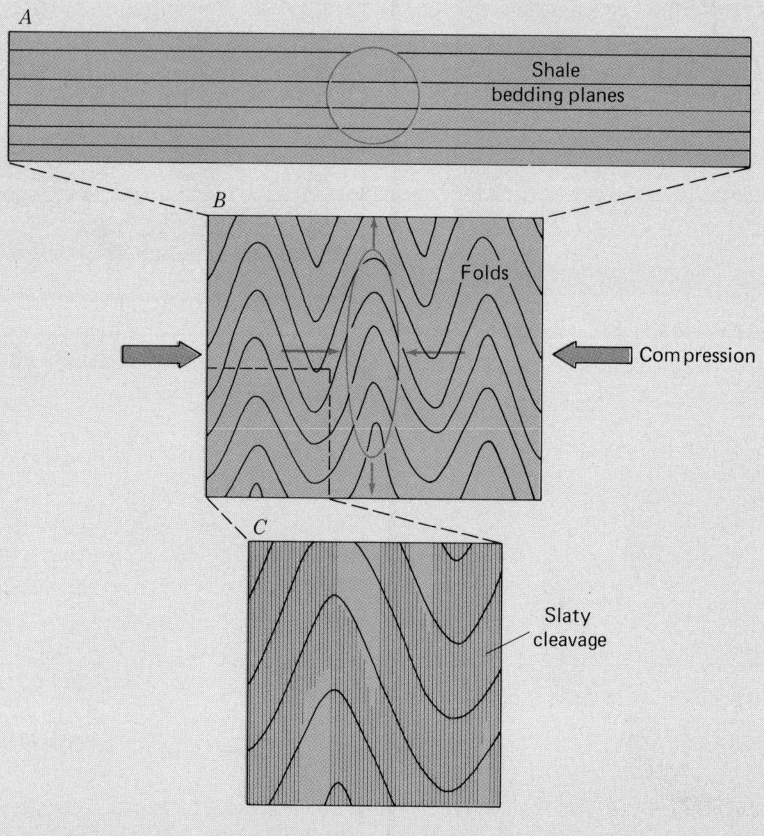

A
Shale bedding planes
B
Folds
Compression
C
Slaty cleavage

into small, tight folds. This tectonic process results from shear stresses that tend to deform a ductile rock mass. We can visualize shale layers originally in a horizontal attitude as being compressed by stresses acting in a horizontal direction (Figure 7.8).

One of the first structures to develop as a result of tight folding in shales is *slaty cleavage* (Figure 7.9). As in mineral cleavage, the word "cleavage" refers to the presence of planes of weakness along which the rock easily splits apart. Slaty cleavage consists of innumerable, closely spaced planes of weakness, formed in an orientation parallel with the long axis of the reference ellipse in Figure 7.8. This is the direction along which the rock has been elongated, or stretched. On a microscopic scale, flakes of clay minerals and mica arrange themselves in parallel layers in the rock, forming planes of weakness in the rock. A fine-grained metamorphic rock derived from shale and exhibiting slaty cleavage is known as *slate*. Because water has been driven out during deformation, slate is dense and hard. It can be split into thin layers that are strong and rigid, as we know from examining roofing shingles made of slate. Layers of slate a few centimeters thick make ideal flooring slabs for patios and walkways.

FIGURE 7.9 Slaty cleavage, represented by closely set slanting planes throughout this outcrop of low-grade schist. Salmon River, Idaho, near Squaw Creek. (W. B. Hamilton, U.S. Geological Survey.)

FIGURE 7.10 Mica schist. This fragment, 15 cm long, shows a glistening, undulating surface of natural parting (*above*). An edgewise view (*below*) shows the thin foliation planes. (A. N. Strahler.)

FIGURE 7.9

FIGURE 7.10

As metamorphism continues to a more intense phase, slate begins to develop a structure known as *foliation*, in which a large percentage of the minerals assume a platelike shape and are assembled in parallel orientation in the rock. Minerals with a strong tendency to platy cleavage—chlorite, muscovite, biotite, talc—concentrate in layers along which the rock splits easily into parallel sheets, or leaves. (The Latin word for leaves is *folia*, hence our word foliation.) Metamorphic rocks with strongly developed foliation are known as *schists*. The foliation typical of schists is referred to as *schistosity*. A freshly broken piece of schist shows glistening or silky surfaces because of the strong reflection of light from the cleavage planes of minerals occupying the foliation planes (Figure 7.10).

As the foliation structure is developing in schists, crystals of certain minerals such as garnet and staurolite are growing in size. Geologists refer to these large crystals as *porphyroblasts*. An example is the large garnet crystal shown in Figure 7.3.

Another structure found in metamorphic rocks is *lineation*, the presence of mineral grains drawn out into long, thin, pencillike objects, all in parallel alignment (Figure 7.11). Lineation is associated with massive types of metamorphic rock and does not produce planes of weakness. The elongated mineral grains may have been formed by lengthening as the rock was stretched along one axis, or may have assumed an alignment parallel with the direction of maximum rock stretching.

Another common structure of metamorphic rocks is *banding*, a rough kind of layering in which minerals of different varieties or groups have become segregated into alternate layers. These layers are usually of different shades—light or dark—so that the banding is conspicuous (Figures 7.12 and 7.27). Metamorphic rock of this description is called *gneiss*. (The word is also applied to some metamorphic rocks showing lineation, but poor banding.) Within individual bands, the rock is coarsely crystalline and strongly bonded. Orientation of grains parallel with the banding is usually present in the banded gneisses. Weakness may exist between individual bands, so that the rock may tend to break apart in layers and can be said to show coarse foliation.

Porphyroblasts are common in many kinds of

banded and lineated gneiss. These large crystals or crystal masses appear on a rock exposure as lumps around which the lineation or foliation is deflected. The lumps thus resemble eyes and the rock is described as *augen gneiss,* from the German word for "eye."

Gneiss is thought to originate in different ways and from different parent rocks. Some banded gneisses are derived from sedimentary rock; others, from igneous rocks. *Granite gneiss* differs little from ordinary granite except that the dark grains of biotite and hornblende show a distinct lineation, as if flowage had slightly affected the granite when it was in a plastic state.

The texture and structure of schists and gneisses show almost infinite variation, and it is not surprising that even a highly trained geologist is often at a loss to reconstruct the history of a particular rock seen in an outcrop. No metamorphic rock can be studied in the environment in which it formed. Moreover, we can see only the final stage in a long series of changes through a changing environment. Detailed laboratory analysis, using sophisticated tools of research, is needed to unravel the mysteries of dynamothermal metamorphic rocks. Chemical analysis of the regional metamorphic rocks sheds a great deal of light on their origin. We therefore turn from metamorphic textures to the assemblages of minerals present in the rock and how they are related to the environment in which metamorphism occurred.

Mineralogical Classification of the Regional Metamorphic Rocks

The regional metamorphic rocks involve large-scale dynamothermal processes. Recrystallization takes place along with wholesale shearing of thick masses of crustal rock. The environments in which these processes act span a wide range of temperatures and pressures. The most important concept relating to these rocks is that they are produced in a well-defined sequence according to increasing temperature and confining pressure. Followed across country, exposures of regional metamorphic rocks show *metamorphic zones,* which reflect the increased temperature that accompanied recrystallization. Geologists found that each zone can be defined by the first appearance of an *index mineral,* not present in zones of lower temperature. A typical sequence of index minerals is shown in Figure 7.13; it runs as follows: chlorite, biotite, almandite, staurolite, kyanite, sillimanite. This sequence would result from metamorphism of a large mass of sedimentary rock made up mostly of shale. The appearance of each mineral can be shown on a map by a line, called an *isograd.* Figure 7.14 is a schematic representation of isograds forming a concentric pattern.

The metamorphic rock is thus zoned into a succession of *metamorphic grades,* each associated with a particular type of metamorphic rock. Figure 7.15 shows typical sequences of metamorphic rock types in terms of pressure and temperature. The broad arrows suggest the sequence of changes that took place as the rock was subjected to increasing temperature. As metamorphism begins at comparatively low temperatures and pressures, a mass of shale develops new minerals of a group called the *zeolites.* We have not referred to the zeolite minerals previously and mention them only briefly at this point. They are hydrous aluminosilicates of calcium and sodium with a rather large water content, and they are soon destroyed as water is driven out of the changing sedimentary clay minerals.

The rock now enters the first major metamorphic grade, *greenschist,* formed under moderate pressure

FIGURE 7.11 Lineation in a gneiss. Individual crystals are drawn out into long, thin, pencillike bodies aligned in the direction in which the rock mass was elongated.

FIGURE 7.12 This banded gneiss of Precambrian age consists of lighter bands of felsic minerals grading into darker bands rich in amphibole and biotite mica. The rock lacks foliation planes and breaks into large blocks. (A. N. Strahler.)

FIGURE 7.11

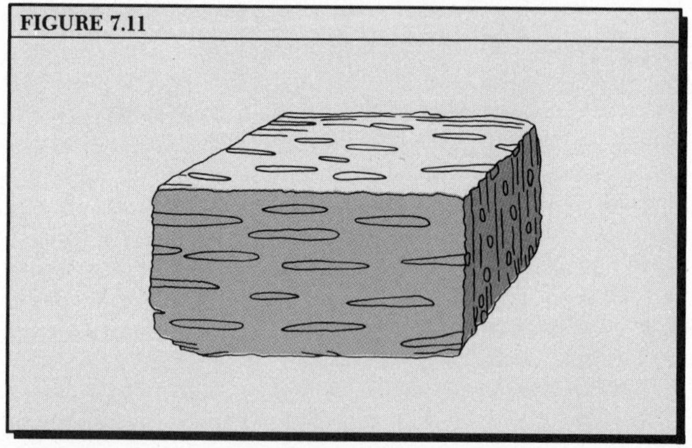

FIGURE 7.12

and fairly low temperature (low-grade metamorphism). Minerals dominant in greenschist are chlorite, muscovite mica, biotite, sodic plagioclase feldspar, and quartz. From this point on, the pressure is assumed to hold about constant, while temperature increases.

The next higher, or intermediate grade is characterized by a rock called *amphibolite,* named for the presence of hornblende (an amphibole). In some places this grade is preceded by *epidote amphibolite,* characterized by the presence of epidote. Amphibolite may also contain quartz, plagioclase feldspar, almandite garnet, and biotite. The rock composition varies according to the original rock composition, whether felsic or mafic.

At still higher temperature, above about 600 °C, a high-grade metamorphic rock, *pyroxene granulite,* is produced. *Granulite* is a term describing the rock texture, which consists of small mineral grains of more or less equal size, developed through shearing action. Grains of quartz and feldspar show a flattening into platelike shapes. Pyroxene is an important new mineral in the rock, while sillimanite appears as the index mineral. With an increase in temperature above the granulite grade, the melting point of the rock would be reached, and new magma could be produced, yielding an igneous rock.

Figure 7.15 shows another possible pathway in the sequence of metamorphism. It is possible for the confining pressure to increase to high levels while temperature remains low. The result is formation of *blueschist,* named for the presence of a blue variety of amphibole (glaucophane).

With an increase in temperature to a middle range (400 °C to 600 °C) and with extremely high pressure (8 to 10 kb or higher), blueschist may become *eclogite.*

FIGURE 7.13 This graph shows how the grades of regional metamorphism are related to pressure and temperature. The arrow shows the typical series of changes from lower to higher grades at a given depth. (Based on data of W. Ernst, 1969, *Earth Materials,* Prentice–Hall, Englewood Cliffs, N.J., Figure 7–9, p. 140.)

FIGURE 7.14 Schematic diagram of isograds of regional metamorphism forming concentric zones around a region of the highest grade of metamorphism. The granite intrusion might represent melting or granitization occurring at a higher temperature.

FIGURE 7.15 A schematic graph of the major types of metamorphic rocks in relation to pressure (depth) and temperature. The arrows show two possible patterns of evolution in the sequence of regional metamorphic rocks. Contact metamorphism is limited to a shallow zone. (Based on data of W. Ernst, 1969, *Earth Materials,* Prentice–Hall, Englewood Cliffs, N.J., Figure 7–10, p. 140.)

FIGURE 7.13

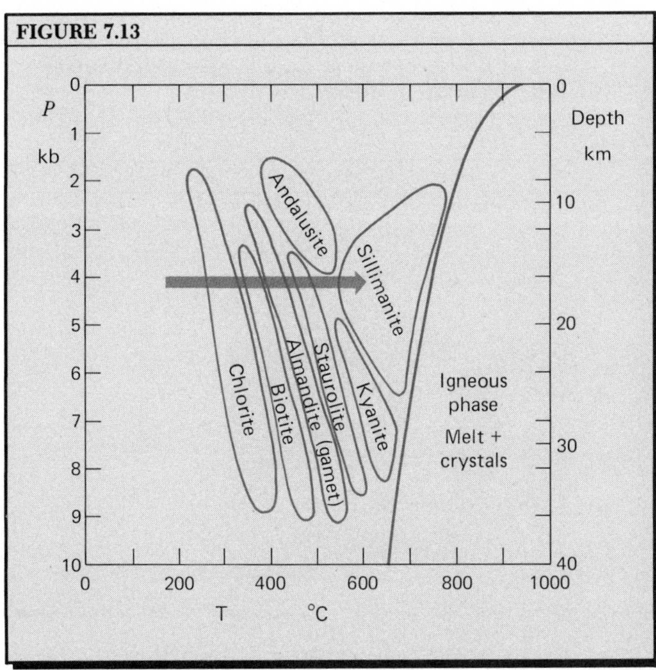

FIGURE 7.14

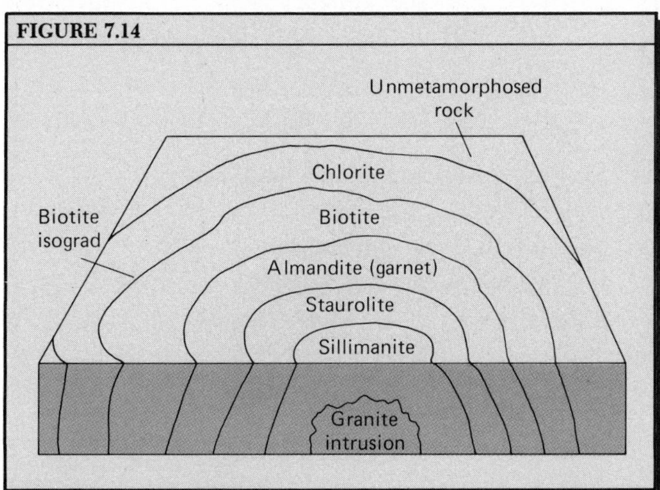

FIGURE 7.15

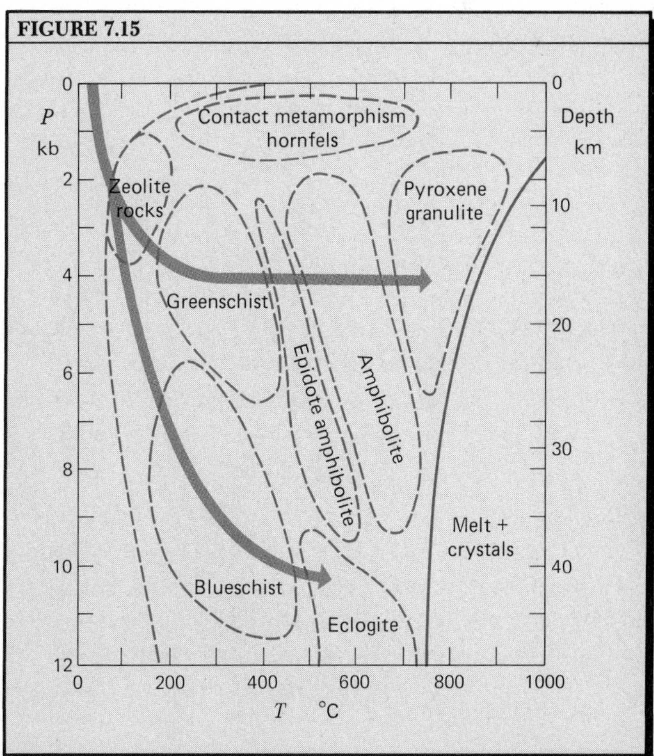

Dominant minerals in eclogite are an unusual green pyroxene rich in sodium and calcium and a red-brown variety of garnet rich in magnesium. The rock grains are coarse in texture. The total chemical composition of eclogite is approximately that of gabbro and basalt. Thus, eclogite may be the end product of dynamothermal metamorphism of mafic igneous rocks at very high confining pressures corresponding to great depths.

Two regional metamorphic rocks that show recrystallization with little or no chemical change are marble and quartzite. *Marble* is recrystallized limestone or dolostone, and may also be formed by contact metamorphism. Under moderate shearing stresses, calcite is easily deformed, because the lattice structure pemits gliding motions to occur easily within crystals. Marble often shows a granular structure such that a freshly broken surface has the appearance of a broken sugar cube. *Quartzite,* formed from quartz sandstone or siltstone, may consist almost entirely of quartz with only minor impurities. It is an extremely hard and durable rock. When struck a hard hammer blow, quartzite shows a conchoidal fracture in which the fracture surface cuts through the quartz grains.

Granitization

Certain gneisses appear to be composed partly of metamorphic rock and partly of igneous rock. These are banded rocks known as ***migmatites*** (Figure 7.16). Between bands that appear to be true metamorphic rock, originally of sedimentary origin, are bands of the composition and texture of granite. In some cases, the granite bands may represent magma that forced its way between layers of metamorphic rock. An alternative explanation is that the granitelike layers represent matter imported by slow injection of fluids and the diffusion of ions from a nearby magma body. These imported substances have replaced the original rock, and many of the original constituents have been exported to other locations. Thus, granitelike layers may have been formed by metasomatism.

Conversion of preexisting rock to a granite by metasomatism is called *granitization*. Melting does not

FIGURE 7.16 Migmatite, a contorted gneiss showing intense deformation in a plastic state. Douglas Channel area, Coast Mountains of British Columbia, Canada. (Geological Survey of Canada, Photo No. 119788.)

FIGURE 7.16

occur during granitization, but the end result is much the same as if the affected rock had been melted and recrystallized by cooling to form a plutonic rock. Since we have already stated that large granite plutons are perhaps best explained by wholesale melting of large masses of crustal rock, this means granite may form both by metasomatism and by melting. The process of granitization is obscure and its importance has been strongly debated for many decades as part of the general question of the origin of granite.

*Hydrothermal Alteration and Serpentinite

In Chapter 4, we described hydrothermal solutions that rise from intrusive magma bodies in the final stages of cooling. These hot-water solutions consist of volatiles and carry a wide variety of mineral-forming ions in solution. Under favorable conditions, hydrothermal solutions are capable of altering the mineral composition of large masses of rock, a process known as *hydrothermal alteration*. Because the temperatures and pressures under which hydrothermal alteration occurs are generally low, the process is often excluded from the scope of metamorphism. Hydrothermal solutions adjacent to shallow magma bodies range in temperature from as low as 60 °C to as high as 500 °C. Thus, in terms of temperature alone, the process can fall well within the range found in regional metamorphism. However, the hydrothermal process usually operates at pressures under 1 kb and at depths of only 1 to 3 km. This is much less than the pressure and depth required for regional metamorphism.

Chemical changes brought about by hydrothermal alteration are those in which water combines with the rock-forming minerals to form new hydrous minerals. Perhaps the most important of these minerals is serpentine, $Mg_3Si_2O_5(OH)_4$ (see Table 7.2). It can be formed through the alteration of olivine, $(Mg,Fe)_2SiO_4$, and other mafic silicate minerals rich in magnesium. The resulting rock is called *serpentinite;* the process by which it is formed is *serpentinization*. Serpentinite is a fine-grained, dark green to black rock with massive structure. Bands of lighter-colored minerals fill joint fractures or are deformed into swirling lines (serpentine patterns), giving the rock an ornamental quality similar to that prized in marble.

Geologists are interested in the origin of large bodies of serpentinite exposed in orogenic belts. The rock is generally considered to have been derived from oceanic crust altered in the presence of water. Evidence is now accumulating to show that hydrothermal activity is intense along spreading plate boundaries. Seawater, which is rich in magnesium, penetrates the rifted crust and is highly heated by the underlying rock. Returning to the surface, the heated water alters the mafic minerals of the crustal basalt to form serpentine-rich basalt. Descending seawater is thought to penetrate the crust to depths on the order of 2 to 3 km, but probably no deeper. For serpentinization to occur deeper in the oceanic crust, affecting the entire crust down to the Moho at a depth of 5 to 6 km beneath the ocean floor, a different source of water would be required.

Professor Harry Hess of Princeton University suggested that the lower part of the oceanic crust is formed from rising peridotite rock of the upper mantle. In the axis of the spreading zone, peridotite is altered to serpentinite on a large scale by the addition of water moving upward from the deeper mantle beneath. The serpentinized mantle rock then spreads laterally to form the oceanic crust. This possibility is to some degree credible, because serpentine and peridotite are often found in adjacent masses at the surface in deeply eroded orogenic belts. These rocks may represent parts of the lower oceanic crust and upper mantle that were finally lifted to the surface by overthrust faulting during orogeny. Evidently, serpentinization may be a process that operates under a wide range of environments, including both low and high confining pressures, and with widely different sources of the water required to form the serpentine.

Radioisotopes and Radioactivity

Although we have now covered the full scope of the rock transformation cycle, two basic problems of geology connected with that cycle require detailed attention: (1) What is the source of the internal energy that drives plate motions and provides heat for igneous and metamorphic processes? (2) How do geologists determine the ages of the crustal rocks and of the earth itself? Both questions are answered through an understanding of radioactivity, a fundamental process of nuclear physics.

Recall from Chapter 2 that for a given element—carbon for example—there exist isotopes. Whereas the atomic number (number of protons in the nucleus) is always the same for a given element, the number of neutrons may vary. Thus, the mass number of a given element differs from one isotope to another. In referring to a given kind of isotope in terms of the composition of its nucleus, physicists use the term *nuclide*. All nuclides with the same atomic number answer to the same element name, but as isotopes of that element, they differ from one another because of differences in mass number.

Protons and neutrons in the atomic nucleus are held together by nuclear forces quite different from any kinds of forces with which we are familiar from everyday experience (e.g., gravitational attraction and electromagnetic forces). The nuclear forces are effective only when protons and neutrons are very close together, as they are in the nucleus of the atom. Some nuclides, with a *stable nucleus* that effectively resists natural forces which might tend to break apart the nucleus, are *stable isotopes*. Other nuclides, with an *unstable nucleus* whose protons and neutrons are not strongly bound and with excess energy, are *unstable isotopes*.

An unstable isotope will undergo a natural process of nuclear disintegration to be transformed into a stable isotope of a different element. This spontaneous process of nuclear disintegration is called *radioactivity*, a term first applied by a French physicist, Henri Becquerel, who observed in 1896 that uranium emitted a mysterious radiant form of energy capable of leaving a photographic image, even under conditions of total darkness. Isotopes that undergo spontaneous disintegration are called *radioisotopes* or *radionuclides*, and the process is referred to as *radioactive decay*.

Radioactive decay can occur in one of three ways: (1) alpha decay, (2) beta decay, and (3) electron capture. In *alpha decay*, the nucleus emits an *alpha particle*, consisting of two protons and two neutrons. (You will recognize this particle as the nucleus of a helium atom.) As a result of alpha particle emission, the mass number is decreased by four; the atomic number by two. An example is shown in Figure 7.17, in which the unstable isotope of uranium, U^{238}, decays to produce an isotope of thorium, Th^{234}. The original radioisotope is referred to as the *parent isotope*, the product of decay is a *daughter isotope*.

In *beta decay*, the nucleus emits a high-speed electron, whose expulsion has the effect of changing one of the neutrons into a proton. As a result, the atomic number is increased by one but the mass number remains unchanged. An example shown in Figure 7.17 is the decay of thorium-234 (Th^{234}) to produce protactinium (Pa^{234}).

In *electron capture*, one of the protons in the nucleus acquires an electron from one of the electron orbitals. The positive charge of the proton is thus neutralized and the protron becomes a neutron. The atomic number decreases by one, but the mass number remains unchanged.

In viewing radioactive decay as an energy-producing mechanism for geologic processes, we must look at the types of emissions associated with the nuclear changes. In alpha decay, the motion of an alpha particle represents a form of kinetic energy. When the alpha particle is lost, the remaining nucleus is in an excited state and becomes stable only by the emission of a *gamma ray*, which is a high-energy photon. Energy contained in both the alpha particle motion and the gamma ray is produced by a very small reduction in the combined mass of the alpha particle and the remaining nucleus. This mass difference is equated to energy by the Einstein formula:

$$E = mc^2$$

where E is energy
m is mass and
c is the speed of light (300,000 km/sec)

FIGURE 7.17 Radioactive decay series of U-238 to Pb-206. (Based on data of P. M. Hurley, 1959, *How Old Is the Earth?*, Doubleday, Garden City, N.Y., p. 62, Figure 9.)

FIGURE 7.17

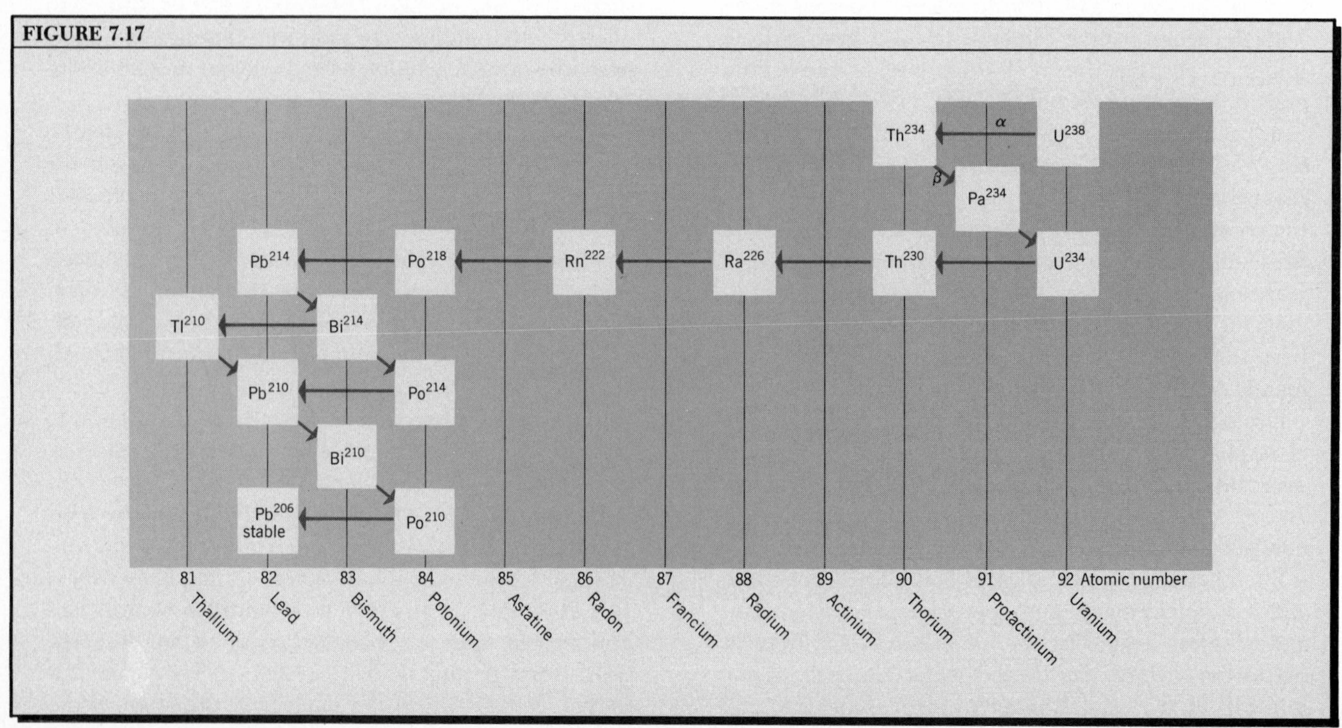

From this equation, we see that an extremely small quantity of mass converts into a comparatively enormous quantity of energy.

In beta decay, another form of energy, the *beta particle*, is produced. It is an electron traveling at high speed, derived from one of the neutrons in the nucleus. In furnishing one electron (one negative unit electrical charge) for the beta particle, the neutron acquires a unit positive charge and becomes a proton. (The neutron also emits a particle called an *antineutrino*, but this is not important in our discussion of energy released by radioactive decay.)

The important point about radioactive decay is that it generates energy which takes the form of heat in the substance surrounding the radioisotope. This form of heat is called radiogenic heat. As alpha particles travel outward through the surrounding matter, they lose energy by interacting with electrons in the orbitals of other atoms and by colliding with other atomic nuclei. This lost energy is transformed into heat and raises the temperature of the substance. Both gamma rays and beta particles interact with electrons in the surrounding matter, also causing a buildup of heat. The total quantity of radiogenic heat produced per unit of time can be exactly calculated for a given quantity of a radioisotope.

We can illustrate radioactive decay and heat production by using the important *decay series* (shown in Figure 7.17) in which the parent radioisotope uranium-238 (U^{238}) eventually ends up as a stable isotope of lead, lead-206 (Pb^{206}). In the figure, arrows show the direction of successive changes. Note that each step in the direction of the arrow to the left is alpha decay; each diagonal step downward to the right is beta decay. Uranium-238 decays to produce thorium-234. This is followed by a succession of isotopes of seven different elements, listed along the bottom of the graph. The disintegration process achieves a steady rate, or equilibrium, with time. In the case of the uranium-238–lead-206 series, each gram of uranium produces 0.71 calories of heat per year.

FIGURE 7.18 Exponential decay and growth curves for K-40, Ca-40, and Ar-40. (Based on data of P. M. Hurley, 1959, *How Old Is the Earth?*, Doubleday, Garden City, N.Y., p. 101, Figure 17.)

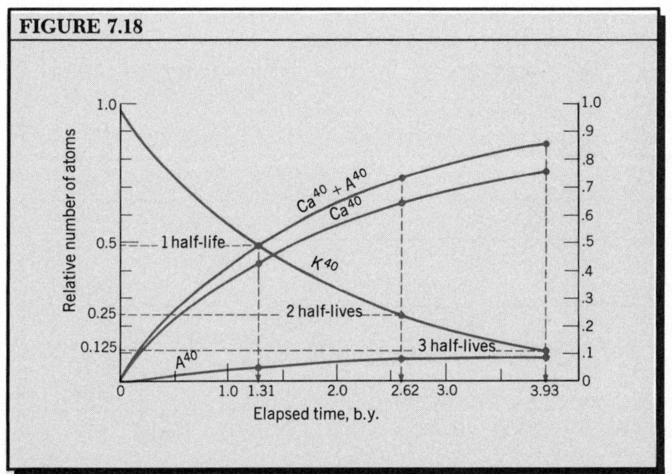

Other important heat-producing decay sequences in the rocks of the earth are those of uranium-235, thorium-232, and potassium-40. Uranium-235 produces 4.3 calories of heat per gram per year, thorium-232 produces 0.20 calorie, and potassium-40 produces only 0.000027 calorie.

Of great importance in both the early history of the earth and the dating of geologic events is a physical law that governs the rate of decay of radioisotopes. Once an equilibrium has been reached in the process of radioactive disintegration, the ratio of decrease in the number of atoms of the parent isotope with each unit of time is a constant.

Take for example, postassium-40, which decays to the stable isotopes calcium-40 and argon-40. We can start at any point in time. Let the number of atoms of potassium-40 at time zero be designated by unity (1.0), as shown at the upper left corner of the graph in Figure 7.18. After 1.31 billion years have elapsed, the number of atoms of potassium-40 will have been reduced to half the initial number, designated as 0.5 on the vertical scale. The span of time of 1.31 billion years is designated as the half-life. In a second elapsed span of 1.31 billion years, the number of atoms of potassium-40 will again be halved, reducing the remaining quantity to 0.25 on the vertical scale. Notice that the ratio of reduction is always the same, i.e., one-half. Such a schedule of decrease in quantity with time is known as *exponential decay*. This schedule applies to all radioactive decay, but the ratio of change, and hence the value of the half-life, is different from one isotope to another.

Figure 7.18 also shows the rate at which the daughter isotopes accumulate. Both are rising curves. Both follow an *exponential increase*, which is the inverse of the exponential decay curve. It should be noted that calcium-40 is produced about 7⅓ times more rapidly than argon-40, hence the curve of calcium-40 rises more steeply. We shall refer to the ratios of parent isotopes to daughter isotopes in our discussion of methods of rock dating.

Table 7.3 gives the half-life of each of the important natural radioisotope series. Notice that there are vast differences in the half-lives of these isotopes. In projecting the production of radiogenic heat back into earliest geologic time, these differences will be highly important. Isotopes with the shorter half-lives were then present in very much larger quantities than today, in contrast with isotopes having extremely long half-lives.

Table 7.3 Half-Lives of Important Radioisotopes

Parent isotope	Stable daughter products	Half-life, billions of years
Uranium-238	Lead-206, plus helium	4.5
Uranium-235	Lead-207, plus helium	0.71
Thorium-232	Lead-208, plus helium	14
Rubidium-84	Strontium-87	51
Potassium-40	Argon-40, calcium-40	1.3

Source: Data from Brian Mason (1966), *Principles of Geochemistry,* 3d ed., New York, John Wiley, p. 9.

Distribution of Radiogenic Heat

Heat flows continuously upward from depths of the earth toward the surface. The increase in temperature with depth, or *geothermal gradient,* is well known from observations in deep mines and bore holes and has a value of about 3 C° per 100 m. The flow of heat because of this thermal gradient averages about 1.4 microcalories (0.0000014 calories) per square centimeter per second. At this rate, the total heat flow in one year is about 50 calories per square centimeter, enough to melt an ice layer 6 mm thick. This quantity of heat is extremely small compared with that received by one square centimeter of the earth's surface from solar radiation. Therefore the earth's heat flow from depth is of no significance in the earth's surface heat balance or in powering the atmospheric and oceanic circulation systems.

Referring to the temperature–depth graph in Figure 4.1, you will notice that the rate of temperature increase with depth falls off very rapidly after the first 200 km or so. The flattening of the temperature curve in the lower mantle and core expresses the very low thermal gradient that is postulated to exist within the deep interior. If the thermal gradient decreases rapidly with depth, a logical interpretation is that the rate of production of radiogenic heat is greatest near the earth's surface and decreases rapidly with depth. It has been concluded that the concentration of radioisotopes is greatest in the rocks of the crust, but falls off rapidly in the mantle rocks and is very small in the lower mantle and core.

Table 7.4 shows both the concentrations and rates of heat production of the radioisotopes of uranium, thorium, and potassium in each of three classes of rocks. These rates are based upon chemical analyses of samples of igneous rocks collected at the earth's surface. If we project these figures to the assumed corresponding rocks of the crust and mantle, as shown in Figure 4.8, we see that the most rapid production of radiogenic heat is by felsic (granitic) rocks of the upper zone of the continental crust. The ultramafic mantle rock produces very little heat per unit of weight. It has been estimated that about one-half of all radiogenic heat is produced above a depth of 35 km in the continental crust.

Some support for the conclusion that the iron core of the earth produces almost no radiogenic heat is found in the analysis of iron meteorites. These fragments of matter are thought to represent the disrupted cores of planetary objects of origin similar to the earth (see Chapter 20), and they show radioactive minerals in only very small quantities.

Early Thermal History of the Earth

Modern hypotheses of the earth's origin, discussed in Chapters 1 and 20, favor the process of accretion of the earth and other planets through the condensation of a hot interstellar cloud of gases and dust as it cooled. The solid particles aggregated through collision and gravitational attraction. Once formed, solid masses would have grown by the infall of solid bodies of many sizes, perhaps including objects similar to the asteroids of our present-day solar system. It is generally supposed that at the time planetary accretion was largely complete, the earth's interior temperature had not risen to the melting point, although local areas may have become molten from the energy of impacts. Modern thinking is thus along quite different lines from that of early scholars, who postulated that hot nebular gases condensed to the molten state and finally cooled to the solid state.

The discovery of natural radioactivity by Henri

Table 7.4 Heat Production in Crustal Rocks

	Concentrations			Heat production, cal/g/yr			
	U ppm	Th ppm	K %	U	Th	K	Total
Felsic (granitic)	4	14	3.5	3	3	1	7
Mafic (basaltic)	0.6	2	1.0	1.5	1.5	0.5	3.5
Ultramafic	0.015	—	0.011	0.01	0.01	<0.001	0.02

Source: Based on data of B. Mason (1966), *Principles of Geochemistry,* 3d ed., New York, John Wiley, Table 11.1; and P. J. Hurley (1959), *How Old Is the Earth?*, New York, Doubleday, p. 64.

Becquerel in 1896, followed by the isolation of radium by Marie and Pierre Curie in 1898, radically altered all scientific thinking about the earth's internal heat. John Joly in 1909 applied the new knowledge of radioactivity to recalculations of the earth's thermal history. Moreover, Joly brought forward the underlying principle that radiogenic heat provides the prime energy source for volcanism, igneous intrusion, and deformation of the earth's crust into mountain belts. Today, radiogenic heat is usually regarded as the basic source of energy for lithospheric plate motions.

We must first accept as a premise that the earth's supply of radioisotopes was furnished, along with all other elements, at the time the earth was formed. It is most unlikely that the earth, at the time of its formation as a planet about 4.5 billion years ago, contained the same quantity and distribution of radioisotopes that we find today. The obvious reason is that radioactive decay progressively reduces the initial store of radioisotopes. We conclude that the total production of radiogenic heat within the earth was at the maximum level at the time of the earth's formation and has diminished ever since. The relative rates of decay of uranium, thorium, and potassium isotopes are not the same, but are well established for each isotope.

Figure 7.19 is a graph in which time is plotted on the horizontal axis starting at an arbitrary zero point at the assumed time of formation of the earth. Total planetary radiogenic heat production per year is given on the vertical scale. Curves have been plotted for

FIGURE 7.19 Rate of production of radiogenic heat projected back 5 billion years. (Based on data of A. P. Vinogradov, 1961, as shown in B. Mason, 1966, *Principles of Geochemistry,* John Wiley & Sons, Inc., New York, p. 61, Figure 3.9.)

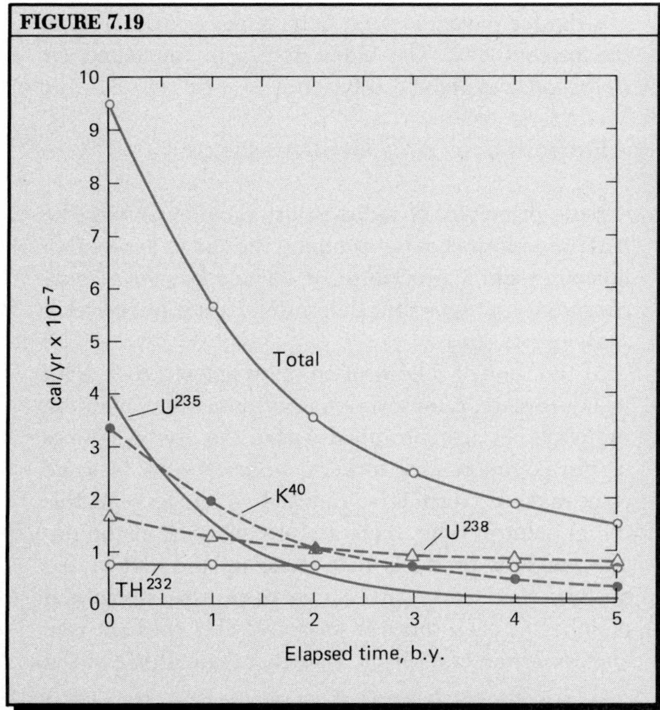

the major radioisotopes of uranium, thorium, and potassium individually, while the total production is shown in a separate curve. Note that uranium-235 and potassium-40 have short half-lives in comparison with these of uranium-238 and thorium-232. Referring to the total curve, it is obvious that total radiogenic heat production was vastly greater when the earth was first formed than it is at present, roughly by a factor of six. The implications of such a history are of great consequence.

First, assume that at the time of the earth's formation by accretion the radioisotopes were uniformly distributed throughout the entire earth. (There is no reason to think otherwise.) Silicate minerals and iron were also more or less uniformly mixed. As radiogenic heat accumulated at great depths, the temperature of the solid rock would have been raised to a level close to the melting point. At a certain point in time, which may have been about one billion years after planetary accretion was completed, the melting point of iron would have been exceeded. This event probably occurred first in a depth zone ranging from 400 to 800 km. The silicate minerals remained crystalline, forming a spongy mass through which droplets of molten iron could filter down under the force of gravity. As molten iron accumulated in the central core region, the silicate minerals were gradually displaced upward.

At this point, a new factor would have come into play to raise the earth's internal temperature. As the iron sank toward the earth's center, its potential energy would have been converted into kinetic energy of molecular motion in the form of sensible heat. It seems likely that the additional heating was sufficient to cause melting or partial melting of a large proportion of the entire earth. At various times and places, melted rock would have risen as magma toward the surface, bringing radioisotopes up with it. Minerals containing the radioactive elements are largely of felsic composition and tend to remain in the liquid state at temperatures lower than the mafic minerals. Cooling and crystallization of the mafic minerals would have been accompanied by a sinking of those mineral crystals (which are denser), leaving the less mafic liquid fraction to solidify closer to the earth's surface. Although such a process of differentiation is speculative, it offers a mechanism for the selective removal of the radioisotopes from the inner earth and their eventual concentration near the surface.

So we see that during this great thermal event, the density layering of the earth came into existence, resulting in the concentration of metallic iron in the core, a less dense ultramafic rock mantle above it, and a mafic–felsic crust at the top. During and after the segregation process, the rate of radiogenic heat production was steadily falling. Consequently, along with redistribution of the heat-generating isotopes, the final episodes of deep melting must have become fewer and eventually ceased. Today the earth is thermally stable, in the sense that melting and movement of magma are limited to an extremely shallow layer compared with the earth's total diameter. The inner core and much of the mantle are no longer subject to melting through the accumulation of excess heat.

It is fortunate, indeed, that the chemistry of the radioactive elements is such that they would tend to rise towards the earth's surface throughout its history. If, on the other hand, they had tended to sink and collect near its center, the heat produced by their concentrated activity would have repeatedly melted the earth. Under such conditions, no planetary stability would have been possible throughout geologic history. As we find conditions today, radiogenic heat production in the core is negligible, while the rate of surfaceward flow of heat from the upper mantle and crust closely balances the rate of heat production. Consequently, the mantle remains for the most part at a temperature lower than its melting point. Only in the soft layer of the mantle (the asthenosphere) is melting on a large scale a likely occurrence.

Determining the Ages of Crustal Rocks

The search for a means by which to establish the absolute age in years of an event in the earth's past history was for decades frustrating and, in retrospect, misleading. Geological estimates of the ages of crustal rocks were based upon two lines of calculation—salinity of the oceans and thickness of accumulated sediments. It was thought that to give the age of the oceans, the total amount of salt in the oceans, a figure subject to rather close estimate, could be divided by the annual increment of salt, which was estimated from chemical analyses of stream waters and estimated annual stream flows. Toward the end of the nineteenth century, the calculation was made that the total weight of sodium in the oceans is about 1.6×10^{16} tons, and that the annual increment of sodium is about 1.6×10^{8} tons. Division yields a figure of roughly 100 million years (10^{8} m.y.). Two major sources of error come immediately to mind. We know now that the salts of the ocean enter sediments and that a given element has a certain residence time, that of sodium being 230 m.y. (Table 5.8). As a result, the salinity of the oceans has probably remained close to its present value for a large part of geologic time. Proof that enormous quantities of sodium have been removed from the ocean lies in the known occurrence of thick salt beds (Chapter 5). A second major source of error lies in the extrapolation of present rates of sodium contribution far into the past. We have good reason to believe that continents stand high today, in comparison with average elevations in much of the past, so that present rates are probably much too high.

The second approach was to total the measured thicknesses of sedimentary strata, taking the thickest known deposit of each age unit of the geologic column. These totaled somewhere in the area of 150 km. Using a value of 1 m per 650 years as an average rate of accumulation, an age of about 100 m.y. was obtained for the start of sedimentation. Allowing for periods of nondeposition by introducing a correction factor of 15 times, the age would come to 1.5 billion years (b.y.). Between uncertainties as to rates of deposition and lengths of periods of nondeposition, this method is scarcely better than a blind guess. A dozen or more estimates made between 1860 and 1909 range from roughly 20 m.y. to 1.5 b.y.

Late in the nineteenth century, the distinguished English physicist, Lord Kelvin, had calculated the age of the earth using the premise that the earth cooled from a molten state and that the cooling rate followed simple laws of radiative and conductive heat loss. On this basis, he concluded that the earth could not be older than 100 m.y. and that an age of 20 to 40 m.y. was a reasonable figure. This figure was disappointingly short to the geologists and also to Charles Darwin and his followers, whose theory of organic evolution by natural selection seemed to require much longer spans of time. Darwin originally estimated that 300 m.y. were needed for only the later stages of evolution. Kelvin had moved his estimate far in the wrong direction, but it was difficult to find any flaw in his application of what were believed to be correct laws of physics.

Dramatically, the dilemma over the age of the earth and the duration of periods of geologic time was solved with the discovery of radioactivity. Using these principles, the first reliable age determinations of rocks were made in 1907 by B. B. Boltwood, a chemist. His figures have required only minor adjustments to the present day. The oldest rock age calculated by Boltwood was about 1.6 b.y.

Radiometric Age Determination

Basic principles of radioactivity provide us with the basis of a method of determining the age in years of an igneous rock, a procedure of science known as *geochronometry.* Ages thus determined are referred to as *radiometric ages*.

At the time of solidification of an igneous rock from its liquid state, minute amounts of minerals containing radioisotopes are entrapped within the crystal lattices of the common rock-forming minerals and in some cases make up distinctive radioactive minerals. At this initial point in time, there are present none of the stable daughter products that make up the end of the decay series. However, as time passes, the stable end member of each series is produced at a constant rate and accumulates in place. Knowing the half-life of the decay system (Table 7.3), it is possible to estimate closely the time elapsed since mineral crystallization occurred. An accurate chemical determination of the ratio between the radioisotope and the stable daughter product must be made. A fairly simple mathematical equation is used to derive the age in years of the mineral under analysis.

Take for example, the uranium–lead series U^{238}–Pb^{206}, which has a half-life of 4.5 b.y. Quantities of both uranium-238 and lead-206 are measured from a sample of uranium-bearing minerals (e.g., uraninite or pitchblende) or from a common mineral (zircon) enclosing the radioactive isotopes. The instrument used for such determinations is the *mass spectrometer*. The ratio of lead to uranium is then entered into the following equation:

$$\text{Age (m.y.)} = 6.50 \times 10^9 \; \text{logarithm} \; (1 + Pb^{206}/U^{238})$$

(The "logarithm" referred to in the equation is the *natural logarithm* of the number within the parentheses and may be found in a set of mathematical tables.)

Similar age determinations can be made using the series U^{235}–Pb^{207}. Because both series of uranium–lead isotopes are normally present in the same mineral sample, analysis of one series can serve as a crosscheck upon the other. Accuracy of the method depends upon the accuracy with which the half-life of the series is known. In this case, accuracy of the half-life of the U^{238}–Pb^{206} series is known to within 1%, and that of the U^{235}–Pb^{207} series to within 2%. It is therefore possible to determine the absolute age of a sample of uranium-bearing mineral to within about 2% of the true value, and in some cases as closely as 1%. But this level of accuracy also assumes that no fraction of the components in the decay series has been lost from the sample. Use of the uranium–lead systems for age determination can be applied to the oldest rocks known, as well as to meteorites. Age of meteorites is close to 4.6 b.y., about 0.8 b.y. older than the oldest rocks of the earth's crust that have thus far been dated.

We mention in passing that the radioactive thorium–lead decay series, Th^{232}–Pb^{208}, listed in Table 7.3 as an important heat-producing system, is in disfavor for age determination because loss of some of the lead tends to occur and to give erroneously low ages.

Of great importance in age determination is the potassium–argon series K^{40}–Ar^{40}, with a half-life of 1.3 b.y. Decay of K^{40} occurs by electron capture. This series is particularly adaptable to use with the micas (specifically muscovite and biotite) and hornblende, which are widely present in igneous rocks. The potassium–argon series gives reliable minimum ages for fine-grained volcanic rocks (lavas) which cannot be dated by other methods. Moreover, the method can be used for relatively young rocks (as young as 1 m.y.), as well as the most ancient rocks. It has been a highly important tool for the geologist, particularly in deciding which of two rock groups is the older.

The rubidium–strontium decay series, Rb^{87}–Sr^{87}, with an extremely long half-life of 47 b.y., is of great value in dating both individual minerals and whole rock samples. It has proved successful in dating metamorphic rocks and thus in dating the tectonic events that produced the metamorphism.

Ideally, the three dating systems—uranium to lead, potassium to argon, and strontium to rubidium—should serve as crosschecks upon one another when all three are applied to mineral samples from the same rock body. In some instances, the ages check out as closely similar, but there are instances in which moderate discrepancies are evident. Despite existing uncertainties, the radiometric ages given for various events in the timetable of the earth's history are now accepted by geologists as valid within small percentages of error. Success of the radiometric age determinations of rocks stands as a striking scientific achievement based upon the application of principles of physical chemistry to geology.

Radiometric age determinations on intrusive and extrusive igneous bodies within a complex arrangement of sedimentary strata can be of great assistance in assigning dates to the rock groups lying above and below unconformities. Figure 7.20 shows a hypothetical example as it might be applied to a case resembling that of rocks of the inner Grand Canyon, shown in Figure 6.2.

Suppose that radiometric ages are found for each of the three different igneous rock bodies, labeled 1, 2, and 3 in the diagram. If the igneous rock 1 has an age of 1.4 b.y., we can say that the adjacent schist, into which the igneous rock was intruded, is older than 1.4 b.y. We can also say that the tilted strata on the right are younger than 1.4 b.y. because they were deposited after igneous body 1 was leveled by erosion.

Igneous body 2 is a sill which was intruded into the tilted strata, but whether before or after they were tilted can only be pure conjecture from the evidence shown. If the age of igneous rock 2 turned out to be 1.0 b.y., it would mean that the tilted strata are at least that old, perhaps much more, but not exceeding the limit of 1.4 b.y. set by igneous rock 1.

If igneous rock 3, a thin vertical dike, yielded an age of 400 m.y., we would know that the Cambrian Period is older than 400 m.y., but that the Mississippian Period is younger than 400 m.y.

Continental Structure

The continental crust, which caps the continental lithosphere, is for the most part composed of rocks of Precambrian age, ranging from close to 3 b.y. to somewhat less than 0.6 b.y. These ancient rocks make up the continental crust that is now stable, meaning that it is unaffected by modern plate-boundary tectonic activity. The stable portion of the continental crust is called a *craton*. In terms of the type of rock exposed at the surface, a craton appears to have two kinds of areas: shields and platforms. A *shield* is an area of craton over which ancient igneous and metamorphic rocks are exposed. A *platform* is a part of a craton covered by sedimentary rocks, most of which are

FIGURE 7.20 Radiometric dating of igneous bodies allows limiting ages to be established for the enclosing and overlying strata. (Refer to Figure 6.2 for names of formations.)

FIGURE 7.21 Generalized world map of continental shields and platforms. The continental nuclei lie within the areas encircled by a solid line. (Based in part on data of R. E. Murphy, 1968, *Annals, Association of American Geographers*, Map Supplement No. 9; P. M. Hurley and J. R. Rand, 1969, *Science*, vol. 164, pp. 1129–1242. Other data sources used. (Goode Base Map copyrighted by the University of Chicago; used by permission of the Geography Department, University of Chicago.)

FIGURE 7.20

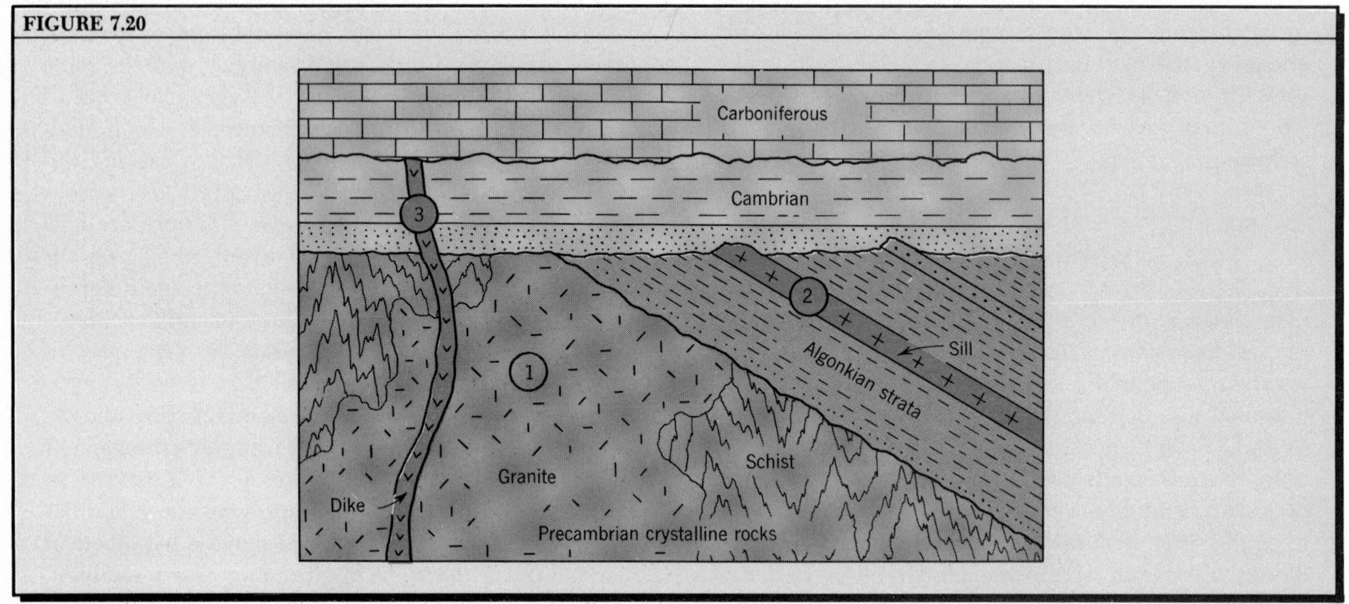

FIGURE 7.21

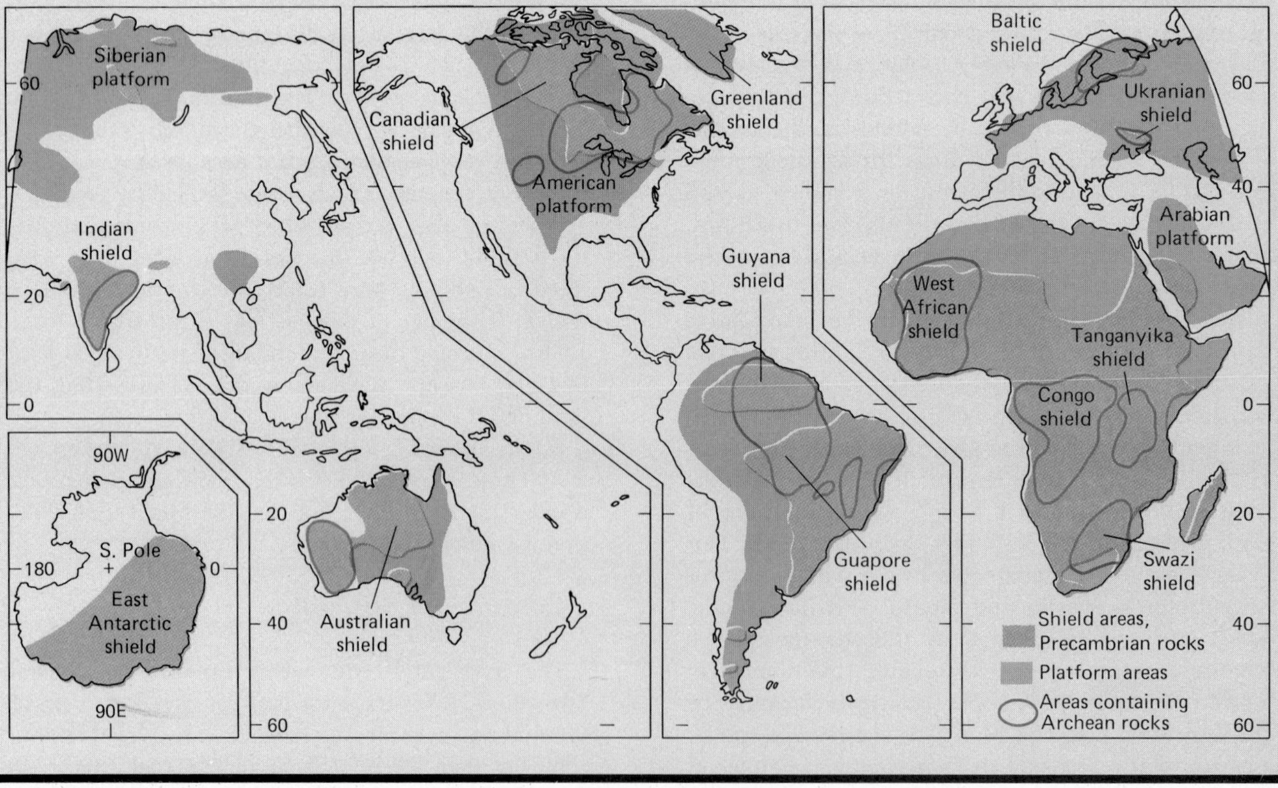

younger than Precambrian. In other words, a platform is a covered shield area. Compared with the thickness of the continental crust, the sedimentary cover represents a very thin skin, but it is effective in obscuring the shield rocks beneath from geological observation. Figure 7.21 is a world map showing the present extent of shields and platforms.

If the cratons have grown in extent throughout geologic time, we should look at the distribution patterns of rock ages in cratons, for they may reveal a growth pattern of diminishing age from a central zone to an outer zone.

The oldest rocks of the cratons, older than about 2.5 b.y., are of the lower Precambrian. They are found only in relatively small shield patches and are called *continental nuclei* (Figure 7.21). Some of the rocks in the continental nuclei are older than 2.7 b.y., although most are in the age range 2.3 to 2.7 b.y.

The continental nuclei are surrounded by or are contiguous with larger areas of shield rock with maximum ages falling in middle and upper Precambrian time. Figure 7.22 is a map of North America showing the crustal rocks zoned into provinces according to ages of the oldest intrusives. Nuclei consist of the Superior (Keewatin), Wyoming, and Slave provinces. The Churchill province (1.3 to 2.3 b.y.) surrounds the nuclei, except on the southeast side, where the Central province (1.3 to 1.7 b.y.) and Grenville province (0.8 to 1.3 b.y.) are contiguous. A great Cordilleran province runs down the west side of the continent with intrusives not older than 440 m.y. These rocks are younger than Precambrian. The Appalachian province is dominated by intrusives under 440 m.y., but some Precambrian ages of the range 0.8 to 1.3 b.y. are scat-

FIGURE 7.22 Geologic provinces of the North American shield. Radiometric ages refer to the oldest granite-forming events in billions of years. Boundaries should be considered largely provisional. (Based on map data of A. E. J. Engel, 1963, *Science*, vol. 140, pp. 143–152; and P. M. Hurley and J. R. Rand, 1969, *Science*, vol. 164, p. 1231, Figure 2.)

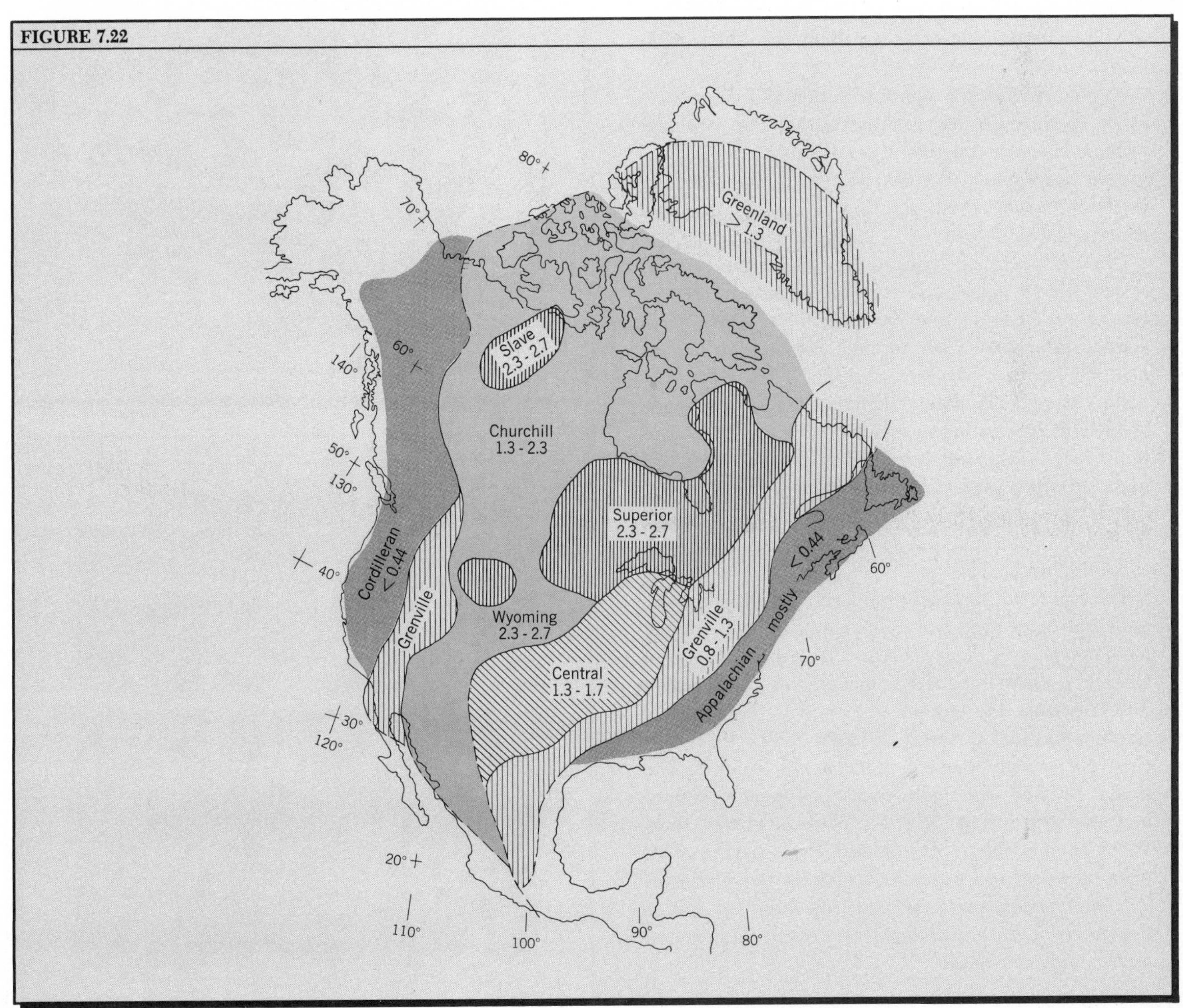

tered throughout. A Greenland province containing rocks older than 1.3 b.y. is perhaps of the same age as the Central province.

Figure 7.23 is a geologic map of portions of Ontario, Manitoba, and Quebec lying north of Lake Superior and between Hudson Bay and Lake Winnipeg, Canada, showing the Superior geologic province, one of the continental nuclei. The patterns on the geologic map represent the several ages and varieties of rocks as they are exposed at the surface. Oldest of these are the Keewatin lavas, over 2.7 billion years old, which may once have been the rocks of an ancient volcanic island chain. Today, these rocks are greenschists. Closely associated with the Keewatin lavas are some sedimentary strata, now greatly metamorphosed, known as the Timiskaming series. These strata were clastic sediments—conglomerates and sandstones—such as those that may have been laid down in a trough near a subduction zone. Other patches of somewhat younger but nevertheless extremely ancient rocks are also found in this province, but most of the area is underlain by intrusive rocks (now gneisses) which invaded later in the Precambrian, largely engulfing the older rocks (Figure 7.24).

As you can see, the age distribution of Precambrian rocks in North America suggests that the continent originated with relatively small nuclei and grew greatly throughout Precambrian and later time by repeated collisions and suturing of continental lithospheric plates.

Cratons made of Precambrian rocks have experienced the removal by denudation of several kilometers of rock during the 600 million years that have elapsed since they were formed. Denudation has repeatedly reduced the surfaces of the shields to low plains, rising only a few hundred meters above sea level. This process has exposed metamorphic and plutonic rocks that were formerly in the deep zone of high pressures and high temperatures. Thus, deep-seated rocks have arrived at the surface environment, as shown in the diagram of the rock transformation cycle, Figure 5.1.

The low shield surfaces produced by continental denudation have experienced many rising and sinking movements over broad areas. As we explained in Chapter 6, these epeirogenic movements are distinctly different from the tectonic movements that severely fracture and fold the crust. When a downward (negative) epeirogenic movement occurred, parts of the shields became vast shallow seas. Sediment deposited in these seas formed a cover of sedimentary rock. When epeirogenic uplift followed, these marine strata were elevated and exposed, becoming the platforms (covered shields) that now make up large sections of the cratons. Let us investigate these epeirogenic movements in greater depth.

FIGURE 7.23 Simplified geologic map of the western portion of the Keewatin geologic province of Canada. Solid-color patches are underlain by Keewatin metamorphic rocks, and pale areas by intrusive igneous rock and gneisses. Diagonally ruled areas are of younger Precambrian rocks. (After J. A. Jacobs, R. D. Russell, and J. T. Wilson, 1959, *Physics and Geology*, McGraw-Hill, New York, p. 325, Figure 15.6.)

FIGURE 7.24 Outcrop of banded gneiss of Precambrian age, east coast of Hudson Bay, south of Povungnituk, Quebec. (F. C. Taylor, Geological Survey of Canada, Ottawa, No. 12522.)

FIGURE 7.23

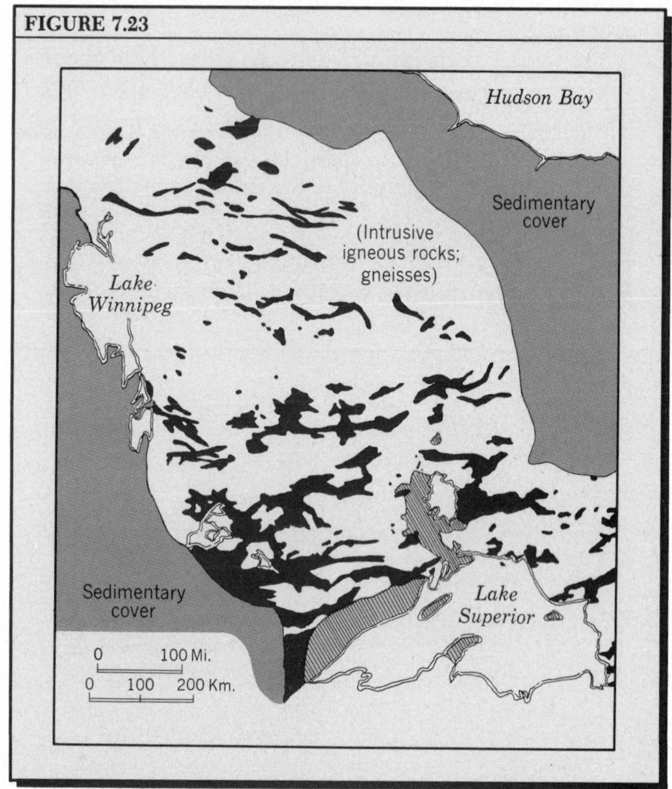

FIGURE 7.24

Isostasy and the Continental Crust

In Chapter 1, we pointed out that the theory of plate tectonics includes the premise that a strong lithosphere rests upon a soft, plastic asthenosphere. In Chapter 4, we explained that the continental crust is considerably thicker than the oceanic crust (see Figure 4.8). Because the asthenosphere is capable of yielding slowly when subjected to unequal stresses, it behaves much like a true fluid. The continental lithosphere is, in fact, floating upon the asthenosphere much as a layer of sea ice floats on the ocean. The continental lithosphere floats because of the lower density of its thick crust. This is the same principle that applies to the buoyancy of all floating objects—a ship's hull, for example.

We are also familiar with the fact that a floating object comes to rest with a certain proportion of its bulk submerged and the remaining portion above the water line. The *Archimedes principle* requires that a floating object displace a volume of liquid having the same mass as the entire floating object. If an object has a density half that of the liquid, it will float at rest with half its volume below the water line and half above.

FIGURE 7.25 A plumb bob will be deflected by a pyramid on a plain.

FIGURE 7.26 The deflection that the Himalaya Range causes for a plumb bob on the Indo-Gangetic plain is not nearly as great as might be expected for so large a mountain mass.

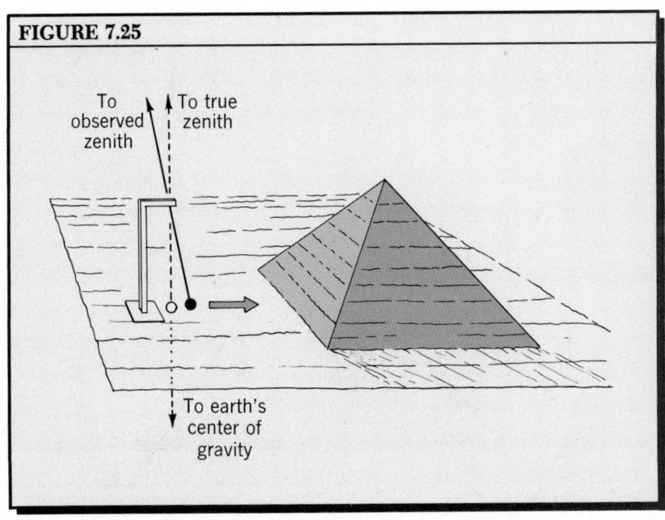

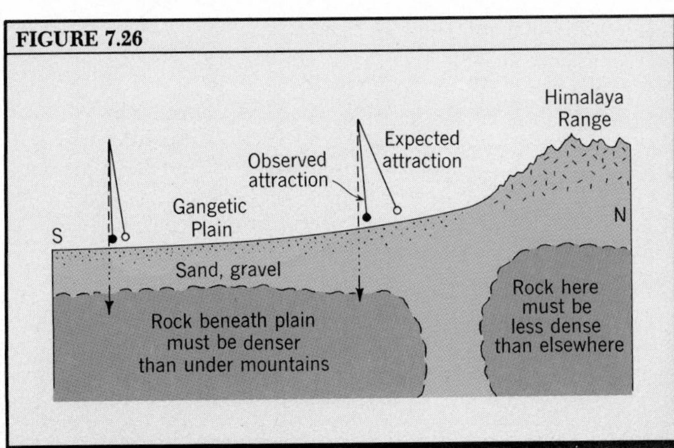

When applied to the lithosphere, the principle of a mass floating at rest in a supporting liquid is referred to as the principle of *isostasy.* This word comes from the Greek words *isos,* equal, and *stasis,* standing still. Isostasy requires that a particular column of the lithosphere must float in an equilibrium state on the underlying asthenosphere. Over most areas of the earth, including the ancient continental cratons and the young ocean floors, the equilibrium required by isostasy exists or is closely approached. In other words, *isostatic equilibrium* prevails.

The circumstances under which the concept of isostasy came into focus are rather interesting and unusual. The story behind it is sometimes called the "Indian mystery." It involves a commonplace instrument called the plumb bob, used by masons to make sure that a brick wall is vertical and by surveyors to make sure their instrument is precisely located over a reference point on the ground. The plumb bob points close to the earth's center of mass. Let us now imagine a smooth, low plain on which an enormous pyramid has been built, as shown in Figure 7.25. The mass of rock in the pyramid will exert a gravitational attraction upon the plumb bob, and it will seem to be pulled toward the pyramid, deflecting slightly from the vertical attitude. Knowing the dimensions of the pyramid and density of the rock of which it is built, we can predict quite precisely the amount of deflection of the plumb bob.

And now, back to the "Indian mystery." Well over a century ago, the British geodesist Sir George Everest was engaged in triangulation surveys involving the precise measurement of the length of a reference line, called a "base line," on the Indo-Gangetic Plain in north India. To the north rises the greatest of continental mountain ranges, the Himalaya, culminating in many peaks over 7 km above sea level. Knowing the dimensions of the range and the average density of rocks exposed in it, Everest computed the amount by which the plumb bob should be deflected toward the range from a given distance away. To his surprise, the plumb bob was attracted far less than the calculated amount (Figure 7.26).

The importance of Everest's discovery was soon realized by two fellow Englishmen, Sir George Airy, Astronomer Royal of England, and J. H. Pratt, Archdeacon of Calcutta. In 1850, they proposed an explanation that has since been one of the most powerful influences in the development of geologic theories. It was immediately obvious to both of these men that if the plumb bob is not deflected toward the mountains in the amount expected, it is because the crustal rock lying deep beneath the mountains is less dense than elsewhere and that this lack of density largely makes up for the additional mass of the mountains.

Airy then proposed a simple concept of the structure of the earth's crust to describe the arrangement of less dense material under mountains and it has since been shown to be in good agreement with certain facts about crustal density and thickness derived from the interpretation of earthquake waves. (This evidence is given in Chapter 8.) Airy's hypothesis is nicely demonstrated by a simple physical model, shown in Figure 7.27. Suppose that we take several blocks, or prisms, of a metal such as copper. Although all prisms have the same dimensions of cross section, they are cut to varying lengths. Because copper is less dense than mercury, the prisms will float in a dish of that liquid metal. If all blocks are floated side by side in the same orientation, the longest block will float with the greatest amount rising above the level of the mercury surface, and the shortest block has its upper surface lowest. With all blocks now floating at rest, it is obvious that the block rising highest also extends to greatest depth.

Airy supposed that the rock of which high mountains are composed extends far down into the earth to form roots composed of less dense rock. This crustal rock protrudes downward into a location normally occupied by denser mantle rock. Under a low-lying plains region, the root of less dense rock will be shallow. The thin oceanic crust consists of mafic (basaltic) rock, resting on ultramafic rock of the mantle. Both because it is a thin layer and is of higher density than the upper felsic zone of the continental crust, its upper surface will take a position lower than the surface of the continental crust. This explains why the ocean floors are the lowest parts of the earth's surface.

Suppose, now, that we cut off a piece of the top of the longest prism in our model. The prism will rise and come to rest in a new position with the top surface lower than before and the bottom at a shallower depth. The ratio of volumes above and below the mercury surface level with remain the same. If, on the other hand, we should add a small prism of copper to the longest block it would sink and come to a new position of rest. The top of the longer prism would now stand higher than before, but the bottom would sink deeper.

The effect of unloading and loading a floating prism is seen in epeirogenic movements of the cratons caused by removal and deposition of rock. Figure 7.28 shows removal by denudation of mountains (left), transport of the sediment, and sediment accumulation on a low plain (right). With each increment of removal and deposition the mountains rise a bit and the adjacent plain sinks a bit, but in each interval of time the mountain does not rise quite as high as formerly, nor does the surface of the plain sink to quite as low a level as formerly. After vastly long spans of time the mountain and plain will reach nearly the same level.

Underneath the mountains, the plastic material of the asthenosphere will flow in slowly to occupy the space vacated by the rising block, just as the mercury in the model flows in beneath the rising copper prism. Of course, as this newly introduced plastic material rises it comes into a region of reduced temperature and cools to become strong—that is, it becomes a part of the lithosphere. Similarly, under the sinking plain the brittle rock mass is forced down into the region of the asthenosphere, but here it is heated and becomes plastic itself, forming a part of the plastic zone. This excess material at depth must slowly flow horizontally to escape. It seems only logical that it will move toward the region under the rising mountains. The level

FIGURE 7.27 Prisms of copper floating in a dish of mercury can be used to illustrate the Airy hypothesis of mountain roots in conformity with the principle of isostasy.

FIGURE 7.28 Isostatic compensation takes place when rock is either removed by denudation or added by sediment deposition.

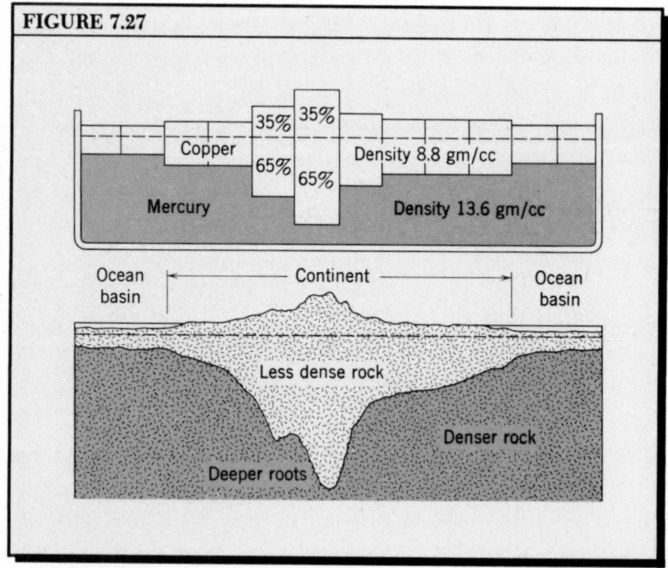

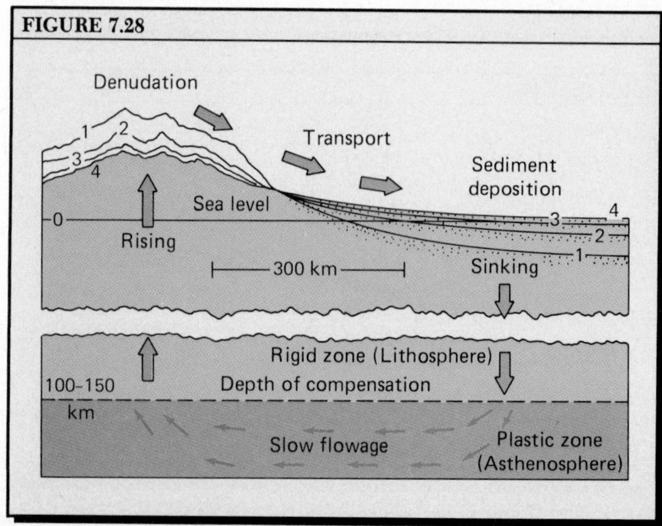

of the base of the lithosphere is therefore referred to as the *depth of compensation,* for it is below this depth that changes in load on the earth's surface are equalized, or compensated for, by slow flowage of the asthenosphere.

*Gravity and the Continental Crust

Extremely precise measurements of gravity at the earth's surface have been of great value in interpreting the structure of the continental crust. The study of place-to-place differences in the value of gravity is a major branch of geophysics and has been intensively pursued by geophysicists the world over for many decades. To make clear how extremely small variations in gravity are used as a research tool, we need to review some basic principles relating to gravity and its measurement. (If necessary, review Newton's law of gravitation and the definition of gravity, given in Chapter 2.)

FIGURE 7.29 The outline of the oblate earth, seen in a cross section through the polar axis, is an ellipse. The elliptical form is greatly exaggerated in this drawing.

FIGURE 7.30 The combined effect of earth rotation and oblate earth form results in an increase in the values of normal sea-level gravity from the equator to the poles. These are standard values used in correcting gravity readings.

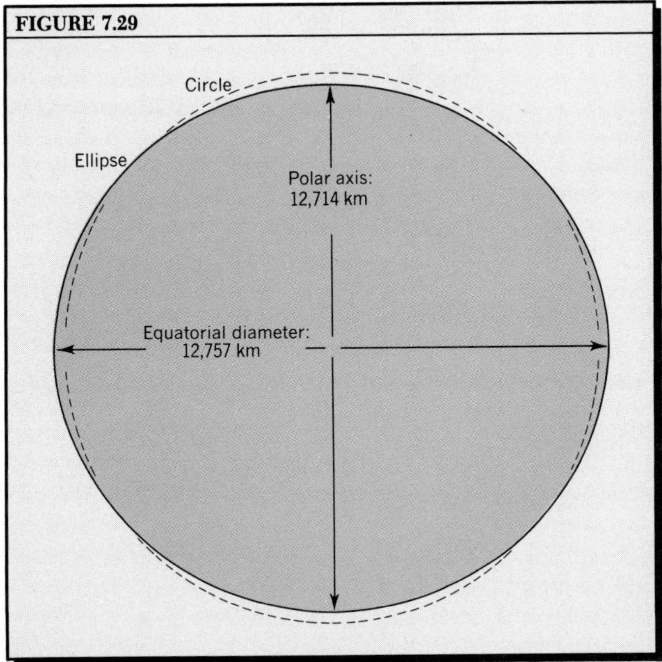

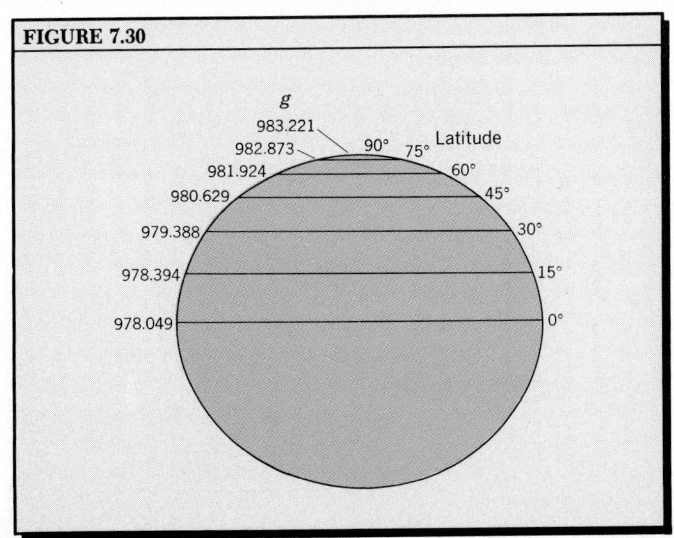

A pendulum clock provides one of the basic mechanisms for the measurement of the acceleration of gravity, *g*, at the earth's surface. The beat-rate, or period, of a pendulum of a given length depends upon the value of gravity. If the earth were a true sphere, perfectly uniform in density throughout, and did not rotate on its axis, gravity would be everywhere the same. A pendulum clock would keep perfect time no matter where on the earth's surface it was located. But, because the earth rotates (spins) on its axis, gravity is affected in two ways.

First, earth rotation sets up an outward or centrifugal force that tends to counteract gravity. A small portion of the gravitational force, acting as a centripetal force, balances the centrifugal, or outward force. This effect is greatest at the equator and decreases to zero at the poles. At the equator the centripetal force is about $1/289$ as great as the gravitational force. This means that an object that now weighs 288 kg at the equator would weigh 289 kg if the earth were not rotating.

A second effect of earth rotation is that the centrifugal force causes a spherical earth to be deformed into an *oblate ellipsoid* (Figure 7.29). As compared to a sphere, the equatorial diameter of an oblate ellipsoid is increased at the same time that the length of polar axis is shortened. On an ellipsoidal earth, points located at the equator are farther from the center of the earth's mass than points near the poles. The ellipsoidal form by itself causes gravity to increase slightly from a minimum value at the equator to a maximum value at the poles. The effect is to diminish the value of gravity at the equator by $1/547$ of the value it would have at the pole.

We must now add the two effects, as follows:

$$\frac{1}{289} + \frac{1}{547} = \frac{1}{189}$$

In practical terms, then, an object that weights 189 kg at the north pole will weigh only 188 kg if taken to a sea-level point on the equator.

Standard Values of Gravity

For use as a research tool of geophysics, gravity must be measured with extreme precision, and this requires a world standard of reference. A world standard station for all gravity measurements was set up at Potsdam, Germany, located at a latitude of about 52½° N. There the value of 981.274 cm/sec^2 was determined for the acceleration of gravity. The unit of gravity, known as the *gal* (named for Galileo), is equal to 1 cm/sec^2. Because gravity differences over the earth's surface are very small, the thousandth part of a gal, termed the *milligal,* is used as the unit for stating differences in gravity measurements. Thus, the standard Potsdam value of gravity is equivalent to 981,274 milligals.

Once the standard value of gravity was established at Potsdam, it was possible to set up a standard system of gravity values according to latitude, from the equator to the poles, taking into account the effects of earth oblateness and rotation. A simple formula is used to determine the standard value at any given latitude. As Figure 7.30 shows, *normal sea-level gravity* ranges from a minimum of 978.049 cm/sec^2 at the equator to 983.221 cm/sec^2 at the poles, the difference being 5.172 cm/sec^2 or 5172 milligals.

Gravity may be measured with either a pendulum apparatus or an instrument called a *gravimeter.* The pendulum apparatus is used at standard stations for absolute gravity measurements. Pendulums may also be used to make relative gravity measurements, that is, to observe differences in gravity from one point to another. In its simplest form, a pendulum apparatus would be a single pendulum so constructed that its length remains absolutely constant. When the length is held constant, differences in period from one place to another reflect differences in gravity.

Actually, one pendulum is not enough in the precision gravity apparatus, because its motion disturbs the housing from which it is hung. To cancel out the effects of such disturbances, two pendulums may be swung simultaneously in opposite phase from one instrument housing. Figure 7.31 shows a modern apparatus consisting of two quartz pendulums hung on precise knife edges.

The gravimeter uses the spring balance as its basic mechanism (Figure 7.32). A weight is hung from a coiled spring. Change in length of the spring is proportional to the change in acceleration of gravity. Slight changes in spring length are amplified by a combination of mechanical, optical, and electrical means. The gravimeter must be calibrated against a series of base stations so that the differences in spring length can be interpreted as gravity values.

Correcting Gravity Readings

Suppose that we have calibrated a portable gravimeter at a base station and have taken the instrument to a distant field location, perhaps in a mountainous region, where we have set it up and taken a reading of gravity. For what reasons might this gravity reading differ from that of the base station? Base station readings are in terms of normal sea-level gravity, taking into account only the earth's ellipsoidal form and its rotation. Normal sea-level gravity thus represents an earth-model in which the land areas are assumed to be

FIGURE 7.31 The Gulf gravity pendulum is shown here with top case removed. At right is a quartz pendulum and the Pyrex flat on which it rests. (Gulf Research & Development Company.)

FIGURE 7.32 Simplified schematic diagram of a spring gravimeter. (Based on data of B. F. Howell, Jr., 1959, *Introduction to Geophysics,* McGraw-Hill, New York, p. 212, Figure 14.3.)

FIGURE 7.31.

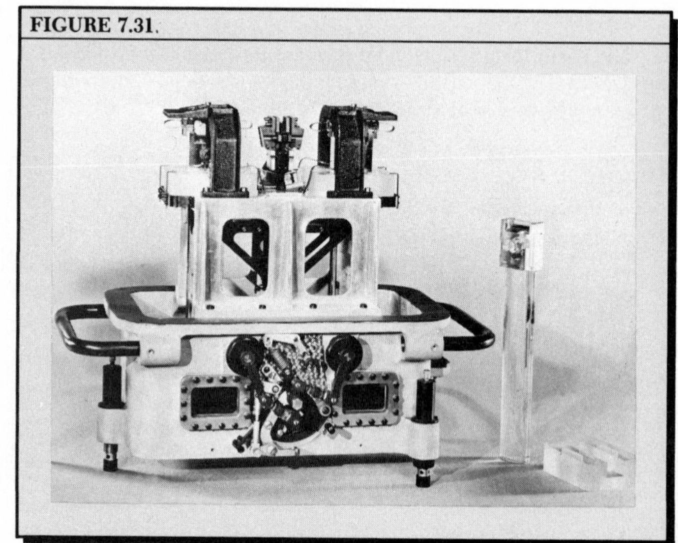

FIGURE 7.32

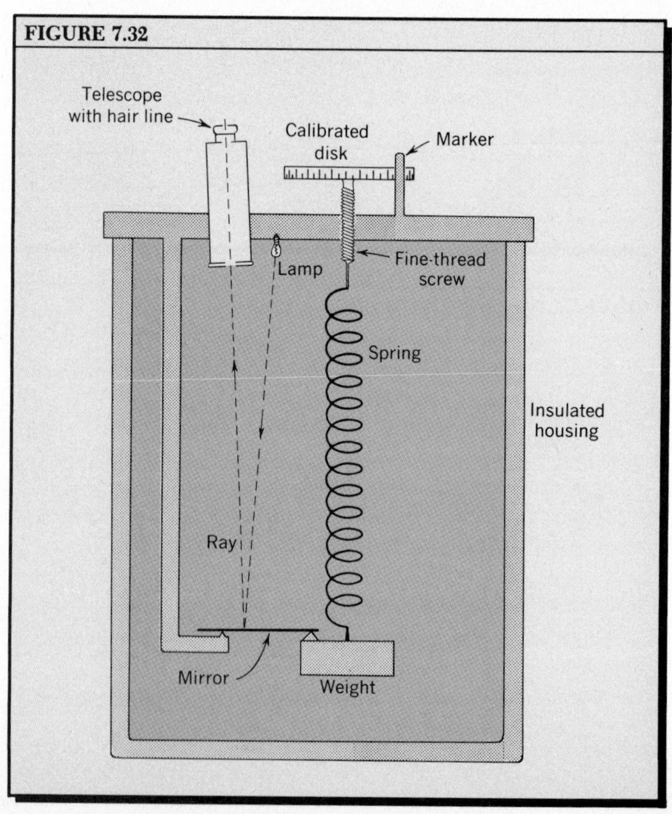

reduced to smooth surfaces at sea leavel while the ocean basins are assumed to be filled to sea level with crustal rock. Moreover, the density of crustal rock is assumed to be equal everywhere over the globe and to increase downward in density uniformly or in a succession of uniform layers.

When our gravimeter has been taken to a higher elevation than that of the base station, it will register a lower value of gravity simply because it is farther from the earth's center. The correction for elevation, known as the *free-air correction*, can be obtained by solving a simple formula. As a rough approximation we can say that the value of gravity decreases about 0.3 milligal for each meter of ascent. This correction is added to the gravimeter reading.

A second correction to be made on our gravity reading is required by the earth's gross surface configuration. Where a high mountain or plateau exists, a large mass of rock lies above sea level. This mass exerts a gravitational pull of its own. Therefore, gravity at the imaginary ellipsoid surface will be somewhat less beneath high continental masses than on an open plain at the ellipsoid surface, all other conditions being the same.

The correction for effects of masses lying above the ellipsoid surface or for deficiencies of mass below that level is termed the *Bouguer correction*. It was named after Pierre Bouguer, a leader of an early eighteenth-century geodetic expedition to Peru. The Bouguer correction for mass distribution with respect to the ellipsoid surface is made with the assumption of a horizontal rock slab of uniform density. Usually, a density of 2.67 g/cc is assumed for the rock slab. To make the correction, we need to know only the thickness of material between the earth's solid surface and the ideal ellipsoid. Using a formula, a subtraction is made from the gravimeter reading obtained on the elevated land surface. The Bouguer correction amounts to about 0.1 milligal per meter of rock thickness.

A third correction takes into account the presence of local topographic features such as a nearby mountain or an adjacent valley. These features will either subtract from or add to the value of the gravimeter reading. A *topographic correction* is therefore made in addition to the Bouguer correction, because the latter assumes that the land surface on which the gravimeter rests is a horizontal surface, i.e., a featureless plain.

After applying all the corrections thus far noted, we may find that the observed value of gravity still does not agree with the normal sea-level gravity. A difference between observed and predicted gravity values is, in general, termed a *gravity anomaly*. In this case, it is termed the *Bouguer anomaly*. When an anomaly is found, a cause other than those already taken into account must be sought to explain the discrepancy.

Isostatic Gravity Anomalies

Suppose you were to join a party of geophysicists in a traverse across the United States from the Pacific to the Atlantic coast. Stopping frequently to take gravity readings with a portable gravimeter, you apply the corrections we have listed, calculating the Bouguer anomaly for each station. After plotting these anomaly values against the line of traverse, you compare the resulting *gravity anomaly profile* with the topographic profile, as shown in Figure 7.33. You would be struck by the fact that strong negative Bouguer anom-

FIGURE 7.33 Topographic and gravity anomaly profiles crossing the United States from west to east. (Based on data of G. P. Woollard, 1966, *The Earth Beneath the Continents, Monograph 10,* American Geophysical Union, p. 560; L. C. Pakiser and I. Zietz, 1965, *Reviews of Geophysics,* vol. 3, pp. 505–520.)

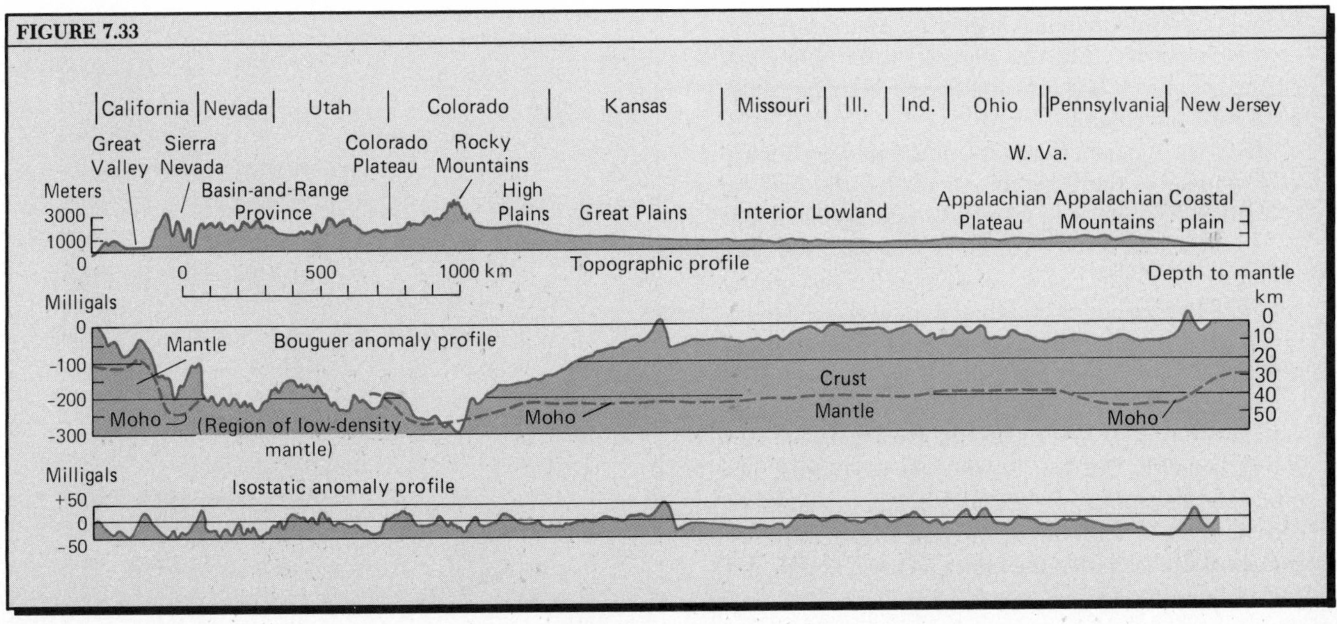

alies coincide with the Sierra Nevada and Rocky Mountains. You also observe that negative anomalies are large in the Basin-and-Range region that lies between them. Over the interior lowland and Appalachian Plateau, the anomaly values are almost always negative and rather uniformly small, but there is a marked increase in the negative value beneath the Appalachian Mountains. If the traverse were carried out over the Atlantic Ocean basin, it would show a switch to a strong, uniform positive Bouguer anomaly.

The significance of the Bouguer anomaly we observe in this profile across a continent and over an ocean basin is simply this: We made a wrong assumption about the way rock density is distributed in the crust and upper mantle. We had been assuming that there exist horizontal rock layers of constant thickness, and that within each layer, density is the same in all horizontal directions. We were also assuming that density increases progressively downward from one layer to the next. Evidently, the assumption about constant thickness of each layer does not hold valid. From what you have already learned in Chapter 4 of crustal structure and thickness and of the depth to the Moho (crust–mantle boundary), you could have predicted what the problem is (see Figure 4.8). When the profile of the Moho is drawn over our gravity anomaly profile (Figure 7.33) deep crustal roots are revealed beneath the Sierras and Rockies. These roots consist of less dense rock than average for this depth. The Moho is somewhat shallower over the interior lowland region, but again shows a deepening beneath the Appalachians. The Moho then rises toward the Atlantic margin and becomes very shallow beneath the oceanic crust (beyond the limit of the profile).

Obviously, a further correction, beyond the Bouguer and topographic corrections, is needed for our gravity readings. To make this correction, we must use evidence from earthquake waves (seismic evidence) allowing us to establish the thickness and density of the crustal layers and mantle (Chapter 8). An *isostatic correction* is needed. For this purpose, a set of vertical crustal columns is laid out, and the correction is calculated for each column.

The Airy model of isostasy, illustrated in Figure 7.27, is used as the basis for calculating the isostatic correction. The first step is to make assumptions about a column whose surface coincides with sea level. For example, we might choose to assume that this crustal column has a depth of 30 km and an average density of 2.9 g/cc (Figure 7.34). An assumed value for density of the mantle rock beneath is 3.3 g/cc. Next, it is assumed that for each increase of 1 km in surface elevation with respect to sea level, the base of the column is lengthened by 7.5 km. Figure 7.34 shows such columns for elevations of 1, 2, and 3 km. A model of oceanic crust is also shown, for comparison. Depth of compensation is double the crustal column length and exceeds 100 km under the longest column.

After the isostatic crustal model is set up, the next step is to calculate a gravity correction based upon the masses of the columns. This isostatic correction is subtracted from the Bouguer correction, leaving a much smaller anomaly known as the ***isostatic gravity anomaly***. If the crustal model is a true one and the condition of isostasy prevails, this anomaly should have zero value. In actual fact, computed isostatic anomalies ranging from −30 to +30 milligals characterize most of the surface of the continents. This small range is seen in the isostatic anomaly profile in Figure 7.33. The anomalies take the form of small "hills" and "depressions." They reflect place-to-place differences in rock density caused by the presence of local crustal rock masses of various dimensions, compositions, and structures. On the whole, however, much of the earth's crust, both continental and oceanic, is thought to be near isostatic equilibrium.

The Continental Crust in Review

In this and the previous four chapters, we have covered some basic areas of geology—minerals and rocks, their structures, and their assembly into the major earth shells. Viewing the rock transformation cycle as an expression of plate tectonics has been especially helpful in understanding how the continental crust

FIGURE 7.34 Simplified Airy isostatic model of crust. (Based on data of G. P. Woollard, 1966, *The Earth Beneath the Continents, Monograph 10*, Washington, D.C., American Geophysical Union, p. 563.)

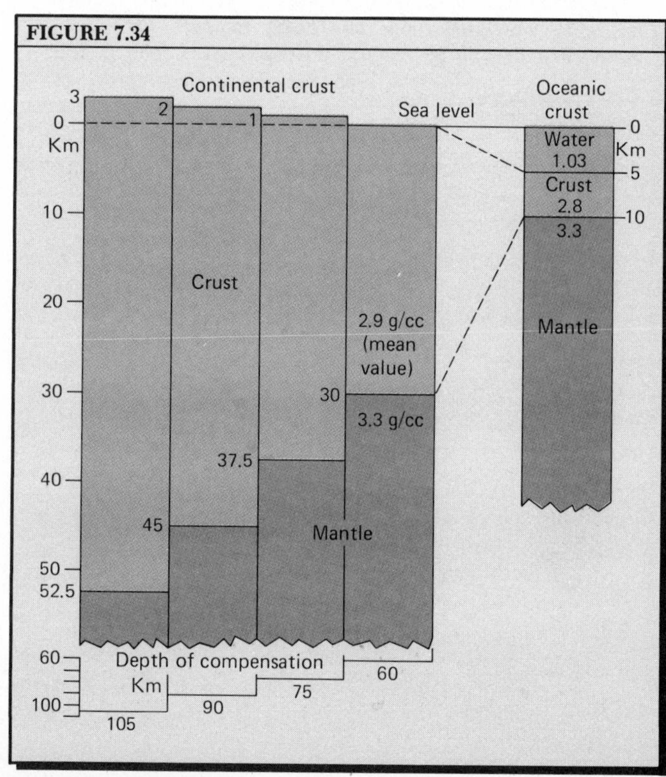

came into existence. We have also covered six other key topics in geology—stratigraphy, rock deformation, radiogenic heat, radiometric age determination, isostasy, and gravity.

Other important topics in geology remain to be examined. One is the nature of earthquakes and earthquake waves and the ways in which their study reveals a great deal about the earth's interior and the processes of plate tectonics. Another new area yet to be visited is the ocean floor, its topography and sediments, and the crust beneath it. As we investigate these topics in the next few chapters, many important details of plate tectonics will emerge as essential parts of the total picture of our earth and its workings.

Key Facts and Concepts

Rock metamorphism *Rock metamorphism* consists of physical and chemical changes in rock in an environment of high temperature, high confining pressure, intense shearing, or some combination of two or three of those factors. Metamorphism may result in the formation of new minerals, new textures, or new structures. Metamorphism may involve *metasomatism,* the import or export of mineral substances.

Classes of metamorphic rocks Major classes are *cataclastic rocks,* those undergoing only *dynamic metamorphism,* and *recrystallized rocks,* those undergoing change in mineral chemistry and crystal lattice structure. Recrystallized rocks include two subclasses: *contact metamorphic rocks,* adjacent to igneous intrusions, and *regional metamorphic rocks* that have undergone *dynamothermal metamorphism.* A third class consists of rocks transitional between metamorphic and igneous rocks.

Metamorphic minerals Certain minerals of igneous rocks persist or reappear in metamorphic rocks, e.g., quartz, biotite mica, pyroxene, amphibole, and feldspar. Important new minerals include aluminosilicates (kyanite, andalusite, sillimanite), a silicate of iron (almandite garnet), a silicate of calcium (wollastonite), hydrous aluminosilicates (staurolite, chlorite, epidote), and hydrous silicates of magnesium (talc, serpentine).

Cataclastic rocks *Cataclastic rocks,* formed under intense shearing that produces grain fragmentation, include *friction breccia* and *mylonite* formed on fault planes.

Contact metamorphic rocks *Contact metamorphic rocks* form close to intrusive magma bodies where temperature is high. An example is *hornfels,* forming an *aureole* that is often zoned in mineral content.

Texture of regional metamorphic rocks New textures are produced by intense shearing action accompanied by recrystallization affecting rock masses on a regional scale. Shale develops *slaty cleavage* and may become *slate.* Further metamorphism alters slate to *schist,* possessing a form of *foliation* known as *schistosity.* Other structures are *lineation* and *banding,* seen in *gneiss.*

Metamorphic zones and grades Dynamothermal metamorphism produces a sequence of *metamorphic zones,* reflecting increased temperature accompanying recrystallization. Zones are identified by *index minerals* in sequence: chlorite, biotite, almandite, staurolite, kyanite, and sillimanite. *Isograds* delineate *metamorphic grades* within zones. In a typical sequence of grades *greenschist* (low-grade metamorphism) gives way to *amphibolite* (intermediate-grade) and ends in *pyroxene granulite* (high-grade). Another common sequence, formed under high confining pressure but low temperature, leads to *blueschist,* followed by *eclogite.* Carbonate rocks are re-formed as *marble;* pure quartz sandstone becomes *quartzite.*

Granitization *Migmatites,* banded rocks containing layers of granitelike rock, are possibly produced by metasomatism. Wholesale conversion of felsic rock to granite in the metasomatic process of *granitization* without melting may account for large masses of crustal granite.

Hydrothermal alteration *Hydrothermal alteration,* caused by hot, watery solution emanating from shallow magma bodies, forms new hydrous minerals. Most important of these is *serpentine,* hydrous magnesium silicate, forming *serpentinite* rock in the process of *serpentinization,* a process of probable major importance along spreading plate boundaries.

Radioisotopes and radioactivity *Nuclides,* all isotopes of one element, may have a stable or an unstable nucleus. *Unstable isotopes* undergo spontaneous disintegration, or *radioactivity,* by a process of *radioactive decay.* The unstable isotopes are called *radioisotopes* or *radionuclides.* Radioactive decay may occur as *alpha decay, beta decay,* or *electron capture,* processes that generate heat energy in the surrounding rock. Heat production of a *decay series* reaches a steady state during *exponential decay* with a specific *half-life.* Accumulation of daughter isotopes correspondingly follows an exponential increase.

Radiogenic heat The *geothermal gradient,* a temperature increase with depth, averages about 3 °C per 100 m for shallow depths and results in slow, upward heat flow. Heat is generated mostly by radioisotopes of uranium, thorium, and potassium, largely concentrated in the felsic upper zone of the continental crust. Little heat is generated in the mantle and almost none in the core.

Early thermal history of the earth Total production of radiogenic heat was at the maximum at the time of the earth's formation and has steadily declined.

Initially uniformly distributed, the radioactive minerals became concentrated near the surface during and following an episode of melting of iron that saw the iron core formed by differentiation from mafic silicate minerals.

Radiometric age determination *Geochronometry* uses radioactivity to derive *radiometric ages* of rocks. Ages determined from ratios of parent to daughter products are measured for the uranium–lead, thorium–lead, potassium–argon, and rubidium–strontium decay series.

Continental structure A *craton,* the ancient stable portion of the continental crust, consists of igneous and metamorphic rocks, ages mostly 0.8 to 2.7 b.y. *Shield* portions of a craton consist of exposed Precambrian rock; *platforms* bear a cover of sedimentary rocks usually younger than Precambrian. Portions of the cratons older than about 2.5 b.y. make up the *continental nuclei;* they are surrounded by younger craton masses forming age provinces. Continental denudation accompanying repeated epeirogenic uplift has exposed deep-seated craton rocks.

Isostasy Under the principle of *isostasy,* lithosphere floats in *isostatic equilibrium* on the asthenosphere. Mountains of low-density felsic rock are underlain by deep felsic roots under the Airy model of isostasy. Loading and unloading by deposition and denudation, respectively, are compensated for by sinking or rising motions that maintain isostatic equilibrium through rock flowage within the *depth of compensation.*

Gravity and the crust Minute variations in gravity at the earth's surface, measured in *milligals* by use of the *gravimeter,* are corrected for elevation, mass excess or deficiency (Bouguer correction), and topography to reveal a *gravity anomaly* (Bouguer anomaly). Correction for variations in crustal density, based on seismic information, allows calculation of an *isostatic gravity anomaly,* which is small over most of the continental crust.

Questions and Problems

1. Examine specimens of schist containing large crystals of such metamorphic minerals as kyanite, staurolite, or garnet. (Photographs in the text can serve in place of specimens.) How is it possible for these remarkably perfect crystals to develop and grow intact in a mass of schist that is undergoing continual shearing along innumerable foliation planes? Would not the shearing action destroy the crystals as rapidly as they could form? Do you suppose that the crystals grew in the rock after it had ceased being formed?
2. How might you distinguish a friction breccia from a sedimentary breccia, assuming that you have available the equipment necessary to determine the rock type and mineral composition of all particles that make up the rock?
3. Suppose that you are examining a large outcrop of granite that may be part of a batholith and the question comes up as to the origin of the granite. By studying the rock in every possible way, including chemical analysis, could you decide whether it was formed by granitization, without melting and involving only metasomatism, or whether it represents a body of magma that intruded the country rock by stoping and assimilation to form a batholith? Would it be helpful to follow the granite body to its limits and observe its contact with older rocks? Explain.
4. If we accept the statement that production of radiogenic heat is much greater in felsic rock of the upper layer of the continental crust than in the lower crust and mantle, how does it happen that mantle rock in the asthenosphere is close to its melting point (an assertion made in Chapter 1)? Would it not be more reasonable to suppose that the rock is hottest where radiogenic heat is most rapidly produced, and hence that rock temperature must become cooler in the downward direction through the crust and into the mantle? What, if anything, is wrong with such reasoning?
5. Would it be possible to use the uranium–lead series U-238 to Pb-206 to determine with high accuracy the age of an igneous rock as young as one million years? Why would the use of the potassium–argon decay series tend to be more accurate for rocks as young as one million years?
6. Apply your elementary knowledge of plate tectonics, as presented in Chapter 1, to the explanation of the age-provinces of the Precambrian Canadian shield. How might each of these parts of the craton have originated? How might they have come to be joined together?
7. We stated near the end of this chapter that the calculated isostatic gravity anomaly is comparatively small across the entire United States, and that this situation is to be expected if isostatic equilibrium exists. Suppose that a large positive isostatic anomaly were discovered to coincide with a high mountain range, such as the European Alps or the Andes. How might that large anomaly be explained in terms of what lies beneath the mountain range?

8

CHAPTER EIGHT

Earthquakes and the Earth's Interior

Determining the composition and physical properties of the earth's interior is one of the most difficult problems to have confronted earth scientists. The deepest mines, such as those of the Rand in South Africa, give us direct views of rock down to depths of about 3600 m—only about 1/2000 of the distance to the earth's center. The deepest boreholes driven in search of petroleum now penetrate more than 6 km, but even this depth is trivial in comparison with the earth's radius.

Some information on what lies at depth comes from the examination of surface rocks that were formerly near the base of the continental crust or in the uppermost mantle. For example, the diamond-bearing rock, kimberlite, in South African mines, may have come up directly through the crust from the upper mantle at a depth of perhaps 35 to 40 km. Metamorphic rocks, such as the blueschists and eclogites described in Chapter 7, may have originated at depths of 40 to 60 km, at most.

Obviously, inferences as to the composition and structure of most of the mantle and all of the core must come indirectly. The methods used draw upon geophysics, which applies principles of physics and mathematics to interpret data obtained from sensitive and precise instruments. Thus, geophysics differs strikingly from the direct-observation methods used by the earlier naturalists and geologists. Interpretations of the findings of geophysics in terms of the rock chemistry of the earth's interior are made by geochemists, who apply principles of chemistry to geologic problems.

Pressure, Temperature, and Density in the Earth's Interior

In Chapter 4, we showed estimates of confining pressure and temperature from the surface to a depth of 700 km (Figure 4.1). Graph *A* of Figure 8.1 shows the increase of pressure to the earth's center, where it reaches about 3500 kb. Matter at the earth's center has no weight, because it is being attracted outward equally in all directions by the mass of the sphere that surrounds it, but the confining pressure is greatest here because the weight of the entire earth mass is directed toward that one point. We have already seen that high values of confining pressure have profound effects upon the physical properties of rocks and their behavior when unequal stresses are applied.

Figure 8.1, Graph *B*, shows the profile of temperature increase with depth, which is rapid in the first 200 km of depth, but the rate of increase falls off sharply below this depth (see Figure 4.1). Through the core, temperature increase is very gradual and reaches a maximum estimated to be about 2500 °C (2775 °K) at the earth's center.

Increase in density, shown in Graph *C* of Figure 8.1, is fairly steady with depth through the mantle, but makes an abrupt jump where the iron core is encountered. Measurement of the average density of the earth as a whole requires a knowledge of both its mass and its volume. The earth's mass was determined in 1798 by the English physicist, Lord Cavendish, who was able to measure the universal gravitational constant by means of a delicate instrument, the torsion balance. This constant allows the gravitational equation of Newton to be solved. The earth's radius and volume were already quite accurately known from geodetic surveys. Thus, we have the two necessary quantities:

Earth's mass:	5.975×10^{27} g	(round to 6×10^{27} g)
Earth's volume:	1.08×10^{27} cc	(round to 1.1×10^{27} cc)

Dividing mass by volume gives us a density of about 5.5 g/cc. This figure is twice the density of granite (2.6 g/cc). Iron, in comparison, has a density of nearly 8 g/cc.

The geophysicist reasons that because the greater part of known rock of the earth's outermost zone is of the density of granite (2.6 g/cc) or of basalt (3.0 g/cc)—which are much less than the global average—density must increase greatly toward the earth's center. The most reasonable conclusion is that there exists a core of high density, in the range of 10 to 12 g/cc. (This reasoning is strongly supported by the astronomer's determination of another physical quantity—the earth's moment of inertia.) The core is usually considered to be composed largely of iron, because if iron were compressed in the earth's center under 2000 to 3000 kb of confining pressure, it would decrease in volume and undergo a density increase from 8.0 to 11.0 g/cc. Independent supporting evidence (see Chapter 20) for an iron core is found in the composition of meteorites, one class of which is composed of iron.

Between the outermost layer and the earth's core, density must range somewhere between 4.0 and 6.0 g/cc. Rocks that fit this requirement are peridotite and dunite, both ultramafic and composed largely of olivine (see Chapter 4). Their density is normally about 3.3 g/cc, but would be increased to an acceptable value under the known confining pressures.

Earthquakes and Faults

The most important evidence about the nature of the earth's interior, and particularly about the actual depths at which the composition and properties change, comes from the science of *seismology*, a branch of geophysics dealing with earthquake waves. To understand how seismologists interpret earthquake waves, we must first consider how earthquakes are generated and how they are recorded.

The *earthquake* is known to humans directly as a trembling or shaking of the ground. Commonly this motion is barely perceptible to the senses, but sometimes it is so violent that it damages or collapses strong buildings, breaks water and gas mains, and causes long cracks in the ground. In cities, structural damage and

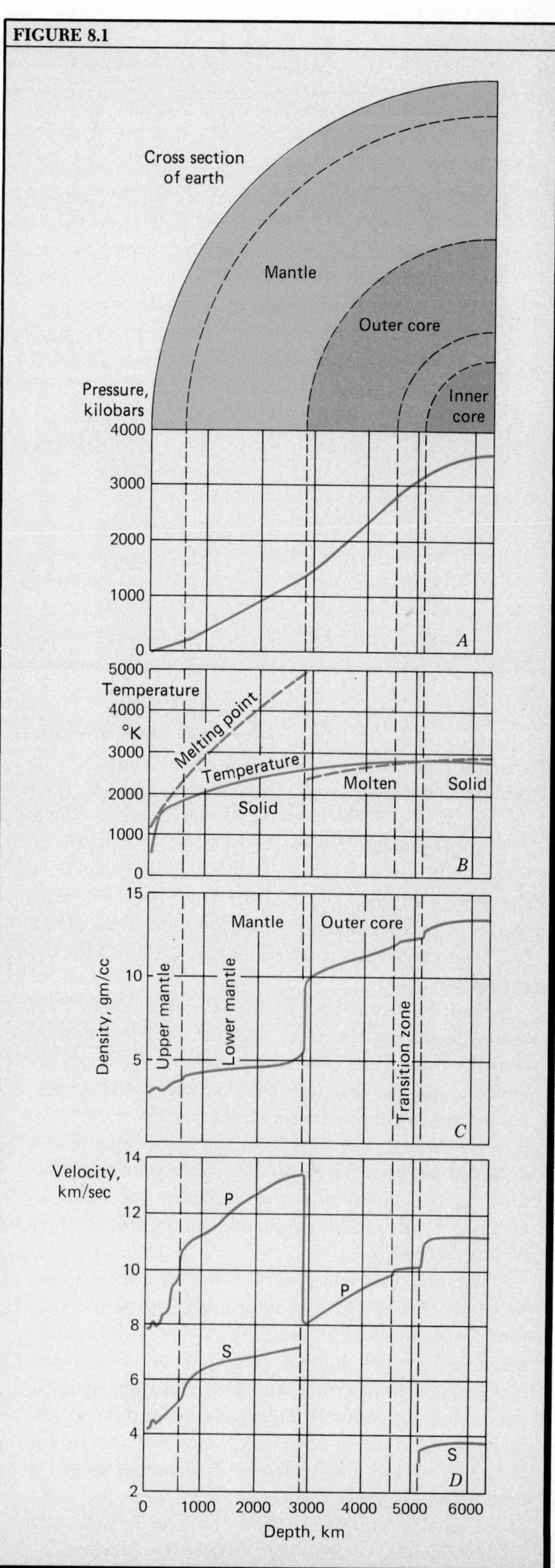

ensuing fire can bring injury and death to large numbers of human beings.

The important problem at hand is the nature of the earthquake waves, or *seismic waves*, as interpreted from the records of the ***seismograph***, a sensitive instrument that can record earthquakes thousands of kilometers distant and so small that their vibrations could not possibly be recognized by the human senses. The records of seismic waves that have passed through the earth's interior provide the evidence we seek about that remote and inaccessible region.

A major class of earthquakes is produced by faulting of brittle crustal rock that bends elastically and thus stores energy as elastic strain. When the elastic limit is exceeded, faulting occurs suddenly and a large quantity of energy is released. A good analogy is the bending of a crossbow, storing energy slowly, but releasing it with the sudden flight of an arrow. Great accumulations of elastic strain build up constantly along active lithospheric plate boundaries, particularly along converging boundaries and active transform faults. The entire lithosphere is capable of generating earthquakes, and many of these occur within downbent slabs of lithospheric plates that are descending deep into the asthenosphere. Slippages of opposed plates in these converging zones release enormous quantities of energy in the form of great earthquakes.

Geologists often describe the mechanism that generates earthquakes along plate boundaries as *stick–slip friction*. Lithospheric plates do not glide smoothly past one another in the motion we see, for example, in the well-lubricated pistons of an automobile engine. Instead, the strong friction that exists between plates causes them to "stick" fast under stress, then to "slip" suddenly, and again to "stick," resulting in a series of jerky motions. Each slip generates an earthquake and releases a certain quantity of energy. You can easily illustrate stick–slip friction with a long, flexible wooden lath or yardstick. Hold the lath in both hands perpendicular to a cement floor, then pull it toward you, applying downward force at the same time. The end of the lath will move over the cement in a series of jerks. The staccato sounds that are emitted represent earthquake waves traveling outward from the point of slip.

FIGURE 8.1 A. Increase in pressure with depth in the earth. *B.* Increase in temperature with depth. *C.* Increase in density with depth. *D.* Velocity of P and S waves. (Data for *A* from K. E. Bullen, 1963, *An Introduction to the Theory of Seismology,* 3d ed., Cambridge Univ. Press, Cambridge. Data for *C* and *D* from B. A. Bolt, 1973, *Scientific American,* vol. 228, no. 3, p. 33.)

FIGURE 8.2 A schematic diagram of the spread of rupture along a fault plane. Rupture begins at the focus and travels outward rapidly, as suggested by the numbered concentric bands. Earthquake waves are not shown here. (Modified from Bruce A. Bolt, *Nuclear Explosions and Earthquakes,* W. H. Freeman and Company, San Francisco. Copyright © 1976.)

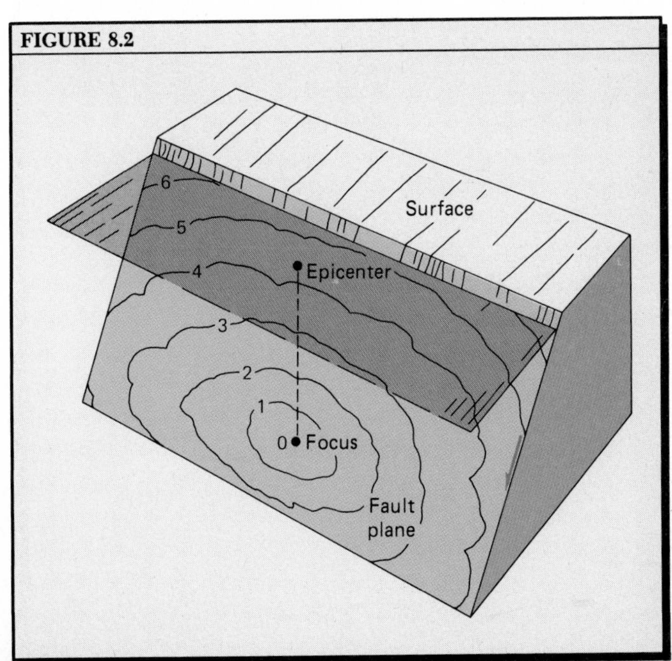

Rupture on a Fault

As long as the frictional force binding the two crustal blocks together along an active fault is sufficiently strong, elastic strain continues to build in the adjacent rock. Rupture begins at a point on the fault plane where the frictional force is first overcome. This initial point is the *earthquake focus* (Figure 8.2). The point on the surface directly above the focus is called the *epicenter*.

Rupture, which rapidly spreads outward along the fault plane from the focus to the ground surface, allows the elastic strain to be relieved by fault slippage. Ultimately, slippage spreads over a horizontal distance that may measure many kilometers or even hundreds of kilometers. The leading edge of the rupture travels at a speed on the order of 3.5 km per second.

Besides the main rupture that produces a moderate-to-large earthquake, many small earthquakes are generated after and sometimes before the main shock. Yielding by slip along the fault surface may begin days or weeks in advance of the main shock in the form of very small movements that cause *foreshocks*. Usually, these foreshocks are too weak to be felt by humans, but they can be recorded by seismographs close to the fault.

Following the main shock, there are usually many secondary shocks, called *aftershocks*, some of which are strong enough to be felt. These occur at irregular intervals over a period of many months. Following a major earthquake, aftershocks of considerable severity usually occur, causing additional damage to structures. Each aftershock relieves a small part of the elastic strain that was not fully relieved by the main shock until, finally, so little strain energy remains unrelieved that the fault becomes stable. Even so, sensitive seismographs record a background seismic activity of very small shocks that goes on more or less continually with foci distributed at various points along the length of the fault.

Normal, reverse, and transcurrent faults affect only the lithosphere, which has brittle properties. An earthquake generated on one of these three types of faults will be a *shallow-focus earthquake*, with a focus less than about 50 km deep. Many earthquakes have foci much deeper, and may be located within the entire thickness of the asthenosphere to a depth of 700 km. These *deep-focus earthquakes* are generated in subduction zones within down-plunging lithospheric plates. We shall have more to say about deep-focus earthquakes later in this chapter and in Chapter 11.

Surface Displacements Accompanying Earthquakes

Where a fault intersects the land surface, ground movements that precede, accompany, and follow a major earthquake can be observed and measured. A simple model of these effects can be made with a strip of tempered steel, such as a coping saw blade, with its ends tightly clamped in wooden blocks (Figure 8.3). If the blocks are forced to move parallel with one another to produce an S-bend in the blade, it can be bent slowly to the breaking point. When the blade snaps, the broken ends whip back into straight pieces, but considerably offset from each other.

This model illustrates the *elastic-rebound effect* that occurs on both sides of a fault following an earthquake. That this movement exists is documented by actual measurements of bending of the ground on either side of a known line of faulting both before and after the great San Francisco earthquake of 1906. Two scientists of the University of California at Berkeley based their measurements on precise geodetic surveys of the exact positions of several triangulation monuments on mountain peaks on both sides of the San Andreas Fault, the transcurrent fault which runs through San Francisco and on which the slip of 1906 took place. The survey data showed that after the earthquake was over the ground lying close to the fault line had moved most, with a gradually decreasing amount of movement away from the fault (Figure 8.4). The maximum measured displacement along the fault was 6.4 m.

The slow bending of surface rock prior to the earthquake showed that stored energy had been accumulating at depth for tens or even hundreds of years before the earthquake occurred. Thus, the rock was already bent almost to its elastic limit when the first triangulation surveys were made, and slow bending was still in

FIGURE 8.3 A steel blade, bent until it snaps, illustrates certain basic features of the earthquake mechanism.

FIGURE 8.4 An earthquake results from the sudden release of elastic strain that has been accumulated in rock over a long period of time.

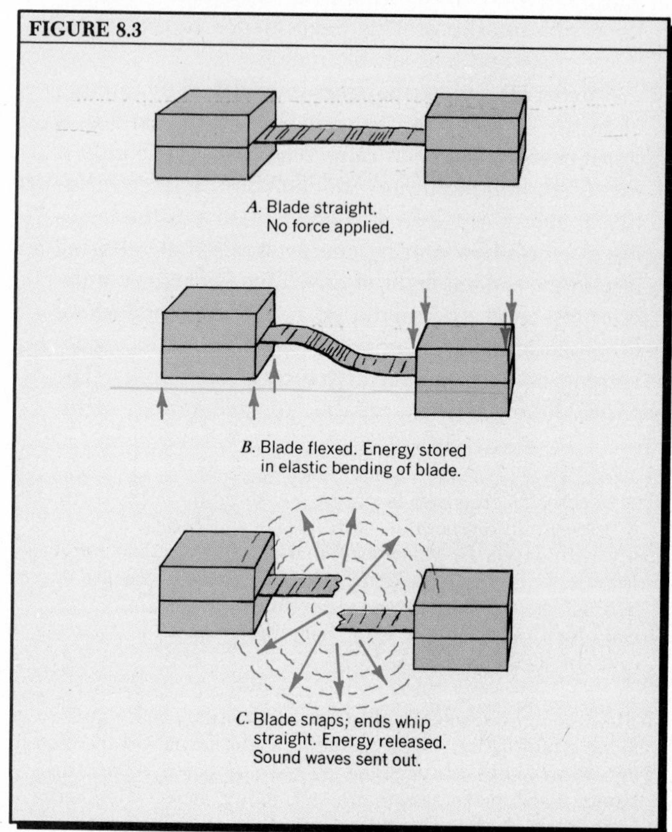

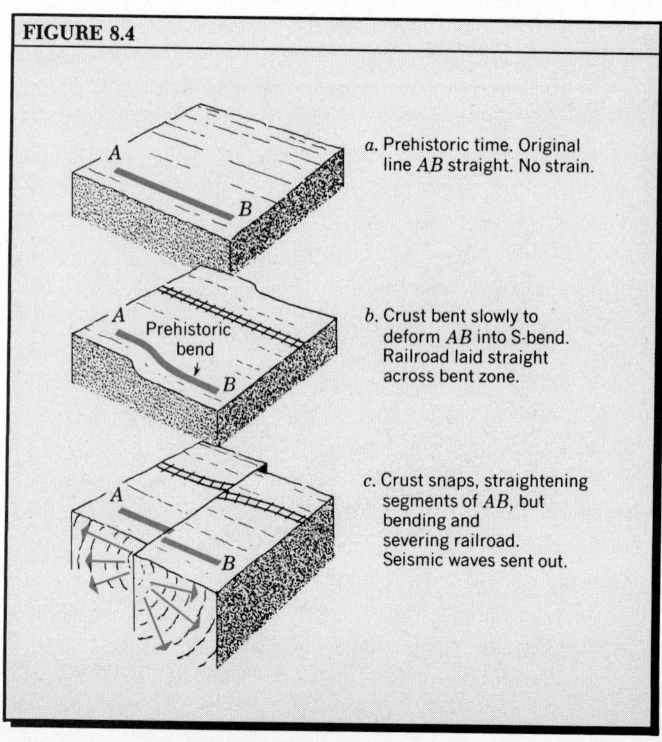

progress between 1851 and 1906 (Figure 8.4). Then, in 1906, the rock rapidly rebounded to a new position.

Transcurrent faults on a large scale are well represented in California, where they form dominant geologic elements. The major California fault systems are described later in this chapter, but the longest and best known is the San Andreas Fault (mentioned above), nearly 950 km long. Horizontal movement occurs locally from time to time along California's transcurrent faults, generating earthquakes of greatly varying intensity. Movement is strikingly illustrated by the offsetting of such linear features as roads, fences, pipelines, and orchard rows (Figure 8.5).

A single movement on a normal fault may produce a fresh fault scarp that runs for many kilometers across the land surface. A good example is a fault scarp produced in soft alluvium (sand and gravel) during the Hebgen Lake earthquake, which occurred in 1959 in Montana (Figure 8.6).

FIGURE 8.5 Evidences of lateral earth movement accompanying the San Francisco earthquake of 1906, Marin County, California. (Photographs by G. K. Gilbert, U.S. Geological Survey.) *A*. A fence offset 2.4 m along the main fault near Woodville. *B*. A road offset 6 m along the main fault near Point Reyes Station. The shear zone is 18 m wide.

FIGURE 8.6 A fault scarp in alluvial materials formed during the Hebgen Lake earthquake. Displacement was about 6 m at the maximum point. The vehicle stands on the upthrown side of the fault. (J. G. Stacy, U.S. Geological Survey.)

FIGURE 8.5 A

FIGURE 8.5 B

The Seismograph

The mechanical problem in recording an earthquake and measuring the directions and amounts of the earth motions involved is that the instrument itself must be resting on the ground and will therefore also move. Because the instrument cannot be physically separated from the earth, the seismograph designer must make use of the principle of inertia to overcome the effect of the attachment. *Inertia,* the tendency of any mass to resist a change in a state of rest or of uniform motion in a straight line, is greater the greater the mass of the object.

To record an earthquake, then, a heavy weight, such as an iron ball, might be suspended from a very thin wire or from a flexible coil spring, as shown in Figure 8.7. When the earth moves back and forth or up and down in earthquake wave motion, the large mass will stay almost motionless because the supporting wire or spring flexes easily and does not transmit the motion through to the weight. If a pen is now attached to the weight, so that the point is just touching a sheet of paper wrapped around a moving drum, the pen will produce a wavy line on the paper. Strong shocks will

FIGURE 8.6

give waves of high *amplitude* (distance from the rest position to a peak or trough), and weak shocks will give waves of low amplitude. When the number of back and forth movements per second (i.e., the *frequency*) is higher, the undulations of the line will be more closely crowded (Figure 8.8).

The seismograph as we have described it is too simple to be workable. In the first place the movement of the ground is so very small that the motion must be greatly magnified if it is to produce a record suitable for study. We know that a small pocket mirror can be used to reflect a spot of sunlight onto the wall of a distant house and that a very slight twist of the hand causes the spot of light to jump several meters. In the seismograph, a mirror may be attached to the heavy weight and a very tiny light beam reflected from it onto photographic paper attached to the slowly moving drum (Figure 8.9). (Of course, either the room is darkened or the instrument is enclosed, so that only the pinpoint of light exposes a line on the photo-

FIGURE 8.7

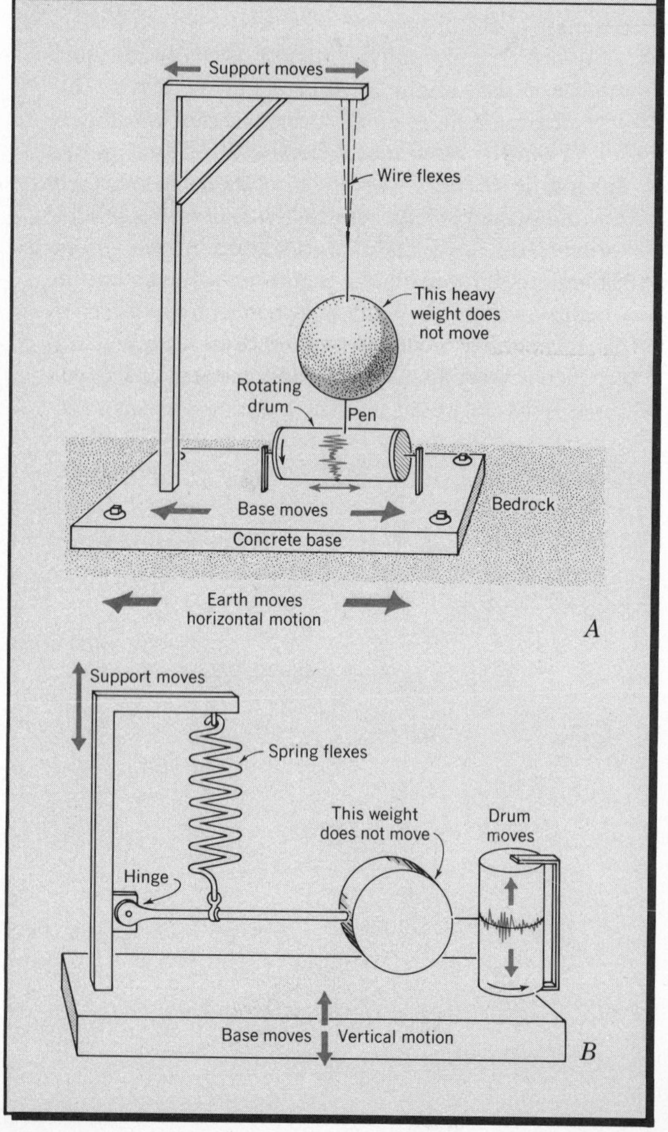

FIGURE 8.7 Inertia of a large mass provides a means of observing seismic waves. Horizontal motions might be detected by the mechanical arrangement shown in *A*, and vertical motions by that shown in *B*. Neither device would actually be useful unless further refined.

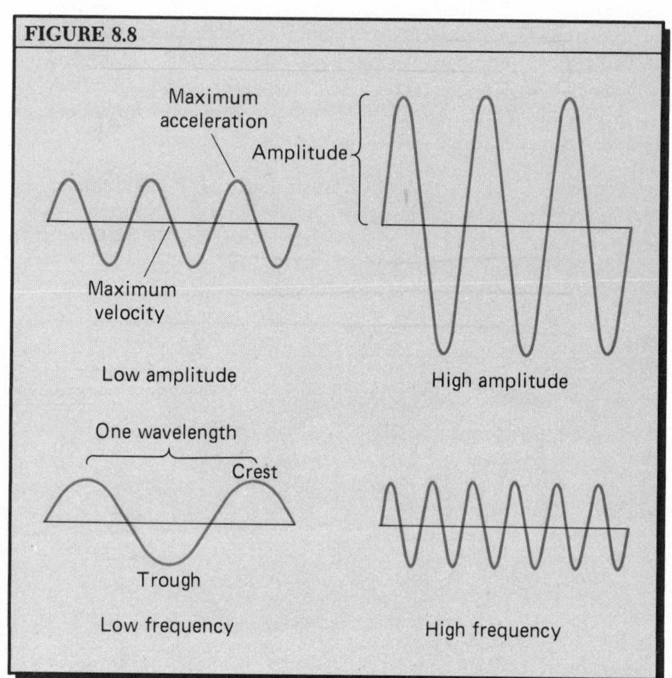

FIGURE 8.8 Amplitude and frequency of seismic waves.

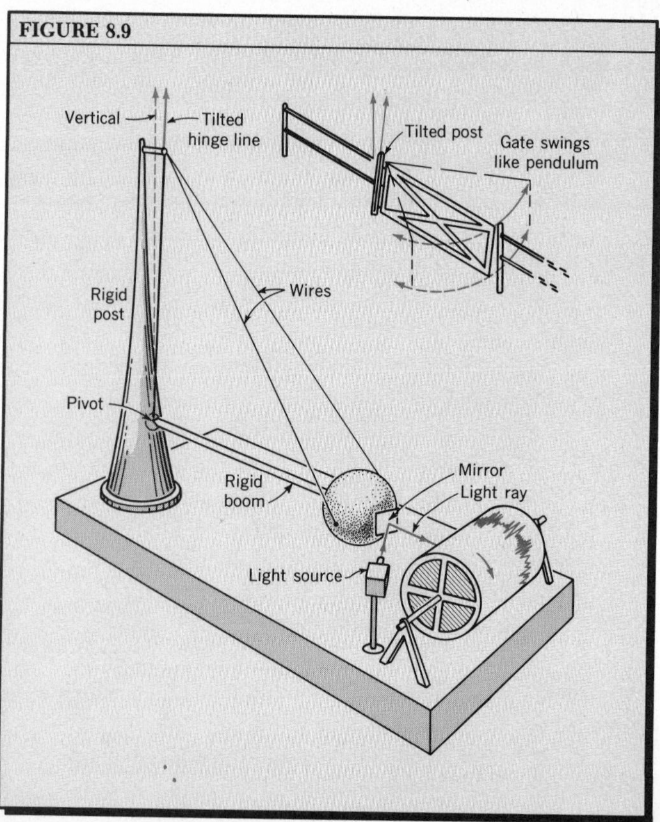

FIGURE 8.9 Principle of a horizontal, hinge-type pendulum seismograph. Earth motions are greatly magnified by use of a light ray reflected from a mirror.

graphic paper.) Later removed and developed, this paper becomes the record of the earthquake—the *seismogram*.

If the apparatus is to have any real scientific value, it will require a highly accurate clock to record the time of occurrence of each event shown on the paper. One way to do this is to install a clockwork mechanism within the drum to control precisely the rate at which it rotates, and at intervals of one minute, to make a small, sharp dent in the recorded line. It is thus possible to tell exactly when a particular train of waves arrives at the observing station.

FIGURE 8.10 *A*. A horizontal seismograph for detecting long-period seismic waves. Mechanism is basically of the design illustrated in Figure 8.9. *B*. A vertical seismometer for measuring long-period seismic waves. (Courtesy of the Lamont-Doherty Geological Observatory of Columbia University.)

FIGURE 8.10 A

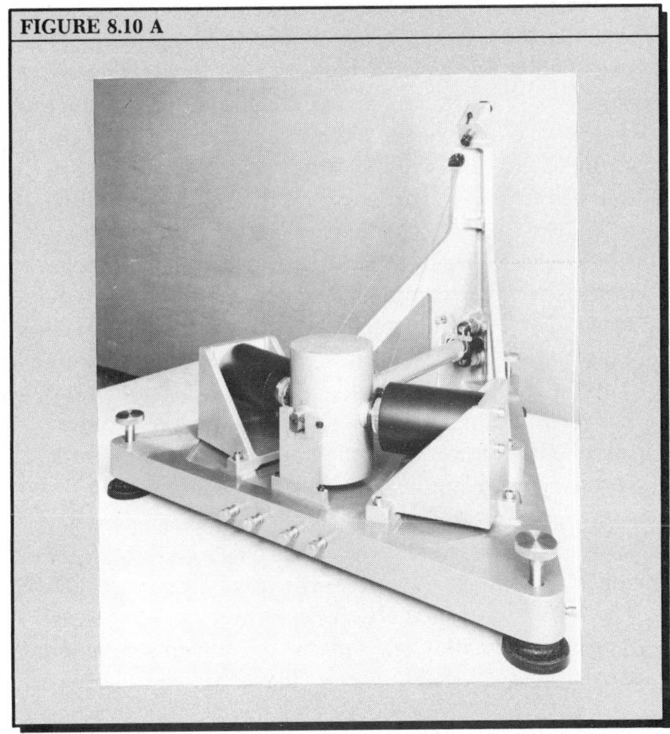

FIGURE 8.10 B

A second difficulty with the simple apparatus first described is that the earthquake waves rise and fall very slowly, each back-and-forth movement of the ground often taking several seconds. The wire from which the heavy weight is suspended would have to be at least 25 m long to keep the weight—acting as a simple pendulum—from swinging at about the same frequency as the waves (or faster). To make a practical seismograph, the weight can be hung from a support hinged like a gate, as shown in Figure 8.9. When the gatepost is tilted slightly, the gate will swing slowly, like a pendulum with a very long wire.

A third difficulty is that any pendulum tends to continue to swing at a natural frequency depending on its length. Because a record of these pendulum movements would obscure the earthquake movements, it is necessary to add to the pendulum a damping mechanism—that is, a device to prevent it from swinging at its own natural frequency.

Modern seismographs make use of magnetic and electronic devices to pick up, amplify, filter, and record the motions of the earth, just as a high-fidelity phonograph uses a magnetic pickup, transistorized amplifier, and magnetic loudspeaker instead of the old-style mechanical pickup head and horn.

To analyze earthquakes adequately, a whole battery of seismographs must be operated simultaneously (Figure 8.10), because each instrument records only the wave motion in one particular line of movement, such as east-west, north-south, or vertically. Then, too, earthquake waves include a wide range of frequencies superimposed in a complex way. Just as with a radio receiver, each seismograph is tuned to receive a particular frequency band, and thus several are needed to register the full range.

Interpreting the Seismogram

Now that we know how a seismograph works, we can turn to the interpretation of the wave record, or seismogram (Figure 8.11). Between earthquakes the seismogram shows a continuous record of waves, usually very small, called *microseisms*. These waves are made by atmospheric disturbances or causes other than those that generate true earthquakes.

The first indication that a severe earthquake has occurred at a distant point is the sudden beginning of a series of larger-than-average waves called the *primary waves* (*P-waves*). These waves die down somewhat for a few minutes; then, a second burst of activity sets in with the arrival of the *secondary waves* (*S-waves*), which are usually somewhat larger in height than the primary waves. The seismogram will then record smooth waves that increase greatly in amplitude to a maximum and then slowly die; these last, high-amplitude waves are the *surface waves*. While the primary and secondary waves have traveled through the earth, the surface waves have traveled along the surface of the ground much as storm swells travel over the sea (Figure 8.12).

For an earthquake occurring one-quarter of the globe's circumference away (that is, 90° of arc distant, or 10,000 km), the primary waves will take about 13 minutes to reach the receiving station, and the secondary waves will begin to arrive about 11 minutes after that.

It was soon apparent to the first seismologists that the farther away the earthquake epicenter, the longer the spread of time between the arrival of the primary and secondary wave groups. Both groups start from the focus at the same instant, but the primary group travels faster. Likewise, the surface waves travel more slowly and thus always come in last. From this discovery, about 1900, came the obvious conclusion that the spread of time between arrival of the wave groups could be used to measure the distance from the epicenter to the seismograph station.

Figure 8.13 is a graph on which are plotted *travel-time curves* for primary, secondary, and surface waves. The difference in times of arrival for the primary and the secondary waves, read from the graph as the vertical separation of the two curves, corresponds to the distance from focus to seismograph, measured along a great circle on the earth's surface. Units of measure are given in both kilometers and degrees of arc.

In the seismogram shown in Figure 8.11, the difference in arrival times of primary and secondary waves is 9.4 minutes. As indicated by a dashed line on Figure 8.12, this time difference corresponds to a distance of 8465 km, or an arc of 76.4°.

Using the figure of distance derived from the travel-time curves, a circle of that radius can be drawn on a globe to show the location of all possible points of origin of the earthquake. When three such circles are drawn from three widely separated observing stations, the earthquake epicenter can be located within the limits of a small triangle of error (Figure 8.14).

Varieties of Earthquake Waves

When an earthquake occurs, several basically different kinds of wave motion are generated. Primary and secondary waves appearing on the seismogram are of a class called *body waves* because they travel through solid rock. Body waves are distinguished from surface waves, which move along the free upper surface of the earth. Of the body waves, those which form the primary group (P-waves) have the same kind of motion as we can observe in sound waves; they are *compressional waves*. As illustrated in Figure 8.15, particles transmitting the compressional wave form move only forward and backward in the direction of wave travel. This motion consists of alternate compression and expansion and is a form of dilatation. We can remember this relationship by thinking of the P-wave as a "push" wave, made easier to remember because the words "primary" and "push" both begin with "P".

In waves of the secondary group (S-waves), particles transmitting the waves move back and forth at right angles to the direction of wave travel (Figure 8.15). The waves are called *shear waves* because a shearing type of deformation takes place. We may also think of shear waves as "shake" waves. The three key words—

FIGURE 8.11 This seismogram shows the record of an earthquake whose epicenter was located at a surface distance of 8460 km from the receiving station, equivalent to 76.4 degrees of arc of the earth's circumference. Figure 8.12 shows the ray paths for this earthquake. (After L. Don Leet, 1950, *Earth Waves*, Harvard Univ. Press, Cambridge, Mass. Copyright © 1950 by the President and Fellows of Harvard College.)

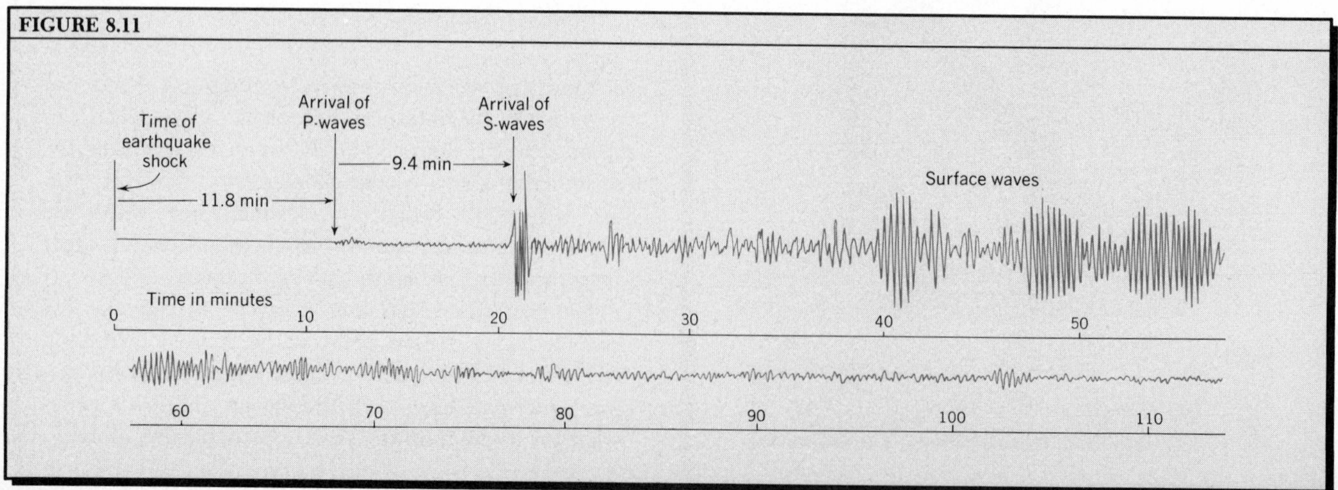

secondary, shear, and shake—begin with "S". P-waves travel approximately 1.7 times more rapidly than S-waves.

Two types of surface waves travel more or less together, but with different motions. One type consists of *Rayleigh waves*, named for the English physicist, Lord Rayleigh, who first gave a mathematical analysis of their behavior. They can be visualized as water waves traveling across the surface of a still pond after a pebble has been tossed into the water. In the Rayleigh wave a rock particle travels in a vertically oriented elliptical orbit that lies in the direction of wave travel (Figure 8.16*A*). When the particle is at the crest of the wave, it is moving opposite to the direction of wave travel. A person standing on the ground and experiencing Rayleigh waves would feel motion forward and backward as well as up and down. Rayleigh waves die out rapidly with depth.

The second type of surface wave is the *Love wave*, named for the physicist A. E. H. Love, who discovered it. Motion in the Love wave is entirely horizontal, at right angles to the direction of wave motion (Figure 8.16*B*), and in this respect, the Love wave resembles

FIGURE 8.12 Cross section of the earth showing diagrammatically the paths of P-waves, S-waves, and surface waves. (After L. Don Leet, 1950, *Earth Waves*, Harvard Univ. Press, Cambridge, Mass. Copyright © 1950 by the President and Fellows of Harvard College.)

FIGURE 8.13 Travel-time curves for earthquakes of 100-km depth of focus. (Based on data of C. F. Richter, 1958, *Elementary Seismology*, W. H. Freeman and Company, San Francisco, Appendix VIII.)

FIGURE 8.14 Circles drawn from three seismological observatories yield the location of an earthquake epicenter.

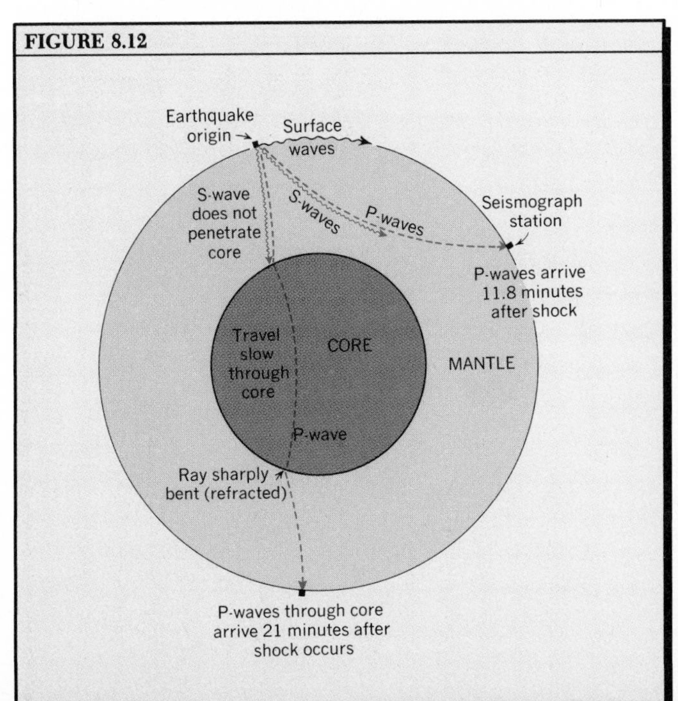

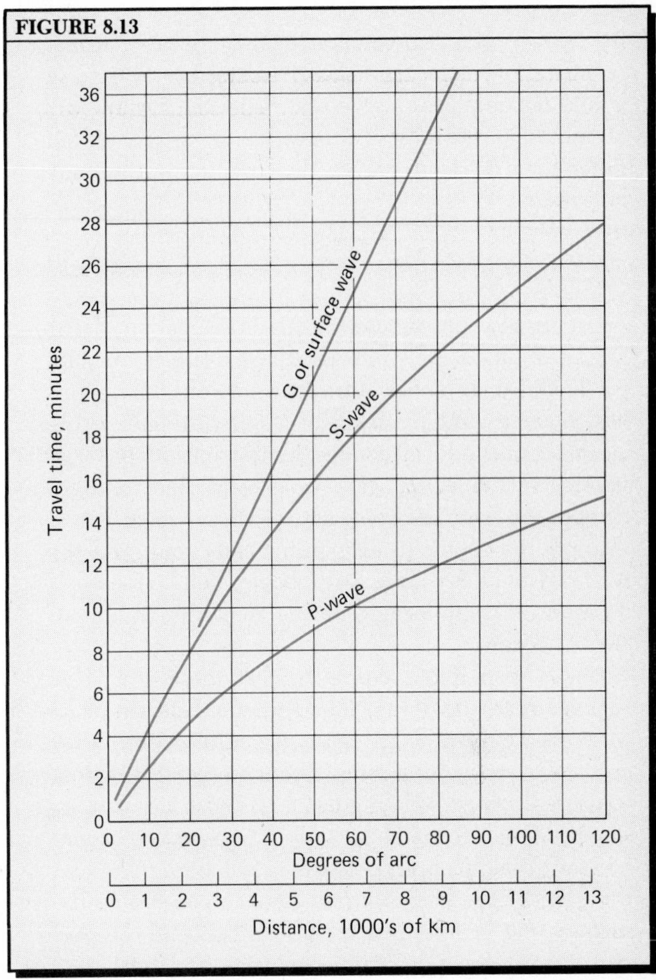

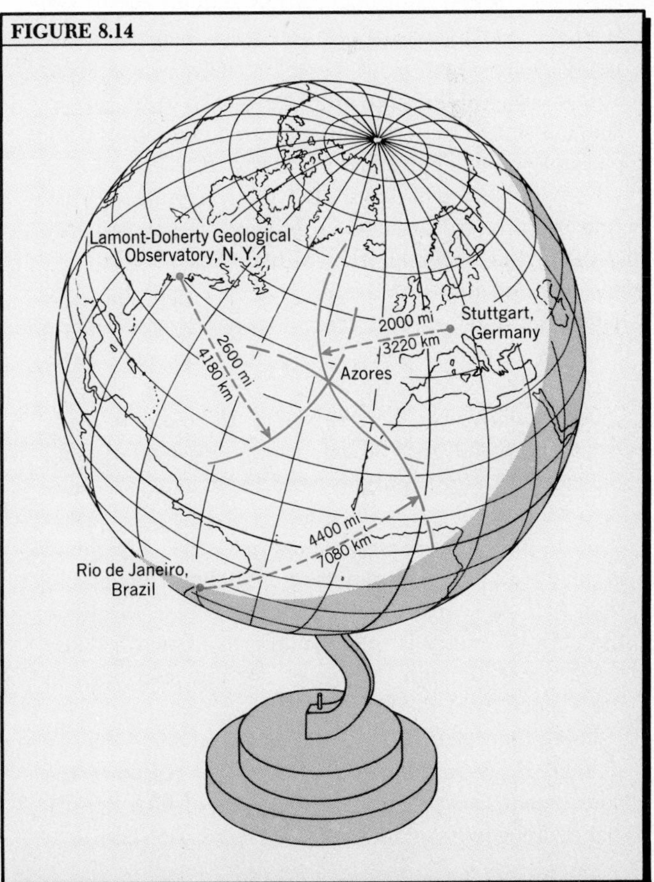

the S-wave. A person standing on the ground would feel the Love wave as a sharp side-to-side motion, of a kind which places a severe stress on the foundations of buildings. Because Love waves travel somewhat faster than Rayleigh waves, their motions are usually felt first.

The Earth's Core

Let us now put to use, in our investigation of the nature of the earth's interior, the principles of body waves and their modes of travel. Study of seismograms has confirmed the existence of a spherical core at the earth's center and has added insight into its physical nature. If the earth were in a solid state entirely throughout, both P-waves and S-waves would travel through the center in all possible directions, and the body waves of any large earthquake could be recorded by a seismograph located on the globe directly opposite the focus.

It was soon found, however, that there is a large region on the side of the globe opposite the earthquake focus where S-waves are not received. Evidently they are prevented from passing through a central region, the core, in the earth (Figures 8.12 and 8.17). Testing shows physicists that shear waves, or S-waves, cannot be sent through a liquid; hence, they have agreed that the earth's core is in a liquid state in contrast to the surrounding mantle, which is solid. "Solid" in this case may mean either crystalline or glassy; it also means here that, when subjected to the sudden twists and bends of earthquake waves, the rock behaves as an elastic solid—that is, it changes shape when shear stresses are applied, but returns exactly to its former shape when those stresses are removed (see Chapter 6).

As shown in Figure 8.17, S-waves are received only within a distance of about 103° of arc from the earthquake source (somewhat more than one hemisphere). Because they bend as they travel through the core, P-waves are not directly received in a zone between 103° and 143° distant from the focus. Neither are S-waves received there, since they do not travel through the core. Only surface waves and complex reflected waves can be received in this *shadow zone*. A zone beyond 143° receives only P-waves passing through the core, complex reflected waves, and surface waves.

From the extent of the shadow zone, the earth's core is calculated to have a radius of 3470 km, a little more than half the earth's total radius (Figure 8.18). That the core boundary is quite definite is known because the P-waves of high frequency are reflected back to the surface from this boundary.

Using our calculation of the earth's average density (5.5 g/cc), we may guess that the outer region of the core is composed largely of liquid iron of high density (10 to 12 g/cc), under enormous pressure, and at a

FIGURE 8.15 Diagrammatic representation of particle motions in compressional and shear waves.

FIGURE 8.16 Forms of surface seismic waves. *A.* Rayleigh waves. *B.* Love waves.

FIGURE 8.17 Diagrammatic representation of many possible ray paths from a single earthquake source. (Based on data of Gutenberg, 1951, *Internal Constitution of the Earth*, Dover, New York.)

FIGURE 8.18 Dimensions of the earth's mantle and core.

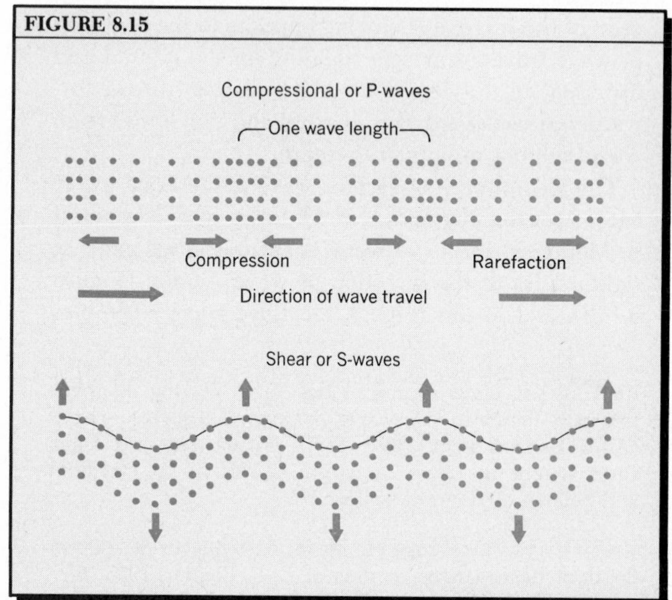

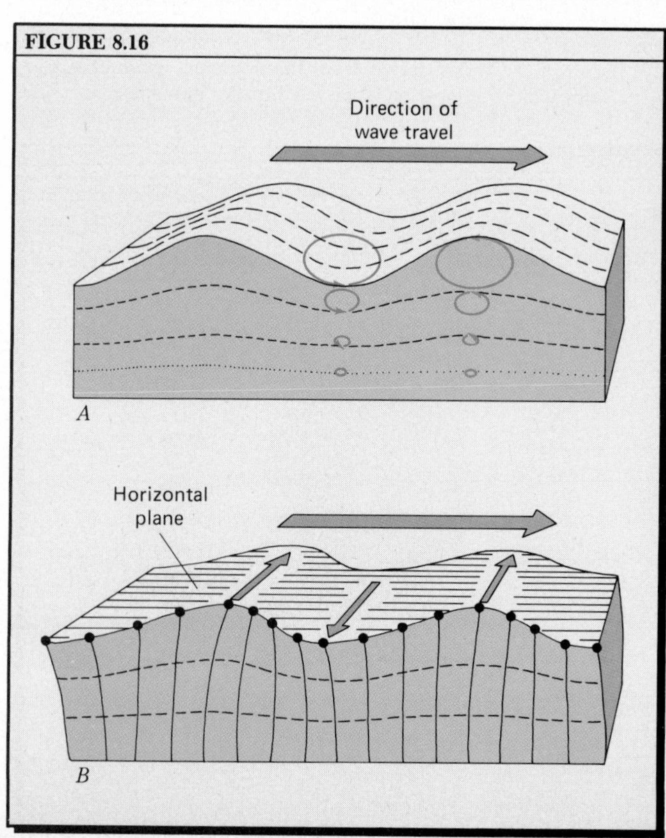

FIGURE 8.17

0°
Both P- and S-waves received here
P- and S-waves
P- and S-waves received here
Ray path
CORE
MANTLE
103°
103°
Shadow zone
Shadow zone
No direct P- and S-waves received here
Surface waves received here
143°
143°
P-waves and surface waves received here

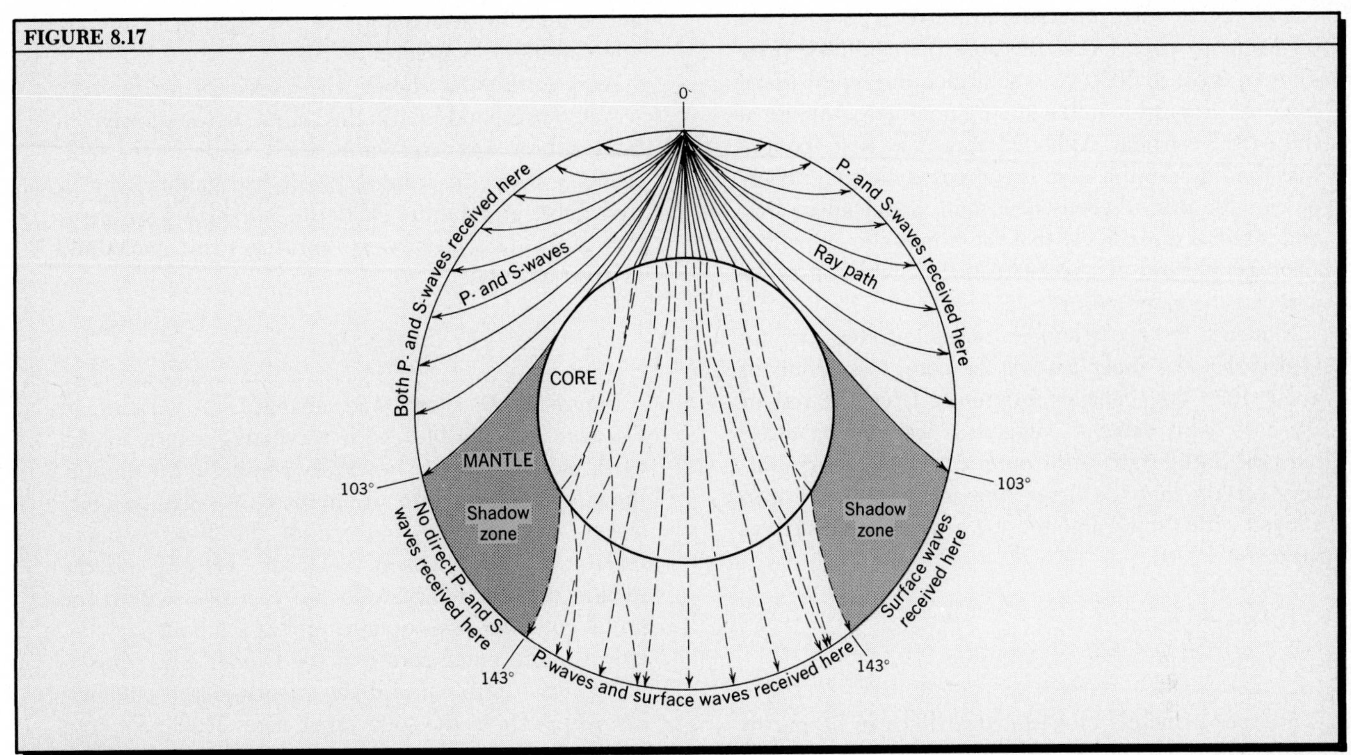

FIGURE 8.18

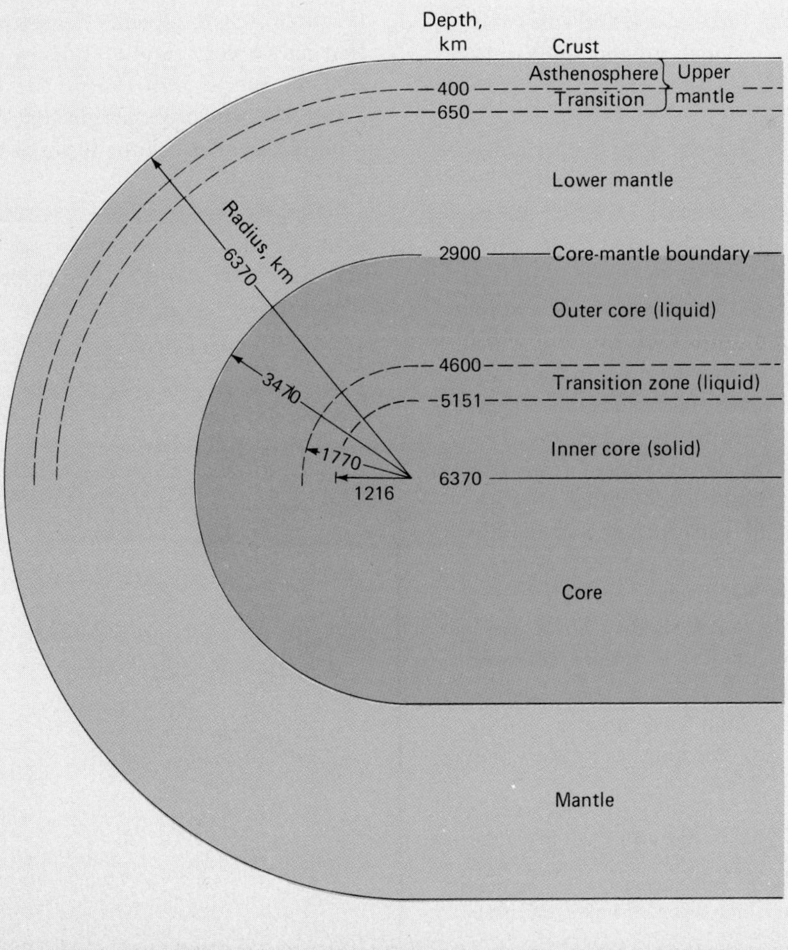

Depth, km
Crust
Asthenosphere
Upper mantle
400
Transition
650
Lower mantle
Radius, km
6370
2900
Core-mantle boundary
Outer core (liquid)
3470
4600
Transition zone (liquid)
5151
Inner core (solid)
1770
1216
6370
Core
Mantle

high temperature. Obviously it is quite impossible for us to imagine what this material is like in terms of our own sensory experiences, because the confining pressures of 2000 to 3000 kb would give the liquid metal physical properties unlike anything we can examine at the earth's surface. Although it was at first thought that the liquid outer core is composed almost entirely of metallic iron, there is now some doubt about this, and it seems more likely that the iron is alloyed with a small proportion of another element, which may be carbon, silicon, or sulfur.

Evidence from earthquake seismology has also revealed that the inner part of the core, to a radius of about 1220 km, behaves differently from the rest of the core. This behavior suggests a solid state, rather than the liquid state of the outer core. It seems reasonably certain that the inner core is composed of iron alloyed with a small amount of a heavier metal such as nickel.

Seismic-Wave Speeds and Rock Materials

Another principle of seismology that can be put to use to reveal the nature of the earth's interior is that P and S types of earthquake waves travel faster through highly rigid material than through less rigid material.[*] The word ***rigidity*** applies only to elastic solids and is their resistance to shearing forces. Steel and rubber are both elastic, but steel has a much greater rigidity because it bends very much less than rubber under the same deforming stress.

Rocks in general have a high degree of rigidity, whether composed of crystallized or glassy mineral matter, but there is quite a marked variation among different rocks. Rigidity can be measured by the physicist in the laboratory, not only under the conditions at the earth's surface, but also under great confining pressures and high temperatures such as those which might be expected many kilometers deep in the earth.

Because the rigidity of rocks determines the velocity of earthquake waves and because the velocity of these waves at various depths can be calculated from seismograms, it is possible to make a good guess concerning the kinds of rock in the earth's crust and mantle. Figure 8.19 lists several rock types and their densities under surface conditions. Sand, clay, and silt layers, made up of loose grains, have low rigidity. Shale, sandstone, and limestone have moderate rigidity, followed by granite and the other felsic igneous rocks. Diabase and gabbro, mafic rocks which have greater density than granite, have about one-third higher rigidity than granite. Next come the ultramafic rocks, pyroxenite and dunite, with about twice the rigidity of granite and considerably greater density.

As Figure 8.19 shows, the speed of earthquake waves corresponds with this same series: slowest in sand, silt, or clay, and increasingly faster in the sedimentary rocks, in granite, basalt and gabbro, and in pyroxenite and dunite. Dolomite, although a sedimentary rock, is exceptional in that it has wave velocities as great as basalt.

Composition of the Mantle

Figure 8.1*D* shows the changes in velocity of P-waves and S-waves with increasing depth in the earth. Both curves are upwardly convex in the mantle, showing that the rate of increase in wave velocity lessens with depth. Notice that the P-wave curve makes an abrupt drop to lower velocity at the mantle–core boundary; velocity then rises within the liquid core and steps up abruptly at the boundary between liquid outer core and solid inner core. The S-wave curve terminates at the mantle–core boundary, but reappears in the solid inner core. This inner-core S-wave is produced by energy transmitted as a P-wave through the liquid zone and reconverted to an S-wave in the solid zone.

Through most of the earth's mantle, nearly 2900 km (Figure 8.18), the speed of earthquake waves is so high that only a very rigid and dense rock, such as pyroxenite or dunite, will satisfy the observed conditions. Thus, it is presently thought that the mantle consists of solid ultramafic rock made up of magnesium–iron silicate minerals.

Based on the behavior of seismic waves, the mantle is sub-divided into two major parts: the ***upper mantle*** and the ***lower mantle***. The upper mantle, extending

FIGURE 8.19 Speed of travel of P-waves in various types of rocks. (Based on data of M. B. Dobrin, 1960, ***Introduction to Geophysical Prospecting***, McGraw-Hill, New York.)

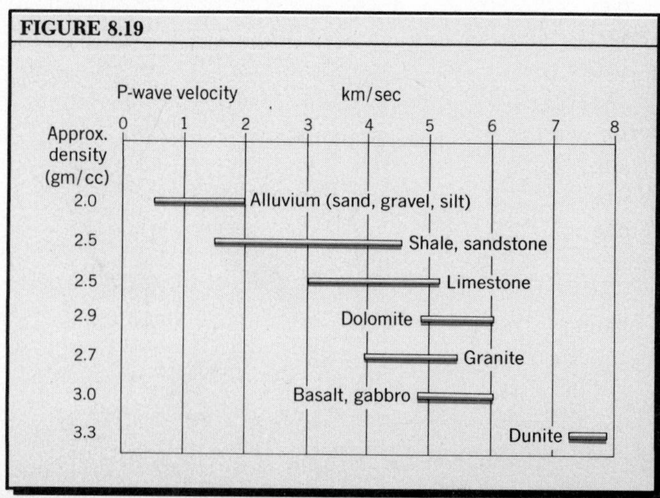

[*]"Rigidity" as used here corresponds with the *shear modulus*, a measure of the resistance of an elastic solid to shearing deformation. P-waves are also influenced by the *bulk modulus*, which measures ratio of compressive stress to resulting volume change by dilatation.

from the crust to a depth of about 650 km, includes the asthenosphere, which occupies the upper 300 to 400 km.

Below this depth, in the range from 400 to 650 km, is a *transition zone*, grading into the lower mantle. At a depth averaging about 400 km a sharp increase in P-wave velocity occurs, suggesting a sudden increase in rock density. Rather than attribute this *discontinuity* to a sudden change in chemical composition of the mantle rock, geochemists postulate that the atoms composing the rock are reordered into a more closely packed arrangement. This kind of physical change in atomic structure is called a *phase transition*. Olivine, which is considered to be the major mineral component of the mantle, is capable of undergoing just such a phase transition, as demonstrated by laboratory experiments conducted under appropriate conditions of pressure and temperature.

Below the 400-km discontinuity, P-wave velocity and density increase uniformly to a depth of about 650 km. Here, another discontinuity occurs, yielding another sharp increase in P-wave velocity. It is suspected that at this critical depth pressure is so great that another phase transition occurs. The change may consist of the breakdown of the densely compacted olivine, formula $(Mg,Fe)_2SiO_4$, into three separate oxides: FeO, MgO, and SiO_2, since a similar breakdown of olivine into oxides has been produced in the laboratory at conditions of pressure and temperature comparable with those probably existing at a depth of about 700 km. Within the lower mantle, deeper than 700 km, seismic wave velocity increases rather smoothly, as Figure 8.1*D* shows, indicating that the lower mantle has somewhat uniform composition.

Physical Properties of the Upper Mantle

Numerous earthquakes with foci in the uppermost 50 to 60 km show that the earth has a brittle outermost layer which breaks by faulting when unequal stresses exceed the elastic limit of the rock. Below the brittle layer, which includes the crust and a part of the upper mantle, rock has low strength and, instead, a plastic quality that enables it to undergo very slow flowage.

As early as 1914, Joseph Barrell, an American geologist, proposed that the plastic zone of the mantle be called the *asthenosphere* (from a Greek word meaning "weak"), to distinguish it from the overlying brittle zone, or *lithosphere*. Both of these terms were introduced in Chapter 1, where we explained that the lower limit of the brittle zone does not represent a change in rock composition, but rather, a change in physical properties.

Recall from Chapter 1 the analogy of the iron bar: If one end of a cast-iron bar is held in a furnace, it becomes white hot and soft; farther out along the bar, the iron is red hot and not so soft; toward the cool end the iron is hard and brittle. In a similar way, the behavior of rock changes gradually upward from within the upper mantle.

Rock in the asthenosphere behaves as both a plastic solid and an elastic solid. We have noted in Chapter 7 that matter possessing these remarkable properties is an elastico–viscous substance—it can be elastic and plastic at the same time, depending on whether the forces that tend to deform it are applied suddenly and released (as if struck a sharp blow) or applied steadily (as under the force of gravity). It is said to be possible to make out of cobbler's wax a tuning fork so rigid that it will vibrate with a musical tone when struck, but if the same fork is left to stand for a long time, it will slowly collapse into a shapeless mass. The asthenosphere does, indeed, have the elastico–viscous properties we see in cobbler's wax, because it can transmit seismic waves, but will also flow slowly like a highly viscous plastic solid, with little strength, when shearing stresses are steadily applied.

A layer within the upper mantle, in depth from 60 to 200 km, is of particular interest because here the mantle rock is at a temperature very close to its melting point. Notice in Figure 8.1*B* that the temperature curve comes close to touching the melting-point curve at a depth of about 160 km. The term *plastic layer*, or *soft layer*, has been applied to this part of the mantle, which is within the asthenosphere.

Figure 8.20 shows details of temperature and strength properties of this part of the upper mantle. The temperature curve gives the difference between the rock temperature and its melting point. Notice that this difference declines almost to zero in the depth range from 150 to 200 km. The curve of rock strength correspondingly declines, with a minimum approaching zero near the 200-km level. Zero strength would represent a true liquid, incapable of supporting any load without immediate yielding. Notice that the curve representing numbers of large earthquakes per year declines to very low values (1 or 2 per year) in the depth zone of low strength, then rises again at increasing depth.

The presence of the soft layer in the mantle was suspected as far back as 1926, when the distinguished seismologist Beno Gutenberg presented evidence from wave amplitude studies that earthquake wave velocities are slowed below 60 km, after first increasing rapidly from the surface to that depth (Figure 8.20). This region, centered about at a depth of 150 km, is referred to as the *low-velocity zone*.

Recently, it has been postulated that a condition of partial melting may exist in the asthenosphere. As we explained in Chapter 4, partial melting of mantle rock is thought to provide the magma needed to produce new basaltic rock of the oceanic lithosphere at spreading plate boundaries. Estimates place the proportion of melt in the asthenosphere as on the order of 1% to 10%. Presence of the melt fraction can explain the lower strength of the asthenosphere, as well as the reduced P-wave velocity.

In the early period of seismology, observation indicated that earthquakes were set off by fault movements at comparatively shallow depths—down to little more than 55 km. Then further study of seismograms showed that many earthquakes have much deeper foci. Earthquakes of intermediate depth originate between 55 and 240 km, whereas deep-focus earthquakes originate mostly between 300 and 650 km, and a few as deep as 720 km.

We have now arrived at a serious scientific dilemma. If the rock of the asthenosphere flows plastically in response to stresses applied over long periods of time, it must be impossible for elastic strain to accumulate in that rock in sufficient quantities to generate earthquakes. The unequal forces produced by lithospheric plate motions must, instead, be continuously relieved by flowage. Yet deep-focus earthquakes of great magnitude are known to be generated adjacent to subduction boundaries in the depth region from 300 to 700 km, which is in the low-strength plastic layer. The nature of this contradiction was fully appreciated by seismologists before the theory of plate tectonics emerged. How it was resolved by plate tectonics will become clear in Chapter 11 in our investigation of seismic activity in subduction zones.

Seismic Waves and Crustal Structure

Most of our information on the structure and composition of the crust has come from observations of seismic-wave behavior. (Refer to Figure 4.8 for details of crustal structure.) The crust is distinguished from the mantle by the presence of a rather abrupt and clearly defined change in the velocity of seismic waves, indicating that there is a corresponding abrupt change in rigidity of the rock from crust to mantle. A change in rigidity indicates in turn an abrupt change in mineral composition or in physical state of the rocks.

Where an earthquake has a focus close to the surface and is located only a few hundred kilometers from the seismograph station, the seismic waves do not penetrate the earth more than about 100 km before they are gradually turned back toward the surface and reach the seismograph. Interpretation of the complex wave records will reveal the velocities at which the waves travel at different depths.

The times and places of natural earthquakes are unfortunately unpredictable except in a very general way. Therefore, the best source of information about the shallow zones of the earth come from human-made shocks. One method is to make use of blasts set off at rock quarries and to record them with portable seismographs at various distances from the blasts. In analyzing shallow zones for possible petroleum-bearing structures, small dynamite explosions are used.

Generally speaking, rigidity of crust and mantle rocks increases with depth. Very simply, two possibilities may be considered for the subsurface structure. Figure 8.21A shows the case of gradual increase in rock rigidity with depth. As the shock wave penetrates this rock, it encounters regions of progressively faster travel. This change results in a continuous bending, or *refraction*, of the wave path, or ray, in such a way as to turn it toward a path parallel with the surface. Continued bending of the path causes the wave to return eventually to the surface, following a curved path. Continuous refraction of this type might indicate that

FIGURE 8.20 Physical properties of the upper mantle. (Based on data of D. L. Anderson, 1962, *Scientific American*, July, pp. 58–59.)

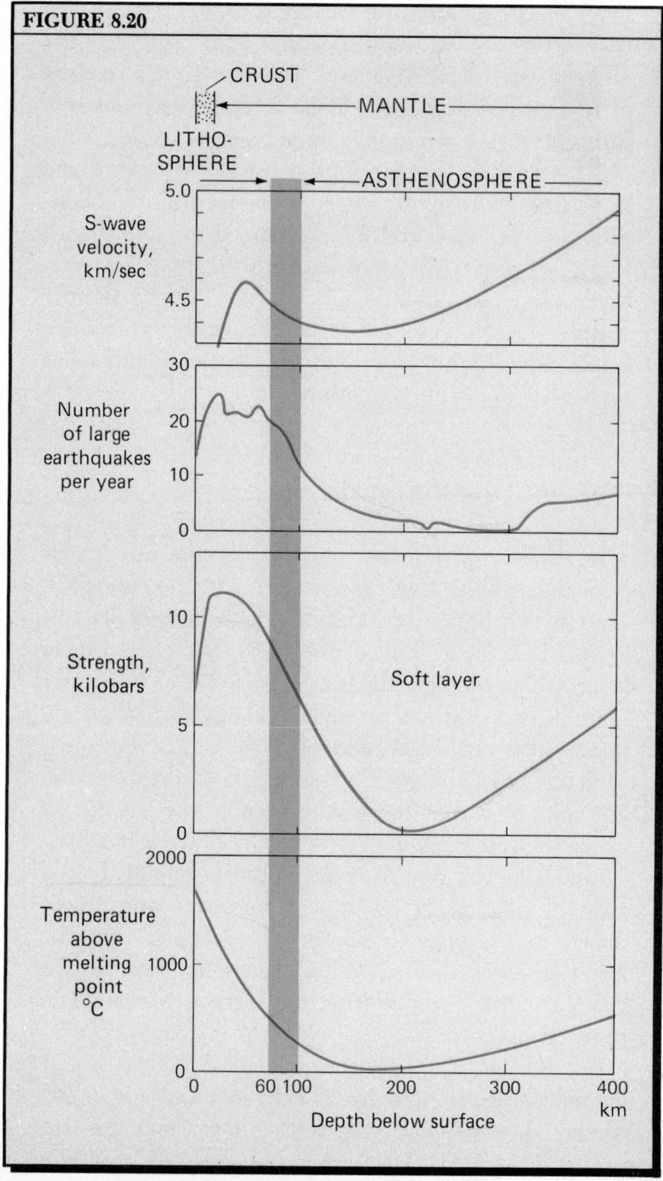

FIGURE 8.21 Bending of seismic waves as they travel through rock layers of differing degrees of rigidity. (After M. B. Dobrin, 1960, *Introduction to Geophysical Prospecting*, McGraw-Hill, New York.)

FIGURE 8.22 Properties of the continental crust and oceanic crust.

FIGURE 8.21

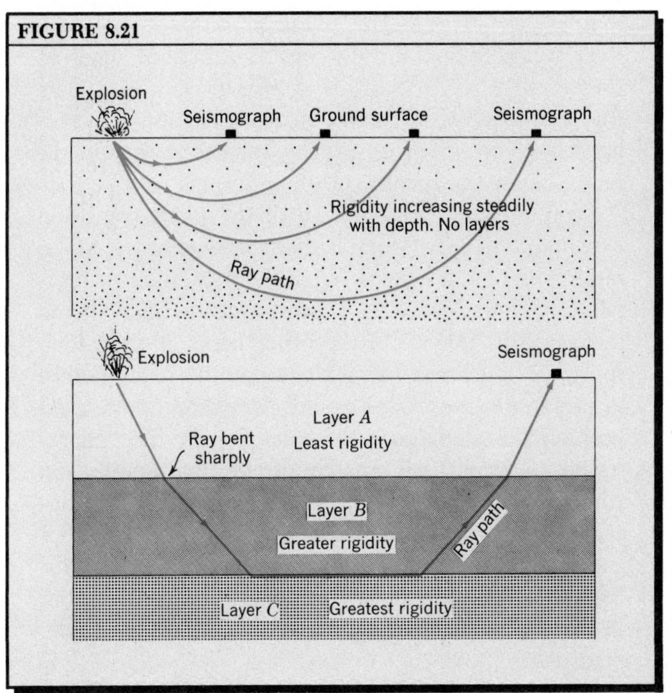

the rock becomes more rigid because of increasing confining pressure at depth. The continuous refraction might also indicate that the rock is changing gradually in composition to denser minerals—that is, from granite (felsic) to basalt (mafic), and then to ultramafic rock without any abrupt change.

Figure 8.21*B* shows the case in which there exist layers of rock, each of uniform rigidity within itself, but with each successive deeper layer changing abruptly to higher rigidity. In this case, a particular ray of the shock wave travels in a straight line through each layer but is refracted sharply as it enters the next layer. When it strikes a new layer at a certain critical angle, the wave travels along the contact between layers for a certain distance and is then turned upward to return to the surface. When the seismograms of several recording stations are compared, the subsurface paths of the seismic waves can be reconstructed and the velocities of wave travel estimated for different depths. This information in turn makes possible the selection of rock varieties whose physical properties fit the observed wave velocities.

Waves of both shallow earthquakes and surface explosions show quite definitely that the continents consist of a platelike crust, averaging about 30 km thick, resting upon quite different rock of the mantle. As indicated in Figure 8.22, the upper part of the continental crust consists largely of felsic (granitic) rock, while the lower part is largely mafic (basaltic) rock.

It is known that the P-waves near the surface travel

FIGURE 8.22

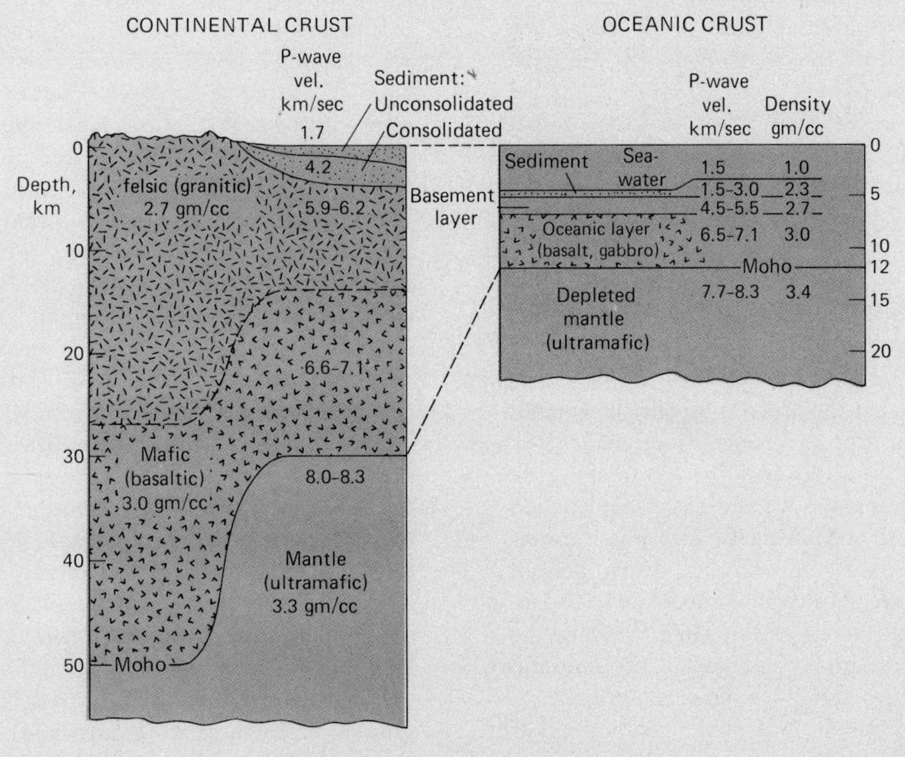

at about 6 km per second, which is expected in granitic rock, and that this velocity increases gradually or abruptly to the base of the crust, where it is about 7 km per second, a velocity expected in basaltic rock at this depth. At about 30 km deep, on the average, the P-wave velocity increases abruptly to more than 8 km per second, a speed to be expected of an ultramafic rock, such as peridotite. (S-waves undergo a corresponding velocity increase with depth.) This surface of sudden increase in wave velocity, which separates the crust above from the mantle below, is the *Mohorovičić discontinuity,* named after the Yugoslav seismologist who first recognized the discontinuity in 1909 from the records of shallow-focus earthquakes. To make things easier, it has become accepted practice to designate this discontinuity as the *Moho,* or simply as the *M-discontinuity.*

Seismic waves reveal a quite different crustal structure beneath the ocean basins. Layers typical of the North Atlantic Ocean basin are shown schematically in Figure 8.22. Beneath a thickness of about 4.5 km of seawater is a thin layer of unconsolidated deep-sea sediment, averaging about 0.5 km. Beneath the sediment is a rock layer about 1.5 km thick with P-wave velocity in the range from 4.5 to 5.5 km/sec. This basement layer might represent either basaltic lava formed at the spreading plate boundary in the mid-oceanic region, or consolidated sediments, or both. Its density is on the order of 2.7 g/cc. The next layer, called the oceanic layer, shows a P-wave velocity of 6.5–7.1 km/sec and a density of 3.0 g/cc. It probably consists of solid basalt or its coarse-grained igneous equivalent, gabbro, also formed during lithospheric plate accretion at the spreading boundary. The Moho is encountered at an average depth of 12 km—much shallower than beneath the continental crust. Here the P-wave velocity jumps to between 7.7 and 8.3, and the density increases to 3.4 g/cc. The rock in the upper mantle here is believed to be peridotite depleted of the mineral constituents that moved upward to form the basaltic crust. This material is sometimes described as *depleted mantle.*

*A Possible Phase Transition at the Moho

We have stated that the Moho is usually interpreted as representing an abrupt change from mafic rock (gabbro or basalt) in the crust to ultramafic rock (peridotite) in the mantle. An alternative explanation for the Moho exists in the concept of a phase transition, such as that we referred to as being present deeper in the mantle. At a depth of about 30 to 35 km, which is about the depth of the Moho beneath the stable continental interior, conditions of pressure and temperature are favorable for a phase transition.

In this phase transition, rock of the composition of gabbro is transformed into eclogite. Recall from Chapter 7 that eclogite found at the earth's surface is a metamorphic rock of total element composition about equivalent to that of gabbro. Eclogite consists of an unusual green variety of pyroxene rich in sodium and calcium and a red-brown variety of garnet rich in magnesium. It is denser than gabbro and would transmit seismic waves at a velocity equivalent to that of peridotite under the same conditions of pressure and temperature. Under the phase-transition hypothesis, the lower continental crust consists of gabbro, whereas the mantle directly below it consists of eclogite. Under the ocean basins, a similar phase transition is postulated, with eclogite forming the uppermost mantle, directly below the basaltic crust.

An interesting consequence of the phase-transition explanation for the Moho is that the level of the Moho might rise or fall because changes of pressure or temperature might cause the phase-transition boundary to move up or down. On the other hand, if the Moho represents a change in rock composition from gabbro to peridotite, the Moho would always occupy a fixed position between those two rock layers. The phase-transition hypothesis has an interesting application which we shall discuss in Chapter 12. Briefly, the concept behind this application is that when a layer of gabbro undergoes a change to eclogite, the density increase in that layer must cause the lithosphere to sink to a lower level. The opposite change would cause the lithosphere to rise.

Earthquake Magnitude and Energy

Interpretation of seismograms has made possible a calculation of the quantities of energy released as wave motion by earthquakes of various magnitudes. In 1935, a leading seismologist, Charles F. Richter, created a rating scale of *earthquake magnitude* that can be related to the energy released at the earthquake focus (Figure 8.23). The *Richter scale* consists of numbers ranging from less than 0 (negative numbers) to more than 8.5. Values are given to the nearest one-tenth, thus: 2.5, 4.9, 6.2, 7.8, 8.5. There is neither a fixed maximum nor a minimum, but the highest-magnitude earthquakes thus far measured have been rated as 8.9 on the Richter scale. Earthquakes of magnitude 2.0 are the smallest normally detected by the human senses, but instruments can detect quakes as small as −3.0. The scale is logarithmic, which is to say that the amplitude of the recorded waves increases tenfold for each integer increase in Richter magnitude. An earthquake of magnitude 5.0 is ten times larger than one of magnitude 4.0. Some information on the meaning of various magnitudes is given in Table 8.1.

Richter magnitude is measured from the seismogram of an earthquake (Figure 8.24). A standard seis-

Table 8.1 Richter Magnitude and Energy Release

Magnitude, Richter scale	Energy release, joules	Comment
2.0	2.5×10^7	Smallest quake normally detected by humans.
2.5–3	10^8–10^9	Quake can be felt if it is nearby. About 100,000 shallow quakes of this magnitude per year.
4.5	10^{11}	Can cause local damage.
5.0	10^{12}	Energy release about equal to first atomic bomb, Alamogordo, N.M., 1945.
6.0	2.5×10^{13}	Destructive in a limited area. About 100 shallow quakes per year of this magnitude.
7.0	10^{15}	Rated a major earthquake above this magnitude. Quake can be recorded over whole earth. About 14 per year this great or greater.
8.25	6.0×10^{16}	San Francisco earthquake of 1906.
8.5	1.5×10^{17}	Chile, 1960; Alaska, 1964. Close to the maximum known.
8.9	8.8×10^{17}	Maximum ever recorded. Only two known: Colombia–Ecuador border, 1906; Japan, 1933.

mograph must be used; it is the Wood–Anderson seismograph, or an equivalent type. Two measurements are made directly from the seismogram: (1) amplitude in mm of the largest (highest) wave recorded; and (2) time in seconds elapsed between arrivals of the S and P waves. This second measurement is equivalent to measuring the distance from the epicenter. In the example illustrated, the amplitude is 23 mm; the S–P time difference is 24 seconds, equivalent to a distance of 210 km. To avoid a difficult calculation from a formula, the nomograph shown in Figure 8.24 provides a quick estimate of the Richter magnitude. A straight line drawn between points on the amplitude and distance scales will intersect the magnitude scale to provide a reading. In this case, the Richter magnitude is 5.0. Using this rapid method, an observatory

FIGURE 8.23 Energy release compared for various Richter magnitudes in terms of the relative volumes of spheres. The upper diagram shows spheres to their true scale relationships for magnitudes 0 through 3. The lower four diagrams give the diameters of spheres on the same relative scale as for magnitudes 0 through 3.

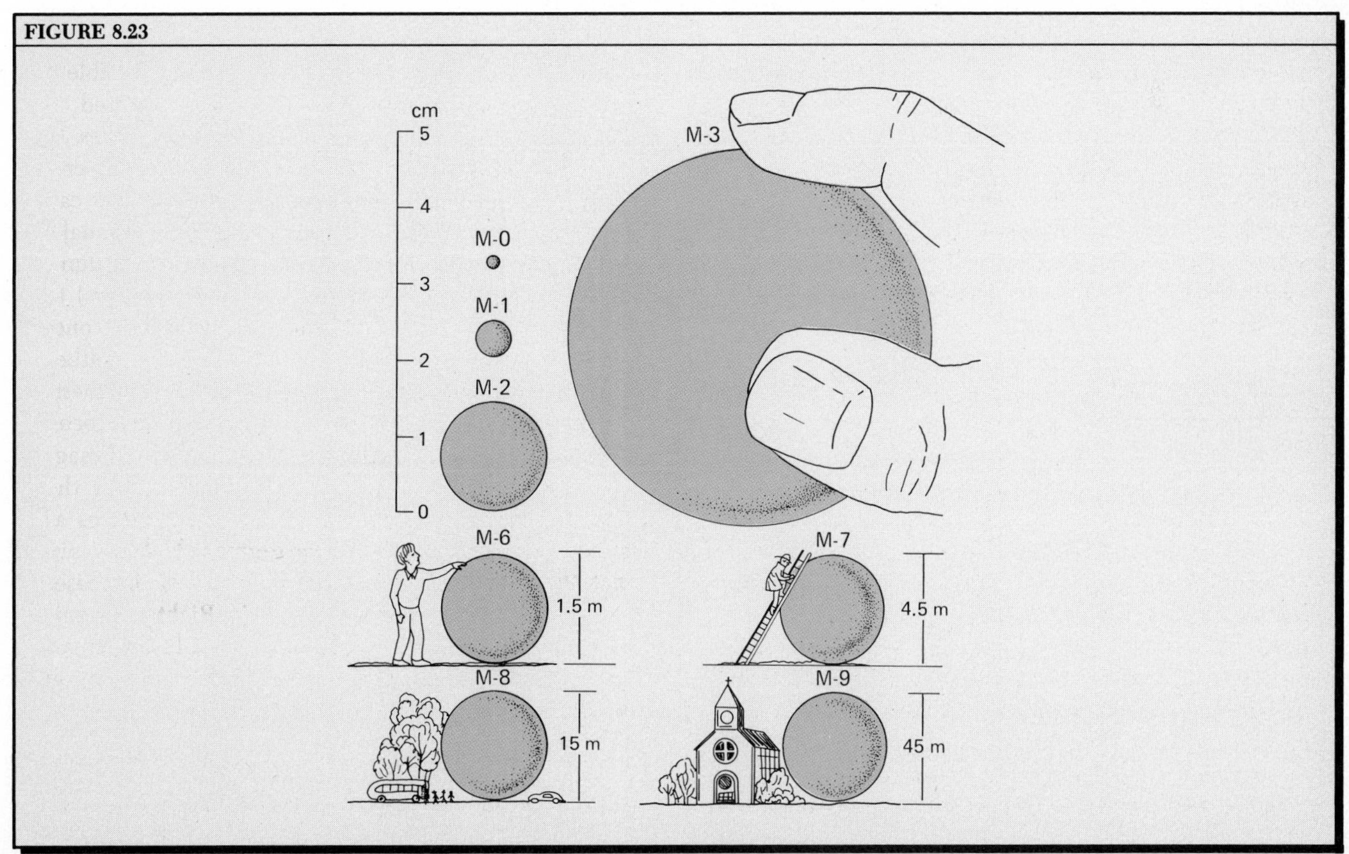

can immediately issue a news report on the occurrence of a potentially damaging earthquake. When two or more stations have made their determinations, the epicenter can be quickly plotted. The method works well for small to moderate shallow-focus earthquakes, but must be modified for use with very large and deep-focus earthquakes.

As we have seen, numbers on the Richter scale relate to the amplitude of the largest earthquake waves. To convert the Richter numbers to tell the actual amount of energy release involves further calculation. For every increase of one integer of Richter magnitude, the quantity of energy increases by a factor of about 32 times. Thus, an earthquake of magnitude 7.0 releases about 32 times as much energy as one of magnitude 6.0. An earthquake of magnitude 8.0 releases about 1000 times as much energy as one of magnitude 6.0 ($32 \times 32 = 1024$). Figure 8.25 is a graph showing energy release (joules) plotted against Richter magnitude. An atomic bomb blast, such as that at Bikini Atoll in 1946, yields about 10^{12} joules, equivalent to an earthquake of magnitude 5.0, which usually causes only minor damage in an urban area. At magnitude 8.9, the energy release is nearly 1,000,000 times greater than from a single atomic bomb.

The total quantity of energy released by all earthquakes of the world in a single year is estimated to be from 10^{18} to 10^{19} joules. Most of this quantity is from a very few earthquakes of Richter magnitude greater than 7.0.

Earthquake Intensity Scales

The actual destructiveness of an earthquake depends upon factors other than the energy release given by Richter magnitude—for example, closeness to the epicenter and nature of the subsurface earth materials. An *earthquake intensity scale*, designed to measure observed earth-shaking effects, is important in engineering aspects of seismology.

An intensity scale used extensively in the United States is the *modified Mercalli scale*. The original

FIGURE 8.24 A graphic method of determining the Richter magnitude of an earthquake, using the actual seismograph trace as recorded on the Wood–Anderson seismograph. (The California Institute of Technology.)

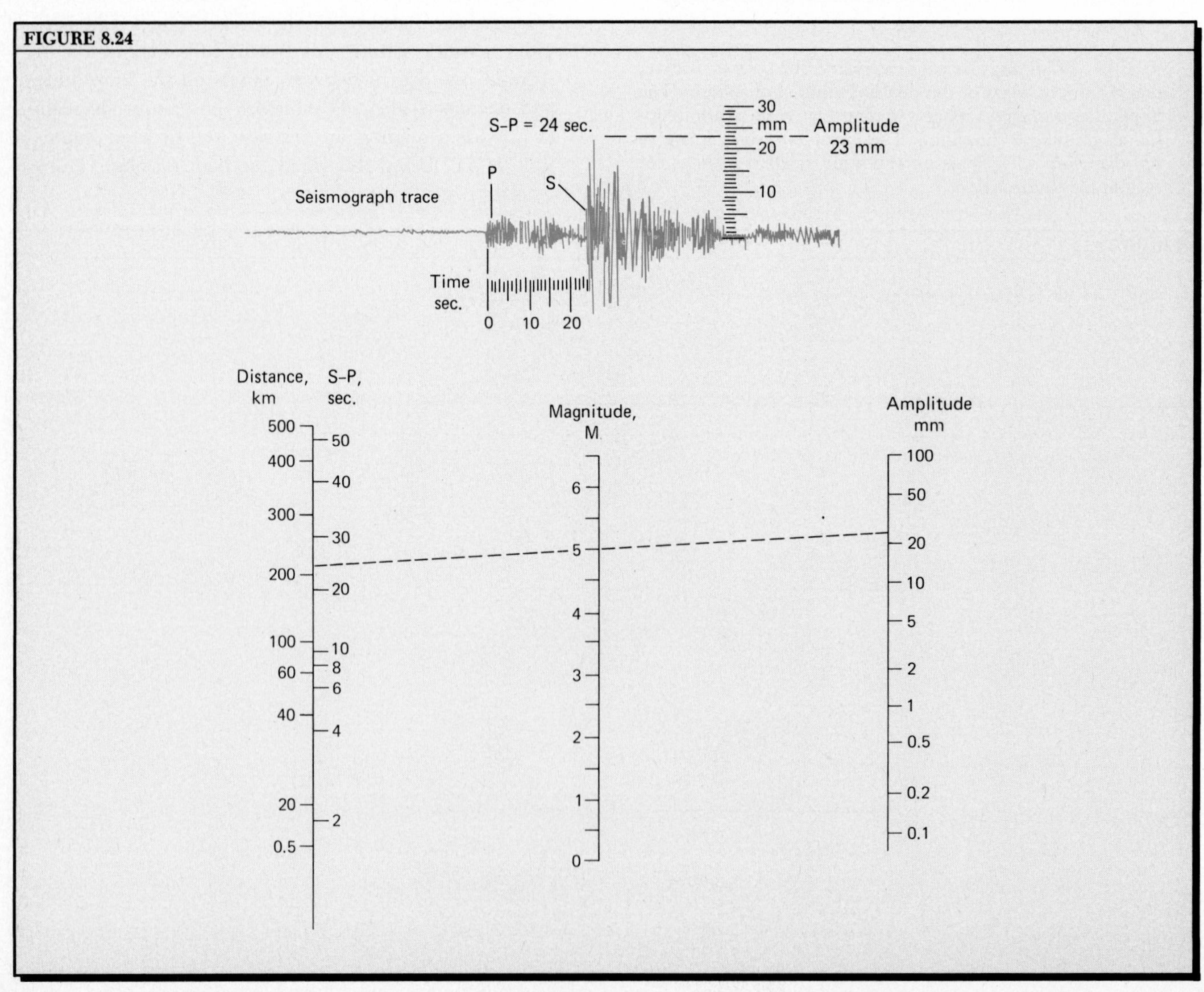

Mercalli scale was prepared by an Italian seismologist of that name in 1902, and was modified in 1956 by Charles Richter to apply to various types of building construction. Previously, the Rossi–Forel intensity scale was in use in this country and remains in use in other parts of the world. The modified Mercalli scale recognizes 12 levels of intensity, designated by Roman numerals I through XII (Table 8.2). Each intensity is described in terms of phenomena that any person might experience. For example, at intensity IV, hanging objects swing, a vibration like that of a passing truck is felt, standing automobiles rock, and windows and dishes rattle. Damage to various classes of masonry is used to establish criteria in the higher numbers of the scale. At an intensity of XII, damage to human-made structures is nearly total and large masses of rock are displaced. On the basis of reports gathered after an earthquake, maps can be prepared to show concentric zones of intensity. The numbered lines are *isoseismals* (Figure 8.26).

Earthquake Ground Motions

Today, special instruments are used to record the strong, damaging ground motions accompanying earthquakes. The ***strong motion seismograph*** records the amplitudes and accelerations of motions so large that they would cause an ordinary seismograph to "go off the scale." Portable instruments can be moved into an area where an earthquake is believed to be imminent. The recording mechanism is triggered by earthquake motions but otherwise remains inoperative.

An index of the severity of ground shaking during an earthquake is found in measurement of the peak acceleration—the rate of change of velocity. In simple

FIGURE 8.25 Earthquake energy release (joules) and Richter magnitude.

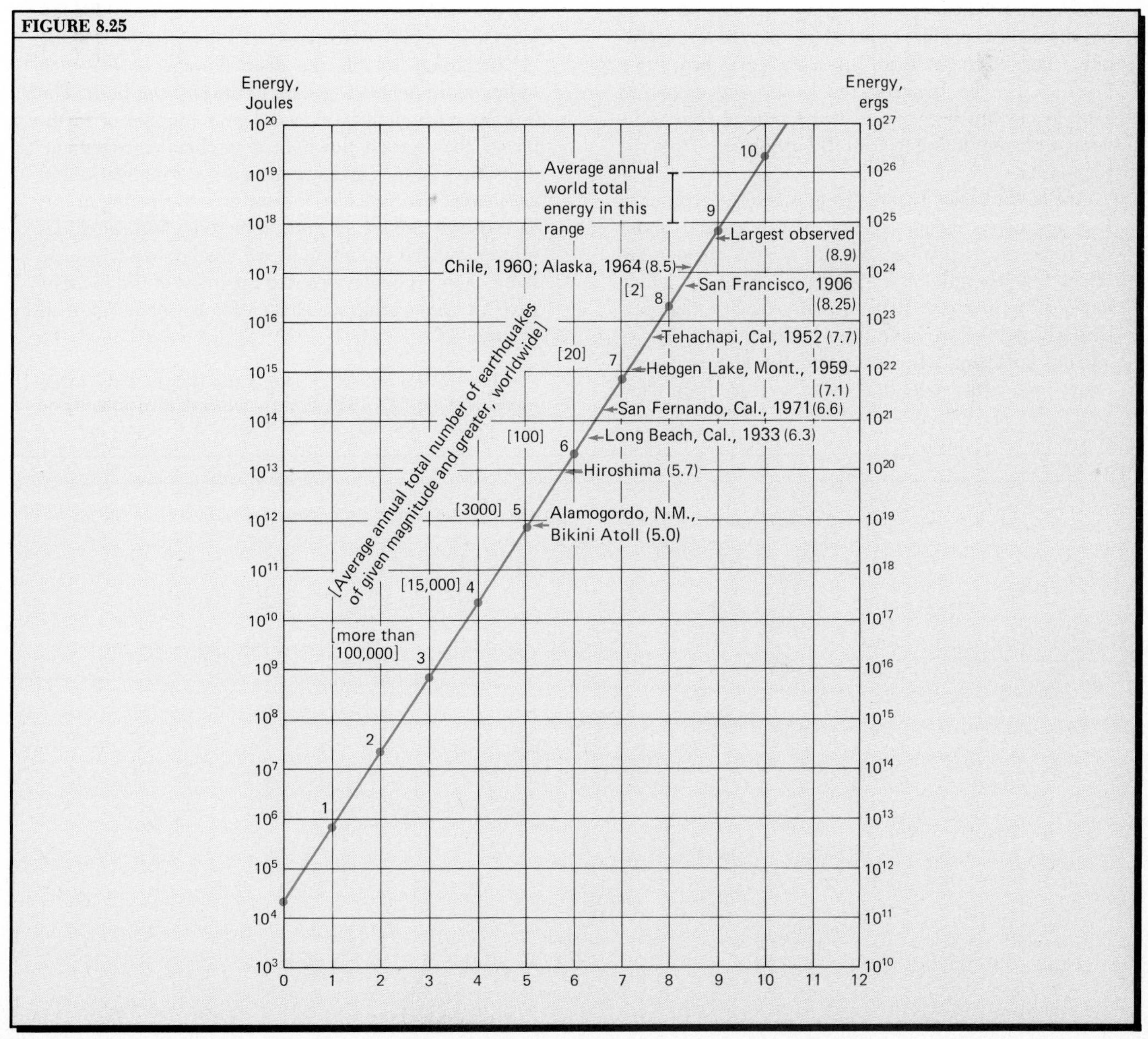

language, the index measures how fast the ground motion speeds up or slows down. Units of acceleration are centimeters-per-second-per-second (cm/sec/sec or cm/sec^2), the same as for the acceleration of gravity, denoted by the symbol g. The value of g is 980 cm/sec^2. Earthquake acceleration is also given in proportional parts of g. Peak ground accelerations occur when the direction of motion reverses itself, producing a whiplash action that causes breakage in structural materials. Maximum accelerations are observed on solid bedrock, as compared with loose or soft sediments, and usually, the horizontal acceleration is greater than the vertical acceleration. High values of acceleration, in the range from 0.5 to 1.0 g, are associated with high velocities of ground motion. The combination of high values of horizontal velocity and acceleration easily topples unreinforced masonry buildings (Figure 8.27). Duration of shaking is also important, since the longer the shaking lasts, the more likely is structural failure. Strong vertical accelerations sometimes produce the interesting effect of throwing objects, such as pebbles or boulders, into the air repeatedly "like peas on a drum." Even objects as heavy as a firetruck can be bounced repeatedly and displaced sideways as much as one or two meters. In the same way a frame building can be displaced from its foundation.

One of the major hazards from a severe earthquake is a secondary effect—collapse of the ground under the force of gravity because the ground shaking reduces the strength of the earth material on which heavy structures rest. Parts of many major cities, particularly port cities, have been built on naturally occurring bodies of soft, unconsolidated clay-rich sediment (such as the delta deposits of a river) or on filled areas in which large amounts of loose earth materials have been dumped to build up the land level. These water-saturated deposits often experience a change in property known as *liquefaction* when shaken by an earthquake. The material loses strength to the degree that it becomes a highly fluid mud, incapable of supporting buildings, which show severe tilting or collapse.

During the great San Francisco earthquake of 1906, the most severe shaking and structural damage occurred on areas bordering San Francisco Bay, where the buildings had been constructed on mud deposits or artificial fill. These areas showed effects associated with a Mercalli intensity of VII, VIII, or higher, whereas areas of the city where buildings rested on solid bedrock showed intensities lower than VII.

Liquefaction of clay-rich sediments during an earthquake can also lead to downhill flowage of the material under the force of gravity. These movements go under such names as landslides and earthflows; they are discussed in Chapter 14. For example, in the Alaskan Good Friday Earthquake of 1964, ground shaking at Anchorage set off the deep flowage of clay-rich sediments on which a section of the city was built. The flowage movement produced a large number of earthblocks that settled down in a steplike arrangement, dislocating houses and streets and breaking water and gas mains (Figure 8.28). Violent ground shaking in this earthquake also set off great snowslides and rockslides in unpopulated mountain areas. Earthquake-triggered slides of this type have caused great loss of life in other parts of the world; we shall refer to them again in Chapter 14.

FIGURE 8.26 An isoseismal map of the Hebgen Lake earthquake of August 17, 1959. Roman numerals give intensity on the modified Mercalli scale. (U.S. Coast & Geodetic Survey.)

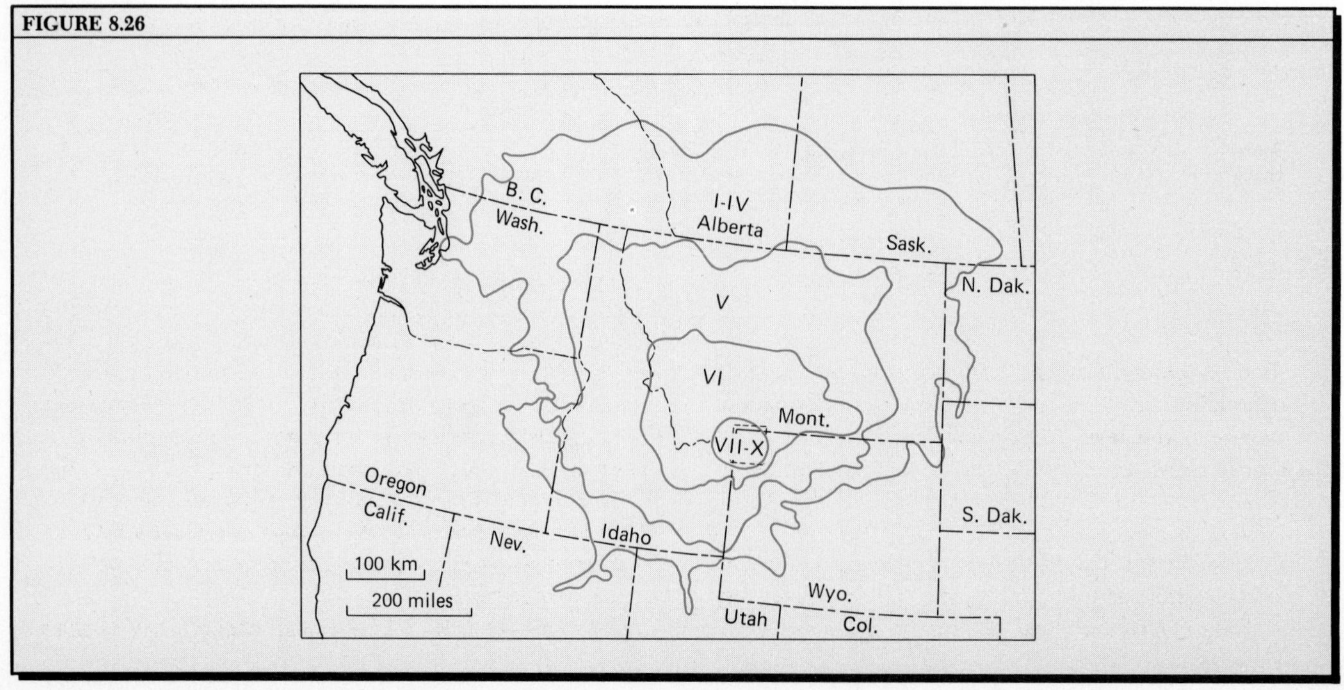

Table 8.2 The Modified Mercalli Intensity Scale (Richter, 1956)

Masonry types:	The quality of masonry, brick or otherwise, is specified by the following letter code:
Masonry A	Good workmanship, mortar, and design; reinforced, especially laterally, and bound together by using steel, concrete, etc.; designed to resist lateral forces.
Masonry B	Good workmanship and mortar; reinforced, but not designed in detail to resist lateral forces.
Masonry C	Ordinary workmanship and mortar; no extreme weaknesses like failing to tie in at corners, but neither reinforced nor designed against horizontal forces.
Masonry D	Weak materials, such as adobe; poor mortar; low standards of workmanship; weak horizontally.

Intensity value	Description
I.	Not felt. Marginal and long-period effects of large earthquakes.
II.	Felt by persons at rest, on upper floors, or favorably placed.
III.	Felt indoors. Hanging objects swing. Vibration like passing of light trucks. Duration estimated. May not be recognized as an earthquake.
IV.	Hanging objects swing. Vibration like passing of heavy trucks; or sensation of a jolt like a heavy ball striking the walls. Standing cars rock. Windows, dishes, doors rattle. Glasses clink. Crockery clashes. In the upper range of IV, wooden walls and frame creak.
V.	Felt outdoors; direction estimated. Sleepers wakened. Liquids disturbed, some spilled. Small unstable objects displaced or upset. Doors swing, close, open. Shutters, pictures move. Pendulum clocks stop, start, change rate.
VI.	Felt by all. Many frightened and run outdoors. Persons walk unsteadily. Windows, dishes, glassware broken. Knickknacks, books, etc. off shelves. Pictures off walls. Furniture moved or overturned. Weak plaster and masonry D cracked. Small bells ring (church, school). Trees, bushes shaken visibly, or heard to rustle.
VII.	Difficult to stand. Noticed by drivers. Hanging objects quiver. Furniture broken. Damage to masonry D, including cracks. Weak chimneys broken at roof line. Fall of plaster, loose bricks, stones, tiles, cornices, also unbraced parapets and architectural ornaments. Some cracks in masonry C. Waves on ponds, water turbid with mud. Small slides and caving in along sand or gravel banks. Large bells ring. Concrete irrigation ditches damaged.
VIII.	Steering of cars affected. Damage to masonry C; partial collapse. Some damage to masonry B; none to masonry A. Fall of stucco and some masonry walls. Twisting, fall of chimneys, factory stacks, monuments, towers, elevated tanks. Frame houses moved on foundations if not bolted down; loose panel walls thrown out. Decayed piling broken off. Branches broken from trees. Changes in flow or temperature of springs and wells. Cracks in wet ground and on steep slopes.
IX.	General panic. Masonry D destroyed; masonry C heavily damaged, sometimes with complete collapse; masonry B seriously damaged. General damage to foundations. Frame structures, if not bolted, shifted off foundations. Frames cracked. Serious damage to reservoirs. Underground pipes broken. Conspicuous cracks in ground. In alluviated areas sand and mud ejected, earthquake fountains, sand craters.
X.	Most masonry and frame structures destroyed with their foundations. Some well-built wooden structures and bridges destroyed. Serious damage to dams, dikes, embankments. Large landslides. Water thrown on banks of canals, rivers, lakes, etc. Sand and mud shifted horizontally on beaches and flat land. Rails bent slightly.
XI.	Rails bent greatly. Underground pipelines completely out of service.
XII.	Damage nearly total. Large rock masses displaced. Lines of sight and level distorted. Objects thrown into the air.

Source: C. F. Richter, (1958), ***Elementary Seismology,*** W. H. Freeman and Company, San Francisco, p. 136–138. Minor editorial changes, following B. A. Bolt, (1978), ***Earthquakes: A Primer,*** W. H. Freeman and Company, San Francisco, Appendix C, p. 204–205.

Changes in Ground Level Accompanying Earthquakes

Besides ground shaking by seismic waves, a major earthquake may be accompanied by permanent changes in the level of the land surface. The Good Friday Earthquake of March 27, 1964, centered about 120 km from the city of Anchorage, Alaska (Figure 8.29) had a magnitude on the Richter scale of 8.4 to 8.6, close to the maximum known. Intensity on the Mercalli scale was probably VII to VIII in Anchorage.

Sudden changes of land level, both up and down, took place at points as far distant as 500 km from the epicenter of the Alaskan earthquake and covered a total area of about 200,000 sq km. A belt of uplift reaching a maximum of 10 m ran parallel with the coast and largely offshore (Figure 8.30). A broad zone of shallow subsidence reaching amounts somewhat more than −2 m lay along the landward side of the uplift zone. The epicenter lay between these zones. Sudden movement of the seafloor produced a train of seismic sea waves, a phenomenon described in following paragraphs.

As we will explain later in this chapter, the Alaskan

earthquake of 1964 was associated with an active subduction zone along which the oceanic lithosphere of the Pacific plate is being forced down beneath continental lithosphere of the Alaskan mainland.

Seismic Sea Waves, or Tsunamis

A special kind of ocean wave not related to wind or tide is the *seismic sea wave*, or *tsunami*, produced by a sudden displacement of the seafloor. The displacement may be caused by a submarine landslide set off by faulting, a sudden rising or sinking of a rock mass when faulting occurs, or a submarine volcanic eruption. The effect is very much like that of dropping a stone into a shallow, quiet pond. A series of simple progressive water waves is sent outward in concentric rings (Figure 8.31).

Seismic sea waves are of great length—100 to 200 km—compared with wind waves and swell, whereas the wave height at sea may be only 0.5 to 1.0 m, and such a wave would not be felt by persons on a ship. Seismic sea waves have periods of 10 to 30 minutes, which is a vastly longer period than even a long swell (20 seconds). To obtain some idea of the velocity of a seismic sea wave, we may imagine it to have a length of 150 km and a period of 20 minutes. Thus, three waves, each 150 km long, will pass a fixed point each hour, which means a velocity of 450 km per hour.

Seismic sea waves are very long in comparison with the depth of water in which they travel. For example, a 200-km wave length is 40 times as great as an ocean depth of 5 km. In such comparatively shallow water, the velocity of travel of the seismic sea wave varies as the square root of the water depth. Therefore, if we know the time at which the wave was sent out (this information is available from earthquake records) and the time at which the wave arrived at a distant coast, we may calculate roughly the depth of ocean water lying between. Just such a procedure was used in 1856 to estimate the average depth of the Pacific Ocean, long before soundings were available to give direct measurements.

FIGURE 8.27 In the Peruvian city of Huaraz many masonry walls built of brick or stone with weak mortar, and lacking in steel reinforcement, crumbled to rubble in the earthquake of May 31, 1970. Measuring 7.7 on the Richter scale, this earthquake took a death toll of 66,000 persons. (U.S. Geological Survey.)

FIGURE 8.28 Slumping and flowage of unconsolidated sediments, resulting in property destruction at Anchorage, Alaska, Good Friday earthquake of March 27, 1964. (U.S. Army Corps of Engineers.)

FIGURE 8.27

FIGURE 8.28

FIGURE 8.29 Map of south-central Alaska showing crustal uplift and subsidence associated with the Good Friday earthquake of March 27, 1964. Contours in meters. Profile and structure section below are drawn through the epicenter along a NW–SE line (AA′). (Redrawn and simplified from G. Plafker, 1964, *Science*, vol. 148, p. 1677, Figure 2, and p. 1681, Figure 6.)

FIGURE 8.30 A geologist determines the elevation above sea level of the zone of barnacles (white crust on marine cliff) to calculate the amount of tectonic uplift (about 3 m) accompanying the Alaska earthquake of 1964. Latouche Island, Prince William Sound, Alaska. (U.S. Geological Survey.)

FIGURE 8.29

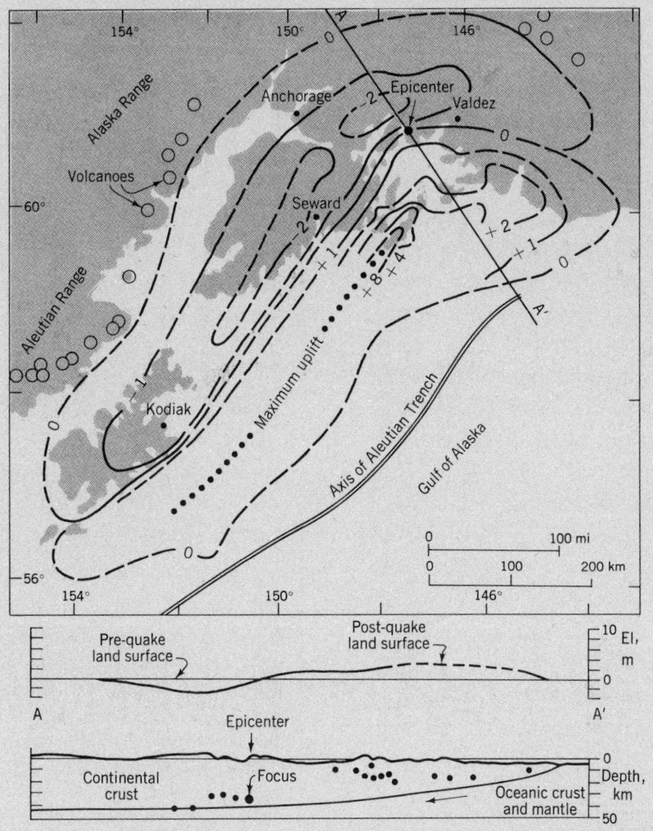

FIGURE 8.30

Upon arriving at a distant shore, the individual wave crest of a tsunami takes the form of a slow rise in water level over a period of 10 to 15 minutes. Superimposed on this rise are ordinary wind waves. These waves break close to shore, producing a destructive surf which demolishes houses and trees.

The first evidence of a tsunami may, however, be the lowering of water level, causing a seaward withdrawal of the water line and exposing the floors of shallow bays. This is what seems to have happened at Lisbon, Portugal, on November 1, 1755, following an earthquake centered off the Portuguese coast. The sight of the exposed seafloor attracted a large number of townspeople, who were drowned when the following wave crest arrived. Several great catastrophes in recorded history seem to have been wrought by tsunamis—for example, the flooding of the Japanese coast in 1703, with a loss of more than 100,000 lives.

A particularly destructive train of seismic sea waves occurred in the Pacific Ocean on March 3, 1933, providing a good example of the phenomenon. The waves were produced by an earthquake centered beneath the ocean floor at a place 500 km northeast of Tokyo (lat. 39° N, long. 144° E). The wave train first reached the Japanese shore in one-half hour, Yokohama in 2 hours, Honolulu in 7.5 hours, San Francisco in 10.3 hours, and Iquique, Chile, in 22 hours. On exposed parts of the Japanese coast, where deep water lies close to shore, the waves rose as high as 10 m, causing destruction and death in low-lying areas.

The Good Friday Alaskan earthquake of 1964 released a tsunami that did great damage at a number of Pacific coastal points. In Alaska, Kodiak Island and Prince William Sound suffered severely from coastal flooding (Figure 8.32). San Francisco Bay experienced a water-level rise of about a meter that damaged small craft moored in marinas within the bay. Very severe effects were felt at Crescent City, in northern California, where damage amounted to about $10 million. Hilo, Hawaii, experienced a steep-walled water wave (a *bore*) with a height of seven meters that rushed landward in a narrow estuary, causing great destruction. In Hilo alone, 83 persons died, and the total death toll for the Hawaiian Islands was over 170. Following the 1964 tsunami, the Seismic Sea Wave Warning System (SSWWS) was set up, with headquarters in Honolulu. Seismologists at the headquarters issue warnings of possible destructive seismic sea waves, using the known depths of the ocean and the known instant of the earthquakes as a basis for computing the time the first waves will reach a given coast (Figure 8.31).

Earthquakes and Plate Tectonics

The relationship between earthquake activity, or *seismicity*, and lithospheric plate boundaries is remarkably strong, although some important earth-

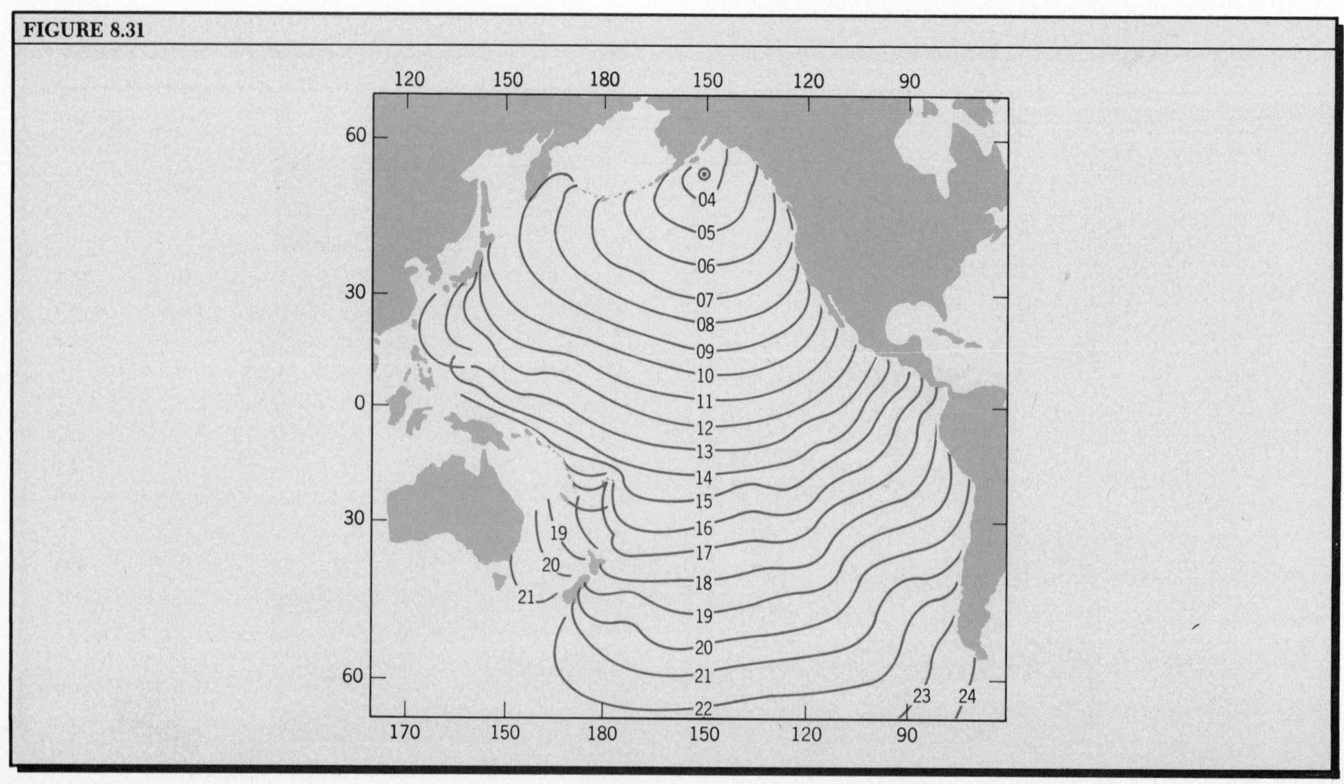

FIGURE 8.31 Map of the Pacific Ocean showing the locations of a tsunami wave front at one-hour intervals, GMT. The wave originated in the Gulf of Alaska as a result of the Good Friday earthquake of March 27, 1964. (After B. W. Wilson and A. Torum, 1968, U.S. Army Corps of Engineers, *Tech Memorandum No. 25*, Coastal Engineering Research Center, Washington, D.C., p. 38, Figure 27.)

quakes have occurred in the interiors of continental plates and along stable continental margins. Figure 8.33 is a world map showing the belts of high seismicity and the locations of many of the largest earthquakes.

Notice particularly the abundance of large earthquakes in a ring surrounding the Pacific Ocean. Most of this *circum-Pacific belt* is a line of intense volcanic activity as well as seismic activity. Comparison with the global map of lithospheric plates, Figure 1.11, will immediately show that the circum-Pacific belt marks the outer margins of the Pacific, Nazca, and Cocos plates, and that most of this plate boundary is one of active subduction.

A second belt of intense seismic activity stretches from the Mediterranean Sea to the Indonesian region. This belt corresponds to the tectonically active boundary between the Eurasian plate and the African and Australian plates that border it on the south. This, too, is a subduction zone throughout part of its length, and part is a zone of recent continental collision extending across southern Asia. We can conclude as a general rule that active subduction and collision zones produce the highest-magnitude earthquakes.

A third important belt of high seismicity runs down the middle of the Atlantic Ocean basin, through the Indian Ocean, and across the southern Pacific basin. Comparison with the global map of lithospheric plates (Figure 1.11) shows that this belt coincides with the active spreading plate boundaries where new oceanic crust is being produced. The earthquakes that are numerous along the spreading plate boundaries are for the most part of moderate to low intensity and are of the shallow-focus type.

FIGURE 8.32 The waterfront area of Seward, Alaska, devastated by the tsunami which accompanied the Good Friday Earthquake of 1964. (U.S. Army Corps of Engineers.)

FIGURE 8.32

Further details of earthquake activity in subduction zones and in spreading plate boundaries are presented in Chapter 10. One feature that will emerge more clearly is that seismicity is found not only in the rift zone, where two oceanic plates are separating by spreading, but also on transform faults that offset those spreading boundaries.

The San Andreas Fault

The San Andreas Fault of California, which we have described as an active transcurrent fault, is now interpreted by most geologists as representing a major plate boundary—the boundary between the Pacific plate and the North American plate (see Figure 1.11). Under this interpretation, as a transform-fault plate boundary, it is unusual, because nearly all other transform plate boundaries are within the oceanic lithosphere and lie submerged beneath deep ocean water.

The San Andreas Fault runs for a distance of about 950 km. It starts in the Salton Sea Basin of southern California, trends in a northwesterly direction, about parallel with the Pacific coastline, to the San Francisco Bay region (Figure 8.34), then follows the northern California coast to Cape Mendocino. There are many other major active faults in California, as shown in Figure 8.34. Some of these appear as branches of the San Andreas Fault, but others, such as the Garlock Fault, intersect the principal trend at an obtuse angle.

Along the San Andreas Fault, motion has consistently been such that the crustal mass on the western (Pacific) side has moved in a northwesterly direction, relative to the eastern block. With this type of motion, the San Andreas Fault is a right-lateral transcurrent fault (Chapter 6). If you stand close to the fault line, facing across the fault, the block on the opposite side moves toward your right. Movement along the San Andreas Fault has been going on for many millions of years. Matching of similar rock masses displaced along the two sides of the fault shows a total lateral movement of about 560 km over the past 150 million years.

Continued right-lateral slippage on the San Andreas Fault has produced a conspicuous trenchlike feature, which in places is nearly straight (Figure 8.35). Locally, the trench is occupied by small lakes, called *sag ponds*, or is followed for short distances by streams. Where a stream has flowed across the fault line for many centuries, repeated fault movements have caused the stream to be offset (Figure 8.36); it has been forced to follow the fault line for a short distance to maintain its flow to the displaced downstream portion of its channel.

Although the average rate of movement on the entire San Andreas Fault has been about 5 cm per year, certain lengthy sections have shown no movement for

FIGURE 8.33 Seismicity of the earth. World maps of earthquake epicenters 1961–1967 and epicenters of great earthquakes. A. Epicenters of earthquakes originating at depths from 0 to 100 km (color dots). Each color dot represents a single epicenter or a cluster of epicenters. Black circles show epicenters of earthquakes of Richter magnitude 8.0 and greater, 1897–1976. B. Epicenters of earthquakes originating at depths from 100 to 700 km. Each color dot represents a single epicenter or a cluster of epicenters. (Epicenter data compiled from ESSA, Coast and Geodetic Survey, in J. Dorman and M. Barazangi, 1969, *Seismological Society of America, Bull.*, vol. 50, No. 1, Plates 2 and 3. Epicenters of great earthquakes from W. Hamilton, 1979, *Professional Paper 1078*, U.S. Geological Survey, Figure 3.)

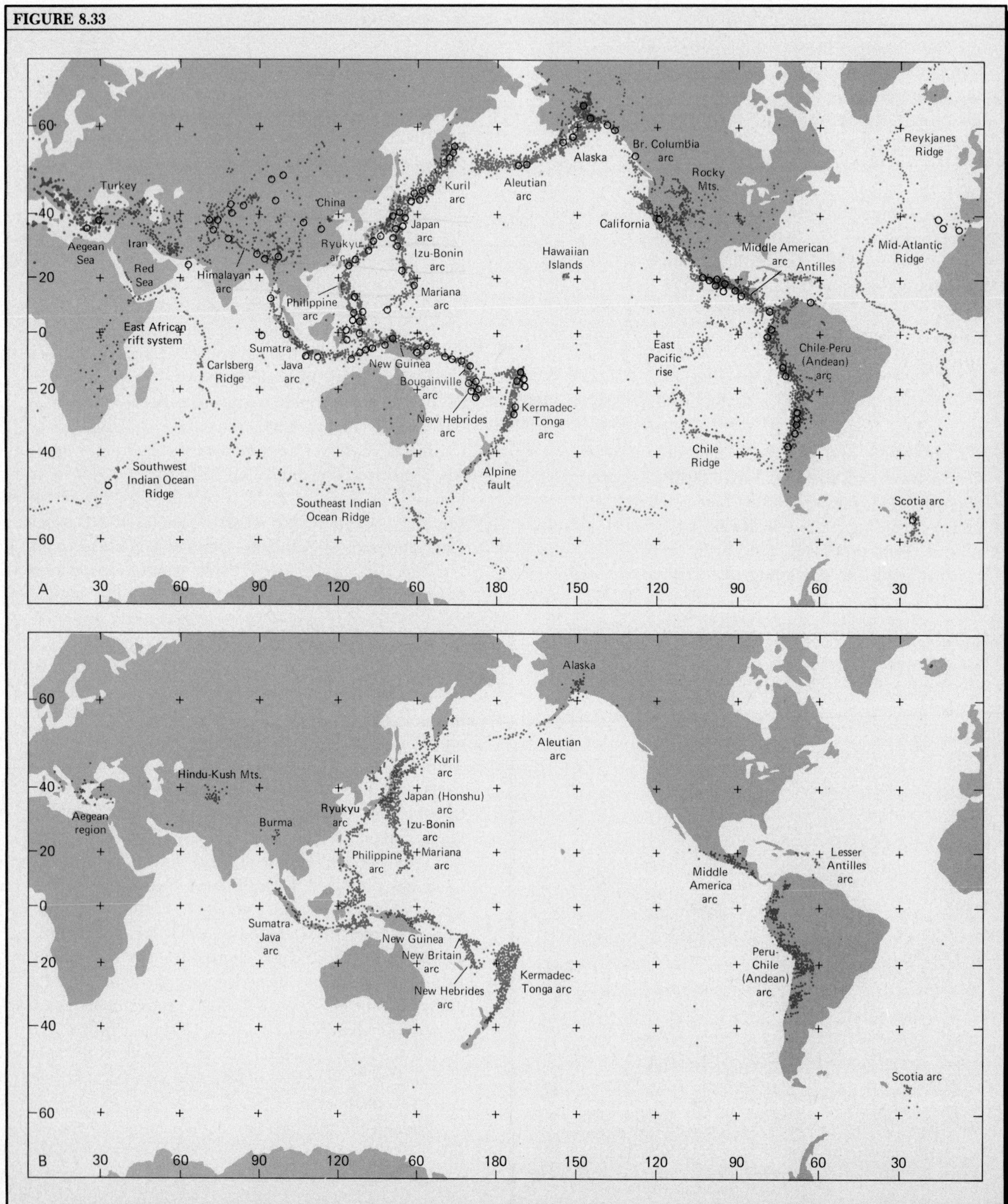

FIGURE 8.34 The San Andreas Fault and associated major faults of California. (Based on data of R. H. Jahns and C. G. Higgins.)

FIGURE 8.35 The San Andreas Fault, seen here in the Temblor Range of central California, takes the form of a straight valley through low hills. The view is to the southeast. In the foreground is Grant Lake, a sag pond. Along the recently active fault, which lies to the left of the road, four successive stream channels are offset along the fault line. (Spence Air Photos.)

FIGURE 8.36 Offsetting of streams along an active transcurrent fault.

FIGURE 8.34

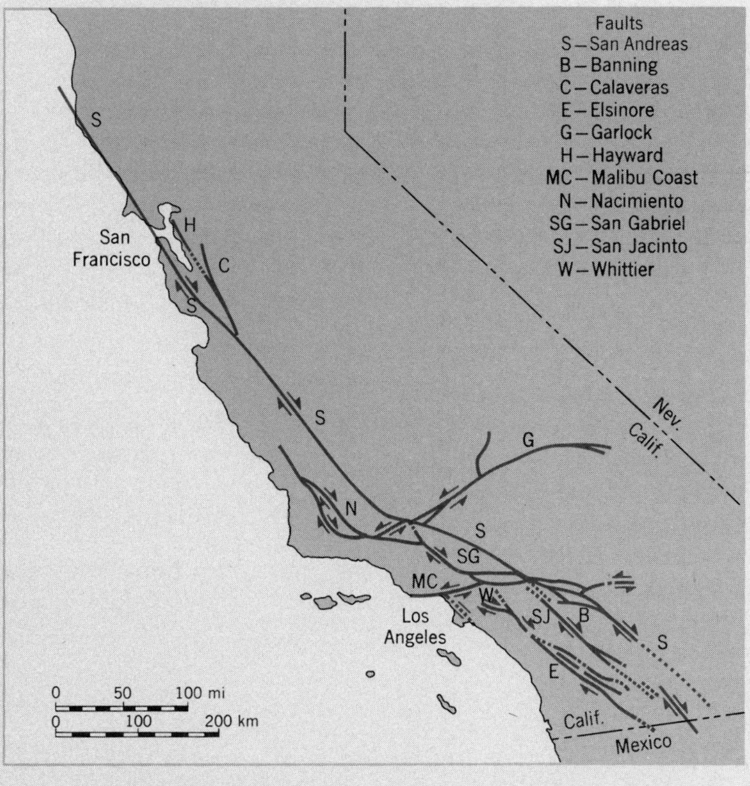

Faults
S—San Andreas
B—Banning
C—Calaveras
E—Elsinore
G—Garlock
H—Hayward
MC—Malibu Coast
N—Nacimiento
SG—San Gabriel
SJ—San Jacinto
W—Whittier

San Francisco
Los Angeles
Nev.
Calif.
Mexico
0 50 100 mi
0 100 200 km

FIGURE 8.35

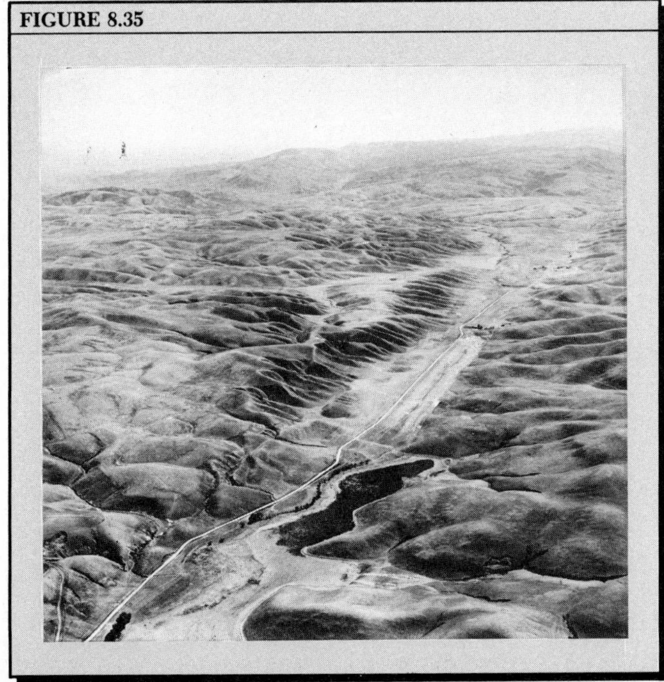

FIGURE 8.36

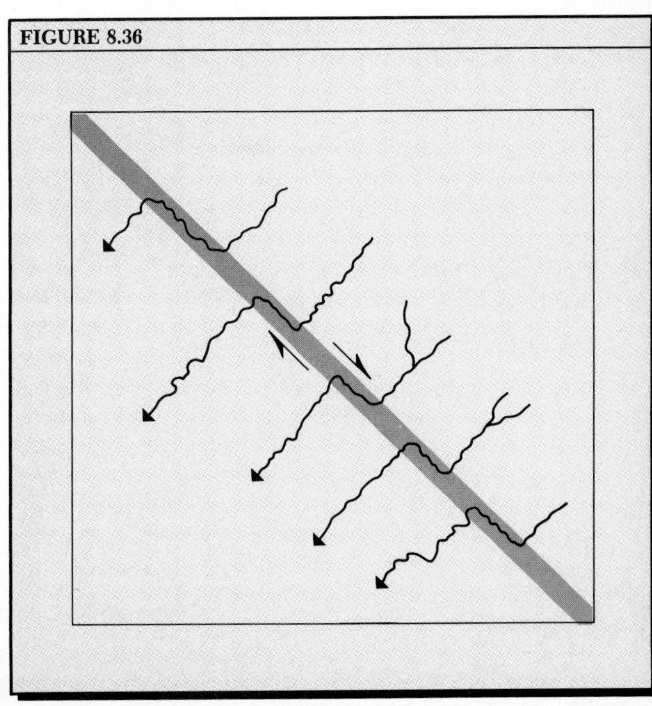

a half-century or more. These seismically inactive sections are referred to as *locked faults;* they have a great potential for generating major earthquakes. In 50 years, at the average rate, the movement should have been about 2.5 m. If enough strain has been accumulated to yield a break of that displacement, the energy release will be enormous when it comes. As shown in Figure 8.37, one locked section runs from near Hollister to Cape Mendocino. There has been no significant fault motion in this section since the great San Francisco earthquake of 1906. A second locked section, lying between San Bernardino and Parkfield, last experienced a major break in 1857. Thus, both the San Francisco and Los Angeles metropolitan areas can anticipate a major earthquake at a point in time that could range from the very near future to a few decades from now.

In contrast to the locked sections, there are other sections of the San Andreas Fault along which a very slow, almost continuous movement is occurring. This motion is referred to as *fault creep* and is monitored closely. Where creep is occurring, there are also generated many earthquakes of small to moderate magnitude. The combination of creep and small slippages is thought to prevent the large buildup of elastic strain that would ultimately generate a major earthquake. (This supposition has recently been questioned.) Figure 8.37 shows the zones of creep and seismic activity along the San Andreas Fault and related branch faults.

*Earthquake Prediction

The great destructiveness of earthquakes occurring in densely populated areas has led seismologists of the United States, Japan, China, and several other countries to test every possible means of predicting the time and place a major earthquake will occur. In general, two approaches have been taken to this problem. One method observes several kinds of physical phenomena that undergo significant change in a short period immediately prior to an earthquake.

A second approach is historical, and long-term seismic histories of known faults are studied. For example, the identification of locked sections along the San Andreas Fault marks them as capable of yielding earthquakes of high magnitude. The two methods can

FIGURE 8.37 Active areas (*color*) and locked sections of the San Andreas Fault. (Based on data of C. R. Allen, 1968, in *Proc., Conference on the Geologic Problems of the San Andreas Fault System*, Stanford Univ. Publ., Univ. Ser. of Geological Science no. 11, p. 70.)

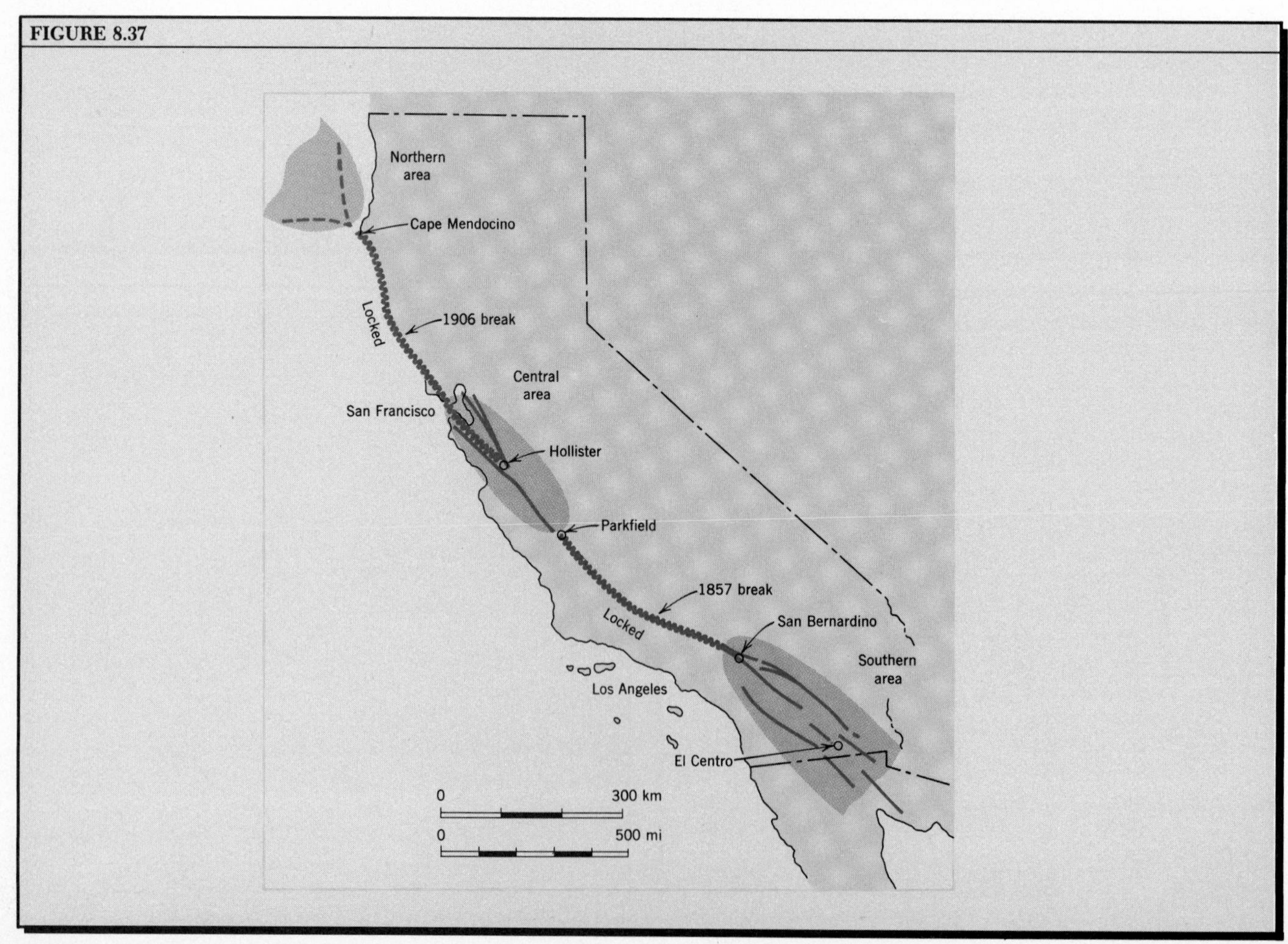

be combined, concentrating intensive instrumental observation where the historical data suggest that earthquake energy may have been stored almost to the breaking point.

Let us take up first the interpretation of changes in physical phenomena prior to an earthquake. These go under the general heading of precursory phenomena, or precursors (from the Latin, meaning "to run before"). A general theory has been proposed to relate five kinds of physical changes that can be expected in a period of weeks or months before an earthquake of substantial magnitude occurs. This theory states that rock strain accompanying the slow movement of crustal masses along a fault zone causes a large volume of rock to become riddled with countless minute cracks or *microcracks*. Cracking increases the volume of the rock, a change referred to as *dilatancy*. In most regions, all rock pores and cracks are filled with water, called ground water (Chapter 15), at a depth below a few tens of meters. After rock strain has caused dilatation of the rock, ground water moves into the newly created microcracks. The presence of the water acts like a great hydraulic jack, with the result that the rock mass swells and the ground surface may experience a doming. The lifting effect of the water also tends to reduce the friction holding the rock masses together on the two sides of the fault, so that eventually the rapid fault-slip motion is triggered.

FIGURE 8.38 A schematic graph of the precursory events and stages included in the dilatancy hypothesis of earthquakes. (Based on data of National Academy of Sciences, 1976, *Predicting Earthquakes*, Figure 2, p. 41.)

Figure 8.38 is a schematic graph showing how several observable physical changes are thought to occur prior to a substantial earthquake. Five time-stages are recognized, and physical changes are as follows:

- **P-wave Velocity** Active faults produce numerous small earthquakes which can be monitored by seismographs placed in strategic positions. Dilatation can be expected to cause a decrease in P-wave velocity, taking place in Stage II. Then, in Stage III, as ground water fills the microcracks, P-wave velocity increases, returning to the normal value immediately before the large earthquake.
- **Ground Uplift** Uplift of the ground surface accompanies dilation and remains high until the earthquake occurs. In an actual case in Japan, the Niigata earthquake of 1964, a gradual increase in ground elevation had occurred near the epicenter through a 60-year period prior to the earthquake. A sudden drop in ground level accompanied the earthquake. This information was not, however, assembled until after the earthquake occurred, so it played no part in prediction.
- **Radon Emission** Monitoring of the emission of radon gas from wells close to active faults has sug-

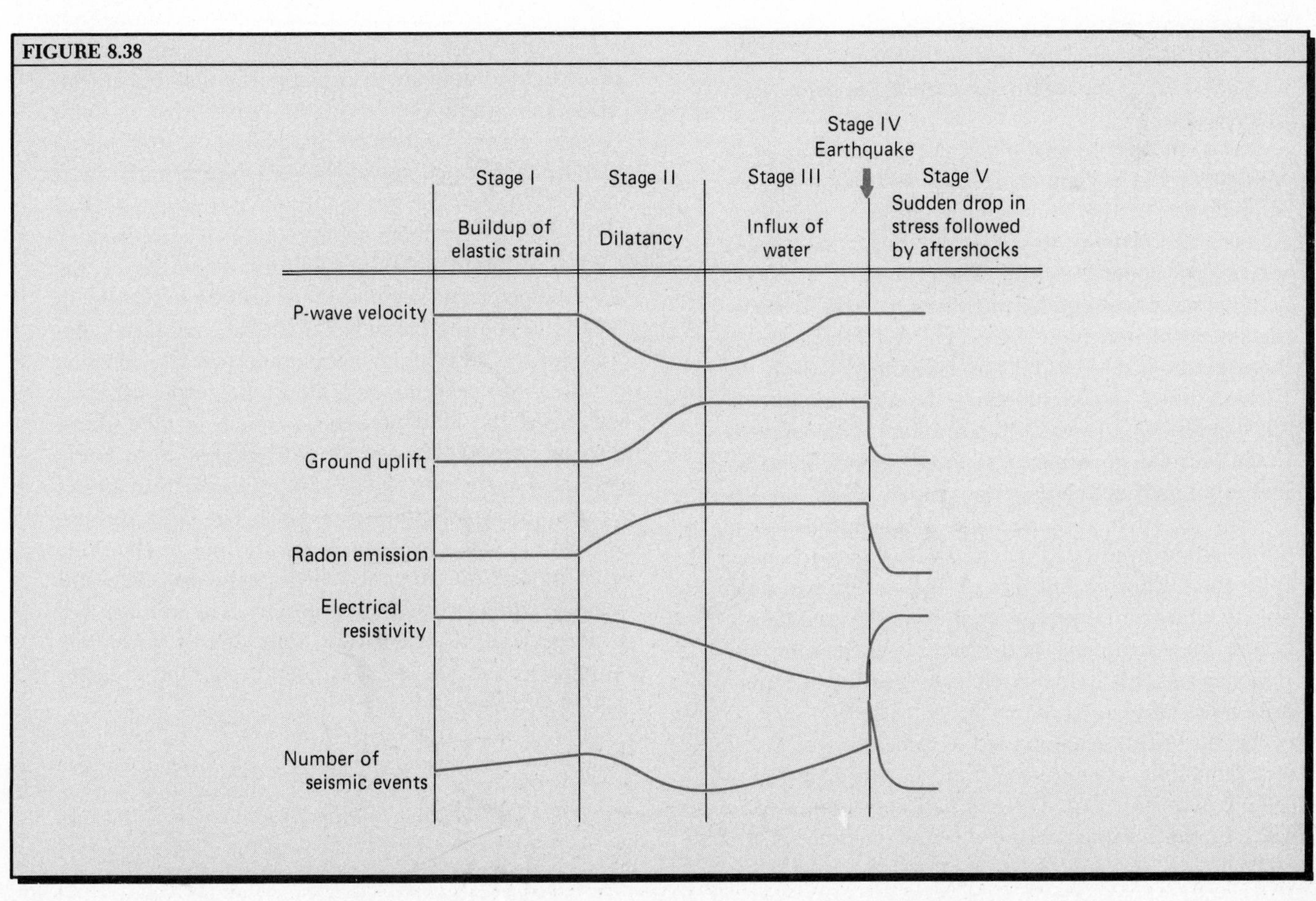

gested that the emission increases just before an earthquake takes place. The significance of this observation is considered questionable.

- **Electrical Resistivity** Electrical resistivity of rock in the vicinity of the fault is thought to decrease during dilatancy. Some field measurements support this decrease, but the relationship to faulting is uncertain.
- **Number of Seismic Events** The frequency of occurrence of small earthquakes has been observed to increase prior to a large earthquake. Many small quakes occurring in a short time span constitute an *earthquake swarm*. In the case of an Italian earthquake in 1976, a sharp increase in minor earthquakes along an active fault prompted officials to give a warning that doubtless saved many lives. On the other hand, many earthquake swarms have been observed that did not presage a major quake.

At present, the dilatancy theory of earthquakes remains a hypothesis under discussion, while the use of precursory phenomena based on the theory has had a mixed record of success in earthquake prediction.

The historical method is based on an assessment of the time that has elapsed since a major earthquake last caused a release of strain along a known active fault. It is a method best applied to the subduction zone that surrounds the Pacific plate. When a careful study is made of the times of occurrence and epicenters of major quakes in this belt, there emerges a pattern in which certain sections of the subduction zone have had no major activity for several decades. These sections are designated *seismic gaps* when no major seismic activity has occurred within the past 30 or more years.

One such seismic gap along the southern coast of Mexico in the region of Oaxaca was identified by seismologists of the University of Texas at Galveston. About 150 km long, this gap had shown no seismic activity for about 50 years. Then, in 1976, a seismic activity stage was noticed in which a series of minor quakes originated in the gap. On the hunch that a large earthquake—as great as magnitude 8.0 on the Richter scale—was in the offing, the same scientists in 1977 published a forecast to that effect. Their request to the Mexican government to install seismic monitoring equipment in the gap was denied. Then, on November 29, 1978, an earthquake of magnitude 7.5 to 7.8 occurred in the gap. Fortunately, a research team from the California Institute of Technology had been able to set up portable seismographs in the area shortly before the earthquake and was able to monitor the minor shocks that immediately preceded the major energy release.

The historical technique led to forecasting of a major earthquake, magnitude 7.5 to 8.0, that occurred in 1979 in a seismic gap. The gap had been identified in 1971 in the Aleutian Island–Alaskan volcanic chain. The last major earthquake in this gap had occurred in 1900 and the gap had been singled out, along with several other gaps in the island chain, as a likely place for intense activity within a few decades. No precursory phenomena had been monitored in this gap, so that the forecast spanned a long period of time in which an earthquake might be expected.

It is hoped that intense monitoring of similar seismic gaps can be undertaken and will lead to some good predictions in which the time of occurrence is narrowed down to a span of a few weeks, rather than of several decades.

Induced Earthquakes

A number of cases have come to light in which human activities have set off earthquakes.* A case that is now almost a classic is that of the Denver, Colorado, region. Here, near the Rocky Mountain Arsenal, hundreds of earthquakes have been recorded since 1962; they seem to be correlated with pumping of fluids under pressure into a disposal well penetrating to a depth of 3600 m. As an explanation, it has been proposed that the increased fluid pressure within the rock caused the release of strain already present within the rock—in other words, the water pressure had a triggering effect.

Naturally, this hypothesis leads us to wonder if it might not be possible to use engineering methods, such as fluid injections, to induce many small fault movements and thus to prevent strain buildup to dangerous levels. This possibility has attracted the attention of scientists, and they are working on the theoretical concepts. You might also be led to reason that if fluids (natural ground water or petroleum in rock pores) were to be pumped out of the rock near a major fault zone, the tendency for the fault to become locked would be increased and seismic inactivity induced.

To investigate these possibilities, scientists of the U.S. Geological Survey undertook a study beginning in 1969 in which the Rangely Oil Field of western Colorado was the guinea pig. They found that the injection of water into deep oil wells (to induce more oil flow) had raised the fluid pressure as much as 60% above normal. During this period, earthquakes were being generated at the rate of 15 to 20 per week from a fault system passing through the oil field. The fluid pressure was then lowered by pumping water out of the same wells for a six-month period. A dramatic drop in earthquake frequency resulted generally; the number fell to none at all near the wells. Pumping of fluid back into the ground has been resumed to find out if earthquakes will then increase.

---

*This section on induced earthquakes has been reprinted from A. N. Strahler and A. H. Strahler, 1974, *Introduction to Environmental Science*, Hamilton Publishing, Santa Barbara, California, pp. 221–224.

It is now suspected that the pumping of fluids into the Inglewood Oil Field to raise the hydrostatic pressure and increase oil recovery was responsible for setting off the earthquake of 1963 which fractured a wall of the Baldwin Hills Reservoir. Water spilling from the reservoir brought an inundation of mud to many homes in an area below the reservoir and resulted in five deaths. The correlation between fluid pumpage and fault movements in this Los Angeles locality is now considered to have been demonstrated. Increased fluid pressure reduces the frictional force across the contact surface of a fault, allowing slippage to occur.

Another situation in which human beings may have been responsible for setting off earthquakes is in connection with the building of large dams on major rivers. In a few cases, the load of water from new lakes impounded behind these dams is thought to be responsible for triggering earthquakes. In a ten-year period following the filling of Lake Mead, behind Hoover Dam in Arizona and Nevada, hundreds of minor earth tremors were observed emanating from the area; they are attributed to loading of the crust by lake water. Another case in point is Lake Kariba, behind the Zambezi River in Zambia, which has been generating earthquakes of even greater magnitude.

Several scientists have been concerned with the possibility that underground nuclear explosions can set off significant earthquakes, and that a hazard may exist in this testing activity. Research thus far has shown that an underground blast does set off a number of small earthquakes close to the site of the blast. Seismic energy of the blast triggers the release of strain along faults in the vicinity, but the radius of the known effects is on the order of 10 to 20 km. These observations have led to the suggestion that underground nuclear blasts can be placed where they will induce strain release and thus prevent buildup of strain to dangerous levels.

FIGURE 8.39 Collapsed buildings of the Veterans Administration Hospital, Sylmar, Los Angeles County, California. San Fernando earthquake of February 1971. (Wide World Photos.)

FIGURE 8.39

Earthquakes and Cities

The impact of an earthquake on a major urban area and its inhabitants is illustrated by the San Fernando earthquake of February 9, 1971. The epicenter was located 40 km north of the center of the city of Los Angeles. The San Fernando Valley, close to the ruptured fault, was hardest hit. The entire Los Angeles community was shocked into renewed awareness of the wide range of damaging effects an earthquake brings to a modern city. This earthquake was only of moderate magnitude, 6.6 on the Richter scale, but in a few places the shaking of the ground was as severe as anything that had been previously measured in an earthquake. Peak accelerations exceeded 1 *g*. Fortunately, the strong shaking was of brief duration—about 15 seconds—and only the most susceptible structures were severely damaged. Also fortunate was the fact that the quake occurred at 6:00 A.M., when few persons were traveling the freeways or occupying public buildings.

Of the major disastrous effects to buildings, one of the most frightening was the partial collapse of the Olive View Hospital in Sylmar, a new structure believed to have been built in conformity with earthquake-resistant standards. At the Veterans Hospital in Sylmar, severe damage with 40 deaths occurred (Figure 8.39). Hospital buildings that remained standing were weakened and have since been demolished. The Van Norman Dam was dangerously cracked, and the water had to be drained from the reservoir behind it to forestall potential flooding of a densely built-up area below the dam. An important converter station in the electrical power transmission system of the Los Angeles area was severely damaged. On the Golden State Freeway, an overpass collapsed, blocking the roadway below, while the freeway pavement itself was badly cracked and dislocated (Figure 8.40).

The San Fernando earthquake was generated along a comparatively minor fault some 25 km from the great San Andreas Fault. What would happen if a major earthquake comparable in intensity with the San Francisco earthquake of 1906 or the Alaskan earthquake of 1964 were to occur along the San Andreas Fault in this area? A joint panel of experts of the National Academy of Sciences and the National Academy of Engineering found it clear that existing building codes do not provide adequate damage control features. The cost in damage of the San Fernando earthquake of 1971 was estimated at on the order of $500 million. In contrast, experts anticipate that an earthquake as great as the San Francisco earthquake of 1906 would cause damage on the order of $20 billion, should it occur today. In coming decades the progress of urbanization in the Los Angeles area will have greatly expanded both the population and the building structures subject to devastation.

°*Assessing Earthquake Hazards*

Geologists have intensified their field studies of active faults in order to map those areas where risk of earthquake damage is high. They are devising criteria by which to rate the level of risk that is incurred by the construction of urban housing and power plants—especially nuclear plants—on or close to faults. Each fault that is mapped is assigned a level of capability of producing an earthquake. One kind of field evidence is the determination of the geologic date when the last significant movement occurred, sometimes by obtaining the age of the youngest bedrock or regolith that is displaced by the fault. Another approach is to set up special seismographs to determine if weak seismic activity (background activity) can be detected along a fault that seems not to have undergone movement in recent geologic time.

Particular importance is being attached to determining the capability of faults close to nuclear power generating plants. The Nuclear Regulatory Commission defines a *capable fault* as one that has exhibited movement at or near the ground surface at least once within the past 35,000 years or movement of a recurring nature within the past 500,000 years.

Geologists and seismologists of the U.S. Geological Survey have prepared maps of the United States showing the level of seismic risk. One type of risk map tells the intensity of ground shaking (peak acceleration) in percentage of the value of the earth's surface gravitational acceleration (g), that has a 10% probability of being exceeded in 50 years (Figure 8.41). One can rephrase this statement to read "a 90% probability of not being exceeded in 50 years."

A study of this risk map shows some interesting things. First, the central and eastern portions of the United States, which lie on the geologically "stable" craton of the North American lithospheric plate, show some areas of moderate earthquake risk, with probable peak accelerations exceeding 10% of g in some fairly large areas. The fact is that several severe earthquakes have been recorded in this supposedly stable area. One was the great Charleston, South Carolina, earthquake of 1886. Another, and perhaps one of the greatest of historical record within the craton, was the New Madrid, Missouri, earthquake of 1811, which affected a large area of the Mississippi River floodplain. A series of violent shocks occurred over a period of many

FIGURE 8.40 Broken pavement and collapsed overpass on the Golden State Freeway at the northern end of the San Fernando Valley, California, resulting from the earthquake of February 1971. (U.S. Geological Survey.)

FIGURE 8.40

months, including three big ones that are estimated at Richter magnitudes 7.5, 7.3, and 7.8. It is said that the seismic waves rang church bells in Boston and that chimneys were felled and plaster walls cracked in Richmond, Virginia. In 1755, a severe quake centered near Cape Ann, Massachusetts, damaged buildings in Boston. The intensity is estimated as VIII on the modified Mercalli scale.

According to the U.S. Geological Survey, the eastern half of the United States has experienced more than 3500 earthquakes since 1700 and most of these produced effects in the range of III to V on the modified Mercalli scale. Earthquakes with epicenters in the North American craton are rarely associated with surface faults along which movement can be observed. The geologic circumstances associated with their occurrence are quite obscure in most cases. In contrast, earthquakes occurring in the western part of the country are rather directly associated with visible faults and produce obvious surface changes.

Seismic Activity in Review

In this chapter, we found that earthquake waves play a varied role in both scientific investigation and the everyday world of humankind. Fault movements generate most earthquakes, which represent sudden releases of great quantities of energy temporarily stored in crustal rock. Earthquake waves serve as scientific probes deep into the crust and mantle, and even through the central core.

Severe earthquakes are a natural hazard to humans, particularly in urban areas near major active faults. Science is making slow progress toward its goal of predicting the time of occurrence of potentially dangerous earthquakes, and urban areas are presently ill equipped to cope with a major earthquake.

Key Facts and Concepts

The earth's interior Pressure, temperature, and density increase from the earth's surface to the center. Density increases abruptly at the mantle–core boundary, reaching 10 to 12 g/cc in the iron core, where pressure is 2000 to 3000 kb.

Seismology and earthquakes *Seismology,* the study of *earthquake waves* (*seismic waves*), interprets the records of the *seismograph.* An earthquake represents the sudden release of energy stored as elastic strain in rock where large rock masses in fault contact move intermittently past one another by the mechanism of *stick-slip friction.*

Fault rupture Rupture, releasing stored energy, begins at a point on the fault plane, the *focus,* above which lies the *epicenter.* Rupture spreads rapidly along the fault plane. *Foreshocks* may precede the main shock, which is usually followed by *aftershocks. Shallow-focus earthquakes,* less than 50 km deep, are generated on active normal, reverse, and transcurrent faults. *Deep-focus earthquakes,* originating in the asthenosphere at depths to 700 km, are generated within a descending lithospheric plate.

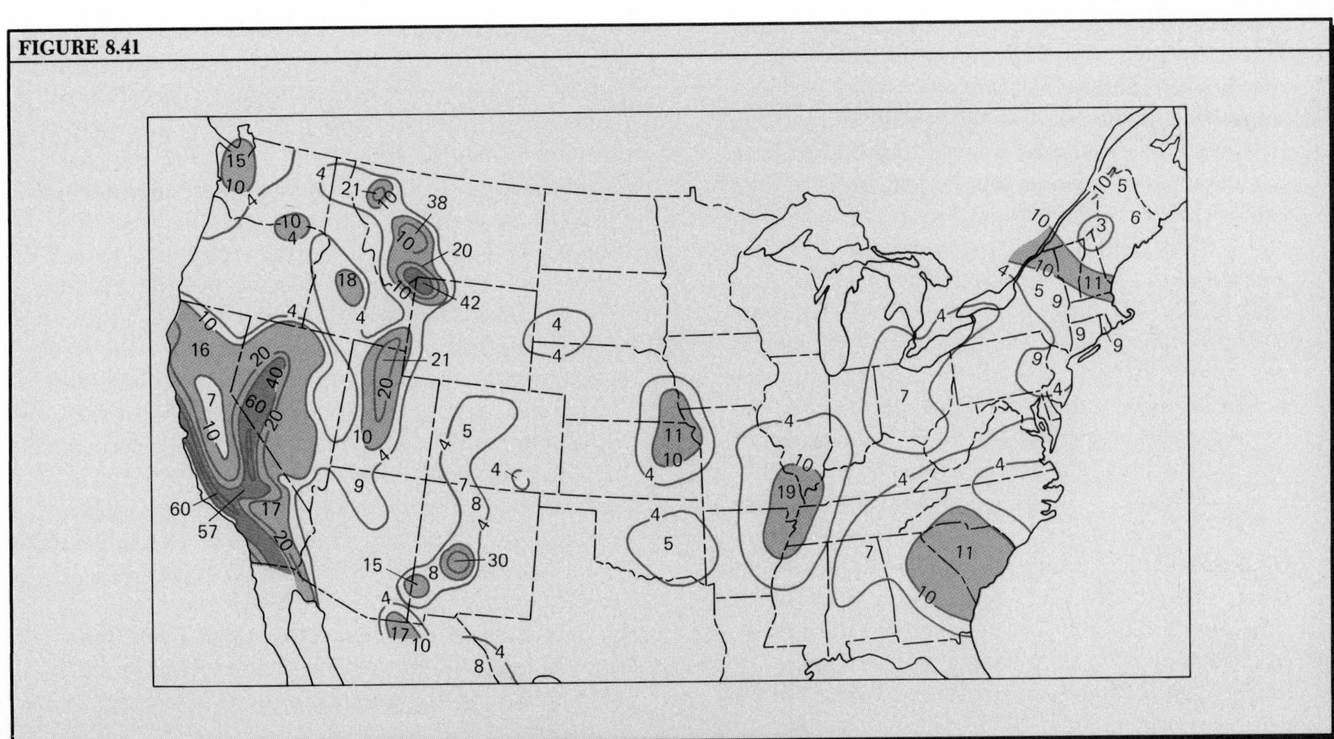

FIGURE 8.41 Seismic risk map of the United States. Numbers give the percentage of the value of the earth's surface gravitational acceleration *(g)* that has a 10% probability of being exceeded in 50 years. (U.S. Geological Survey.)

Elastic-rebound effect Transcurrent faults show an *elastic-rebound effect,* in which horizontal offsetting of reference lines crossing the fault accompanies and follows an earthquake, relieving elastic strain built up prior to the earthquake. Movement on a normal fault produces a fresh fault scarp.

The seismograph The seismograph uses the principle of *inertia* to register ground motion on a *seismogram,* showing earthquake waves of varying *amplitude* and *frequency.* A battery of seismographs is required to register several directional components and frequency bands.

Earthquake waves Wave motion recorded by a seismograph consists of *primary, secondary,* and *surface waves.* Difference in arrival times of primary and secondary waves indicates distance between earthquake focus and seismograph. Primary and secondary waves are *body waves,* traveling through solid rock. Primary waves (P-waves) are *compressional waves;* secondary waves (S-waves) are *shear waves.* Surface waves are *Rayleigh waves,* with vertical orbital motion, or *Love waves,* with horizontal ground motion.

The earth's core Simple S-waves do not pass through the core, a fact indicating the core to be in the liquid state. Because of outward bending (refraction), P-waves passing through the core fail to reach a *shadow zone,* lying more distant than 143° of arc from the epicenter. Distance to the core can be measured from P-waves reflected back to the surface from the sharp core–mantle boundary. An inner core region is in the solid state.

Travel speeds of seismic waves Speed of travel of P- and S-waves increases with increasing *rigidity* of the rock medium. Speeds are slow in unconsolidated sediment, moderate in most sedimentary rocks (dolomite excepted), and high in igneous rocks. Highest speeds are in ultramafic rock (pyroxenite, dunite).

Composition of the mantle Speeds of P- and S-waves in the mantle indicate solid ultramafic rock of magnesium–iron silicate composition, as in peridotite or dunite. A discontinuity in P-wave speed at 650 km distinguishes the *upper mantle* from the *lower mantle.* Upper mantle includes the asthenosphere (300–400 km) and a *transition zone* (400–650 km). A discontinuity at 400 km may represent a *phase transition* involving olivine. The discontinuity at 650 km may represent another phase transition, involving the breakdown of olivine into oxides of iron, magnesium, and silicon.

The upper mantle The asthenosphere of the upper mantle shows a *plastic layer,* or *soft layer,* between 60 and 200 km, in which rock is close to its melting point, strength reaches a minimum value close to zero, and numbers of large earthquakes are fewest. In this *low-velocity zone,* seismic wave velocities are slower than in the layers above and below. Partial melting may occur in the asthenosphere, causing strength loss.

Seismic waves and crustal structure Seismic waves generated by shallow earthquakes or shocks made by explosives are observed to travel by *refraction* and reflection, following paths that return to the surface. Seismograms can thus reveal rock layers of differing density. The crust–mantle boundary is shown by a seismic discontinuity called the *Mohorovičić discontinuity* (*Moho,* or *M-discontinuity*), with abrupt increase in P-wave velocity. The Moho beneath the continental crust averages 30 km in depth, representing change from a gabbroic lower crust to an ultramafic mantle of peridotite composition. Beneath the oceanic crust, the Moho lies at about 12 km depth, representing change from gabbro to peridotite. The Moho beneath the continents may represent a phase transition in which gabbroic rock changes to eclogite of similar element composition.

Earthquake magnitude and energy *Earthquake magnitude* in terms of energy release is rated by the logarithmic *Richter scale* of numbers. An integer increase in Richter number represents an energy increase by a factor of 32 times. Greatest known earthquakes are of magnitude about 8.9. Most energy release on a global basis is from a few earthquakes of magnitude over 7.0.

Earthquake intensity scales Earth-shaking effects are measured by an *earthquake intensity scale,* such as the *modified Mercalli scale* with 12 intensity levels. Maps show concentric intensity zones by means of *isoseismals.*

Earthquake ground motions The *strong-motion seismograph* records large amplitudes and peak accelerations produced by an earthquake hazardous to structures. *Liquefaction* of water-saturated sediments can lead to structural collapse in urban areas and to earth flowage. Sudden permanent changes in ground level often accompany a major earthquake.

Seismic sea waves The *seismic sea wave,* or *tsunami,* is often generated by an earthquake that displaces the ocean floor. Trains of sea waves of long period and length travel across the oceans to strike distant shores with damaging high water and surf.

Earthquakes and plate tectonics *Seismicity,* the frequency of earthquake activity, is largely concentrated at active lithospheric plate boundaries. Subduction zones and lines of recent continental collision are highly seismic, with large earthquakes and deep-focus earthquakes. Low-intensity, shallow earthquakes are numerous along the spreading plate boundaries and transform faults that offset spreading boundaries.

The San Andreas Fault An active right-lateral transcurrent fault, the San Andreas Fault of California is interpreted as the transform-fault boundary between American and Pacific plates. Lateral movement has totaled 560 km in 150 m.y. Certain sections of the fault are locked, with high potential for generating major earthquakes; other sections show continuous slow creep.

Earthquake prediction Using precursory phenomena and historical records, attempts are made to predict the places and times of major earthquakes. Under the theory of *dilatancy,* formation of *microcracks* precedes rupture and can be detected by changes in ground water level, ground surface level, radon gas emissions in wells, and electrical conductivity of rock. Precursory *earthquake swarms* may occur. The historical method detects *seismic gaps,* which show no major seismic activity within the preceding 30 or more years.

Induced earthquakes Earthquakes possibly induced by human activity have been attributed to pumping of fluids into wells and to loading effects of dams and reservoirs.

Earthquake hazards Urbanization greatly increases earthquake hazards to human life and property in the vicinity of a known active fault. Rating of faults as *capable faults* can improve risk assessment associated with construction of nuclear power plants. Seismic risk maps are useful in general planning.

Questions and Problems

1. Figure 8.1 shows increases of pressure, temperature, and density from the earth's surface to the center. Of the three graph curves, which one is probably the most speculative and uncertain? Explain. How would you go about making a rough calculation of the confining pressure at the center of the earth, using an average earth density of 5.5 g/cc? What is the basic source of information on the value of density at various depths in the earth? Specifically, what clue might exist that there is a sudden jump in density at a depth of about 2900 km?
2. Imagine that during an earthquake you are standing very close to the initial rupture point at shallow depth on a fault plane. What initial ground motion would you feel if the fault were a normal fault? A transcurrent fault? Gaping soil cracks are often observed to have formed after movement on a transcurrent fault. Make a sketch map of a number of such cracks showing their position and orientation with respect to a line representing the fault. Use arrows to show the relative fault motions.
3. Features called "sand cones" or "sand craters" sometimes form on a low-lying, flat land surface close to a fault on which rupture has occurred. The sand cone resembles in miniature a low volcanic cone with a small crater in the center. It has formed by an eruption of water carrying sand upward through a small vent. What conditions would favor the development of a sand cone? What physical process has occurred in the underlying material, and how is it related to the earthquake?
4. A simple toy can be used to demonstrate the distinction between the P-wave and S-wave; it is a coil spring sold under the trade name of "Slinky." Stretch this spring out gently on a smooth hard floor. The person at one end can manipulate the spring with a rhythmic motion that propagates a wave toward the other end of the spring. Try this demonstration, figuring out for yourself how to obtain the necessary wave motions.
5. We have alluded to a serious scientific dilemma that troubled seismologists for many years prior to the maturing of the theory of plate tectonics. How can intermediate-focus and deep-focus earthquakes originate in the soft zone of the mantle, if the rock in that zone can easily adjust itself to stresses by slow flowage? What solution can you offer for this dilemma in terms of plate tectonics?
6. How would you go about obtaining the data for construction of isoseismals showing the distribution of earthquake intensities according to the modified Mercalli scale? Could isoseismals be constructed from historical data of an earthquake that occurred many decades ago—say in 1880? What records would you consult?
7. Imagine yourself placed in charge of a committee that is empowered to issue a warning of an impending, potentially severe earthquake that seems to be in store for your city in a matter of days to weeks. What reaction might you expect from the public? Suppose that the earthquake did not materialize in the forecast period. What would be the economic and political consequences? Suppose that you deliberately withheld the forecast from the public and the earthquake actually occurred as predicted. Should you be subject to criminal prosecution and to financial liability for damages?

9

CHAPTER NINE

The Ocean Basins and Their Sediments

Built into the theory of plate tectonics is the concept that the ocean basins are young in contrast to the continental cratons, which are very old. Oceanic lithosphere forms quite rapidly, geologically speaking, along the spreading mid-oceanic boundaries, but is rapidly destroyed as it dives beneath the continental lithosphere in subduction zones. Thus, while oceanic crust cannot be very old, it can certainly be very young—so young, in fact, that some of it is even younger than you or I. If we calculate on the basis of seafloor spreading at a rate of 5 cm per year and assume that new oceanic crust must form at the same rate in order to keep the gap between the plates filled, a belt of new crust 1 m wide will have formed since the birth of a person who today is 20 years old.

Perhaps because the ocean basins are the young portions of the earth, it is logical to examine them in detail before the continents. In this chapter, we investigate the ocean basins in terms of their broad configuration, their relief features, their oceans, and the sediments that form within them. With this information in hand, we can turn in the following chapter to those details of plate tectonics affecting the oceanic lithosphere and its plate boundaries.

Ocean Basins and Continents

The first-order relief features of the earth are the continents and ocean basins. A detailed relief map of the earth and its ocean floors shows immediately that the natural limits of the continents are much larger and more regular than a conventional map showing only the oceanic shoreline. Broad shelves and platforms border the continents in many places. These shallow border zones typically show an abrupt outer edge and a steep drop to deep ocean floors.

To get a generalized picture of the earth's solid surface form we can make a depth-graph such as that shown in Figure 9.1. The graph shows the proportion of surface lying between equal units of vertical distance, or elevation, each unit being a kilometer. The length of each horizontal bar is proportional to the percentage of surface area found within each one-kilometer elevation zone.

The graph shows most of the surface as concentrated in two general zones: (1) on the continents between sea level and 1 km elevation, and (2) on the floors of the ocean basins from about 3 to 6 km below sea level. This graph also tells us that in a general way the continents are broad, tablelike areas whose edges slope away rapidly to the deep ocean floor. Although the floor does not lie entirely within one elevation zone, vast areas lie at approximately the same depth.

We must accept the idea that the ocean basins are full to the brim with water, so that the oceans overlap considerable areas of the continental margins to produce shallow seas bordering the shores. A true picture of the continents in relation to the ocean basins emerges when the ocean level is imagined to be dropped some 180 m to uncover these shallow continental shelves and inland seas. When the area lying above the 180-m submarine contour is included with the lands, it will be found that the continents make up about 35% of the total earth's surface area and the ocean basins, about 65%.

The reasons for a two-level form of the earth's solid surface have already been explained in Chapters 7 and 8. Perhaps the most generally satisfactory definition of *continents* and *ocean basins* is that the former are high-standing features underlain by felsic continental crust; the latter are low-lying features underlain by mafic oceanic crust.

Charting the Ocean Floor

Few persons get a chance to see the deep ocean floor at close range, but many undersea photographs, taken by a special type of camera lowered to the seafloor from a ship (Figure 9.2), are available for examination. Direct observation and bottom sampling are now becoming commonplace, thanks to a fleet of *manned submersibles*—undersea vessels capable of withstanding the enormous confining pressures at depths greater than 3 km. In a single major expedition to the floor of the Mid-Atlantic rift in 1974, three submersibles from the United States and France made 47 dives to study an active seafloor spreading zone. The findings of this international study, project FAMOUS, are reviewed in Chapter 10. Scientists in the submersibles took many photographs of the features they saw and collected many samples of rocks and sediments. *Alvin*, a manned submersible operated by the Woods Hole Oceanographic Institution, was one of the vessels taking part in project FAMOUS. In the same year, *Alvin* made dives in the North Atlantic to investigate a chain of submarine mountain peaks (seamounts) (Figure 9.3).

Even the deepest points of the ocean floor—the bottoms of trenches above subduction zones—have been visited by manned diving vessels called ***bathyscaphes***, suspended from cables. Two French bathyscaphes, the *Trieste* and the *Archimede*, have carried out many such deep dives. In 1960, the *Trieste* reached the deepest point humans have yet gone on the ocean floor, 11,033 m in the Marianas Trench of the western Pacific Ocean.

Making detailed maps of the ocean floor lagged far behind the mapping of the continental surfaces. In the early decades of oceanographic research, sounding of the ocean bottom had to be done by lowering a heavy weight on a thin steel cable until the weight reached bottom, when the depth could be determined by the length of cable. Because this was a slow and costly process, our knowledge of the seafloor was scanty until World War II, when naval vessels began to use a continuously recording sounding apparatus.

The *precision depth recorder* (*PDR*) makes use of a sound-emitting device attached to the bottom of the ship. Its pulses of sound are sent down through the water from the ship's hull and are reflected from the ocean floor back to the ship, where they are picked up by a microphone. The method, called *echo sounding*, uses an automatic recording device to indicate the time required for sound waves to reach the bottom and return. Reflections are plotted continuously by a writing instrument to give a line representing the profile of the ocean bottom (Figure 9.4). By allowing the precision depth recorder to operate continuously while the ship travels, a profile across the seafloor is obtained, but this information can only be used to make seafloor maps if the exact position of the ship is known at all times. With navigational satellites available, precise positioning of a vessel is no problem. Echo sounding enormously increased our knowledge of the con-

FIGURE 9.1 The length of each bar represents the percentage of surface area in each 1-km elevation zone of the earth's solid surface.

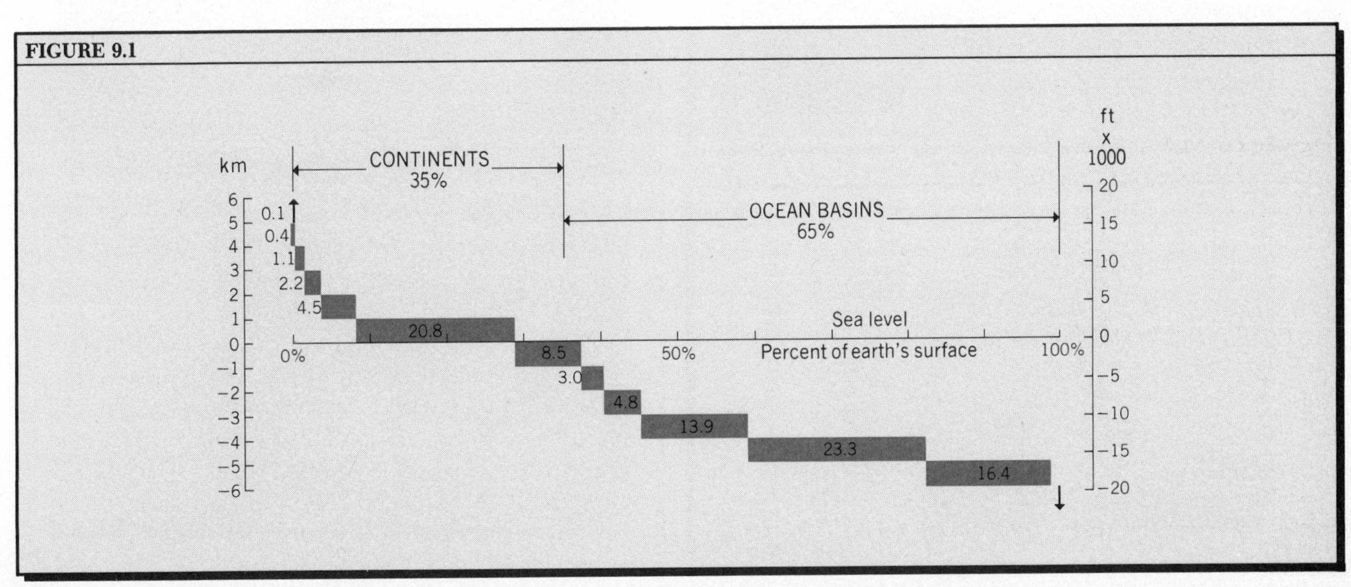

FIGURE 9.2 Submarine photography. *A.* A research scientist of the Lamont–Doherty Geological Observatory staff prepares to lower an undersea camera over the side of the Research Vessel *Vema.* (National Academy of Sciences, IGY.) *B.* This photograph of the ocean floor at a depth of 2000 m shows an outcrop of bedrock on the side slopes of Ampere Seamount. Location is lat. 35° N, long. 13° W. (Lamont–Doherty Geological Observatory of Columbia University.) *C.* A starfish (*left*) and a sea spider (*right*) seen on a mud bottom at a depth of 1800 m on the continental slope of the eastern United States. (D. M. Owen, Woods Hole Oceanographic Institution.)

FIGURE 9.2 A

FIGURE 9.2 B

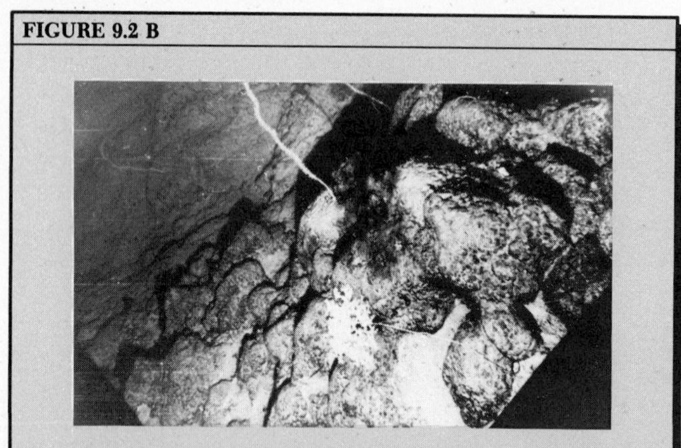

FIGURE 9.2 C

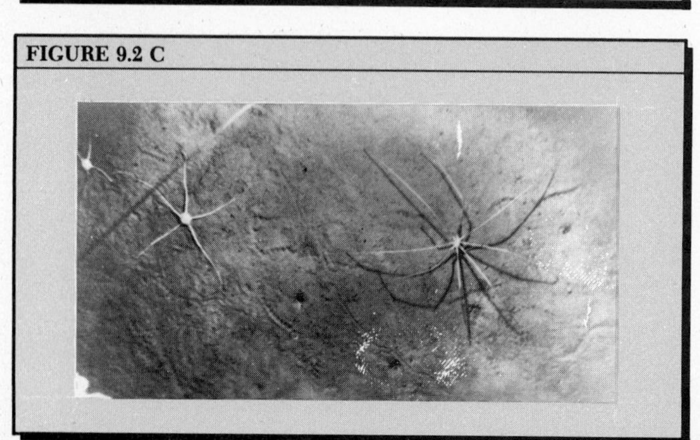

FIGURE 9.3 *Alvin,* a self-propelled submersible that has participated in many important scientific expeditions to the ocean floor. *A. Alvin* raised for inspection. Near the bow is one of the two lateral viewing ports that resemble great fish-eyes. *B. Alvin* submerged. An articulated arm, used for collection of bottom samples, is folded against the bow; below it is a rack holding photographic apparatus. (John Porteous, Woods Hole Oceanographic Institution.)

FIGURE 9.3 A

FIGURE 9.3 B

figuration of the ocean floors within a span of only two decades, and thus the mid-twentieth century can truly be viewed as a golden age of undersea exploration.

Topographic Divisions of the Ocean Basins

Modern study of the topography of the ocean floors was greatly advanced through the work of Professor Bruce C. Heezen of the Lamont–Doherty Geological Observatory at Columbia University. In scientific collaboration with Marie Tharp, Heezen compiled seafloor topographic data as rapidly as they became available, and prepared many detailed maps of the ocean floors. In 1959, Heezen set up a system of submarine landform classification. (*Landform* refers to the geometrical configuration of the solid earth's surface, whether exposed to the atmosphere or covered by water.)

According to the Heezen classification, the topographic features of the ocean basins fall into three major divisions: (1) the continental margins, (2) the ocean-basin floors, and (3) the mid-oceanic ridge. As the terms themselves make clear, the continental margins lie in belts directly adjacent to the continents, and the mid-oceanic ridge divides the basin roughly in half, with one part of the ocean basin on either side of the ridge. Figure 9.5 and the accompanying profile show these major topographic divisions as they apply to the North Atlantic basin. Figure 9.6 is a schematic block diagram idealizing the major features.

The explanation for symmetry of the ocean-basin topography is familiar through our model of continental rifting and the opening up of an ocean basin through seafloor spreading (see Figure 1.10).

The Continental Margins

Perhaps the best known and most easily studied of the units within the continental margins are the *continental shelves*, which fringe the continents in widths from a few kilometers to more than 300. The shelves have very smooth and gently sloping floors, and are for the most part less than 180 m deep. A particularly fine example is the continental shelf of the eastern coast of the United States (Figure 9.7).

The *epicontinental marginal seas*, bodies of water lying well within the continental blocks, form parts of the continents. Generally more than 180 m deep, the epicontinental seas not only are deeper than the continental shelves, but have appreciably greater relief as well. Examples are the Gulf of Maine and the Gulf of St. Lawrence.

Along their seaward margins the continental shelves give away to the *continental slopes*. Although the actual inclination of the slopes with respect to the horizontal is only 3° to 6°, this is exceptionally steep for submarine relief features and appears quite precipitous on highly exaggerated profiles (Figure 9.8).

Notching the continental slope are *submarine canyons*, which may be visualized as akin to gullies cut by erosion in the side of a hill, but on an enormous scale (Figure 9.7). The continental slope drops from the sharply defined brink of the shelf to depths of 1400 to 3200 m. Here the slope lessens rapidly, though not overly abruptly, and is replaced by the *continental rise*, a surface of much gentler slope decreasing in steepness toward the ocean-basin floor. Ranging in width from perhaps 200 to 500 km, the continental rise has generally moderate to low relief. (*Relief* refers to the local differences in elevation from point to

FIGURE 9.4 Photograph of the actual trace made by a precision depth recorder. Depth in meters is given by figures at the right. The entire profile spans about 16 km. An abyssal hill is shown flanked by the very flat surface of the Pernambuco Abyssal Plain. Location is about lat. 13° S, long. 28° W. (Lamont–Doherty Geological Observatory of Columbia University.)

FIGURE 9.4

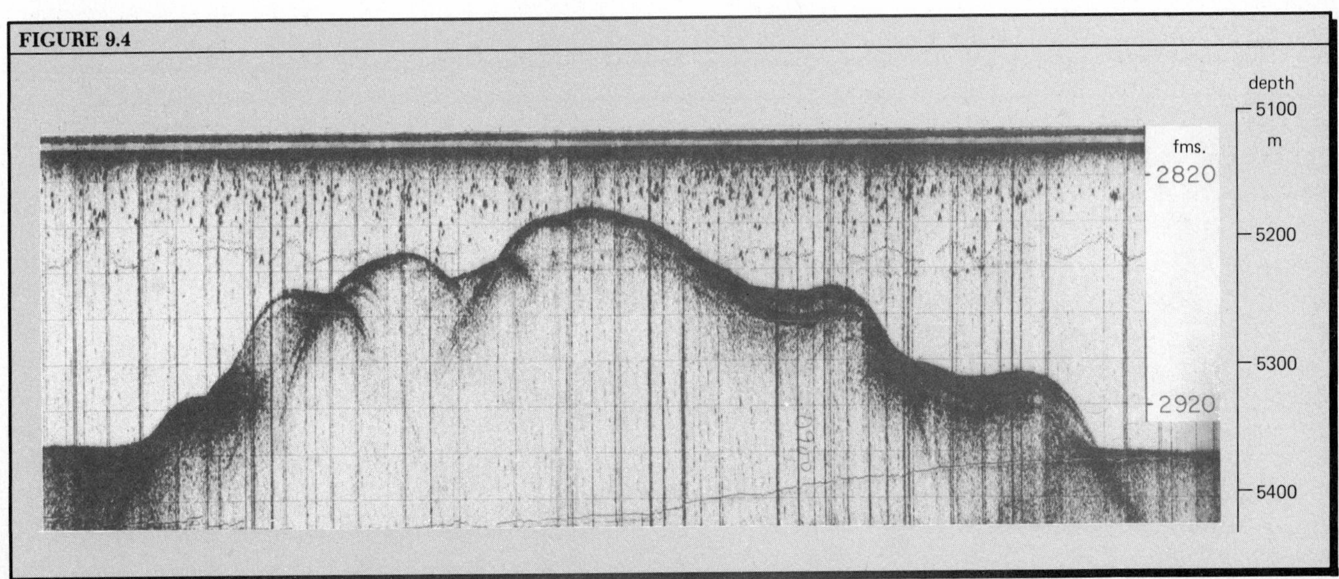

point.) At its outer margin, the continental rise reaches a depth of 5000 m, where it may be in direct contact with the deep floor of the ocean basin.

The Ocean-Basin Floor

Second of the major topographic divisions of the ocean basins is the extensive region of basin floor, generally lying in the depth range of 4500 to 5500 m. The ocean-basin floor contains three classes of forms: abyssal plains and hills, oceanic rises, and seamounts.

An *abyssal plain* is an area of the deep ocean floor having a flat bottom with the very faint slope of less than one part in 1000 (a drop of one meter in a horizontal distance of 1000 m). Characteristically situated at the foot of the continental rise, the abyssal plain is present in all ocean basins. Examples are the Hatteras and Nares Abyssal plains at depths of roughly 5500 m (Figure 9.7). The only reasonable explanation for such nearly perfect flatness is that the abyssal plains are surfaces formed by long-continued deposition of very fine sediment.

Abyssal hills are small hills rising to heights of a few tens of meters to a few hundred meters above the

FIGURE 9.5 Outline map of the major divisions of the North Atlantic Ocean basin, with representative profile from New England to the Sahara coast of Africa. Vertical exaggeration about 40 times. (After B. C. Heezen, M. Tharp, and M. Ewing, 1959, *The Floors of the Oceans*, Geological Society of America Spec. Paper 65, p. 16, Figure 9.)

FIGURE 9.6 This idealized block diagram shows the major units of the North Atlantic Ocean basin as symmetrically placed on both sides of the central ridge axis.

FIGURE 9.5

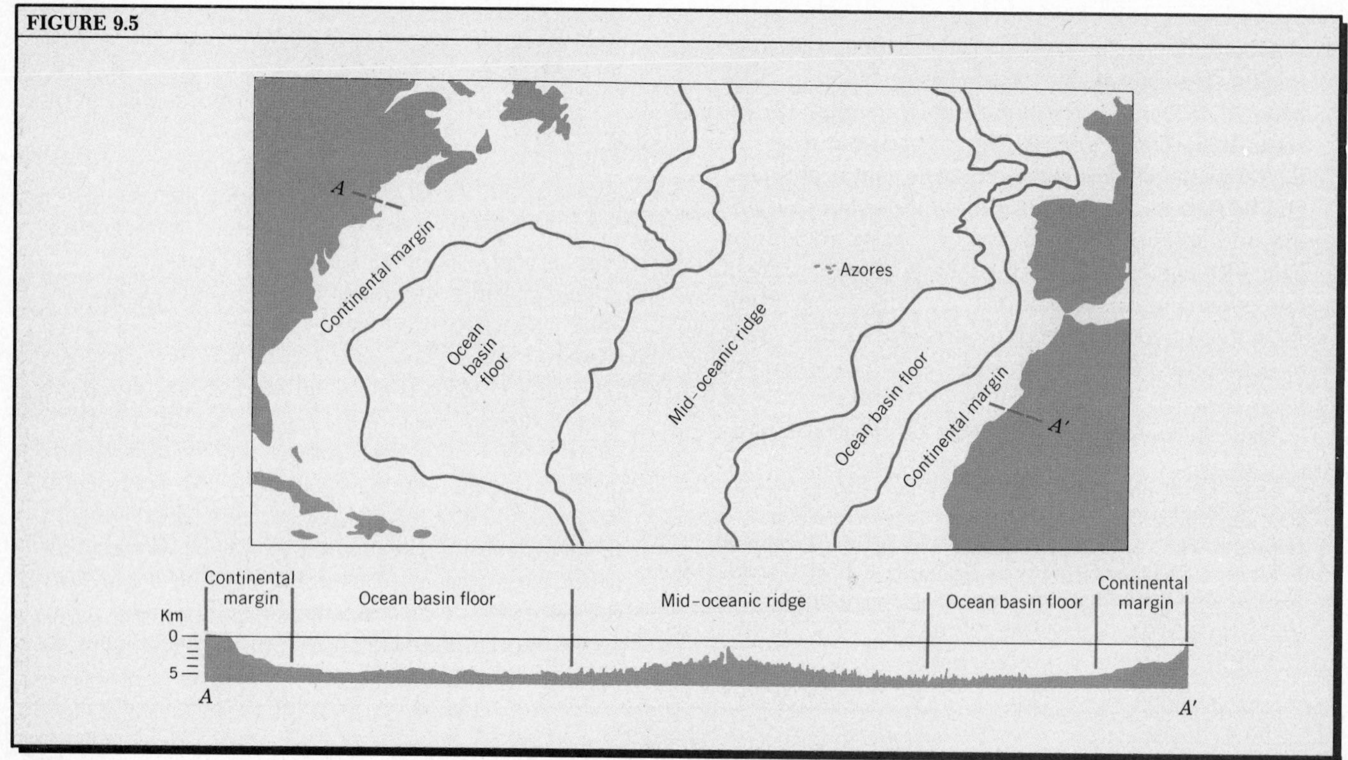

FIGURE 9.6

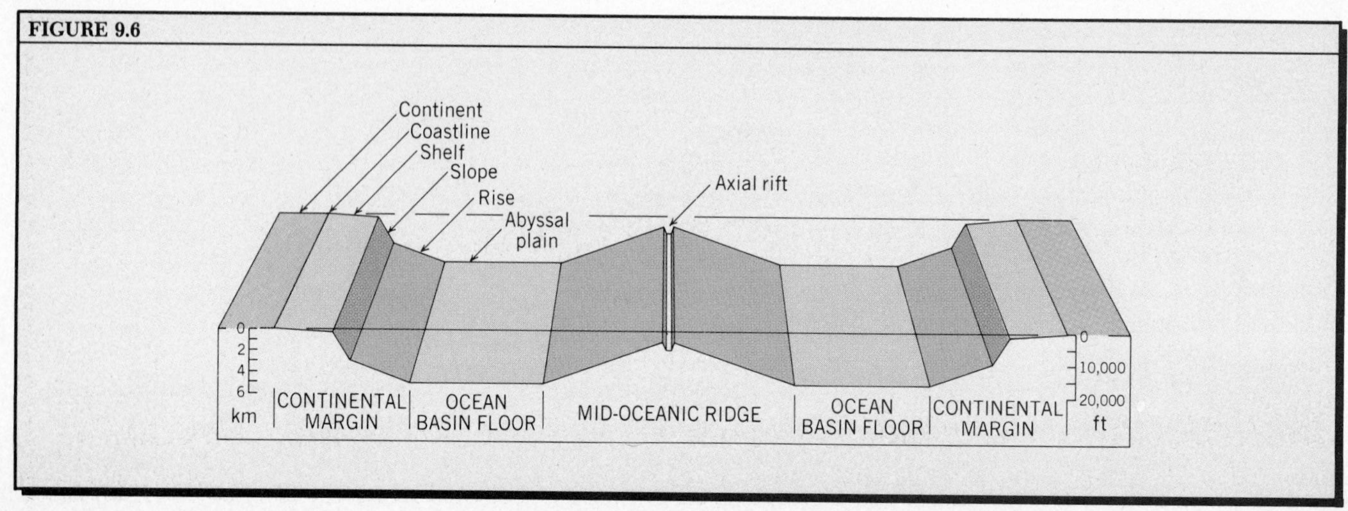

ocean-basin floor and may be so numerous as to occupy nearly all the floor. Isolated abyssal hills are well developed in the North Atlantic basin along the mid-ocean side of the great abyssal plains.

Yet another characteristic topographic unit of the ocean-basin floor is the *oceanic rise,* an area hundreds of kilometers in breadth over which the surface rises several hundred meters above the surrounding abyssal plains. Within the rise, relief may range from subdued to very rugged. An example is the Bermuda Rise, shown in both profiles of Figure 9.8 and in the lower right corner of Figure 9.7. In places, this rise consists of hills 40 to 100 m high and 3 to 16 km wide. Along the eastern edge, the rise is broken by a series of scarps

FIGURE 9.7 Relief features of the North Atlantic continental margin and ocean-basin floor off the northeastern United States and the Maritime Provinces of Canada. Depths in kilometers. (Drawn by Marie Tharp. Portion of *Physiographic Diagram of the North Atlantic Ocean* (revised 1972), by B. C. Heezen and M. Tharp. Copyright © 1968 by Bruce C. Heezen and Marie Tharp.)

FIGURE 9.8 Two profiles across parts of the North Atlantic Ocean basin show several characteristic features of the ocean floors. (From B. C. Heezen, M. Tharp, and M. Ewing, 1959, *The Floors of the Oceans,* Geological Society of America, Spec. Paper 65, Plate 27.)

FIGURE 9.7

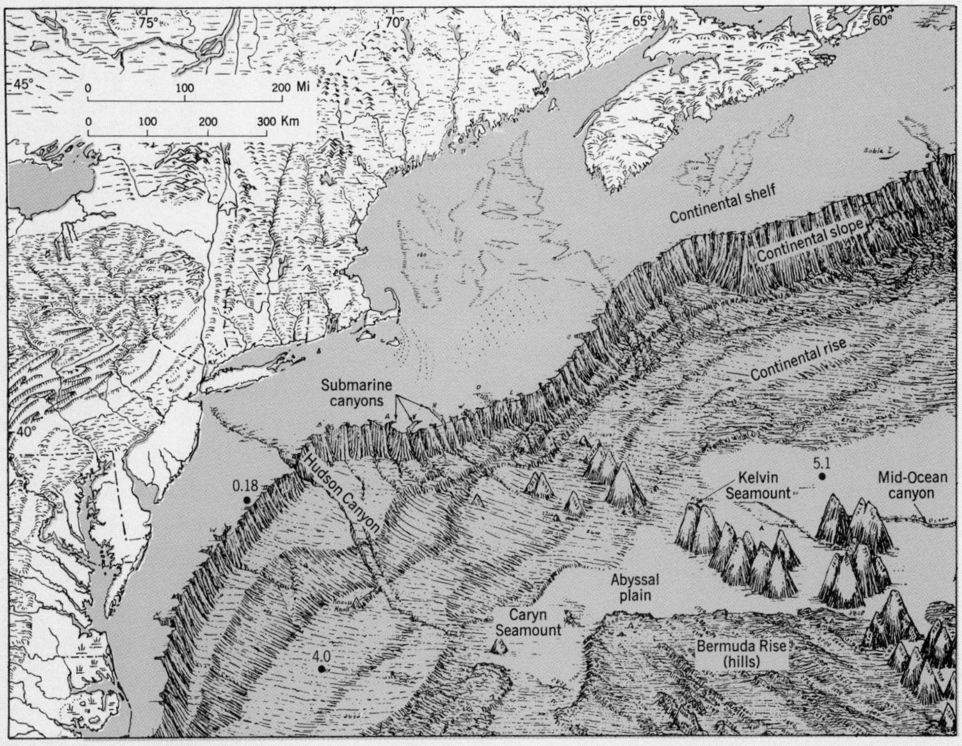

75°
70°
65°
60°
45°
40°
0 100 200 Mi
0 100 200 300 Km
Continental shelf
Continental slope
Continental rise
Submarine canyons
Hudson Canyon
0.18
5.1
Kelvin Seamount
Mid-Ocean canyon
Abyssal plain
Caryn Seamount
Bermuda Rise (hills)
4.0

FIGURE 9.8

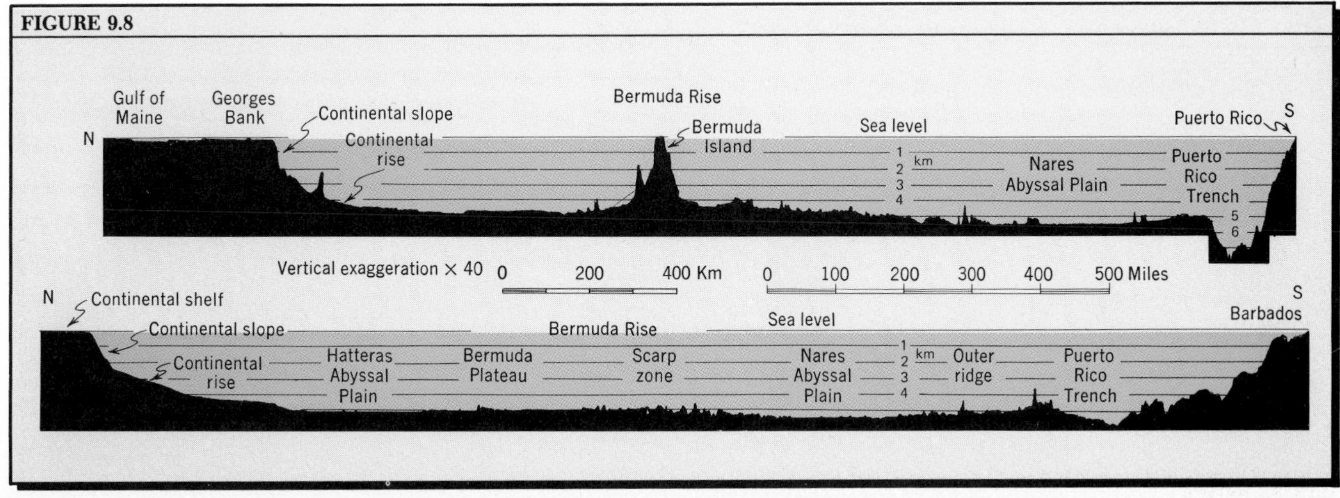

550 to 1600 m high. Located near the center of the Bermuda Rise is a pedestal or platform 80 by 130 km at its base, upon which the islands of Bermuda are situated.

Perhaps the most fascinating of the strange features of the ocean basins are the *seamounts*, isolated peaks rising 1000 m or more above the floor. Although seamounts also occur on the continental rises, they are most conspicuous on the ocean-basin floors. A good example from the Atlantic basin is the Kelvin Seamount Group, forming a row of conical peaks extending across the continental rise for 1000 km southeastward toward the Bermuda Rise (Figure 9.7). Shown in natural-scale profile (Figure 9.9), even the most striking seamount of this group may not appear particularly impressive (few mountains do when thus presented), but its bulk is apparent when we note that the seamount rises almost 3300 m above the abyssal plain and is 40 km wide at the base.

Altogether, several hundred seamounts have been found in the Pacific Ocean, a number vastly greater than in the Atlantic. Many of the Pacific seamounts are conspicuously flat topped and extremely steep sided (Figure 9.10). This kind of seamount is named a *guyot*, in honor of a geologist of the nineteenth century. The origin of guyots is explained in Chapter 10 and illustrated in Figure 10.29.

The Mid-Oceanic Ridge

We turn now to the third of the major divisions of the ocean basins, the *mid-oceanic ridge*. One of the most remarkable of the major discoveries coming out of oceanographic explorations of the mid-twentieth century was the charting of a great submarine mountain chain extending for a total length of some 64,000 km (Figure 9.11). The ridge runs down the middle of the North and South Atlantic ocean basins, into the Indian Ocean basin, then passes between Australia and Antarctica to enter the South Pacific basin. Turning north along the eastern side of the Pacific basin, where it is named the East Pacific Rise, the ridge contacts the North American continent along the coast of Mexico. The mid-oceanic ridge also extends into the Arctic Ocean basin.

Details of the mid-oceanic ridge are well illustrated by the Mid-Atlantic Ridge, seen in the profiles of Figure 9.12. The ridge in its entirety is a belt, 2000 to 2400 km wide, in which the surface rises through a series of steps from abyssal plains on both sides toward the central, or median, line, where the ridge assumes mountainous proportions. Since the higher points lie at depths of 1800 to 2700 m, the Mid-Atlantic Ridge has a height of roughly 3700 m.

A distinctive feature of the principal continuous ridge is that, instead of having a single high crest line, as many narrow continental mountain chains have, there is a characteristic trenchlike depression, or ***axial rift***, running precisely down the mid-line of the highest part of the ridge. This rift shows well on the upper profile of Figure 9.12. Along with other parallel scarps and steplike rises on both sides, the rift strongly suggested to geologists, when it was first discovered, that the crust has been pulled apart. Today, we recognize the axial rift as a spreading plate boundary along which new oceanic lithosphere is being formed. At the time when details of the mid-oceanic ridge were just being gathered, the theory of plate tectonics had not yet emerged in its present form. Despite the suggestive appearance of the axial rift and the symmetrical steps on either side, proof had not then been marshalled to demonstrate active seafloor spreading.

A particularly significant feature of the axial rift is that it is broken into many segments, the ends of which appear to be offset along transverse *fracture zones* (Figure 9.13). This arrangement of offset segments is particularly striking in the equatorial zone of the Atlantic Ocean, where a single offset displaces the main axial rift by as much as 600 km. It might seem obvious by inspection that the transverse fracture zones are transcurrent faults.

Today, we recognize the fracture zones between offset ends of the axial ridge as transform faults marking sections of the boundary between lithospheric plates. However, the fracture zones often extend many hundreds of kilometers out on either side of the offset ends of the axial rift. For example, in the eastern North Pacific basin, fracture zones over 5000 km long extend

FIGURE 9.9 Kelvin Seamount shown in true-scale profile. (After B. C. Heezen, M. Tharp, and M. Ewing, 1959, *The Floors of the Oceans*, Geological Society of America Spec. Paper 65, p. 77, Figure 34.)

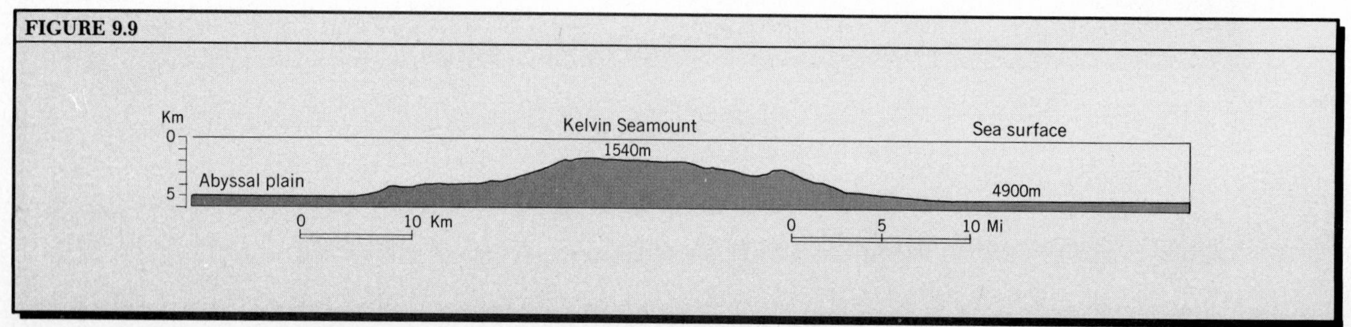

FIGURE 9.10

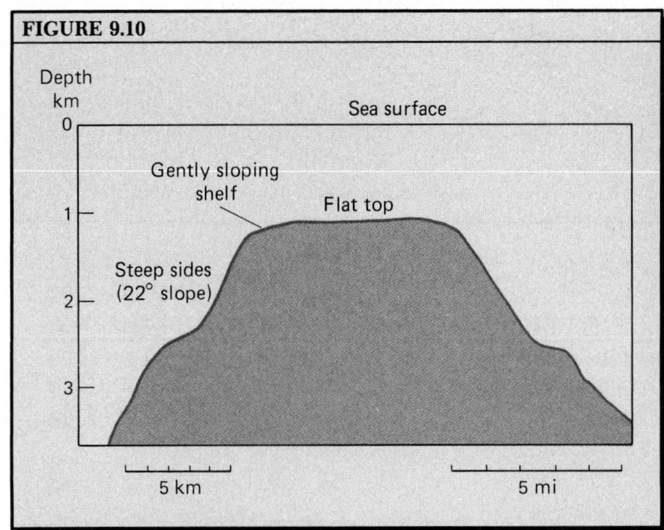

FIGURE 9.10 Profile of the first guyot to be discovered in the Pacific Ocean basin (1944). The location of this flat-topped seamount is about lat. 9° N, long. 163° E. (Based on data of H. H. Hess, 1946, *American Journal of Science*, vol. 244.)

FIGURE 9.11 The mid-oceanic ridge system and related fracture zones. (See Figure 10.3 for data source.)

FIGURE 9.12 These two profiles across the Mid-Atlantic Ridge show two somewhat different sets of features associated with the ridge. The profiles are based upon soundings continuously recorded by echo sounder on the Research Vessel *Atlantis*. (From B. C. Heezen, M. Tharp, and M. Ewing, 1959, *The Floors of the Oceans*, Geological Society of America Spec. Paper 65, Plate 22.)

FIGURE 9.11

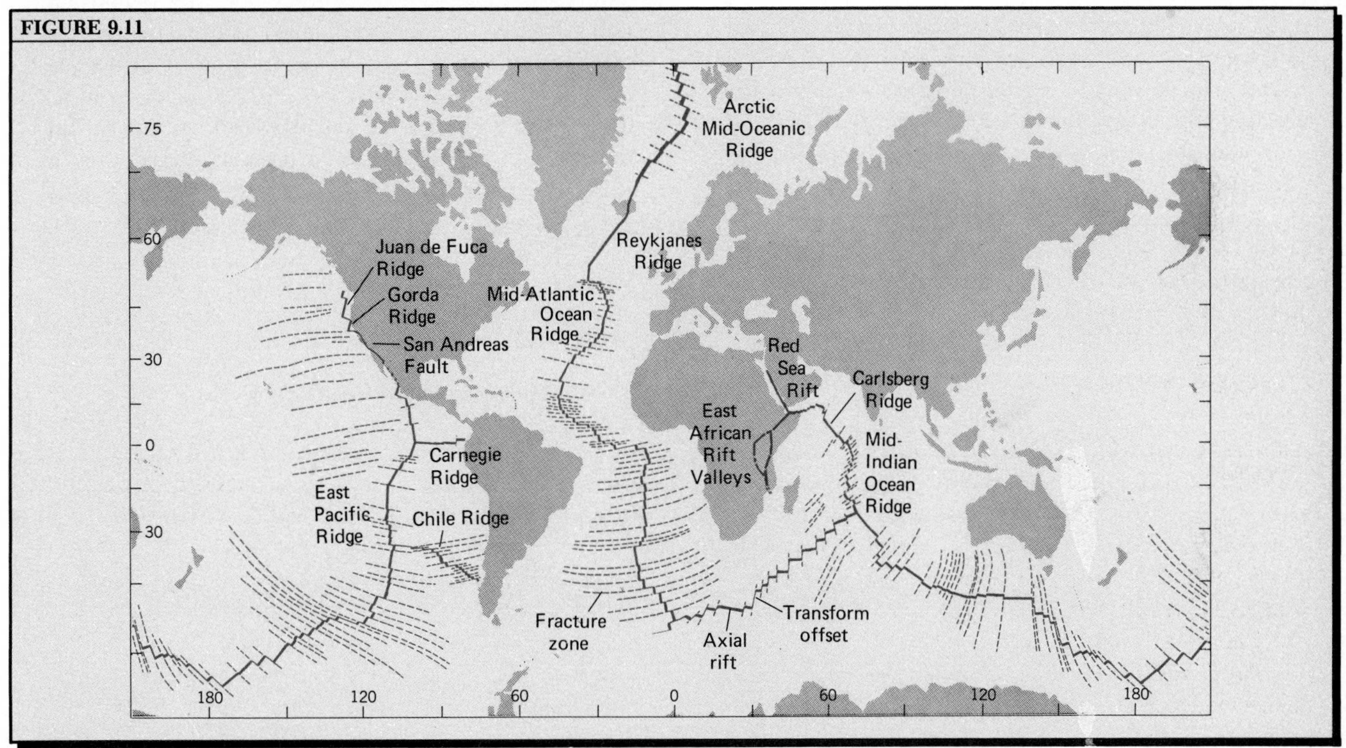

FIGURE 9.12

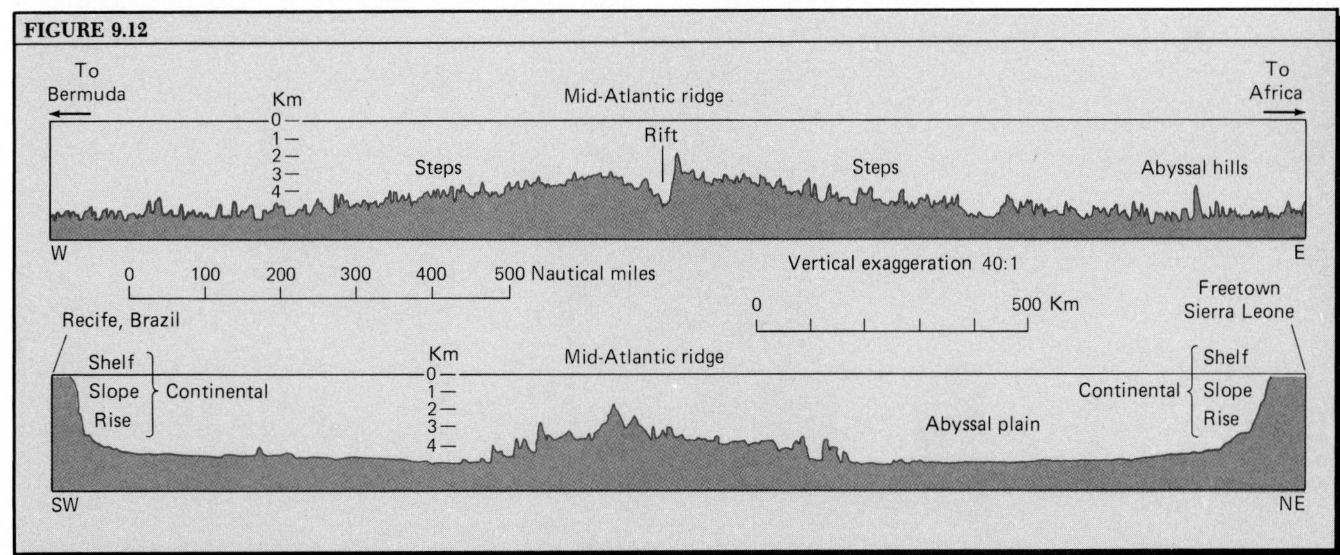

from points close to the west coast of North America to as far west as the Hawaiian islands (see Figure 10.3). In Chapter 10, we interpret these fracture zones as crustal scars formed by long-continued activity on the transform faults.

Trenches and Island Arcs

So far, our description of landforms of the ocean floors has been limited to basins bounded by stable continental margins. Another class of ocean-floor landforms is associated with active subduction zones, where plates are being consumed as they dive into the asthenosphere.

The deepest points on the ocean floors occur in long, narrow *trenches*, also referred to as *foredeeps*, commonly with maximum depths of 7500 to 9000 m. Almost invariably, the trenches lie immediately adjacent to and on the oceanward side of submarine ridges, the *island arcs*, or close to coastal mountain ranges of the continental margins, the *mountain arcs*. Both kinds of arcs were included in the term volcanic arcs, used in Chapter 4.

The sole example of a deep trench of the North Atlantic Ocean is the Puerto Rico Trench, running along the north side of that island, and its extension to the west in the Cayman Trench, lying south of Cuba (Figure 9.14). A profile of the Puerto Rico Trench is shown in the right-hand side of the upper part of Figure 9.8. The deepest point in this trench is more than 8400 m below sea level.

Trenches of the western North Pacific Ocean are particularly striking (Figure 9.15). Deepest of all may be the Marianas Trench, where a record depth of 11,033 m has been measured. At least five other Pacific trenches have depths over 10,000 m.

Trenches have widths of 40 to 120 km and lengths of 500 to over 4500 km. Longest of all is the Peru–Chile Trench, off the west coast of South America, extending for 5900 km (Figure 9.16). This trench is particularly striking because of the proximity and great height of the Andes range, a volcanic mountain arc that lies only a short distance to the east.

In Chapter 4 we explained volcanic island arcs and mountain arcs as built by magma rising from sources close to the upper surfaces of downbent plate margins plunging into the asthenosphere. One good example of a volcanic island arc comes from the Caribbean–Atlantic boundary—the Lesser Antilles, a chain of islands with a number of active volcanoes, but by far the greatest development of volcanic island arcs is along the northern and western sides of the Pacific

FIGURE 9.13 Relief features of the Mid-Atlantic Ridge in the vicinity of the Azores Islands. (Drawn by Marie Tharp. Portion of *Physiographic Diagram of the North Atlantic Ocean* (revised, 1972), by B. C. Heezen and M. Tharp. Copyright © 1968 by Bruce C. Heezen and Marie Tharp.)

FIGURE 9.13

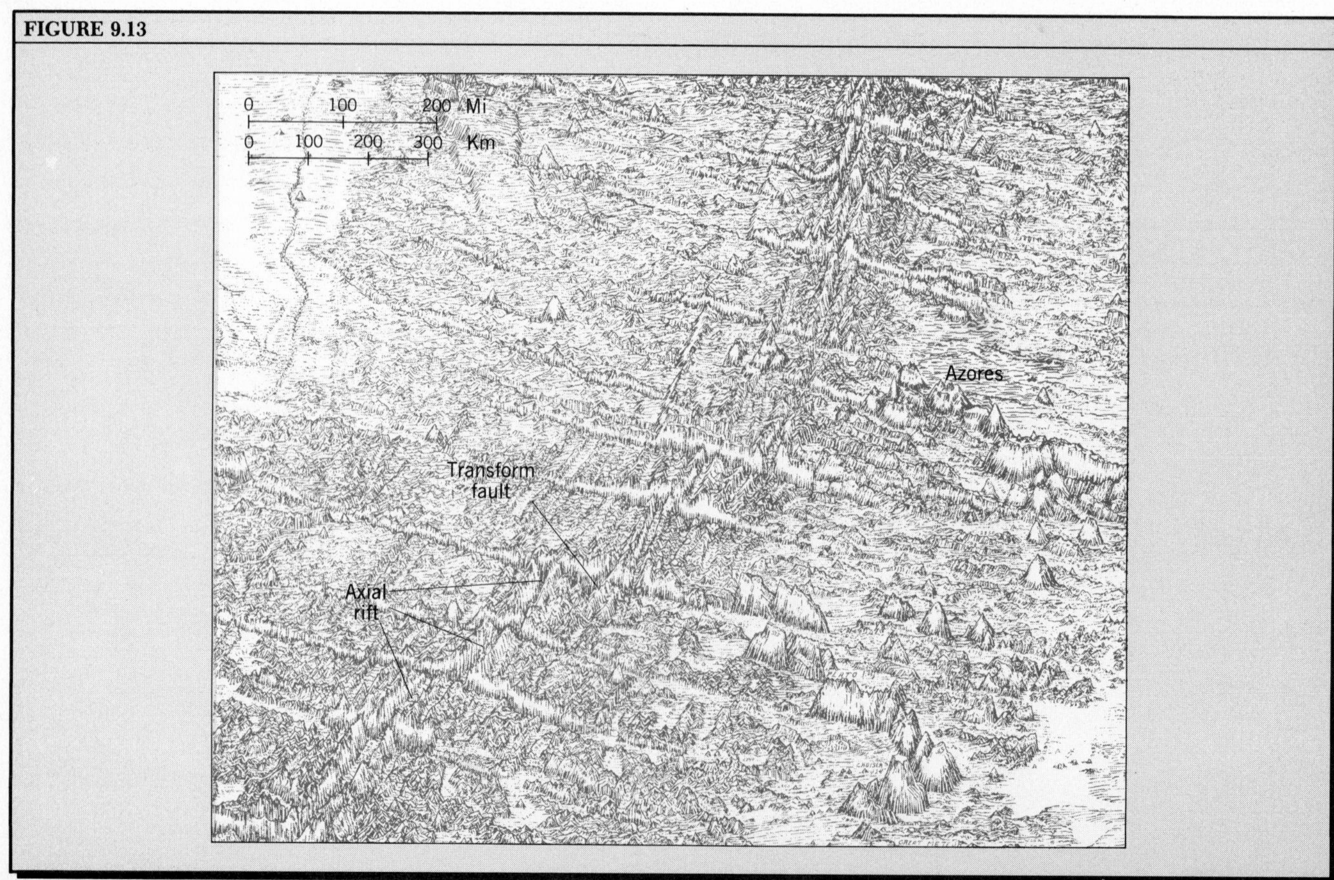

Ocean basin (Figure 9.15). The Aleutian Islands form a great volcanic arc adjacent to the Aleutian Trench. Multiple arcs characterize the western Pacific basin between Japan and New Guinea. Yet another great arc–trench system runs through the Indonesian islands of Sumatra and Java, bordered on the south by the curving Java Trench.

In the western Pacific region of multiple arcs, small but deep ocean basins lie between a given arc and the continental mainland, or between two island arcs. These small ocean basins are called *backarc basins*. One example is the Sea of Japan, between the Japan arc and the Asiatic mainland; another is the basin lying west of the Marianas arc but east of the Ryukyu arc (Figure 9.15).

The wide range of classes of landforms that make up the floors of the oceans implies varied formative processes. Some features are obviously related to active

FIGURE 9.14 The Puerto Rico and Cayman trenches. Depths in kilometers. (Drawn by Marie Tharp. Portion of *Physiographic Diagram of the North Atlantic Ocean* (revised, 1972), by B. C. Heezen and M. Tharp. Copyright © 1968 by Bruce C. Heezen and Marie Tharp.)

FIGURE 9.15 Generalized map of the major arcs and trenches of the Pacific Ocean region. (Redrawn, by permission of the publisher, from A. N. Strahler and A. H. Strahler, *Elements of Physical Geography*, Figure 17.4. Copyright © 1976 by John Wiley & Sons, Inc.)

FIGURE 9.14

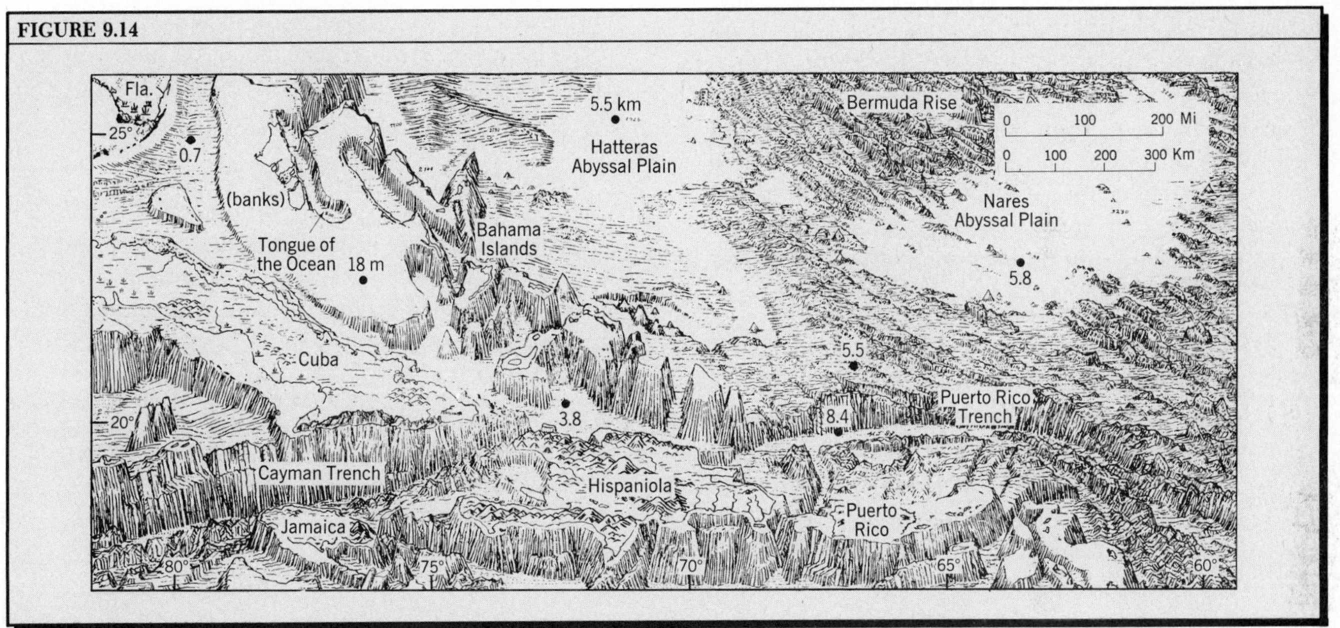

FIGURE 9.15

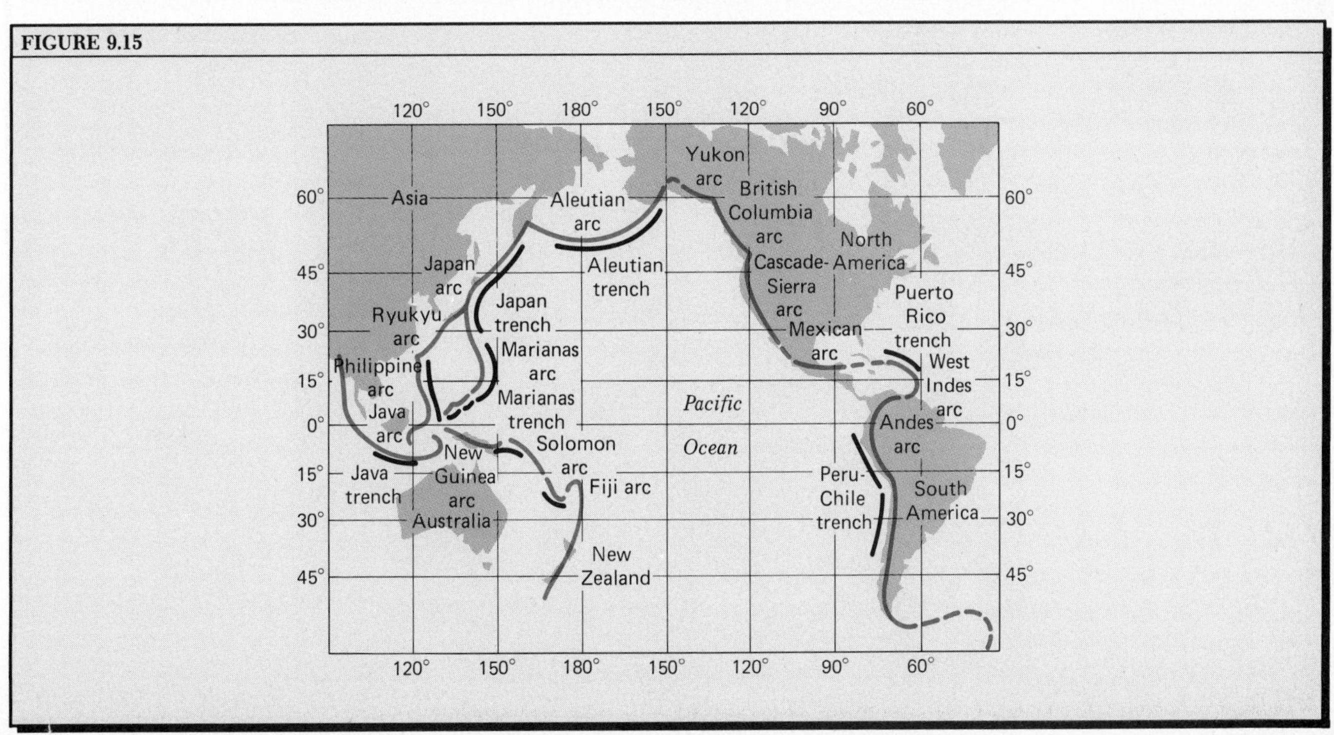

plate boundaries and are either tectonic or volcanic in origin. Others are related to the erosion of the seafloor or the accumulation of sediments, but before we turn to these latter processes, we will review briefly the nature of the oceans themselves as bodies of water in continual circulation.

Ocean Science and Geology

Oceanography, the science of the oceans, is closely allied with geology in many ways. The ocean waters are an enormous reservoir for mineral substances that participate in the rock transformation cycle. Salts of seawater have been derived in large part from the rocks of the continents, and those salts are eventually returned to the solid earth as sedimentary rocks. Thus, marine sedimentary rocks owe their character to the ocean environment.

Seawater supports countless organisms that secrete mineral matter to form their shells and skeletons, and this mineral matter makes an important contribution to marine sediments. Geologists who specialize in *marine geology* (or *submarine geology*) must know a great deal about the chemical properties of seawater and the physical processes that affect the ocean.

The World Ocean

The *world ocean,* meaning the combined oceans of the earth, occupies about 71% of the earth's surface and has a mean depth of about 3800 m, including shallow seas in addition to the main basins. The round figure of 4000 m applies quite well to the average depth of the main portions of the Atlantic, Pacific, and Indian oceans. Volume of the world ocean is about 1.4 billion cubic kilometers (1.37×10^9 cu km), which constitutes 97.2% of the world's free water. Most of the remaining 2.8% is locked up in glaciers.

If we compare masses of the atmosphere and world ocean with the total earth mass (including atmosphere and oceans), the following estimates apply:

Atmosphere	0.0052×10^{21} kg
World ocean	1.43×10^{21} kg
Entire earth	5983×10^{21} kg

On a fractional basis, these figures show that the world ocean has a mass about 1/4000 that of the entire earth, while the atmosphere has a mass about 1/275 that of the world ocean.

*Temperature Structure of the Oceans

Water temperature is one factor determining water density, and also influences the abundance of various species of sediment-producing marine organisms that live suspended in the upper water layers.

FIGURE 9.16 The Peru-Chile Trench, off the west coast of South America. (Drawn by Marie Tharp. Portion of *Physiographic Diagram of the South Atlantic Ocean.* Copyright © 1961 by Bruce C. Heezen and Marie Tharp.)

As a general statement, the temperature structure of the oceans over middle and low latitudes can be described as a three-layer system (Figures 9.17 and 9.18). In those latitudes, the surface water is subjected to intense solar radiation—all year in low latitudes and in summer in middle latitudes. The heated water takes the form of an upper layer of quite uniform temperature, a result of mixing within the layer. This warm layer may attain a thickness of 500 m and a temperature of 20 °C to 25 °C or higher in equatorial latitudes.

Immediately below the warm layer, water temperatures drop sharply downward. This layer of rapid temperature change, which may be 500 to 1000 m thick, is known as the *thermocline*. In low latitudes, temperatures decline gradually from about 5 °C immediately below the thermocline to about 1 °C close to the bottom at depths of around 4000 m. In arctic and antarctic latitudes, surface water temperatures are close to 0 °C and so the temperature changes with increasing depth are very slight.

FIGURE 9.17 Typical changes in temperature, salinity, and dissolved oxygen are shown for oceans in low and middle latitudes. (After W. E. Yasso, 1965, *Oceanography*, Holt, Rinehart & Winston, New York, Figure 2–4; modified by data of A. Defant, 1961, *Physical Oceanography*, Pergamon, New York, vol. 1, chap. 4.)

FIGURE 9.18 A schematic north–south cross section of the world ocean shows that the warm surface water layer disappears in arctic latitudes, where very cold water lies at the surface. (Redrawn, by permission of the publisher, from A. N. Strahler and A. H. Strahler, *Elements of Physical Geography*, Figure 2.8. Copyright © 1976 by John Wiley & Sons, Inc.)

*Salinity of Ocean Water

Seawater is a *brine*, that is, a solution of salts. One way of describing the chemical composition of seawater is to list the ingredients that one would need to

FIGURE 9.17

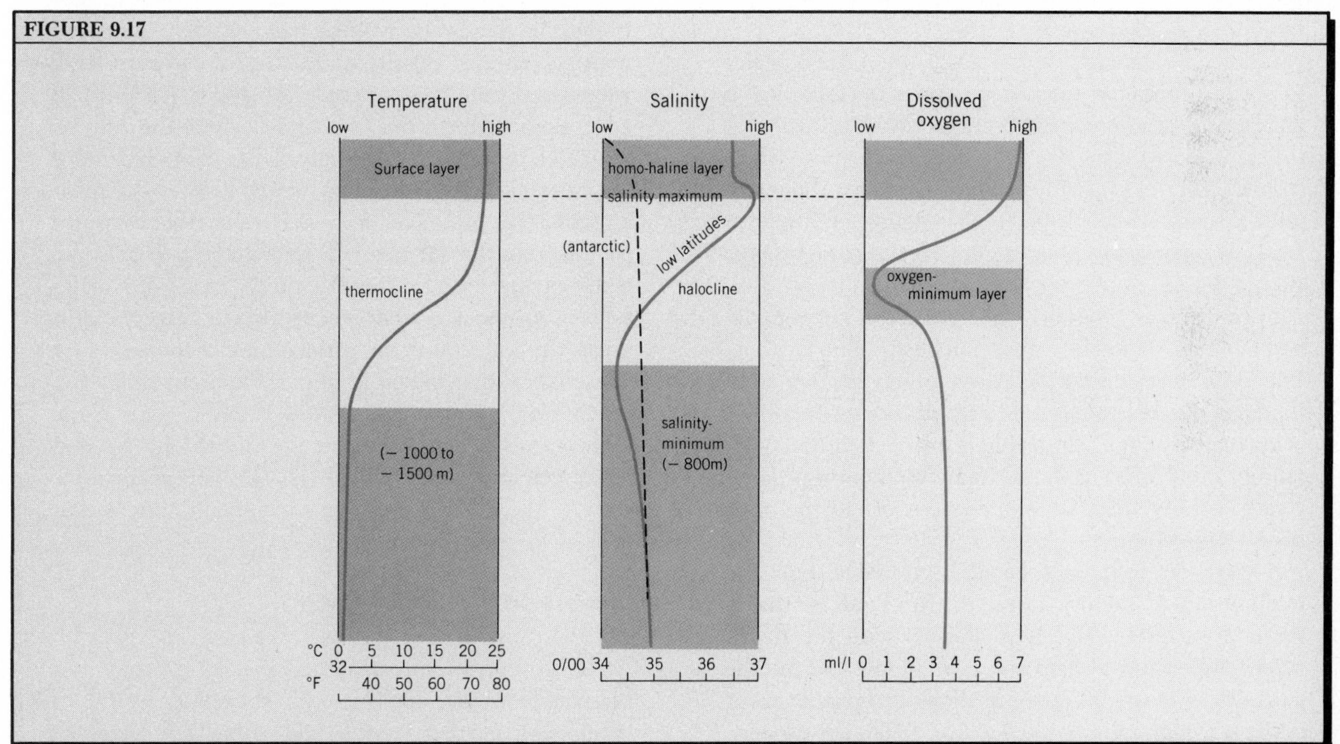

FIGURE 9.18

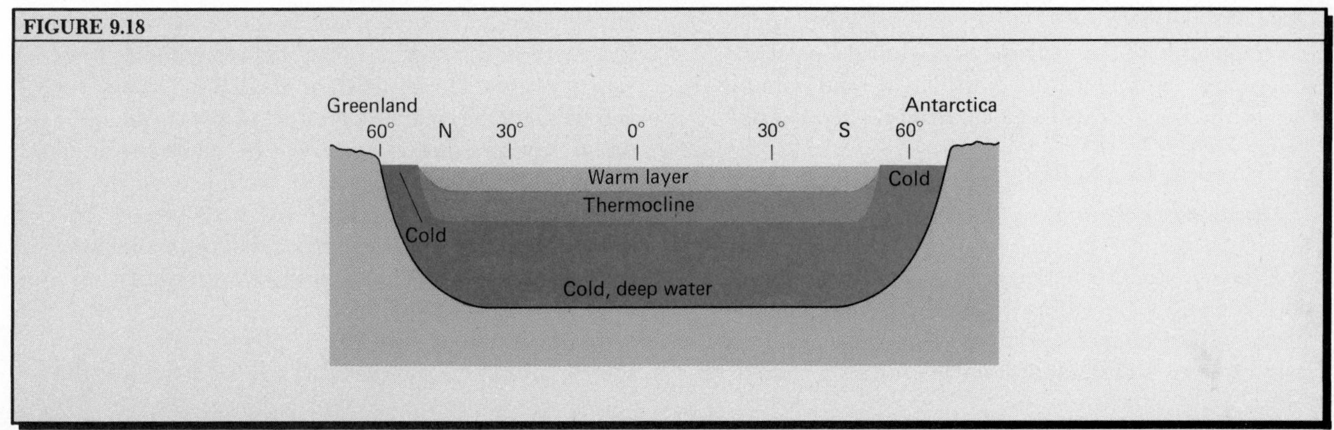

add to pure water to make an artificial brine closely resembling seawater. Table 9.1 lists the five most important salts required.

In Chapter 5, we described the ion composition of seawater and the residence times of various elements. We made a comparison between the dissolved solids composition of seawater and that of stream water. We found that the dominant base cations of seawater (sodium, magnesium, calcium, and potassium), along with aluminum and silicon, are derived largely from rock weathering, whereas outgassing of volatiles explains elements of the dominant anions in seawater—chlorine, sulfate, carbonic acid, and bromine.

Although the proportions in which the various chemical elements are present in relation to one another in seawater are remarkably constant throughout all oceans, the *salinity* of the water, which is the total weight of dissolved solids to weight of water, is a variable quantity, differing in value from place to place over the oceans and at various depths. Salinity is measured from water samples taken at any desired depth by a device called a water bottle.

Salinity is commonly expressed as the ratio of solids to water. Thus, the total value given in Table 9.1 is 34.5 grams of salt per 1000 grams of water, which is the same as a salinity of 3.45%. Because the percentage is small, salinity is usually stated in parts per thousand, with a special symbol: 0/00. The figure 34.5 0/00 is read "34.5 parts per thousand" and is the same proportion as 3.45%.

In the oceans generally, salinity ranges between 33 and 37 0/00, depending upon the geographic location. In certain bays or arms of the sea largely shut off from the open ocean and located in tropical deserts, salinity is abnormally high. An example is the Red Sea, with salinity of 40 0/00 or more. Near the mouths of large rivers, salinity may be low because of mixing with fresh river water.

If one were to make a general statement about the relationship of salinity to depth, it would be that a three-layer system closely analogous with the three-layer temperature system is typical in low and middle latitudes (Figure 9.17). A shallow surface layer of uniformly high salinity (35.0 to 36.5 0/00) corresponds with the uniformly warm layer. Below this layer is a zone of rapid decrease in salinity, the *halocline,* which corresponds with the thermocline. Below the halocline differences in salinity are very small, and salinity lies in the range of 34.6 to 34.9 0/00 for most of the ocean body.

*Dissolved Oxygen

Although the distribution of dissolved oxygen is not well known for the entire world ocean, certain generalizations can be made (Figure 9.17). An oxygen-rich surface layer is characteristic of the oceans generally, because atmospheric oxygen is dissolved in the seawater to the saturation point and phytoplankton (oxygen-releasing plants) live in surface layers. It is also known that, for the vast bulk of water at great depth, the oxygen content is about on the order of half the surface value. This reduction is attributed to the consumption of oxygen through biological activity. Over large areas, a very strongly defined layer of minimum oxygen content is found immediately below the oxygen-rich surface layer. Here, at depths on the order of 500 to 1000 m, the oxygen content falls to values less than one-tenth of the surface-layer value.

*Density of Seawater

Density of seawater is defined as the mass of a unit volume of water. One cubic centimeter of pure fresh water at 4 °C, which is the temperature of greatest density of pure water, has a mass of 1 gram and therefore has a density of 1.00 g/cc.

Because of the presence of dissolved solids, seawater is slightly denser than pure fresh water. Compared with an assumed density of 1.000 g/cc for pure fresh water, seawater has a density of about 1.026–1.028 g/cc. Both temperature and salinity affect the density. Seawater becomes increasingly dense as it cools until the freezing point is reached, at about −2 °C. This is an important principle, because it means that seawater cooled near the surface will tend to sink, displacing water of less density. Density also becomes greater as salinity increases, so that where surface evaporation is great, the water near the surface may become slightly denser than that below it, and it therefore sinks to a lower level. Because temperature is the stronger of the two controls of density, the densest seawater is formed in the cold arctic and polar seas. This very cold water sinks to the bottom and tends to remain close to the floor of the deep ocean basins.

*Pressures Within the Oceans

The principle of hydrostatic pressure, or confining pressure in a fluid, applies within the oceans as it does within the earth's mantle and core. The pressure which the water exerts equally in all directions upon any exposed surface increases in direct proportion to the depth in the ocean. Taking the atmospheric pressure at sea level to be 1 bar, or about 1 kg/sq cm, we can estimate the confining pressure at any depth in the ocean. The pressure increase will be about one gram per square centimeter for each centimeter of depth, and will amount to 1 kg in 10 m, or 100 kg per km. Thus, a water layer 10 m thick produces a confining pressure about equal to atmospheric pressure at sea level (1 bar).

Oceanographers state ocean pressure after atmospheric pressure is subtracted. Thus, by definition, the

Table 9.1 Principal Constituents of Seawater

Name of salt	Chemical formula	Grams of salt per 1000 grams of water
Sodium chloride	$NaCl$	23
Magnesium chloride	$MgCl_2$	5
Sodium sulfate	Na_2SO_4	4
Calcium chloride	$CaCl_2$	1
Potassium chloride	KCl	0.7
With other minor ingredients to total		34.5

pressure has a value of zero at the sea surface. Also, a specialized unit of pressure—the *decibar,* defined as 100,000 dynes per square centimeter—is used. One scientific advantage of using this unit, apart from its conformity to the metric system, is that the water pressure in decibars is approximately equal to the water depth in meters.

In contrast to the gases of the atmosphere, ocean water can be loosely described as "incompressible," but that is not strictly the case. All liquids and crystalline solids contract somewhat in volume under an increase in confining pressure, even though the amount may be scarcely measurable. Under enormous confining pressures near the bottom of the deep ocean basins, the density of seawater is measurably increased by the pressure. Consider, for example, a small mass of surface water with a salinity of 35 0/00 at 0 °C and a density of 1.028. If forced to descend to a depth of 4000 m, this same water would increase in density to a value over 1.048. Thus, we see that three variable quantities determine the density of seawater: temperature, salinity, and pressure.

Ocean Currents

An *ocean current* is any predominantly horizontal water movement persistently in one direction. Ocean currents thus differ from tidal currents (Chapter 19), which alternate their direction of flow in a rhythmic manner. Currents can range in scale from oceanwide flow systems to local currents of small extent, and can be generated by several mechanisms, but the most important mechanisms for the great ocean currents are: (1) the drag of winds over the ocean surface, and (2) unequal forces set up by differences in water density. Current speeds are given in meters per second, while the direction of the current is given as the direction in which the water is moving.

Despite their low speeds, ocean currents move enormous volumes of water in a given time because their cross-sectional area is great. For example, the Gulf Stream, in the region opposite Chesapeake Bay, has a total water transport of from 75 to 90 million cubic meters per second. We could try to visualize this flow as consisting of 75 to 90 million cubes of water, each one meter on a side, passing each second through an imaginary vertical plane placed across the current. The Gulf Stream extends to a depth of not more than about 1.5 km and across a width of roughly 240 km. In comparison, the Mississippi River averages no more than 15 m deep and about 0.8 km wide and is discharging water at about 20,000 cubic meters per second.

Air flowing over a water surface exerts a dragging force upon that surface, setting the surface water layer in motion. Prevailing winds, such as the tropical easterlies (trade winds), blowing from east to west, and the middle-latitude westerlies, blowing from west to east, exert a one-way drag on the sea surface over vast expanses of oceans. This drag produces a system of *drift currents.* Because of the Coriolis effect—the effect of the earth's rotation—the direction of water drift in the northern hemisphere is in a compass direction about 45° to the right of the direction of prevailing wind.

Differences in water density can also set currents in motion, and these are described as *thermohaline currents.* Whether because it is warmer or lower in salinity, a surface water layer in one place may be less dense than an adjacent water layer that is either colder or has higher salinity. The water surface then slopes gently from the region of less dense water to the region of denser water. Water begins to move down the gradient of this slope, but because of the Coriolis effect, is turned horizontally through 90° until it flows along the contour of the sea-surface slope.

The Atlantic and Pacific Oceans have a simple and orderly pattern of surface-water movement, which can be generalized in a schematic diagram (Figure 9.19). Both oceans extend unbroken from high latitudes north and south across the equator and are bounded by continents on both sides. The dominant features are two great circular water motions called *gyres,* each of which is centered on a subtropical region of high barometric pressure. Within each gyre, the water turns clockwise in the northern hemisphere and counterclockwise in the southern hemisphere. The westward drifts of water in equatorial latitudes are referred to as the *north equatorial current* and the *south equatorial current* (Figure 9.20). These are separated by the *equatorial countercurrent* that moves eastward and is caused by the return of lighter surface water which has been piled up on the western side of the ocean basin by the equatorial currents.

An essential feature of the great gyres is that the current is greatly narrowed and intensified in velocity on the western side of the ocean, whereas on the eastern side the current is diffused into a slow drift spread over a broad zone. This asymmetry results from the Coriolis effect: On an eastward-rotating earth the gyre is deflected westward and forced to be compressed against the western boundary of the ocean.

The intensification of poleward currents on the western sides of the North Atlantic and North Pacific Oceans results in two powerful warm currents, the Gulf Stream in the Atlantic and the Kuroshio Current in the Pacific. Speeds up to 2.5 m/sec are developed in these narrow streams.

Upon turning eastward, each warm poleward current becomes part of the *west-wind drift*, produced by the highly variable prevailing westerly winds in lat. 40° to 65°. The gyre is then completed by a cold coastal current moving equatorward close to the west coast of the bordering continent. The four most important examples are the California Current, the Canary Current, the Peru or Humboldt Current, and the Benguela Current, shown in Figure 9.20.

The Southern (Antarctic) Ocean, which is simply the southern part of the Pacific, Atlantic, and Indian oceans in the region of the fiftieth and sixtieth parallels, forms a continuous circular ribbon of ocean, scarcely interrupted by land. Here the *antarctic circumpolar current* flows eastward in an uninterrupted path following the parallels of latitude (Figure 9.20*B*).

Surface currents are important to marine geologists who study the sediments that rain down upon the ocean floor. Currents of both wind and water transport many kinds of fine sediment far from their places of origin on the lands. Minute organisms that live suspended in the shallow upper layer of the ocean are also transported by currents.

In addition to the horizontal movement described above, vertical motions of ocean water are also important in the total circulation of the oceans. In areas where surface water layers tend to move closer together (convergence), sinking of the surface water takes place (Figure 9.21). Where surface water layers are separating and moving apart (divergence), there takes place replacement by rising of water from below, a movement called *upwelling*. Causes of sinking and upwelling are several; they relate to wind action, evaporation of surface water, addition of surface water by rainfall, and changes in density due to cooling or warming of the surface layer. Upwelling is concentrated along narrow zones off the west coasts of continents in tropical latitudes and takes place in the cool currents that move toward the equator, such as in the Peru Current, off the west coast of South America.

The most important cause of sinking of ocean water on a large scale is cooling of surface water by loss of heat to the overlying atmosphere in high latitudes. Here, during long winters, much more heat is lost to space than is gained by solar radiation. Relatively warm surface waters brought poleward by ocean currents are chilled and increase in density. This water, which may be close to the freezing point, sinks to the deep ocean floor. A surface *convergence zone* is thus created in both arctic and antarctic latitudes.

In the North Atlantic, the arctic convergence zone lies to the south and east of Greenland in an area into which the relatively more saline water arriving from tropical latitudes via the Gulf Stream is cooled and becomes unusually dense. The antarctic convergence zone, roughly concentric about the continent of Antarctica, is the most important zone of sinking of cold water, from a global standpoint. The principal convergence lies in the latitude belt between 50° S and 60° S and forms a relatively narrow and well-defined line. Closer to Antarctica, chilling of the surface water is even more severe and results in sinking of dense water that flows down the continental slope of that landmass.

*Water Masses and the Deep Ocean Circulation

By recognizing unique combinations of temperature and salinity, oceanographers distinguish large water bodies, known as *water masses*, from one another. Identification of a particular water mass cannot be made on the basis of density alone, because density depends simultaneously on both temperature and salinity. Water masses move sluggishly and tend to spread out into stable layers, one above the other, ar-

FIGURE 9.19 Schematic diagram of an ocean with an idealized system of surface currents.

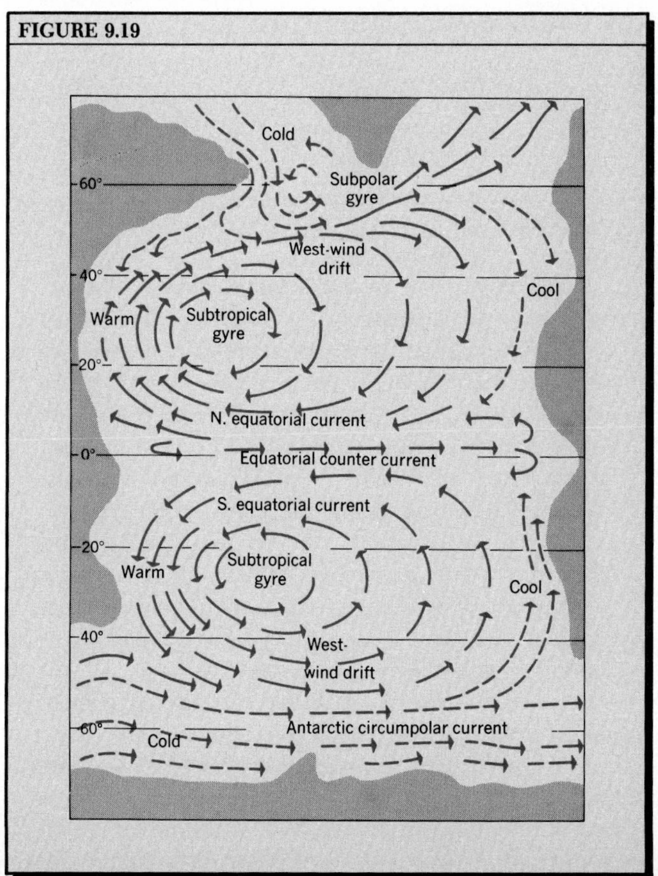

ranged in order of density. For this reason, water masses are described according to position from the surface downward as upper, intermediate, deep, and bottom water masses. In addition, water masses are classified in terms of latitude and the particular ocean

FIGURE 9.20 *A*. World map of average surface drifts and currents of the oceans. (From U.S. Navy Hydrographic Office, *Stream Drift Chart of the World*.) *B*. Streamlines of surface water motion about Antarctica. (After H. U. Sverdrup, 1942, *Oceanography for Meteorologists*, Prentice-Hall, Englewood Cliffs, New Jersey, p. 206, Figure 57. Used by permission of Prentice-Hall, Inc.)

FIGURE 9.20 A

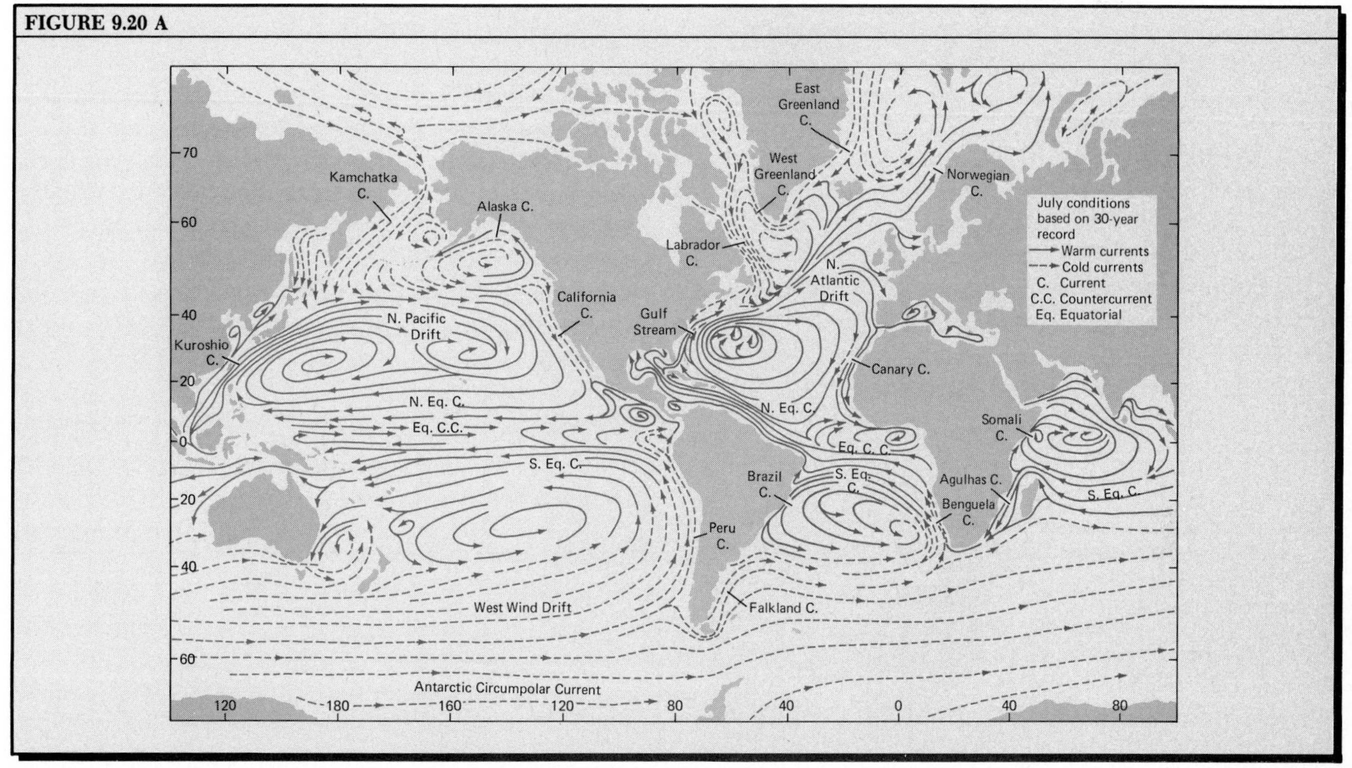

FIGURE 9.20 B

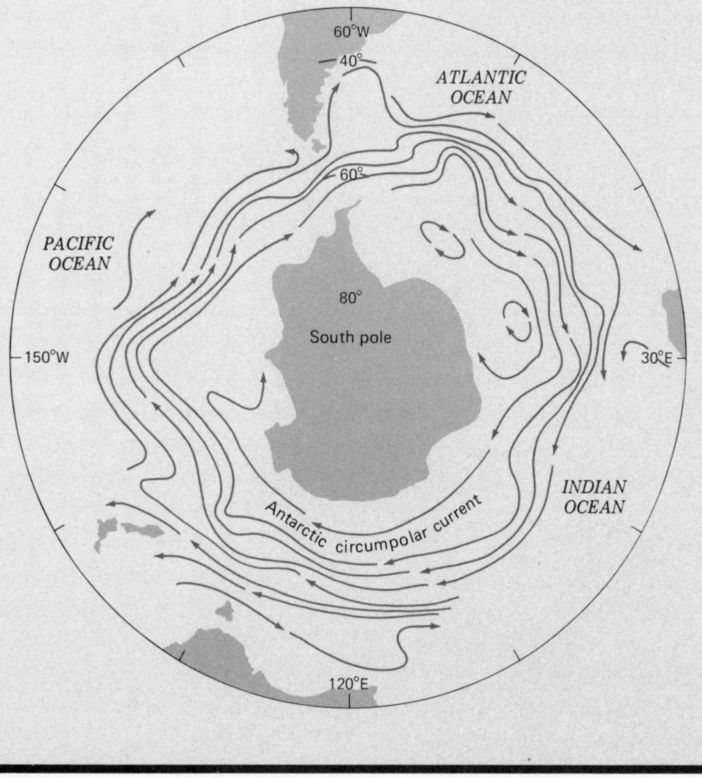

basin they occupy. In order from equator to poles there are, for example, equatorial, central (in subtropical latitudes), and arctic and antarctic water masses.

Figure 9.22 is a schematic diagram showing the meridional (north–south) flow of deep water in the Atlantic Ocean. The Atlantic central water mass is produced at the surface in the subtropical region of convergence, where it sinks and moves northward toward the equator. Sinking of cold water in the antarctic convergence produces the antarctic intermediate water mass, which moves northward as it sinks, passing beneath the central water layer, then reversing to southward flow as it encounters and mixes with North Atlantic deep water and North Atlantic bottom water. The resulting mixture is a circumpolar water mass surrounding the antarctic continent. Very cold water produced close to the coast of Antarctica moves down the continental slope and spreads out upon the ocean floor as antarctic bottom water. Because it is colder, and therefore denser, than the water mass produced in arctic regions of the North Atlantic Ocean, antarctic bottom water travels across the equator and far into the northern hemisphere, reaching to 35° N to 40° N. Above this bottom layer, North Atlantic deep water flows in the opposite direction (southward) to arrive in the region of circumpolar water, at about 60° S. (No deep water is formed in the North Pacific Ocean.)

On the average, speed of flow of deep water is extremely slow in comparison with surface currents above the thermocline and amounts to less than a few centimeters per second. Because a water mass develops its characteristics at the ocean surface, the surface provides a convenient starting point for tracing the flow path of an average water particle. The elapsed travel time from this initial point is termed the *water age*. Estimates of water age for deep masses have been made by subjecting water samples to the radiocarbon method (see Chapter 18). Ages of from 750 to 1000 years are obtained for Atlantic bottom water, while ages of Pacific bottom water run higher—1500 to 2000 years. On this evidence, a complete cycle of movement of a water particle from surface to deep ocean and back to the surface might require a time span on the order of some 2000 to 4000 years.

Deep Bottom Currents

The movement of cold, deep water is not everywhere sluggish, as the average flow rates would seem to imply. Locally, this cold bottom water moves swiftly in narrow currents paralleling the continental margins on the western sides of the ocean basins. Two such *bottom currents* have been identified and are

FIGURE 9.21 Convergence and sinking; divergence and upwelling.

FIGURE 9.22 Meridional cross section of the Atlantic Ocean showing water masses and their movement. (After J. Williams, *Oceanography,* Copyright © 1962, Little, Brown & Company, Boston, p. 121, Figure 9–9.)

FIGURE 9.21

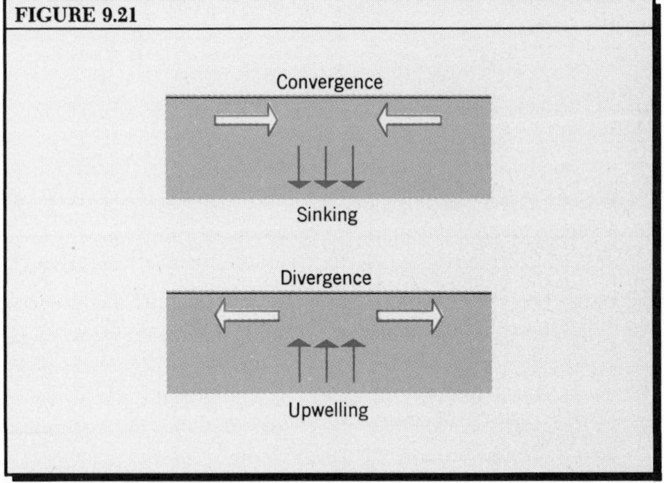

FIGURE 9.22

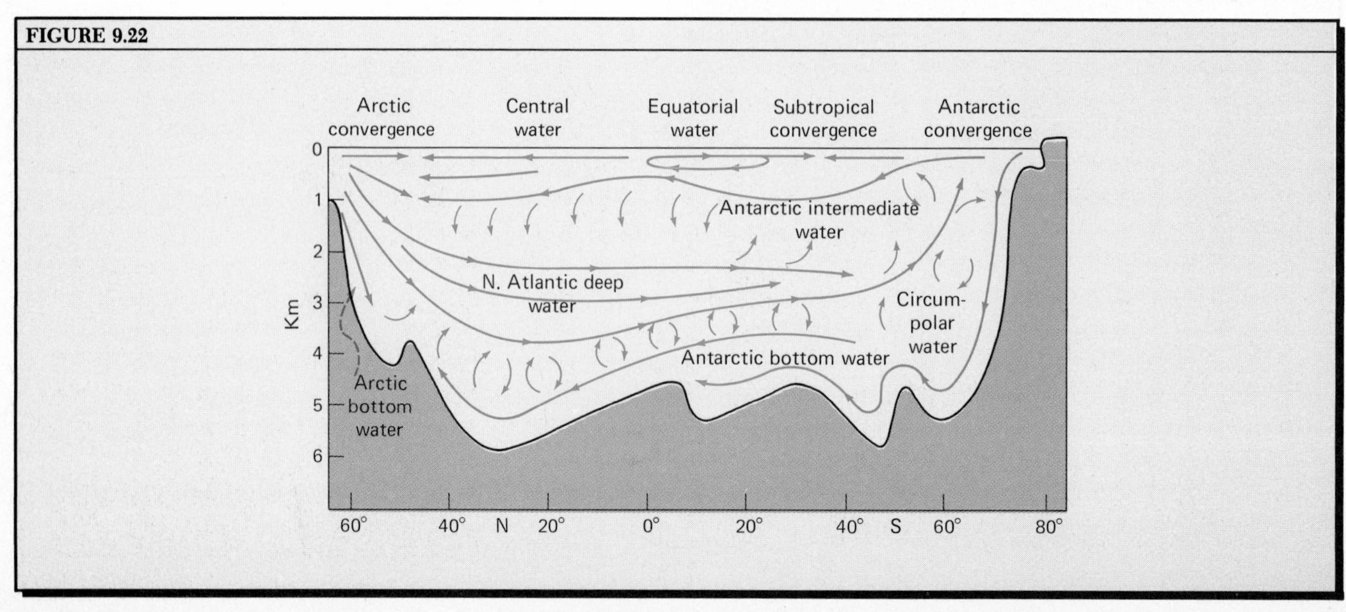

thought to be of great importance in transporting bottom sediment and shaping the configuration of the seafloor. North Atlantic deep water, flowing southward, is concentrated in a narrow current paralleling the contour of the continental rise. Referred to as a *contour current*, it flows at a depth of 2400 to 3200 m and its measured speed is as high as 18 cm/sec. A second current lies beneath the northward-flowing antarctic bottom water mass; it flows over the outer continental rise at a depth of 4800 to 5500 m (see Figure 9.32).

Probing the Ocean Floor

Since the 1870s, oceanographers have systematically taken samples of materials from the ocean floors. At first this could be done only by means of dredges that scraped off a thin layer and brought it to the surface for examination.

By the 1930s, information about the sediment layer itself began to be obtained by the process of *coring*, which is simply vertical penetration by a long section of pipe that cuts a cylindrical sample, or *core*. Brought to the surface, the core is extruded and gives a complete cross section of the layer (Figure 9.23). Cores up to 30 m long are readily obtained, then cut in half longitudinally, revealing the bedded structure and permitting small interior samples to be taken for microscopic examination and chemical and physical analysis.

A great advance in deep sampling of the sediment of the ocean floor came in the late 1960s through the use of oil-well drilling methods. The drilling equipment is mounted on a special 10,000-ton vessel, the *Glomar Challenger*, built with the capability of drilling into the ocean floor to a depth of 750 m, in water depths as great as 7000 m (Figure 9.24). Not only can the drill pass through a sediment layer, but it can also obtain cores of the solid crustal rock beneath.

Glomar Challenger looks like an oil-drilling rig, mounted on an ordinary ship's hull. The drilling rig is like other land-based oil rigs, but the hull is unlike that of any commercial vessel. The difference lies in an elaborate positioning mechanism, by means of which the ship can maintain its exact position during the laborious job of drilling. Four motors stabilize the vessel by directing water jets from tunnels in the hull. They are computer-controlled and respond instantly to the pitching and rolling motions the vessel experiences in heavy seas. The flexible drill string is capable of accommodating a considerable motion of the platform. It is even possible for the drill to be withdrawn, the bit replaced, and the hole again found and penetrated.

FIGURE 9.23 A

FIGURE 9.23 Cores from the deep ocean floor. *A*. Aboard the Research Vessel *Vema*, crew members bring alongside a piston corer with its sample of ocean-floor sediment. A heavy weight, seen at the top of the photograph, drives the 30-m pipe into the soft sediment. *B*. At Lamont–Doherty Geological Observatory, two research scientists examine the contents of a deep-sea core. The Observatory has a collection of over 9000 cores and is also a repository for drilled cores obtained by the *Glomar Challenger*. (Photographs by courtesy of Lamont–Doherty Geological Observatory of Columbia University.)

FIGURE 9.23 B

Called the most successful scientific expedition in the history of science, the long voyage of the *Glomar Challenger* that began in 1968 enabled scientists to probe the floor of every ocean except the Arctic. Hundreds of holes have been drilled and over 50,000 m of core samples recovered. In these cores are found a historical record of the sediments of the ocean basins undreamed of a few decades ago.

The voyage of the *Glomar Challenger* was at first under the supervision of the Deep Sea Drilling Project (DSDP) of the National Science Foundation. The costs ran to some $10 million per year, but the scientific results made the outlay seem a bargain. After seven years of operation, the *Glomar Challenger* entered a new phase. Although international participation had always been included in DSDP, 1976 saw the beginning of the International Phase of Ocean Drilling (IPOD), with regular staff representations from the Soviet Union, France, Great Britain, Japan, and West Germany. Each nation pledged to contribute $1 million annually to the IPOD operating budget.

With the change to an international scientific program, the drilling program was changed to concentrate upon deep holes penetrating up to 600 m into the basaltic crust—a remarkable feat where the ocean water is 4000 m deep.

Seismic Reflection Profiles

A different type of information about the sediments of the ocean floors comes from application of the *seismic reflection* principle. In a manner somewhat like that of the precision depth recorder, the impulse from a small explosion is sent to the ocean bottom and reflected from the bottom and from layered structures below the bottom. As in the case of the echo sounder, the returning waves are picked up by a hydrophone and recorded on paper. In a *seismic profiler traverse*, explosions or other forms of sudden energy release are made at ten-second intervals along the line of the ship's course, and a profile results showing the contact of sediment with bedrock, as well as certain reflecting horizons within the sediment. The method was developed by Professor Maurice Ewing, John E. Ewing, and co-workers of the staff of the Lamont–Doherty Geological Observatory. Seismic reflection profiling has yielded a remarkably clear picture of the distribution and thickness of the sediment layer over the ocean floors.

One type of sediment accumulation, seen in the mid-oceanic ridge with its strong relief, is *ponding* of sediment in isolated topographic basins (Figure 9.25). Notice that sediment is lacking near the ridge axis but thickens outward toward the flanks, where sediment thicknesses up to 500 to 600 m are present.

A different type of sediment accumulation is shown in Figure 9.26. Here, the ocean floor is a smooth abyssal plain at a depth of over 5000 m, grading eastward into the rising flank of the Mid-Atlantic Ridge. The sediment accumulation, over 3000 m thick in places, forms a continuous blanket.

Examples of seismic reflection profiles as they are actually recorded are shown in Figure 9.27. Keep in mind that the vertical scale is about 18 times greater than the horizontal scale, greatly exaggerating the heights and thicknesses of the features shown. Profile

FIGURE 9.24 A. The *Glomar Challenger* is 120 m long and weighs 10,000 metric tons. Her drilling derrick, which rises 60 m above the waterline, has a capability of raising a weight of 500 metric tons. The pipe sections used in drilling lie on a rack forward of the derrick. Altogether, 7,000 m of 13-cm pipe is stored here. (Deep Sea Drilling Project.) *B.* Two scientists and the cruise operations manager examine the bottom of a core barrel filled with sediment taken from the bottom of the North Atlantic. (Deep Sea Drilling Project.)

FIGURE 9.24 A

FIGURE 9.24 B

A shows the ponding of sediments in depressions of the Carlsberg Ridge (an extension of the mid-oceanic ridge in the northern Indian Ocean). At the right is a portion of the India Abyssal Plain, its floor at a depth of over 4 km. Profile *B* shows a gently sloping abyssal plain that extends southward from the submarine cone of the Indus delta. Sediments have completely buried many hills of the Carlsberg Ridge. Profile *C* shows a third kind of sediment accumulation, explored by seismic reflection in oceanic trenches. Partial filling of the Java Trench is shown at the extreme right. Many reflective layers give a strongly bedded appearance to this wedgelike accumulation, which is more than one-half kilometer thick.

Terrestrial Sources of Deep-Sea Sediments

There are several possible terrestrial sources of deep-ocean sediments, such as those revealed by the seismic profiler method. "Terrestrial" refers to sources on the continental land surfaces or along continental coastlines. We will omit from consideration the thick continental-shelf sediment accumulations obviously derived from the continents through direct transportation by streams, waves, and shallow-water currents. The continental shelf wedge rests on the continental lithosphere.

Atmospheric circulation provides an important transport mechanism for extremely fine particles from lands to the deep oceans. Mineral particles are raised high into the atmosphere by dust storms in the tropical deserts in latitudes 10° to 30° north and south. High-level tropical easterly winds carry these particles far westward over the adjacent oceans, where they may settle to the ocean surface or may be carried down in raindrops. Other important sources of atmospheric dusts are volcanic eruptions, emitting minute shards of volcanic glass, and the vaporization of meteors in the upper atmosphere.

Transport by surface ocean currents is an obvious means for the wide distribution of very fine suspended particles derived from sources close to the continental margins. Suspended particles follow the patterns of oceanic circulation, particularly the great gyres and the Antarctic circumpolar current system.

A related mechanism is transport by icebergs, which break off land-based glaciers and float far out to sea and melt, dropping mineral fragments of many sizes. By this means, even huge boulders may reach positions hundreds of kilometers from the nearest land.

The great bulk of the thick sediment layers found beneath abyssal plains and in trenches requires transport mechanisms of far greater capability than those so far mentioned.

FIGURE 9.25 Tracing of seismic reflection profile obliquely crossing the Mid-Atlantic Ridge at about 40° N. Sediment deposits shown in solid color. Basement rock (bedrock) lies beneath. (After J. Ewing and M. Ewing, 1967, *Science,* vol. 156, p. 1591, Figure 2.)

FIGURE 9.26 Sketch of seismic reflection profile of the Argentine Basin of the South Atlantic, off Buenos Aires at latitude 36°–38° S. (After M. Ewing, W. J. Ludwig, and J. I. Ewing, 1964, *Journal of Geophysical Res.*, vol. 69, p. 2011, Figure 6.)

FIGURE 9.25

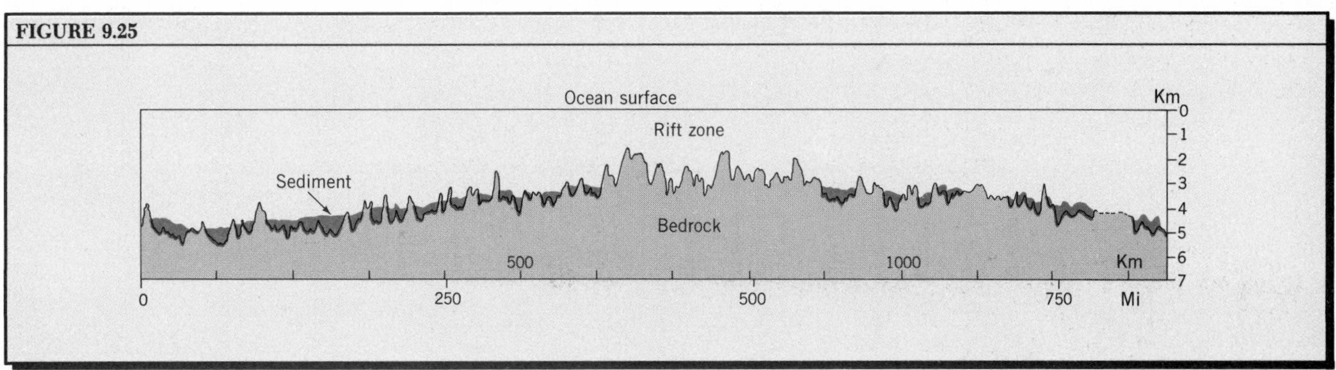

FIGURE 9.26

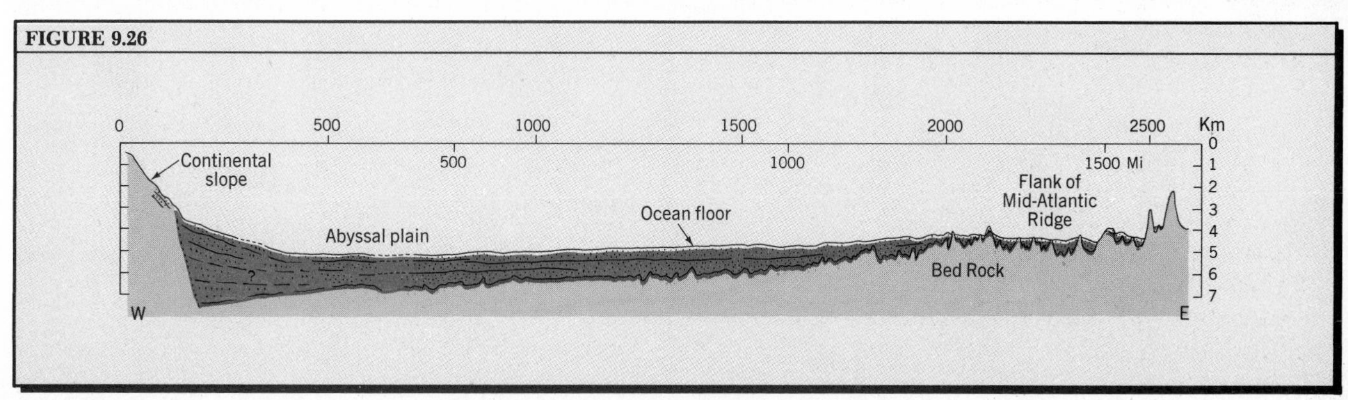

FIGURE 9.27 Seismic reflection profiles along three different tracks in the Arabian Sea region of the Indian Ocean. (Courtesy of Allen Lowrie, Geology and Geophysics Branch, Naval Oceanographic Office.)

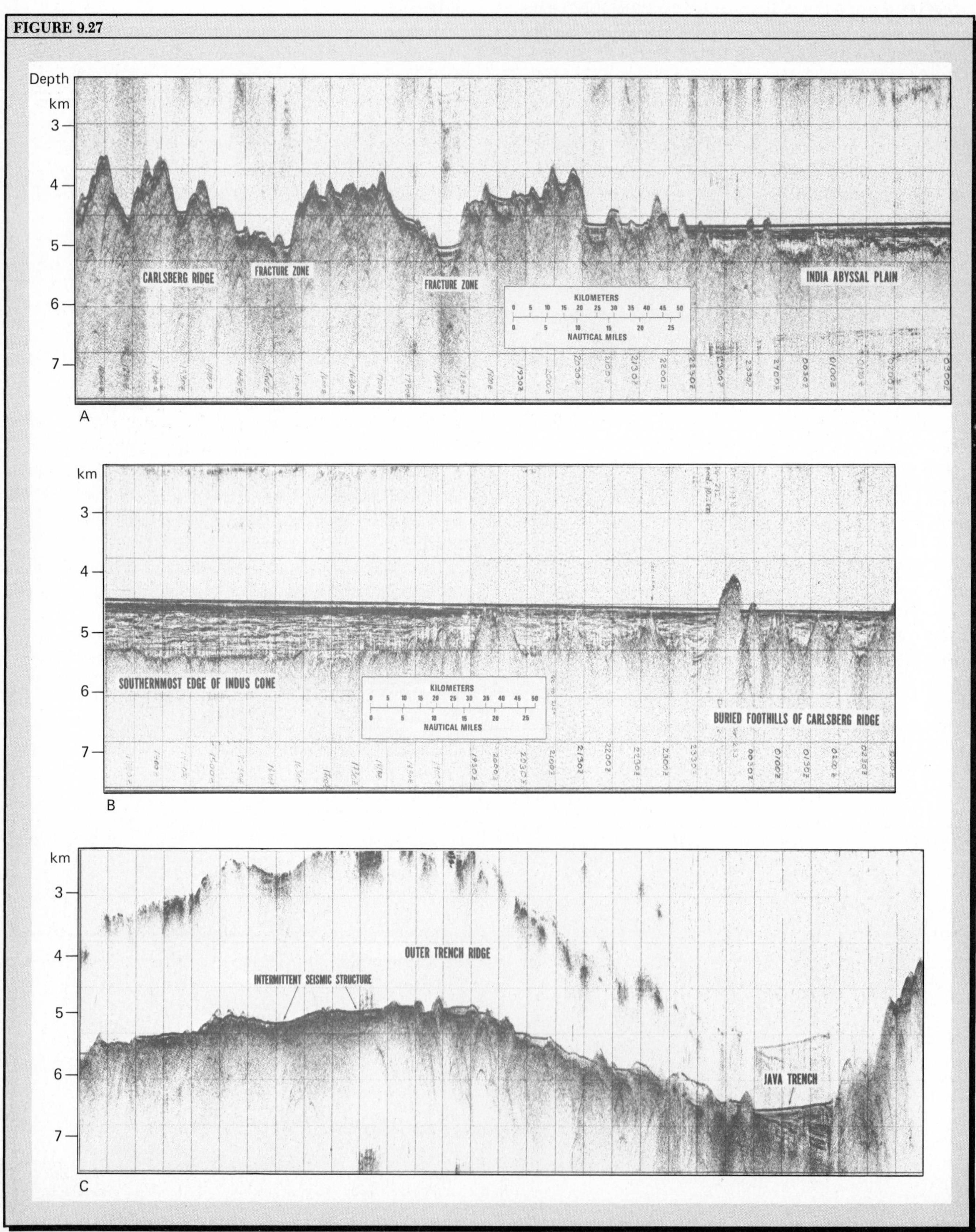

Turbidity Currents and Turbidites

By far the most capable sediment-transporting mechanism operating in many parts of the ocean basins is a type of current flow powered by gravity and flowing directly down the gradient of sloping portions of the ocean floor. Called *turbidity currents,* these flows consist of highly turbid (muddy) water moving swiftly in long, narrow tongues. A turbidity current requires a source mass of soft sediment occupying a high position at the brink of a steep submarine slope. Collapse of a mass of sediment causes it to become a turbid liquid (process of liquefaction) with a density greater than the surrounding clear seawater. Under the force of gravity, the denser mixture moves directly downslope, as a tongue, to reach an adjacent seafloor plain or depression where the turbid water spreads out into a thin sheet and comes to rest.

The turbidity current is just one variety of *density current,* a class that includes any gravity-powered current of a denser fluid flowing to lower levels beneath a less dense fluid. Turbidity currents repeatedly occur from the same source areas and their deposits accumulate in numerous thin sheets to reach great total thicknesses. In the accumulation process, hills or other topographic irregularities of the bedrock surface are gradually buried until ultimately an abyssal plain is formed. The sediments thus accumulated are classified as *turbidites.*

The deposit of a single turbidity-current flow is represented by a distinctive unit layer of sediment called a *turbidite.* Within this turbidite is an arrangement of layers changing in texture from coarse at the bottom to fine at the top. Typically, the basal zone of the turbidite also shows a distinctive particle size arrangement known as *graded bedding,* characterized by a continuous upward gradation of sizes from coarse sand or gravel to fine sand. The arrangement is illustrated in Figure 9.28. You can simulate graded bedding by placing a clean mixture of fine pebbles, coarse and fine sand, and coarse silt in a tall glass cylinder filled with water. Upend the cylinder and shake it vigorously, then allow it to stand upright. The rain of particles reaching the base of the container will be assorted in the same manner as in graded bedding. Above the basal graded zone, the turbidite consists of sand in parallel and rippled laminations. Close to the top is fine sandy to silty clay. The topmost layer is made up of clay-size particles.

As a turbidity current advances, it passes over the soft clay layer that forms the top of the previous turbidity current deposit. The force of the fresh turbulent current digs into the soft clay bed, creating distinctive *scour marks.* On a bedding surface, these appear as elongate cuplike or troughlike indentations. The scour marks are later filled by coarse sand of the bottom zone of the next turbidity-current deposit. Much later, after the entire turbidite sequence has become lithified and finally exposed as an outcrop on the continental surface, the molds of scour marks appear on the undersides of the coarse sandy beds as *sole marks* (Figure 9.29). Geologists use scour marks and

FIGURE 9.28 A schematic cross section through a single turbidite showing the typical structure. Above the basal layer of graded bedding are laminated sand and silt layers, topped by a clay layer.

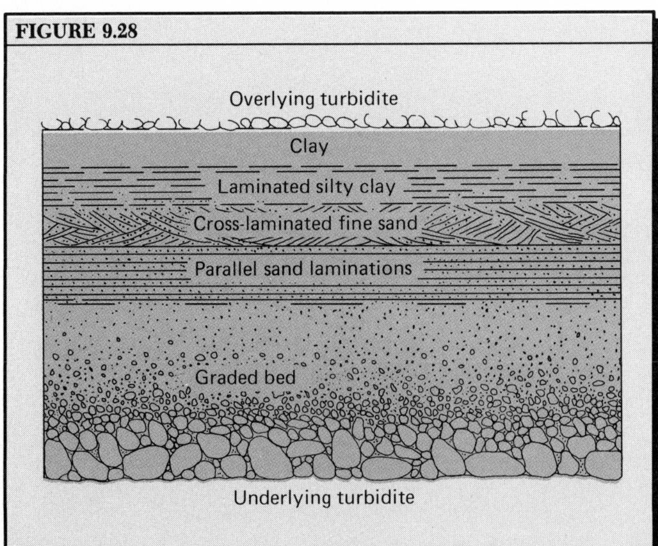

FIGURE 9.29 These sole marks, seen on the underside of a sandstone slab found in the Carpathian Mountains of Poland, are the molds (counterparts) of scour marks made by currents flowing over a bed of cohesive sediment on the ocean floor. The current flowed from left to right (*arrow*). The long, narrow features are the impressions of flutes. (From S. Dzulynski and J. E. Sanders, 1962, *Transactions of the Connecticut Academy of Arts and Sciences,* vol. 42, Plate IV. Courtesy of John E. Sanders.)

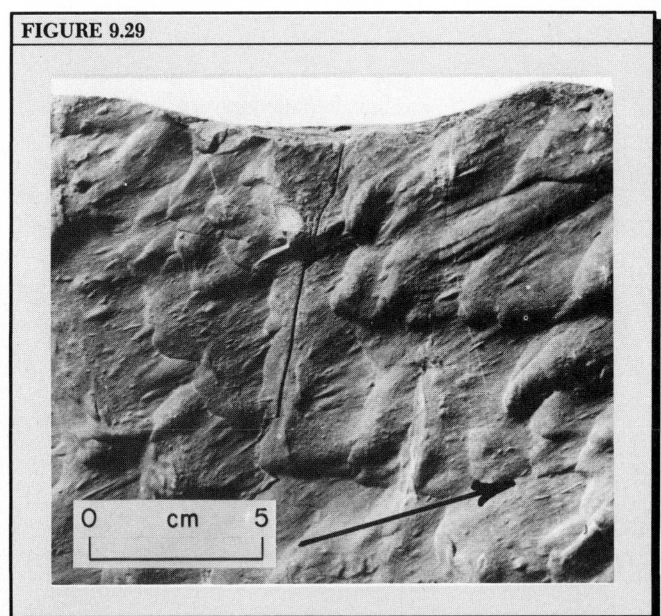

related current-formed features to determine the direction of the flow of ancient turbidity currents.

Turbidity currents were first recognized in connection with submarine canyons of the Atlantic continental slope of North America. These remarkable gully-like features score the continental slope by the hundreds and have attracted the curiosity of geologists for decades. Many consider it quite unlikely that these features could have been carved by terrestrial streams, so they have sought an explanation in processes taking place within the oceans. Although turbidity currents had been proposed as a mechanism of submarine canyon erosion, it remained for a chance earthquake to set off a chain of events that could document a single turbidity current in action.

In 1929, a turbidity current crossing the continental slope and rise off the Grand Banks of Newfoundland was sufficiently powerful to break in succession several transatlantic cables lying in its path. From a knowledge of the exact time each cable broke, the velocity of the turbidity current was determined, and it was found to have decreased from 100 km per hour where the bottom gradient was 1 in 170, to 23 km per hour where the gradient was only 1 in 2000. The turbidity current was set off by the earthquake shock, causing unconsolidated sediments of the continental slope to lose strength and to mix with seawater to produce a highly turbid suspension. The turbid tongue ultimately spread out upon the abyssal plain, forming a layer of sediment averaging 1 m thickness over an area of perhaps 200,000 sq km.

Deltas of major rivers (Chapter 19) contribute exceptionally large amounts of sediment to the continental shelf on which they emerge. A major submarine canyon is often present, representing a seaward extension of the river channel across the outer shelf. For example, both the Hudson River and the Congo River have major submarine canyons. Sediment carried by turbidity currents from the steep outer slopes of the continental shelf in the vicinity of the river delta build a *submarine fan* (or *cone*) that slopes gently away from the base of the continental slope for many hundreds of kilometers (Figure 9.30). At the outer limit of the fan may lie an abyssal plain. The two greatest submarine fans are those of the Indus and Ganges–Brahmaputra rivers. Also important are the Mississippi River fan, built out upon the floor of the Gulf of Mexico, and the Amazon River fan. Turbidite deposits of submarine fans are from several hundreds to a few thousand meters thick.

Contourites

The deep bottom currents that flow parallel with the western continental margin of the North Atlantic Ocean basin (and possibly the Pacific Ocean basin) are also important transporters of sediment at depths of 3000 to 6000 m. Contour currents that flow over the continental rise at speeds of 10 to 20 cm/sec carry silt and clay and build important sediment accumulations. Sediment transported by these currents is probably brought down the continental slope and rise by turbidity currents. The bottom currents are capable of winnowing out fine particles from sand of medium and coarse grades originally deposited by turbidity currents. Evidence of current transportation is seen in *ripple marks*, photographed in many places on the ocean floor where the bottom currents operate (Figure 9.31). One type of coastal ripple marking exposed at

FIGURE 9.30 Block diagram of a deep-sea fan formed at the base of a continental slope and extending far out upon the deep ocean floor. (Drawn by A. N. Strahler.)

FIGURE 9.31 These short-crested ripples on the ocean floor attest to the action of strong bottom currents of the Scotia Sea at a depth of about 4000 m, lat. 56° S, long. 63° W. (Charles D. Hollister, Woods Hole Oceanographic Institution.)

FIGURE 9.30

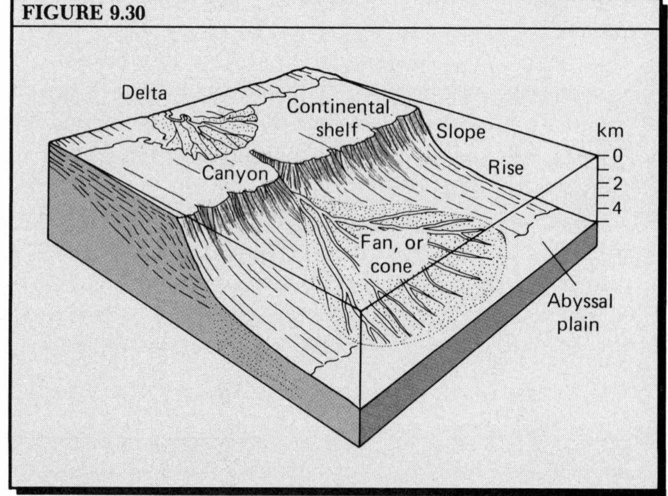

FIGURE 9.31

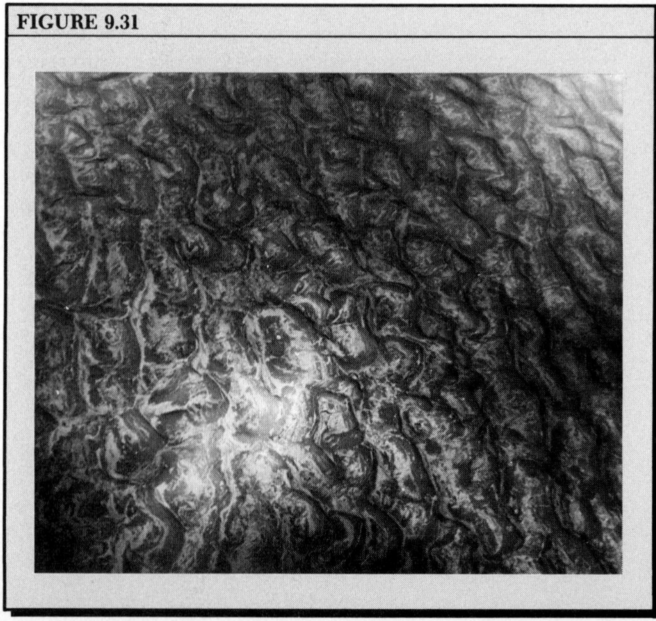

low tide shows clearly the action of swift tidal currents (see Figure 19.53). Similar ripple marks occur on the deep ocean floor where bottom currents have been measured by current meters. The direction of water flow indicated by the ripple marks is known to coincide with the measured direction of the bottom current.

Sediments deposited by contour currents are called *contourites,* to set them apart from turbidites. In contrast to turbidites, which show a large range in particle grade size and a distinctive graded bedding, contourites are formed of well-sorted fine sediment. Figure 9.32 is a schematic diagram showing a lenslike body of contourites underlying the continental rise. The thickness of the deposit is on the order of 500 m. The diagram also shows a layer of sediment deposited beneath the antarctic bottom current. This kind of contourite deposit forms on the very gentle outermost slope of the continental rise, where it grades into the abyssal plain.

Classification and Composition of Deep-Sea Sediments

Four main classes of deep-sea sediments are shown in Table 9.2. The first is the *biogenic–pelagic sediments,* which consist principally of calcareous or siliceous mineral matter secreted by organisms. The word *biogenic* is used in the same sense as in Chapter 5. The word *pelagic* (from the Greek word *pelagikos,* for "sea") simply means "originating in, or derived from, the ocean." The pelagic organisms of most importance in furnishing deep-sea sediment are *plankton,* the very small floating plants and animals growing in vast numbers in the shallow, well-oxygenated surface layer of the ocean. These organisms secrete hard structures referred to as *tests* which, upon the death of the organism and the destruction of the organic matter, sink down to great depths. If the tests are not dissolved away as they sink, they will reach the ocean floor.

Table 9.2 Classification of Deep-Sea Sediments

I	Biogenic-Pelagic Sediments
	Oozes
	Calcareous ooze
	Siliceous ooze
	Organic compounds
II	Pelagic-Detrital Sediments
	Brown clay (Red clay)
	Glacial-marine sediment
	Volcanic ash
III	Bottom-Transported Detrital Sediments
	Turbidites
	Contourites
	Terrigenous muds
IV	Hydrogenic Sediments
	Montmorillonite
	Zeolites (Phillipsite)
	Manganese nodules

Source: Based in part on a classification by K. Turekian (1968), *Oceans,* Englewood Cliffs, N.J., Prentice-Hall, p. 34, Table 3-1.

Accumulated sediment formed of 30% or more of tests is classified as *deep-sea ooze* and is further subdivided according to whether the tests are of calcareous or siliceous composition. *Calcareous ooze* is composed of the tests of foraminifera, pteropods, or coccoliths, all of calcareous composition. *Foraminifera* are one-celled animals, of which the genus *Globigerina* is particularly important (Figure 9.33). The term *globigerina ooze* is applied to sediment rich in these tests. Commonly present along with foraminifera are tiny molluscs, known as *pteropods,* which secrete tests of aragonite. *Coccoliths* are fragile calcite tests of a type of algae (microscopic plants) and are an abundant constituent of the very-fine grained calcareous oozes.

At a depth of about 3500 to 4000 m lies the *compensation depth,* at which calcium carbonate begins to dissolve in seawater (Figure 9.34). Tests can reach bot-

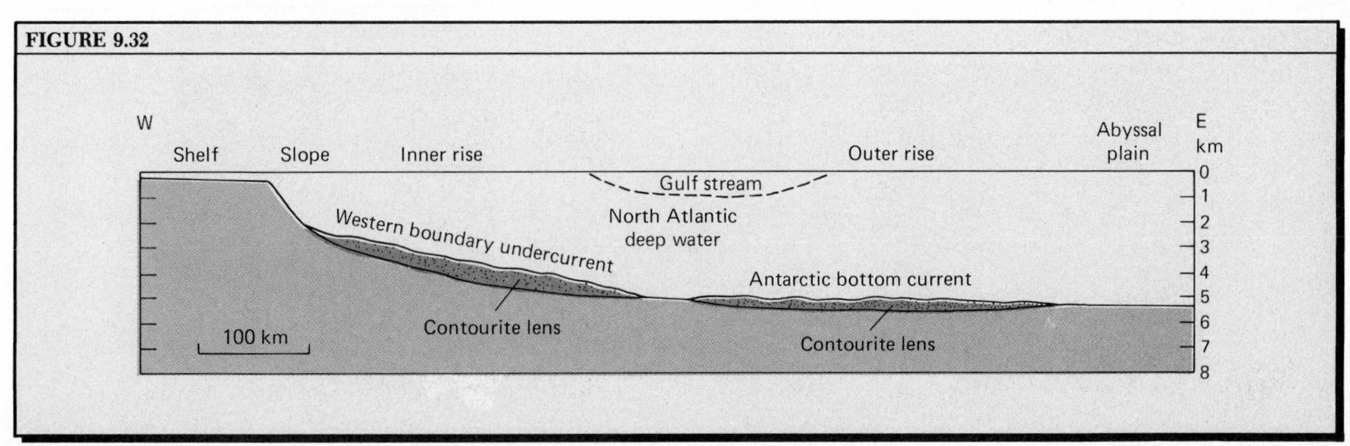

FIGURE 9.32 Cross section of the continental slope and rise, showing lenses of contourites formed by bottom currents. (After B. C. Heezen, C. D. Hollister, and W. F. Ruddiman, 1966, *Science,* vol. 152, Figure 3, p. 505.)

FIGURE 9.33 Calcareous ooze of the ocean floor. These shells and shell fragments were separated by seiving a sample of globigerina ooze, the bulk of which is composed of fine clay material. The sample is from a core obtained by scientists of the Research Vessel *Vema* from a depth of 3000 m in the South Atlantic Ocean. Enlargement about 12 times. (Andrew McIntyre, Lamont–Doherty Geological Observatory of Columbia University.)

FIGURE 9.34 A schematic diagram showing that carbonate tests begin to dissolve at the compensation depth and will fail to reach the ocean floor below about 4 km.

FIGURE 9.35 Plankton seen in high magnification under the electron scanning microscope. Magnifications indicated on individual photographs. These specimens were collected in plankton nets from within the uppermost 200 m in the western North Atlantic Ocean. (*A*, *B*, and *D*) Diatoms. (*C* and *E*) Radiolaria. (*F*) Silicoflagellate. (Photographs by courtesy of Allan W. H. Bé, Lamont–Doherty Geological Observatory of Columbia University.)

FIGURE 9.33

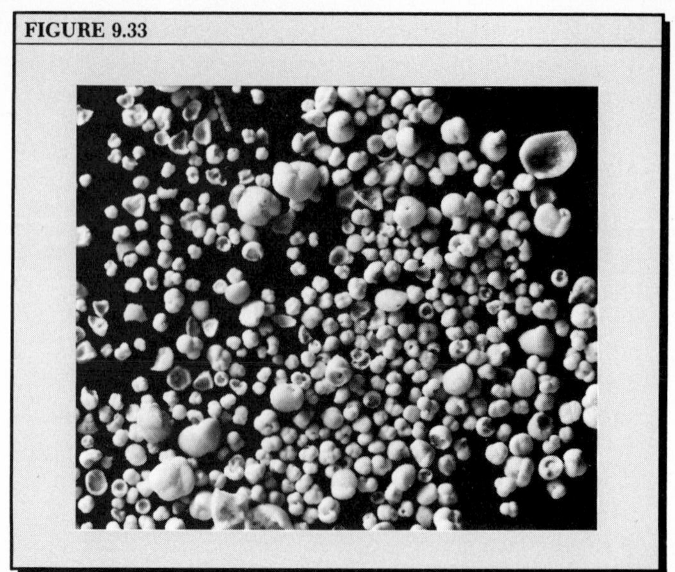

FIGURE 9.34

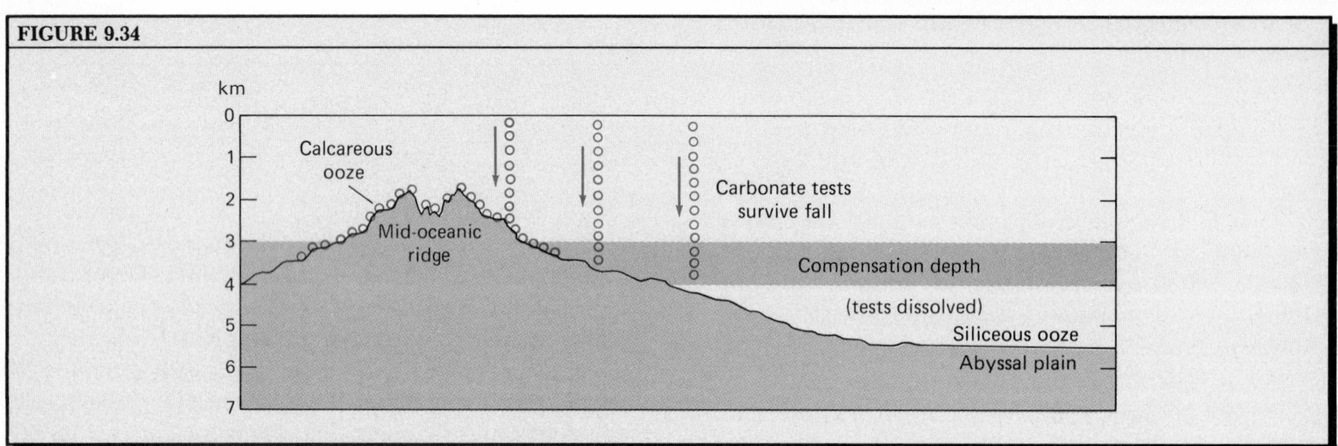

FIGURE 9.35

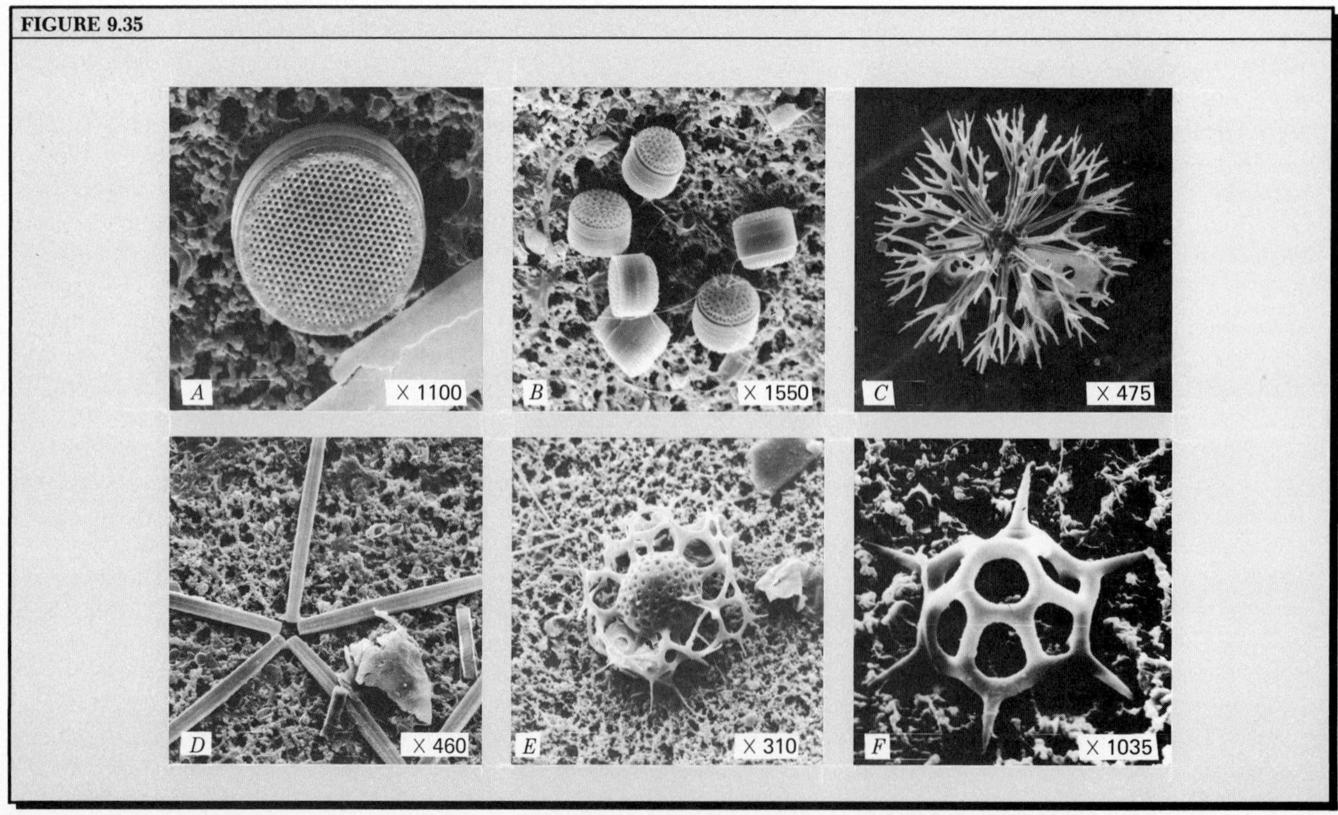

tom only in depths less than 4500 m, at most. Consequently, bottom areas with high proportions of calcium carbonate are quite closely correlated with topographically high areas of the ocean basins. In the Atlantic Ocean, concentration of calcium carbonate is high along the axis of the Mid-Atlantic Ridge, but is relatively low on the abyssal basins on either side. Rate of biological productivity, which tends to be high in areas of warm ocean currents and in zones of upwelling, also influences the richness of calcium carbonate in bottom sediments.

Siliceous ooze, consisting of 30% or more of siliceous tests, is derived from a number of organisms, of which the diatoms and radiolaria are most important. *Diatoms* are one-celled plants, and *radiolaria* are microscopic animals, both of which secrete ornate siliceous tests with radial symmetry (Figure 9.35). Siliceous oozes are found largely in two oceanic zones—between latitudes 45°–60° in both hemispheres, and in limited portions of the equatorial Pacific Ocean.

Rate of accumulation of oozes has been estimated as 1 to 5 cm per 1000 years. Both calcareous and siliceous oozes usually contain substantial amounts of inorganic clays.

Under the heading of biogenic–pelagic sediments, we can also include organic matter that escapes decomposition and becomes incorporated into muds on the floors of deep basins, where anaerobic conditions (the absence of oxygen) prevail. This environment, which is comparatively rare today, is described in Chapter 5 in connection with the origin of petroleum. An example of such a basin is the Black Sea, over 2200 m deep and almost completely cut off from the Mediterranean Sea. Here, under stagnant bottom conditions, oxygen is depleted and a black mud rich in organic matter is deposited. Over the oceans generally, water circulation is adequate to bring oxygen to even the deepest places, allowing destruction of organic matter as it sinks or upon its arrival at the bottom.

Pelagic–detrital sediments are particles of nonbiogenic matter that have settled to the bottom from the near-surface layer above. (Refer to Chapter 5 for an explanation of "detrital.") As we have already explained, such particles may be brought from continental locations as suspended matter in surface ocean currents or in icebergs drifting with those currents. Volcanic and terrestrial dusts carried by winds also furnish detrital matter to the ocean surface, as does the vaporization of meteors in the overlying outer atmosphere. It is obvious that the mineral composition of these detrital sediments can be quite complex, with the proportions of the components depending upon geologic nature of the sources and distances from those sources.

The most widespread of pelagic–detrital sediments is *brown clay*, which is a soft plastic material with a greasy feel. This clay typically is low in calcium carbonate (less than 30%) and consists for the most part of clay minerals, among them illite, chlorite, and kaolinite derived from continental sources. (These minerals are described in Chapter 5.) Montmorillonite may be present and is believed to have been derived by alteration of volcanic materials after their deposition. Quartz in minute grains is abundant in some brown clays, and there may be minor amounts of feldspars and micas, all derived from continental surfaces. Rate of accumulation of brown clay is extremely slow and has been estimated at 0.1 to 1 mm per 1000 years.

The relative abundances of kaolinite and chlorite in brown clay change with geographical position in a manner that reflects conditions of origin of those minerals. Kaolinite, which is produced by silicate rock weathering in warm, humid climates of low latitudes, is 5 to 15 times more abundant than chlorite in the Atlantic Ocean in the latitude range 20° N to 20° S. Chlorite, which is easily destroyed in a warm climate, becomes two to four times more abundant than kaolinite in latitudes poleward of about 45°. Illite, produced from the alteration of micas, shows its greatest abundances near the continents, from which it is derived. Quartz shows concentrations in the lee of tropical deserts, from which it is brought by easterly winds, and in middle to high latitude locations, where it can be attributed to an abundance of freshly pulverized rock produced by Pleistocene continental glaciation.

On deep ocean floors in both Arctic and Antarctic waters, we find sediments that appear to have been brought by icebergs from continental glaciers. Designated as *glacial–marine sediments*, they consist of silt with some clay and are formed of finely ground fresh rock. Consequently, they show little oxidation or mineral alteration. Glacial–marine sediments were evidently far more widespread in area of deposition during stages of glaciation, for they have been found buried beneath globigerina ooze forming today.

As we have already stated, fine volcanic ash travels widely as dust in the atmosphere. Dust from a single great volcanic explosion achieves global distribution and can be expected to produce a very thin pelagic sediment layer simultaneously in many parts of the world ocean. Horizons of volcanic ash have been found in many sediment cores, and there is no doubt that a definable layer up to several hundreds of kilometers in extent can be associated with a single volcanic eruption. Ash horizons are represented in deep-sea sediment by concentrations of glass shards, highly susceptible to mineral alteration.

Bottom-transported detrital sediments have been described earlier and include thick accumulations of turbidites and contourites. Bordering the continents, in a zone along the upper and middle continental slope, are found *terrigenous muds*, apparently brought to the deep ocean floor by bottom currents. *Terrigenous* means "originating on land" and is synonymous with "terrestrial." The terrigenous muds are silty, but finer grained than most turbidites, and their silt and lack of complete oxidation set them apart from brown clay. Colors of the terrigenous muds may be blue, green, black, or red. Blue and green colors result from the presence of ferrous iron oxide and reflect a deficiency of oxygen, or a lack of time for oxidation to have occurred. Red muds, colored by ferric iron oxide, show complete oxidation of iron, but this probably occurred during transport on the lands. Black muds, as we have already noted, have a relatively high organic content and show a depositional environment of stagnant water with little oxygen present.

Hydrogenic sediments of the deep ocean floors include minerals formed by alteration in place or reformed from other minerals. Perhaps the most important alteration product is the clay mineral montmorillonite, described in Chapter 5. This mineral is derived from volcanic materials, including volcanic ash and basaltic rock exposed on the ocean floor. A second alteration product is ***phillipsite***, a silicate mineral of the zeolite group, also derived from volcanic materials of basaltic composition. Phillipsite, a hydrous aluminosilicate of calcium, sodium, and potassium, forms in minute needlelike crystals that may constitute as much as 50% or more of the bottom sediment in parts of the central Pacific Ocean basin, where basaltic volcanic rocks are abundant. This mineral is rare near the bordering continents and is not found in the other oceans.

A great deal of interest centers around the finding of abundant nodules of hydrous manganese and iron oxides exposed on the surface of the deep ocean floors in many places. Referred to as *manganese nodules*, these objects often prove to be thick mineral coatings surrounding nuclei of volcanic rock (Figure 9.36). Manganese nodules are believed to be formed from manganese and iron derived either from detrital sediments of continental origin, or from volcanic rocks of the ocean floor. Although widely distributed over the deep ocean floors, manganese nodules seem not to be present in areas where sediment rich in calcium carbonate is accumulating in abundance.

The deep-sea sediments we have described here in some detail are of great importance for the interpretation of global environmental conditions. Cores of these deposits show alternations of various types of sediment, each reflecting certain physical and chemical conditions of the ocean waters or a certain set of climatic conditions prevailing in the atmosphere. For example, a thick accumulation of turbidites could reflect a glacial stage in which vast amounts of sediment were being brought to the brink of the continental shelf. Particular species of foraminifera are associated with water temperature of a given range. By studying species variations in layers of calcareous ooze, we can derive a record of changing atmospheric temperatures (see Chapter 18).

Coral Reefs and Reef Deposits

Coral reefs are massive rock structures of biogenic origin built close to sea level along coasts situated in warm waters of low latitudes. We inquire into this subject as a part of an investigation of the ocean basins because coral reefs form on the shores of many volcanic islands rising sharply from the deep ocean floor. Here detrital fragments of coral rock contribute sediment to the adjacent ocean floor. Because coral reefs form only in very shallow water, they are important indicators of changes in sea level. Rising or sinking of the crust beneath the ocean floor can be interpreted from reef deposits found today far above or below the level at which they were formed.

The framework of reefs is built by the calcium carbonate secretions of both corals and algae growing vigorously in the surf zone. The living animals and plants build new structures upon old, extending reefs seaward into deeper water or upward to the surface. Wave attack pulverizes exposed coral structures to form a calcareous sediment, which may be deposited in cavities in the reef or spread out on the seaward slope.

Coral reefs are largely limited to the latitude zone between 30° N and 25° S, where water temperatures are at least as high as 20 °C, and usually between 25 °C and 30 °C. Vigorous growth of reef-building corals is limited to the surface-water zone, less than 45 m in depth. Corals thrive where the water is highly agitated as well as free of suspended sediment, but turbid water issuing from the mouths of streams inhibits or prevents reef development. As a result, reef growth is most rapid along exposed coastal positions—as off headlands—and along sides of islands facing into the prevailing direction of wave approach.

FIGURE 9.36 Manganese nodules resting undisturbed on the mid-Pacific ocean floor at a depth of 4500 m. Their average diameter is about 10 cm. (Deepsea Ventures, Inc.)

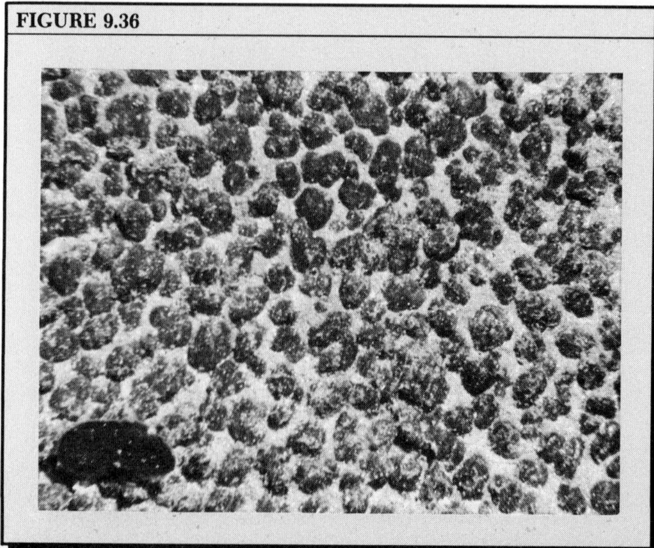

Forms of Coral Reefs

Coral reefs take three basic forms. First and simplest is the *fringing reef*, a shelflike attachment to the land varying in width from 0.5 to 1.0 km or more when well developed (Figure 9.37). As we noted above, fringing reefs are best developed along exposed headlands and may be absent in bays into which fresh water is being brought by streams. The reef surface is remarkably flat and lies at a level about one-third of the tide range below mean high water. The reef surface is thus exposed at low tide, but covered by surf at high tide.

FIGURE 9.37 A fringing reef off the south coast of Java forms a broad bench between the surf zone (*left*) and a white coral-sand beach. Inland is dense rainforest. (Luchtvaart-Afdeeling, Ned. Ind. Leger, Bandoeng.)

FIGURE 9.38 Barrier coral reef, reef islands, and lagoon, Tahaa Island, Society Islands. (U.S. Navy.)

FIGURE 9.37

FIGURE 9.38

A second form is the *barrier reef*, a long, narrow coral embankment lying offshore and enclosing a lagoon between reef and mainland (Figure 9.38). The lagoon may be up to 15 km wide or wider, and the barrier reef up to 1000 m wide. The lagoon is normally 35 to 75 m deep and is flat-floored, but it has numerous stalklike columns of coral that may reach up to the water surface. On the seaward side of the barrier, the submarine surface slopes steeply away into deep water, and coral fragments are spread upon this slope. At intervals along the barrier reef, there are gaps (passes) through which excess water brought into the lagoon by breaking waves is returned to the sea.

A third reef form is the *atoll*, a ringlike reef of coral enclosing only a lagoon of open water (Figure 9.39). The reef and lagoon of an atoll are similar in form and development to the barrier reef and its lagoon. Here and there on the atoll are low islands of coral sand sufficiently large and high to be habitable but vulnerable to inundation in tropical storms. Atolls appear in isolated groups far from any islands of noncoralline rock in the vast expanses of the western Pacific Ocean.

Theories of Coral Reef Development

Of several plausible but divergent theories of the origin of coral reefs, particularly of the fringing reefs and atolls, which have been proposed and debated, we will discuss two here. Earliest and one of the most successful is the *subsidence theory* proposed by Charles Darwin in 1842 as a result of that great naturalist's observations during the voyage of H.M.S. *Beagle*. Darwin's theory was again taken up and strongly supported by the geomorphologist W. M. Davis in the 1920s.

According to Darwin and Davis, a fringing reef is first formed during the slow subsidence of a volcanic island (Figure 9.40). Coral growth, continuing uninterruptedly during subsidence, builds the reef upward, maintaining the reef surface at or close to sea level. Because the reef is built directly upward while the island shore is gradually being inundated, a lagoon is formed and widens as subsidence continues. With continued subsidence, the volcanic island is diminished to a small remnant, then it finally disappears, leaving an atoll lagoon in its place.

More recently, additional support of the subsidence theory has come from Bikini Atoll in connection with seismic refraction studies carried out there after World War II. As shown in Figure 9.41, there is strong indication of the presence of more than 1500 m of calcareous deposits beneath the atoll. A drill hole penetrated 760 m of reef materials identified as formed in shallow water, beneath which is a possible volcanic core. If the seismic data are being correctly interpreted, there is no escape from the conclusion that reef growth kept pace with slow subsidence over a long period. If the reef had been rapidly brought down to depths below 60 m or so, the reef corals would have died and continued subsidence would have resulted in a flat-topped seamount, or guyot.

Holes drilled in 1952 into Eniwetok Atoll penetrated over 1200 m of reef rock to reach a basement of basalt. All of the reef rock is identified as a shallow-water deposit and clearly represents the upbuilding of a submerging reef. Age of the reef rock was found to be progressively greater with depth—that below 850 m was of Eocene age, 36 million years old or more. Sinking of the oceanic crust of this magnitude and duration is of great importance in interpreting the development of the ocean basins, and we shall discuss this subject in Chapter 10.

A second important explanation of reefs is the *glacial-control theory* developed largely by the geologist Reginald A. Daly in the 1930s and 1940s. The theory is based upon an observation made in 1894 to the effect that the surprising uniformity of water depth in coral lagoons, approximately 75 m, might be explained by a low stand of ocean level during worldwide glaciations. Daly added to this point the postulate that the lowering of water temperature greatly reduced, or entirely stopped, coral-reef growth during these low-level stages, permitting wave action to carve broad rock platforms along the coasts and in some cases to bevel small islands completely. Then, as the oceans rapidly warmed, coral growth set in at the margins of the eroded platforms. Reef upbuilding was maintained during the postglacial rise in sea level to reach the present level.

That the worldwide lowering of sea level and reduction of water temperatures occurred during Pleistocene glaciations is inescapable in the light of independent evidence. Whether wave erosion was sufficiently rapid to produce the broad rock platforms required by the glacial-control theory is, however, subject to serious question. Platforms as broad as those required for the larger atolls could not have been eroded in basaltic rock in the short spans of time available. The mechanisms of slow crustal subsidence and glacially controlled changes of sea level must both be taken into account and applied to the degree that evidence dictates in explaining the outward form and internal structure of coral reefs in any given locality.

Oceanic Carbonate Platforms

Related in some respects to atolls of the Pacific Ocean is the Bahamas Platform, a great mass of carbonate rock on which rest the numerous low-lying islands of the Bahama Banks, lying east of the Florida peninsula (Figure 9.42). The Bahama Banks are a group of shallow-water areas bounded by the 200-m depth contour. They extend over a total area about 400 km wide and 1300 km long. Around each bank,

FIGURE 9.39 Rongelap Atoll, Marshall Islands, Pacific Ocean, photographed by astronauts aboard *Gemini V* spacecraft at an altitude of about 240 km. (NASA.)

FIGURE 9.40 Stages in the development of an atoll, according to the subsidence theory. (From W. M. Davis, 1916, *Scientific Monthly*, vol. 2, No. 5, p. 23, Figure 13.)

FIGURE 9.41 Block diagram of Bikini Atoll, Pacific Ocean. (Based on data of M. Dobrin and others.)

FIGURE 9.39

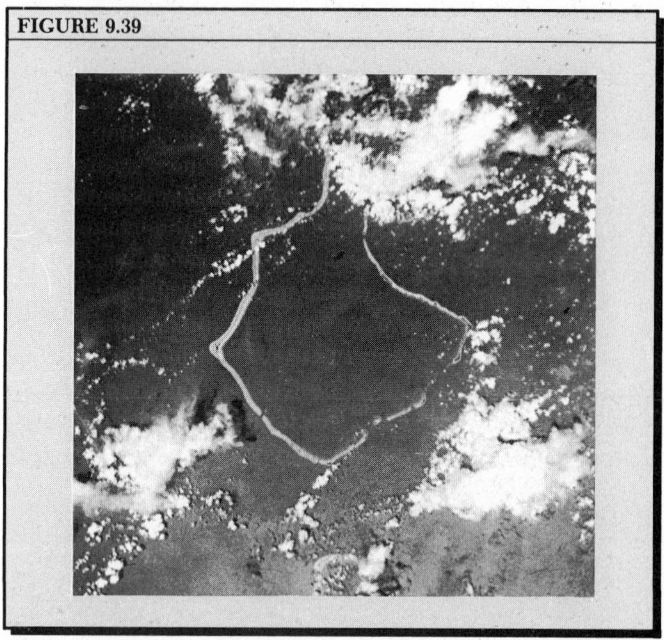

FIGURE 9.40

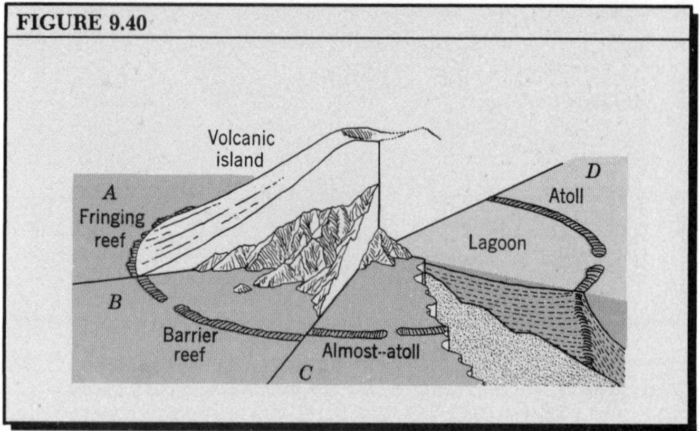

FIGURE 9.41

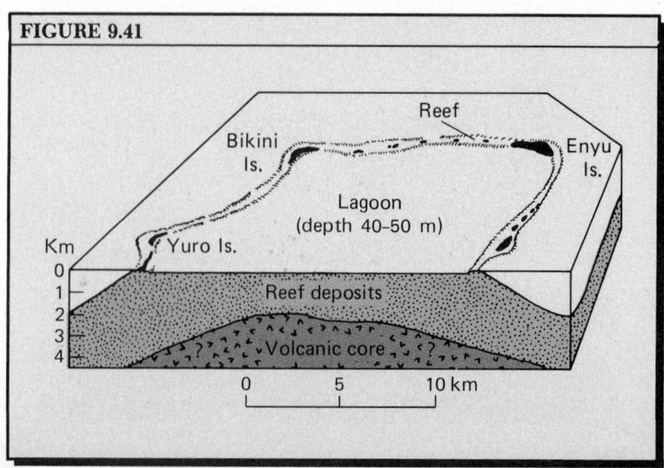

the sea bottom falls off rapidly to the deep ocean floor. On the eastern side, abyssal depths lie close to the banks.

Geologists have been greatly interested in the origin of the Bahama Banks and have studied their composition and structure intensively. Calcareous sediments including lime muds, oolite sands (see Chapter 5), and reefs, are being formed over the banks today. The coral–algal reefs flourish along the eastern sides of the banks where vigorous surf is generated by prevailing easterly winds (trade winds). The modern carbonate sediment layer is only a thin veneer on the banks overlying a thickness of at least 5700 m of carbonate sediments. From deep borings and analysis of core samples, it is known that the oldest carbonate rocks, those at the base of the explored sequence, are of early Cretaceous age (about −130 m.y.). Seismic evidence suggests that carbonate strata may extend down as far as 10 km. It seems an inescapable conclusion that the *carbonate platform* underlying the Bahama Banks has been accumulating for almost as long a time as the Atlantic Ocean basin has been in existence following continental rifting. Because all the sediments were formed in shallow water, there must have been a steady subsidence of the crust as the layers accumulated over a period of some 130 m.y., and the total subsidence would equal the total thickness of carbonate sediments—from 6 to 10 km. (Crustal subsidence of the Atlantic margin is explained in Chapter 13.)

FIGURE 9.42 A generalized relief diagram of the Bahama Banks. (Based on data of Bruce C. Heezen and Marie Tharp.)

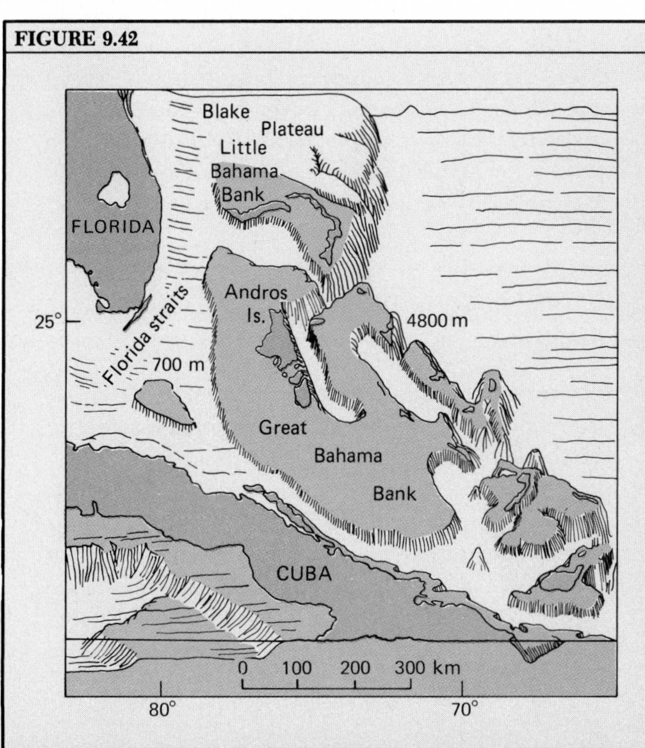

The Rise of Marine Geology

Much of the scientific data we have covered in this chapter is the product of modern research in a 40-year period. Prior to about 1940, most of the ocean floor was *terra incognita*—a vast unexplored region, cold and dark, covering over two-thirds of the global surface. In 1940, plate tectonics was yet to be developed as a theory; it was represented only by an unpopular hypothesis stating that the supercontinent of Pangaea had broken apart about 130 million years ago and that the fragments had somehow drifted to their present positions.

The first outlines of the true configuration of the topography of the ocean floor began to emerge in the late 1940s. Rapidly, this information snowballed into an enormous accumulation of details. At the same time, the sediments of the ocean floor were being explored by coring, then later by seismic profiling, and finally by deep-sea drilling. While this growing body of information on the configuration and sediments of the seafloor added a great deal of evidence to support the emerging theory of plate tectonics, that revolution in geology was decided by means of geophysical information about the crust beneath the ocean floors. In the next chapter, we fill in many details about plate tectonics of the oceanic crust and lithosphere; we will review the crucial evidence on which that theory rests.

Key Facts and Concepts

Ocean basins and continents Ocean basins and continents are the earth's first-order relief features. Area is concentrated in a tablelike continental surface under 1 km elevation, underlain by felsic crust, and over the ocean floors between −3 and −6 km, underlain by mafic crust.

Ocean floor investigation Direct observation and sampling of the ocean floor is carried out by *manned submersibles* and diving vessels (*bathyscaphes*). The *precision depth recorder* produces topographic profiles by *echo sounding.*

Divisions of the ocean floor Major topographic divisions of the ocean floors are (1) continental margins, (2) ocean-basin floors, and (3) the mid-oceanic ridge.

Continental margins The typical passive continental margin consists of a *continental shelf,* bounded by a *continental slope* with *submarine canyons,* grading into a *continental rise* in contact with the deep ocean floor.

Ocean-basin floor The ocean-basin floor includes *abyssal plains* with flat floors, *abyssal hills,* areas of *oceanic rise,* and isolated *seamounts,* some of them flat-topped *guyots.*

The mid-oceanic ridge The *mid-oceanic ridge* extends continuously through the global ocean basins as an elevated zone of rugged relief rising symmetrically to a central *axial rift*, offset by numerous transverse *fracture zones*. The mid-oceanic ridge is an expression of active spreading plate boundaries in oceanic lithosphere.

Trenches and island arcs Long, narrow *trenches*, reaching depths of 7.5 to 9.0 km, lie oceanward of ridgelike volcanic *island arcs* and marginal *mountain arcs*. Trench-arc zones are expressions of converging plate margins with subduction in progress. *Backarc basins* of the Pacific Ocean lie between island arcs and continental margins.

The oceans *Oceanography*, the science of the oceans, is linked to geology through *marine geology*. The *world ocean*, occupying 71% of the earth's surface with mean depth of 4 km throughout the main basins, holds over 97% of the world's free water.

Physical properties of seawater A three-layer system characterizes the temperature structure of the oceans, with a shallow, warm surface layer separated from the deep cold layer by a *thermocline*. Seawater is a *brine* of dissolved mineral solids. *Salinity*, the ratio of dissolved solids to weight of water, ranges generally between 33 and 37 parts per thousand. Dissolved oxygen is high in a surface layer. Seawater density depends on both temperature and salinity, and hydrostatic pressure increases downward in direct proportion to depth.

Ocean currents *Currents* are horizontal water motions produced largely by wind drag over the sea surface and by place-to-place differences in water density. Winds set in motion broad, slow *drift currents;* differences in water density produce *thermohaline currents*. Flow direction is subject to the Coriolis effect, resulting in great *gyres* centered on the subtropical ocean regions. Equatorial currents and the west-wind drift form the latitudinal flows of the gyres while Coriolis causes concentrated poleward flows along the western basin margins. The *antarctic circumpolar current* of the Southern Ocean encircles Antarctica. Converging water motion leads to sinking; divergence to *upwelling*. *Water masses* with distinctive combinations of temperature and salinity form a layered ocean structure. Deep *bottom currents* include *contour currents* flowing over the outer continental rise.

Ocean floor coring and drilling *Coring* obtains sample *cores* of soft ocean-bottom sediment. Deep drilling by the *Glomar Challenger* samples sediments and rock to depths of several hundred meters.

Seismic reflection profiling Seismic reflection is used to obtain profiles showing sediment layering and contacts of sediment with rock. *Ponding* of sediment is revealed in bedrock basins. Trenches show wedgelike sediment masses.

Terrestrial sediment sources Sediment from terrestrial sources reaches the deep ocean floor after transport by currents and winds. Wind-transported sediment consists of mineral dusts from soil and volcanic dust. Icebergs transport assorted sizes of rock debris to deep-ocean locations.

Turbidites and contourites *Turbidity currents*, a form of *density current*, carry terrestrial sediment down the continental slope and rise to reach abyssal depths, forming *turbidites* that show *graded bedding*. Submarine canyons are scoured by turbidity currents that may construct large *submarine fans* (*cones*) located opposite the mouths of major rivers. Contour currents paralleling the continental margins deposit sandy *contourites* formed by winnowing of turbidites.

Deep-sea sediments Four main classes of deep-sea sediments are *biogenic–pelagic*, *pelagic–detrital*, *bottom-transported detrital*, and *hydrogenic*. *Plankton* living in shallow depths furnish biogenic–pelagic sediment in the form of *tests*, accumulating as *deep-sea ooze*. *Calcareous ooze* formed of carbonate tests accumulates in depths usually less than 3.5 to 4 km; *siliceous ooze* consisting of *diatoms* and *radiolaria* accumulates largely in middle-latitude zones. Pelagic–detrital sediments include *brown clay* rich in clay minerals, *glacial-marine sediments* of relatively unaltered minerals, and volcanic ash. Bottom-transported detrital sediments include turbidites, contourites, and *terrigenous muds*. Hydrogenic sediments, formed in place by chemical precipitation and alteration, include zeolites and *manganese nodules*.

Coral reefs Coral reefs form in warm, shallow water close to shorelines of low latitudes. Coral and algae construct the reef framework, taking the form of *fringing reefs*, *barrier reefs*, and *atolls*. The *subsidence theory* of upbuilding accompanying crustal subsidence explains atolls developed from volcanic seamounts. *Carbonate platforms* have accumulated during long-continued ocean-floor subsidence accompanied by deposition of calcareous sediments.

Questions and Problems

1. The idealized model of an ocean basin with its mid-oceanic ridge almost perfectly centered between the continental margins on both sides applies best to the North and South Atlantic Ocean basins. How does plate tectonics explain this remarkable basin symmetry? In contrast, in the Pacific Ocean basin, the mid-oceanic ridge (East Pacific Rise) lies far over toward the eastern side, off South America. How does plate tectonics explain this situation? Why is there no mid-oceanic ridge between South America and Antarctica?

2. Calculate the confining pressure of seawater at the ocean floor at a depth of 5 km. State the answer in kilopascals. (Hint: As a rule of thumb, the confining pressure in decibars is approximately equal to the water depth in meters. One *bar* is defined as 100,000 pascals.)
3. Investigate the Coriolis effect. Toward which direction (right or left) is a fluid in horizontal motion turned in the northern hemisphere? Where on the globe is the Coriolis effect greatest? Where least? Where zero? How does the Coriolis effect influence the oceanic gyres? Explain.
4. Suppose that you are preparing to examine a submarine core that has just been obtained from a location on the flank of the mid-oceanic ridge. You know that the sample point lies in a basinlike depression bounded by high hills or scarps. What class or type of sediment might you expect to find in the core? Name some kinds of deep-sea sediment that might accumulate at such a location.
5. In an interval of one million years, would the layer of turbidites deposited on a continental rise be thicker or thinner than the layer of brown clay deposited in the same time span at a location far from any continent? Would the difference in thickness be great or small? Explain.
6. Using information on seismic exploration described in Chapter 8, how might a scientist investigate the structure and origin of an atoll? Would it be possible to distinguish a contact of reef limestone with an underlying basalt layer? If the reef limestone deep in the atoll has been altered to dolomite, could it be readily distinguished from basalt by seismic investigation? Explain.

10

CHAPTER TEN

Tectonics of the Oceanic Lithosphere

Beginning in Chapter 1, we sketched the broad outlines of plate tectonics. This widely accepted general theory brings together a great variety of geologic forms and processes into a unified framework. In subsequent chapters, we introduced new facets of plate tectonics, and we elaborated upon them to show their value in explaining rock types, rock structures, and tectonic processes. But many details remain to be examined, and a great deal of scientific evidence remains to be presented. In this chapter and the three which follow, we present those details as they apply to the oceanic and continental lithosphere.

A study of the oceanic lithosphere is a study in present-day processes of geology; it is a study of new features of the solid earth in the process of being formed. In contrast, a study of rocks and structures of the continental lithosphere is largely a study of what has already happened through the activity of plate tectonics during all of recorded geologic time.

Geologists have long been influenced by the concept, explained in Chapter 6, that "the present is the key to the past." In keeping with this concept, we can say that the present-day tectonic activity of the oceanic lithosphere and its plate boundaries is the key to the geologic past—that past being represented by the continental lithosphere. Many rock features that puzzled geologists for decades have been solved by the key of the present. For example, ancient turbidites found on the continents defied explanation until geologic research on the ocean basins revealed the conditions under which turbidites are now being deposited.

Plate Boundaries and Their Junctions

As presented in Chapter 1, the boundaries of lithospheric plates fall into three types: (1) spreading plate boundaries, (2) converging boundaries where subduction is in progress, and (3) transform-fault boundaries. Each plate is an independent fragment of a spherical earth shell, completely surrounded by other plates. Apparently, the nature of the forces that drive plates is such that several major plates are required to be present on the globe simultaneously. Their sizes and outlines are quite varied, and they change in both size and shape as time passes. Entire plates can disappear by subduction and new plates can form by rifting processes.

To familiarize yourself with the geometry of spherical plate systems on a globe (and have some fun at the same time), get hold of a half-dozen well-worn tennis balls. With a very soft lead pencil, experiment with drawing various shapes and numbers of plates that might conceivably exist on a hypothetical globe. One rule you must follow is that every boundary shared by two plates (a plate-boundary segment) terminates in a common triple junction with two other boundary segments. A quadruple or quintuple junction is not allowed. For the time being, draw all plate boundaries as single lines, without any distinction as to the type of boundary each represents. Later, after you have become acquainted with the possible kinds of boundary junctions, you may wish to mark the boundaries as to their types.

Figure 10.1 illustrates some results. As you can see from these samples, models with as few as two, three, and four symmetrical plates can be imagined. The four-plate model is based on the geometry of a tetrahedron, while the six-plate model consisting of four-sided plates is based on the geometry of the cube or rhombohedron. The 12-plate system with pentagonal plates is based on a common crystal form known as a pyritohedron (commonly seen in the mineral pyrite).

The Global Plate System

The first global map of lithospheric plates appeared in published form in 1968, and there have since been many minor changes and revisions in the map. Boundaries have been relocated and new plates identified; differences have appeared in the naming of plates, as well. Today, however, a fairly good consensus exists in the geologic community as to the numbers and names of the major plates, the nature of their boundaries, and their relative motions. Differences of interpretation persist in many boundary details. Also, a few sections of certain plate boundaries are of uncertain classification or location.

For a particular lithospheric plate to be identified and named, its boundaries should all be active. In other words, there must be good evidence of present or recent relative motion between the plate and all its contiguous (adjoining) plates. A plate is uniquely defined, or set apart, by its boundary segments and triple junctions. A *plate boundary segment* is the continuous line of separation between two plates and is terminated at both ends by a triple junction. A *triple junction* is the common point of meeting of three boundary segments, or the common point at which three

FIGURE 10.1 Six ideal symmetrical plate systems inscribed on a sphere. See Figure 10.4 for explanation of symbols. (Drawn by A. N. Strahler.)

FIGURE 10.1

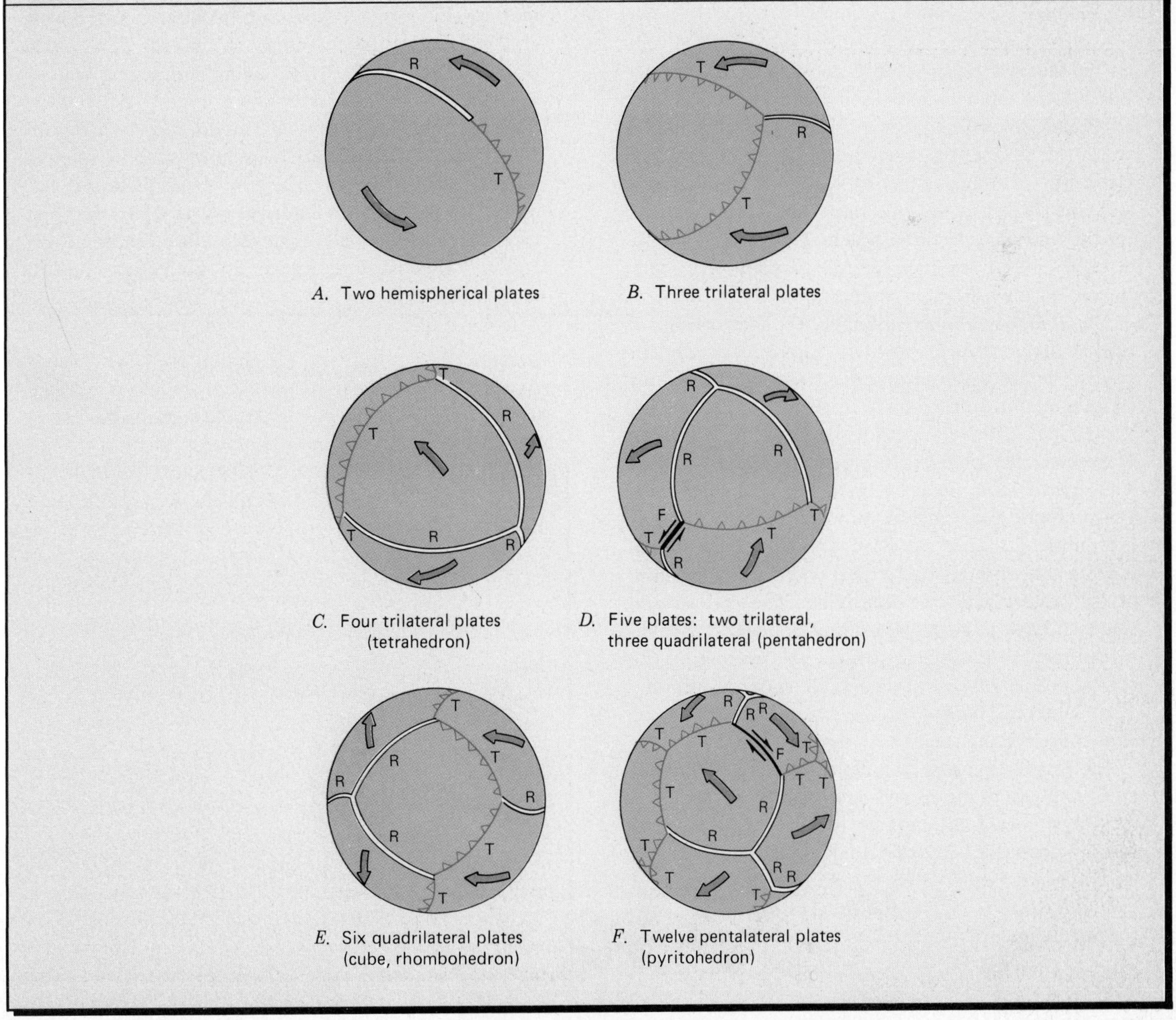

A. Two hemispherical plates

B. Three trilateral plates

C. Four trilateral plates (tetrahedron)

D. Five plates: two trilateral, three quadrilateral (pentahedron)

E. Six quadrilateral plates (cube, rhombohedron)

F. Twelve pentalateral plates (pyritohedron)

plates are in contact. It follows that for a given plate the number of boundary segments must equal the number of triple junctions. Figure 10.2 shows six forms of plates in order of increasing numbers of boundary segments and triple junctions, from the simplest, or lenticular type, with only two boundary segments, to the complex type, combining one or more lenticular forms with a simple polyhedral form.

While the lenticular type, with its two boundary segments, sets the lower limit of simplicity, there seems to be no upper limit to the number of boundary segments. The two largest plates (Pacific, American) each have 10 boundary segments and 10 triple junctions. Table 10.1 lists 12 global plates in order of numbers of boundary segments. The median number is five, and this might lead us to guess that there is a tendency for an "ideal" plate to be pentagonal. This speculation would seem to favor a simple global model of 12 pentagonal plates, as shown in Figure 10.1*F*.

Another requirement of an active lithospheric plate is that it must move as a unit; the surface track of its motion over the spherical globe must be an arc of a circle. This is a subject we will elaborate upon later in the chapter.

A given boundary segment need not consist of only one boundary type (spreading, converging, transform). Instead, the segment may be made up of two boundary types joined in sequence. Many plate boundaries consist of alternating short sections of spreading and transform boundary meeting at about right angles, and this pattern is the rule rather than the exception for the mid-oceanic ridge system. Converging boundaries are often strongly curved in plan-form (map trace) and may show serpentine bends.

The global system of lithospheric plates consists of twelve major plates, shown in Figure 10.3. Of the twelve, six are plates of enormous size; the remaining six range from intermediate in size to comparatively small. Geologists have identified and named a number of even smaller plates within the twelve major plates. The great Pacific plate occupies much of the Pacific Ocean Basin and consists almost entirely of oceanic lithosphere. Its relative motion is northwesterly, so that it has a converging (subduction) boundary along most of its western and northern edge. The eastern and southern edge is mostly spreading boundary. A sliver of continental lithosphere is included, making up the coastal portion of California and all of Baja California. The California portion of the plate boundary is the San Andreas Fault, an active transform fault.

The American plate includes most of the continental lithosphere of North and South America, as well as the entire oceanic lithosphere lying west of the mid-oceanic ridge (Mid-Atlantic Ridge) that divides the Atlantic Ocean Basin down the middle. For the most part, the western edge of the American plate is a converging boundary with active subduction extending from Alaska through Central America to southernmost South America. Some geologists recognize a North American plate and a South American plate, with the boundary running east-west at about latitude 15° N. That boundary is not, however, tectonically active and is simply an arbitrary line.

The Eurasian plate is largely continental lithosphere, but is fringed on the east and north by a belt of oceanic lithosphere. The African plate can be visualized as having a central core of continental lithosphere nearly surrounded by oceanic lithosphere. The Austral-Indian plate (also called the Australian plate) takes the form of an elongate rectangle. It is mostly oceanic lithosphere, but contains two cores of continental lithosphere—Australia and peninsular India. The Antarctic plate has an elliptical outline and is almost completely enclosed by a spreading boundary. The continent of Antarctica forms a central core of continental lithosphere completely surrounded by oceanic lithosphere.

Of the remaining six plates, the Nazca and Cocos plates of the eastern Pacific are rather simple fragments of oceanic lithosphere bounded by the Pacific mid-oceanic ridge, a spreading boundary, on the west and by a converging boundary on the east. The Philippine plate is noteworthy as having converging boundaries on both eastern and western edges. The Arabian plate resembles the "sunroof" model shown in Figure 1.9; it has two transform boundaries and its relative motion is northeasterly. The Caribbean plate also has important transform boundaries on parallel sides. The tiny Juan de Fuca plate is steadily diminishing in size and will eventually disappear by subduction beneath the American plate.

FIGURE 10.2 Six plate outlines with their triple junctions. Except for the complex type (*F*) the number of triple junctions is equal to the number of plate boundary segments.

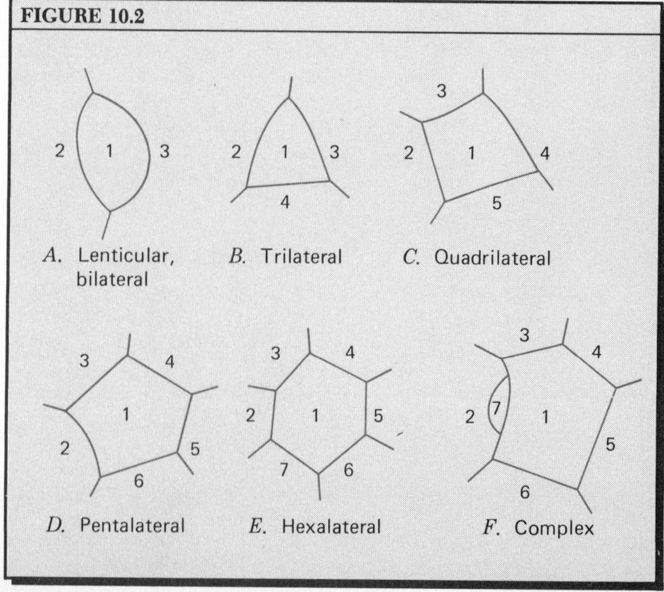

Table 10.1 The Lithospheric Plates

	Plate name	Number of boundary segments and triple junctions
The Great Plates	Pacific	10
	American	10
	Eurasian	7
	Persian subplate	
	Southeast Asian, or China subplate	
	African	5
	Somalian subplate	
	Austral–Indian	5
	Antarctic	5
	Nazca	4
	Cocos	5
	Philippine	2
	Caribbean	2
	Arabian	4
	Juan de Fuca	2
	Total	61

Geologists recognize one or more subplates within certain of the major plates. A *subplate* is a plate of secondary importance set apart from the main plate by a boundary that is uncertain or questionable, either as to its true nature or the level of its activity. An example is the Somalian subplate of the African plate. It is bounded by the East African Rift Valley system and there is good reason to think that this portion of the African plate is beginning to split apart and will become an independent plate (see Chapter 13).

Triple Junctions

Given all possible combinations of three boundary types, irrespective of the order in which they are listed, triple junctions are limited in theory to ten basic varieties, shown in Figure 10.4. The three boundary types are assigned a letter: R (Rift)—spreading boundary; T (Trench)—converging, subduction, or consuming boundary; F (Fault)—transform-fault boundary. In stating the three-letter code for a given triple junction, we always take the letters in the order R–T–F. Of the ten possible combinations, one must be ruled out because the nature of transform faulting requires that plates must move in the same direction as the fault line. Thus, combination FFF is eliminated, leaving nine possible triple junctions. Six examples are shown in Figure 10.5. Note that a triple junction can take the form of either a "Y" or a "T," since the only requirement is that three plates meet at a common point. The "T" form is the more common of the two.

Triple junctions, like the plates that form them, can move over the lower mantle and disappear by subduction beneath adjacent lithospheric plates. New triple junctions come into existence by rifting or by the formation of a new subduction boundary. A triple junction of one type may evolve into a triple junction of another type.

Table 10.2 lists the 20 triple junctions that form the common corners of the 12 plates we recognized in Table 10.1 and on our plate maps. For convenience, each junction is assigned a reference number, which appears on the maps of Figure 10.3. The numbers have no significance except to help you keep track of the junctions and correlate one map with another.

Table 10.2

Arbitrary I.D. number	Variety	Plates meeting at junction
1	RRF or RFF	Pacific, Antarctic, Austral–Indian
2	RFF (?)	Pacific, Philippine, Austral–Indian
3	TFF (?)	Eurasian, Philippine, Austral–Indian
4	TTT	Pacific, Eurasian, Philippine
5	Undetermined	Pacific, Eurasian, American
6	RTF	Pacific, American, Juan de Fuca
7	TFF	Pacific, American, Juan de Fuca
8	RTF	Pacific, American, Cocos
9	RRR	Pacific, Cocos, Nazca
10	RRF or RFF	Pacific, Nazca, Antarctic
11	TTF	American, Cocos, Caribbean
12	TTF (?)	American, Nazca, Caribbean
13	TTF	Cocos, Nazca, Caribbean
14	RTT	American, Nazca, Antarctic
15	RFF	American, African, Antarctic
16	RRF or RFF	American, African, Eurasian
17	TFF	African, Eurasian, Arabian
18	RFF	African, Arabian, Austral–Indian
19	TFF	Arabian, Eurasian, Austral–Indian
20	RRR	African, Antarctic, Austral–Indian

FIGURE 10.3 A

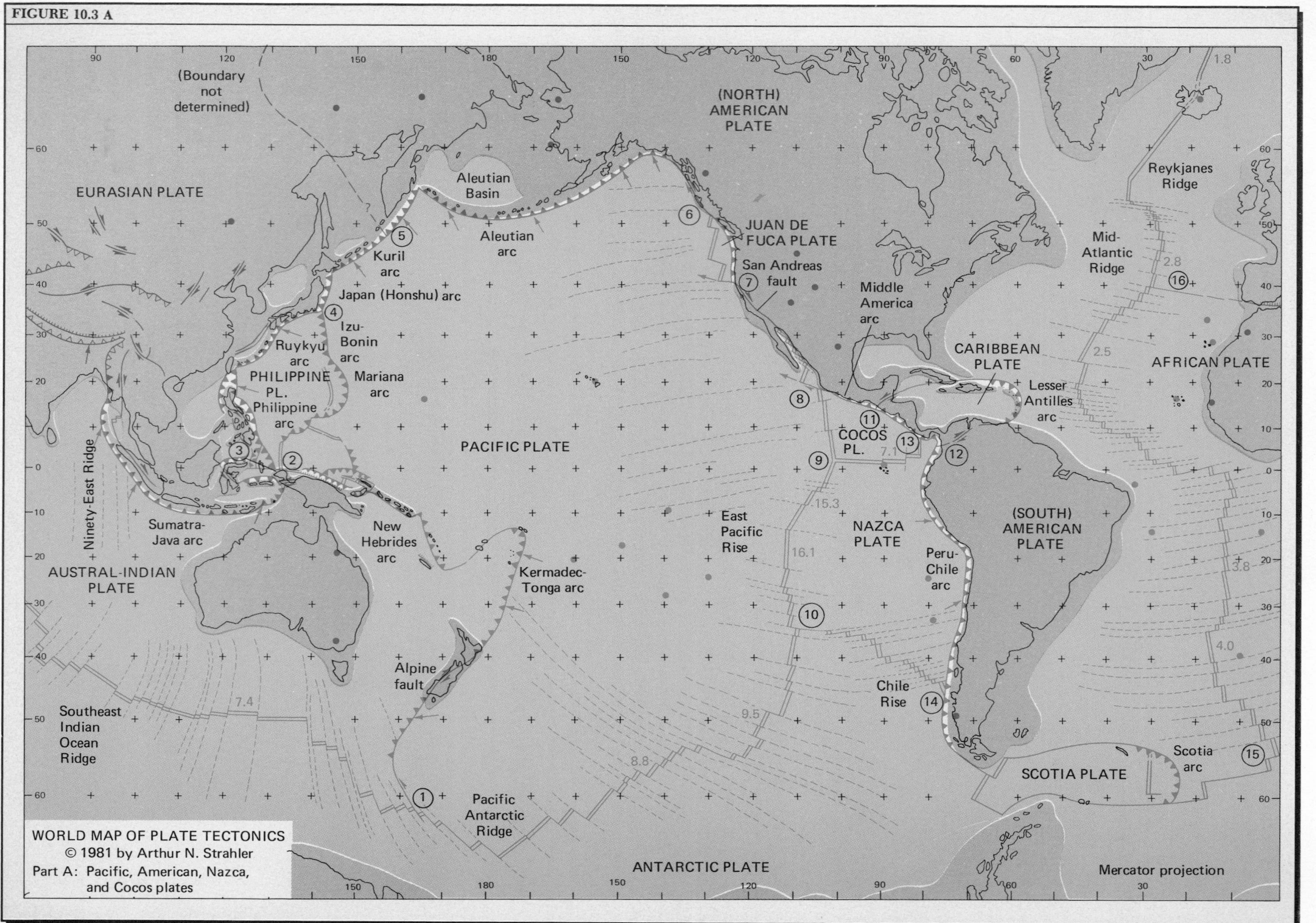
(Boundary not determined)
EURASIAN PLATE
(NORTH) AMERICAN PLATE
Aleutian Basin
Aleutian arc
Kuril arc
Japan (Honshu) arc
Izu-Bonin arc
Ruykyu arc
PHILIPPINE PL.
Philippine arc
Mariana arc
JUAN DE FUCA PLATE
San Andreas fault
Middle America arc
CARIBBEAN PLATE
Lesser Antilles arc
Mid-Atlantic Ridge
Reykjanes Ridge
AFRICAN PLATE
PACIFIC PLATE
COCOS PL.
NAZCA PLATE
East Pacific Rise
(SOUTH) AMERICAN PLATE
Peru-Chile arc
Chile Rise
SCOTIA PLATE
Scotia arc
Ninety-East Ridge
Sumatra-Java arc
AUSTRAL-INDIAN PLATE
New Hebrides arc
Kermadec-Tonga arc
Alpine fault
Southeast Indian Ocean Ridge
Pacific Antarctic Ridge
ANTARCTIC PLATE
Mercator projection
WORLD MAP OF PLATE TECTONICS
© 1981 by Arthur N. Strahler
Part A: Pacific, American, Nazca, and Cocos plates

FIGURE 10.3 B

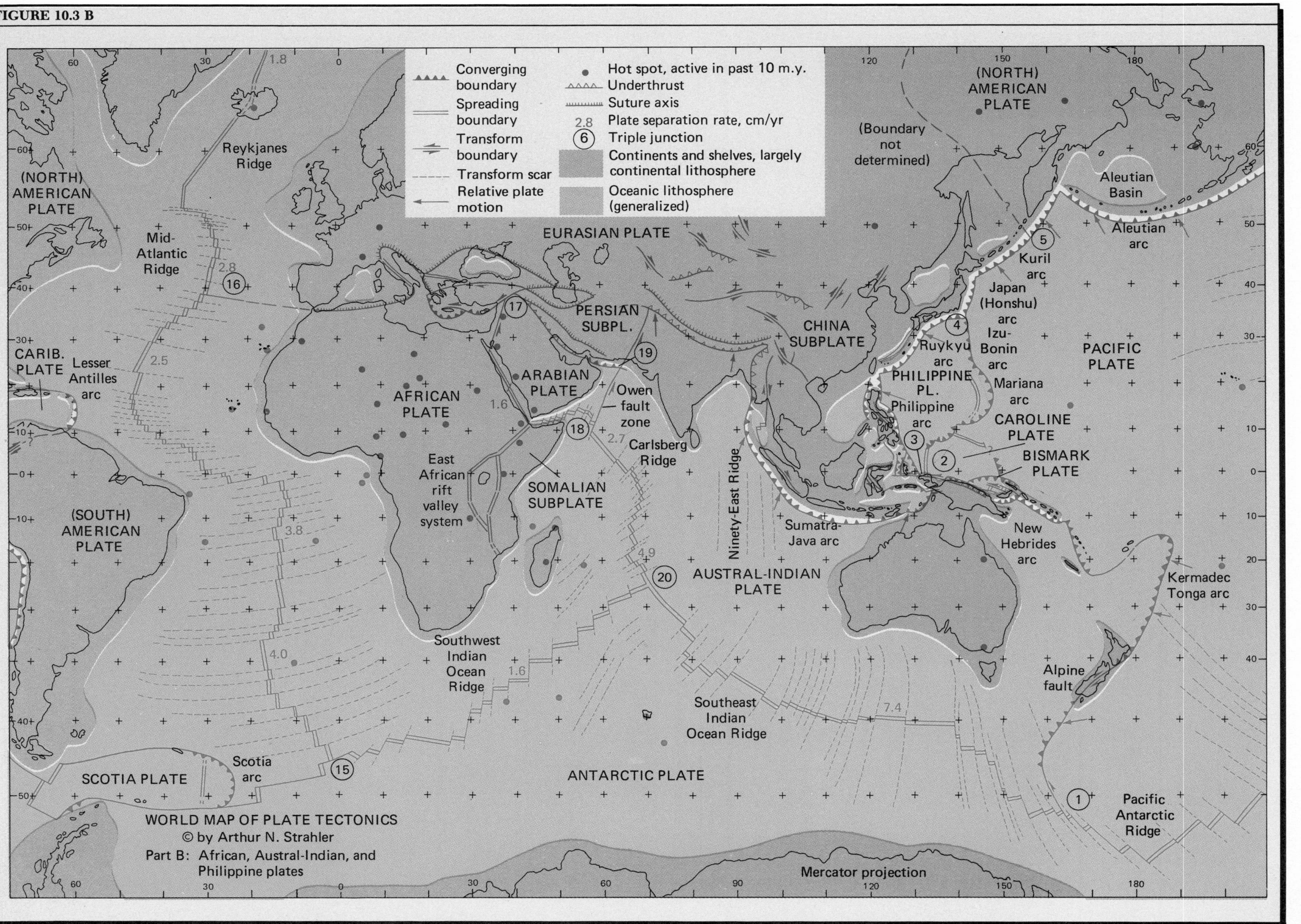

Converging boundary
Spreading boundary
Transform boundary
Transform scar
Relative plate motion
Hot spot, active in past 10 m.y.
Underthrust
Suture axis
2.8 Plate separation rate, cm/yr
6 Triple junction
Continents and shelves, largely continental lithosphere
Oceanic lithosphere (generalized)
(NORTH) AMERICAN PLATE
Reykjanes Ridge
Mid-Atlantic Ridge
CARIB. PLATE
Lesser Antilles arc
(SOUTH) AMERICAN PLATE
SCOTIA PLATE
Scotia arc
EURASIAN PLATE
PERSIAN SUBPL.
CHINA SUBPLATE
AFRICAN PLATE
ARABIAN PLATE
Owen fault zone
Carlsberg Ridge
East African rift valley system
SOMALIAN SUBPLATE
Southwest Indian Ocean Ridge
Ninety-East Ridge
Sumatra-Java arc
AUSTRAL-INDIAN PLATE
Southeast Indian Ocean Ridge
ANTARCTIC PLATE
(Boundary not determined)
Kuril arc
Japan (Honshu) arc
Ruykyu arc
Izu-Bonin arc
Mariana arc
PHILIPPINE PL.
Philippine arc
CAROLINE PLATE
BISMARK PLATE
PACIFIC PLATE
Aleutian Basin
Aleutian arc
New Hebrides arc
Kermadec Tonga arc
Alpine fault
Pacific Antarctic Ridge
Mercator projection
WORLD MAP OF PLATE TECTONICS
© by Arthur N. Strahler
Part B: African, Austral-Indian, and Philippine plates

FIGURE 10.3 C

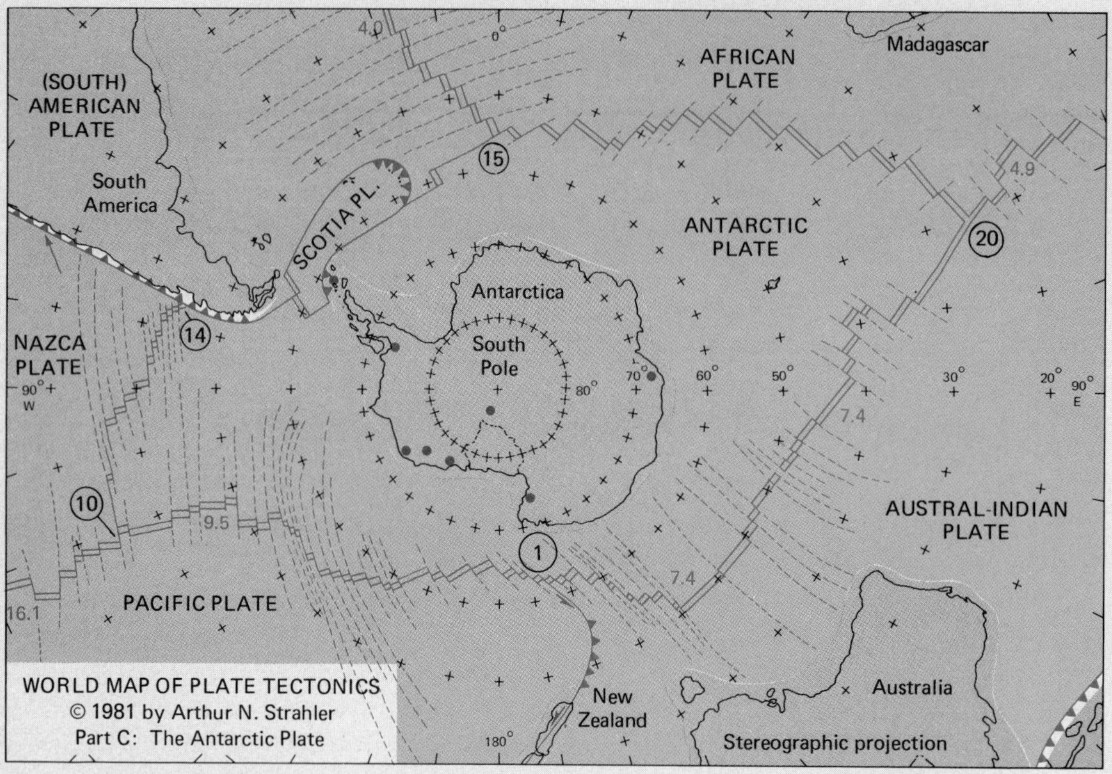

(SOUTH) AMERICAN PLATE
South America
SCOTIA PL.
AFRICAN PLATE
Madagascar
ANTARCTIC PLATE
Antarctica
South Pole
NAZCA PLATE
PACIFIC PLATE
AUSTRAL-INDIAN PLATE
Australia
New Zealand

WORLD MAP OF PLATE TECTONICS
© 1981 by Arthur N. Strahler
Part C: The Antarctic Plate

Stereographic projection

FIGURE 10.3 D

Europe
Mid-Atlantic Ridge
Iceland
Reykjanes Ridge
North Atlantic Ocean
EURASIAN PLATE
Barents Shelf
Greenland
Labrador
Nansen Basin
Nansen-Gakell Ridge
North Pole
Fram Basin
Lomonosov Ridge
Makarov Basin
Alpha Ridge
Siberia
Verkhoyansk Mts.
Canada Basin
Canada
Siberian Shelf
(NORTH) AMERICAN PLATE
(Boundary not determined)
Alaska
Bering Sea
Juan de Fuca plate
Kuril arc
Aleutian arc
PACIFIC PLATE

WORLD MAP OF PLATE TECTONICS
© 1981 by Arthur N. Strahler
PART D: The Arctic Ocean Basin
Stereographic projection

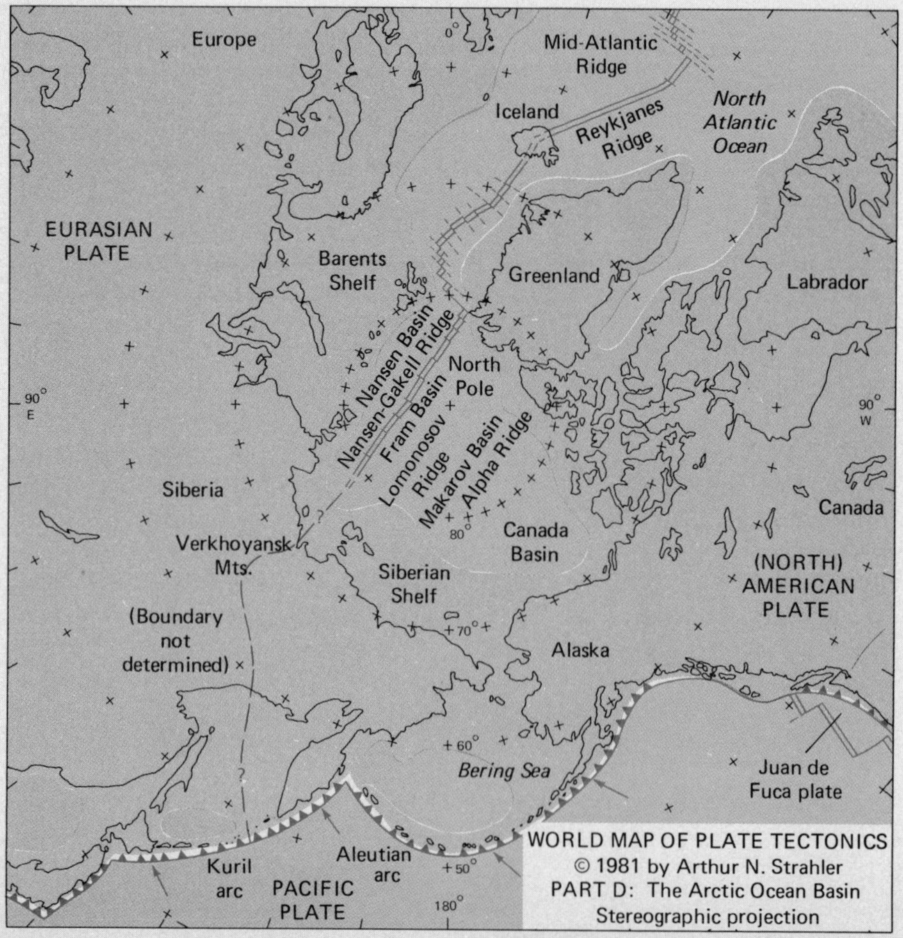

Three of the junctions (1, 10, and 16) are listed as "RRF or RFF." Choice of one of these two possibilities depends on how much importance is given to the possible presence of a short transform fault offsetting a spreading boundary. Junction 5 is said to be "undetermined"; it cannot be identified by code because one of the three boundaries is of uncertain position and classification. The uncertain boundary, which separates the Eurasian and American plates, may cut through eastern Siberia, but there is no clear line of seismic activity by which it can be located. As we shall explain, active plate boundaries are located and defined by their seismic activity, indicating stick-slip movement between adjoining plates.

FIGURE 10.3 Maps of lithospheric plates. (Based on data of W. C. Pitman III, and others, 1974, *Magnetic Lineations of the Oceans,* copyright © 1974, by the Geological Society of America. Many other sources also used.)

FIGURE 10.4 The triple junctions. Alternate forms, (*a*) and (*b*), are given for some of the types.

The Geometry of Transform Faults

The concept of a transform fault eluded geologists during the early period of discovery that finally led to a full-blown theory of plate tectonics. Perhaps the concept is also elusive to many geology students, so we must make a special effort to show how transform faults work. The simplest concept of a transform fault is one we presented in Chapter 1, modeled after the sliding "sunroof" or the rolltop desk (Figure 1.9). Let us designate transform faults like this as ***Type 1.*** A neat example seems to exist in the case of the Arabian plate (Figure 10.3). This plate is being formed by accretion on a spreading boundary located in the Red Sea and Gulf of Aden, while at the same time it is being consumed by subduction beneath the Eurasian plate. Two Type 1 transform faults make up the side boundaries of the plate.

A second type of transform fault, ***Type 2,*** occurs between the offset ends of a spreading boundary and is illustrated in Figure 10.6*B*. We have already called

FIGURE 10.4

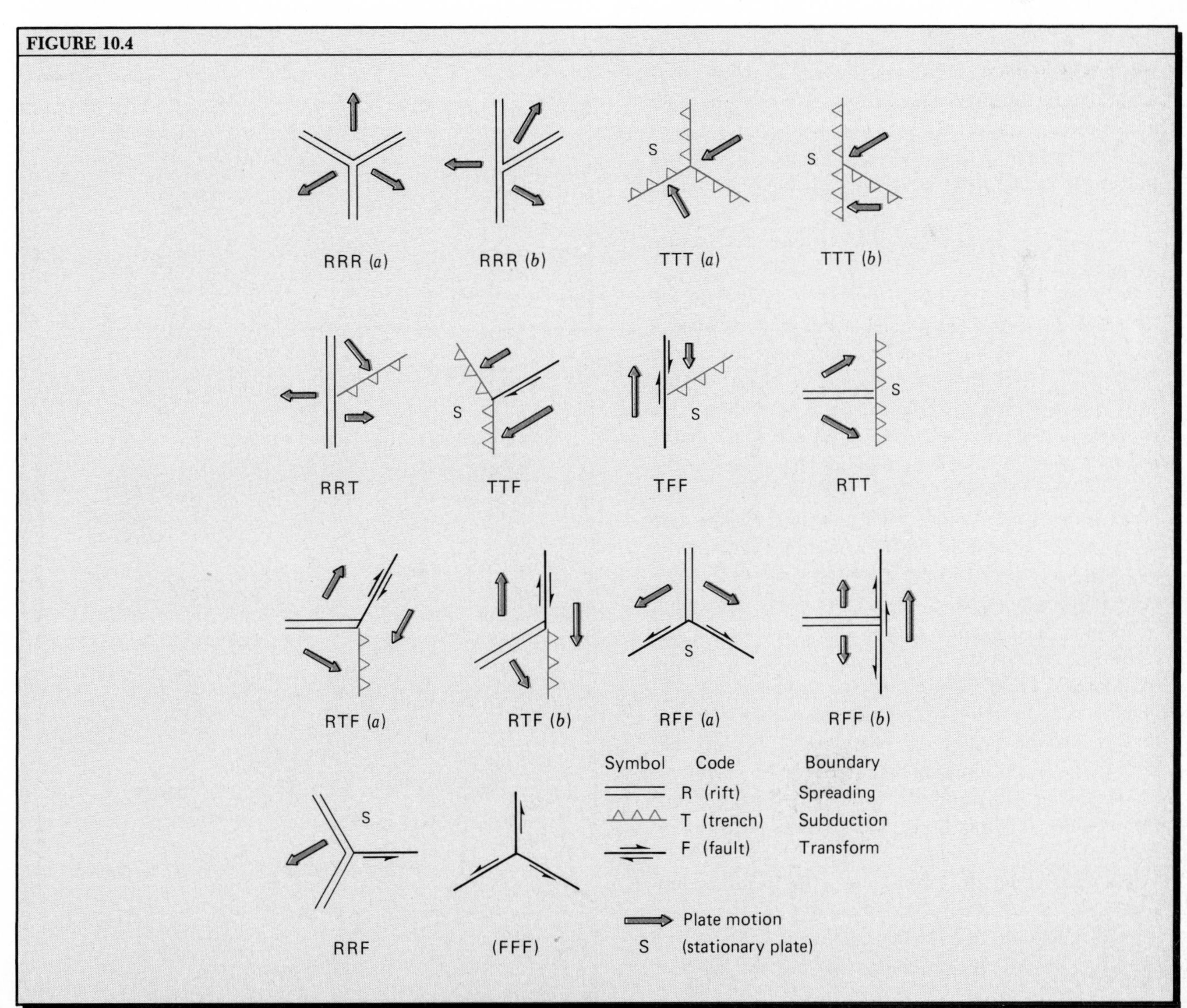

attention to the numerous transverse fracture zones that cut across the mid-oceanic ridge throughout much of its length. By inspection of a map showing the offset of the ridge at each transverse fracture, many geologists jumped to an obvious conclusion: "We have here a lot of simple transcurrent faults." In Chapter 6, we defined a transcurrent fault in strict geometrical terms as a fault in which all motion is horizontal and in the direction of the strike of the fault plane. If the transverse fractures of the mid-oceanic ridge are indeed simple transcurrent faults, the direction of relative movement on the fault will be as shown by arrows in Figure 10.6*A*. The entire plate south of the fault moves toward the right; the entire plate north of the fault moves toward the left (i.e., it is a left-lateral transcurrent fault).

While this simple interpretation of transverse fractures may have seemed adequate, it had some unacceptable consequences. What happens at the two ends of the fault? This question aroused the curiosity of a Canadian geologist and geophysicist, Professor J. Tuzo Wilson. If the fault movement is continuous along the entire fracture line, as it must be between rigid plates, how can the fracture line simply end in a point of zero relative movement? You can get the point of Wilson's argument by taking a sheet of paper and cutting a slit across the middle part. If you try to move one part of the sheet parallel with the other along the slit, you will find it impossible to do so without buckling or tearing the paper. Clearly, the simple transcurrent fault model is not workable in the case of the transverse fracture zones.

Professor Wilson in 1965 found the answer to the puzzle of the transverse fracture zones by postulating a new type of fault, which he named the transform fault. Its relative motion is shown in Figure 10.6*B*. First, we postulate that active plate spreading is in progress on the ridge axis both north and south of the fracture zone. A single lithospheric plate lies to the right of the ridge axis on both sides of the fault; another plate is on the left side, as shown by the two colors in the diagram. Shearing motion between the two plates occurs only on segment *b–c* of the fracture zone lying between the offset ends of the spreading boundary. As the small arrows show, motion on the transform fault is the opposite of that in the simple transcurrent fault. There is no movement along segments *a–b* and *c–d* of the fracture zones. Keep in mind that along the spreading boundaries new oceanic lithosphere is continuously being formed by accretion. Motion along the transform fault is necessary to accommodate the plate motion required by plate accretion.

But what about the extensions of the fracture line across the ocean floor on both sides of the offset spreading boundaries? If you visualize the progress of spreading at point *b*, you realize that the line segment *a–b* is a trail or scar left by the end of the transform fault and impressed on the surface of the plate. We can call these extensions of the transverse fractures *healed transform faults* or *transform scars*. (They have also been called *extinct transforms*.) They appear as narrow ridges or scarps on the ocean floor because the topographic features on the two sides on the line are not matched. Also, it seems likely that basaltic magma, rising in dikes along the active transform fault, builds up the ocean floor close to the fault, giving a ridgelike form to the healed fault.

As we suggested in Chapter 8, the proof of Professor Wilson's transform fault hypothesis was to be looked for in seismic activity. Under the older, transcurrent-fault hypothesis, seismic activity should be observed along the entire length of the fracture zone. Under the transform-fault hypothesis, seismic activity will be limited to the transform segment only, and, of course, to the axial rift of the spreading boundary segments. We will present the seismic evidence later in this chapter.

FIGURE 10.5 Examples of triple junctions. (See Figure 10.4 for explanation of symbols.)

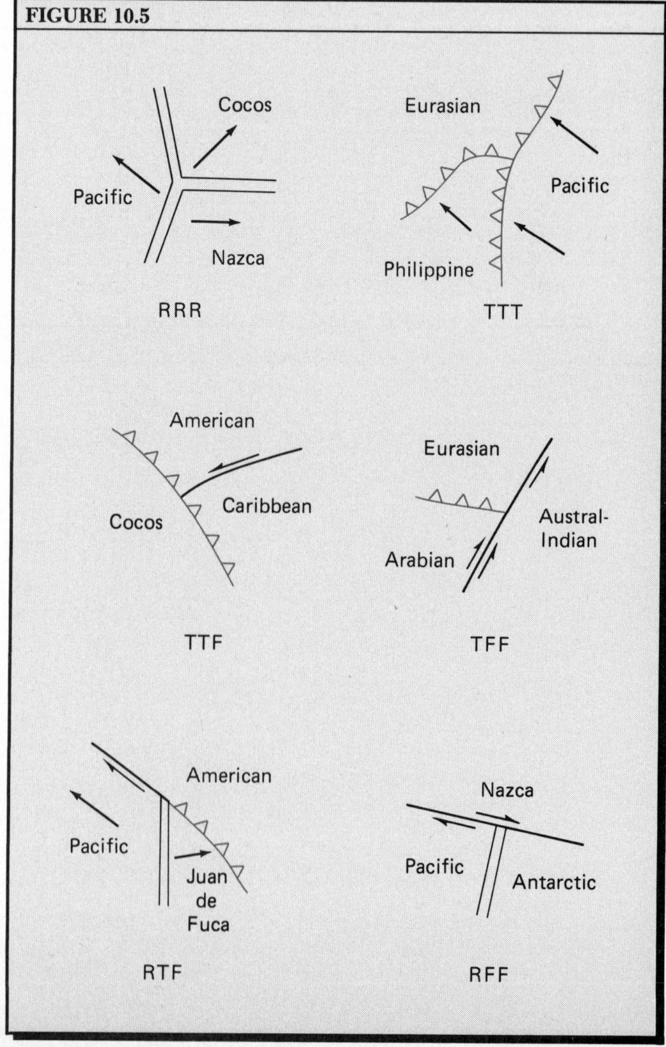

The transform fault that occurs between offset ends of a spreading boundary can be designated as *Type 2*. What then is *Type 3*? A similar situation can exist between offset ends of a converging plate boundary—a subduction zone—as shown in Figure 10.6*C* and *D*. In this case, there are no continuations of the fault into the adjacent plates so that no submarine fracture zones would be formed. The subducting plate edges may both belong to the same plate and dip in the same direction, or both plates may undergo subduction, with downbending plate margins dipping in opposite directions.

Of the two transform types related to offset plate boundaries, Type 2, the offsetting of spreading boundaries, is by far the more common. On the eastern boundary of the Pacific plate, no less than 60 offsets can be counted, each with an active transform fault (see Figure 10.3).

Plate Rotation

A lithospheric plate is a fragment of a spherical shell that moves as a unit over a complete spherical surface. If you cut apart an old tennis ball, making small pieces of it, you can slide these pieces over the surface of a second ball. In so doing, you will observe that the center point of the small piece must follow a circular path as it moves (Figure 10.7). You might then attach a thread to a point near the center of the small piece and anchor the other end of the thread to a pin driven into the complete ball. As you move the small piece, the taut thread serves as a radius controlling the circular track of the small piece. The pin in this model represents the *pole of rotation* of the plate. Note that the "pole" described here is in no way related to the pole of earth rotation, although the two might by barest chance happen to coincide.

To make our model more nearly complete, we should use a pin long enough to pass completely through the ball along its true diameter. When this is done, the pin represents the *axis of rotation* of the lithospheric plate. When we give the exact latitude and longitude of the pole of rotation of a given plate and describe the angular distance, or arc, through which it turns per time unit, we have exactly described its motion. The radius of plate motion is always a *great circle*, whereas the point of reference on the plate produces a track or path that is a *small circle* (Figure 10.8). Other points of reference on the same plate produce other small circles, but all of the small circles are parallel and concentric; they share the same pole and axis (Figure 10.7). (A point lying 90° from the pole of rotation would produce a great circle, but this is a special case.) We can visualize that points on the plate that are closer to the pole of rotation move shorter distances than those farther from the pole, although all points cover the same arc (degrees) in the same unit of time. This is the same principle that applies to the grooves of a rotating phonograph record or to a person riding on a carousel: Points nearer the outer edge move with greater speed and cover more distance per unit time.

Next, consider a transform fault between two lithospheric plates (Figure 10.9). In order for one plate to slide past the other without any gap opening up and without one plate overriding the other, the transform

FIGURE 10.6 Transform faults. *A*. A simple transcurrent fault. *B*. A transform fault between offset ends of a spreading boundary. *C*., *D*. Transform faults between the offset ends of subduction boundaries.

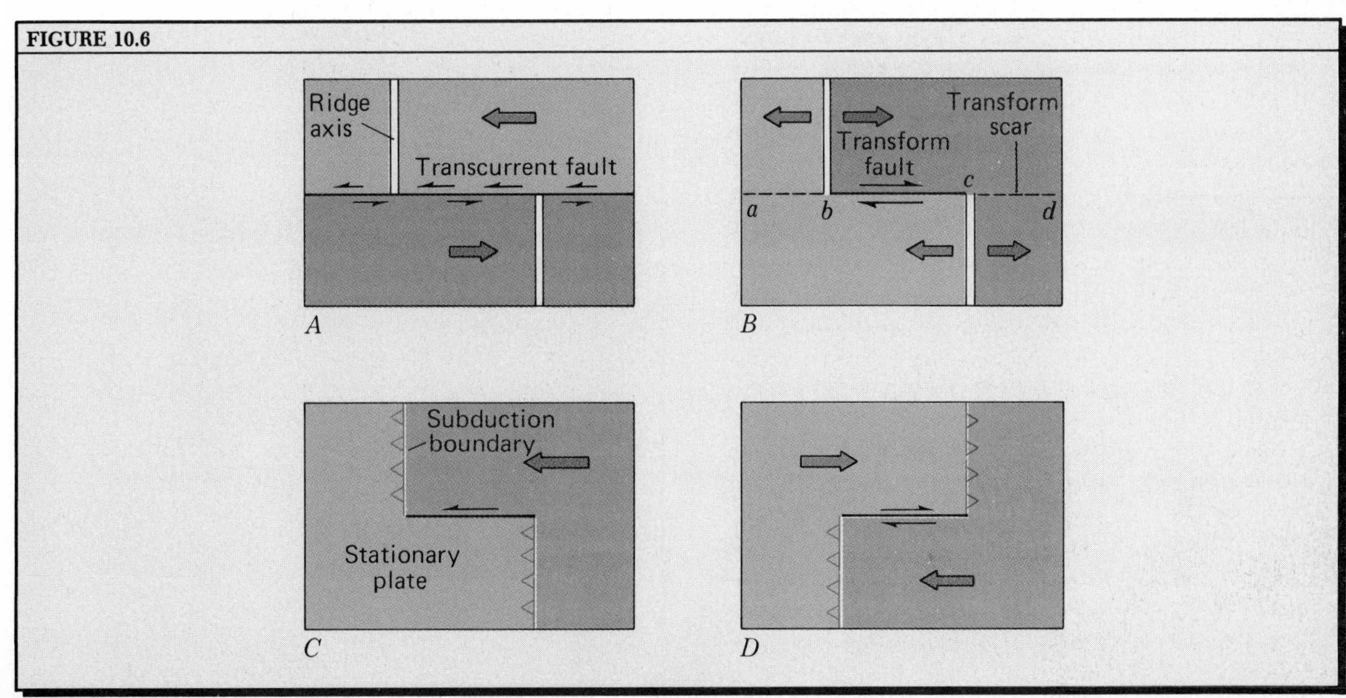

fault must itself be an arc of a small circle conforming to the single pole of rotation that describes the *relative plate motion* between Plates *A* and *B*. These plates share the same pole and axis. It is important to use the adjective "relative" for plate motions with respect to this common pole of rotation.

When we introduce a third plate (Plate *C*), sharing a plate boundary with Plate *B*, another pole of relative rotation (Pole *B–C*) must be plotted, as shown in Figure 10.9. Pole *B–C* describes the relative motions between Plates *B* and *C*. Keep in mind that when we speak of plate motion, we usually mean only the motion of one plate relative to another. Looking at the broad arrows placed at the two plate boundaries in Figure 10.9, it may seem that Plate *B* is moving in a different direction at the spreading boundary with Plate *A* than at the converging boundary with Plate *C*. How can a single plate be moving in two different directions at the same time? To resolve this difficulty, imagine that Plate *B* is not moving at all, but that Plate *A* is moving away from it in a westward direction while at the same time Plate *C* is plunging beneath it from the southeast.

The actual motion, or *absolute plate motion*, would be its motion with respect to the earth's entire inner mantle and core. Absolute motion is very difficult, if not impossible, to measure. We will have more to say about this subject later in the chapter.

The segments of the spreading boundary shown in Figure 10.9 are shown as parts of great circles and lie on radii connecting the boundary segments with the pole of rotation. The segments of spreading boundaries are offset by transform faults. As distance from the pole increases (i.e., as radius of rotation increases), (a) the rate of plate spreading increases and (b) the rate of slip on the transform faults increases. (Rate refers to centimeters of movement per year.) Therefore, new oceanic lithospheres will be produced more rapidly at a greater distance from the pole of relative rotation. For the same reasons, for Plates *B* and *C*, sharing the same pole but joined along a subduction boundary, the rate of plate consumption must increase with increasing distance from the pole of relative rotation. The increase in rate of spreading or consumption takes place only up to a distance of 90 degrees of arc from the pole of rotation. Beyond this distance the rate will decrease because the radii of the small circles will diminish.

One task that geologists have performed is to calculate the pole of relative rotation of each pair of adjoining lithospheric plates. Of course, the pole of relative rotation between two given plates can change its position through time. When such a change occurs, the track of a reference point undergoes an abrupt change in direction, following a new small-circle track required by the new pole position. Changes in pole positions play an important role in the geologic history of the earth's crust.

FIGURE 10.7 As a lithospheric plate moves over the asthenosphere, a point on the plate follows a small circle. The radius of rotation with respect to the pole of rotation is an arc of a great circle.

FIGURE 10.8 Great circles are formed by planes slicing through the center of a sphere; circle diameter is the same as sphere diameter. Small circles are formed when planes slice through a sphere without passing through the center.

FIGURE 10.7

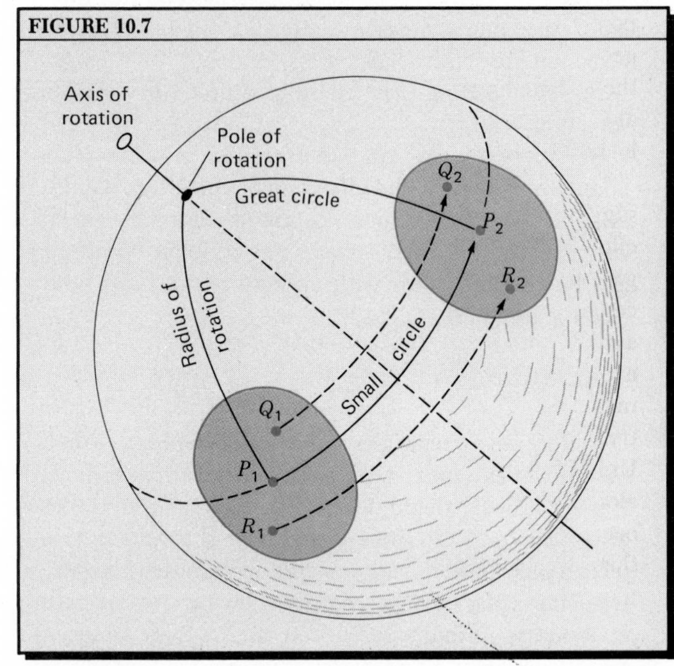

FIGURE 10.8

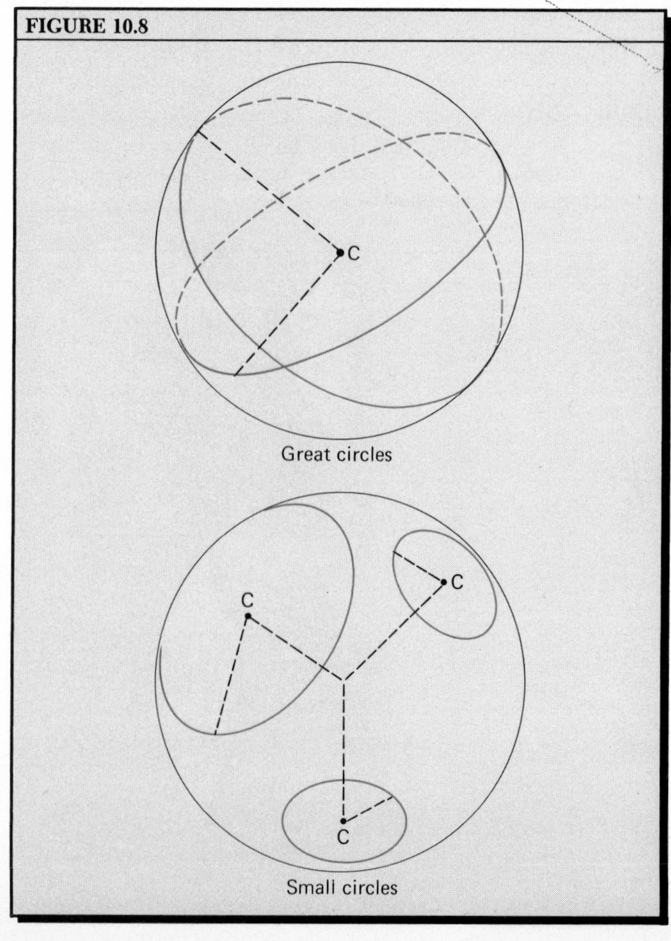

Seafloor Spreading—Where Is the Proof?

We have now completed an overview of the spherical geometry of lithospheric plates, their boundaries, and their motions. If you have a healthy skepticism, as a good scientist should, you will want to be shown some proof that plates actually do move as we have postulated. Until the middle 1960s, a large number of geologists were very skeptical about the proposal that the crust of the ocean floor is actually spreading apart along the axial rift of the mid-oceanic ridge. Let us look into the history of development of this hypothesis.

The concept of seafloor spreading had been vaguely suggested as early as 1889, and it had been quite specifically presented by a highly respected British geologist, Professor Arthur Holmes, as early as 1931. Of course, very little was then known about the geology and topography of the ocean floors. By 1956, seafloor exploration had revealed the enormous length of the mid-oceanic ridge as well as its continuity and symmetry. In 1960, Professor Harry H. Hess of Princeton University proposed that seafloor spreading was active along the axis of the mid-oceanic ridge and that new oceanic crust was being produced continuously along this line by magma rising from the mantle. At very nearly the same time, another American scientist, Robert S. Dietz, outlined a very similar mechanism and added the concept that spreading involves a thick lithosphere, extending much deeper into the mantle than previously postulated.

FIGURE 10.9 The relative motion of two plates in contact along a spreading boundary with transform faults is described by a common pole of relative rotation. The rate of spreading increases with distance from the pole.

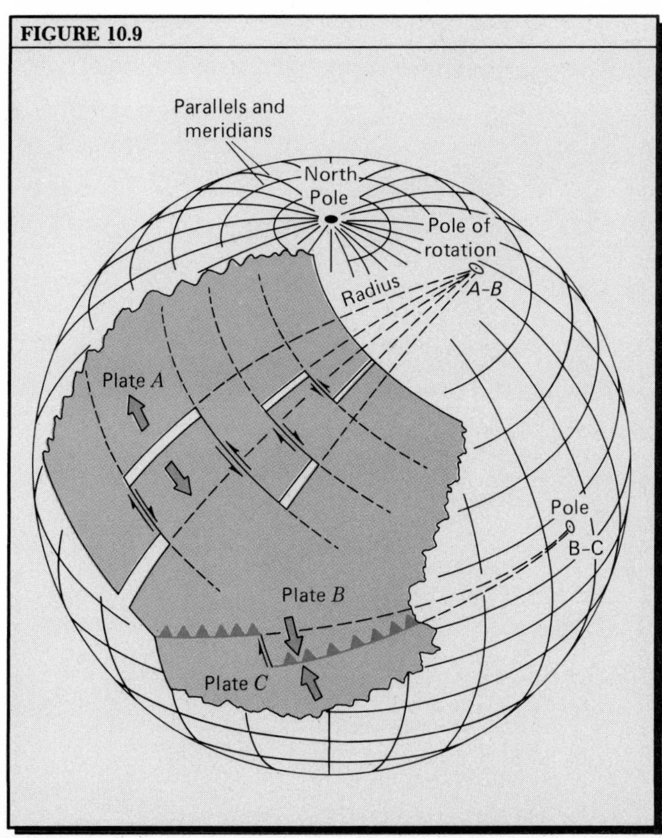

But one major aspect of modern plate tectonics remained incomplete in the early 1960s—namely, the consumption of oceanic crust by subduction along the lines of the great trenches and volcanic arcs. In the absence of this explanation for the disposal of crust produced in the spreading zones, a few scientists had proposed that the earth must be expanding rapidly. This was a rather drastic proposal, but it seemed necessary to accommodate a steadily increasing expanse of oceanic crust.

Actually, the idea of a return of crustal rock to the mantle along zones of convergence seems to have been quite clearly presented by Professor Holmes in diagrams appearing in the 1944 edition of his textbook on physical geology (Figure 10.10). We reproduce them here so that you can judge for yourself. Holmes envisioned a system of slow rock flowage in the upper mantle, called a *convection system,* and consisting of rising and sinking currents connected by horizontal currents which exert a strong drag on the crust. His diagrams show that the horizontal drag force of the moving mantle layer has caused the rupture of a continent and is pulling it apart to produce a new ocean basin. This particular drag mechanism of plate motion is no longer accepted (see Chapter 11), but the general scheme of lithospheric motion conforms with modern plate tectonics. Holmes shows the horizontal flow currents as converging and turning down beneath oceanic trenches at the continental margins. His illustration carries the germ of the idea of subduction as a means of disposal of crust, but it shows the continental crust to be bent down, along with the oceanic crust. Holmes explained in his text that the buoyant continental crust would not descend into the mantle, but would form a deep mountain root. He does, however, show basaltic crust to be converging from two directions to descend into the mantle. The mechanism of a single downbent, rigid slab undergoing subduction is clearly missing from his diagrams and explanation. We must, however, give Holmes credit for a rather simple conveyor-belt model that ingeniously provides for disposal of old basaltic crust as fast as new crust is produced. Thus, you can understand why many geologists credit Holmes with laying the foundations for the theory of plate tectonics.

Scientific proof of seafloor spreading came with tremendous impact in the early 1960s, first through the use of the principles of magnetism of rocks, followed by independent confirming evidence offered by seismology. We will take up each of these subjects in turn, then follow up with more supporting evidence from the record of deep-sea sediment layers.

The Earth as a Magnet

Human awareness that the earth acts as a great magnet may have been reached during the eleventh century A.D., when mariners' compasses using lodestone were put into use by Arabs and Persians. *Lodestone* is a naturally magnetic variety of the iron mineral magnetite (Fe_3O_4) (see Figure 5.11). It was discovered that a piece of lodestone could be floated upon a piece of wood or a cork to serve as a magnetic compass (Figure 10.11). Similarly, an iron needle could be magnetized by contact with lodestone and floated on water. That lodestone possessed the property of attracting iron was known to the Greeks as early as the seventh century B.C., for this fact is stated in the writing of Thales (640–546 B.C.).

In the middle of the thirteenth century, Petrus Peregrinus, a Frenchman, experimentally investigated the properties of lodestone. He can be credited with the discovery that when a piece of lodestone is broken into many smaller fragments, each piece becomes a magnet. Using a small magnetized needle, he was able to demonstrate the existence of a magnetic axis within a spherical piece of lodestone, finding that the needle was oriented perpendicular to the surface of the lodestone over the polar position.

The first really scientific analysis of the earth's magnetism is contained in a treatise entitled *De Magnete*, published in 1600 by Sir William Gilbert, physician to Queen Elizabeth. In trying to explain why the north-seeking end of a compass needle, if free to rotate on a vertical axis, points downward into the earth's surface in the northern hemisphere, but outward in the southern hemisphere, Gilbert carried out some of the first truly scientific experiments ever recorded. He made a sphere of lodestone and observed the orientation of a magnetized needle when held at various points with respect to the sphere (Figure 10.12). He was thus able to construct the lines of magnetic force. Gilbert found that the magnetized needle assumed an orientation with respect to the lodestone sphere much like that which the compass needle assumes on the globe, from which he concluded that the earth acts as a great magnet.

The earth's magnetism is called *geomagnetism*. In its simplest aspect, the earth's magnetic field resembles that of a bar magnet located at the earth's center (Figure 10.13). The axis of the imaginary bar magnet coincides roughly with the earth's geographic axis. At the points where the projected line of the magnetic axis, or *geomagnetic axis*, emerges from the earth's surface are the *magnetic poles*. Note that the earth's magnetic axis forms an angle of about 20° with re-

FIGURE 10.10 Published in 1944, this drawing illustrates Professor Arthur Holmes's hypothesis of mantle convection currents as the driving force for continental drift and the opening up of a new ocean basin. The general plan comes remarkably close to modern plate tectonics, but lacks essential features of spreading and subduction boundaries. (From Arthur Holmes, 1944, *Principles of Geology.* Reproduced by permission of Thomas Nelson and Sons, Ltd.)

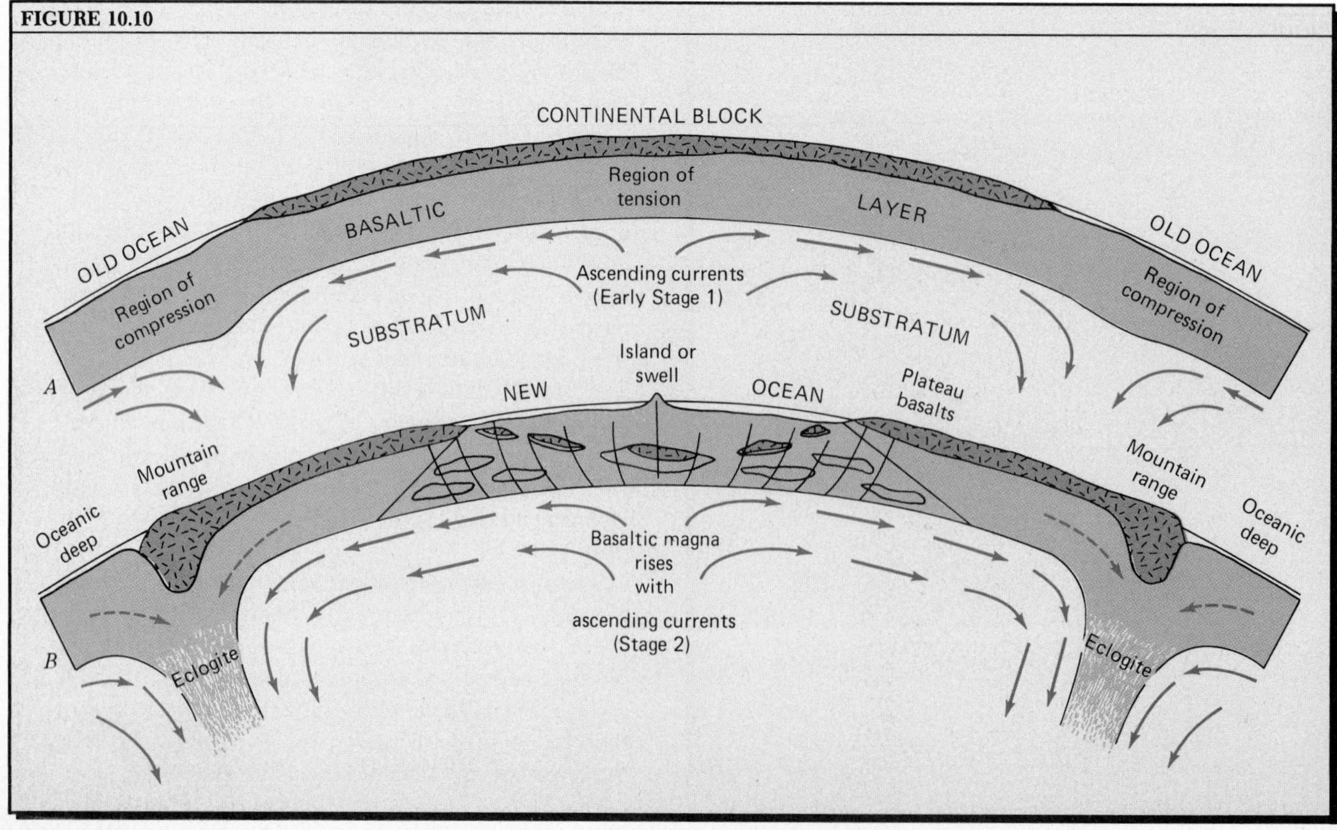

spect to the geographic axis. As a result, the magnetic poles do not coincide with the geographic poles.

Figure 10.13 shows lines of force of the earth's magnetic field in relation to the earth's core. The force lines pass through a common point close to the earth's center. The magnetic axis is oriented vertically in this diagram. There exists a *magnetic equator,* lying in a plane at right angles to the geomagnetic axis and encircling the earth's surface approximately in the region of the geographic equator. Visualized in three dimensions, the lines of force of the earth's magnetic field form a succession of doughnutlike rings, suggested in Figure 10.13. The small arrows show the attitude that would be assumed by a small compass needle, free to orient itself parallel with the force lines close to the earth's surface. The vertical angle between

FIGURE 10.11 A medieval floating compass. Stars mark the poles of the piece of lodestone. (From Athanasius Kircher, 1643.)

FIGURE 10.12 Gilbert's diagram of his terella with small magnets showing inclination. (From W. Gilbert, 1600, *De Magnete.*)

FIGURE 10.13 Lines of force in the earth's magnetic field are shown here in a cross section passing through the magnetic axis. Letter *M* designates *magnetic,* and *G, geographic.* Arrows at the surface of the earth show the orientation of a magnetized needle.

FIGURE 10.11

FIGURE 10.12

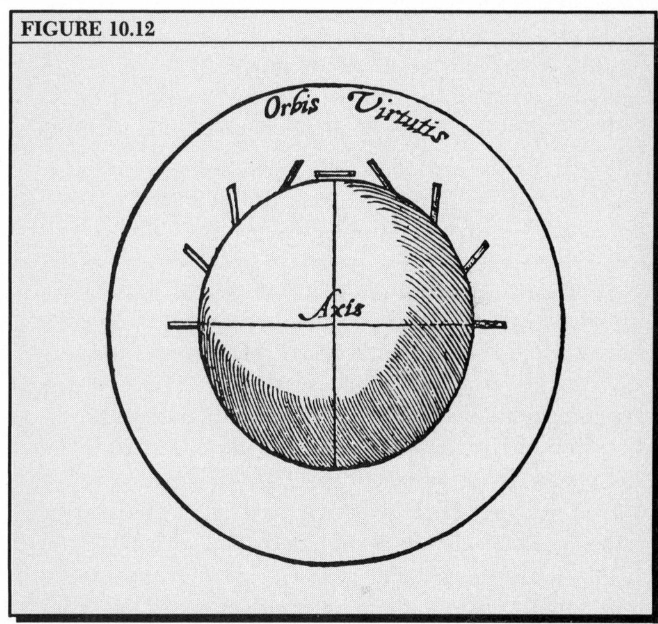

FIGURE 10.13

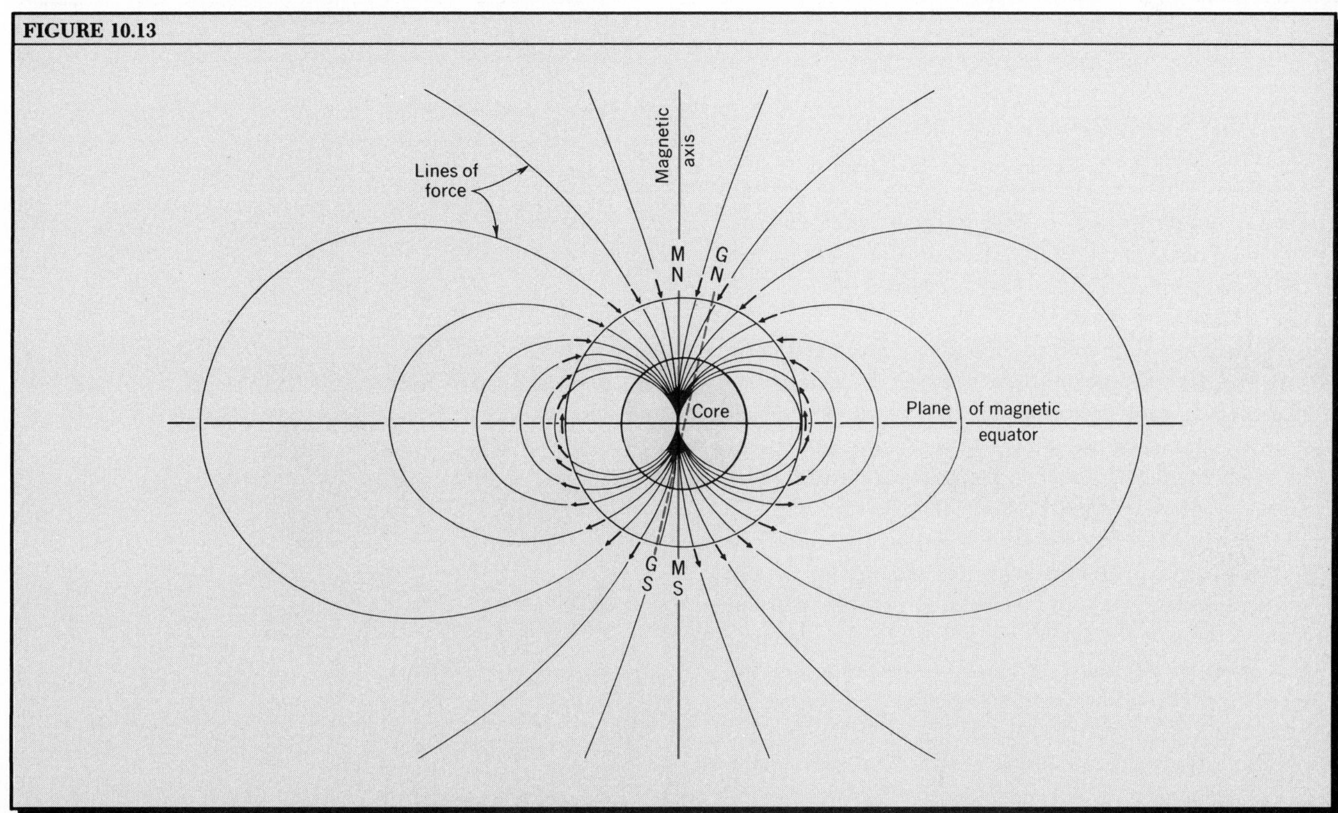

the lines of force and the horizontal plane is called the *magnetic inclination*. Inclination is nearly zero over the region of the geomagnetic equator but nearly 90° over the area of the geomagnetic poles. Inclination is measured by means of a *dip needle*, a magnetized needle mounted on a horizontal axis. Force lines extend far out into space surrounding the earth. This entire field of magnetic effect is the *magnetosphere*.

Earth magnetism is explained by the *dynamo theory*. This theory postulates that the liquid iron of the core is in slow rotary motion with respect to the solid mantle that surrounds it. It can be shown that such motion will cause the core to act as a great dynamo, generating electrical currents. These currents at the same time set up a magnetic field (Figure 10.14). A single, symmetrical current system can thus explain the magnetic field as essentially resembling a simple bar magnet.

One of the most remarkable scientific discoveries of recent decades has been that the earth's magnetic field has undergone repeated changes in polarity. In other words, the magnetic north pole and south pole have switched places, but with the axis unchanged in position. The phenomenon is called *magnetic polarity reversal*. Reversal occurs by a weakening (decay) of the geomagnetic field of force while the polarity and direction of the force lines remain the same. After the strength of the magnetic field has fallen to zero, it begins to build back to normal, but with the polarity in the opposite direction—i.e., reversed polarity. The time intervals between polarity reversals seem to be irregular in length; the sequence does not follow a cyclic or rhythmic pattern.

Paleomagnetism

Basaltic lavas contain, in addition to the abundant silicate minerals, minor amounts of oxides of iron and titanium. These oxides typically occur as magnetite (Fe_3O_4) intergrown in a solid solution with titanium oxide ($FeTiO_4$ or $FeTiO_3$). Magnetite and other iron minerals susceptible to becoming magnetized are called *magnetic minerals*.

At the high temperatures of a magma, the magnetic minerals have no natural magnetism. As cooling sets in, however, each crystallized magnetic mineral passes a critical temperature known as the *Curie point*, below which the mineral becomes magnetized. The Curie point is reached between 600 °C and 400 °C. Because the magma crystallizes within the influence of the earth's magnetic field, the magnetism of the mineral particles assumes a direction parallel with the lines of force of the earth's field. With further cooling, this *thermal remnant magnetism* becomes permanently locked within the solidified rock and thereafter serves as an enduring record of the earth's magnetic field at the time the magma cooled. The permanent magnetism that is acquired at various points in geologic time is called *paleomagnetism*.

In the study of igneous rock magnetism, a sample of rock is removed from the surrounding bedrock. Orientation of the specimen core is carefully documented in terms of geographic north and horizontality. The specimen is then placed in a *magnetometer*, a sensitive instrument that measures the direction and intensity of the permanent magnetism within the rock. Measurement is also made of the inclination of the magnetic axis of the specimen. After a number of samples have been obtained from a single lava flow, the magnetic readings are compared for consistency and averaged. The direction and inclination of the paleomagnetism can be compared with present conditions and with the magnetic field at other locations and at different times in the geologic past. Two other forms of paleomagnetism are useful in working out past geologic events. One is a permanent magnetism acquired by a hydrogenic mineral as it grows by chemical precipitation. Hematite, an oxide of iron (Fe_2O_3), is one such mineral that acquires permanent magnetism as it forms from weathering processes involving the oxidation of magnetite. More important, however, is *detrital remnant magnetism*, which is found in detrital sediments laid down in quiet water of lakes or the oceans. Particles of magnetic minerals, previously magnetized, come to rest in a position of orientation in which their magnetic axes are parallel with the earth's magnetic force lines. In other words, the particles act

FIGURE 10.14 Electric currents, shown as color lines on the earth's core, are believed capable of producing the earth's magnetic field.

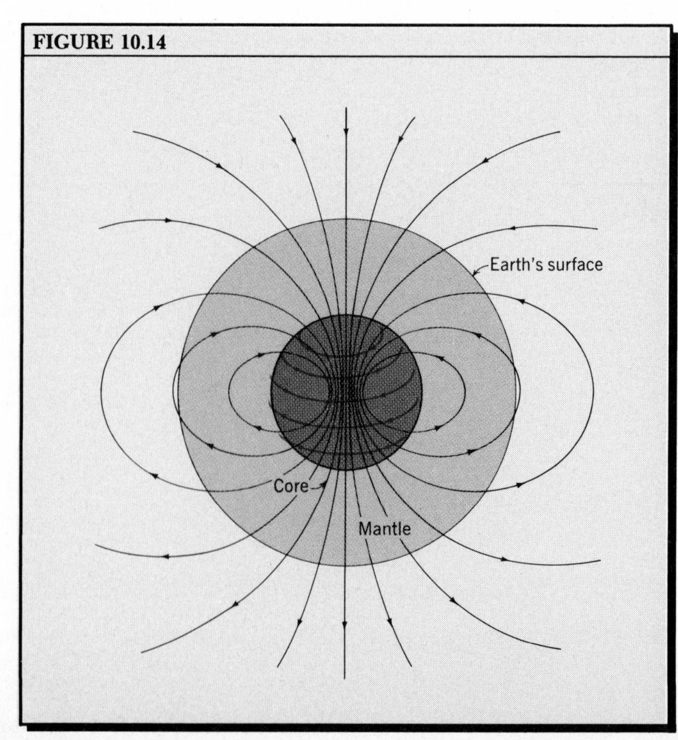

like compass needles that swing into line with the force field surrounding them. Needless to say, detrital remnant magnetism is of great value in determining the paleomagnetic history of layers of pelagic–detrital sediment that have accumulated on the deep ocean floor.

Geomagnetic Polarity Reversals

As early as 1906, Bernard Brunhes, a French physicist, had observed that the magnetic polarity of some samples of lavas is exactly the reverse of present conditions. He concluded that the earth's magnetic poles must have been reversed at the time the lava solidified. You might wish to propose as an alternative hypothesis that the rock magnetism itself has undergone a change in polarity, and this possibility has been given careful study, along with other possible alternatives. In recent years, there has been general agreement among members of the scientific community that the rock magnetism is a permanent and reliable indicator of the former states of the earth's magnetic field.

In addition to establishing the magnetic data of the lava, it is essential to determine the age of the rock—i.e., the date the magma solidified. Such information is available through radiometric methods (Chapter 7). Extensive determinations of both magnetic data and rock age soon revealed that there have been at least ten *geomagnetic polarity reversals* of the earth's magnetic field in the last 3.5 million years (m.y.) of geologic time. Figure 10.15 shows the time scale of magnetic events. Polarity such as that existing today is referred to as a *normal epoch,* opposite polarity as a *reversed epoch.* Each epoch was named for an individual or a locality. For example, the pioneer work of Bernard Brunhes is recognized in assigning his name to the present *Brunhes normal epoch,* which began about 700,000 years ago. An epoch of reversal, named for the Japanese scientist Motonori Matuyama, extends to −2.5 m.y. (two and one-half million years ago). The *Matuyama reversed epoch* contains three shorter periods of normal polarity classified as *events:* They are Jarmillo, Gilsa, and Olduvai normal events. Still older is the *Gauss normal epoch,* named in honor of the mathematician Karl Gauss (1777–1855); it carries back the paleomagnetic record to about −3.4 m.y. It contains one major reversal, the Mammoth reversed event. Oldest of the reversed epochs shown on the diagram is the *Gilbert reversed epoch,* named after Sir William Gilbert.

The time scale of geomagnetic polarity reversals was then carried back to about −7 m.y. by use of the potassium–argon dating method applied to lava layers. A particularly valuable find was a succession of lava layers in Iceland totaling 3500 meters in thickness. There, three additional magnetic polarity epochs were found to extend beyond the Gilbert epoch, and these were simply designated as epochs 5, 6, and 7.

Using the detrital remnant magnetism of deep-sea sediments, it has been found possible to identify the normal and reversed magnetic epochs in samples of soft sediment obtained from the ocean floor by piston-coring. Here epochs are encountered in sequence from top to bottom within the core.

Core orientation is not known in terms of north and south, since they are turned frequently during extraction and handling, but top and bottom are known. Therefore paleomagnetic analysis can be based upon magnetic inclination. Small specimens are cut from the core and subjected to measurement of magnetic inclination. At high latitudes, inclination is a high angle with respect to the horizontal.

FIGURE 10.15 Time scale of magnetic polarity reversals. The graph of geomagnetic declination fluctuations is schematic. (After A. Cox, R. R. Doell, and G. B. Dalrymple, 1964, *Science,* vol. 144, p. 1541, Figure 3, and other sources.)

FIGURE 10.15

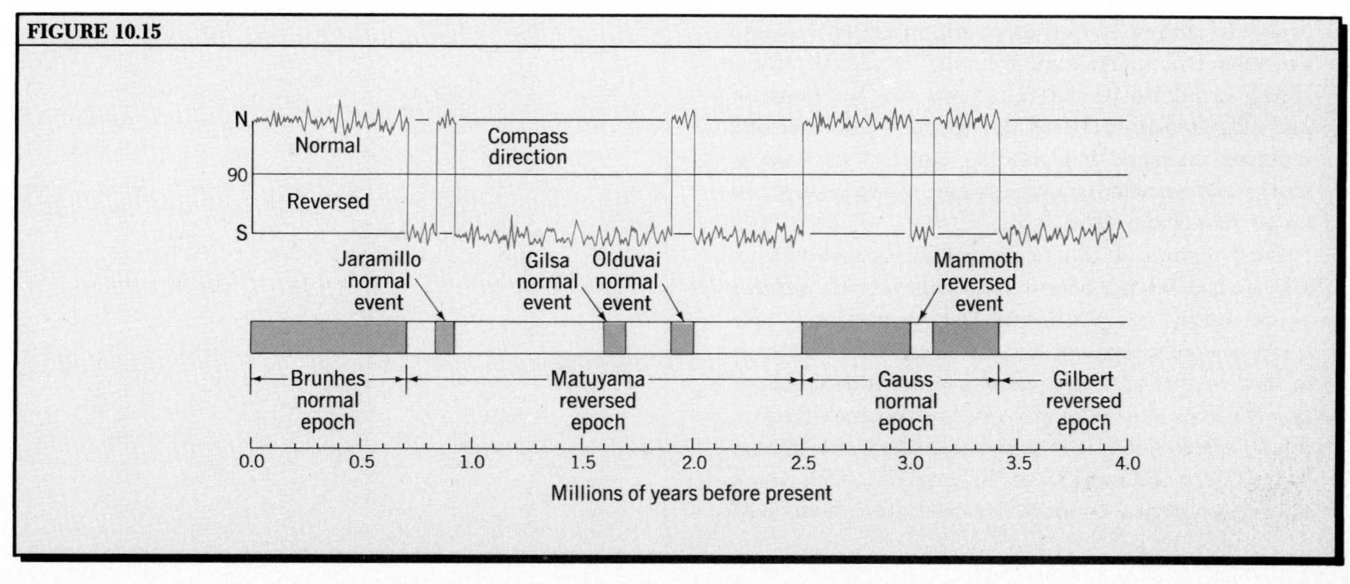

Figure 10.16 shows actual data from a core taken from North Pacific waters, each dot representing a sample from the core. Depth in core is given on the vertical axis, and inclination in degrees on the horizontal axis. A negative sign means that the north-seeking end of a dip needle would point down (normal field); a positive sign means that the same end of the needle would point up into the air (reversed field). Along the right of the graph is the history of polarity reversals, with time in millions of years before the present.

Because of the established time scale of polarity reversals, it is possible to establish the rate of accumulation of deep-sea sediment at the locations sampled by the cores. For North Pacific cores, this rate was found to differ from core to core and ranged between 0.3 and 1.1 cm per year. Sediment cores reveal older epochs of polarity reversals dating back tens of millions of years.

Paleomagnetism and Seafloor Spreading

We are now prepared to return to the subject of seafloor spreading along the mid-oceanic ridge. If the axial rift valley is a line of upwelling basaltic lavas, and if crustal spreading is a steady continuous process, the lava flows that have poured out in the vicinity of the axial rift will be slowly moved away from the rift. The solid lava, being brittle, is easily fractured by extensional stress. Cracks, or fissures, form continuously in the floor of the axial rift. The newly formed ribbon of lava is continuously divided into two halves, one half adhering to the crust on one side of the rift and the other half adhering to the opposite side. A similar process of division takes place in the underlying igneous rock of the newly formed crust. Thus the lithospheric plates on the two sides of the spreading boundary undergo accretion at equal rates. As a result of this continuous process, the oceanic crust produced in a given interval of time—say one million years—forms two parallel bands or stripes of equal width, one on each side of the rift. As time passes, stripes of the same age-span will increase in distance of separation, as shown in Figure 10.17*A*. The lavas can be identified and classified in terms of the epochs of normal and reversed magnetic field; in the figure, these epochs will be represented by symmetrical striped patterns on either side of the rift.

Confirmation of the symmetrical magnetic stripes was gained in the course of oceanographic surveys made during the mid-1960s. Geologists cannot take oriented core samples of lavas from the ocean floors. It is, however, possible to operate a sensitive magnetometer during a ship's traverse of the mid-oceanic ridge. When this is done, it is found that there are minute variations in the strength of the magnetic field. These departures from a constant normal value are referred to as *magnetic anomalies*.

When several parallel lines of magnetometer surveys have been run across the mid-oceanic ridge, the magnetic anomalies can be resolved into a pattern, such as that shown in Figure 10.17*B*. Figure 10.18 is a map showing the striped pattern of magnetic anomalies near Iceland. Notice the mirror symmetry of the striped pattern on either side of the axis of the ridge. From a study of the anomaly pattern, it is possible to identify the normal and reversed epochs, as we have done in Figure 10.17*B*.

Magnetic anomaly patterns clearly reveal the presence of transform faults and transform scars, which show as abrupt offsets of the magnetic stripes. Figure 10.19 is a set of map-diagrams showing how the stripes can become offset during their growth.

Finding the magnetic stripes on the ocean floor proved to be the key to the revolution in geology. There followed rapidly a series of magnetic surveys along many sections of the mid-oceanic ridge, all revealing similar striped patterns in mirror image. Magnetic evidence not only made a virtual certainty of seafloor spreading, but also allowed the rates and total distances to be estimated as well.

Take, for example, the case of the anomaly pattern

FIGURE 10.16 Changes in magnetic inclination with depth in sediments of a core taken at about lat. 45° N in the Pacific Ocean by the Research Vessel *Vema*. (After Neil D. Opdyke, in Robert A. Phinney, Ed. The History of the Earth's Crust: A Symposium. Copyright 1968 by Princeton University Press. Reprinted by permission of Princeton University Press.

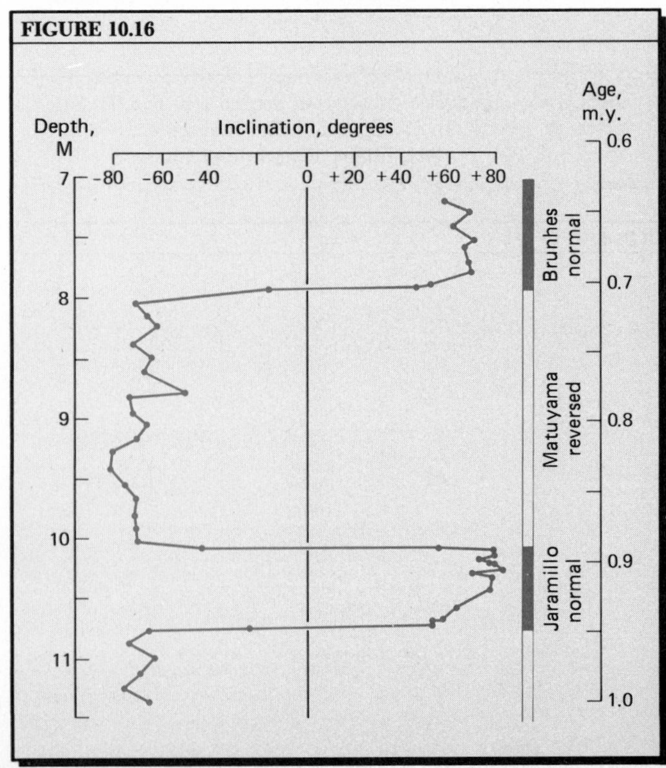

shown in Figure 10.18. Here the width of the anomaly zone is 200 km, which represents the total distance of crustal separation in about 10 m.y. The average spreading rate (rate of horizontal motion of one plate) during this time has been about 1 cm/yr, which means that the rate of separation of the two plates (rate of rift widening) is double this value, or 2 cm/yr.

From one ocean region to another, the magnetic stripe of a given epoch can have different widths, and this can only mean that the rate of spreading has been different in one place from another. Figure 10.20 is a graph on which the distance from the axial rift to the line of each polarity reversal is plotted against time in millions of years, using the time scale established by radiometric methods. The scale extends only to −4 m.y., which is about halfway into the Pliocene Epoch. The steeper the plotted line, the faster the rate of spreading. The highest observed separation rate—9 cm/yr—is along the East Pacific Rise, which is the spreading boundary between the Pacific plate and the Nazca plate. The rate for the North Atlantic basin, to which we have already referred, is the lowest observed for a major plate boundary.

Throughout the late 1960s and well into the 1970s, anomaly patterns were mapped over large parts of all of the ocean basins. After the magnetic anomaly pat-

FIGURE 10.17 *A*. Schematic block diagram of development of symmetrical pattern of magnetic polarity belts in oceanic basalts during crustal spreading. (See Figure 10.15 for time scale.) *B*. The observed profile of magnetic intensity along a traverse of the mid-oceanic ridge at about lat. 60° S in the South Pacific. Below it is the theoretical magnetic profile correlated with the time scale of magnetic polarity reversals. (After W. C. Pitman, III and J. R. Heirtzler, 1966, *Science*, vol. 154, p. 1166, Figure 3. Copyright 1966 by the American Association for the Advancement of Science.)

FIGURE 10.18 Magnetic anomaly pattern for Reykjanes Ridge, located on the Mid-Atlantic Ridge southwest of Iceland, with approximate rock ages in millions of years. (After J. R. Heirtzler, X. Le Pichon, and J. G. Baron, 1966, *Deep-Sea Res.*, vol. 13, p. 427.)

FIGURE 10.17 A

Magnetic polarity epochs and events
Gil Mam Gau Old Mat Jar Bru Bru Jar Mat Old Gau Mam Gil
Ridge axis
Basalt lava
Spreading Magma Spreading

FIGURE 10.18

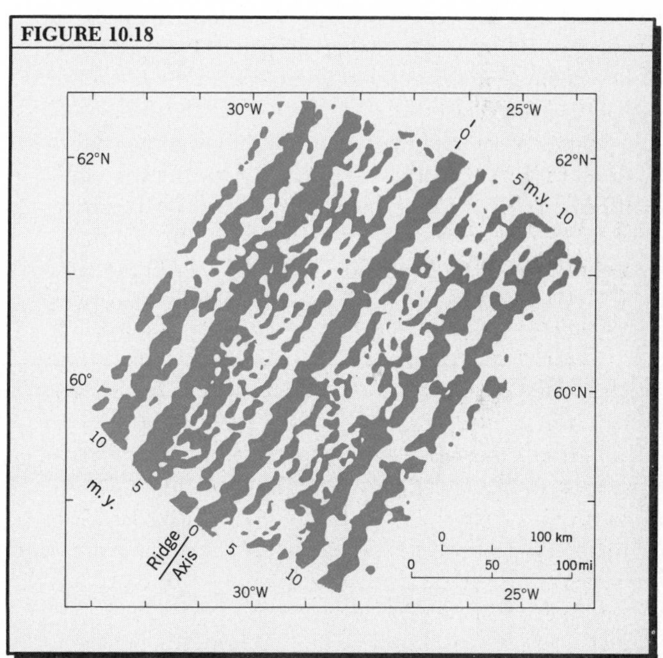

FIGURE 10.17 B

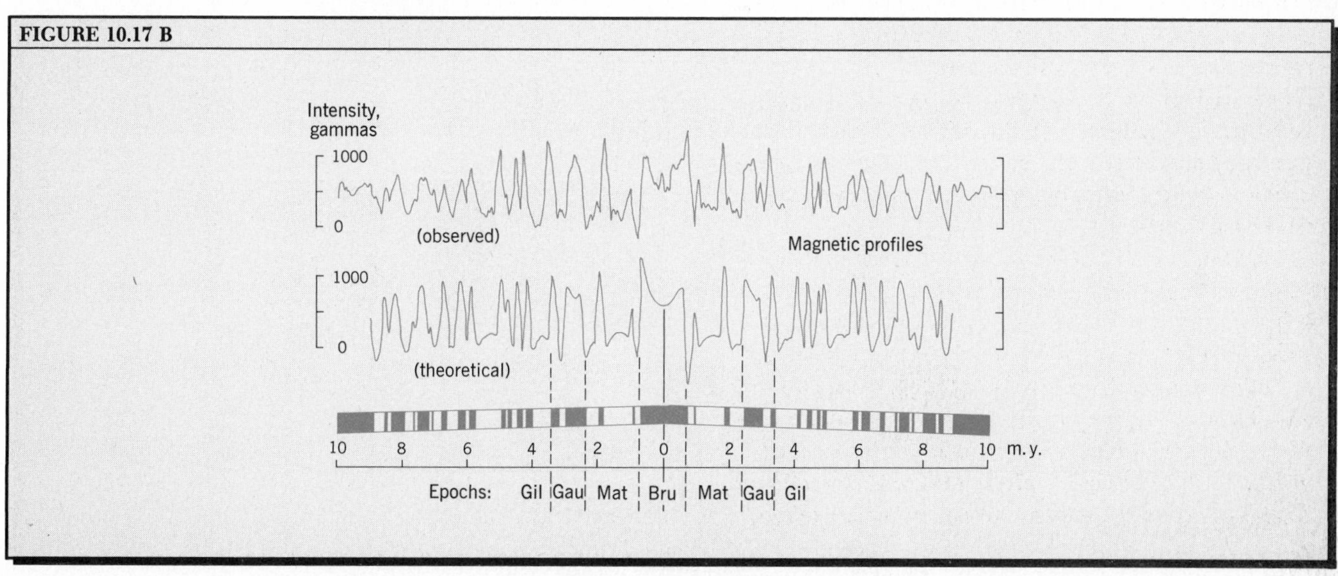

tern was mapped in a particular area, the magnetic reversals were dated on the basis of rock ages obtained by seafloor drilling. Cores obtained by the *Glomar Challenger* were used in this way. Cores that penetrated the basalt were dated by the radiometric method, establishing the point in time at which a reversal occurred. The result is a continuous chronology of magnetic polarity reversals dating back to the late Cretaceous Period, about to −83 m.y. (Figure 10.21). In this total time span, approximately 80 polarity reversals have been counted, and these have been grouped into 34 numbered magnetic anomalies.

Starting about −84 m.y. and continuing to about −110 m.y. there was a long interval within the Cretaceous Period when normal polarity seems to have persisted without reversal. This blank period in the reversal record is referred to as a *magnetic quiet interval*. Then, in the early Cretaceous time, about −110 m.y., polarity reversals again appear and these extend into the Upper Jurassic Period, as far back as about −150 m.y. Altogether, 22 polarity reversals are documented in this activity group, called the *Mesozoic polarity reversals*.

Magnetic anomaly patterns have allowed geologists to make a map of the ocean-basin floors; it was published in 1974 (Color Plate 1.1). Colors on the map show the ages of the basalt bedrock that lies immediately below the sedimentary layer. Of course, the ages of sediment layers lying above the bedrock floor will include all geologic periods younger than the bedrock age, right through to the present. Thus, the map shows the age of the ocean floor as it was formed by seafloor spreading and the accretion of new oceanic crust along spreading boundaries.

The geologic map is not complete, but it tells us a great deal about the history of the ocean basins. For example, the oldest rock floor mapped in the North Atlantic basin is of Jurassic age and occurs in two bands close to the continents. From this pattern, we infer that the North Atlantic Ocean basin began to open up by continental rifting at some point during the Jurassic Period. Looking at the Pacific plate, we note that there is a large area of Jurassic bedrock in the western part of the plate. This area terminates abruptly along the trenches, since it is on this line that the plate is being consumed in a subduction boundary. Notice on the eastern margin of the Pacific plate that the spreading boundary has two triple junctions; these mark corners of the Nazca and Cocos plates.

While the evidence of the magnetic polarity reversals might seem to have proved beyond any possible doubt the reality of seafloor spreading, there remained

FIGURE 10.19 This set of idealized map-diagrams shows how the offsetting of magnetic stripes can take place during continued movement on a transform fault. The maps are drawn at intervals of one million years. We have assumed that the spreading rate holds constant. The offset in the magnetic polarity stripes is propagated as a lengthening transform scar.

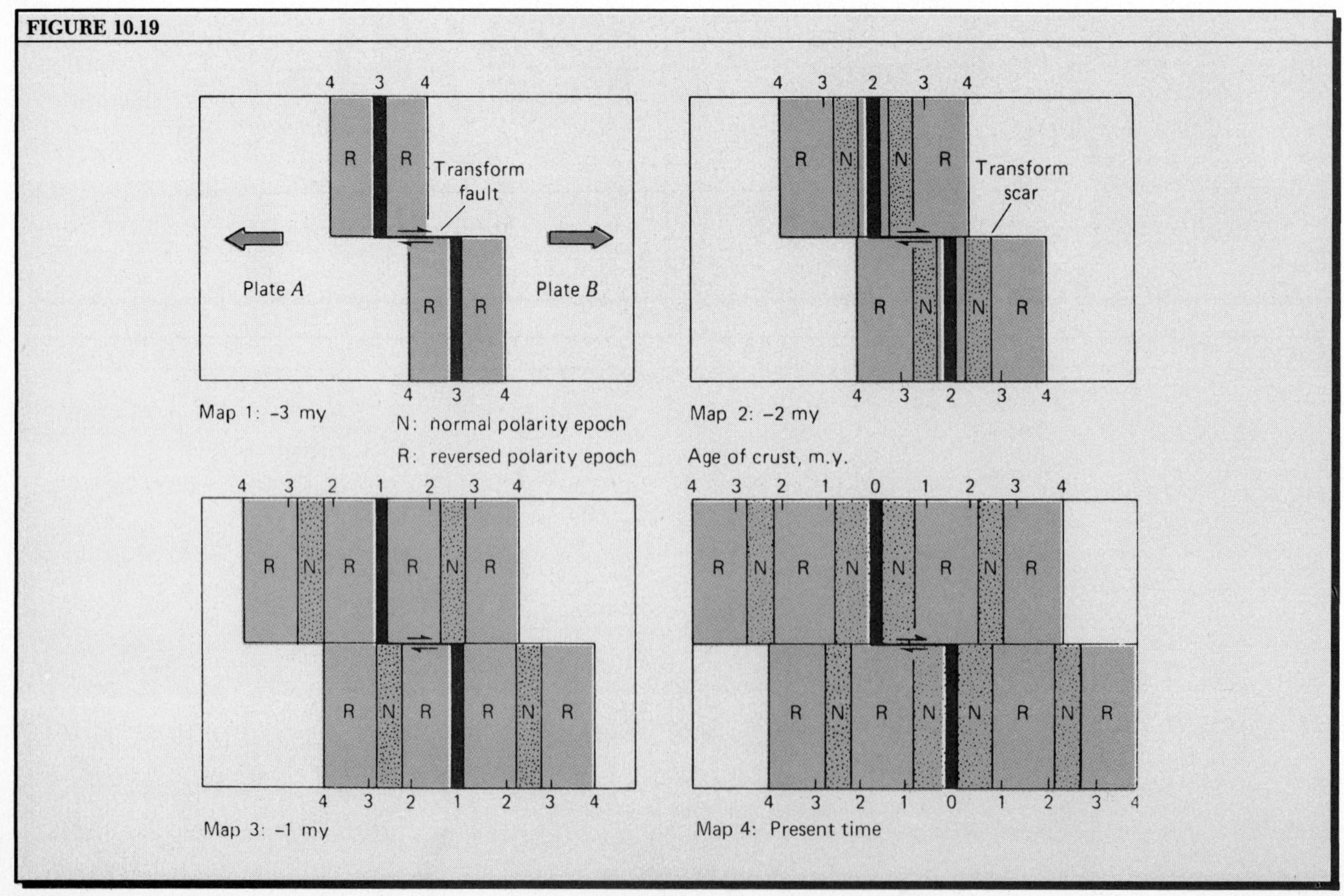

skeptics within the geological community. We must, therefore, continue our search for independent evidence of an entirely different sort to confirm plate motion along spreading boundaries and transform faults.

Seismic Evidence of Transform-Fault Motion

If the offsets of the spreading boundary between two plates are indeed transform faults, active slip movement must be limited to the fault line between the offset ends. If so, seismic maps should show that shallow earthquake epicenters are limited to those active portions. A close look at a detailed map of the mid-oceanic ridge will show this to be the case. Figure 10.22 shows a portion of the Mid-Atlantic Ridge. There are many closely spaced offsets of the ridge axis in this region. Nearly all the shallow-focus earthquake epicenters lie along either the ridge axis or the connecting portion of the transverse fracture zones. Few, if any, epicenters lie on extensions of the fracture zones beyond the ridge axis. Many other examples of the same seismic pattern could be offered.

While the evidence of the epicenters might seem strong proof that the transform segments are active faults, we need more precise evidence in the form of the relative direction of the fault movement. This evidence comes from analysis of seismograms of earthquakes originating along a supposed transform fault. Confirmation came through the work of Professor Lynn R. Sykes, then a seismologist on the research staff of the Lamont–Doherty Geological Observatory

FIGURE 10.20 As the rate of plate spreading increases, the distance between the ridge axis and a given magnetic stripe increases in direct proportion. The half-spreading rate refers to the rate at which one plate moves away from the axis. Full-spreading rate of plate separation by rift widening is double the half-spreading rate. (Based on data of F. J. Vine, J. R. Heirtzler, and others.)

FIGURE 10.21 A magnetic polarity time scale extending back into the late Cretaceous Period. This 1977 version was prepared by scientists of the Lamont–Doherty Geological Observatory of Columbia University. Anomaly No. 14 has been deleted. (From J. L. LaBreque, D. V. Kent, and S. C. Cande, 1977, *Geology*, vol. 5, pp. 332–333, Figures 1, 2, and 3. Copyright © by the Geological Society of America.)

FIGURE 10.20

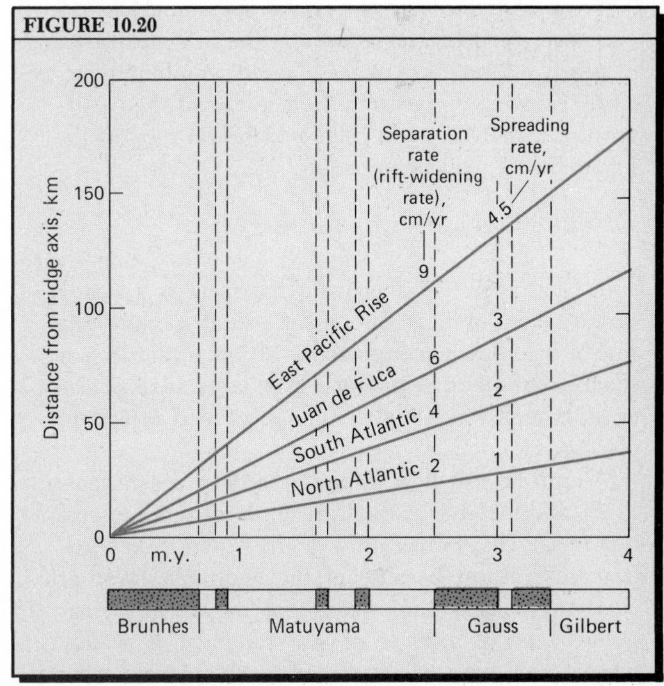

FIGURE 10.21

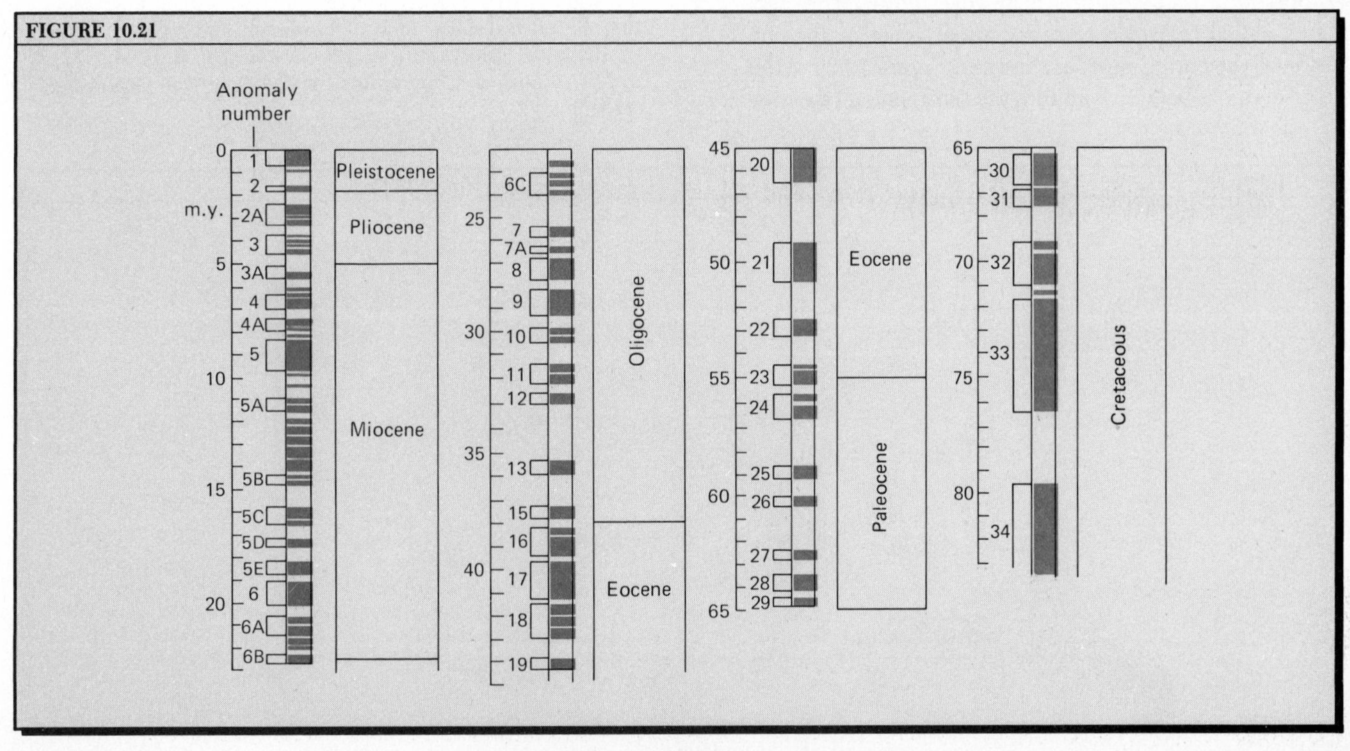

of Columbia University. Examination of a seismogram will reveal the direction of *first motion;* it is the initial impulse that generates a compressional wave (P-wave). Sykes was able to report in 1966 that the first motion along transverse faults between the offset ridge ends is just what would be expected of a transform fault, as J. Tuzo Wilson had envisioned only a short time previously (Figure 10.22). First motions of the faults with epicenters along the axial rift showed them to be normal faults, such as we would expect of movement on crustal blocks breaking off on the two sides of the widening rift and settling down into the bottom of the trough.

It was at this point that crustal spreading and the existence of transform faults won the full acceptance of nearly all members of the scientific community actively carrying on research on tectonics of the earth's crust. But there is more good evidence to support the hypothesis of seafloor spreading.

*Heat Flow and Spreading Plates

In Chapter 7, we explained that there is a steady upward flow of heat through the earth's crust. The rate of heat flow averages about 1.6 microcalories per square centimeter per second over large areas of the crust, though much higher rates are found in certain places.

Measuring heat flow through the deep ocean floors is not as difficult as it might seem, largely because of two facts. The temperature of the ocean floor is extremely uniform because of the enormous layer of overlying cold bottom water that moves very sluggishly. Second, a layer of soft sediment is widely present and is easily penetrated by the piston coring apparatus. To determine heat flow, temperature sensors, called *thermister probes*, are attached to the outside of the coring pipe, as shown in Figure 10.23. After the core has been driven into the sediment, a few minutes are allowed for the thermisters to respond to the temperature of the surrounding ooze. Differences of temperature from one thermister to the next are recorded, allowing the heat flow rate to be calculated.

Some very high rates of heat flow are observed along the active spreading boundaries, represented by the axial rift of the mid-oceanic ridge. Generally, heat flow along the axial rift is over two microcalories per square centimeter per second and in some spots as high as eight. The high rate of heat flow is just what we would predict for a zone of lava extrusion, with a chain of magma chambers probably present at shallow depth.

If, as we postulate, new oceanic lithosphere is being formed along the spreading boundary and the plates are moving apart, that new lithosphere will become cooler as it moves away from the ridge axis. Scientists of the Lamont–Doherty Geological Observatory have developed a model of the ideal distribution of heat in the oceanic lithosphere on the two sides of a spreading boundary. The lithosphere is assumed to be 100 km thick.

Figure 10.24*A* shows the results of their computations in the form of a vertical cross section on which are drawn isotherms (lines of equal temperature). Temperature near the rift is 1300 °C, close to the melting point of basalt. To read this kind of diagram, you should follow a selected horizontal depth line across the graph, noting the decrease in value of the

FIGURE 10.22 Sketch map of the Atlantic Ocean basin between Africa and South America showing the locations of earthquake epicenters (open circles) with respect to the axial rift of the Mid-Atlantic Ridge and its offsets on transform faults. The solid circles with arrows are epicenters for which direction of first motion was determined, establishing that they lie on transform faults. (Simplified and stylized from L. R. Sykes, 1967, *Journal of Geophysical Res.*, vol. 72, p. 2137, Figure 4. Copyrighted by The American Geophysical Union.)

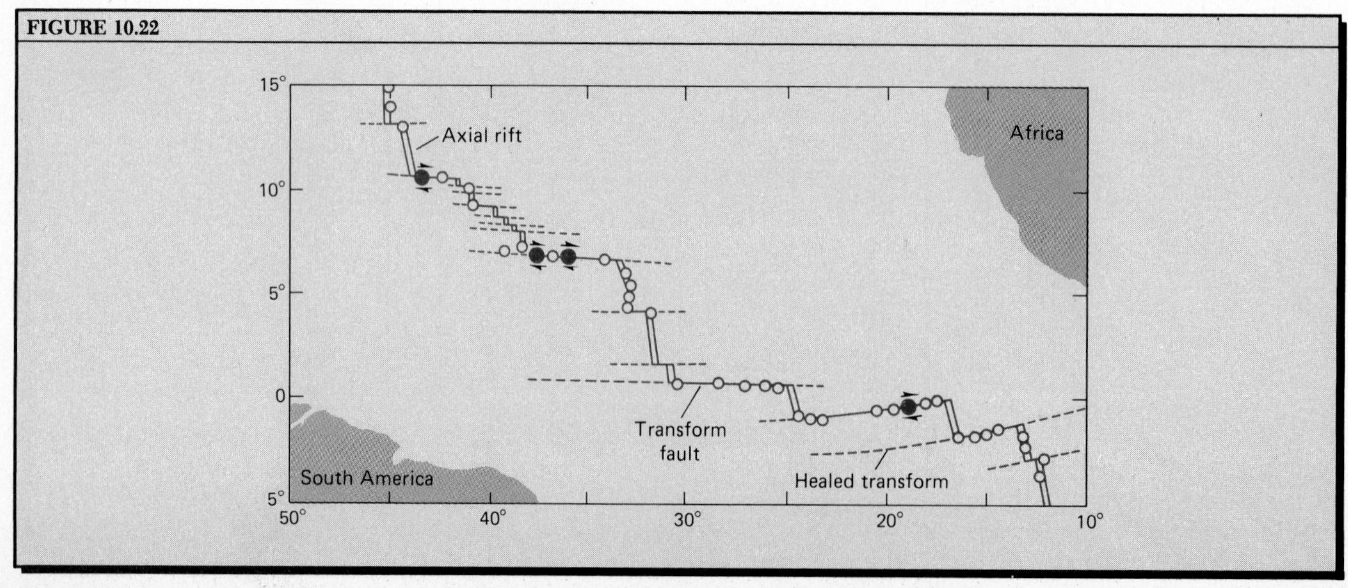

isotherms that cross the line. As the plate becomes cooler, the volume of the rock contracts and its density increases. The plate then sinks lower into the asthenosphere, with the result that the depth to bedrock seafloor increases away from the spreading axis. Figure 10.24*B* shows the increase in depth with distance actually observed in the northwestern Atlantic Ocean. The theoretical curve of deepening provided by the model is shown as a continuous curve on the same graph. The agreement of the observed depths with the predicted depths is remarkably good.

Because the plate becomes cooler with increasing distance from the spreading axis, the rate of heat flow through the ocean bottom should also decrease with distance. Figure 10.24*C* shows the heat-flow curve predicted by the model; dots show heat-flow values measured in seafloor sediment in the northwestern Atlantic Ocean. Although the observed values are highest near the axis and decrease with distance, the rates are much lower than predicted by the model in the first 1000 km. Scientists of the Lamont–Doherty Observatory have offered an explanation for the lower observed heat-flow values.

Cold ocean water penetrates deep into the crustal rock near the spreading axis, becomes highly heated, and returns to the ocean floor by other flow paths. The water emerges in submarine springs located in areas where bedrock is exposed on the seafloor and sediment cover is absent. Thus, by a convection process water circulation effectively carries out a great deal of the residual heat from the crust, and explains why heat-flow measurements taken by thermister probes in seafloor sediments show values lower than the model predicts. The same relationships between observed and predicted heat-flow values have been found near

FIGURE 10.23 The apparatus used to measure heat flow in sediments of the ocean floor consists of thermister probes attached to the pipe used to obtain piston core samples. The instrument for recording the temperature of the probes is located in the lead weight. (Courtesy of Lamont–Doherty Geological Observatory of Columbia University.)

FIGURE 10.24 Temperature, depth, and heat flow in the vicinity of a spreading plate boundary. *A*. Ideal distribution of temperatures in the lithosphere, as computed by model. *B*. Observed and predicted depths of the ocean floor. *C*. Observed and predicted rates of heat flow through the ocean floor. (Redrawn from M. G. Langseth, Lamont–Doherty Geological Observatory of Columbia University, *Yearbook 1977*, vol. 4, p. 42. Used by permission.)

FIGURE 10.23

OCEAN SURFACE

Wire rope

Wire scope

I

II

III

Tripping release

Lead weight

Thermistor probes

Trigger weight

OCEAN BOTTOM

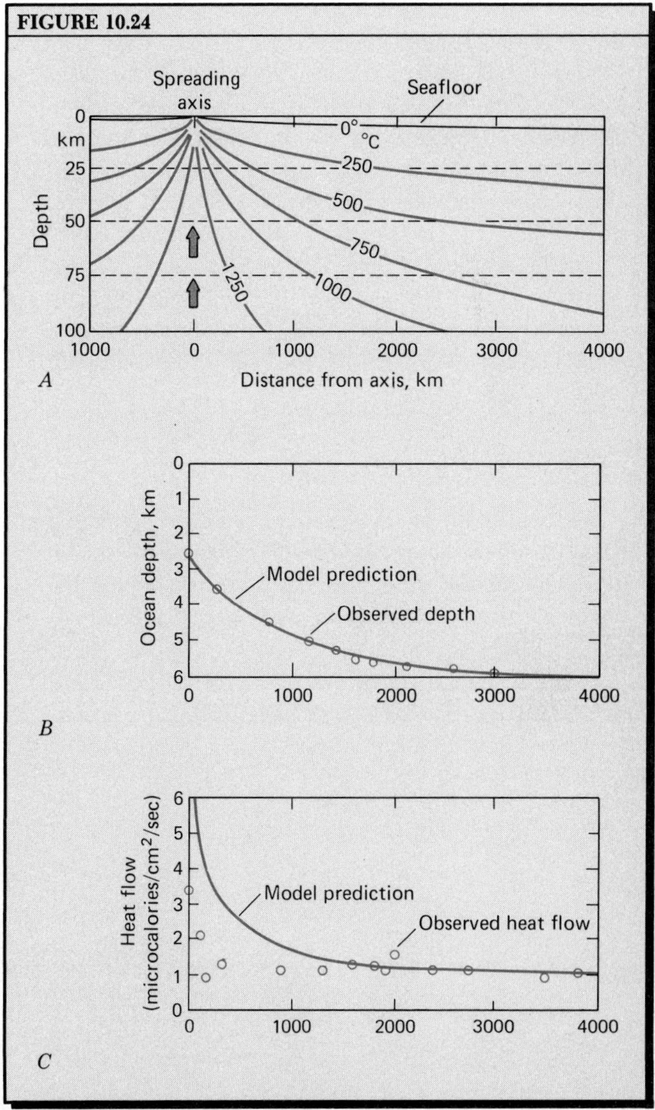

the spreading plate boundary in the North Pacific Ocean.

In Chapter 13, we shall refer to the rise of heated water from fissures in the spreading axis in the Red Sea. The same phenomenon has been found in spreading axis locations in the Galapagos Islands and the Gulf of California.

*Seafloor Spreading and the Pelagic Sediment Cover

If we accept the observed data of increase in depth of seafloor in proportion to its age, it follows that the thickness of pelagic sediment should increase away from the ridge axis and that the character of that sediment should show a change from calcareous ooze to siliceous ooze beyond a line where the carbonate compensation depth is exceeded. (Refer to Chapter 9 for an explanation of pelagic sediments and the carbonate compensation depth.) The predicted increase in thickness of the sediment layer overlying the basaltic floor is very evident from the numerous deep-sea sediment cores extracted by scientists aboard the *Glomar Challenger.* Figure 10.25 is a schematic diagram showing the increase in sediment thickness and the change in sediment type with increasing distance from the ridge axis. Until the compensation depth is reached, carbonate sediment thickens rapidly, but thereafter remains the same thickness as it becomes buried under pelagic clay, which increases gradually in thickness. In cores drilled into the oldest parts of the western North Pacific Ocean floor, where bedrock is of Lower Cretaceous or Upper Jurassic age, the thickness of clays equals or exceeds the thickness of the underlying carbonate layers, while the total thickness of the sediment layer exceeds 500 m.

Project FAMOUS

We turn now to evidence of seafloor spreading based on direct observation of the axial rift by geologists. One of the most remarkable recent submarine discovery efforts has been the probing of the axial rift of the mid-oceanic ridge of the North Atlantic Ocean at a point about 650 km southwest of the Azores Islands. You can find this area on the topographic diagram, Figure 9.13; it lies in the middle of the map, just below the center, in a short length of ridge axis offset between two transform faults. Here, starting in 1971, began the French-American Mid-Ocean Undersea Study known as Project FAMOUS. The project culminated in the summer of 1974 in a series of 42 dives by three submersibles, each carrying a crew of three and capable of reaching the seafloor, here at a depth of 2000 to 3000 m, for direct observation and sampling. Thousands of photographs were taken, and a detailed map was made of a small area of the floor.

As expected, the narrow floor of the axial rift zone proved to be underlain with fresh basaltic lava flows, taking the form of pillow lavas. As the lava is extruded from narrow cracks and comes in contact with the seawater, it solidifies into strange bulbous and tubelike shapes, shown in Figure 10.26. Although no volcanic activity was in progress at the time of the explorations, the very high rate of heat flow upward through the ocean floor suggested that a magma chamber is present not more than about 2 km below the surface.

Another feature of great significance is the presence of numerous open fractures, or ***fissures***, up to 10 m wide, in the basalt lava floor. One is illustrated in Figure 10.26*C*. More than 400 cracks, some only hairline fractures, were counted in an area of about 6 sq km. Most of the fractures were oriented parallel with the ridge axis, and they are regarded as conclusive evidence of a pulling apart of the ocean crust.

Researchers have since turned their attention to the spreading boundaries of the eastern Pacific Ocean basin. Since 1977, *Alvin* has made many dives to permit observation of the floor of these rifts in two general localities. One is along the Galapagos rift, which is the east-west plate boundary between the Cocos and

FIGURE 10.25 A schematic cross section showing the thickening of pelagic seafloor sediment away from a spreading plate boundary. The vertical scale (thickness) of the sediment layers is enormously exaggerated.

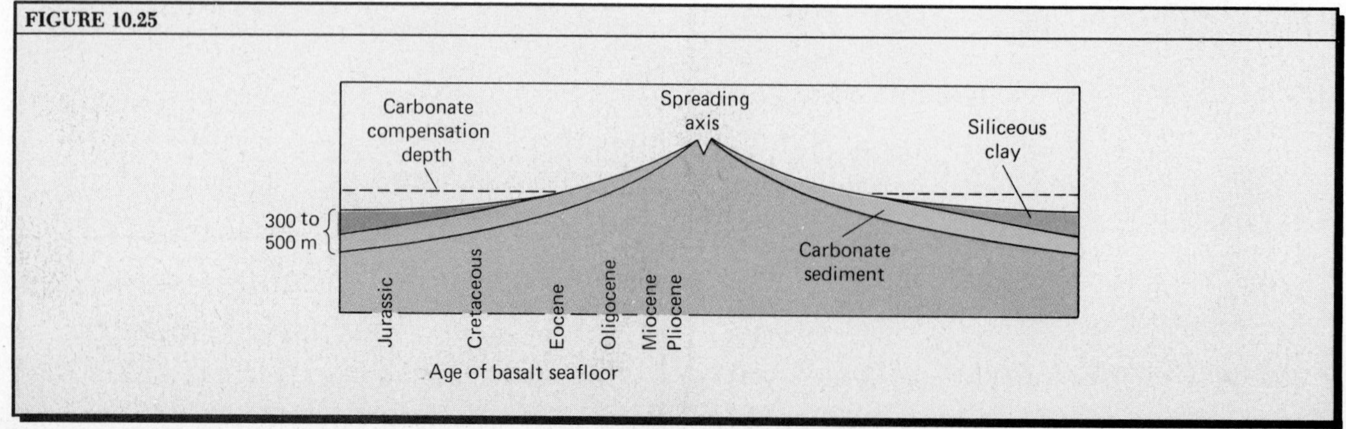

Nazca plates, approximately on the equator and not far from the Galapagos Islands. The second locality is on the East Pacific Rise off the Gulf of California, in the rift boundary between the Pacific plate and the northern end of the Cocos plate. (This part of the Cocos plate is also recognized as a small independent plate, the Rivera plate.) Spreading rates on these plate boundaries are about 6 cm/yr, which is much faster than on the Mid-Atlantic Ridge axis, where the rate is only 2 cm/yr.

Perhaps as a result of faster spreading, the eastern Pacific rifts are low in topographic relief; they lack the mountainous flanks seen in the Mid-Atlantic rift. Fresh lavas on the rift floors also differ in surface appearance. Whereas pillow lavas dominate the Mid-Atlantic rift floor, smooth-surfaced sheet lavas are extensive in the eastern Pacific rift floors (Figure 10.27*A*). The sheet lavas have flow features somewhat like the pahoehoe (ropy) lava surfaces of the Hawaiian basalt flows. The eastern Pacific rifts also have pillow lavas that form hill-like volcanic masses.

The most newsworthy discovery in the eastern Pacific rift floors was of numerous thermal springs issuing from vents in the pillow lavas. In the Galapagos rift these are warm springs that support unique colonies of sea worms 2.5 m long and large clams, as well as shrimp, crabs, and fish. In the East Pacific Rise were found strong jets of hot water (temperature over 350 °C). These jets are densely turbid and appear either white or black; they have been called "smokers" (Figure 10.27*B*). The hot water carries high concentrations of sulfides of copper, iron, and zinc, which are precipitated around the vent, building up a solid stalk-like "chimney" of mineral matter. The hot water jets represent seawater that has penetrated the permeable pillow lavas, become heated by the surrounding rock, heavily mineralized, and returned to the ocean floor under high pressure. This hydrothermal process not

FIGURE 10.26 A

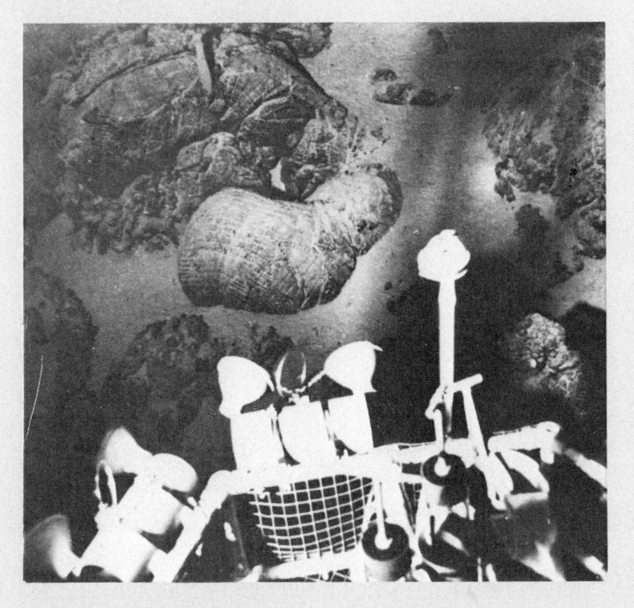

FIGURE 10.26 B

FIGURE 10.26 Project FAMOUS investigates the spreading rift of the North Atlantic Ocean basin. *A.* Pillow lavas are examined at close range by scientists in the deep-diving submersible *Alvin*. Sediment occupies the low places. *B.* Extruded like toothpaste from a tube, this freshly formed lava is scored by sharp edges of the vent from which it emerged. *Alvin*'s arm is reaching for a sample. *C. Alvin* hovers over an open fissure in basalt, proof that seafloor spreading is in progress. A dusting of sediment lies like snow over the basalt. (Woods Hole Oceanographic Institution.)

FIGURE 10.26 C

only transports a large quantity of heat out of the crust beneath the rift, but appears to supply a large quantity of manganese to the ocean, along with other chemical components of seawater.

Sheeted Dikes and Ophiolite Suites

The North Atlantic island of Iceland lies squarely upon the axis of the mid-oceanic ridge. Here, geologists can see a surface exposure of the spreading axis and examine the rocks in detail. The spreading axis is marked by a great troughlike rift running through the center of Iceland. Several points of current and recent volcanic activity occur along this rift zone. In deeply eroded canyons, carved by glaciers on older parts of Iceland, the geologist can examine the structure of the spreading zone. Particularly striking is the presence of innumerable vertical basalt dikes, stacked side by side like a deck of cards in an arrangement called *sheeted dikes*.

Combining the seafloor observations of Project FAMOUS, which showed pillow lavas at the surface, with the observations of sheeted dikes on Iceland, we can reconstruct an idealized section of the rock structure in new oceanic crust formed in spreading zones (Figure 10.28). At the top of the figure are pillow lavas, fed by dikes, above a zone of sheeted dikes. Still farther down we infer the presence of a layer of gabbro, a plutonic mafic rock of the same composition as basalt. The gabbro would represent magma bodies that cooled very slowly as they moved away from the spreading axis. Below the gabbro we would expect to find the Moho, with a rather abrupt transition into ultramafic rock (peridotite) of the mantle.

The layered sequence of mafic and ultramafic rocks we have described has been identified in rock exposures on the continental crust where it is called an *ophiolite suite*. We now realize that these occurrences represent slabs or slices of oceanic crust that have somehow become caught in continental collisions and lifted along faults, eventually becoming part of the continental crust (Chapter 12).

Mantle Plumes and Hot Spots

In describing the process of plate accretion along spreading zones, we have referred to the rise of magma along the entire line of the plate boundary. This motion has long been regarded as part of a great mantle convection system, which we previously referred to in describing the early hypothesis offered by Arthur Holmes. In Chapter 11, we will discuss the various models of convection systems that have been envisioned in connection with plate tectonics.

A related convection phenomenon is the inferred rise of a stalklike column of heated mantle rock—a mantle plume (Chapter 4). The concept is hypothetical because we have no way of directly observing or verifying the existence of such a phenomenon. It has been proposed that plumes rise from near the base of the mantle, penetrating the entire mantle as narrow columns that reach to the base of the lithosphere. Here, in the asthenosphere, mantle rock spreads radially outward, moving horizontally in all directions. A very slow sinking motion distributed through the entire mantle replaces the material that rises in numerous active plumes.

FIGURE 10.27 A. Ropy lava surface of a sheet flow on the floor of the Galapagos rift. B. A "black smoker" in the rift axis of the East Pacific Rise. (Dudley Foster, Woods Hole Oceanographic Institution.)

FIGURE 10.27 A

FIGURE 10.27 B

Although there is no direct evidence of the existence of mantle plumes, there are real features that may represent the effects of plumes on the lithosphere. These are hot spots, small centers of past or present volcanic activity, occurring within lithospheric plates and often located far from any active plate boundary (Chapter 4). Hot spots may be represented by isolated volcanoes or volcanic groups on both continental and oceanic lithosphere. They are also seen in long chains of volcanoes rising from the deep ocean floors, within an oceanic lithospheric plate. Other hot spots are situated on a spreading plate boundary, but we exclude from the hot spot category the volcanoes of island and mountain arcs formed over subduction zones.

FIGURE 10.28 Composition and structure of oceanic crust, formed at an accreting plate margin in a zone of seafloor spreading. The sediment is deposited later, after the crust has moved away from the spreading zone.

FIGURE 10.29 Schematic diagram of a hot spot producing a chain of volcanic islands as the oceanic plate moves away from a spreading zone. Beveled extinct volcanoes are submerged to become guyots. Ocean depth and height of volcanoes are enormously exaggerated.

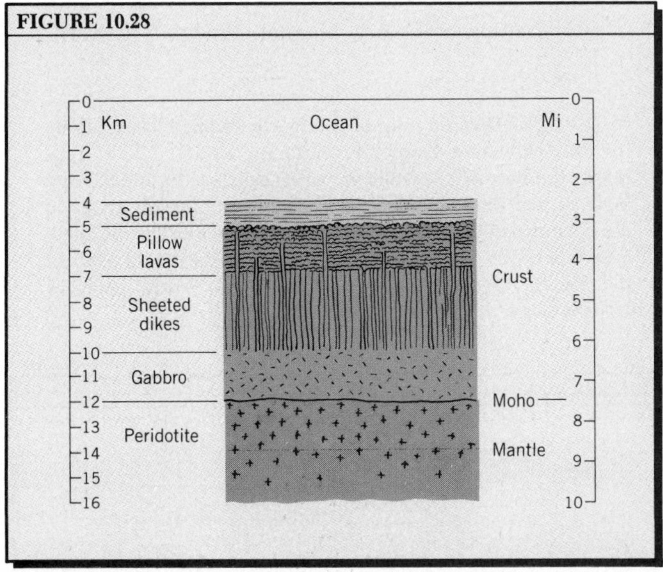

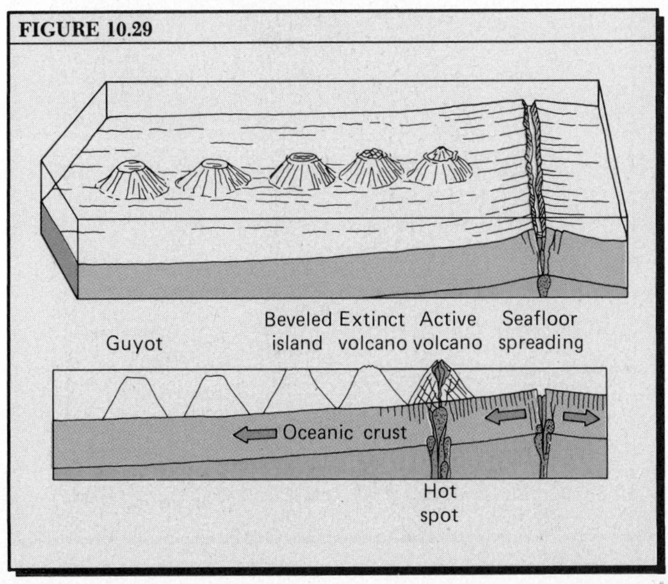

The concept of hot spots formed over magma sources in the upper mantle was first suggested by J. Tuzo Wilson in 1963, well before he explained the significance of the transform fault. The important point that Professor Wilson made is that the magma source in the mantle remains fixed in position, while the lithospheric plate above it moves steadily over the source. In this way, volcanoes are formed over a hot spot but then move away from the magma source, becoming extinct. These extinct volcanoes form a chain that is a record of the plate motion (Figure 10.29).

An extinct volcano may be worn down by wave action and this, combined with the subsidence of the moving plate beneath it, submerges the beveled island beneath the sea surface and creates a guyot (Chapter 9).

The hypothesis of a deep mantle plume beneath each hot spot was proposed by W. Jason Morgan in 1971; it was not included in Wilson's original presentation, but has served to reinforce the hot spot hypothesis by providing a causative mechanism.

An important class of hot spots consists of those occurring along spreading plate boundaries. Iceland is perhaps the best example of this type. Of another class, the active hot spots rising from a position within an oceanic lithospheric plate, the best known is the island of Hawaii, on which the great basaltic domes of Kilauea and Mauna Loa are active. A third class, hot spots on the continental lithosphere, is represented by the Yellowstone Park thermal area where hot springs and geysers are active and young lavas have been extruded.

Of a list of 122 global hot spots identified by J. Tuzo Wilson in 1975, 53 are on oceanic crust and 69 on continental crust. Of those on oceanic crust, 15 are located on the axial rift of the mid-oceanic ridge and 9 others are not far from the axis. Of the major lithospheric plates, the African plate has more hot spots than any other—43 in all. Of these, 25 are on continental crust and the remaining 18 are on oceanic crust.

Professor Wilson's major contribution to plate tectonics through the interpretation of hot spots was to show how the volcanic chains, or lines, formed over hot spots and how they can be used to determine the direction and rate of motion of plates. Consider, first, the hot spots of the Pacific plate. There are three major Pacific volcanic island chains, represented by active volcanoes and lines of seamounts that are interpreted as inactive volcanoes. Consider some details of the Hawaiian–Emperor chain (Figure 10.30), which begins with active volcanoes on the island of Hawaii, easternmost of the Hawaiian Islands. Maui, the next island to the west, has extinct volcanoes, only partially destroyed by erosion. Volcanoes on islands farther west are long extinct and even more deeply eroded. A chain of seamounts continues far westward, rising steeply from a broad submarine ridge. At the end of this straight chain lies Midway, some 2400 km from Hawaii.

Ages of Hawaiian basalts have been determined, using the potassium–argon method. In agreement with Wilson's hypothesis, the ages increase consistently from Hawaii (recent to −0.5 m.y.) westward to Kauai (−4.5 to −5.6 m.y.). The small island of Necker, located much farther west, has a rock age of −11 m.y. Using this evidence we can calculate that the Pacific plate has moved west–northwest at the rate of about 10 cm per year. At a point west–northwest of Midway, another chain of seamounts begins—the Emperor Seamounts. They run in a direction just west of north for a distance of about 2400 km. Ages of lavas of the Emperor Seamounts start with about −40 m.y. at the southern end and reach −75 m.y. at the northern end, where the chain meets the Aleutian Trench. The interpretation seems clear enough: From about −75 m.y. (or earlier) to −40 m.y., the Pacific plate moved almost due north, then changed direction rather abruptly to move west–northwest. Two other seamount and island chains, both in the South Pacific, show the same west–northwest direction of Pacific plate motion as does the Hawaiian chain.

Hot spots of the Atlantic Ocean basin show a different type of behavior from those of the Pacific plate. The Atlantic hot spots seem to have originated on or near the early axial rift that developed when the American plate broke free from contact with the African–Eurasian plate, beginning the opening up of the Atlantic basin. (See Chapter 13 for an explanation.) In the South Atlantic basin, a hot spot exists today in the active volcanic island group of Tristan de Cunha (Figure 10.31). Extending from this point northeast to the African continent is a submarine ridge called Walvis Ridge. Extending northwest to South America is another such topographic feature, the Rio Grande Plateau. Both ridges can be interpreted as basaltic lava accumulations built originally on the same site as the present hot spot, Tristan de Cunha, but gradually transported away from the spreading axis by diverging plate motions. The evidence favoring this interpretation of paired "volcanic tracks" is not as strong as might be hoped for, but it is suggestive.

In Chapter 12, we shall have more to say about hot spots on the continental lithosphere, and particularly their relationship to such volcanic features as the great outpourings of flood basalts that have occurred at certain places. In Chapter 13, we shall develop a hypothesis to the effect that the rifting of a continental lithospheric plate begins through the upward impacts of mantle plumes, cracking the plate to generate triple fractures that eventually become connected into a major spreading boundary.

*Relative Plate Motions

Relative direction and rate of motion of one lithospheric plate with respect to an adjacent plate are, as we have shown earlier in this chapter, formally stated in terms of the angular motion with respect to their common pole of relative rotation. We can also state the direction and rate of motion of a particular reference point on a plate in terms of its compass direction in degrees with respect to the earth's meridians and parallels and its speed in centimeters per year relative to another plate. An arrow on the map will display this information.

Plate motion can be estimated by various means. Direction is indicated by movement on transform faults, confirmed by determination of first motion of P-waves. Seismic evidence will also show the direction of motion of the downbent edge of a lithospheric plate in a subduction boundary. Magnetic anomaly patterns

FIGURE 10.30 Sketch map of the North Pacific Ocean, showing the Hawaiian–Emperor seamount chains. Color limit marks the base of the volcanic piles; dots are summits. Geologic ages of the volcanic rock of the seamounts are given on the accompanying scale. Present motion of the Pacific plate is indicated by arrows. (Based on data of Geologic Map of the Pacific Ocean, Sheet 20, Geological World Atlas of UNESCO.)

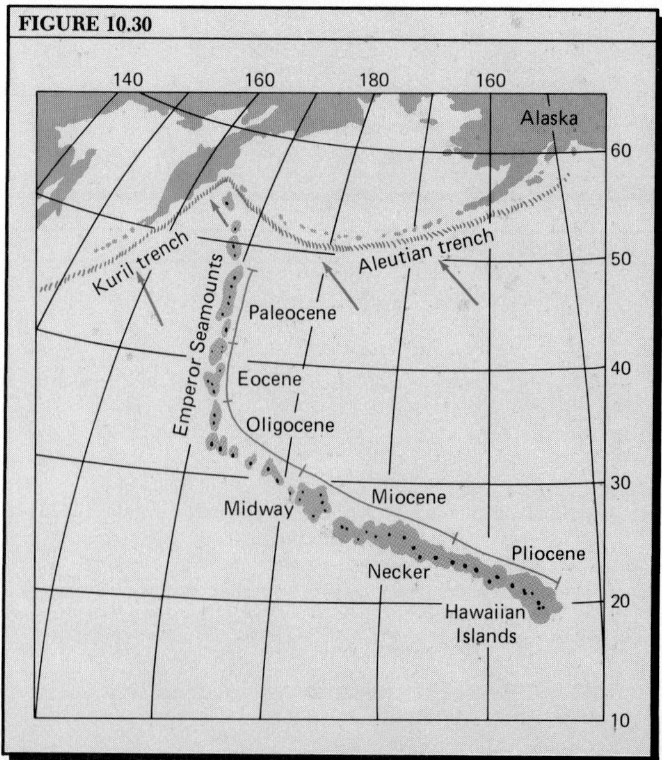

and accessory data on age of the basaltic bedrock floor of the oceans establish the rate of plate motion. Finally, we have seen that chains of volcanic islands and seamounts can be used to establish both direction and rate of plate motion.

The problem in showing plate motions on a world map, such as that of Figure 10.3, is that we can establish only the relative motion of one plate with respect to another that adjoins it on a plate boundary segment, or the relative motions of three plates with respect to one another where they meet in a triple junction. For practical purposes, one major plate can be assumed to be fixed in position, meaning that it is assumed to have no motion with respect to the lower mantle. The motions of all other plates are then shown with respect to the fixed plate.

Some investigators have selected the African plate as a fixed plate because some of the hot spots on the African continent contain lavas of a wide range of geologic ages stacked up one upon the other in a small area. This evidence suggests that the lithospheric plate has been fixed for a long time with respect to the source of magma, a mantle plume, in the upper mantle. Another reason one could almost anticipate the choice of the African plate as the plate of reference is the fact that some five-sixths of the length of its perimeter is a spreading boundary. Although there is a possible convergent boundary along the northern (Mediterranean) side of the African plate, much of it seems to be inactive, since seismic activity that indicates subduction is weak except in the Aegean region.

For the same reasons, the Antarctic plate is also a candidate for use as a fixed plate, as it is bounded almost entirely by a spreading boundary (see Figure 10.3). Only a very short arc of plate subduction is included in the perimeter of the Antarctic plate: the South Sandwich Trench. But if the African plate is fixed, the Antarctic plate must be moving away from it in a southerly or southeasterly direction, and, further, all other plates that bound the Antarctic plate must be moving away from it at a rate sufficient to accommodate its growth on all spreading boundaries. We are assuming, of course, that along a spreading boundary both plates are growing at about the same rate, and this assumption seems to be generally accepted as a requirement of plate tectonics.

We can expect that plates with about half their perimeter made up of consuming plate boundary and half of spreading boundary will be moving rapidly over the asthenosphere. Such is the case with the Pacific, Cocos, Nazca, and Austral–Indian plates. On the other hand, neither the Eurasian plate nor the Antarctic plate seem to have boundary conditions that require rapid motion over the asthenosphere. The Eurasian plate has a spreading boundary only in the North Atlantic and Arctic Ocean basins, but no boundary along which the plate is being consumed. The rate of spreading in the North Atlantic is quite slow and may decrease to nearly zero at the polar end of that boundary.

World maps showing plate boundaries and plate

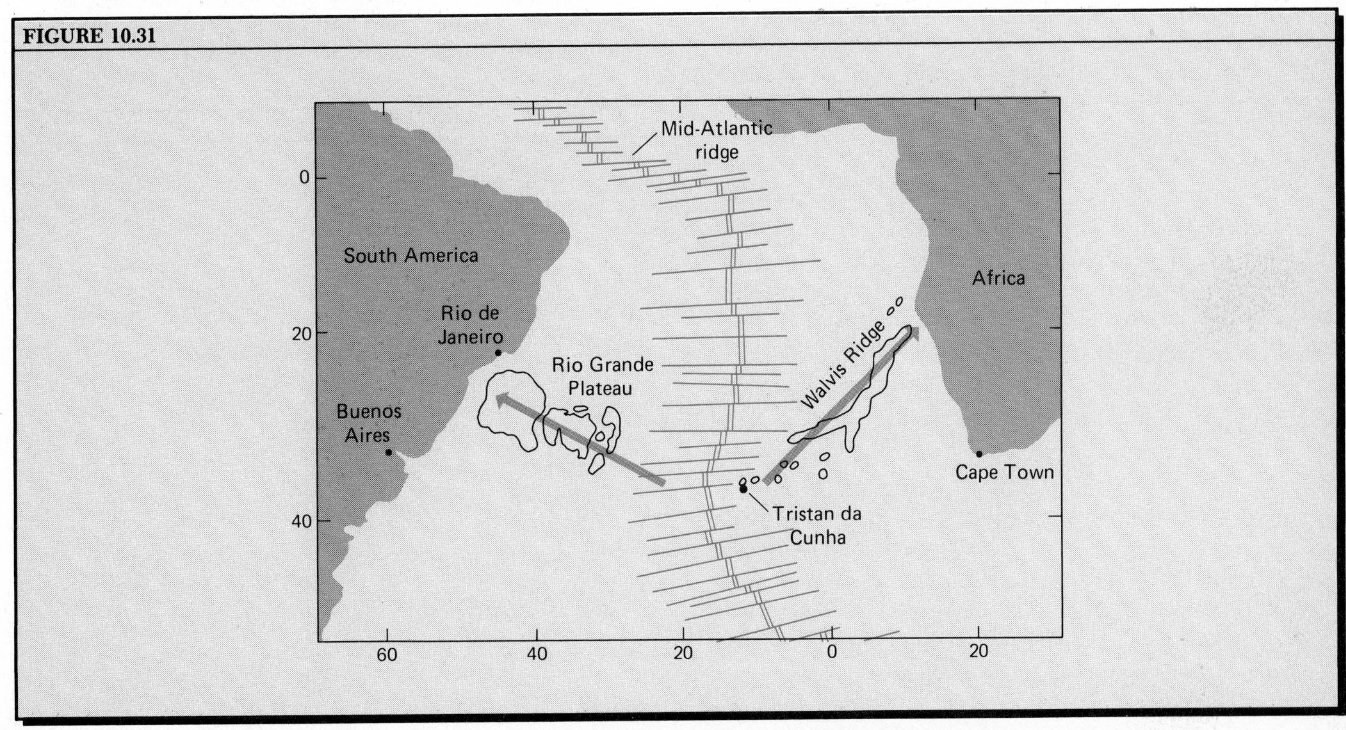

FIGURE 10.31 Sketch map of the South Atlantic Ocean, showing the paired volcanic submarine ridges that may have originated from a single hot spot now located beneath the active volcano of Tristan da Cunha.

motions are almost always drawn on the Mercator map projection (Figure 10.3*A*), which virtually eliminates the two polar regions from consideration. For this reason, we rarely see a map (Figure 10.3*C*) that shows the relative motion of the Antarctic plate with respect to its neighbors, or one that shows the relative motions on the boundary between the Eurasian plate and American plate in the polar regions (Figure 10.3*D*).

*Magnetic Puzzles of the Pacific Plate

The Pacific plate displays two mysterious features that require explanation. One is a curious, abrupt bend in the trend of magnetic stripes in the Gulf of Alaska. Discovered in 1966 and 1967, this feature has been named the Great Magnetic Bight (Figure 10.32). Another puzzle lies in the Mesozoic magnetic anomalies, for which the three groups of stripes show conflicting trends.

The abrupt change in trend of magnetic stripes in the Gulf of Alaska can be interpreted in terms of what happens at a triple junction of the RRR type (Rift–Rift–Rift) (see Figures 10.4 and 10.5). Figure 10.33 is a schematic map of an RRR triple junction with the magnetic anomaly patterns it would generate, allowing the Great Magnetic Bight to fit into a portion of the Pacific plate. If this explanation is valid, there should have been two other plates (Plate "*X*" and Plate "*Y*") to complete the triple junction, and where these other two plates are, or were, is the mystery.

An answer was proposed in 1968 by Walter C. Pitman and Dennis E. Hayes of Lamont–Doherty Geological Observatory. Their reconstruction of the tectonic history of the North Pacific region is shown in Figure 10.34. During the late Cretaceous Period (−75 m.y.), the predicted triple junction existed. One of the two presently missing plates was called the Kula plate, the other, the Farallon plate. The triple junction migrated northeastward, producing the pattern of stripes we see today in the Great Magnetic Bight. When the western branch of the triple junction reached the subduction zone of the Aleutian Trench, the entire Kula plate simply disappeared, "down the drain" so to speak, and the spreading zone disappeared with it. The east-moving Farallon plate headed toward a subduction boundary along the western side of the American plate where it, too, was largely consumed, but, as we shall explain later in this chapter, two small portions of it remain today (the Juan de Fuca plate and the Cocos plate).

The problem with the Mesozoic magnetic anomalies of the western Pacific (Figure 10.35) lies in the fact that the northern set of stripes, called the Japanese lineations, runs about at right angles to a second set, the Hawaiian lineations, while a third set, the Phoenix lineations, runs about parallel with the Japanese lineations. Separating the three sets of stripes are wide gaps. The line of Emperor Seamounts, resting on a seafloor region without stripes, separates the Mesozoic anomalies from those of the Great Magnetic Bight. As we explained earlier in this chapter, areas without stripes

FIGURE 10.32 Magnetic anomaly map of the northeastern Pacific Ocean. Magnetic anomalies are represented by numbered lines and dotted lines. Healed transform faults (transform scars) offset the anomaly pattern as transverse fault zones. The numbered anomalies are approximately on the time boundaries between Cenozoic epochs and the Cretaceous Period. (Based on data of W. C. Pitman, R. L. Larson, and E. M. Herron, 1974, *Magnetic Lineations of the Oceans* (Map), Geological Society of America.)

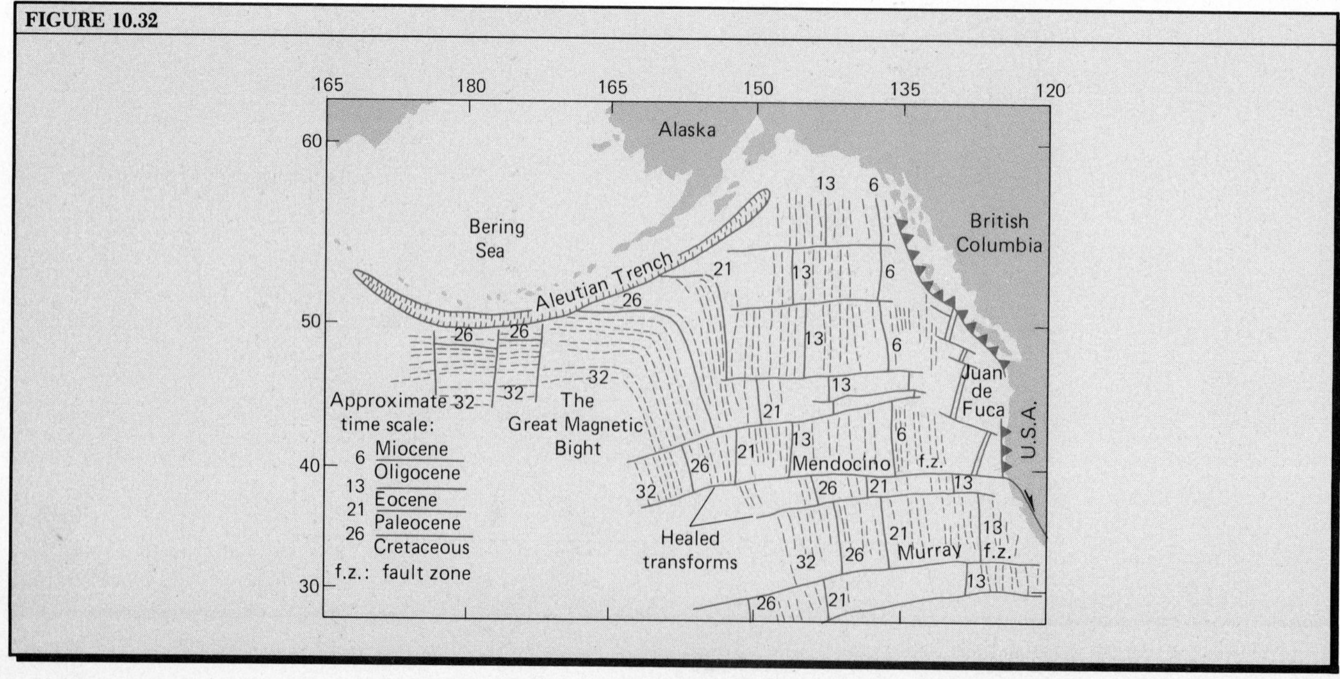

FIGURE 10.33 Schematic map of a triple rift plate junction with its magnetic anomaly pattern (dashed lines).

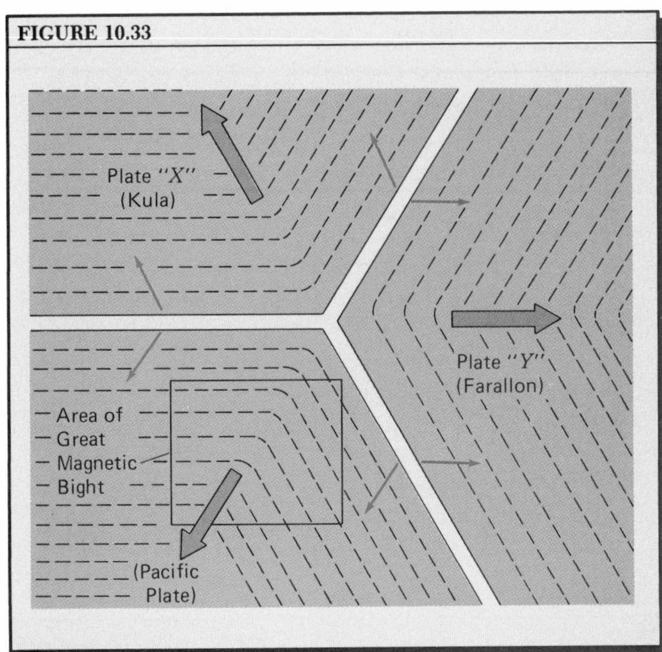

represent periods during which crust was being produced by seafloor spreading, but no reversals were taking place by means of which to produce a record of anomalies.

A full account of the events that produced the Mesozoic anomalies would be long, involved, and perhaps difficult to follow. We will, however, show how a pattern of stripes such as those we observe could have come about by seafloor spreading and plate accretion. Figure 10.36 is a schematic diagram making use of two triple junctions of the RRR type. One of these (that at the top of the figure) is the same RRR junction shown in Figures 10.33 and 10.34, except that we have now regressed in time another 40 m.y. to things as they might have been earlier in the Cretaceous Period,

FIGURE 10.34 Schematic maps of the northeastern Pacific Ocean basin showing the migrations of spreading plate boundaries and the disappearance of the Kula plate. (From W. C. Pitman and D. E. Hayes, 1968, *Jour. Geophysical Research*, vol. 73, p. 6577, Figure 4. Copyrighted by the American Geophysical Union.)

FIGURE 10.34

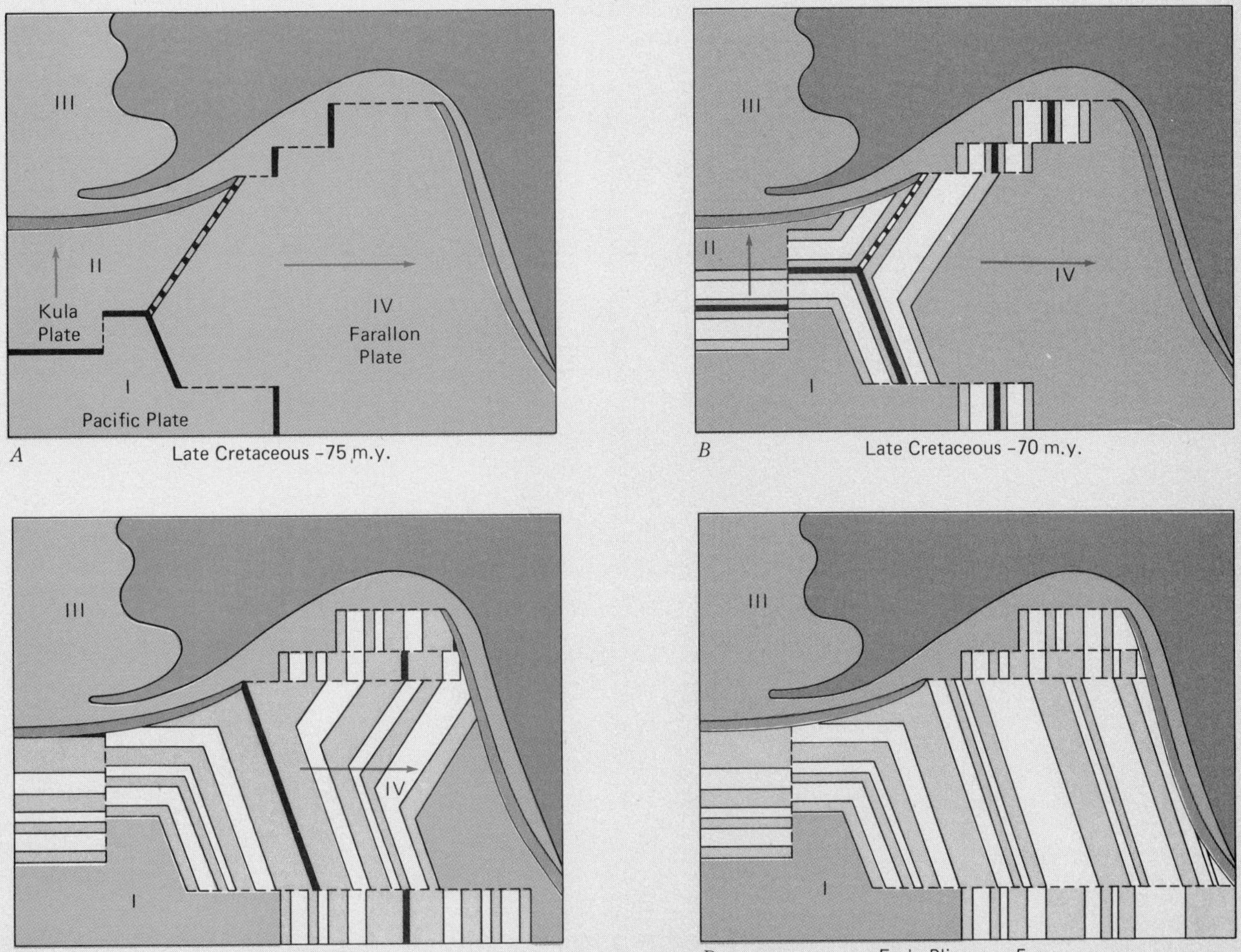

A Late Cretaceous -75 m.y.

B Late Cretaceous -70 m.y.

C Early Paleocene -60 m.y.

D Early Pliocene -5 m.y.

at about −110 m.y. The second triple junction lay to the southeast and involved another plate, the Phoenix plate, which formed a triple junction with the ancestral Pacific plate and the Farallon plate. We can consider the Pacific plate to be stationary with reference to the other three plates. If so, a threefold pattern of stripes emerges that resembles the Mesozoic anomaly pattern. We have outlined the relative position of three groups of stripes, from which you can see that the northern and southern groups have stripes about parallel in direction but arranged in opposite order of appearance from north to south. A third, intermediate, group of stripes is oriented about at right angles to the other two.

FIGURE 10.35 Map of the western Pacific Ocean basin showing the Mesozoic magnetic anomalies. (Based on data of W. C. Pitman, R. L. Larson, and E. M. Herron, 1974, *Magnetic Lineations of the Oceans* (Map), Geological Society of America.)

°*Evolution of the San Andreas Transform Boundary*

Returning to the history of the Farallon plate, let us look into the events that may have led to the development of the San Andreas Fault as a transform plate boundary. Although not all geologists agree, the San Andreas Fault—a right-lateral transcurrent fault—is thought by most to form the boundary between the Pacific plate and the American plate from the head of the Gulf of California to a point on the northern California coast near Cape Mendocino. (The San Andreas Fault is described in Chapter 8; see Figure 8.37.) During late Cretaceous and early Cenozoic, the east-moving Farallon plate had been undergoing subduction along the entire North American plate boundary from as far north as British Columbia to as far south as central Mexico (see Figure 10.34). As subduction continued, the North American plate and the spreading

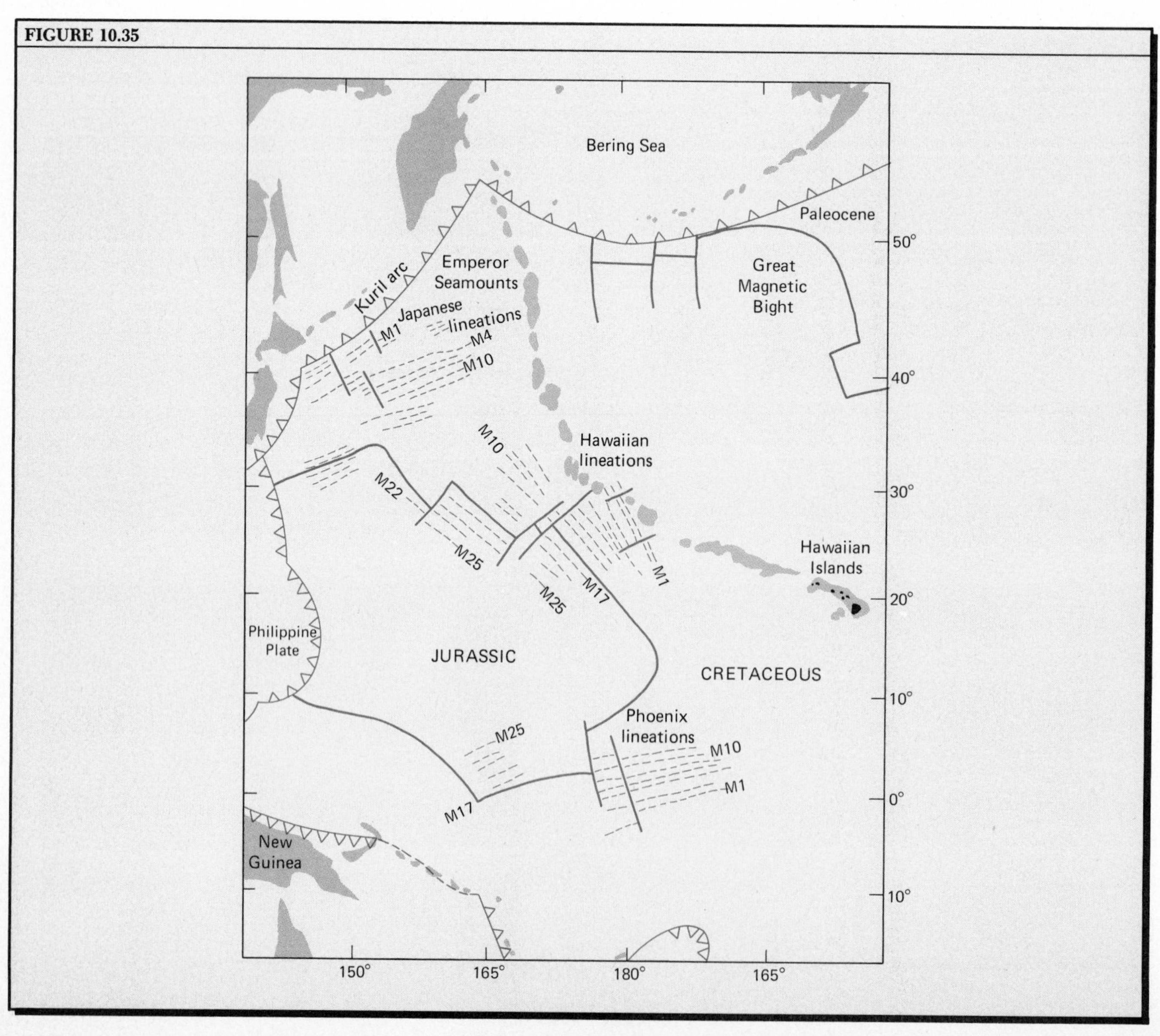

FIGURE 10.36 A schematic map showing how the Mesozoic magnetic anomalies fit into a scheme of four lithospheric plates separated by spreading boundaries.

FIGURE 10.37 Sketch maps showing the consumption of the Farallon plate and the formation of the San Andreas transform fault. (Based on data of T. Atwater, 1970, Implications of plate tectonics for the Cenozoic tectonic evolution of western North America, *Bulletin, Geol. Soc. Amer.*, vol. 81, pp. 3513–3536, Figures 5 and 18.)

FIGURE 10.36

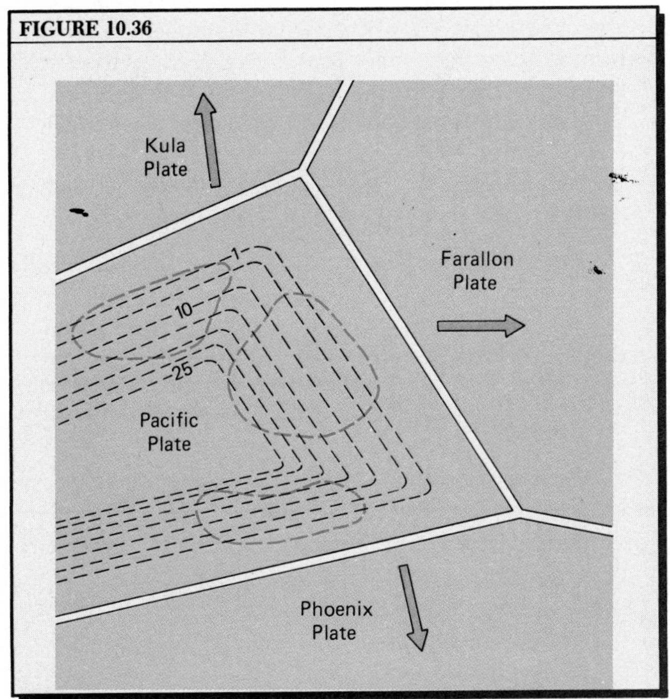

boundary between the Farallon and Pacific plates were coming closer together. Thus, the exposed surface area of the Farallon plate was shrinking in width.

Figure 10.37 is a series of maps showing the subduction of the Farallon plate beneath the American plate, starting with conditions in late Eocene, about −40 m.y. Notice particularly that the American plate is shown in Map *A* to be moving in a southeasterly direction, about parallel with the subduction boundary. The spreading axis between the Pacific and Farallon plates was offset by long transform faults, with the result that a salient (a promontory) of the Pacific plate first made contact with the American plate at a single point, shown in Map *B* to be located not far from the present-day position of Guyamas on the Mexican mainland. But the Pacific plate had a northwesterly motion with respect to that of the American plate and, as a result, the plate boundary here became a transform fault (Map *C*), and further subduction ceased along the transform boundary. As subduction of the Farallon plate continued both to the north and to the south, the new transform boundary increased in length (Map *D*). If we plot the locations of Los Angeles and San Francisco on the American plate and observe their locations on successive maps, we see that the transform boundary is extended in a relative northwesterly direction, passing first Los Angeles, then San Francisco. In Map *E*, showing present conditions, two remnants of the Farallon plate persist today. The northern remnant is the Juan de Fuca plate; the southern remnant is the extreme northwest corner of the Cocos plate. (This

FIGURE 10.37

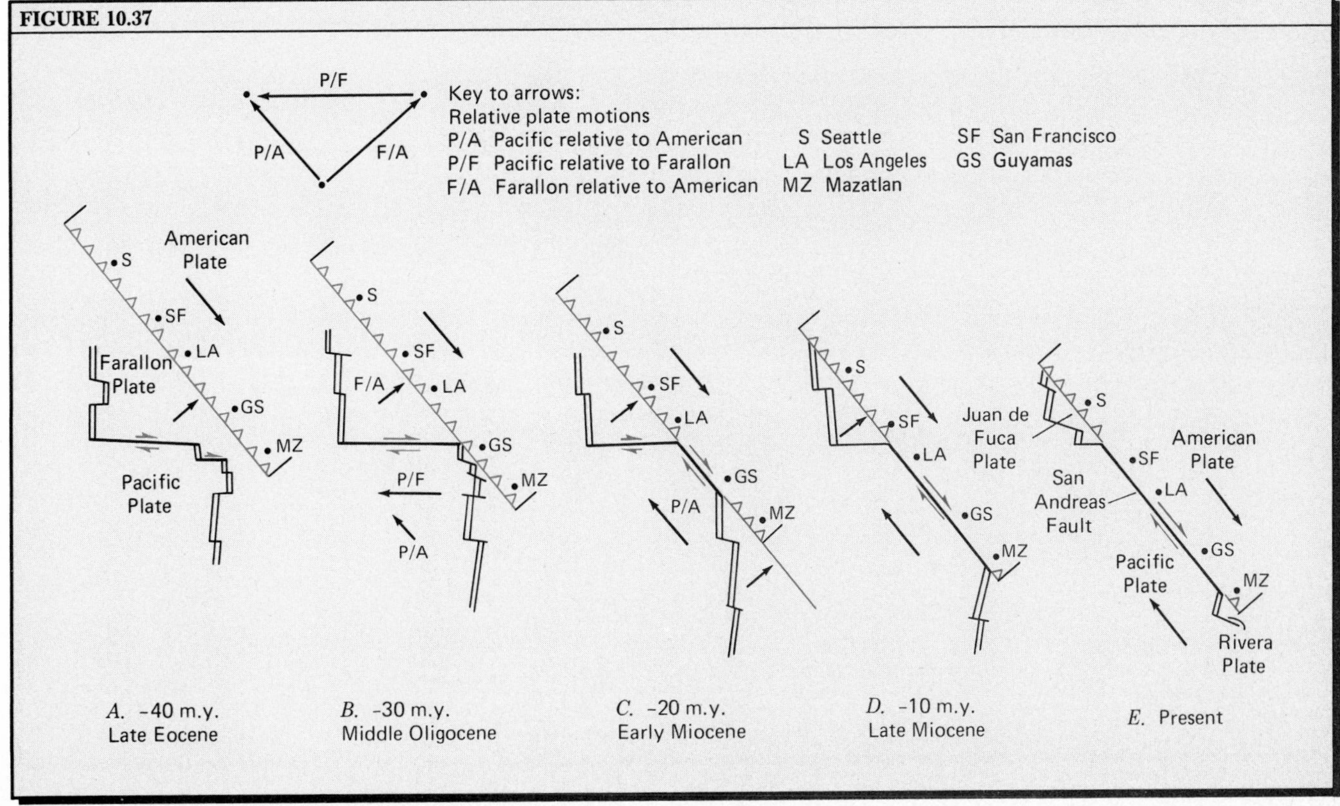

small plate remnant is sometimes considered to be a separate plate called the Rivera plate.)

Figure 10.38 shows by cross sections how subduction ceased when the Pacific plate reached the American plate.

At this point, you may have thought of a good question to ask about this hypothesis. The transform boundary shown to evolve in Maps *C, D,* and *E* of Figure 10.37 must lie along the contact between oceanic lithosphere and continental lithosphere. Instead, the San Andreas Fault, which is the present-day plate boundary, runs through continental lithosphere from the head of the Gulf of California to a point north of San Francisco. How do geologists explain the discrepancy? The only answer thus far proposed to deal with this problem is to postulate that the transform boundary made a sudden "jump" to a new position within the continental lithosphere. This jump is shown to have taken place between cross-sections *C* and *D* of Figure 10.38. The suggestion has been made that the continental lithosphere is weaker than the contact zone between continental and oceanic lithosphere, so that the transform boundary would not remain for long in its original location.

Subduction and the Oceanic Lithosphere

The break between this chapter and the next is quite arbitrary. We have not completed our study of the oceanic lithosphere because the consumption of that kind of lithosphere in subduction boundaries remains to be considered. Viewed from a different perspective, however, subduction plate boundaries are also those lines along which continental lithosphere meets oceanic lithosphere. New continental lithosphere is being formed along these same boundaries, both by the rise of magma that penetrates the continental crust and by addition of new belts of metamorphic rock which form above a downmoving plate margin or by collision. In Chapter 11, we resume the topic of plate tectonics with a study of trenches and island arcs.

FIGURE 10.38 Schematic cross sections of the Pacific, Farallon, and American plates showing the formation of the San Andreas Fault as a transform plate boundary. (Based on data of T. Atwater, 1970, Figure 15 of same reference as for Figure 10.37.)

FIGURE 10.38

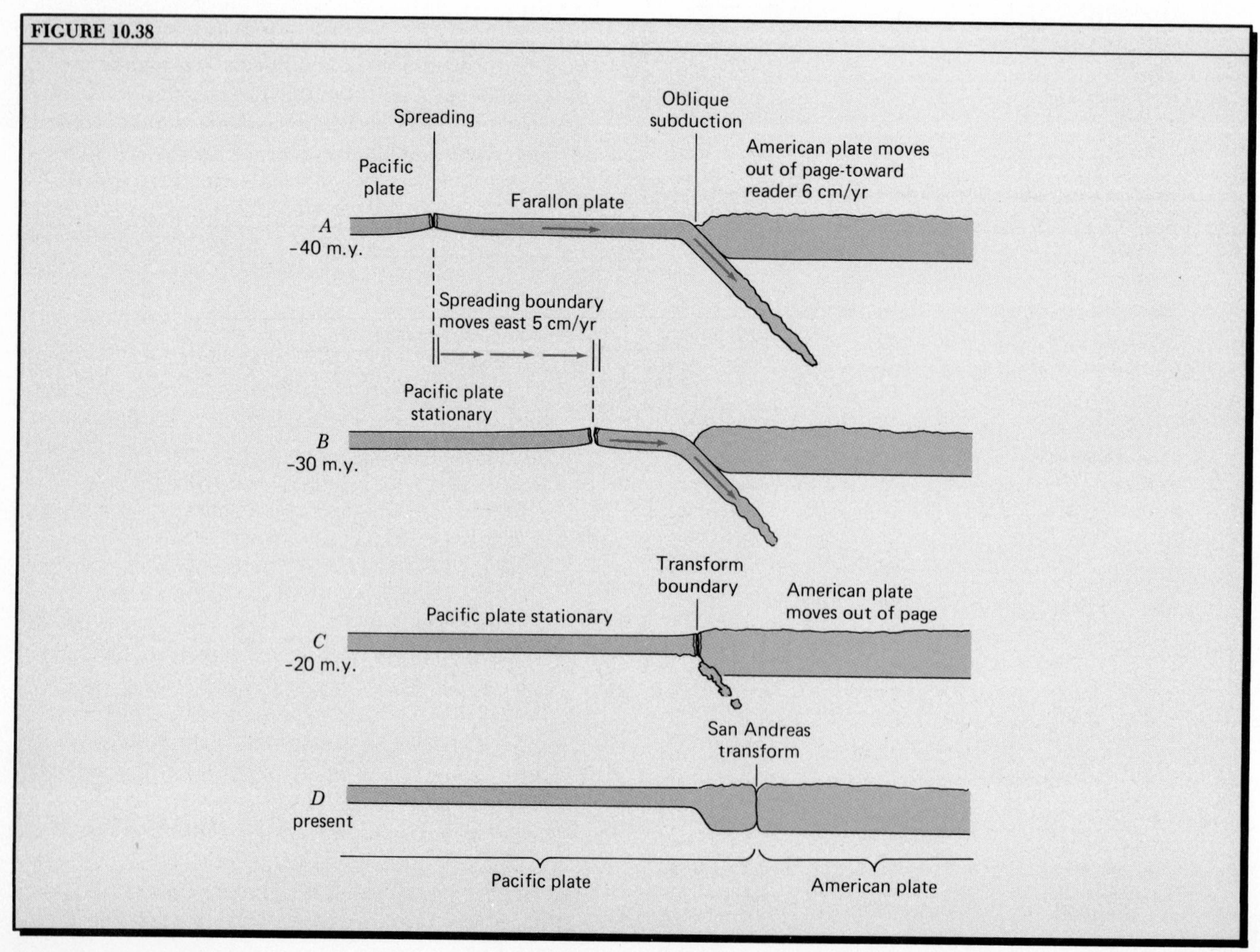

Key Facts and Concepts

Plate boundaries and junctions Lithospheric plate boundaries are of three types: spreading, converging, and transform. Every boundary shared by two plates—a *plate boundary segment*—terminates in a common *triple junction* with two other boundary segments. For a given plate, the number of plate boundary segments equals the number of triple junctions. In general, a given plate is recognized on the criterion that all of its boundaries are or have recently been tectonically active. Subplates can be recognized as incipient independent plates.

Triple junctions Triple junctions, limited in practice to nine varieties, are described by a letter code stating the combination of boundary types involved.

Transform faults One type of transform fault is an active plate boundary segment extending between spreading and converging boundaries of the same plate. A second type lies between offset ends of a spreading boundary and does not terminate in triple junctions. A third type similarly connects the offset ends of a converging plate boundary. Motion on a transform fault of the second type is opposite to that of a simple transcurrent fault suggested by the ridge offsetting. Continued movement on offsetting transform faults leads to the formation of *healed transform faults,* or *transform scars,* that were earlier described as fracture zones.

Plate rotation Motion of a lithospheric plate relative to another plate is described by a *pole of rotation* through which passes an *axis of rotation*. The radius of plate motion is an arc of a great circle; the path of a reference point on the plate describes a small circle. Two plates sharing a common transform boundary also share the same pole of rotation. *Absolute plate motion* is described in relation to the inner core and mantle.

Geomagnetism *Geomagnetism,* the earth's magnetism, takes the form of a magnetic field with a *geomagnetic axis* passing through the *magnetic poles*. Force lines pass through the earth's center and emerge at the surface. *Magnetic inclination* (dip), which measures the vertical angle between force lines and the horizontal plane, is zero in the vicinity of the *magnetic equator*. The entire external magnetic field is the *magnetosphere*. The *dynamo theory* explains geomagnetism through rotation of the liquid core with respect to the solid mantle.

Paleomagnetism *Magnetic minerals* are susceptible to natural magnetization. During cooling of magma, *thermal remnant magnetism* sets in below the *Curie point* and becomes permanent as *paleomagnetism* in the solidified magma. Using the *magnetometer,* direction and intensity of paleomagnetism can be measured from a rock specimen. *Detrital remnant magnetism* is acquired by waterlaid sediments during deposition.

Geomagnetic polarity reversals *Geomagnetic polarity* measured from rocks and sediments shows numerous past *polarity reversals* alternating between *normal epochs* and *reversed epochs*, with shorter reversal *events* within the epochs. Polarity reversal in sediments is determined by sign of magnetic inclination. A continuous chronology of polarity reversals dates back to the late Cretaceous Period (−83 m.y.) and includes 80 reversals.

Paleomagnetism and seafloor spreading *Magnetic anomalies* are recorded on magnetometer traverse lines crossing the mid-oceanic ridge, enabling polarity reversals to be recognized and mapped in parallel bands, symmetrical on both sides of the spreading boundary. Mirror symmetry of the magnetic anomaly stripes demonstrates seafloor spreading and allows spreading rates to be determined. Age of basaltic basement, determined from sample cores, is correlated with the magnetic anomaly pattern to map the age of the bedrock ocean floor.

Seismic evidence of transform-fault motion Shallow-focus earthquake epicenters lie along the spreading ridge axis or the connecting transform faults, but do not extend out along the transform scars. First motion studies verify the direction of transform fault movement.

Heat flow and spreading plates Heat flow through the ocean floors makes use of *thermister probes* into deep-sea sediment. Rates of heat flow are high along spreading plate boundaries, decreasing with distance away from the spreading axis because of plate cooling. Some heat is conducted out of the basaltic crust by circulating ocean water returning to the surface through fissures.

Seafloor spreading and pelagic sediments Thickness of pelagic sediment increases gradually with distance away from an oceanic spreading boundary. Carbonate sedimentation changes from calcareous ooze to siliceous ooze as floor depth exceeds the carbonate compensation depth. Thick pelagic clays dominate the oldest areas of ocean floors.

Crustal rock of the axial rift zone Direct observation shows fresh pillow lava with extension fissures in the floor of the axial rift zone of an oceanic spreading boundary. As seen exposed in Iceland, *sheeted dikes* are intruded below the zone of pillow lavas. Gabbro, formed by magma crystallization, lies below the sheeted dikes. This sequence, which includes ultramafic mantle rock (peridotite) below the crustal gabbro, constitutes an *ophiolite suite*.

Mantle plumes and hot spots A rising column of heated mantle rock, called a mantle plume, has been postulated to lie beneath a hot spot over which an isolated volcano or a volcanic group is extruded through the ocean floors or continental crust. Oceanic volcanoes form in chains over relatively stationary hot spots, giving evidence of the direction and relative rate of plate movement—for example, the Hawaiian–Emperor chain. Hot spots in Africa seem to have remained stationary on the continental lithosphere, suggesting that the African plate may be fixed relative to the asthenosphere.

Pacific plate magnetic anomalies Abrupt changes in trends of magnetic anomaly stripes in the Pacific basin are interpreted in terms of migrating triple plate junctions, leading to subduction and disappearance of a plate. Older magnetic lineations of Mesozoic age can be similarly interpreted.

Evolution of the San Andreas transform boundary The San Andreas transform boundary may have come into existence when the spreading boundary between two oceanic plates reached the converging boundary of the North American plate, replacing that boundary with a transform boundary.

Questions and Problems

1. Referring to the world map of lithospheric plates, Figure 10.3, identify two lenticular plates, each with only two boundary segments and two triple junctions. Describe the boundary segments and triple junctions of the Arabian plate. Would it be possible to have a single plate with no converging boundary along any part of its perimeter? What plate comes closest to fulfilling this condition?
2. Is it possible for the plates on the two sides of a spreading boundary to have relative motion in a direction other than at right angles to the boundary? Make a sketch map of such a situation, showing plate motions by arrows. Is it necessary that a plate undergoing subduction must always move at right angles (perpendicular) to the boundary line? Illustrate with a sketch map. If so, locate a place where diagonal motion is occurring.
3. Cut a single straight slit across the center area of a sheet of paper, being sure not to carry the slit to the edge of the sheet. Prove Professor Wilson's argument that motion of a pure transcurrent fault cannot be demonstrated without buckling or tearing the paper. Is it possible to demonstrate a normal fault with this same device? What happens to a normal fault when it is traced horizontally to its limits?
4. Would it be possible for a set of magnetic anomaly stripes to form on only one side of a spreading plate boundary? If so, explain how it would happen. If not, why not?
5. During a long magnetic quiet interval, magnetic anomaly stripes are not formed during seafloor spreading. Is there any other way in which to measure the rate of spreading that applied during a long quiet interval?
6. Of the plates listed in Table 10.1, which one do you nominate as the candidate for "first to disappear"? Give some specific information in support of your answer.

11

CHAPTER ELEVEN

Tectonics of the Active Continental Margins

The *continental margins*, broadly defined, are those long, narrow zones where oceanic lithosphere meets continental lithosphere. In recent years, the continental margins have been the object of an enormous research effort by a large number of geologists and geophysicists. There is, of course, a keen desire on the part of many research geologists—particularly those on the faculties of colleges and universities—to investigate the continental margins as storehouses of untapped scientific information relating to many fields of geology, but there is another compelling reason for concentration of research on the continental margins: Present and former continental margins hold most of the world's petroleum resources. A large number of professional geologists in private industry direct their research to basic problems of petroleum accumulation. Their work is complemented by research conducted by governmental geologic and oceanographic organizations, such as the U.S. Geological Survey and the U.S. Naval Oceanographic Office. Small wonder, then, that we have witnessed in the past decade an enormous increase in scientific knowledge of the continental margins.

Passive and Active Continental Margins

As we have defined the term, the continental margins fall largely into two great classes. First are the *passive margins*, where the meeting line of continental and oceanic lithosphere falls within a single lithospheric plate. This inactive type of margin was illustrated in Figures 1.14 and 1.15; it will be the subject of our detailed investigation in Chapters 12 and 13. The passive margins were formed by the rifting of a continental lithospheric plate and the growth of new oceanic lithosphere between the two separating continental plates. At present, because the greatest extent of the world's passive continental margins is along the borders of the Atlantic Ocean basin on both western and eastern sides, the passive margins are sometimes called Atlantic-type margins. They are also found on other margins of the continental plates that moved apart following the breakup of Pangaea, so they border the Indian Ocean basin and the Arctic Ocean basin as well.

The second major class of continental margins consists of *active margins* associated with plate boundaries that are presently or recently tectonically active. While the active continental margins can be located along any one of the three basic kinds of plate boundaries, by far the most important in total extent are the converging plate boundaries where subduction is in progress. Here lithospheric plates bend down to become dipping slabs that plunge deep into the asthenosphere. Subduction boundaries ring the Pacific Ocean basin, so it is these Pacific-type continental margins that largely concern us in this chapter.

Figure 11.1 is a special map of the Pacific Ocean basin and its enclosing continental margins. The map

FIGURE 11.1

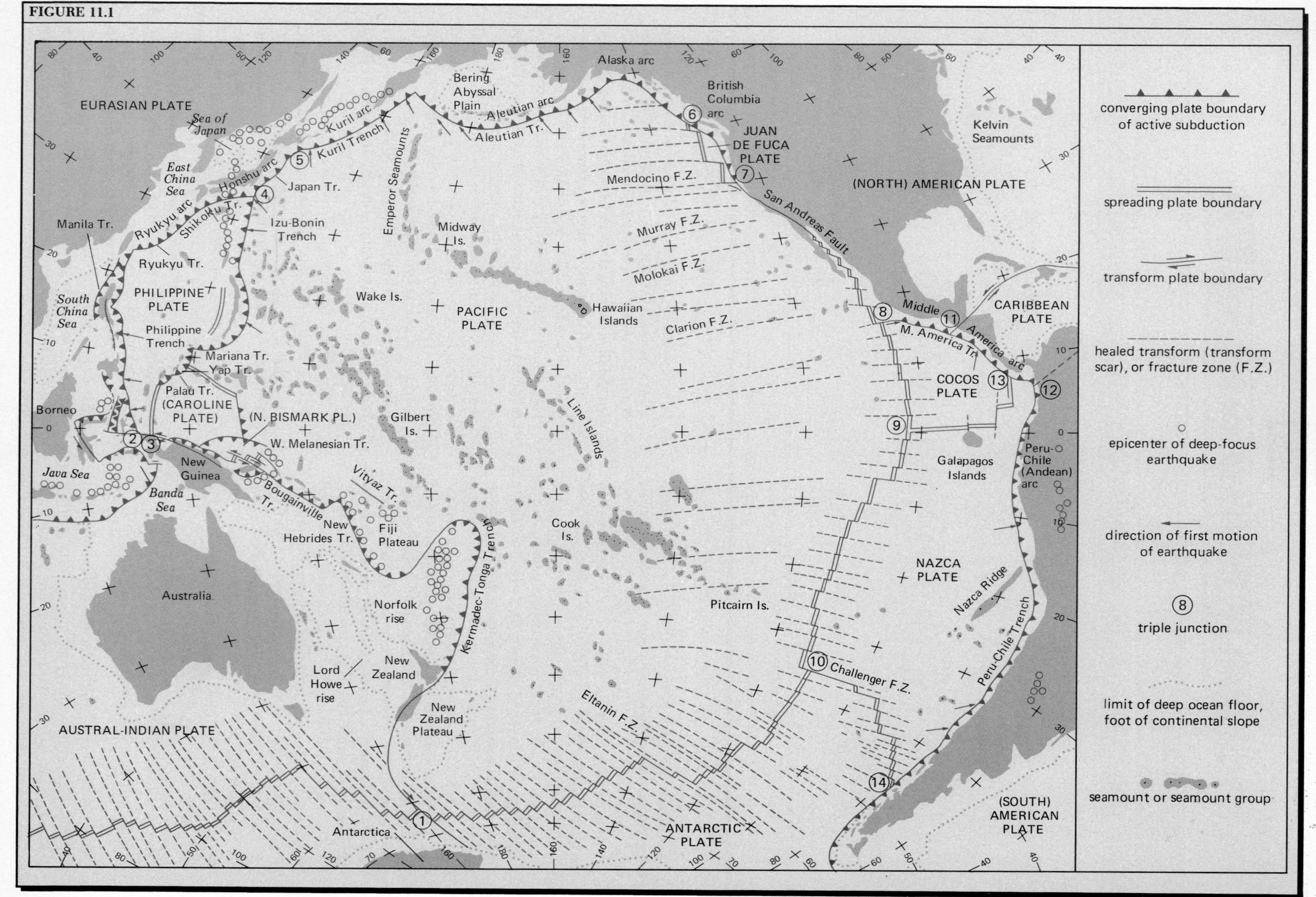

EURASIAN PLATE
Sea of Japan
East China Sea
Manila Tr.
Ryukyu arc
Shikoku Tr.
Honshu arc
Ryukyu Tr.
PHILIPPINE PLATE
South China Sea
Philippine Trench
Mariana Tr.
Yap Tr.
Palau Tr.
(CAROLINE PLATE)
Borneo
Java Sea
Banda Sea
New Guinea
(N. BISMARK PL.)
W. Melanesian Tr.
Bougainville Tr.
New Hebrides Tr.
Vityaz Tr.
Fiji Plateau
Australia
Norfolk rise
Lord Howe rise
New Zealand
New Zealand Plateau
AUSTRAL-INDIAN PLATE
Antarctica
Kuril arc
Kuril Trench
Japan Tr.
Izu-Bonin Trench
Emperor Seamounts
Bering Abyssal Plain
Aleutian arc
Aleutian Tr.
Midway Is.
Wake Is.
Gilbert Is.
PACIFIC PLATE
Hawaiian Islands
Line Islands
Cook Is.
Kermadec-Tonga Trench
Alaska arc
British Columbia arc
JUAN DE FUCA PLATE
Mendocino F.Z.
Murray F.Z.
Molokai F.Z.
Clarion F.Z.
Pitcairn Is.
Eltanin F.Z.
ANTARCTIC PLATE
San Andreas Fault
(NORTH) AMERICAN PLATE
Kelvin Seamounts
CARIBBEAN PLATE
Middle America arc
M. America Tr.
COCOS PLATE
Galapagos Islands
Peru-Chile (Andean) arc
NAZCA PLATE
Nazca Ridge
Peru-Chile Trench
Challenger F.Z.
(SOUTH) AMERICAN PLATE
converging plate boundary of active subduction
spreading plate boundary
transform plate boundary
healed transform (transform scar), or fracture zone (F.Z.)
epicenter of deep-focus earthquake
direction of first motion of earthquake
8 triple junction
limit of deep ocean floor, foot of continental slope
seamount or seamount group

is special in the sense that the gridwork of parallels and meridians on which it is drawn is centered at a point on the equator in mid-Pacific. The grid is symmetrical in its scale properties outward in all directions from the center. The Mercator projection used as a base for our world map of lithospheric plates (Figure 10.3A) greatly expands the map scale in high latitudes, such as the Aleutian–Alaskan region. In contrast, the Pacific-centered map has only small scale changes over its entire area of coverage. There is some distortion of the outline of the continental shorelines, but you can correct for this problem by noting the trend of the parallels and meridians shown on the map. Other important active margins, not shown on the Pacific map, are the Sumatra–Java arc of Indonesia, the Lesser Antilles volcanic arc of the eastern Caribbean Sea, the Scotia arc in the South Atlantic, and a subduction zone in the eastern Mediterranean Ocean basin (for these, see Figure 10.3).

The active continental margins that are subduction zones fall into two broad classes. One consists of mountain arcs along the borders of the large continental plates. Here the trench that marks the line of subduction lies a short distance off the coast, with only a narrow shelf zone. Examples are the Alaskan arc, the Peru–Chile arc of South America, and the Sumatra–Java arc of Indonesia. A second class consists of island arcs separated from the main bodies of continental lithosphere by small ocean basins called backarc basins, described in Chapter 9. These basins are floored by oceanic crust. The island arcs are usually bowed convexly outward from the continents toward the central ocean basin. One important task that faces us in this chapter is to explain the island arcs and their backarc basins.

The active continental margins as a class also include major transform plate boundaries that lie close to the contacts of oceanic with continental lithosphere. Also included are young zones of continental rifting where a spreading boundary lies in the axis of a narrow gulf, such as the Red Sea or the Gulf of California. This rifted type of margin will be a subject of special investigation in Chapter 13.

Deep-Focus Earthquakes and Plate Subduction

If we accept the evidence of continued accretion of oceanic lithosphere at spreading plate boundaries, we must look for equally strong evidence that oceanic lithosphere is being consumed at converging plate boundaries. Otherwise, the material flow cycle of plate tectonics could not form a complete circuit capable of being sustained for hundreds of millions of years.

The concept of disposal or disappearance of oceanic crust into trenches appeared long before plate tectonics was fully developed. The task of establishing the phenomenon of plate subduction fell to the seismologists and they put on a brilliant scientific performance. Proof that lithospheric plates do, indeed, plunge deeply into the asthenosphere came slowly at first, as the locations and depths of earthquake foci were plotted on the map in increasing numbers and the points began to fall into significant global patterns. These early data led to cautious inferences, but progress was slowed by a seeming contradiction that arose to confront the researchers. Finally, modern seismology applied its sharpest tools to a great fund of new data and revealed remarkable details of plate subduction and of the forces acting within a descending plate.

To understand the nature of the seismic evidence of plate subduction, consider the case of the island arcs made up of northern Japan and the Kuril Islands (Figure 11.2). Adjoining these arcs on the east is the Japan Trench, and to the west lie the Sea of Japan and the Sea of Okhotsk, which are backarc basins. Earthquakes are frequent in this region, and their epicenters have been plotted in great numbers. Shallow quakes are abundant in a zone immediately adjacent to the trench on the landward (western) side; those of intermediate depth occur under the island belt and under the Sea of Japan. Those of deep focus, however, are centered under the margin of the Asiatic mainland. When we plot these centers on a vertical cross section, as is done on the face of the block diagram in Figure 11.2, it is obvious that they define a slanting zone along which internal fracturing has been occurring.

If you now examine closely the map of the Pacific Ocean basin, Figure 11.1, you will find that the same relationship of shallow and deep earthquakes to trenches is present both in the Kermadec (Tonga) Trench zone of the southwestern Pacific Ocean and along the western side of South America. The slanting plane of earthquake foci in these belts was recognized as early as 1930 by a Japanese seismologist, K. Wadati, and shortly thereafter by an American seismologist, H. Benioff. In the scientific publications of western seismologists, the feature became known as the *Benioff zone*, but it is now proper to refer to it as the *Wadati–Benioff zone*.

The existence of the Wadati–Benioff zone posed some very difficult problems. If, as seismologists hold, earthquakes are generated by stick-slip movements on fractures in brittle rock, the occurrence of the deep earthquakes with foci far down in the mantle seems particularly puzzling. The seismologists were assuming, of course, that at temperatures prevailing in the mantle, the rock there would not have the strength

FIGURE 11.1 Tectonic map of the Pacific Ocean basin and its surrounding continental margins. (Based on data of B. C. Heezen and D. J. Fornari, Geological Map of the Pacific Ocean, *World Geological Atlas*, UNESCO, Paris, France, and other sources.)

and brittle properties necessary to produce earthquakes. Instead, the soft mantle rock would yield by slow flowage and would continually relieve any unequal stresses that might tend to build up. On the basis of this reasoning, it appeared quite unlikely that a slanting fault zone could be responsible for the deep earthquakes.

When the concept of a descending lithospheric plate was introduced, the problem seemed largely solved. Here we have a rigid and comparatively cold rock slab penetrating deep into the mantle and providing the right kind of material to yield by the fracturing mechanism that generates earthquakes. But there remained some unanswered questions.

First, as the descending plate moves into the hot asthenosphere, will it not be rapidly heated and become too soft to produce earthquakes? The answer to this question seems to lie in the rapid rate of descent of the cold slab. If it moves down rapidly enough, it will remain cool and brittle, at least in the inner part of the plate. Cracking of the brittle zone could then continue, even when the plate has reached a depth as great as 600 to 700 km.

The next question was this: How do we know that the descending lithospheric plate remains cold enough to generate earthquakes? Some independent evidence of its internal temperature is needed.

This evidence was also supplied by seismologists who studied the way in which seismic waves generated by deep earthquakes are transmitted along the sloping Wadati–Benioff zone. They discovered that something about the property of the rock along this travel path causes wave velocity to be faster and to undergo less weakening by absorption of energy than waves transmitted directly upward through the mantle. Now, it is known that seismic wave velocity is greater and energy absorption is less for a given kind of rock when the rock temperature is lower. This seismic evidence strongly supports the hypothesis that the descending slab is cooler than the surrounding mantle.

Figure 11.3 is a cross section of a descending lithospheric plate showing by means of isotherms (lines of equal temperature) the temperature conditions postulated for a plate descending at a rate of 8 cm/yr after subduction has been in progress for about 6.5 m.y. The plate is outlined and the foci of earthquakes are shown schematically. This idealized model seems to fit conditions known to exist under Japan and in the subduction zone of the Tonga Trench in the southwestern Pacific basin.

Some further support for the hypothesis of a cold lithospheric slab descending into the asthenosphere is found in observations of heat flow at the surface. Generally, heat flow is lower than average in the floors of the trenches, the value being less than one heat-flow unit. This is what we would expect, because the cool mass of the oceanic lithospheric plate has been carried to a greater depth. Heat flow is greater than average (more than two units) in the surface belt that lies directly above the middle and lower part of the descending slab. These conditions are shown by a profile line in Figure 11.3. The higher rate of heat flow over the deeper part of the slab does not seem to be expected by this hypothesis and will require some further dis-

FIGURE 11.2 Block diagram of the Japan–Kuril arc showing how earthquake foci are distributed in the crust and mantle beneath. (Based on data of B. Gutenberg and C. F. Richter, 1949, *Seismicity of the Earth*, Princeton Univ. Press, Princeton, N.J.)

FIGURE 11.3 Schematic cross section of a descending lithospheric plate. Isotherms are drawn for each 400 C°. A profile of heat flow is shown above. (From M. Nafi Toksöz, 1972, *Technology Review*, vol. 75, No. 2, p. 28. Copyright © 1972 the Alumni Association of the Massachusetts Institute of Technology.)

FIGURE 11.2

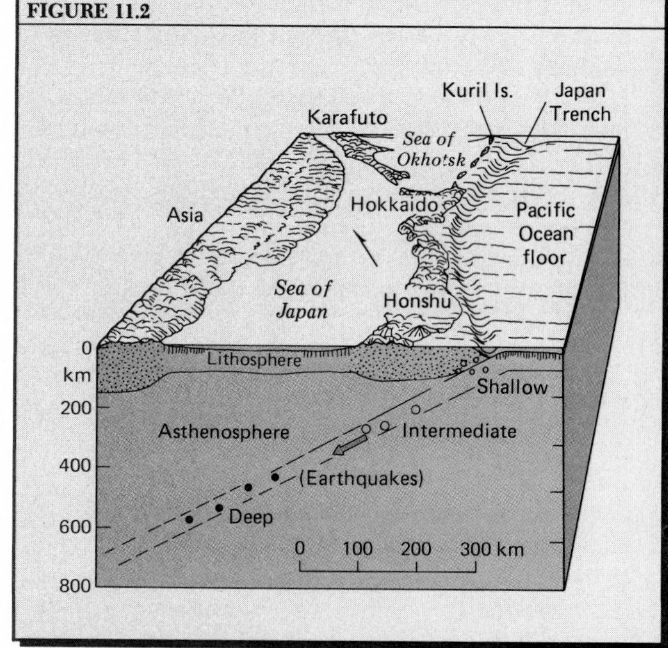

FIGURE 11.3

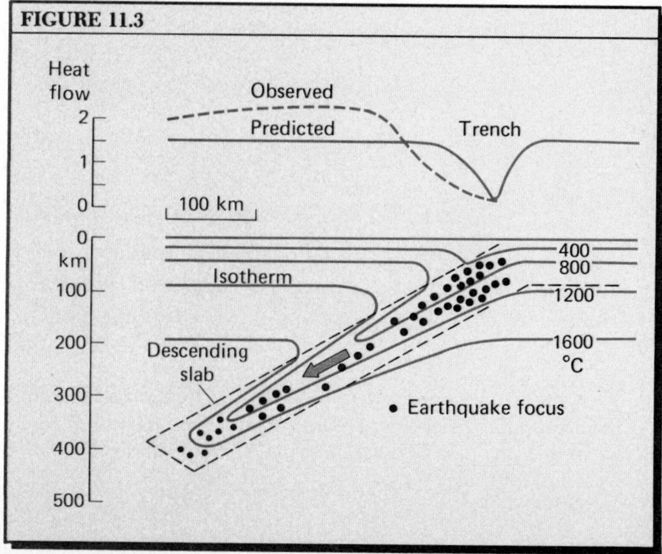

cussion. The same conditions of rates of heat flow apply also to the Peru Trench, in which the rate is low, and the adjacent Andes volcanic range, where the rate is high. The presence of chains of active volcanoes adjacent to a trench seems to account for the high rate of heat flow because magma is rising from depth. The puzzling part is that a cold descending slab should furnish the surplus heat needed for the formation of magmas. One simple explanation you might think of is that the heat is generated by friction between the descending slab and the surrounding mantle. We will discuss this problem later in the chapter and find that another, more workable explanation can be offered for the formation of magma above the descending plate.

*Stress Directions Within a Descending Plate

Seismology has more to offer on the subject of subduction of lithospheric plates. To present this additional material, we need to look further into the nature of the stresses that produce an earthquake. In Chapter 8, while explaining the production of earthquakes along a fault surface separating two blocks, we made use of a pair of opposite-facing but parallel arrows suggesting that the forces causing the fault slip act parallel with the fault plane, as shown in Figure 11.4*A*. The simple relationship of two opposite but parallel forces is called a *force couple*. In the 1930s, a Japanese seismologist, H. Honda, and his colleagues showed that the complete picture of forces that produce an earthquake is a *double couple*, in which two force couples are perpendicular to one another, as shown in Figure 11.4*B*. Now, if we substitute for each pair of crossing arrows a single arrow that is the vector sum of the pair, we produce two different pairs of arrows, as shown in Diagram *C*. One pair, meeting head-on, represents an axis of compressional stress (see Chapter 6 and Figure 6.12), and will tend to compress rock acted upon by these opposed forces. The pair of arrows pointing in opposite directions represents an axis of tensional stress, which will tend to pull apart or elongate rock subject to it. Earthquakes can be produced by both kinds of stress—one represents a yielding to compressional stress that exceeds the strength of the rock, the second represents a yielding of the rock to tensional stress that exceeds its strength.

FIGURE 11.4 Force couples and their resolution into axes of compression and tension. (After S. Uyeda, 1978, *The New View of the Earth*, p. 138, Figure 5–7. Copyright © 1971 by S. Uyeda; © 1978 by W. H. Freeman and Company. Used by permission of author and publisher.)

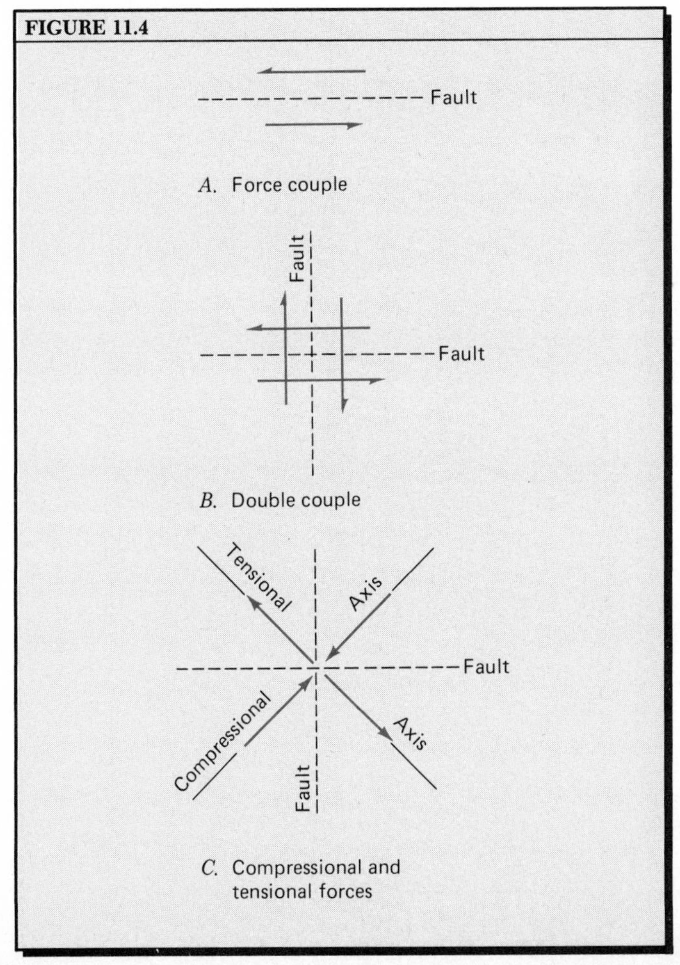

To distinguish whether an earthquake has been generated by compressional or tensional stress requires study of the first motion of S-waves. This operation proved to be very difficult, but a method was finally perfected. Japanese seismologists began to determine the type of stress and its direction for intermediate and deep earthquake foci beneath the Japanese islands. They found that the axis of compressional stress, when plotted on a map, is consistently at right angles to the trend of the trench and volcanic arc associated with it (Figure 11.5). Seen in cross section, the axis is also consistently tilted so as to be parallel with the plane of the earthquake foci. This evidence strongly supported the hypothesis of a lithospheric plate being forced to descend, but it also showed that the earthquake-producing force acts parallel with the sloping plate and is not a horizontal force. Why is the stress axis parallel with the plate? The answer seems to lie in the resistance offered by the surrounding mantle to the downward motion of the rigid plate. As the leading edge of the slab passes through the asthenosphere and begins to penetrate the stronger mantle that lies deeper, resistance to further penetration begins to build. The force of resistance is thus passed back up the rigid slab as a compressional stress.

Later, the same kind of seismic analysis revealed that in many subduction zones stresses causing intermediate-depth earthquakes near the top of the downplunging slab are of the tensional type (Figure 11.6). Apparently, the upper part of the slab is under tensional stress at the same time that the deep portion is under compressional stress. Why is the slab being "stretched" in its upper part? The answer offered to this question is that the slab is sinking under the force of gravity. As it moves into the low-strength zone of the asthenosphere, resistance to downsinking is diminished, and the plate tends to sink faster, causing a stretching force in the plate in this upper region. Simultaneously, the leading edge of the same plate may be under compressional stress because it is penetrating stronger mantle material.

The interesting implication of these stress-axis data is that they require us to discount the idea that a plate is being forced to descend by horizontal compressive stress exerted by horizontal plate motion on either side of the subduction zone. Instead, it appears that the plate is being pulled down by gravitational force be-

FIGURE 11.5 Short lines on this map show the directions of greatest horizontal compressional stress at the foci of intermediate and deep earthquakes beneath the islands and the Sea of Japan. The smaller diagram shows the axes of compressional stress in a vertical cross section of the map zone A–B. (Data of H. Honda and M. Ichikawa. From S. Uyeda, 1978, *The New View of the Earth*, pp. 139 and 140, Figures 5–8 and 5–9. Copyright © 1971 by S. Uyeda; © 1978 by W. H. Freeman and Company. Used by permission of author and publisher.)

FIGURE 11.6 Schematic diagrams of a lithospheric plate plunging into the asthenosphere. Locations of tensional and compressional stresses are shown by closed and open circles, respectively. (After B. Isacks and P. Molnar, 1969, *Nature*, vol. 223, p. 1121.)

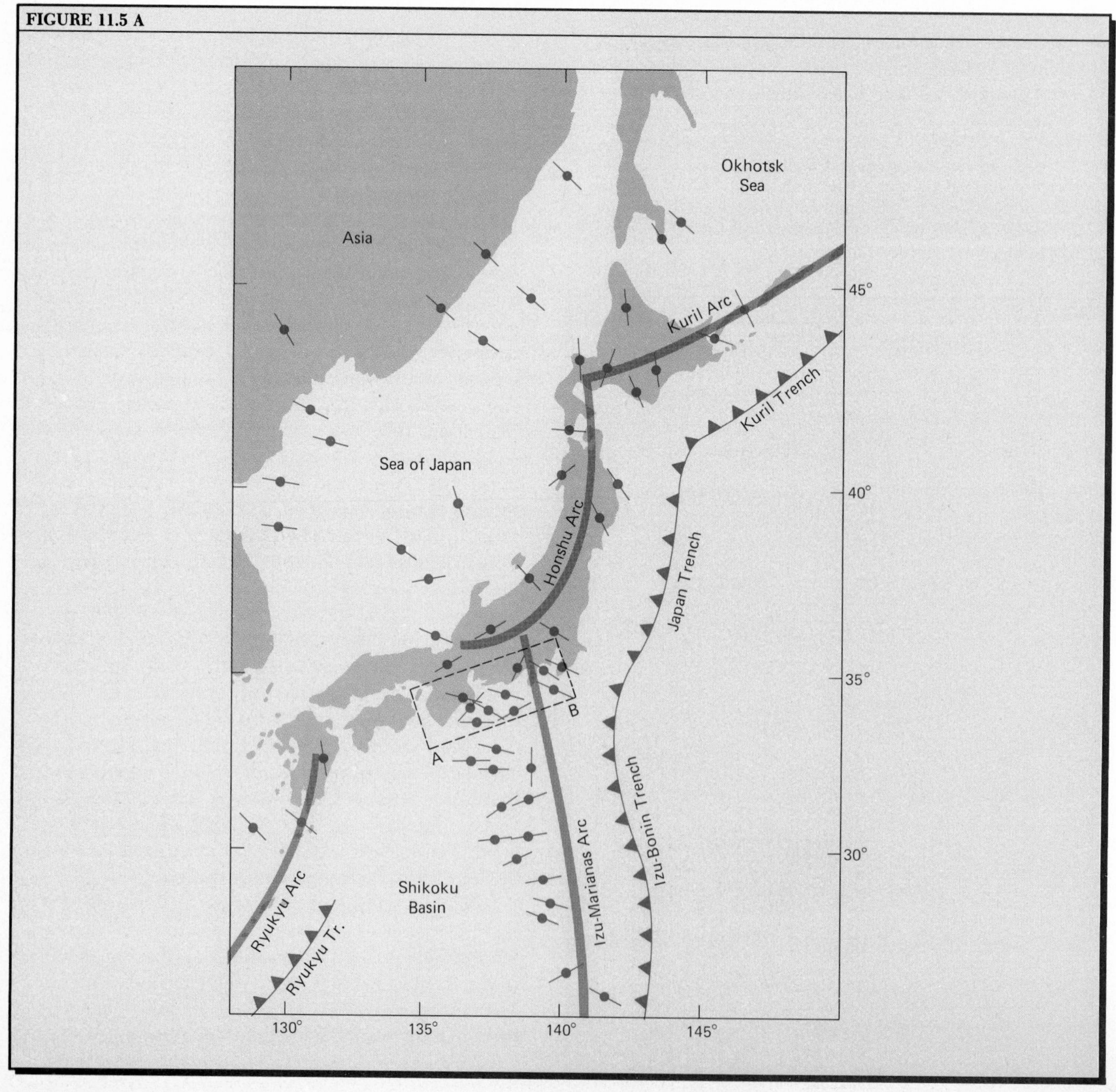

cause it is much colder, and therefore denser than the surrounding high-temperature asthenosphere. We shall refer again to this sinking force as we review the various possible forces that drive plate motions. The idea that the downbent edge of an oceanic lithospheric plate, once it breaks free from contact with a continental plate, can rapidly "founder of its own weight" seems, at first thought, to contradict the doctrine of isostasy, which holds that the lithosphere is floating at rest on the denser asthenosphere that lies beneath it. We will provide an explanation for this seeming paradox later.

Once the hypothesis of a descending cold lithospheric plate is tentatively accepted on the basis of the seismic evidence thus far presented, there comes an opportunity to fit another bit of seismic activity into the picture. Recall from Chapter 8 that in the case of major earthquakes observed both in Japan and Alaska, changes in land surface elevation preceded and followed the earthquake. In the Japanese earthquake, the coastal zone adjacent to the trench was observed to have gradually subsided prior to the quake, then suddenly to have risen following the quake. The same type of postquake rise in ground level was observed for the Good Friday Earthquake near Anchorage, Alaska. The explanation of these land movements may lie in the effect of downward motion, or *underthrusting*, upon the continental lithospheric plate. As shown in Figure 11.7, the steadily sinking plate tends to pull down with it the edge of the overlying continental plate. This downbending occurs prior to the earthquake, when elastic strain is stored in the bent continental plate. When the earthquake occurs, the bent plate rebounds upward. The mechanism of elastic rebound thus seems to hold for shallow earthquakes in subduction zones, but the direction of fault motion is quite different from that on a transcurrent fault.

FIGURE 11.5 B

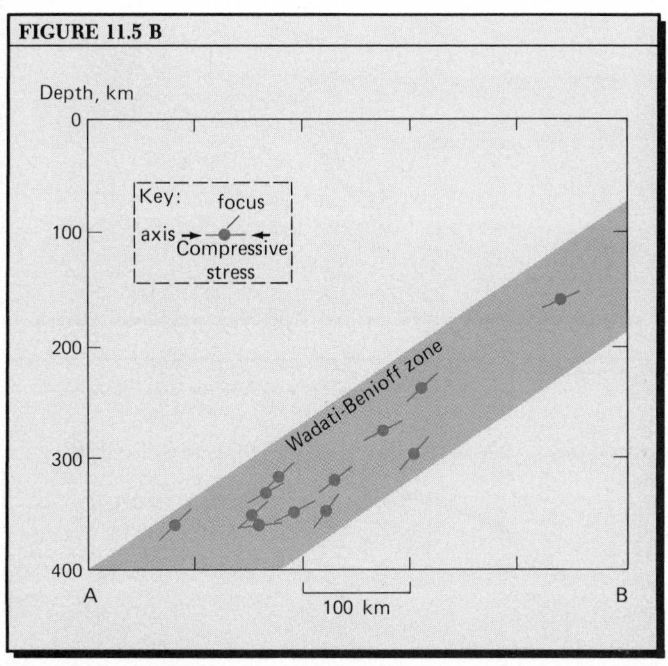

Mantle Convection Systems

Now that we have presented some rather detailed evidence of the entire cycle of lithospheric plate development from accretion to consumption, we need to focus upon the mechanisms and forces causing plates to move. Throughout the early stages in development of plate tectonics as a general theory, scientific thought about the forces sustaining motion of the lithosphere was strongly centered on the mechanism of *mantle convection*. In Chapter 10, we introduced an early model of mantle convection as suggested by Arthur Holmes (see Figure 10.10).

Convection, as the term is used in physical science, simply means the spontaneous rising and sinking of fluid masses because of density differences from one place to another. In most cases, convection develops in a fluid body because of the need to transport heat

FIGURE 11.6

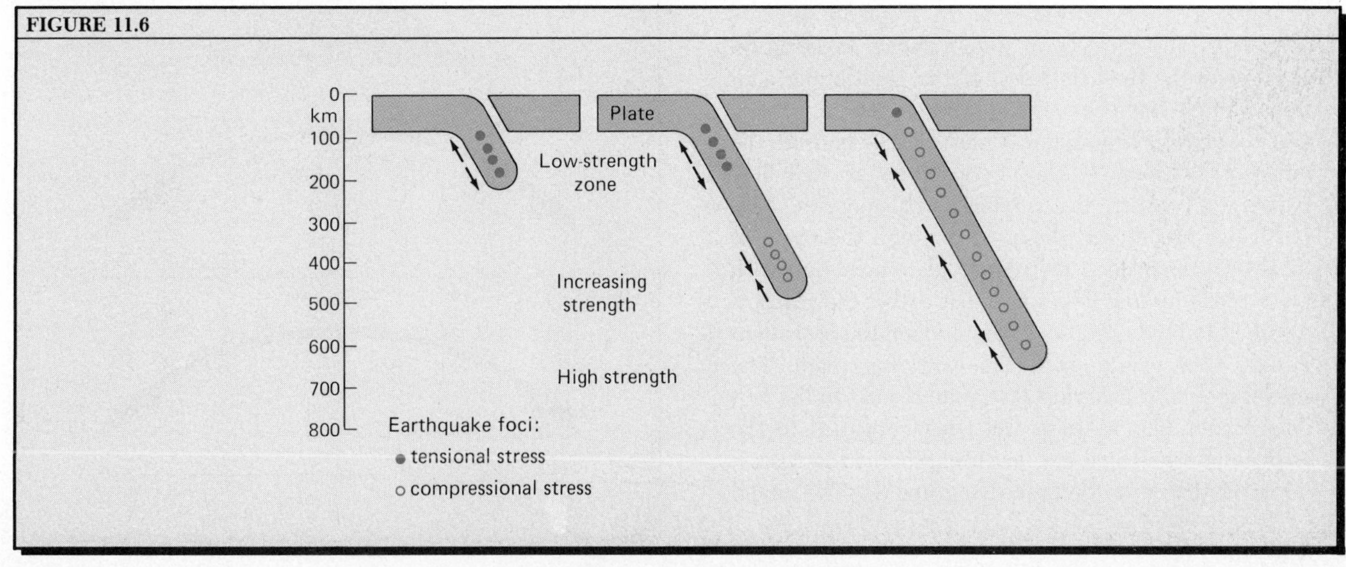

from one point in the system to another. Simple convection can be illustrated by a container of water placed over a flame. Application of intense heat at the bottom of the water body produces a warm water mass of low density, which rises to the top where heat is disposed of. Sinking currents of cooler water replace the rising warm water. Carefully controlled laboratory experiments with various fluids, such as melted paraffin, show that sustained convection takes the form of *convection cells*. The heated fluid rises within the center of the cell and descends in the boundary zone between cells. The cells are typically of hexagonal outline and more or less equal in size.

When applying the concept of convection to plate tectonics, it is important to keep in mind that the lithosphere must be included within the convection system. Suppose that we have devised a model convection system using a waxlike substance that quickly solidifies when exposed at the free upper surface of the liquid. We could thus produce a miniature replica of a lithospheric plate. The plate would come into existence by solidification of the molten wax as it emerges from a rising convection cell and spreads laterally. Our plate of solid wax would then move with the underlying fluid wax until it came to a descending convection zone. Here the wax plate would bend down and begin its descent, but the edge would quickly melt as it absorbed heat from the surrounding molten wax. The point is that the wax plate is an integral part of the matter that makes up the convection system. It is a part of the convecting mass that changes from a fluid state to a solid state and back to a fluid state because of cooling and heating. In the same way, an oceanic lithospheric plate in motion represents the top part of the convection system, which is not merely confined to the soft asthenosphere below the plate. With this point clear, we can continue with a discussion of convection systems that may operate in conjunction with lithospheric plate movements.

Three of the models of proposed mantle convection systems are shown in Figure 11.8. All three share in common a rising current beneath a spreading plate boundary. The convection system shown in Diagram *A* involves the total thickness of the mantle and contains a single large convection cell beneath each major plate of oceanic lithosphere. Other cells lie beneath the adjacent continental lithospheric plates, so that flow converges beneath the subduction boundaries. This particular model was designed to fulfill the cycle of mass transfer needed to balance plate accretion with plate consumption. The force that drives the plates is assumed to be the frictional drag which the horizontal mantle flow exerts upon the overlying plate. The model shown in Diagram *B* is quite similar to the first one, except that it limits the convection cell to the asthenosphere, which has low strength and low viscosity as compared to the underlying mantle. The model

FIGURE 11.7 A greatly exaggerated cross section of alternate elastic bending and rebound of the edge of the continental plate at a subduction boundary. The actual amount of deformation is much too small to be shown to true scale here.

FIGURE 11.8 Three models of mantle convection systems shown in schematic cross section.

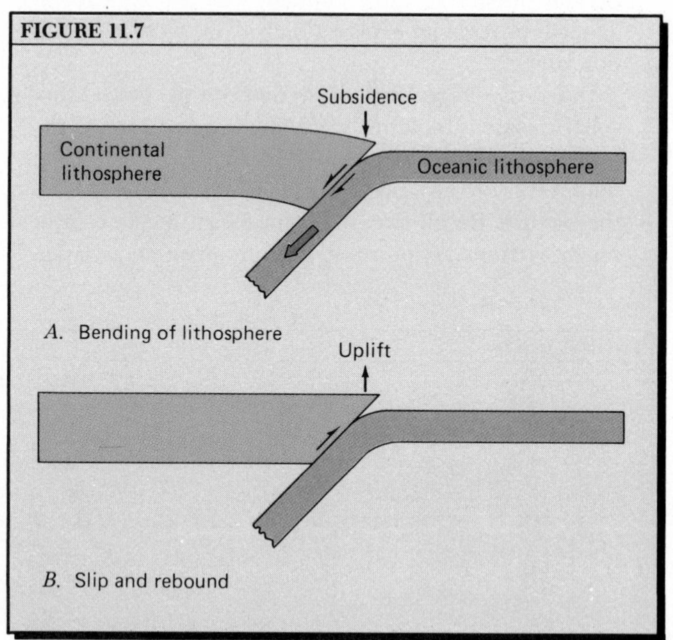

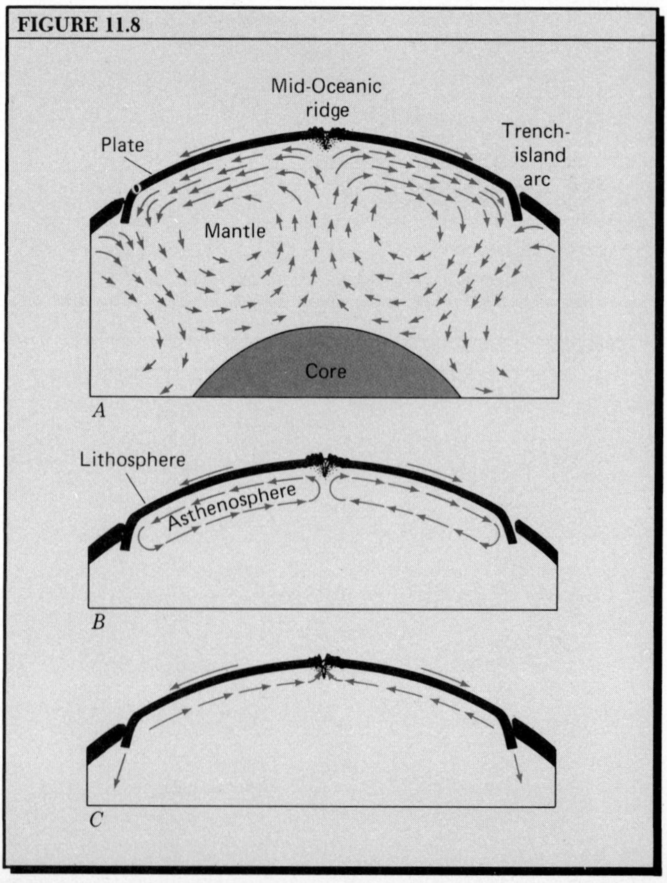

shown in Diagram *C* is quite different from the other two in that flow in the asthenosphere, immediately below the moving plate, is opposite in direction to plate motion, and the plate itself represents the upper part of the convection system. Because mantle flow opposes plate motion, an entirely different force is required to move the plate.

In setting up a workable model of mantle convection, two assumptions had to be made in accordance with what was known about the earth's interior. First, viscosity of the mantle is much lower in the asthenosphere than in the deep mantle. This assumption leads to the requirement that flowage of rock must be much faster in the relatively thin asthenosphere. Second, we assume that the convection system is driven by internal heat supplied largely by radioactivity and that most of the radioactive isotopes are concentrated in the earth's crust. In other words, the heat source is at the top of the system rather than at the bottom.

In line with the first assumption, we have devised the convection model shown in Diagram *A* of Figure 11.8 to make the rate of motion fast in the asthenosphere and very slow in the remaining mantle. Alternatively, we can favor the second and third models in which all convection is limited to the asthenosphere.

A deep mantle plume, rising from the base of the mantle in a narrow column and spreading out like a mushroom in the asthenosphere, is a special case of a convection model involving the entire mantle. It requires a convection cell of small horizontal area—much smaller than even a medium-sized lithospheric plate, and it therefore requires that many cells exist under a single large plate. Because a plume involves horizontal spreading of the mantle rock in all directions from the rising center, it would not provide a force for moving a single plate in one direction. The African plate, with many hot spots, might be used as an example of a plate situated over numerous mantle plumes but experiencing little or no motion relative to the mantle as a whole.

Geophysicists who studied the various aspects of hypothetical mantle convection systems were not satisfied with the first two models shown in Figure 11.8. Some strong theoretical arguments were made that the dragging force which a horizontal current in the upper asthenosphere could exert on the undersurface of a lithospheric plate would be much too weak to be effective in causing plate motion. In the late 1960s, as the theory of plate tectonics entered the final stages of assembly as a working hypothesis, the point was made that it is much more likely that the motion of a lithospheric plate governs the flow of convection currents in the asthenosphere, rather than the reverse. Thus, the third model shown in Figure 11.8 became the favored interpretation.

The change in viewpoint seems to have occurred at the same time that it was realized that a subducting plate is not being forced by some external pressure to descend into the asthenosphere, but rather that it sinks spontaneously because it is denser than the surrounding asthenosphere. In the latter case, the descending plate drags along the adjacent soft mantle rock of the asthenosphere, impelling it downward and forcing a horizontal flow within the asthenosphere in a direction opposite to that of the horizontal plate above it. This return flow transports mantle rock to a position beneath the spreading plate boundary, where the soft, hot rock rises and completes the flow cycle.

*Forces that Drive Lithospheric Plates

Let us now summarize the various possible forces that may affect plate motion to come up with a rough evaluation of these forces and to make a tentative statement as to which are most likely to be directly responsible for plate motion.

All forces involved in plate motion fall into two categories: (1) forces that tend to produce plate motion relative to the mantle, and (2) forces that tend to resist or oppose such motion. Figure 11.9 shows the major forces involved and places them into three classes.

Class I: Forces related directly to mantle flowage. A *mantle-drag force,* F_{MD}, is exerted by a current in the asthenosphere and will tend to move a horizontal plate in a horizontal direction. It is met by a *resisting force, RF.* Mantle-drag force may act on both oceanic and continental lithospheric plates; it may also act on the upper or lower side of a descending plate.

Class II: Gravitational forces. Because the mid-oceanic ridge is an elevated zone, the oceanic plates slope away from the spreading axis. Thus, a small part of the force of gravity acts in a direction parallel with the base of the plate, tending to cause the plate to move away from the spreading axis; this is the *ridge-push force,* F_{RP}. If the asthenosphere is not moving, or is moving more slowly than the plate, a resisting force opposes the ridge-push force. A second gravitational force is the *slab-pull force,* F_{SP}, acting within the cold descending slab; it is opposed by a resisting force acting on its upper and lower boundaries.

Class III: Resistance forces between plates. The forces that tend to produce plate motion are met by forces between moving plates along two kinds of plate boundaries. In subduction zones, a *colliding resistance force,* F_{CR}, is set up at the plate contact. Along transform faults, a *transform resistance force,* F_{TR}, is set up between plates. Of these two forces, we might expect the colliding resistance force to be much greater in magnitude and to produce important tectonic effects. Resistance forces between plates are continually dissipated by rock fracture in a succession of earthquakes.

Other forces have been recognized as affecting plate motion, but we have limited the list to those which particularly relate to the crucial issue: Are plates directly driven by a mantle drag force or by the direct pull of gravity? The most recent speculations on this question seem to favor the direct action of gravitational force as responsible for rapid motion of oceanic lithosphere and its subduction. At the same time, convection currents are by no means ruled out as unimportant in the total system. Rising mantle currents under the mid-oceanic ridge can provide the heat energy that forms rising magma and lifts the plates to a higher elevation at the spreading boundary. This uplift provides the gradient that allows gravity to generate the ridge-push force. Cooling of the oceanic plate as it moves toward the continent is part of the total convection system, since it represents heat loss that must occur if the system is to function. Cooling of the plate results in volume contraction and an increase in density. The slab-pull force is proportional to the difference in density between the plate and the surrounding mantle. Thus, old oceanic lithosphere, which is relatively cold and dense, will tend to sink faster than new oceanic lithosphere, which is comparatively warm and less dense.

The total picture we have drawn seems to show that the oceanic lithosphere supplies the main activity of plate tectonics. The continental lithosphere tends to remain intact because its thick capping crust of low-density rock gives it buoyancy. The great depth to which the base of the continental lithosphere extends in the mantle is also a factor that tends to reduce plate activity, because the asthenosphere becomes more viscous and less capable of flowage as the intermediate layer of the stronger mantle below it is approached.

Some geologists have envisioned that continental lithospheric plates will eventually become "grounded" on the deeper mantle and will not be able to move freely, if at all. As the rate of heat production by radioactivity slowly declines, the asthenosphere can be expected to become generally thinner. Thus, the freedom of slabs to plunge deeply into the asthenosphere will gradually be reduced and both tectonic and volcanic activity will gradually diminish. By the same reasoning, looking back into geologic time when the asthenosphere was hotter and thicker, plate motions may have been even more rapid, causing more intensive tectonic and volcanic activity than occurs today.

*Why Does a Lithospheric Slab Sink?

We have stated that a cold slab of oceanic lithosphere sinks through the asthenosphere because it is denser than the soft, hot rock of the surrounding asthenosphere. From this, we might infer that rock of the horizontal lithosphere of the ocean basins is also denser than rock of the asthenosphere below it. Evidence from seismology seems to confirm this inference.

In 1970, Professor Frank Press, a leading American seismologist, published the results of an elaborate investigation in which a computer was programmed to generate a large number of models of the possible change of rock density with depth in the earth. The results, which were in good general agreement, are shown in a simplified version in Figure 11.10. Density increases rapidly in the first 100 km but levels off in the depth range of 100 to 200 km. Then, below about

FIGURE 11.9 This schematic block diagram shows three classes of forces thought to be important in the motions of lithospheric plates.

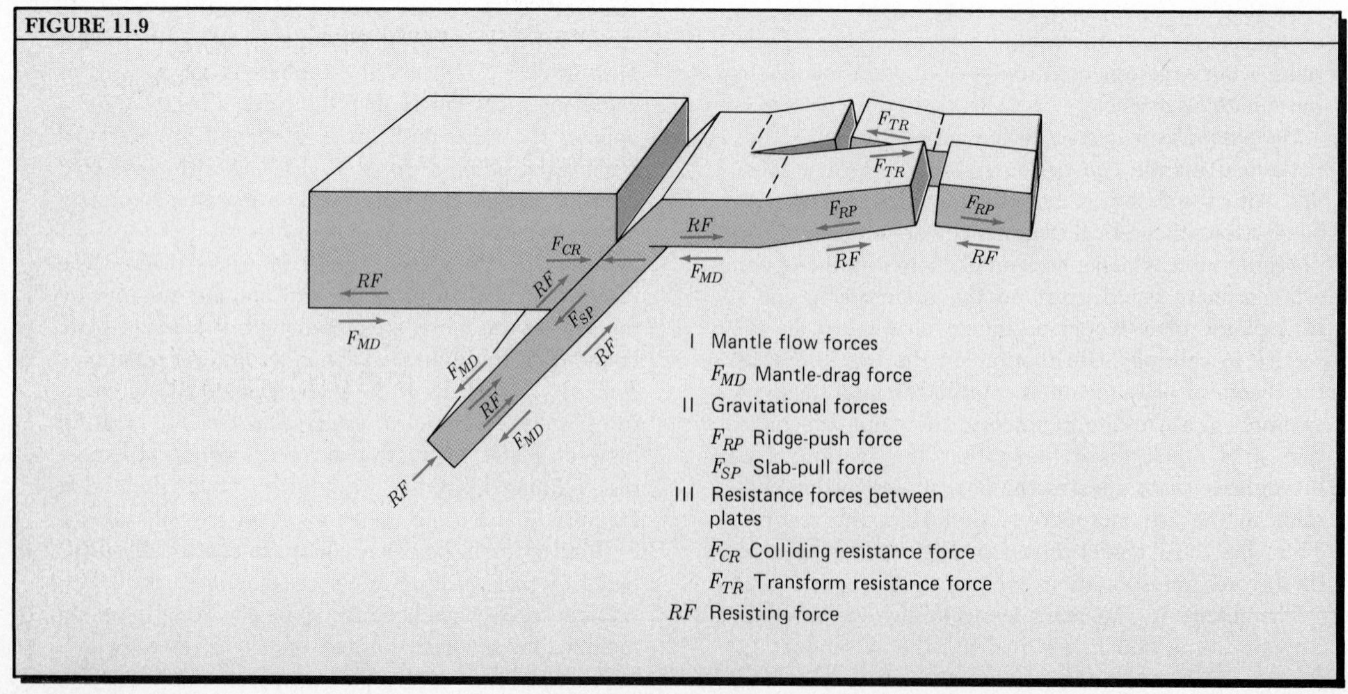

200 km, density begins to decrease. In the general depth range of 200 to 400 km, we see a zone of reduced density values. Below about 400 km, density increases rapidly into the mantle. Figure 11.10 also shows the results of this computer study with respect to the S-wave velocities. The low-velocity zone of the asthenosphere (described in Chapter 8) appears very distinctly on the graph with a minimum value in the depth range of about 150 to 200 km. This depth zone is the "softest" part of the asthenosphere.

The important message from these results comes from the reversal of the density curve within the asthenosphere. The reversal indicates a condition of potential instability favoring the possibility that a lithospheric plate will sink of its own weight once it begins to descend into the asthenosphere. Now, a new physical principle needs to be taken into account. When a dipping slab of oceanic lithosphere begins to enter the asthenosphere, density conditions change because the slab is continually moving into a region of increasing confining pressure. The slab contracts slightly in volume under this confining pressure and becomes denser than it was at the surface. Under this principle, at any given level in the asthenosphere, rock of the cold slab must actually be denser than mantle rock of the surrounding asthenosphere. Remember that most of the oceanic lithosphere is composed of ultramafic rock, probably of peridotite composition, and that it is probably of the same chemical composition as the asthenosphere. Because the chemical compositions of the two materials are alike, differences in temperature are very important in causing density differences, where the two materials are under the same confining pressure. We can also postulate that, as oceanic crust is subducted and enters a region of higher pressure, the basalt and gabbro of which it is composed undergo a phase change and turn into eclogite, which has a density about equal to that of peridotite. In any case, the oceanic crust is only a very thin layer (about 5 km thick) in comparison with the oceanic lithosphere, which is on the order of 50 to 70 km thick over large areas. Because the oceanic crust is so thin, it could have little buoyant effect on a subducting slab.

If our argument is sound, in order for subduction to start along some particular line in a plate of otherwise continuous horizontal oceanic lithosphere requires an initial rupturing force which is not the slab-pull force. That initial force must be capable of breaking the lithospheric plate and forcing one edge down into the asthenosphere. The ridge-push force is one possible candidate for causing initial plate rupture along a new subduction zone. We may imagine that if the plate is lifted high enough by rising mantle rock along a potential spreading axis, the ridge-push force will be transmitted to distant points on the plate. Then, at some place where a structural weakness exists, the plate ruptures along a shear plane in a reverse-fault movement. The shear plane then grows rapidly in both directions and a new subduction boundary comes into existence.

Island Arcs and Backarc Basins

Earlier in this chapter, we took note of the characteristic arrangement of island arcs and backarc basins in the western Pacific Ocean (Figure 11.1). The Sea of Japan and the Sea of Okhotsk, already mentioned, occupy two such backarc basins. The Bering Sea, lying north of the Aleutian volcanic arc, is another. A narrow but deep backarc basin forms the extreme eastern part of the East China Sea, lying between the Ryukyu arc and the Asiatic mainland. A deep elongate basin, the Marianas Trough, lies west of the Marianas Trench, within the Philippine plate. West of the Philippine Islands and Celebes are three deep basins. Depths of the backarc basins range from moderately deep (2 km) to abyssal (5 km). Although small areas of abyssal plain are present in the deeper basins, many

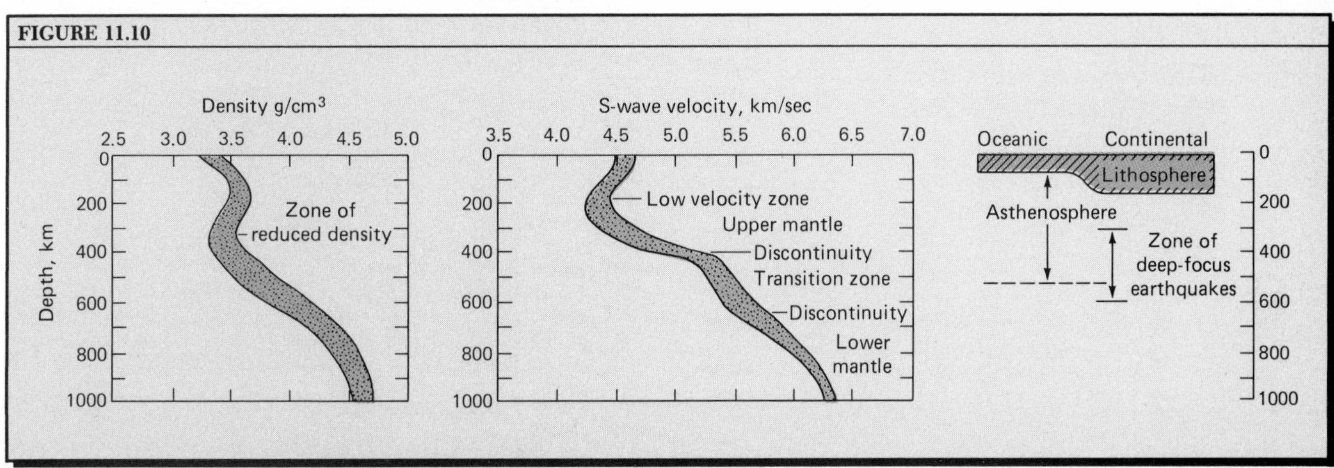

FIGURE 11.10 Graphs showing computer-derived changes of density and S-wave velocity with depth. The descending band in each graph shows the limits within which numerous model curves fell. (Data of F. Press, 1970, as shown in J. A. Jacobs, R. D. Russell, and J. T. Wilson, 1974, *Physics and Geology*, 2d ed., McGraw-Hill, New York, pp. 45, 46.)

topographic irregularities are also present, taking varied forms such as rises, abyssal hills, and broad submarine ridges.

The backarc basins have been a real puzzle to geologists who seek to apply the theory of plate tectonics to all major crustal features. What kind of crust lies beneath these basins? Is it oceanic or continental crust? How did these basins come into existence? Despite their enigmatic quality, the backarc basins have divulged scientific data on which to base some firm statements and at least two strong hypotheses of origin.

First, examine a cross section of the crust along a line drawn from northwest to southeast across the Japanese island of Honshu (Figure 11.11). The crust beneath the island of Honshu seems best classified as continental crust, because it is thick and the seismic P-wave velocities suggest an upper felsic layer and a lower mafic layer. Continental crust is again encountered on the Asiatic mainland. Under the Sea of Japan, seismic wave evidence shows a thin crust, with the Moho situated at a depth of about 15 km. Seismic velocity layers within this portion of crust match rather closely those of the Pacific oceanic crust east of the Japan Trench. There seems to be general agreement that this backarc basin, and perhaps others, are underlain by oceanic crust and oceanic lithosphere. The question is how the oceanic crust came to be located in the backarc basin. Two very different mechanisms have been proposed to answer this question.

The *entrapment hypothesis* explains backarc basins as areas of oceanic lithosphere that have been cut off from a large oceanic lithospheric plate by the formation of a new subduction zone. Figure 11.12 is a schematic diagram illustrating this hypothesis. In Diagram *A*, continental and oceanic lithosphere meet in a stable contact zone and form a single plate. In Diagram *B*, a new subduction zone has formed within the oceanic lithosphere. Much later, as shown in Diagram *C*, an island arc of continental crust has been formed adjacent to the subduction zone. The backarc basin now lies between the island arc and the continental mainland. This entrapment hypothesis has been proposed to explain the Bering Sea, which is a large, deep basin.

A good test of the entrapment hypothesis would be to look for magnetic anomaly patterns in the trapped portion of oceanic crust. If these are present and can be identified as belonging to patterns found on the floor oceanward of the island arc, the hypothesis would be acceptable. In the case of the Bering Sea, a magnetic anomaly pattern has been found and can be interpreted as a part of the Pacific system of Mesozoic magnetic anomalies, which we described in Chapter 10. The Aleutian Islands volcanic chain is composed mostly of rocks of Cenozoic age and is therefore a younger feature than the backarc basin floor. Thus, the evidence seems to favor the entrapment hypothesis for the Bering backarc basin.

An objection to the entrapment hypothesis in other cases is that it requires the oceanic crust of the backarc basin to be older than the crust of the island arc. In the Sea of Japan–Honshu case, however, some metamorphic rocks of the island arc are as old as the Paleozoic Era—older than 250 m.y. If the Sea of Japan had been entrapped at such an early date, it would contain a large thickness of marginal sediments derived from the mainland. Instead, the marginal sediment accumulation of the Sea of Japan is no more than about 2 km thick, and this is considered much too thin for a very old basin. Thus, in the case of the Sea of Japan, the entrapment hypothesis has been rejected.

The second hypothesis to explain a backarc basin is that of *lateral drift* of the island arc, illustrated in Figure 11.13. There first exists a subduction boundary between an oceanic plate and a continental plate (Diagram *A*). A narrow zone of continental plate then begins to separate from the main continental plate along a rift (Diagram *B*). New oceanic crust forms in the

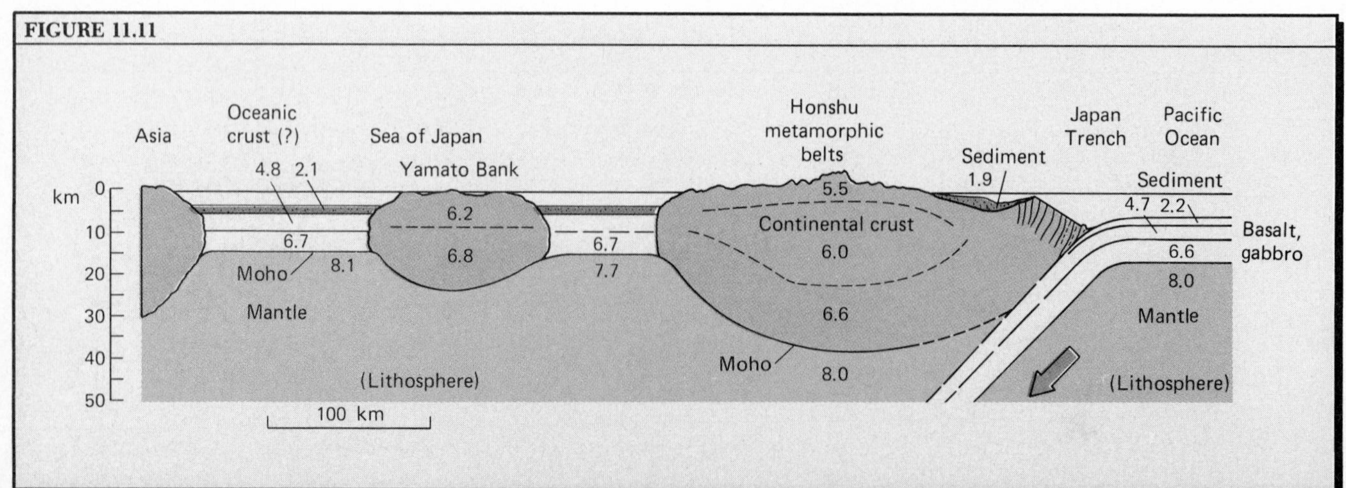

FIGURE 11.11 Details of the continental and oceanic crust in a cross section through Japan and the Sea of Japan. (Based on data of A. Miyashiro, Y. Matsuda, and S. Uyeda, as shown in J. F. Dewey and J. M. Bird, 1970, *Journal of Geophysical Res.*, vol. 75, p. 2634, Figure 7.)

widening rift, which becomes a backarc basin (Diagram *C*). In this scenario, the oceanic crust of the backarc basin is quite young—much younger than the continental crust of the island arc.

Supporters of the lateral drift hypothesis use the strongly curved plan of the island arc and trench as a feature favoring this origin of the backarc basin. As the island arc moves away from the continental plate (or vice versa), the arc becomes increasingly curved or bowed out. A much more convincing line of evidence of lateral drift is in the high rate of heat flow observed over the Japanese backarc basin. Recall that we questioned whether friction of the downplunging oceanic plate would be capable of supplying the heat needed to melt magmas and furnish the observed high rate of heat flow. Under the lateral drift hypothesis, a convection mechanism is available to supply the necessary heat. As shown in Figure 11.14, lateral drift causes replacement of thick continental lithosphere with thin oceanic lithosphere. As seafloor spreading continues in the backarc basin, a convection current system is generated in the underlying asthenosphere. A rising current brings up high-temperature mantle rock to shallow depths, furnishing the heat needed to form magmas beneath the island arc, but above the descending plate.

FIGURE 11.12 Three stages in the formation of a backarc basin according to the entrapment hypothesis.

FIGURE 11.13 Stages in the formation of a backarc basin according to the hypothesis of lateral drift of an island arc.

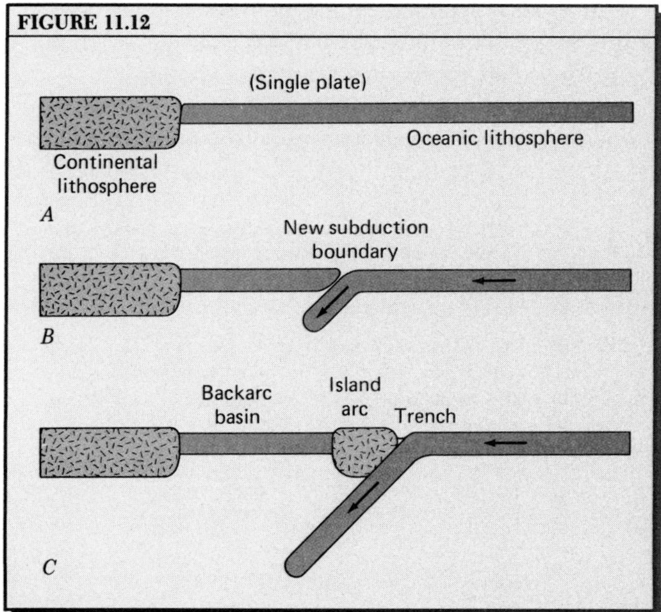

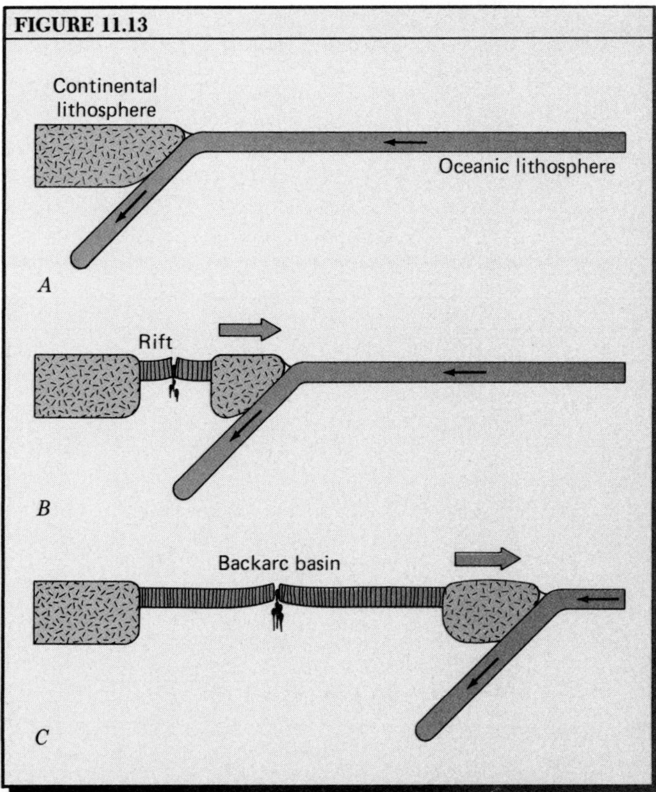

Further details of a convection mechanism operating above a descending lithospheric plate have been worked out on the basis of studies of the volcanic arc that borders the Aleutian Trench on the north and separates the Aleutian Islands from the Bering Sea. The central and western part of the arc, which trends about east–west, consists of a rather simple line of active volcanoes formed of a lava type intermediate between andesite and basalt. Immediately north of the volcanic arc lies a deep backarc basin floored by the Bering abyssal plain. We mentioned earlier that this backarc basin seems to have been formed by entrapment; if so, it is a new subduction zone formed in the midst of an area of oceanic lithosphere. In any case, the volcanic arc seems to be a relatively young feature, as compared with the complex Japan and Kuril arcs, which contain quite old rocks. The Aleutian volcanic arc has been studied for whatever light it may shed on the origin of the magma that rises to build the volcanic chain. The active volcanic centers are quite uniformly spaced along the chain, and it appears that comparatively small bodies of magma are being produced at regular intervals at a position just above the downgoing plate and near the base of the stationary oceanic plate that lies above it. As we noted in connection with the Honshu arc of Japan, the source of enough heat to cause melting has been a scientific problem difficult to solve. The descending slab is relatively cold, and calculations show that friction along the upper boundary of the slab is not a sufficient source of heat to cause melting. However, as Figure 11.15 shows, the magma seems to be formed as a thin ribbon lying close to the descending plate.

A recent hypothesis proposed to explain the location of the narrow magma ribbon states that the downmoving plate exerts a drag on the asthenosphere above the plate, setting in motion a strong current in the asthenosphere, as shown by the arrows in Figure

11.15. To complete the flow, asthenosphere is drawn into the narrow V-shaped opening between the two plates, bringing heat to this location. Coming into contact with the top of the descending plate, the hot asthenosphere supplies enough heat to melt the top of that slab and produce the narrow magma ribbon. From time to time, small magma bodies ascend through the overlying lithosphere.

This model is called a *corner-flow model* because of the sharp bend of the flow lines as they enter and leave the narrow corner formed by the two plates. The general flow pattern is in agreement with the larger convection system depicted in Figure 11.14 as responsible for the widening of the backarc basin. Thus, the convection system seems to be generated by the downmoving plate and is the effect of plate motion, rather than the cause.

The corner-flow model of island-arc volcanism can also be applied to continental volcanic arcs, such as the Andes Range of South America. It explains the continuous volcanic activity limited to a narrow zone paralleling a trench. As long as subduction continues, magma will be produced. The model represents a valuable refinement of our general statement, made in earlier chapters, that magma produced by melting of the downgoing plate rises to form a chain of volcanoes above a subduction zone.

At this point you may have thought of an expected consequence of the lateral drift hypothesis that should be looked for. If seafloor spreading has formed a backarc basin, a new magnetic anomaly pattern should be imprinted on its bedrock floor. No clear pattern of symmetrical anomaly stripes has been found in the Sea of Japan, although a vaguely defined linear pattern has been mapped. Those who have supported the lateral drift hypothesis for the Sea of Japan have visualized multiple rifts, so that new oceanic crust formed simultaneously along several roughly parallel lines. This mechanism would produce irregular, roughly aligned topographic features on the ocean floor but with no clear symmetrical magnetic anomaly pattern. It is thought that the Sea of Japan is no longer expanding, so no active rift zones are to be expected.

Recently, the characteristic mirror-image pattern of anomaly stripes has been recognized in two other

FIGURE 11.14 Schematic cross section through a backarc basin and island arc showing how convection currents in the asthenosphere might furnish a high rate of heat flow beneath the arc.

FIGURE 11.15 Schematic cross section showing the sharp bend in flow lines postulated under the corner-flow model of currents in the asthenosphere. A volcanic arc lies above the line where a magma ribbon is formed.

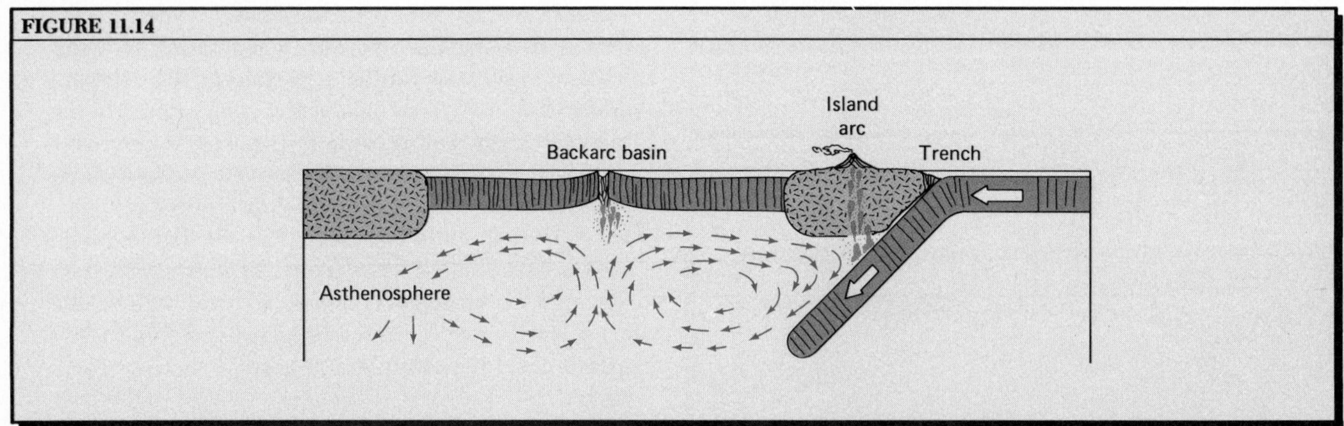

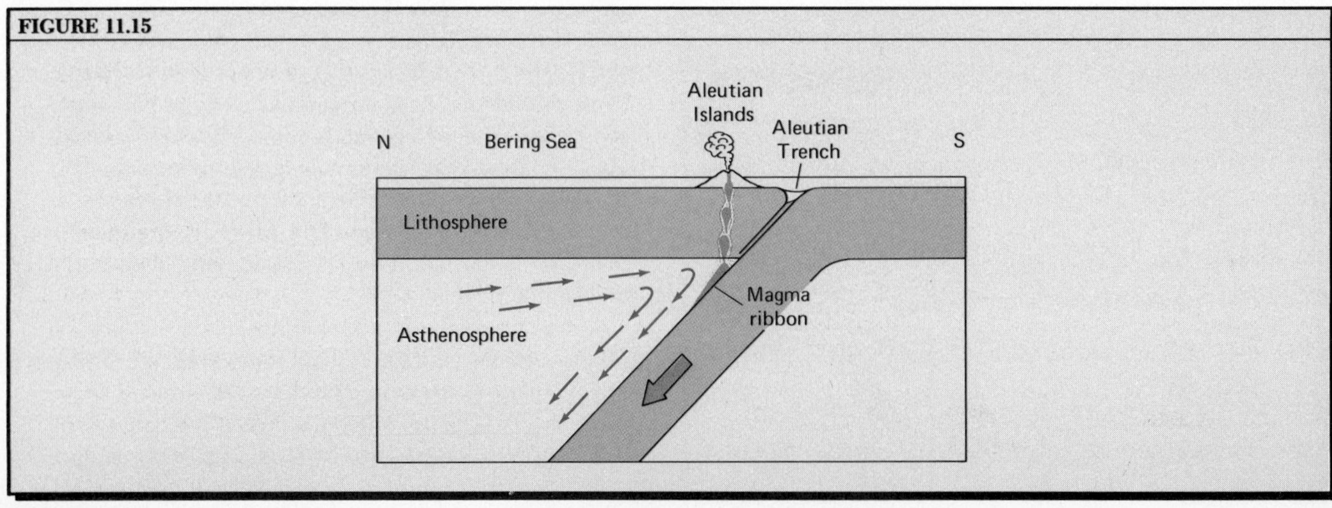

backarc basins: the Lau Basin west of Tonga, and the Shikoku Basin of the eastern Philippine Sea. This evidence strongly supports the lateral drift hypothesis as applied at least to those two basins.

Basic Modes of Plate Subduction

From a study of seismic activity of subduction zones—most of which ring the Pacific Ocean basin—it has been concluded that at least two basically different styles or modes of subduction are taking place. They were described and analyzed in 1979 by Japanese seismologists Seiya Uyeda and H. Kanamori, who studied the distribution of great earthquakes along subduction zones and uncovered a rather striking fact. Over 90% of seismic energy release on a global basis is from subduction zones lacking an active backarc basin. Earthquakes substantially larger than magnitude 8.0 on a revised scale (designed by Kanamori especially for large earthquakes) have occurred almost entirely along subduction zones that lack an active backarc basin. Uyeda and Kanamori reasoned that the explanation for the presence or absence of great earthquakes must be tied in with the mechanics of plate subduction.

Before going further with this account, we should take stock of the major varieties of subduction segments, or arcs, and place them in one of two classes: (I) those without active backarc basins, and (II) those with active backarc basins.

	Examples
I. Subduction arcs without active backarc basins	
A. Continental arcs	Peru–Chile arc Alaska arc
B. Arcs with inactive backarc basins	Kuril arc Japan arc Shikoku arc Aleutian arc
II. Subduction arcs with active (spreading) backarc basins	Mariana arc Scotia arc Lau Basin arc (Tonga Trench)

It appears that subduction zones of the first class of arcs have a different mode of subduction from those of the second class. The principal differences are shown schematically in Figure 11.16. Uyeda and Kanamori have applied the name *Chilean type* to the mode of subduction associated with arcs of the first class; *Mariana type* for the mode associated with the second class. The Chilean-type mode is characterized by a low dip angle for the subducting slab, which presses strongly against the opposing, overlying plate. The plate shows a low upward bulge oceanward of the trench. The strong compressive stress is responsible for the occurrence of great earthquakes along shear zones. A strong scraping action is exerted upon the descending plate by the overlying plate.

The Mariana-type subduction mode is characterized by a steeply dipping slab and the presence of an actively spreading backarc basin. Because the spreading is accomplished by a retreating motion of the continental plate, the new oceanic lithosphere beneath the backarc basin is continually subjected to tension. Thus, the subduction zone is relieved of compression and great earthquakes do not occur. Lacking strong compression, the arc is not strongly uplifted and may furnish little sediment to the trench, because the arc consists of a chain of volcanic islands. The trench tends to be deep and barren of sediment. It is also possible that slices of the edge of the upper plate can be broken off and carried down with the subducting slab, a process called *tectonic erosion*. This description of the two modes of subduction is quite generalized and does not necessarily fit well in all cases.

Uyeda and Kanamori show that their interpretation of the two subduction modes fits rather well with relative plate motions. For the Chilean-type arcs, the upper plate has a relative motion of 1 to 2 cm/yr toward the line of downbending of the subducting plate, generating strong compressive stress. For the Mariana-type arcs, the upper plate has a rather rapid speed (6 to 8 cm/yr), but in a direction away from the line of downbending. This plate motion allows the extension of the backarc basin and tends to relieve the subduction zone of compressive stress.

In both subduction modes, we can envision convection within the asthenosphere, following the corner-flow model shown in Figure 11.15. However, in the case of the Chilean-type mode with the low-dipping slab, conditions would not be as favorable for deep overturn of the mantle that would generate a high rate of heat flow at the surface. On the other hand, the combination of a steeply dipping slab and continual extension of the backarc basin would favor the rise of heated mantle rock from great depths to supply the observed high rate of heat flow at the surface.

Uyeda and Kanamori have gone further in attempting to explain why two modes of subduction should occur. One possibility is that the oceanic plate associated with the Chilean-type mode is comparatively young in age of formation. As such, the plate is relatively warm and of relatively low density, as compared with an old plate, and thus sinks less rapidly into the mantle. The Mariana-type mode may be associated with old lithosphere that is relatively cold and dense and therefore sinks rapidly and steeply into the asthenosphere. The slab may become anchored to the relatively strong mantle below the asthenosphere, a factor in preventing it from pressing strongly against the overlying plate.

You should view this analysis of modes of plate subduction and their causes as a working hypothesis that will require a great deal of further research. On the other hand, it appears to be a good working hypothesis because it brings together in an orderly way some quite diverse lines of evidence, such as seismic activity, heat flow, plate geometry, plate motions, and plate stresses. A working hypothesis must be tested against evidence coming from many different branches of a science. As each test is passed, by showing that the evidence conforms with the needs of the hypothesis, it becomes a stronger hypothesis. On a larger scale, the grand hypothesis of plate tectonics has been subjected to many difficult tests and has grown stronger in successful encounters with new forms of evidence.

Sediment Deformation in Subduction Zones

In Chapter 1, we described the trenches of subduction zones as places where large quantities of terrestrial sediment can accumulate. A high-standing volcanic arc can supply large quantities of detrital sediment to the coastline, from which it is swept across

FIGURE 11.16 Two modes of plate subduction. The Chilean-type subduction zone is one of strong compression, causing great earthquakes. The Mariana-type subduction zone is associated with plate extension, and is subject to little or no compressional stress; it is associated with an active backarc basin. (Based on data of S. Uyeda, 1979, *Oceanus*, vol. 22, no. 3, pp. 53–62.)

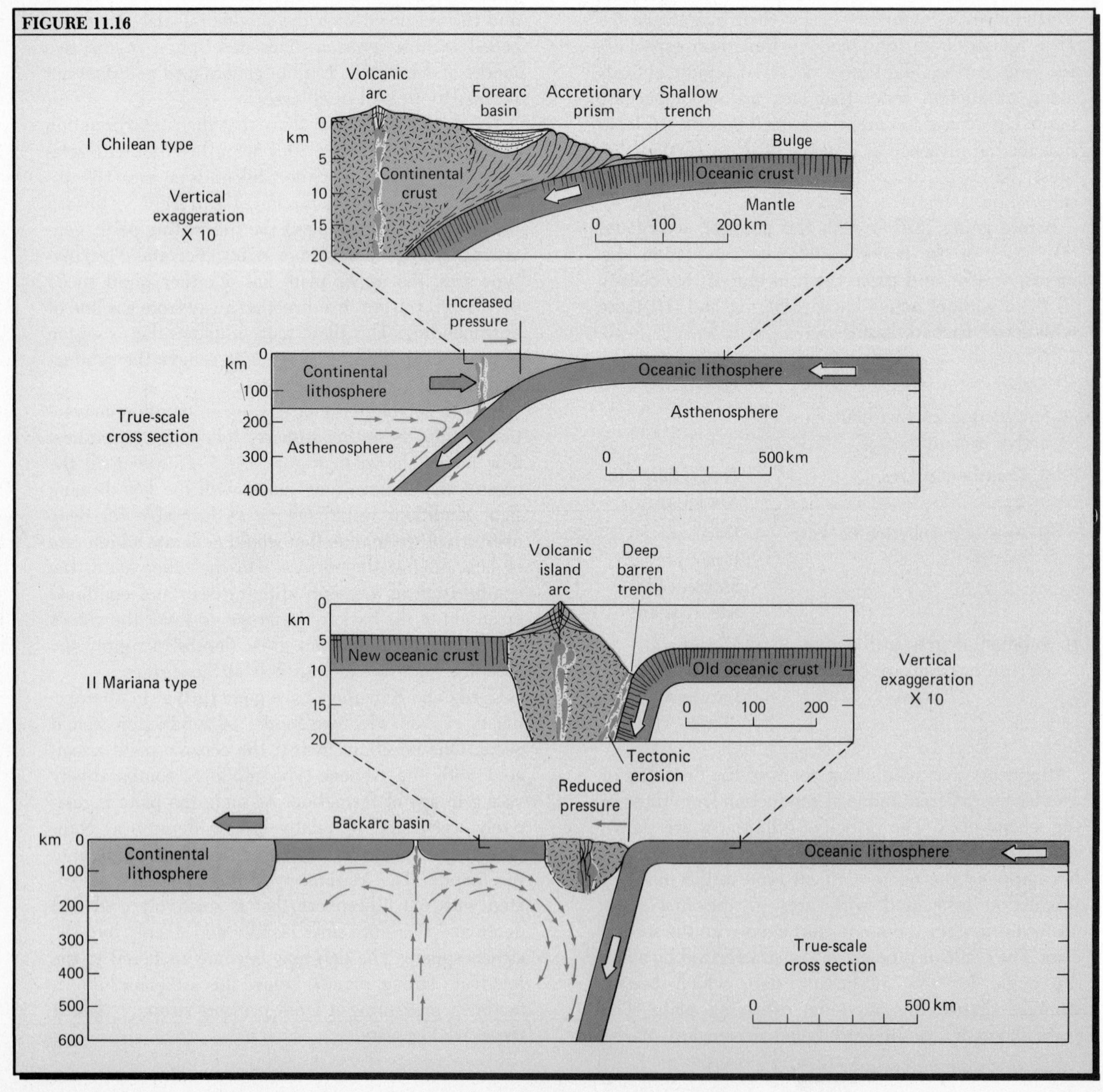

the continental shelf by currents. At the outer edge of the shelf, turbidity currents are formed and carry the sediment in tonguelike surges down canyons in the steep inner slope of the trench to reach the trench floor. The movement and accumulation of terrestrial sediments were pictured in Figure 1.13, which also shows pelagic sediment of the ocean floor to be carried toward the trench axis on the moving plate.

Arriving at the trench floor, terrestrial and pelagic sediments are subjected to intensely disruptive tectonic activity. The fixed edge of the continental plate acts as a gigantic scraper for sediments dragged against it by the downgoing plate. The activity is often described as *offscraping,* and is pictured as a process somewhat like the action of the blade of a snowplow. Although some of the snow at the bottom of the layer passes beneath the blade, most is pushed along in front of the blade, where it piles up in a succession of crumpled wedges. Every now and then, the blade may strike a bump in the pavement beneath, breaking loose fragments of the pavement, and these become mixed in with the snow. The pavement in this model might be represented in reality by the basaltic oceanic crust.

Figure 11.17 is a series of schematic cross sections showing the growth of a succession of thrust slices of crumpled trench sediments to form a rising pile. The process has been called *understuffing.* The imbricate thrusts are formed at a low angle, but are pushed upward by newer wedges and steepen in dip. Figure 11.18 is a cross section showing the imbricate structure in detail, based on seismic reflection profiling. Notice that, as the wedges are pushed upward, they are also being buried by newer sediment layers. The newer sediments in turn become folded between the upper edges of the wedges. In this way, numerous small catchment basins are formed and filled with sediment on the inner trench slope.

The result of the understuffing process is an *accretionary prism* of deformed sediments. (As an alternate term, *subduction complex* may be used.) The density of the material in the prism is quite low and it tends to rise to form a *structural high,* the crest of which tends to form a barrier to terrestrial sediment arriving from the coastline, trapping part of it in a shallow depres-

FIGURE 11.17 Stages in the construction of an accretional prism by offscraping and understuffing. (Based on data of W. R. Dickinson and D. R. Seely, 1979, *Amer. Assn. of Petroleum Geologists, Bull.,* vol. 63, no. 1, pp. 2–31.)

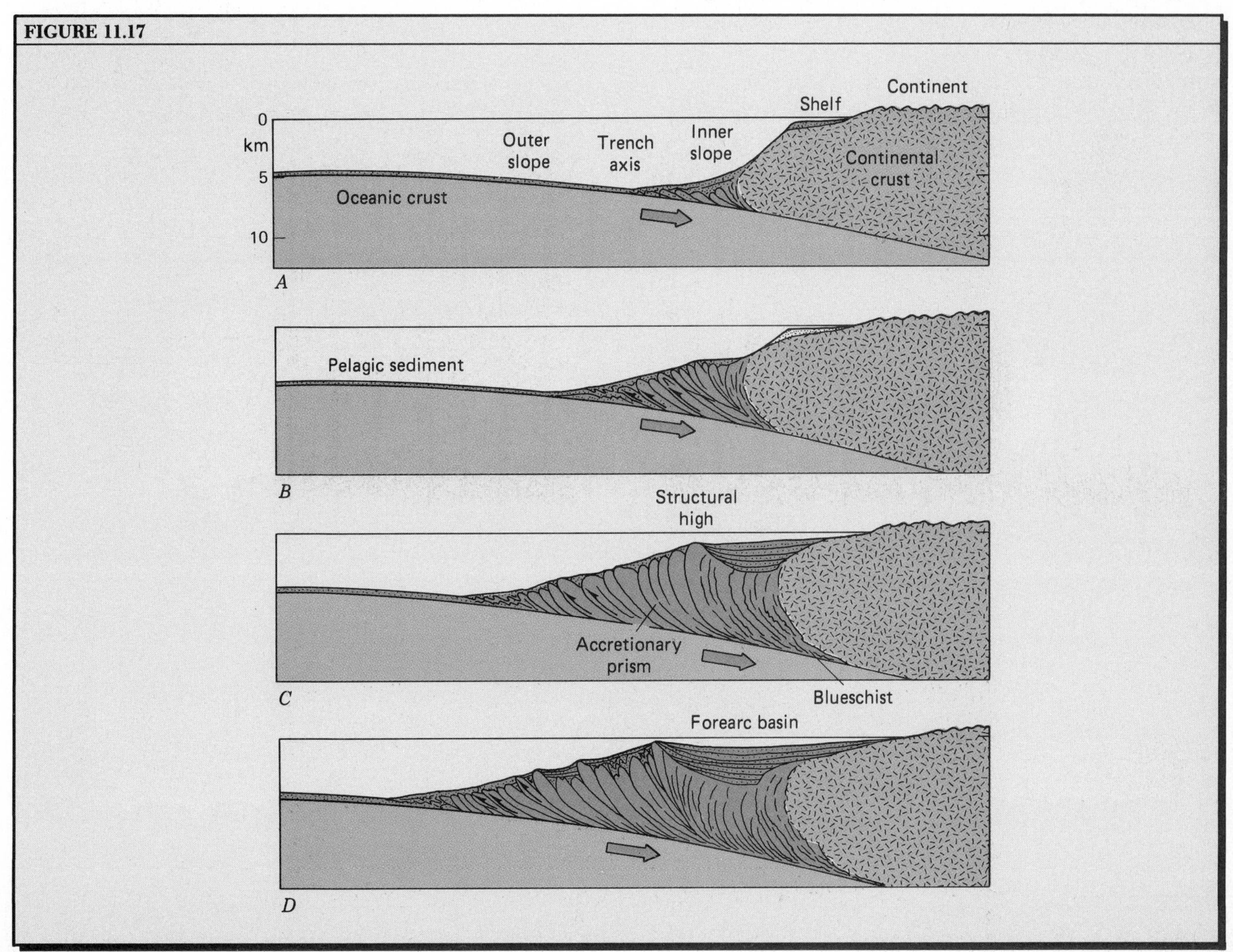

sion which we may call a *forearc trough*. Here sediment layers accumulate and the base of the deposit subsides because the combined load of the accretionary prism and the forearc trough sediments causes the moving plate to sink somewhat lower. At the same time, the accretionary prism grows in width, extending farther and farther out upon the moving plate. The process of growth of the accretionary prism and forearc trough can continue more or less interruptedly for tens of millions of years. In some cases, it continues as long as 100 million years, or about the duration of a major geologic period such as the Cretaceous Period.

A somewhat different mode of evolution of the forearc basin and accretionary prism occurs when a new subduction boundary forms by breakage within the oceanic plate some distance out from the margin of the continental plate. Figure 11.19 shows stages in this form of evolution. The lithosphere breaks along a low-dipping fault and subduction begins. Understuffing then takes place beneath the tapered edge of the broken crustal segment, forcing it upward. The upturned edge of the crust serves as a structural high, forming a forearc trough which receives sediment. As the stack of imbricate wedges rises higher, the crustal wedge and perhaps some of the peridotite mantle below it is elevated further. At the same time, the accumulation of sediments in the forearc trough thickens. So far as the surface forms are concerned, the end result is much the same as shown in Figure 11.17.

A particularly interesting feature of this scenario is that during an ensuing orogeny, portions of the oceanic crust of the upraised wedge may be further faulted and uplifted, eventually becoming exposed at the surface on the continental margin. This is an event we will discuss in Chapter 12.

Whatever may be the details of development of the accretionary prism, it gives an opportunity for the formation of metamorphic rocks. Deep in the subduction zone (deeper than 20 km), pressures are very high (greater than 6 kb). Here the environment is favorable for the formation of blueschist from turbidites. Blueschist requires high pressure combined with relatively low temperature (see Figure 7.15). Temperatures close to the descending plate are comparatively low (200 °C to 400 °C) at this depth because the plate itself is cool in comparison to the surrounding mantle. At still greater depth, with increased pressure, conditions may be favorable for the formation of eclogite.

The metamorphic rock is added to the continental crust and thus the continental margin is gradually extended. Denudation of the continental margin is accompanied by a rise of the continental crust, so that eventually these deep-seated metamorphic rocks may appear at the surface. For example, on the Japanese island of Honshu, a narrow belt of blueschist-

FIGURE 11.18 Structure sections of imbricate thrusts and sediment accumulation of an accretionary prism on the lower trench slope of the Sunda arc, Indonesia. This interpretation is based on a seismic reflection profile. Horizontal and vertical scales are the same. (From G. F. Moore and D. E. Karig, 1976, *Geology*, vol. 4, p. 696, Figure 5. Copyright © 1976 by the Geological Society of America.)

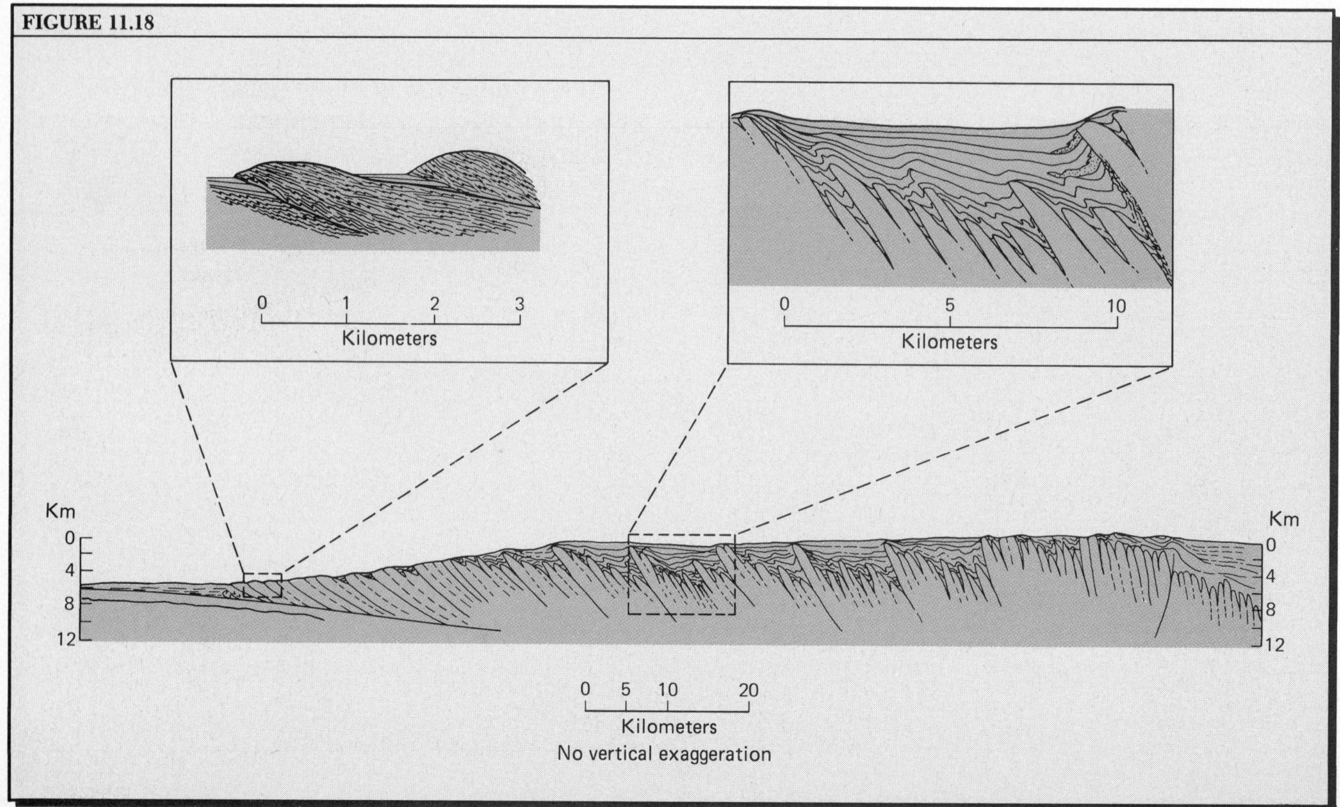

FIGURE 11.19 Stages in the construction of an accretionary prism by understuffing beneath an uplifted wedge of oceanic crust. (Based on data of W. R. Dickinson and D. R. Seely, 1979, *Amer. Assn. of Petroleum Geologists, Bull.*, vol. 63, no. 1, pp. 2–31.)

FIGURE 11.20 Mélange of early Cenozoic age from subduction complexes in the Indonesian region. *A.* A turbidite sandstone block surrounded by dark-gray scaly clay that has been strongly sheared. The knife is 9 cm long. Sabah, northern Borneo. *B.* Lenses of oceanic crustal basalt enclosed by strongly foliated scaly clay. In the extreme upper left is a block of gabbro. Near Port Moresby, Papua, New Guinea. (Both photos by Warren Hamilton, U.S. Geological Survey.)

FIGURE 11.19

Km 0 5 10 20 30
Ocean
Oceanic crust
Continental crust
Mantle
A
B
C
Km 0 10 20 30 40
Forearc trough
D

type metamorphic rocks lies exposed along the southeastern coastal zone, paralleling the Japan Trench. This rock is of Jurassic and Cretaceous ages; it may have formed in the lower wedges of a subduction complex generated in the Mesozoic Era, well before the backarc basin between Japan and the mainland of Asia was opened out.

Mélange

The intense tectonic activity that goes on within a growing subduction complex affects a wide variety of sediment types as well as oceanic crust. These diverse ingredients become mixed together in a remarkable kind of rock called *mélange*. As you know, the French word "mélange" refers to an incongruous mixture. Here and there, among the deformed rocks of ancient mountain ranges, occur exposures of rock that would best be described as a very coarse breccia. What is remarkable about this breccia is not only the great size of some of the blocks—but also the diversity of their composition and origin. Some of the blocks appear to have been torn off the downmoving plate and consist of basaltic oceanic crust; these are *exotic blocks*. Other blocks are fragments of brittle layers that were formed within the deformed trench sediments; they are *native blocks*. Enclosing the blocks is the *matrix*, consisting of intensely deformed plastic (ductile) sediments—a complex mixture of turbidites and pelagic sediment. If the mélange is carried down to greater depths in the subduction zone, it may be subjected to high pressure, transforming the mélange into a schist (Figure 11.20).

Subduction and Plate Tectonics

In this chapter, we have investigated the evidence for active subduction on converging plate boundaries. Geophysical evidence of subduction based on seismol-

FIGURE 11.20 A

FIGURE 11.20 B

ogy has proved to be a decisive factor in completing the theory of plate tectonics, complementing the magnetic and seismic evidence accumulated from seafloor spreading boundaries.

Now the cycle of oceanic plate accretion and consumption stands complete. In the next chapter, we turn to the continental lithosphere, which carries on its crust the record of a long and complex history of tectonic and igneous activity begun in early Precambrian time and still in progress along plate margins in many parts of the world.

Key Facts and Concepts

The continental margins *Continental margins* are of two basic types, passive and active. Along tectonically *passive margins,* oceanic and continental lithosphere are joined within a single plate. These inactive margins resulted from continental rifting and ocean-basin opening. *Active margins* coincide with plate boundaries presently or recently tectonically active. Most active margins are on converging plate boundaries with active subduction, including both mountain arcs and island arcs of backarc basins.

Deep-focus earthquakes and subduction At converging boundaries, foci of deep-focus earthquakes define the *Wadati–Benioff zone,* a slanting plane extending from an oceanic trench into the asthenosphere. Earthquakes produced in this zone originate in the cold, brittle interior of a subducting plate, as demonstrated by seismic wave velocities within the slab. Low heat flow in trenches supports the cold-slab hypothesis.

Stress directions in a descending plate Seismic methods distinguish compressional stress from tensional stress at earthquake foci. Axes of compressional stress for intermediate and deep foci are consistently at right angles to the trend of the trench-arc and are tilted to parallelism with the Wadati–Benioff plane, indicating resistance to descent within the slab. Tensional stress is found near the top of the slab, indicating downpulling of the slab by gravitational force. Severe earthquakes of shallow focus result from *underthrusting* of the subducting plate, causing downbending of the overlying plate followed by elastic rebound.

Mantle convection systems *Mantle convection* consists of slow plastic flowage within the asthenosphere and deeper mantle; it may take the form of a *convection cell,* of which the lithosphere is an integral part. A model in which mantle drag causes motion of a lithospheric plate is rejected on physical grounds. More plausible is the hypothesis that drag exerted by lithospheric plate motion sets mantle convection in motion, forcing deeper flow in the asthenosphere opposite to that of the overlying plate, with upwelling beneath the oceanic spreading boundary.

Plate-driving forces Forces that tend to produce or to resist plate motion fall into three classes. Class I forces include *mantle-drag force* and a *resisting force.* Class II forces are gravitational and consist of *ridge-push force* and *slab-pull force,* opposed by resisting forces. Class III forces are resistance forces between plates and include *colliding resistance force* and *transform resistance force.* Forces of Class II appear to dominate oceanic plate motion.

Sinking of a lithospheric slab In theory, a slab of oceanic lithosphere will be denser than the asthenosphere and will sink spontaneously under gravitational slab-pull force, once the leading edge has penetrated the asthenosphere. Phase change in the downgoing slab may contribute to an increase in plate density.

Island arcs and backarc basins Backarc basins of the western Pacific are underlain by oceanic crust and lithosphere. The *entrapment hypothesis* involves cutting off of an area of oceanic lithosphere by formation of a new subduction zone. The hypothesis of *lateral drift* involves the separation of a narrow belt of marginal continental lithosphere along a widening rift, causing new oceanic crust to form the backarc basin. Occurrence of symmetrical magnetic anomaly patterns in a backarc basin favor the lateral drift hypothesis. Under the *corner-flow model,* subduction may set in motion a strong asthenospheric current that brings heat to the upper surface of the descending slab, causing melting and magma production.

Modes of plate subduction Two basic modes of plate subduction have been recognized. The Chilean-type mode, associated with subduction arcs lacking active backarc basins, has low-angle slab dip with strong compressive stress on the overlying plate. The Mariana-type mode, associated with actively spreading backarc basins, has a steeply dipping slab with the subduction zone relieved of compression, allowing *tectonic erosion.*

Sediment deformation in subduction zones *Off-scraping* of terrestrial sediments (turbidites), pelagic sediments, and mafic crustal fragments from the downgoing plate constructs a succession of wedges along imbricate thrusts in the process of *understuffing* and forms an *accretionary prism (subduction complex).* The low-density prism rises to form a *structural high* and a *forearc trough,* which may accumulate sediment. Metamorphism affects the accretionary prism at depth producing blueschist and eclogite, which are added to the continental crust. Within the subduction complex are masses of *mélange* consisting of *exotic blocks* and *native blocks* enclosed in a matrix of deformed ductile sediments.

Questions and Problems

1. Do you favor the use of public funds to subsidize basic research on the geology of the continental margins? What benefits are to be derived from such research? Should public support be concentrated only on those margins now believed to contain important petroleum accumulations? Should all research on petroleum resources be left to private corporations?
2. Does the rate at which a subducting slab moves down through the asthenosphere have anything to do with the frequency and magnitude of deep-focus earthquakes generated within the slab? What reasoning do you apply to reach your answer?
3. Both Japan and Chile experience great earthquakes. Japan is part of an island arc with a backarc basin. Chile is part of the continental mainland. How do you explain the similarity in level of seismic intensity in two such geologically unlike regions?
4. If the oceanic lithosphere has a tendency to sink into the asthenosphere, how can it remain intact over vast areas of the ocean basins?
5. Why have geophysicists rejected the hypothesis that the drag of mantle convection currents causes lithospheric plates to move? Can you devise a simple working model, using a slablike object floating on a water surface, to illustrate your answer?
6. Use the principle of isostasy to explain the formation of a structural high above an accretionary prism. How does isostasy affect the accumulation of sediment in a forearc trough? How is isostasy involved in the exposure of blueschists at the surface of the continental crust, where subduction has been in progress for many tens of millions of years?

CHAPTER TWELVE

Geosynclines, Orogens, and Continental Structure

In this third chapter on global plate tectonics, we concentrate on the continental lithosphere. Because most of that lithosphere is very old, our investigation will include reference to events in geologic history as far back as Precambrian time. One task is to describe the different classes of geologic structures that make up the continental crust and to relate these features to plate tectonics. In so doing, we can gain a better understanding of the recent and ancient mountain systems of North America and Eurasia.

Structural Components of the Continental Crust

The crust of the continents, which is merely the uppermost layer of the continental lithosphere, can be said to consist of a variety of structural components, which Figure 12.1 shows in schematic form. Eleven types of structural components are designated by capital letters, but the letters are for convenience only. You need attach no particular significance to the letter sequence. A brief description of each structural component is as follows:

A. **Exposed shields.** Shields are stable craton regions of exposed metamorphosed igneous and sedimentary rocks produced by plate subduction and collision throughout Precambrian time. (See Chapter 7 for a description of the exposed shields and their history.)

B. **Sedimentary platforms.** These consist of strata of nearly horizontal attitude resting on Precambrian basement rocks of the cratons. The strata range from Paleozoic through Cenozoic in age and were largely deposited in shallow seas, although continental sediments may be present. Included with the platforms are *sedimentary basins* and low arches or domes.

C. **Geoclines of the passive continental margins.** These are *miogeoclines* (continental-shelf wedges) undeformed by tectonic stresses. Sediments range in age from mid-Mesozoic through Cenozoic. Adjacent *eugeoclines* (turbidite wedges) can be studied in conjunction with miogeoclines, but are classed as features of the oceanic lithosphere.

D. **Geosynclines (sedimentary troughs) and arcs of the active continental margins.** Active geosynclines include forearc troughs, backarc troughs, and foreland troughs. They are long, narrow belts of thick sediments of late Mesozoic through Cenozoic age, largely undeformed by compressional stresses. Positive, ridgelike features (geanticlines) adjacent to or between troughs are tectonic arcs and volcanic arcs. (Trench deposits and accretionary prisms are excluded.)

E. **Cordilleran-type orogens.** These are orogenic structures produced by subduction tectonics and volcanism active through late Mesozoic and Cenozoic time. Structures include deformed accretionary prisms of metamorphic rock, volcanic arc accumulations, plutons, and belts of thrust sheets, with associated foreland sedimentary accumulations.

F. **Eurasian-type orogens.** These are orogenic structures produced by plate collision throughout late Mesozoic and Cenozoic time. They consist of alpine-type structure, including thrust slices with ophiolites, nappes, flysch, and molasse. Foreland folds are included. Underthrusting of continental lithosphere may be present.

G. **Tectonic uplifts and domes of the continental forelands.** These tectonic structures consist of platform and geosynclinal strata uparched or downbent by foreland basement deformation.

H. **Orogenic root systems of Paleozoic subduction and collision.** These are lower roots of orogens produced during Caledonian, Hercynian, and Appalachian orogenies associated with ocean closings.

I. **Block-faulted and rift-valley structures.** These are tectonic forms produced by plate extension and thinning in Cenozoic time and active today. They may represent incipient plate breakup leading to new spreading boundaries. Included are transform faults representing active plate boundaries that cut through continental lithosphere.

FIGURE 12.1 Schematic cross sections of the principal structural components of the continental crust. (Drawn by A. N. Strahler.)

J. **Inactive downfaulted and rifted crust.** These structures represent failed arms of triple junctions with aulacogens. Many are associated with Mesozoic breakup of Pangaea, but others are related to older Paleozoic or even Precambrian rift events.

K. **Isolated masses of extrusive igneous rock.** These volcanic structures represent hot spot and plume activity beneath the continental plates, as distinct from volcanism above subducting oceanic plates. They include flood basalts and isolated volcano groups.

Sedimentary Platforms of the Cratons

Over large areas of the continental cratons, a layer of nearly horizontal sedimentary rocks now covers the ancient shield rock, most of which is of Precambrian age (see Figure 7.21). This sedimentary cover of Paleozoic and younger strata constitutes a *continental platform*. Typically, the platform strata are of marine origin, deposited in shallow water, and consist of limestones, shales, well-sorted quartz sandstones, and conglomerates. The maximum total thickness of platform strata at any one place averages some 1000 to 2000 m, for the rate of platform sediment deposition is comparatively slow. The Paleozoic strata of Grand Can-

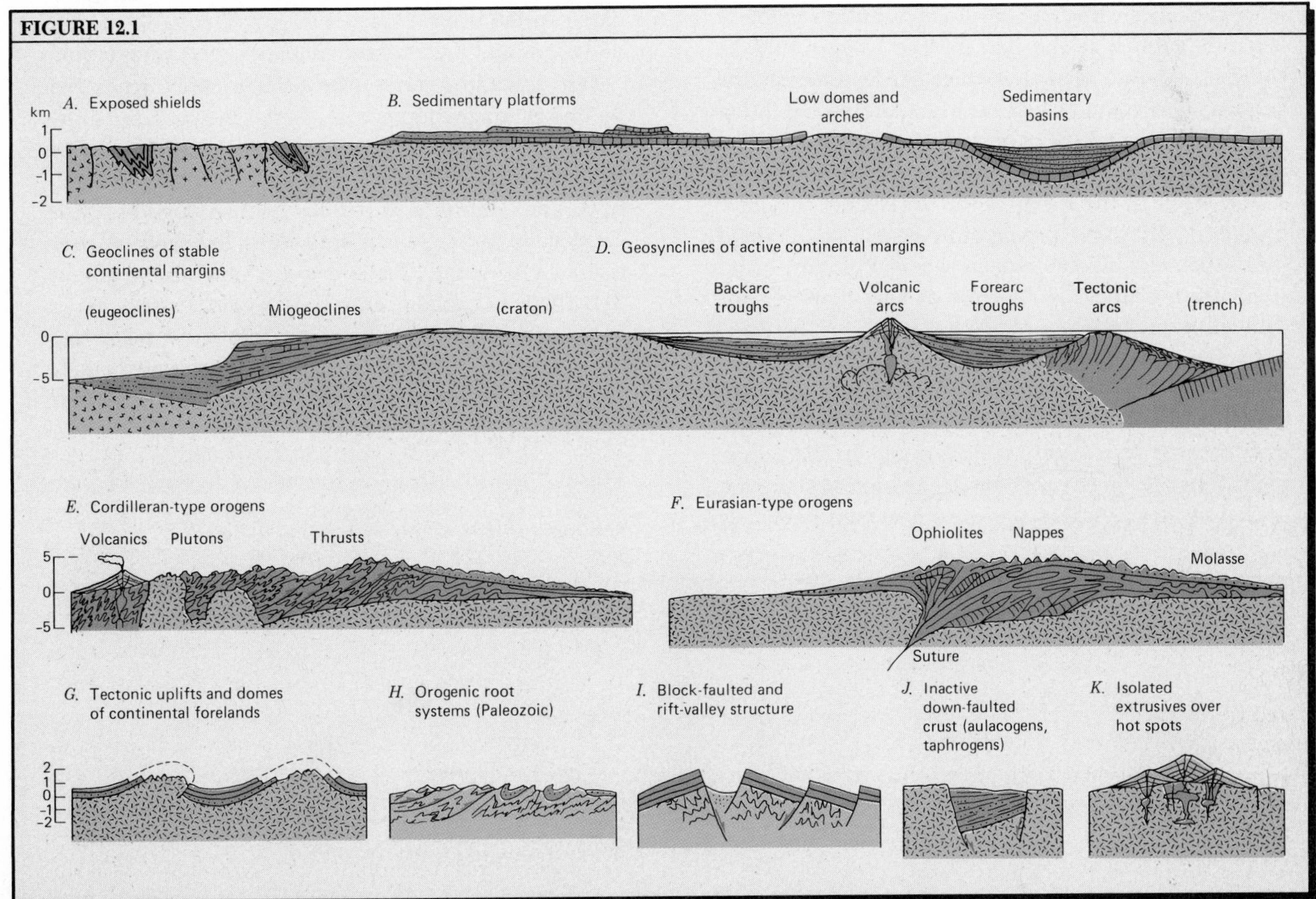

yon, described in Chapter 6, are an example of the continental platform.

In Chapter 6, we explained that slow negative (downward) epeirogenic movements of the cratons allowed shallow seas to invade the continental interiors. Positive epeirogenic movements reversed this process and brought the same areas above sea level. Many alternations of positive and negative epeirogenic movements affected the cratons throughout the Paleozoic, Mesozoic, and Cenozoic eras. Platform strata of a wide range of ages are thus found in sequence. During periods of time when the craton surface stood above sea level, platform strata deposited in earlier periods were subject to erosion by running water. During these long episodes of denudation, part or all of the older platform sequence was removed. The detrital sediment was transported by streams to the nearest ocean shoreline, from which it was swept by currents into an adjacent oceanic trench or across a continental shelf. An advance of the marine shoreline over the craton is called *transgression,* and shoreline retreat is called *regression*.

Epeirogenic movements of the stable cratons are difficult to explain, but a working hypothesis has been advanced. Recall from Chapter 8 that the Moho has been interpreted as marking the phase transition from gabbro to eclogite which occurs at a particular combination of pressure and temperature. Suppose, now, that the temperature just above the Moho should be increased above the critical phase-transition point. A layer of gabbro at the base of the crust would be transformed into eclogite, which is denser than the gabbro. The result would be a sinking of the lithosphere to a lower level in order to reach isostatic equilibrium. The opposite change, from eclogite to gabbro, would result in the rising of the lithosphere to a new level of equilibrium. Temperature changes that would cause such epeirogenic movements of the crust could result from changes in the rate of heat flow upward through the mantle.

Another possibility to consider in explaining platforms is that marine inundation of a low, plainlike craton might occur because of a rise of sea level, rather than because of a negative epeirogenic crustal movement. Changes in the position of the marine shoreline caused by rise or fall of the ocean level are known as *eustatic changes*. For example, a popular prediction sometimes featured in the news media is the unpleasant prospect that the great ice sheet of Antarctica may rapidly melt, causing sea level to rise and many of the great coastal cities of the world to be drowned. Conceivably, the alternate growth and melting of great ice sheets in the geologic past may have caused eustatic changes of sea level, but because such ice ages were rare and comparatively brief, it does not seem likely that they would have significantly affected the alternate deposition and erosion of platform strata. Another, more positive, mechanism linked to plate tectonics has been proposed as effective in causing important eustatic changes in sea level.

As we explained in Chapter 10, oceanic lithosphere recently formed along a spreading axis stands high in elevation. As the lithosphere moves away from the spreading plate boundary, the plate becomes cooler and denser, subsiding gradually to a lower level. Suppose that the rate of spreading should be greatly increased—say, from 2 cm/yr to 10 cm/yr. In that case, the width of the two belts of high-standing lithosphere adjacent to the axis will be increased, displacing a certain volume of water from the ocean basin. As shown in Figure 12.2, world sea level will be raised and this positive eustatic change will cause a transgression of the ocean over the low cratons. When the spreading rate then slows, the former profile will be resumed, and a certain volume of water can again be accommodated in the ocean basin. A negative eustatic change will occur and the cratons will experience a regression.

Past spreading rates can be estimated from the widths of magnetic anomaly belts. In the case of the Cretaceous Period, a transgression that is known to have occurred between 110 million years ago (−110 m.y.) and −85 m.y. is matched by an episode of exceptionally rapid seafloor spreading, followed by a return to a lower rate. Plate tectonics thus offers a potent hypothesis to explain transgressions and regressions.

An alternative hypothesis also based on plate tectonics uses the possibility that a spreading plate boundary can drift toward a subduction boundary and that

FIGURE 12.2 An increase in spreading rate along the mid-oceanic plate boundary may be the cause of sea-level rise and transgression of the ocean upon the craton.

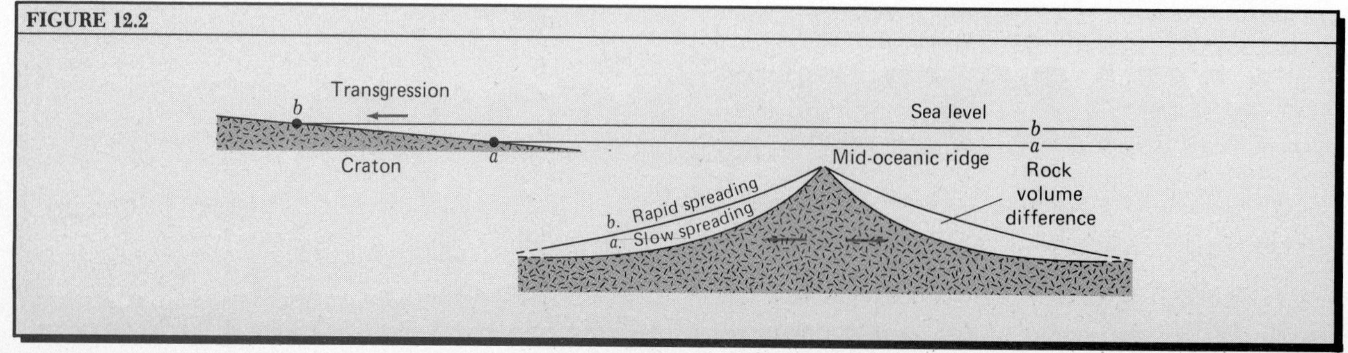

eventually the entire mid-oceanic ridge is consumed by subduction. During this process, the oceanic lithosphere that is being consumed is progressively younger and more high-standing, which means that the average age of unsubducted lithosphere elsewhere must increase, and so must the average depth of the ocean basins. The result will be eustatic lowering of sea level and regression. Later, as new spreading boundaries appear and an increased amount of new oceanic lithosphere is generated, average depth of the ocean basins will again decrease and a positive eustatic movement will bring on transgression.

Carrying this last speculation a bit further, consider the effect of the opening up of a new backarc basin by extension, following the model explained in Chapter 11. During this process new hot lithosphere within the backarc basin replaces in areal extent the old, cold lithosphere lost by subduction in the trench on the outward side of the laterally moving island arc. The substitution of new lithosphere for old would decrease the average depth of the ocean basins, as required for transgression.

Sedimentary Basins within the Cratons

From time to time throughout the Paleozoic and later eras, elliptical or circular *sedimentary basins* were formed within the interior region of the continental cratons. Marine sediments deposited in shallow water in these subsiding basins reached thicknesses on the order of 1000 m during a single geologic period lasting 30 to 50 m.y. This thickness is some four to five times greater than platform strata laid down in the same time period.

A remarkable example of a sedimentary basin of early Paleozoic time is the Michigan Basin, so named because it happens to coincide quite closely in geographic position and extent with lower Michigan, though it also includes adjacent parts of Lake Michigan and Lake Huron (Figure 12.3). During the Cambrian and Ordovician periods, this region received platform deposits that covered the Precambrian basement rocks of the craton. Then, in the Silurian Period, starting about −430 m.y., the crust began to subside in a nearly circular basin with a diameter of about 350 km. The basin lay within a shallow sea in which coral reefs formed surrounding shallow banks. The reefs nearly enclosed the subsiding basin, but two or three passes connected the basin with the outer ocean.

The situation was ideal for the deposition of salt beds, as we described the process in Chapter 5 (see Figure 5.18). The climate was evidently arid, and evaporation of seawater was rapid in the stagnant waters of the basin. Water lost by evaporation was replaced by water inflow through the passes. Over 1200 m of salt beds and interbedded limestone strata formed in the central part of the basin, as shown in the geologic cross section of Figure 12.4. Subsidence of the basin continued through the ensuing Devonian, Mississippian, and Pennsylvanian periods, during which time another 2000 m of sediments accumulated. Today, the Precambrian floor in the center of the basin lies about 3700 m below sea level, and since we know some of the uppermost strata of the sequence were later eroded away, the total basin deposit was probably more than 4000 m thick.

The cause of subsidence of a craton basin such as the Michigan Basin is not known, and we can only speculate on what mechanism may have operated—for example we might propose that a downsinking column in the mantle persisted beneath the lithosphere at this place. However, whatever the mechanism was, it must have operated for nearly 150 m.y., which is much longer than the life span of most geosynclines.

Along with sedimentary basins of various sizes and depths, the craton has shown from time to time low arches separating shallow basins. Over the arches, seas were very shallow or the arches appeared as low belts of land. Thus, the record of the strata over these positive belts tends to be incomplete.

The Geosyncline Concept in Classical Geology

In the 1850s, a pioneer American geologist, James Hall, State Geologist of New York, was studying the strata of the northern Appalachian Mountains. Using the principles of superposition and continuity that we explained in Chapter 6, Hall reconstructed the total succession of strata deposited in that region in each period of the Paleozoic Era. In this way, he was able to determine at what locations the strata were thickest and how their thickness changed along a line of traverse. He was led to conclude that for a given geologic period marine strata were deposited in a slowly subsiding trough. While the floor of the trough was subsiding, the depth of ocean water remained shallow, because the rate of sediment deposition kept pace with the rate of crustal subsidence. This is essentially the same scenario as that for subsiding craton basins, such as the Michigan Basin, but the subsiding feature observed by Hall was a long, narrow trough—many times longer than it was wide.

In 1873, James D. Dana, a professor of geology at Yale University, confirmed Hall's interpretation and gave the name "geosynclinal" to the subsiding trough. Later, this name was changed to *geosyncline*. This term is appropriate because, when sediment deposition has been completed, the strata are warped down into a shallow syncline. The prefix "geo" signifies that the syncline is of very large dimensions as compared with synclines produced by compressional deformation of strata. The dip of strata in a geosyncline is visualized as extremely gentle before any folding occurs.

Let us examine a particular geosyncline in order to follow the interpretive method used by a stratigrapher. Figure 12.5 is a graphic presentation of all strata of the Devonian Period in southern New York State. Diagram *A* shows an east–west profile of the land sur-

FIGURE 12.3 Maps of the Michigan Basin. *A*. Geologic map showing the surface exposure of strata of Paleozoic age within the basin. *B*. Configuration of the Precambrian basement rock surface that lies beneath the Paleozoic sedimentary strata, shown by contour lines. *C*. Paleogeographic map showing the area of salt accumulation in relation to surrounding reef banks and inlets. (From M. Kay and E. H. Colbert, 1965, *Stratigraphy and Life History*. Copyright © 1965 by John Wiley & Sons, Inc., New York. Reproduced by permission.)

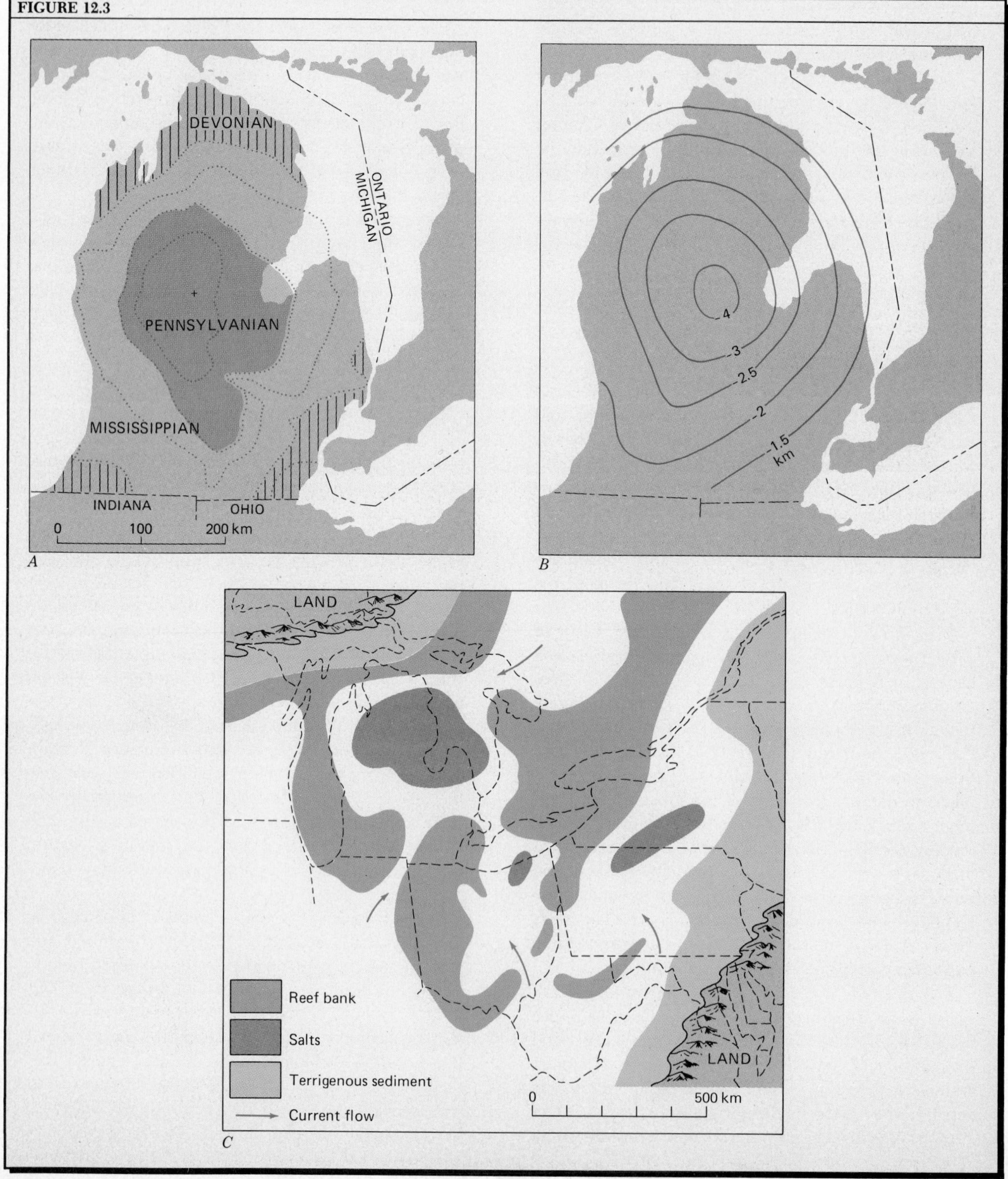

face extending from the Hudson Valley on the east to about Erie, Pennsylvania, on the west, and, below the profile, a *stratigraphic cross section* of the rocks present today along the line of profile. The vertical scale is greatly exaggerated—about 40 times greater than the horizontal scale. This form of exaggeration is standard practice to focus attention on the details of the stratigraphy at any given point, but it greatly increases the apparent dip of strata. For example, the limestone strata at the extreme eastern end of the section appear to have a dip of about 70°, whereas in fact the dip is somewhat less than one degree!

The strata actually show very gentle folding that occurred after they were deposited. (Strong folding affected strata of the same age to the east of this locality.) The stratigraphic section shows each kind of sedimentary rock by a different pattern. The thin strata at the base of the section are marine limestones, but all the rest are clastics and range from coarse-textured red sandstones and conglomerates (red beds),

FIGURE 12.4 A restored stratigraphic section of the Silurian strata of the Michigan Basin. (From M. Kay and E. H. Colbert, 1965, *Stratigraphy and Life History.* Copyright © 1965 by John Wiley & Sons, Inc., New York. Reproduced by permission.)

FIGURE 12.5 A geosyncline of Devonian age in western New York State. *A.* East–west surface profile and stratigraphic cross section of the Devonian strata as they are found today. *B.* Restored stratigraphic section showing conditions at the close of Devonian sedimentation. (*Section A* after J. G. Broughton et al., 1962, New York State Museum and Science Service, Chart Series No. 5. *Section B* from M. Kay and E. H. Colbert, 1965, *Stratigraphy and Life History,* p. 218, Figure 11-9. Copyright © 1965 by John Wiley & Sons, Inc., New York. Reproduced by permission.)

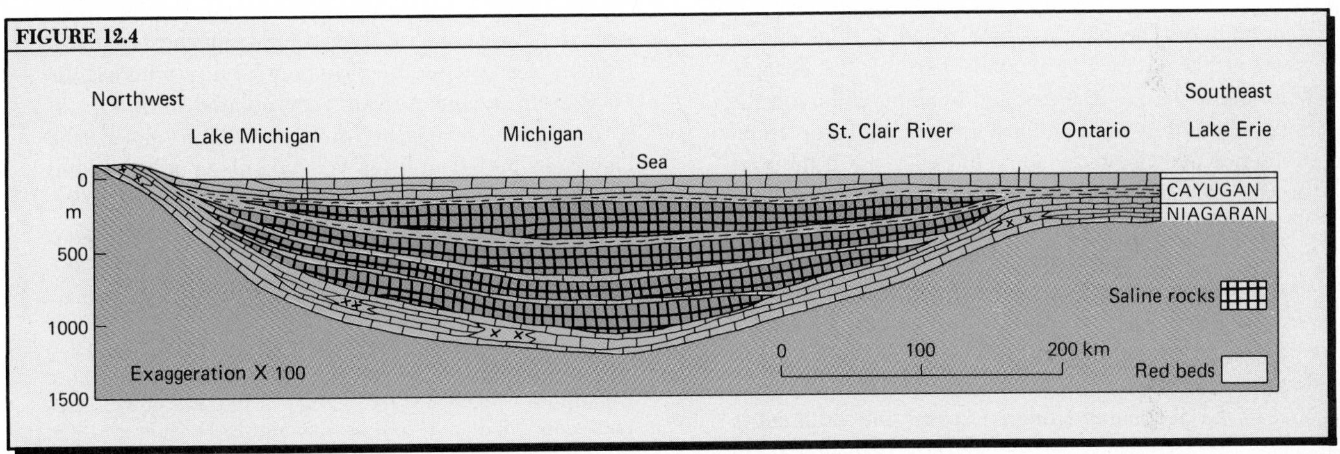

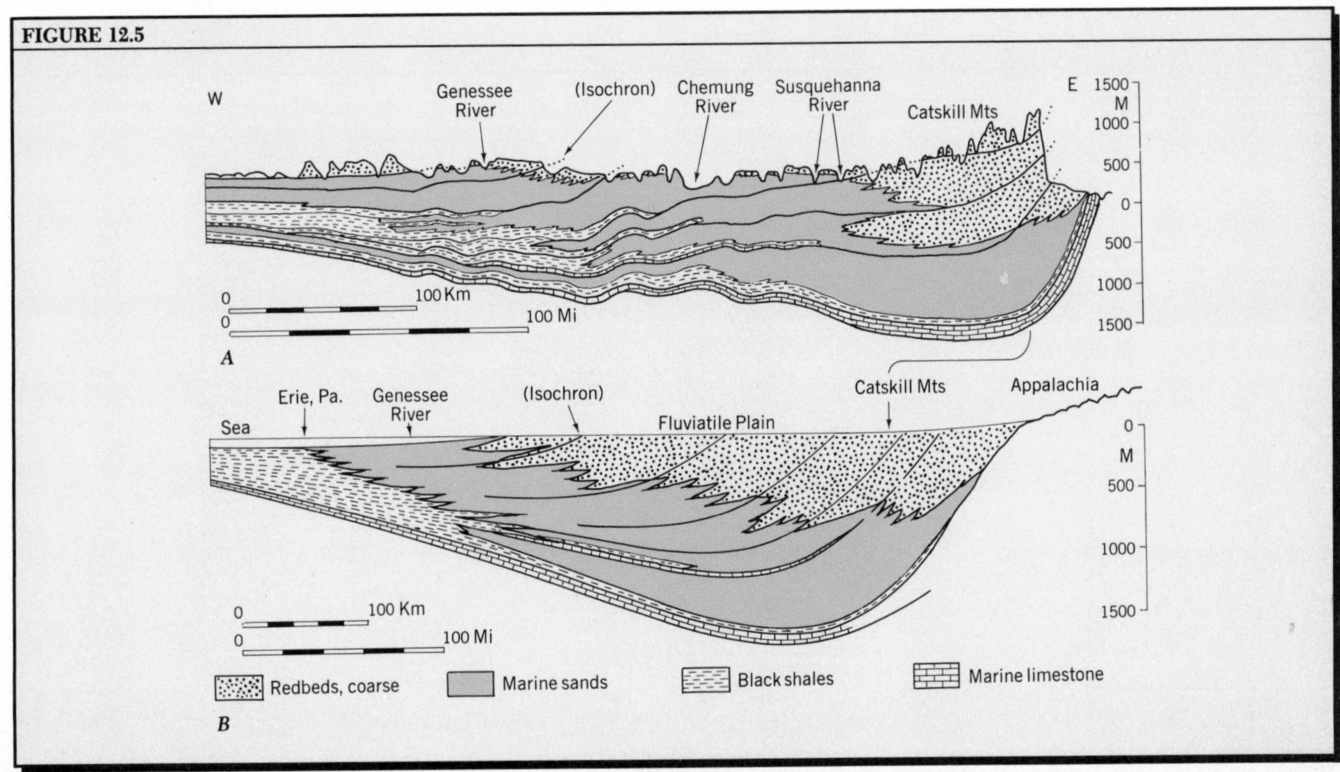

through sandstones, to shales. Undulating solid lines running through the section are time lines, or isochrones, explained in Chapter 6. The layer of strata between any two successive isochrones was deposited within the same time span. A stratigraphic section of this type is usually constructed from the study of rock cores obtained from boreholes drilled into the rock, since most of the rock mass lies far beneath the present land surface.

Diagram *B* of Figure 12.5 is a *restored stratigraphic section*. It takes the information in the upper stratigraphic section and replots the vertical dimension with respect to a horizontal upper reference line. This reference line represents sea level at the end of the period of sediment deposition—in this case, the end of the Devonian Period. All undulations caused by later folding are removed, and the isochrones become smooth curves.

Let us now interpret the restored stratigraphic section. First, we see the broad troughlike outline of a geosyncline. The clastic strata reach a thickness of about 1500 m in the deepest part of the trough, thinning to zero on the east and to about 250 m on the west where they are evidently typical platform strata. Evidence that the geosyncline did not exist at the start of the Devonian Period lies in the nearly uniform limestone layer at the base of the section. It must have been deposited in shallow water over a submerged craton or platform that was present early in Devonian time.

Then, for some reason we do not know, a trough began to form, and sediment from the east was deposited in the deepening trough. Because the sediment is coarse in texture along the eastern margin of the trough and grades into finer texture to the west, we postulate that a high mountain range, called Appalachia, had become uplifted along the eastern side of the geosyncline. It seems reasonable that the coarse clastic sediment was part of a delta system formed by streams carrying sediment down from the mountain range. The delta surface was a low plain, only slightly above sea level. Finer sediment was spread as a sand deposit in shallow ocean water lying to the west, while clay particles were carried still farther west into the inland sea. As time passed, the coarse delta deposits spread farther toward the west. The Devonian Period ended with a tectonic event—an orogeny—that uplifted and gently deformed the strata of the geosyncline.

The Devonian strata of New York State make up only one major layer of a complete geosyncline. Deposition continued during the ensuing Mississippian Period in the Appalachian region. Figure 12.6 is a restored stratigraphic section along a line running across Virginia and West Virginia. This portion of the Appalachian geosyncline received a total thickness of about 8000 m of clastic sediment in only 20 m.y. whereas the 1500 m of Devonian strata accumulated in 50 m.y. It seems that the nearby orogeny which closed the Devonian Period uplifted the axis of Appalachia and resulted in a much faster rate of sediment deposition in the Mississippian Period than in the Devonian Period.

FIGURE 12.6 Restored stratigraphic section of a geosyncline of Lower Carboniferous age in the central Appalachians. Conditions are shown as they existed at the close of Mississippian sedimentation. (From M. Kay and E. H. Colbert, 1965, *Stratigraphy and Life History*. Copyright © 1965 by John Wiley & Sons, Inc., New York. Reproduced by permission.)

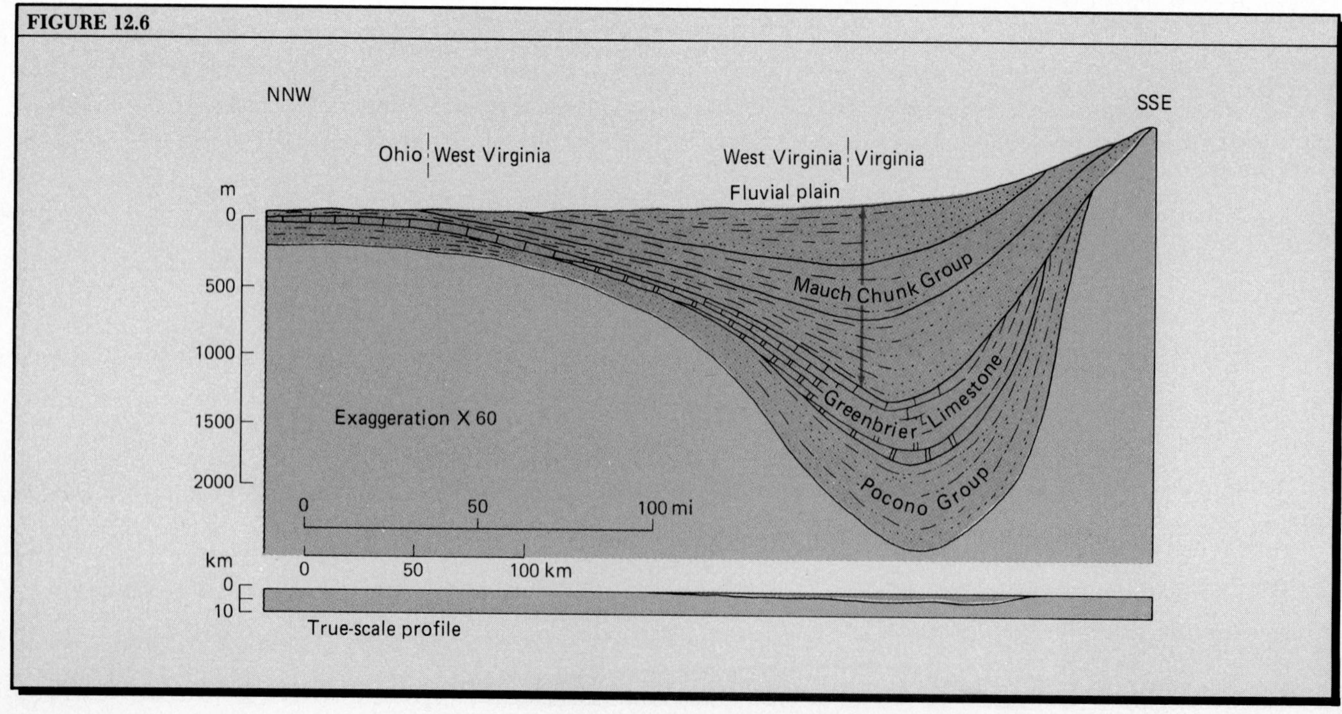

From the time of Hall and Dana until about the late 1930s and early 1940s, studies of geosynclines were based almost entirely on ancient sedimentary and metamorphic rocks seen exposed on the continents, rather than on present-day geosynclines in which sedimentation is now taking place. Keep in mind that seismic refraction and reflection methods of subsurface exploration had not been fully developed or widely used. True, oil wells had been drilled into the geologically young strata of the Gulf Coast region, and it was understood that a great wedge of undeformed sediments, dating back to the Cretaceous Period, lay along the continental margin in this region. But this wedge of sediments did not fit the early or classical concept of a geosyncline, because there was no crustal trough bounded by highlands. Instead, the sediment wedge ended in a steepened continental slope.

The classical geosyncline was described as consisting largely of marine sediments deposited in shallow water. It was inferred that the rate of crustal subsidence of the trough closely matched the rate of sediment accumulation (or vice versa). The geosyncline was always underlain by older basement rocks of the continents, and the idea of thick sediment deposition on the deep ocean floors was totally unknown. The classical geosyncline was usually pictured as having a mountain range along one side. This mountain range furnished sediment to the subsiding trough, as did a lowland area of the craton on the opposite side. Two parallel geosynclines were separated by an uparched land belt called a *geanticline* (see Figure 12.7).

One of the fundamental tenets of the classical model of geosynclines was that the period of sediment deposition always ended in orogeny, destroying the geosyncline and creating a belt of folds and overthrust faults. This conceptual linkage between geosynclines and orogenies was very strong—so strong, in fact, that every belt of folded strata was assumed to have originated as a geosyncline. The implication was that orogeny occurs only where geosynclinal deposition has taken place.

The arrival of the great geologic revolution of the 1960s played havoc with the classical concept of the geosyncline. Some geologists tried to accommodate plate tectonics to the established varieties and forms of geosynclines in ways such as suggested by Figure 12.7. They retained the term "geosyncline" and many of the terms applied to varieties of geosynclines. Other geologists were ready to scrap the classical model entirely, even to abandoning the name "geosyncline." This entire question was discussed at a notable conference of geologists held in 1969 at Asilomar, California, under the auspices of the Geological Society of America. By this time, plate tectonics had been widely accepted and a great deal was known about the sediment deposits of the continental margins and subduction zones. The consensus reached by the conference participants seemed to be that the entire subject of geosynclines needed to be revised to include new knowledge of contemporary deposits of thick sediments along both active and passive continental margins. Revision has since been largely accomplished, and little remains of the classical model. What geologists have done is to follow the principle of uniformity, which implies that "the present is the key to the past." They have looked closely at contemporary sediment deposits of the continental margins, both stable and tectonically active. Refraction and reflection profiling has allowed them to probe these thick sediment deposits and to discern their internal structure and the nature of the rock floors on which they rest.

As we turn to the subject of geosynclines and their tectonic evolution, keep in mind that we are trying to incorporate principles of plate tectonics as well as new information on ocean-floor sediment accumulations.

FIGURE 12.7 Schematic block diagram to show how the classical concept of the geosyncline might be adapted to plate tectonics. At the left is a geosyncline adjacent to the tectonic arc formed at a subduction boundary. A volcanic arc runs along the middle of the geosyncline. At the right is a geosyncline bounded on one side by a broad geanticline. (Drawn by A. N. Strahler.)

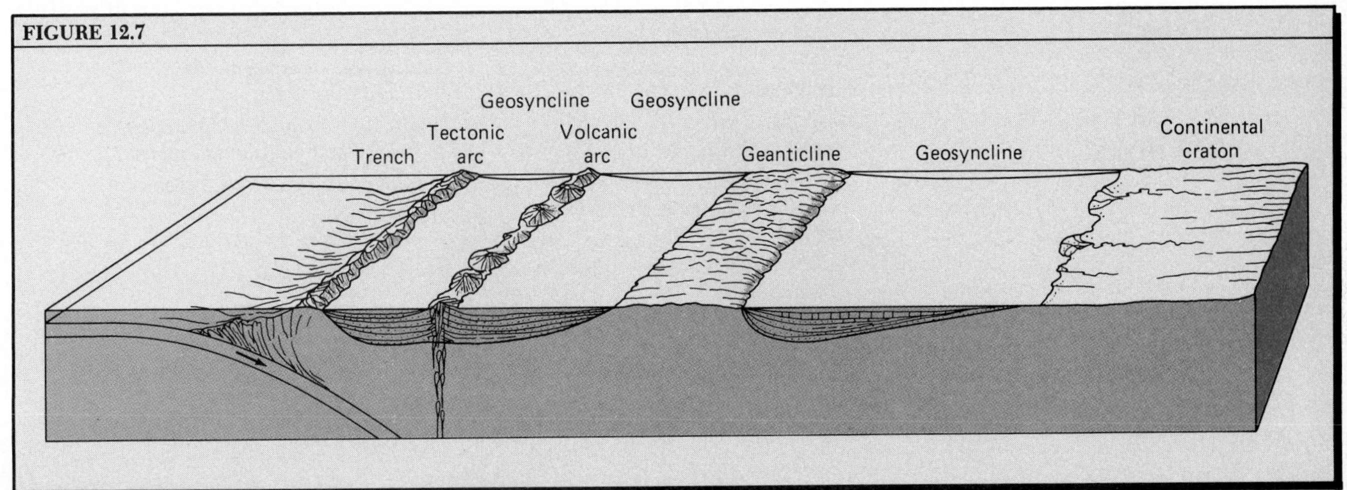

This is a subject difficult even for research geologists to view clearly, because of its complexity and because data are still fragmentary. We will make many generalized and sweeping statements in an effort to keep the picture simple, consistent, and meaningful.

Contemporary Geosynclines and Tectonic Activity

Our definition of geosyncline is broadly worded: A *geosyncline* is a thick, rapidly accumulating body of sediment formed within a long, narrow belt usually paralleling the margin of the continental lithosphere. A geosyncline may accumulate in a trough or trench with rising slopes on either side, or it may accumulate on a gently descending slope leading down or up to the contact of continental lithosphere with oceanic lithosphere. Sediments of a geosyncline can be deposited either in shallow marine water or on the deep-ocean floor, or they may accumulate as continental deposits on a land surface above sea level. The continental plate margin adjacent to or beneath a geosyncline may be either an active plate boundary, where subduction is in progress, or a passive contact between continental and oceanic lithosphere, far from an active plate boundary. Geosynclines can, and often are, controlled by tectonic activity, but the reverse is not true: Geosynclinal deposition does not set off tectonic activity. Because ocean basins open and close more or less continually through geologic time, however, it is almost inevitable that a geosyncline will become caught up in an orogeny and its strata will undergo deformation, and it is possible for a geosyncline to be formed in an environment of ongoing tectonic activity. In other words, sediment deposition and tectonic activity can take place simultaneously in the same narrow zone. With these general principles in mind, we can turn to a classification of geosynclines. (Those who prefer to abandon the term "geosyncline" may substitute another term, such as "sediment trough.")

Table 12.1 and Figure 12.8 show the various classes and kinds of geosynclines and related belts of rapid sediment deposition that can be identified as active today. We recognize four major classes of geosynclines, covering the common or important types.

I. **Atlantic Type.** There are two kinds of geosynclines in this class: miogeocline and eugeocline. They are wedge-deposits on passive continental margins, accumulating on the subsiding contact between a continental plate and an oceanic plate. Best described as *geoclines*, they accumulate during the opening of an ocean and are not involved in tectonic activity during later stages of their formation.

II. **Indonesian Type.** Named for their occurrence in the Indonesian region, the three kinds of geosynclines listed here are associated with tectonic and volcanic activity of a subduction plate boundary. The *trench wedge* forms directly over the plate boundary. The *forearc trough deposit* lies between an outer tectonic arc and an inner volcanic arc and is thus located over the descending plate. The *backarc trough deposit* lies between the volcanic arc and the craton.

III. **Eurasian Type.** This geosyncline is a *foreland trough deposit;* it may accumulate after a continental collision has formed a suture. It rests on continental lithosphere on either side of the high mountain belt of the suture zone and consists of coarse clastics laid down either in a shallow seaway or on a sloping piedmont plain. The foreland trough deposit may develop while suturing is still in progress or after its completion.

Table 12.1 Contemporary Sedimentary Troughs (geosynclines)

		Name	Sedimentary type	Position
I	Atlantic Type (Passive)	Miogeocline	Shallow-water marine sediments of cratonic origin	Shelf on continental margin, inactive
		Eugeocline	Deep-ocean sediments; turbidites, contourites	Continental rise over oceanic lithosphere, inactive
II	Indonesian Type (Active)	Trench wedge	Deep-ocean sediments; turbidites, pelagic sediments, mélange	Subduction boundary; contact of oceanic and continental plates
		Forearc trough deposit	Deep-water and shallow-water marine sediments; shales, lithic sands	Continental slope at subduction boundary
		Backarc trough deposit	Shallow-water marine sediments; clastics	Behind volcanic arc
III	Eurasian Type (Active)	Foreland trough deposit	Continental sediments; coarse clastics (molasse)	Adjacent to collision suture
IV	African Type (Active)	Taphrogen or Aulacogen	Continental sediments; coarse clastics, red beds	In rift valley belt or at failed arm of triple junction

IV. African Type. This kind of geosyncline may be a *taphrogen* or an *aulacogen;* it represents thick sediment deposition in a downfaulted basin. It has been interpreted as related to the rifting of a continental plate preceding the opening of a new ocean basin. In Chapter 13, we explain that the aulacogen is thought to be the failed arm of a triple junction developed above a mantle plume.

Atlantic-Type Geoclines of the Passive Continental Margins

In contrast to geosynclines that form along tectonically active continental margins are those thick sediment wedges that form on passive continental margins far from active plate boundaries. A sediment wedge is named a *geocline* to make clear that the strata dip in one direction only (Figure 12.9). The continental-shelf wedge has been named a *miogeocline;* the turbidite wedge next to it is a *eugeocline.* Both these sediment wedges are formed during long, slow subsidence of the lithosphere.

The miogeocline receives its sediments from streams that are eroding older rocks of the adjacent stable continental craton. The miogeocline also loses sediment at its outer edge, where masses slump off the edge of the continental shelf and travel as turbidity currents down the steep continental slope. The thick outer edge of a miogeocline butts directly against the thick edge of the eugeocline formed in deep water of the continental rise. The eugeocline may include turbidites with graded bedding and well-sorted sandy contourites. The eugeocline will rarely, if ever, include important quantities of volcanic materials except where a hot spot has generated an isolated volcanic seamount.

Fortunately, there is no lack of examples of geoclines being formed at the present time. Stable continental margins exist on both sides of the Atlantic Ocean basin, both along the American margins on the

FIGURE 12.8 Four basic classes of geosynclines. See Table 12.1 for further data. (Drawn by A. N. Strahler.)

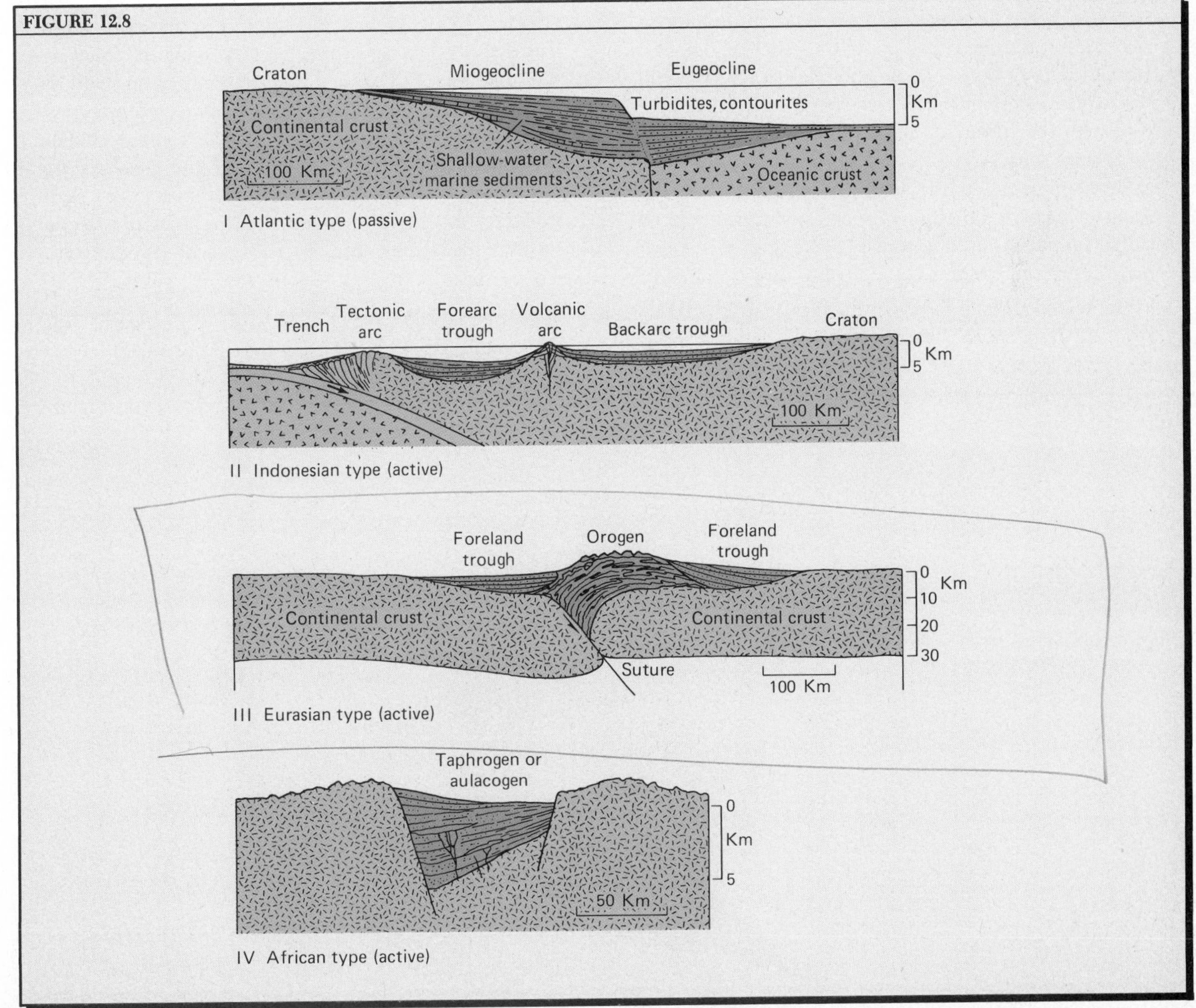

west and the western European and African margins on the east. The Atlantic Ocean basin came into existence by continental rifting that began in late Triassic time. Thus, continental shelf wedges and turbidite wedges have been accumulating for some 200 m.y., and it is not surprising that thick geoclines border the continental coasts for thousands of kilometers. We need look no farther than our North American continent for prime examples: the East Coast Geocline and the Gulf Coast Geocline. These geoclines join in Florida and have had essentially the same history of development. They include no volcanic materials of importance and lack such structures as folds that might indicate tectonic compression.

Of the two geoclines, the Gulf Coast Geocline is the thicker, with a maximum thickness of 20,000 m in the continental shelf wedge, or miogeocline (Figure 12.10). Strata of Jurassic age form the base of this sediment wedge, but most are of Cretaceous and Cenozoic ages. In deeper water of the eugeocline that lies beneath the floor of the Gulf of Mexico, the deep-sea sediment accumulation reaches a thickness of 10,000 m. An interesting feature of the Gulf Coast Geocline is the presence of a thick salt formation in the oldest strata near the base of the geocline. The salt formation may be as old as late Triassic age. We will discuss the origin of the salt in Chapter 21, in connection with the formation of salt domes that are major petroleum sources.

Strata making up the Gulf Coast and East Coast miogeoclines are limestones, shales, and sandstones. A single formation usually shows thickening in the seaward direction, while in the landward direction it may thin to the point of disappearance. This geometry is what we would expect of deposition in shallow coastal waters that receive sediment from the mouths of streams. The sediment is spread seaward by currents and comes to rest in deeper water offshore. Strata in the upper part of the geocline, which are of Miocene, Pliocene, and Pleistocene ages, are mostly unconsolidated and consist of sand and mud layers, including calcareous muds.

Even the oldest strata of a miogeocline, those at the bottom of the sequence, are typically shallow-water deposits, and this may be characteristic of the entire accumulation of strata. This fact means that the marine shoreline shifted back and forth many times over a broad zone, sometimes invading far into the continental interior, and that the rate of sediment accumulation must have closely matched the rate of crustal sinking. This relationship was recognized many decades ago by geologists who interpreted thick geosynclines of marine strata in various mountain belts.

The principle of isostasy can be applied to the development of a geocline. As we explained in Chapter 7, the loading of the crust by accumulating sediment layers will be met with an isostatic response: The crust and entire lithosphere will sink beneath the load (see Figure 7.29). At first thought, this principle might seem to explain geoclines, because once sediment accumulation set in along a shallow continental shelf, the trough would deepen steadily as more sediments arrive, and the water depth would remain shallow throughout the entire formation of the geocline. But on second thought, this conclusion appears false. Sedimentary layers have lower density than the deeper crustal rocks they displace to a lower level. Even a

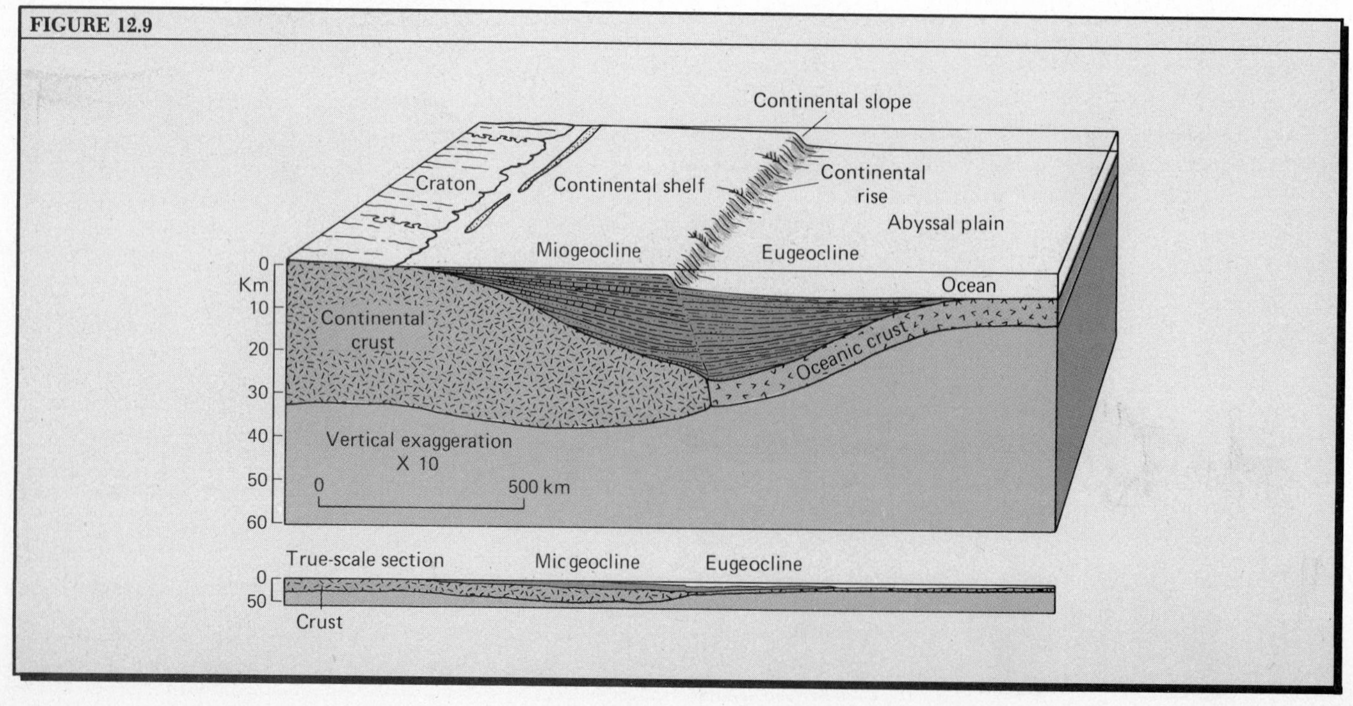

FIGURE 12.9 Schematic block diagram of geoclines on a stable continental margin. (Drawn by A. N. Strahler.)

dense sedimentary rock, such as limestone, is not as dense as mafic rock of the lower crust, so the amount of sinking by isostatic response will not be as great as the thickness of the sedimentary layer that accumulates. It can be shown that a downsinking of only 5000 m will follow the accumulation of 6000 m of compacted sediment—a ratio of five to six.

And, if the lithosphere were affected only by the isostatic force of buoyancy, the shallow sea present at the onset of sediment accumulation would soon be filled and deposition would then change from marine to continental sediments. The surface of deposition would continue to rise in elevation above sea level, or the shelf would simply be built farther out into the ocean basin. Clearly, an independent mechanism of sinking must be sought to explain a geocline.

One such mechanism might be a convection current system in the mantle. Such a flow converging along a downsinking axis beneath the lithosphere might exert a downpull, creating a trough. An objection to this mechanism is that it would tend to compress the plate, causing faulting and seismic activity, and the presence of a strong negative gravity anomaly over a geocline would be expected, but the geoclines of the Atlantic continental margins show very little seismic activity and only small isostatic gravity anomalies.

A more plausible explanation lies in plate cooling accompanying the widening of the Atlantic Ocean basin. At the time of rifting, the continental plate would be relatively warm because of the rise of mantle rock. As the stable continental plate margins retreated from the spreading axis, the plate would have steadily cooled, increasing its density, and therefore slowly sinking. Added to the sinking required by isostasy as the sediment wedges thickened, the total effect would have been to maintain a shallow continental shelf for nearly 200 million years.

FIGURE 12.10 A cross section of the Gulf Coast Geocline. Data for depths below 6 km are from seismic refraction exploration. Structure beneath the Gulf of Mexico is conjectural. (Based on data of G. C. Hardin, Jr.)

Pacific-Type Geosynclines of Subduction Boundaries

Tectonic activity in subduction zones was covered in Chapter 11, where we explained the process of upbuilding of the accretionary prism and the formation of a structural high and forearc trough. Let us now investigate in more detail the troughs of sediment deposition associated with a subduction zone in the Indonesian region where sediment accumulations have recently been investigated by seismic reflection profiling.

Figure 12.11 is a map of the Sumatra–Java tectonic and volcanic arcs, bordered on the south by the deep Java (Sunda) Trench. From seismic evidence, we can be sure that subduction is an active process here. Figure 12.12 shows four transverse cross sections of the troughs of contemporary sediment deposition and the tectonic and volcanic arcs. The first feature that can be classified as a kind of geosyncline is an accumulation of sediments in the trench itself. This *trench wedge* (TW) represents sediments that have yet to be scraped off the moving plate. The floor of the Java Trench lies at a depth of about 6000 m. Turbidites and pelagic sediments are deformed into mélange at the line of the tectonic front. A succession of imbricate thrusts develops in the soft sediments, producing an accretionary prism (AP). The deeper portions are changed into metamorphic rock. The prism is slowly forced upward to form an elevated zone—a low tectonic ridge—which we previously have called a structural high. Here we designate it as a *tectonic arc* (TA), which one might classify as a kind of geanticline, using that term rather broadly. In the Java region, the tectonic arc lies submerged at a depth about 1000 to 2000 m for much of its length, but to the west, off the coast of Sumatra, it rises above sea level as a chain of islands, the Mentawai Islands (Section A).

Adjacent to the tectonic arc is a shallow *forearc trough* (FT), a geosyncline that receives sediment from the volcanic arc of the islands of Java and

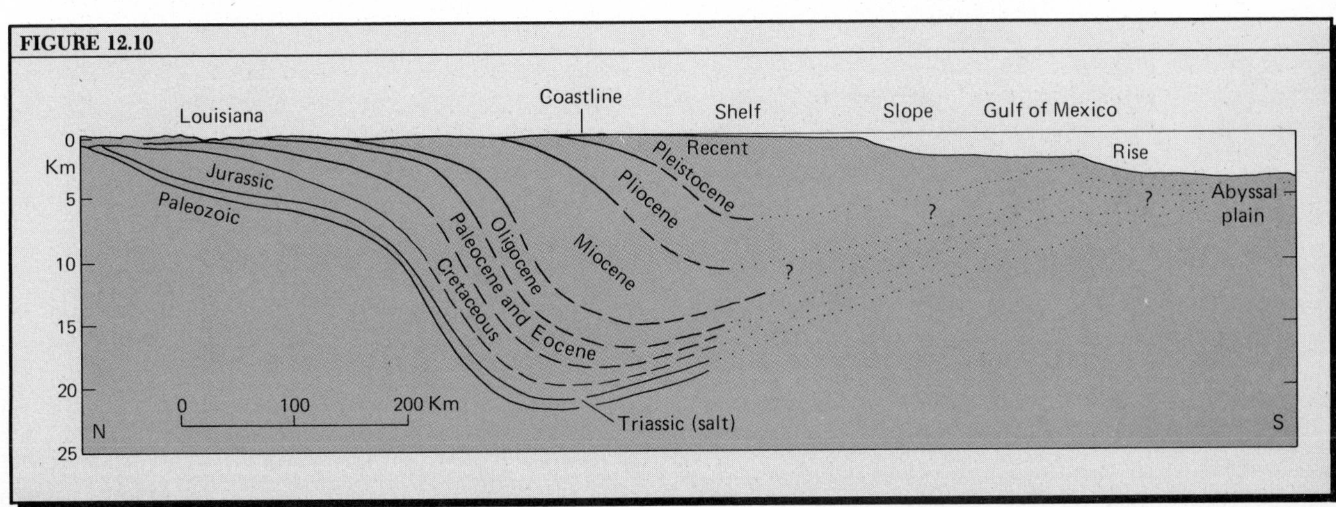

Sumatra. Sediments of the forearc trough are over 4000 m thick off Sumatra (Section A and Figure 12.13). These sediments may be largely clays, muds, and lithic sands eroded from the steep slopes of the volcanic mountain range. They are transported as turbidites where the floor of the trough is deep and the bottom slope is sufficiently steep. Because the forearc trough acts as a catchment basin or trap for sediment from the volcanic arc, the trench may become deprived of much of its sediment. Later, the tectonic ridge may rise above sea level to form a narrow land ridge. When this happens, a sediment supply will again be available. This seems to have happened off Sumatra, where the Mentawai Islands now furnish sediment to the trench of the adjacent subduction zone.

A narrow axial mountain chain on Sumatra and Java represents the volcanic arc (VA) of the system. It can be classed as a geanticline because Mesozoic and older basement rocks lie beneath the volcanic rocks along this arc, and it thus represents continental crust that has been uparched. North of the volcanic arc lies a *backarc trough* (BT) 100 to 200 km wide, representing the third type of geosyncline of the system. In Sumatra, this zone lies above sea level and is a low coastal plain, underlain by Cenozoic sediments previously deposited in the geosyncline (Section A). Traced eastward to the region of the Flores Islands, the backarc trough deepens to become the Flores Deep, with bottom depths over 3000 m (Section C). Still farther east, the volcanic arc descends below sea level with only a few small volcanic islands showing above sea level (Section D). Here the forearc and backarc trough form a more or less continuous single geosyncline with only a median dividing ridge. Clastic sediments accumulating close to the volcano chain may perhaps be interbedded with volcanic rocks—volcanic ash beds and lavas. Looking back to Figure 12.7, you will notice that we depicted a similar composite geosyncline in which a volcanic arc lies along the median line of the trough. Somewhat similar geosynclines of early Paleozoic age have been reconstructed from examination of folded strata in orogenic

FIGURE 12.11 Map of the Sumatra–Java tectonic–volcanic arc. (Geologic interpretation based on data of J. Aubouin, as presented in P. J. Wyllie, 1971, *The Dynamic Earth*, John Wiley & Sons, Inc., New York, p. 227, Figure 9–7, and W. Hamilton, Subduction in the Indonesian Region, pp. 15–31 in *Island Arcs, Deep-Sea Trenches and Back-Arc Basins*, 1977, Maurice Ewing Series 1, M. Talwani and W. C. Pitman III, eds., American Geophysical Union, Washington, D.C.)

FIGURE 12.11

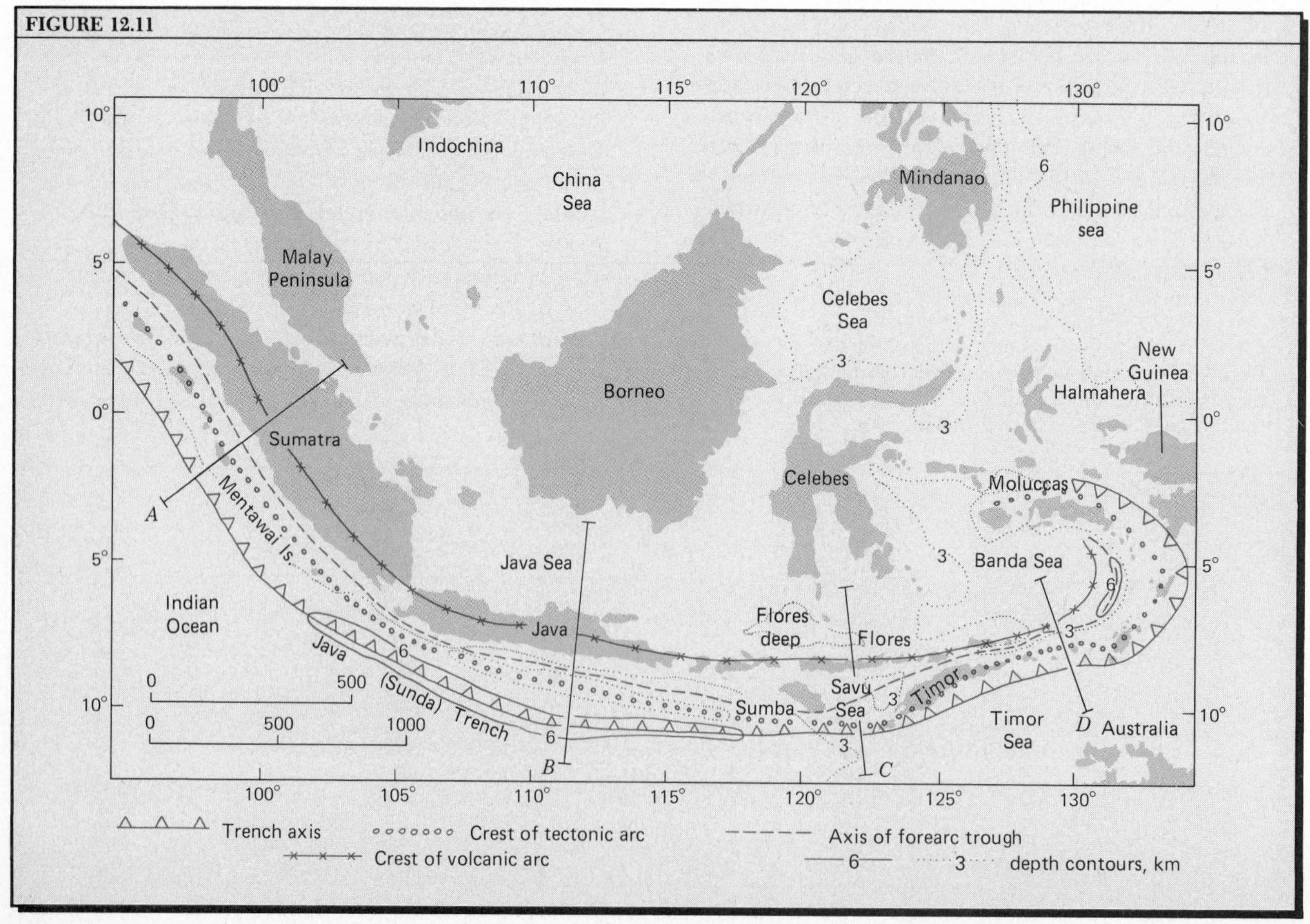

belts of North America and Europe.

North of the backarc trough lies continental crust with rocks of Paleozoic and Late Triassic age exposed in many places. Thus, the Malay Peninsula and the island of Borneo represent exposures of the *continental foreland* (CF). We use the term "foreland" to refer to any part of the stable continental crust located adjacent to a tectonic belt.

The Franciscan Complex—An Ancient Mélange?

Underlying much of northern California in a belt that reaches nearly 200 km from the Pacific shores almost to the western foot of the Sierra Nevada is a remarkable assemblage of rocks that was long a mystery to geologists. This assemblage is roughly in two halves, a western half consisting of the metamorphic Franciscan Complex and an eastern half consisting of sedimentary strata of the Great Valley sequence (see Figure 12.14*D*). The rocks of both halves have an age span of about 100 m.y., the oldest being of Late Jurassic age (−150 m.y.) and the youngest of Eocene age (−40 m.y.). It appears that the rocks in both halves were formed at the same time and more or less continuously throughout this 100 m.y. time period, which includes all of the Cretaceous Period (duration 70 m.y.). Otherwise, in terms of kind of rock that was formed, the two halves are as different as day and night.

From the standpoint of making a geologic interpretation, "day" applies to the Great Valley sequence. Any college geology major would immediately identify it as a succession of clastic sedimentary rocks, sandstones, and shales of marine origin, with a basin-like (synclinal) shape. To label the Franciscan Complex as the "night" side is quite appropriate. Much of the Franciscan Complex is an incredibly mixed-up assemblage of small bodies of such rocks as schist (blueschist, glaucophane schist, amphibolite, and eclogite), pillow-lava basalt, peridotite, serpentinite, and sedimentary rock (turbidites, dark shales, and thin-bedded cherts). The entire mass is strongly sheared, and the various rock types usually occur in lenslike masses separated by steeply dipping thrust faults. This incongruous mixture is a mélange, some of its ingredients originating as pelagic sediments on the deep ocean floor, others as turbidites of terrestrial origin, and the remainder as parts of the oceanic crust

FIGURE 12.12 Cross sections to accompany Figure 12.11. (Drawn by A. N. Strahler. Interpreted from data in references cited in Figure 12.11.)

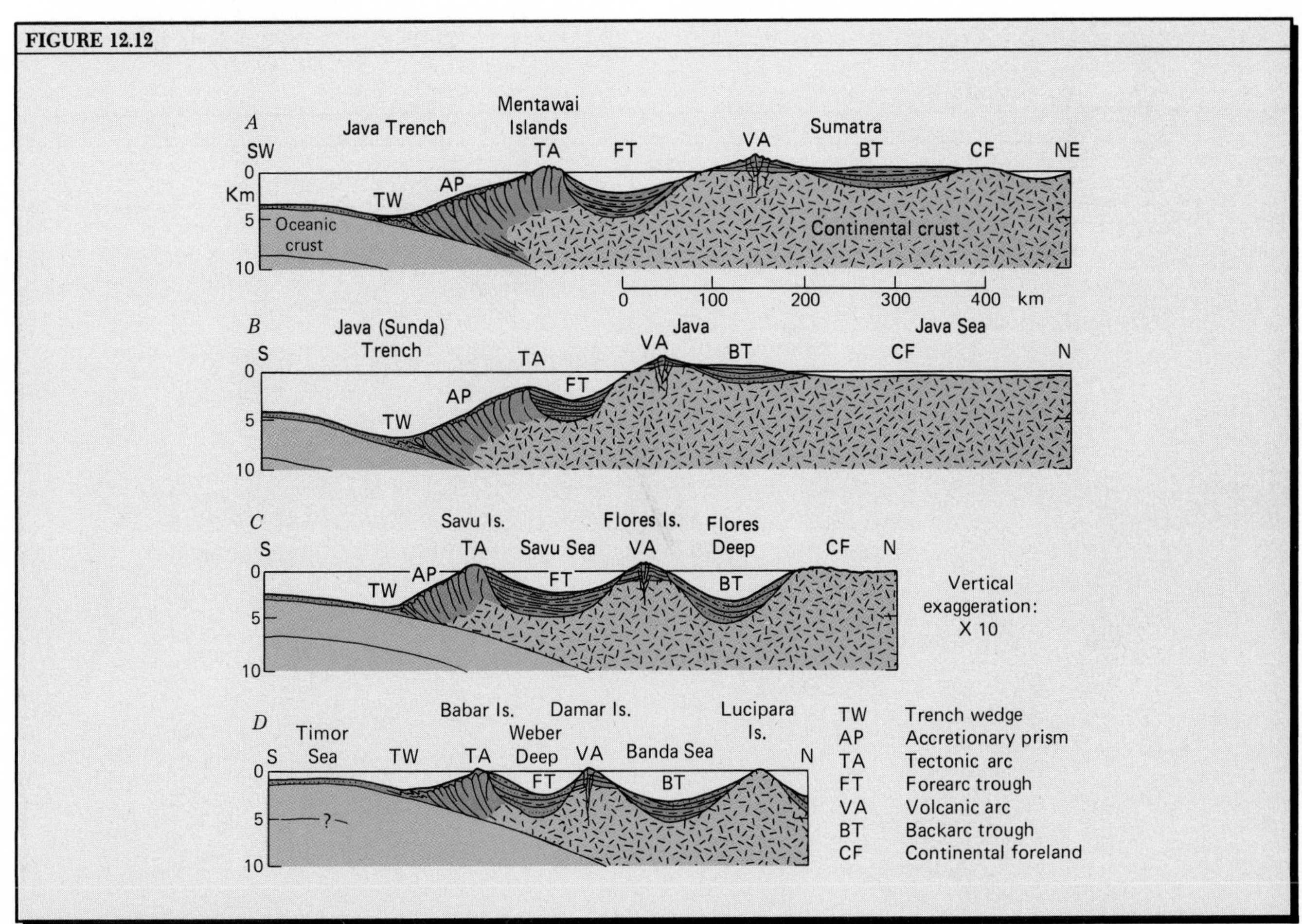

and even the mantle below the crust.

In light of what is now known of the geology of the tectonic arc and forearc trough of the Indonesian region and of other subduction zones of the same type, geologists can now say with considerable confidence that the Franciscan Complex is what remains of a massive accretionary prism, formed more or less continually through 100 m.y. The Great Valley sequence can be neatly fitted into the picture as the accumulation within a forearc trough that persisted at the same time, lying between a structural high—the crestline of a tectonic arc—and a volcanic arc that lay to the east. Figure 12.14 shows by a series of cross sections the development of this continental margin. It makes use of the subduction model shown in Figure 11.18, in which a wedge of oceanic crust is lifted by understuffing and is bent upward to form a forearc trough. As the accretionary prism was growing and rising, the forearc trough was receiving sediment and slowly subsiding. The end to this long span of subduction came rather abruptly about in the middle of the Cenozoic Era (Oligocene or Miocene) when folding and overthrusting affected both the accretionary prism and the forearc trough. This tectonic activity greatly complicated the structure of the Franciscan rocks. About this time, too, the subduction boundary between the Farallon and American plates was replaced by the San Andreas transform plate boundary between the Pacific and American plates. (See Chapter 10 for details.)

Cordilleran-Type Orogens

While tectonic and volcanic activity have been continuous through Cenozoic time in the trench of a subduction zone such as the Sumatra–Java arc, we do not consider this activity as being an orogeny. Orogeny involves intense compressional deformation of geosynclines, resulting in folding and overthrusting and the intrusion of magma to produce plutons. Deformation limited to turbidites of a trench wedge would not fall within this definition of an orogeny. According to our classification of continental structural elements, two types of orogeny are recognized. One, the Cordilleran type, results from an episode of rapid subduction in which geosynclines of the continental margin are strongly affected. The other, or Eurasian type, results from continental collision. Although both types produce deformation—folds and

FIGURE 12.13 A seismic-reflection profile across the Sumatra-Java trench and tectonic arc. The profile was obtained by scientists of the Lamont—Doherty Geological Observatory of Columbia University during a cruise of the research vessel *Robert Conrad*. The profile is located off the southern coast of Sumatra, and is approximately equivalent to Section *A* of Figure 12.12. Vertical exaggeration is about ×25. Layers of sediment are clearly visible in the trench floor at the left and in the forearc trough at the right. Imbricate wedges of the accretionary prism appear to be vertical because of the great vertical exaggeration. (Courtesy of Warren Hamilton, U.S. Geological Survey.)

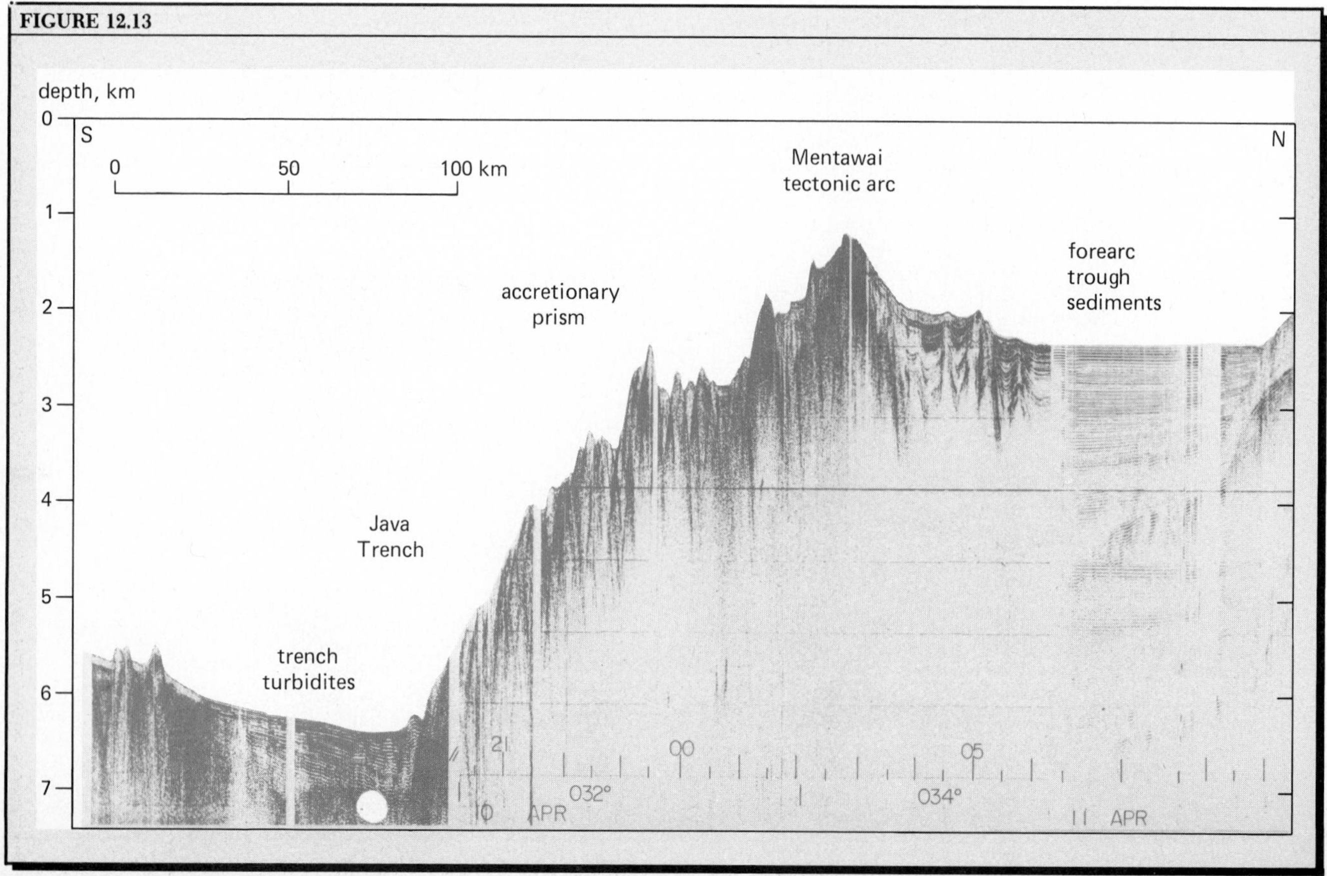

thrust sheets that appear to be quite similar—there are important differences in their total structure. We will find it convenient to use the term *orogen* to mean the total mass of crustal rock deformed during an orogeny.

A reconstruction of the Cordilleran-type orogeny was published in 1970 by John F. Dewey and John M. Bird. We present this model of an orogeny in somewhat simplified form (Figure 12.15). The scenario opens with a passive continental margin bearing geoclines of the Atlantic types (Diagram *A*). Orogeny is initiated by the formation of a new subduction zone, as shown in Diagram *B*. The oceanic crust is broken through some distance away from the continental plate margin, not at the contact with the continental lithosphere. The location of the break may vary from one case to another. As a result, a broad belt of oceanic crust beneath the eugocline (turbidite wedge) remains attached to the continental plate. When the new subduction margin is formed, the oceanic plate is first arched upward, then broken into a number of overthrust slices. Melting of the upper surface of the downplunging plate produces basaltic magmas that rise behind the uparched oceanic crust, appearing as submarine volcanoes.

As subduction continues, a deep core of magma forms in the continental plate above the descending oceanic plate. Called a *mobile core,* it consists of magma of mafic (gabbro) and intermediate (diorite) composition, along with some magma of more felsic composition (granodiorite). The rising mobile core causes *thermal uparching* of the earlier volcanic rocks and the strata of the outer turbidite wedge. When the arch has risen above the ocean surface, it becomes a highland belt and begins to furnish clastic sediments to the trough that lies between the arch and the continental shelf wedge and to the oceanward slope of the

FIGURE 12.14 Evolution of the Franciscan Complex and Great Valley Sequence in northern California. Horizontal and vertical scales are the same. (Redrawn and simplified from W. R. Dickinson and D. R. Seely, 1979, *American Assn. of Petroleum Geologists, Bull.*, vol. 63, p. 25, Figure 11. Used by permission.)

FIGURE 12.14

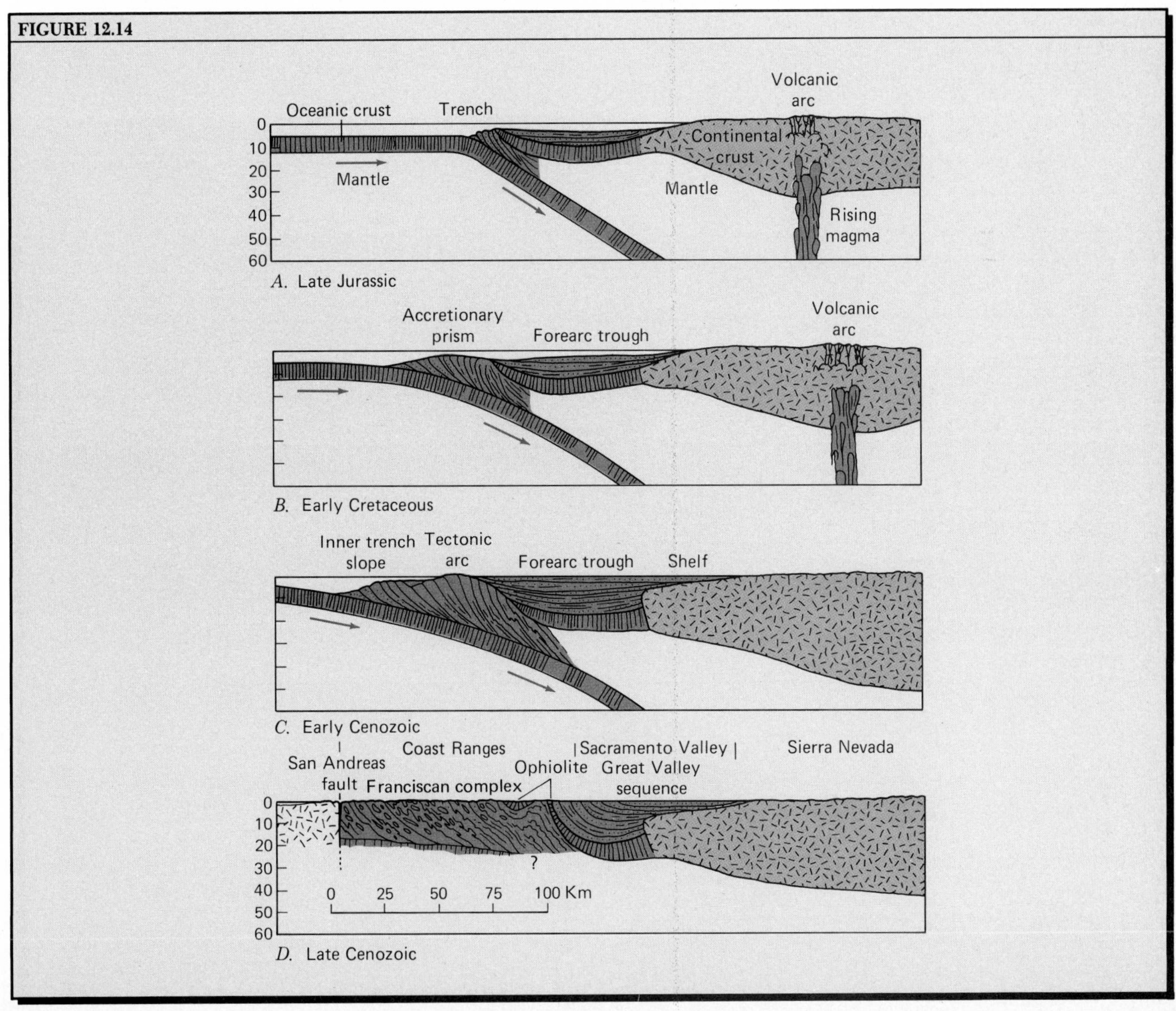

arch. These detrital sediments consist of dark shales and lithic sandstones that are mostly deep-water turbidites. We might, if we choose to, regard the accumulation as a special variety of geosyncline. If so, it would need to be added to our list of geosynclines given earlier in this chapter. Below the new sediment accumulation, sediments of the original turbidite wedge are now being subjected to intense heating and to horizontal compression caused by the rising magma core. These sediments are then metamorphosed into schists, while the lower portions may be converted to plutonic rocks (migmatites). Meantime, back at the arch above the mobile core, magma of granite composition is rising in increasing quantities and forming batholiths in the upper part of the arch (Diagram *D*).

The final event in the Cordilleran-type orogeny is intense folding and overthrusting of the continental shelf wedge (former miogeocline) and its overlying younger sediments (Diagram *D*). More batholiths are being formed, and magma also reaches the surface to form volcanoes and lava flows of andesite and rhyolite

FIGURE 12.15 Reconstruction of imagined stages in a Cordilleran-type orogeny. *A*. The stable continental margin prior to orogeny. *B*. A new subduction boundary is formed; a volcanic arc is constructed. *C*. Rise of a mobile core forms an arch. Volcanics and sediments are deformed and metamorphosed; at depth, migmatites are produced. *D*. A final tectonic paroxysm results in emplacement of granite plutons and formation of imbricate thrusts. (Drawn by A. N. Strahler. Modified and simplified from J. F. Dewey and J. M. Bird, 1970, Mountain belts and the new global tectonics, *Journal of Geophysical Res.*, vol. 75, p. 2638, Figure 10.)

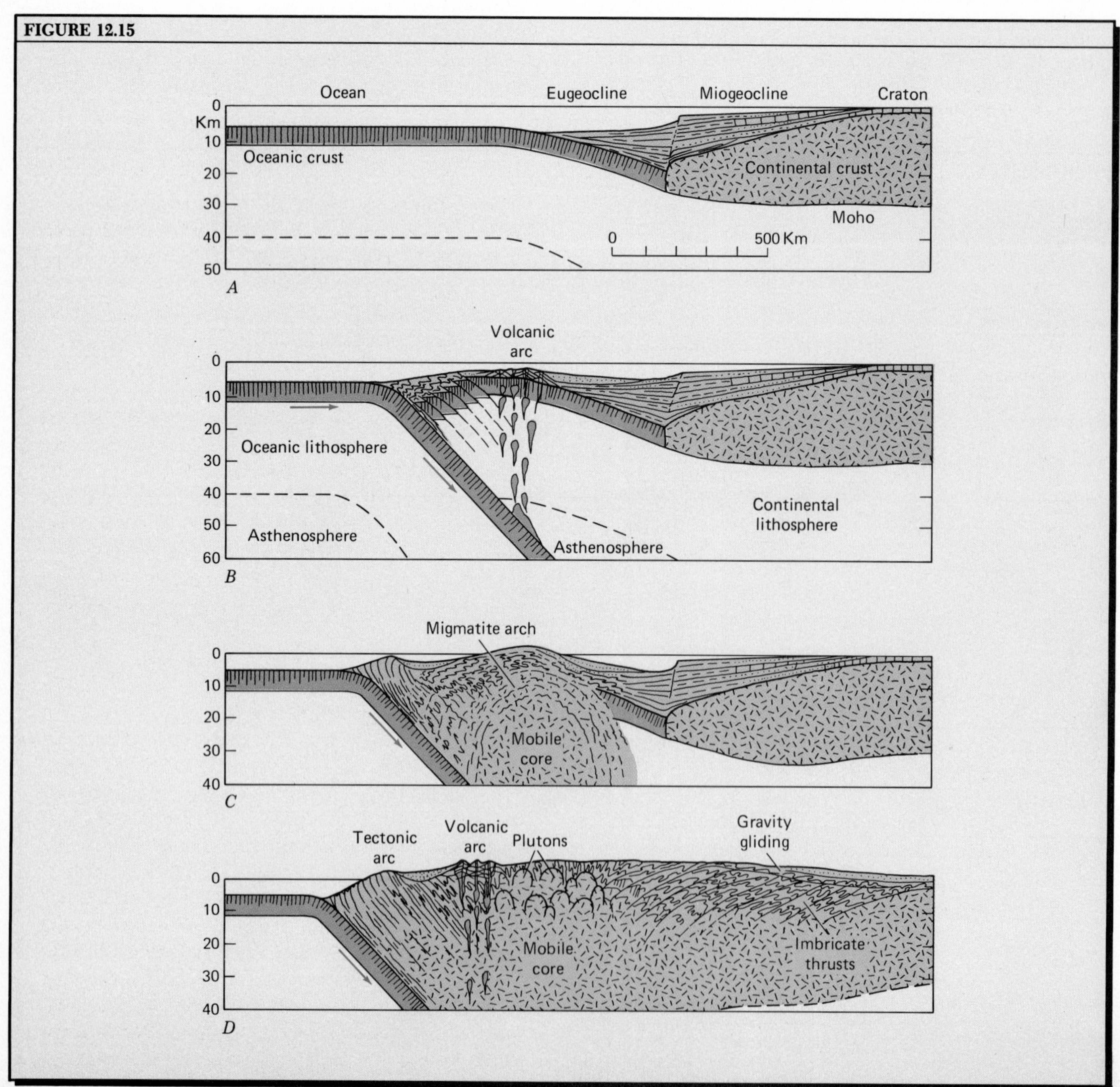

composition. From these volcanic piles, coarse clastic sediment is transported by streams to the forearc trough. Clastic sediments are also being swept toward the foreland, forming a new deposit of coarse sands and gravels spread by streams on a sloping plain. The new sediment wedge, which is classed as a foreland deposit, may experience some folding before the orogeny ends.

The orogen is now complete, and activity ceases, perhaps because by now the rate of plate subduction has fallen off rapidly, and there is no longer an abundant supply of magma to maintain the mobile core and continue its uparching. Later erosion exposes plutons, thrust structures, and metamorphic rocks.

The Cordilleran Orogeny in North America

An example of a Cordilleran-type orogen can be found (as the name implies) in the Cordilleran ranges of North America. The orogeny is called the Cordilleran Orogeny; it began near the close of the Cretaceous Period and continued into the Cenozoic Era. The area involved in this orogeny lay to the east of the northern California region illustrated in Figure 12.14. However, the subduction that was responsible for the accumulation of an accretionary prism and forearc trough, shown in Figure 12.14, seems also to have been responsible for the uprising of a mobile core above the deeper portion of the descending slab farther east. One feature of the Cordilleran Orogeny was the intrusion of numerous batholiths in late Cretaceous time, among them the Sierra Nevada batholith and the Boulder batholith (Figure 12.16). Intrusion of granite continued into the Cenozoic Era in what is now the northern Rocky Mountains in Idaho and eastern Washington. Farther east, over what is now Alberta, Montana, Idaho, Wyoming, Utah, and Colorado, lay the Cordilleran (or Rocky Mountain) geosyncline, which had been receiving sediment throughout middle and late Mesozoic time. Intense low-angle overthrust faulting of the geosyncline marked the final stages of the orogeny and affected a great belt of older rocks extending from western Wyoming across Utah into southern Nevada. This intense overthrusting affected the Cretaceous strata of the Cordilleran geosyncline along what is now the eastern front range of the Rockies in Alberta (Figure 12.17). Many extrusions of rhyolite and andesite also occurred at this time.

FIGURE 12.16 Major Cretaceous batholiths of western North America (solid color). Trends of overthrusts are indicated by a line pattern. (Batholith data after C. O. Dunbar and K. M. Waage, *Historical Geology.* Copyright © 1969 by John Wiley & Sons, Inc., New York. Used by permission.)

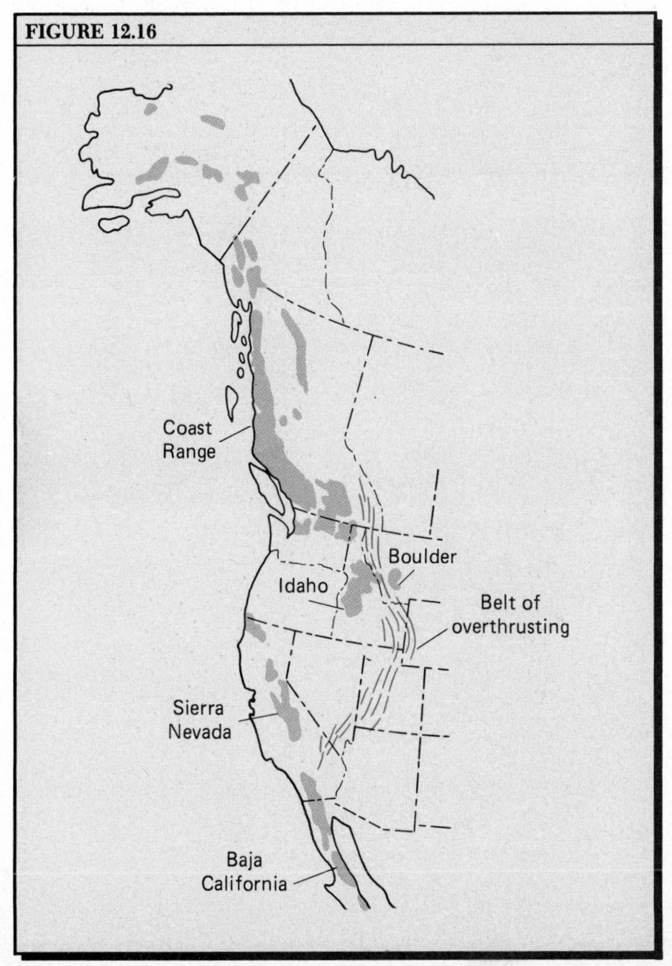

***Tectonic Domes and Uplifts of the Continental Foreland**

The Cordilleran Orogeny produced some unusual tectonic features in the central and southern Rocky Mountain region of the United States, east of the limits of the intense overthrusting to which we have just referred. These are elliptical, domelike structures pushed up locally in the craton that formed the foreland with respect to the belt of orogeny. Platform and geosynclinal strata that had been previously laid down in a horizontal cover over the Precambrian basement were forced upward into dome or arch forms and now dip outward from the summit or axis of each uplift. Intrusion of plutons does not seem to have occurred beneath these *structural uplifts,* which are purely tectonic in origin.

The large structural uplifts are particularly numerous in a belt extending from Montana south through Wyoming and Colorado into New Mexico. Individual uplifts in this group are from 50 to 75 km wide and 200 to 400 km long (Figure 12.18). The vertical movements amount to several thousand meters along the crest region of the larger uplifts. Denudation has long since removed the top portions of the uplifts, exposing cores of Precambrian shield rock, but the upturned strata can be seen today along the lower flanks of the uplifts (Figure 12.19). The harder, more resistant

FIGURE 12.17

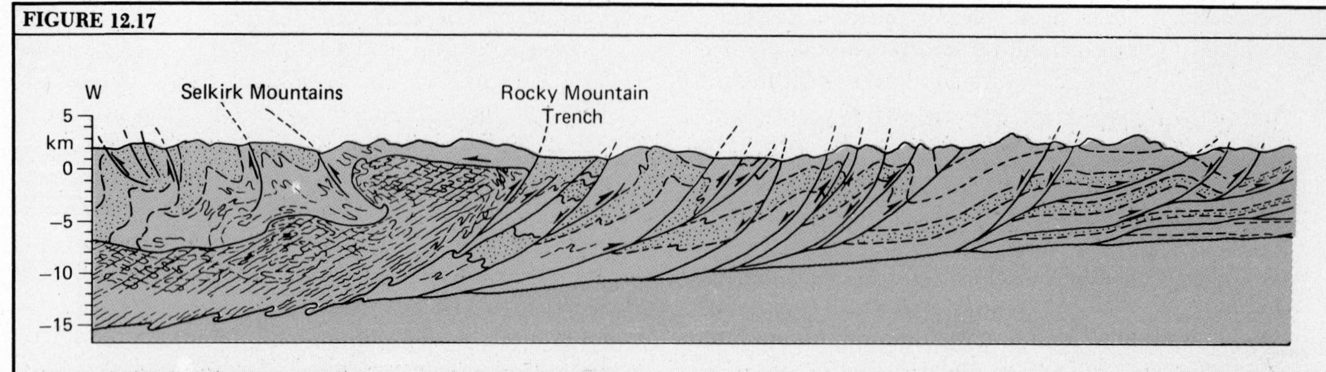

FIGURE 12.17 A structural cross section through the Canadian Rockies between the Bow and Athabasca rivers in Alberta and British Columbia. Horizontal and vertical scales are the same. Notice that the dip of the thrust planes diminishes with depth. (Prepared by R. A. Price from data of the Geological Survey of Canada. Geological Association of Canada, Special Paper No. 6, 1970.)

FIGURE 12.18 Outline map of the major structural uplifts and basins of the Middle and Southern Rocky Mountains. (Based on data of *Tectonic Map of the United States,* 1944, American Assn. of Petroleum Geologists and National Research Council.)

FIGURE 12.18

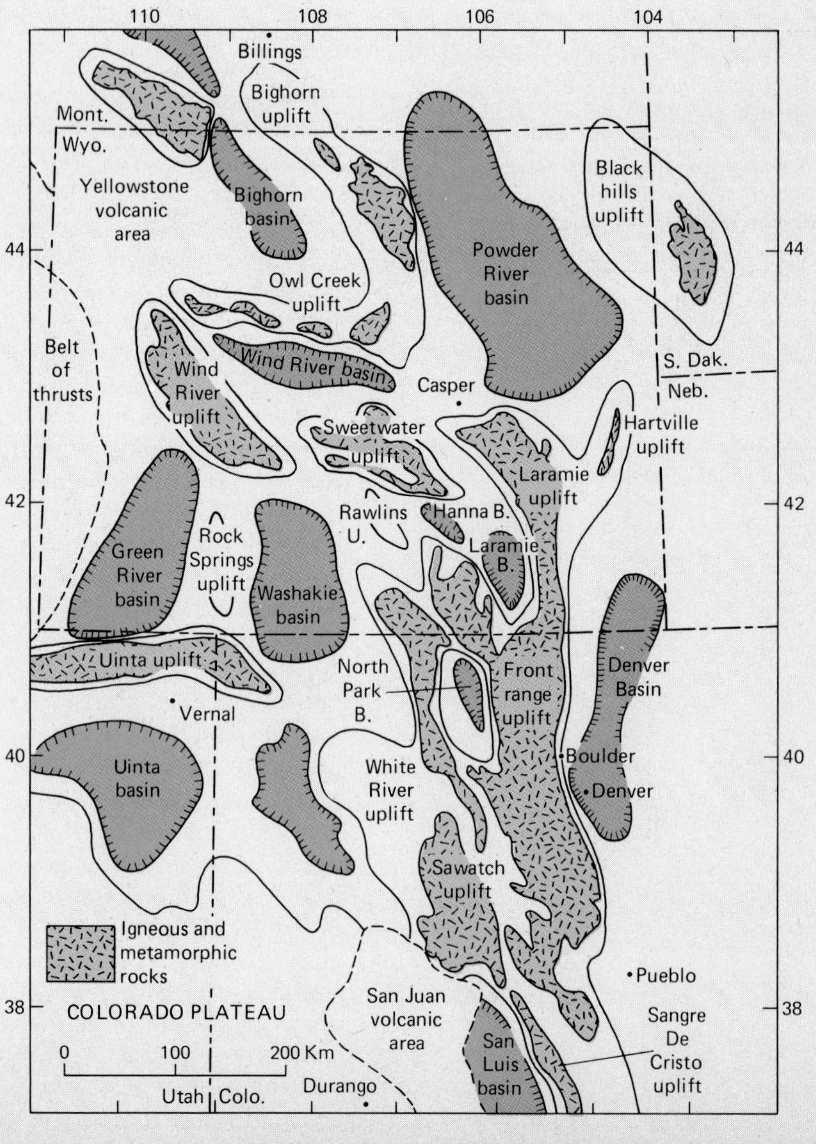

Rocky Mountains
foothills
plains
E
5
km
0
−5
−10
−15
crystalline basement rocks
0 5 10 15 20 25 km
thrust faults
stratification

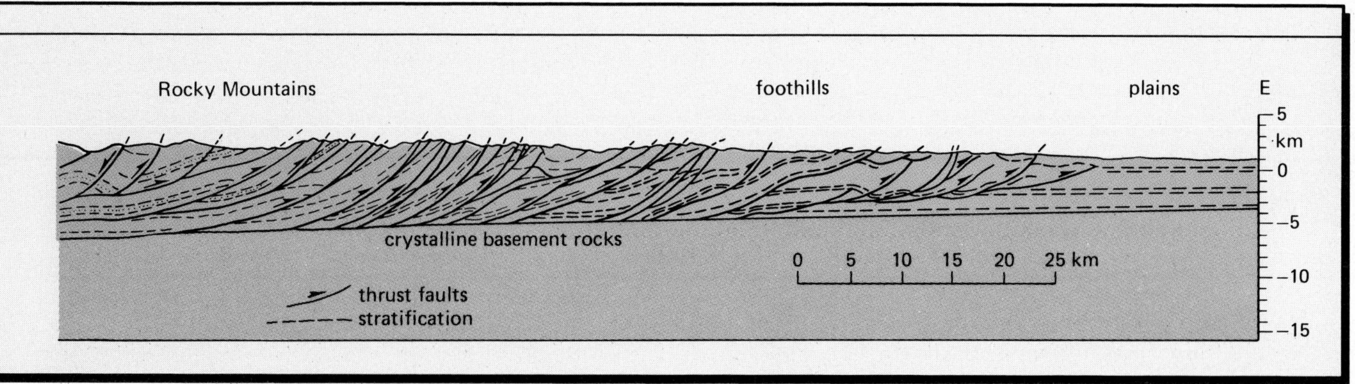

FIGURE 12.19 Stylized and generalized cross sections of two structural uplifts of the Rocky Mountains. (Drawn by A. N. Strahler.) *A*. Colorado Front Range, north of Denver. The restored arch structure is purely imaginary, because erosion would have removed the strata as the uparching occurred. (Based on data of W. T. Lee, U.S. Geological Survey.) *B*. Sheep Mountain uplift, Wyoming. Note that this is a much smaller structure than the Colorado Front Range. (Based on data of R. H. Beckwith, 1938, *Geological Society of America, Bull.*, vol. 49, Plate 1.)

FIGURE 12.19

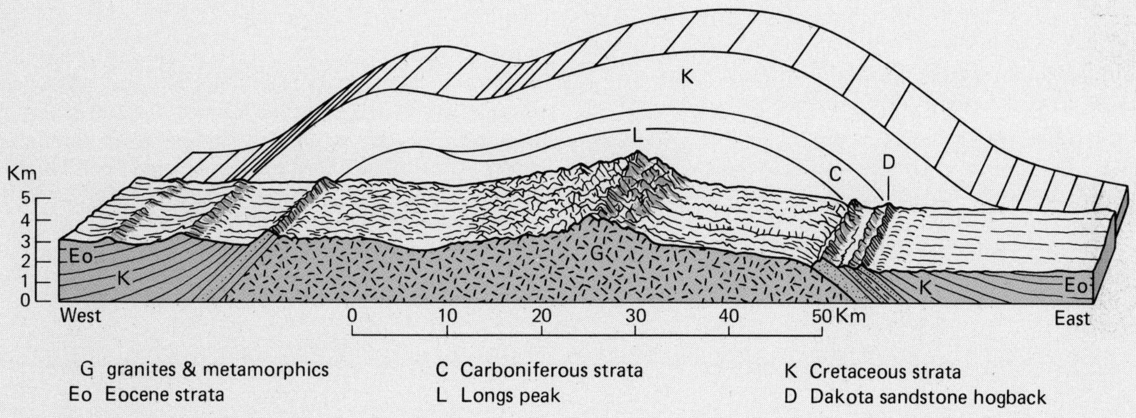

K
L
C
D
Km
5
4
3
2
1
0
Eo
K
G
K
Eo
West
0 10 20 30 40 50 Km
East
G granites & metamorphics
Eo Eocene strata
C Carboniferous strata
L Longs peak
K Cretaceous strata
D Dakota sandstone hogback

A

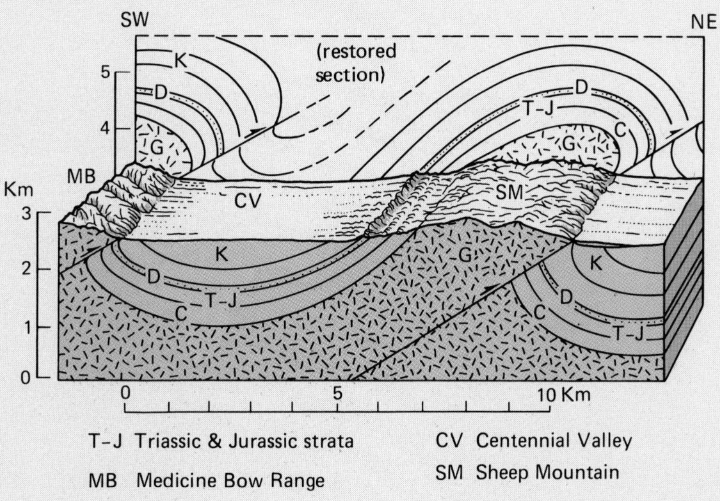

SW
NE
(restored section)
K
D
G
D
T-J
G
C
MB
Km
CV
SM
K
D
T-J
C
G
K
D
C
T-J
0 5 10 Km
T-J Triassic & Jurassic strata
MB Medicine Bow Range
CV Centennial Valley
SM Sheep Mountain

strata form sharp-crested hogbacks (Figure 12.20).

Between the uplifts are basins in which the strata were warped and which, once formed, received large amounts of sediment from the eroding uplifts. This accumulation of sediment continued throughout much of the Cenozoic Era. The sediments consist of coarse clastic materials deposited by streams, as well as fine-textured water-laid clays, silts, and lime muds laid down in lakes that occupied the basins from time to time.

Some of the smaller uplifts, particularly those in Wyoming, show overturning of strata on their flanks. The overturned strata are cut through in places by thrust faults or reverse faults, as shown in Diagram *B* of Figure 12.19. But other uplifts, such as the Black Hills dome and the Colorado Front Range, show only moderate dips on both flanks and are lacking in thrust faults. The crustal deformation that occurred here has been called *foreland basement deformation*.

Platform strata of the Colorado Plateau region were also affected by foreland basement deformation. The crust beneath the sedimentary platform was broken by several long north–south normal faults. These faults did not reach the surface, but instead, the strata bent sharply down in a monocline above each fault. Another form of foreland basement deformation that occurred in this region consists of a number of broad low domes and arches, similar to those of the Rocky Mountain region, but not as strongly uplifted (see Color Plate E.2).

The Rocky Mountain structural uplifts are thought to have been produced by compressive stresses exerted on the continental plate by subduction along the western continental margin. It has also been suggested that the descending oceanic plate had a very low inclination and passed beneath this foreland region, then became detached and inactive. As it warmed to the temperature of the surrounding mantle, its buoyant force pushed the overlying continental plate upward, supplying the vertical stresses needed to produce the uplifts.

According to the idealized model of a Cordilleran-type orogeny, as we presented it in Figure 12.15, the Cordilleran Orogeny in western North America should have ended when low-angle overthrusting ceased. Final deposition of the coarse foreland sediments would be a postorogenic event. Using this criterion, we can say that the Cordilleran Orogeny ended in late middle Miocene time, about −12 m.y. Thick clastic sediments deposited during the closing stages and following orogeny are found in structural basins between the Rocky Mountain uplifts and over the high plains region that lay to the east.

FIGURE 12.20 View north along the eastern base of the Colorado Front Range, near Morrison, Colorado. The prominent sharp-crested hogback in the center is formed of the Dakota sandstone. Farther to the left massive Carboniferous sandstone forms tilted plates of light-colored rock lying at the base of rising slopes of granite. Compare with Figure 12.19*A* (T. S. Lovering, U.S. Geological Survey.)

FIGURE 12.20

*Basin-and-Range Block Faulting

As the Cordilleran Orogeny came to a close, another and possibly related major tectonic event began in the western United States and northern Mexico. Over a large area, the continental crust began to break up into fault blocks, bounded by normal faults. (Normal faults are explained in Chapter 6 and illustrated in Figures 6.16 through 6.21.) Normal faulting on a large scale can produce mountain blocks many kilometers wide and many tens of kilometers long. As shown in Figure 12.21, a mountain block may be tilted so that one side is upfaulted and the other downfaulted, or it may be lifted as a horst.

Block faulting began to occur in middle Miocene times (−15 m.y.) in Nevada, western Utah, southeastern California, southwestern Arizona, northern Mexico, and in parts of several neighboring states. The

FIGURE 12.21

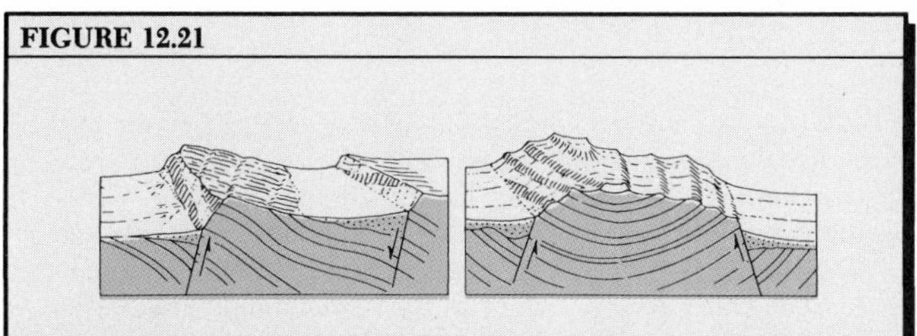

FIGURE 12.21 Tilted and lifted mountain blocks. (From W. M. Davis, 1903, *Bull. Mus. Comparative Zoology,* vol. 42, Figures 113 and 114.)

FIGURE 12.22 The Basin-and-Range Province. (From A. K. Lobeck, 1948, *Physiographic Diagram of North America,* Hammond Incorporated, New York. Reproduced by permission of Hammond Incorporated.)

FIGURE 12.22

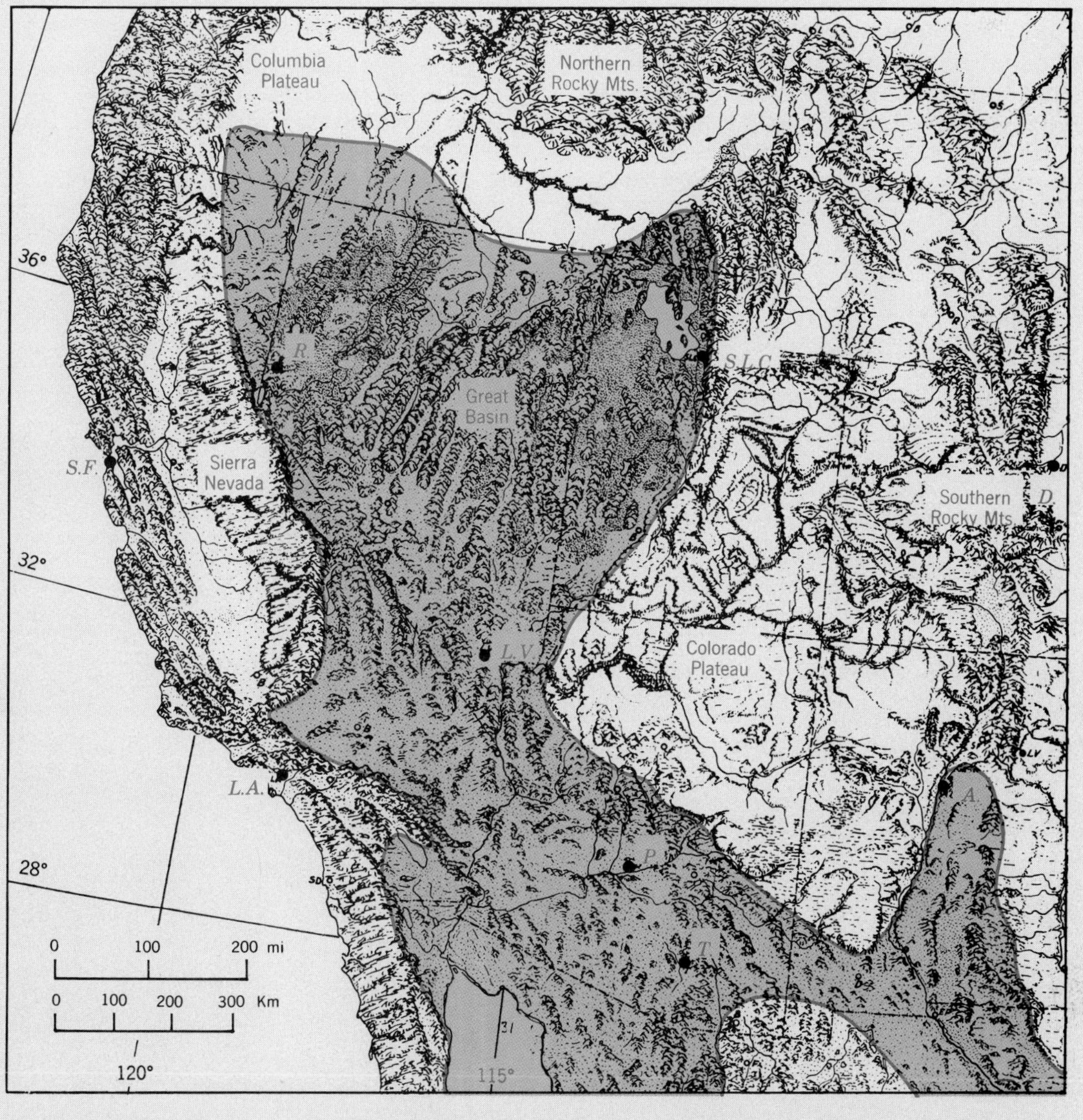

entire region is called the Basin-and-Range Province (Figure 12.22). Much of this area had previously experienced overthrust faulting and volcanism in the closing stages of the Cordilleran Orogeny. The normal faults cut across the older structures along new lines, lifting or dropping crustal blocks through vertical distances as great as 3000 m. The downdropped blocks formed deep structural basins, which received large volumes of coarse sediment from the adjacent high-standing blocks that were undergoing rapid erosion (Figure 12.23). Much of this region today lies in a desert climate (see Color Plate D.1).

Block faulting of the Basin-and-Range region is interpreted as the result of a pulling-apart, or extension, of the continental lithosphere. The faulting began about the time that the Farallon plate was becoming converted into a transform plate boundary (Chapter 10), and plate subduction had come to a halt along much of the western boundary of the American plate immediately west of the Basin-and-Range region. It has been suggested that a large rising mantle convection current or plume appeared beneath this region, lifting the lithosphere and at the same time spreading it apart so that the crust collapsed and became deeply fractured (Figure 12.24). This hypothesis is strengthened by observations that the rate of heat flow here is high (about two heat-flow units) compared with the low rate typical of the stable continental lithosphere of the craton (about one unit).

*Columbia Plateau Flood Basalts

A special event that occurred in the closing phase of the Cordilleran Orogeny was the outpouring of flood basalts of the Columbia Plateau region. We described these basalts in Chapter 4. Figure 4.42 shows the extent of the flood basalts; they covered an area of

FIGURE 12.23 Block mountains of the western Basin and Range Province in California. The view is westward from the crest of the Panamint Range, across Panamint Valley (downfaulted), to the Argus Range (upfaulted block), and beyond the Owens Valley to the distant Sierra Nevada. Notice the flattened top surface of the Argus Range, suggesting uplift as a horst. (Warren Hamilton, U.S. Geological Survey.)

FIGURE 12.24 Block faulting of the Basin-and-Range Province. *A*. Structures of the Cordilleran orogeny exposed by erosion. *B*. Extension of the crust causes crustal fracturing to produce fault blocks. (Drawn by A. N. Strahler.)

FIGURE 12.25 Mount Shasta, a partially dissected volcano in the Cascade Range. The bulge on the left-hand slope is a more recent subsidiary cone, called Shastina. (Infrared photograph by Eliot Blackwelder.)

FIGURE 12.23

130,000 sq km with a thickness of 600 to 1200 m of basalt. We explained the Columbia Plateau and Deccan flood basalts as having been formed over hot spots beneath a continental plate, possibly created by a mantle plume. If the hot spot explanation is correct, the outpouring of the Columbia Plateau basalts is not necessarily related to the Cordilleran Orogeny. On the other hand, it is possible that the basalt originated in the oceanic plate (Farallon plate) that was descending beneath the continental plate during this time. If no mantle plume were active, some other special explanation would be needed for the enormous quantity of basalt—250,000 cu km—brought up through the crust in fissures in this particular area.

*Late Cenozoic Tectonic and Volcanic Events

Some tectonic and volcanic events that took place in the western United States in late Cenozoic time were of major importance in shaping the geologic structure and modern landscape. In the Pliocene Epoch, about −5 m.y., intense volcanic activity began in the Cascade region of northern California, Oregon, and southern Washington. Great extrusions of andesite and rhyolite lavas resulted in the building of the Cascade Range, with its chain of lofty composite cones (Figure 12.25). Quite possibly, this volcanic activity, which continues today, was caused by subduction of the Juan

FIGURE 12.24

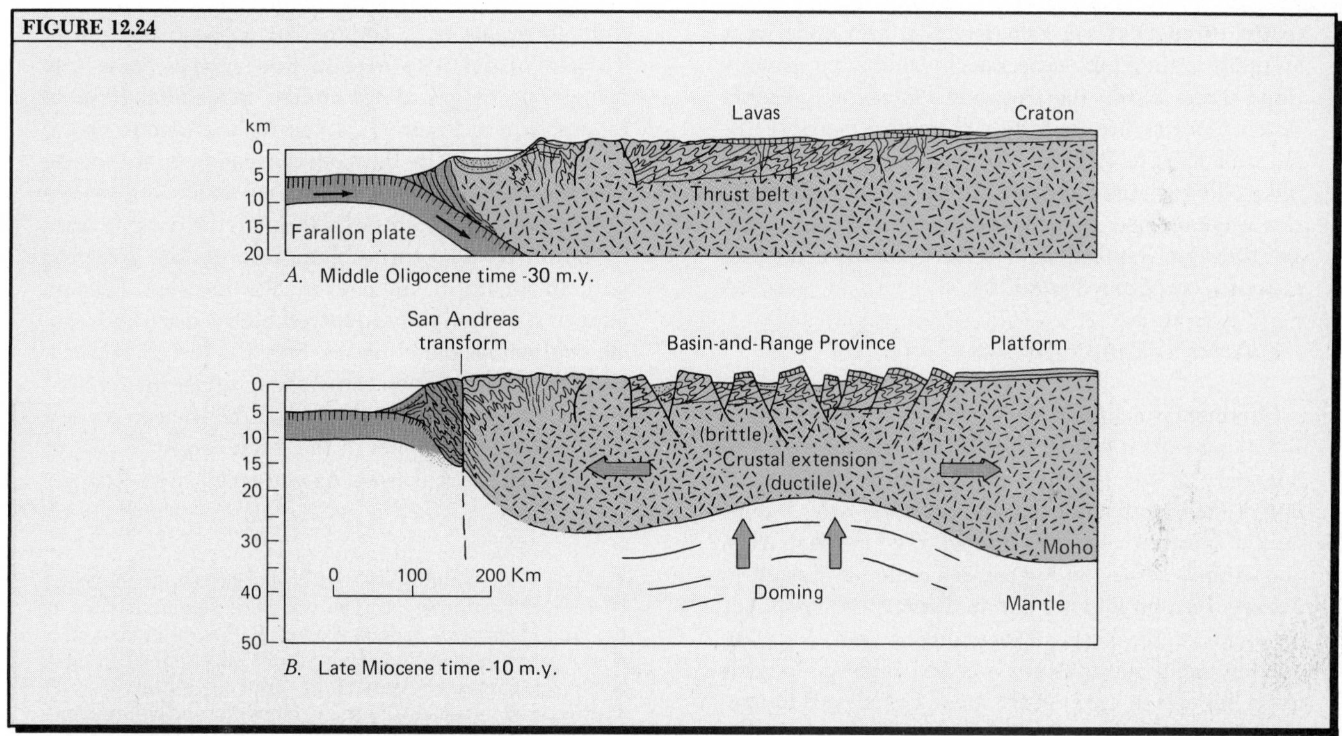

FIGURE 12.25

de Fuca plate (a remnant of the Farallon plate).

A second event of major importance was the uplift of the Sierra Nevada in California. This lofty range is an enormous tilted fault block composed in part of a granite batholith intruded early in the Cordilleran Orogeny (Figure 12.26). The block is thought to have risen as recently as the early Pleistocene Epoch, beginning about −2 m.y. A great system of normal faults bounds the Sierra Nevada on the eastern side, while the western side is downdropped, forming the Great Valley of California. The cause of the uplift of the Sierra Nevada is not understood, but it may have been related to the thermal uplift accompanied by plate extension in the adjacent Basin-and-Range Province.

The Colorado Plateau, which we used earlier in this chapter as an example of a platform on the stable craton throughout the Paleozoic Era, also underwent an uplift in the Pleistocene Epoch. Unlike the strongly tilted Sierra block, the Colorado Plateau was simply raised more or less uniformly through a vertical distance of 2500 to 3000 m. The tectonic significance of this uplift remains a mystery. Uplift of platform strata was accompanied by deep canyon cutting by rivers, and thus the Grand Canyon of the Colorado came into existence (see Color Plate E.1).

Eurasian-Type Orogens

Continental collision has occurred in the Cenozoic Era along a great tectonic line that marks the southern boundary of the Eurasian plate (Figure 12.27). This line of collision tectonics begins with the Atlas Mountains of North Africa, runs through the European Alps, and extends across the Aegean Sea region into western Turkey. Beyond a major gap in Turkey, the line takes up again in the Zagros Mountains of Iran. Jumping another gap in southeastern Iran and Pakistan, the collision line sets in again in the great Himalayan Range. Thus, we are dealing with three collision segments: European, Persian, and Himalayan. Each collision segment represents the collision of a different north-moving plate against the single and relatively immobile Eurasian plate. The European segment represents collision of the African plate with the Eurasian plate in the Mediterranean region. The Persian segment represents the collision of the Arabian plate with the Persian subplate of the Eurasian plate. The Himalayan segment represents the collision of the Austral–Indian plate with the Eurasian plate. Between the three segments lie major transform faults; these are the transform boundaries of the Arabian plate. The tectonic structures of the three collision segments differ markedly, and we shall discuss each in turn.

Figure 12.28 is a series of cross sections in which the tectonic events of a continental collision are reconstructed in detail to explain how nappes come into being. The diagrams are similar in general form to those shown in Figure 1.14, but focus attention on the structural details in the collision zone. As the ocean between the converging continents narrows, a succession of overthrust faults cuts through the oceanic crust (Diagram *B*). The thrust slices ride up, one over the next, in an imbricate pattern. As the slices become more and more tightly squeezed between the converging continental plate masses, they are forced upward. The upper part of each thrust sheet, under the force of gravity, bends over to a horizontal position to form a nappe. The final nappes in the series consist in part of the slices of oceanic crust. As you recall from Chapter

FIGURE 12.26 A highly stylized block diagram of the Sierra Nevada of California. *A*. The imagined sloping surface of the fault block prior to deep erosion. *B*. Present mountainous landscape with glaciated High Sierra, deep glacial troughs, and river gorges. (Suggested by drawings of Francois E. Matthes, U.S. Geological Survey. Drawn by A. N. Strahler.)

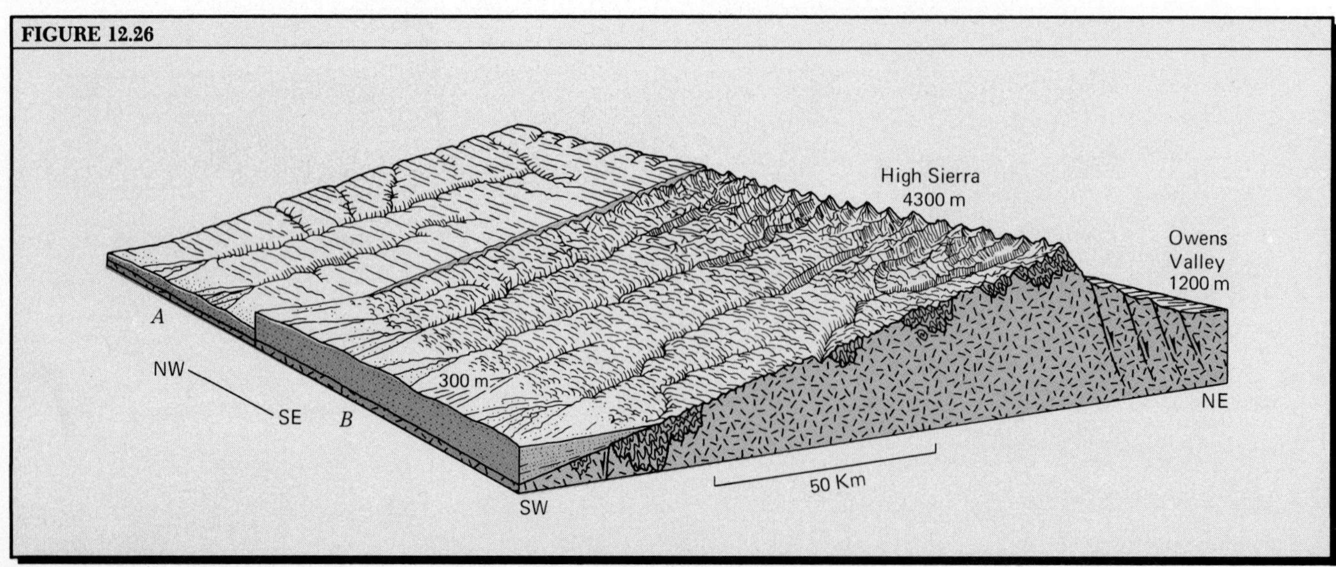

10, the oceanic crust sequence formed at a spreading plate boundary is an ophiolite suite, with rocks ranging from ultramafic periodotite at the base to mafic basalt dikes and lavas at the top. When a slice of this oceanic crust is carried up and squeezed to form a nappe, it is metamorphosed into a rock known as an *ophiolite*. It is now generally agreed that a narrow zone of ophiolites marks the line of a continental suture. Ophiolites are found at intervals along the entire collision boundary of southern Eurasia.

Continual erosion by streams and glaciers reduces the upper surfaces of the rising pile of nappes. This sediment is swept to lower levels, accumulating as a younger sedimentary deposit. Deposits of this type have been found in the European Alps, and two major types have been identified. Sediments accumulating in deep, narrow ocean troughs during early stages of the collision are of a type called *flysch* by European geologists. It consists of shale and fine-grained lithic sandstone (graywacke). Flysch becomes involved in later stages of overthrusting and may be incorporated into younger nappes. As the collision reaches its final stage, a high-standing mountain mass results, and the ocean troughs disappear. At this time, the sediment produced by land erosion becomes coarser in texture and accumulates as a thick sediment wedge called *molasse*, which forms adjacent to the new mountain mass. The molasse is deposited on an exposed land surface or in very shallow coastal water of expanding deltas.

FIGURE 12.27 Sketch map of the Eurasian collision segments.

*European Collision Segment

Tectonic history and structure of the European collision segment is extremely complex—so complex, in fact, that only a geologist specializing in this region could form a comprehensive and cohesive picture of what took place here through time.

The tectonic history of the European collision segments goes back to the late Triassic, at which time Pangaea remained intact, but was soon to begin its breakup (see Figure 13.15). The Eurasian and African plates were then separated by the Tethys Sea, which is pictured as a V-shaped embayment or gulf of *Panthalassa*, the enormous ancestor of the Pacific Ocean. In earliest Jurassic time, about −205 m.y., the breakup of Pangaea had started. While the North Atlantic began to open up by continental rifting, the Tethys Sea had begun to close, a process which can be visualized as the closing of a nutcracker, the apex or pivot being about where the Strait of Gibraltar lies today. In the European segment, the closing was complicated by the presence of several small lithospheric blocks, or subplates, which were moved about in various ways.

*Persian Collision Segment

The Persian collision segment is represented by the Zagros Mountains of southwestern Iran. This mountain range is remarkable in that it is a broad belt of open folds of sedimentary strata. Penetrating the Zagros fold belt in many places are great salt columns, called

FIGURE 12.27

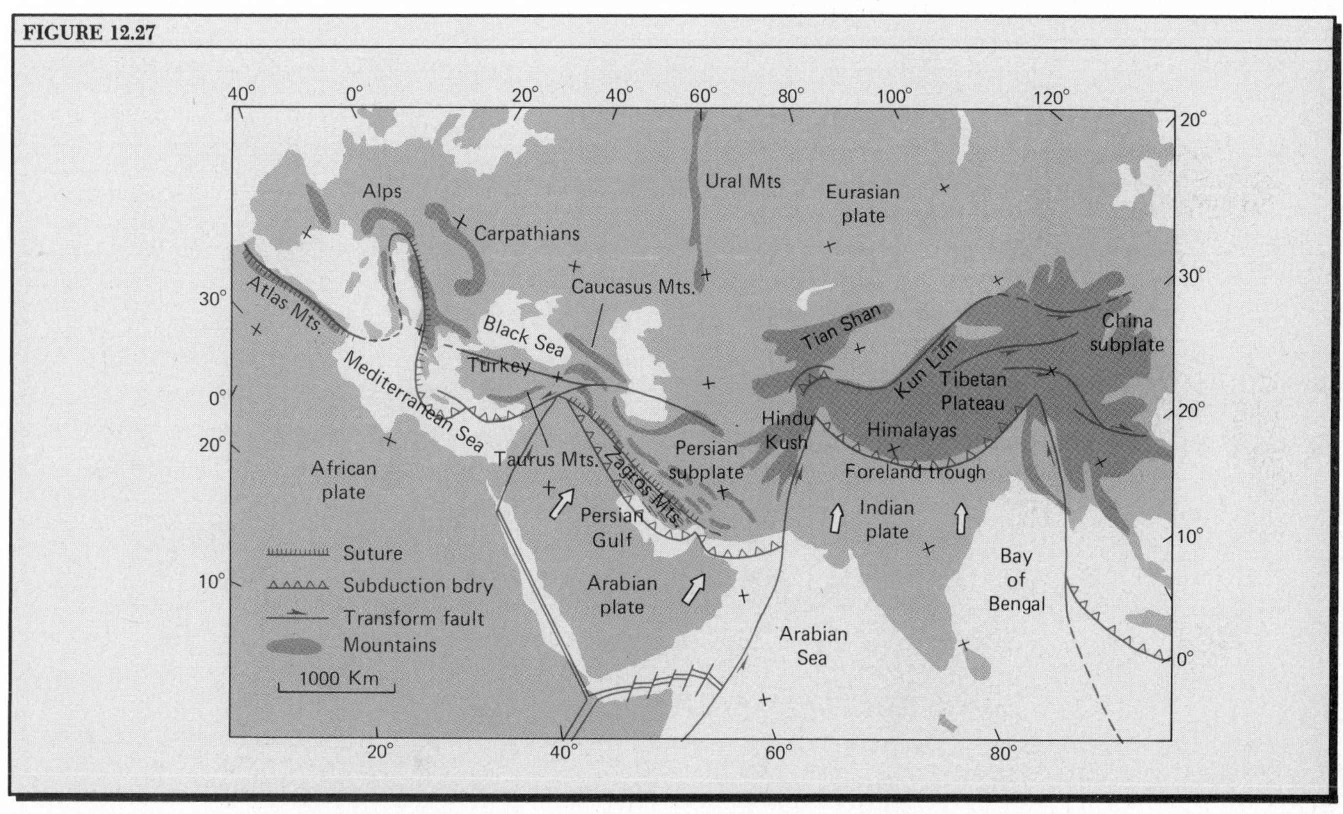

salt domes or *salt plugs*. We shall discuss these features in Chapter 21 in connection with their petroleum accumulations. The salt originally formed as a thick bed of evaporites, largely halite, near the bottom of a geosyncline. After folding occurred, the salt began to rise in tall stalklike columns, forcing its way up through the overlying strata to reach the surface. The salt of a single column may form an entire hilltop or mountainside.

Geologists have attempted to reconstruct the Zagros fold belt in terms of collision of the Arabian plate with the Persian subplate, as shown in Figure 12.29. In Diagram *A*, ocean closing is nearly complete. Thrust slices that include oceanic crust with trench deposits have already been formed as the oceanic lithosphere was subducted beneath the Persian subplate. A miogeocline lies on the margin of the Arabian plate; its sediments include carbonate strata (limestones) and red shale beds, and a thick salt formation lies near the base of the series of strata. In Diagram *B*, continental collision has brought the Arabian continental plate in contact with the thrust slices that mark the edge of the Persian subplate. Compression of the miogeocline has thrown the strata into open folds, through which rising columns of salt have since forced their way.

The Zagros Mountains are seismically active today, with numerous earthquake epicenters distributed over the entire belt. This activity suggests that the Arabian plate is strongly pressing upon the Eurasian plate and that many faults are active throughout the crust beneath the Zagros folds.

*Himalayan Collision Segment

The Himalayan collision segment is of particular interest to geologists because the great alpine-type Himalayan arc is bordered on the north by the Tibetan Plateau, an enormous elevated crustal mass unequaled in size anywhere on earth. Seismic activity is concentrated in the Himalayan arc and on two narrow zones nearly at right angles on either side of it—one in Pakistan, the other in Burma. This activity suggests that the boundary between the Indian plate and the Eurasian plate is tectonically active today, although the major collision has already occurred. According to one interpretation, the lateral zones of seismic activity are located on transform faults that bound the Indian plate. Thick alluvial sediments fill the deep Indo-Gangetic Trough, a foreland trough which lies adja-

FIGURE 12.28 Schematic cross sections showing continental collision and the formation of nappes. *A*. An ocean basin is closing, bringing two continental margins closer together. On one side (*left*) is a stable continental margin with geoclines; on the other side (*right*) is an active subduction boundary. *B*. A series of imbricate thrusts breaks through the oceanic crust and causes folding of sediments. Flysch is produced by erosion of the updomed mass. *C*. Final suturing produces nappes with ophiolites and metamorphosed sediments. Molasse is deposited in foreland troughs. Note that the continental crust on the left has been forced beneath the edge of the opposing continental crust along a surface of underthrusting. (Drawn by A. N. Strahler. Based in part on illustrations in J. F. Dewey and J. M. Bird, 1970, *Journal of Geophysical Res.*, vol. 75, pp. 2625–2647.)

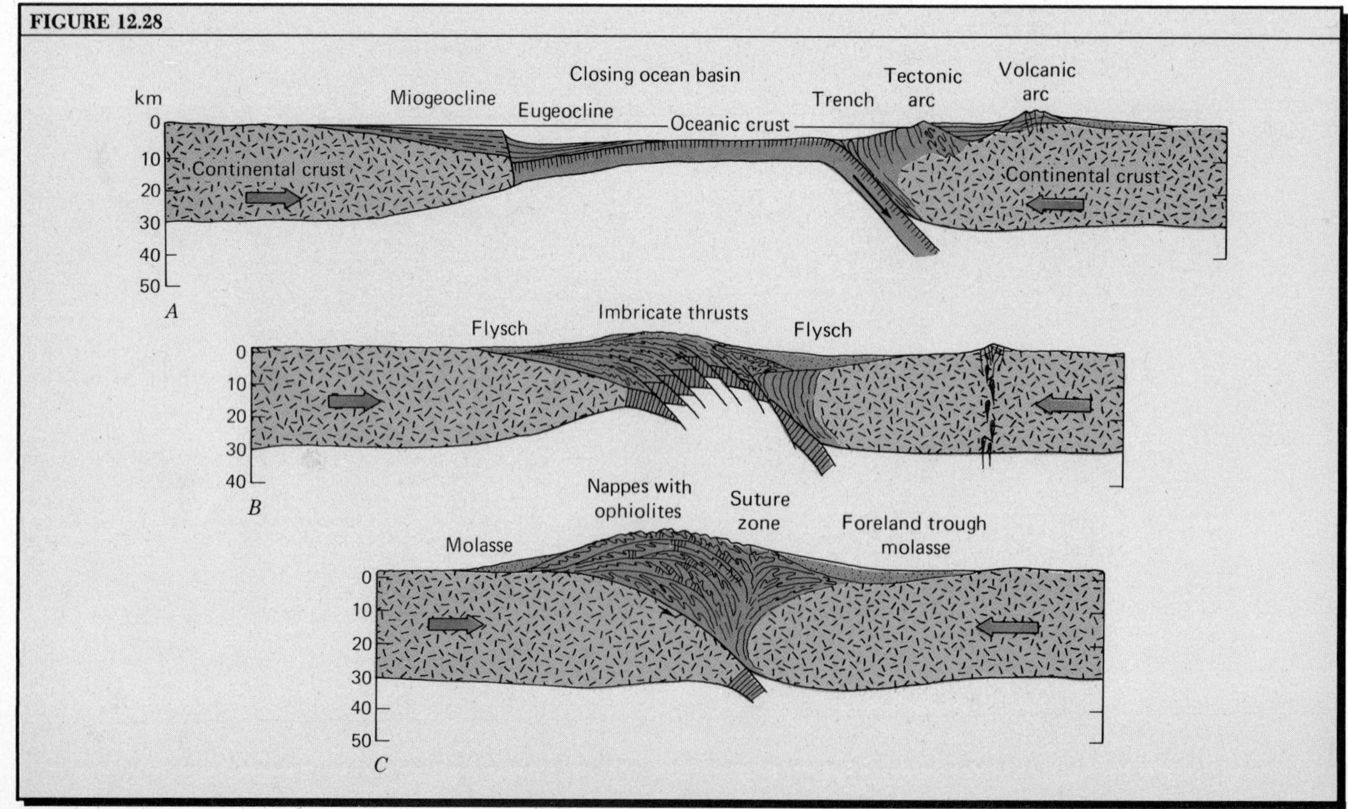

PLATE A

1. FOLDED STRATA IN NORTHEASTERN MEXICO. Folds trending east-west in the lower right (southeastern) portion of the image involve strata of Jurassic and Cretaceous age; they are part of the Sierra Madre Orientale in the State of Nuevo Léon. The plunging anticlines are deeply eroded. Farther to the west the folds are more open and many take the form of elliptical domes. Sediments of Cenozoic age fill the broader valleys. The whiteness of alluvial surfaces indicates deposition of evaporites in an arid climate. (NASA 1130-16440, 30 November 1972.)

2. EAST AFRICA RIFT VALLEY SYSTEM IN SOUTHERN KENYA. The grabenlike rift valley floor (shades of blue, violet, and green), about 40 km wide, is bounded by a series of fault steps leading up to forested highlands on either side. The red color indicates green foliage, which becomes denser with increasing elevation. Fault scarps show as nearly straight contacts between belts of different color. In the floor of the rift valley lies Lake Naivasha (black); south of it are two volcanoes, Longonot and Suswa (violet). (NASA 1048-07172, 9 September 1972).

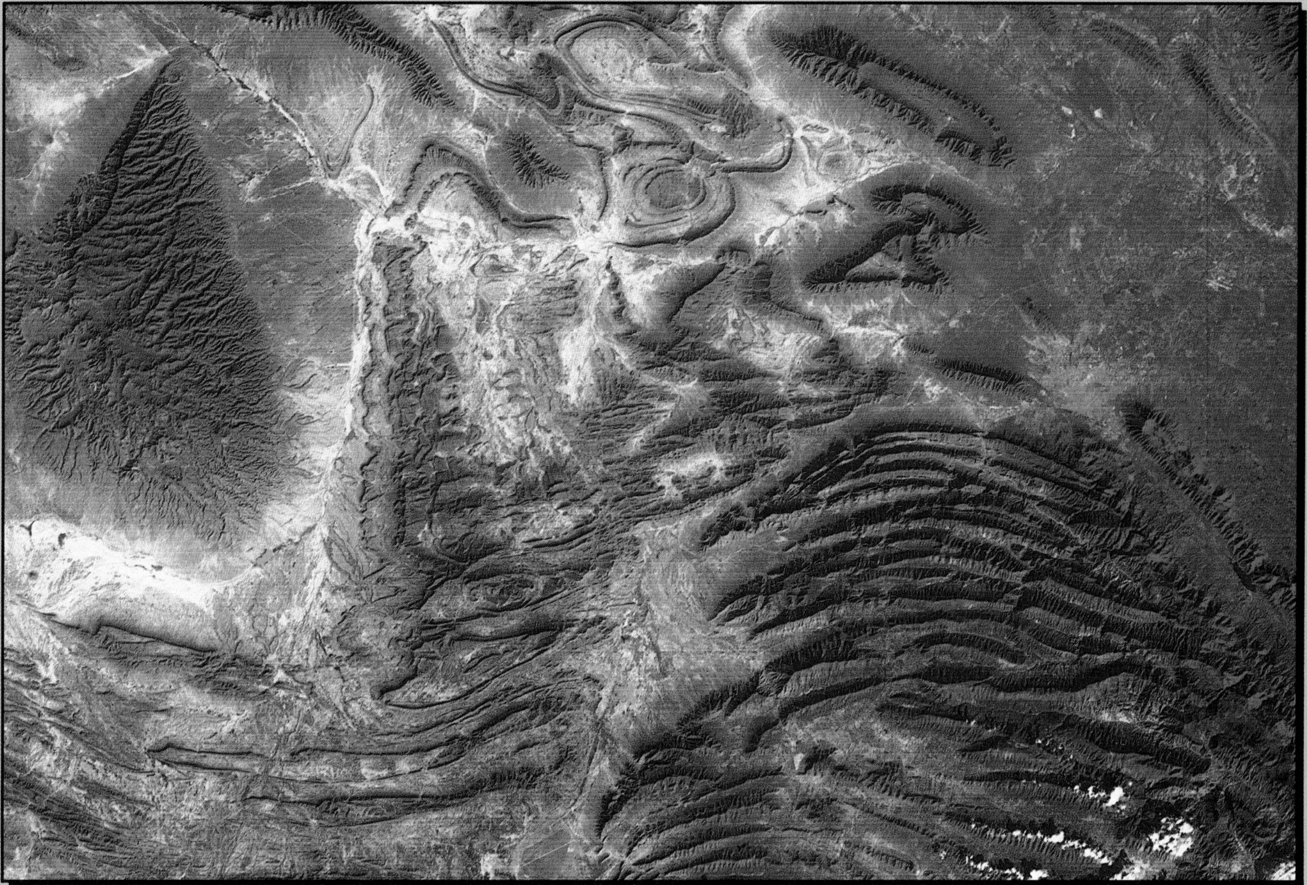

PLATE B

LANDSAT IMAGES OF THE EARTH'S SURFACE. Shown on Plates A, C, D, E, and F are images of the earth's surface produced from data collected by instruments mounted on an orbiting satellite, originally called the Earth Resources Technology Satellite (ERTS) and since renamed Landsat. From a height of 570 km Landsat photographs a strip of the earth's surface 185 km wide, completing 14 earth orbits each day. Landsat carries a special remote sensing system called the Multispectral Scanning System (MSS); it simultaneously gathers data from four spectral bands in the range from about 0.4 to 1.1 microns. A scanner feeds the reflected ground radiation to detectors and the information is digitized for continuous transmission to a ground receiving station. Each of three bands is then reproduced in a different color and the three colors are superimposed in a single color print, as seen on these pages. The colors derived from the MSS imagery are thus "false" colors. Healthy, dense vegetation appears deep red. Mature crops and dried vegetation appear yellow or brown. Shallow water areas appear blue; deep water, dark blue to blue-black. Snow and salt deposits appear white. Shading of relief features represents true

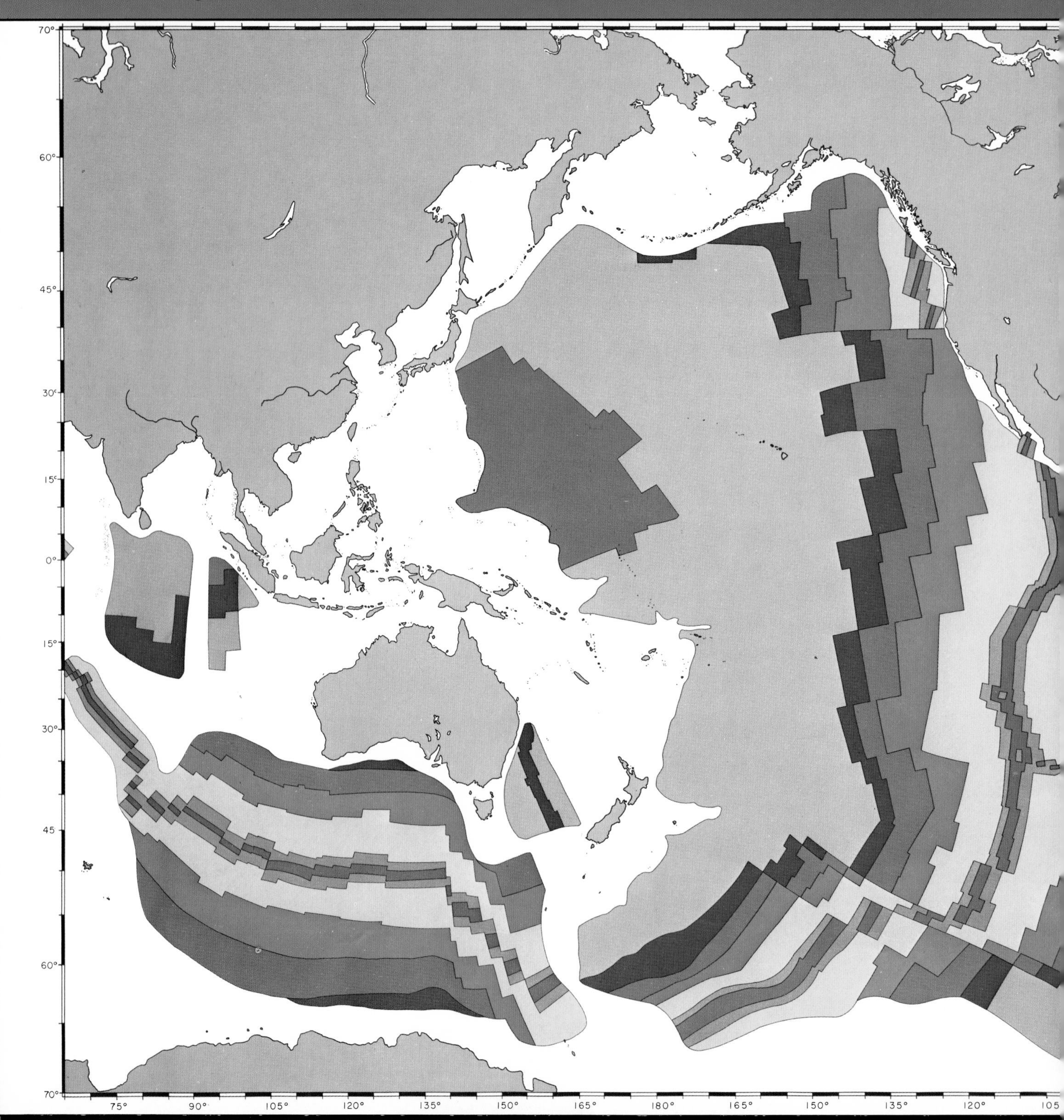

PLATE B

shadows resulting from the oblique angle of the sun's rays, which provide the natural illumination of the earth's surface that is reflected to the satellite overhead. In some cases, the impression of relief will be strengthened by viewing the plates in the inverted position (turned upside down); for example, Plate C.2.

All images reproduced on these pages are to the same scale, 1:1 million, on which 1 centimeter represents 10 kilometers. To determine distances on the images, simply use a centimeter scale, multiplying the number of centimeters by 10 to obtain distance in kilometers. The images are supplied by NASA and are available to the public on order through the U.S. Geological Survey. Identification numbers appear with each legend, along with the date the image was obtained.

The images were selected from a color atlas of Landsat images: N. M. Short and others, 1976, *Mission to Earth: Landsat Views the World*, NASA SP-360, 459 pp., National Aeronautics and Space Administration, Washington, D.C. Detailed information on each image can be found in this atlas, along with nearly 400 additional examples of Landsat imagery.

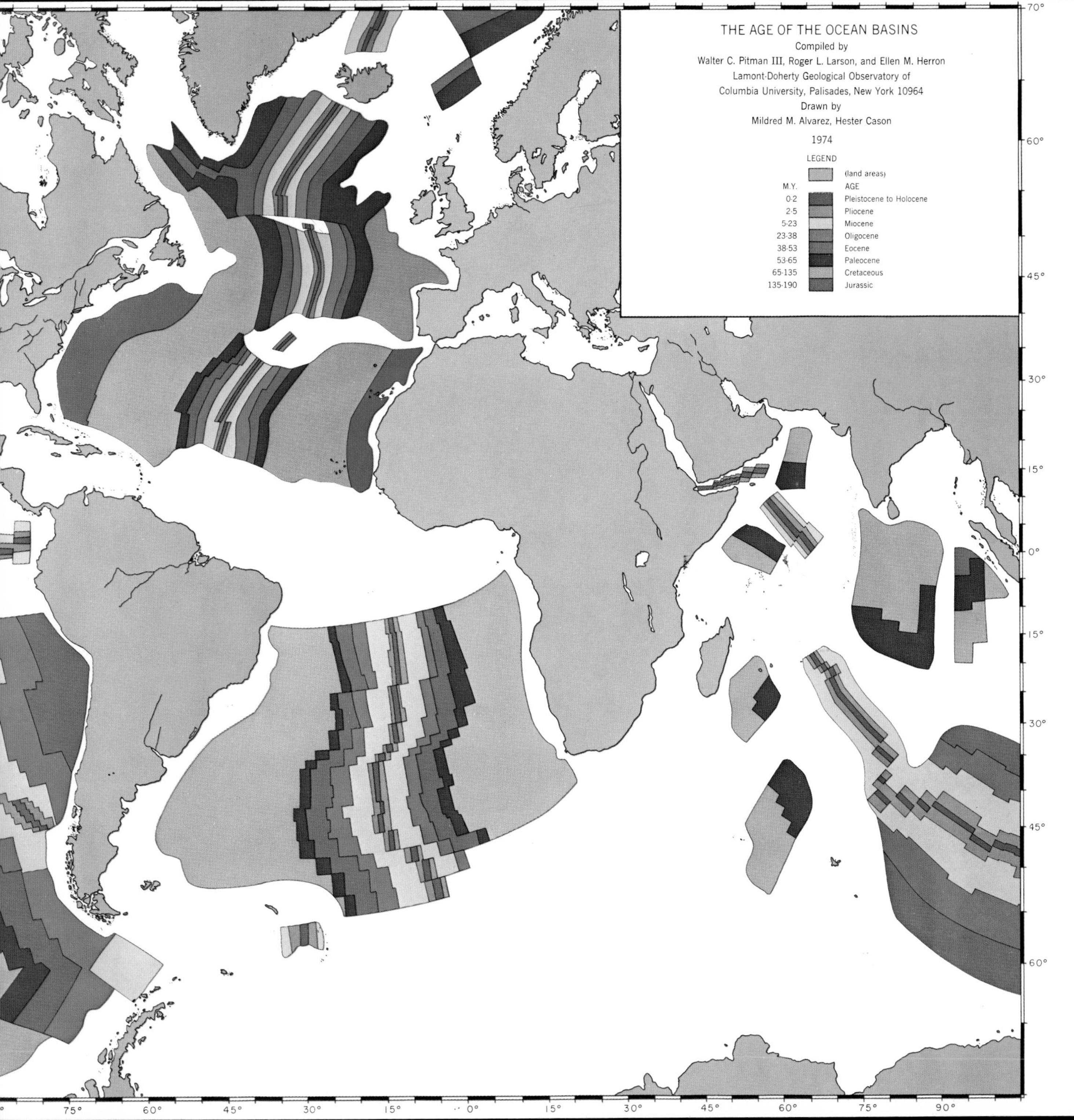

PLATE C

1. THE HIMALAYAN CONTINENTAL SUTURE IN NEPAL. Along the southern border of the image lie the high plains of the Ganges lowland. The first lines of low forested hills are the Siwalik Hills, formed of dissected Pliocene and Miocene sediments deposited as gravels brought from the higher mountains to the north. About in the center of the image are the Lesser Himalayas, with summits rising to over 3000 m. Major underthrusts are expressed as valleys and ridge crests trending east-west. Along the upper border of the image are the snow- and glacier-covered peaks of the main Himalayan Range, with many summits reaching elevations of 6000 to 8800 m. Mount Everest (8848 m) lies in the extreme upper right-hand corner. (NASA 1144-04174, 14 December 1972.)

2. ERODED FOLDS OF THE NEWER APPALACHIANS IN CENTRAL PENNSYLVANIA. Paleozoic strata ranging in age from Cambrian through Pennsylvanian were deformed into open foreland folds during the final collision that closed the ancestral Atlantic Ocean basin in late Carboniferous time. Deep erosion of plunging folds has brought into relief sharp-crested ridges of quartzitic sandstone and conglomerate — the Tuscarora (Silurian), Pocono (Mississippian), and Pottsville (Pennsylvanian) formations — which make zigzag patterns over much of the area of the image. The Allegheny Plateau lies to the west; the Great Valley and Blue Ridge occupy the southeastern corner of the image. (NASA 1495-15222, 30 November, 1973.)

3. PRECAMBRIAN SHIELD OF THE LABRADOR FOLD BELT IN QUEBEC AND NEWFOUNDLAND. Complex folding and faulting, possibly occurring during continental collision and suturing, has affected metamorphosed sedimentary rocks of Precambrian age. Hematitic iron ores occur within the metamorphic sequence. Long periods of fluvial denudation throughout the past half-billion years reduced the shield crust to a peneplain, which was heavily abraded by Pleistocene ice sheets. Numerous lakes (black) occupy the elongate ice-scoured valleys that follow the strike of weaker rock bands and fault lines. (NASA 1483-15013, 4 October 1973.)

PLATE C

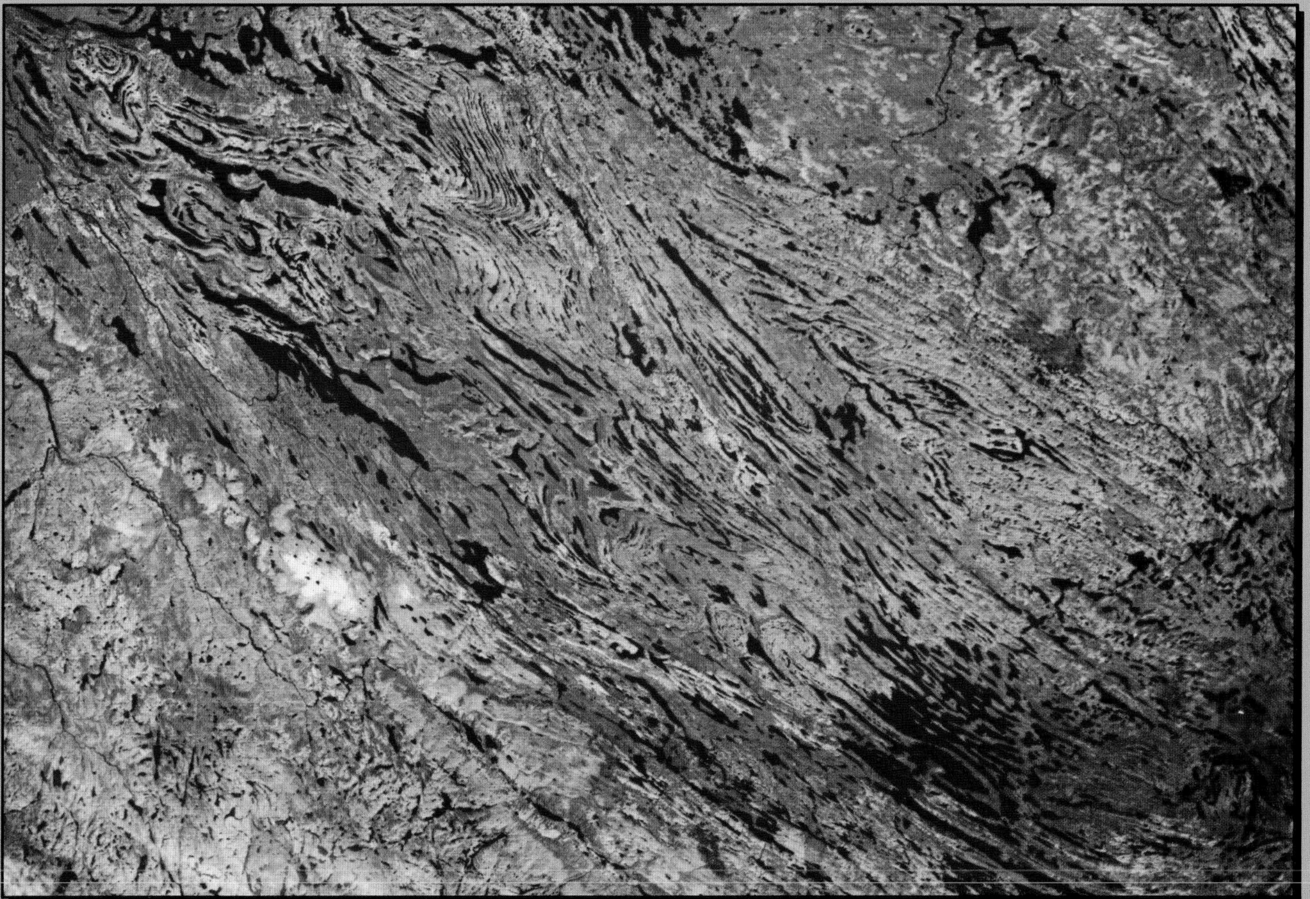

PLATE D

1. Basin and range region in central Nevada. While exploring this mountainous desert region in the 1870s, Major C. E. Dutton described the individual fault blocks, depicted by hachures on his crude map of the region, as resembling an "army of caterpillars crawling northward out of Mexico." The uplifted and tilted fault blocks have been deeply eroded, while alluvium has accumulated in the downfaulted areas, partially burying the mountain fringes. Mountain summits rise to elevations 1000 to 1500 m above the valley floors. The white areas are playas with saline deposits; they occupy closed basins with internal drainage. Small red patches are irrigated fields. (NASA 1342-17594, 30 June 1973.)

2. San francisco bay and san andreas fault. The San Andreas Fault can be identified on the San Francisco peninsula by long narrow lakes (black) located about midway between the southwestern shore of San Francisco Bay and the Pacific Ocean shoreline. Its faint trace can be followed southeasterly to the Hollister area, exiting at the lower edge of the image. The Pacific plate, represented by the Santa Cruz Mountains (solid red) has moved toward the northwest relative to the American plate. True north can be determined from the right-angle grid of roads and from boundaries of farm parcels in the Sacramento Valley, occupying the northeastern corner of the image. (NASA 1075-18173, 6 October 1972.)

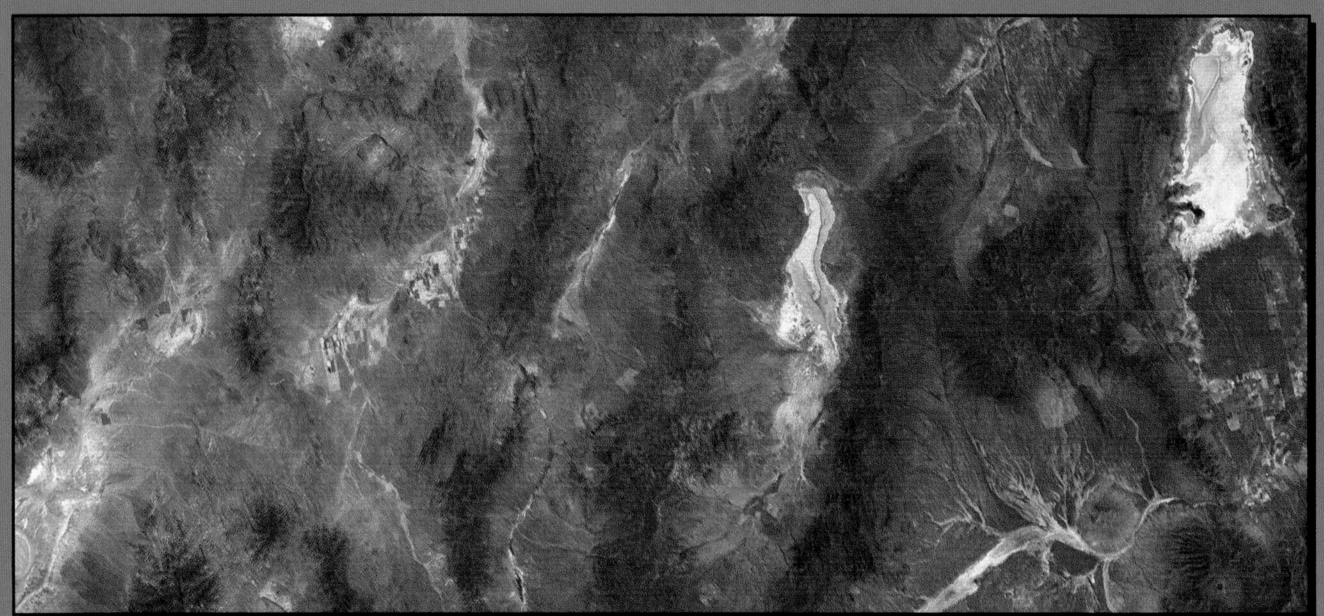

PLATE E

1. GRAND CANYON OF THE COLORADO RIVER, NORTHERN ARIZONA. The winding gorge of the Colorado River, with its numerous tributary side canyons, is deepest where it crosses uparched strata of the high Kaibab Plateau (white snow cover). The Kaibab uplift is bounded on the east by a monocline and on the west by normal faults. Several minor faults can be identified as narrow lines crossing the plateau surfaces east and south of Grand Canyon. Far to the left is the Unikaret volcanic field with late Cenozoic basaltic lava flows and cinder cones. (NASA 1284-17384, 3 May 1973.)

2. ARABIAN DUNES AND DISSECTED PLATFORM STRATA. North is to the right in this image of a portion of the southern Arabian Peninsula. The pale yellow bands are longitudinal sand dunes that roughly parallel the direction of prevailing northeasterly winds. The left half of the image shows stream-dissected Cenozoic strata — limestones and calcareous shales — lying almost horizontal. A dendritic drainage pattern is well developed. (NASA 1186-06381, 20 January 1973.)

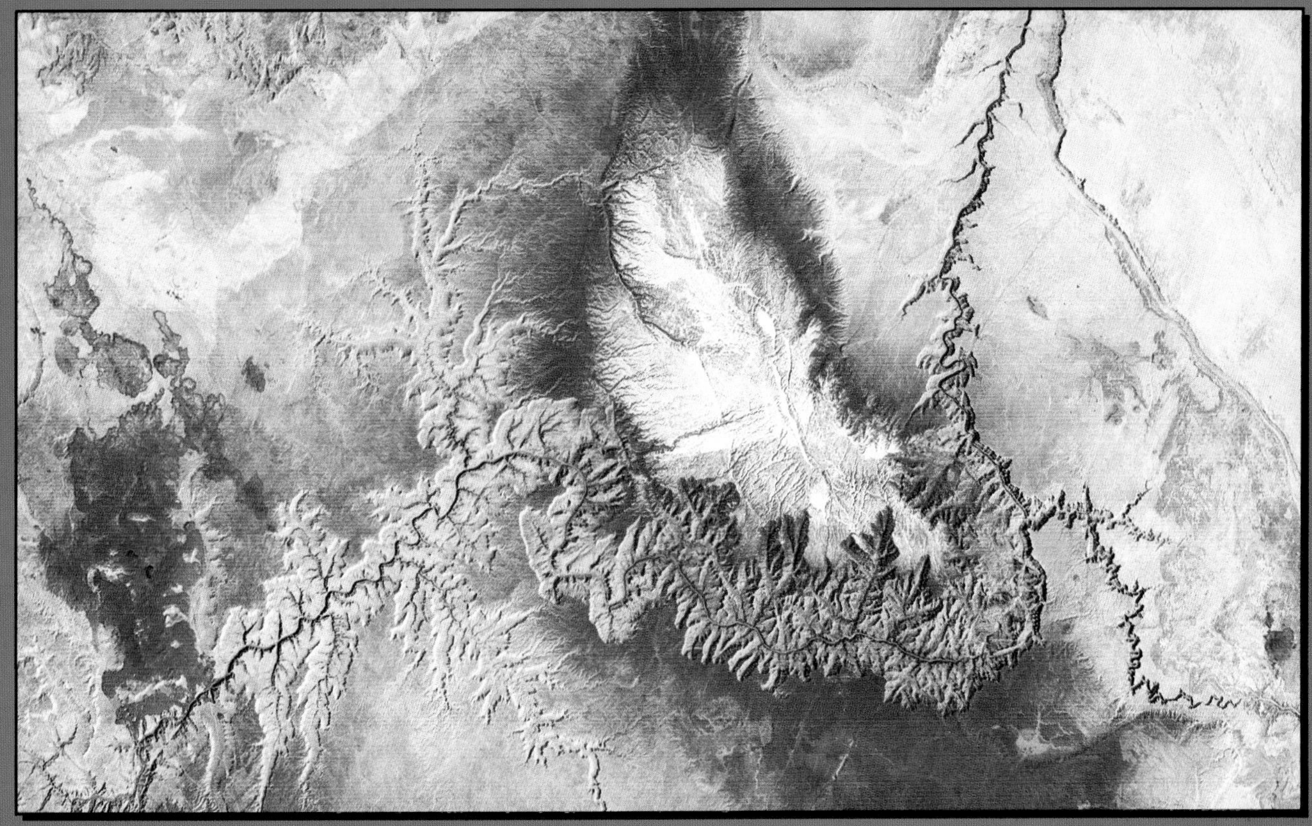

PLATE F

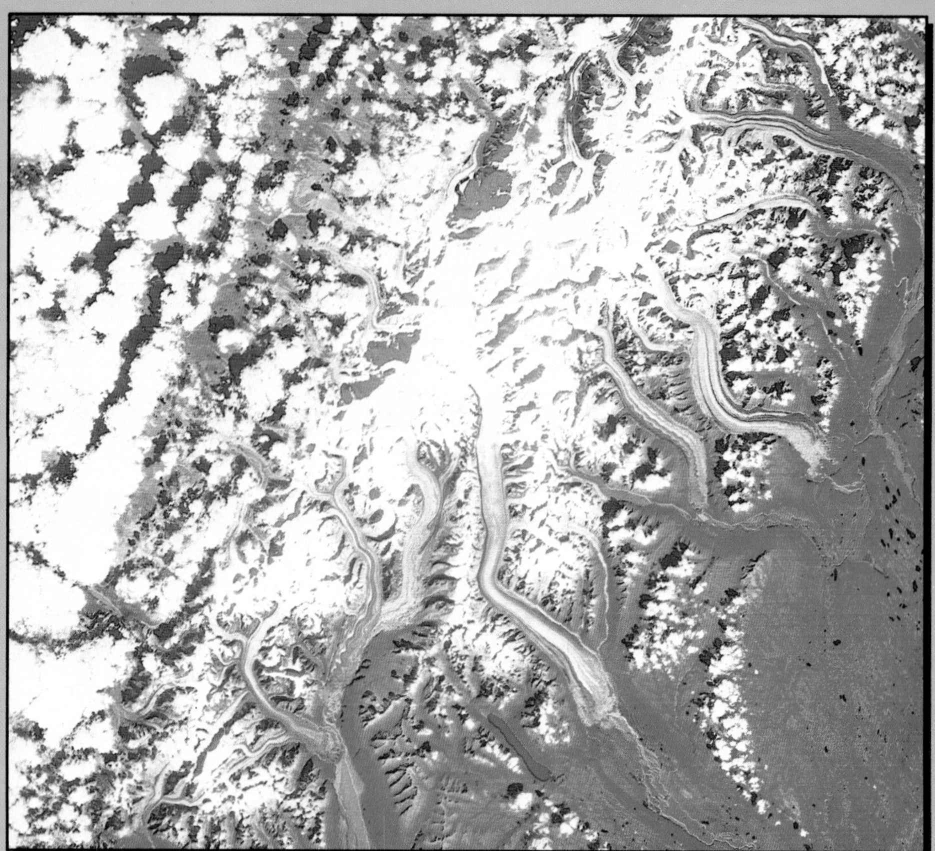

1. **Glaciers of south central Alaska.** Glaciers of the Alaska Range appear as blue curving bands, emerging from high firn (granular snow) fields on a snow-covered mountain axis trending from northeast to southwest across the center of the image. (The white patches at the left and right are clouds.) Darker lines running down the length of a glacier are medial moraines, formed of debris on the ice surface. Where these moraines have been distorted into a sinuous pattern the glacier has experienced surging at rates up to one meter per hour. A glacier with moraines smoothly paralleling the banks is experiencing slow uniform flow throughout its entire length. The group of mountain peaks in the upper central part of the image includes Mount McKinley, highest point in North America. (NASA 1033-21020, 25 August 1972.)

2. **Delta of the Mississippi River.** The modern Mississippi delta is of the bird-foot type, with multiple branching distributaries. Natural levees of the distributaries appear as lacelike filaments in a great pool of turbid river water. West and north of the active modern delta are the remains of older deltas. The Chandeleur Islands, a system of arcuate barrier islands (upper right) have been built of material from the subsided remains of these older deltas. Part of the city of New Orleans can be seen at the extreme upper left, occupying natural levees of the lower Mississippi River. A number of salt domes show as circular spots on salt marsh close to the modern delta. (NASA 1177-16023, 16 January 1973.)

cent to the Himalayan arc on the south. This accumulation can be regarded as a type of geosyncline—a foreland trough (see Table 12.1). The sediment has been derived from the Himalaya and Hindu Kush ranges by intense erosion of streams and glaciers (see Color Plate C.1).

A problem that greatly interests geologists is to explain the significance of the enormous Tibetan Plateau, covering nearly 2 million sq km and mostly over 5000 m in surface elevation. The low-density continental crust must be extremely thick beneath the plateau for it to stand so high. A thickness of continental crust of 60 km is indicated by seismic evidence—nearly double that of most continental cratons. Several interpretations of this thick crust have been made, one of which involves a unique concept in collision tectonics. This unusual hypothesis proposes that the entire Tibetan Plateau is underlain by a double thickness of continental lithospheric plate.

As shown in Figure 12.30, collision did not end in an edge-to-edge suture between the Indian and Eurasian plates. Instead, the Indian plate was forced beneath the Eurasian plate. Because of the buoyancy of the continental lithosphere, the Indian plate simply slid northward under the base of the Eurasian plate. This hypothesis of *continental underthrusting* has three points in its favor: (1) It explains the present seismic activity beneath the entire Tibetan Plateau. (2) It calls for an active underthrust fault at the southern foot of the Himalayas. (3) It explains the persistence of the Indo–Gangetic Trough as a sharply depressed crustal feature similar to an oceanic trench.

A major difficulty with the continental underthrusting hypothesis lies in details of the way in which the lower slab made its way horizontally beneath the upper slab. If the entire Eurasian lithospheric plate remained intact, a thick layer of ultramafic mantle rock must remain sandwiched between the two plates, as in Diagram *A* of Figure 12.30. An alternative explanation suggests that the Indian plate split the Eurasian plate horizontally, "peeling off" the mantle portion of the lithosphere along the plane of the Moho, as in Diagram *B*.

Other hypotheses explain the thickening of the crust beneath the Tibetan Plateau by processes that do not involve plate underthrusting and thus avoid the difficulty we have mentioned. For example, it has been proposed that following the initial collision between the Indian and Eurasian plates, continued pressure of the Indian plate caused the Eurasian crust to be thickened under lateral compression, as in Diagram *C*. As the Eurasian crust contracted from north to south, the ductile lower part of the crust yielded by plastic flowage; the upper, more brittle, part yielded by fracturing.

The hypothesis of compression and thickening of the Tibetan crust is strengthened by the presence of some very long transcurrent faults that lie north of the plateau in central Asia. These are shown in Figure 12.27. The direction of crustal movement along these faults gives the impression that great east–west strips of the Eurasian plate are being forced to move eastward, to "get out of the way" of the north-pushing Tibetan plate section. If this interpretation is correct, the entire plate area that lies beneath China is being moved eastward as a unit, or subplate. The region in

FIGURE 12.29 Simplified and largely schematic cross sections of the Zagros Mountains collision zone. *A*. Final closing of ocean basin is accompanied by imbricate thrust faulting of oceanic crust. *B*. Structure as it is today, inferred from surface geology and data of exploration for petroleum accumulations. (Drawn by A. N. Strahler. Suggested by data of S. J. Haynes and H. McQuillan, 1974, *Geological Society of America, Bull.*, vol. 85, p. 742, Figure 2.)

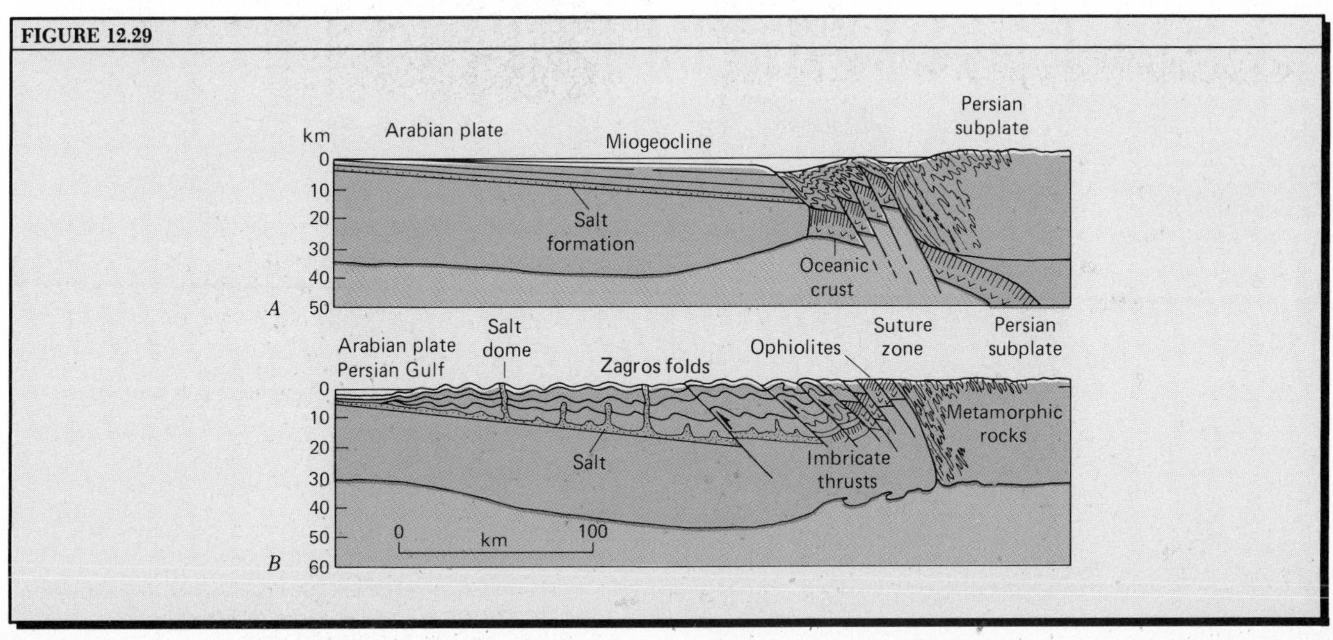

which these faults lie is remote and few geologic details are known to western geologists. They do, however, have the use of remote-sensing images that reveal remarkably fine details of the terrain. The fault system is largely interpreted from such images.

Structure of the Continental Lithosphere

The length of this chapter and the complexity of its topics reflect quite directly the long and complex history of the continental lithosphere. The oceanic lithosphere and its structural features are quite simple by comparison, because the oceanic crust is new and has formed rapidly at the seafloor spreading boundaries.

We have by no means finished a full account of the continental lithosphere. There remains for investigation the rifting of continental plates, followed by the opening of new ocean basins. We will have to go back to the original controversy that attended the once-disreputable hypothesis of continental drift as it was presented by Alfred Wegener and his supporters in the 1920s and 1930s.

Key Facts and Concepts

The continental crust Structural components of the continental crust are: exposed shields; sedimentary platforms; geoclines of passive margins; geosynclines and arcs of active margins; Cordilleran-type orogens; Eurasian-type orogens; tectonic uplifts of the continental foreland; orogenic root systems; block-faulted structures (active); inactive faulted and rifted crust; and isolated extrusives.

Sedimentary platforms A *continental platform* bears marine strata deposited during a *transgression* and followed by a *regression*. Phase change near the Moho may explain epeirogenic movements. *Eustatic changes* accompanying changes in rates of crustal spreading may also produce transgressions and regressions.

Sedimentary basins Within cratons, *sedimentary basins* of circular or elliptical outline have been sites of persistent subsidence and sedimentation over long time spans, sometimes with conditions favorable to accumulation of thick salt beds.

FIGURE 12.30 Schematic cross sections of the Tibetan Plateau showing three hypotheses of origin of a thickened crust. Vertical and horizontal scales are the same. *A*. Underthrusting of the Indian plate beneath the Eurasian plate. *B*. Underthrusting with mantle peeling, doubling the thickness of the crust. *C*. Crustal thickening under lateral compression following collision. (Drawn by A. N. Strahler. Suggested by compiled data of C. M. Powell and P. J. Conaghan, 1975, *Geology*, vol. 3, p. 729, Figure 2.)

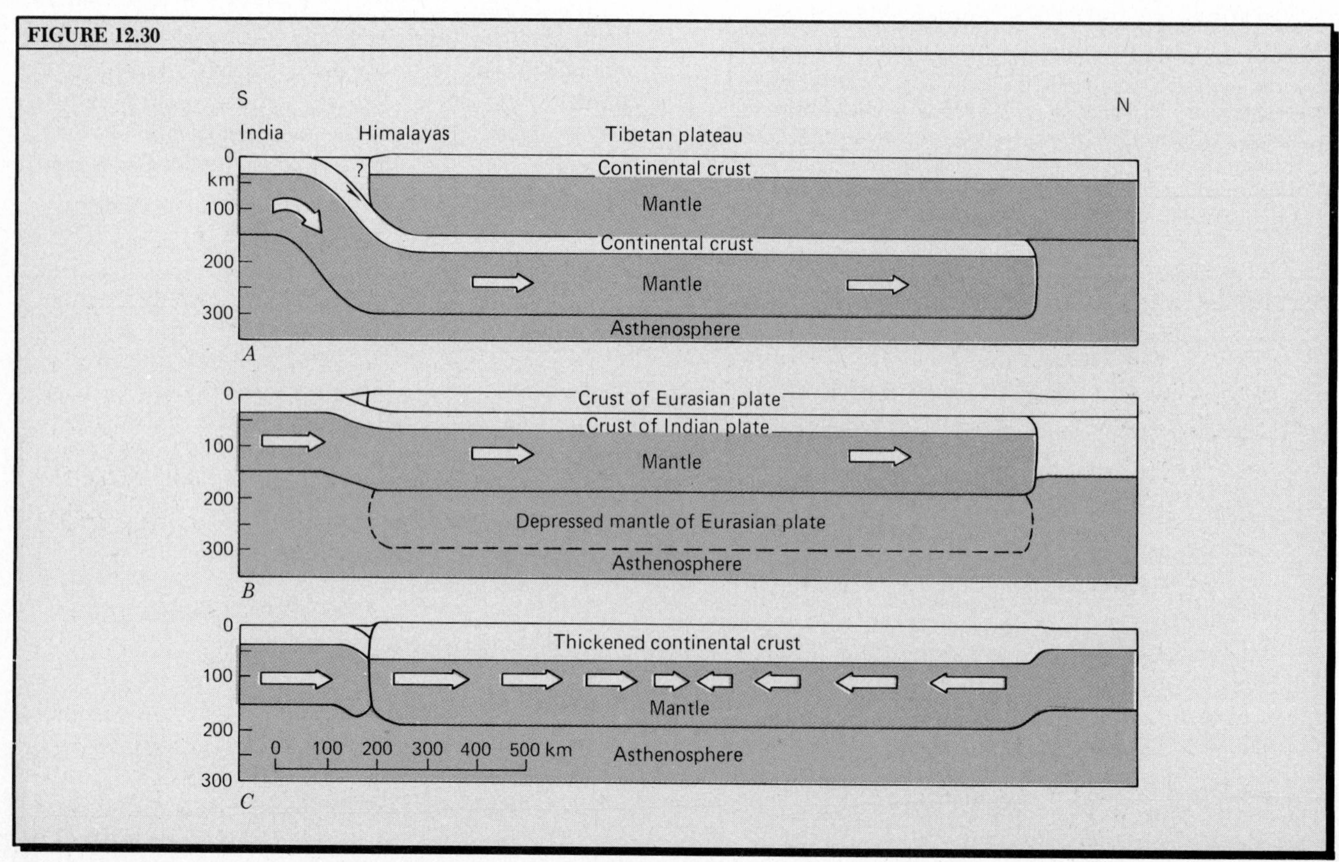

The geosyncline concept Under the classic concept, the *geosyncline,* a trough-shaped accumulation of thick, largely marine sediments, was thought to subside at a rate equaling sedimentation and to receive detrital sediment from a bordering mountain range or a *geanticline*. Geosynclinal deposition invariably terminated in orogeny. Geosynclines are presented graphically by means of a *stratigraphic cross section* with *isochrons;* it can be replotted as a *restored stratigraphic section*.

Contemporary geosynclines In modern terms required by plate tectonics, a geosyncline is any thick, rapidly accumulating body of sediment formed within a long, narrow belt usually paralleling the margin of a continent. Geosynclines may show control by tectonic activity. Four major classes of geosynclines are: I. Atlantic type, II. Indonesian type, III. Eurasian type, and IV. African type.

Atlantic-type geoclines Along passive continental margins are formed wedgelike *geoclines*. Typically, a *miogeocline,* or continental-shelf wedge, lies landward of and adjacent to a *eugeocline,* or turbidite wedge. Geoclines are found along both eastern and western margins of the Atlantic Ocean basin, with sediments ranging in age from Late Triassic to Recent. Subsidence of geoclines may result from continued plate cooling following opening of a broad ocean basin by rifting.

Pacific-type geosynclines Troughs of sediment deposition in subduction zones include a *trench wedge* separated by a *tectonic arc* (geanticline) from a *forearc trough* (geosyncline) bounded by a volcanic arc (geanticline). Landward of the volcanic arc may lie a *backarc trough* (geosyncline) bounded by the *continental foreland* of stable continental crust.

The Franciscan Complex The Franciscan Complex and Great Valley Sequence of northern California are interpreted as the accretionary prism and forearc basin, respectively, of an active continental margin in continuous subduction for nearly 100 m.y. A wedge of oceanic crust, lifted during understuffing, now forms an ophiolite zone between the two rock sequences.

Cordilleran-type orogens The Cordilleran-type orogen results from formation of a new subduction plate boundary close to a passive continental margin bearing thick geoclines, which are lifted in *thermal uparching* over a *mobile core*. Magma intrusion and extrusion follow. Newer turbidites of an accretionary prism are metamorphosed adjacent to the subduction boundary, while pluton intrusion continues. Finally, intense folding and overthrusting of the miogeocline and younger sediments, along with continued volcanic activity and batholith intrusion, terminates the orogeny. The Cordilleran Orogeny in North America, beginning late in Cretaceous time and continuing into Cenozoic time, produced many batholiths and a belt of low-angle imbricate overthrusts in the Northern Rockies.

Tectonic domes and uplifts Domelike *structural uplifts* were produced by *foreland basement deformation* during the Cordilleran orogeny in the Rocky Mountain region. These have been eroded to reveal cores of Precambrian shield rocks, bordered by upturned strata eroded into hogbacks.

Flood basalts Cenozoic extrusion of flood basalts in the Columbia Plateau region may have originated from a hot spot over a mantle plume.

Basin-and-Range faulting Block faulting beginning in middle Miocene time lifted and dropped crustal blocks of the Basin-and-Range region. Faulting may have resulted from crustal extension over a rising mantle arch or plume.

Cascade volcanism and Sierra Nevada uplift Pliocene volcanic activity in the Cascade region may have been related to subduction of the Juan de Fuca plate. Uplift of the tilted Sierra Nevada fault block occurred in early Pleistocene time.

Eurasian-type orogens A line of collision orogens, or sutures, marking the southern boundary of the Eurasian plate, consists of three segments: European, Persian, and Himalayan. Each segment represents collision with a different north-moving plate—African, Arabian, and Indian, respectively. Continental collision resulted in formation of nappes from imbricate thrust sheets in a rising pile. Final nappes include ophiolites that mark the suture line. *Flysch* and *molasse* accumulate during orogeny and are deformed in later stages. *Continental underthrusting* may have carried the Indian plate beneath the Eurasian plate, forming the Tibetan Plateau.

Questions and Problems

1. Imagine that you are examining a 10-m thickness of flat-lying platform strata exposed in the wall of a quarry in the Middle West. At the base is a 1-m thick bed of pebble conglomerate, above which are 2 m of thin beds of quartz sandstone. Above is a shale bed 3 m thick with many marine fossils. The uppermost 5 m consists of thin-bedded limestone rich in marine fossils. Does this sequence represent marine transgression or regression? Explain your interpretation.
2. The concept of paired geoclines—a miogeocline joined with a eugeocline—is quite modern. Deformed and uplifted strata that we now interpret as deep-water turbidites of a eugeocline were studied by geologists for decades and usually interpreted as sediments shed from a rising island arc on the oceanward side. Why do you suppose the eugeo-

cline concept eluded geologists until the early 1960s, when Robert S. Dietz proposed that a thick turbidite wedge lies below the continental rise?

3. Turbidites of a eugeocline are rarely associated with volcanic (extrusive) rocks. Why is this so? Are turbidites of a trench wedge commonly interbedded with volcanic rocks? Explain.
4. Does the exposure of ophiolites in an orogen always indicate the presence of a suture produced by continental collision? If not, how else might ophiolites come to be located at the surface in deformed rocks of the continental crust?
5. The Sierra Nevada, a great tilted fault block, is believed to have been elevated to its present height as recently as 2 m.y. ago. What geologic evidence might you seek in proof of this comparatively young age of uplift and tilting?
6. Both the Zagros Mountains of Iran and the Gulf Coast geocline of North America have numerous salt domes (salt plugs). The Zagros Mountains are made up of folded strata; the Gulf Coast strata are not affected by tectonic activity. What do these two geologic regions have in common that explains the existence of numerous salt domes?

13

CHAPTER THIRTEEN

The Opening and Closing of Ocean Basins

Our final chapter in this four-part series on global plate tectonics deals with continental rifting and the opening and closing of ocean basins. As we did in the last chapter, we must go back into geologic time to review major events of plate tectonics that shaped the continents. We shall first make a retrospective journey into the history of science to review the lines of evidence used by Alfred Wegener and his followers in the 1920s and 1930s to support their contention that a united continent of Pangaea existed in the early Mesozoic. Why did so many English-speaking geologists, particularly those of North America, reject Wegener's evidence so vigorously and so outspokenly?

Wegener's Pangaea

Alfred Wegener's 1924 map of Pangaea is shown in Figure 1.12. He had worked on the problem for several years prior to that date and had published maps of Pangaea in earlier editions of his major work, *The Origin of Continents and Oceans*. Figure 13.1 is a reproduction of a map in the first edition of that book, published in 1915, showing the American continents fitted to Europe and Africa. Wegener chose the outer limits of the present continental shelves as his boundaries, using the 200-meter depth contour as a guide. In so doing, he was on the right track because we know today from seismic evidence that the limit of the continental crust coincides about with the outer edge of the shelf.

Pangaea has traditionally been divided into two parts, or subcontinents. A southern subcontinent, *Gondwana* (or *Gondwanaland*), consisted of Africa, Madagascar, India, Australia and New Zealand, and Antarctica. A northern subcontinent, called *Laurasia*, consisted of North America, Greenland, and Eurasia.

A strong supporter and admirer of Wegener was the South African geologist Alexander du Toit who, in 1937, published a revised map of Gondwana (Figures 13.2 and 13.8). Du Toit lined up geosynclines of the same age in South America, southernmost Africa, Antarctica, and Australia to form a continuous trough of sedimentation. Two more recent versions of Gondwana are also shown in Figure 13.2.

Positioning of the southern continents, particularly Australia and Antarctica, proved a difficult puzzle, but with the passing years new data and computers have been used to assist in this operation. In earlier years, great importance was attached to geological fitting of the continental margins, and the geologic ages of the rocks, together with tectonic patterns, were a major source of information. As radiometric ages of rocks became more readily available, this method took on increased accuracy. Figure 13.3 is a modern reconstruction of Laurasia and Gondwana showing ages of shield rocks. In the 1977 reconstruction shown in Figure 13.2, information from magnetic anomaly patterns and other paleomagnetic studies was combined with geological information to achieve the best possible fit, and the 2000-meter depth contour was used as the arbitrary boundary of a continental plate. We shall probably see newer and more refined reconstructions of Gondwana and Laurasia in the near future.

Not much is known about the outline and position of the Asian continent, particularly its extreme eastern part. Wegener showed Asia complete with India connected to it and closely nested in with the southern continents (Figure 1.12). In later reconstructions, India is always shown as an independent continental fragment separated from Asia by the Tethys Sea (Figure 13.3). Other fragments of Asia have been added to Gondwana in some reconstructions. As we suggested in Chapter 12, the collision of India with Asia probably produced profound changes in the Asiatic continental structure, perhaps pushing the China subplate far eastward of its earlier position.

Wegener in Difficulty

When we compare Wegener's reconstruction of Pangaea with the newer versions that have almost universal acceptance, we cannot help but wonder why his hypothesis of continental drift fared so badly at the hands of the geological establishment. This is a question to keep in mind as we cover the geologic evidence, point by point, that Pangaea existed and that it broke apart. In retrospect, it seems likely that the un-

FIGURE 13.1 Alfred Wegener's 1915 map fitting together the continents that today border the Atlantic Ocean basin. The sets of dashed lines show the fit of Paleozoic tectonic structures between Europe and North America and between southernmost Africa and South America. (From A. Wegener, 1915, *De Entstehung der Kontinente und Ozeane*, F. Vieweg, Braunschweig.)

FIGURE 13.2 Three reconstructions of Pangaea. (*A.* Alexander du Toit, 1937, *Our Wandering Continents*, Oliver & Boyd, Edinburgh. *B.* A. G. Smith, and A. Hallam, 1970, The fit of the southern continents, *Nature*, vol. 225, p. 139. Copyright by Macmillan (Journals) Limited. *C.* E. J. Barron, C. G. A. Harrison, and W. W. Hay, 1977, *EOS*, vol. 58:9, p. 844, Figure 8. Copyrighted by the American Geophysical Union. All reproduced by permission of the respective publishers.)

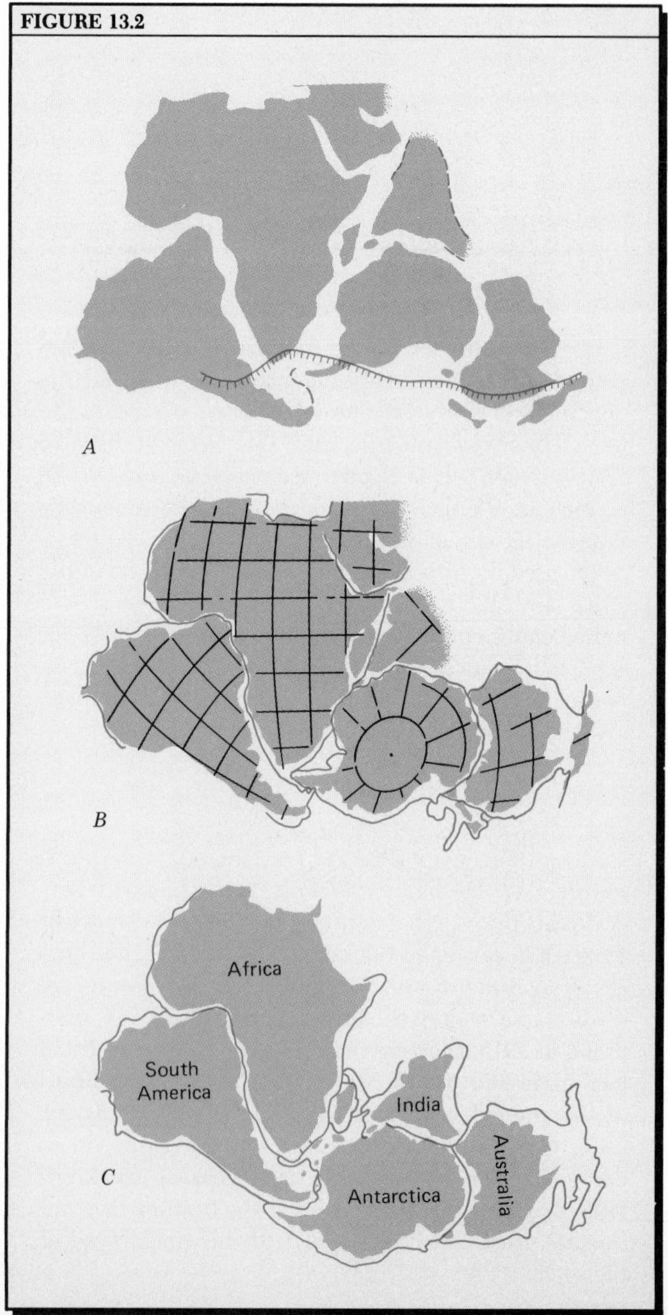

relenting opposition grew out of Wegener's failure to accompany his scenario of events with an acceptable physical mechanism for the drifting apart of continents.

In the period from 1915 to 1929, when Wegener published his advocacy of continental drift, ideas then current about the earth's crust and mantle were quite primitive, but in some respects basically sound. It was understood that the continents consisted of a layer of relatively low-density rock of the approximate composition of granite. This material was called *sial,* a word coined by combining the chemical symbols for silicon (Si) and aluminum (Al). The ocean floor was understood to be underlain by a crust of denser rock of basaltic composition, called *sima,* from two syllables in silicon and magnesium. It was supposed that the layer of sima also passed beneath the sial of the continents. The concept of isostasy was then fully accepted, and it was understood that vertical motions of the sial and sima could take place by slow yielding of rock beneath those crustal layers. Wegener argued that if such vertical movements could take place, it would be equally possible for the sial of the continents to move horizontally with respect to the sima, provided the horizontal forces to propel the sial existed. He seems to have envisioned the continents as great rafts or barges of sial moving with almost imperceptible slowness through a "sea" of sima. Wegener speculated at some length on what the propelling force might be. One force that might be adequate, he wrote, was related to the tide-producing force generated by the rotation of the earth while deformed by mutual attraction between the earth and moon.

FIGURE 13.3 Patrick M. Hurley's 1974 reconstruction of Pangaea showing the full Eurasian continent. Extensive exposures of older crystalline basement rocks are shown by stippled pattern. Paleozoic tectonic trends are shown by dashes. (From P. M. Hurley, 1974, *Geology,* vol. 2, p. 373, Figure 1. Copyright © 1974 by the Geological Society of America.)

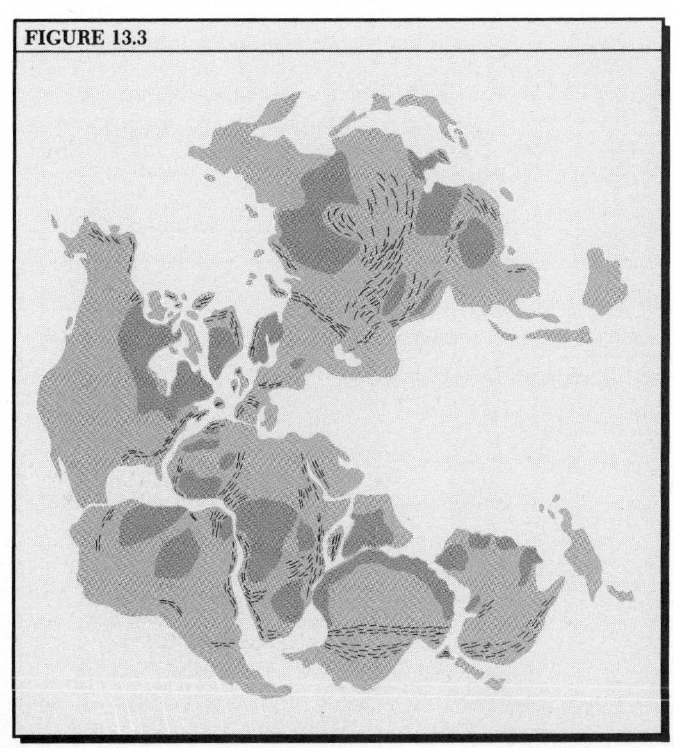

Perhaps more than any other critic, the distinguished Cambridge geophysicist Harold Jeffreys was responsible for bringing Wegener's entire hypothesis into disrepute. In the late 1920s Jeffreys attacked the physical mechanisms in Wegener's hypothesis. He showed that the tidal force invoked by Wegener was much too feeble to produce horizontal motion of the sial, but Jeffrey's major coup was to demonstrate that the strength of the sima was so great that it would be impossible for the sial to move through it. Jeffreys was, of course, correct in his physical analysis, and he was supported without hesitation by his colleagues. What seems to have ensued was a *non sequitur* (does not follow)—namely, that, because the physical model was in error, all the geologic evidence that favored a united continent must also be in error.

On the American side of the Atlantic one of the most merciless critics of Wegener was Professor Rollin T. Chamberlin of the University of Chicago. This leading geologist imperiously attacked not only continental drift, but its author as well. We quote one of Chamberlin's blasts: "Wegener's hypothesis in general is of the foot-loose type, in that it takes considerable liberty with our globe, and is less bound by restrictions or tied down by awkward, ugly facts than most of its rival theories. Its appeal seems to lie in the fact that it plays a game in which there are few restrictive rules and no sharply drawn code of conduct."

But Wegener had strong supporters, few as they were, even in the dismal years of the 1920s and in the period that followed his untimely accidental death on the Greenland Ice Sheet in 1930. One was his early admirer and supporter, Alexander du Toit. Equally strong in a supportive role was Professor Arthur Holmes, whose early insights into the scheme of plate motions we reviewed in Chapter 10. Like Jeffreys, Holmes rejected Wegener's model of the moving raft of sial, but, unlike Jeffreys, Holmes searched for an alternative model that was acceptable in terms of principles of physics. The result was a system of mantle convection, first introduced in 1929. Holmes showed flowage to occur below both the sial and the sima, in a deeper mantle layer (Figure 13.4). Mantle currents moving in opposite directions would exert a strong tensional force on the overlying sima, which would be stretched horizontally and thinned in the process. The overlying sial of a single large continent might then be fractured into large pieces, and these would separate from one another, exposing sima between them and thus creating new ocean basins. Holmes was then held in high esteem in the geological profession, and many now regard him as the greatest British geologist of the twentieth century. Nevertheless, his efforts to support Wegener's hypothesis failed to check the rain of skepticism and contempt showered upon Wegener and his ideas by a majority of geologists and geophysicists through the period between the First and Second World Wars.

Polar Wandering and Continental Drift

One of the strongest lines of evidence that the continents were formerly united in Pangaea came from paleomagnetism, but not until long after Wegener's death. Recall from Chapter 10 that rock samples which retain permanent magnetism acquired at the time of their origin yield two kinds of information. One is the compass direction of the lines of force; it tells the direction one would follow on a great circle to reach the former north (or south) magnetic pole (Figure 13.5). The other item of information is the inclination of the former magnetic lines of force at the point where the sample is taken. Looking back to Figure 10.13, it is obvious that the magnetic inclination, as registered by the angle of a dip needle at the earth's surface, will vary according to the magnetic latitude. Thus, the amount of inclination read from the sample tells the distance to the former magnetic north pole. "Distance" in this case is the number of degrees of arc along the great circle to the former pole (Figure 13.5). The paleomagnetic data tell us the position of the former north magnetic pole in terms of present latitude and longitude, stated in our conventional system of geographic coordinates. But, as you can see from Figure 13.5, the sample point could have been located anywhere on the former geomagnetic parallel of latitude. In other words, while we know the former geomagnetic latitude of the sample, we do not know the former geomagnetic longitude.

At the present time, the geomagnetic axis does not quite coincide with the earth's axis of rotation. One axis is inclined relative to the other by an angle of about 11.5°. As a result, the geomagnetic north pole is located at about latitude 70° N, which is at a distance of about 2000 km from the geographic north pole. The north-seeking end of a magnetic compass needle points to the geomagnetic north pole. The horizontal angle between geographic (true) north and geomagnetic north indicated by the compass is called the *compass declination*. Records of compass declination have been kept at both Paris and London since about the year 1600. In London, between 1600 and 1820, compass declination changed in a westerly direction through a total angle of nearly 35°; since 1820, the direction of change has reversed itself and shifted eastward about 20°. From 1600 to 1915 in Paris, inclination of the compass needle underwent a change amounting to 10° of latitude. Patterns of change at the two localities are quite similar, so it is reasonable to think that these changes follow a cyclic pattern and that, over a time span of millions of years, the geomagnetic axis has, on the average, coincided very closely with the axis of rotation. For any given sample point in past geologic time, however, there is a good chance that an error of several degrees is made when we assume that the geomagnetic axis coincided with the rotation axis.

Another source of error in paleomagnetic data lies in the place-to-place irregularities in the lines of magnetic force. These irregularities exist today and are known to be changing through time. There is no way that the pattern of magnetic irregularities can be reconstructed for periods in the geologic past. Thus, ev-

FIGURE 13.4 A model of mantle convection introduced by 1929 by Arthur Holmes. Extension of the continental crust is shown to have resulted in the opening of new ocean basins. (From A. Holmes, 1929, Radioactivity and earth movements, *Trans. Geological Society of Glasgow*, vol. 18, pp. 559–606, Figures 2 and 3.)

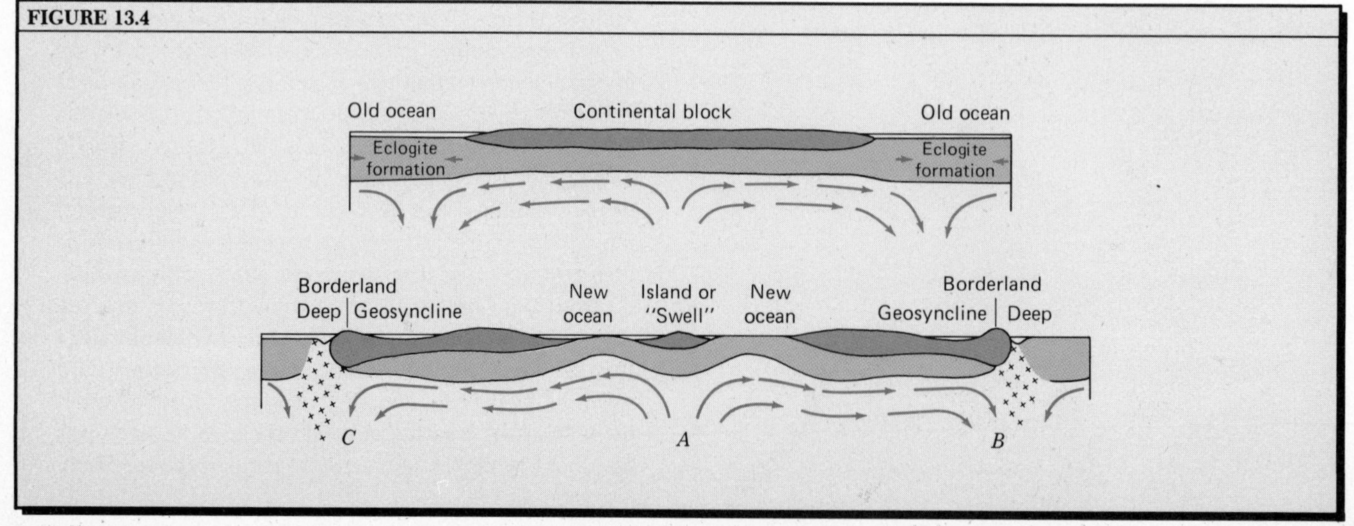

ery determination of a former position of the earth's pole of rotation is subject to an error on the order of five degrees of latitude and longitude. Even so, paleomagnetism provides us with some very strong evidence of past movements of the continents relative to one another.

Former pole positions have been determined for rocks of all geologic periods and eras, even far back into the Precambrian, and thousands of such determinations now exist. Taking a continent that behaved as a single rigid plate, we can plot on a map or globe the positions of the paleomagnetic pole through geologic time. The plotted positions are then connected by a smooth line, called the *polar wandering path*.

Figure 13.6 shows the polar wandering paths for North America and Europe for the past 400 m.y. Both paths end at the present location of the magnetic north pole, but going back into time, the two paths separate, with the distance of separation increasing rapidly to about −200 m.y. (Separation is measured in degrees of longitude.)

At this point, you can try an experiment. Using a small piece of tracing paper, trace the North American curve. Poke the point of a pin through the tracing paper and into the page at the pole position. Now rotate the tracing paper slowly counterclockwise until the two curves coincide. Coincidence is almost perfect between −200 and −400 m.y., but the curve portions between 0 and 100 m.y. become widely separated.

When we rotated the tracing paper, we also moved North America and Greenland closer to Europe. Using the present north pole as the pivot, we rotated the North American pole path through about 45° of longitude. This same amount of rotation brings Newfoundland about in contact with Scotland. The interpretation of the two polar wandering paths thus leads to the conclusion that, prior to about −200 m.y., North America and Europe were close together and moved as a single lithospheric plate over the asthenosphere. Subsequent to −200 m.y., which is in early Jurassic time, the single plate broke up into two plates, which gradually separated from one another.

Figure 13.7 shows how paleomagnetic data can be used to fit two continents together. The map shows polar wandering paths of Africa and South America in their present positions. The color overlay shows South America moved eastward and rotated so that the curves are superimposed. In rearranging the continents to a new fit, the latest and most reliable paleomagnetic data are used in this manner. Because of the range of error that exists in the method, some flexibility is allowed in order to match the continental plate boundaries in the most suitable manner and to align various structural trends in the bedrock geology.

A note of historical interest is appropriate here. Polar wandering paths were first plotted in the middle 1950s, and by the early 1960s the basic results were quite clear, as we have presented them. A pioneer in this field was S. K. Runcorn of Cambridge University. Plate tectonics had not been formulated as a working hypothesis at that time, but the debate about continental drift was beginning to enter a new and rather heated phase. In 1960, an important international

FIGURE 13.5 Present and past geomagnetic coordinates. *A*. Present geomagnetic pole, equator, and meridian. Paleomagnetism of sample at Point P indicates where the former magnetic pole was located. *B*. Former geomagnetic coordinate system based on the former magnetic pole.

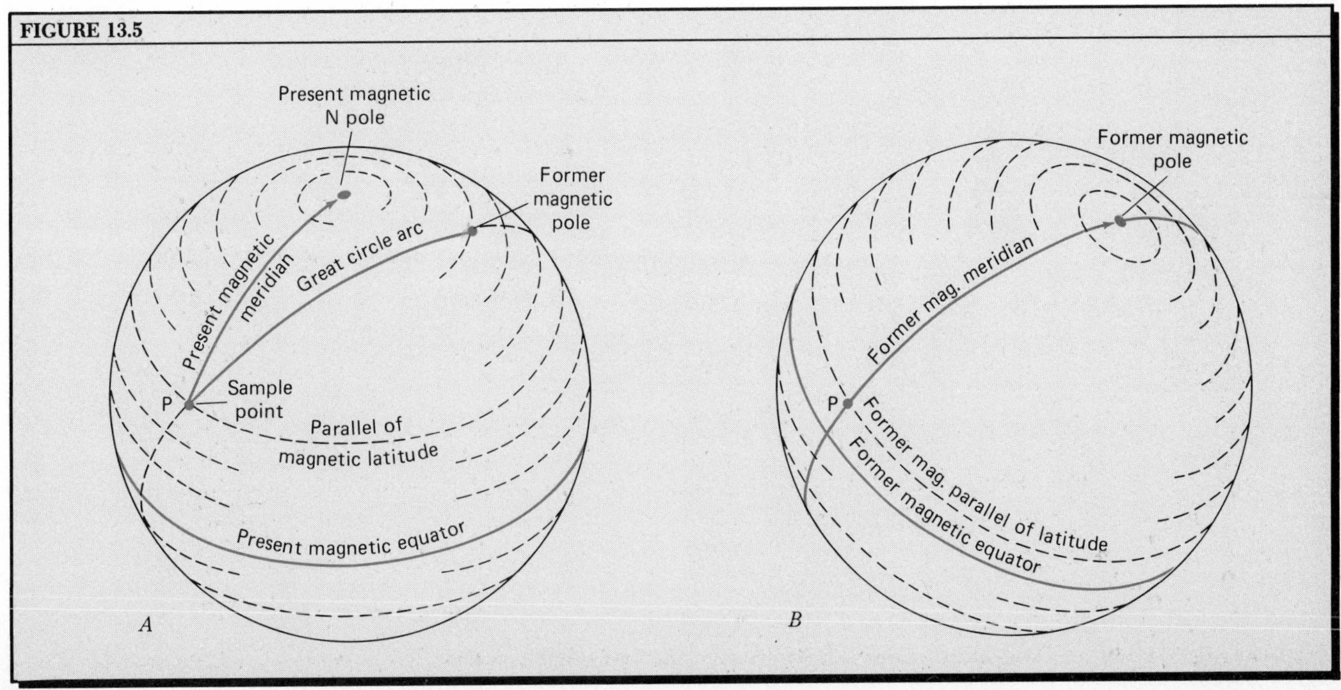

symposium was held to discuss the impact of the new paleomagnetic discoveries on the Wegener concept of continental drift. By then seismic evidence of the Wadati–Benioff zone was beginning to be interpreted as evidence of deep underthrusting at the active Pacific continental margins. Seafloor spreading was becoming an inescapable interpretation from the configuration and seismic activity of the mid-oceanic ridge. New geologic studies of Antarctica were beginning to point strongly to a former positioning of that continent close to Australia and Africa. The high point of the 1960 symposium was the imposing display of polar wandering paths, which left little doubt in the minds of many that the continents had indeed been united in Pangaea and that they had separated. Although paleomagnetism had finally won a victory for Alfred Wegener, many geologists still failed to see it that way. The nonbelievers were not going to give their approval until an acceptable physical model could be produced to explain how continents could drift apart. Several years were to elapse before plate tectonics provided that physical model.

FIGURE 13.6 Polar wandering paths for Eurasia and North America. (Based on data of C. K. Seyfert and L. A. Sirkin, 1979, *Earth History and Plate Tectonics*, 2d ed., Harper & Row, New York, p. 123, Figure 7.12.)

Geological Evidence of Pangaea

Matching of rock ages and structural patterns across the fitted edges of the nested continents provides evidence in favor of the unity of Pangaea. Wegener showed this kind of matching on his early map, reproduced in Figure 13.1. He matched a system of east–west folds at the southern tip of Africa with similar folds of the same age and rock sequence in eastern Argentina, and he noted in his book that shield rocks in eastern Brazil are remarkably similar to those in

FIGURE 13.7 South-polar wandering paths during Paleozoic time for South America and Africa. To superimpose the curves, it is necessary to move the continents close together and to rotate South America in a counterclockwise direction. Although these curves cannot be made to coincide in trend when the continental coastlines are closely fitted, former unity of the two continents is strongly indicated. (Geographic coordinates of average paleomagnetic poles are from C. K. Seyfert and L. A. Sirkin, 1979, *Earth History and Plate Tectonics*, 2d ed., Harper & Row, New York, pp. 120–121, Table 7.1.)

FIGURE 13.8 Alexander du Toit's 1937 reconstruction of Pangaea showing the position of the Samfrau geosyncline. (A. du Toit, 1937, *Our Wandering Continents*, Oliver & Boyd, Edinburgh, Figure 7. Reproduced by permission.)

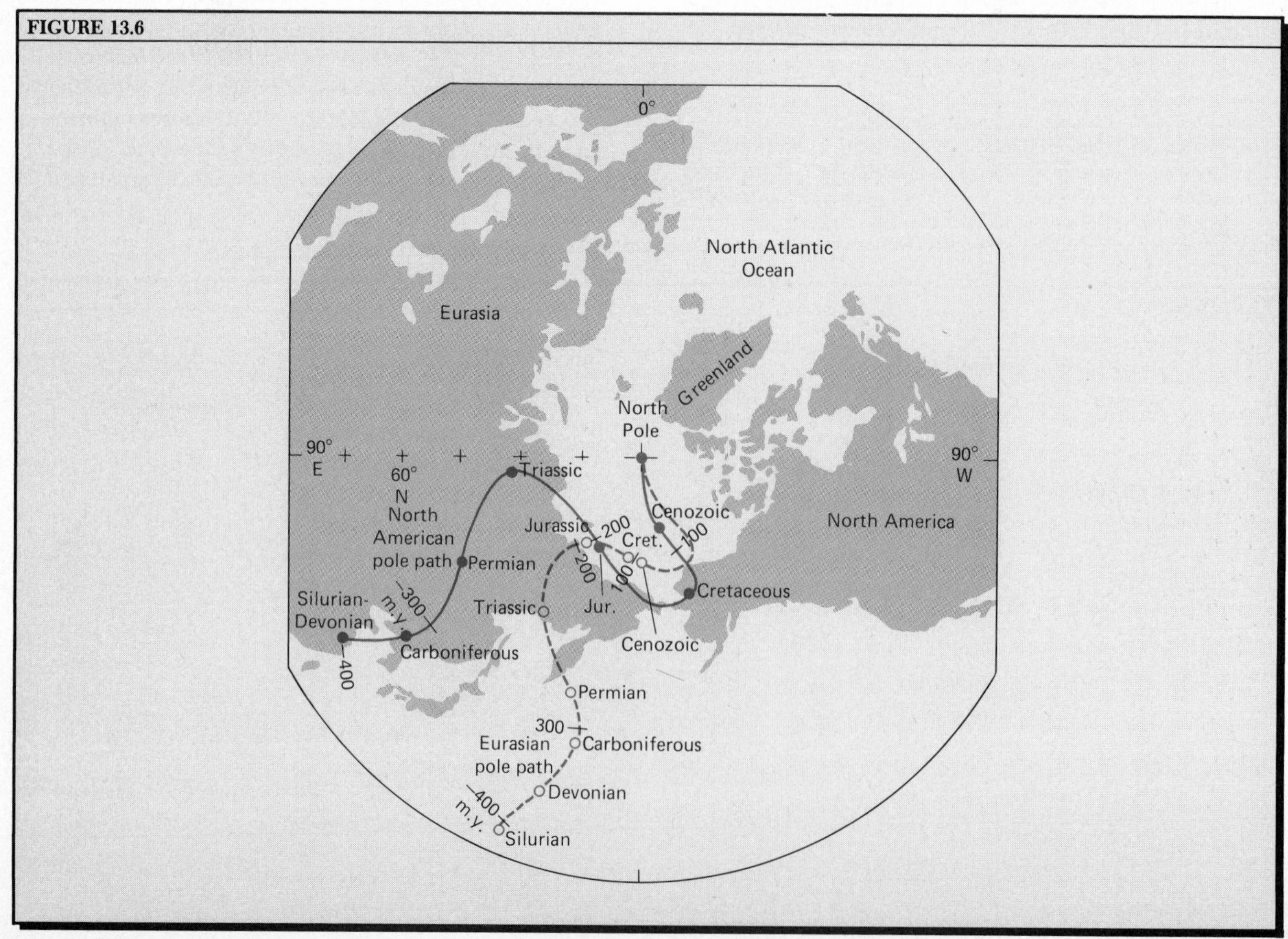

western Africa. His early map shows that he matched the Appalachian orogenic structures of North America with those of comparable age in the British Isles and France. Opponents of Wegener argued that the matching of these features was merely coincidence or, if real, that an intervening crustal block had since dropped down, leaving an ocean gap between.

Supporters of Wegener continued to document the similarities of rock types, ages, and structural patterns along fitted continental boundaries. One of these was Alexander du Toit, who made a careful reconstruction of Gondwana. He was able to trace a single geosyncline, of Paleozoic age and now folded, from South America, through South Africa, across the Antarctic Peninsula, and through eastern Australia (Figure 13.8). For this feature he coined the name *Samfrau geosyncline*, using letters in each of the names of the three landmasses it crossed (South *AM*erica, *AFR*ica,

FIGURE 13.7

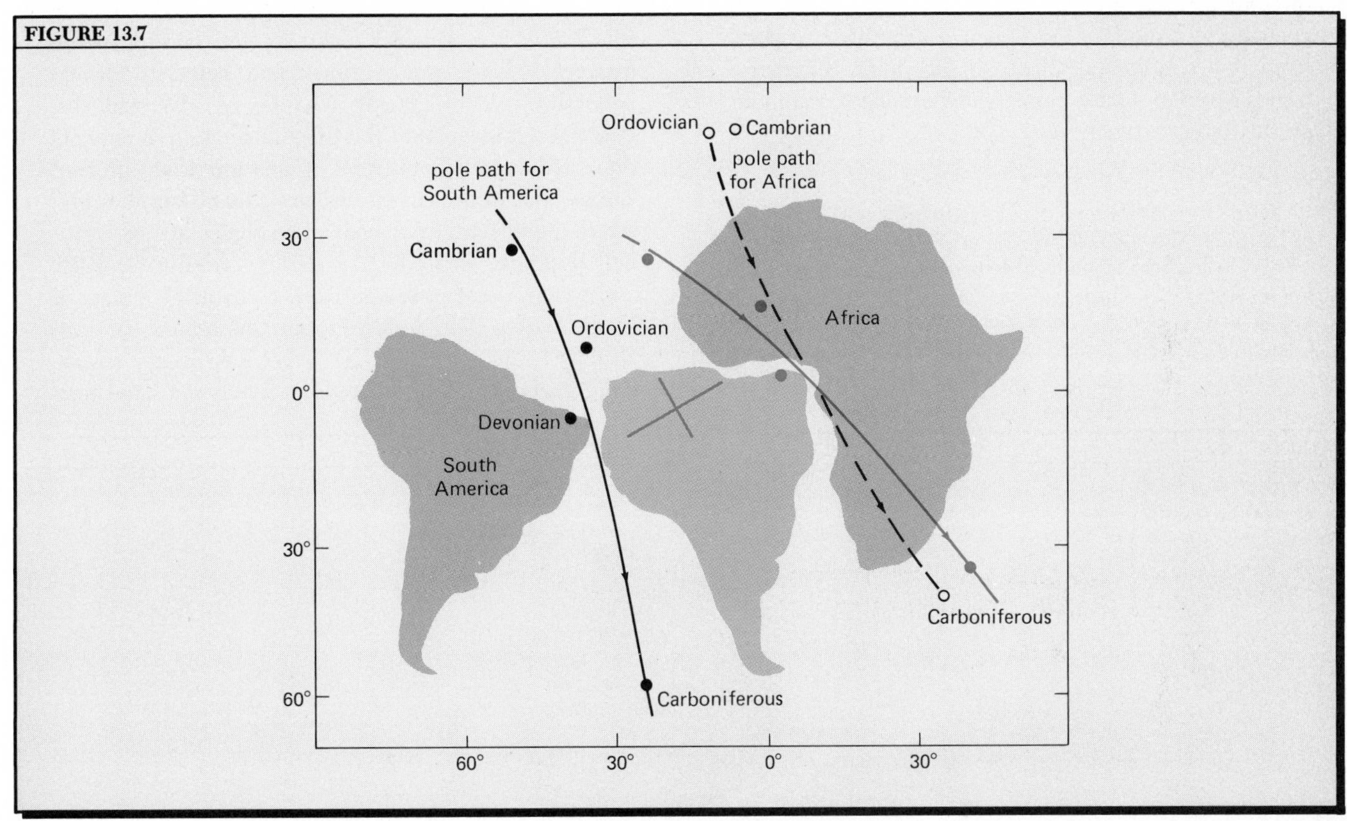

FIGURE 13.8

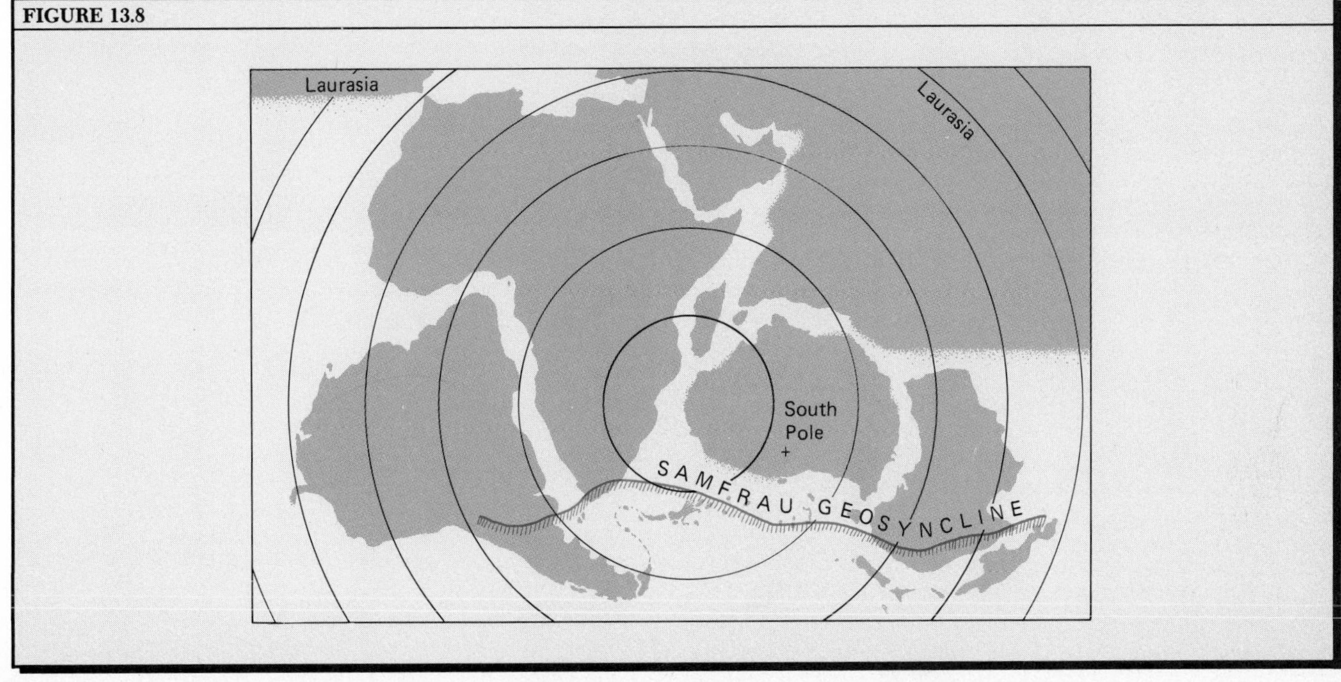

AUstralia.) Du Toit also pointed out the similarities of strata of comparable ages forming shield platforms on the landmasses of Gondwana.

As the unified theory of plate tectonics made its appearance in the late 1960s, a major research effort was launched by Professor Patrick M. Hurley and colleagues at the Massachusetts Institute of Technology to make a detailed plot of rock types, ages, and orogenic structures of the continents. An example of the results of this investigation is shown in Figure 13.9. When those areas composed of rock older than −2 b.y. are enclosed by boundary lines, they clearly show overlaps from one continental shield to the other. Structural trends based on fold axes also show excellent matching on the two continents.

FIGURE 13.9 South America and Africa prior to the breakup of Pangaea. Areas in color consist of shield rocks older than −2 b.y. Trends of tectonic structures are shown in dashed lines. (From P. M. Hurley, 1968, The confirmation of continental drift, *Scientific American*, vol. 218, No. 4, p. 60. Copyright © 1968 by Scientific American, Inc. All rights reserved.)

Evidence from Paleontology and Stratigraphy

One of Wegener's strongest points in favor of the united continent was a close similarity in the platform strata lying on the cratons of the Gondwana landmasses. The strata he correlated span the range from the Carboniferous Period through the Mesozoic Era. In the upper Carboniferous sequence of all the Gondwana landmasses is found a sedimentary formation known as a *tillite*, consisting of lithified glacial till. *Till* is the general term for the heterogeneous mixture of clay, sand, gravel, and boulders formed beneath a moving glacier (Chapter 18). We shall discuss the significance of the tillite formation in connection with a Carboniferous glaciation that affected Gondwana. Above the tillites lie shales of Permian age, interbedded with coal beds. The shales are of continental origin, deposited in shallow freshwater lakes and swamps; they contain fossil remains of a class of small freshwater reptiles, the mesosaurs (Figure 13.10). These creatures had a bone structure resembling a salamander, and would probably not have been

FIGURE 13.9

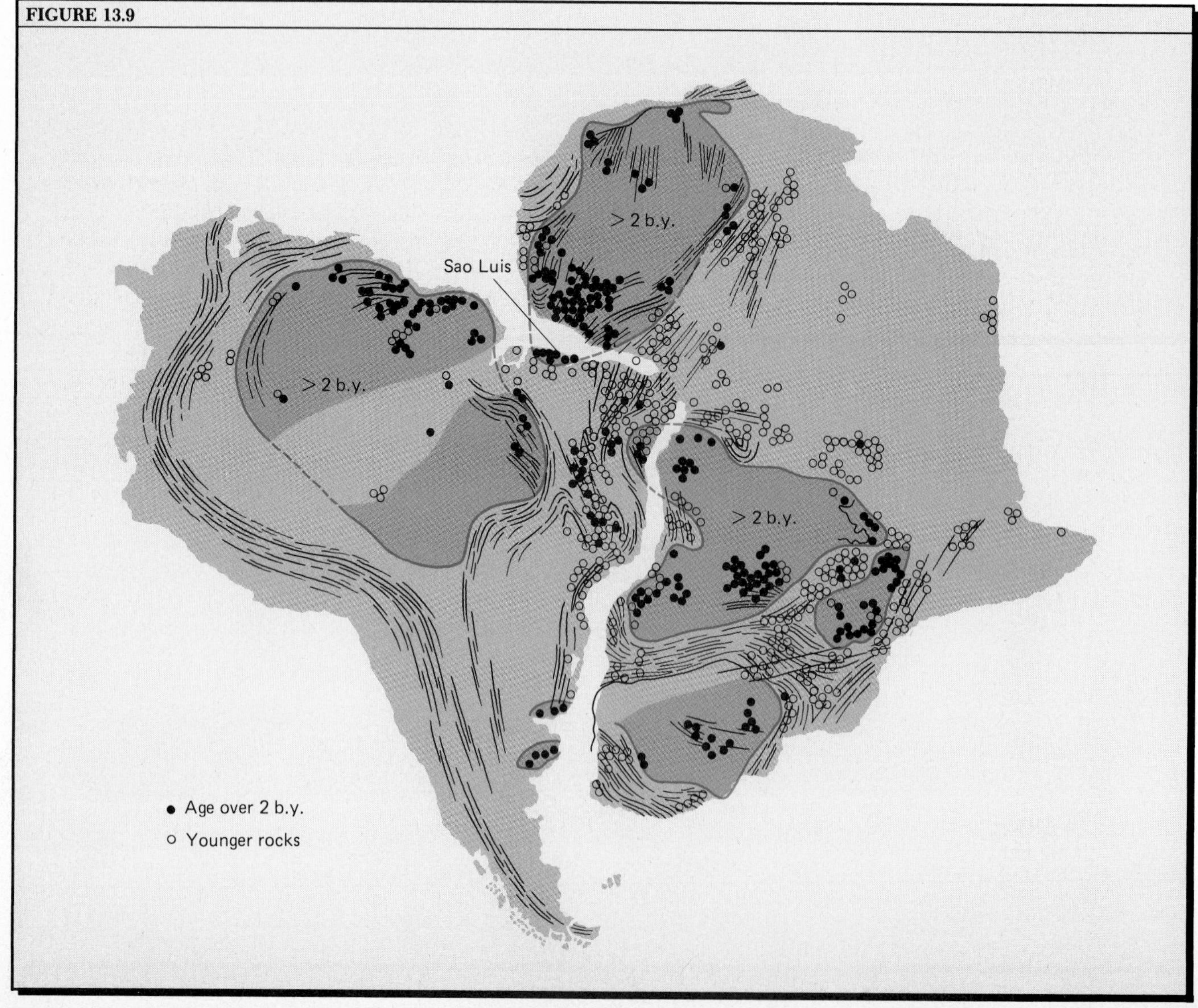

capable of swimming a wide ocean. Wegener was quick to realize that the finding of mesosaurs in the Permian strata of the several Gondwana landmasses was strong evidence that they were formerly connected. On the other hand, it was argued by his opponents that land bridges then connected the Gondwana continents, enabling mesosaurs to move freely from one to the other. Today the land-bridge argument has little force, because we know from seismic data that there are no large sunken areas of continental crust beneath the ocean floors. Wegener also noted that the Permian strata of all Gondwana landmasses contain a similar assemblage of plant fossils, seed ferns belonging to the *Glossopteris* flora. The wide distribution of this flora seemed to strengthen the case for a single continent.

The Triassic strata of the Gondwana landmasses consist of clastic rocks—sandstones and shales—containing fossil remains of a mammallike reptile of the genus *Lystrosaurus* (Figure 13.11), a small animal with massive, wide-set legs. Fossil remains of *Lystrosaurus* are abundant in Triassic strata of southern Africa and also are found in India, Russia, and China. In the late 1960s, as the case for plate tectonics gained strength, paleontologists began an intensive search for *Lystrosaurus* fossils in Antarctica, for it was considered impossible for this animal to have migrated to Antarctica across the broad, deep ocean basin that now separates Antarctica from the other continents. The search met with success in December of 1969 when remains of *Lystrosaurus* were found in the Transantarctic Mountains about 650 km from the south pole. This fossil find was hailed as one of the most significant in modern times, for it threw paleontologic evidence strongly in favor of the existence of a single continent of Gondwana as late as the Triassic Period.

FIGURE 13.10 A reconstruction of *Mesosaurus*, a small aquatic reptile that lived in southern Africa during late Carboniferous and early Permian times. (From E. H. Colbert, *Evolution of the Vertebrates*. Copyright © 1955 by John Wiley & Sons, Inc., New York. Reprinted by permission of John Wiley & Sons, Inc.)

Another line of stratigraphic evidence was developed by Wegener through the interpretation of global climate zones prevailing at the time of deposition of platform sediments. *Paleoclimates*, which are climates of past geologic periods, are thought to have followed global patterns similar to those on the earth today. An equatorial belt of warm, wet climate is flanked by two great subtropical desert belts in the latitude range from 15° to 30°. Wegener reasoned that coal layers of Permian age would have been formed in the equatorial zone, where lush swamp forests would have grown. He supposed that Permian strata consisting in part of salt beds (halite) and gypsum would have been formed in the belt of the subtropical deserts. Sandstones representing lithified dunes of desert sand would also occur in this zone. When Wegener plotted the locations of these various kinds of sedimentary deposits on his map of Pangaea, he was able to show that they fell approximately into the expected latitude zones, but only if the position of the earth's polar axis and equator were rotated through about 30° latitude (Figure 13.12). The position of the south pole would then have fallen in the southern part of Gondwana, where a great glaciation was in progress through late Carboniferous and early Permian time.

The Glaciation of Gondwana

One of the most remarkable geological displays to be seen anywhere is the Dwyka tillite of South Africa, a formation of late Carboniferous age which is a lithified glacial till. It contains boulders that clearly show the effects of strong abrasion beneath moving ice, because they display scratches, called ***striations***, a distinctive mark of glacial action. Beneath the tillite is a floor of igneous rock that shows strong abrasion by the overriding ice (Figure 13.13). In Wegener's time, the evidences of a great glaciation in the Carboniferous and the Permian periods were known from other parts of Gondwana, including South America, India, and Australia. You will notice that on Wegener's world map (Figure 13.12) the places where this evidence was

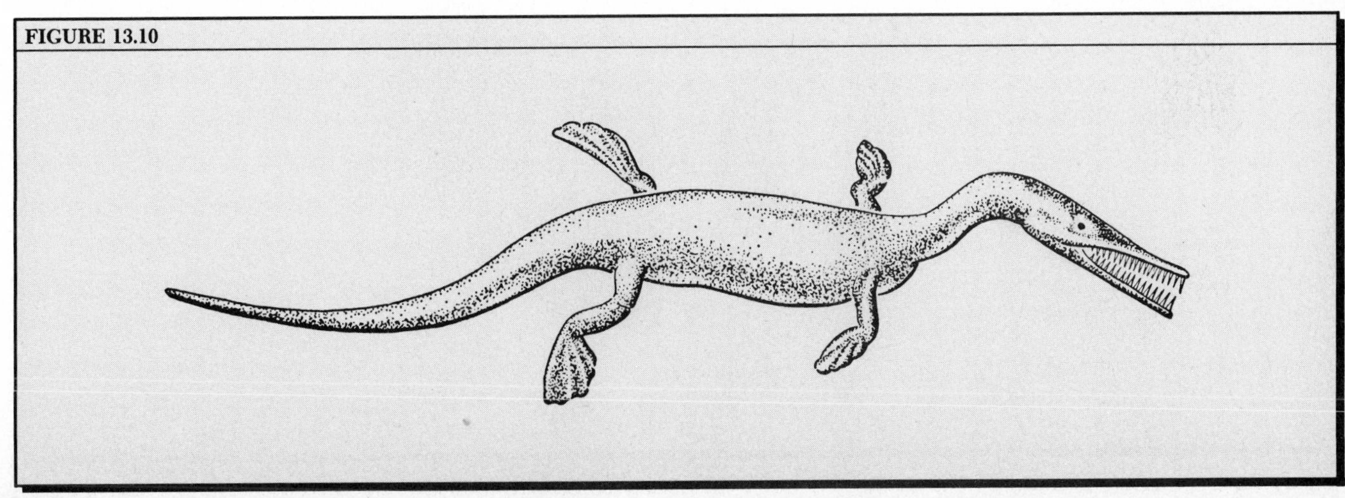

found are shown as clustered around the position of the south pole in late Carboniferous time.

The glacial argument for a united continent was strong, because it was difficult to explain how great ice sheets of subcontinental size could have formed near the margins of widely separated fragments of Gondwana. We know that a great ice sheet, several thousand meters thick at its center, requires a large landmass situated at high latitude, and these conditions were satisfied by a united Gondwana, but not by widely separated continents near their present latitudes.

Figure 13.14 is a reconstruction of Gondwana and Laurasia in the Carboniferous Period, published in 1979. Shown on this map are localities where evidence of glacial action has been found, along with arrows to show the inferred directions of ice movements as interpreted from striations on the bedrock floor on which the tillites rest.

We have also shown the polar wandering path as calculated from paleomagnetic data. Notice that the

FIGURE 13.11 This partial skeleton of *Lystrosaurus*, a mammallike reptile of Triassic age, measures about 1 m in length. (Sketched from a photograph.)

FIGURE 13.12 Wegener's maps showing climate and sedimentary deposits of the united landmass of Pangaea during the Carboniferous and Permian Periods. (From A. Wegener, 1929, *The Origin of Continents and Oceans*, translated from the 4th revised German ed., 1929, by J. Biram, Methuen, London, Figures 35 and 36.)

FIGURE 13.11

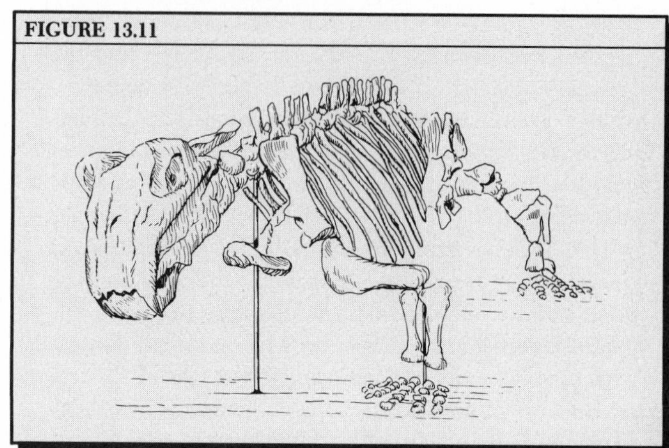

FIGURE 13.12

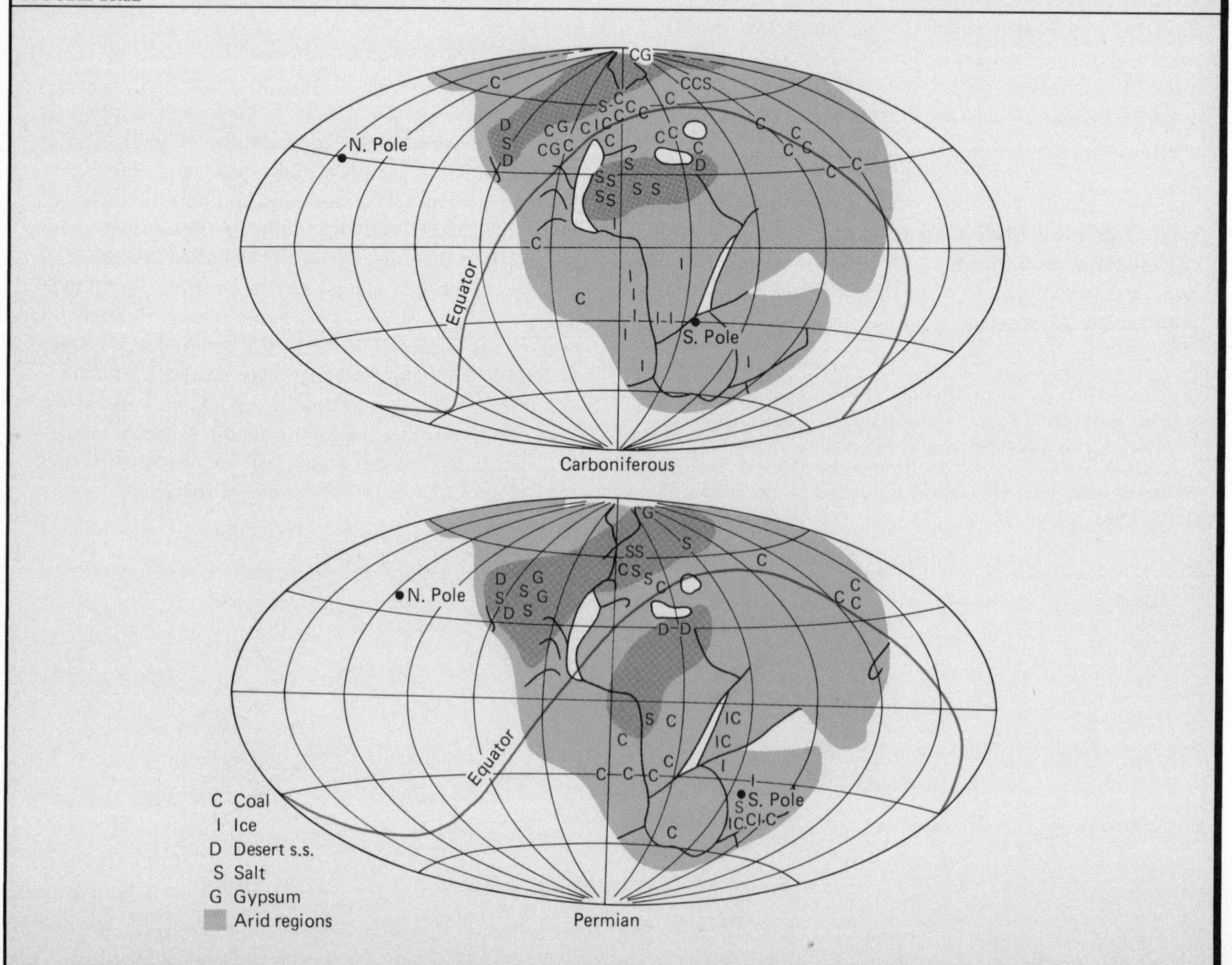

FIGURE 13.13 This surface of mafic igneous rock shows the scorings made by a glacier of Carboniferous time in South Africa. The ice moved in a direction away from the observer, riding up the sloping surface. The deep intersecting cracks are joint fractures enlarged by weathering. (Douglas Johnson.)

FIGURE 13.13

pole was located in western Africa in Ordovician time, moved to a position near the southern tip of Africa in the Devonian, and by the Carboniferous had reached Antarctica. Thus, the data agree remarkably well to reveal a large land area near the pole at the time the glacial till was deposited.

If the pole lay in northwest Africa in Ordovician time, there might be evidence of glaciation in that area in rocks of early Paleozoic age. As a matter of fact, in 1961, geologists had found in the Sahara Desert the distinctive markings of a glaciation—striations and grooves—in rocks of Upper Ordovician age. In 1970, an international expedition traveled to the

FIGURE 13.14 Restoration of the nested continents of Gondwana and Laurasia in Carboniferous time, showing the areas probably covered by an ice sheet, inferred directions of ice motion, and the locations of coal seams, red beds, and evaporites. (From *Earth History and Plate Tectonics* by C. K. Seyfert and L. A. Sirkin. Copyright © 1979 by Carl K. Seyfert and Leslie A. Sirkin. By permission of Harper & Row, Publishers, Inc.)

FIGURE 13.14

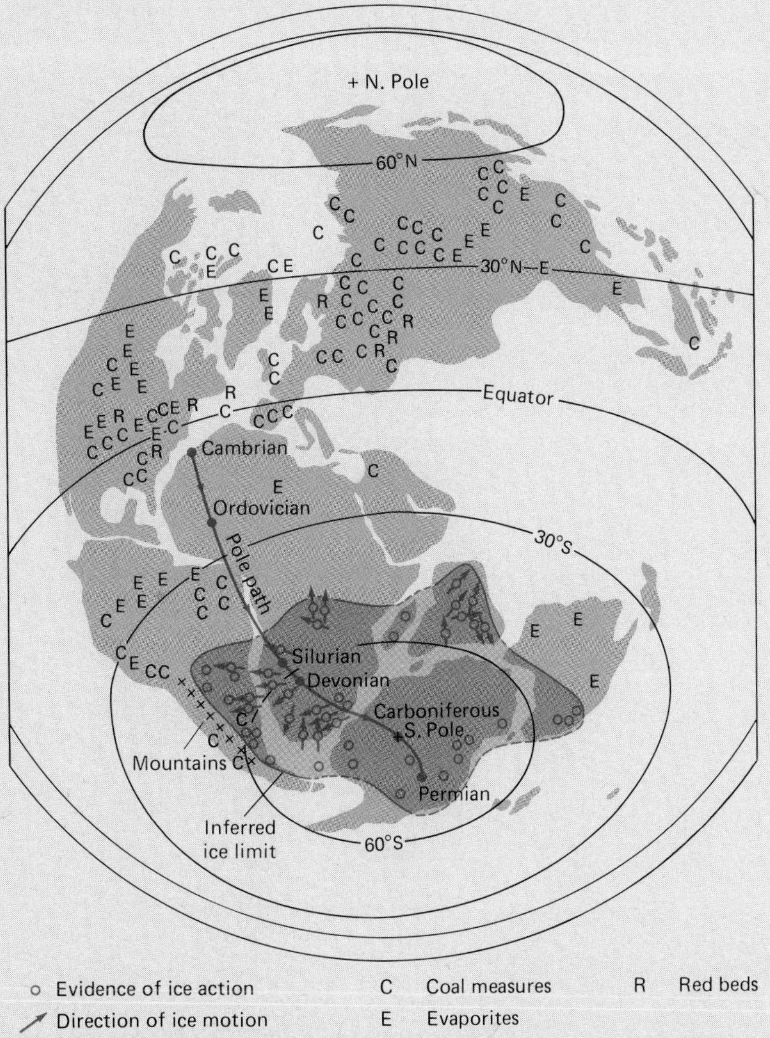

o Evidence of ice action
↗ Direction of ice motion
C Coal measures
E Evaporites
R Red beds

central Sahara to evaluate this evidence. The visitors were convinced that a great glaciation had, indeed, occurred here, for not only were the glacial striations found, but also tillites and many other forms typical of such a glaciation.

Although the pole position was also favorable for glaciation of Gondwana in the Silurian and Devonian periods, there is no evidence of it in rocks of that age. For reasons not known to geologists, glaciation did not recur until the Carboniferous Period.

The Breakup of Pangaea

The breakup of Pangaea probably began in earliest Jurassic time, about −205 m.y. One reason to use this date is that the polar wandering paths of Africa and North America begin to diverge just after this time. Figure 13.15 shows the breakup of Pangaea and the changing pattern of separated continents at intervals to the present. Laurasia began to separate from Gondwana in the earliest phase of the opening of the North Atlantic Ocean basin. South America and Africa remained in contact through the Triassic Period, but India, Antarctica, and Australia were beginning to separate in an early phase of the opening of the Indian Ocean basin. The Tethys Sea was beginning to close by

FIGURE 13.15 Five stages in the breakup of Pangaea to form the modern continents. Arrows indicate the directions of motion of lithospheric plates. The continents are delimited by the 1800-m submarine contour in order to show the true extent of continental lithosphere. (Redrawn and simplified from maps by R. S. Dietz and J. C. Holden, 1970, *Journal of Geophysical Res.*, vol. 75, pp. 4943–4951, Figures 2–6. Copyrighted by the American Geophysical Union.)

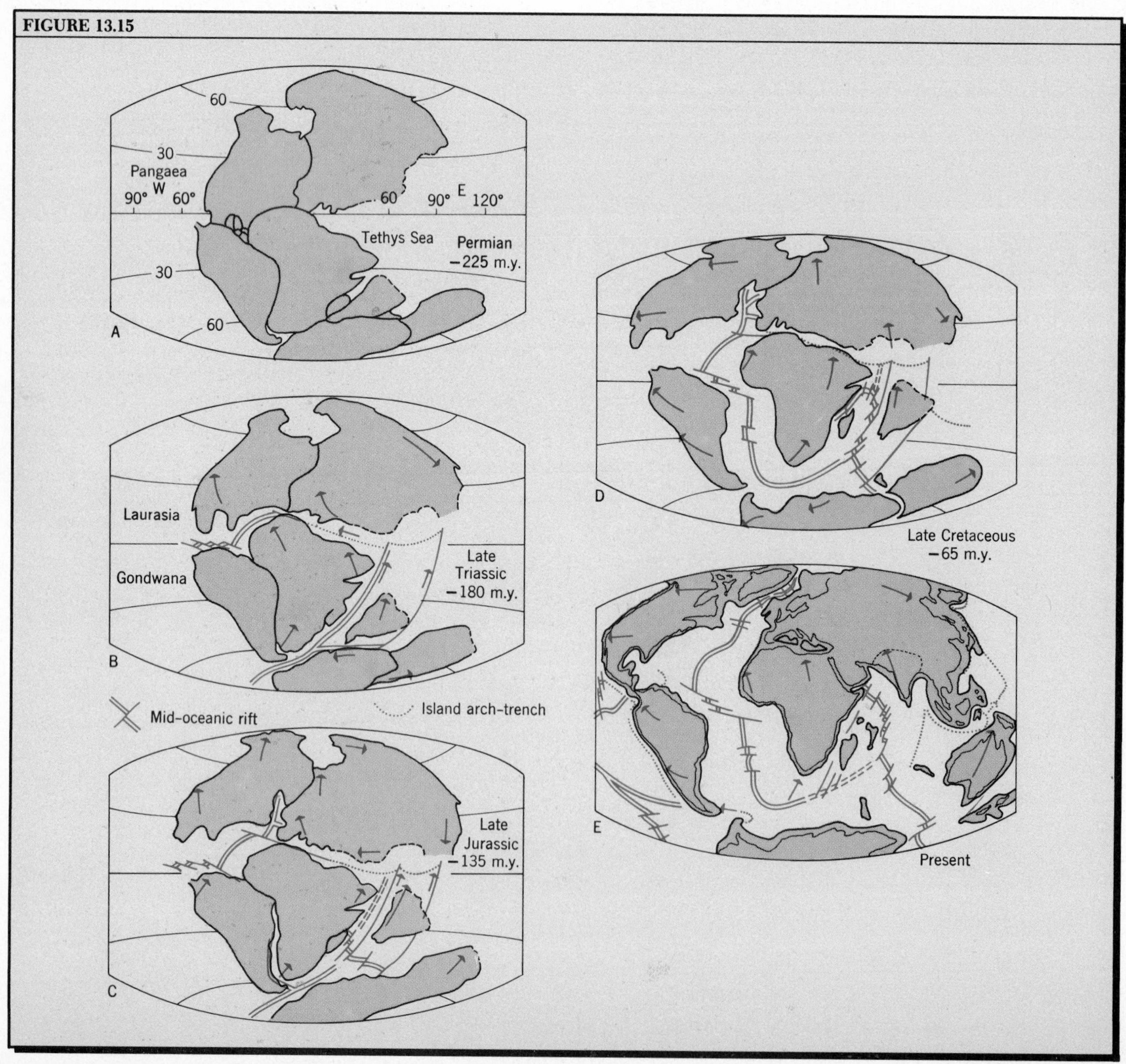

a rotating motion of Eurasia, pivoting on the point of contact of Eurasia and Africa in the western Mediterranean.

By the end of Jurassic time, −135 m.y., the North Atlantic basin was rather broadly opened up, and India was making its way rapidly northward toward Asia. The Indian continental plate seems to have moved independently of the Australian plate, although the two are now usually regarded as one major plate. Two major transform faults bounded the sides of the Indian plate as it traveled rapidly northward away from the spreading boundary of the widening Indian Ocean basin. The trace of these transform boundaries is seen today in two long, narrow features of the Indian Ocean—the Owen Fracture Zone and the Ninety East Ridge. Subduction of the oceanic portion of the plate north of India was continually taking place, so that an enormous volume of oceanic lithosphere was consumed before the Indian continental fragment finally collided with the Eurasian plate. In the late Jurassic, separation of South America and Africa had just begun, starting at the south and working north. By the end of Cretaceous time, −65 m.y., South America was well separated from Africa; the wide opening between them formed the South Atlantic basin. Important events of the Cenozoic Era have been reviewed in Chapter 12. These included the collision of Africa and India with Eurasia and the Cordilleran Orogeny that affected western North America.

Now that we have completed our investigation of Pangaea and its breakup, we can turn to some basic questions of plate tectonics. Why did Pangaea break apart? What mechanisms were involved in this process?

FIGURE 13.16 Domes formed by lifting of the lithosphere over mantle plumes are fractured into three-armed tension rifts. (Drawn by A. N. Strahler.)

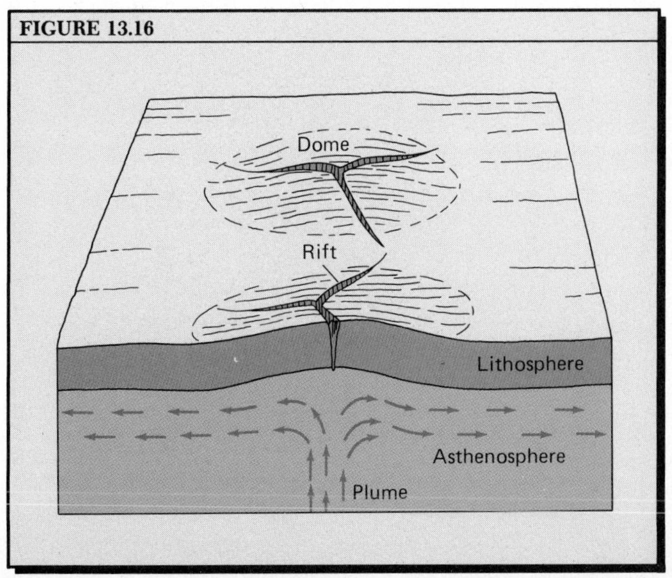

The Mechanics of Continental Rifting

What causes a single continental plate to break apart? Any brittle plate that is very thin in comparison with its horizontal dimensions will yield to a hammer blow by fracturing, so we might envision impacts being delivered from above by the infall of solid objects of the solar system. But in the case of lithospheric plates floating on a plastic asthenosphere, the most commonly used working hypothesis for continental plate breakup involves rising mantle plumes that lift the lithosphere in a doming action. Doming of the plate causes tension that expresses itself in widening fractures, or rifts.

As shown in Figure 13.16, the typical fracture pattern of a dome is a three-branched tension rift system. A four-branched fracture system is thought to have occurred in some instances, but the triple fracture is considered the most common type. Geologists envision the simultaneous formation of an irregular line of domes (Figure 13.17). One rift extending out from a dome is approximately aligned with a rift from the next dome in the chain, so that the rifts become connected as a single continuous rift line that zigzags across the plate.

Because the plate has been lifted along the continuous rift system, ridge-push forces are set up (Chapter 11 and Figure 11.9). These act on both sides of the rift in opposite directions. For the two plate sections to begin moving apart, subduction must occur at the far side of at least one of the two plate sections. Perhaps a new subduction zone forms in oceanic lithosphere, and horizontal plate motion is started.

As we showed in Figure 1.10, the continuous rift now begins to widen and becomes a deep trench, or grabenlike feature, often called a *rift valley.* Basaltic lava rises in the floor of the rift valley, appearing in the form of basaltic volcanoes and lava flows. Basalt dikes fill vertical fractures in the crust below, and other dikes radiate outward from the dome center. Sections of crust that form the sides of the rift valley fail and collapse along normal faults paralleling the trend of the rift. The fault blocks sink down and rotate to lower positions in the widening rift valley.

As plate-spreading continues, ocean water invades the widening rift from one end or the other, forming a narrow ocean. In early stages, however, parts of the rift may remain blocked off from ingress of ocean water. Stages in widening of the new ocean basin are depicted in Figure 1.10. As the two margins of the continental plates recede from the axis of spreading, the plate is cooled and gradually sinks to a lower elevation. In Chapter 12, we explained that this slow subsidence of the plate margins allows the thickening of miogeoclinal and eugeoclinal sediment wedges that develop on both sides of the ocean basin.

An alternate hypothesis to continental rifting by rising mantle plumes switches cause and effect completely around. It envisions the rise of domelike mantle protrusions at hot spots as the effect, rather than the cause, of splitting of a continental lithospheric plate. Careful analysis of the history of continental rifting of Pangaea shows that a particular rift began to appear at the perimeter of the supercontinent. As the rift widened, it was propagated (extended) into the heart of the continent, eventually meeting another rift that had penetrated from a different marginal point.

We can imagine that as a rift penetrates the continent, it abruptly changes direction to the right or left because of the presence of vertical zones of weakness in the lithosphere that trend at various angles to the lengthening rift. In this way, the lengthening rift might follow a zigzag path. At each point of direction change, a secondary rift might be expected, forming a triple rift. Updoming would be induced by the widening triple rift, since magma must rise to fill the gap, and this rise would set off an upward movement in the asthenosphere below, perhaps through the triggering of a deep-seated phase change in the mantle rock. While this hypothesis seems to explain the triple rifts and their domes, it leaves unexplained the cause and mechanism that forced the rifting to begin in the first place.

Failed Arms, Deltas, and Aulacogens

Going back to the initial rifting that sets off plate separation, we note that the connected rift system makes use of only two of the three radial fractures formed at each dome. The third fracture is not used, but simply extends back into the continental plate for some distance. This unused fracture is called the *failed arm* of the triple rift junction. As the plates recede, the failed arm becomes the site of a downdropped crustal block that may be either a graben with two boundary faults, or a downtilted fault block with a major normal fault on only one side. In either case, it forms a trenchlike feature that leads out from the continental interior to the new coastline of the continent, thus making the failed arm a natural site for a major river.

The river mouth occupies a V-shaped indentation in the plan of the coastline where the sediment carried by the river accumulates in a delta. (Deltas are described in Chapter 16.) But because sediment also accumulates in the floor of the failed arm, downsinking

FIGURE 13.17 Schematic maps of the opening of an ocean basin. *A*. Chain of lithospheric domes with three-armed rifts. Rifts are connected in a zigzag line. *B*. Plate spreading and formation of a new ocean basin. Failed arms become downfaulted troughs (aulacogens) with thick sediment accumulations and the construction of large deltas. (Drawn by A. N. Strahler.)

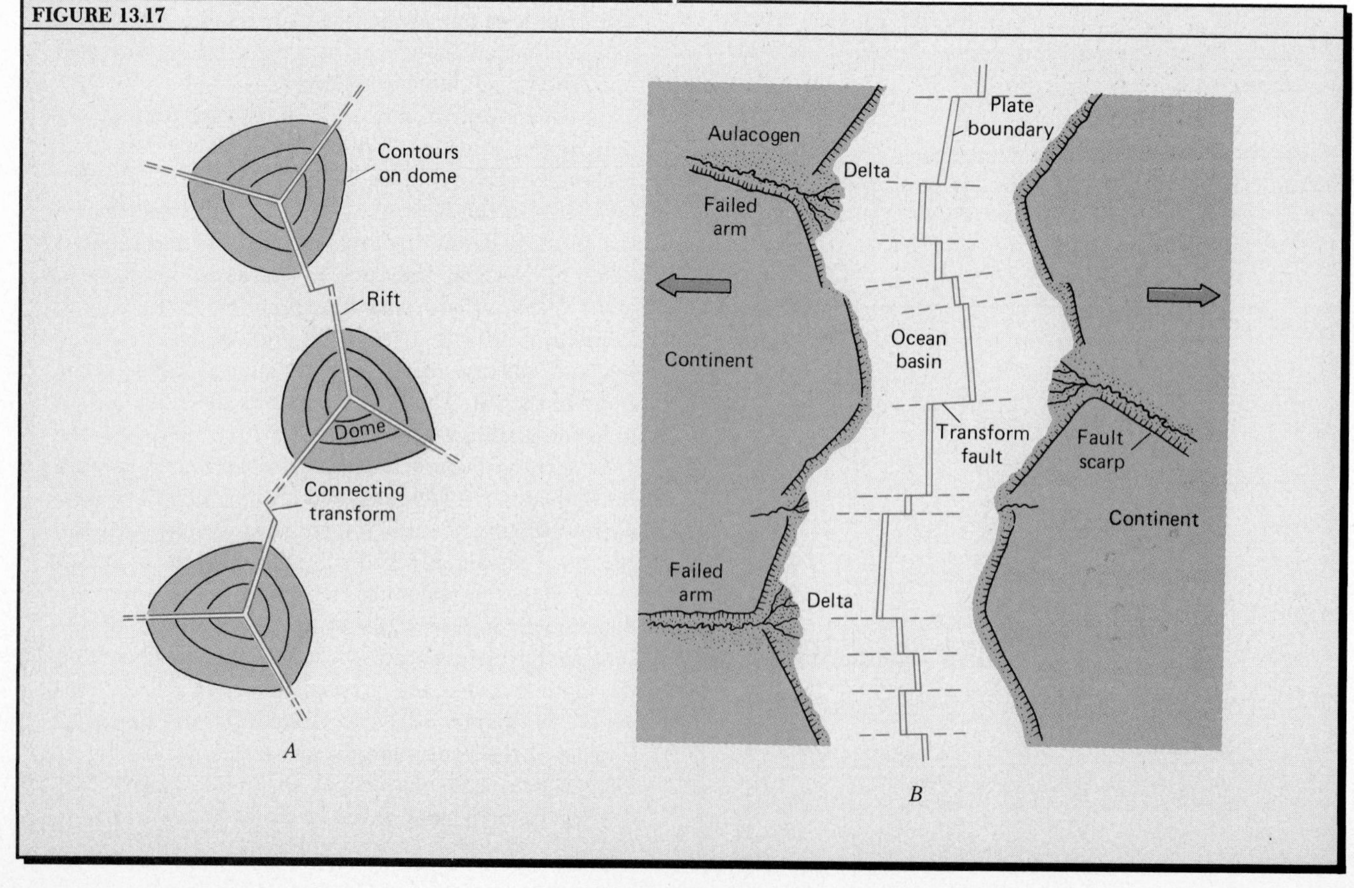

of the floor of the failed arm allows more and more sediments to accumulate until they may reach a thickness of several kilometers. This deposit is a type of geosyncline that has been given the name *aulacogen* (see Table 12.1); it was recognized as a special form of sedimentary trough of the continental crust as early as the early 1940s. The name had been coined from Greek words that mean "born of furrows," but the origin of the "furrow," or trough, was not understood at that time in the framework of plate tectonics.

Stages of Ocean-Basin Opening and Closing—The Wilson Cycle

Professor J. Tuzo Wilson, whose pioneer contributions to the concepts of hot spots and transform faults we described in earlier chapters, recognized a series of stages in the rifting of a continental plate followed by opening of an ocean basin. He followed this series with stages in the closing of the same ocean basin, leading to continental collision and orogeny (Figure 13.18). The full cycle of opening and closing of an ocean basin has since been named the *Wilson cycle,* honoring the contribution made to plate tectonics by this distinguished Canadian geologist.

Wilson recognized the following stages in his tectonic cycle:

- **Stage 1:** Rift-valley system, the embryonic stage in continental rifting (examples: the East African rift-valley system)
- **Stage 2:** Narrow ocean gulf, the formation of a young ocean basin (example: the Gulf of Aden)
- **Stage 3:** Wide ocean basin, fully developed and mature (example: the Atlantic Ocean basin between Europe and North America)

FIGURE 13.18 Schematic diagrams of stages in the Wilson cycle of opening and closing of an ocean basin. *Stage 1.* Continental rifting begins, forming a rift valley that is the embryonic ocean basin. *Stage 2.* Young stage; a narrow ocean gulf. *Stage 3.* Mature stage; a wide ocean basin with passive continental margins on both sides. *Stage 4.* (*a*) Closing is initiated by the formation of new subduction boundaries in the oceanic lithosphere, close to the continental margins. (*b*) Cordilleran-type orogens are formed at the continental margins as the ocean basin narrows. *Stage 5.* As subduction continues and the basin becomes narrower, accretionary prisms are added to the continental plates. *Stage 6.* Continental collision produces a Eurasian-type orogen above a suture, ending the cycle. (Drawn by A. N. Strahler. Interpreted from descriptions in J. A. Jacobs, R. D. Russell, and J. T. Wilson, 1974, Chapter 15, The life cycle of ocean basins, *Physics and Geology,* 2d ed., McGraw-Hill, New York, pp. 397–470.)

FIGURE 13.18

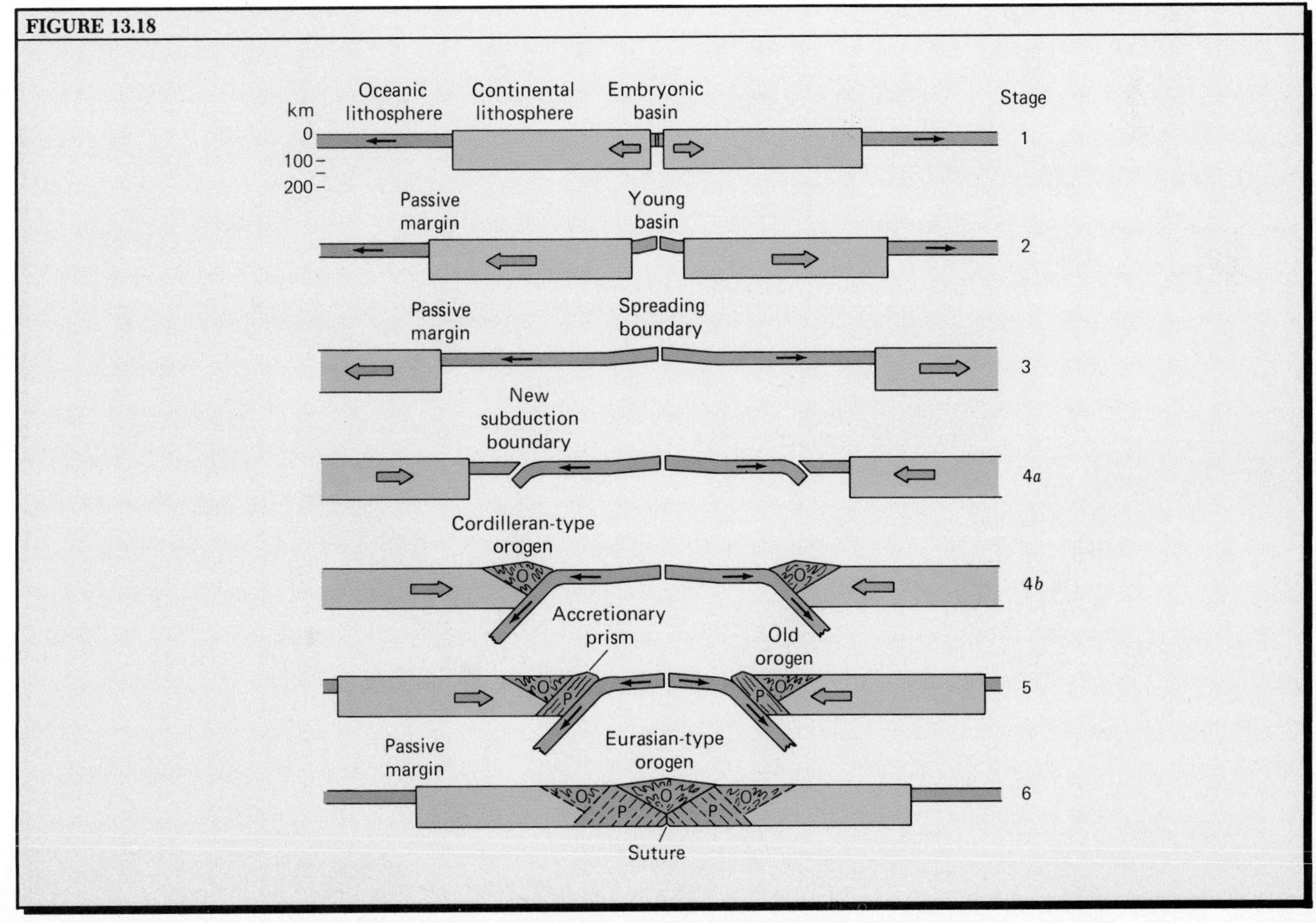

- **Stage 4:** Shrinking ocean basin, the beginning of ocean-basin closing (example: the Pacific Ocean basin)
- **Stage 5:** Narrow ocean basin, the late stage of ocean-basin closing (example: the Mediterranean basin)
- **Stage 6:** Continental collision, ocean-basin closing complete (example: the Himalayan arc and Tibetan Plateau)

Stages 4, 5, and 6 are familiar, as we have already covered them in Chapters 10, 11, and 12. These are the waning stages in the Wilson cycle, involving subduction boundaries and leading to orogeny. We will, however, need to cover in detail the first two stages, and mention the third stage briefly.

*The East African Rift-Valley System (Stage 1)

The East African rift-valley system has attracted the attention of geologists since the early 1900s. The term *rift valley* was first used in 1920 to describe the structural elements of this faulted area in eastern Africa (Figure 13.19). The overall length of the system, which runs from the Red Sea on the north to the Zambezi River on the south, is a full 3000 km. The system consists of a number of grabenlike troughs, each a separate rift valley ranging in width from 30 to 60 km. As geologists had noted in earlier field surveys of this system, the rift valleys are like keystone blocks of a masonry arch that have slipped down between neighboring blocks because the arch has spread apart somewhat. Thus, the floors of the rift valleys are above the elevation of most of the African continental surface, even though some of the valley floors are occupied by long, deep lakes and by major rivers (Figure 13.20). The sides of the rift valleys may consist of multiple fault steps (Figure 13.21 and Color Plate A.2).

The rift-valley system consists of a number of domelike swells in the crust, the highest of which forms the Ethiopian Highlands on the north. Basalt lavas have risen from fissures in the floors of the rift valleys and from the flanks of the domes. Sediments,

FIGURE 13.19 Present and future tectonics of eastern Africa. *A.* Sketch map of the East African rift-valley system and its relationship to the Red Sea graben. *B.* Plate tectonics of the same region. Continental rifting is just beginning in East Africa, where the rift-valley system is shown by a spreading boundary symbol. *C.* A prediction of the tectonic map 50 m.y. from now. The Somalia region of Africa has become an independent tectonic plate, moving north. A new ocean basin is opening up in East Africa, while the Red Sea has widened considerably. (Based on data of R. S. Dietz and J. C. Holden, 1970, *Scientific American*, vol. 223, no. 4.)

FIGURE 13.19

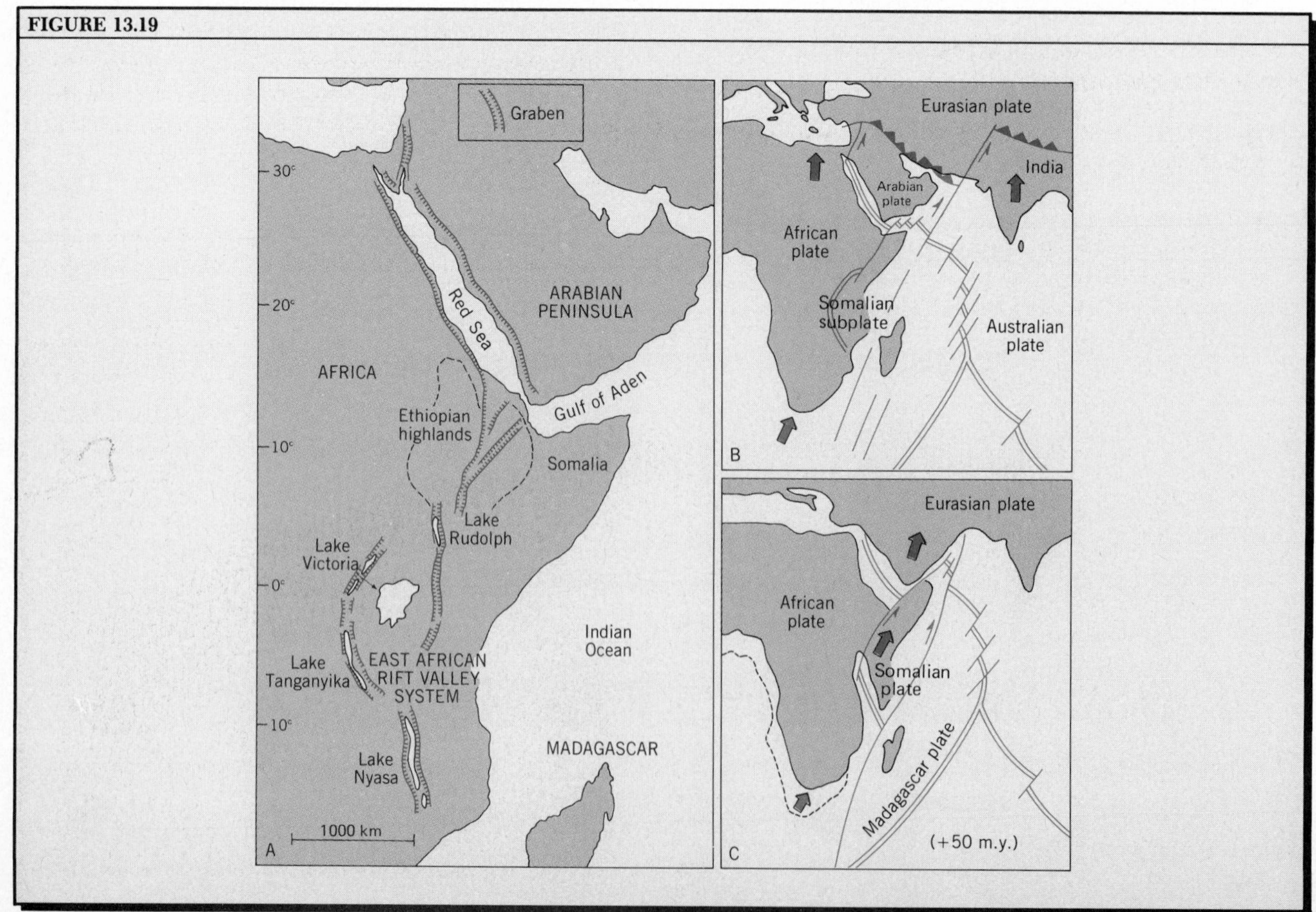

derived from the high plateaus that form the flanks of troughs, make thick fills in the floors of the valleys. Lake Victoria is flanked by two rift valleys, which join south of the lake. A single rift valley extends southward from this junction, and thus there may be a triple junction at this point, as in our idealized model of a three-armed rift.

Tectonic and volcanic activity that produced the modern rift system began in this region in Eocene time and has continued sporadically ever since. The speculative suggestion has been made that the rift system will eventually become the boundary of a detached lithospheric plate, the Somalian plate. According to the interpretation of future relationships shown in Figure 13.19, a spreading plate boundary will open up along the Lake Nyasa rift valley, making a new ocean basin, while a transform boundary will follow the northern part of the rift system to the Gulf of Aden. The entire plate will travel northeastward, moving past the Arabian plate.

FIGURE 13.20 Development of a typical rift valley in East Africa. The diagrams are schematic, combining elements found in several localities. Width of the area shown is about 150 km. A. Late Miocene and early Pliocene. Normal faulting has produced a tilted fault block on the left. Crust at right is deformed into a broad monocline with a cap of lava. B. Late Pliocene. Renewed normal faulting has broken the valley floor into narrow blocks and raised the eastern side. The rift valley is now a graben structure. Lava flows have filled the valley floor. C. Pleistocene and Holocene. After another episode of minor faulting, extrusive activity has built volcanoes in the rift valley and on the flank of the uplift. (Drawn by A. N. Strahler. Based on data of B. H. Baker, in *East Africa Rift System*, 1965, UNESCO Seminar, University College, Nairobi, p. 82.)

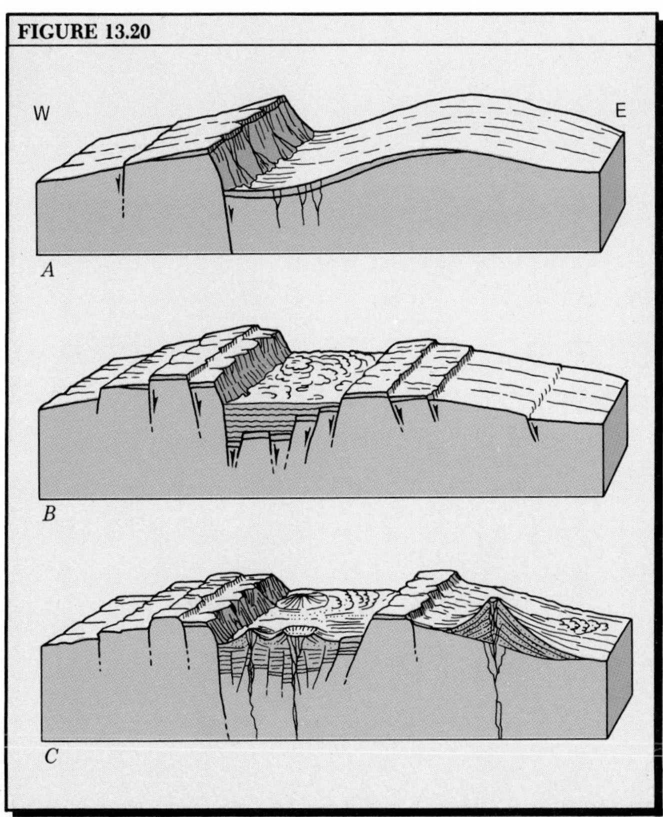

The Gulf of Aden and the Red Sea (Stage 2)

Lying between the Arabian Peninsula and the mainland of Africa are two narrow ocean troughs, the Gulf of Aden and the Red Sea (Figure 13.22). The two troughs are oriented almost exactly at right angles to one another, forming a great "L." Take particular note of a triangular lowland area on the mainland of Africa at the point of junction of the two troughs; it is the Afar Triangle. The steep western wall of the Red Sea Trough and the steep northern wall of the Gulf of Aden Trough bound the Afar Triangle on two sides, meeting in a right-angle junction just east of Addis Ababa. The nearly circular dome shape of the Ethiopian Highlands is thus missing its northeastern quarter. Looking across to the Arabian Peninsula, you will find the missing sector—it is represented by the highlands of Yemen. Thus, on the basis of topography alone, the rifting of a continental plate to form an L-shaped oceanic trough is most convincing.

Fortunately, there is abundant scientific evidence that the Gulf of Aden and the Red Sea are indeed new ocean basins. Seismic data show that the floors of the troughs are made up of oceanic crust. The troughs have a median axis along which earthquake epicenters are concentrated. The Gulf of Aden shows a symmetrical magnetic anomaly pattern and several faults offsetting the spreading axis; analysis of first motion on the faults shows them to be true transform faults. A high

FIGURE 13.21 Rift valley wall in Ethiopia. Multiple fault scarps give the landscape a stepped appearance. (George Gerster/Photo Researchers, Inc.)

rate of heat flow characterizes the floor of the troughs—on the Red Sea floor at a depth over 1800 m, about opposite the city of Mecca, local pockets of hot brine have been found. These pockets show exceptionally high salinity, nearly seven times greater than that of normal seawater.

Sediments directly beneath the hot brines were also found to be saturated with the brine and to contain remarkably high concentrations of sulfide minerals rich in iron, zinc, and copper, with smaller amounts of lead and silver, and even a minute quantity of gold. The hot brine, with temperatures as high as 60 °C, is interpreted as a hydrothermal solution rising from highly heated rock above a magma chamber under the axial rift. The hydrothermal solution represents seawater that has penetrated the ocean floor, moving down through permeable sediment and volcanic rock to reach the heat source. The heated water then returns to the surface charged with ions dissolved from the altered mafic igneous rock.

There seems to be little doubt that the Gulf of Aden and Red Sea Troughs are young ocean basins in the process of widening as the Arabian plate moves northward. In Chapter 12, we described the collision boundary between the Arabian plate and the Eurasian plate in the region of the Zagros Mountains. The L-shaped junction of the two troughs at the Afar Triangle can be interpreted as a triple junction of three plates: Arabian, African, and Somalian. The failed arm runs southwest across the Ethiopian Highland as a narrow rift, but may in the future become an active plate boundary. The Afar Triangle contains thick salt beds and basalt lavas that represent materials filling the failed arm. These layered rocks are broken by innumerable normal faults (Figure 13.23).

At the northern end of the Red Sea lies the triangular Sinai Peninsula, bounded by two fault trenches. The eastern fault zone extends from the Gulf of Aqaba northward through the Dead Sea depression, and has been interpreted as a transform boundary of the Arabian plate. Displacement along this transform fault is estimated to be 105 km.

*The Atlantic and Indian Ocean Basins (Stage 3)

The Atlantic and Indian Ocean basins represent the mature stage of the Wilson cycle. The rifted plate segments of Pangaea have separated by a distance of roughly 3500 to 4000 km in the 200 m.y. that have elapsed since middle Triassic time, when rifting be-

FIGURE 13.22 Sketch map of the Afar Triangle and surrounding region. (Based on data of F. Barberi and J. Varet, 1977, *Geological Society of America, Bull.*, vol. 88, pp. 1251–1266.)

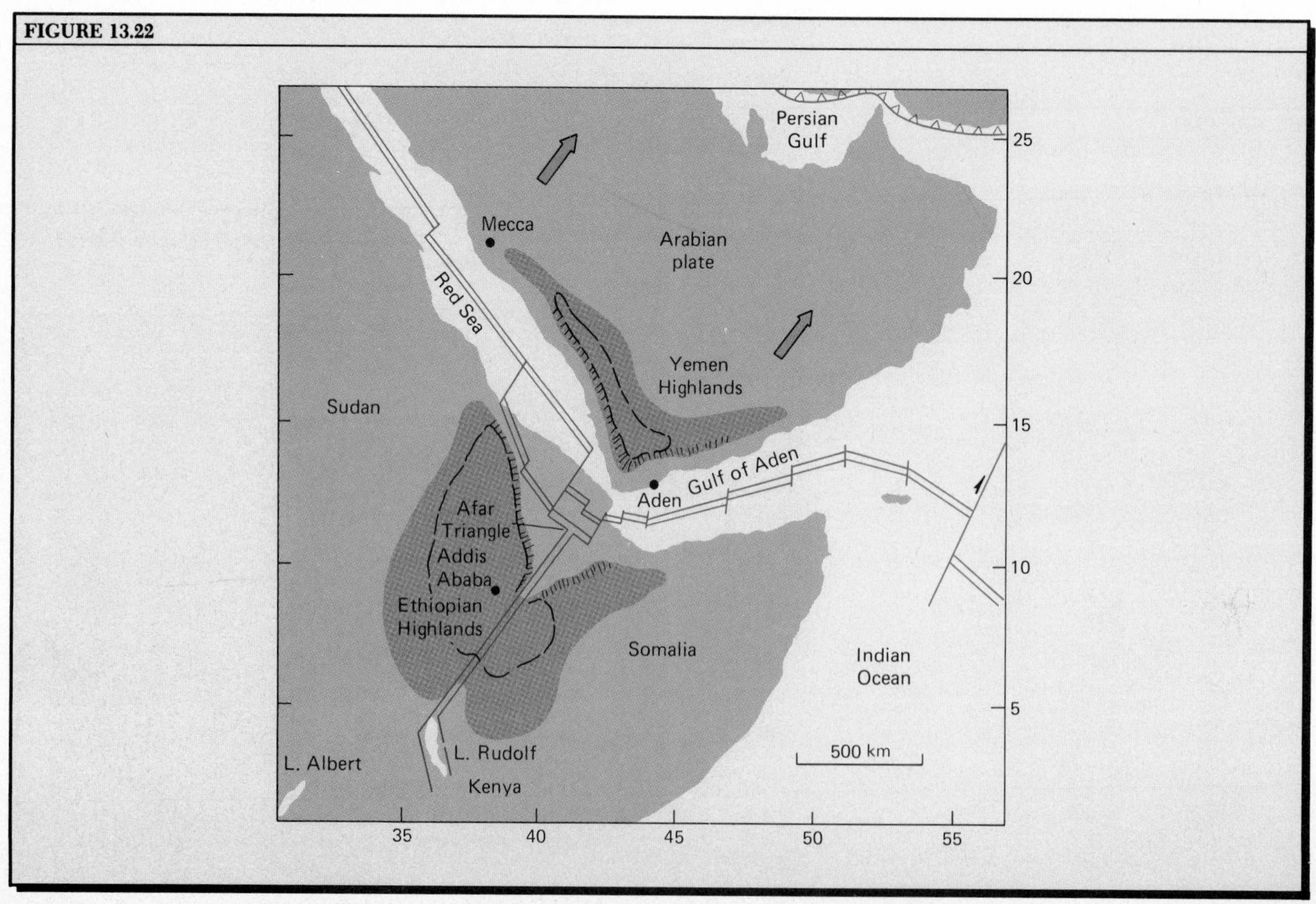

gan. The stable continental margins are, for the most part, now well endowed with thick geoclines made up of strata dating back as early as late Triassic or early Jurassic time. In Chapter 10, we described the geological character of these mature ocean basins. A feature of particular interest has been the formation of seamount chains and submarine ridges that are the tracks of hot spots over which the oceanic plates moved as spreading continued.

Closing of an Ocean Basin

The final stages of the Wilson cycle call for narrowing and closing of an ocean basin. In Stage 4, shown in Figure 13.18, new subduction zones have formed near the continental margins, and subduction of oceanic plates occurs at a more rapid rate than production of new oceanic lithosphere at the spreading boundary. As a result, the expanse of oceanic lithosphere diminishes and the continents begin to draw together. By Stage 5, only a narrow ocean basin remains. Actually, two opposed subduction boundaries are not necessary to accomplish the closing of the basin; one subduction zone is enough.

Stage 6, collision and suturing, is an event we have covered in some detail in Chapter 12. What we shall investigate here is an ancient opening and closing of an ocean basin that occurred in Paleozoic time. In Chapter 12, we listed as one component of continental structure the orogenic root systems produced by Paleozoic subduction and collision. Root systems of this type exist in eastern North America and western Europe where they have been intensively studied and mapped by geologists since the time of James D. Dana and James Hall, who interpreted these orogens as deformed geosynclines.

FIGURE 13.23 Multiple tilted fault blocks comprise the floor of a large graben structure in the Afar Triangle, between the Red Sea and the Ethiopean Plateau. These fault forms indicate that the crust has been stretched and thinned during the opening of the Red Sea as a new ocean basin. (Drawn by A. N. Strahler from a photograph.)

Caledonian and Hercynian Orogens

Long before plate tectonics came on the scene, and even long before Wegener proposed that a Pangaea existed, geologists had recognized two major orogenies that occurred in the Paleozoic Era, each preceded by one or more lesser orogenies. Each major orogeny was preceded by deposition of thick sediments in geosynclines, and each caused folding and thrust-faulting of the strata of a geosyncline.

The earlier of the two major orogenies, named the *Caledonian Orogeny,* occurred about the close of the Silurian Period in western Europe and produced a belt of close folds and thrusts seen today in Ireland, Scotland, and Scandinavia. Geologists refer to this tectonic belt as the *Caledonides.* In North America, the corresponding orogen runs through Newfoundland, Nova Scotia, and New England. Figure 13.24 shows the two orogenic belts as a single continuous belt at the close of the Paleozoic Era, when the continents were united in Pangaea.

The second of the two great orogenies is called the *Hercynian Orogeny* in Europe and the *Alleghenian Orogeny* in North America. In Europe, the Hercynian orogen runs from west to east through southern Ireland and Wales, northern France, Belgium, and Germany, trending at an angle to the Caledonian orogen, which runs in a more northerly direction. In North America, the Alleghenian orogen is a fold belt that can be traced through the Appalachian Mountains from southern New England, southwestward through Pennsylvania, Maryland, the Virginias, Tennessee, and Ala-

FIGURE 13.23

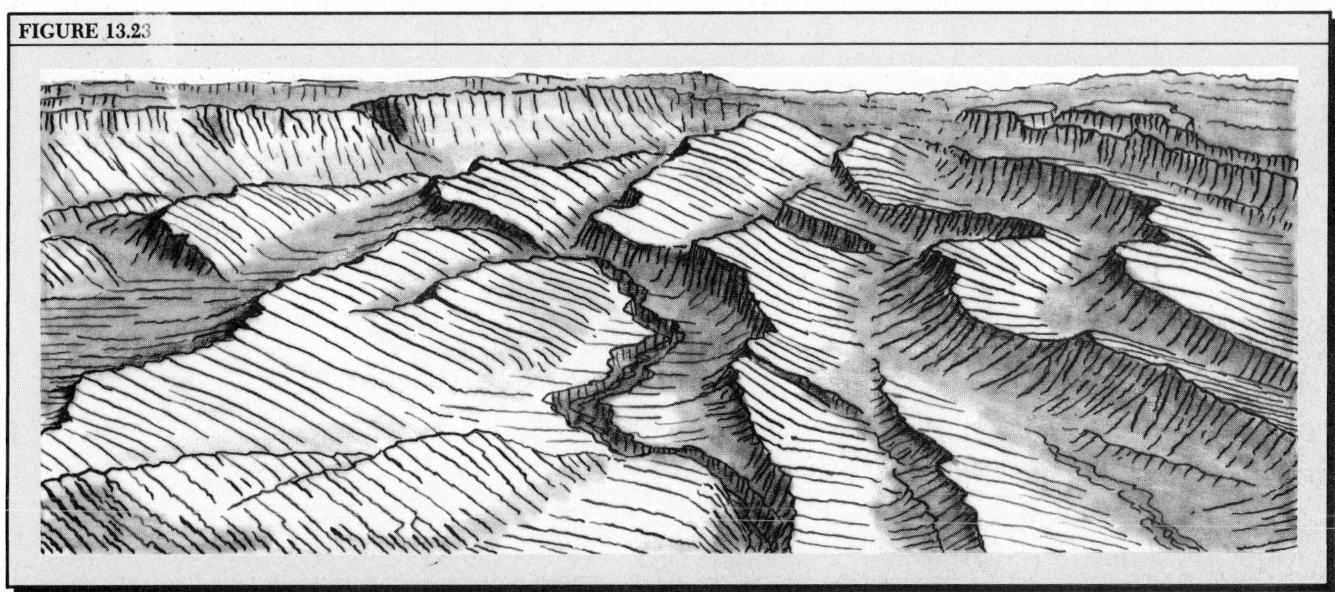

bama. Of the same age is the fold belt of the Ouachita Mountains in Arkansas and Oklahoma. (The gap between these two fold belts is concealed by Cretaceous and younger strata of the Gulf Coast miogeocline.) Strata of Paleozoic age involved in the Hercynian–Alleghenian Orogeny are as young as early Permian age, so the orogeny is placed in Permian time and closed the Paleozoic Era.

The two orogens we have described are clearly the result of two different tectonic events separated in time by about 150 m.y. In Chapter 12, we identified two basic kinds of orogenies: (a) the Cordilleran type, produced when a new subduction zone develops close to a stable continental plate margin, and (b) the Eurasian type, produced by collision of two continental plate margins. How can we interpret the Paleozoic orogenies in this framework of plate tectonics? Both kinds of orogenies require an oceanic plate and at least one active subduction boundary, as shown in Stages 4, 5, and 6 of Figure 13.18.

One conclusion is inescapable: Pangaea was not always a single continent. Instead, in Paleozoic time, it must have been divided in such a way that an ocean basin lay between what are now North America and Eurasia and another between those northern continents and the Gondwana continents of South America and Africa (Figure 13.25*A*). This ocean basin must have been in existence from the very beginning of the Paleozoic Era, which starts with Cambrian time, −570 m.y. Geologists have given the name *Iapetus Ocean* to the Paleozoic ocean between North America and Eurasia; it is thought to have opened up in late Precambrian time (about −700 m.y.) and to have begun closing early in the Paleozoic Era. During the early closing stages, new subduction boundaries developed along one or both margins of the separated continental plates. Geosynclines were formed as forearc and backarc troughs, with tectonic and volcanic arcs to provide sediment. By late Silurian time, the Iapetus Ocean had closed with a Eurasian-type collision in the northern section that is now the region of Greenland and Norway. Collision extended southward in the Devonian Period and by the close of Devonian time was complete, welding the two continental plates together (Figure 13.25*B*) in what we have called the Caledonian Orogeny.

This scenario of opening and closing of the Iapetus Ocean is grossly oversimplified, of course, but gives you a general idea of the way in which a great orogen—the Caledonides—came into existence in eastern North America and western Europe. What we see today in the Caledonide belt are only the root structures of a great alpine-type mountain range that must have existed in late Silurian and early Devonian time (Figure 13.26).

Closure of the Iapetus Ocean welded western Europe to northeastern North America to form the Laurasian continent. There remained, however, an ocean basin between Laurasia and Gondwana (Figure 13.25*B*). This ocean has received no special name, but the geosynclines that bordered it on continental margins are referred to as the *Appalachian Geosyncline* (in what is now North America) and the *Hercynian Geosyncline* (in what is now central Europe). Followed eastward, this ocean connected with a narrow

FIGURE 13.24 Sketch map of the Caledonian and Hercynian–Alleghenian orogens. The continents are shown as joined in Pangaea at the beginning of the Mesozoic Era. (Continental fitting according to Sir Edward Bullard, 1965.)

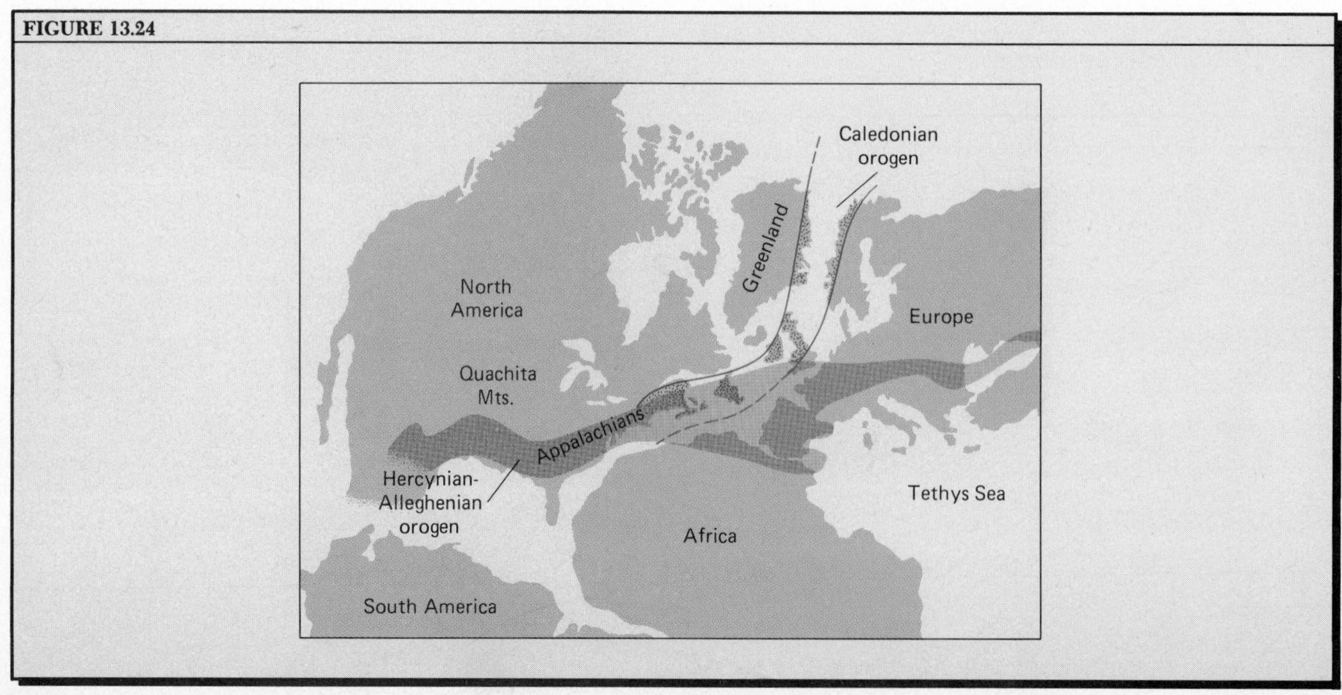

western gulf of the Tethys Ocean. An active subduction zone was probably present on one or both margins of the Appalachian–Hercynian ocean basin throughout nearly all of the Paleozoic Era. At intervals, a Cordilleran-type orogeny deformed strata of the bordering geosynclines, which were forearc and backarc troughs and miogeoclines. Then, toward the end of the Paleozoic Era, closing of the ocean basin began to take place (Figure 13.25*C*).

The terminal continental collision began in Carboniferous time and was completed in the Permian Period; it was the Alleghenian–Hercynian Orogeny. The roots of nappes produced in this collision have been identified in the Piedmont region of South Carolina (Figure 13.27). Foreland folds were produced in Paleozoic strata that had accumulated on the margin of the craton and in the inner zone of a miogeocline. Prior to the Carboniferous collision, these strata had been little affected by earlier orogenies, but now they were thrown into large open folds. The foreland folds

FIGURE 13.25 A schematic map–diagram showing the closings of Paleozoic oceans, bringing together Laurasia and Gondwana into the single continent of Pangaea.

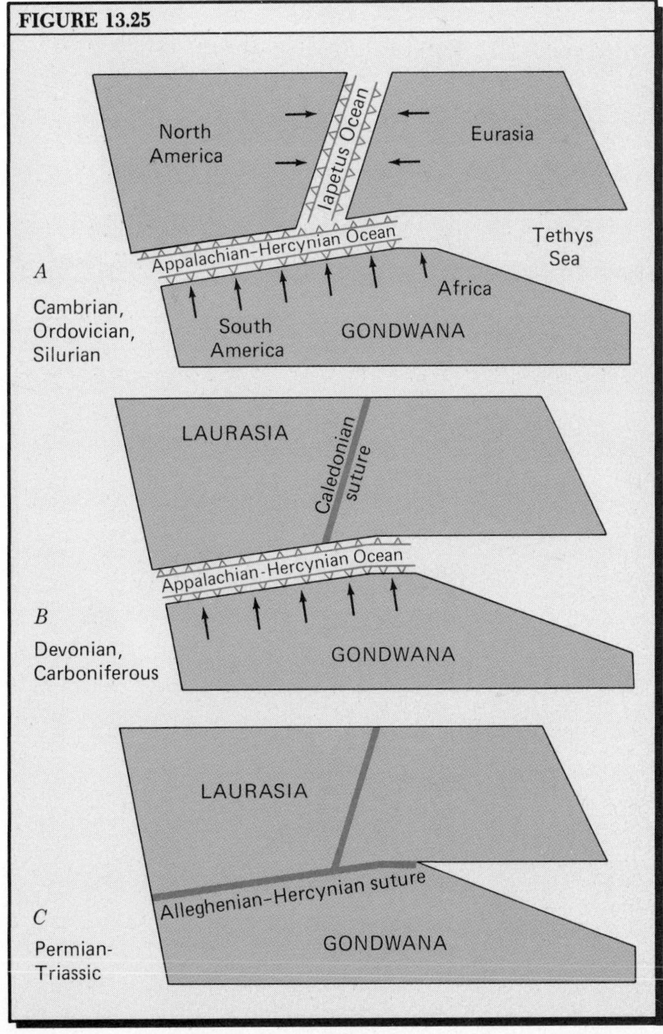

die out toward the continental interior. Today the fold belt is known as the Newer Appalachians (or Folded Appalachians); it extends from Pennsylvania to Alabama and is continued farther west as the Ouachita Mountains. Plunging anticlines and synclines of Carboniferous and older strata are now eroded into sharp-crested ridges that zigzag across the landscape (Figure 17.32 and Color Plate C.2). In the Blue Ridge Mountains, metamorphic rocks are exposed, and these are broken by numerous low-angle overthrust faults, while still farther east are many exposed plutons.

Opening of the Atlantic Basin

An interesting feature of the geologic history of the continental margins of North America and Europe has to do with details of the final rifting that began in Triassic time. Consider first that when the two continents had been joined by collisions in the Silurian–Devonian and Carboniferous Periods strong sutures were formed. The question is this: Did the new rift that opened up in Triassic time follow these sutures? Suppose that you glue two blocks of soft balsa wood together, then later pull them apart to form a new break. The chances are that the new break will not follow the glued contact very closely. Instead, some splinters will be pulled off one block and become a part of the other block, and vice versa. This is exactly what seems to have happened when the final opening of the Atlantic began in Triassic time. Figure 13.28 shows this in a highly schematic way.

Without going into details, which are very complex, we can point out a very strange thing that resulted from the principle shown in Figure 13.28. When the Atlantic Ocean opened up in late Triassic time, a substantial strip of what was formerly the North American continent on the western side of Iapetus Ocean adhered to the European plate and is now northern Ireland and the Highlands of Scotland (Figure 13.29). At the same time, a strip of what was formerly a part of the European continent on the east side of Iapetus Ocean adhered to the American plate and is now part of Newfoundland, Nova Scotia, and New England. We leave it to you to decide whether this was a fair trade of real estate.

**An Ancient Aulacogen on North America*

If our general scenario of continental rifting is accurate, the zigzag line of the rift, connecting one three-arm rift with another, we should expect to find examples of failed arms of triple rifts. The first rifting, which took place in late Precambrian time to form the Iapetus Ocean and the Appalachian–Hercynian ocean, might have left a failed arm in which Paleozoic sediments of an aulacogen accumulated.

What appears to be a remarkable example of what we are looking for turns up in western Oklahoma, almost due west of Oklahoma City and directly beneath the city of Sayre. Geologists call this aulacogen the Anadarko Basin. You would see nothing of it in the landscape, which is quite uninspiring, but wells drilled here show Paleozoic sediments nearly 10,000 m thick in a narrow, one-sided fault trough. A cross section of the Anadarko Basin is shown in Figure 13.30. The sediment trough is bounded on the southwest side by an ancient fault cutting the Precambrian shield rocks.

What seems to have happened is illustrated schematically in Figure 13.31. Following the formation of a triple rift over a mantle plume and hot spot in late Precambrian time (Figure 13.31*A*), the Appalachian–Hercynian ocean opened fully. In early Paleozoic time (Figure 13.31*B*), marine strata of Cambrian, Ordovician, and Silurian ages were deposited in the Anadarko Basin, which was then an arm of the ocean. In Carboniferous (early Pennsylvanian) time, as ocean closing was occurring, a massive tectonic uplift formed across the mouth of the aulacogen. Streams eroding this uplift brought great quantities of coarse clastic sediment into the Anadarko Basin. About 7000 m of conglomeratic strata accumulated during this period. This material can perhaps be identified as a foreland trough in our model of Cordilleran-type orogeny (Figure 12.15). Then, in late Pennsylvanian time, the final collision occurred, forming a belt of Appalachian-type folds that are today the Ouachita Mountains (Figure 13.31*D*). Molasse from this orogen poured into the aulacogen, and as it accumulated, the floor of the aulacogen subsided. This subsidence was perhaps largely due to an isostatic adjustment to added load, but we can perhaps attribute the continued sinking of the lithosphere to steady cooling following the eruption of the hot spot in late Precambrian time.

FIGURE 13.26

FIGURE 13.26 Steeply tilted coarse sandstone beds along the west coast of Wales illustrate the tectonic effects of the Caledonian Orogeny. (NERC Copyright. Reproduced by permission of the Director, Institute of Geological Sciences, London.)

FIGURE 13.27

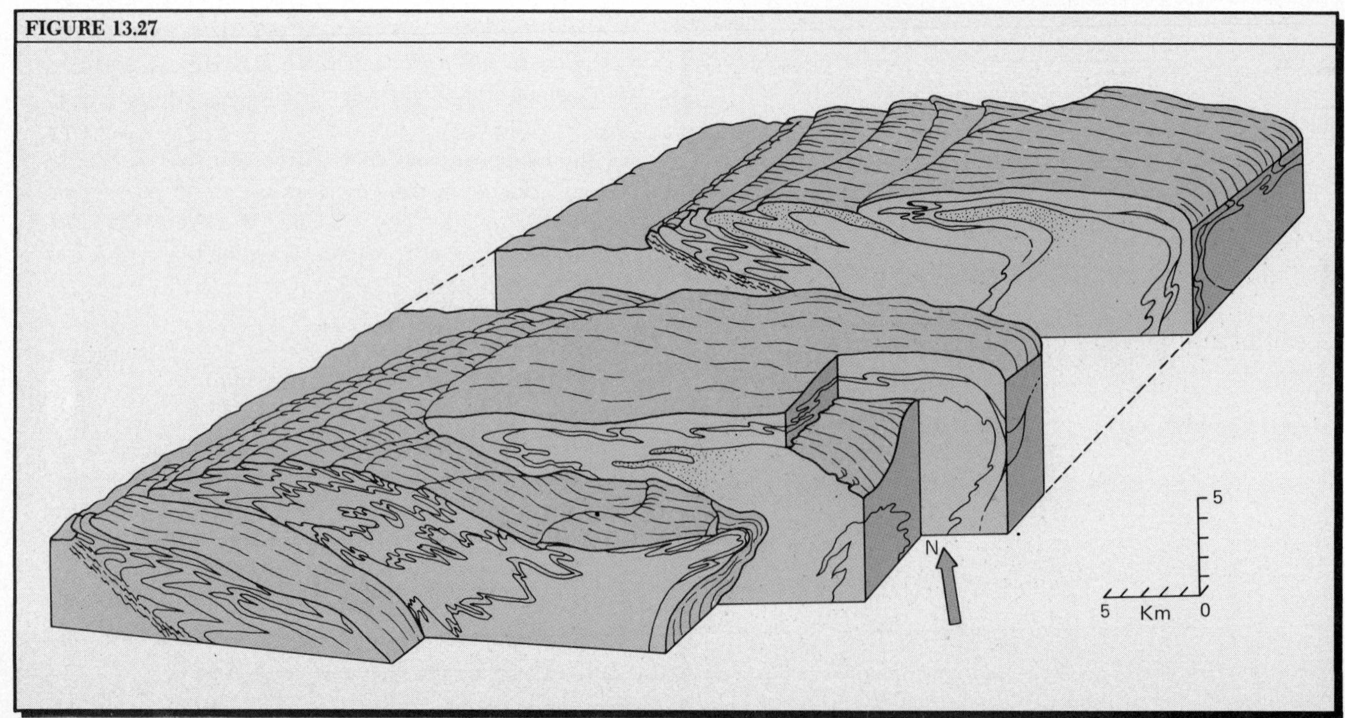

FIGURE 13.27 Block diagram showing reconstructed nappes of the Piedmont region in South Carolina. These structures were produced by continental collision during the Alleghenian Orogeny. (From V. S. Griffin, 1974, *Geological Society of America, Bull.*, vol. 85, p. 1123, Figure 13. Copyright © 1974 by the Geological Society of America.)

The Evolution of Continental Margins

In Chapter 1, we described the opening of an ocean basin such as the Atlantic Basin illustrated by a series of block diagrams (Figure 1.10). Now, from a study of the East African rift valley system, we realize that the early stage of continental rifting produces many tilted fault blocks over a wide zone.

In recent years, it has become evident from studies of deep crustal structure along the Atlantic margin of North America that the continental crust was stretched and faulted just prior to the appearance of the early Atlantic Ocean basin. There were also intrusions and extrusions of magma within and upon the faulted continental crust, and thick sediments filled the downfaulted troughs and grabens. As a result, a broad zone—perhaps 100 to 150 km wide—was changed into crust of intermediate geologic properties between the old continental crust and new oceanic crust on either side.

A series of schematic cross sections, Figure 13.32, shows how a belt of transitional crust is formed. When continental rifting starts, as in East Africa, the lithosphere is highly heated by the rise of mantle material (Section *A*). Heating is accompanied by softening of the lower lithosphere, which develops a ductile property, making possible horizontal stretching of the lithosphere. Block faulting represents the fracturing of the brittle crust overlying the stretched zone. Dipping normal faults allow the crustal extension and thinning to occur. The downfaulted basins are filled by terrestrial sediments, mostly red beds—conglomerates, red sandstones, and red shales (Section *B*). Basaltic lava

FIGURE 13.28

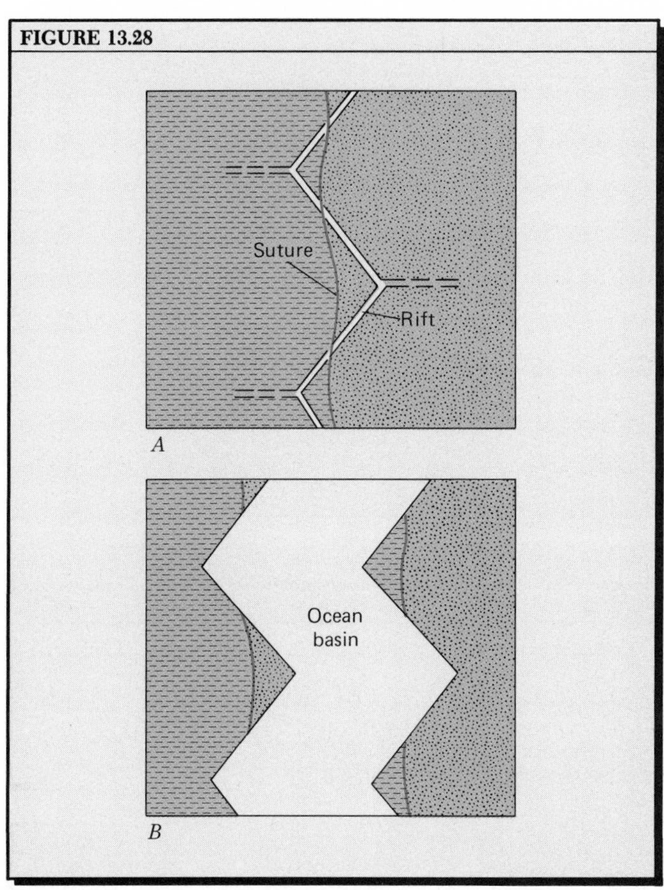

FIGURE 13.28 Schematic maps showing how a new continental rift system may be superimposed on an older continental suture. *A*. A zigzag rift system lies across a suture. *B*. When the continents separate, portions of the former continent on the left adhere to the continent on the right, and vice versa.

FIGURE 13.29

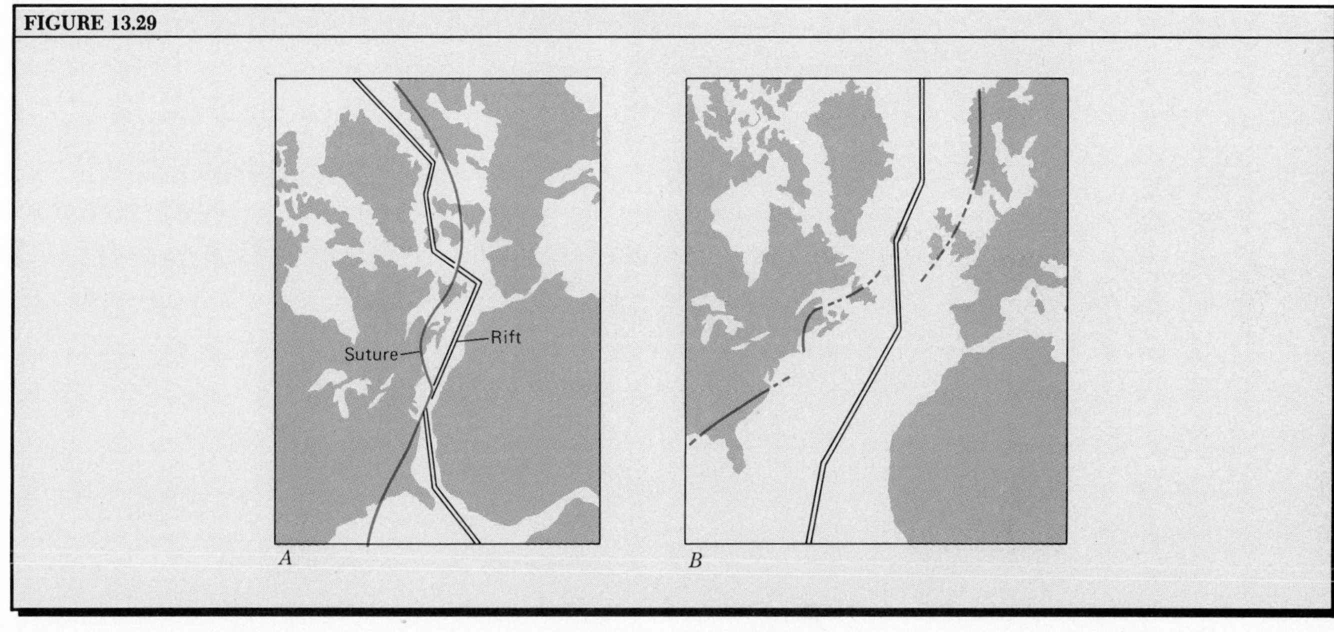

FIGURE 13.29 Maps showing conditions before and after the opening of the North Atlantic ocean basin. *A*. Early Triassic time, immediately prior to rifting. The suture shown by a bold line was formed during closing of the Iapetus Ocean basin during Paleozoic time. *B*. Later Mesozoic time. As the new ocean basin widens, the suture is broken into segments, which become widely separated.

FIGURE 13.30

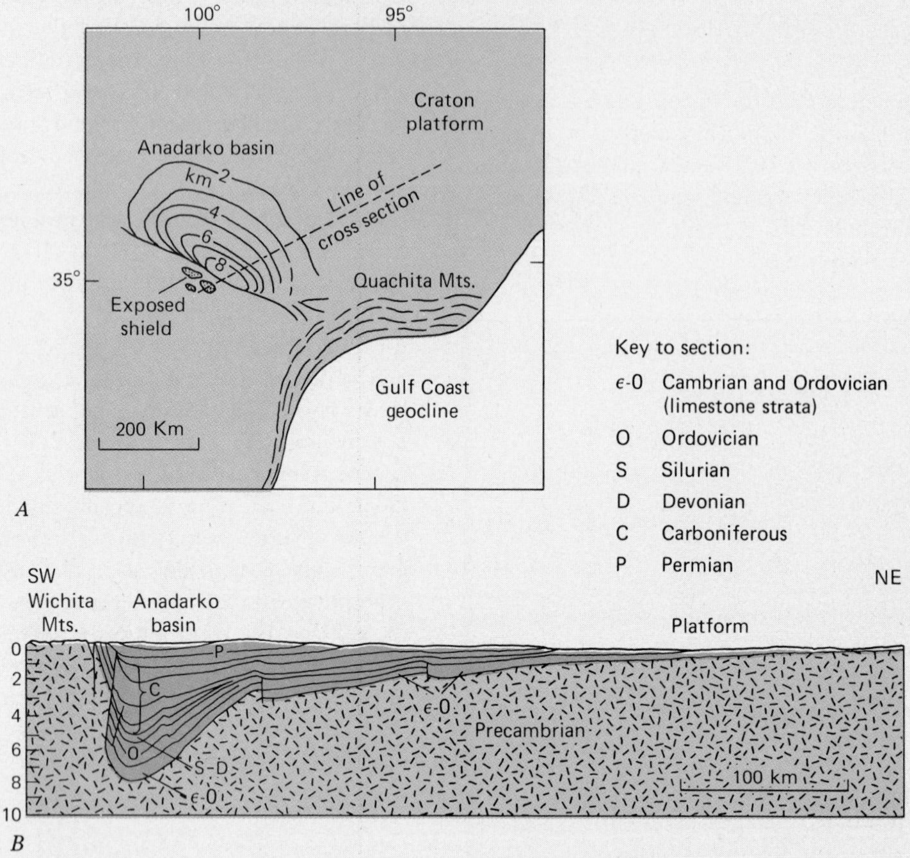

100°
95°
Craton platform
Anadarko basin
km 2
4
6
8
Line of cross section
35°
Exposed shield
Quachita Mts.
Gulf Coast geocline
200 Km

A

Key to section:

Є-0	Cambrian and Ordovician (limestone strata)
O	Ordovician
S	Silurian
D	Devonian
C	Carboniferous
P	Permian

SW
Wichita Mts.
Anadarko basin
NE
Platform
0
2
4
6
8
10
P
C
O
S-D
Є-0
Є-0
Precambrian
100 km

B

FIGURE 13.31

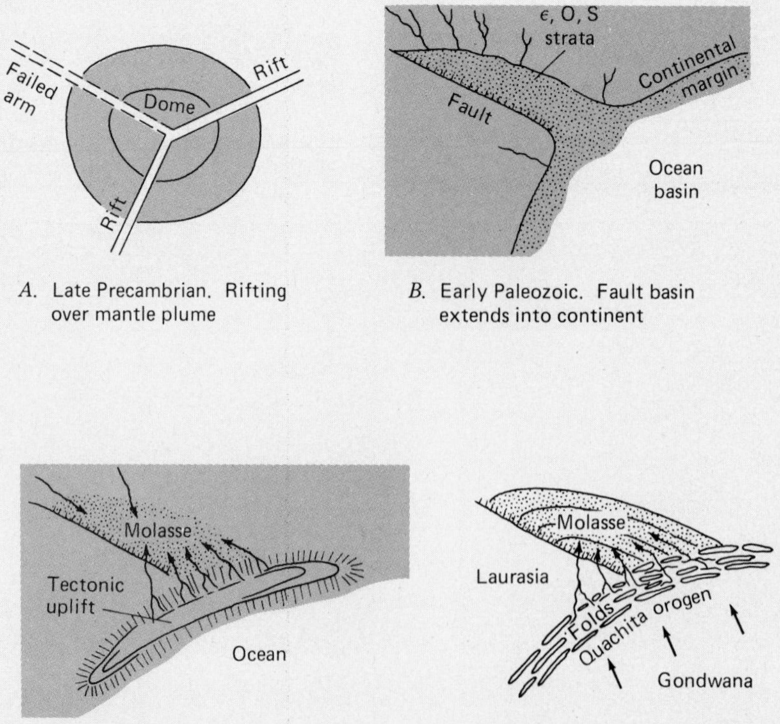

Failed arm
Rift
Dome
Rift

A. Late Precambrian. Rifting over mantle plume

Є, O, S strata
Continental margin
Fault
Ocean basin

B. Early Paleozoic. Fault basin extends into continent

Molasse
Tectonic uplift
Ocean

C. Early Pennsylvanian. Tectonic uplift of continental margin

Molasse
Laurasia
Folds
Quachita orogen
Gondwana

D. Late Pennsylvanian. Continental collision forms Quachita orogen flows are poured out on the basin floors and numerous basalt dikes are injected into the crust below. Basaltic sills may also form within the red beds. A transitional type of crust has been formed.

When the period of crustal stretching is over, the now passive continental margin is cooled and subsides (Section *C*). There begins a period of deposition of terrigenous sediment and carbonate strata, burying the faulted crust and its sediment basins. Gradually, a eugeocline and miogeocline come into existence and continue to thicken as time goes on (Section *D*).

FIGURE 13.30 Map and cross section of the Anadarko Basin, an aulacogen of Paleozoic age. Map contours show depth to the Precambrian basement. Vertical exaggeration of the cross section is about 10 times. (Based on data of P. B. King and others, *Tectonic Map of the United States, 1944*, American Assn. of Petroleum Geologists.)

FIGURE 13.31 Hypothetical stages in the evolution of the Anadarko Basin as an aulacogen.

FIGURE 13.32 Schematic cross sections showing stretching and faulting of marginal continental crust to produce transitional crust. (Based on data of K. Burke, 1979, *Oceanus*, vol. 33, no. 3, pp. 3–9.)

*Triassic Basins of Eastern North America

If our scenario of marginal stretching, faulting, and subsidence is correct, we should find some of the downfaulted sedimentary basins today along the eastern margin of North America. Indeed, they are present and they contain rocks of Upper Triassic age, which is correct for the early stage of opening of the North Atlantic basin. Wedgelike bodies of Triassic sedimentary strata, occupying downfaulted basins, occur from New England to North Carolina in the Piedmont geologic province. In Chapter 12, we gave the name *taphrogen* to this sedimentary accumulation, classifying it as a variety of geosyncline. The basins are surrounded by older gneisses and schists, which are root structures of the Caledonian and Hercynian orogens. Extensive erosion has removed all but the lower parts of the fault basins.

The largest of the basins is the Newark Basin of New Jersey and eastern Pennsylvania. Figure 13.33 shows how sediments accumulated in the basin as downfaulting took place. Successive basalt flows were buried and tilted with the sediments (Sections *A* and *B*), then a massive sill of gabbro was intruded into the red beds

FIGURE 13.32

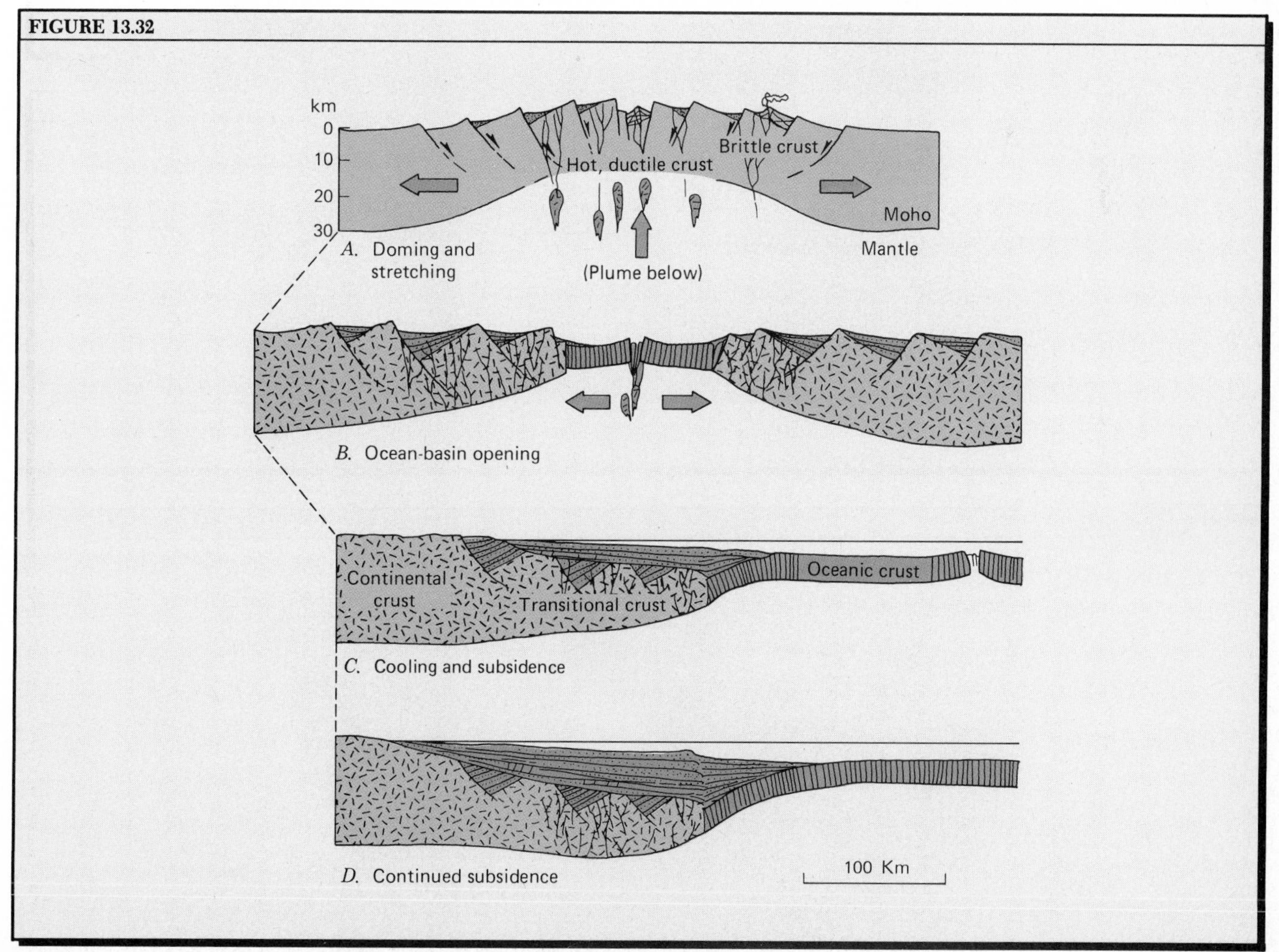

(Section *C*). Today, the edge of the sill is exposed to view as the Palisades of the Hudson River (Figure 13.34), which owes its ribbed cliff form to massive columnar jointing in the gabbro (see Figure 14.31).

The Triassic basins exposed along the Piedmont belt represent only the westernmost of the fault basins that formed in late Triassic time by stretching of the heated continental crust. Other basins lie concealed beneath Cretaceous and Cenozoic strata of the coastal plain sediment wedge (the modern miogeocline). They have been detected by means of detailed seismic exploration of the continental shelf. Both reflection and refraction seismology have contributed to unraveling the details of crustal structure and sedimentary deposits beneath the Cenozoic wedges.

*Structure of the Atlantic Continental Margin

Figure 13.35 is a set of block diagrams showing steps in the formation of buried structures and sedimentary units that probably underlie the Cenozoic geoclines in the Baltimore Canyon region of the continental shelf. This is the part of the Atlantic continental shelf offshore from New Jersey, Delaware, Maryland, and Vir-

FIGURE 13.33 The Newark Basin of downfaulted Triassic sediments and lava flows.

FIGURE 13.34 A schematic diagram of the Palisades of the Hudson River, New Jersey and New York. Relief is greatly exaggerated. (Drawn by A. N. Strahler.)

FIGURE 13.33

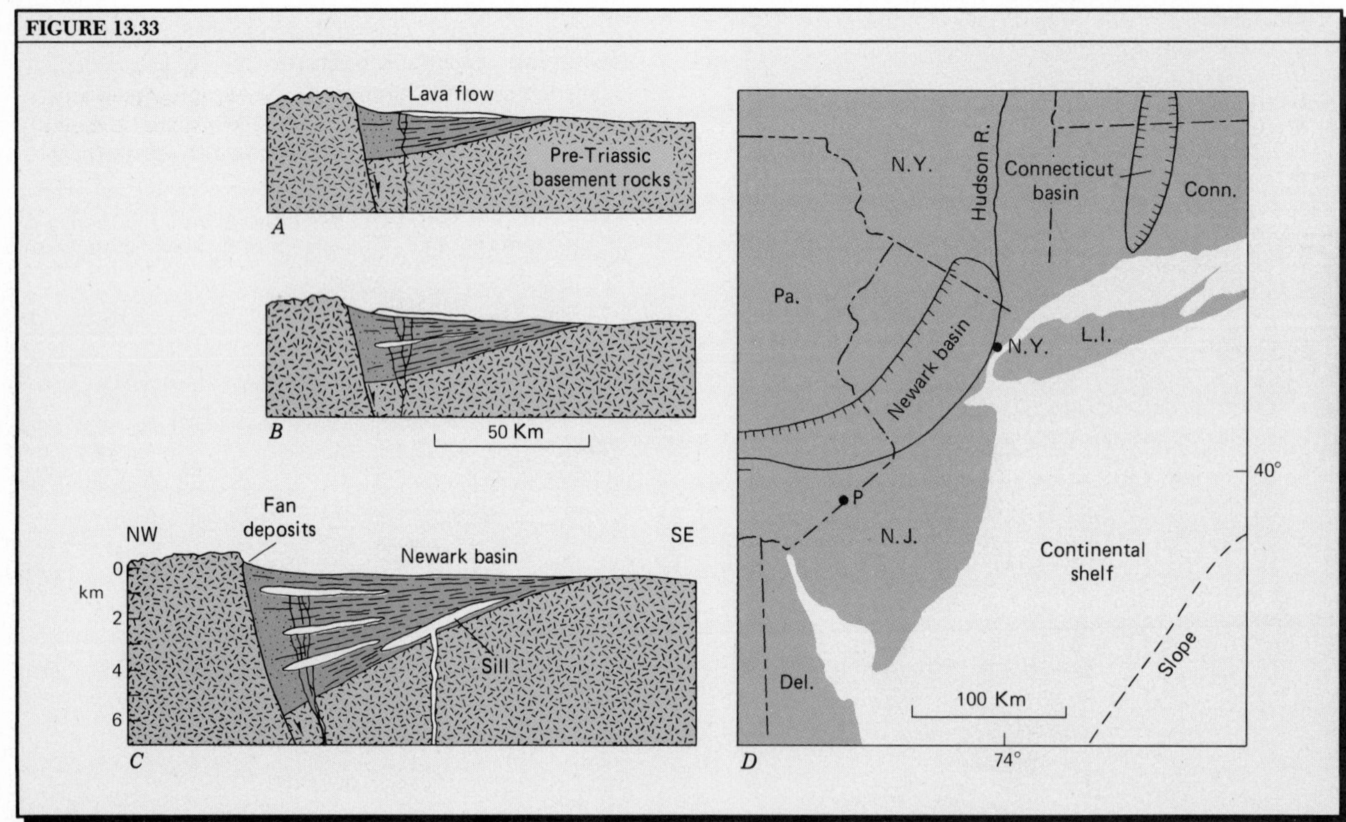

FIGURE 13.34

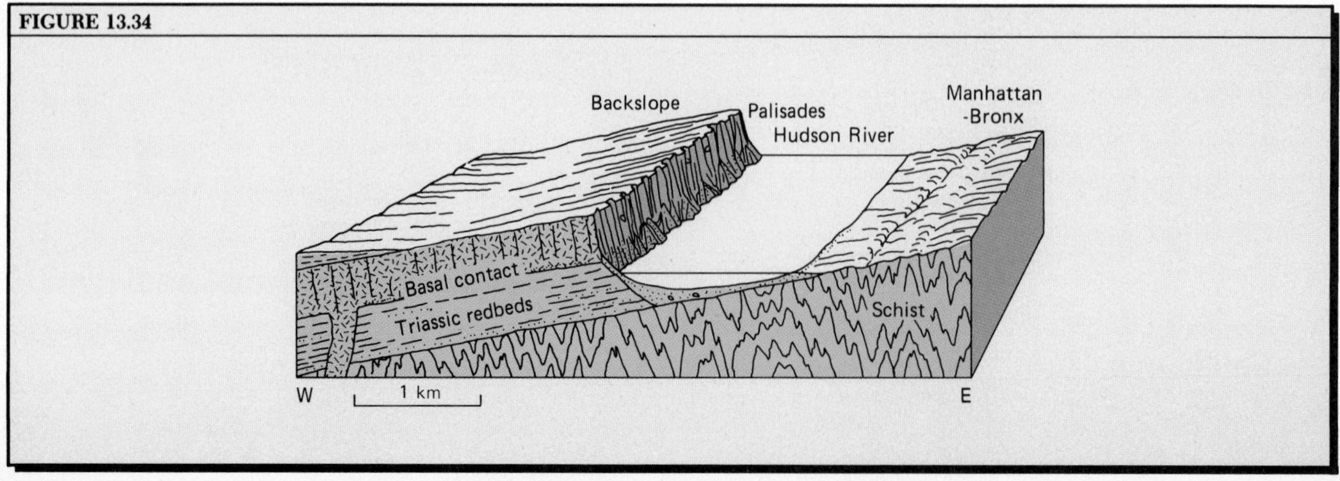

ginia. Block *A* shows the situation in the late Triassic. The Triassic fault basins have been filled by red beds. Block *B* shows conditions in Jurassic time, when the Triassic basins have been partly removed by erosion and are being overlapped by a wedge of Jurassic marine strata. It seems that conditions were right for the growth of massive coral reefs in a shallow water zone paralleling the coast. These reefs continued to maintain themselves as the continental margin subsided, forming a great rock barrier that separated deep ocean water on the east from a shallow marginal sea on the west. Rapid evaporation of seawater resulted in the deposition of thick salt beds landward of the reef barrier, while at the same time clastic sediment was being brought into the shallow sea from the mainland to the west. Altogether eight to ten kilometers of clastic sediments and salt beds of Jurassic age accumulated in this marginal geosyncline. Seaward of the reef barrier a deep-water wedge of sediments was also accumulating on the continental slope and rise.

In the Cretaceous Period, the reef barrier ceased to grow upward and was buried under Cretaceous strata, as shown in Block *C*. These were clastic sediments and limestones of the shelf miogeocline and deep-water turbidite clastics of the eugeocline. Slow subsidence continued through the Cenozoic Era, with deposition

FIGURE 13.35 Evolution of the Atlantic continental margin. Vertical exaggeration is 5 times. Surface features are greatly exaggerated. (Drawn by A. N. Strahler. Based on data of J. S. Schlee et al., 1979, *Oceanus,* vol. 22, no. 3, pp. 40–47; J. A. Grow et al., 1979, *Memoir 29, American Assn. of Petroleum Geologists,* Tulsa, Okla., pp. 65–83). *A.* Late Triassic time. Horst and graben structure has been produced by crustal extension. Red beds have been deposited in grabens. (Volcanic rocks and igneous intrusions are not shown in these diagrams.) *B.* Jurassic time. After extensive denudation of the Triassic horsts, the sea has encroached upon a beveled surface of Triassic and older rocks. A barrier reef and lagoon dominate the shelf sedimentation. Evaporites are shown by a pattern of small squares. *C.* Present time. Continued subsidence has been accompanied by geoclinal sedimentation through Cretaceous (*K*) and Cenozoic (*C*) times. A diapir (salt plug) is shown rising from the evaporite deposits and projecting through overlying sediments.

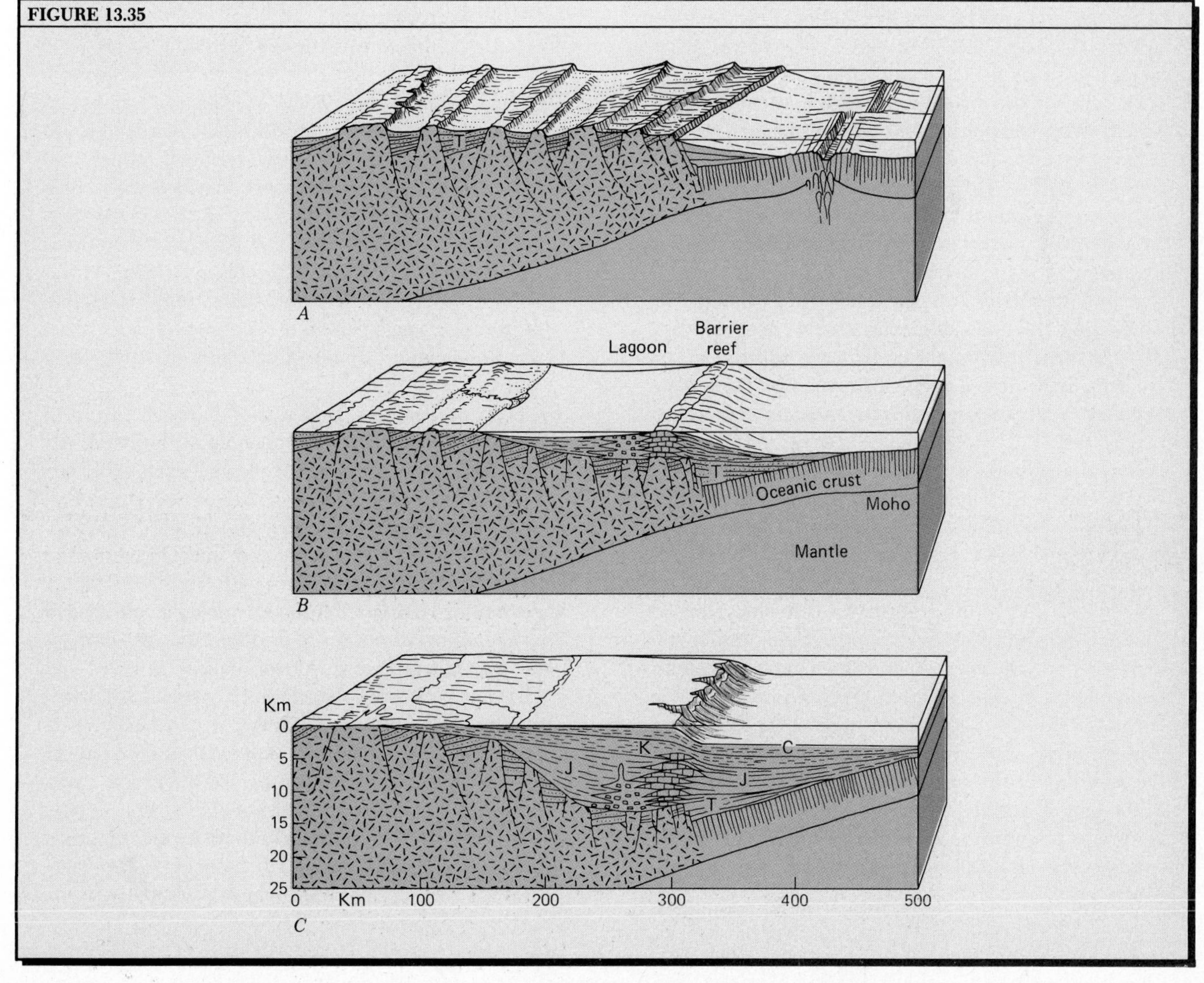

of more miogeoclinal beds on the shelf and turbidites in the eugeocline.

Although this history of evolution of the Atlantic continental margin seems complex and perhaps difficult to follow in detail, it is a topic of great interest to exploration geologists. The prospect of discovering large accumulations of petroleum in the thick sediments of this continental margin has been a powerful motivating force for the intensive research given to the structures and sediments that lie concealed here.

Global Plate Tectonics in Review

In four chapters, we have investigated all of the important aspects of plate tectonics as they relate to earth dynamics and to structures produced by plate motions. We hope that the many lines of geological and geophysical evidence we have followed through have made a convincing case for the validity of plate tectonics as a central theory in geology. While doing this, we have made a special attempt to interpret the major structural elements of North America, so that you can have an understanding of the geologic history of the region in which you live and of contrasting regions you may visit. We have, of course, oversimplified many geologic structures and events.

Perhaps we have not given enough attention to alternative explanations, based on the earlier theory of stationary continents. Perhaps, as some say we have done, those of us who formerly supported a concept of stationary continents and treated Wegener's hypothesis with disdain have, in becoming converts to the new paradigm of plate tectonics, lost part of our sense of scientific objectivity and balance. We are willing to let time bring the final judgment, but we are confident that the enormous weight of evidence will remain on our side. Some serious problems remain in fitting plate tectonics to all the observed facts; those problems must be resolved.

Continuing into the chapters that follow, we will turn to the earth's land surfaces to examine geologic processes powered by the sun's energy. These external processes attack all exposed rock, altering it physically and chemically and transporting those products to lower levels and to the marginal seas for recycling in subduction boundaries and for reprocessing into metamorphic and igneous rock during orogeny and continental collision. The external processes do not come last in the cycle of rock transformation, for there is no "last place" in that cycle. In another sense, however, the sculpting of the surfaces of the continents belongs at the end of a study of physical geology because the features of a landscape are typically and as a group the youngest of all geologic products one sees on the continents.

Key Facts and Concepts

Pangaea *Pangaea,* the supercontinent that Alfred Wegener postulated to exist in Carboniferous time, consisted of a southern subcontinent, *Gondwana,* and a northern subcontinent, *Laurasia.* Modern reconstructions of Pangaea are based on radiometric and paleomagnetic data.

Continental drift Wegener's hypothesis of breakup of Pangaea and drifting apart of the continental fragments required horizontal motion of crustal plates of *sial* through a layer of *sima.* The mechanism was declared physically impossible by geophysicists. Mantle convection was proposed by Arthur Holmes as a more plausible mechanism.

Polar wandering Paleomagnetic data allow the former position of the geomagnetic pole to be determined. Geomagnetic latitude of a sample can be determined, but not geomagnetic longitude. Pole positions plotted for long spans of geologic time form a *polar wandering path.* Drifting apart of two continents is shown by the separation of paths plotted for each continent.

Geologic evidence of Pangaea Matching of orogenic structural trends and rock ages strongly suggests a former unified continent of Pangaea. Intercontinental similarities in stratigraphic sequences, including tillites and coal beds, reptilian faunas, and the *Glossopteris* flora support the hypothesis of a single continent. Climatic zones, reconstructed from coal formations, salt beds, and gypsum, form a unified pattern on the reconstructed continent. Glaciation, reconstructed from tillites and striations, requires a pole-centered continental ice sheet on Gondwana through Carboniferous and early Permian time.

Breakup of Pangaea Breakup of Pangaea began in the early Jurassic with the opening of the North Atlantic Ocean basin. There followed opening of the South Atlantic and Indian oceans basins with subsequent rapid separation of the Gondwana fragments. The Tethys Sea was later closed, and ultimately the Indian subcontinent collided with Southern Asia.

Continental rifting Continental rifting is envisioned as beginning with the rise of domes in the continental lithosphere, possibly over mantle plumes, and production of three-branched rift systems. After formation of a continuous zigzag rift, plate separation produced a widening *rift valley.* Rise of basalt in dikes and fissures was accompanied by normal faulting of the rift margins. Stretching of the ductile deep crust caused fault blocks to rotate and separate. Invasion of the deepening rift by ocean water initiated the new ocean basin, which widened with the formation of new oceanic crust and lithosphere.

Failed arms, deltas, and aulacogens The *failed arm* of a triple rift may develop into a downfaulted crustal block occupied by an *aulacogen,* a form of geosyncline in which thick sediments accumulate. Major deltas tend to form in the coastal reentrants of the failed arms.

The Wilson cycle A full cycle of opening and closing of an ocean basin—the *Wilson cycle*—consists of six stages: (1) rift-valley system; (2) narrow ocean gulf with young ocean basin; (3) wide, mature ocean basin; (4) shrinking ocean basin (closing); (5) narrow ocean basin; (6) continental collision and suturing.

East African rift-valley system The *rift-valley* system of East Africa consists of domelike swells with axial grabens containing basaltic lavas and sediment fills. Continued rifting may lead to separation of the Somalian subplate.

Gulf of Aden and Red Sea Stage 2 of the Wilson cycle is illustrated by the Gulf of Aden and Red Sea, meeting in the Afar Triangle formed in the rifted dome of the Ethiopian and Yemen highlands. High heat flow rates and hot brines are found on the Red Sea floor, indicating hydrothermal activity over a magma chamber.

Caledonian and Hercynian orogens In earliest Paleozoic time, the *Iapetus Ocean* basin separated North American and Eurasian plates. Closing of that basin began in the present North Atlantic region and ended with collision in the Caledonian Orogeny. There followed a late Paleozoic closing of an ocean between Laurasia and Gondwana, ending in the Alleghenian–Hercynian Orogeny. Mesozoic rifting did not follow the earlier suture lines, but left fragments of the Paleozoic North American crust attached to the European plate and fragments of the European Paleozoic crust attached to the North American plate. The Anadarko Basin, a Paleozoic aulacogen, was strongly affected by the late Pennsylvanian collision.

Evolution of continental margins During continental rifting and opening of new ocean basins, crustal stretching and thinning strongly affects the margins of the continental lithospheric plate, with block faulting taking place over a widening ductile lower crust. A zone of transitional crust is produced between continent and ocean basin. Sedimentation in deepening fault troughs forms *taphrogens* (variety of geosyncline) with red beds and accompanying basalt extrusion. Subsidence may then lead to crustal submergence and deposition of reef sediments and salt beds. Continued submergence leads to deposition of a continental shelf sediment wedge (miogeocline) and a turbidite wedge (eugeocline). This sequence is illustrated by the North Atlantic continental margin of North America.

Questions and Problems

1. Put yourself in the place of a conservative western geologist of the 1920s and 1930s, reading Alfred Wegener's treatise on continental drift. You react with hostility to his proposal and set about to refute his geological arguments favoring a single supercontinent. How would you explain the structural, stratigraphic, and paleontologic facts under a system of separate continents, as they are today, with intervening ocean basins?

2. Do you consider the paleomagnetic evidence of polar wandering paths presented in the early 1960s as convincing proof of the separation of continents over the past 200 m.y.? What other evidence favorable to the original scenario of continental drift was beginning to appear about that time? On which side would you have stood in the great debate?

3. We have suggested the alternate possibility that in the initial phase of continental rifting, the rise of mantle plumes would have been the result (rather than the cause) of a number of rifts penetrating the interior of a continental plate. Which sequence of events do you consider most likely? Why would a number of mantle plumes begin to rise at about the same time beneath an unbroken continental lithospheric plate? On the other hand, what forces would cause rifts to penetrate the continental lithosphere in the absence of mantle plumes?

4. At present, the Atlantic Ocean basin is continuing to open, representing Stage 3 in the Wilson cycle. Imagine how the geology of this ocean basin would change if opening were to cease and closing were to begin. Describe the changes that might be expected in tectonic and topographic features of the ocean floor. What changes would you expect in seismic activity? In volcanic activity?

14

CHAPTER FOURTEEN

Landforms of Weathering and Mass Wasting

With this chapter, we enter a new area of geology to study *landforms*, the varied relief features of the land surface we see about us at all times. The study of the origin and evolution of landforms is a branch of geology known as *geomorphology*. In the study of landforms, the various landscape features are sorted out according to processes of origin.

Examples of landforms are hills and valleys, plains, and mountains. Volcanoes and fault scarps, which we studied in earlier chapters, are also landforms. Volcanoes and fault scarps are created by *internal earth processes*, driven by energy sources deep within the earth: volcanism and tectonic activity. In contrast, the landforms under examination in this and later chapters are shaped by *external earth processes*. External processes, driven by solar energy, act through the atmosphere and oceans, where air and water come in contact with the lithosphere.

You have already been introduced to chemical processes of rock weathering as one phase of the interaction between the atmosphere and the lithosphere. Weathering in the form of alteration of silicate minerals produces the clay minerals (Chapter 5), and the production of sediments and sedimentary rocks depends upon mineral alteration, which is part of the overall process of rock weathering.

Weathering is a passive process, in the sense that the products of rock decay and decomposition tend to remain where formed, although these same products, and in some cases bedrock as well, often yield to the force of gravity and move downslope by rolling, sliding, or flowage to lower levels. These spontaneous movements under gravity are referred to collectively as *mass wasting*, which is also a passive process.

The land surfaces are also subject to a group of active processes by which weathered rock is transported, often for long distances, and eventually deposited to form new sedimentary accumulations. The active processes are carried out by several agents: running water, glacial ice, waves and their associated currents, and wind. These *active agents* are all, strictly speaking, fluids. As we explained in Chapter 6, a fluid is a substance that flows easily when subjected to unbalanced stresses, no matter how weak those stresses are. In moving over the land surface, fluids perform *erosion*, which is simply the forceful removal of mineral matter from the parent mass of soil or rock. Erosion is always accompanied by transportation and must always eventually end in deposition.

Denudation

With a few exceptions, erosion, transportation, and deposition by the active agents persistently carry mineral matter from higher places on the continents to lower places of accumulation. The overall process of lowering of the lands is called denudation, a term we introduced in Chapter 5. It is a very convenient term to cover at once all of the kinds of work done by both active and passive agents.

Allowed to go on for millions of years, denudation would eventually lower the surface of a continent almost to sea level. Waves and currents of the oceans—also agents of denudation—might be expected to complete the job, and the continent would end up as a shallow submarine platform. This has been the end result over some large continental areas at certain points in geologic time, but in most places tectonic and volcanic activity have periodically elevated the crust and thwarted the progress of denudation.

Human beings and all other forms of terrestrial life evolved upon continental surfaces of denudation, extremely varied from place to place in terms of physical environmental qualities. In combination with climatic factors of available heat and water, the nature of a given type of land surface determines the capacity of that surface to support life. Because denudation never ceases on the lithospheric surface, there has been a continual change in life environments through geologic time. These long-term environmental changes have influenced the course of organic evolution of terrestrial life forms.

Initial and Sequential Landforms

Using the concepts offered in previous paragraphs, we find that all landforms fall into two great groups. Those formed directly by volcanic and tectonic processes belong to the group known as *initial landforms*. Those landforms shaped by the agents of denudation are *sequential landforms*, from the meaning of sequential, "following after," and these landforms must be derived from the initial landforms. Figure 14.1*A* shows an uplifted fault block—a horst—produced by internal earth forces. This enormous mass of bedrock has been elevated by forces active in moving and dislocating lithospheric plates; it is an initial landform, subsequently carved into sequential landforms by the active agents (Figure 14.1*B*).

FIGURE 14.1 Two great classes of landforms. *A*. Initial. *B*. Sequential. (Redrawn, by permission of the publisher, from A. N. Strahler, *Physical Geography*, 4th ed., Figure 24.1. Copyright © 1975 by John Wiley & Sons, Inc.)

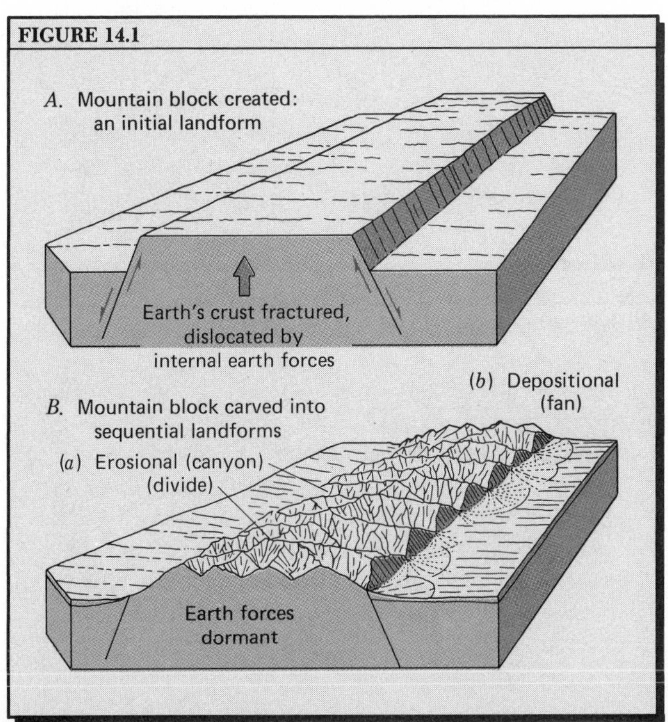

The sequential landforms can in turn be subdivided into two varieties: erosional and depositional. *Erosional landforms* are those resulting from the progressive removal of earth materials; *depositional landforms* are those resulting from the accumulation of the products of erosion and transportation. In the example shown in Figure 14.1, erosion by water flowing on hillslopes and in streams has carved a host of erosional landforms, consisting of canyons and the intervening divides and peaks. Each agent of erosion produces a characteristic assemblage of erosional and depositional landforms. For example, deposition of sediment by streams has at the same time been forming a type of depositional landform known as an "alluvial fan."

All landscapes of the continents reflect an unending conflict between internal and external processes. Where internal processes have been active recently along lithospheric plate boundaries, there exist rugged alpine mountain chains and high plateaus. Within the stable cratons of continental lithosphere, external processes have been given an opportunity to operate with little disturbance for vast spans of time. Here, the land surfaces have been reduced to low plains.

Flow Systems of Energy and Matter in Geomorphology

Our approach to geomorphology is a dynamic one, in which we focus on the action of forces that disrupt and transport masses of regolith and bedrock. Force acting through distance constitutes work in the mechanical sense, and energy must be transformed when work is done. Each of the agents of landscape development is involved in an energy system coupled with a material flow system. (See Chapter 2 for an explanation of flow systems of energy and matter.) All of these energy systems are open systems, and they are subsystems of the great solar radiation energy system that powers the external processes.

The internal processes are responsible for raising large masses of rock above sea level. In so doing, they furnish every particle of the elevated mass with a supply of potential energy. Work is required to achieve the mechanical and chemical breakdown of this rock into regolith and sediment. As individual sediment particles move from high to low levels, the potential energy is transformed into kinetic energy, which in turn is dissipated through friction as heat.

The external processes operate on solar energy. Some of this energy is utilized directly when solar heating and evaporation occur under direct impact of the sun's rays. But most of the external processes are powered indirectly through motions of the atmosphere and oceans and processes of water condensation. Thus, streams and glaciers depend upon precipitation to deposit water at high elevations on land where the water has an initial store of potential energy to be expended in flow to lower levels.

Wind is a fluid flow induced by inequalities in barometric pressure from place to place in the lower atmosphere. Kinetic energy of moving air powers two active agents of erosion and transportation. First, the direct frictional drag of air over the ground surface, called wind action, results in certain forms of erosion and deposition. Second, winds blowing over ocean surfaces generate waves. Energy transferred to waves ultimately is transported to the shores of the continents, where it is absorbed in erosion and transportation of mineral matter.

To sum up, the external processes fall into three groups; (a) those sustained by direct input of solar radiation, (b) those that are simple gravity-flow systems (mass wasting, streams, and glaciers), and (c) those which derive energy from atmospheric motion (wind action and wave action).

Landforms of Mechanical Weathering

The processes of rock weathering, both mechanical and chemical, were explained in some detail in Chapter 5. There, we emphasized weathering as the preparation of sediment prior to its transport to distant sites of accumulation. As weathering takes place, it produces a great variety of landforms, some of which are features of exposed bedrock, and others of which are configurations of the regolith. One aspect of weathering important in landform development is the continual agitation of the soil and regolith as moisture content increases and decreases seasonally and as soil temperatures rise and fall both daily and seasonally. Drying and wetting, freezing and thawing, growth and decay of plant roots, and the burrowing and trampling of the regolith by animals continually agitate the regolith. Such disturbances affect the regolith long after the mineral matter has been reduced to minute particles and the principal chemical changes have largely occurred.

Weathering is indeed a complex natural phenomenon. Looking at its total nature, we see that certain one-way, or irreversible, changes occur in the mineral matter. Superimposed on the long-term change process is a pattern of rhythmic fluctuations in physical and chemical state.

The effects of frost action are most conspicuous above the timberline in high mountains and at lower levels in arctic latitudes. The ground may be covered with large angular blocks of fresh rock in an accumulation known as a *felsenmeer* (literally, a rock sea), or *boulder field* (Figure 14.2). Frost shattering is particularly active on the steep rock walls rising above alpine glaciers, for here the meltwater produced in the warmth of the summer days percolates into joint cracks to refreeze at night, a process repeated many

FIGURE 14.2 Quartzite blocks above timberline, making up a felsenmeer at 3700 m elevation on the summit of Medicine Bow Peak, Snowy Range, Wyoming. (A. N. Strahler.)

FIGURE 14.3 Stone polygons grading into stone stripes on an arctic slope. (After C. F. S. Sharpe, 1938, *Landslides and Related Phenomena*, Columbia Univ. Press, New York.)

FIGURE 14.4 Stone rings near Thule, Greenland. (A. E. Corte, Geology Department, Universidad Nacional del Sur, Bahia Blanca, Argentina.)

FIGURE 14.5 A V-shaped ice wedge surrounded by alluvial silt is seen here exposed in the banks of a stream near Livengood, Alaska. (Troy L. Péwé, U.S. Geological Survey.)

FIGURE 14.2

FIGURE 14.3

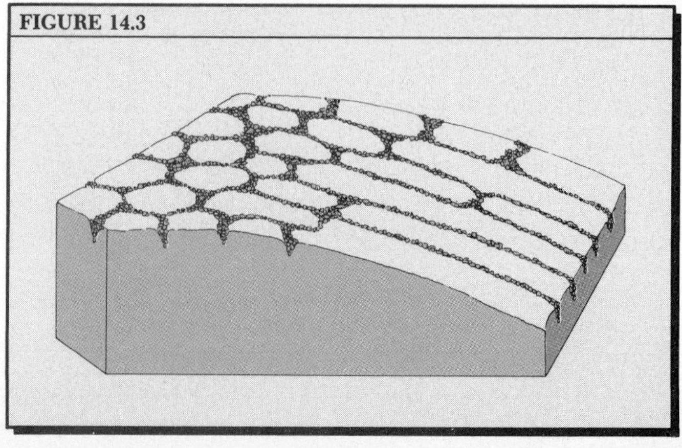

times in each season.

Formation of ice bands and ice layers in the soil is a widespread occurrence in winter in the colder climates. Where the soil is rich in fine-grade silt and clay, soil water tends to freeze in the form of horizontal ice layers that consist of densely crowded ice crystals perpendicular to the surface. As the ice layer thickens, a strong upward pressure is exerted upon the overlying soil layer, thus lifting or heaving the soil. Because of the irregularity of growth of the ice layers, *frost heaving* is uneven and produces mounds of soil.

FIGURE 14.4

FIGURE 14.5

An interesting effect of ice-crystal growth in soils (*ground ice*) is the moving of larger fragments of rock upward toward the surface or laterally to form rows. These fragments are in the size range of pebbles, cobbles, or small boulders. Where a rock fragment lies close to the surface, soil heat is conducted more rapidly to the surface, causing the growth of ice under the rock. Continued thickening of the ice layer heaves the rock fragment upward, causing it to rise to the surface, or even to be lifted above the surrounding ground surface. In arctic environments, particularly in the tundra climates, the larger fragments tend to be moved sidewise as well as upward by such frost action and to be sorted out into narrow bands. The bands intersect to form a netlike pattern consisting of *stone polygons* (Figures 14.3 and 14.4).

Throughout vast arctic regions of North America and Siberia, the average annual temperature is below freezing. This region, the treeless arctic tundra, consists of continually frozen ground except for a shallow surface layer in which soil moisture thaws during the short summer. The condition of perennially frozen ground is called *permafrost;* it also extends southward into forested regions of subarctic latitudes, but in discontinuous patches. Ice bodies are found in various shapes within alluvial silt deposits. Particularly interesting are the *ice wedges*, perpendicular in orientation, embedded in silty materials (Figure 14.5). Ice wedges are linked into polygonal systems, much like stone polygons in their pattern (Figure 14.6).

The arctic tundra is extremely vulnerable to human-made disturbances, because these can easily lead to deep melting of the permafrost (Figure 14.7). Stripping away of tundra vegetation and forest exposes the mineral soil to summer thawing. Ground ice layers and wedges melt, producing a soft, water-saturated mud, into which highways and buildings can sink. A major concern of environmentalists was that construction and operation of the Trans-Alaska Pipeline would cause serious disturbance to the permafrost areas which it crosses. Some feared that the hot oil within the pipeline might melt the surrounding permafrost, leading to breakages in the line and serious oil spills. To avoid such problems, the pipeline was engineered to be elevated where crossing ground areas that might produce mud when thawed, or buried in gravels that would not be softened by warming.

Granular disintegration of rocks by salt crystal growth in dry climates is often a conspicuous process. Sandstones are particularly affected. Water may emerge gradually near the base of a sandstone cliff, supplying water and dissolved salts for continual evaporation (Figure 14.8). As the rock disintegrates and the sand particles are blown away or washed out in rainstorms, the rock wall recedes to produce a *niche,* or in some cases a shallow cave, or even a *rock arch* (Figure 14.9). Such well-protected rock recesses were used by

Indians of the arid southwestern United States as sites for dwellings.

The simple process of wetting and drying of regolith and rock can result in forces capable of agitating regolith and disintegrating rock. Where clay minerals—particularly the expanding clays—are present, rock or regolith will swell greatly when permitted to take up water. Certain varieties of shale and those siltstones and sandstones containing clay particles tend to disintegrate by moisture absorption on exposed surfaces. Some shales actually disintegrate spontaneously into a mass of tiny chips, a process known as *slaking*. Clay-rich regolith—swelling when wet, contracting when dry—is continually affected by changes in moisture content. The contraction that occurs in dry seasons causes deep cracks to occur in the soil surface.

Important in creating many large and bold landscape features is the process of spontaneous expansion of plutonic and metamorphic rock that was formerly deeply buried in the crust and has arrived at the surface environment after millions of years of continental denudation. We explained this process in Chapter 6 under the heading of dilatation of rock, a form of rock deformation. Spontaneous rock expansion results in the development of sheeting structure in massive, joint-free rocks. Where great monolithic bodies of granitic rock are subject to formation of sheeting, domelike mountain summits, known as *exfoliation domes*, are produced. Fine examples are seen in the Yosemite National Park, where individual rock shells are 5 to 15 m thick (Figure 14.10). Sheeting in granite of a coastline will result in bedrock slabs dipping seaward everywhere along the shore.

Geomorphic Effects of Chemical Weathering

Chemical decay of joint blocks of igneous rock takes two forms. Granular disintegration commonly affects the coarse-grained igneous rocks, and tends to produce rounded, egg-shaped boulders (Figure 14.11). The products of disintegration, in the form of a coarse sand or gravel of individual mineral crystals, are swept away by wind or water to become the sediment load of streams. The finer-grained igneous rocks commonly show *spheroidal weathering*, a form of exfoliation in which the joint blocks are modified into spherical cores surrounded by shells of decayed rock (Figure 14.12).

In the warm, humid climates, chemical decay of igneous and metamorphic rocks extends to depths as great as 100 m. Here, the residual regolith is a thick layer of soft, clay-rich material known as *saprolite*. Examples are common throughout the Piedmont and Appalachian regions of the southeastern United States. Saprolite is easily removed by power shovels and bulldozers, with little or no blasting required. Presence of

FIGURE 14.6 Ice-wedge polygons in fine-textured floodplain silts near Barrow, Alaska. Dark areas within polygons are lakes. In the middle distance is a meandering river channel. (R. K. Haugen, U.S. Army Cold Regions Research & Engineering Laboratory.)

FIGURE 14.7 After one season of thaw, this vehicular winter trail through the Alaskan arctic forest had suffered severe thermal and water erosion. (R. K. Haugen, U.S. Army Cold Regions Research & Engineering Laboratory.)

FIGURE 14.8 Sandstone has disintegrated at the base of a high cliff, resulting in a niche and cliff overhang. (Redrawn, by permission of the publisher, from A. N. Strahler, *Physical Geography*, 4th ed., Figure 24.6. Copyright © 1975 by John Wiley & Sons, Inc.)

FIGURE 14.9 Rock arches resulting from granular disintegration of sandstone, Arches National Monument, Utah. The rock of which the arch is composed is the Entrada sandstone formation. (A. N. Strahler.)

FIGURE 14.10 North Dome (*left of center*) and Basket Dome (*right of center*), two great exfoliation domes of Yosemite National Park, California. (Douglas Johnson.)

FIGURE 14.6

FIGURE 14.7

FIGURE 14.8

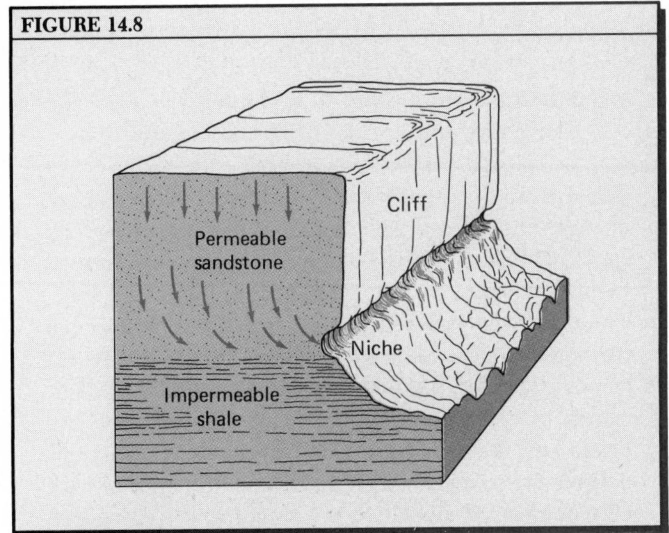

FIGURE 14.9

FIGURE 14.10

abundant clay minerals with plastic behavior greatly reduces the ability of the saprolite to support heavy structures.

Carbonic-acid action, called carbonation for brevity, plays an important role in the decomposition of many mineral and rock varieties. Its effects are most striking in the weathering of the carbonate rocks—limestone, dolomite, and marble. Carbonic acid combines readily with calcium carbonate to produce a highly soluble salt, calcium bicarbonate, which is carried away in streams. Limestone surfaces commonly show elaborate pits, grooves, and cup-shaped hollows on exposed surfaces (Figure 14.13), the effects of etching by carbonic acid in rainwater and soil water. Deep below the surface, carbonic acid acts upon limestone strata to produce cavern systems (Chapter 15).

The Soil Layer

The uppermost layer of the regolith is the *soil*, a natural zone containing living matter and supporting or capable of supporting plants. Our definition of "soil" is that preferred by soil scientists, but you should be alert to finding the word used by civil engineers to refer to any regolith or soft bedrock that can be easily moved about without blasting. Living matter in the true soil consists of plant roots, partly decomposed plant matter (humus), and many varieties of living organisms. When fully developed over long spans of time, soils have *soil horizons*—distinctive horizontal layers set apart from other soil zones or layers by differences in physical and chemical composition. Soil horizons are formed by the interactions of climate with living organisms over long periods of time. Configuration of the land surface, covered by the general term *relief*, is also a factor in the development of soil horizons. The display of soil horizons as exposed in the side of a freshly dug test pit is known as a *soil profile*.

**Soil Colloids and Cation Exchange*

From the standpoint of geology, the soil layer is important as a place where mineral matter is changed and stored in various ways. Soil processes involving mineral matter are part of the total rock cycle. We may think of the soil layer as a processing place and way station for mineral matter produced by chemical decomposition and physical disintegration of bedrock. Ultimately, soil particles are removed from the soil and carried as sediment to other locations, where that sediment can become sedimentary rock. In fact, under certain favorable conditions, hard rocklike layers are formed by accumulation of mineral matter in the soil itself. To a casual observer, a layer of this type might easily be mistaken for bedrock.

The mineral fraction of the soil consists of two classes of inorganic material. The *skeletal minerals* are the coarser particles. They are fragmented minerals and rocks of the grade size of sand and silt which provide the bulk of most soils but play no important chemical role in development of the soil. (Definitions of particle grade sizes are given in Chapter 5.) The second material class consists of the clay minerals, together with associated oxides and hydroxides; they are formed by chemical weathering of silicate minerals.

The clay minerals are particularly important because of their colloidal dimensions and platelike particle forms. Figure 14.14 is a schematic diagram of a single clay particle. In the lattice layer structure of the clay minerals, oxygen ions, negatively charged, are located nearest to the upper and lower surfaces of the particle. These charges are shown in Figure 14.14 as negative signs on the clay particle. Positively charged ions—cations—that may be present in the soil water surrounding the colloidal particle are attracted by the negative charges and held as a surface layer. Among the common cations found in the soil-water solution are hydrogen (H^+), aluminum (Al^{3+}), sodium (Na^+), potassium (K^+), calcium (Ca^{2+}), and magnesium (Mg^{2+}). You will, of course, recognize at once that the metallic cations in this list are readily available from regolith that is the parent material of the soil and that they are derived by chemical decay of the aluminosilicate minerals of the bedrock beneath. The hydrogen ion, as we have already explained in Chapter 5, is present in carbonic acid and is universally available in moist soils. All of these cations, in varying proportions, will be found on clay particle surfaces. Moreover, one cation may replace another in a process called *cation exchange*.

The balance between two classes of cations determines the pH of the soil solution, whether acid (pH below 7) or alkaline (pH above 7). Hydrogen and aluminum ions are the principal acid-generating cations in most soils. When these ions are present in large proportions, the soil is acid. The *base cations* are Ca^{2+}, Mg^{2+}, K^+, and Na^+; when they are present in dominant proportions the soil is alkaline.

An important property of a given soil is its *cation-exchange capacity* (CEC), which is the capacity of a given quantity of soil to hold and to exchange cations. A high value of CEC is a general indication of a high degree of chemical activity within a soil; a low CEC indicates a low level of chemical activity. The value of the CEC of a given soil is largely related to the kind of material that makes up the colloidal fraction of the soil. Among the clay minerals we described in Chapter 5, montmorillonite has a high CEC, illite an intermediate CEC, and kaolinite a low CEC. Finely divided organic matter, known collectively as *humus*, has a high CEC. The hydrous sesquioxides of iron and aluminum have a very low CEC.

Of what importance is the cation-exchange capacity of a soil? To answer this question, let us arrange the colloidal minerals in descending order of their CEC:

montmorillonite → illite → kaolinite →
hydrous sesquioxides of aluminum and iron

The arrows are significant in terms of a sequence of changes that affects soils of moist climates over long spans of time. A soil developed from recently formed regolith tends to be rich in montmorillonite and illite. Then, as time passes and the clay minerals undergo further changes, kaolinite becomes dominant. In extremely old soils, there remains little except the sesquioxides of aluminum and iron. In general, as time passes and the soil of a given geographical region undergoes the sequence of changes from minerals of high CEC to those of low CEC, the chemical activity of the

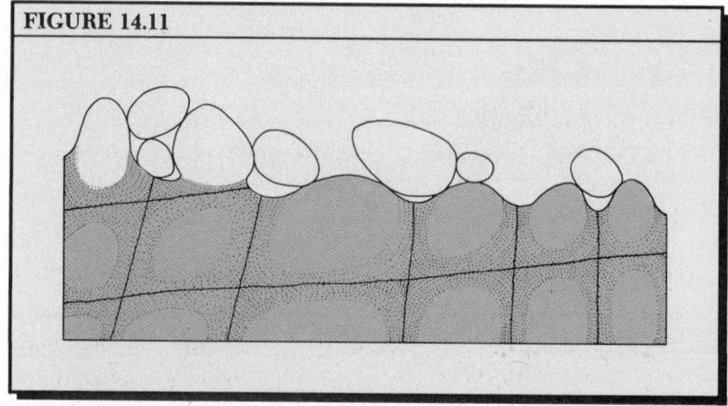

FIGURE 14.11 Rounded boulders produced by granular disintegration of rectangular joint blocks of granite. (From W. M. Davis, 1938, *Bull. Geol. Soc. of Amer.*, vol. 49, Figure 14.11.)

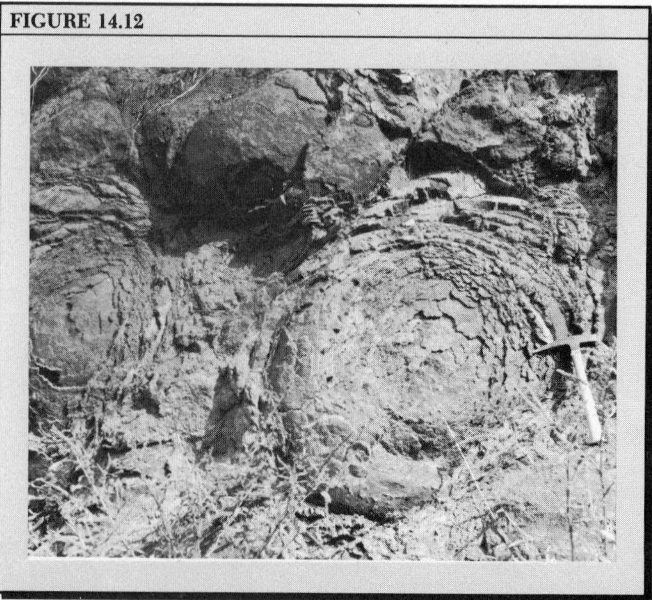

FIGURE 14.12 Spheroidal weathering, shown here, has produced many thin concentric shells in a basaltic igneous rock. (U.S. Geological Survey.)

soil diminishes. This change is highly significant in terms of soil fertility, which depends upon the richness of the holding of the base cations by soil colloids. Fertility, judged by the ability of a soil to produce grass-like crops, including all cereals and grains, may be initially high in a young soil, but will ultimately decline to low levels in very old soils. Keep in mind that this principle applies to soils of moist climates in the middle and low latitudes, but not to soils of dry climates. Soils of high cation-exchange capacity that have a high proportion of base cations are said to have *high base status;* they are the naturally rich agricultural soils. Soils of low cation-exchange capacity with a low proportion of base cations are said to have *low base status;* they are the naturally poor agricultural soils.

*Examples of Soil Orders

On a global basis, soils are classified into major classes known as *soil orders*. The modern classification system recognizes ten soil orders. In this brief overview of soils, we will select four of the orders that represent a broad spectrum in terms of environments of formation, age, and base status. All four are important in terms of the great areas they occupy and their significance to humankind as being either good or poor agricultural soils. Two of these orders include soils of high base status, two are of low base status. Three of the orders are typical of moist climates, the fourth occurs in a dry climate.

FIGURE 14.13 Deeply pitted surface of limestone, near Fremantle, Western Australia. (Douglas Johnson.)

FIGURE 14.14 A schematic diagram of a clay particle with surrounding cations.

FIGURE 14.13

FIGURE 14.14

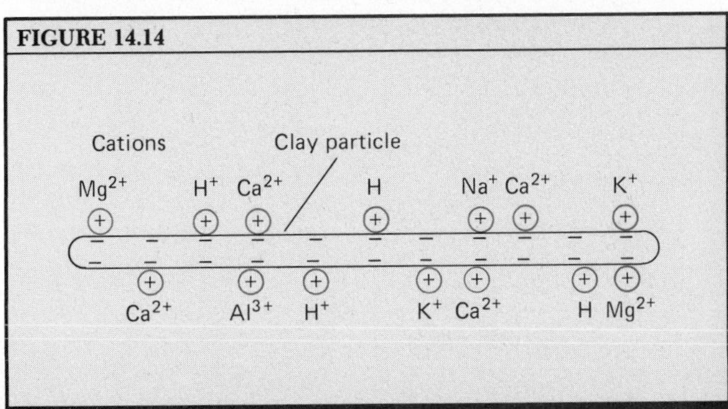

Soils of the order *Spodosols* occur mostly in a middle-latitude climate that is both cool and moist. The order is widespread in North America and northern Europe, where the natural plant cover is a dense needleleaf forest including such trees as pine, spruce, and fir, and where winters are long and bitterly cold. Soil moisture is frozen for several consecutive winter months, and soil remains moist throughout the summer. There is a large surplus of precipitation, so that water from melting snows and rains passes down through the soil profile in rather large quantities. The Spodosols are best developed on a soil layer consisting mostly of quartz sand, with little content of clay minerals. The Spodosol profile, shown in Figures 14.15 and 14.16, has strongly developed horizons that catch the eye. At the top is a dark, almost black, layer composed largely of humus. Immediately below the dark layer is a layer with a very pale color—light gray to almost white. This layer, called the *albic horizon*, often consists almost entirely of quartz grains of sand or silt texture. The albic horizon has been depleted of clay minerals, humus, iron oxide, and base cations by the leaching action of soil water, which is weakly acid. The acids in the soil solution are carbonic acid and various complex organic acids produced in the organic layer at the surface.

Below the albic layer lies a dark-colored horizon, called the *spodic horizon*. It is hard and dense, in some cases with a rocklike quality, and it is composed mostly of colloidal oxides of iron and aluminum, along with some dense organic colloids. These substances have been brought down through the albic layer and have accumulated through time to form the spodic horizon, often given a reddish color by the presence of iron oxide. Below the spodic horizon lies the parent material, which is typically a sandy regolith. Spodosols are acid soils with low cation-exchange capacity and a large proportion of hydrogen and aluminum ions. Base cations have been largely carried downward through the soil by leaching action and have been removed from the region by streams. Thus, the base status of the Spodosols is low and they are poor agricultural soils unless heavily treated with lime (to neutralize the acidity) and fertilizers (to furnish the necessary base cations). Spodosols of the cold northern forests are young soils, formed on sandy deposits left by ice sheets that disappeared only about 10,000 years ago. Leaching has been rapid on the sandy regolith, but few clay minerals were present to begin with.

We have selected the soil order of *Alfisols* because their environment of formation is not greatly different from that of the Spodosols, yet they are fertile soils of high base status. We are referring now to Alfisols that are widespread in middle latitudes in both central North America and in western and central Europe. They lie in a region just to the south of the Spodosols, where the climate is moist, with cold winters. The natural forest cover is dominated by deciduous trees that shed their leaves in winter. The Alfisols have a soil profile similar to that of the Spodosols, but with horizons less intensely developed. There is a pale horizon of leaching corresponding to the albic horizon of the Spodosols. There is a dense horizon of clay accumulation corresponding to the spodic horizon of the Spodosols.

The important difference between these two orders is that the Alfisols have a good supply of base cations derived from clay minerals such as montmorillonite and illite. Thus, the Alfisols have moderately high base status. The northern Alfisols are comparatively young soils—they date from the period following the last advance of the ice sheet, or from earlier ice advances. The materials left by the glacial ice are well endowed with alumino–silicate minerals derived from pulverized bedrock over which the ice moved. In some areas of Alfisols, a surface layer of wind-deposited silt, rich in the silicates that produce clay minerals, forms the parent matter of the soil. The Alfisols are rich agricultural soils and require rather modest inputs of lime and fertilizers.

In low latitudes, where the climate is warm and wet throughout the year or shows a strong wet season, we find the *Oxisols*. This order illustrates a group of very old soils, as their age runs into the hundreds of thousands of years. Typical of the Oxisols is their reddish color, a reflection of the abundance of sesquioxides of iron. Chemical weathering has proceeded to an end point in which there remain largely sesquioxides of iron and aluminum, some kaolinite, and various forms of silica (Figure 14.15). Distinct horizons are usually lacking in the Oxisols. Because of the mineral composition, the cation-exchange capacity of the Oxisols is low and the soils have very low base status. The native plant cover is typically a rainforest of tall, broad-leaved trees which recycle a meager store of base cations held close to the soil surface.

An interesting feature of the Oxisols is the presence of a deep horizon of mottled appearance in which sesquioxides are concentrated. Called ***plinthite***, this material is soft when formed as a deep soil horizon. When brought to the surface and exposed to the air and to repeated wetting and drying, plinthite becomes a hard, rocklike material resembling building brick, and called ***laterite***. Laterite forms over wide areas where the overlying soil horizons have been eroded away by natural processes, exposing the plinthite. Humans have made use of plinthite to produce building blocks of laterite that are extremely durable in the warm, moist tropical climate.

FIGURE 14.15 Schematic diagrams of soil profiles of Spodosols and Oxisols. (Redrawn, by permission of the publisher, from A. N. Strahler, *Physical Geography*, 4th ed., Figure 18.8. Copyright © 1975 by John Wiley & Sons, Inc.)

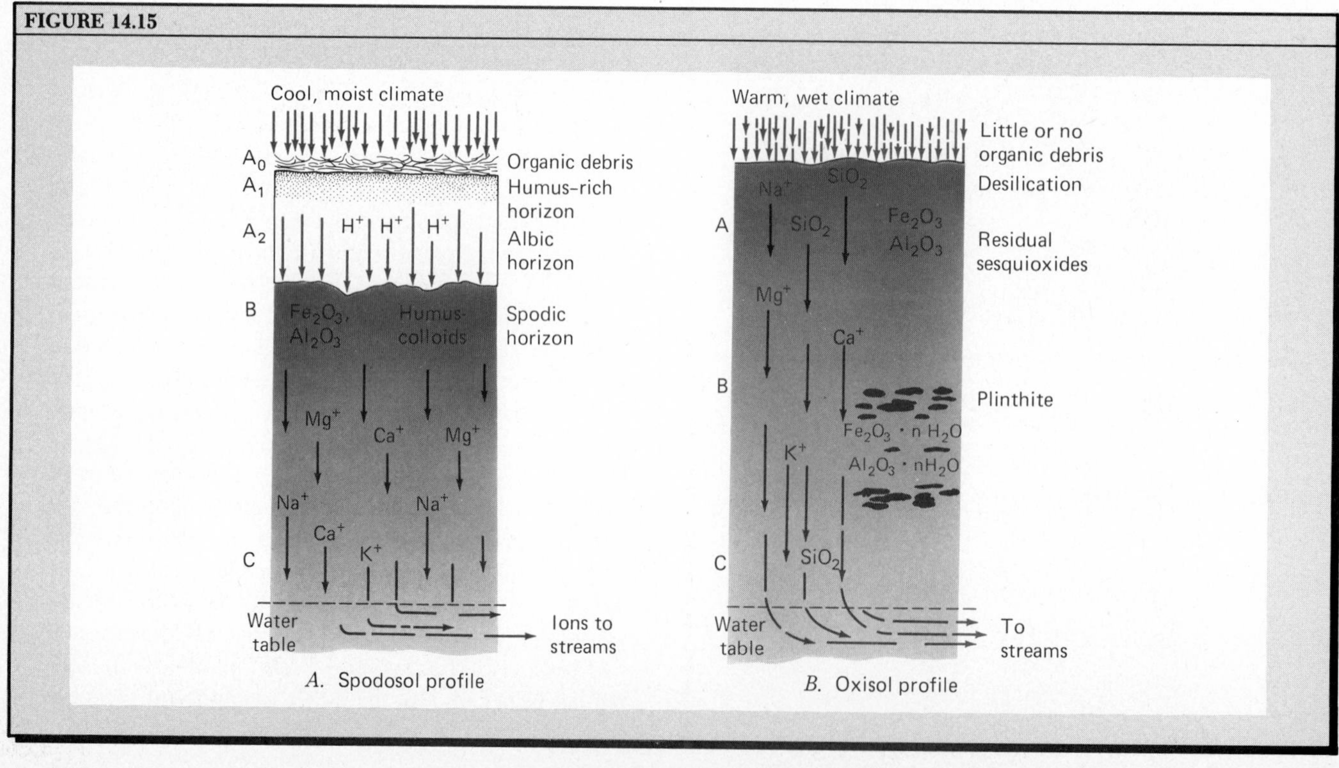

Our final example is the soil order of *Mollisols*, found in middle latitudes under a climate that ranges from semiarid to arid. Great areas of Mollisols occur in the western half of North America. Mollisols extend in an enormous belt from eastern Europe (the Ukraine) across Siberia to northern China. In this region, summers are warm to hot with scanty rainfall, while winters are cold and tend to be dry. The soil is dry throughout much of the summer, and the natural plant cover of grasses and shrubs becomes dormant except when revived by a brief rain period. Because of the aridity of climate, rainfall tends to be held in the soil layer, where it later evaporates. Base cations dissolved by rainwater may be carried down through the upper soil layer, only to be redeposited in a lower horizon. One base cation in particular, calcium (Ca^{2+}), is abundant in the Mollisols and occurs in the solid state as calcium carbonate ($CaCO_3$). Sodium, potassium, and magnesium ions are also abundant.

The Mollisols are characterized by their thick brown horizons (Figure 14.17). An upper dark brown horizon more than 25 cm thick owes its color to minute particles of humus, derived from the growth and decay of roots of grasses and annual herbs that penetrate the entire horizon (Figure 14.18). Below the upper horizon lies another dark brown horizon, but this layer contains calcium carbonate in the solid mineral state, taking form of small nodules or lumps. In some places, where the climate is particularly dry—bordering on the desert climate—calcium carbonate forms a dense white layer immediately below the soil. In the southwestern United States, this material is locally called *caliche* (Figure 14.19). The general name is *calcrete*. In eroded areas, calcrete often forms sharp cliffs and table-topped uplands (mesas). The Mollisols have very high base status and are extremely fertile soils for the production of grains. Large exports of wheat come out of regions of Mollisols.

This sample of five soil orders gives you some insight into the way in which climate, acting through time, influences the soil-forming processes and generates distinctive soil horizons. Because of the close association of soil horizons with climate, the geologist has been able to derive information about changes of climate that may have occurred throughout the past few tens of thousands of years. This information is based on the study of soil horizons that were buried and pre-

FIGURE 14.16 Profile of a Spodosol formed on sandy regolith in Maine. (Soil Conservation Service.)

FIGURE 14.17 Profile of a Mollisol in North Dakota. The horizon labeled *A* is the mollic epipedon. The *B* horizon is also dark in color but contains particles of calcium carbonate. (Soil Conservation Service.)

FIGURE 14.16

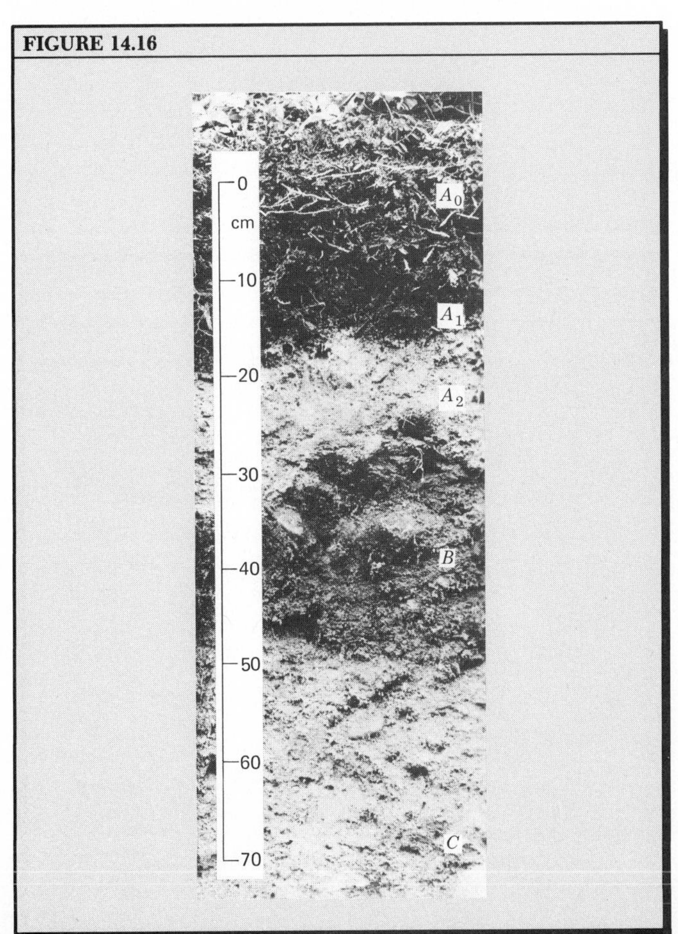

FIGURE 14.17

served under new regolith. Called *paleosols*, these buried soils often show that the climate existing at the time of their formation was different from the climate acting today upon the soil at the surface.

Mass Wasting

The force of gravity acts constantly upon all soil, regolith, and bedrock. In most places, the internal strength of these materials is sufficient to keep them in place. Consequently, we rarely see soil or rock moving spontaneously except when carried by an active agent of erosion. Wherever the ground surface is sloping, a proportion of the force of gravity is directed downslope parallel with the surface. Every particle has at least some tendency to roll or slide downhill and will do so whenever the downslope force exceeds the resisting forces of friction and cohesion that tend to bind the particle to the rest of the mass.

The forms of mass wasting range from the catastrophic slides in alpine mountains, involving millions of cubic meters of rock and capable of wiping out a whole town, down to the small flows of water-saturated soil seen commonly along the highways in early spring. But extremely slow movement of soil, imperceptible from one year to the next, also acts on almost every hillside.

Soil Creep

Careful inspection of a hillside often discloses evidence that the soil has been very slowly moving downslope rather steadily over a long period of time, a phenomenon termed *soil creep* (Figure 14.20). Where a distinctive type of rock outcrops high up on a hillside, perhaps as a vein or dike, you may find that the larger joint blocks have moved away from their original locations and that smaller fragments of the rock have been carried far down the slope in the soil mass. Yet it is unlikely that these particles have at any time slid or rolled rapidly.

Where steeply dipping, layered rocks such as slates or shales underlie a hillside, the upper edges of the layers are commonly turned downhill as if bent. This phenomenon is the result of shear distributed along countless joint fractures and bedding or cleavage surfaces in the rock (Figure 14.21). Trees, posts, poles, and monuments may be found tilted downhill, suggesting rotation as the soil has crept downslope, the surface layers moving more rapidly than those at depth. Masonry retaining walls paralleling the slope are often found to be tipped over and broken, yielding to the pressure of soil creep.

The mechanism of soil creep is a combination of the various weathering processes that agitate the soil, acting in concert with the force of gravity. Whatever mechanism disturbs the soil induces downslope movement of the particles, because gravity exerts an influence on the motions and its influence is in the downslope direction.

Consider, as an illustration, that you pour some dry sand into a conical pile on a table. Once the motions of the grains have ceased, there is no further change of the slope of the sand surface because the forces of friction exceed the downslope component of the force of gravity. Now suppose you tap the table repeatedly, sending a series of shock waves through the sand. With

FIGURE 14.18 A schematic profile of a Mollisol. Compare with Figure 14.17. (Redrawn, by permission of the publisher, from A. N. Strahler, *Physical Geography*, 4th ed., Figure 18.8. Copyright © 1975 by John Wiley & Sons, Inc.)

FIGURE 14.19 White, rocklike slabs and nodules of calcium carbonate, exposed by removal of the brown upper horizon of a Mollisol. The material is known locally as caliche. Pecos Plains of New Mexico. (A. N. Strahler.)

FIGURE 14.18

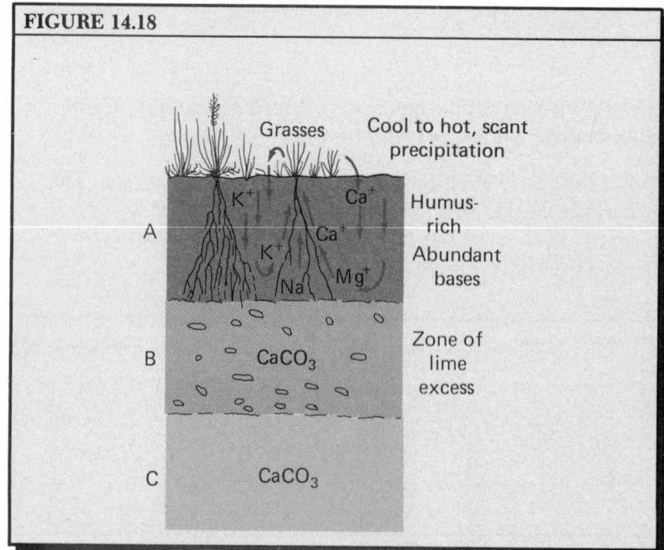

FIGURE 14.19

each tap the sand slope reclines and consequently the conical pile lowers and widens. Each mechanical shock momentarily reduces the friction between sand grains, and at that instant the force of gravity moves the grains a very slight distance downslope.

In nature, disturbances of soil result from many causes: Growth and melting of ice crystals, drying and wetting accompanied by shrinking and swelling of the soil, and the volume expansion and contraction (dilatation) from temperature changes are important, as is the growth of plant roots, pushing aside the soil, with later collapse of root cavities after the roots have decayed. Burrowing by many forms of animal life, with later closing of the cavities, is another biological cause. Trampling of slopes by large animals—such as deer, bison, and cattle—may also cause a great deal of downslope soil creep.

FIGURE 14.20 Commonplace evidences of the almost imperceptible downslope creep of soil and weathered rock. (Redrawn, by permission of the publisher, from A. N. Strahler, *Physical Geography,* 4th ed., Figure 24.17. Copyright © 1975 by John Wiley & Sons, Inc.)

FIGURE 14.21 Downbend by creep of the upper edges of thin-bedded sandstones. (Ward's Natural Science Establishment, Inc., Rochester, N.Y.)

FIGURE 14.20

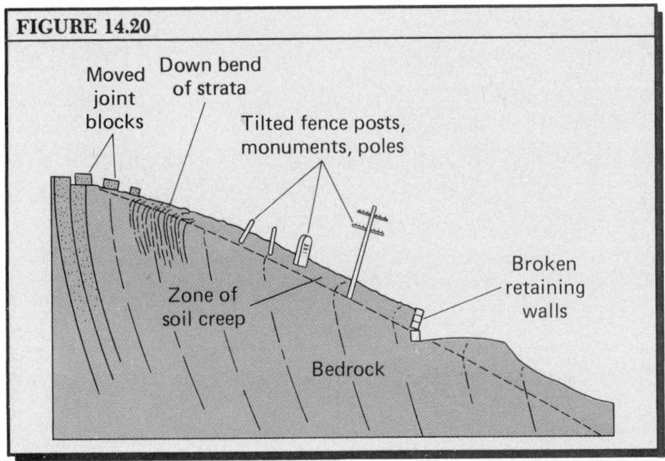

FIGURE 14.21

Earthflow

In hilly and mountainous regions of humid climate, yielding of water-saturated soil and regolith rich in clay minerals takes the form of an *earthflow.* This is a tonguelike mass that has flowed a limited distance down the hillside, perhaps coming to rest before reaching the base, or in some cases turning and flowing downvalley for a short distance. At its upper end, the earthflow leaves a depression bounded on the uphill side by a curved scarp (Figure 14.22). At its lower end, the earthflow bulges convexly downslope in a toe. Where the hillside flattens to a broad valley floor, the toe spreads out into a broad, rounded mass resembling a pancake. Where the valley is narrow, the toe forms a dam, sometimes creating a lake.

Earthflow is a form of mass wasting in which behavior of the earth material is that of a *plastic solid.* (Deformation of solids by plastic shear was explained in Chapter 6.) Any mixture of solid particles (sand, silt, or clay, for example) with a liquid such as water or oil forms a variety of plastic solid characterized by a certain degree of strength. Such a mixture will resist flowage up to a given limit, the *yield stress,* but above this it will flow much as a true liquid. If the quantity of water is small relative to the amount of solid matter, the material will resist flowage to the point that it will come to rest on a relatively steep slope, as we see in the earthflow. The presence of clay particles tends to reduce the resistance to plastic flowage by providing a form of lubrication. Clays tend to absorb and hold water, thus aiding in the accumulation of the necessary water to give a mixture of soft consistency.

Some earthflows, particularly those large flows affecting weak bedrock, start out with a very slow flowage but increase in rate because the mixing of the water with the mineral particles reduces the internal resistance to flow. In such cases, the flow quickly develops a fluidlike behavior and may travel a considerable distance downvalley as a long narrow tongue. Rate of motion is probably on the order of several meters per hour. Shallow earthflows in regolith tend to become stiffer in consistency after a few meters of travel because of drainage of the soil water. These small, shallow flows may come to rest on the hillside.

Solifluction is an arctic variety of earth flowage important in the treeless tundra where the permanently frozen subsoil (permafrost) acts as a barrier to downward percolation of water released in the spring by the melting of the snow cover and ice in the surface layer of the soil. Unable to escape by drainage, moisture builds up until the thawed soil is saturated, resulting in slow flowage of a shallow layer, producing a succession of *solifluction lobes* (Figure 14.23). Motion is on the order of a few meters per year, with a speed of a few centimeters per day at most.

*Induced Earthflows

Earthflows are an environmental hazard in terms of property damage to highways, railroads, and structures of all kinds. Because the flowage is slow, these movements are rarely a threat to human life. In many cases, destructive earthflows are aggravated or induced by human activity that involves the grading or removal of regolith required to produce a level roadbed or building site.

Examples of both small and large earthflows induced or aggravated by urbanization are numerous in central and southern California. A good place to study these effects is the Palos Verdes Hills, a blunt coastal peninsula in Los Angeles County. Largest of the earthflows of this area is known in the news media as the Portuguese Bend Landslide (though use of the term "landslide" is incorrect). It affected an area of about 1.6 sq km. The total earth motion over a three-year period was about 20 m. This is extremely slow motion and is best described as an almost imperceptible creep of the underlying mass, but as blocklike masses subsided along deep cracks, houses and driveways were broken apart. The total damage during the active life of the earthflow was about $10 million.

This earthflow has been attributed by geologists to the downsinking of waste water from cesspools and of irrigation water applied to lawns and gardens. The 150 homes located on the earthflow discharged over 115,000 liters of water per day. Much of this water percolated slowly down into the mass of weak shale beneath, increasing its weight and causing a loss of internal strength.

Earthflows in Quick Clays

A special form of earthflow has proved to be a major environmental hazard in parts of Norway and Sweden and along the St. Lawrence River and its tributaries in Quebec Province of Canada. In all these areas, the flowage involves horizontally layered clays, sands, and silts of late Pleistocene age that form low, flat-topped terraces adjacent to rivers or lakes. Over a large area, which may be 500 to 1000 m broad, a layer of silt and sand 5 to 10 m thick begins to move toward the river, sliding on a layer of soft clay that has spontaneously turned into a near-liquid state. The moving mass also settles downward and breaks into steplike masses. Carrying along houses or farms, the layer ultimately reaches the river, into which it pours as a great, disordered mass of mud.

Figure 14.24 is a block diagram of an earthflow of this type that occurred in 1898 in Quebec, along the Rivière Blanche. Beyond is the scar of a much older earthflow of the same type. The Rivière Blanche earthflow involved about 3 million cu m of material and required three to four hours to move into the river

*The text under this heading is reprinted from A. N. and A. H. Strahler, *Modern Physical Geography.* Copyright © 1978 by John Wiley & Sons, Inc. Reproduced by permission of the publisher.

FIGURE 14.22 An earthflow with slump features well developed in the upper part. (Redrawn, by permission of the publisher, from A. N. and A. H. Strahler, *Modern Physical Geography,* Figure 17.13. Copyright © 1978 by John Wiley & Sons, Inc.)

FIGURE 14.23 Solifluction on this Alaskan tundra slope has produced lobelike masses of soil, locally called "earth runs." (U.S. Geological Survey.)

FIGURE 14.22

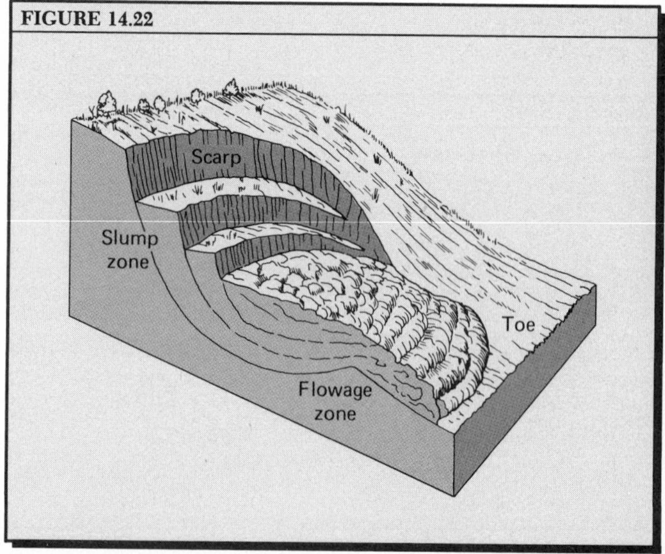

FIGURE 14.23

through a narrow, bottleneck passage.

Disastrous earthflows have occurred here a number of times since the occupation of Quebec by Europeans. A particularly spectacular example was the Nicolet earthflow of 1955, which carried a large piece of the town into the Nicolet River (Figure 14.25). Fortunately only three lives were lost, but the damage to buildings and a bridge ran into the millions of dollars.

Clays that spontaneously change from a solid condition to a near-liquid condition, called *quick clays,* are said to undergo *spontaneous liquefaction.* A sudden shock or disturbance will often cause a layer of quick clay to begin to liquefy, and once begun, the process cannot be stopped. A good example comes from the city of Anchorage, Alaska (Chapter 8). Severe ground shaking by the Good Friday earthquake of 1964 set off the liquefaction of quick clays underlying an extensive flat area, or terrace, which was the site of a housing development. Clay flowage allowed the overlying sediment layer to subside and break into blocks (see Figure 8.28).

A study of quick clays in North America and Europe has led to an explanation of their strange behavior. The clays are marine glacial deposits, meaning that they are sediments laid down in shallow saltwater estuaries during the Pleistocene Epoch. The thin, platelike particles of clay accumulated in a heterogeneous arrangement, often described as a "house of cards" structure. There is a very large proportion of water-filled void space between clay particles. It is thought that the original salt water saturating the clay acted as an electrolyte to bind the particles together, giving the clay layer strength. Some areas where quick clays occur have experienced a crustal uplift since they were laid down. The sediment deposit has been raised from immersion in salt water and the salt solution has been gradually replaced by fresh ground water. Now the clay is no longer bound by the electrolyte action and becomes sensitive to the sort of mechanical shock that causes the house-of-cards structure to collapse. Because a large volume of water is present (from 45 to 80% water content by volume), the mixture behaves as a liquid, with almost no strength remaining.

FIGURE 14.24 Block diagram of the 1898 earthflow near St. Thuribe, Quebec. (Redrawn, by permission of the publisher, from A. N. and A. H. Strahler, *Modern Physical Geography,* Figure 17.15. Copyright © 1978 by John Wiley & Sons, Inc.)

FIGURE 14.25 A portion of the city of Nicolet, Quebec, Canada, was carried into the channel of the Nicolet River by an earthflow that occurred in November 1955. The flow moved downvalley (right), passing beneath the bridge. (Raymond Drouin.)

FIGURE 14.24

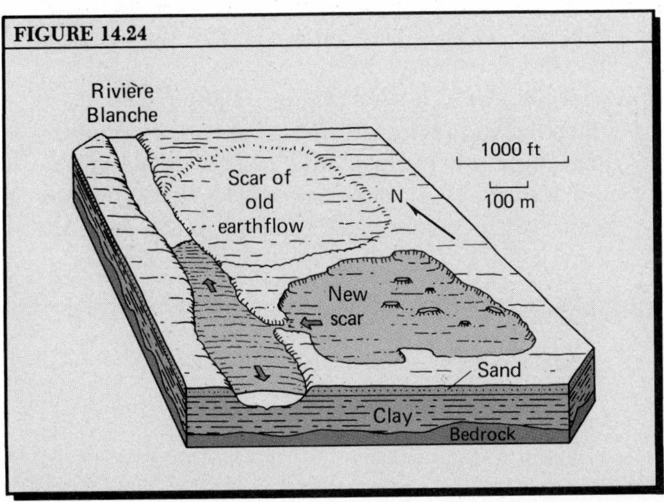

FIGURE 14.25

Mudflows

Where the proportion of water to mineral matter is large, mass wasting takes the form of a rather fluid mixture, termed a *mudflow,* capable of traveling rapidly in streamlike masses down the channels of streams. In nature one finds all gradations between earthflow and mudflow, and it is not practical to try to draw a precise line of distinction.

One type of mudflow, common in arid regions, originates in the watersheds of streams high in a mountain range. Here, torrential thunderstorm rains sometimes wash large quantities of loose soil and regolith down steep mountain slopes into the adjacent canyons. As the heavily laden streams progress downvalley, the loss of water by seepage and the increasing proportion of solid matter picked up from channel floors and banks cause the stream to thicken into a mudflow. The mudflow may attain the consistency of ready-mix concrete and will continue downvalley, where it commonly spreads out upon the plain at the foot of the mountain range (Figure 14.26). Eventually, thickening of the mud by loss of water causes flowage to cease.

Mudflows are also produced on the slopes of active volcanoes where torrential rains saturate freshly fallen volcanic ash. Herculaneum, a Roman village at the western base of Mount Vesuvius, was buried by mudflows in the eruption of A.D. 79.

Another type of mudflow originates high in alpine mountains, where melting of winter snows saturates deeply weathered rocks rich in clay minerals. As a mass of weathered rock on a steep slope begins to yield, it undergoes a spontaneous strength loss to form mud of high fluidity. Mudflows of this type occur seasonally and flow intermittently for weeks at a time.

Both earthflows and mudflows are common on the flanks of spoil heaps built from coal mines, including strip mines. In recent times a major disaster occurred at Aberfan, Wales, when a great waste heap 180 m high, built during coal-mining operations, spontaneously lost strength and moved quickly downslope. The tongue of debris overwhelmed buildings in the town below, causing the loss of over 150 lives (Figure 14.27).

The highly fluid type of mudflow, which resembles a water stream in flood, is called a *debris flood* in the far western United States. It is particularly prevalent in southern California, where it occurs commonly and with disastrous effects. Throughout urban areas of the coastal hills and mountains of California, housing developments have been extended to very steep hillsides by the process of bulldozing roads and homesites out of deeply weathered regolith and weak forms of bedrock. The excavated material is pushed into adjacent embankments, where its unstable mass poses a serious hazard to slopes and stream channels below. When saturated by winter rains, these embankments may give way, producing earthflows, mudflows, and debris floods that travel far down the canyon floors and spread out over lowland surfaces. When this happens, streets and yards are buried in bouldery mud and homes are severely damaged. Even where the mountain slopes have not been disturbed by grading operations, heavy winter rains produce debris floods, and these are particularly severe in areas where the vegetation has been burned off by brush fires in the preceding dry summer.

Landslides

Although the term *landslide* is often used in the news media and popular writing in reference to any form of rapid mass wasting, including earthflow, it is correctly limited to the rapid sliding of large masses of bedrock without plastic flowage in the early stages. As most landslides travel, they undergo a disintegration of the rock mass into a debris of assorted sizes that may travel with a gross flowage motion and produce a tonguelike body of rubble. In the early stages, however, the landslide moves with one of two basic types of motion: rockslide and slump. In a *rockslide*, a single block slides on its lower surface on a bedding plane, joint plane, or fault plane. In a *slump*, a block slips on a curved fracture plane rotating backward upon a horizontal axis as it sinks (Figure 14.28).

Rockslides are found in high mountain ranges of the alpine type where steep rock walls have previously been formed by glacial erosion (Chapter 20). Many great rockslides have been reported in the Alps, the

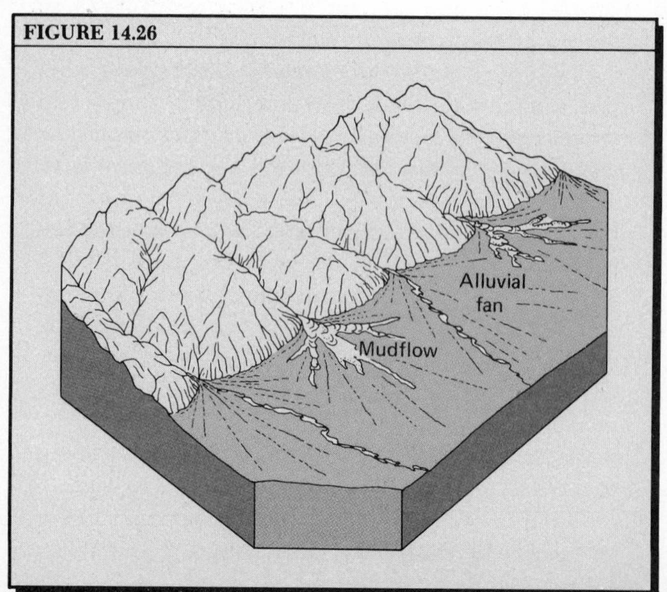

FIGURE 14.26 Mudflows issuing from the mouths of canyons in a semiarid region. (Redrawn, by permission of the publisher, from A. N. Strahler, *Physical Geography*, 4th ed., Figure 24.22. Copyright © 1975 by John Wiley & Sons, Inc.)

FIGURE 14.27 A destructive mudflow at Aberfan, Wales. (Sketched from a photograph.)

Canadian Rockies, and the mountains of Norway. A single slide may involve many millions of cubic meters of rock and can travel with the speed of a freely falling object. Towns, highways, and railroad lines in the path of the slide are obliterated. Immediate causes of rockslides are not often evident, and the time of their occurrence cannot be predicted. In some cases, hydrostatic pressure of water in the rock interstices may have pried the blocks loose.

An earthquake shock may set off a rockslide, an example being the Madison Canyon landslide caused by the Hebgen Lake earthquake of 1959 in Montana (see Figure 8.6). The Madison Slide involved the motion of 28 million cu m of rock which formed a part of the south wall of the canyon of the Madison River (Figure 14.29). The rock mass, which measured over 600 m in length and 300 m in height, descended 0.5 km to the Madison River at a speed later estimated to have exceeded 150 km/hr. The rock mass quickly disintegrated into boulders and pulverized rock, which crossed the canyon floor. Momentum of the moving mass carried the debris over 120 m in vertical distance up the opposite canyon wall. Acting as a huge natural dam, the slide debris blocked the Madison River and a new lake quickly began to form; within three weeks' time, it was 60 m deep. The lake outlet has since been stabilized and the lake is permanent; it is named Earthquake Lake. At least 26 persons died beneath the slide and their bodies have never been recovered.

The amazing speed and freedom with which rockslide rubble travels down a mountainside has been explained through the presence of a layer of compressed air trapped between the slide and the ground surface. The air layer reduces frictional resistance to nearly zero and may keep the rubble from making contact with the surface beneath.

When a slump block forms, a new plane of fracture is created which is almost vertical where it emerges at

FIGURE 14.28

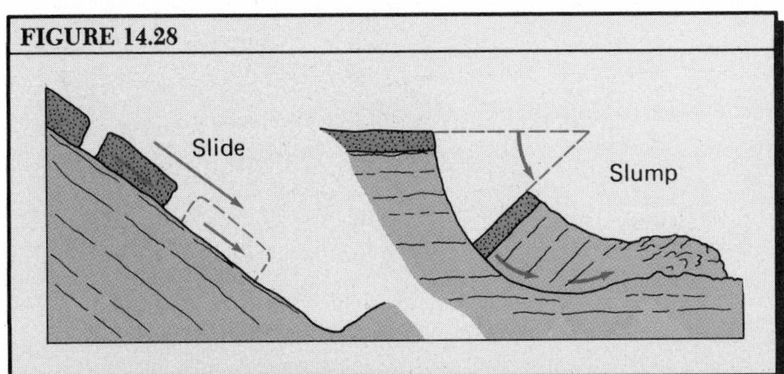

FIGURE 14.28 Rockslide and slump are two basic forms of landsliding. (Redrawn, by permission of the publisher, from A. N. Strahler, *Physical Geography*, 4th ed., Figure 24.23. Copyright © 1975 by John Wiley & Sons, Inc.)

FIGURE 14.29

FIGURE 14.29 Seen from the air, the Madison Slide forms a great dam of rubble across the Madison River Canyon. (U.S. Geological Survey.)

the upper surface. At depth, the fracture plane becomes upwardly concave and may emerge at the base of the slope with a reverse dip. Slumping of bedrock masses on a vast scale is found along cliffs of sedimentary strata or lavas (Figure 14.30). Blocks as long as 1 km or more and with a thickness of 300 m are known. Steepening of the cliff by running-water erosion of shale formations outcropping in the cliff base precedes the slump. Hundreds of such large slump blocks are known in the semiarid regions of the Colorado Plateau and the Columbia Plateau. Nearly all appear to be at least some thousands of years old and may have occurred under climatic conditions perhaps moister than exist there today.

The slumping of small masses of soil and regolith is seen on a small scale in the banks of streams or along cliffs cut into by wave action. The upper parts of many earthflows show the characteristic blocklike and stepped appearance of slumping.

Rockfall and Talus

On a near-vertical rock cliff, fragments are continually being pried free by the processes of physical weathering. The fall of any such fragment, whether it be merely a boulder or a huge mass weighing hundreds of tons, is described as *rockfall.* Although the fragment may bounce or roll down the cliff face, its descent is approximately at the speed of a freely falling body accelerating under the force of gravity. Most large falling rock masses shatter into many smaller fragments, strewing the base of the slope (Figure 14.31).

Rockfall that continues from a cliff face over many decades and centuries eventually builds a *talus slope* of loose rock fragments at the cliff base (Figure 14.32). Most cliffs have narrow, V-shaped recesses and ravines that funnel the fragments into a chutelike exit, causing the pile of fragments to take the shape of a *talus cone* (Figure 14.33). The fragmental material making up a talus deposit is called *slide rock.* The surface slope of a talus cone is remarkably constant at an angle of close to 35° with the horizontal.

The maximum angle of slope that can be held by a pile of loose coarse grains—whether of slide rock, sand, or gravel—is termed the *angle of repose.* Although this angle increases slightly as the size of fragments decreases, it is rarely less than 34° or higher than 37° in natural slopes composed of mixtures of sizes. Increased angularity of the particles and greater roughness of their surface texture yield the somewhat steeper repose angles.

In simple talus cones, particles tend to be finer near the apex, coarser near the base, because the greater momentum and radius of the larger particles permit them to travel farther. Very large fragments may roll well beyond the base of the talus cone.

FIGURE 14.30 A slump block descending from a cliff of horizontal strata rotates backwards on an axis parallel with the cliff. (Redrawn, by permission of the publisher, from A. N. Strahler, *Physical Geography,* 4th ed., Figure 24.26. Copyright © 1975 by John Wiley & Sons, Inc.)

FIGURE 14.31 This rockfall on the Palisades of the Hudson River, just north of the George Washington Bridge, took place in 1955. About 1200 metric tons of broken rock fragments resulted from the fall of a large columnar joint block of gabbro. The cliff is an exposure of an igneous sill of Triassic age. (Bergen Evening Record.)

FIGURE 14.32 Talus cones of quartzite fragments, Snowy Range, Wyoming. (A. N. Strahler.)

FIGURE 14.33 Talus cones. (Redrawn, by permission of the publisher, from A. N. Strahler, *Physical Geography,* 4th ed., Figure 24.16. Copyright © 1975 by John Wiley & Sons, Inc.)

FIGURE 14.30

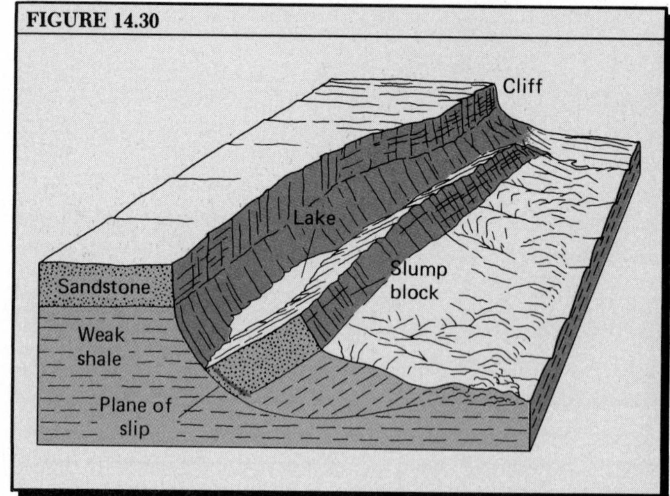

FIGURE 14.31

Weathering and Mass Wasting in Review

The passive processes of weathering and mass wasting act continuously on the land surfaces of the earth. Usually inconspicuous, these contributors to denudation perhaps rank first in importance to life on earth, in comparison with all of the agents involved in denudation. Without weathering there would be no soil, without soil no plant life, and without plant life no terrestrial animal life. Even the life of the oceans owes its existence to weathering, because the essential nutrients of plant life found in seawater are ions derived from rock weathering on the lands. Soil creep assists in continual renewal of the soil profile, allowing new parent matter to become exposed. In this way the supply of soil nutrients is replenished.

The more rapid forms of mass wasting are often destructive and cataclysmic in action. Not only do many forms of mass wasting—earthflows, mudflows, and rockslides—constitute an environmental hazard to humankind and its structures, but they are often in-

FIGURE 14.32

FIGURE 14.33

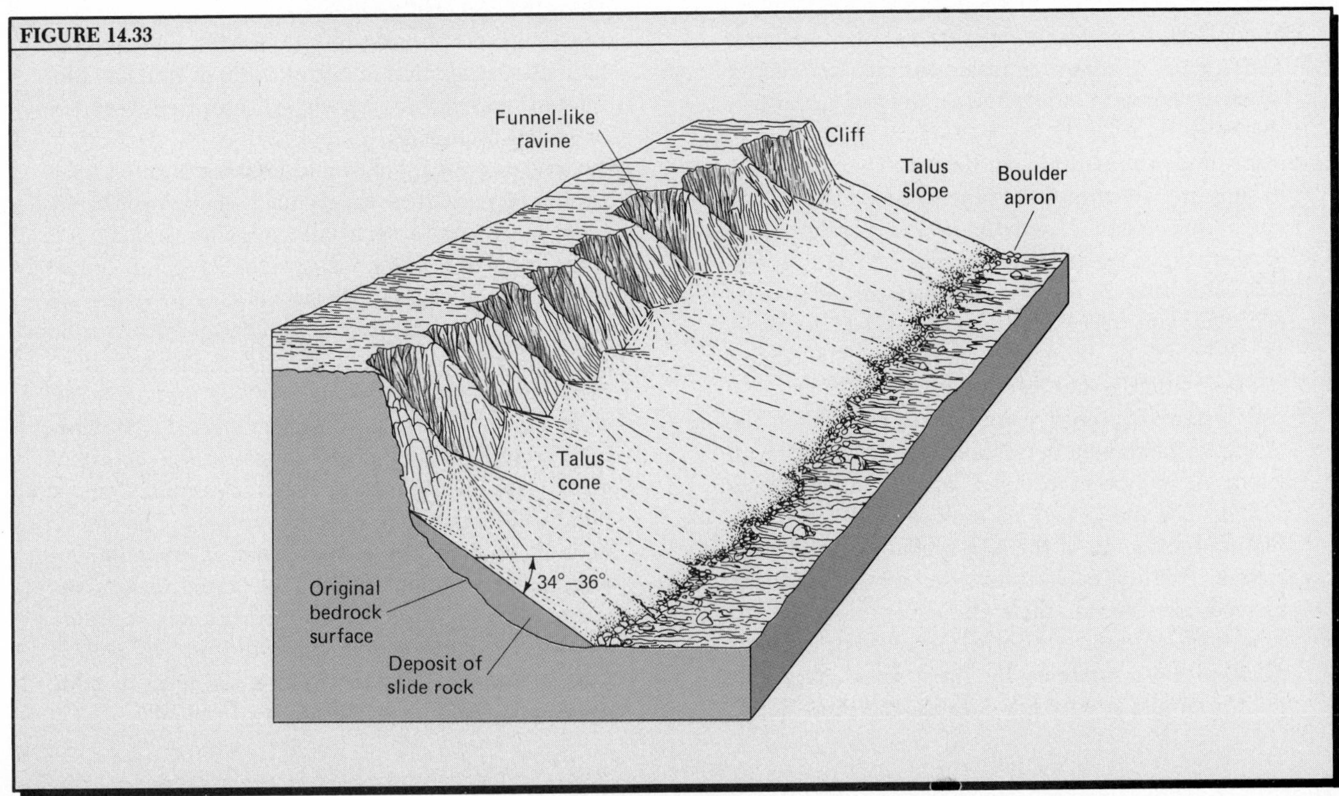

duced by human activities. Both the removal of support by excavation and the piling up of rock debris contribute to instability of slopes and lead to unwanted mass movements of regolith and soil.

The thin soil layer which supports plant life develops its unique properties in part by weathering processes and in part by organic processes. Soils differ greatly in natural fertility for agricultural crops, and much of this variation in fertility is a consequence of regional differences in forms and intensities of weathering. Soils of high productivity are limited in global extent; most are highly susceptible to deterioration and destruction under the impact of our expanding crop agriculture and urbanization.

Repeatedly in this chapter we have referred to the role of water. Water is essential to chemical weathering and to many forms of physical weathering and mass wasting. Water is, of course, essential to formation of mature soils and their horizons. In the next chapter we will investigate the geologic role of water in the soil, regolith, and bedrock beneath the land surface.

Key Facts and Concepts

Landforms *Geomorphology* is the study of *landforms*, which are land-surface features shaped by both internal and external earth processes. *Internal processes* include volcanic and tectonic activity. *External processes*, powered by solar energy, include weathering, *mass wasting*, and the *active agents* (fluid agents) of *erosion:* running water, glacial ice, waves and currents, and wind.

Denudation *Denudation* consists of the combined activity of weathering, mass wasting, and active agents, tending to reduce the continental surfaces to sea level.

Initial and sequential landforms Two landform groups are: (1) *initial landforms*, produced by volcanic and tectonic activity; (2) *sequential landforms*, shaped by denudation processes from the initial landforms. Sequential landforms are classed as *erosional landforms* and *depositional landforms*. A given landscape represents the status of a contest between internal and external processes.

Flow systems of energy and matter Each of the agents of landscape development is involved in an energy-flow system coupled with a material-flow system. The energy systems are open systems, representing subsystems of the total global solar radiation system. Within each subsystem energy is transformed and stored, then ultimately dissipated as heat. Three groups of external processes can be recognized: those sustained by direct solar energy input, by simple gravity-flow sytems, and those deriving energy from atmospheric motion.

Landforms of mechanical weathering Physical weathering processes not only cause rock disintegration (an irreversible change), but continue to cause rhythmic changes in the physical and chemical state of the mineral matter of regolith. Products of frost action are the *felsenmeer* (boulder field), *frost heaving* forms produced by *ground ice,* and features of *permafrost* such as *stone polygons* and *ice wedges*. Salt crystal growth forms *niches* and *rock arches* in sandstone. Shales disintegrate by *slaking*. Unloading produces *exfoliation domes* and sheeting structure in monolithic rock masses.

Effects of chemical weathering Chemical weathering of rock by oxidation, hydrolysis, and acid reaction results in granular disintegration, *spheroidal weathering*, and production of *saprolite*. Carbonation acts rapidly to dissolve carbonate rocks, producing pitted and grooved rock surfaces.

The soil layer The *soil*, uppermost zone of the regolith, is a natural zone containing living matter and supporting plants. Fully developed soils have *soil horizons*, seen in the *soil profile*, produced by interactions of climate with living organisms and affected by *relief* of the land surface. The mineral fraction of the soil consists of coarse *skeletal minerals* and colloidal clay minerals. The clay colloids are active in *cation exchange* involving acid-generating cations and *base cations*. Soils with high *cation-exchange capacity* (CEC) have *high base status* (fertile); those of low CEC have *low base status* (infertile).

Soil orders Major classes of soils are the *soil orders*. Examples are acid *Spodosols* of cool, moist climates, fertile *Alfisols* of moist forest climates, infertile low-latitude *Oxisols* rich in sesquioxides of iron and aluminum, and fertile, brown, calcium-rich *Mollisols* of semiarid climates.

Mass wasting All regolith and bedrock is acted upon by gravity, tending to produce downslope movement in various forms of mass wasting. *Soil creep* is a very slow downslope movement of regolith caused by continuous regolith disturbances. *Earthflows* are produced by rapid downslope flowage of saturated regolith. The motion is one of *plastic flowage* above the limit of the *yield stress*. In the arctic tundra, *solifluction* forms *solifluction lobes*. Earthflows may be formed by the *spontaneous liquefaction* of *quick clays*. *Mudflows* are rapid streamlike flows of highly fluid mud on steep mountains. The *debris flood* is an environmental hazard in mountainous arid regions. *Landslide* is the rapid downslope movement of a large mass of bedrock as a *rockslide* or with *slump* movement. *Rockfall* from cliffs builds a *talus slope* of *slide rock* taking the form of *talus cones* with surfaces at the *angle of repose*.

Questions and Problems

1. The rate at which solar energy flows as kinetic energy of the atmosphere and oceans is about ten times the rate of outflow of internal heat by conduction through the earth's surface. Does this mean that a geomorphic process such as wave action (generated by winds) expends energy at a much greater rate than internal energy is expended in all tectonic and volcanic processes combined? Should we conclude that wave action must bevel off all continents to sea level as fast as the crust is raised by tectonic and volcanic activity? Why must such a conclusion be false?
2. The maximum concentration of a natural solution of carbonic acid in rainwater or lake water is higher for cold water than for warm water. (Saturation of CO_2 in pure fresh water occurs at 0.8 ppm at 5 °C; at 0.5 ppm at 20 °C.) Does it follow that limestone succumbs to carbonic acid action faster in a cold climate than in a warm climate?
3. The Oxisols as a major soil group are found over vast areas of the tropical and equatorial latitude zones of the continents. Where rainfall is abundant the natural plant cover of the Oxisols is rainforest. Would it be wise to attempt to clear these rainforest areas and plant instead cereal crops such as corn or wheat? Explain your answer in terms of base status of soils.
4. The downturn of bedding in shales because of soil creep is often seen in road cuts along highways in mountainous areas. How might you go about distinguishing this downturn from true folding caused by tectonic forces?
5. In southern California, a serious environmental hazard is earthflow in unconsolidated strata and regolith rich in montmorillonite (smectite). Strangely enough, flowage may not commence until some weeks after winter rains have ceased and the ground has begun to dry out. Why does this time lag occur? What permanent engineering measures would you recommend to reduce the possibility of earthflow beneath homesites on a steep hillside susceptible to earthflows?

15

CHAPTER FIFTEEN

Ground Water and Its Geologic Activity

The total plan of movement, exchange, and storage of the earth's water is called the *hydrologic cycle*. Water moves from the world ocean to the lands and back, following various paths (Figure 15.1). This water can move and be stored in all three of its natural states: liquid, water vapor, and ice. Most of the world's water, 97%, is held in the oceans. Ice sheets and mountain glaciers hold about 2%, and water held on the continents amounts to only about 0.6%. The amount of water present as vapor in the atmosphere is an extremely tiny fraction of the total, or about 0.001%.

Some idea of the quantities of water passing through the hydrologic cycle each year can be had from Figure 15.2. As the principal reservoir of the earth's water, the oceans form a convenient point at which to start. An estimated 455,000 cu km of water evaporates annually from the ocean surface, and about 62,000 cu km evaporates from the lands, including lakes and marshes. Thus, a total of 517,000 cu km of water evaporates, and an equal amount must be returned to the earth's surface annually by precipitation. Of this, about 108,000 cu km falls as rain or snow upon the land surfaces. These figures show that the quantity precipitated upon the lands is some 73% greater than the amount of water returned to the atmosphere by evaporation from the lands. We conclude that the remaining precipitation on the lands, 46,000 cu km or about 46% is returned annually to the oceans in *runoff*, the liquid or glacial flow over and beneath the ground. The most obvious part of this return flow is, of course, by streams emptying into the oceans, but some water seeps into the ground and travels beneath the lands into the coastal ocean waters.

The study of the earth's water and its motions through the hydrologic cycle makes up the science of *hydrology*, which, like geology, is one of the geosciences. Geologists work closely with hydrologists in studying water on and beneath the lands. The hydrologist is primarily interested in "where water goes," the geologist in "what water does." The U.S. Geological Survey is responsible for measuring the flow of streams and investigating the movement and storage of all water lying beneath the ground surface. Water on and beneath the lands is a geologic agent, performing geologic work.

Surface Water and Subsurface Water

Water that flows in streams on the land surface or lies stagnant with an exposed upper surface in lakes and marshes is *surface water*. Water that lies beneath the land surface—enclosed in pores of the regolith or bedrock—is *subsurface water*.

Most soil surfaces can absorb the water from light to moderate rains and transmit it downward to the underlying regolith or bedrock by a process called *infiltration*. Natural passageways are available between individual soil grains and between the larger aggregates of soil. These passages may be soil cracks caused by previous drying; borings made by worms and burrowing animals; openings left by the decay of plant roots, or created by the alternate growth and melting of ice crystals. Such openings tend to be kept clear by the protective mat of decaying leaves and plant stems, which also acts to break the force of falling raindrops. When rain falls too rapidly to escape downward through the soil passages, the excess quantity escapes as a surface layer of water following the slope of the ground. This escaping surface water, known as *overland flow*, is one of the forms of runoff.

We have a choice now as to whether to trace first the paths of surface water in overland flow and in streams, or the underground paths. Because water emerges from the ground in many places to feed into streams, it will be best to begin with subsurface water.

Zones of Subsurface Water

Water is held in regolith and the underlying bedrock in different ways. For simplicity, let us refer to an idealized body of mineral matter beneath the ground surface. A good model will be a densely packed mass of pure sand (such as beach sand or dune sand) extending downward indefinitely. In such sand, about 35% of the bulk volume is open space; the voids are fully interconnected and are large enough to permit movement of water through the mass.

Under typical conditions of a moist climate with ample precipitation, there exists an upper zone of regolith and rock in which water is held in the form of small films or droplets clinging to mineral surfaces with a force stronger than the force of gravity. Because the adhesive force is called capillary force, this water is described as *capillary water.* Air occupies the remaining volume of the pore spaces, either as a connected network of air spaces or as separate air bubbles. This subsurface region of capillary water and air is known as the *unsaturated zone* (Figure 15.3), and it is geologically significant as a zone in which oxidation of mineral matter can take place. Hydrologists also recognize a *soil water zone*, a shallow layer holding capillary water within reach of plant roots. Water is returned to the atmosphere from this zone as water vapor, both by direct evaporation from the soil surface and by a process called *transpiration*, in which plants return water to the atmosphere by evaporation from leaf pores. Water that has percolated to a depth greater than that of the deepest plant roots remains in the unsaturated zone.

Below the unsaturated zone lies the *saturated zone*, in which all pore space is occupied by water called *ground water*, which moves slowly in response to gravity. The upper surface of this zone is the *water table*.

The capillary state of water above the water table is temporarily destroyed whenever heavy rain or rapidly melting snow allows large amounts of water to infiltrate. At such times, the soil openings are fully saturated, and the water moves down under the influence of gravity to reach the water table. In this way, the ground-water body is replenished, a process called *recharge*.

The depth to the water table at a given place can be

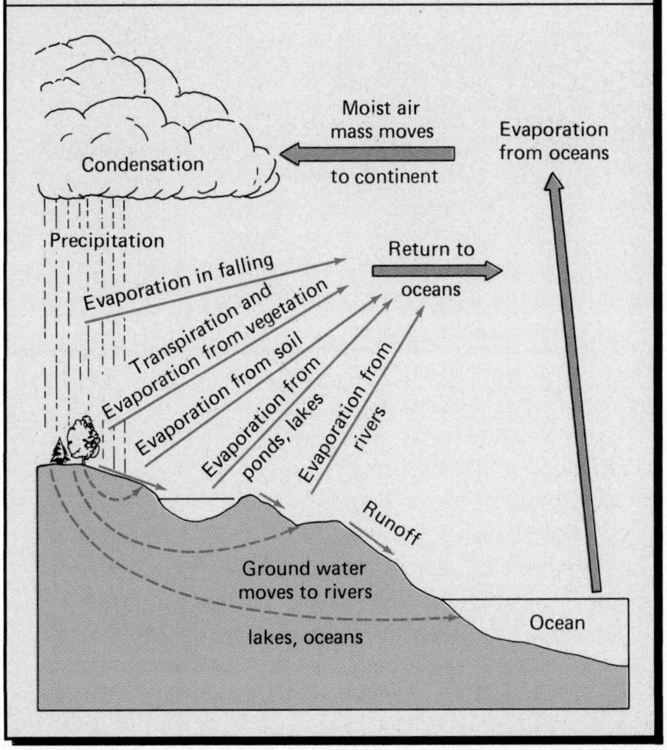

FIGURE 15.1 The hydrologic cycle traces the movement of water through the atmosphere and oceans and its flow over and beneath the land surfaces. (Redrawn, by permission of the publisher, from A. N. Strahler, *Physical Geography*, 4th ed., Figure 12.1. Copyright © 1975 by John Wiley & Sons, Inc.)

FIGURE 15.2 A schematic flow diagram of the hydrologic cycle showing the estimated quantities of water that flow through the principal pathways each year. Units are thousands of cubic kilometers. (Data of M. I. Budyko, 1971.)

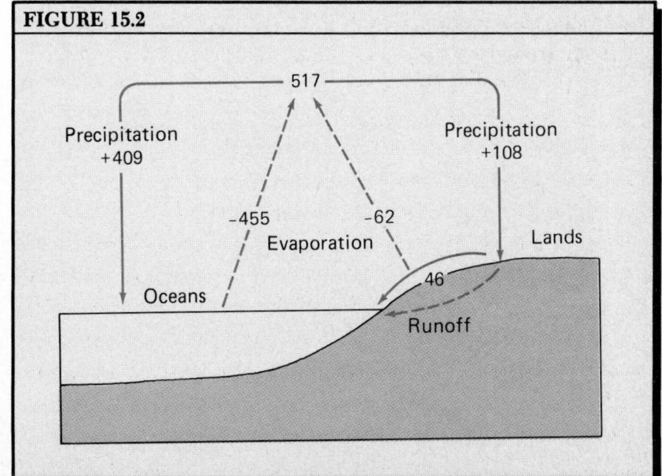

determined by noting the level at which water stands in a well of large diameter. In the surrounding regolith or bedrock, however, capillary force draws water upward in minute cracks and connected pore spaces. Actually, then, the rock can be saturated (part of the saturated zone) to a higher level than the standing water surface in the well. This additional layer of saturation, held up by capillary force and thus called the *capillary fringe* (Figure 15.3), may be no more than 2 or 3 cm in thickness in coarse, sandy materials, but can be more than 50 cm in thickness where the material consists of fine, silty material. Because the water table is usually determined from measurements of water height in wells, the capillary fringe is usually disregarded. Beneath flat, low-lying land surfaces, however, the capillary fringe may reach to the soil surface, where water continually evaporates. Steady evaporation in dry weather leaves behind various salts, and these often accumulate in the soil to form a salt crust. Many "salt flats" seen in desert regions originated in this way.

Ground Water Movement

Where many wells are closely spaced over an area, the configuration of the water table can be shown by connecting the levels of standing water in the wells (Figure 15.4). The water table will usually be highest in elevation under the hill summits. From these high points, it slopes toward the nearest valleys, intersecting the surface in the channels of streams or at the shores of lakes and marshes.

Water in an open body, such as a lake, assumes a horizontal surface because there is little resistance to flowage. In the saturated zone, however, gravity movement of ground water is through the very tiny spaces between mineral grains and along thin cracks in bedrock. These narrow spaces impede flowage. Consequently, percolating water reaching the water table under the hill summits cannot readily escape. The ground water tends to accumulate, raising the water table to higher levels than at the streams, where water is escaping. The difference in level between the water table under a summit and at the low point of a valley sets up an unbalanced pressure, under which the water flows very slowly within the ground-water body, following curved paths as shown in Figure 15.5. A particular molecule of water, if it could be traced, might follow a curving path that carries it deep into the rock and returns it by upward flow to the line of a stream channel, where the water escapes into the stream as surface runoff.

Speed of flow of ground water has actually been measured in many places, using dyes, salts, or radioactive substances as tracers. The normal range is between 1 m per year and 1 m per day, but the flow can be much slower or much faster, depending upon the nature of the material through which the water moves, and rates as high as 30 m per day have been measured.

FIGURE 15.3

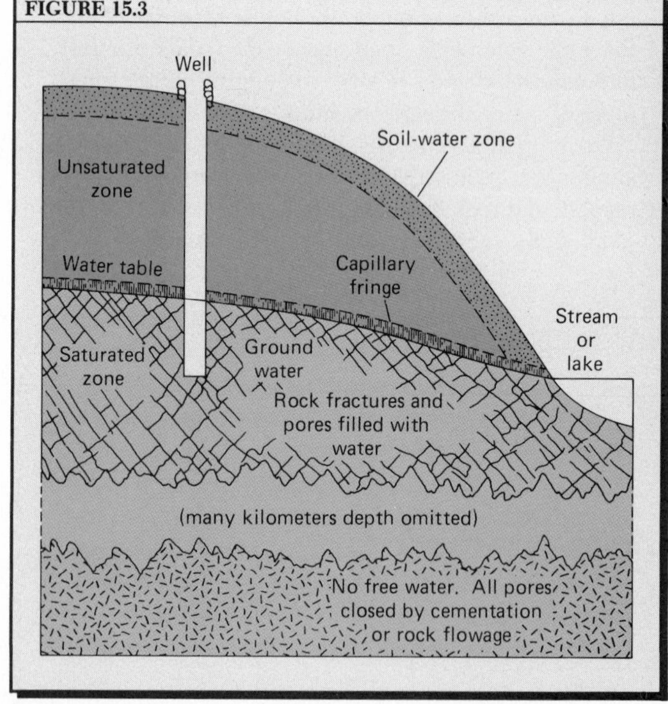

FIGURE 15.3 A schematic cross section through the various zones of subsurface water. (Based on a diagram by Ackerman, Colman, and Agorsky, U.S. Dept. of Agriculture.)

FIGURE 15.4 Position of the water table can be determined by the height of standing water in wells. (Redrawn, by permission of the publisher, from A. N. Strahler, *Physical Geography*, 4th ed., Figure 13.3. Copyright © 1975 by John Wiley & Sons, Inc.)

FIGURE 15.4

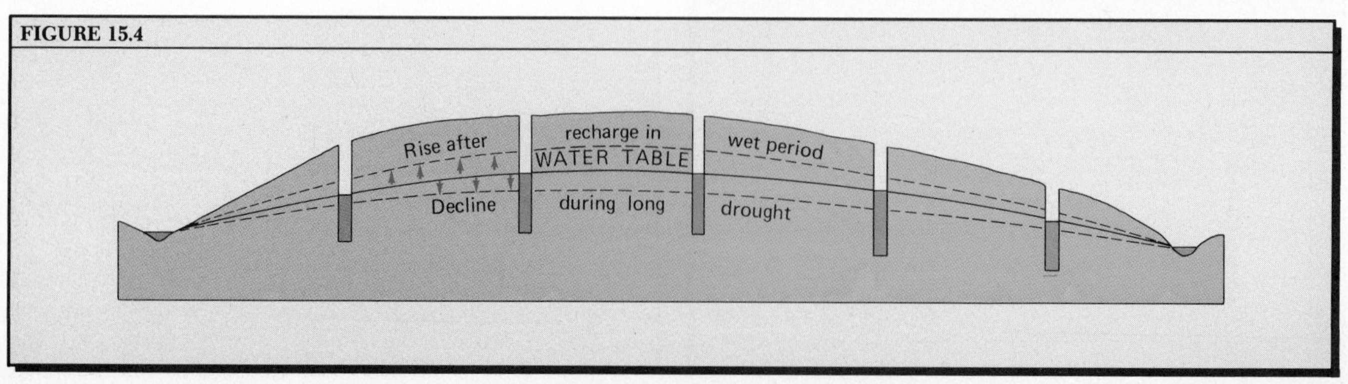

With a generally constant climate, the water table will become approximately fixed in position. The rate of recharge will, on the average, balance the rate at which water is returned to the surface flow by seepage in streams, lakes, and marshes. When there is a period of unusually dry years, the water table slowly falls, while during a period of unusually wet years, it will gradually rise (Figure 15.4).

Smaller seasonal fluctuations in the level of the water table are normal in certain climates. A seasonal decline in the water table results from the cutting off of recharge when a dry season occurs or when soil moisture is solidly frozen. A seasonal rise results from recharge by the percolation of excess rainfall or snowmelt through the unsaturated zone.

FIGURE 15.5 Ground water follows strongly curved paths of travel where the subsurface material is uniform throughout. (Redrawn, by permission of the publisher, from A. N. Strahler, *Physical Geography*, 4th ed., Figure 13.2. Copyright © 1975 by John Wiley & Sons, Inc.)

FIGURE 15.6 Graph of changing water level in an observation well on Cape Cod, Massachusetts. Recharge usually is greatest in the late winter and early spring, whereas little recharge occurs after midsummer. (Redrawn, by permission of the publisher, from A. N. Strahler, *Physical Geography*, 4th ed., Figure 13.4. Copyright © 1975 by John Wiley & Sons, Inc. Data of U.S. Geological Survey.)

FIGURE 15.5

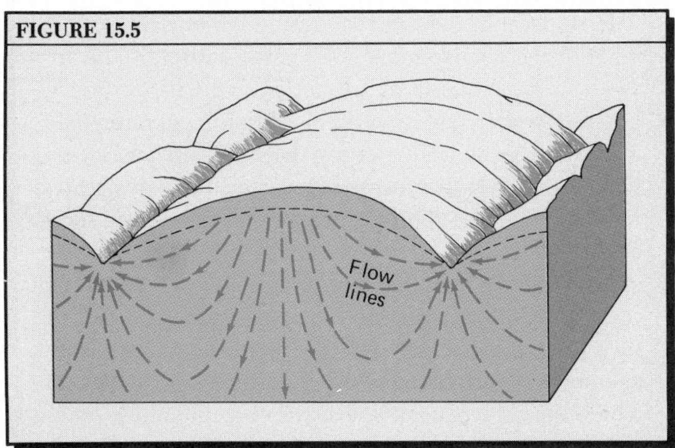

FIGURE 15.6

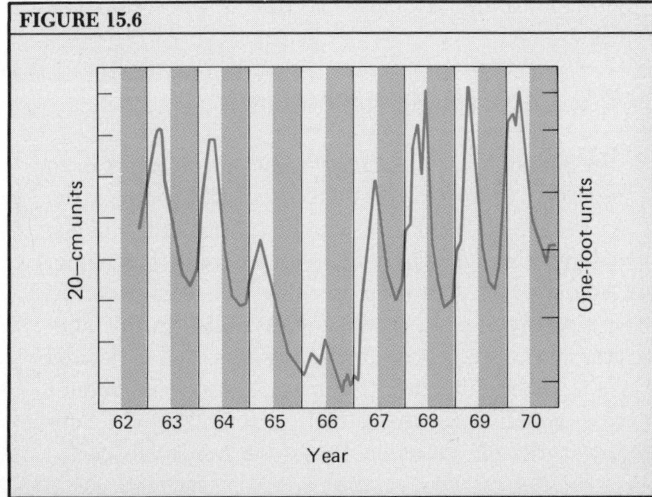

The seasonal cycle of fluctuation in water table height is illustrated by the record of a well on Cape Cod, Massachusetts (Figure 15.6). To obtain data such as these a special observation well is used. No water is withdrawn from the observation well and it is placed far away from wells from which water is being pumped in large quantities. The graph in Figure 15.6 shows the effects of a major drought in 1965 and 1966. Because recharge was small in these years, the average level declined severely.

*Darcy's Law

The speed of flow of ground water can be estimated quite accurately by using a simple formula. It was first stated in 1856 by a French hydraulic engineer, Henry Darcy, and has since been known as *Darcy's law*. A simplified version of Darcy's law is as follows:

$$V \propto \frac{H}{L}$$

where V is the average speed (velocity) of flow of the ground water along its path from a high point to a low point

$\propto$ means "proportional to"

H is the vertical distance between the two points on the flow path (the *hydraulic head*)

and L is the length of the flow path

Figure 15.7 illustrates some terms in this equation. The flow path lies between the points P_1 and P_2. The ratio H/L is called the *hydraulic gradient;* it is a measure of the average steepness of the flow path.

To express Darcy's law completely, we would need to introduce another term that tells whether the connected openings within the rock mass are small or large. We shall say something about this subject in a later paragraph. For the moment, we assume that the material with which we are dealing is always the same in physical properties, for example, a well-sorted, fine-

FIGURE 15.7 Diagram showing the meaning of terms used in Darcy's law.

FIGURE 15.7

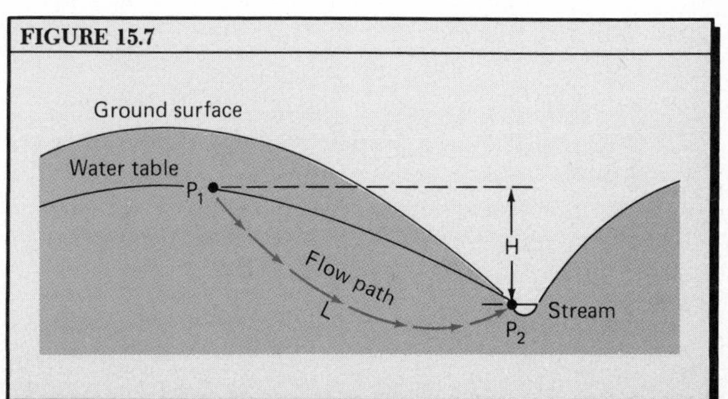

grained quartz sand.

From Darcy's law, we see that the average speed of flow of ground water in the very deep paths shown in Figure 15.5 must be very slow, because the path length, L, is very great. Flow will be faster in the shallow paths, where L is relatively short.

Darcy's law tells us only the average speed of water flow between the entering and exit points, P_1 and P_2. Actually, the speed may increase or decrease along the flow path. For example, speed normally increases near the exit point, P_2, where the flow paths become closely crowded together (as shown in Figure 15.5).

Porosity and Permeability

Ground water can occupy openings in any type of geologic material, whether it be bedrock or regolith. The bedrock may be of any variety of igneous, sedimentary, or metamorphic rock, and the regolith may range from dense clays to coarse gravels. Consequently, the speed of flow of ground water and the quantity which can be held in the rock are subject to wide variations. It is such variations that concern ground-water geologists, who must apply the laws of flow of fluids to varied and complex geologic conditions.

The total volume of pore space within a given volume of rock is termed the *porosity.* It gives an indication of the capacity of a rock to hold a fluid in storage. Clastic sedimentary rocks, such as sandstone and conglomerate, can have high porosity because of the relatively large openings possible between the well-sorted, rounded grains (see Figure 5.10*A*). Transported regolith of stream-laid gravels and sands or beach deposits of hard, well-rounded coarse mineral grains have a high porosity. Where sands are mixed with silts, the porosity is reduced, because the fine particles fit into the openings between the large grains (see Figure 5.10*B*).

Soft clays and muds have a high original porosity, but when compacted into shale they have extremely low porosity. Some rock masses have very large openings—for example, limestone, which may have cavernous openings resulting from solution. Scoriaceous lavas have numerous bubblelike cavities formed by expanding gases.

Certain of the very dense rocks, such as igneous plutons and metamorphic varieties, have negligible pore space in the fresh, unweathered state because the mineral crystals are tightly intergrown. If fractured severely, however, the porosity of these rocks can be moderately great.

Porosity is measured in percentage—the ratio of open void space to total volume of the rock mass. Porosity of well-sorted, well-rounded grains in the size range from sand to gravel usually ranges between 30% and 50%. In a sandstone consisting of the same grade sizes, the porosity is usually much less—10% to 20%—because cementing mineral matter fills much of the void space that was originally present. Dense shale and limestone have porosities in the range from 1% to 10%. Porosity is a property of great importance in petroleum geology, as well as in ground-water geology, because it determines in large part the ability of a given rock formation to hold in storage liquid petroleum or natural gas.

Even though a rock has a high porosity, water cannot move freely through the rock mass unless the openings are interconnected and of sufficient diameter to permit flow. The property of *permeability* is the relative ease with which water will move through the rock under unequal pressure. Permeability is of primary importance in determining the rate of ground-water movement and the amount of water that can be withdrawn by pumping from wells.

Permeability of unconsolidated sands and gravels is often extremely high, while that of dense clays and shales is extremely low. In fact, clay and shale can be described as impermeable rocks for all practical purposes. Permeability of sandstone formations is commonly high but may be greatly reduced if the pores are closed by cementation of mineral matter. Igneous and metamorphic rocks, where they are broken by numerous joint fractures and faults, may have high permeability.

How deep in the crust is ground water held, and at what depth does its movement cease entirely? The answers to these questions are not known with certainty, but few wells yield any water at depths greater than 3 km and most wells deeper than 0.8 km yield very little water, which means that permeability is greatly reduced at these depths. Perhaps this is because most of the pore spaces have been filled with mineral water or have been forced to close under the high confining pressure. Complete closure of openings in all kinds of rock takes place under pressure in a zone of rock flowage in which rock exhibits ductility. The depth at which the zone of rock flowage is encountered is probably at least 15 km, and this is well within the continental crust and far above the base of the continental lithosphere.

*The Coefficient of Permeability

In our discussion of Darcy's law, we noted that an important term is omitted from our statement of the equation. This term measures the permeability of the material through which the ground water is moving. Designated K, the term is called the *coefficient of permeability.* It is a quantity determined in the laboratory, using rock or sediment samples, and is defined as the volume of water that passes through a standard cross section of the given rock material under a standard hydraulic gradient. When we insert this coefficient into our earlier equation, Darcy's law becomes

$$V = K\frac{H}{L}$$

If we let the value of the coefficient of permeability be represented by the number 1 (unity) for fine sand, the value would be as high as 10^6 for clean gravel, but as low as 10^{-4} for dense unweathered clays consisting of such minerals as illite or montmorillonite. This means that the total range of permeability spans ten orders of magnitude. By the same token, the speed of flow of ground water can range through ten orders of magnitude for a single value of hydraulic gradient.

Aquifers and Aquicludes

Layered rocks can offer strongly contrasting zones of permeability, particularly where sedimentary strata of varying types are interlayered. In this way, the geology of an area exerts a strong control on the movement of ground water. Figure 15.8 shows a thick layer of sandstone with a shale formation below it.

A sandstone layer, when high in porosity and permeability, can hold and transmit large quantities of water; it is designated as an *aquifer*. A shale layer may be almost impermeable, preventing or greatly retarding flow of ground water through itself, and such a layer is called an *aquiclude*.

As shown in Figure 15.8A, the main water table lies within the sandstone aquifer, whose lower part is occupied by the upper portion of the saturated zone. The shale aquiclude is also in the saturated zone, but the water it holds is, for all practical purposes, stagnant, while ground water moves rather rapidly through the aquifer. The direction of this motion must necessarily be almost horizontal, and it is in the direction of downward slope (gradient) of the water table.

The diagram also shows a lenslike body of shale within the upper part of the sandstone formation. Acting as an aquiclude, this lens blocks the downward percolation of water through the unsaturated zone. As a result, a lenslike body of ground water collects above the shale lens. The upper surface of this ground-water lens is known as a *perched water table*. Ground water in this lens moves toward the edges of the lens, from which it trickles down through the lower part of the aquifer and eventually reaches the main water table. The aquifer shown in Figure 15.8 is described as an *unconfined aquifer*, because the water table is free to receive recharge from above and can rise or fall freely within the aquifer in response to changes in the amount of recharge received.

Diagram *B* of Figure 15.8 shows how aquicludes can influence the way in which ground water emerges at the surface. Using the same geological setup as in Diagram *A*, we have shown a deep river valley cut through the strata. The shale lens is now exposed in the upper wall of the valley. Ground water emerges from the perched water table in a line of springs. A *spring* is any flow of ground water emerging naturally from the solid surface of the earth. (Springs may also emerge from the floors of streams, lakes, or the ocean.) A second line of springs is formed in the lower wall of the valley, where the main water table intersects the surface. Springs of emerging ground water occur under a wide variety of geological conditions, and we have shown only one simple case.

Springs are usually insignificant features, going unnoticed because of a concealing cover of vegetation, and many small springs cease to flow in summer, but some large springs yield copious, sustained flows of water. An impressive example is the Thousand Springs of the Snake River Canyon in Idaho, where a large volume of flow emerges from highly permeable layers of scoriaceous basalt (Figure 15.9).

FIGURE 15.8 Unconfined ground water in relation to aquifers and aquicludes. (Drawn by A. N. Strahler.)

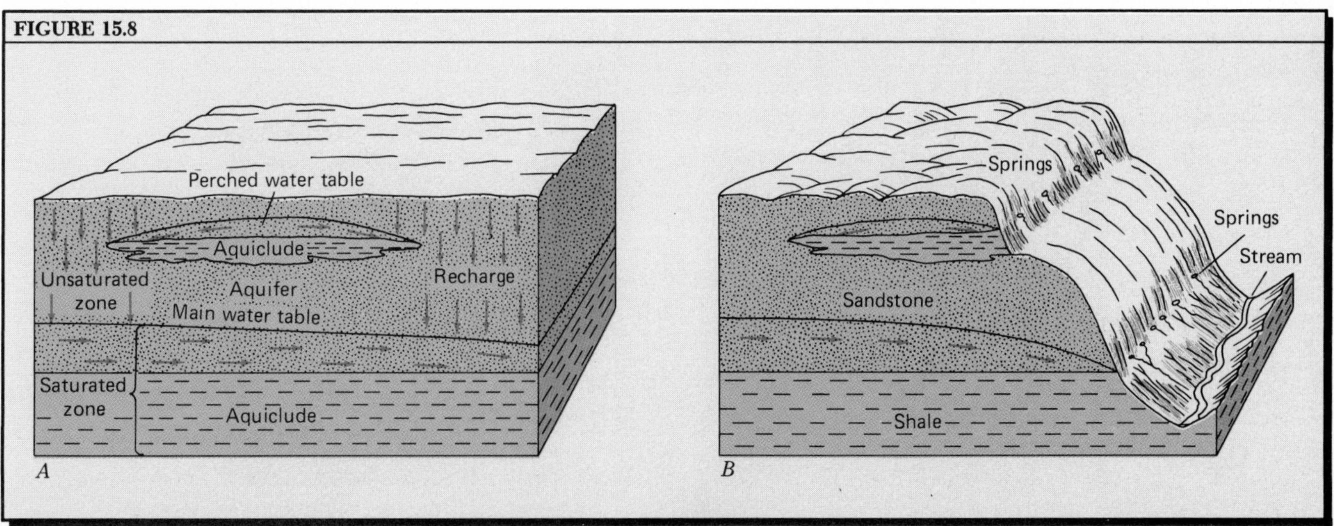

Gravity Wells

To supply fresh water for human needs, ground water in unconfined aquifers must be reached by wells and lifted to the surface against the pull of gravity. Wells of this type are classed as ***gravity wells***. In preindustrial times, the small gravity well needed to supply domestic and livestock water needs of a home or farmstead was actually dug by hand as a large cylindrical hole and lined with masonry where required. Dug wells are still in wide use today in the less-developed nations.

In contrast, the modern gravity well is drilled with powerful rotary drilling equipment capable of penetrating hard bedrock as well as unconsolidated regolith. These drilled wells are often called ***tube wells*** because a steel pipe, or casing, is lowered into the hole to provide a barrier against impure near-surface water and to prevent clogging by particles broken from the sides of the drill hole. A perforated section of casing at the bottom of the well admits water from the aquifer. Small domestic wells range in diameter from 5 to 10 cm, whereas large wells that supply industrial and irrigation water may have diameters of 30 to 50 cm or larger. While the small domestic well can yield up to 200 liters of water per minute, the large industrial wells can furnish several million liters per day where the aquifer is highly permeable.

Pumping of water from a well commonly exceeds the rate at which water can enter the well, so the water level drops progressively (Figure 15.10). As this is done, the water table is drawn down into a conical form, or *cone of depression*, surrounding the well. The *drawdown* is the difference in height between the surrounding water table and the water level in the well. Formation of the cone of depression actually increases the rate at which water flows into the well, and so increases the yield of the well, but this effect is limited. Beyond a critical limit of drawdown, the yield no longer increases.

The cone of depression may extend several kilometers from a large well, and the combined effect of closely spaced wells is to depress the water table generally. Continued heavy pumping can eventually lead to a serious decline in water yield. In arid regions, particularly, the ground water stored in valley deposits can be withdrawn much faster than it is restored by seepage from the beds of surface streams. In humid regions, careful regulation of pumping and the artificial recharge of the ground-water table by waste water, pumped down into ***recharge wells***, can bring about a balance in the rates of withdrawal and recharge.

Artesian Ground Water

Another type of water well is the ***artesian well***, in which water is forced upward under natural pressure. Geologic conditions necessary to produce an artesian well are illustrated in Figure 15.11. The aquifer consists of an inclined sand or sandstone layer, receiving water along its exposed outcrop at a relatively high position on a ridge or mountain. Ground water moves through the aquifer and is confined below an aquiclude of clay or shale. Hydrostatic pressure in this ***confined aquifer*** is sufficient to raise water in wells to levels higher than the upper surface of the aquifer. The upward water movement is called ***artesian flow***; its upper limit of rise is the ***hydrostatic pressure surface***, shown by a sloping line in Figure 15.11. The elevation of this surface declines with increasing dis-

FIGURE 15.9 Thousand Springs, Idaho. Located on the north side of the Snake River Canyon, nearly opposite the mouth of the Salmon River, these great springs extend for 0.8 km along the edge of a layer of scoriaceous basalt. The copious discharge is nearly constant. (U.S. Geological Survey.)

FIGURE 15.9

tance from the intake area because energy is expended in the flow of ground water through the aquifer. As the diagram shows, a flowing artesian well occurs where the ground surface and wellhead lie below the pressure surface. Where the wellhead lies above the pressure surface, the water will rise only to the level of that surface; it will not emerge as a flowing well. Natural artesian springs can occur where faulting or solution removal has produced a passageway for water to make its way upward through the confining aquiclude.

The word "artesian" is taken from a French word ***artêsien***, meaning "pertaining to the province of Artois." In about 1750 in this northernmost province of France, deep wells were first drilled to reach confined aquifers and obtain artesian flow. Since then, many regions have furnished large amounts of artesian water.

One of the most extensive areas producing artesian ground water is the Great Plains region, covering parts of Kansas, Nebraska, and eastern Colorado and extending northward into the region of the Black Hills of South Dakota. Here the Dakota sandstone underlies the entire area, overlain by the Pierre shale, and has furnished artesian water in many places. In eroded stratigraphic domes, such as the Black Hills uplift, the Dakota sandstone is exposed in high, steeply dipping hogback ridges, which serve as intake surfaces for precipitation. Numerous irrigation wells were drilled into the Dakota formation early in this century, and artesian water was withdrawn at a rate that greatly exceeded the rate of recharge. Consequently, few of these wells produce surface flow today and most must be pumped to furnish water.

Another region highly favorable to artesian water is the Coastal Plain of the Atlantic and Gulf coasts of the United States (Figure 15.12). Here the seaward-dipping strata of the miogeocline contain many aquifers of sand and sandstone, confined beneath impermeable clays and shales. Most of this coastal belt has ample precipitation to recharge the aquifers.

*Ground Water Pollution and Contamination

Ground-water supplies are subject to pollution by various kinds of human-made wastes that infiltrate the ground and become a part of ground-water recharge. Discharge of sewage and liquid industrial wastes into infiltrating basins forms one potential pollution source, while landfill waste disposal sites (dumps) are another. In both cases, because of the abnormally high rate of infiltration, the water table is built up into a mound. From this high point, the infiltrating fluid moves out-

FIGURE 15.10

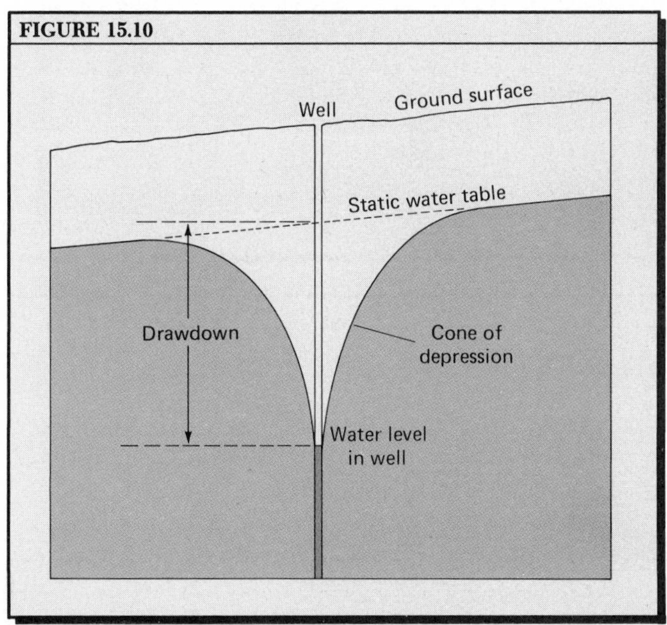

FIGURE 15.10 Rapid pumping of a gravity well produces a cone of depression in the water table. (Redrawn, by permission of the publisher, from A. N. Strahler, *Physical Geography*, 4th ed., Figure 13.6. Copyright © 1975 by John Wiley & Sons, Inc.)

FIGURE 15.11 Schematic cross section of a confined aquifer, producing artesian flow in wells.

FIGURE 15.11

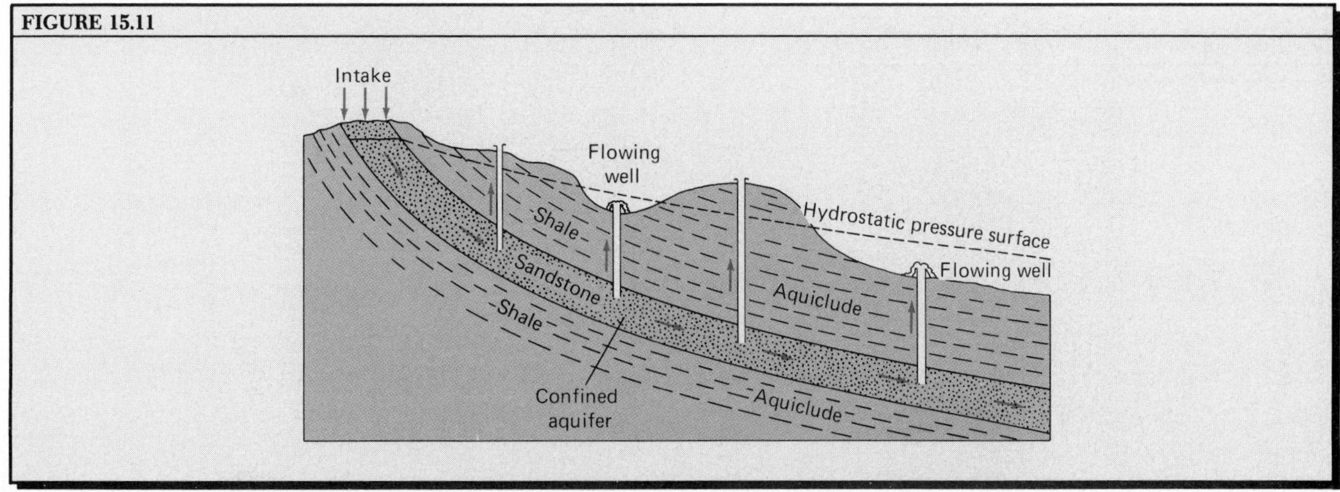

ward to surrounding parts of the ground-water body, where it may emerge in streams or lakes or may be drawn into nearby wells (Figure 15.13).

Another common ground-water pollutant is road salt, which is used on highways and city streets in winter and can pollute shallow wells close to the roadways. Spills of petroleum fuels and chemicals from highway accidents and from leaking storage tanks are yet another source of ground-water pollution.

An important form of contamination of fresh ground water is by salt water drawn into coastal water supply wells. For this reason, the relation of fresh ground water to salt ground water in coastal zones and beneath islands is most important in the management of water supplies. Figure 15.14 is a highly diagrammatic cross section of the ground-water relations under an island or in a long, narrow peninsula. Fresh water, being less dense than salt water, forms a ground-water body resembling a huge lens, with convex surfaces above and below.

The fresh-water body actually floats upon the salt water, much as the hull of a ship displaces water and floats at rest. Normally the densities of fresh water to salt water are in the ratio of 40:41. Consequently, the elevation of the upper surface of the water table above sea level is one-fortieth as great as the depth of the base of the fresh-water body below sea level. For example, in approximate figures, if the bottom of the fresh-water lens lies 40 m below sea level, the ground-water table will rise 1 m above sea level.

FIGURE 15.12 A relic of the past. This gushing artesian well was a common sight a half century ago. Because of excessive water use, most such wells no longer bring water to the surface under natural pressure. Oneco, Florida. (Ewing Galloway.)

FIGURE 15.13 Pollution from a solid-waste disposal site can move within the ground water beneath to reach a well or a stream.

FIGURE 15.14 Relation of fresh to salt ground water under an island or peninsula. (Redrawn, by permission of the publisher, from A. N. Strahler, *Physical Geography*, 4th ed., Figure 13.8. Copyright © 1975 by John Wiley & Sons, Inc.)

FIGURE 15.12

FIGURE 15.13

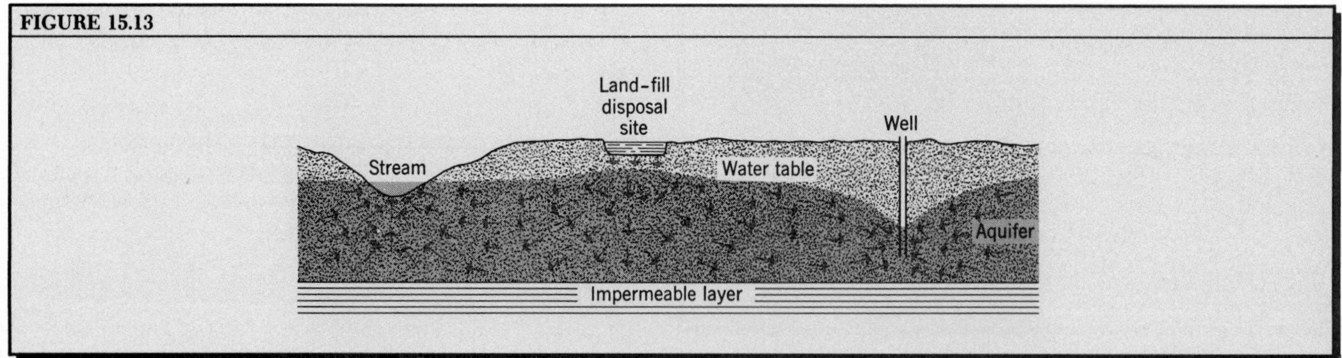

FIGURE 15.14

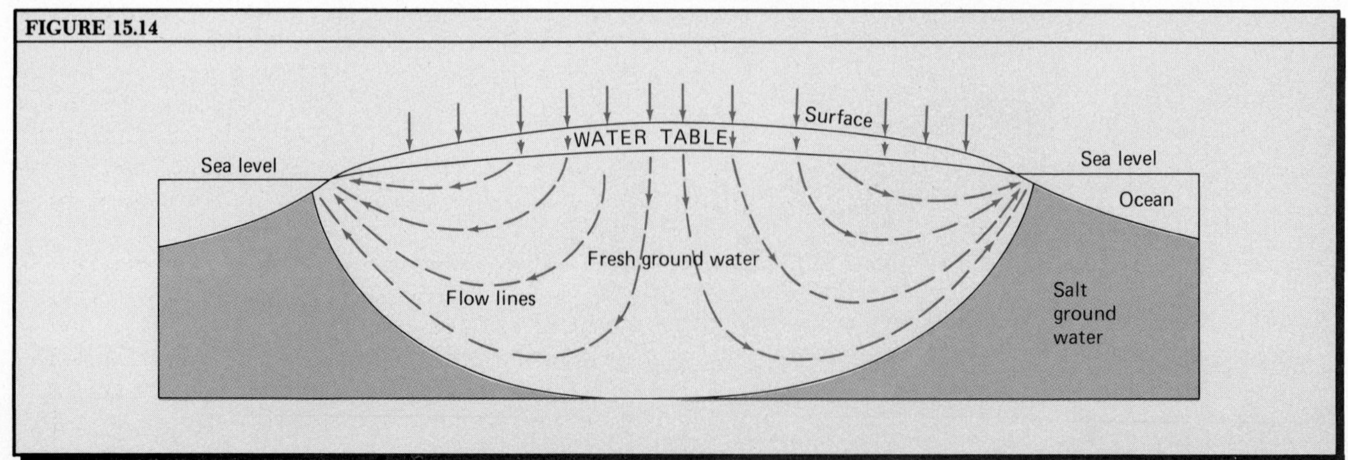

As the shoreline is approached, the contact between fresh and salt water rises in elevation and emerges on the sea floor along a line close to the land. As the arrows in Figure 15.14 indicate, the fresh ground water follows deeply curving paths. These paths turn upward and seaward to emerge under the ocean close to the coast. The salt ground water remains more or less stagnant beneath the fresh. Some mixing of the salt and fresh water will occur through the effects of tidal changes in ocean level.

A serious environmental problem is that the rapid pumping of water from wells located close to a coastline will draw the salt-water–fresh-water contact landward, and there will come a time when the salt water begins to be drawn into the well, contaminating the fresh-water supply (Figure 15.15). If pumping is stopped for a long period, the salt water will again be pushed seaward. This process can be hastened by pumping fresh water down into the contaminated wells.

*Ground-Water Withdrawal and Land Subsidence

One serious environmental effect of excessive ground-water withdrawal is land subsidence, which is particularly severe where water is pumped from thick beds of unconsolidated sediment. These are beds of gravel, sand, and silt lying beneath broad valley floors. As the water table declines, the silty sediments which come into the unsaturated zone become compacted under the load of overlying layers.

A now-classic example of subsidence is that of the San Joaquin Valley in California. Over a period of many years, the water table beneath the valley has been drawn down over 30 m. Land subsidence has been as much as 3 m in a 35-year period and has resulted in damage to wells in the area. Water withdrawal beneath the city of Houston, Texas, has resulted in subsidence of from 0.3 to 1 m over an area 50 km in diameter, with serious damage to pavements, building foundations, airport runways, and flood-control works. Minor faults have been activated by the subsidence and can be traced through the city. Mexico City has suffered even greater subsidence—from 4 to 7 m in the period between 1891 and 1959—as a result of ground-water withdrawals from aquifers beneath the city.

FIGURE 15.15 Salt-water intrusion into a well situated near the ocean.

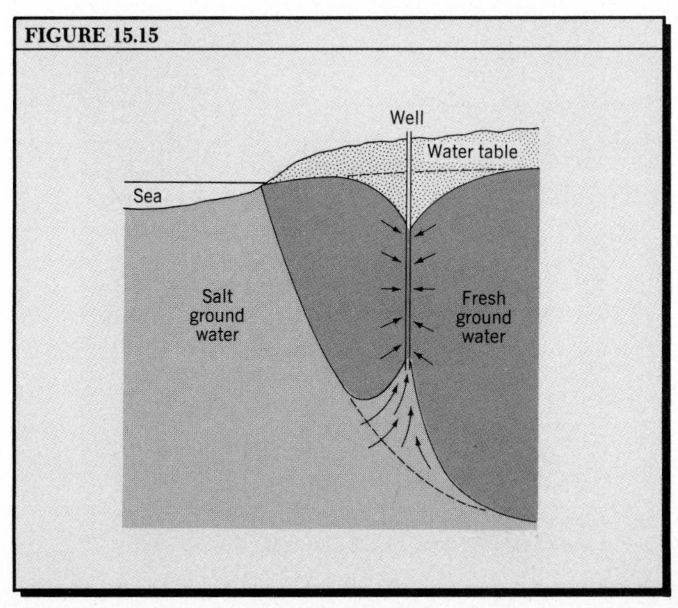

Subsurface Water as a Geologic Agent

Subsurface water in motion is an important geologic agent despite its extremely slow rate of flowage, and its action in the unsaturated zone above the water table is quite different from that in the saturated zone below. All natural subsurface water is capable of chemical activity because it contains various ions in solution as well as dissolved molecular gases of the atmosphere. Perhaps the most important chemical reagent in subsurface water is carbonic acid (H_2CO_3). In Chapter 5, we explained how carbonic acid is formed as atmospheric carbon dioxide is dissolved in water. Another acid commonly present in subsurface water is sulfuric acid. It can be formed from sulfur compounds, abundant in organic sediments, and usually present in coal, lignite, peat, and liquid petroleum. Various complex organic acids are generated in the soil layer by plant decay. Thus, weak acids in the water droplets and films that occupy the unsaturated zone can react with mineral surfaces. Because air is also present in the unsaturated zone, oxidation of minerals is an important natural process. Ions released by these reactions are carried down to the ground-water body, where they circulate through permeable rocks.

Circulating ground water can take part in any one of three kinds of chemical activity affecting mineral matter: solution removal, precipitation, and replacement. Acids in ground water can remove mineral matter in solution, as they do in the unsaturated zone. Under a different set of conditions, precipitation of mineral matter is carried out by circulating ground water. Precipitation of silica and carbonates of calcium and magnesium cement loose grains of sand, silt, and gravel in sedimentary strata, causing them to be hardened (lithified) into strong rocks.

Mineral replacement by ground water action consists of the slow removal—one at a time—of ions, atoms, or molecules of organic or mineral matter and their simultaneous substitution by mineral ions carried by the ground water. Perhaps the example best known to amateur mineral and rock collectors is petrified wood, formed as silica (SiO_2) gradually replaces the organic molecules of logs that have been buried in sediments. In many cases, the replacement has left intact fine details of the cell structure of the tree trunk. Fossil shells also undergo replacement of the original mineral by others, leaving all minute details perfectly preserved.

Another commonplace example of the geologic action of ground water is the *geode*, a rock mass with a spherical cavity lined with mineral crystals (Figure 15.16). Geodes commonly occur in limestones and shales, and can be explained by a two-stage development. First, the cavity is excavated by removal of rock in solution in ground water. Then the cavity is filled by mineral matter carried in by ground water. These minerals—commonly quartz, calcite, or fluorite—accumulate very slowly and form perfect crystals, pointing inward toward the center of the geode.

The chemical action of moving subsurface water has played a major role in forming valuable ore deposits in which compounds of various rare metals have become greatly concentrated. This is a topic we will cover in detail in Chapter 21.

Limestone Caverns and Karst

Limestone, composed of carbonates of calcium and magnesium (calcite and varying proportions of dolomite), is highly susceptible to the action of carbonic acid in rainwater and subsurface water. The carbonic acid reaction with calcite is given in Chapter 5 and should be reviewed now. Commonly called carbonation, it results in the production of bicarbonate ions (HCO_3^-) and calcium ions (Ca^{2+}), and these are removed from the region by the outflow of surplus ground water to streams. In Chapter 14, we noted the effects of carbonation on exposed surfaces of limestone (see Figure 14.13). Lowering of the land surface by carbonation is a comparatively rapid process in areas underlain by limestone or marble, and where rainfall is abundant. As a result, carbonate rocks typically form valleys and lowlands in humid climates.

Limestone caverns consist of subterranean passageways and rooms forming complex interconnected systems. Patterns of joints in the limestone and variations in composition among individual limestone beds exert a strong control over the size and spacing of the network of passages (Figure 15.17).

It is now generally agreed by geologists that many, if not all, cavern systems are excavated in the ground-water zone. As suggested in Stage 1 of Figure 15.18, water percolating downward in the unsaturated zone reaches the water table, below which it moves as ground water along curved paths indicated by the arrows. Carbonic acid reaction with limestones is believed to be concentrated in the uppermost part of the saturated zone. Here passageways are enlarged, and the calcium bicarbonate is carried into streams and out of the region.

In Stage 2 of Figure 15.18, the stream has deepened its valley and the water table has been lowered accordingly. New caverns are being formed at the lower level, while those previously formed are now in the unsaturated zone where percolating water is exposed to the air on ceilings and walls of the caverns.

The carbonic acid reaction is reversible. Under favorable conditions that permit some of the dissolved carbon dioxide in the solution to escape into the air as a gas, calcite is precipitated as a mineral deposit:

$$\underset{\text{carbonic acid}}{H_2CO_3} + \underbrace{Ca^{2+} + 2(HCO_3)^-}_{\text{calcium and bicarbonate ions in solution}}$$

$$\rightarrow \underset{\text{carbon dioxide gas}}{2CO_2} + \underset{\text{water}}{2H_2O} + \underset{\text{calcite precipitate}}{CaCO_3}$$

Precipitation of calcium carbonate on the inner surfaces of caverns occurs as *travertine*, a banded form of calcite. Encrustations forming where water drips from cave ceilings are called *dripstone*, while encrustations made in moving water of pools and streams on cave floors form *flowstone*. Dripstone and flowstone accumulate as elaborate encrustations, known popularly as "formations," that give many caverns their great beauty.

Certain scenic features of caverns are illustrated in

FIGURE 15.16 This geode has a rim of agate and is lined with quartz crystals. (Ward's Natural Science Establishment, Inc., Rochester, N.Y.)

FIGURE 15.16

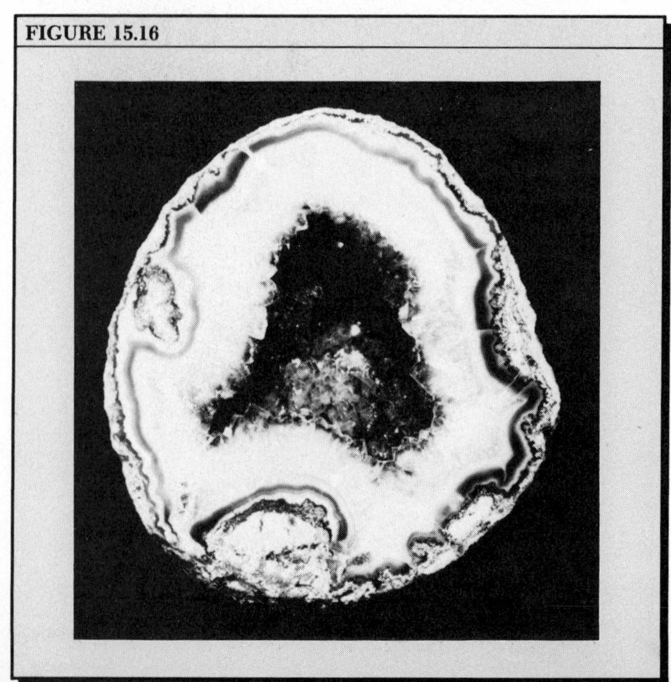

Figure 15.19. From ceiling points where the slow drip of water takes place, spikelike forms known as *stalactites* are built downward, while from points on the cavern floor upon which there falls a steady drip of water, postlike columns termed *stalagmites* are built upward. Stalactites and stalagmites may join into *columns*, and the columns forming under a single joint crack may fuse into solid walls. Growth below a joint crack may produce a drip curtain. Blocks of limestone fallen from the ceiling pond the runoff along the cavern floor, making pools in which *travertine terraces* are formed.

FIGURE 15.17 Map of a portion of Anvil Cave, Alabama. Empty cavern space is in white. (Courtesy of W. W. Varnedoe and the Huntsville Grotto of the National Speleological Society.)

FIGURE 15.18 Stages in cavern development.

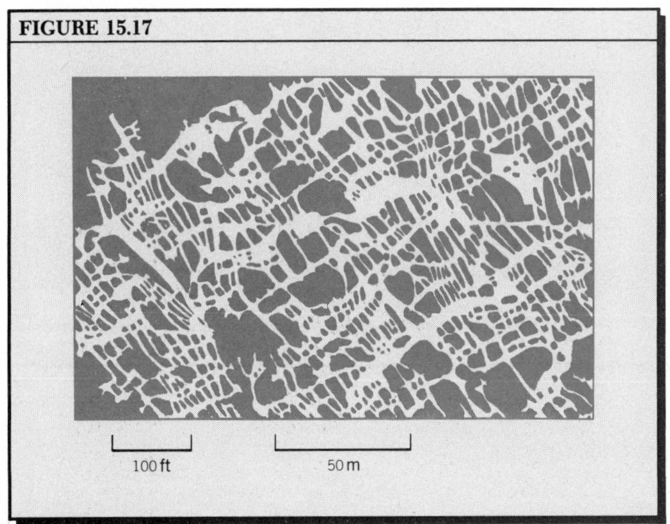

After cavern development has been in progress for a long period of time, the land surface above is deeply pocked with depressions, which are parts of the old cavern system lying close to the surface. These depressions are termed *sinks*, or *sinkholes* (Figure 15.20). They may be partly filled with clay soil and can hold ponds or marshes, or may be cultivated if the soil is well drained. Some sinks are gaping holes or fissures leading down to open caverns beneath.

The name *karst* is given to the landscape of a region of carbonate rocks bearing a well-developed subterranean drainage system and largely devoid of surface streams. The name comes from a region in the Dalmation coastal belt of Yugoslavia, where such landscapes are plentiful. A number of distinctive landforms of karst regions are shown in Figure 15.21. Diagram *A* shows numerous sinkholes over the land surface, and while there are no flowing surface streams, underground streams may be found in the floors of the deeper caverns. In Diagram *B*, a more advanced stage of development is shown. The ceilings of large caverns have collapsed, leaving natural rock arches, and streams flowing on impermeable strata beneath the limestone formation are now exposed in many places. At a later time, when most of the limestone formation has been removed, masses of limestone will stand as spires or isolated, steep-sided hills, riddled with solution passages (Figure 15.22).

Regions of karst impose some unique environmental hazards to humans, particularly in urban areas where heavy structures are built on cavernous limestone. Collapse of the limestone occasionally occurs, damaging buildings with little warning. Because ground water moves rapidly from place to place in large cavern passageways, pollution and contamination of water supplies is a serious threat.

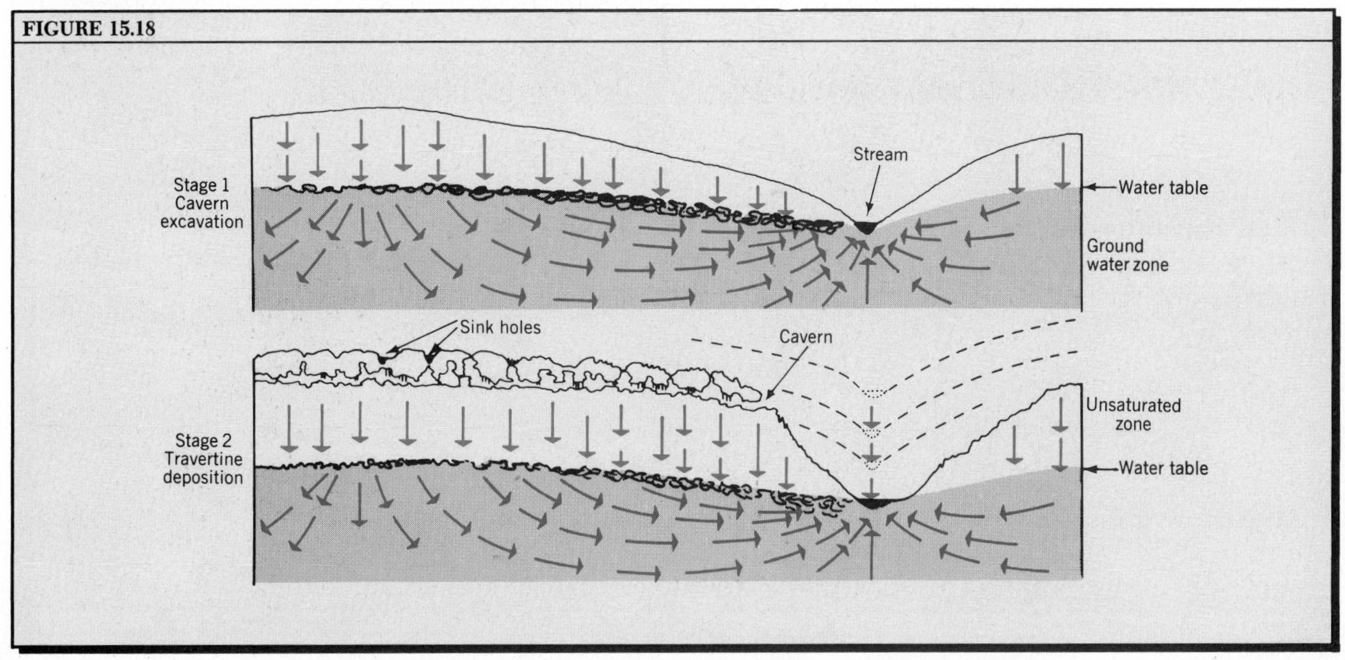

FIGURE 15.19 The Jenolan Caves of New South Wales, Australia, have remarkably fine displays of cavern deposits. *A*. A massive stalagmite. *B*. A drip curtain formed beneath a joint in the ceiling rock. *C*. Travertine terraces with small stalagmites and stalactites. (New South Wales Government Printer.)

FIGURE 15.19 A

FIGURE 15.19 B

FIGURE 15.19 C

FIGURE 15.20 A sinkhole about 20 m deep on the Kaibab Plateau, Arizona, elevation 2500 m. Horizontal limestone strata outcrop on the far wall of the depression. (A. N. Strahler.)

FIGURE 15.20

FIGURE 15.21 Block diagrams showing some features of the karst landscape of the Dalmatian coastal region. (Drawn by E. Raisz. (Redrawn, by permission of the publisher, from A. N. Strahler, *Physical Geography*, 4th ed., Figure 28.25. Copyright © 1975 by John Wiley & Sons, Inc.)

FIGURE 15.22 Called "The Stone Forest," this unusual form of karst topography is a popular tourist attraction in eastern Yunnan Province, China. The grooved spires are found over an area of 260 sq km. (Courtesy of Zhang Zhiyi, Chinese Academy of Sciences, Beijing.)

FIGURE 15.21

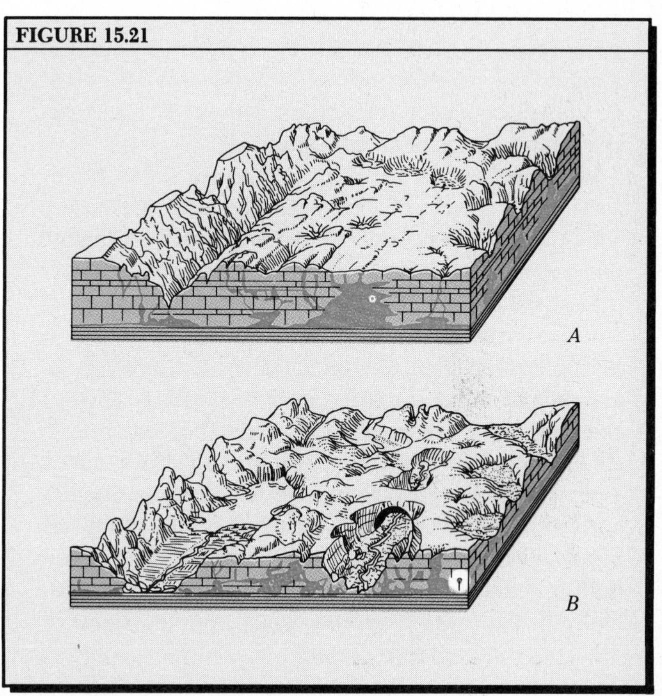

FIGURE 15.22

Hot Springs, Geysers, and Fumaroles

At a number of widely separated *geothermal localities*, ground water emerges in *hot springs* at temperatures not far below the boiling point of water, which is 100 °C at sea-level pressure. At some of these localities, periodic jetlike emissions of steam and hot water occur from small vents and are known as *geysers*. The heated water emitted by hot springs and geysers is largely ground water that has been heated by contact with rock, in turn heated by conduction from a magma body that lies below. Probably only 1% or less of the water in hot springs and geysers is water derived from the magma beneath.

Some geothermal localities have vents, called *fumaroles*, from which only superheated steam is emitted. The superheated steam of fumaroles provides valuable sources of natural heat with which to operate power plants (see Figure 21.11).

Because the heated ground water of hot springs is highly active chemically, it dissolves unusually large amounts of mineral matter, most of which is silica or calcium carbonate, depending upon the composition of the bedrock through which the solutions move. Upon reaching the surface, the hot water is rapidly cooled and must precipitate much of the dissolved mineral matter. Encrustations of mineral matter are built up close to the springs and gradually spread laterally in flat-topped terraces which may be stepped, one above the other (Figure 15.23). Terraces formed of silica, known as *siliceous sinter*, are typical of hot springs in which water rises through igneous rocks. Cones of siliceous sinter are built around the orifices of geysers. Where limestone bedrock furnishes calcium carbonate, the deposits are of travertine, as in the case of Mammoth Hot Springs, pictured in Figure 15.23. Certain species of algae thrive in the pools of hot water, and these may also precipitate calcium carbonate.

Geyser action consists of the occasional or periodic emission of a column of steam and water droplets under high pressure from a small vent (Figure 15.24). This action can be explained by the *Bunsen theory*, which makes use of the principle that the boiling point of water is raised with increasing hydrostatic pressure at increasing depth. For example, at a depth of 150 m the boiling point is about 200 °C. The geyser is thought to consist of a long, narrow tube with tortuous shape and numerous constrictions (Figure 15.25), which fills with ground water entering from passageways in the surrounding rock. Gradually the water temperature rises by conduction of heat from the enclosing rock until the boiling point is approached throughout a large part of its length. Boiling point is first reached near the base of the tube, converting

FIGURE 15.23 This early photograph of Mammoth Hot Springs, Yellowstone National Park, Wyoming, was taken by the pioneer photographer, W. H. Jackson, in 1870. (U.S. Geological Survey.)

FIGURE 15.23

water to steam, which lifts the entire water column. Displaced water pours out of the geyser vent and the hydrostatic pressure, and along with it the boiling point, are lowered simultaneously at all points in the column. With sufficient lift, the entire column passes the boiling point and conversion to steam takes place suddenly throughout the entire tube. Under tremendous pressure, the entire column of water and steam is ejected within a few minutes, after which eruption ceases. In 1969, a successful human-made geyser was put into operation in a geothermal locality in Oregon. It was made by boring a hole 15 cm in diameter to a depth of 35 m, and it erupted once in every eight to ten hours.

Almost all of the world's geysers are located in three places: Yellowstone Park in Wyoming, Iceland, and New Zealand. Old Faithful, in Yellowstone National Park, perhaps the best-known geyser, erupts for about four minutes at intervals averaging close to one hour. With each eruption, about 40,000 liters of water are discharged in a column that may rise to a height of 45 m.

In addition to the hot springs associated with magmatic activity, there are countless springs of warm water known simply as *warm springs* or *thermal springs*. For example, at Warm Springs in Georgia, water emerges at a temperature of 31 °C. This water is ground water that has followed a permeable rock layer downward to a depth of almost 1200 m, where it is heated according to the normal geothermal gradient. A favorable geologic structure enables the water to return to the surface as a warm spring.

FIGURE 15.24 Waikite Geyser, Rotorua, North Island, New Zealand. (New Zealand Tourist Bureau.)

FIGURE 15.24

Ground Water as Both a Resource and a Geologic Agent

Subsurface water plays divergent roles in geoscience, as we have found in this chapter. As a reservoir of fresh water, ground water is a resource of enormous value to human beings. Unfortunately, the supply of ground water can too easily be withdrawn more rapidly than it can be recharged.

In arid lands, particularly, ground water is best described as a nonrenewable mineral resource being "mined" with reckless abandon for irrigation and urban use. Rapidly falling water tables in these areas of overdraft are a recorded fact, which goes largely unheeded because people cannot see the water table. Ultimately, these depleted ground-water supplies will fail.

In humid regions of abundant rainfall, ground water is copiously recharged by nature, and a substantial proportion can be withdrawn on a year-to-year basis with relatively little environmental damage. In these regions, ground water is a renewable resource, when its use is carefully regulated and its quality protected.

The geologic role of ground water as a cementer of rock and a former of ore deposits is an inconspicuous activity, carried out with incredible slowness over long periods of geologic time. As a subterranean sculptor of grotesque and beautiful cavern forms, subsurface water plays out a still different role. Finally, emerging waters of hot springs and geysers deposit their mineral

FIGURE 15.25 Schematic diagram of the Bunsen geyser theory.

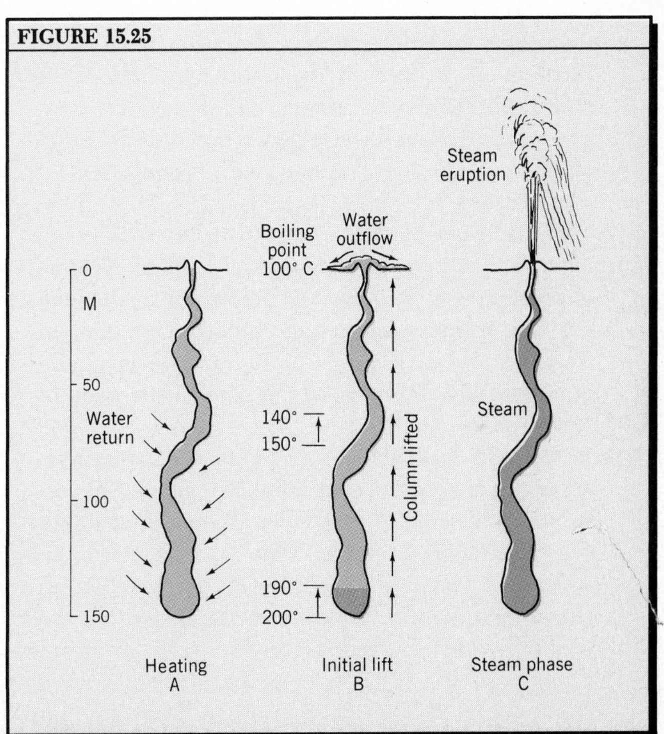

load to construct surface forms unlike anything else seen in nature.

In the next chapter, we return to surface water. We will be particularly interested in the geologic role of streams, as agents of erosion, transportation, and deposition.

Key Facts and Concepts

The hydrologic cycle The *hydrologic cycle* involves the total system of movement, exchange, and storage of the earth's water in liquid, vapor, and solid states. Precipitation falling on the lands is returned to the atmosphere by evaporation and to the oceans by *runoff*. *Hydrology* is the science of the earth's water balance.

Surface water and subsurface water *Surface water* is exposed in streams, lakes, and marshes. *Subsurface water* occupies pores in regolith and bedrock. Water enters the subsurface water zone by *infiltration*. Excess water escapes by *overland flow*, a form of runoff.

Zones of subsurface water In an upper *unsaturated zone*, subsurface water is held as *capillary water* in pores. Capillary water in the *soil water zone* is within reach of plant roots and may return to the atmosphere by direct evaporation or by *transpiration* from plant foliage. In the *saturated zone* beneath the *water table*, *ground water* moves slowly in response to gravity. The water table receives *recharge* by infiltration. A *capillary fringe* lies above the water table and adds height to the saturated zone.

Ground water movement The water table, observed in wells, slopes from high areas and divides toward streams or lakes at lower levels. Ground water movement follows the direction of the water table slope. Fluctuations in level of the water table follow the annual climatic cycle. *Darcy's law* states that average speed of ground water flow is equal to the product of the *hydraulic gradient* and the *coefficient of permeability*.

Porosity and permeability Ground-water storage and movement depend on *porosity*, the available volume of pore space in the rock, and *permeability*, the relative ease of motion of ground water through pores or fractures in a rock mass. Porosity and permeability diminish with depth, reaching zero in the zone of rock flowage.

Aquifers and aquicludes A rock mass or stratum having high porosity and permeability is an *aquifer* (example: sandstone). A rock layer of low permeability is an *aquiclude* (example: shale). A *perched water table* may form above a lenslike aquiclude. In an *unconfined aquifer*, the water table lies within the aquifer, which is free to receive recharge from above by infiltration. Ground water may emerge as a *spring* from the intersection of the water table with the surface.

Gravity wells *Gravity wells*, placed below the water table, yield water only when pumped. Pumping produces a *cone of depression* with *drawdown* of the water table. *Recharge wells* return used water to the water table.

Artesian ground water A dipping *confined aquifer* can be tapped to produce *artesian flow*, which rises to the level of the *hydrostatic pressure surface*. Water in an *artesian well* or an artesian spring may reach the surface.

Ground-water pollution Ground-water supplies are vulnerable to pollution from sewage, road salt, and solid wastes. In coastal zones, salt ground water of greater density lies beneath fresh ground water. Excessive withdrawal of fresh water from wells can result in salt-water intrusion and salt contamination.

Ground-water withdrawal and land subsidence Excessive ground-water withdrawal from sediments can result in ground surface subsidence with damage to surface structures.

Subsurface water as a geologic agent Moving subsurface water causes chemical changes in minerals of the unsaturated zone. Circulating ground water can cause chemical changes through solution removal, precipitation, and mineral replacement. Crystal growth in cavities produces a *geode*.

Limestone caverns and karst Carbonic acid action on carbonate rocks within the upper ground-water zone, close to the water table, results in the opening out of *limestone caverns*. Following later uplift with a drop in the water table, caverns enter the unsaturated zone where partial filling with *travertine deposits* may occur. *Sinkholes* are common surface features of *karst regions*, riddled by caverns.

Hot springs, geysers, and fumaroles In *geothermal localities*, hot ground water emerges in *hot springs;* steam and hot water in *geysers;* superheated steam in *fumaroles*. Deposits of *siliceous sinter* or travertine accumulate around geysers and hot springs. Geyser action is explained by the *Bunsen theory*. *Warm springs* (thermal springs) occur where deep ground water heated by the normal thermal gradient emerges at the surface.

Questions and Problems

1. Explain the capillary force that causes films and droplets of water to cling to surfaces despite the downward pull of gravity. (See Chapter 2, molecular structure of water.) Is it possible that capillary force might draw water down into the soil when rain falls on the soil surface? Does this process oper-

ate when a lawn or field is irrigated after a spell of dry weather?

2. If it is true that the water table is usually highest in elevation under hill summits forming divides between streams, does it follow that the height of a summit has a direct physical influence on the height of the water table? Explain.
3. Imagine a situation in which a drought begins abruptly in a region of moist climate with a high water table. The drought is total and lasts for ten years. You record the height of the water table in an observation well, making one reading on the same date each year. Will the recorded drop in water table be the same in each of the ten years? Justify your answer in terms of Darcy's law.
4. The rural water "dowser" uses a forked stick to locate "veins" of subsurface water. The stick, held firmly with both hands in a horizontal attitude, is pulled down by a mysterious force when the dowser reaches a point over the water "vein." Does this phenomenon have an acceptable explanation in terms of laws of physical science? Do you believe that some classes of physical phenomena we observe are beyond explanation by use of the scientific method?
5. Draw a structural cross section to show how a thermal spring might occur in steeply dipping strata. Would you expect the water of a natural artesian spring to be warm or cold? Explain.

CHAPTER SIXTEEN

The Geologic Work of Running Water

Running water on the surface of the lands is undoubtedly the most important of the active agents of denudation. Running water is most conspicuous in the form of large streams or rivers that transport enormous volumes of surplus precipitation from the lands to the oceans. To a geologist, however, streams are much more than just mechanisms for the disposal of runoff from the lands; they are major agents of land sculpture. Streams create a vast array of erosional and depositional landforms. Streams also transport sediment and deposit it in basins and shallow seas—geoclines and geosynclines—where it can be transformed into sedimentary rock.

Stream action, in combination with weathering, mass wasting, and overland flow, is responsible for a total process we call *fluvial denudation*, which creates most landscapes we see on the surfaces of the continents (Figure 16.1). True, glacial ice is a dominant agent in high mountains, wind creates conspicuous forms in a few desert and coastal localities, and wave action shapes the shorelines. But from the standpoint of total surface area affected, fluvial denudation is the predominant sculptor of landscapes.

Overland Flow and Channel Flow

Running water as a geologic agent acts in two basic forms. First is *overland flow*, the movement of runoff downhill on the ground surface in a more or less broadly distributed sheet or film. Second is *channel flow*, or *stream flow*, in which water moves to lower levels in a long, narrow, troughlike feature called a *stream channel*, bounded on both sides by rising slopes, called *banks;* these contain the flow. The *stream* itself is a body of water flowing within a channel, and includes all sediment carried by the water. Overland flow starts near the hill summits and converges upon the stream channels, supplying them with both water and sediment (Figure 16.2).

When overland flow is taking place during a heavy rainstorm, or when snow is melting rapidly, the sheet of water moving downhill exerts a drag force over the ground surface. Soil and other forms of soft or unconsolidated regolith or bedrock are susceptible to the drag force. Progressive removal of mineral grains by this force, together with downhill transport, is called *sheet erosion*.

A good plant cover, particularly a grass sod, breaks the force of falling raindrops and absorbs the energy of the overland flow, reducing the rate of sheet erosion. Even under heavy and prolonged rains, a thickly vegetated hillside yields very small quantities of mineral solids. In contrast, the barren slopes of a desert landscape or the unprotected surface of a cultivated field will produce large quantities of sediment with each rainstorm.

Steepness of the ground surface also strongly influences the rate of sediment removal by overland flow. On steep hillsides the flow moves swiftly, and its power to erode and transport is much greater than on gentle slopes. Mineral particles removed in sheet erosion are carried to the base of the hillslope, where some of the coarser particles accumulate in thin layers to form a deposit known as *colluvium*. Some particles will, of course, be carried into stream channels, to be deposited on the valley floor as *alluvium*, a broad term for any stream-laid sediment deposit.

Stream Channels and Stream Flow

Stream channels range in size from insignificant brooks one can take a single step across to the trenches of great rivers several hundred meters wide. Many channels of desert streams are dry most of the time, yet they bear the unmistakable markings of rapidly flowing water and are occupied by raging torrents during rare floods. In the semidesert region shown in Figure 16.1, runoff is strongly seasonal—copious in winter months, but absent in the dry summer. Even in a humid region, small channels normally become dry in the summer season, when the water table falls so low that no ground water can seep into the stream.

An essential feature of every stream is that it descends to lower elevations along its length. The rate of this descent is the *stream gradient*. Gradient is stated as the vertical drop in m/km of horizontal distance, or as the percentage of grade, the ratio of vertical drop to horizontal distance in the same units. Thus, a gradient of 0.05, or 5% means a vertical drop of 5 m for each 100 m of horizontal distance. Of course, when you examine a stream bed in detail, it will usually prove to have deeper places, called *pools*, separated by shallower places, called *riffles*. Because of these irregularities, the bed itself can locally rise in the downstream direction, but the water surface must have a downstream gradient at all points, or flow will not be possible.

A stream moves downgrade in its channel because a fraction of the force of gravity acts in a downstream direction, parallel with the bed. The water responds freely to this force, each layer moving over the layer beneath it in fluid shear (Chapter 6). *Stream velocity* is the speed of downstream water flow, as measured at any selected point above the bed. Because of friction with the bed and banks, stream velocity varies from zero in contact with the bed to a maximum in midstream, some distance above the bed.

Arrows in the lower part of Figure 16.3 show the tracks that would be taken by water particles starting out together on a vertical line. We see that velocity increases very rapidly from the bed upward, then decreases, but less rapidly, so that near the midline of the stream the maximum velocity is found at a point about a third of the distance from the surface. The upper part of the figure shows that, on the stream surface, velocity increases from zero at the banks to a maximum near the center line.

Turbulence in Streams

Statements we made in previous paragraphs imply that each particle of water moves downsteam in a direct simple path. That would be the case in true

FIGURE 16.1

FIGURE 16.1 This California landscape is almost completely dominated by the processes of fluvial denudation. Sheet flow, along with soil creep, carries sediment from the hill summits and hillsides to countless stream channels. The small channels in turn converge into larger ones, more efficient in transporting the sediment to lower levels and ultimately to the sea. (Spence Air Photos.)

FIGURE 16.2

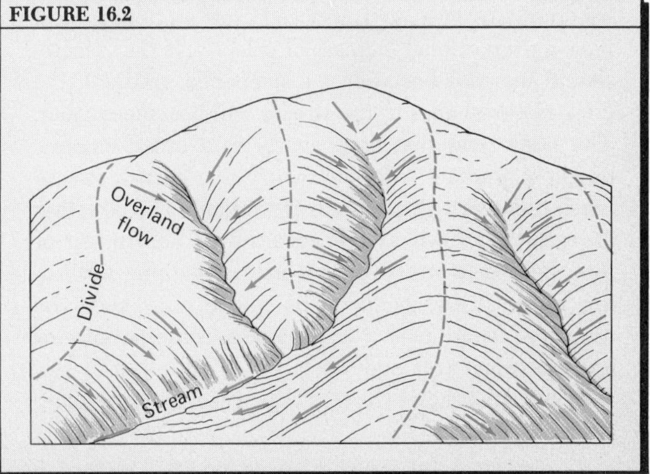

FIGURE 16.2 Stream channels are fed by overland flow. (Redrawn, by permission of the publisher, from A. N. and A. H. Strahler, *Modern Physical Geography*, Figure 18.18. Copyright © 1978 by John Wiley & Sons, Inc.)

streamline flow (or *laminar flow*), which occurs in fluids whose motion is very slow. In most forms of runoff, including most overland flow and nearly all stream-channel flow, the water particles follow highly irregular paths in twisted corkscrew motions that include sideways and vertical movements. Such fluid motion is called *turbulent flow,* and consists of innumerable eddies of various sizes and intensities continually forming and dissolving. Velocity, as we referred to it earlier, and the simple paths of flow shown by the parallel arrows in Figure 16.3, are merely the average velocities and average paths of the particles at given levels in the stream.

Turbulent flow in fluids is of vital importance in the geologic activity of streams, because the transportation of fine particles held suspended in the fluid depends upon the upward movement of currents in turbulence to support the particles. Without turbulence, particles could only be rolled or dragged upon the bed, or lifted a short distance above it.

Discharge, Mean Velocity, and Cross-Sectional Area

Because of the differences in average flow velocity from point to point in a stream, we need a single statement of velocity to apply to the stream as a whole. Called the *mean velocity,* it is approximately equivalent to six-tenths of the maximum velocity.

A most important characteristic of the flow from the standpoint of describing the magnitude of the stream is the *discharge,* the quantity of water that flows through the cross section of a stream channel in a given period of time. Commonly, discharge, Q, is given in cubic meters per second, and is computed by multiplying the mean velocity, V, times a cross-sectional area, A, in the formula $Q = AV$.

If a long stream channel is to conduct a given discharge through its entire course, the discharge must be constant at all cross sections. If this were not so, water would accumulate by ponding. If the stream is neither gaining nor losing discharge, the product of cross-sectional area and mean velocity must be constant in all cross sections along the stream. Where a stream becomes narrower, with reduced cross section, it must have a proportional increase of velocity. If the velocity should increase because of a steepened gradient, the cross-sectional area of the stream will become smaller. The same river that flows slowly in a broad channel on a low gradient will flow swiftly in a narrow stream when it enters a gorge of steep gradient. The equation $Q = AV$ is known as the *equation of continuity* of flow, because a stream that is neither gaining nor losing water at any point on its course must keep the discharge constant by appropriate combinations of cross-sectional area and velocity.

Figure 16.4 shows these relationships in a schematic way. What we see is a side view of a vertical slice down the middle of the channel. Where the gradient (S) is steep over a rapid, mean velocity (V) is high but cross-sectional area (A) is small. In the pool between rapids, the gradient of the water surface is very low, the cross-sectional area is large, and the velocity is very low.

If we were to set up a measuring station at some point along a large stream (a gauging station) and make a continuous record of the flow for several hours, we would probably find that the discharge and mean velocity would hold nearly constant for that rather short period. Why doesn't the flow velocity become faster and faster, as in the case of an object falling in a vacuum? The answer is that the stream encounters resistance at the boundary between the fluid water and the solid channel. This resistance to flow is a form of friction which drains away energy from the stream flow. The supply of energy available to a stream is potential energy it possesses because of its elevation. As the stream descends, potential energy is converted to kinetic energy, which in turn is converted to heat energy. That heat energy leaves the stream by conduction through the channel walls and upper surface. When the rates of energy conversion become

FIGURE 16.3 Speed of water flow in a stream channel is fastest near the center. Velocity is proportional to the length of the arrow.

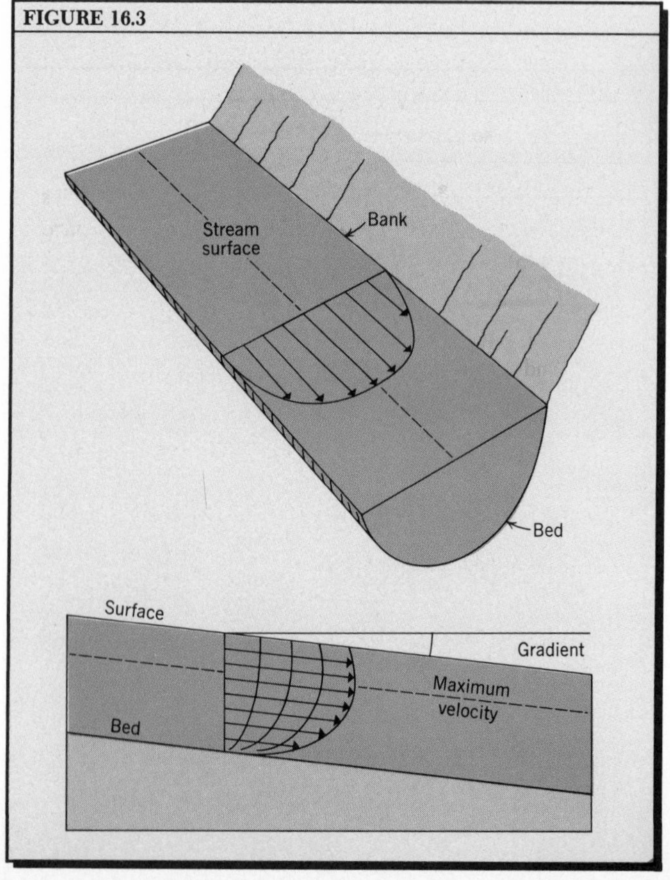

constant, we say that the stream has attained its *terminal velocity.* We see many examples of terminal velocity in moving objects. On a pavement of uniform gradient, a person on a skateboard quickly attains a constant terminal velocity. The key to the phenomenon is that the frictional resistance to movement increases as the speed increases, so that a point is quickly reached at which the energy wasted in overcoming the friction balances the rate at which the kinetic energy is being derived from potential energy.

*Stream Velocity in Relation to Gradient and Depth

What controls the mean velocity of a stream? For over two centuries, hydraulic engineers have tried to find a formula that relates velocity to the three most important controlling factors. Common sense guides us to suppose that gradient is one important factor—the steeper the gradient, the faster the flow. Less obvious is the factor of water depth. Given the same gradient, a deep stream flows faster than a shallow stream, because the deep stream loses a smaller proportion of its total kinetic energy to friction with the stream bed.

As early as 1775, a French engineer formulated a simple equation to relate mean velocity to gradient and depth. We can state his formula or equation as follows:

$$V \propto \sqrt{S} \times \sqrt{D}$$

where V is mean velocity
$\propto$ means "proportional to"
S is the gradient in percentage
and D is the average depth of the cross section

In words, the formula states that mean velocity varies both as the square root of gradient and as the square root of water depth. Although this early formula has since been somewhat modified, it is generally correct.

A third important factor in controlling a stream's velocity is the nature of the channel bed over which it flows. This channel factor goes under the name of *bed roughness*. Channels that are rough because there are many cobbles or boulders on the bed set up more resistance to flow than channels with smooth surfaces of clay or polished bedrock. Engineers use a scale of numbers to describe bed roughness, and when the appropriate number for a particular channel is inserted into the gradient-depth formula, a fairly accurate working formula results.

FIGURE 16.4 Relations among cross-sectional area, mean velocity, and gradient in a stream of uniform discharge.

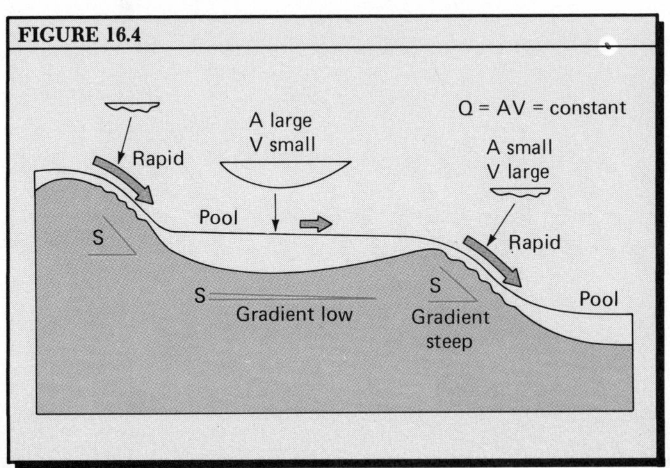

Drainage Systems and Channel Networks

So far, we have thought of a stream channel in terms of a short length of narrow trough which corresponds to what we would see of a channel from a single vantage point on the stream bank or on a bridge, looking upstream and downstream. Of course, if we traced the stream channel in the upstream direction, we would soon come to a point where the channel divides into two channels of about equal size, or where a small channel comes in to join the main channel from one side or the other. Should we choose to follow the channel downstream, we would soon encounter a junction where this and another channel join to form a single channel of larger size.

Obviously, stream channels are organized into branching *channel networks*. Together with the land surface slopes that contribute runoff by overland flow, the channel network forms a *drainage system* (or *drainage basin*). The entire drainage system is bounded by a *drainage divide*, which is a line following a chain of continuous ridge crests. The line includes all of the surface, or *watershed*, that slopes toward the channels of the system. Thus, all runoff within the limits of the divide feeds into the channel system. Of course, a drainage system has only one exit point, or outlet. Here a single channel—the trunk channel of the network—carries out all runoff as well as all sediment that moves with the water.

Figure 16.5 is a map of a stream network and its drainage divide. Within a drainage system the slope of the ground surface converges toward the head of each of the smallest, or fingertip, channels. Here, enough runoff collects to scour a permanent channel and to maintain it against the filling action of soil creep, colluvium, and vegetative growth. During the long period of time within which the landscape evolves by fluvial-denudation processes, the runoff of all the available land surface becomes apportioned to the individual channels in such a manner that each channel obtains just the amount of runoff needed to sustain it. The surface area that feeds a single fingertip channel is a unit cell of the drainage system (Figure 16.5). Furthermore, the gradients of both valley sides and channels become adjusted between themselves to permit a generally uniform disposal of rock waste.

As channels join each other, at angles that are usually acute, the discharge of water and sediment load is funneled into exit channels of progressively larger dimensions. The flow paths are not necessarily the shortest possible. Instead, they are the most economical paths under the requirement that each unit of channel length be provided with a proportionate surface area to furnish runoff.

*An Ordered System Within a Channel Network

A complete stream network, as represented by a set of lines on a flat map, can be subjected to mathematical analysis to reveal a rather remarkable degree of orderliness of form. Using an idealized network map, such as that shown in Figure 16.6, we can study the geometry of a channel network. The first step is to recognize that the entire network consists of individual *channel segments*, or links, each of which lies between two channel junctions—one at the upstream end, the other at the downstream end. Next, we designate by integer numbers the segments in terms of orders of magnitude within a hierarchy. Every fingertip channel, from its point of origin to its first point of junction, is designated as a channel segment of the first order. The junction of any two first-order channels produces a segment of the second order, the junction of any two second-order segments produces a segment of the third order, and so forth. The junction of a single first-order segment with a second-order or higher-order channel does not, however, produce any change in the order of the segment it joins.

When this ordering system is applied throughout the entire drainage network, it will be found that a single trunk-stream segment bears the highest-order designation. Where a large stream network is taken into consideration, the order of any segment will, on the average, reflect the magnitude of the channel in terms of its channel dimensions, discharge, and contributing watershed area. A relatively large sample must be taken to reveal consistent relations, because individual segments may be much longer or shorter than the average and there will be many chance distortions in the pattern of the network.

Without going into detail, we can suggest the nature of the orderliness, or constancy, of relationships that exists from one stream order to the next within most stream networks. Figure 16.7 shows three graphs on which data are plotted for a single large drainage basin—that of the Allegheny River in the Appalachian Plateau region of Pennsylvania. On the horizontal scale of each graph are stream orders from one to seven, represented by equally spaced vertical lines. (There are no intermediate values of order—only the seven integers.) The vertical scales of all three graphs are logarithmic scales. (Powers of ten form equal vertical units.) Graph *A* shows the numbers of stream segments of each order, plotted against order. There are nearly 6000 segments of the first order, but only one segment of the seventh order. A straight line fits the seven dots rather well and suggests that the logarithm of the number of segments is directly proportional to

FIGURE 16.5 A map of the stream network and drainage basin of Pole Canyon, Spanish Fork Peak Quadrangle, Utah. Elevations are given in meters. (Data of U.S. Geological Survey and Mark A. Melton.)

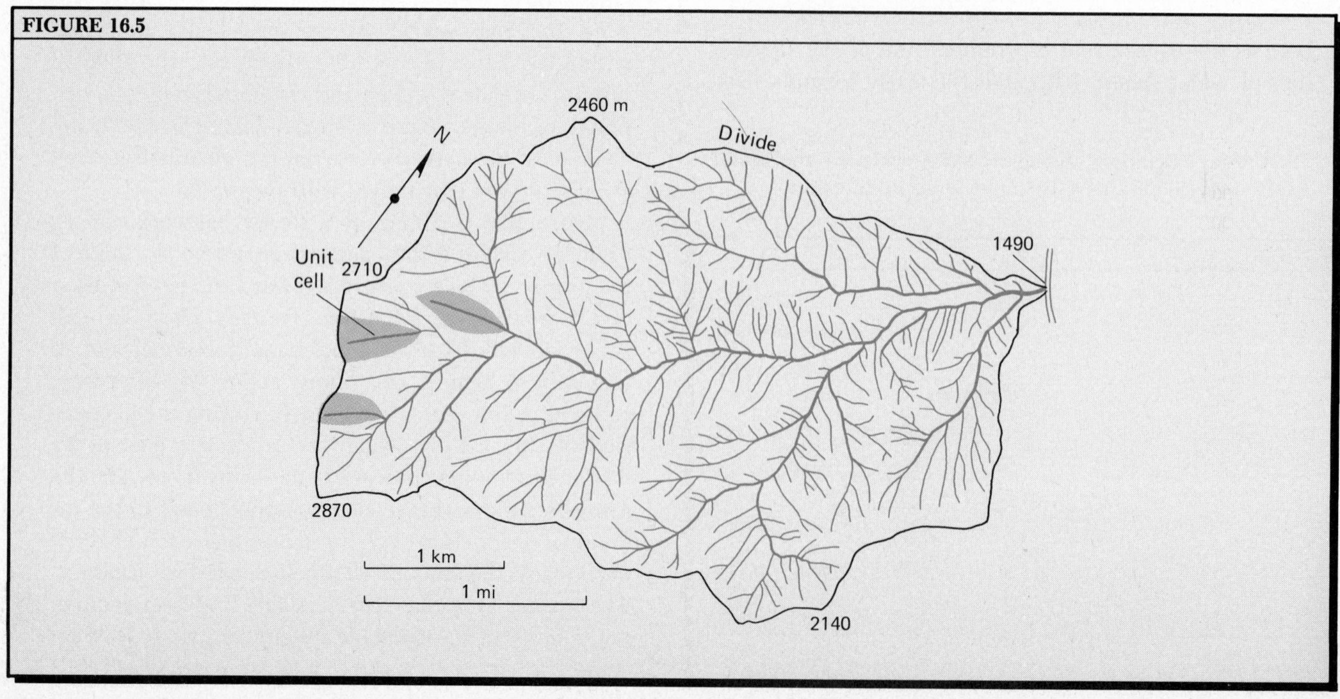

FIGURE 16.6 A system of assigning orders to the stream segments within a drainage system. (Redrawn, by permission of the publisher, from A. N. Strahler, *Physical Geography,* 4th ed., Figure 27.1. Copyright © 1975 by John Wiley & Sons, Inc.)

FIGURE 16.7 Graphs showing the semilogarithmic relationship of stream segment order to numbers of segments, channel length, and contributing runoff area. (Data of Marie E. Morisawa, 1962.)

FIGURE 16.6

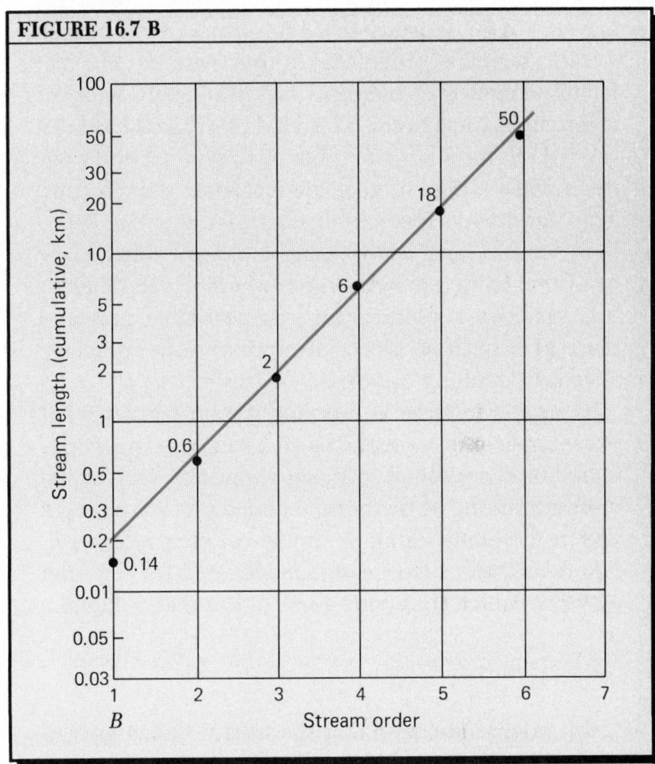

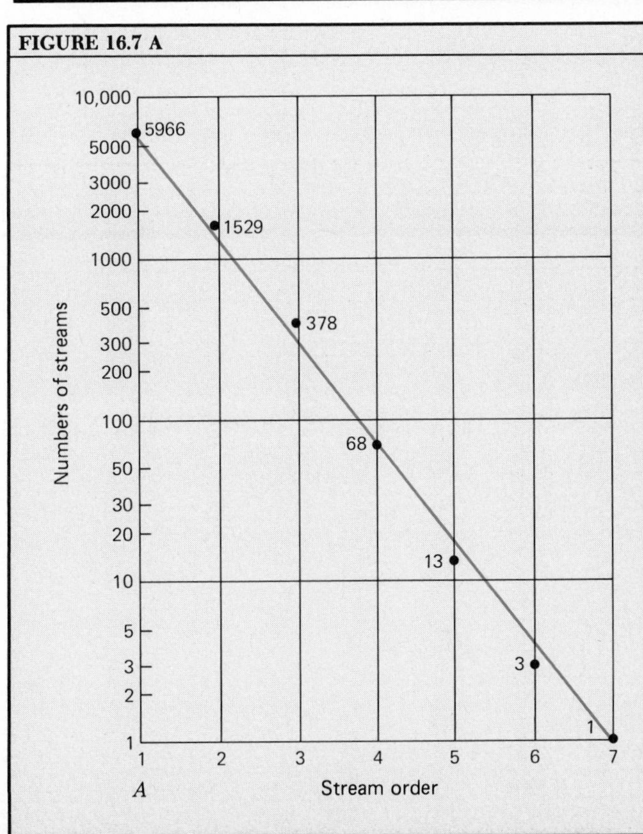

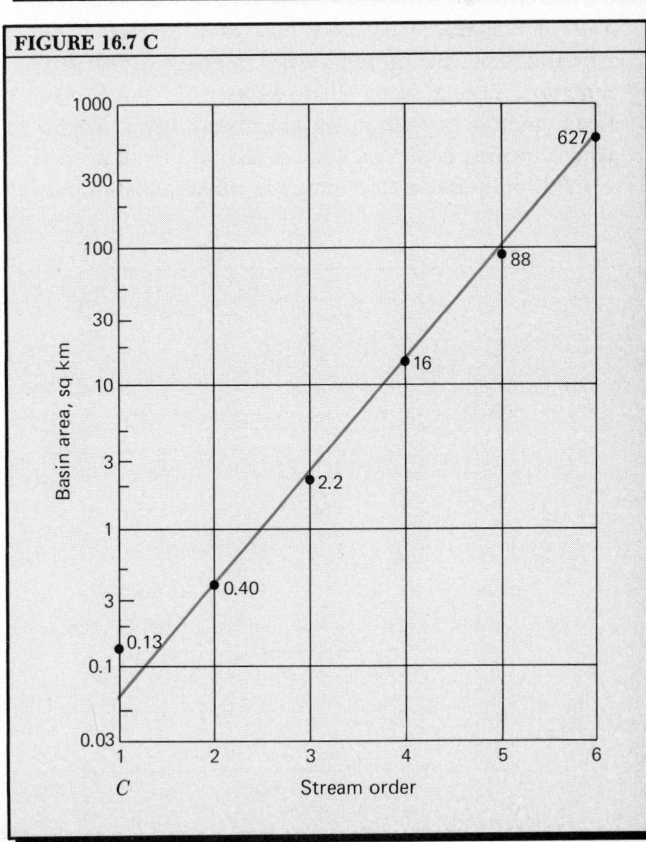

order. This is a semilogarithmic relationship.

Graph *B* suggests that a similar semilogarithmic relationship exists between the average length of segments of each order and the order itself. (In this case, the length of segments of a given order includes the lengths of segments of all lower orders.) Note, however, that the slant of the fitted line in Graph *B* is opposite to that for Graph *A*. Graph *C* shows the average area of land surface that contributes runoff to each stream segment order. As in the case of channel length, runoff area increases logarithmically with increase in channel order. This kind of analysis has been carried out for a large number of stream networks under a wide variety of geologic materials and climates. The logarithmic relationships between segment numbers, lengths, and contributing areas are remarkably consistent from region to region where fluvial denudation has been in progress for long periods of time and there have been no recent disruptions of the drainage system by faulting or volcanic extrusion.

The same method of ordering stream segments has been applied to other kinds of branching networks, including those found in living organisms. One of the most interesting of these is the network of air passages within the human lung. A similar set of semilogarithmic relationships has been demonstrated for the lung network, which transports a gas rather than a liquid.

Stream Discharge and Basin Area

You might anticipate that the stream discharge will tend to increase with increasing basin area. This is a reasonable guess, since the discharge is derived from overland flow and ground-water seepage from precipitation falling upon that watershed. When the mean annual discharge, as calculated from stream gauge records, is plotted against the total area of watershed lying above that gauge, a simple relationship of discharge to basin area is revealed (Figure 16.8). Although the data are plotted on a double-logarithmic graph, the straight line of best fit is so inclined that the discharge increases in direct proportion to the increase in drainage area. Once the existing relationship has been established for a given watershed, it is possible to

FIGURE 16.8 Plot of mean annual stream discharge against drainage basin area. (Data of John T. Hack, 1957, *U.S. Geological Survey Professional Paper 294-B*, p. 54, Figure 15.)

FIGURE 16.9 Longitudinal profiles of the Arkansas and Canadian Rivers. The middle and lower parts of the profiles are for the most part smoothly graded, whereas the poorly graded upper parts reflect rock inequalities and glacial modifications within the Rocky Mountains. (Redrawn, by permission of the publisher, from A. N. Strahler, ***Physical Geography***, 4th ed., Figure 25.14. Copyright © 1975 by John Wiley & Sons, Inc. Data of Gannett, U.S. Geological Survey.)

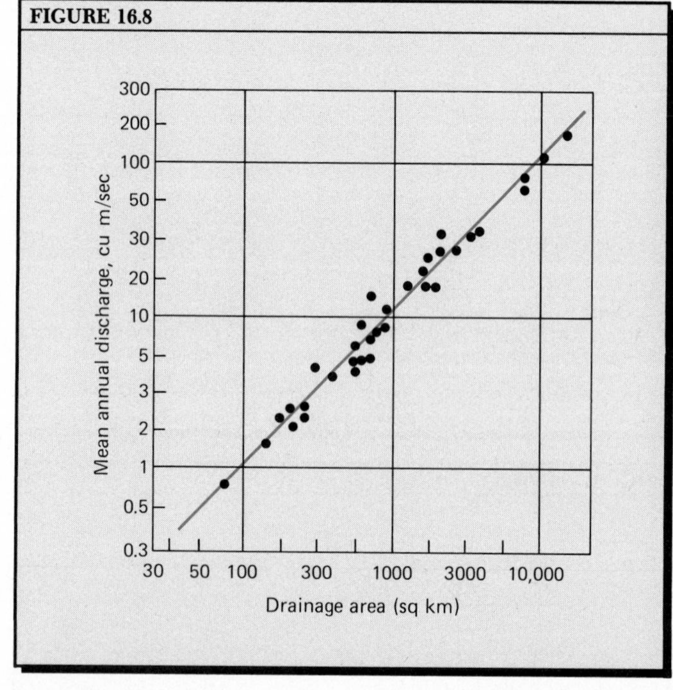

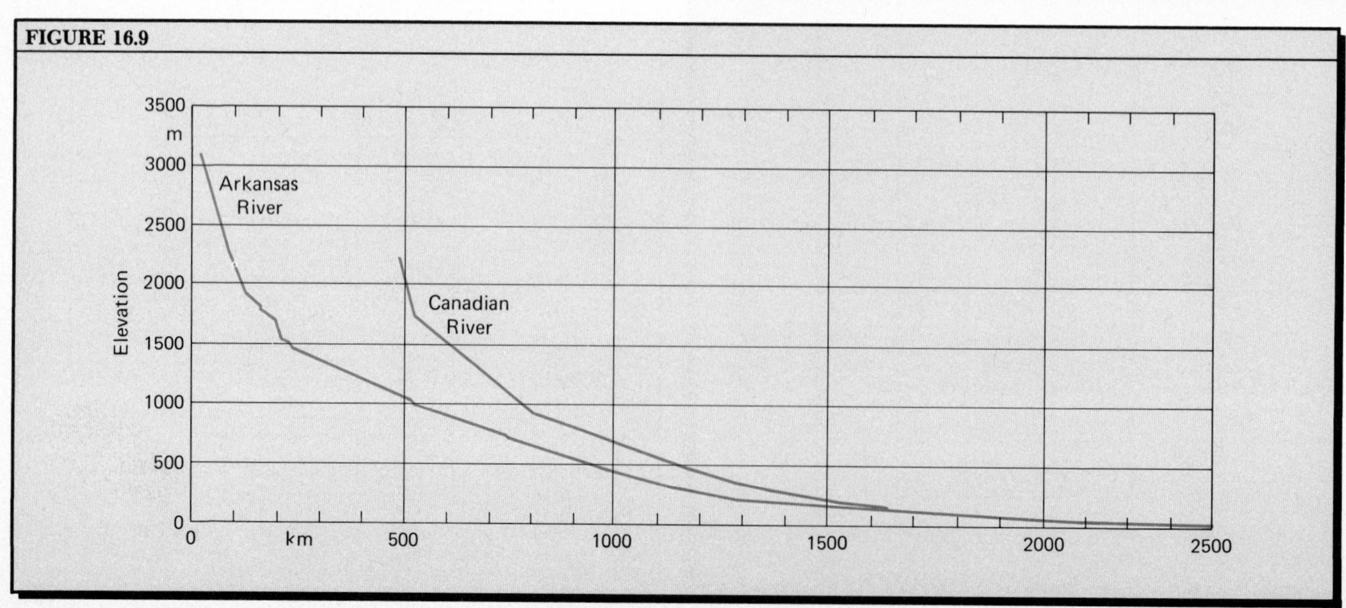

make good estimates of the mean annual discharge at any give point on a trunk stream by merely measuring the watershed area lying above that point.

The Longitudinal Stream Profile

As a general rule, the trunk stream of a drainage system shows systematic changes in channel gradient, width, depth, and mean velocity of flow as it is followed downstream to lower elevations. Changes in gradient can be examined by plotting the *longitudinal profile* of the stream as in Figure 16.9. Distance is plotted on the horizontal scale; elevation above sea level on the vertical scale. In general, the typical longitudinal profile is upwardly concave and the gradient flattens more gradually as the stream mouth is approached. In detail, the longitudinal profile usually shows abrupt decreases in gradient wherever a large tributary enters. In the upstream portions, various irregularities—rapids and falls—are commonly present.

FIGURE 16.10 Average gradient of stream segments of each stream order in the drainage basin of Home Creek, Ohio. *A*. Uniform (arithmetic) vertical scale. *B*. Logarithmic vertical scale. (Data of Marie E. Morisawa, 1959.)

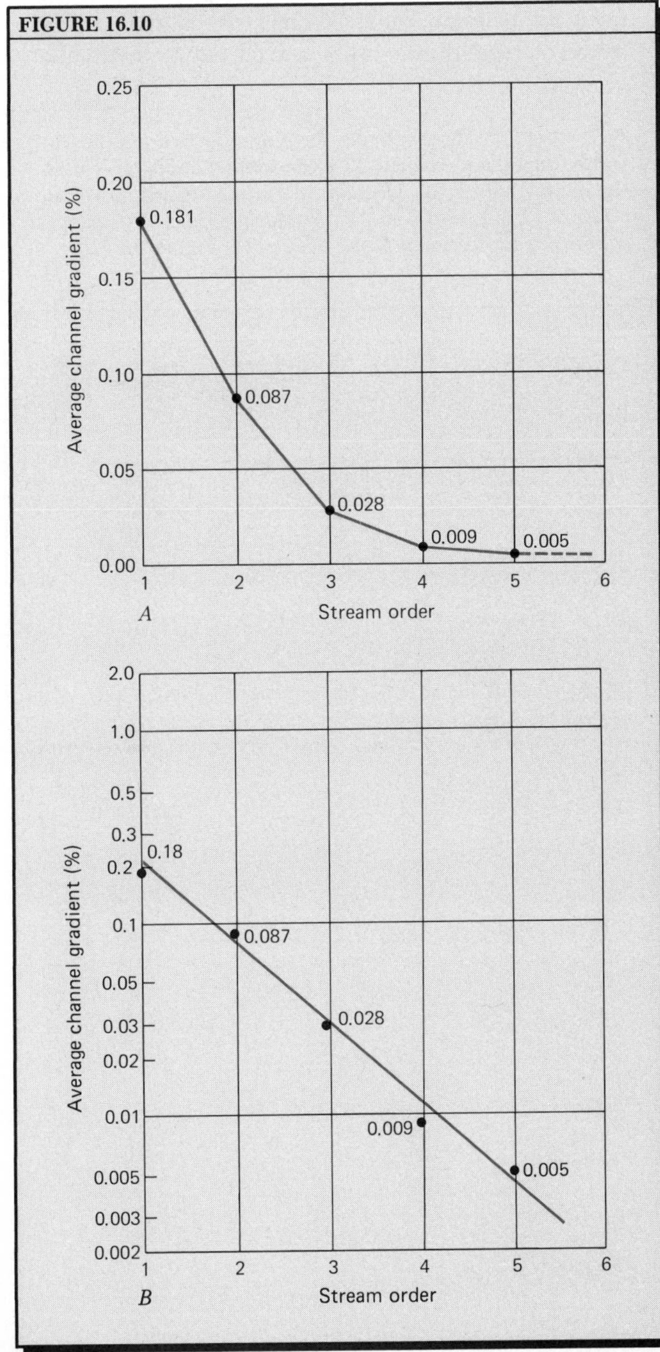

*Downstream Channel Changes

Another way to look at channel gradient is in terms of the increasing order of channel segments in the total stream network. First, we compute the average gradient for all segments of the first order, then for segments of the second order, and so on. The data are then plotted as points on a graph. Figure 16.10 is such a graph in two forms for a stream network in Ohio. Graph *A* shows the points plotted on a uniform (arithmetic) vertical scale of elevation. When the points are connected, an average gradient profile results and is upwardly concave, as in the simple longitudinal profile of Figure 16.9. The lower graph uses a logarithmic vertical scale, and the points lie close to a straight line. From this, we conclude that the relationship of gradient to order is semilogarithmic.

At numerous gauging stations located at intervals on American streams, both small and large, continuous records are kept of the discharge and of the height of the water surface, or *stage*. Average velocity is determined when discharge is measured, and periodic measurements are made of the width of the stream and of the depth of the bed from bank to bank. As we showed in Figure 16.8, the mean annual discharge increases in the downstream direction because of the increasing area of the watershed that supplies runoff above each gauging station.

Graphs of Figure 16.11 put together three sets of data in relation to increase in discharge in the downstream direction: depth, width, and velocity. In each case, the mean annual discharge is plotted on the horizontal scale, which is logarithmic, as are the vertical scales. Although there is a moderate scatter of the points, each of which represents a gauging station, the trends are clearly indicated and can be emphasized by straight lines fitted to the points. As we would expect, both width and depth increase with discharge and downstream distance. The product of width and average depth gives the cross-sectional area of the stream, which is the term A in the equation of continuity, $Q = A \times V$. Greater discharge is largely accommodated by increased cross-sectional area. Mean velocity shows only a small downstream increase, on the average, as reflected in the low upward slant of the fitted line. You may wonder why the velocity does not increase much more rapidly, considering the strong increase in depth. But then you realize that the gradient is steadily decreasing downstream, and the decrease in gradient largely cancels out the effect of increase in depth.

The data we have presented on stream channels and stream networks show that a remarkable degree of orderliness prevails in a drainage system. We can conclude that a drainage system gradually evolves its configuration in such a way as to carry out its work most efficiently. That work consists not only of carrying off surplus water, but of transporting sediment as well. To learn more about the workings of streams, we must turn next to their geologic activities of eroding, transporting, and depositing mineral matter.

Stream Erosion

Streams perform three closely interrelated forms of geologic work—erosion, transportation, and deposition. *Stream erosion* is the progressive removal of mineral matter from the surfaces of a stream channel which itself may consist of bedrock or regolith. *Stream transportation* is the movement of eroded particles in chemical solution, in turbulent suspension, or by rolling and dragging along the bed. *Stream deposition* consists of the accumulation of any transported particles on the stream bed, on the adjoining floodplain, or on the floor of a body of standing water into which the stream empties. These phases of geologic work cannot be separated from each other, because where erosion occurs, there must be some transportation, and eventually the transported particles must come to rest.

The nature of stream erosion depends upon the materials of which the channel is composed and the means of erosion available to the stream. One simple means of erosion is *hydraulic action*, the pressure and drag of flowing water exerted upon grains projecting from the bed and banks. Weak bedrock and various forms of regolith are easily carved out by hydraulic

FIGURE 16.11 Three graphs showing the relationships of width, depth, and velocity to mean annual discharge within the main trunk of the Missouri and lower Mississippi rivers. (Data of L. B. Leopold and T. Maddock, 1953, *U.S. Geological Survey Professional Paper 252*, p. 13, Figure 8.)

FIGURE 16.11 A

Width, m

Depth, m

Velocity, m/sec

Discharge, 1000s of cu m/sec

A

FIGURE 16.11 B

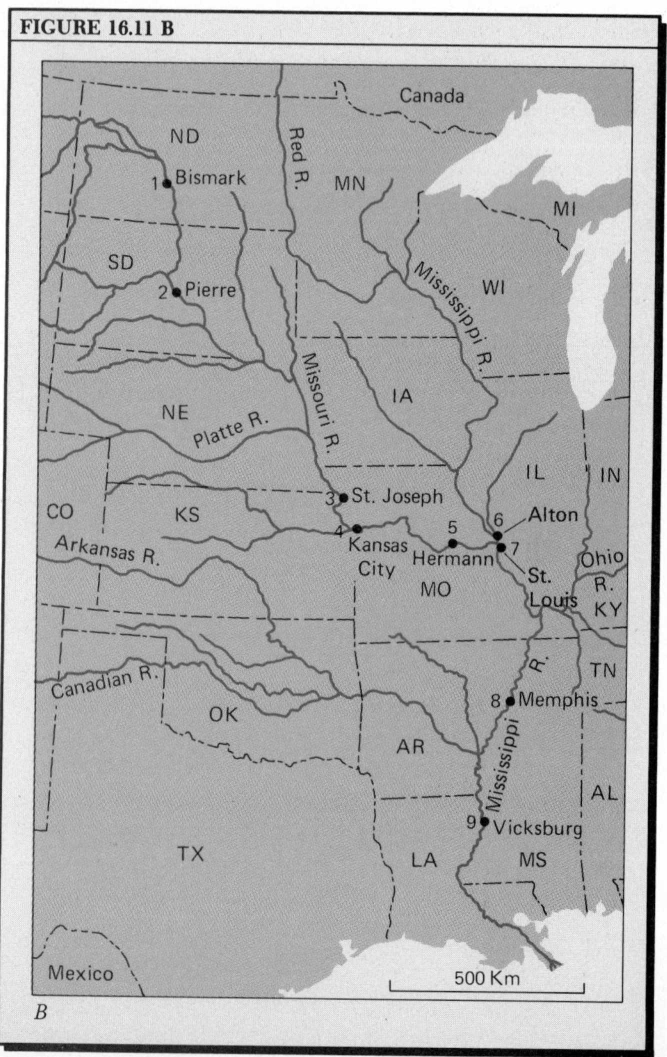

action alone, but the process has little effect on strongly bonded bedrock.

Hydraulic action is the dominant process of stream erosion in weak alluvial deposits. In flood stage, the swift, highly turbulent flow undermines the channel wall, causing masses of sand, gravel, silt, or clay to slump and slide into the channel. This activity is *bank caving*. Huge volumes of sediment are incorporated into the stream flow in times of high water, and the channel can shift laterally by many meters in a single flood (Figure 16.12).

Mechanical wear, termed *abrasion*, occurs when rock particles carried in the current strike against the exposed bedrock of the channel surfaces. Small particles are further reduced by crushing and grinding when caught between larger cobbles and boulders. Chemical reactions between ions carried in solution in stream water and exposed mineral surfaces result in a form of erosion called *solution*. It is essentially the same as chemical rock weathering.

Abrasion of hard-rock channels yields a variety of minor erosional forms such as chutes, plunge pools, and a type of cylindrical pit known as a *pothole* (Figure 16.13). The pothole is deepened by a spherical or discus-shaped stone, a grinder, rotated by the force of corkscrew water currents in the cylinder.

Stream Transportation

We can distinguish three forms of stream transportation of mineral matter. First, solution yields *dissolved solids*, which can travel downstream indefinitely and may reach the ocean. They do not affect the mechanical behavior of the stream.

Second, particles of clay, silt, and sometimes fine sand are carried in *suspension*. In this form of transport, the upward currents in eddies of turbulent flow are capable of holding the particles indefinitely in the body of the stream. Material carried in suspension is referred to as the *suspended load* and constitutes a large share of the total load of most streams. Clay particles, once lifted into suspension, are so readily carried that they travel long distances. The finest colloidal particles remain in suspension until they reach the ocean. Silts settle rapidly when turbulence subsides. Coarse sands are rarely transported in suspension except in floods. As a result, suspension provides a means of separating solid particles of various sizes and carrying each size category to a different location, a process known as *sorting*.

Third of the modes of transportation is that of roll-

FIGURE 16.12 Severe erosion occurred here during the 1976 flood of the Big Thompson River, a mountain stream of steep gradient near Denver, Colorado. Bank caving removed the foot of an earlier landslide, setting off renewed sliding of bouldery debris. A road, which formerly ran in front of the overhanging house at the left, was totally removed by the floodwaters. (U.S. Geological Survey.)

FIGURE 16.12

ing or sliding of grains along the stream bed. These dragging motions can be conveniently included in the term *traction*, a result of both the direct pressure of the water flow against the upstream face of a grain and the dragging action of the water as it flows over the grain surface. Fragments moved in traction are referred to collectively as the *bed load* of the stream (Figure 16.14). In bed-load movement, individual particles roll, slide, or take low leaps downstream, then come to rest among other grains. Leaping of grains is called *saltation;* it is relatively unimportant in the movement of bed load in a stream, but is the dominant process of transport of sand grains by wind (Chapter 19).

*Erosion Velocities and Settling Velocities

The key to understanding how sediments are sorted according to size grades during transport by water currents lies in the relative ease with which particles are set in motion as compared with the ease with which they are held in suspension. Consider, first, what levels of current velocity are needed to set in motion the particles on the bed of a stream. Many observations have been made, both in natural streams and in laboratory flumes, of the minimum velocity required to set in motion particles of a given grade size. One of the pioneer investigators of this problem was G. K. Gilbert, who is the early 1900s experimented with traction of coarse materials in flumes. He was seeking an answer to the problem of choking of stream beds with debris as a result of uncontrolled hydraulic mining of gold-bearing gravels in the Sierra Nevada of California.

When stream velocity steadily increases, there comes a critical point, designated the *erosion velocity,* at which particles of a given size begin to roll or slide on the stream bed or are lifted into suspension. You might think—and correctly so—that erosion velocity must increase as the size of particles is increased, since the larger the grains, the greater is the stress required to set them in motion. But contrary to what you might expect, the erosion velocity of densely compacted colloidal clay is quite high—about the same as for coarse pebbles—while the lowest erosion velocity is for medium sand. The shaded band on the graph in Figure 16.15 shows erosion velocities in relation to particle size.

Evidently two factors control erosion velocity. Cohesion, which is greatest in fine clays and diminishes to nothing in sand, tends to increase resistance to erosion. Particle size directly affects erosion velocity beginning with sand grades. So we see that the most easily entrained bed material is sand, for it lacks cohesion and is least impeded by particle size resistance.

The area of the graph above the shaded band is labeled "zone of erosion"; it represents the velocities at which specific particle grades will be eroded from the stream bed. Below the shaded band is a zone on the graph labeled "zone of transportation" where we encounter the principle that particles, once lifted into suspension or set to rolling on the stream bed, will continue in motion at lower velocities than were required to set them in motion.

This principle requires some knowledge of *settling velocity,* which is the terminal velocity with which a particle falls through still water. Obviously, the smaller the particle, the slower will be its rate of settling. A solid line on the graph shows how settling velocity is related to particle size.

Particles of clay and silt, once entrained, rise quickly

FIGURE 16.13 These potholes were carved into granite in the channel of the James River, Henrico County, Virginia. (U.S. Geological Survey.)

FIGURE 16.14 This bouldery debris represents the bed load of a desert stream, swept downvalley on the rare occasions when the narrow channel is in flood, Gonzales Pass Canyon, Arizona. (Mark A. Melton.)

FIGURE 16.13

FIGURE 16.14

into the body of the stream and are carried in suspension. As the left-hand side of the graph shows, the settling velocities of the finest particles are extremely low and are easily exceeded by the upward components of motion in the turbulent eddies of the stream. Consequently the fine clays continue in transportation almost indefinitely, while the coarser silts settle out when stream velocity has dropped to moderate values. In this way, clay is carried to the sea, where it contacts salt water and undergoes clotting into larger particles, a process known as *flocculation*. Silts settle out in sluggish waters of inundated floodplains and over deltas in standing water.

Particles of medium to coarse sand and larger particles travel as bed load in a stream. These particles cease to move at velocities only slightly less than the erosion velocities required to entrain them. Consequently, as shown on the right-hand half of the graph, the region of bed load transportation is a narrow zone.

The principle that emerges is as follows: When a mixture of clay, silt, and coarse particles (sand grains and larger) has once been set in motion by increasing velocity of a stream, the reverse process of deposition during falling velocity quickly affects the coarse grades, which cease to move. The silts and clays, however, continue to be transported in suspension and do not settle out until velocity has dropped to comparatively low levels. Therein lies the answer to the sorting of sediment by streams.

FIGURE 16.15 Graph showing how erosion velocity and settling velocity are related to particle size. Both scales are logarithmic. (Based on data of F. Hjulström, 1935, *Bull. of Geological Instit. of Uppsala,* vol. 25, p. 295, Table 7, and p. 298, Figure 18; and W. W. Rubey, 1933, *American Journal of Science,* vol. 24, p. 334, Figure 1.)

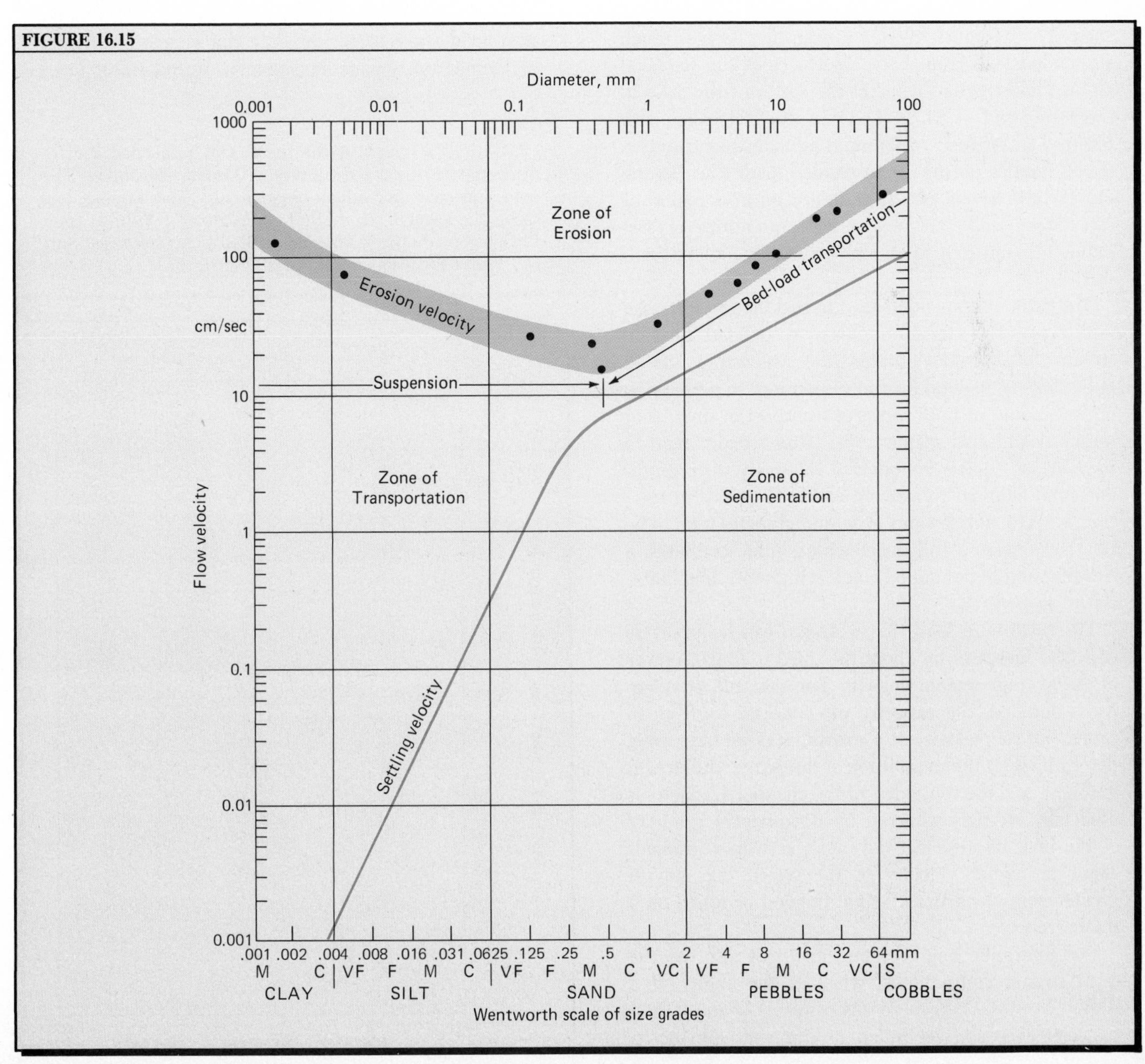

Load of Streams

We measure the solid load of a stream in terms of the weight of sediment moved past a fixed cross section in a unit of time: for example, metric tons of sediment per day. Geologists refer to the maximum load a stream can carry as the *stream capacity*. The increase in suspended load with increase in discharge is very striking. In a typical case, shown by the graph in Figure 16.16, a tenfold increase in discharge brings a 100-fold increase in load. Increasing discharge increases the stream velocity; an increase in velocity increases the capacity. Obviously, the great bulk of stream transportation occurs at high stages, including floods, while little is moved at low stages. Many rivers which are very turbid (murky) at high stage are quite clear at low stage, with only a small amount of sand being dragged along the bed.

Rivers differ greatly in their typical suspended loads, depending upon the environment of the watershed. The Missouri River, for example, derives much suspended load from barren surfaces of the semiarid Great Plains region, whereas the eastern tributaries to the Mississippi—such as the Ohio and Tennessee, with forested watersheds in a humid climate—contribute a much smaller portion of suspended load. The Hwang Ho (Yellow River) of China drains an arid region of silts (loess) and can receive vast quantities of suspended matter to give an extraordinarily high concentration.

The ratio of suspended load to bed load in a stream will range from predominantly suspended load in streams of humid climates (the Mississippi carries about 90% of its total load in suspension) to perhaps an equal amount of both forms of transport in streams of semiarid and arid regions. The latter streams tend to have broad shallow channels of relatively steep gradients well adapted to moving coarse materials in traction, whereas the streams of humid climates tend to be relatively narrow and deep, with gentler gradients, a combination better suited to carrying more fine material in suspension.

The maximum load that a stream can transport as bed load increases by about the third to fourth power of the average stream velocity. For example, if velocity is doubled, the capacity can increase by 8 to 16 times. But the velocity of a stream is, as we have seen, determined by the magnitude of discharge, the stream gradient, and the roughness of the channel. Capacity is especially strongly affected by downstream gradient of the bed, because not only does a steeper gradient result in higher velocity of the water, but also in greater ease of particles being dragged or rolled on a steeper slope.

The hydraulic engineer and geologist also use the term *stream competence*, which is the ability of a stream to move bed materials in terms of the largest particles that can be rolled or dragged. Obviously, when a stream increases in velocity it can move larger particles, because more pressure and dragging force can be exerted upon them. From laboratory experiments it has been observed that the weight of the largest particle that a stream can move on its bed varies about as the sixth power of the mean velocity. Just as in the case of capacity, competence is strongly affected by the gradient of the stream. Huge boulders can be rolled down the channel of a mountain stream of very steep gradient, whereas the same boulders would be immovable in the bed of a large river of low gradient, a principle illustrated by the Colorado River in the Grand Canyon. Boulders brought down in floods from steep side canyons form great accumulations, making rapids in the channel of the river (Figure 16.17).

Channel Changes in Flood

Persons watching a river rise to the top of its banks as a flood wave passes see only the increase in height of the stream surface, because the turbid water hides

FIGURE 16.16 Graph of the relation of suspended load to discharge for the Powder River at Arvada, Wyoming. The dots represent individual observations. The sloping line shows the average trend of all observations. (Modified from L. B. Leopold and T. Maddock, 1953, *U.S. Geological Survey Professional Paper 252*, p. 20, Figure 13.)

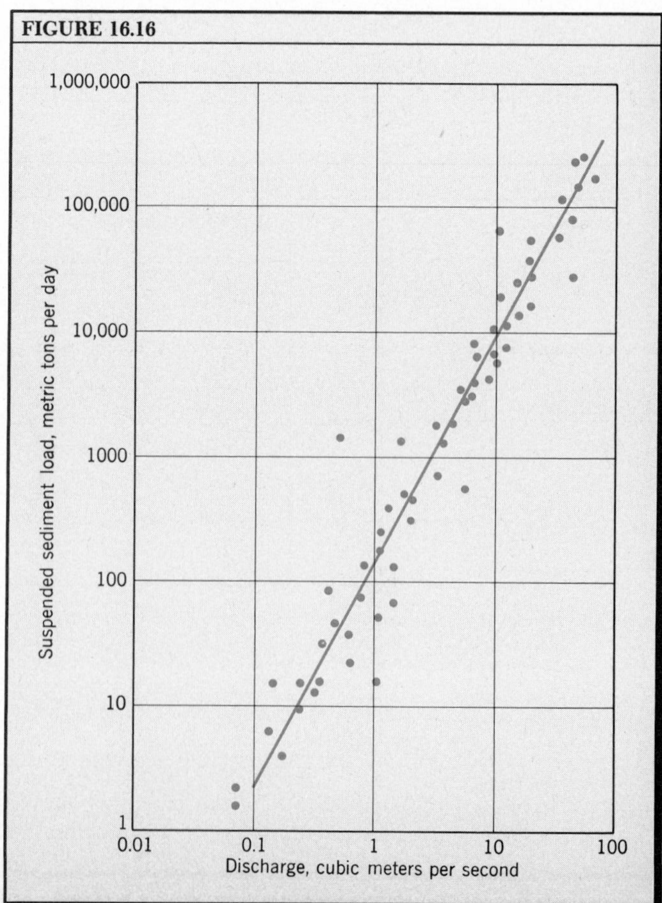

what goes on beneath. By means of stream gauging at various stages of discharge, the hydraulic engineer can record important changes in the level of the stream bed during periods of floods.

Figure 16.18 shows a typical set of changes in channel cross section during rise and fall of river stage over a period of a few weeks. As discharge increases, cross-sectional area increases. At first, as the stream surface rises, the bed remains at about the same level or may actually be raised by deposition, because the initial rise in discharge is accompanied by a sharp increase in bed load. Then, as discharge continues to increase, the increasing velocity of the stream greatly raises its transporting power and the bed is strongly scoured. In this phase, the channel floor is deepened by several meters; later, as discharge falls, bed materials cease to move and are deposited in layers on the stream bed. Thus, deposition restores the channel approximately to its previous depth.

A stream is subjected almost constantly to either an increasing or a decreasing discharge. We can infer that the channel is almost continuously being either scoured or filled. In this way, alluvial deposits are being continuously reworked by a stream, often through a depth of several meters.

How Streams Become Graded

The way in which a stream shapes its landforms is nicely illustrated by a model life history of stream development. We can imagine that this history starts with a newly formed block of land, which we will call

FIGURE 16.17

FIGURE 16.17 Boulders, carried into the Colorado River by Tapeats Creek (channel entering from left), have formed an obstruction (far right) with rapids. Grand Canyon, Arizona. (L. R. Freeman, U.S. Geological Survey.)

FIGURE 16.18

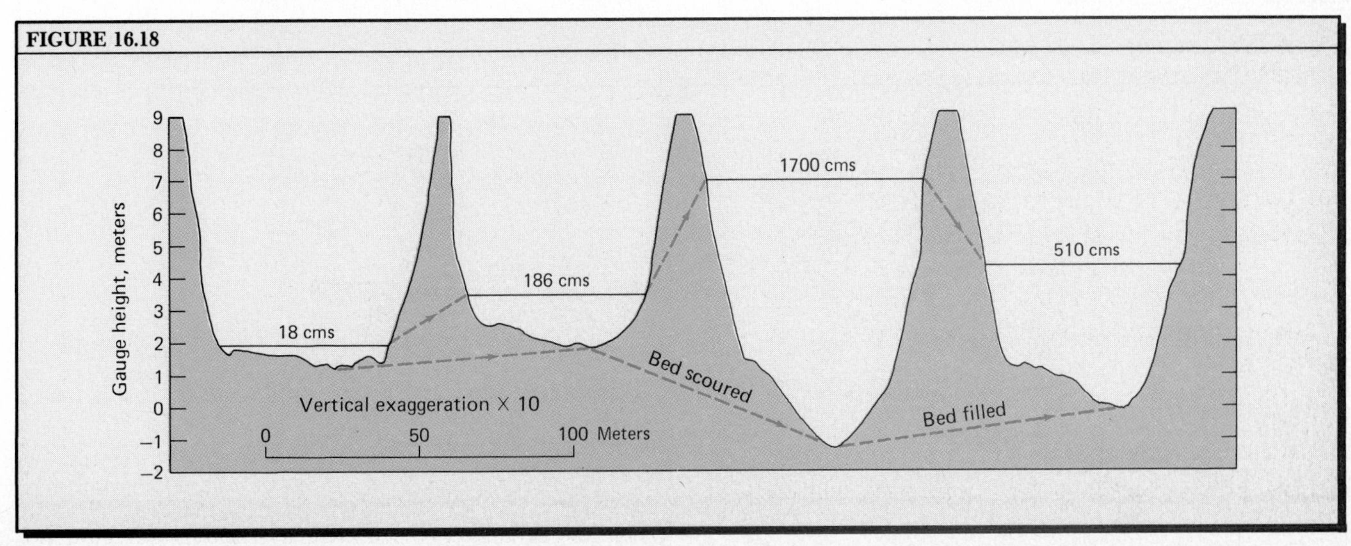

FIGURE 16.18 These cross sections of the San Juan River show great changes as the river first rises in flood stage, then falls. Based on data of L. B. Leopold and T. Maddock, 1953, *U.S. Geological Survey Professional Paper 252*, p. 32, Figure 22. (Redrawn, by permission of the publisher, from A. N. Strahler, *Physical Geography*, 4th ed., Figure 25.8. Copyright © 1975 by John Wiley & Sons, Inc.)

a *landmass*. For example, a section of the smooth continental shelf, rapidly upraised from beneath the sea and faulted into blocks, makes a good starting landmass for our purpose. As shown in the upper surface of Figure 16.19, this landmass consists of basinlike depressions and steplike alternations of steep and gentle slopes.

Surface runoff immediately organizes itself into a crude and inefficient drainage system. The flow path consists of a series of lakes connected by narrow streams passing over the steep fault steps in a succession of falls and rapids.

Intense channel abrasion is concentrated at those points where the steam passes over steep *waterfalls*. Here, narrow trenchlike *gorges*, or *canyons*, are quickly formed. As the rock barriers are reduced, lakes are lowered and finally drained. Although the stream now has a continuous, narrow channel, it has many rapids along its course. The transporting capacity of the stream exceeds the sediment load available, and so the stream channel consists largely of exposed bedrock. Abrasion continues rapidly and the channel is lowered, causing the steep-walled rock gorge to be deepened. The result may be a canyon of spectacular proportions (Figure 16.20). Weathering, sheet erosion, and mass wasting act upon the rock walls to widen the canyon. Rock fragments that roll, slide, or are washed down to the stream are swept away as suspended and bed load.

As time passes, many new stream branches are formed and feed into the main stream. As these branches cut their valleys, new sources of sediment are created, and the total load of the main stream steadily increases. At the same time, the gradient of the main stream is becoming less steep, so that the stream's capacity to transport load is lessening comparably. Obviously it is only a matter of time before the increasing supply of coarse sediment being fed into the stream matches the stream's capacity for bed-load transportation.

This point in the life history of the stream is a most important one, for when the stream is receiving and transporting sediment to the limit of its capacity, the period of rapid channel downcutting comes to an end, and a *graded channel* results. The stream has now become a *graded stream* (level 3 of Figure 16.19). Rapids will have been removed by abrasion, and the channel will have formed a smooth gradient throughout its length. A layer of alluvium will normally cover the channel floor and will be continually reworked as the stream stage rises and falls.

Once graded, the stream begins to produce a new landform. This feature is a ***floodplain***, a flat strip of land subjected to flooding about once a year (Figure 16.21). The floodplain is formed as the stream cuts horizontally on the outsides of the stream bends. This activity, termed *lateral planation*, resembles the action of a saw turned on its side. Planation takes place in times of large discharges, when the stream has great energy and scours both the banks and the bed. On the insides of the bends, bed load is deposited in the form of sand and gravel bars, creating a widening belt of low, nearly flat ground. When extreme floods occur, this low ground is inundated and is the site of deposition of silt that settles out from the turbulent water. The silt layers thus accumulate to produce a flat, fertile floodplain.

In stage *D* of Figure 16.21, the main stream is shown to be graded and to be producing a floodplain. As the floodplain is widened by further lateral planation, the valley walls are less rapidly undermined and decline

FIGURE 16.19 The irregular profile of a stream becomes graded into a smoothly upconcave form after the original falls and rapids have been removed. With the passage of time, the profile is lowered and gradually approaches base-level. (Drawn by A. N. Strahler.)

FIGURE 16.20 The Grand Canyon of the Yellowstone River, viewed from over Inspiration Point. The canyon is carved into the surface of a lava plateau. (U.S. Army Air Service.)

FIGURE 16.21 During stages *A*, *B*, and *C*, the stream deepens its valley and its profile is graded. In stages *D* and *E*, the graded stream carves a floodplain and begins to develop meanders. (From W. M. Davis, 1912, *The Explanatory Description of Landforms*, Teubener, Leipzig.)

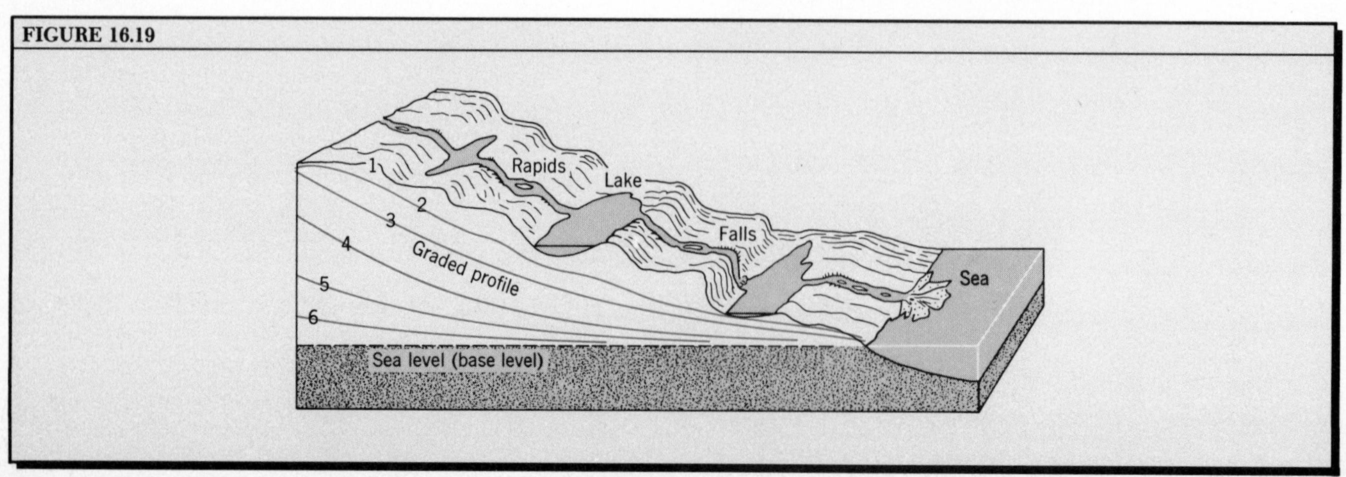

in steepness through the action of weathering and mass wasting. In stage *E*, the floodplain has widened to the degree that wavelike channel bends are able to form. Widening of the valley occurs only where the outside of a bend impinges upon the valley wall.

The longitudinal profile of a graded stream is known as the *graded profile*. It is upwardly concave and shows a decreasing gradient from head to mouth.

The stream mouth enters a body of standing water at a very low gradient. The level of the body of standing water effectively limits the downcutting of the stream and constitutes the *baselevel* of the stream. Sea level, projected inland beneath the stream system, forms the baselevel for fluvial denudation of the lands. The level we are referring to here is the upper surface of the stream, since for large rivers the channel floor is carved below sea level in lower reaches of the river.

Successive profiles numbered in Figure 16.19 show how the graded profile is gradually lowered and flattened. The surrounding land surfaces are also being reduced in height and steepness, supplying less and less load to the stream. The whole process can be expected to take from 1 to 3 million years, or more.

FIGURE 16.20

FIGURE 16.21

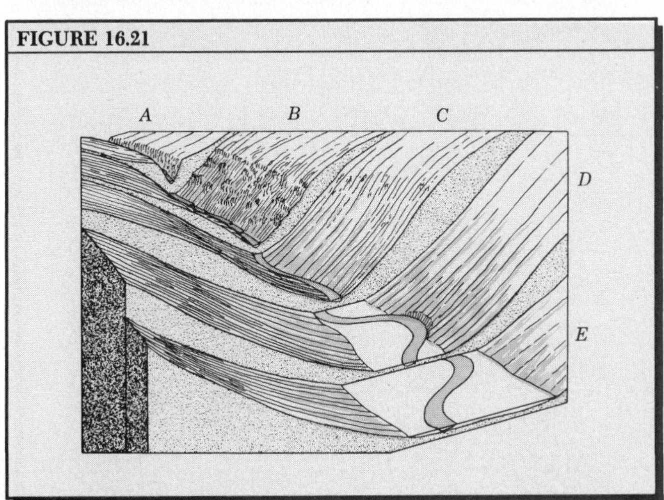

Why is the longitudinal profile of a graded stream upwardly concave? In other words, why does the slope of a graded stream become progressively less downstream? Consider first that in most stream systems, particularly those of humid lands, stream discharge increases progressively downstream. Discharge increases because tributaries enter the main channel, bringing the runoff of an increasingly large watershed area into the main channel. There is also a downstream increase in the load of the main stream because each tributary contributes a share.

One of the fundamental laws of streams was stated in 1877 by the distinguished American geologist G. K. Gilbert to explain the downstream decrease in stream gradient. Gilbert reasoned that as a stream grows larger (that is, as the magnitude of discharge increases downstream) the stream becomes more efficient mechanically, losing a smaller proportion of its energy in friction with the channel floor and sides. Because proportionately less energy is lost in friction, the larger stream can maintain sufficient velocity to carry its load on a gentler gradient.

It is important to keep in mind that the graded stream is the product of a long period of adjustment through progressive downcutting (or upbuilding, in some cases). During this long period, the stream has so adjusted its gradient that throughout its entire length the gradient is just right to permit the load to be carried through the system.

In explaining why a stream profile is upwardly concave, we must also take into account that in most river systems the particles making up the stream bed become finer, on the average, from the head to the mouth. Geologists conclude that a steeper gradient is required near the headwaters to transport the coarser particles and that the gradient diminishes progressively downstream in close relation to the decrease in particle size.

Alluvial Rivers

Many of the world's graded rivers occupy broad floodplain belts. (A river is simply a large stream. We use the word river here because it agrees with popular usage.) Called *alluvial rivers* by hydraulic engineers, these streams flow with very low gradients on thick deposits of alluvium; they have extremely sinuous bends known as *alluvial meanders* (Figures 16.22 and 16.23).

Meanders originate from the enlargement of bends in the path of flow of the stream. For example, the growth of a sand bar along the side of a straight channel will deflect the lines of flow toward the opposite bank, where undercutting takes place and a bend begins to form (Figure 16.24*A*). Material from the undercut bank is carried a short distance downchannel, forming another bar, which in turn deflects the flow to the opposite side to develop a second bend. Once a bend is produced, centrifugal force continues to thrust the flow toward the outside of the bend, and undermining continues the enlargement of the bend until a meander loop is formed.

On the inside of the bend, a series of curved sand and gravel bars accumulates to produce a *point-bar deposit* (Figure 16.24*B*). A downvalley shift, or *meander sweep*, of the entire system of bends continues because of the gradient of the alluvial valley in the direction of the stream's mouth (Figure 16.24*C*).

A meander may become constricted, creating a narrow *meander neck*. As shown in Figure 16.22, the neck may be cut through by bank caving or by overflow in the time of flood, permitting the stream to bypass the bend and to produce a *cutoff*, whose meander bend is quickly sealed off from the main stream by silt deposits and becomes an *oxbow lake*. Gradual filling of the lake results in an *oxbow swamp* (Figure 16.23).

Most alluvial rivers have a yearly flood of such proportions that the water can no longer be contained within the channel and spreads out upon the floodplain (Figure 16.25). Such *overbank flooding* permits fine-grained sediment (silts and clays) to be deposited from suspension in the relatively slow-moving water covering the floodplain. The sediment is laid down in layers, which are called *overbank deposits*.

Adjacent to the main channel, in which flow is relatively swift because of greater depth, the coarsest sediment—sand and coarse silt—is deposited in two bordering belts. Repeated floods build up lateral zones of somewhat higher ground, called *natural levees* (Figure 16.26). The highest points on the levees lie close to the river, with a gentle slope away from the river down to the low-lying marshy areas of floodplain some distance away. In a big flood, such as that pictured in Figure 16.25, the natural levees can be identified by two tree belts, one on each side of the submerged channel.

Between the natural levees and the sharply rising slopes (bluffs) that bound the floodplain lies the *backswamp* (Figure 16.22). Here, floodwaters accumulate and remain ponded long after the flood has subsided in the main channel. Fine sediment—largely clay—slowly settles out of the stagnant floodwater, forming clay layers on the floodplain. A tributary stream that enters upon the floodplain cannot flow directly into the main river channel because the natural levee forms a barrier. The tributary stream must turn downvalley, following the backswamp for a long distance before a junction is possible (Figure 16.22). Because the Yazoo River, a tributary to the lower Mississippi River, shows this type of downvalley extension, streams like it have been called *yazoo streams*.

FIGURE 16.22 A meandering alluvial river produces many interesting landforms on its floodplain. (Drawn by A. N. Strahler.)

FIGURE 16.23 Seen from an altitude of about 6000 m, the Hay River in Alberta is replete with meanders, cutoffs, oxbow lakes, and oxbow marshes. (National Air Photo Library, Surveys and Mapping Branch, Canada Dept. of Energy, Mines, and Resources.)

FIGURE 16.22

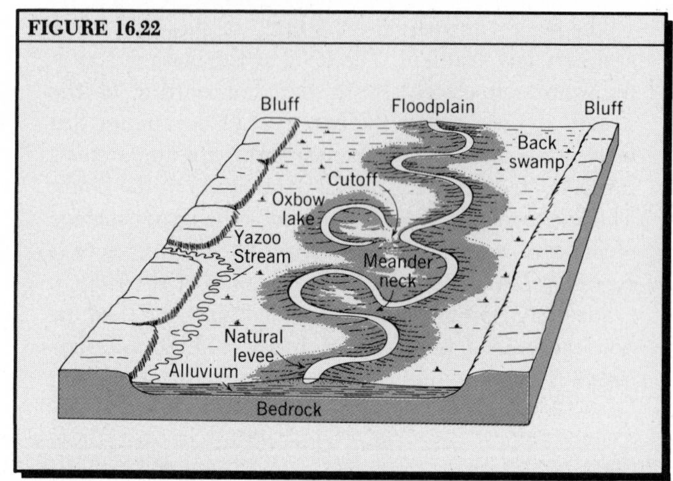

FIGURE 16.23

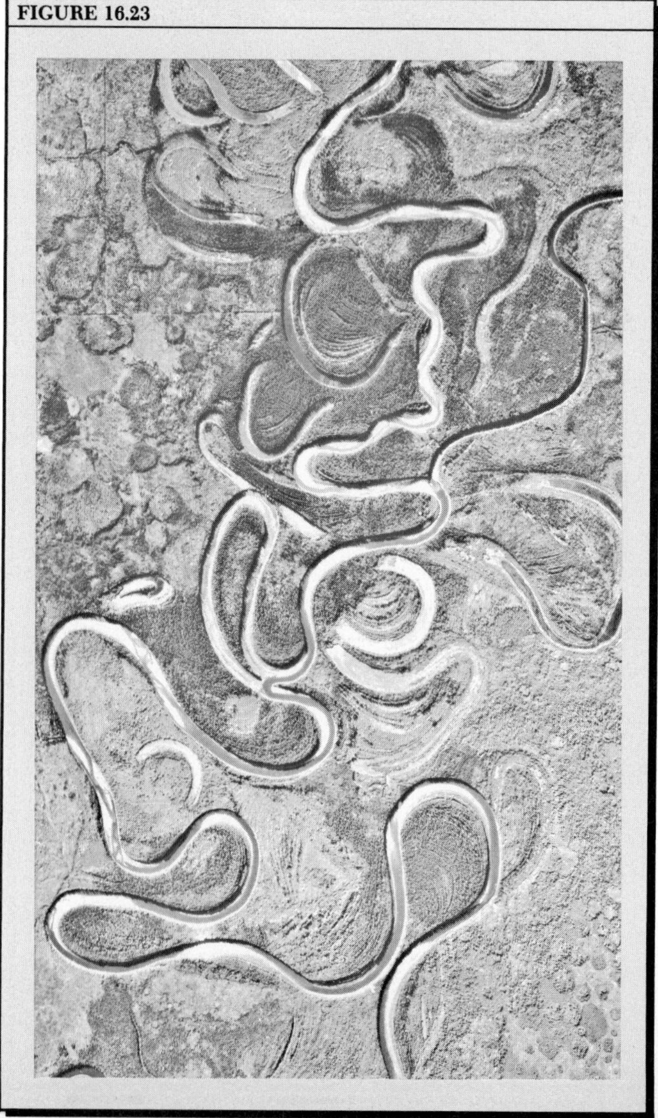

*Flood Hazards and River Control

Alluvial rivers have long posed an environmental problem because of the overbank flooding that is normal on an annual basis and the much higher floods that occur a number of years apart. A conflict of interest arises over the fact that overbank flooding adds a layer of silt rich in plant nutrients. The silt maintains fertility of the floodplain soils, particularly in regions having heavy rainfall that tends to leach out these nutrients. Counterbalancing this beneficial effect is the destruction of life and property caused by floodplain inundation. Flood-control engineering has been practiced for decades in hopes of reducing to the minimum the hazards of overbank flooding.

In general, reduction of flood-peak discharges can be brought about by two forms of control. First, storm runoff on slopes of the smaller tributary basins can be detained and delayed, thus passing the flow to downstream parts of the system more gradually. Second, the lower reaches of larger streams can be improved in efficiency or can be provided with protective structures to confine peak discharges to the natural channel.

Under the first program, watershed slopes can be treated by reforestation and by crop planting in contour belts and terraces so as to increase the capacity of the surface to absorb rainfall and to reduce the rate of runoff. To these measures may be added the construction of many small dams, usually of compacted earth, to store floodwaters temporarily and distribute the downstream discharge over a long period of time. Watershed treatment also has the beneficial effects of reducing soil erosion and of increasing recharge to the ground-water zone.

Control of inundation of floodplains is based on one of two principles. First, a system of *artificial levees,* or dikes, can be built adjacent to the channel to contain

FIGURE 16.24 Development of simple alluvial meanders from an initially straight reach of a stream. (Based on data of G. H. Matthes, 1941, *Trans. American Geophysical Union,* vol. 22.)

FIGURE 16.25 The floodplain of the Washita River, near Davis, Oklahoma, was almost completely inundated in this flood of 1950. A meander bend of the river channel is marked by a double line of trees growing on the natural levees. The channel carries most of the discharge because of its much greater water depth and velocity of flow, whereas water spread over the floodplain is moving very slowly. (Soil Conservation Service, U.S. Dept. of Agriculture.)

FIGURE 16.26 A transverse profile, greatly exaggerated, showing relation of natural levee to river channel.

FIGURE 16.24

Initial straight channel
Bank erosion
(1) Bar deposition
(2) (3) (4)
A INITIAL STAGE
Point-bar deposit
B DEVELOPING STAGE
Bar deposition
Deep channel
Shallow crossing
SWEEP
Undercutting of bank
C EQUILIBRIUM STAGE

FIGURE 16.25

FIGURE 16.26

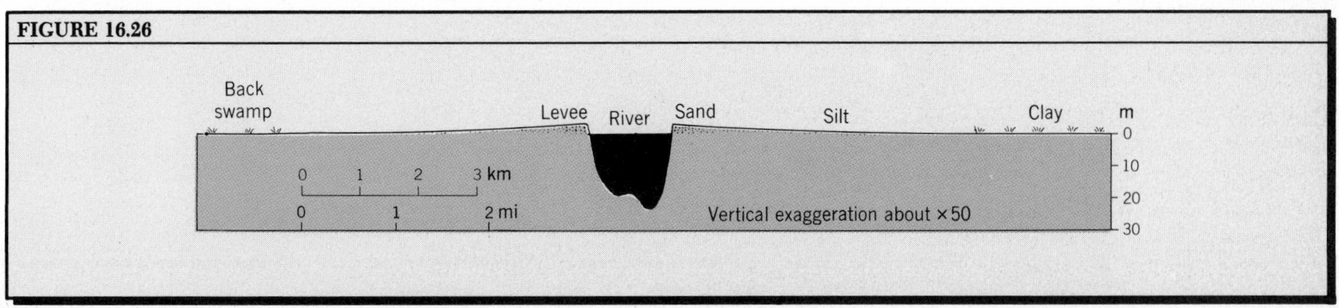

FIGURE 16.27 This air view shows a breach in the artificial levee adjacent to the Missouri River in western Iowa. Water is spilling from the high river level at right to the lower floodplain level at left. (U.S. Dept. of Agriculture.)

FIGURE 16.28 Air photograph of the Mississippi River looking north. The view shows three artificial meander cutoffs made by the Corps of Engineers to reduce river length. Taken in 1937, this photograph shows sediment plugs blocking the ends of Bachelor Bend, which formerly flowed past the city of Greenville, Mississippi. (U.S. Army Corps of Engineers.)

FIGURE 16.27

FIGURE 16.28

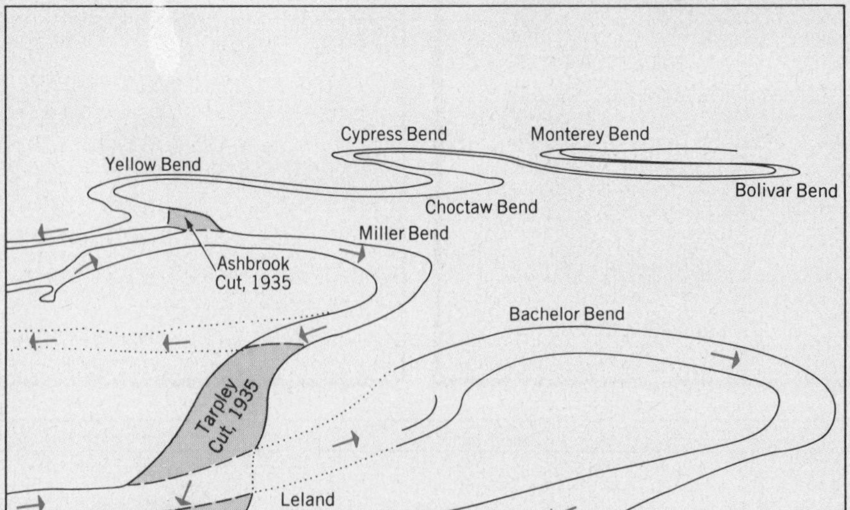

Cypress Bend
Monterey Bend
Yellow Bend
Bolivar Bend
Choctaw Bend
Miller Bend
Ashbrook Cut, 1935
Bachelor Bend
Tarpley Cut, 1935
Leland Neck
Leland Cut, 1933
Greenville, Miss.

overbank flow in stages at which the water would otherwise overspread the floodplain (Figure 16.27). Most levees are broad earth embankments, although flood walls of reinforced concrete can be used where a city lies close to the river channel. If overtopped by an unusual flood discharge, the levee may be broken at a low point. Overflow erodes rapidly to produce a deep breach through which the floodwater will quickly flow to inundate the floodplain (Figure 16.27).

The lower Mississippi River, under the supervision of the Mississippi River Commission from 1897 to 1934, was controlled by a vast levee system designed to protect the floodplain from inundation. The system has been continuously improved and now includes over 4000 km of levees, in places up to 10 m high.

FIGURE 16.29 This braided stream derives its coarse bed load from a melting glacier far up the valley, Peters Creek, Chugach Mountains, Alaska. (Alaska Pictorial Service, Steve and Dolores McCutcheon.)

FIGURE 16.29

A second principle of river control is that of channel modifications designed to shorten the length of the stream and so to steepen the gradient. The increase in velocity of flow reduces the area of cross section, resulting in a lowering of the height of water at flood crest. Shortening is accomplished by the artificial cutoff of meander bends.

Figure 16.28 shows such engineering changes shortly after they were completed. The program was successful in reducing the river-surface height in flood, but considerable difficulty has been experienced in preventing the river from reforming its meanders and returning to the previous condition. Against the advantages of increased security from floods gained by such measures we must also weigh the losses of natural landscapes offering scenic beauty, wildlife habitats, and recreational uses.

Perhaps the most common cause of aggradation in stream channels is the combination of arid climate and mountainous relief, as we find in the southwestern United States. Barren, steep mountain slopes shed large quantities of coarse debris when eroded by runoff of torrential rains. Floods in the mountain valleys are characterized by a large proportion of coarse bed load; this is carried downvalley on steep gradients (see Figure 16.14). Where a canyon emerges upon a valley floor of gentle slope, aggradation occurs because the stream is not able to transport its load on a sharply reduced gradient. Shifting from side to side as aggradation occurs, the stream spreads its excess load in the form of an *alluvial fan* (Figures 16.30 and 16.31).

Aggrading Streams; Alluvial Fans

Coarse rock waste is sometimes supplied to a stream by its tributaries and by runoff from adjacent hillsides in greater quantity than the stream is capable of transporting. When this happens, the excess load is spread along the channel floor, raising the level of the entire channel, a process termed *aggradation*.

A stream channel in which aggradation is in progress is typically broad and shallow. Aggradation takes the form of deposition of long, narrow bars of sand and gravel, which tend to divide the flow into two or more lines. The stream subdivides and rejoins in a manner suggesting braided cords and is thus called a *braided stream* (Figure 16.29). Where the braided stream flows between narrow confining walls, aggradation steadily raises the level of the floodplain.

The alluvial fan takes the form of a low, upwardly concave sector of a cone, steepening in gradient toward an apex situated at the canyon mouth. At its outer edge, the fan slope grades imperceptibly into the flatter plain. As one might suspect, the size of rock particles making up the fan is greatest near the apex, where much bouldery material accumulates. The alluvium grades to progressively finer particles toward the base. Large fans of mountainous deserts may be several kilometers in radius from apex to outer edge.

Alluvial fans hold ground water beneath confined aquifers, layers of sand and gravel deposited by streams which built the fan (Figure 16.31). The water-bearing layers receive recharge near the head of the alluvial fan, where water enters the permeable layers by seepage from the channels of streams originating in mountain watersheds. The water in the aquifers is retained under aquicludes, the relatively impermeable layers of mud spread over the fan surface during its construction at times when mudflows issue from the

FIGURE 16.30

FIGURE 16.31

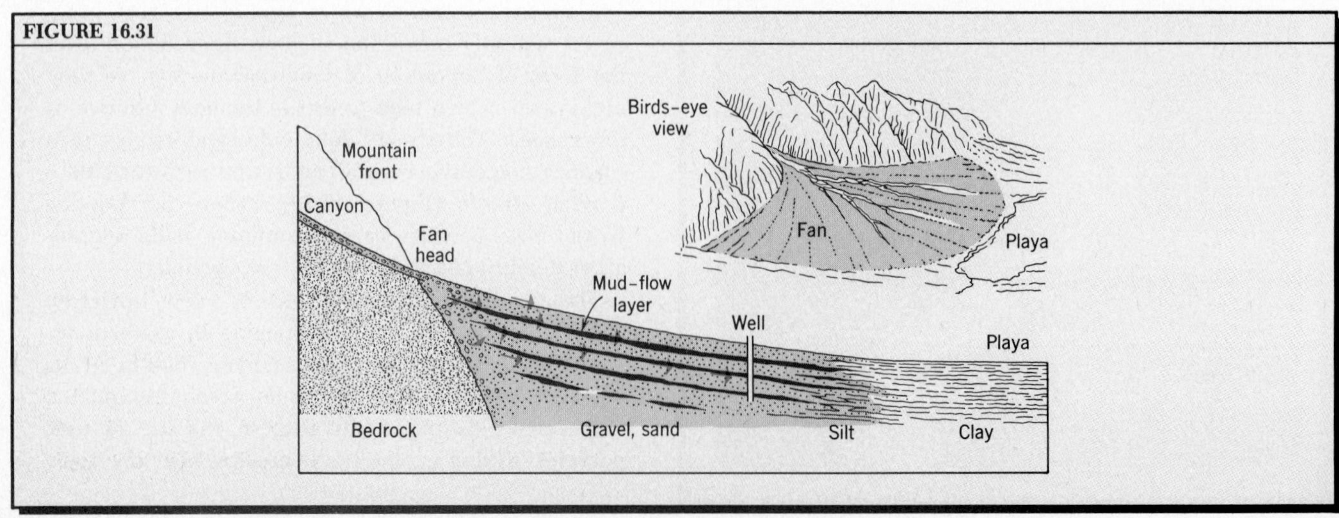

canyon mouths.

Now, because the sediment layers of an alluvial fan slope downward to lower elevations from the apex of the fan, water that enters the aquifers makes its way to lower levels and is confined under pressure beneath the aquicludes; this water possesses a hydraulic head. When a well is driven into the periphery of the alluvial fan, water will spontaneously rise, often forming a flowing artesian well.

Alluvial fans are the most important of all groundwater reservoirs in the arid southwestern United States, but as we noted in the last chapter, water is being withdrawn from these alluvial deposits at a much faster rate than the natural recharge can replenish it. Water tables have fallen drastically in many places, and the supplies may soon be exhausted.

FIGURE 16.30 A great alluvial fan in Death Valley, built of debris swept out of a large canyon. Notice the many braided stream channels. (Spence Air Photos.)

FIGURE 16.31 Idealized cross section of an alluvial fan showing the relation of mudflow layers (aquicludes) to sand layers (aquifers). (Drawn by A. N. Strahler.)

FIGURE 16.32 Formation of alluvial terraces. The letter *R* in Diagram *C* refers to a point where a terrace is defended by a rock outcrop. (Redrawn, by permission of the publisher, from A. N. Strahler, *Physical Geography*, 4th ed., Figure 25.23. Copyright © 1975 by John Wiley & Sons, Inc.)

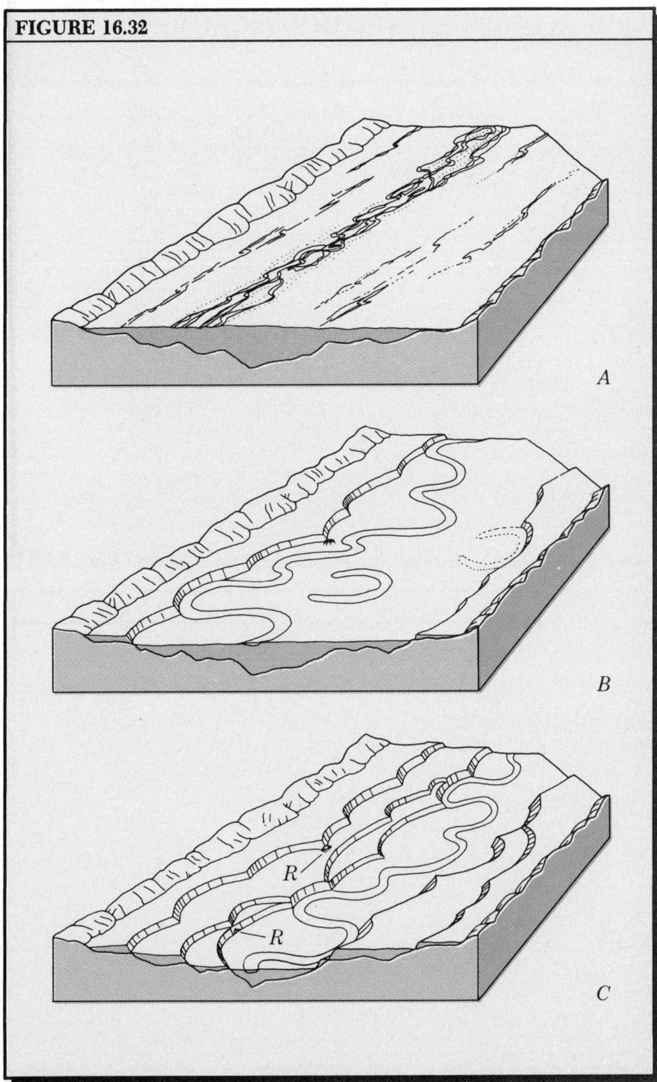

Alluvial Terraces

A *terrace* is a steplike landform, bounded by steeply rising slopes on one side and by descending slopes on the other. There are many kinds of terraces. The *alluvial terrace* is one important type, carved out of alluvial deposits previously deposited in a valley.

Steps in the evolution of alluvial terraces are illustrated in Figure 16.32. Diagram *A* shows a broad, flat valley filled with alluvium by an aggrading stream. Aggradation can result from a change to a more arid climate or from the greatly increased load derived from a melting glacier in the upstream region.

In Diagram *B*, we see that the stream has changed its role from one of aggradation to one of downcutting, or *degradation*. The stream has begun to cut down into the alluvial deposit, at the same time shifting laterally and carving out its floodplain in the easily eroded material. Now a single terrace level appears, its surface the remnant edge of the original floodplain.

The cause of degradation might be found in a change to a more humid climate. In that case, the growth of a denser vegetative cover would tend to hold back the coarser debris from the streams and so reduce both the quantity and the size of particles in the stream loads. A similar effect follows the disappearance of glaciers (see Chapter 18).

In Diagram *C*, degradation has proceeded further. Now we have a series of alluvial terraces, resembling a flight of broad stairs on the valley wall (Figure 16.33).

Stream Trenching and Entrenched Meanders

At any time in the life of a graded stream, sea level may drop, or the earth's crust may rise. Either change causes a drop in baselevel, and the establishment of a lower baselevel has a profound effect upon the graded stream; it scours the bed deeply, making a trench.

Channel trenching forms a series of rapids. Degradation extends rapidly upstream, and a narrow, steep-walled gorge or canyon is produced (Figure 16.34). On either side is a *rock terrace*, representing the former broad-floored valley (Figure 16.35).

In an extreme case of trenching, the floodplain meanders of an alluvial river become deeply carved into the underlying bedrock, producing *entrenched meanders* (Figures 16.36 and 16.37), which form a winding river gorge. Cutoff sometimes occurs at the narrow meander neck, leaving a rock arch, or natural bridge.

FIGURE 16.33

FIGURE 16.34

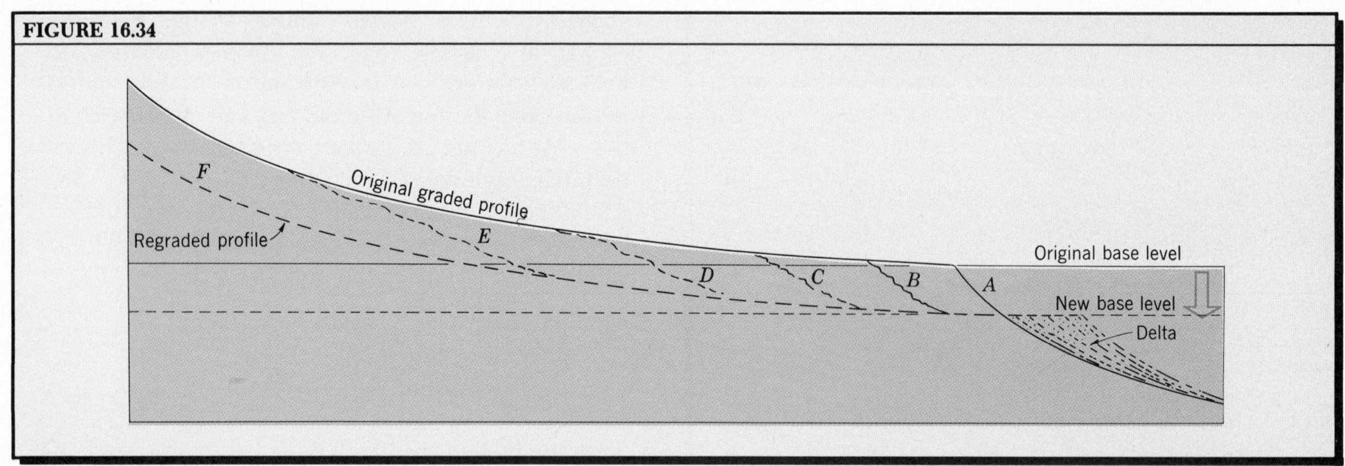

FIGURE 16.35

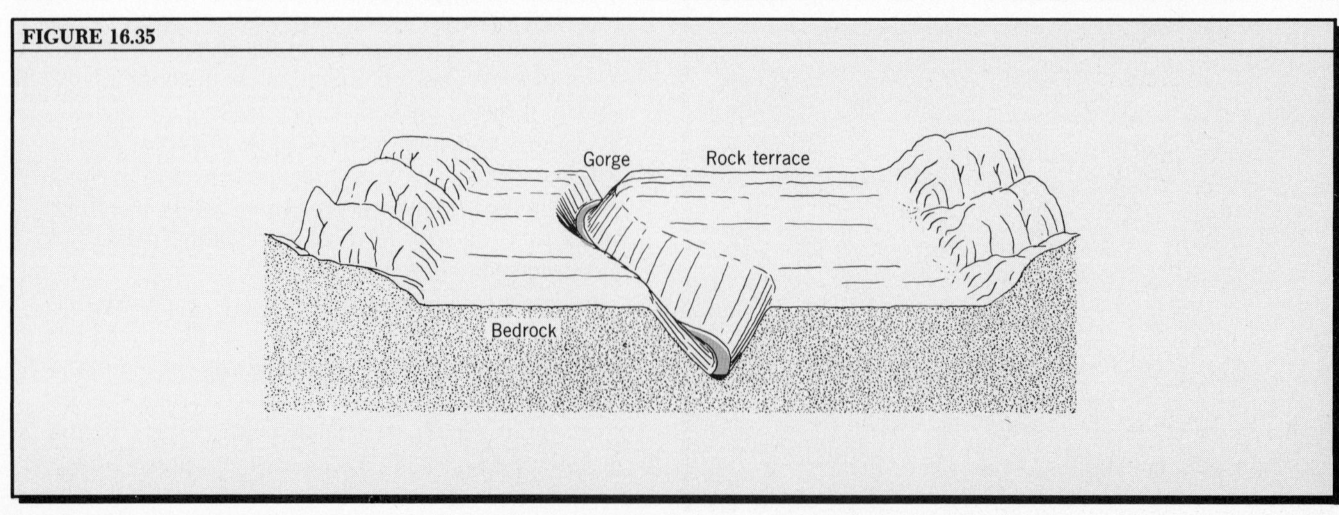

FIGURE 16.33 Alluvial terraces of the Esk River Valley, Canterbury, New Zealand. The uppermost terrace level at the right bears alluvial fans, now deeply trenched by the tributary streams that formed them. (Copyright by V. C. Browne & Sons, Christchurch.)

FIGURE 16.34 A drop in baselevel sets off the regrading process leading to completion of a regraded profile at a lower level. Changes in the stream profile progress headward from *A* through *F*.

FIGURE 16.35 A rock terrace, formerly a floodplain belt, borders a steep-walled inner gorge resulting from stream trenching. (Drawn by A. N. Strahler.)

FIGURE 16.36 Entrenched meanders with cutoffs and a natural bridge. (Drawn by E. Raisz. Redrawn, by permission of the publisher, from A. N. Strahler, *Physical Geography*, 4th ed., Figure 25.28. Copyright © 1975 by John Wiley & Sons, Inc.)

FIGURE 16.37 These entrenched meanders, carved into horizontal strata by the San Juan River in Utah, are known locally as the Goose-Necks. (Spence Air Photos.)

FIGURE 16.36

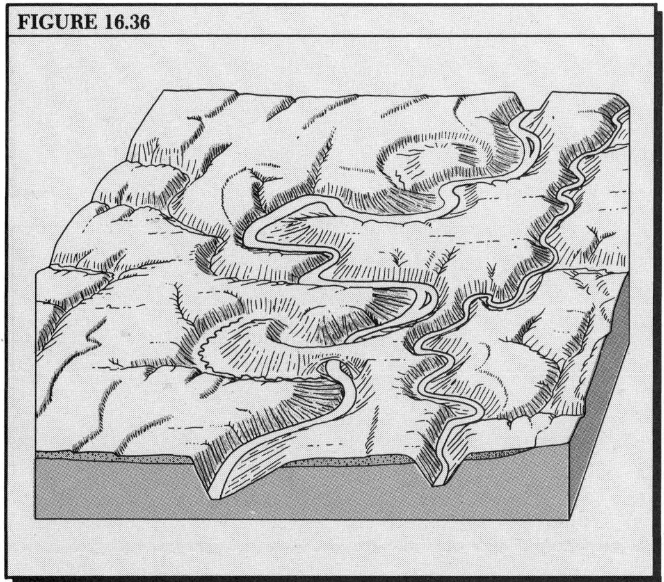

FIGURE 16.37

*How Large Dams Disturb River Channels

Dams are built across the courses of large trunk rivers for production of electric power and to divert water for major downstream irrigation systems. These big dams are among the great engineering achievements of humankind. On the other hand, what was originally conceived as a benevolent giant can turn into a monster when long-term effects begin to spread upstream and downstream from the new structure and its great reservoir of water.

A large dam offers an interesting exercise in the principles of stream aggradation and degradation. Consider, first, the upstream changes that such a dam initiates (Figure 16.38). The reservoir behind the dam is a body of standing water of zero gradient. Bed load carried by the stream along with much of the suspended load comes to rest in the reservoir as a delta and its extension into a layer of fine silt and clay over the lake floor. One can easily see, and in fact predict rather closely, that the days of the reservoir as a holder of water are numbered. Both for generating electric power and for water supply, a substantial reservoir of water is needed to regulate the distribution of flow. Water must be stored in times of high stream flow and released at time of low stream flow. Thus, whatever its purpose, the dam has a limited future.

Next, as the delta is built out into the reservoir, the stream channel begins to aggrade. A wedge of coarse alluvium is spread over the valley floor and begins to spread upvalley (Figure 16.38). Inundation of fertile floodplain soils, and along with it roads, railroads, and towns, can ensue.

A case in point is the valley of the Rio Grande, upstream from the Elephant Butte Reservoir in New Mexico. Thirty years after completion of this dam, aggradation had reached a depth of 3 m at the head of the reservoir, burying the village of San Marcial. At Albuquerque, 160 km upstream, aggradation had reached a depth of over 1 m. So long as the delta deposition continues in the reservoir, aggradation will spread upvalley, for the stream must maintain a gradient for the transportation of its bed load.

Looking in the downvalley direction below a large dam, we must anticipate the effects of release of water practically free of load of any kind. Here it flows over a bed previously adjusted to transport of a large quantity of coarse bed load (Figure 16.38).

In the case of Hoover Dam in the Colorado River, the release of a large flow of clear water caused rapid scouring of the river bed, and this effect has spread far downvalley. In the first 40 km deepening ceased when a residual layer of boulders on the stream bed prevented further erosion. At Yuma, 560 km below the dam, permanent channel changes included a lowering of average position of the stream bed by about 3 m.

The load carried past this point is only about one-fifteenth of its original value, since all load must now come from tributary surfaces and streams entering below the dam. Stream depth has about doubled, but because of downcutting, height of the river in its high stages is not appreciably lower than formerly. One harmful effect of the overall downcutting has been to lower the entire stream cross section several meters below its former position with respect to natural levees on floodplain portions of its course. This change rendered unfit for use a system of gravity-flow irrigation that had previously been installed. Now, pumping must be used to lift the river water into the irrigation ditches—a costly solution in a time of energy shortages.

FIGURE 16.38 Schematic profile of a river showing the effects of a large dam and its reservoir upon the river channel both upstream and downstream.

FIGURE 16.39 Stages in the development of a simple delta built into a lake in which wave action is slight. Based on data of G. K. Gilbert. (Redrawn, by permission of the publisher, from A. N. Strahler, *Physical Geography,* 4th ed., Figure 25.25. Copyright © 1975 by John Wiley & Sons, Inc.)

FIGURE 16.40 Oblique air view of the Kander River delta, Lake Thun, Switzerland. Note the jet of sediment-laden water being projected into the lake. (Swissair photograph.)

FIGURE 16.38

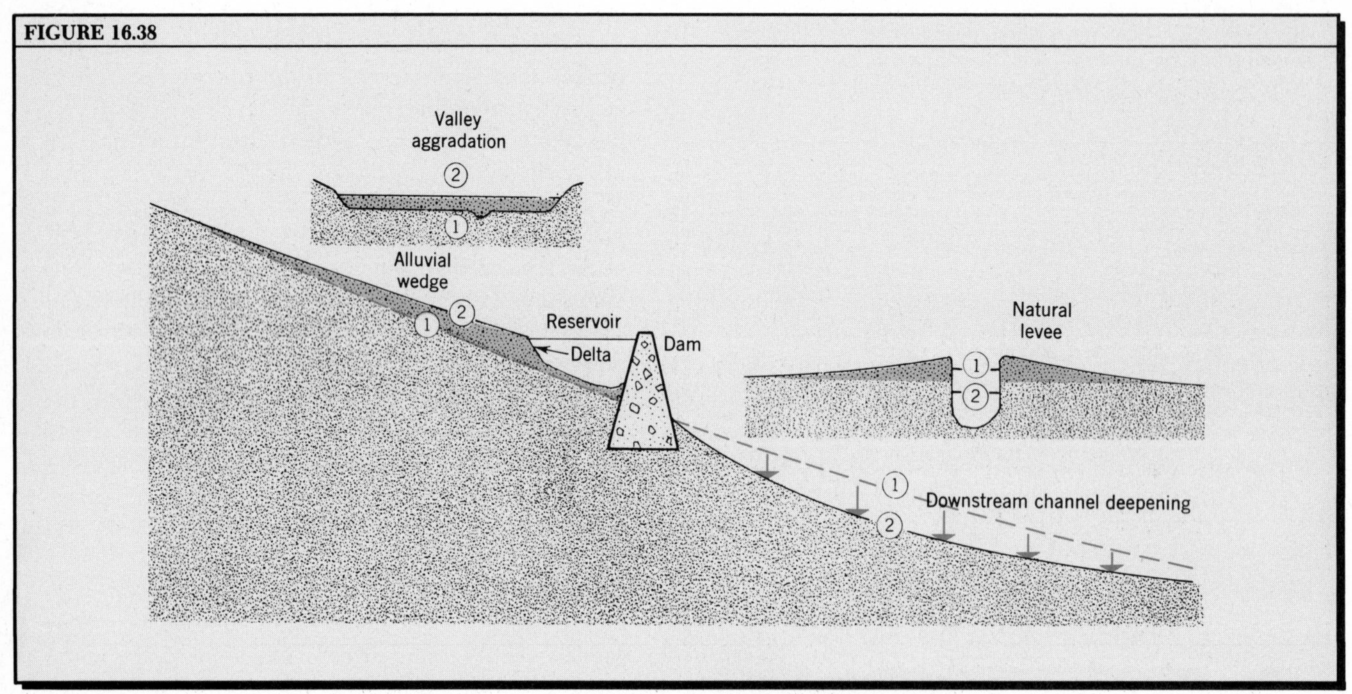

FIGURE 16.39

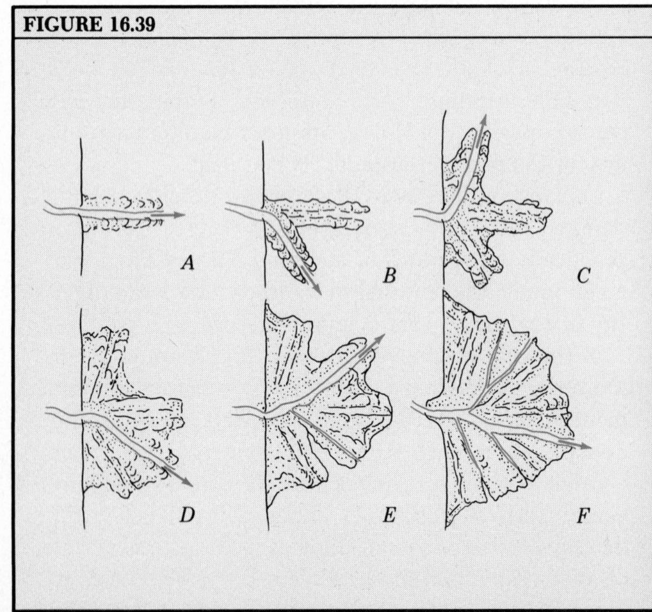

FIGURE 16.40

Deltas

A stream reaching a body of standing water, whether a lake or the ocean, builds a deposit, the *delta*, composed of the stream's solid load. The growth of a simple delta can be followed in stages, shown in Figure 16.39. For simplicity, imagine that the water body is not appreciably affected by waves and tides. The stream enters the standing water body as a jet, but its velocity is rapidly checked (Figure 16.40). The stream channel is extended into the open water by sediment deposited in lateral embankments in zones of less turbulence on either side of the jet. The stream repeatedly breaks through the embankments to occupy different radial positions. In time it produces a deposit of semicircular form, much like the alluvial fan (which is in a sense a terrestrial delta).

In cross section, the simple delta consists largely of steeply sloping layers of sands, termed *foreset beds* (Figure 16.41), which grade outward into thin layers of silt and clay, the *bottomset beds*. As the delta grows, the stream will aggrade slightly and spread new layers of alluvium, the *topset beds*. An important factor in causing the finer suspended particles to settle close to the stream mouth is the presence in seawater

FIGURE 16.41 Internal structure of a simple delta built into a lake. (Redrawn, by permission of the publisher, from A. N. Strahler, *Physical Geography*, 4th ed., Figure 25.26. Copyright © 1975 by John Wiley & Sons, Inc.)

FIGURE 16.42 The Nile delta has an arcuate shoreline strongly shaped by wave and current action. (Redrawn, by permission of the publisher, from A. N. Strahler, *Physical Geography*, 4th ed., Figure 25.27. Copyright © 1975 by John Wiley & Sons, Inc.)

FIGURE 16.41

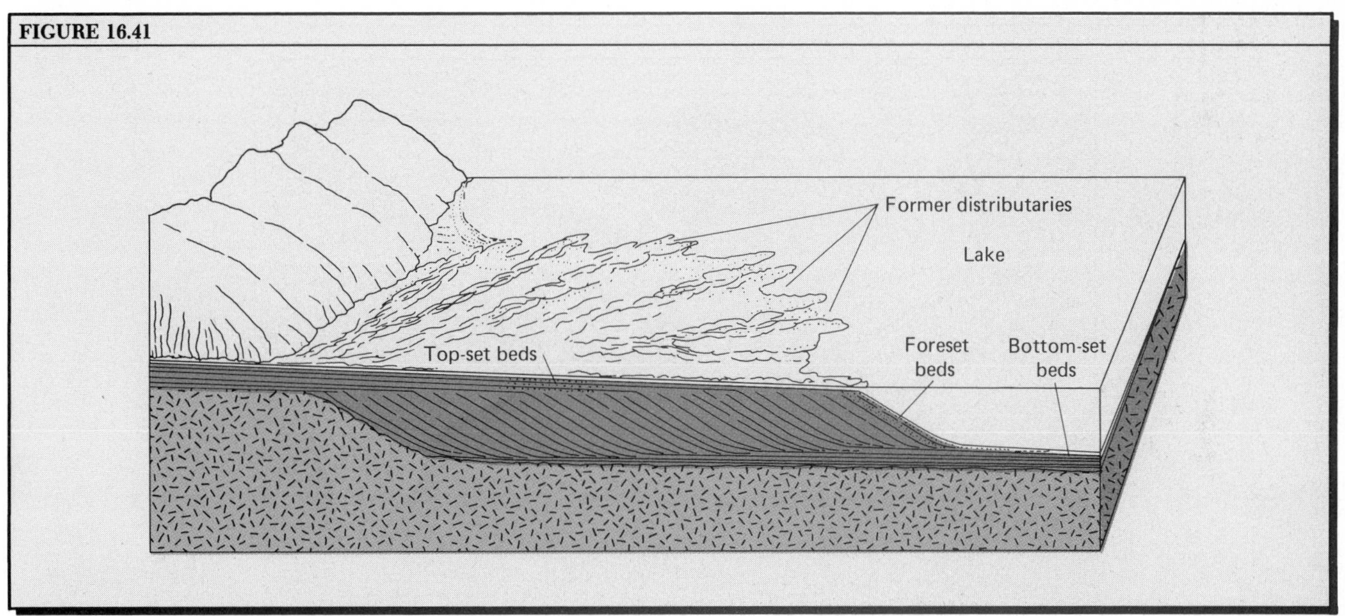

FIGURE 16.42

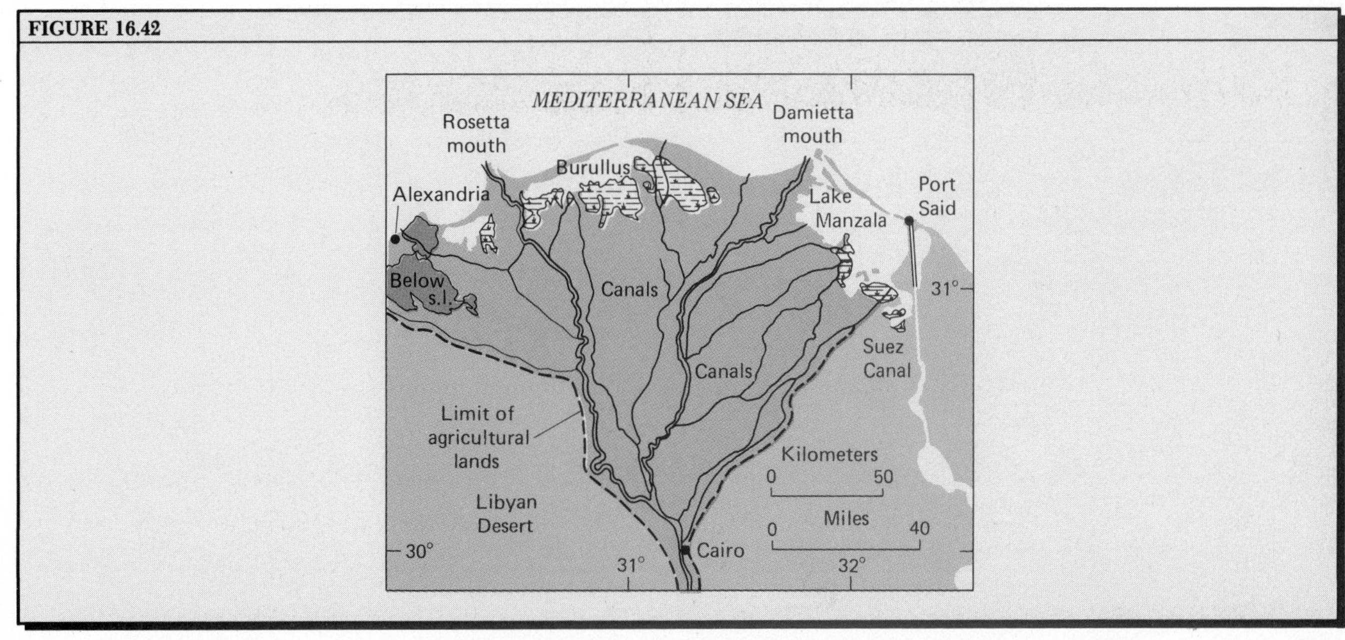

of dissolved salts. These act to cause the particles to clot, or flocculate, into aggregates of larger size that readily sink to the bottom.

The form and structure of a large marine delta is strongly influenced by waves and tides, changes of sea level, and gradual subsidence of the delta. The Nile delta takes the triangular shape of the capital Greek letter *delta*, from which the landform was originally named (Figure 16.42). From an apex at Cairo, *distributary channels* branch in a radial arrangement. Sediment reaching the Mediterranean Sea from the principal distributary mouths is swept along the coast by shallow currents, which shape curved bars enclosing shallow lagoons. Because of wave and current action, the delta shoreline is arcuate in plan, bowed convexly outward.

FIGURE 16.43 The deltaic plain of the Mississippi River. *Upper map*—Abandoned river courses and distributaries are shown by bold lines. *Lower map*—Seven deltas, numbered in order from oldest to youngest. The lower map covers a larger area than the upper map. (Redrawn and simplified from maps by C. R. Kolb and J. R. Van Lopik, 1966, in *Deltas in Their Geologic Framework*, M. L. Shirley, ed., Houston, Texas, Houston Geological Society, p. 22, Figure 2, and p. 31, Figure 8.)

FIGURE 16.44 Block diagram showing structure and sedimentary units within the modern bird-foot delta of the Mississippi River. (Simplified from a diagram by H. N. Fisk, E. McFarlan, Jr., C. R. Kolb, and L. J. Wilbert, Jr., 1954, *Journal of Sedimentary Petrology*, vol. 24, p. 77, Figure 1.)

FIGURE 16.45 Map and structure section of the modern delta of the Mississippi River. (Simplified from H. N. Fisk, E. McFarlan, Jr., C. R. Kolb, and L. J. Wilbert, Jr., 1954, *Journal of Sedimentary Petrology*, vol. 24, p. 87, Figure 8, and p. 92, Figure 12.)

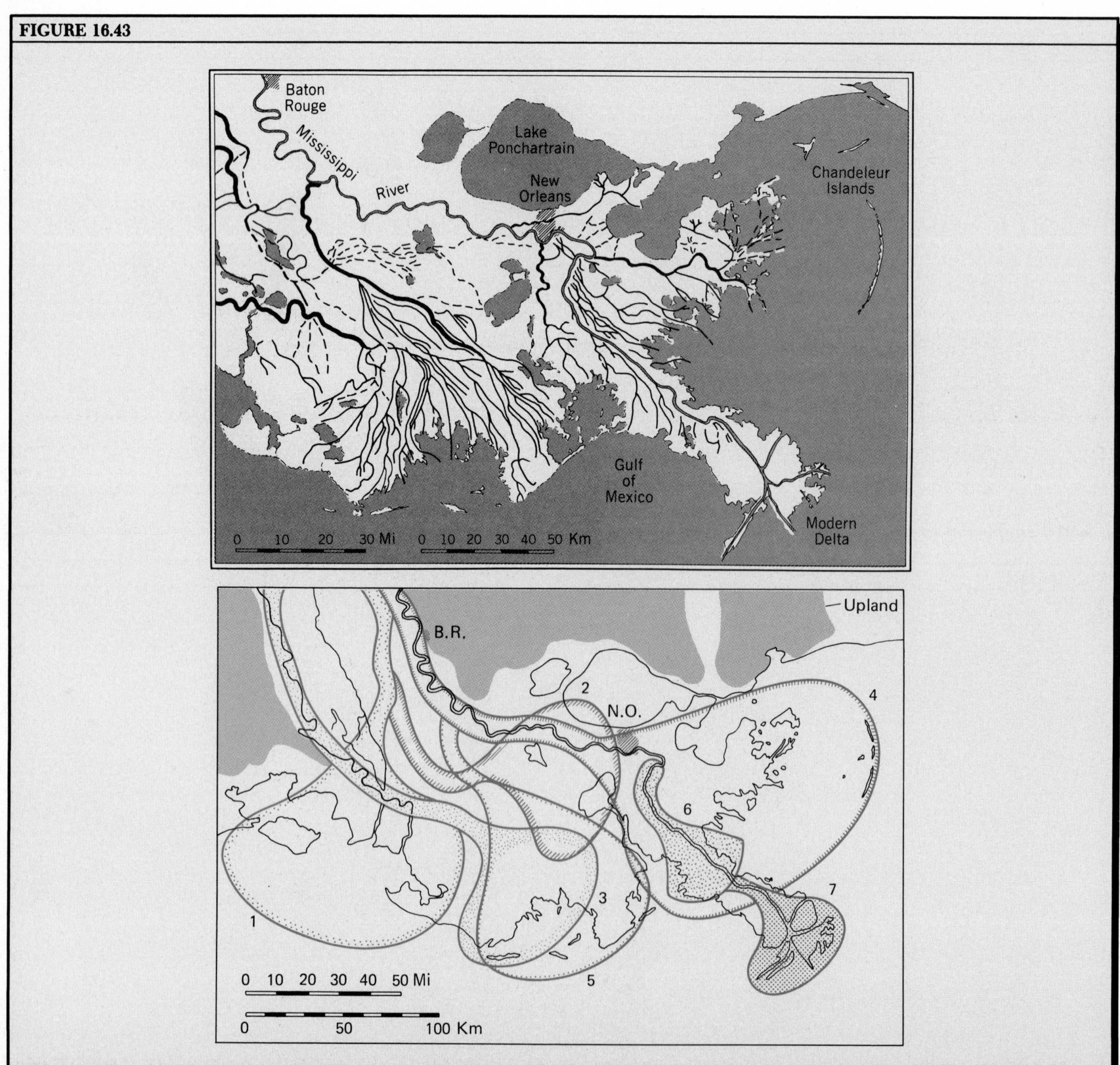

*Geology of the Mississippi Delta

The delta of the Mississippi River has been intensively studied by geologists. Numerous borings have revealed the composition and structure of both the modern delta and of several older deltas which together make up an enormous *deltaic plain*. Figure 16.43 consists of two maps of the Mississippi delta region. The upper map shows the patterns of abandoned channels and distributaries as well as the present channel. The lower map shows the outlines of six older delta lobes and the modern lobe, which constitutes a *subdelta* of the entire complex. Starting over 5000 years ago, each of the older deltas in succession built a fan-shaped deposit, each somewhat like the Nile delta, in the form of radiating distributary channels. Diversions of the main river channel upstream from the delta shifted the zone of accumulation alternately eastward and westward, as the numbered sequence on the map shows. Recent subsidence has resulted in partial inundation of the outer fringes of the older delta plains (see Color Plate F.2).

The modern Mississippi River *bird-foot delta* differs from the earlier deltas in having broadly branched distributaries. Moreover, the modern delta is being constructed in deeper water—over 100 m—than its predecessors. Figure 16.44 shows the structure of the

FIGURE 16.44

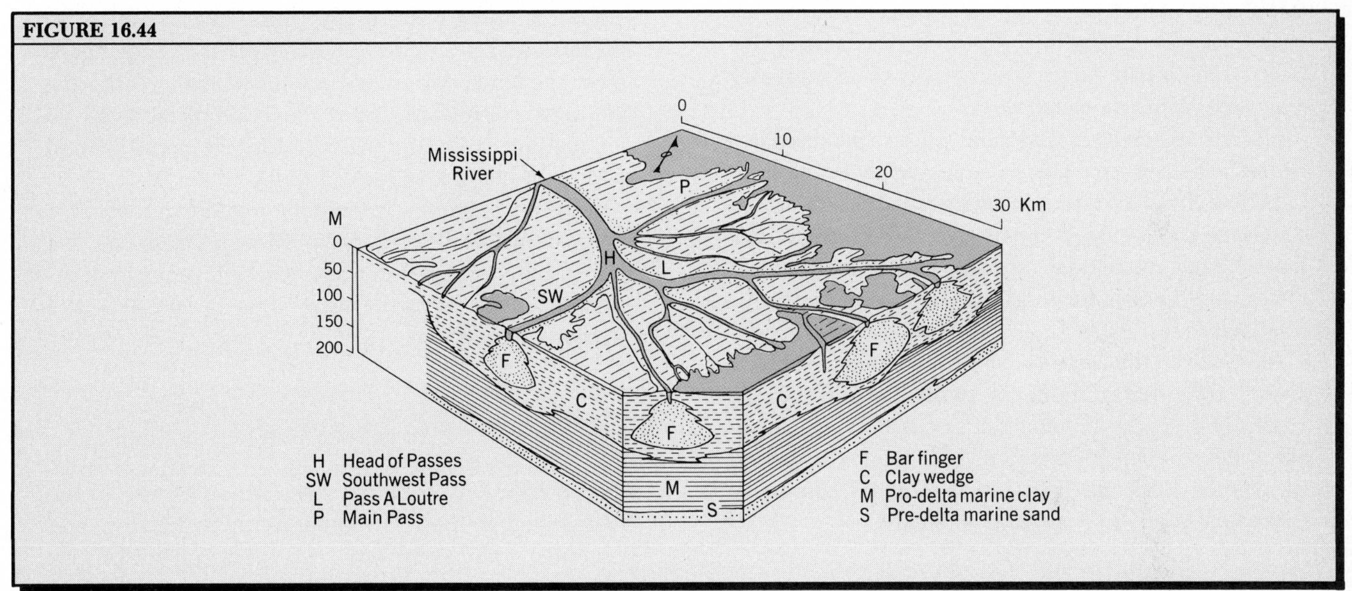

FIGURE 16.45

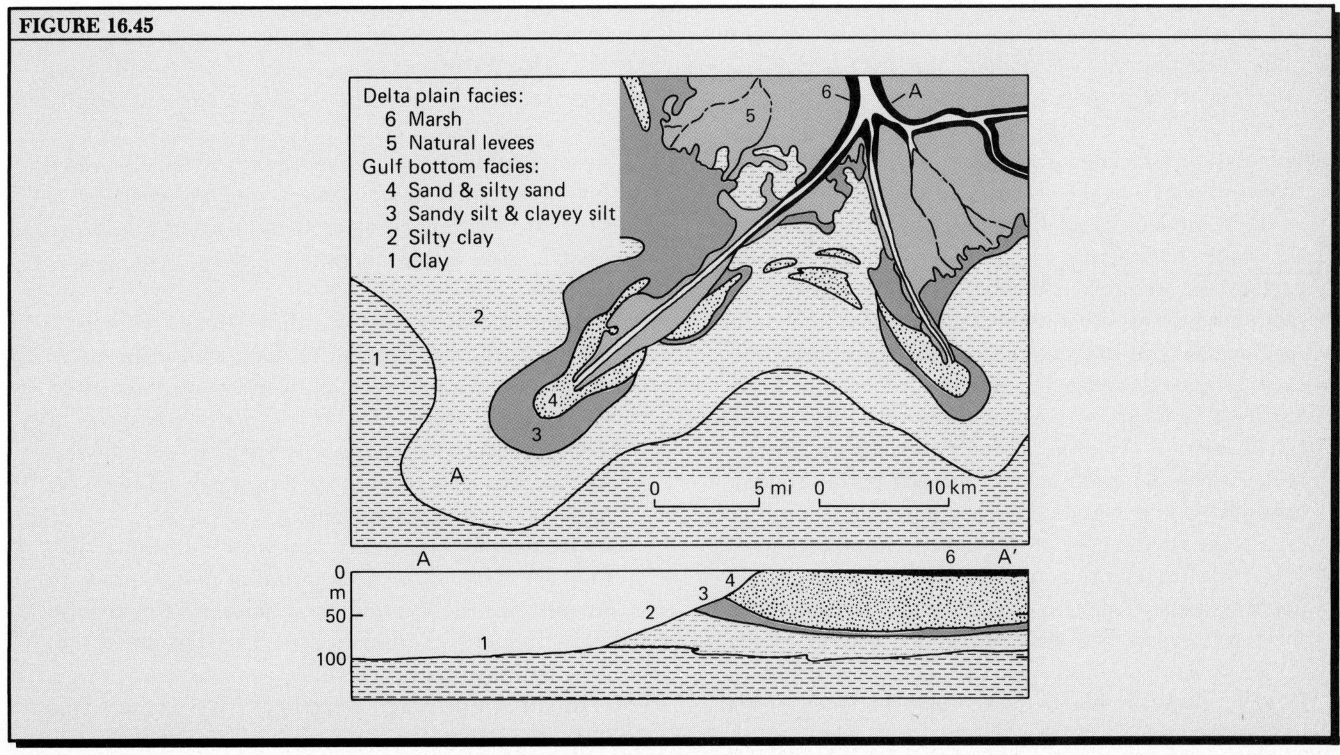

modern delta by means of a block diagram. Advancing distributary mouths have built deep but narrow *bar fingers* of sand upon earlier deposits of marine clays and upon a thin layer of pro-delta clays and silts that were laid down seaward of the advancing sands. As the bar fingers grew, clays and silts were deposited between the fingers.

Sediment types (sedimentary facies) of the modern bird-foot delta platform are shown in Figure 16.45. Notice the succession of deposits encountered at increasing depth outward from the ends of the distributaries (passes). These are foreset beds and grade outward and downward from sand, through silt mixtures, to clay (bottomset beds) spread over the ocean floor at a depth of about 100 m. The topset strata, which form the present delta plain lying close to sea level, are natural levees bordering the distributary channels. Between the distributaries are deposits of organic-rich clay and silt in salt marshes.

Not all sediment is deposited by stream discharge through the principal passes. Here and there the levee is broken by a *crevasse* through which flood waters discharge sediment. A crevasse, a fanlike accumulation of sand upon tidal deposits, is eventually abandoned, and its deposits subside to be covered again by tidal waters.

Altogether, the modern bird-foot delta contains about 113 cu km of sediment, deposited in about 450 years. The yearly increment is about 0.25 cu km, or about 455 million metric tons. This total stream load (suspended load and bed load combined) consists of about 25% sand, 30% silt, and 45% clay.

Deltas and Plate Tectonics

The great deltas of the world can be tied in with plate tectonics. As we pointed out in Chapter 13, rifting apart of a continental plate tends to favor the growth of a large marine delta at the apex of a coastal indentation associated with a triple rift junction (see Figure 13.17*B*). As the ocean basin widens, deltas such as these produce great thicknesses of sediment along the stable continental margins, where they become part of the marginal geocline. In many cases, delta sediment is transported by turbidity currents down the continental slope in front of the delta to accumulate as an enormous deep-sea fan or cone (see Figure 9.30). Good examples are seen today in the deltas and fans of the Mississippi, Amazon, and Congo rivers on the stable continental margins of the Atlantic basin. Another example is the Ganges–Brahmaputra delta and its great cone on the floor of the Indian Ocean.

Many geosynclines consist in part of ancient delta deposits. Going back to Chapter 12, reexamine the structure section of a geosyncline that existed during the Devonian period in New York State (Figure 12.5). Rapidly rising highlands to the east provided rivers with abundant sediment, which was spread as a series of deltas across what is now the Catskill Mountains region. These sediments were red beds of coarse, sandy texture. Westward, the delta sediments graded into finer sands and then into marine black shales. As the restored section shows, the deltaic sediments spread progressively farther westward, overlapping the marine strata. Behind the spreading delta lay an expanding deltaic plain, shown by lines of equal time (isochrones).

Running Water as a Geologic Agent

In this chapter, we have investigated the geologic work of running water on the lands. Overland flow on hillslopes and stream flow in channels work together to lower the lands. We should not forget that weathering and mass wasting are part of the total process of fluvial denudation, preparing the rock for transportation and helping to bring it to lower levels.

In the next chapter, we will put together a model for fluvial denudation of a large continental landmass. We shall also investigate the ways in which geologic structures influence the patterns of streams and of entire stream networks.

Key Facts and Concepts

Fluvial denudation *Fluvial denudation* consists of the combined actions of weathering, mass wasting, and running water of streams. It is the dominant process in shaping continental landforms.

Overland flow and channel flow Running water acts in two forms: (1) *overland flow,* spread broadly over the ground surface; (2) *channel flow* (*stream flow*), concentrated between *banks* in a narrow, troughlike *stream channel.* The *stream* is the body of flowing water, including all sediment it carries. Overland flow may cause *sheet erosion,* yielding particles of sediment that accumulate at the base of a hillslope as *colluvium* or are deposited in stream channels as *alluvium.*

Stream channels and stream flow *Stream gradient* is the vertical drop per unit of horizontal distance. In detail, the channel typically consists of *pools* alternating with *riffles. Stream velocity* is the speed of downstream flow parallel with the bed and banks; it increases rapidly from the bed upward and from the banks to the midstream line.

Stream turbulence At very low velocities, water flow may be *streamline flow* (*laminar flow*). Flow in streams is usually *turbulent flow,* with irregular eddy motions. Turbulence enables the stream to hold sediment in suspension.

Stream discharge *Mean velocity* (V), the average flow rate of the entire stream cross section, is measured directly. Multiplied by cross-sectional area (A) mean velocity yields *discharge* (Q), stated as the *equation of continuity,* $Q = AV$. In a given stream of constant discharge, high velocity is associated with small cross-sectional area; low velocity, with large cross-sectional area. A stream flows at the *terminal velocity,* dissipating energy as frictional resistance is overcome.

Stream velocity, gradient, and depth Empirical formulas state that mean velocity is roughly proportional to square root of both gradient and depth. *Bed roughness* also affects velocity.

Drainage systems and channel networks Stream channels are organized into branching *channel networks,* each forming a *drainage system* bounded by a *drainage divide* that encloses the *watershed.* Stream networks can be analyzed as systems of *channel segments* of increasing order of magnitude. Stream order is logarithmically related to segment numbers, segment lengths, drainage areas, and channel segment gradients. Stream discharge is observed to increase logarithmically with drainage basin area.

The longitudinal stream profile The *longitudinal profile* of a stream, plotting altitude against downstream distance, is upwardly concave, with flattening gradient downstream. In the downstream direction, depth, width, and mean velocity increase logarithmically with increasing discharge.

Stream erosion Streams perform three closely interrelated types of geologic work: (1) *erosion,* (2) *transportation,* and (3) *deposition.* Stream erosion occurs by *hydraulic action* (causing *bank caving*), by *abrasion* (mechanical wear), and by *solution* (chemical reaction). Abrasion produces *potholes.*

Stream transportation Stream transportation occurs in *suspension* as *suspended load* and in *traction* as *bed load.* Transportation results in *sorting* during deposition. The critical *erosion velocity* required to set particles of a given diameter in bed load motion is lowest for medium grade sand. Because of their low *settling velocity,* fine particles remain in suspension for long distances of transport. Clays may settle only after *flocculation* in contact with seawater.

Load of streams The maximum load a stream can carry is the *stream capacity.* Capacity increases sharply with increase in discharge and velocity, at a rate of about the fourth power of the velocity. *Stream competence,* a measure of the largest size of particle a stream can move on the bed, increases with both velocity and gradient. In rising stage, a stream deepens its bed by scour; in falling stage, the bed is built up by deposition; the process is one of reworking of alluvium.

Stream gradation Stream gradation acts upon the available *landmass* above sea level. A stream with initially irregular gradient rapidly erodes *waterfalls* and rapids, draining lakes and carving a *gorge* or *canyon.* In time, a *graded channel* is formed. The *graded stream* then produces a *floodplain* by *lateral planation.* The longitudinal profile of the graded stream is a *graded profile* for which the stream mouth at sea level serves as the *baselevel* for fluvial denudation. Upward concavity of the graded profile is a response to increasing stream efficiency as its discharge increases downstream. The profile is gradually lowered in gradient, while long-term equilibrium is preserved.

Alluvial rivers Graded rivers flowing on broad floodplains are known as *alluvial rivers;* they show well-developed *alluvial meanders.* Meanders originate by growth of bends, creating *point-bar deposits.* Fully developed meanders move downvalley by *meander sweep.* A meander becomes constricted to result in cutoff through the narrow *meander neck,* leaving an *oxbow lake* or *oxbow swamp. Overbank flooding* leaves fine-textured *overbank deposits* and builds *natural levees.*

Flood hazards and river control Flood peaks can be reduced by (1) upstream controls to reduce storm runoff, or (2) downstream channel engineering works to confine peak discharges. *Artificial levees* (dikes) can be built to confine the flow to the channel but are subject to breaching. Shortening of the channel by artificial cutoffs can reduce peak floods.

Aggrading streams; alluvial fans Building up of the stream bed by *aggradation* produces a *braided stream.* Aggradation below the mouth of a canyon in dry lands builds an *alluvial fan,* a structure which holds ground water. Permeable gravel layers serve as aquifers, mudflow layers as aquicludes.

Alluvial terraces Valley aggradation, followed by *degradation,* leads to carving of *alluvial terraces,* forming a succession of steps in the valley sides.

Stream trenching and entrenched meanders Degradation followed by deep trenching into bedrock leaves a *rock terrace.* The gorge may exhibit *entrenched meanders.*

Deltas The accumulation of stream-transported sediment in a body of standing water at the stream mouth is a *delta* with *distributary channels.* Delta structure includes foreset, bottomset, and topset beds.

Questions and Problems

1. What remedial measures might be taken to minimize the loss of soil by sheet erosion on cultivated lands? Do you favor government payments to compensate owners of land kept out of cultivation because the land is subject to severe erosion?
2. Analyze a flowing stream as an open energy system powered by gravity. In what form is the energy at

the point where water enters the stream? How is energy transformed and temporarily stored as it moves downstream? How does energy leave the system?

3. Suppose you were to dig a trench into the alluvium of a dry stream bed, making a careful study of the gradation of the bed materials according to size. Would the deposits of a single episode of falling stream stage show a size decrease or an increase from bottom to top? Explain.
4. If you were to sample the sediment forming a natural levee, taking samples at short intervals along a line extending out at right angles to the river bank, what grade sizes would you encounter, and in what order? Explain.
5. It has been argued that measures to delay storm runoff in upstream areas are ineffective in making any important reduction in the size of a major flood reaching the lower trunk portion of a major river on its floodplain. What reasoning is behind this conclusion? Do you agree? If so, would you abandon programs designed to reduce overland flow on watershed slopes?
6. A geologist is examining an exposure of Triassic strata which prove to be clastic sediments of very coarse textures. The geologist concludes that the strata were deposited as an ancient alluvial fan. What physical features of the rock would suggest such a conclusion?

17

CHAPTER SEVENTEEN

Denudation and Rock Structure

The last three chapters have given us an insight into the workings of the fluvial denudation process, consisting of the action of weathering, mass wasting, and running water. In this chapter, we will develop the total picture of fluvial denudation through long spans of geologic time, using idealized models of landform evolution. We will take a large landmass, uplifted above sea level by a positive tectonic or epeirogenic movement of the crust, and follow its changes as the land surface is lowered toward a general baselevel. Because climate—whether moist or dry—has a great deal to do with the way fluvial processes act, we will need to take into account the special features of the world's deserts, where running water works at a different pace than it does in moist regions.

After viewing landmass denudation on the vast scale of a continent or subcontinent, we will turn to a close examination of landforms controlled by varied structures of the underlying bedrock. Folding and faulting has brought rocks of quite different physical properties into contact in various attitudes, with the result that stream valleys, hillslopes, and ridge crests take on distinctive patterns and profiles.

Concept of a Denudation System

Denudation deals with that portion of the continental crust which lies above baselevel. As we define it here, ***baselevel*** is an imaginary surface of zero elevation (sea level, that is) extended beneath the land surface. The layer of rock lying above the baselevel surface is the *available landmass*, the crustal mass available for removal by denudation. This might seem simple enough and perhaps too obvious to need discussion, until we consider the principle of isostasy. In Chapter 7, we explained that removal of rock by denudation is compensated for by an upward isostatic movement, whereas the deposition of sediment in another place is accompanied by downward isostatic movement under the added load (see Figure 7.29).

To the available landmass as it exists at any one moment must be added the crustal mass that will rise under isostasy to replace part of the rock mass removed in the denudation process. Estimates vary as to what factor to allow for isostatic replacement, but ratios on the order of 3:4 or 4:5 are considered realistic. Using the second of these ratios, the removal of 5000 m of rock would be accompanied by uplift of 4000 m and would result in a net lowering of the land surface of only 1000 m. Thus, the available landmass includes the crustal mass furnished by isostatic replacement.

The available landmass provides potential energy for a denudation system activated by the external agents powered by solar energy. Two forms of work must be done within this system. First, energy must be expended to reduce the strong, dense bedrock to a clastic state or weaken it greatly so that it can be moved to lower levels by mass wasting or by processes of overland flow. Weathering performs the preliminary physical and chemical changes essential to denudation. Energy for weathering processes comes from external sources—from atmospheric heat and heat stored in water in its liquid and vapor states, and to some degree directly from solar radiation.

The second form of work is that of abrasion and transportation by overland flow and channel flow. This fluid flow, as we have seen, is a system of mass transport and energy transformation powered by gravity. Potential energy that water and rock particles possess because of their elevated position is transformed into kinetic energy of flow, which is ultimately dissipated as heat through frictional resistance to flow. Water is brought to high elevations on land by atmospheric processes, which are powered by solar energy. Rock is brought to high elevations by internal forces which are powered by the same sources of energy that cause lithospheric plates to move.

Rates of Orogenic Uplift

Let us take up first the case of an active orogenic belt along a continental margin. Disregarding the horizontal movements involved in folding and overthrust faulting, consider for the moment only the lifting of the crustal mass. Rates of uplift during orogeny can be estimated from the ages of late Cenozoic strata found today at high positions in the mountain zone. Of course, these estimates deal with long time spans within which uplift rates may have been unevenly distributed.

Some rough values of maximum rates of uplift by orogenic processes can be obtained from the data of geodetic surveys. Precise determination of ground elevations has been repeated in a number of instances along survey traverse lines crossing active mountain blocks. A number of such measurements in California showed rates of uplift ranging from 4 to 12 m per 1000 years, with an average of about 8 m per 1000 years. Comparable rates have been measured in orogenic belts in such widely separated places as in the Persian Gulf area and in Japan. In contrast, epeirogenic uplift of certain parts of the stable continental cratons is estimated to be on the order of 1 m per 1000 years.

Using the rate of uplift of 6 m per 1000 years, a mountain summit might rise from sea level to an elevation of 6 km in about 1 million years (m.y.), assuming none of its mass to be removed by denudation processes. This figuring suggests that a full-sized mountain range might be created in a time span less than one-half the duration of the Pleistocene Epoch, which was about 2.5 m.y. Considering the long durations of even the individual epochs of the Cenozoic Era, orogenic uplift appears to be an extremely rapid process in comparison with the long, uninterrupted spans of time in which denudation operates. Even the towering Himalayas might well have been raised through a large part of its height in 2 million years, although the total uplift probably took much longer because of slower rates operating during much of the orogeny.

Observed Denudation Rates

Geologists and hydrologists have made a number of attempts to estimate the rates of continental denudation by using long-term measurements of suspended and dissolved solids in major streams. For example, in 1967, Professor Sheldon Judson of Princeton University published the following estimates of denudation rates in three large continental areas:

	cm/1000 yr	m/1000 yr
Amazon River basin	4.7	0.047
Congo River basin	2.0	0.02
United States	3.0	0.03
Weighted mean	3.6	0.036

An unknown factor in such calculations is the effect of human activity in increasing the rates of soil erosion through deforestation and agriculture. Judson made a generous allowance for this factor. A large part of all three regions consists of a low-lying, stable craton, and the rates are extremely small in comparison with our estimate of rapid tectonic uplift.

Denudation rates are strongly influenced by elevation of the surface of a landmass above baselevel. The higher a mountain block rises, the steeper will be the gradients, on the average, of streams carving into that mass. Both stream abrasion and transportation ability (both capacity and competence) increase strongly as gradients become steeper. For this reason, the rate of denudation will be highest for the most highly elevated crustal masses and will diminish as elevations become lower. We must also take into account the increased rate of rock breakup with increasing altitude because of frost action.

Based upon measured volumes of sediment brought by streams from small watersheds in mountainous areas, denudation rates of 1 to 1.5 m per 1000 years are about the maximum that can be expected as average values for high mountain masses. These rates are very much greater than Judson's figures for large continental areas of moderate to low elevation.

We can conclude that rates of uplift of a landmass during orogeny are faster, by a factor of perhaps five to ten times, than the maximum rates of denudation of the uplifted masses. This difference in rates easily accounts for the presence of great alpine ranges such as the Himalayas, the Alps, and the Andes.

A Model Denudation System

Using certain reasonable assumptions, we can devise a model of the denudation process and from it perhaps obtain some idea of the order of magnitude of time spans involved in reduction of a mountain mass to a low, undulating plain called a *peneplain*.

First, we assume that during orogeny a substantial crustal mass is arched up, or lifted as a block along boundary faults (Figure 17.1). Arbitrarily, we assign a width of 100 km to the uplifted mass, for this is about the order of magnitude of width of a number of present-day ranges, including the Alps, the Carpathians, the Pyrenees, the Caucasus, the Alaska Range, the Sierra Nevada, the Rockies, and the Appalachians. Length of the uplift is not important in this analysis—a segment some tens of kilometers long will do.

The uplifted mass is bordered by low areas, at or below sea level, which can serve as receptors of detritus. An initial surface of reference, close to sea level, is raised to a summit elevation of, say, 6 km. In Figure 17.1, a dashed line shows how this reference surface has been deformed by the orogenic uplift. Figure 17.2 is a graph on which elevation is plotted against time. Orogeny is shown by the steeply rising line. This tectonic uplift is given a span of 5 m.y., but most of the rise in elevation occurs within 2 m.y. Uplift tapers off, then ceases at zero reference time.

Denudation has been in progress during the uplift, increasing in intensity as elevation increases. The elevated mass has been carved into a maze of steep-walled gorges organized into a fluvial system of steep-gradient streams. The profile of the rugged mountain mass and the main stream system are suggested in greatly exaggerated scale in Stage *A* of Figure 17.1. Let us assume that at time zero the average elevation of the eroded surface lies at 5 km. Thus, about 1 km of rock has been removed during tectonic uplift.

Starting at time zero, a denudation rate of 1 m per 1000 yr is assumed for the entire surface. However, as average elevation declines, the rate of denudation itself diminishes in such a constant ratio that one-half of the available landmass is removed in each 15 m.y. period. We may call this time unit the half-life of the available mass. We make an additional assumption that isostatic restoration occurs constantly in the ratio of 4:5. The initial rate of net lowering of the surface will be only one-fifth of the denudation rate, or 0.02 m per 1000 yr. In million-year units, this net denudation rate is 200 m/m.y. at time zero. As shown by labels on the descending curve of the graph, when the elevation is reduced to 2.5 km at the end of 15 m.y., the net rate of lowering will have fallen to about 100 m/m.y. At the end of 30 m.y. (two half-lives), average elevation and denudation rate are again reduced by one-half, to one-quarter of the initial values. Now three-quarters of the mass is gone and the average altitude is down to 1.25 km.

FIGURE 17.1

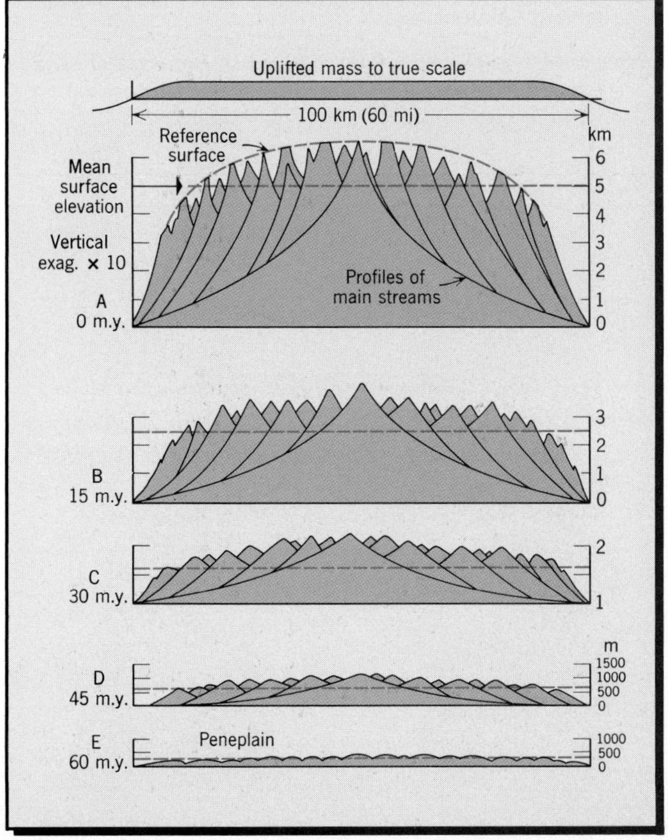

FIGURE 17.1 Schematic diagram of landmass denudation. In this model, the average surface elevation is reduced by one-half every 15 million years.

FIGURE 17.2 Graph of decrease in average surface elevation with time, as shown in Figure 17.1.

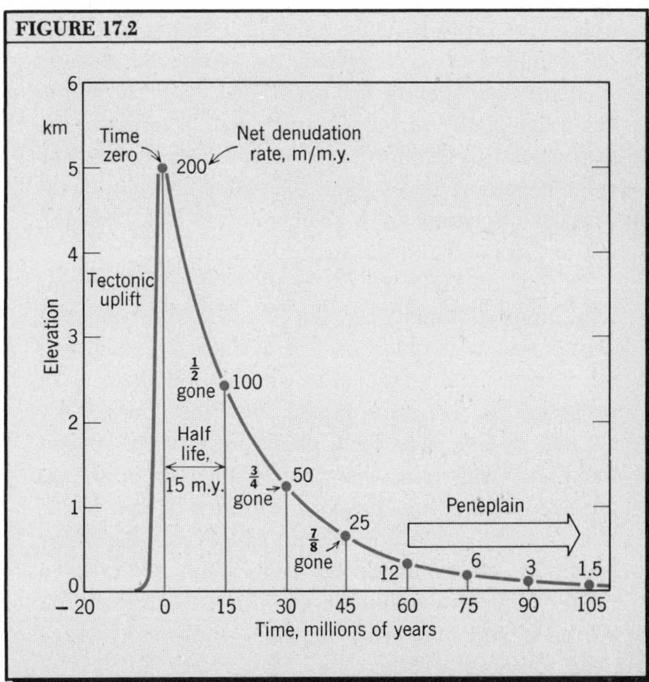

What we are describing here is a negative-exponential decay process not unlike the mass rate of decay of radioactive isotopes (Chapter 7). The curve of elevation flattens with the passage of time. Rates of denudation comparable to those observed today in the central and eastern United States are attained when the average elevation is about 0.3 km (300 m), after a lapse of some 45 m.y. Thereafter, further decline in elevation is extremely gradual.

The exponential decay rate, as applied to the denudation of a landmass, simply reflects the principle that the energy available to perform the work of denudation is proportionate to the height of a mass above sea level. Therefore, as average elevation declines, the intensity of change is proportionately diminished.

In our model denudation program, an average elevation of about 300 m is reached after about 60 m.y. By now, the rate of net lowering has fallen to 12 m/m.y. or 1.2 cm per 1000 yr. The land surface can now be described as a peneplain. No exact definition of a peneplain exists in terms of average elevation or relief. We can only say that hillslopes are very gentle, divides are very broadly rounded, and the region has the aspect of an undulating plain. Peneplain reduction continues to follow the exponential decay schedule, which we have shown carried out to seven half-lives, or 105 m.y. Most likely, crustal stability would not endure nearly this long, and the simple denudation curve would be terminated abruptly by another tectonic event. Another important possibility is that eustatic rise of sea level would cause the sea to encroach upon the continent, submerging and further eroding the peneplain.

Denudation and Plate Tectonics

As fluvial denudation takes place, beginning with an episode of rapid tectonic uplift and ending with the development of a peneplain, the landscape changes gradually but greatly in appearance. In early stages, the relief of the surface is great. *Relief* refers to the difference in elevation between the highest points—ridge crests and peaks—and the valley bottoms. Great relief is associated with steep gradients of both the ground surfaces and the stream channels.

To find examples of extremely rugged mountains with high relief and steep slopes in the early stages of denudation, we would turn first to the active continental margins and island arcs where subduction is in progress. For example, Japan, the Philippines, New Guinea, or Indonesia show many examples of rugged mountainous terrain where recent tectonic uplift has occurred or is still in progress (Figure 17.3). In North America, the western mountain ranges provide many examples of mountainous relief—the Cordilleran ranges of British Columbia and Alaska or the Andes are such mountains. The Eurasian suture belt, where continental collision has recently occurred, is another mountain belt of high relief and steep slopes. Of course, mountain glaciers have also been very active agents of erosion in these mountains. In middle and high latitudes the fluvial landforms in the higher elevations have been largely replaced by glacial landforms (Chapter 18). California shows many rugged coastal ranges that have been recently uplifted as fault blocks (Figure 17.4). The semiarid climate supports only a sparse cover of shrubs and grasses, so that the smallest details of the ridges and valleys are sharply outlined.

For subdued mountain ranges of moderate relief, we would look to stable continental cratons where the last major tectonic event was a Paleozoic orogeny associated with the closing of the ancestral Atlantic Ocean basin (Iapetus Ocean). For example, the Appalachian Mountains of the southeastern United States represent the roots of the Alleghenian–Hercynian Orogeny that closed the Paleozoic Era (Chapter 13), and the Blue Ridge Mountains would be a specific example. Although this is a bold mountain mass, with summit elevations rising 1000 to 1500 m above sea level, the Blue Ridge topography is subdued in form, with many rounded summits (Figure 17.5).

FIGURE 17.3 On the Island of Okinawa, in the Ryukyu Arc between Japan and Taiwan, steep mountainsides are extensively terraced for agriculture. Similar scenes can be found in Japan and the Philippines. (Stephen S. Winters.)

FIGURE 17.4 The San Gabriel Mountains, near Montrose and Altadena, California. (A. N. Strahler.)

FIGURE 17.5 The Blue Ridge Upland in North Carolina, a subdued mountain range formed by prolonged denudation of a stable continental margin. (F. J. Wright.)

FIGURE 17.3

It is difficult to identify an example of a peneplain that is still in its formative stage and lies close to sea level. The Amazon–Orinoco lowland on the craton of northern South America may well represent a present-day peneplain. Much of the area is a lowland in which the bedrock is covered by thick regolith, including stream-deposited alluvium. This cover makes it difficult for the geologist to determine the nature of the concealed bedrock surface.

Landscape Rejuvenation

Peneplains that have been uplifted several hundred meters above present baselevel are easily identified as upland surfaces of low relief. Following epeirogenic uplift, the streams that flow across a peneplain quickly begin to degrade their channels and develop deep, steep-walled rock gorges (Figure 17.6). In Chapter 16, we explained the process of stream trenching and showed that in some cases entrenched meanders are produced. The process of uplift of a peneplain followed by trenching of the major streams is called *rejuvenation*.

Figure 17.7 is a series of block diagrams that illustrate rejuvenation of a peneplain formed on an exposed shield area of a craton close to a stable continental margin. Block *A* shows the peneplain immediately after an uplift of several hundred meters. In Block *B*, major streams have deeply trenched the peneplain, which is now seen as an upland surface between steep-walled stream valleys or gorges. As the stream valleys are deepened and widened, the remaining peneplain surface is reduced in area and is finally totally destroyed. Block *C* shows the region in a rugged condition with steep slopes and narrow divides. From this point on, relief declines and slopes become more gentle (Block *D*). The major streams have become graded with respect to the new baselevel. Block *E* shows that a

FIGURE 17.4

FIGURE 17.5

second peneplain has been formed. Now, if epeirogenic uplift again occurs, the region will revert to the conditions shown in Block *A*, and the denudation cycle will be repeated. The question now arises: What mechanism of plate tectonics can explain repeated epeirogenic uplifts of a stable craton?

*Denudation and Spasmodic Uplift

To explain repeated rejuvenations in the advanced stage of fluvial denudation, our simple model of denudation, shown in Figures 17.1 and 17.2, needs an important modification. We have assumed that isostatic compensation occurs uniformly and constantly as denudation proceeds, but this is probably not the case. It is more reasonable to suppose that isostatic compensation occurs spasmodically. Because the lithosphere is a massive, strong layer, it resists being lifted until a certain minimum thickness of rock has been removed. Then, beyond the critical point of strength, any further removal triggers a rapid uplift, restoring the equilibrium.

We must therefore build a program of spasmodic isostatic compensation into our denudation model. As each critical thickness of mass is removed, an isostatic uplift occurs. Figure 17.8 shows this program, using a vertical line for the instantaneous uplift, followed by denudation on the exponential decay schedule. The result is a sawtooth curve, plotted on the assumption that full isostatic compensation occurs abruptly after 300 m of rock has been removed. The uplift, however, amounts to only 80% of the amount removed, or 240 m.

Our revised model shows that isostatic uplifts are triggered at short intervals during the early stages of denudation, when gradients are steep and relief is great. Because the abrupt changes in elevation are only a small relative fraction of the total elevation, there will be no observable changes in the profiles of

FIGURE 17.6 The St. John peneplain, southeast of Barranquitas, Puerto Rico, is represented by a rolling upland surface at an elevation of about 600 m. The peneplain, of Miocene age, is now deeply trenched by the Rio Usabón flowing in the steep-walled Canyon de San Cristobal. (R. P. Briggs, U.S. Geological Survey.)

FIGURE 17.7 A denudation cycle caused by uplift of a peneplain and a rejuvenation of the stream system. An uplift of the peneplain shown in Block *E* will bring a return to conditions shown in Block *A* and the cycle will be repeated. (Redrawn, by permission of the publisher, from A. N. and A. H. Strahler, *Modern Physical Geography*, Figure 20.5. Copyright © 1978 by John Wiley & Sons, Inc.)

FIGURE 17.6

FIGURE 17.7

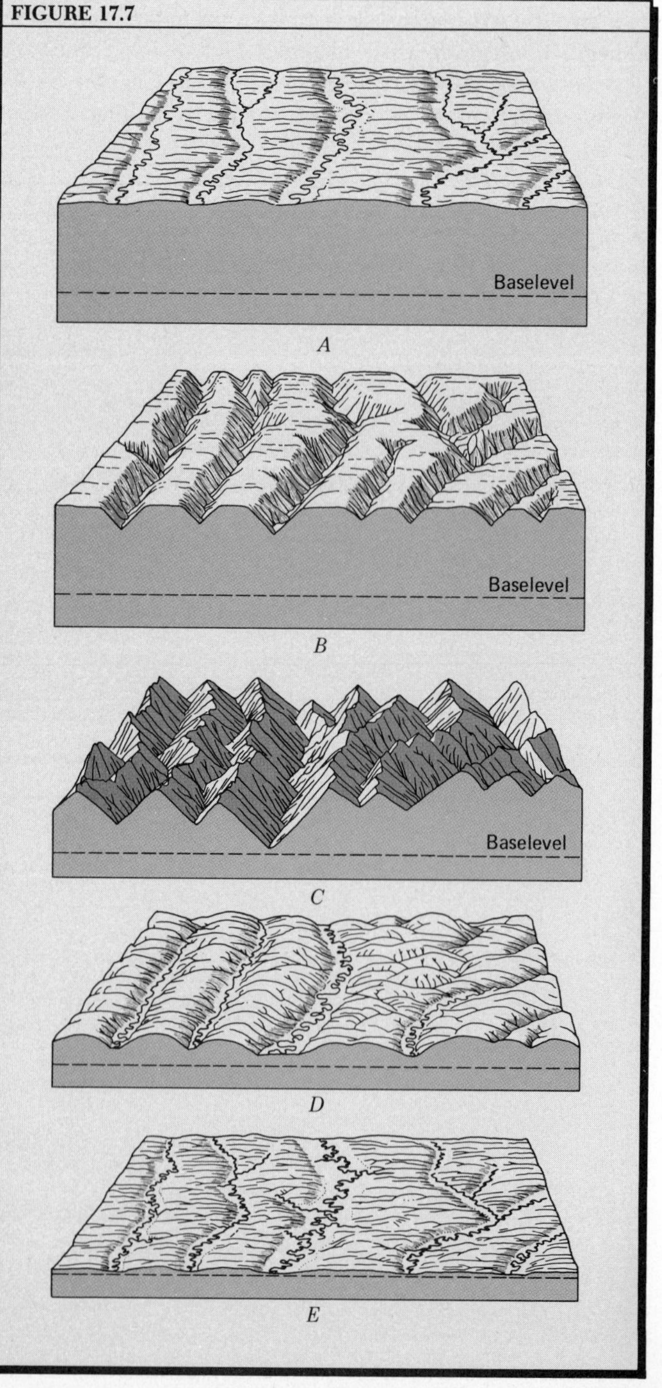

the slopes and stream channels. In contrast, the effects of spasmodic uplift are very conspicuous in the part of the graph representing the peneplain (area shown in Graph *B*). Each sudden isostatic uplift brings a relatively great increase in the available landmass. Depending on the average elevation at which this uplift occurs, the available landmass may be suddenly doubled or tripled in thickness. Following such an uplift, a long period of steady denudation follows, often lasting 5 to 10 m.y.

Other causes of epeirogenic uplift are not excluded by our model of spasmodic isostatic uplift. For reasons that are obscure, epeirogenic movements, both up and down, can be expected as a result of large-scale motions of a continental lithospheric plate. The suggestion has been made that the underlying asthenosphere has a somewhat undulating upper surface, so that when a plate is moving over the crest of a "bump" on the asthenosphere, it undergoes epeirogenic upwarping. Irregularities in the upper surface of the asthenosphere might result from differences in rock temperature and density from place to place in the upper mantle. Perhaps the presence of a large mantle plume is responsible for an upward bulge in the asthenosphere.

As we explained in Chapter 7, a negative isostatic anomaly can be interpreted as meaning that the lithosphere is being forcibly held down against a tendency to rise. Now, we find that negative isostatic anomalies on the order of 20 to 40 milligals exist over wide areas of the stable continental lithosphere of central and eastern North America (see Figure 7.33). There are also many small areas of positive anomalies in these regions, but the tendency toward negative values is quite marked.

George P. Woollard, an authority on gravity measurements and their interpretation, stated that "changes in crustal thickness due to surface erosion (are) going on at a faster rate than can be compensated through crustal buoyancy."* He went on to say that all mountain ranges show a history of reduction to a peneplain followed by epeirogenic uplift. In old mountain ranges, such as the Appalachians, this sequence has been repeated several times. In the Rockies, uplift has occurred at least once since the range was reduced to low relief following its rise during the Cordilleran Orogeny of Cretaceous time. Dr. Woollard predicted that the isostatic anomalies will be about zero when a mountain range is originally formed, but that negative anomalies will develop after reduction to a low surface of erosion. Positive anomalies might be expected following isostatic uplift if the momentum of crustal uplift carried the base of the crust to a level above that required for perfect equilibrium.

Here we have a good example of the interrelationship between geomorphology and geophysics. It serves as a reminder that a field of specialization within geology can be investigated more effectively by calling upon other branches of specialization for supporting information.

FIGURE 17.8 Repeated sudden uplifts because of isostatic compensation modify the curve shown in Figure 17.2.

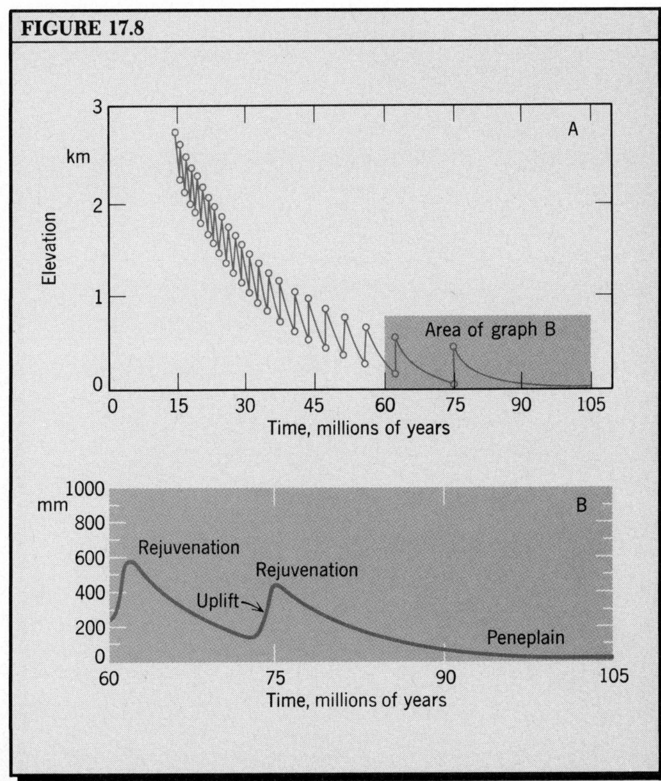

Denudation in an Arid Climate

Desert regions differ strikingly in appearance from humid regions because of differences in landforms and vegetation. Many of the most arid deserts experience heavy rains only at intervals of several years. Despite this fact, the evidences of the action of running water are clearly seen in the presence of numerous dry channels. On the rare occasions when heavy downpours produce copious runoff, the eroding and transporting power of overland flow and channel flow is fully equal to that in any humid region, especially in the absence of a vegetative cover. Without the protective effect of vegetation, there is nothing to stop overland flow from sweeping great quantities of coarse particles of weathered rock down the slopes and into adjacent channels. Charged with a heavy load of debris, the raging torrent sweeps down steep canyon grades and spreads out upon the adjacent valley floor.

Much desert rainfall, especially that of mountainous

*From G. P. Woollard, 1966, *The Earth Beneath the Continents*, Geophysical Monograph 10, Amer. Geophys. Union, Washington, D.C., p. 584.

FIGURE 17.9 Fluvial denudation in a mountainous desert region. *A*. Maximum tectonic relief. *B*. Mountain blocks deeply dissected and basins partly filled. *C*. Small residual mountain masses, vast alluvial surfaces, and playa floors. (Redrawn, by permission of the publisher, from A. N. Strahler, *Physical Geography*, 4th ed., Figure 26.13. Copyright © 1975 by John Wiley & Sons, Inc.)

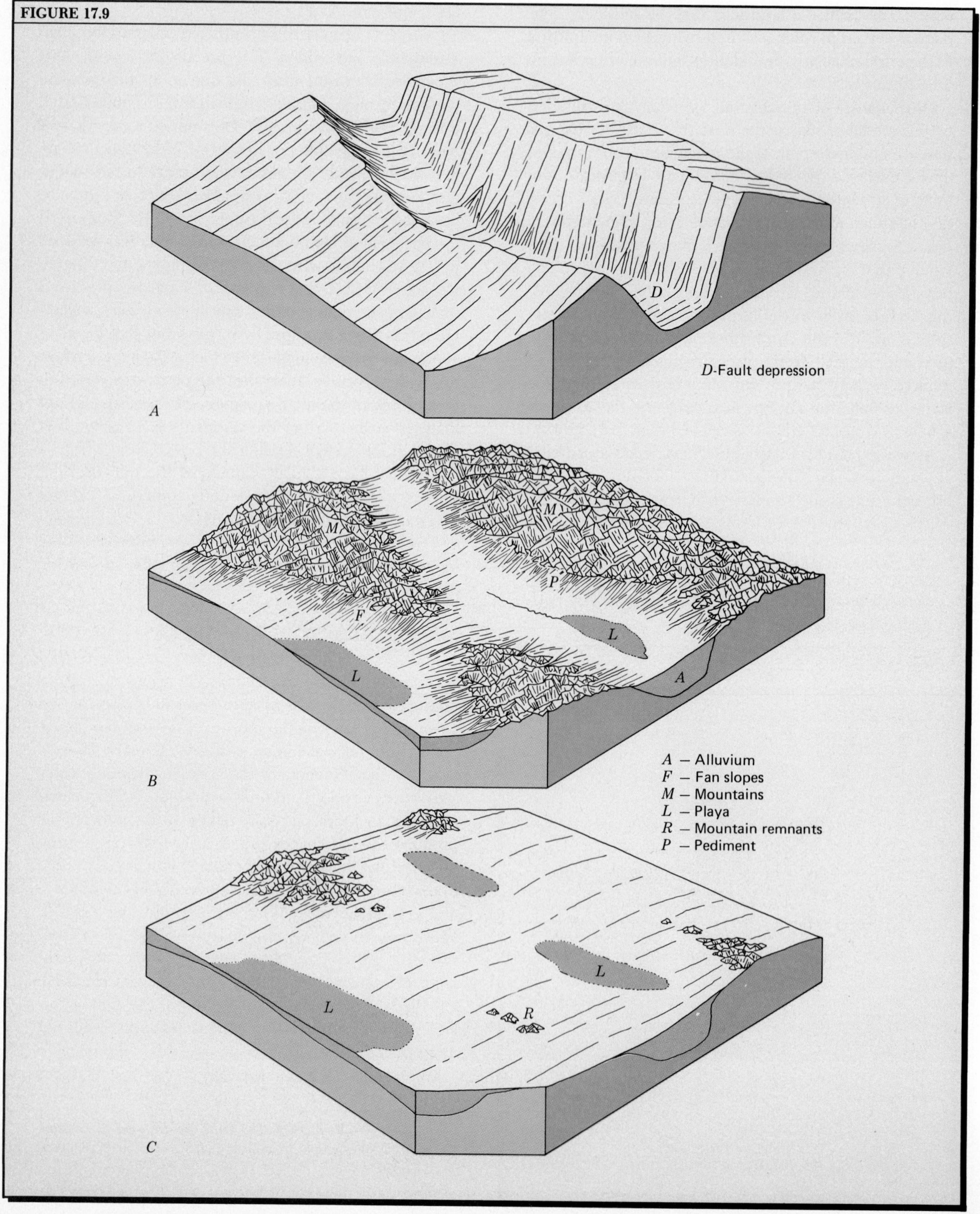

deserts, is from local convectional storms (thunderstorms). These tend to cover only small areas, so that runoff may be limited to a single small watershed. Unlike the widespread regional floods of river systems in humid lands, desert floods affect only short reaches of stream systems. The floodwater is rapidly lost by evaporation and by seepage into the permeable channel floor. Thus, the debris load is carried only short distances—usually from a mountain range to the adjacent valley floor, where it accumulates in alluvial fans.

An idealized model of fluvial denudation in a mountainous desert region is illustrated by the block diagrams in Figure 17.9. Block faulting on a large scale has produced two great tilted fault blocks, with a downfaulted depression between them. Summit elevations may range from 1500 to 3000 m, while the valley floor may lie below sea level, as at *D* in Block *A*. Processes of fluvial denudation are at work from the very beginning, carving up the rising masses and filling the depressions with alluvium, even as they are being formed. However, for purposes of illustration, we have shown the fault blocks as almost unmodified during rapid tectonic uplift. (Recently uplifted fault blocks are shown in Figure 12.22.)

The fault depression shown in Block *A* receives sediment from adjacent mountain slopes, but it does not fill with water to form a lake because evaporation greatly exceeds precipitation in the arid climate. Block *B* shows the effects of prolonged denudation and sedimentation: The mountain blocks have been carved into an intricate system of valleys and ridges. The basins are now filled with hundreds of meters of alluvium, which is spreading over upon the ragged mountain borders. Far out upon the alluvial valley are ***playas,*** flat surfaces underlain by fine silts and clays. At times, the playas are inundated by shallow lakes. Evaporation of the lake water leaves behind dissolved salts, which accumulate upon the playa surface (Figure 17.10).

As denudation goes on, the mountain ranges are reduced both in elevation and in extent, shrinking along the margins at the expense of the rising and expanding surfaces of alluvial deposition. Figure 17.11 shows a desert landscape similar to that depicted in Block *B*.

In an advanced stage of denudation, shown in Block *C*, mountain areas shrink to small remnants with many isolated fragments becoming separated from the main mass by the surrounding sea of alluvium.

As the mountain base is worn back, there is often formed a sloping platform of bedrock, called a ***pediment,*** thinly covered by layers of fan alluvium (Figure 17.12). Zones of rock pediment are shown fringing the mountain bases in Block *B* at points indicated by the letter *P*. Pediments become broader until the intervening mountain mass is entirely consumed, leaving a thinly veneered pediment surface in its place, as in Block *C*.

Completion of denudation results in a ***pediplain,*** a surface formed in part of the eroded bases of the mountain blocks and in part of thick alluvial deposits.

The cycle of fluvial denudation in a desert climate is well illustrated in the Basin-and-Range region of southern Oregon, Nevada, western Utah, and southeastern California, and in the Mojave and Sonoran

FIGURE 17.10 Salt flats on the floor of a playa in Death Valley, California. The surface elevation here is about 70 meters below sea level. (Warren Hamilton, U.S. Geological Survey.)

FIGURE 17.10

deserts extending farther south and east (refer to Chapter 12 and Figures 12.22, and 12.23). Recently uplifted fault blocks are characteristic of northern Nevada and southern Oregon; advanced denudation is illustrated in central and southern Nevada and in the Death Valley region (see Color Plate D.1); and pediplain development can be found in southwestern Arizona.

The profiles of mountainsides seem to evolve differently under an arid climate than under a humid climate. Figure 17.13 is a schematic diagram showing the changes in profiles of mountain slopes as denudation lowers the landmass, and relief declines. Under a humid climate (*left*), divides are usually rounded and the entire profile is covered by a layer of residual regolith. As denudation continues, the divides become

FIGURE 17.11 View southeast up Death Valley and the valley of the Amargosa River, California. (Spence Air Photos.)

FIGURE 17.12 A rock pediment, now bared by erosional removal of the thin alluvial cover, western side of Dragoon Mountains, near Benson, Arizona. (Douglas Johnson.)

FIGURE 17.11

FIGURE 17.12

more broadly rounded and the slopes become gentler, ending in the formation of a peneplain. In an arid climate (*right*), the divides are sharp and remain so throughout the entire denudation sequence. At the same time, a pediment surface begins to form at the base of the mountain slope. The pediment widens as the mountain slope retreats, but a sharp contact is maintained between the two surfaces. Note that the mountain slope maintains a uniformly steep angle, so that it retreats in a set of parallel planes as the pediment widens. Ultimately, the mountain mass is entirely eliminated and the pediment extends across the divide as a low arch.

Landforms and Rock Structure

Our models of fluvial denudation have been based upon the simplest possible landmass—one in which the rock is uniform throughout in its physical and chemical properties. In this respect, the imaginary rock in our models resembles a sugar cube—you can turn any side up and see no difference in the structure or texture. A rock mass answering this description is said to have ***homogeneous structure.***

Truly homogeneous rock masses are a rarity in nature. Some large plutonic masses may qualify, and some thick shale formations behave as if homogeneous. However, most continental crustal rock comes in layers, which may be inclined or folded, or in blocklike masses of unlike rocks faulted one against the other. Intrusion of magma in the form of stocks, dikes, and sills creates still another set of arrangements of rock masses. Some rock bodies are shattered by countless closely set joints; others are almost free of such joints.

Denudation is a selective process, rapidly removing weak rock to form low places and leaving harder rock masses to stand high as hills, ridges, or mountains. Streams seek out the weaker belts of rock, occupying them in preference to the harder rock masses. In this way, the landscape features come to reflect the rock inequalities, or rock structure of the landmass.

In the remainder of this chapter, we will investigate some distinctive landforms controlled by rock structure. These landforms lend variety and beauty to the scenery of the continents; they make one region distinctively different from another.

FIGURE 17.13 Schematic diagram of the evolution of mountainside slopes in humid and arid climates.

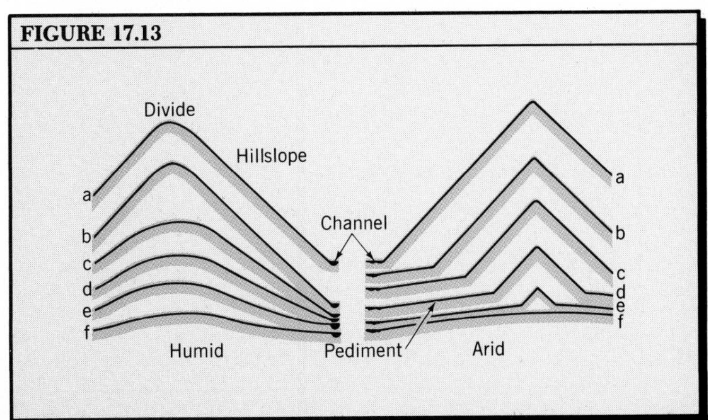

Drainage Patterns

Particularly sensitive to rock structures are ***drainage patterns,*** the distinctive forms of complete stream channel networks viewed on a flat map. The total pattern consists not only of the main stream, but also of all its branches, traced to their very tips. The drainage pattern associated with homogeneous structure is described as a ***dendritic pattern*** (Figure 17.14). The branching is often described as treelike. The dendritic pattern shows no grain; the individual segments of channel between junctions do not favor any single trend.

The dendritic pattern is inherited from the early stages of denudation, in which streams competed for space on a new landmass. In homogeneous structure, no single direction was favored over another, so that the smaller channels came to be oriented randomly.

FIGURE 17.14 Dendritic drainage pattern on igneous rock of the Idaho batholith. (Traced from U.S. Geological Survey topographic map.)

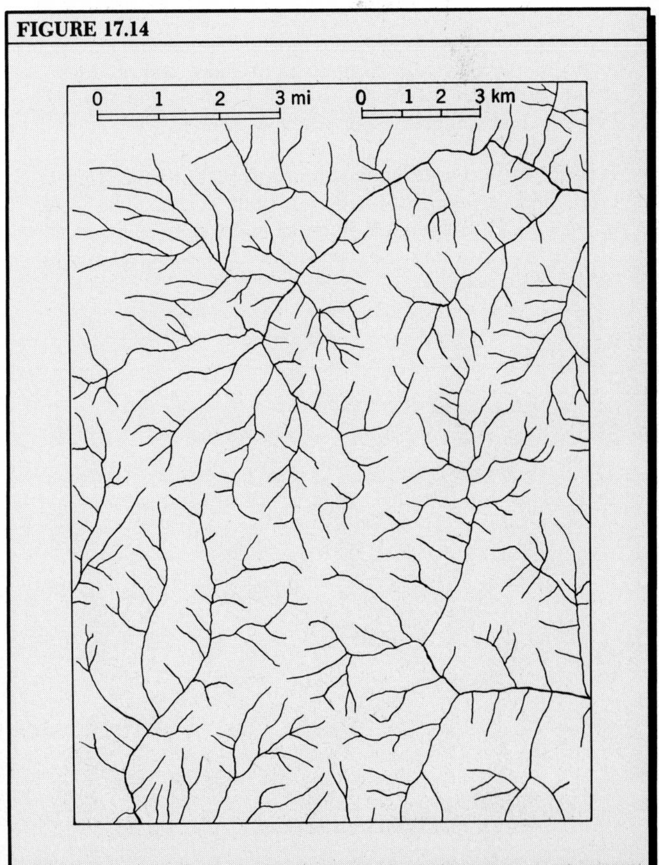

Figure 17.14 is the stream network of the deeply dissected granite batholith pictured in Figure 17.15. The granite is nearly homogeneous, and the pattern is dendritic. Although this rock is broken by countless joints and minor faults, they do not noticeably affect the pattern.

Some areas of plutonic rock are broken by major faults. The rock is crushed along the fault plane and is easily eroded, as streams seek out the fault lines. Figure 17.16 is the drainage pattern of a faulted plutonic mass in which the faults intersect at about right angles. A stream following a fault runs straight for a long distance, then makes a sharp right-angle bend to follow another fault. Any stream controlled by a zone or belt of weak rock is classed as a *subsequent stream;* streams following major faults are one variety. The resulting network pattern is called a *rectangular pattern.*

Erosion Forms of Horizontal Layers

Vast areas of the continental cratons are platforms of marine sedimentary strata (Chapter 12). These beds were laid down upon a basement of ancient igneous and metamorphic rocks that had been reduced to a peneplain and submerged to become the floor of a shallow sea. The thickness of these strata does not exceed a few thousand meters, which is relatively thin in comparison with the strata accumulated in subsiding structural basins and geosynclinal troughs.

Over large parts of the continental platforms, strata of Paleozoic, Mesozoic, and Cenozoic ages have been raised in epeirogenic crustal uplifts involving little or no tilting, folding, or faulting. A particular sandstone layer, for example, may have been raised to an elevation of a few hundred meters above sea level with so little disturbance that it dips only a fraction of a degree from the horizontal; the platform landmass consists of *horizontal strata.*

In the United States, much of the platform region lying between the Rockies and the Appalachians is underlain by nearly horizontal strata. Another large region is that of the Colorado Plateau of Arizona, Utah, New Mexico, and Colorado from which we draw examples of the landforms developed in horizontal strata.

The plateau basalts, described in Chapter 4, also make up large expanses of near-horizontal layered rocks. An example is the Columbia Plateau region of Washington, Oregon, and Idaho (see Figures 4.41 and 4.42).

Details of landform development in horizontal strata are illustrated in Figures 17.17 and 17.18. A cap rock of resistant sandstone maintains a nearly flat surface called a *plateau.* The exposed edges of this hard layer stand as nearly vertical *cliffs,* kept in sharp definition by constant undermining of weak shale beds beneath. Continued erosion causes a portion of the plateau to become detached and results in a flat-topped mountain, bounded on all sides by steep cliffs, called a

FIGURE 17.15 The Idaho batholith of Cretaceous age, exposed in the Sawtooth Mountains of south central Idaho. (From *Geology Illustrated* by John S. Shelton. W. H. Freeman and Company, Copyright © 1966.)

FIGURE 17.16 A rectangular drainage pattern consisting of subsequent streams located on fault zones. The rock underlying this region is a pluton of anorthosite, Adirondack Mountains, New York. (Traced from U.S. Geological Survey topographic map.)

FIGURE 17.15

FIGURE 17.16

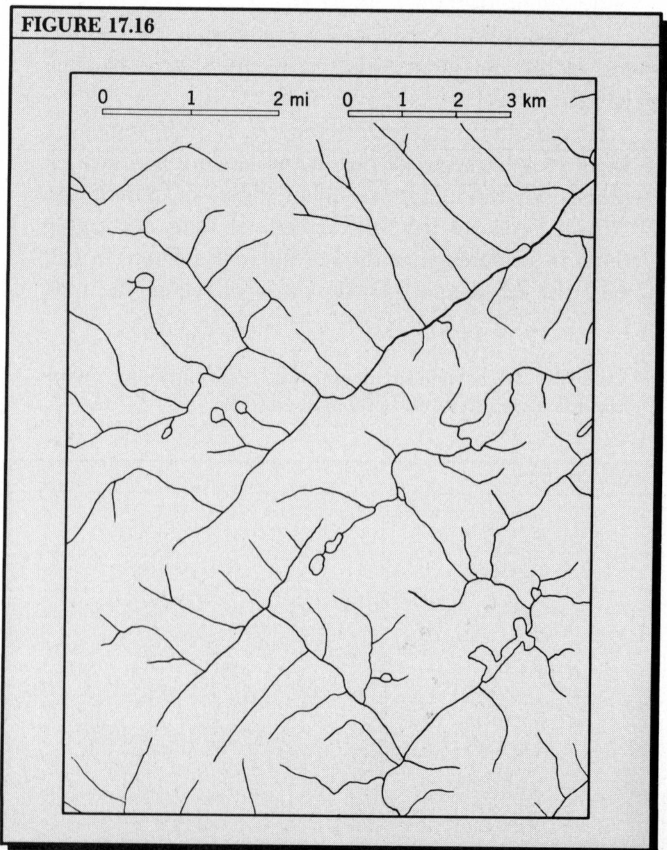

mesa. As a mesa shrinks by wastage of its surrounding cliffs, it assumes the form of a small, flat-topped hill, a *butte* (Figure 17.19).

Where several weak and resistant strata are found in alternation, the wasting back of a canyon wall produces a composite slope consisting of a succession of cliffs and intervening gentle slopes (Figure 17.20). A wide bench may be formed on the upper surface of a particularly thick, massive layer.

Drainage patterns developed upon horizontally layered rocks are dendritic, and a given layer is homogeneous in all horizontal directions, so that it does not appreciably control the direction taken by the growing fingertip streams in the early stages of denudation.

FIGURE 17.17 Landforms of horizontal strata evolving in an arid climate. (Redrawn, by permission of the publisher, from A. N. Strahler, *Physical Geography*, 4th ed., Figure 28.16. Copyright © 1975 by John Wiley & Sons, Inc.)

FIGURE 17.18 This panoramic drawing by the noted geologist-artist, W. H. Holmes, published in 1882, shows the Grand Canyon at the mouth of the Toroweap. In this part of the canyon, a broad bench called The Esplanade is well developed. (From C. E. Dutton, *Atlas to accompany Monograph II*, U.S. Geological Survey.)

FIGURE 17.19 Buttes of massive sandstone, Monument Valley, Arizona. They are known locally as The Mitten Buttes. (A. N. Strahler.)

FIGURE 17.19

FIGURE 17.17

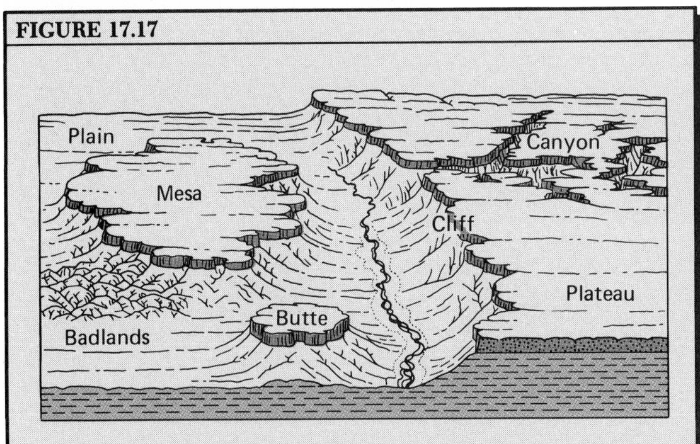

FIGURE 17.18

Gently Dipping Strata—Coastal Plains

Marine sedimentary strata of the U.S. Atlantic and Gulf coastal regions were deposited upon a sloping continental shelf during Mesozoic and Cenozoic time. These strata represent the inner edge of the present miogeocline (see Figure 12.9) and became exposed by crustal uplift in late Cenozoic time. The strata have a persistent seaward dip of perhaps 1° or 2° at most. Such strata form a *coastal plain* (Figure 17.21). Because the deposition of the strata took place to the accompaniment of a series of marine invasions and retreats, individual formations tend to thin to a featherlike edge toward the land and to thicken seaward.

When brought above sea level for the last time, the coastal plain presents a very smooth, sloping surface, across which flow streams bringing runoff from the former mainland. Any stream that comes into existence upon a newly formed land surface by following the direction of the initial slope is a *consequent stream*. Figure 17.21 shows two consequent streams flowing to the new shoreline.

After fluvial denudation has begun to act upon the poorly consolidated clays and sands of the coastal plain, there emerge belts of low hills called *cuestas*,

FIGURE 17.20 This map of a portion of Grand Canyon shows characteristic benching of horizontal strata. The area is about 10 km across; it occupies the northwest quarter of the Bright Angel Quadrangle. The isolated butte to right of center is Shiva Temple (*S*). The Granite Gorge of the Colorado River crosses the bottom of the map. The Tonto Platform (*T*) lies adjacent to the Granite Gorge. (From C. E. Dutton, 1882, *Tertiary History of the Grand Canyon District*, U.S. Geol. Survey, Mono. 2, U.S. Government Printing Office, Washington, D.C., Plate 42.)

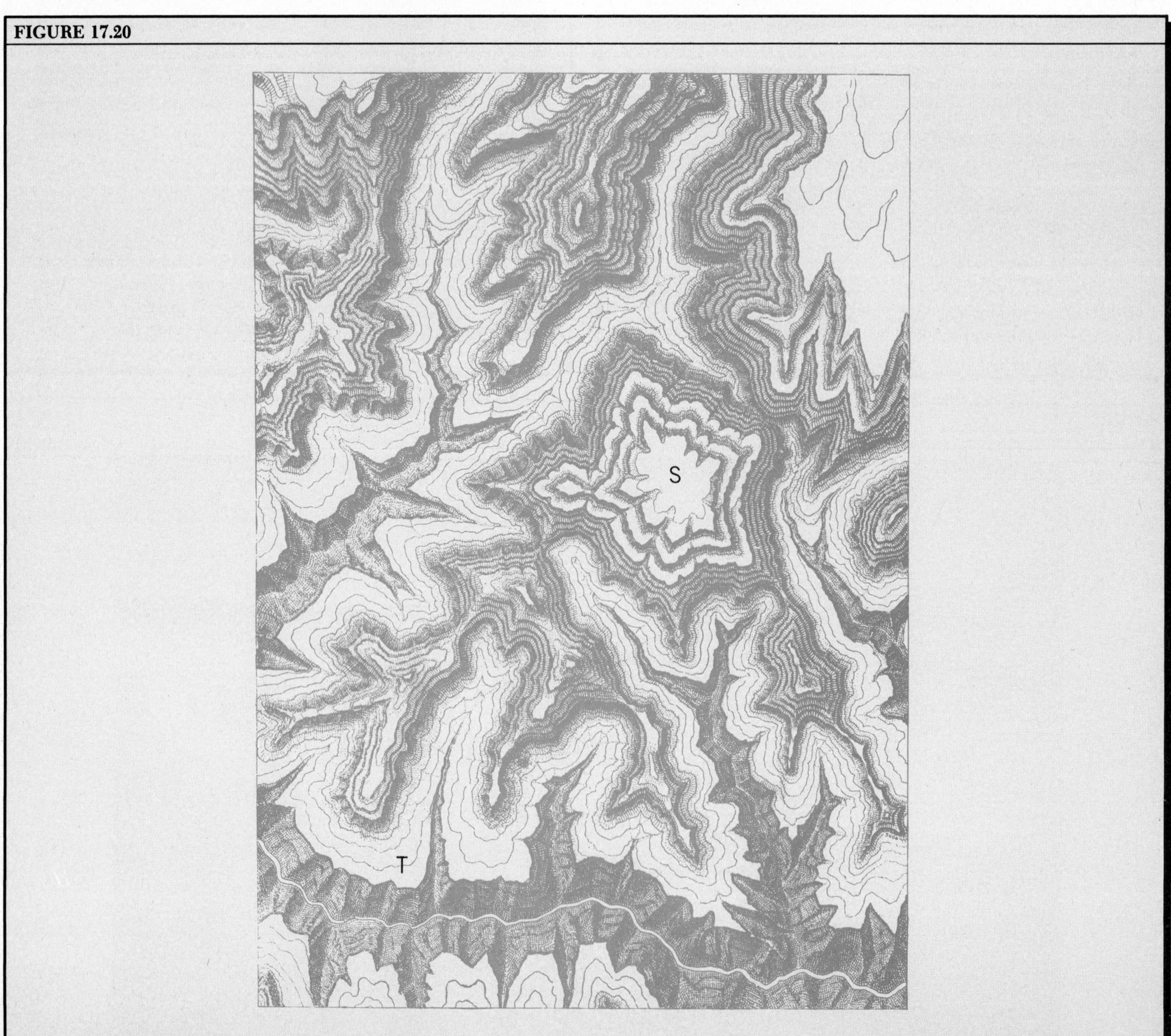

which represent exposed sand layers. Although the sand is not lithified, it resists erosion because of its high capacity to absorb precipitation, for even heavy rains produce little overland flow on sand. Separating the hill belts are broad, low valleys—called *lowlands*—where the surface is underlain by layers of soft clay.

As the lowlands are excavated, a new set of streams develops, following the lowlands and joining the consequent streams at right angles. These subsequent streams are formed by extending themselves headward along a belt of weaker rock.

The drainage pattern of a dissected coastal plain is a *trellis pattern,* in which the most important elements are the consequent and subsequent streams (Figure 17.22). The trellis form is completed by short tributaries which drain the cuestas and join the subsequent streams.

The gently dipping strata of coastal plains form ideal structures for the accumulation of artesian ground-water supplies (Chapter 15). Sandy strata outcropping in cuestas serve as intake areas for precipitation. Many kilometers down the dip, artesian flow is obtained from the aquifer, where it is buried deeply beneath impermeable clays (Figure 17.23). Coastal plain aquifers are a major source of water along much of the Atlantic and Gulf coastal belt.

*An Old Cuesta and a New Waterfall

Not all gently dipping layered rocks are in coastal plains. Any platform area within the continent may be warped into broad arches and saddles to produce zones of low dip. After long denudation, there will emerge a series of cuestas and lowlands. A good example of a cuesta of Paleozoic strata is the Niagara cuesta, of massive Silurian limestone. It runs across western New York State, through southern Ontario, then between Lake Huron and Georgian Bay, and finally curves south to form the Door Peninsula in Wisconsin.

Niagara Falls is formed at the point where the north-flowing Niagara River crosses the Niagara cuesta (Figure 17.24). The falls illustrate a basic principle of the control of rock structure upon the gradient of a large stream. The resistant Lockport dolomite sustains the channel of the river upstream from the falls. This rock threshold acts as a *local baselevel* for the upstream portion of the river. Weak shales beneath the dolomite are continually undermined, perpetuating the abrupt (50-m) drop of the river (Figure 17.25). The falls have retreated about 10 km from the line of the Niagara escarpment, where they originated some 20,000 to 35,000 years ago.

The Niagara River has a large discharge—some 17,000 cubic meters per second. This volume, combined with the great drop in level, has been used to generate electricity under the Niagara Power Project, in which water is withdrawn above the falls. The capability of this hydroelectric project is 2400 megawatts of power, and it is the largest single producer of electricity in the western hemisphere.

FIGURE 17.21 Erosional development of a coastal plain. The upper block shows the plain recently emerged. Consequent streams flow directly to the new shoreline. The lower block shows a later stage, in which erosion has produced lowlands and cuestas. Subsequent streams occupy the lowland. (Redrawn, by permission of the publisher, from A. N. Strahler, *Physical Geography,* 4th ed., Figure 28.5. Copyright © 1975 by John Wiley & Sons, Inc.)

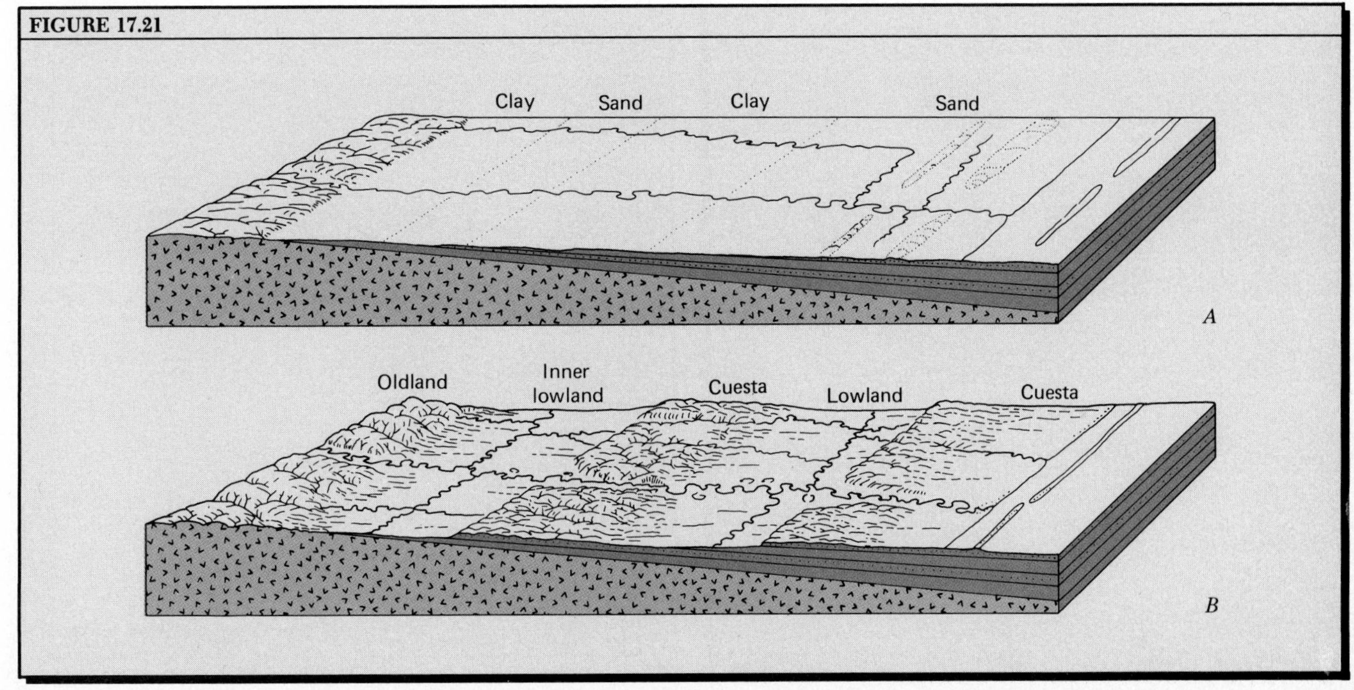

Erosion Forms of Stratigraphic Domes

In Chapter 6, we described the stratigraphic dome, in which platform strata are abruptly raised into a circular or elliptical dome structure (see Figure 6.37). Sedimentary domes may be from 10 to 150 km across the base, around which the sedimentary strata are sharply flexed upward, and the central part is elevated to heights of several hundred or even a few thousand meters. On the dome flanks, the strata dip at angles from 30° to 60°, or even more.

As denudation progresses, strata are removed from the center of the dome and are eroded outward to

FIGURE 17.22

FIGURE 17.22 The trellis drainage pattern of a coastal plain. The major streams are consequent (*C*) or subsequent (*S*).

FIGURE 17.23

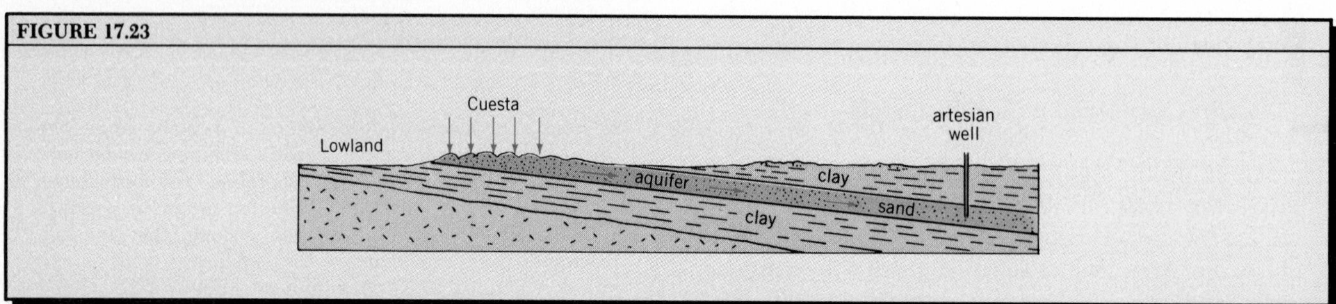

FIGURE 17.23 Artesian ground-water conditions beneath a coastal plain.

FIGURE 17.24

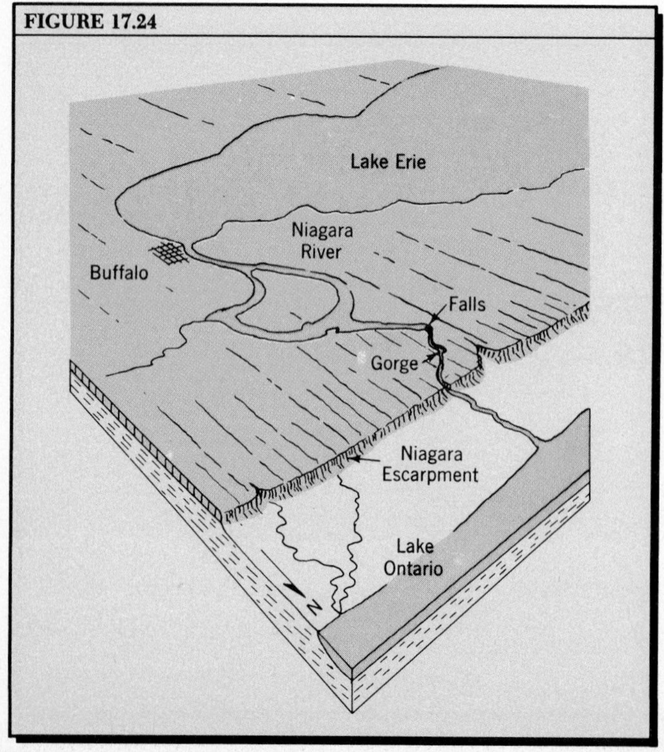

FIGURE 17.24 Bird's eye view of the Niagara Escarpment and Niagara River, looking southwest from a point over Lake Ontario. (Redrawn from a sketch by G. K. Gilbert, 1896.)

FIGURE 17.25

FIGURE 17.25 Cross section and sketch of Niagara Falls, showing cap rock of Lockport dolomite and underlying weak shales. (Drawn by Erwin Raisz and based on an illustration by G. K. Gilbert, 1896. Redrawn, by permission of the publisher, from A. N. Strahler, *Physical Geography,* 4th ed., Figure 25.11. Copyright © 1975 by John Wiley & Sons, Inc.)

FIGURE 17.26 This hogback of steeply dipping sandstone was photographed in the early 1900s by a geologist, whose horse and buggy are on the roadside. Today, Interstate Highway 40 occupies the same gap, a few miles east of Gallup, New Mexico. (U.S. Geological Survey.)

FIGURE 17.27 Landforms of a mountainous dome. (*Upper*) Early stage in removal of sedimentary cover. (*Lower*) Fully dissected dome with exposed core. *S:* subsequent stream. *R:* stripped sandstone formation. *P:* stripped limestone formation. *H:* horizontal strata. *F:* flatiron. *M:* crystalline rocks. (Redrawn, by permission of the publisher, from A. N. Strahler, *Physical Geography,* 4th ed., Figure 28.28. Copyright © 1975 by John Wiley & Sons, Inc.)

FIGURE 17.26

FIGURE 17.27

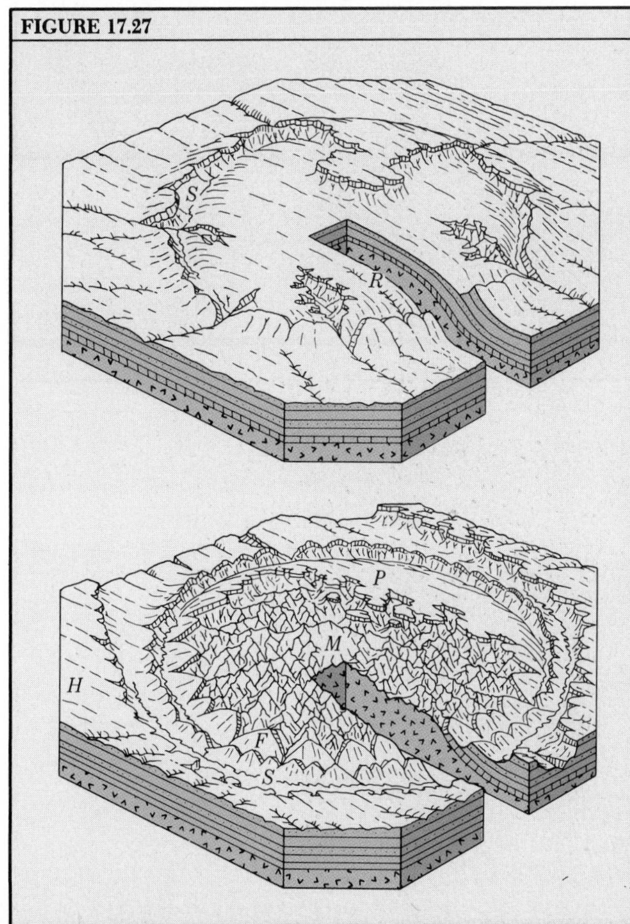

become sharp-crested ridges called *hogbacks* (Figure 17.26; see also Figure 12.19).

When a mountainous dome is deeply dissected, the central region may have been completely stripped of sedimentary strata, leaving exposed a mass of older igneous or metamorphic rock—this central mass is the core of the dome (Figure 17.27). A sandstone layer resting directly upon the older rock now develops a series of triangular-shaped plates known as *flatirons*. Beyond these are hogback ridges separated by narrow circular valleys occupied by subsequent streams.

The drainage pattern of a mountainous dome consists of *radial streams*, which are consequent streams, and *annular streams*, forming arcs of concentric circles. The latter are subsequent streams. The total effect is that of a trellis pattern bent into a circular form (Figure 17.28).

Erosion Forms of Open Folds

Figure 17.29 illustrates erosional landforms developed on a series of simple open folds. As shown in the upper block, a very massive, resistant formation of sandstone has been almost stripped clean of overlying weak formations of shale. Bold *anticlinal mountains* are formed by the anticlinal arches, and deep *synclinal valleys* coincide with the synclinal troughs. Major streams that originally crossed the anticlines may maintain their transverse courses by cutting deep, steep-walled *water gaps* across the anticlinal mountains.

Next, the crests of the anticlines become breached by narrow valleys that grow along the mountain crests. These *anticlinal valleys* are occupied by subsequent streams and continue to lengthen and deepen until the

FIGURE 17.28 Radial and annular drainage on a dissected dome. (Redrawn, by permission of the publisher, from A. N. Strahler, *Physical Geography,* 4th ed., Figure 28.30. Copyright © 1975 by John Wiley & Sons, Inc.)

FIGURE 17.28

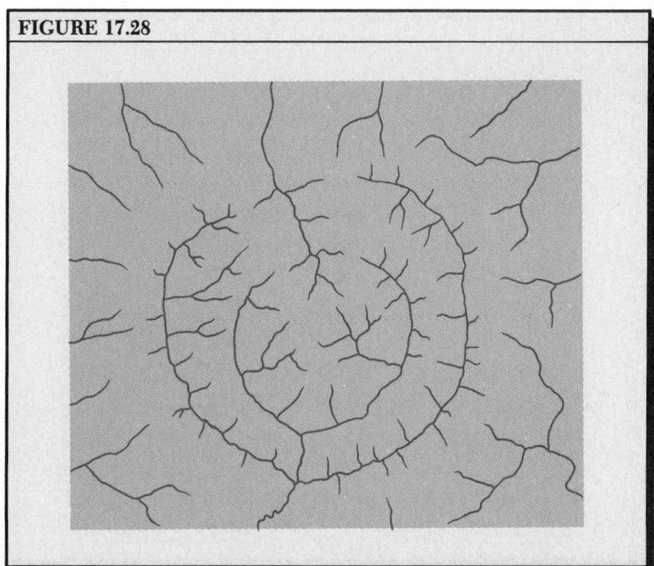

FIGURE 17.29 Landforms on open, parallel folds. (*Upper diagram*) Strong relief, corresponding to a single resistant sandstone formation. *AV:* anticlinal valley. *SV:* synclinal valley. *AM:* anticlinal mountain. *WG:* water gap. (*Lower diagram*) Relief increasing in complexity on two resistant formations. *HV:* homoclinal valley. *SM:* synclinal mountain. *HM:* homoclinal mountain. *S:* subsequent stream. (Redrawn, by permission of the publisher, from A. N. Strahler, *Physical Geography,* 4th ed., Figure 29.1. Copyright © 1975 by John Wiley & Sons, Inc.)

FIGURE 17.30 Trellis drainage pattern on dissected folds. (Traced from U.S. Geological Survey topographic map.)

FIGURE 17.31 Zigzag ridges developed on plunging folds. (Drawn by E. Raisz. Redrawn, by permission of the publisher, from A. N. Strahler, *Physical Geography,* 4th ed., Figure 29.3. Copyright © 1975 by John Wiley & Sons, Inc.)

FIGURE 17.32 A zigzag ridge in plunging folds, south of Hollidaysburg, Pennsylvania. The nearer bend is a plunging syncline while the more distant bend (at right) is a plunging anticline enclosing a cove. The resistant beds underlying the ridge crest make up the Tuscarora formation, a quartzite sequence of Silurian age. (John S. Shelton.)

FIGURE 17.29

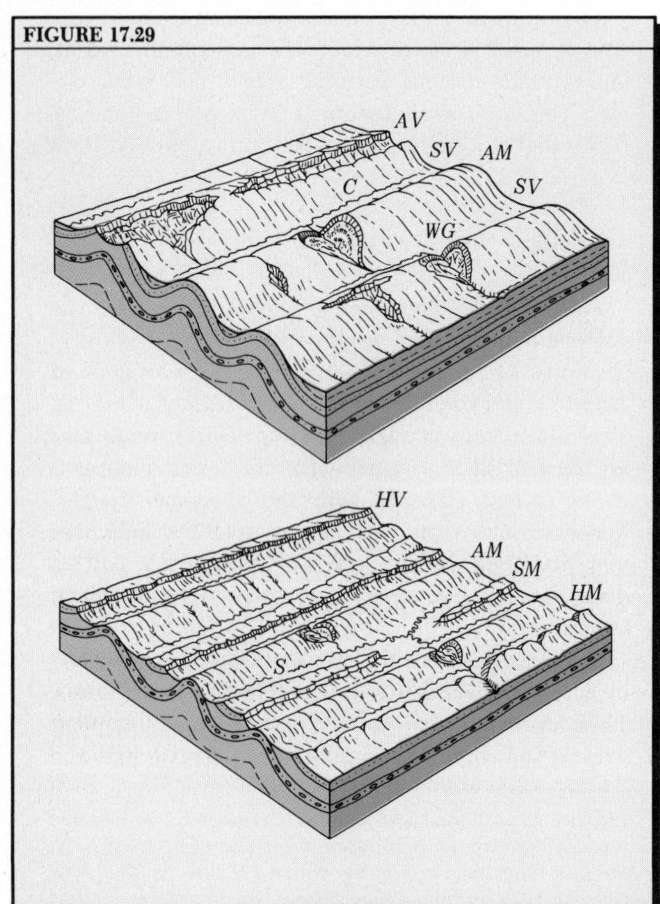

FIGURE 17.30

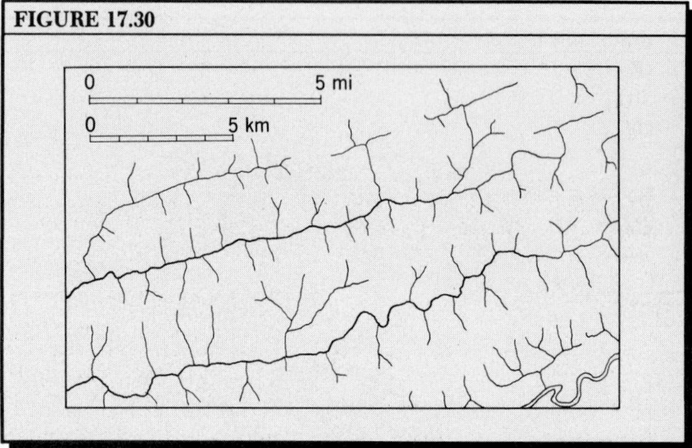

FIGURE 17.31

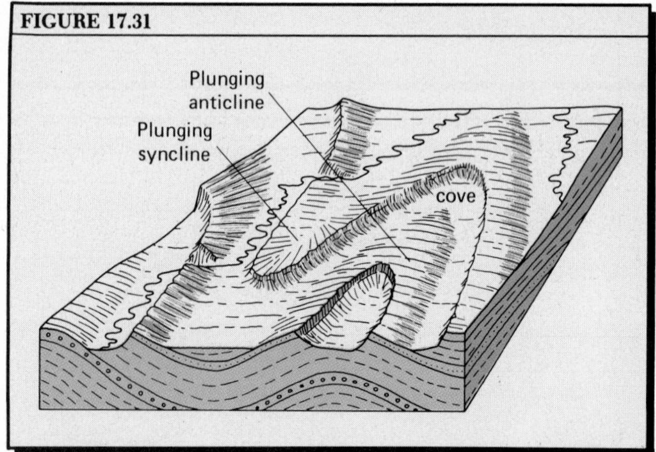

FIGURE 17.32

anticline is completely opened out along its length. On the flanks of the anticline, the sandstone formation now presents its upturned edges to form *homoclinal mountains*, which are essentially the same features as hogback ridges.

The more deeply eroded folds are shown in the lower block of Figure 17.29. The sandstone formation is entirely eroded from above the anticlines but persists along the synclinal axes, where it forms long, narrow, flat-topped *synclinal mountains*. These features are like long mesas, but with shallow summit valleys extending along the center line of the mountain. A trellis drainage pattern is associated with dissected folds (Figure 17.30).

FIGURE 17.33 Geologic map of anthracite coal basins of eastern Pennsylvania. Dark areas are synclinal troughs. (Redrawn, by permission of the publisher, from A. N. Strahler, *Physical Geography,* 4th ed., Figure 29.5. Copyright © 1975 by John Wiley & Sons, Inc.)

FIGURE 17.34 A monocline, breached by erosion. (Redrawn, by permission of the publisher, from A. N. Strahler, *Physical Geography,* 4th ed., Figure 29.9. Copyright © 1975 by John Wiley & Sons, Inc.)

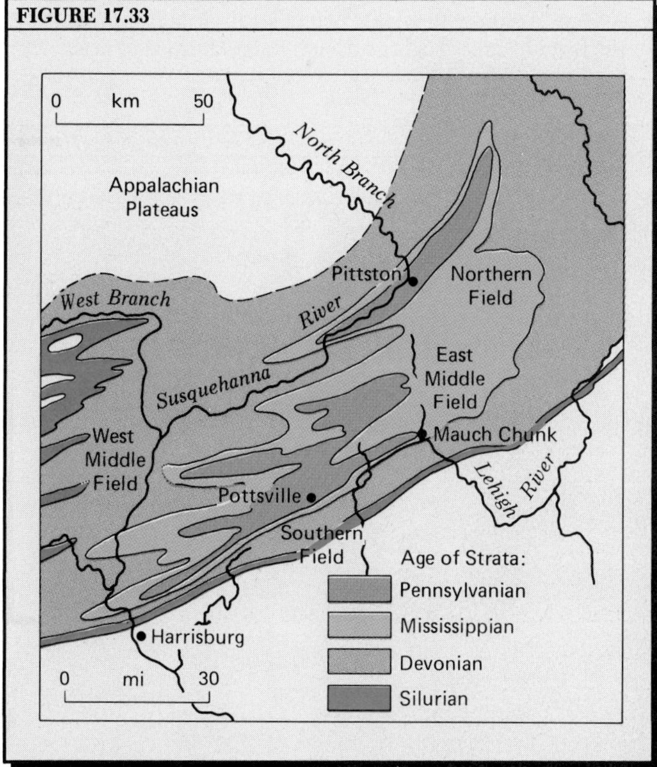

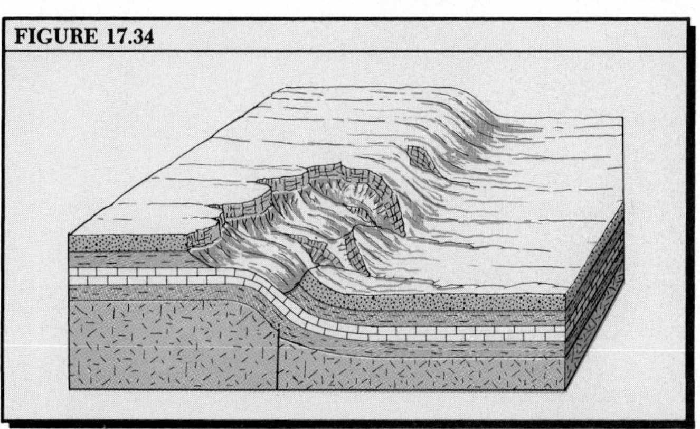

The folds in Figure 17.29 idealize the crests of the anticlines and troughs of the synclines as maintaining horizontality. Most folds crests and troughs, when followed for long distances, alternately descend and then rise again in a series of undulations explained in Chapter 6 as plunging folds (see Figure 6.26).

Where plunging folds are eroded, the homoclinal mountains form *zigzag ridges* (Figure 17.31). Where a ridge crosses a plunging anticline, a steep-walled cove is formed. Where a syncline is crossed, we find a long, spadelike mountain with steep walls around the end and sides. Good examples of zigzag ridges are seen in the Appalachians of central Pennsylvania (Figure 17.32 and Color Plate C.2).

In deeply eroded, folded strata of Pennsylvanian age are layers of anthracite coal, but these remain only in the synclinal troughs (Figure 17.33).

Monoclines

In certain regions of horizontal platform strata, there occur sharp downflexings of strata in the form of monoclines (see Figure 6.30*F*), which seem to be closely associated with normal faults. Traced along its length, a monocline may merge with a normal fault so that both flexing and faulting of strata take up the total displacement, and it may be represented at depth by a normal fault (Figure 17.34). Several prominent monoclines are found in the Colorado Plateau (Figure 17.35). These monoclines flex the same series of sedimentary strata that we described earlier in this chapter as exposed in the walls of Grand Canyon. Erosion bares the upturned strata, which form flatirons and hogbacks resembling those of an eroded stratigraphic dome.

Erosion Forms of Metamorphic Rocks and Thrust Sheets

Much of the area of the continental shields is underlain by metamorphic rocks—gneisses, schists, slates, quartzites, and marbles. A common attribute of metamorphic rocks is that they tend to lie in roughly parallel belts and zones. For example, belts of gneiss or schist run parallel with belts of slates, marbles, and quartzites (Figure 17.36). The result of such an arrangement is that the topography consists of valleys, hill belts, and mountain ridges in parallel zones. Thrust faults commonly separate rocks of one belt from those of another and may bring together rocks of contrasting resistance to erosion, producing sharply defined escarpments. The trellis pattern develops on some belts of metamorphic rocks, but it is lacking the striking parallelism of major streams found in belts of open folds.

In a humid climate, we find that quartzite strongly resists weathering and erosion to form bold ridges much like the homoclinal ridges of a belt of open folds. In contrast, marble is susceptible to solution removal and forms the lowest valleys. Slate is of intermediate resistance and produces belts of hills. Gneisses are often highly resistant to denudation and produce mountainous belts.

Topography of metamorphic rocks is illustrated in its many forms throughout New England and the older Appalachians, including the Blue Ridge Mountains and the Piedmont Plateau.

Recall from Chapter 6 the development of recumbent folds and low-angle overthrust faults in close alpine folding (Figure 6.39). After a large thrust sheet has been deeply eroded, some interesting landforms are produced. As shown in Figure 17.37, a valley carved deeply into the thrust sheet can uncover the younger formations beneath in a feature called a *window.* After much of the thrust sheet is eroded away, an isolated portion often remains, surrounded by younger rock. This feature is called an *outlier;* it may outwardly resemble a mesa.

Erosion Forms Developed on Normal Faults

Many of the straight, steep scarps located along the line of a fault are not produced directly by fault movements of recent geologic date. They are, instead, *fault-line scarps* produced by erosional removal of weaker rock from one side of an older fault plane (Figure 17.38). The fault itself may date far back in geologic time. Stage *A* represents a recently formed fault scarp in weak sedimentary strata. After extensive erosion, shale is stripped away from the resistant basement

FIGURE 17.35 The Comb Ridge monocline in southern Utah flexes strata ranging in age from Permian (*left*) to Cretaceous (*right*). The maximum dip is about 50°, while the total displacement amounts to about 1700 m. Comb Wash, a subsequent stream occupying a belt of weak shale beds, follows the foot of a great hogback (Comb Ridge) formed of massive sandstone. (John S. Shelton.)

FIGURE 17.35

rock, causing the appearance of a fault-line scarp (stage *B*). This landform takes the general position of the original fault scarp but is formed of rock that was deeply buried at the time of the fault movement.

Fault-line scarps are common features of the ancient crystalline rocks of the continental shields (Figure 17.39). Fault structures can persist for hundreds of millions of years, despite deep erosion, because the fault plane extends far down into the crust.

FIGURE 17.36 Relation of metamorphic rock belts to topography. (Redrawn, by permission of the publisher, from A. N. Strahler, *Physical Geography,* 4th ed., Figure 28.3. Copyright © 1975 by John Wiley & Sons, Inc.)

FIGURE 17.37 A deeply eroded thrust sheet with erosional outlier and window. (Drawn by A. N. Strahler.)

FIGURE 17.38 *A*. Fault scarp resulting from recent faulting. *B*. Fault-line scarp resulting from erosion along an inactive fault. (Drawn by A. N. Strahler.)

FIGURE 17.39 A fault-line scarp in ancient rocks of the Canadian Shield. The body of water resting against the scarp is MacDonald Lake, located near Great Slave Lake, Northwest Territories. (Royal Canadian Air Force photograph No. A5120-105R.)

FIGURE 17.40 A deeply eroded fault-block mountain often shows a row of triangular facets along the mountain base. These are eroded remnants of the fault scarp. (Drawn by A. N. Strahler.)

FIGURE 17.36

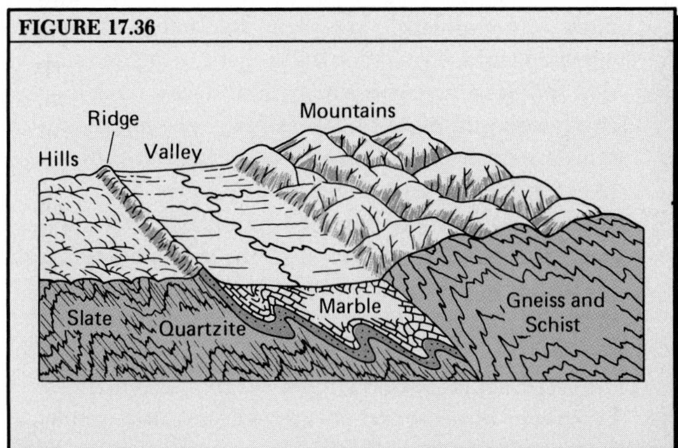

FIGURE 17.37

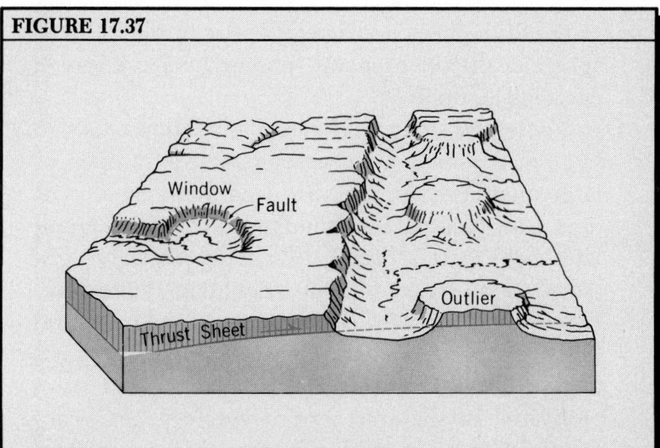

FIGURE 17.38

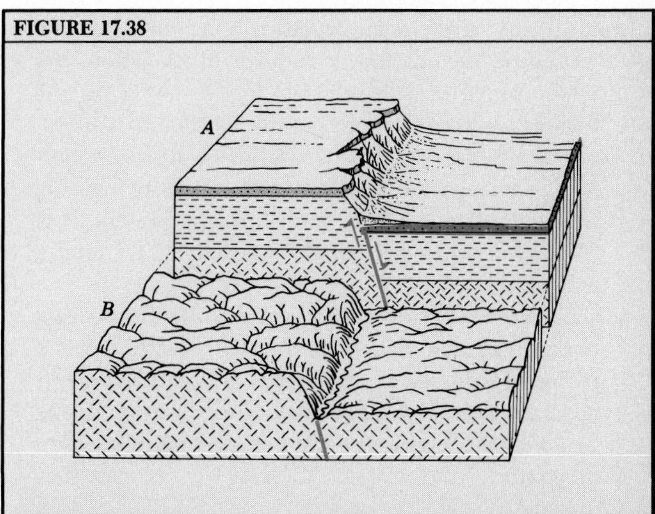

FIGURE 17.39

FIGURE 17.40

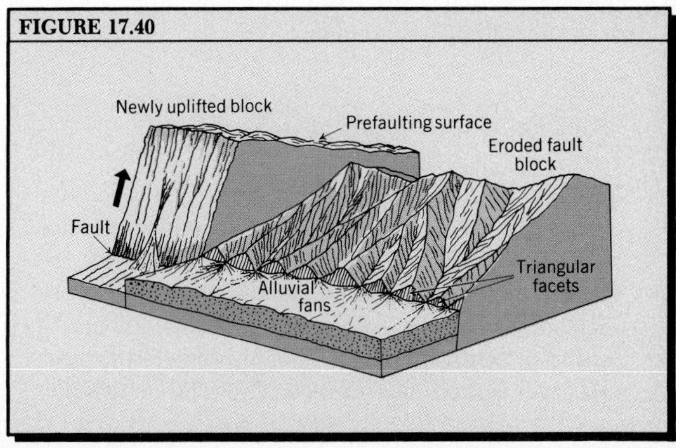

In Chapter 12, we described the lifted and tilted fault-block mountains of the Basin-and-Range province of the western United States (see Figures 12.21, 12.22, and 12.23). Fault-block mountains were used earlier in this chapter as the initial landforms for a denudation system in an arid climate (Figure 17.9). A tilted block is bounded on one side by a major normal fault (Figure 17.40). Because fault movements occur at intervals over many thousands of years, erosion continually attacks the exposed fault scarp, wearing the face of the block down to a lower slope than that of the fault plane. A portion of fresh fault scarp is often visible at the base of the block. Much later, after the block has been carved into a system of deep canyons and ridges, the only remaining vestiges of the fault scarp are the snubbed ends of the descending ridges between canyon mouths. These features, called *triangular facets,* will eventually become buried beneath alluvium of a series of fans formed along the mountain base.

Fluvial Denudation in Review

In this chapter, we first followed the idealized stages in denudation of a model landmass. While tectonic activity lifts a landmass rapidly, the process of denudation is very slow. The final stage, development of the peneplain, drags on for a very long time. The denudation cycle in an arid climate is different mainly because streams evaporate within the landmass, leaving their loads in alluvial basins. The pediplain that results is not graded with respect to sea level as the baselevel.

Rock structure exerts strong control over landforms and drainage patterns shaped by fluvial denudation. Horizontal strata, coastal plains, domes, and folds are dissected into a remarkable variety of landforms as the weaker strata are etched from between resistant strata. Faults provide another class of structurally controlled landforms. Landforms produced by erosion of volcanoes are yet another class of structurally controlled landscape elements; they are described in Chapter 4 and include volcanic necks with radial dikes. Landforms controlled by structure are among the finest scenic features of our land; many are set aside as national parks and national monuments.

Key Facts and Concepts

Concept of a denudation system The *available landmass* lies above *baselevel*, an imaginary surface of zero elevation. As fluvial denudation occurs, the crust rises in isostatic compensation. For every 5000 m of rock removed, a rise of about 4000 m will occur in compensation. Ultimate reduction to near sea level is thus long delayed. The total denudation system uses solar energy in various forms, including gravity flow systems of water.

Orogenic uplift and denudation Rates of uplift by orogenic processes are comparatively rapid—several meters per 1000 years in some cases—and capable of elevating a large mountain range in one or two million years. Denudation rates are moderately rapid for elevated masses with steep slopes (1 m per 1000 yr), but decline to low rates (0.02 m per 1000 yr) as *relief* diminishes and the average elevation is reduced. After a long span of uninterrupted denudation, the land surface is reduced to a *peneplain*—an undulating low plain. Reduction of average land surface elevation and relief during denudation follows a negative exponential decay model.

Denudation and plate tectonics High, rugged mountains are typical of mountain and island arcs of active subduction zones and suture zones of recent continental collision. Subdued mountains, hills, and undulating plains are typical of the stable continental shield interior and passive margins experiencing little tectonic activity after the close of the Mesozoic Era.

Landscape rejuvenation *Rejuvenation*, accompanied by stream incision and an increase in relief, results from epeirogenic uplift that may occur spasmodically as delayed isostatic compensation. Movement of the lithosphere over irregularities in the asthenosphere or the rise of mantle plumes may also lead to epeirogenic uplift.

Denudation in an arid climate Denudation processes in an arid climate are characterized by transport of large quantities of coarse bed load, building alluvial fans with incorporated mudflows. Debris accumulates close to the sources. Mountain blocks are dissected as alluvium fills intermountain depressions. *Playas* are underlain by fine-textured sediment and saline deposits. *Pediments* form as mountain masses recede, eventually leading to development of a *pediplain*. Parallel retreat of mountain slopes is considered typical of arid climates; reclining retreat typical of humid climates.

Landforms and rock structure On *homogeneous structure*, denudation is uniform in its action, but where weak rocks lie adjacent to resistant rocks, the former are eroded more rapidly, producing valleys.

Drainage patterns A treelike *dendritic drainage pattern* develops on homogeneous structure and on horizontal strata. Where fault lines are present, *subsequent streams* occupy the fault lines, forming a *rectangular pattern*.

Erosion forms of horizontal layers *Horizontal strata,* undergoing denudation in an arid climate, develop *plateaus* bordered by steep *cliffs*. Plateau fragments become detached as *mesas* and *buttes*. Strong benching is found in the walls of canyons in horizontal strata.

Gently inclined strata—coastal plains Recently emerged *coastal plains* gradually develop *cuestas* and *lowlands* with a *trellis drainage pattern* formed of *consequent streams* and subsequent streams. Artesian ground water occurs in coastal plains.

Niagara Cuesta and Niagara Falls An ancient cuesta of Silurian limestone forms the brink of Niagara Falls. Undermining of weak shales has maintained the falls during retreat since they were formed in postglacial time.

Domes in layered rock Erosion of a *sedimentary dome* results in formation of *hogbacks* and *flatirons.* The drainage pattern consists of *radial streams* and *annular streams.*

Erosion of folded strata Erosion of open folds brings into relief *anticlinal, synclinal,* and *homoclinal mountains* (often with *water gaps*) and intervening *synclinal* and *anticlinal valleys.* A trellis drainage pattern results. Plunge of folds is expressed by *zigzag ridges.* Anthracite is found deep in remnants of synclinal folds in the central Appalachians. Erosion of monoclines produces a line of hogbacks and flatirons.

Erosion forms on metamorphic rocks and thrust sheets Metamorphic rocks occur as mountain roots exposed by erosion of continental sutures. Belted topography is typical. Erosion of extensive thrust sheets reveals underlying younger rock in *windows,* while remnants of the thrust sheet form isolated *outliers.*

Erosion forms on normal faults Deep erosion along ancient, inactive faults develops persistent *fault-line scarps,* commonly found in shield areas. At the base of a large uplifted *fault block mountain,* the fault scarp persists as a row of *triangular facets.*

Questions and Problems

1. Does the concept of a denudation system in which orogeny ceases and is followed by denudation apply to an active continental margin where subduction is continuous for, say, 100 m.y.? Take into account the formation of an accretionary prism with its rising tectonic arc and an adjacent volcanic arc. Do you think that a tectonic arc might be lowered by denudation at a rate equal to its vertical uplift?
2. Is tectonic uplift still in progress in suture orogens such as the European Alps and the Himalayan range? How would you define "tectonic uplift" in the case of continental suturing? Does the isostatic gravity anomaly give any clue as to whether tectonic activity is in progress (see Chapter 7)?
3. Under an exponential decay model of denudation, is it theoretically possible for the surface of a continent to be reduced to a true horizontal surface of sea-level elevation? The word "peneplain" was coined from the two words *penultimate* (next-to-the-last) and *plain.* In what sense is this word suitable for an exponential decay model?
4. How might the dendritic drainage pattern of eroded horizontal strata be distinguished from the dendritic pattern formed on a batholith, assuming that only drainage network maps are used for study? How can the trellis pattern of eroded folds be distinguished from the drainage pattern formed on an eroded monocline? How can the drainage pattern of an eroded stratigraphic dome be distinguished from the pattern of streams on an eroded composite volcano?
5. You are examining what appears to be a fault-line scarp. The block on one side (upthrown block) consists of Precambrian metamorphic rocks; Triassic red beds lie on the opposite (downthrown) side. Does this feature pose any serious threat to the safety of a nuclear power plant constructed close to the fault line? How might recent faulting (if any has occurred) be detected along this fault line?

18

CHAPTER EIGHTEEN

Glaciers and the Pleistocene Glaciations

Because most of us have dealt with ice only in small pieces or thin layers, we think of it as a brittle solid. Ice fractures easily and often shows crystalline structure, for it is a true mineral. But where an ice layer has accumulated on land to a depth of about 100 m and rests on a sloping surface, the basal part of the ice becomes plastic, yielding by slow internal flowage. The entire ice mass is carried downslope and becomes a glacier (Figure 18.1). A *glacier* is defined quite broadly as any large natural accumulation of land ice affected by motion which may be occurring today or may have ceased.

How Glacier Ice Is Formed

For glaciers to form, the quantity of incoming snowfall—on the average, year in and year out—must exceed the average quantity lost yearly by melting and evaporation, two processes which geologists combine under the single term *ablation*.

Freshly fallen snow has a very low density, with as much as 90% of its volume consisting of air-filled openings. Changes quickly set in, however, and the elaborate hexagonal snow crystals change into more rounded, smaller particles. The mass becomes greatly compacted into old granular snow, in which the air spaces make up less than 50% of the volume. Under the load of newer snow layers and with the aid of some melting and refreezing, the old snow compacts further, reaching a stage given the name *firn*. Density of firn is greater than about 0.4 g/cc, or four-tenths that of water. As the older firn is buried still more deeply under new layers of firn and snow, it compacts further, finally becoming glacier ice. Although most of the air has by this time been expelled, some is enclosed in bubble holes and the ice density may not be greater than about five-sixths that of water.

Geologists have studied glacier ice as if it were rock. The ice is sawed into thin layers and placed under a polarizing microscope to reveal its crystalline structure. Glacier ice closely resembles a coarse-grained igneous rock, such as granite, as far as its crystal units are concerned. Individual ice crystals range in diameter from 1 mm near the head of a glacier to 2 cm or more in the much older ice near the lower end. A simpler method of examining the crystals is to make a pencil rubbing on paper laid over an ice surface that has been slightly etched by melting. The pencil rubbing will reveal the crystal contacts (Figure 18.2).

Kinds of Glaciers

Glaciers formed in high, steep-walled mountain ranges and occupying previously carved stream valleys are shaped into long, narrow ice streams. They have many of the basic elements of the fluvial drainage system, such as tributary channels leading downgrade to a trunk stream, and are thus classified as *alpine glaciers* (Figure 18.3). In the high interior area of a large landmass in the high latitudes, glacier ice accumulates in platelike bodies called *icecaps*. Small icecaps a few tens of kilometers across and roughly elliptical or circular in outline are found today on summit areas of arctic islands such as Iceland, Baffin Island, and Ellesmere Island. Today alpine glaciers and small icecaps are sustained on high mountains and plateaus, where air temperatures are low and there is abundant snowfall. The west coasts of continents in high latitudes are favored by abundant snowfall, and here glaciers flourish. Even in the equatorial zone of South America, we find glaciers thriving on the Andes, because of their very high altitude. Another tropical belt of large alpine glaciers is within the Himalayas of southern Asia, the loftiest of the earth's mountain chains.

Vastly greater ice masses of continental proportions, such as those of Greenland and Antarctica, are *continental glaciers*, or simply *ice sheets*. A great ice sheet may reach a thickness of several thousand meters and may cover an area of several million square kilometers, spreading far beyond the limits of the highlands where it originated. Many ice sheets of the past covered low plains that otherwise would not have sustained glaciers.

FIGURE 18.1 Two famous glaciers of the European Alps; they were studied by Louis Agassiz and other glaciologists in the mid-1800s. Deep cirques can be seen in the distance. *A*. Glacier d'Argentière, with a great ice fall in the lower portion. *B*. Mer de Glace, with a prominent medial moraine. (Swissair Photo.)

FIGURE 18.1 A

FIGURE 18.1 B

Features of Alpine Glaciers

The form of a typical alpine glacier is illustrated in Figure 18.4 by longitudinal and transverse cross sections. Snow accumulates at the upper end of the glacier in a bowl-shaped depression, the *cirque*. Here we find a broad expanse of smooth glacier surface, the *firn field*. By slow flowage at depth, the glacier ice moves toward the exit of the cirque and may pass over a steepened gradient, called a *rock step*, in the floor of the valley. Here the rate of flow is increased and results in an *ice fall*, in which the rigid surface ice is deeply broken by gaping fractures, termed *crevasses*.

Where the gradient lessens, the crevasses tend to close, but the surface may be extremely rough because of ablation of surface ice (see Figure 18.1). Near the *glacier terminus*, the lower end, the rigid ice is thrust forward over its own deposits of rock debris, and here low-angle shear planes, resembling overthrust faults, can be seen in the ice.

The rate of flow of alpine glaciers has long been a subject of interest to *glaciologists*, those scientists specializing in the study of glaciers. Pioneers in this field—Louis Agassiz, the Swiss-born naturalist, and J. D. Forbes, an English physicist—made many observations on glaciers of the Swiss and French Alps in the middle 1800s. They planted rows of stakes across the glacier surface and surveyed the movements of the stakes (Figure 18.5). Both men came independently to the conclusion that surface motion is most rapid near the center, decreasing toward the sides. If the line of stakes were to be mapped at equal intervals of time, it would be found to be bent convexly downvalley into a series of parabolic curves, as suggested by lines in Figure 18.5.

Slow surface motion of this type normally goes on at

FIGURE 18.2 Outlines of individual crystals of glacial ice in a wide range of sizes.

FIGURE 18.3 These schematic maps compare the form of an alpine glacier with that of an icecap.

FIGURE 18.4 Idealized longitudinal and transverse sections show the anatomy of a simple alpine glacier. (Drawn by A. N. Strahler.)

FIGURE 18.2

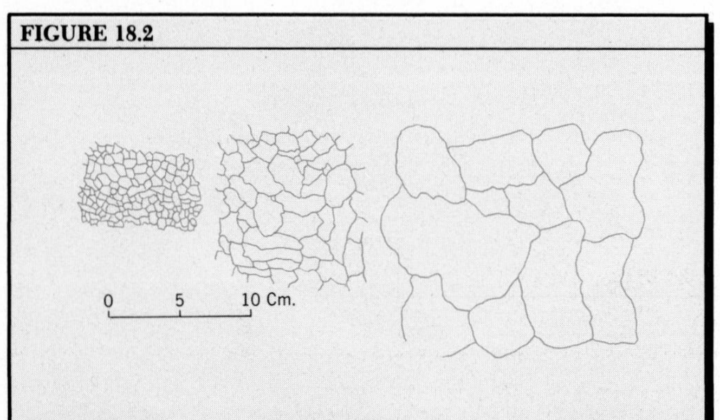

FIGURE 18.3

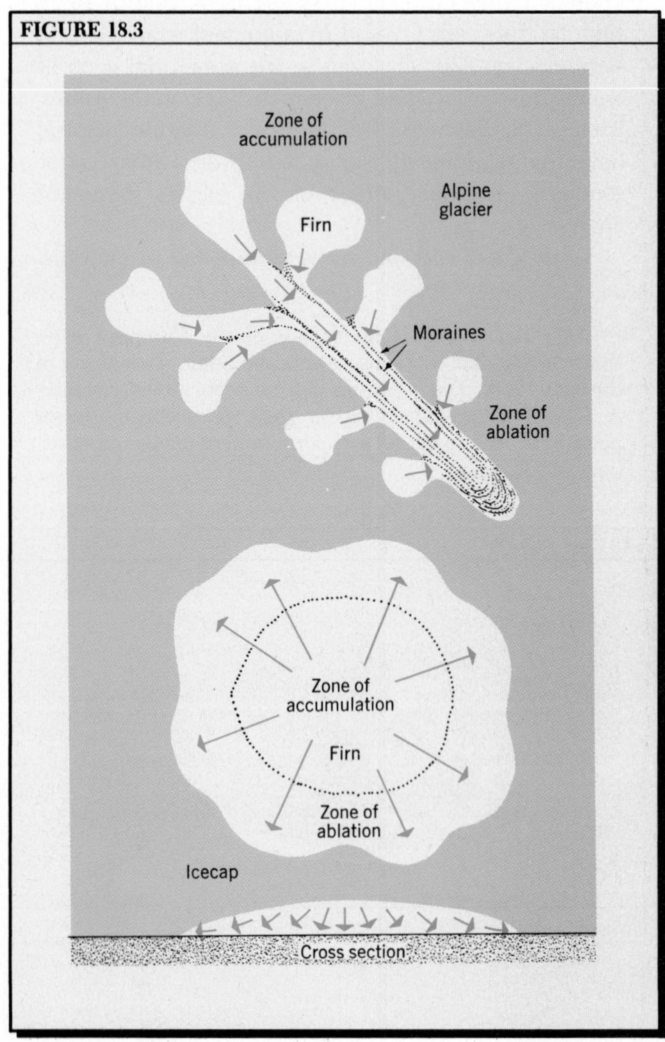

FIGURE 18.4

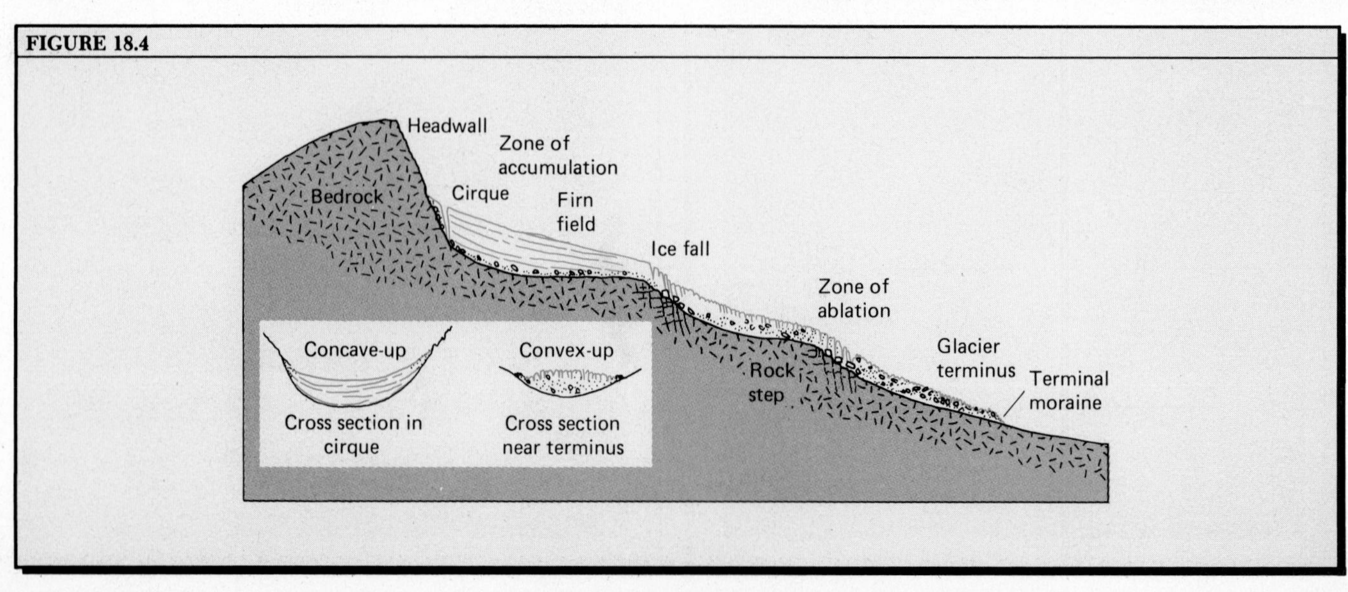

the rate of several centimeters to 1 m/day in an alpine glacier. Agassiz found that his fastest-moving stakes had traveled over 75 m downstream in one year. A large boulder resting on the ice moved almost 150 m in two years. It should be remembered, of course, that the upper zone of the glacier is composed of brittle ice in the rigid zone. The relative movements of surface points, although resembling flowage of a fluid on a large scale, actually occur by slippage between small blocks formed by the fracturing of the brittle ice. Crevasses are produced in this way. Below the rigid zone is the plastic zone, in which the ice deforms by plastic shear. The rate of flow decreases with increasing depth, as indicated in Figure 18.5.

Accompanying this internal deformation, the entire glacier is thrust forward as a block and apparently is undergoing a slipping movement over its bed, a phenomenon termed *basal slip*. The movement is caused by strong push from upstream sections of the ice.

Besides showing the continual slow internal flowage we have described, some glaciers experience episodes of very rapid flowage movement, called *surges*. Surging may result from an accumulation of water beneath the glacier, removing much of the frictional resistance to sliding. When surging occurs, the glacier commonly develops a sinuous pattern of movement, not unlike the open bends of an alluvial river (see Color Plate F.1). A surging glacier (sometimes called a "galloping glacier" by the news media) may travel several kilometers downvalley in a few months.

FIGURE 18.5 The rate of downvalley ice motion is greatest at the glacier surface and along the midline of the glacier.

FIGURE 18.6 A schematic cross section of an alpine glacier as an open system of flow of matter.

FIGURE 18.5

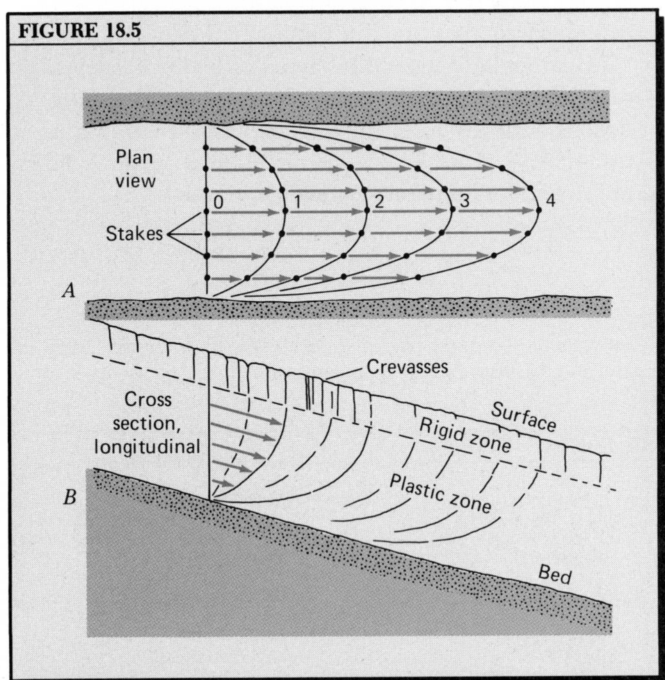

Glaciers as Open Systems

In Chapter 2, we used a simple alpine glacier as an example of an open system of flow of matter (see Figure 2.7). As open systems, all glaciers have much in common regardless of their form or size, and the concept is nicely illustrated by an alpine glacier (Figure 18.6). Matter, in the form of snow, is received upon the upper surface in the zone of highest elevation where the rate of loss of snow by ablation in summer is, on the average, less than the rate at which the snow is received. This region of net gain is called the *zone of accumulation*. In the lower part of the glacier, loss

FIGURE 18.6

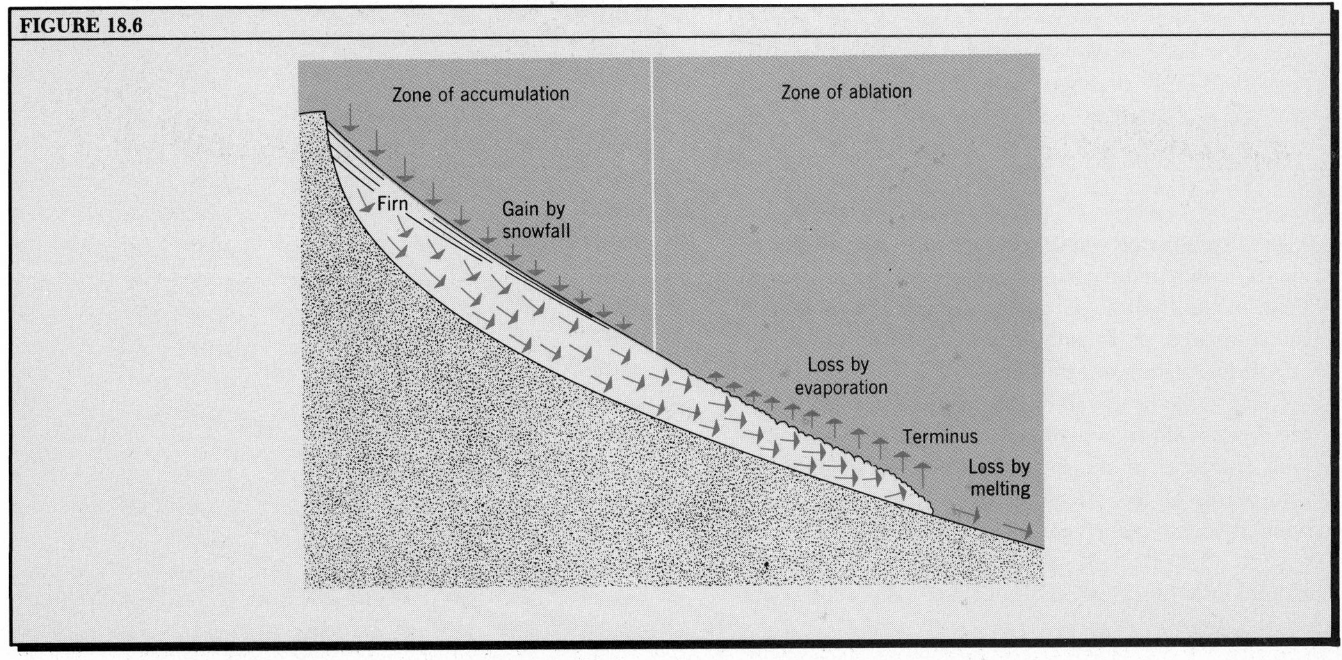

by ablation exceeds the rate at which snow accumulates, and the imbalance is greater as we follow the glacier to lower, warmer elevations. Consequently, the ice disappears at the terminus. This region of net loss is the exit boundary of the system and is known as the *zone of ablation*.

In the glacier flow system, the rate of motion of ice from the zone of accumulation to the zone of ablation depends largely upon the ice thickness, which in turn depends upon rates of nourishment and removal. A glacier easily adjusts its form and dimensions to reach a condition of dynamic balance known as a *steady state*. In the steady state, the terminus can remain essentially fixed in position for long periods of time, a condition also known as *glacier equilibrium*.

Glacier equilibrium is easily upset. Suppose, for example, that the rate of nourishment by snowfall is sharply increased. The ice of the upper end of the glacier thickens and consequently flows more rapidly downgrade. This action moves the glacier terminus farther downvalley, in what is described as a *glacier advance*. But the advance brings the glacier terminus to lower levels, where temperatures are higher and ablation more rapid, so soon the rate of ablation balances the rate of ice advance, and the terminus again becomes stabilized.

Suppose, instead, that the rate of glacier nourishment decreases, as it might if summers become warmer or snowfall lessens. The thinning glacier will flow more slowly and ice will not be brought as rapidly to the terminus. Now the rate of ablation will be excessive, and the glacier terminus and surface will melt away rapidly, causing the terminus to retreat upvalley in a condition called *glacier recession*. But as the glacier terminus recedes upvalley, less and less of its surface lies in warmer climatic zones; as a result, ablation is reduced. Again equilibrium will be established, with the terminus stabilized at a higher level.

Because of their sensitivity to climatic changes, glaciers have been intensively studied for the information they can supply as to past periods of colder or warmer climate, or of wetter or drier periods.

Glacier Erosion and Transportation

Like streams of water, glaciers represent to the geologist much more than mere systems of water disposal in the hydrologic cycle. Glaciers are molders of the landscape, performing erosion, transportation, and deposition of mineral matter.

Close to the headwall of the cirque, meltwater pouring down from snowbanks above the glacier enters the joint fractures in the headwall rock. Here it freezes into seams of ice, breaking loose many joint blocks which become incorporated into the upper end of the glacier. Beneath the glacier, ice flows plastically around joint blocks, then drags them loose when a sudden basal slip occurs. This activity is *glacial plucking*. Blocks of rock being carried within the glacier ice are scraped and dragged along the rock floor, gouging and grooving the bedrock and chipping out fragments of rock in an abrasive process called *grinding*.

The valley floor formerly beneath a glacier shows a number of interesting erosional features resulting from grinding and plucking. Rock surfaces are generally of hard, fresh rock and bear numerous fine scratches called *glacial striations* marking the lines where sharp corners of rock fragments have scraped the surface. Where pressure was strongly applied, the impinging boulders created curved fractures in the bedrock. The most common fracture type is the *chatter mark*, bent concavely toward the downstream direction of ice flow (Figure 18.7). (Another fracture type is the crescentic gouge, illustrated in Figure 18.9.)

In particularly susceptible types of bedrock, such as some limestones, glacial abrasion may produce long, deep *glacial grooves* (Figures 18.8 and 18.9). These troughlike features follow the direction of ice flow and are scored by numerous parallel striations.

FIGURE 18.7 These sets of chatter marks indicate that ice movement was away from the observer. Elmer Creek, Sierra Nevada, California. (D. L. Babenroth.)

FIGURE 18.7

A common small landform produced beneath a glacier is the *glaciated rock knob.* It is a hill of bedrock strongly shaped by abrasion and plucking (Figure 18.9). The side of the knob facing upstream with respect to ice flow is smoothly rounded by ice abrasion; it is called the *stoss side.* The downstream, or *lee side,* is strongly plucked.

Still another source of glacier load is the rolling and sliding of rock fragments down the steep sides of the cirque and the valley walls adjacent to the ice stream. Such slide rock takes the form of talus cones and sheets (Figure 18.10). Rock fragments reaching the glacier margin are dragged along by the moving ice, forming marginal embankments of debris called *lateral mo-*

FIGURE 18.8 Large, unusually deep glacial grooves in massive limestone, Kelley's Island, near the south shore of Lake Erie. (State of Ohio, Dept. of Industrial and Economic Development.)

FIGURE 18.9 Sketch of a glacially rounded and plucked rock knob. Arrows indicate the direction of ice flow. (Drawn by A. N. Strahler.)

FIGURE 18.8

FIGURE 18.9

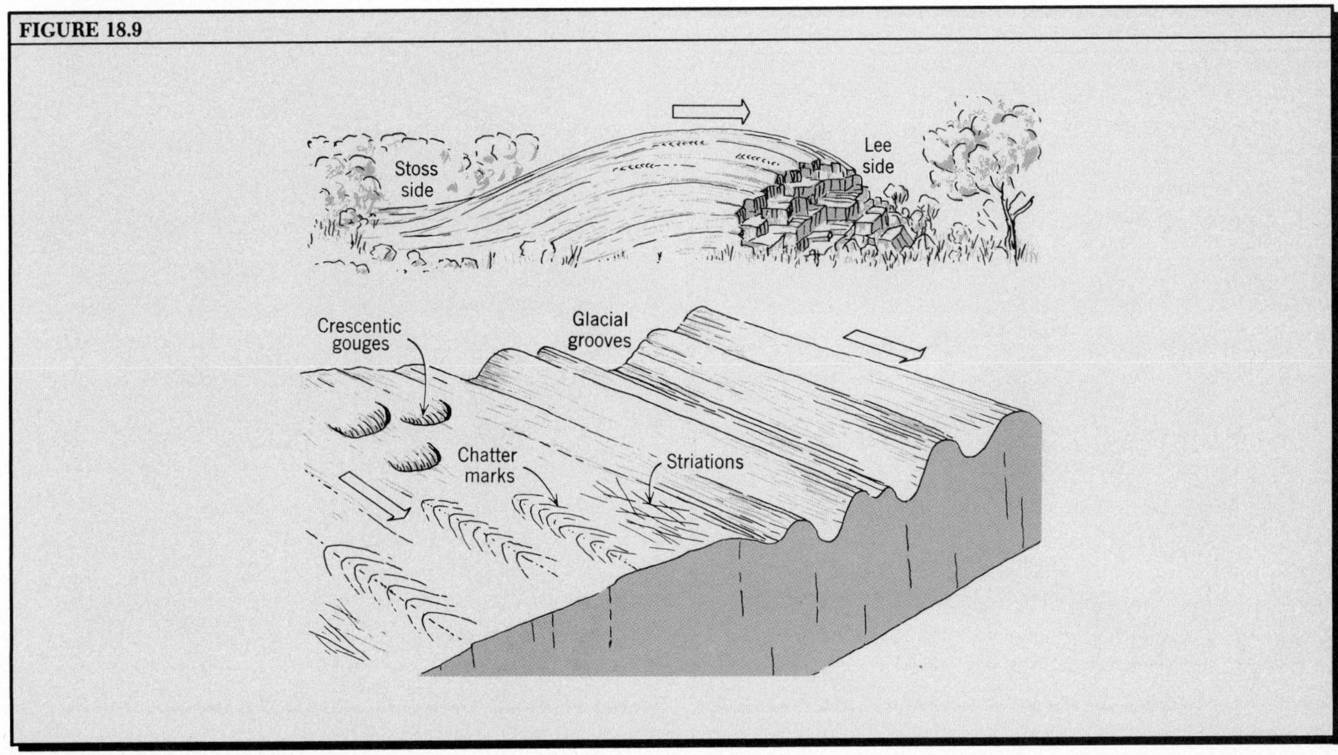

raines. After the glacier has disappeared, these moraines form ridges parallel with the valley walls (Figure 18.10).

Where two ice streams join, the debris of the inner lateral moraines is dragged out into the middle of the combined ice streams to form a long narrow line of debris, a *medial moraine*. Several of these features are visible in Figures 18.3 and 18.11, appearing as parallel lines dividing the glacier into narrow bands. Debris supplied from marginal slopes remains largely on the glacier surface.

Near the glacier terminus, the proportion of solid load to ice increases greatly, until at the very end there is more solid debris than ice. This residual mass of rock debris constitutes the *terminal moraine* of the glacier (Figure 18.10).

Seeing the huge boulders composing glacial moraines, you may not realize that much rock is also ground by the glacier into fine silt and clay called *rock flour*. It gives to meltwater streams issuing from a glacier a characteristic milky appearance (Figure 18.12). Settling out in lakes beyond the glacier limit, the rock flour forms layers of silt and clay.

FIGURE 18.10

FIGURE 18.10 The shrunken remnant of the Black Glacier is almost buried in its own morainal debris. Talus cones have been built from the valley walls. The terminal moraine at the lower left extends upvalley in ridgelike lateral moraines. Bishop Range, Selkirk Mountains, British Columbia. (H. Palmer, Geological Survey of Canada.)

FIGURE 18.11

FIGURE 18.11 This branching alpine glacier flows westward along the northern edge of the Juneau Ice Field in Alaska. The dark bands are medial moraines. (U.S. Army Air Force.)

FIGURE 18.12

FIGURE 18.12 Meltwater issues from a tunnel in stagnant, debris-laden ice at the terminus of the Franz Josef Glacier, South Island, New Zealand. The bouldery debris in the foreground is part of the terminal moraine. (Official New Zealand Government Photograph.)

Landforms Carved by Alpine Glaciation

Many spectacular landforms are shaped by alpine glacier erosion (Figure 18.13). Frame *A* shows a mountain region shaped by processes of fluvial denudation in a period of milder climate preceding glaciation. Thick accumulations of soil and regolith sheathe the mountain slopes; divides are broad and somewhat subdued in appearance.

Frame *B* shows the same mountain mass occupied by valley glaciers that have been in action for many thousands of years. The higher central summit area has been carved into steep-walled cirques which meet in sharp, knife-edge crests, culminating in toothlike peaks, called *horns*. Where two cirques are arranged back to back on opposite sides of a divide, the intervening rock wall may be cut through to form a deep pass. Notice that the somewhat lower mountain summit at the far right of Frame *B* is only partly consumed by cirque development and that the preglacial mountain surface remains intact over the summit.

FIGURE 18.13 Landscape evolution under alpine glaciation. *A*. Preglacial topography formed by fluvial denudation. *B*. Stage of maximum alpine glaciation. *C*. Postglacial landscape, following disappearance of all glacial ice. (Redrawn, by permission of the publisher, from A. N. Strahler, *Physical Geography*, 4th ed., Figure 31.3. Copyright © 1975 by John Wiley & Sons, Inc.)

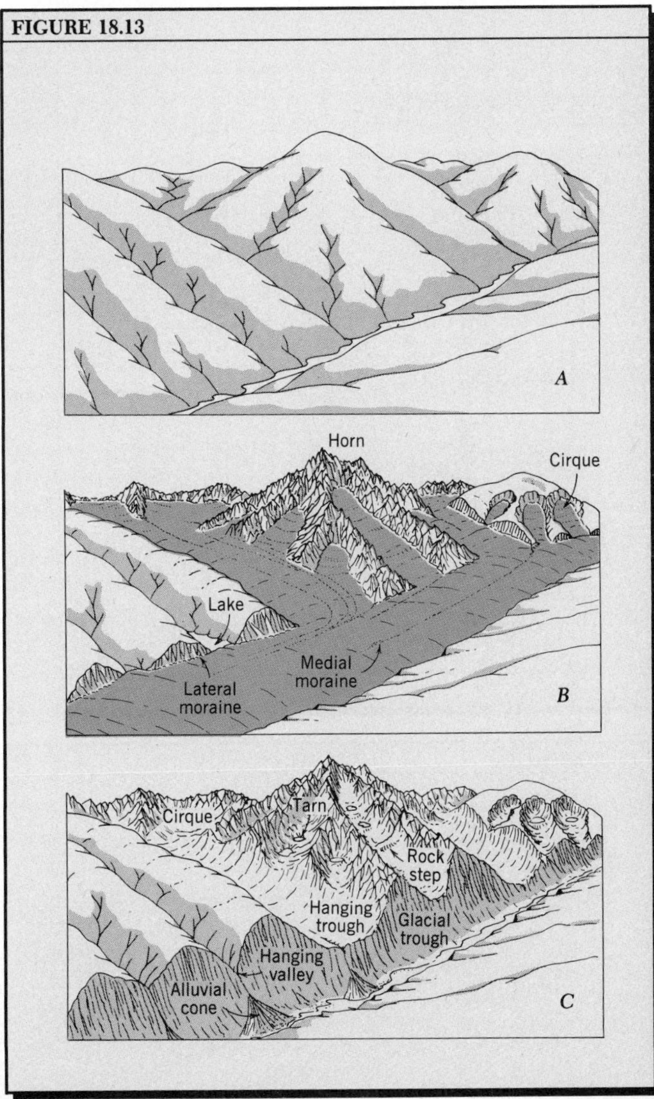

Tributaries enter the main glacier with smooth ice-surface junctions. Abrasion of the valley walls has planed away projecting spurs of the preglacial stream valleys. Where the main glacier blocks a stream valley at low altitude, discharge is impounded in temporary lakes.

Frame *C* shows this same region after the glaciers have entirely disappeared because of a general warming. This condition is found today in many high mountain ranges of the middle latitudes. The trunk glacier eroded a deep, U-shaped *glacial trough*. Its smaller tributaries also carved troughs, but because these ice streams were of smaller cross section, their floors were not so deeply cut. These tributary troughs now enter at levels high above the main trough floor, and so are called *hanging troughs*. In the upper reaches of the troughs are many irregularities of gradient, giving a succession of rock steps and *rock basins;* the latter often hold lakes, as will depressions in the floors of cirques. These lakes are called *tarns*. Torrential streams build steep alluvial cones against the trough walls.

Grandest of all glacially carved landforms are the glacial troughs, which may be a thousand meters deep and tens of kilometers long. Troughs cut below sea level in a coastal mountain range are now invaded by the sea to become *fiords*, characterized by the steepness of their walls and the great depth of their floors (Figure 18.14). Fiords are widely found along mountainous coasts of arctic and subarctic latitudes. Important fiord coasts occur in Alaska and British Columbia, southern Chile, Scotland, New Zealand, and Norway.

Greenland and Antarctic Ice Sheets

From the small, streamlike alpine glaciers, we turn our attention to the two enormous ice masses of subcontinental size that exist today on Greenland and Antarctica.

The Greenland Ice Sheet occupies some 1,740,000 sq km, which is 80% of the area of the island of Greenland where, in fact, ice covers all but narrow land fringes (Figure 18.15). The ice sheet contains almost 3 million cu km of ice. In a general way, the ice forms a single, broadly arched, doubly convex lens. It is generally smooth surfaced on the upper side and considerably rougher and less strongly curved on the underside. The mountainous terrain of the coast passes inland beneath the ice, but gives way to a central lowland area close to sea level in elevation. The ice surface is characterized by wind-eroded and drifted features (Figure 18.16).

The ice thickness measures close to 3 km at its greatest. It is not surprising that the center of Greenland is actually depressed under such a load, in conformity with the principle of isostasy, since 3 km of glacial ice is roughly equivalent to a rock layer at least 1 km thick.

Because the Greenland ice surface slopes seaward, the ice creeps slowly downward and outward toward the margins, where it discharges by glacier tongues. These are *outlet glaciers*—closely resembling alpine glaciers, but fed from a vast ice sheet rather than from a cirque. Greenland's outlet glaciers reach the sea in fiords and are the source of North Atlantic icebergs.

Like Greenland, the Antarctic continent is almost entirely buried beneath glacial ice, in an area of just over 13 million sq km, or about 1.5 times the total area of the contiguous 48 U.S. states (Figure 18.17). Ice volume is about 25 million cu km, which is over 90% of the total volume of the earth's glacier ice. (In comparison, the Greenland Ice Sheet has about 8%.)

The Antarctic Ice Sheet reaches its highest elevations, just over 4 km, in a broadly rounded summit. Surface slope is gradual to within 300 km of the edge of the continent, where a marked steepening occurs. Although, in general, the ice is thickest (4 km) where surface elevation is highest, there are important exceptions to this statement. There is a great subglacial channel, or valley, named the Byrd Basin (see profile in Figure 18.17). Here an ice thickness of over 4 km has been measured, and the rock floor lies some 2 km below sea level.

The subglacial topography of Antarctica is partly mountainous. Mountain peaks and ranges rise above

FIGURE 18.14 This Norwegian fiord has the steep rock walls of a deep glacial trough. (Mittet and Co.)

FIGURE 18.14

FIGURE 18.15 Greenland and its ice sheet. (Redrawn, by permission of the publisher, from A. N. Strahler, *Physical Geography*, 4th ed., Figure 31.9. Copyright © 1975 by John Wiley & Sons, Inc.)

FIGURE 18.16 Blizzard winds have eroded the packed snow surface of the Greenland Ice Sheet. (L. H. Nobles.)

FIGURE 18.15

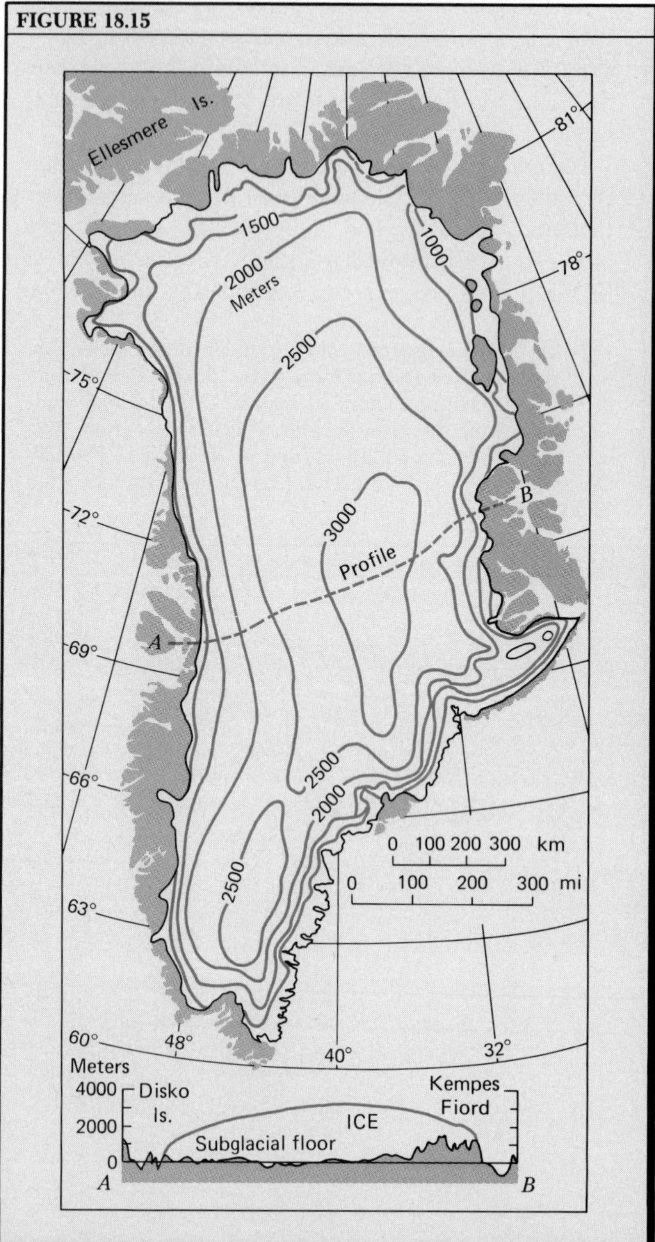

FIGURE 18.16

the ice in several belts, mostly close to the continental margins. Here the ice moves as outlet glaciers from the interior polar plateau to reach the coast (Figure 18.18).

A characteristic feature of the Antarctic coast is the presence of numerous *ice shelves*, which are great plates of partially floating ice attached to the land (Figure 18.17). Largest is the Ross Ice Shelf, about 500,000 sq km in area, with its surface at an average elevation of about 70 m (Figure 18.19). Almost as large is the Filchner Ice Shelf. Smaller ice shelves occupy most of the bays of the Antarctic coast and in places form a continuous but narrow ice fringe. The ice shelves represent those parts of the ice sheet that have been pushed seaward into water of sufficient depth to float the ice off the bottom. The ice shelves, largely maintained by snow accumulation on their surfaces, are the source of enormous tablelike icebergs of the Antarctic Ocean.

The Antarctic Ice Sheet moves outward near its margin at a rate estimated between 25 and 50 m per

FIGURE 18.17 The Antarctic Ice Sheet and its ice shelves. (Based on data of Amer. Geophysical Union and Amer. Geographical Soc.)

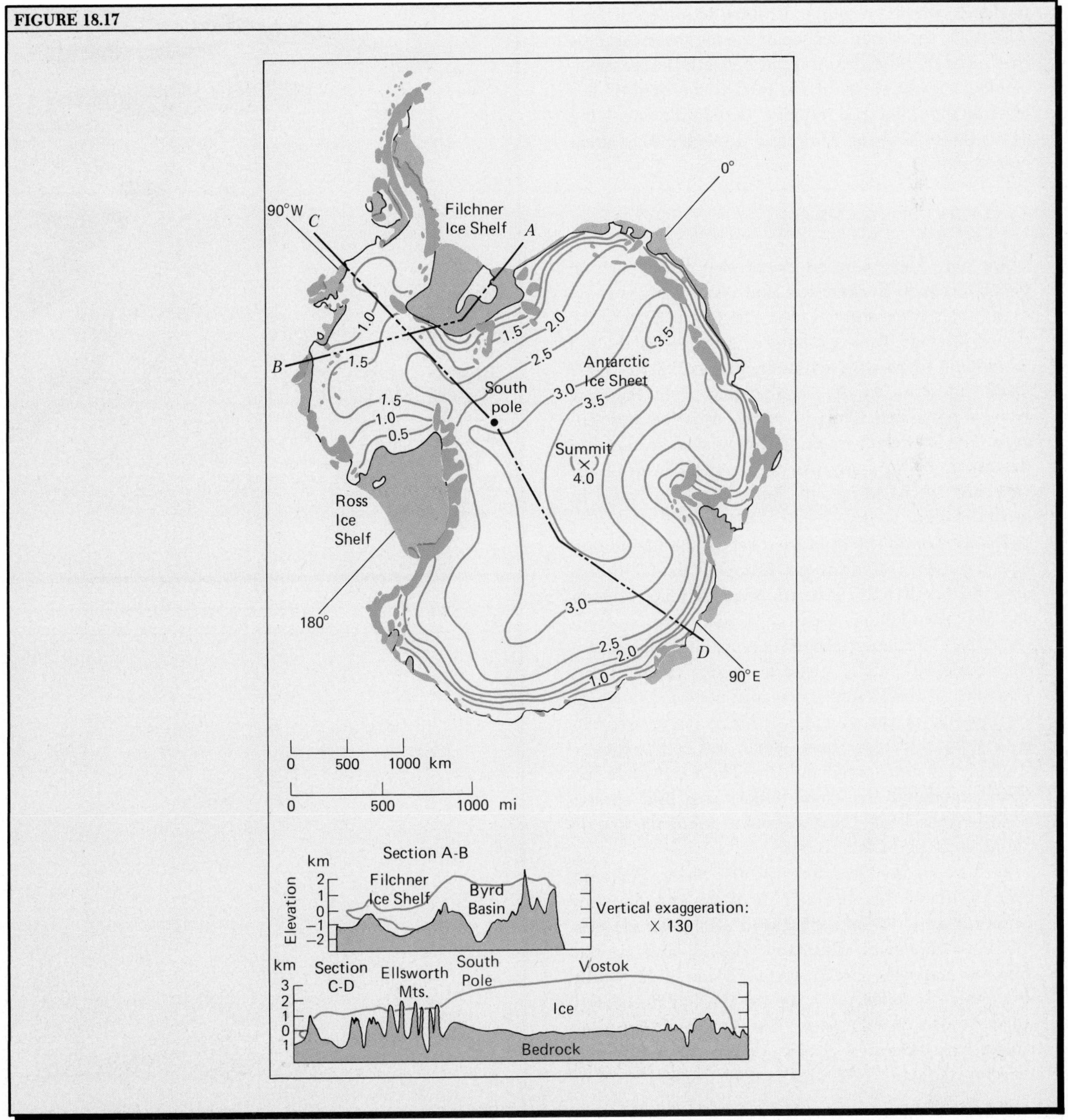

year. Rate of movement of outlet glaciers is much faster, estimated at 400 m per year on the average. For the ice shelves, rate of outward movement is even more rapid, on the order of 1000 to 1200 m per year for the Ross and Filchner shelves. From these estimated flowage rates, scientists infer that a particle of ice following the longest possible path of flow might remain in the ice sheet for as long as 100,000 years.

The Antarctic Ice Sheet seems to be very closely balanced in terms of the net rate of accumulation and the rate of loss of ice to the atmosphere and surrounding ocean. Actually, the figures show a small rate of growth to be in progress, but this may be a short-term effect. The enormous volume of water presently stored in the ice sheet is a source of worry to some scientists who study the water balance of the ice sheet. Suppose that rapid melting should occur beneath the ice sheet, causing a great surge of ice toward the oceans? Sea level might then rise rapidly, inundating low-lying coasts throughout the world and drowning most great coastal cities.

Pleistocene Ice Sheets

We now know beyond doubt that large parts of North America, Eurasia, and South America were covered by great ice sheets in the Pleistocene Epoch, the unit of geologic time spanning approximately the last 2.5 million years. Only within the last 10,000 to 15,000 years did these ice sheets disappear from the now heavily populated lands of North America and Europe. The landforms resulting from glacial erosion and deposition are in many places extremely fresh in appearance, so the former ice limits can be mapped in great detail.

The maximum extent of ice sheets of the Pleistocene Epoch in North America and Europe is shown in Figures 18.20 and 18.21. In North America, all of Canada and the mountainous areas of Alaska were covered. Over the Cordilleran ranges, icecaps and alpine glaciers coalesced into a single ice body which spread westward to the Pacific shores and eastward down to the foothills of the mountains. Much larger was the great Laurentide Ice Sheet, which was centered over Hudson Bay and spread radially. The Laurentide Ice Sheet inundated the Great Lakes area and spread south into the United States about as far as the line of the Missouri and Ohio rivers.

In Europe, the Scandinavian Ice Sheet, centered over the Baltic Sea, covered all of the Scandinavian peninsula and reached southward and eastward into the Low Countries, Germany, Poland, and Russia. This ice mass also spread westward across the North Sea, where it joined with an ice sheet that covered much of the British Isles. The Alps and Pyrenees ranges bore extensive glacier systems formed by the merging of many individual valley glaciers. As you

FIGURE 18.18 In this air view of the head of Shackleton Glacier, Antarctica, the polar ice plateau is seen in the distance. (U.S. Geological Survey.)

FIGURE 18.19 The Ross Ice Shelf, Antarctica. The steep ice cliff, from 15 to 45 m high, presents a formidable barrier. (Official U.S. Coast Guard Photograph.)

FIGURE 18.18

FIGURE 18.19

would expect, glacial activity in all the world's high mountain ranges was greatly intensified. Valley glaciers were then much larger and extended into lower altitudes than they do today.

In Siberia, icecaps formed upon the uplands east of the Ural Mountains. Mountain ranges of northeastern Siberia supported large icecaps formed by the joining together of complex systems of smaller icecaps and valley glaciers. Smaller ice complexes also existed over a number of the higher mountain and plateau areas of central and eastern Asia.

In South America, the only large ice sheet of the southern hemisphere (exclusive of Antarctica) was formed over the Andean Range of Chile and Argentina, largely southward of lat. 40°. At its maximum extent, this ice sheet reached to the Pacific Ocean on the west and spread eastward upon the pampas (piedmont plains) of Patagonia for a distance of 160 km or so beyond the base of the Andes.

The growth and outward spreading of a great ice sheet is known as a *glaciation*. We can safely assume that a glaciation is associated with a general cooling of average air temperatures over the region where the ice sheet originated. At the same time, ample snowfall must have persisted over the growth area to allow the ice mass to increase in volume. The opposite kind of change—a shrinkage of the ice sheet in depth and volume—would result in the ice margins receding toward the central highland areas, a process called *deglaciation*, and the ice sheet would eventually disappear. Following a deglaciation, but preceding the next glaciation is a period of time in which a mild climate prevails; it is called an *interglaciation*.

FIGURE 18.20 Maximum extent of Pleistocene ice sheets of North America. (Redrawn, by permission of the publisher, from A. N. Strahler, *Physical Geography*, 4th ed., Figure 31.11. Copyright © 1975 by John Wiley & Sons, Inc.)

FIGURE 18.20

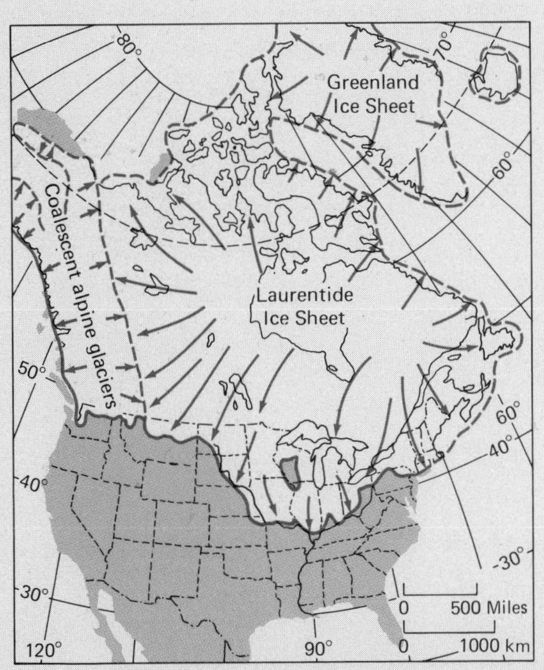

From the middle 1800s until about 1950, the record of glaciations, deglaciations, and interglaciations was almost entirely interpreted from continental deposits left by former ice sheets. During this early period of research, there emerged a history of four distinct North American glaciations in the Pleistocene Epoch. A similar and possibly equivalent four-glaciation history was also established for Europe on the basis of studies in the Alps. Names of the four established glaciations and interglaciations of North America are given in Table 18.1.

Maximum southern extent of ice in each glaciation in the north central United States is shown in Figure 18.22. The ice fronts advanced southward in great lobes. Notice that an area in southwestern Wisconsin escaped inundation by Pleistocene ice sheets. Known as the Driftless Area, it was apparently bypassed by glacial lobes moving on either side.

Later in this chapter, we will investigate new information on the complete global record of glaciations and interglaciations. Newer methods of research have revealed a complex history of Pleistocene and earlier climate changes possibly capable of causing numerous glacial cycles. At this point, our attention turns to landforms and sediment deposits made by ice sheets. Most of these features are the direct result of the last

FIGURE 18.21 Limit of glacial ice of Europe in the last glaciation is shown by a solid line; maximum extent in the entire Pleistocene Epoch is shown by a dashed line. (Redrawn, by permission of the publisher, from A. N. Strahler, *Physical Geography*, 4th ed., Figure 31.12. Copyright © 1975 by John Wiley & Sons, Inc.)

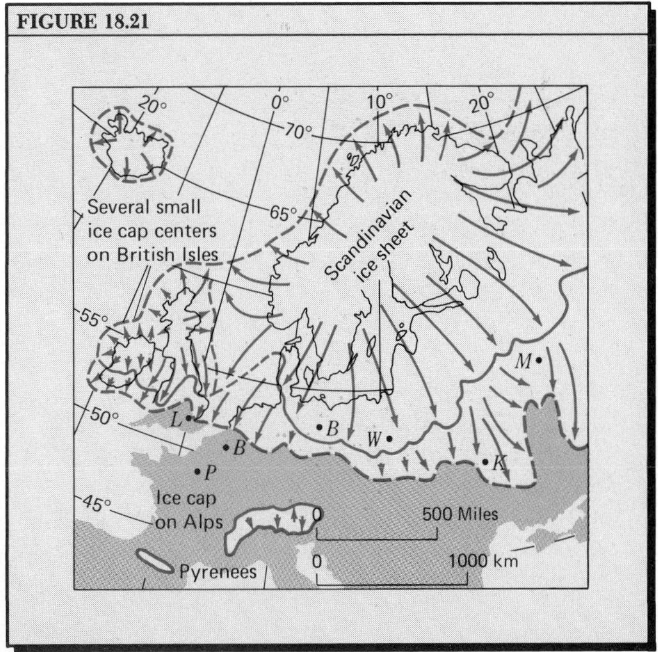

glaciation, which is named the *Wisconsinan Glaciation*, and of the deglaciation which followed. The Wisconsinan ice sheet began to recede quite rapidly, starting about 15,000 years before the present (−15,000 y.), and it had largely disappeared from North America and Europe by about −8000 y. Landforms shaped and abandoned by the Wisconsinan ice sheet are today little changed by weathering, mass wasting, and running water. It is these sharply defined features we now describe.

Glacial Drift

The term *glacial drift* applies to all varieties of rock debris deposited in close association with Pleistocene ice sheets. Drift consists of two major classes of materials: (1) *Stratified drift* is made up of sorted and layered clay, silt, sand, or gravel deposited as bed load or delta sediment from streams of meltwater, or settled from suspension into bodies of quiet water adjoining the ice. (2) *Till* is an unsorted mixture of rock and mineral fragments of a wide range of sizes—from clay to large boulders—deposited directly from glacial ice. One form of till consists of material dragged along beneath the moving ice and plastered upon the bedrock or upon other glacial deposits. Another form consists of debris held within the ice but dropped in place as the ice wasted away. Large rock fragments within glacial till, some of which have been transported hundreds of kilometers, are called *erratic boulders* (Figure 18.23).

Within the area of the United States once covered

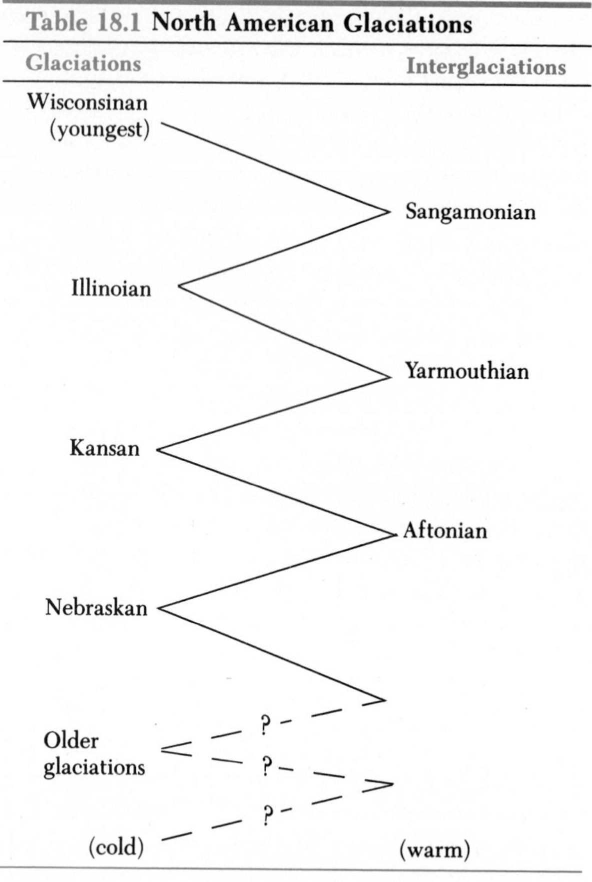

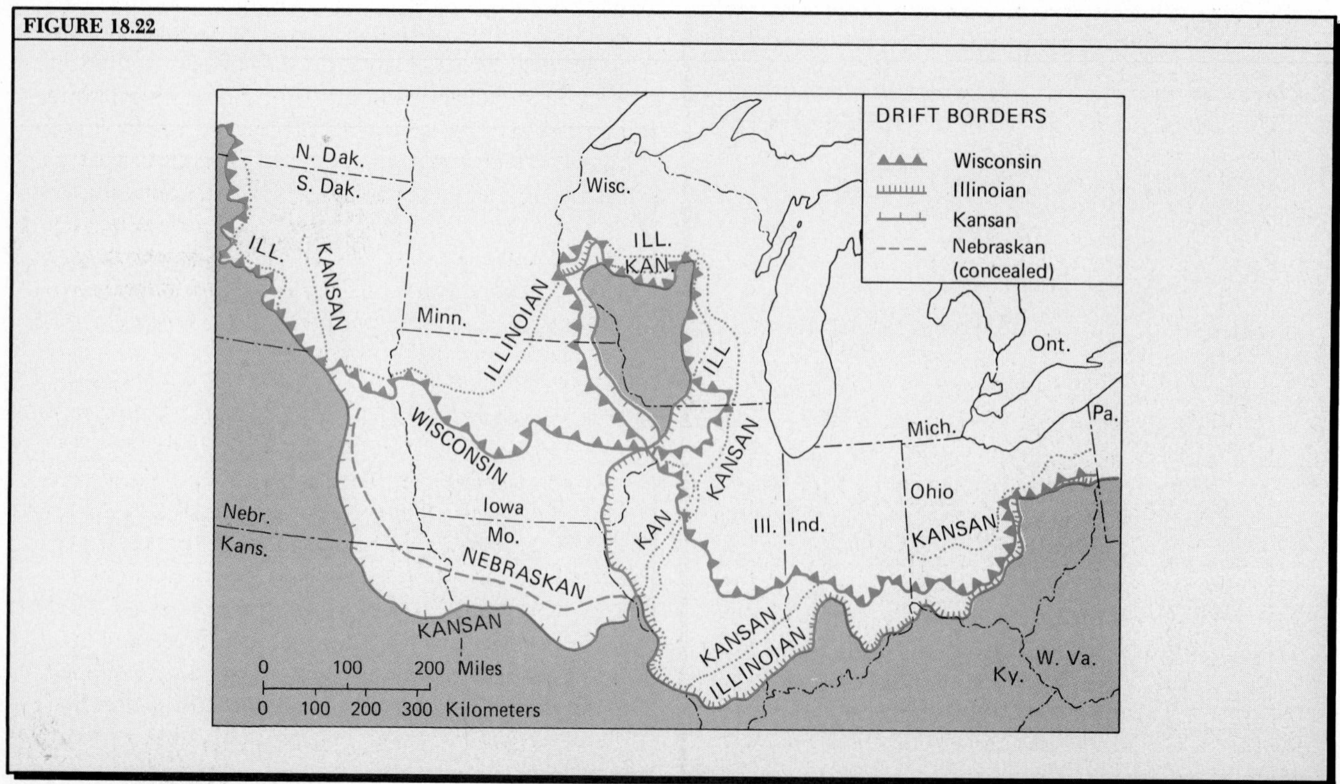

FIGURE 18.22 Ice limits in the north central United States for the four glaciations. (Redrawn, by permission of the publisher, from A. N. Strahler, *Physical Geography*, 4th ed., Figure 31.13. Copyright © 1975 by John Wiley & Sons, Inc.)

by Pleistocene ice sheets, glacial drift has a thickness averaging from 5 m over the mountainous terrain of the northeast to perhaps 15 m or more over the plains of the north-central states. Drift is generally 15 to 60 m thick in Iowa and over 30 m in Illinois. As you would expect, the greatest thicknesses occur in preglacial valleys and lowlands, while the thinnest drift lies over plateaus and hill summits. Consequently, the gross effect of continental glaciation has been to subdue the landscape to a more plainlike aspect.

Landforms of Continental Glaciation

The advance and wastage of a great ice sheet leaves a host of distinctive minor landforms that can best be understood and described with the aid of two block diagrams (Figure 18.24). The upper block shows the ice sheet in place; the lower one shows the landscape after the ice has disappeared. In the upper block, the ice has reached a stable line of advance, though previous forward movement of the ice has created a terminal moraine at the ice margin. Farther back, beneath the ice, is a till layer of variable thickness, the *ground moraine*.

The ice has now become almost stagnant. Meltwater issues from the ice front in numerous streams, some of which are discharging from tubes within and beneath the ice, while others are flowing down the ice surface itself. Sand and gravel, carried as bed load in the meltwater streams, are spread in sheets in a zone in front of the ice, forming alluvial fans that coalesce into a continuous alluvial sheet, the *outwash plain*. A series of ice blocks, left behind during a previous episode of advance and rapid wastage, are surrounded and perhaps buried in the outwash layers. At the rear right-hand side of the block, the ice sheet has dammed the runoff system to produce a temporary *marginal lake*.

FIGURE 18.23 In 1916, a group of New York City geologists visited this enormous erratic boulder on Long Island. The boulder had been carried by the ice sheet from what is now the mainland of New England. (Armin K. Lobeck.)

FIGURE 18.23

Disappearance of the ice sheet reveals more landforms shaped beneath the ice (lower block in Figure 18.24). The terminal moraine now appears as a belt of hilly ground with many deep, closed depressions. This type of terrain is called *knob and kettle* (Figure 18.25). Sloping away from the terminal moraine is the smooth surface of the outwash plain, pitted here and there by steep-sided depressions, called *kettles*, formed where ice blocks were buried. Lakes and ponds are often held in the kettles (Figure 18.26).

Behind the moraine is an expanse of poorly drained, marshy surface underlain by ground moraine. Rising from this low ground are smoothly rounded hills, called *drumlins*, usually occurring in groups. They are oval in outline and composed of till shaped by the ice. The long axes of the hills are roughly parallel to one another and at right angles to the trend of the terminal moraine (Figure 18.27).

Disappearance of the ice also uncovers long, narrow ridges of coarse sands and gravels which extend in a sinuous course for many kilometers, roughly parallel with the direction of ice movement (Figure 18.28). These features are *eskers*, the bed-load deposits of meltwater streams that emerged from tunnels at the ice margin. Some eskers are traceable for tens of kilometers with few interruptions and may receive tributary eskers.

Most large lobes of an ice sheet, during the period of ice recession, underwent temporary halts and perhaps minor readvances. These halts resulted in the formation of additional moraines, termed *recessional moraines*. Between adjacent ice lobes are *interlobate moraines* (Figure 18.24).

A number of distinctive deposits and landforms are found in and around the basins of temporary marginal lakes that drained away following disappearance of the ice (Figure 18.24). Marginal lakes were the sites of deposition of sediment brought by meltwater issuing from the adjacent ice. These lake sediments show a succession of alternating light and dark bands known as *varves*. A single varve consists of a light band of silt below, grading upward into a dark band of fine clay (Figure 18.29). Each varve represents the sediment deposit of a single year. The lighter material consists of silt (rock flour) that settled out of suspension in the summer, when turbid meltwater was pouring into the lake. In winter, when the lake was sealed beneath an ice cover and the water became still, the finest clay particles gradually settled out to form the dark layer.

Meltwater streams emerging from the ice also built deltas in the marginal lakes. After both ice and lake disappeared, deltas were left standing as flat-topped *delta kames*. A *kame* is any steep hill of well-sorted glacial sand and gravel (Figure 18.30). Delta kames contain the steeply sloping foreset beds typical of small, simple deltas (Figure 18.31). Where an aggrading stream ran between the ice and the valley wall, a *kame terrace* was formed, resembling in some respects the alluvial terrace (Chapter 16), but commonly pitted with kettles (Figure 18.30*B*).

FIGURE 18.24 Landforms produced near the margin of an ice sheet. (*Upper*) Ice margin in an almost stagnant condition. (*Lower*) Ice entirely gone, exposing subglacial forms. (Redrawn, by permission of the publisher, from A. N. Strahler, *Physical Geography,* 4th ed., Figure 31.18. Copyright © 1975 by John Wiley & Sons, Inc.)

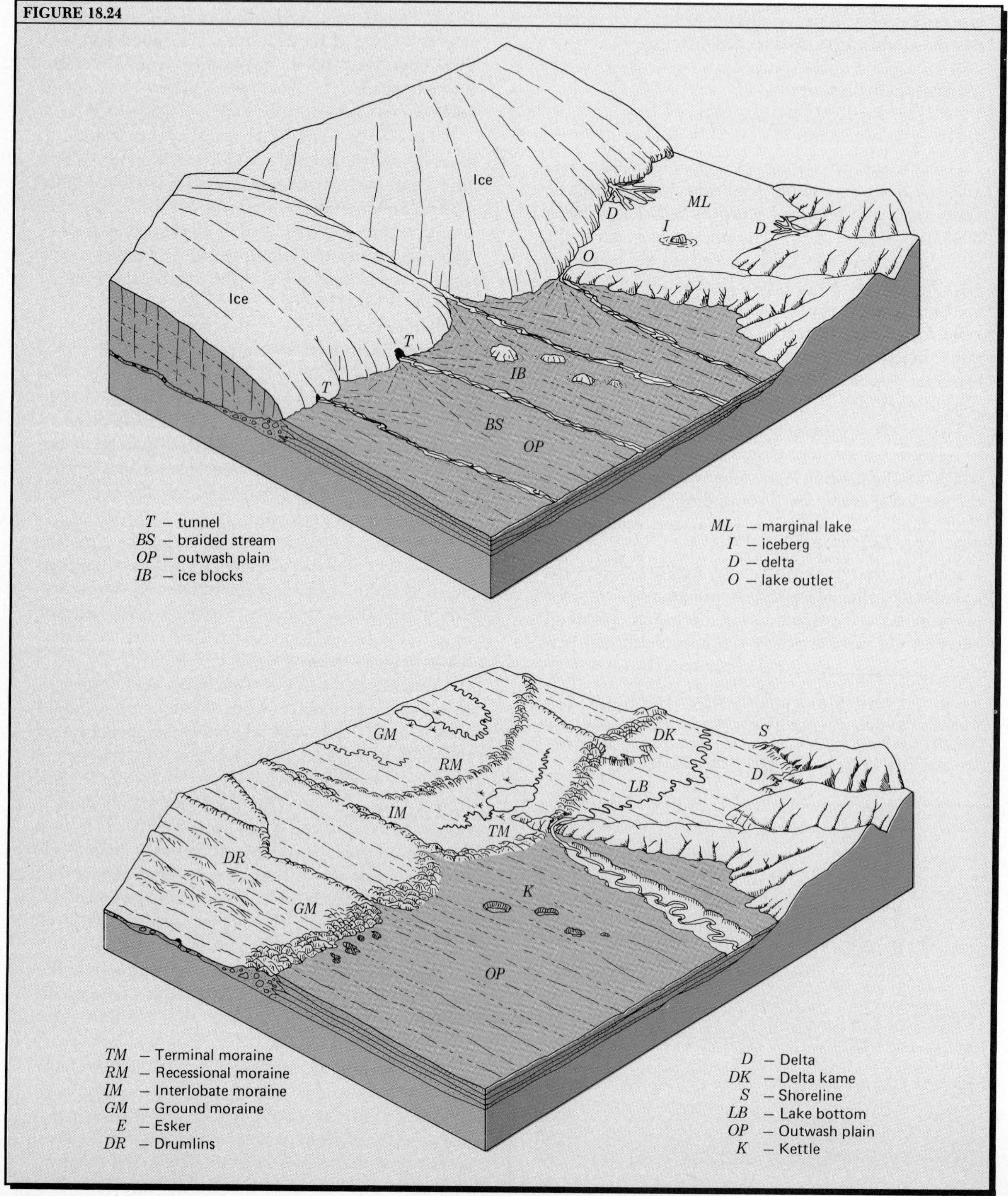

FIGURE 18.25 Rugged topography of small knobs and kettles characterizes this interlobate moraine in Sheboygan County, Michigan. Many erratic boulders litter the surface. (U.S. Geological Survey.)

FIGURE 18.26 A deep kettle pond on Cape Cod. (A. N. Strahler.)

FIGURE 18.27 Drumlins of glacial till are seen in this vertical air photograph. Ice moved from upper right to lower left. The area shown is about 3 km wide. The locality is in British Columbia. (U.S. Army Air Force.)

FIGURE 18.28 This esker, near Boyd Lake in Canada, crosses irregular hills of glacially eroded bedrock. (Canadian Dept. of Mines, Geological Survey.)

FIGURE 18.29 Varved clays from lake deposits near New York City. Pins mark the divisions between each varve. The sample columns are about 25 cm high. (C. A. Reeds, American Museum of Natural History.)

FIGURE 18.25

FIGURE 18.26

FIGURE 18.27

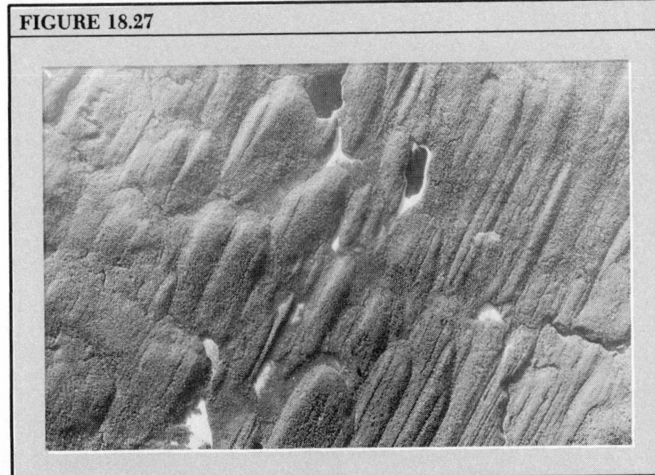

FIGURE 18.28

FIGURE 18.29

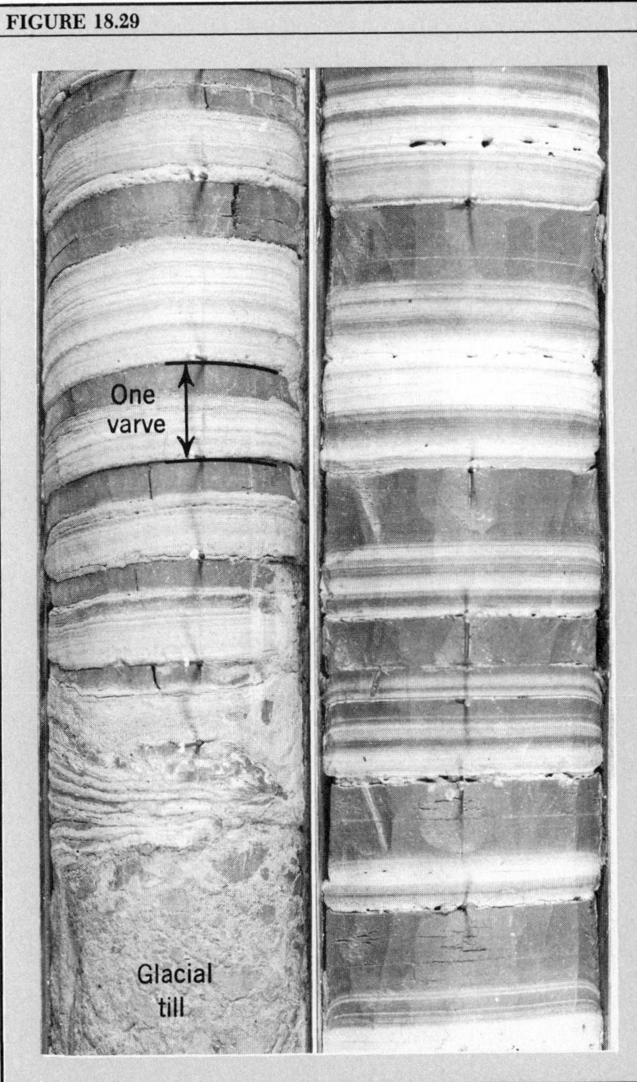

*Origin of the Great Lakes

In North America, a major event of Pleistocene time was the formation of the Great Lakes, five great interconnected inland bodies of fresh water. In preglacial time, the area of the present Great Lakes was occupied by broad, stream-eroded lowlands. Except for the basin of Lake Superior, which is entirely within the Canadian Shield, these lowlands were eroded in weak formations of Paleozoic strata separated by cuestas of more resistant formations. One prominent cuesta is the Niagara Cuesta, mentioned in Chapter 17 and pictured in Figure 17.24. The preglacial lowlands were strongly scoured and considerably deepened by erosive action of successive Pleistocene ice sheets. Debris from the basins provided much of the drift now found in moraine and outwash-plain belts situated south of the Great Lakes.

The Great Lakes as we know them came into existence during recessional substages of the Wisconsinan Glaciation as ice evacuated basins between the ice front and higher ground elevations to the south. A complicating factor in the history of the Great Lakes has been a downwarping of the earth's crust under the load of glacial ice and a subsequent upwarping following removal of ice load.

History of the Great Lakes during ice recession is extremely complex. We will simply pick out certain representative and interesting points in time to illustrate the nature of the process. A series of maps in Figure 18.32 shows conditions at these selected times.

Map *A* shows conditions between −14,000 and −15,000 y. in recessional phases of the Woodfordian Substage. Ice lobes in the basins of Lake Michigan and Lake Erie had receded sufficiently far to form two glacial lakes, named Chicago and Maumee, respectively. The former drained south by means of the Desplaines River, and the latter drained into the Wabash River. Continued recession (Map *B*) enlarged these lakes, lowering the level of Lake Maumee and

FIGURE 18.31

FIGURE 18.30 Glacial deposits are associated with a mass of stagnant ice in the axis of a broad valley. (Redrawn, by permission of the publisher, from A. N. Strahler, *Physical Geography,* 4th ed., Figure 31.23. Copyright © 1975 by John Wiley & Sons, Inc.)

FIGURE 18.31 Sloping foreset beds of well-sorted sand in a delta kame near North Haven, Connecticut. This deposit and most like it have since disappeared, consumed for use in concrete in urban expansion. (R. J. Lougee.)

FIGURE 18.30

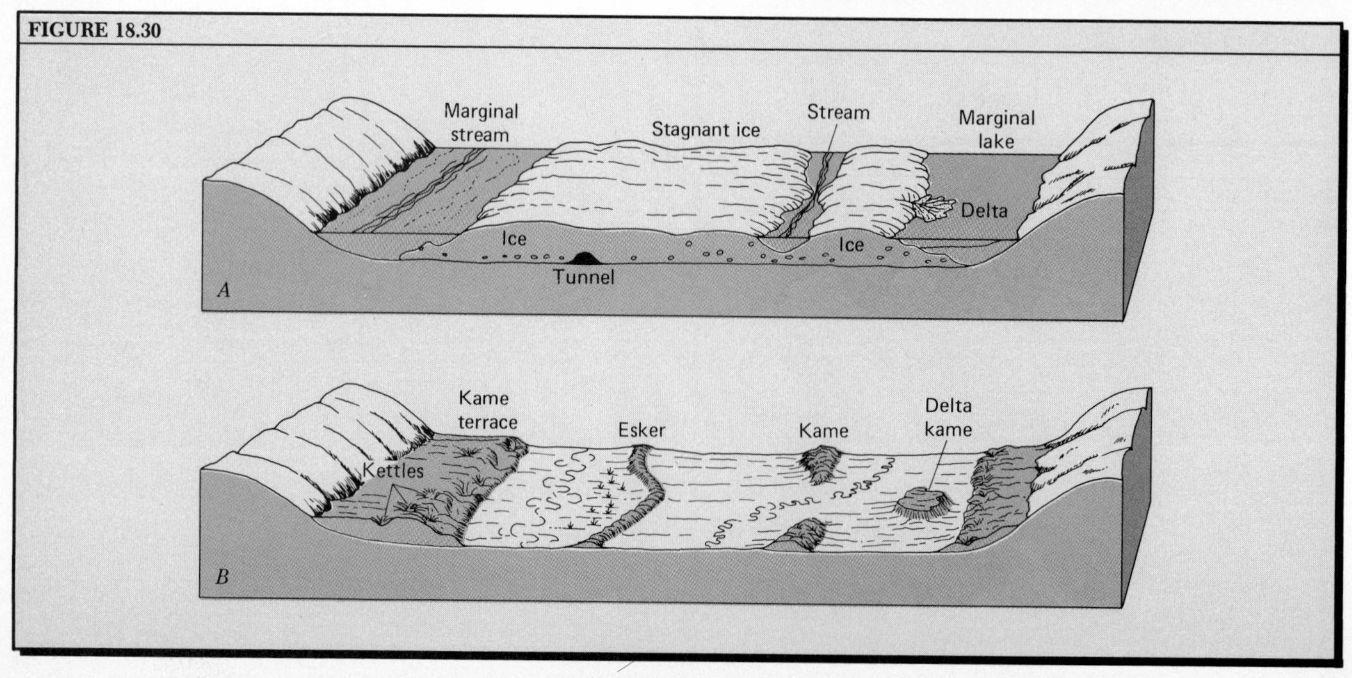

allowing it to drain west along the ice front into Lake Chicago.

There followed an extensive ice recession interval (not shown), followed by a readvance. At about −13,000 y., the Erie basin was occupied by Lake Whittlesey (lower than its predecessor, Lake Maumee), which drained into Lake Saginaw, situated in what is now Saginaw Bay of Lake Huron (Map *C*). Lake Saginaw in turn drained into Lake Chicago. Then followed a major ice recession, not shown in the map series. Much of the area of the present lake basins was occupied by water and probably drained eastward into the St. Lawrence or Lake Champlain Estuaries.

Again the ice advanced. This was the Valderan advance of about −12,000 y., but it did not reach as far as previous advances. (Though the name Valderan has recently been changed to Greatlakean, we will continue to use it here.) Conditions shown in Map *C* were essentially resumed at the time of maximum Valderan advance. Final recession of the ice front now was under way. Map *D* shows conditions later in Valderan

FIGURE 18.32 Selected stages in the development of the Great Lakes. (After J. L. Hough, 1958, *Geology of the Great Lakes*, Univ. of Illinois Press, Urbana, pp. 284–296. © 1958 by the Board of Trustees of the University of Illinois.)

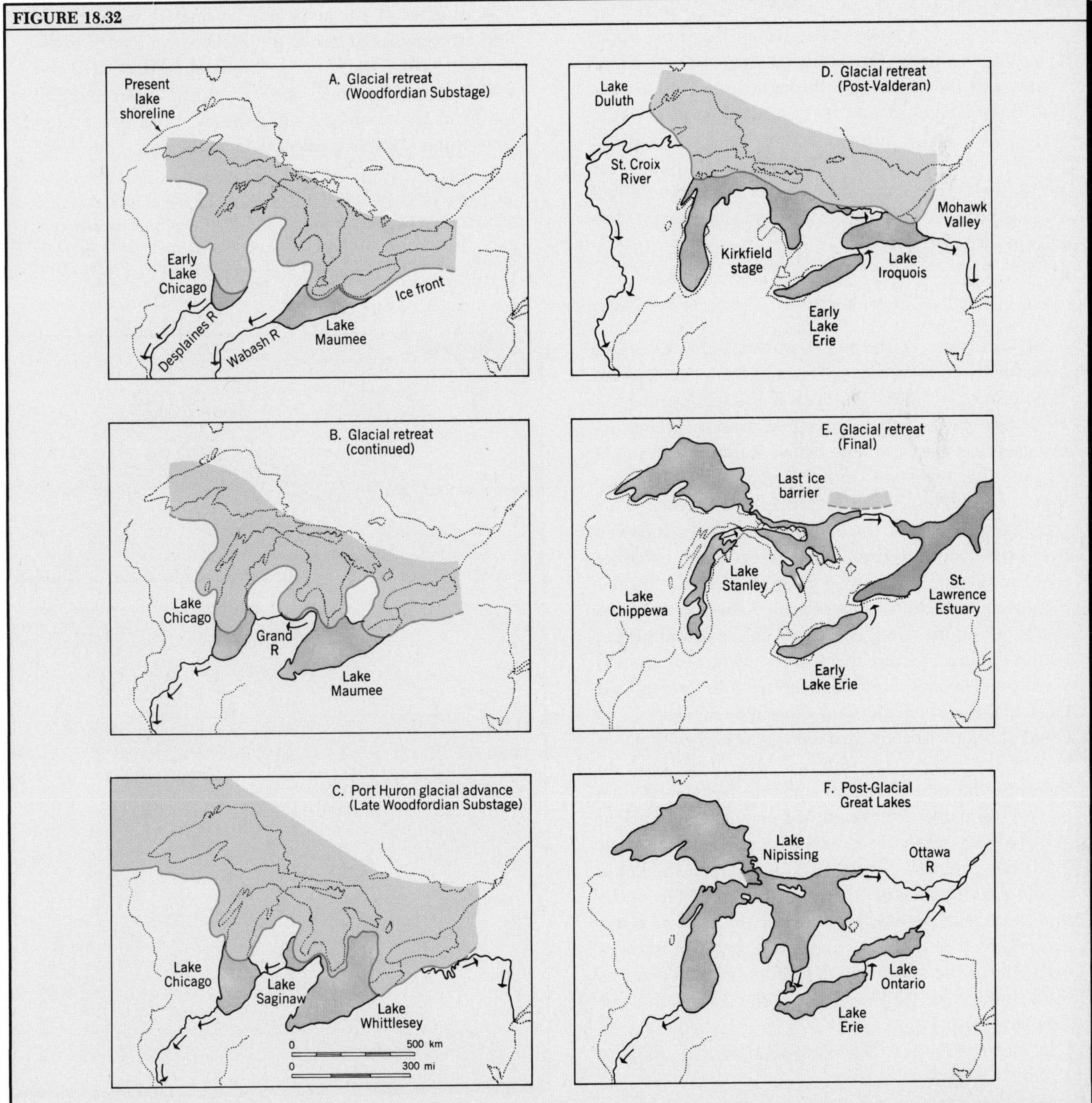

time. Ice had receded from the western end of the Lake Superior basin, allowing the formation of Lake Duluth, which drained south by way of the St. Croix River. Lakes of the Michigan, Huron, Erie, and Ontario basins were now connected, and drainage was eastward through the Mohawk Valley. Note that Huron drained directly into Ontario by means of a channel along the ice front.

At about −10,000 y., or somewhat earlier, ice recession had opened a northern outlet channel from Georgian Bay of Lake Huron, allowing eastward drainage by way of Lake Nipissing and the Ottawa River into the St. Lawrence Estuary (Map *E*). The opening of this channel initiated an extremely low stage of lake levels, but this was reversed as crustal upwarping raised the northern outlet. At about −4,000 y., as shown in Map *F*, lake levels had risen again, fusing the three upper lakes into one body (the Nipissing Great Lakes), which discharged simultaneously through three outlets—one into Lake Erie, a second by reoccupation of the Chicago outlet, and a third by way of the Ottawa River. Subsequently, because of increased crustal uplift in the north, the northern outlet by way of the Ottawa River was abandoned. Lake levels fell, and the Chicago outlet was also abandoned.

Ice Sheets and Isostasy

We took note of the fact that the bedrock surface beneath the Greenland and Antarctic ice sheets has been depressed under the load of the ice layer. Actually, some parts of the continental crustal surface under each ice sheet are now below sea level. We might suppose that the same kind of isostatic crustal depression occurred under the Pleistocene ice sheets. If so, the disappearance of the ice sheets would be followed by isostatic crustal rise. There is excellent evidence that crustal uplift is now in progress over the former European and North American ice sheet centers. At numerous points along the coasts of the Scandinavian peninsula and Finland there are ancient beaches and wave-cut benches now high above the present sea level. Some of these elevated shoreline markers can be dated through various artifacts associated with them and by the radiocarbon method (described later). In this way, the rate of crustal uplift in postglacial time can be estimated. Precise measurements of elevations of surveying reference points also show clearly that uplift is in progress. Figure 18.33 is a map of the Baltic region showing lines of equal rates of uplift. The maximum rate is in the northern end of the Gulf of Bothnia, where it is about 1 m per century. Not surprisingly, this is the approximate location of the center of the Scandinavian Ice Sheet, where the ice was thickest (compare with Figure 18.21).

In North America, the postglacial crustal uplift is nicely documented by a great succession of uplifted beach ridges bordering the present coastlines of Hudson Bay and the Arctic Ocean (Figure 18.34). Each beach ridge consists of boulders rearranged by breaking ocean waves into a low embankment paralleling the water line. The Laurentide Ice Sheet, which was centered about over Hudson Bay, disappeared between −6000 and −8000 y. The crust in this area has risen 70 to 100 meters since the ice disappeared, for an average rate of uplift on the order of 1 m per century. This rate compares favorably with the maximum rate of uplift in the Baltic region.

*Pluvial Lakes of the Pleistocene

A nonglacial phenomenon correlated with glacial and interglacial stages of the Pleistocene Epoch is the rise and fall of water levels in inland lakes of arid and semiarid regions. Such lakes occupied intermountain basins of the western United States. With one exception, these lakes had no outlets to the sea. They have

FIGURE 18.33 Present rate of uplift of the Baltic region is shown here by lines of equal uplift in centimeters per century. (Based on data by B. Gutenberg, as shown in J. A. Jacobs, R. D. Russell, and J. T. Wilson, 1959, *Physics and Geology*, New York, McGraw-Hill, p. 98, Figure 4–5.)

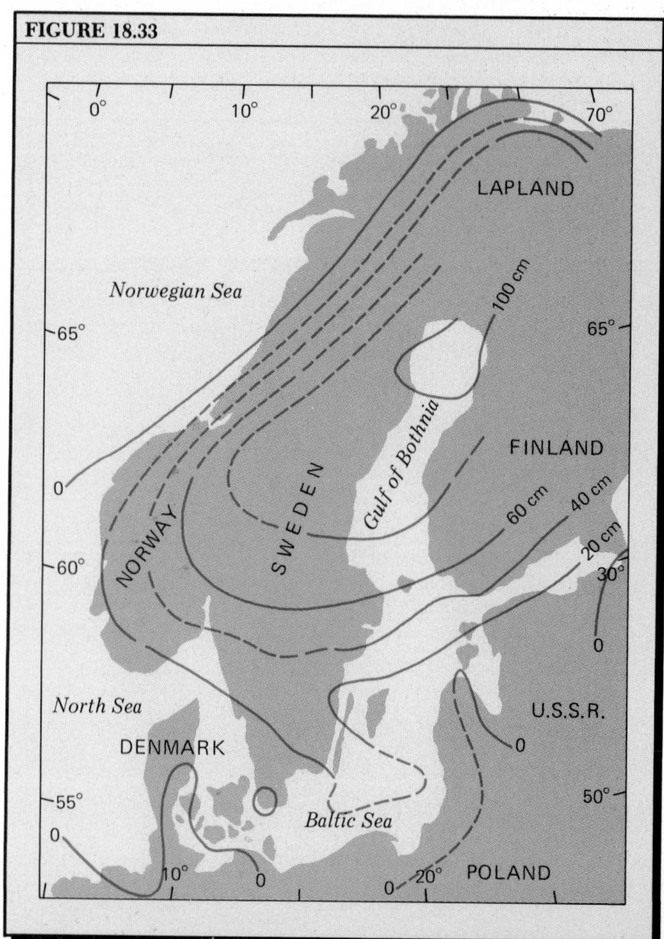

been named *pluvial lakes,* suggesting that the increase in depth and extent of the lakes was a result of increased precipitation and reduced evaporation. Figure 18.35 is a map of pluvial lakes of the western United States showing their shorelines when at the maximum stages. Today these basins contain greatly shrunken lakes or are completely dry playas (see Chapter 17). Altogether, about 120 pluvial lakes were in existence in optimum periods of the Illinoian and Wisconsinan glaciations. Many of these overflowed into neighboring lakes of lower elevation.

Largest of the western pluvial lakes was *Lake Bonneville,* of which the present-day shrunken remnant is the Great Salt Lake. At its maximum extent, Lake Bonneville was almost as large in area (52,000 sq km) as present Lake Michigan and attained a maximum depth of 330 m. The lake expanded and contracted at least ten times in the Pleistocene Epoch. Each expansion is correlated with the advance of local alpine glaciers in valleys of the neighboring Wasatch Mountains. Each time that the lake reached its maximum surface elevation and held steady, the action of breaking waves carved a notch in the mountainsides that contained the lake. A number of these former shorelines can be seen today (Figure 18.36).

FIGURE 18.34 Upraised marine shorelines bordering the shoreline of Hudson Bay. Snow bands lie in swales between beach ridges. The white area at right is the ice-covered surface of the bay. (Canadian Government Dept. of Energy, Mines, and Resources; National Air Photo Library.)

FIGURE 18.34

Radiocarbon Dating of Pleistocene Events

Prior to 1950, geologists could only make educated guesses as to the age of the glaciations and interglaciations of the Pleistocene Epoch. The radiometric methods then in use to date rock samples could not be used on the soft, sedimentary materials and organic compounds of which glacial drift is composed. Then, about in 1950, came a great breakthrough when a new radiometric method was developed by Willard F. Libby of the Institute for Nuclear Studies of the University of Chicago. His work won him a Nobel Prize.

Libby's method made use of a radioactive isotope of carbon, called *carbon-14.* (Carbon-14 is one of the three isotopes of carbon described in Chapter 2 and illustrated in Figure 2.9.) Carbon-14 originates in the earth's upper atmosphere, at levels above 16 km where atoms of ordinary nitrogen (nitrogen-14) are subject to bombardment by neutrons created by highly energetic cosmic particles (cosmic rays) penetrating the atmosphere from outer space. Upon being struck, an atom of nitrogen-14 absorbs the impacting neutron and emits a proton. The nitrogen atom is thus transformed into carbon-14, which quickly combines with oxygen to form carbon dioxide (CO_2). Carbon-14 is radioactive and decays back to nitrogen-14. The half-life of carbon-14 is 5730 ± 40 years. (Refer to Chapter 7 for explanation of radioactive decay and half-life, and methods of radiometric age determination.)

The rate of production of carbon-14 in the upper atmosphere is assumed to be constant. Therefore atmospheric carbon dioxide that is taken up by plants and animals will contain a fixed proportion of carbon-14 relative to the total amount of ordinary carbon (carbon-12). From an initial point in time marked by the death of the organism, the proportion of carbon-14 in the organic structure declines steadily, following the exponential curve of decline. By making precision measurements of the extremely small amounts of carbon-14 in a sample of organic matter, the age in years of that matter can be estimated to within a fairly small percentage of error. The very short half-life of carbon-14 makes it an excellent tool for age determinations in the last few tens of thousands of years. On the other hand, the uncertainty of measurement increases at such a rate that the present limit of usefulness is about −40,000 y.

Libby began by making age determinations of such materials as charcoal, shells, wood, and peat derived from archaeological sites and glacial deposits. Materials whose age was documented from other historical records served as a check upon the accuracy of the method. By 1952, Libby's laboratory had made age determinations of a large number of carefully selected samples. Other laboratories were soon set up, and the radiocarbon method became established as one of the most important research tools in geological and archaeological research.

A good example of the use of the radiocarbon method was to determine the ages of some tree trunks felled by the rapid advance of the ice sheet at a locality in Wisconsin. A forest had grown up in a mild period in late Wisconsinan time, but the advancing ice broke off the tree trunks and incorporated them into the dense red clay of a moraine. Radiocarbon analysis of the logs gave an age close to −12,000 y. This was a record of the last readvance of ice in the Wisconsinan Glaciation, before it finally disappeared from the region.

As the years passed, discrepancies began to appear in the radiocarbon dates when they were compared with dates arrived at by other means. An alternate method of dating makes use of tree rings exposed in a sample cut at right angles to the tree trunk. Each growth ring represents one calendar year, and the age of a tree is obtained by simply counting the rings. A given tree trunk shows distinctive sequences of wide and narrow rings that are controlled by variations in

FIGURE 18.35 Pluvial lakes of the western United States. Dotted lines are overflow channels. (Based on data of R. F. Flint, 1957, *Glacial and Pleistocene Geology*, John Wiley & Sons, Inc., New York, p. 227, Figure 13.2.)

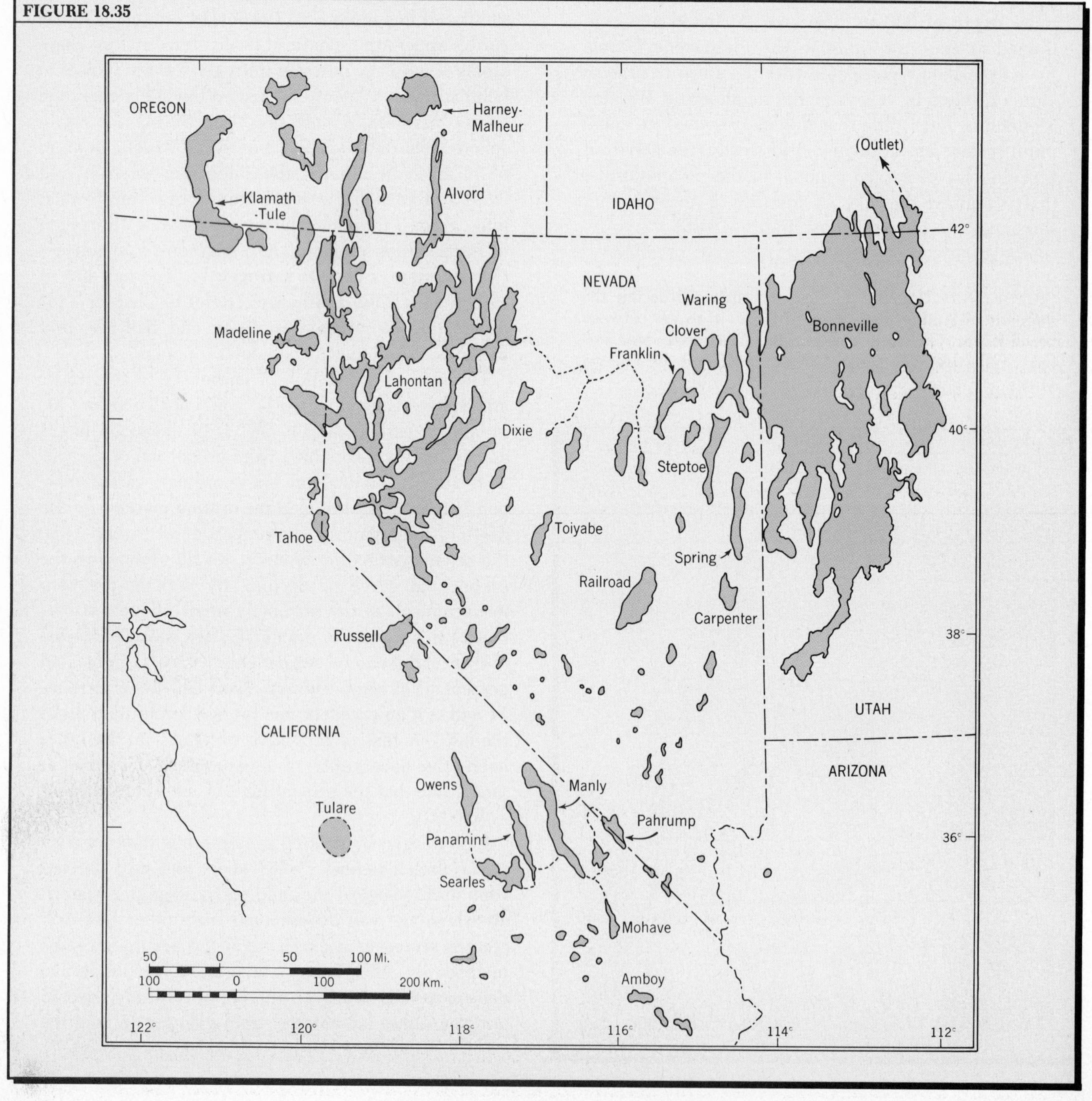

climate from year to year. Trees whose life spans overlapped in time can be correlated by matching the distinctive ring sequences. In this way, logs and timbers used in ancient dwellings can be dated accurately. The method of precise tree-ring dating is called *dendrochronology,* developed by an American astronomer, A. E. Douglass, and extended by a number of collaborators and successors. Another approach has been to count the growth rings of long-lived species of trees and compare the ring counts with carbon-14 dates of the same wood. One tree in particular, the bristlecone pine (*Pinus aristata*), has an extremely long life span, some specimens living to an age of nearly 4600 years.

By the mid-1970s, a full record of tree-ring ages had been developed to −7500 y. The record showed that the carbon-14 ages deviated from the true tree-ring ages by as much as 600 years. While the tree-ring data allow us to make excellent corrections of the carbon-14 ages back nearly 6000 years, this is only a small fraction of the 40,000-year time span over which the carbon-14 ages can be extended.

It seems likely that there have been fluctuations in the error of radiocarbon dates for as far back as the method can be used, and the probable absolute error is thought to be at least 5%.

The dating error seems to be the result of a rhythmic fluctuation in the rate at which carbon-14 is produced in the upper atmosphere. A number of causes for fluctuation have been suggested. One possible cause that interests geologists is related to changes in the strength of the earth's external magnetic field (Chapter 10). Greater strength of the magnetic field tends to reduce the rate of penetration of cosmic particles into the upper atmosphere, whereas reduced strength of the field increases such penetration. It is possible that the strength of the magnetic field fluctuates gradually in a regular time cycle. Another possibility is that magnetic polarity reversals, which seem to have occurred suddenly and would have been accompanied by a greatly weakened magnetic field, allowed a much greater rate of production of carbon-14. Two such reversals are thought to have occurred during the past 35,000 years but their effects on the production of carbon-14 cannot be judged at this time.

The Pleistocene Record on the Ocean Floors

We have seen that the record of Pleistocene events as read from glacial deposits on the continents is, at best, fragmentary. Weathering and erosion have altered or partially removed the older glacial deposits, while the deposits of successive ice advances have partly or largely covered those of earlier advances. So we turn to the one environment in which a continuous and undisturbed depositional record of the Pleistocene can be found—the floors of the deep oceans. It is from the evidence of deep-sea sediment cores that a series of cold and warm episodes can be identified, dated, and perhaps correlated with glaciations and interglaciations. Determination of the age of the Pliocene–Pleistocene boundary seems to have been finally settled from the evidence of the deep-sea cores, bringing to a close a long period of speculation in which a wide range of ages had been offered.

To be successful, a program of Pleistocene research based on deep-sea sediment must develop two systems

FIGURE 18.36 The horizontal step seen at the mountain base is a wave-cut bench with associated gravel deposits. It represents a high stand of ancient Lake Bonneville, which occupied the Great Salt Lake basin during the Pleistocene Epoch. (Hal Rumel, Utah Travel Council.)

FIGURE 18.36

of information: (1) a system of establishing the absolute age of the sediment, and (2) a system of climate indicators within the sediment. Let us look into each of the information systems.

Absolute age of a sediment sample can be estimated by radiometric methods, within the time limits to which they apply. The radiocarbon method, as we have seen, has a limit of about −40,000 y., although recent refinements may push this limit somewhat farther back. Even so, the radiocarbon ages go no farther back than the middle of the last glaciation—the Wisconsinan. Other radioisotope methods are available; they include use of thorium–protactinium (Th^{230}–Pa^{231}) and Th^{230}–Th^{232} ratios. These methods can carry the record back over a range of several hundred thousand years, but errors of unknown magnitude may be introduced by contamination of the sediment samples.

Paleomagnetic age determination has been the mainstay of Pleistocene research based on ocean sediment cores. In Chapter 10, we described the establishment of a continuous chronology of magnetic polarity reversals dating back as far as late Cretaceous time. Looking back at Figure 10.15, you can see that two magnetic epochs—the Brunhes normal and the Matuyama reversed—make up the record back to 2.5 million years ago (−2.5 m.y.). These magnetic epochs, then, encompass the full range of the Holocene and the Pleistocene. For a given deep-sea core, the points at which magnetic polarity reversal occur serve as rather precise fixes on age at those points. Between the reversal points, age is estimated in proportion to depth in the core and is subject to some error.

Interpreting Glacial Cycles from Foraminifera

Cores that are suitable for determining Pleistocene chronology consist of biogenic–pelagic and pelagic–detrital sediments which have accumulated very slowly in locations far from the continents and free from disturbance by bottom currents (see Chapter 9). The materials in these cores consist partly of terrigenous matter that has settled out from the overlying water body, and partly of the tests (hard parts) of planktonic microorganisms that lived in the near-surface zone of the overlying ocean.

In the case of cores taken in low and middle latitudes, the microfossils are almost entirely foraminifera with calcareous tests. The tests of these organisms are coarser than about 76 microns diameter, whereas the terrigenous mineral particles are smaller. Consequently, a small sample of the core can be washed on a sieve to remove the fine mineral particles, leaving only the tests for study. Two methods of using foraminifera have been followed in attempts to interpret water temperatures in the near-surface zone. The inferred seawater temperatures, in turn, are correlated with periods of colder and warmer atmospheric temperatures thought to be associated with glaciations and interglaciations.

Under the first method, percentages of the various species of foraminifera present in a core sample are determined by counting. From this information, it can be decided whether the plankton that lived in the surface water over the site of the core belonged to a cold-water or a warm-water fauna. A cold-water fauna is assumed to be associated with a glaciation; a warm-water fauna, with an interglaciation. For the waters in which these organisms lived, a sea-surface temperature of 21 °C is considered "cold"; while 28 °C is considered "warm." A total range of about 5 to 6 C° is a reasonable maximum sea-surface temperature range from the coldest to the warmest periods during the Pleistocene Epoch.

Data of several cores must be averaged, because we expect that there were many local variations due to controls other than sea-surface temperature. Figure 18.37 shows a generalized curve of water temperature interpreted in 1968 by scientists of the Lamont–Doherty Geological Observatory of Columbia University, using deep-sea cores from the equatorial and tropical zones of the Atlantic Ocean. In this case, three foraminifera subspecies of the species *Globorotalia*

FIGURE 18.37 Generalized curve of ocean-water temperatures interpreted from data of foraminifera in deep-sea cores. (Based on data of D. B. Ericson and G. Wollin, 1968, *Science*, vol. 162, p. 233, Figure 7.)

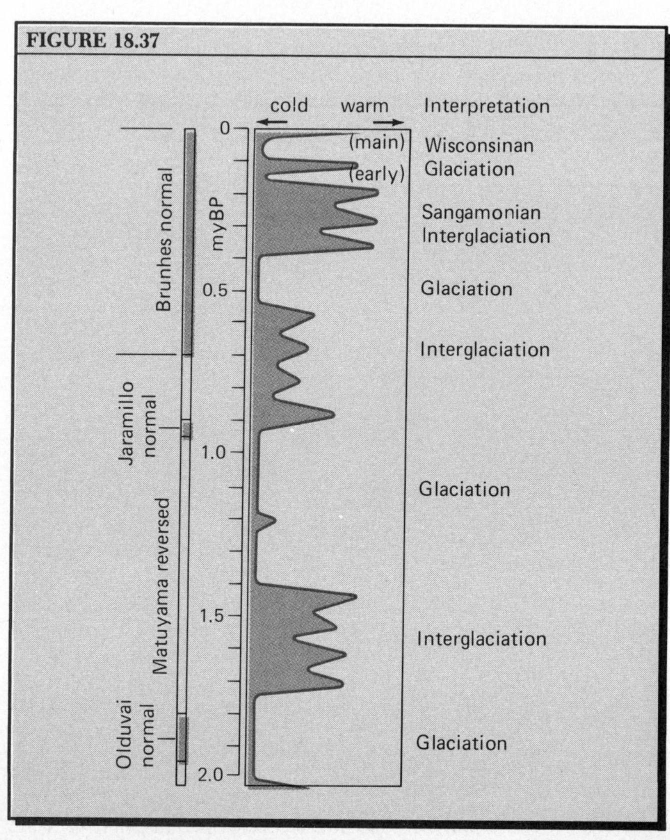

menardii were counted as a group in relation to the total plankton content of the sample. A low count was interpreted as indicating cold water and possible glaciation, a high count as indicating warm water and a possible interglaciation. Also shown are the magnetic polarity scale and the corresponding age scale in millions of years before present.

A second method makes use of a single species of foraminifera in a rather remarkable way. The test of this species is coiled about an axis (Figure 18.38). Some of the tests show a left-hand direction of coiling and others a right-hand direction. It has been established that left-coiling tests are dominant in periods of cold water, while right-coiling tests are dominant in periods of warm water. Data for use in the two methods of foraminifera analysis can be gathered from the same deep-sea core and should show a good degree of correspondence within the same core.

In other core studies, the abundance of assemblages of radiolarian species or of a single radiolarian species have been used to produce a curve of sea-surface temperature. The lower curve shown in Figure 18.39 is an interpretation of summer sea-surface temperature based on a statistical analysis of assemblages of radiolarians. The data come from two cores taken at about lat. 45 °S at points located about intermediate between Africa, Australia, and Antarctica.

Interpreting Glacial Cycles from Oxygen Isotopes

Independent evidence of cycles of glaciation comes from an analysis of the ratio of abundances of isotopes of oxygen. In addition to the common form, oxygen-16, there exist two heavier oxygen isotopes, oxygen-17 and oxygen-18. In 1947, Nobel Prize chemist Harold C. Urey noted that the ratio of oxygen-18 to oxygen-16 (O^{18}/O^{16}) in ocean water depends partly upon water temperature. He then reasoned that the ratio of those isotopes in the carbonate shell matter of marine organisms should reflect the surrounding water temperature at the time that matter was secreted. Thus, changes in water temperature should be reflected in changes in the oxygen-isotope ratio. Through improvement in laboratory techniques, it became possible in ensuing years to measure very small differences in oxygen-isotope ratios and to interpret these differences.

FIGURE 18.38 A species of foraminifera called *Globorotalia truncatulinoides*. Left-coiling test is at the left; right-coiling test at the right. Enlargement about 40 times. (Allan W. H. Bé, Lamont–Doherty Geological Observatory of Columbia University.)

FIGURE 18.38

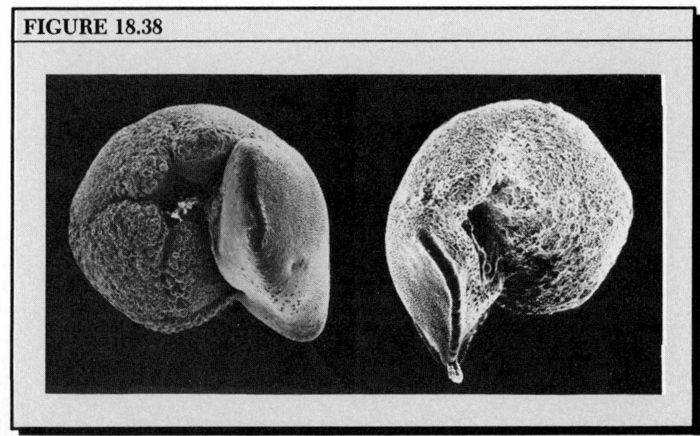

The oxygen-isotope method was applied in the late 1950s to foraminifera tests by Professor Cesare Emiliani, who had begun his research under Professor Urey's direction. Emiliani translated the scale of oxygen-isotope ratios directly into a scale of water temperatures, called *paleotemperatures*. The result of his study was to reveal a paleotemperature curve with a large number of peaks and valleys—far more than would fit the classical four established continental glaciations. Emiliani first estimated that about eight climatic cycles, each representing a glaciation and an interglaciation, occurred following the last reversal of magnetic polarity—i.e., back to about −700,000 y.

Emiliani's paleotemperature analysis came under criticism of other scientists who pointed out that water temperature is only a small factor in determining the oxygen-isotope ratio of seawater. The true significance of variations of the isotope ratio in the tests of plankton, such as the foraminifera, lies in a more complex but nevertheless quite straightforward chain of events involving the hydrologic cycle. As Professor Urey had pointed out in 1946, when water evaporates from the sea surface, there is a tendency for a larger proportion of those water molecules containing the lighter isotope of oxygen, O^{16}, to enter the vapor state than of the molecules containing the heavier isotope, O^{18}. This selective process tends to leave the remaining seawater richer in O^{18}; in other words, the ratio O^{18}/O^{16} tends to become larger. On the other hand, when the hydrologic cycle functions in a state of balance, as we described it in Chapter 15, the precipitation of water on the continents and oceans and the runoff of water from continents to oceans assures that the rate of return of lighter isotopes to the oceans as liquid water will balance the quantity leaving by evaporation. Under this state of global water balance the quantity of water stored on the lands in both liquid and ice forms will hold constant, and so will the oxygen-isotope ratio of the seawater.

Suppose, now, that the total quantity of glacier ice begins to increase because of a climate change that favors the expansion of glaciers. As ice volume grows, more of the lighter oxygen isotopes are withheld in the ice and prevented from entering the return flow of the hydrologic cycle, resulting in an increase in the value of the ratio O^{18}/O^{16} of seawater. On the other hand, when a climate change results in a decrease in the total quantity of glacier ice, more of the lighter isotopes are released as runoff, and the isotope ratio of seawater becomes smaller.

The oxygen-isotope curve determined from carbonate matter in deep-ocean cores is now regarded as a reliable indicator of the total volume of glacier ice present on the earth at the time the plankton secreted their tests. For this reason, the isotope-ratio curve can be called a *paleoglaciation curve*. Low points on the curve are associated with glaciations, while the high points are associated with interglaciations.

The oxygen-isotope method has been applied to many deep-sea cores, and the result has been a recognition of a succession of *isotope-ratio stages* extending back as far as about −2 m.y., and perhaps even to −3 m.y. Stages within the 2-m.y. time span are numbered from 1 to 41. An example is an isotope-ratio curve, or paleoglaciation curve, published in 1976 by a team of scientists representing several universities (Figure 18.39). It covers 500,000 years and is based on two deep-sea cores from the subantarctic ocean in a location between Africa, Australia, and Antarctica. The record shows the first 13 isotope stages. If you smooth out the smaller wiggles in both curves, perhaps you will conclude that there were five glaciations in the 450,000 years of record, for an average of about one glaciation every 90,000 years.

It now appears that glaciations may have been occurring as far back as −3 m.y., or well into the Pliocene Epoch, and that the total number of Cenozoic glaciations may have numbered more than 30, spaced at intervals of about 90,000 years.

*Evidence of Earlier Cenozoic Glacial Ice

Deep-sea cores carry yet another kind of evidence of the presence or absence of great ice sheets on the continents. When a large ice sheet is present, sediment is carried out over the adjacent deep ocean by icebergs, which melt and allow the mineral fragments they carry to be dropped to the ocean floor. In the 1960s, cores taken in the southern Indian Ocean showed the presence of glacial-marine sediments as old as −4 m.y., and it was then realized that the Antarctic continent bore an ice sheet long before the Pleistocene Epoch began. In 1973, drill cores taken from the *Glomar Challenger* near the Antarctic coast were found to contain ice-rafted materials dated as about −20 m.y., which is early in the Miocene Epoch. More recently, ice-rafted sediments in cores in that area have been dated at about −38 m.y., which is at the start of the Oligocene Epoch. These earliest ice-rafted materials may have come from alpine glaciers on Antarctica, rather than from an ice sheet. It has been guessed that the Antarctic Ice Sheet began to build up between −11 and −14 m.y. and reached a large size by about −4 to −5 m.y. The Greenland Ice Sheet may have built up to its approximate present size by about −3.5 m.y. During the Miocene Epoch, the Antarctic Ice Sheet apparently underwent many changes in volume and may even have largely disappeared in periods of exceptionally mild climate. As we shall see, drastic changes in the volume of ice held on Antarctica would have caused large swings in worldwide sea level, alternately inundating and exposing low coastal areas of all continents.

Pleistocene Changes of Sea Level

One of the most important phenomena of the Pleistocene and Holocene Epochs has been the changing of sea level with respect to the continents. These sea-level

FIGURE 18.39 Curves of oxygen-isotope ratios (*above*) and inferred summer sea-surface temperatures (*below*) from two deep-sea cores in the subantarctic ocean. (Based on data of J. D. Hays, J. Imbrie, and N. J. Shackleton, 1976, *Science*, vol. 194, p. 1130, Figure 9.)

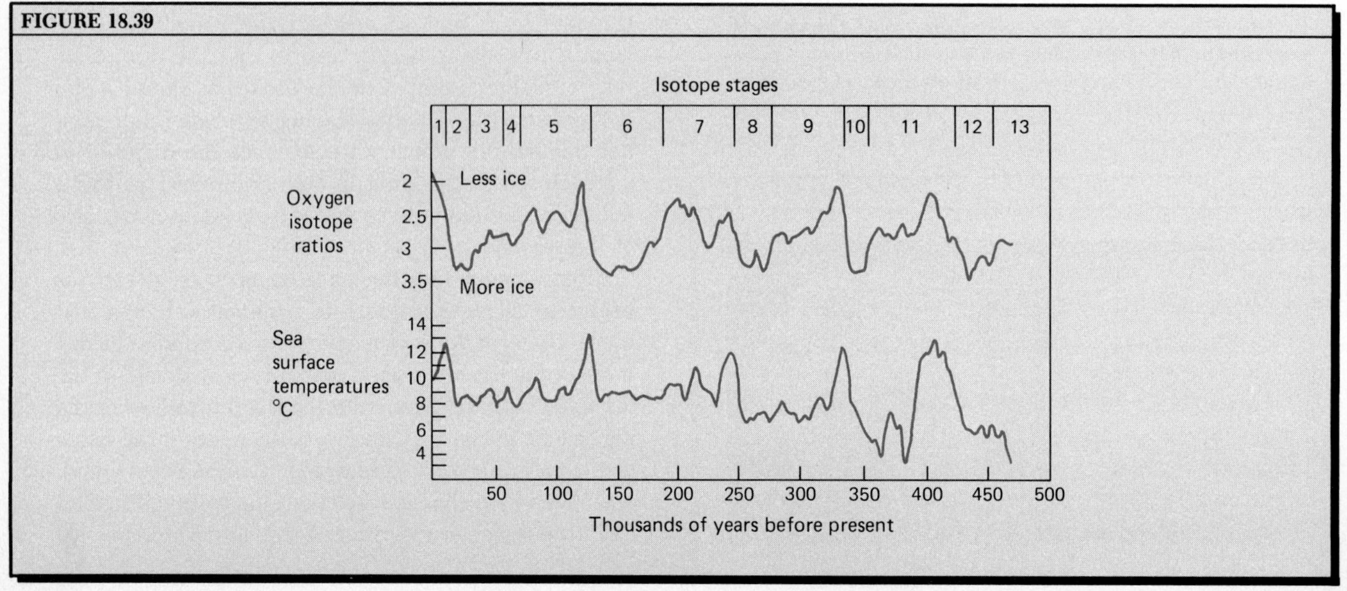

changes caused the alternate exposure and inundation of continental margins and also had many secondary effects relating to denudation of the continents and sedimentation upon the ocean floors.

The changes of sea level we are concerned with here are eustatic changes. In Chapter 12, we considered the possibility that changes in rates of seafloor spreading might have been responsible for positive and negative eustatic changes of sea level. That scenario involves a change in the volume of the ocean basins—that is, a change in their capacity to hold water. In the case of glaciations and interglaciations, we are dealing with changes in the volume of the ocean water itself. These changes result from the accumulation of ocean water in storage on land as glacial ice, followed by its release to the world ocean by melting.

The amounts of sea-level change from this cause can be calculated from known volumes of existing ice masses and from estimates of volumes of former ice sheets. The Antarctic Ice Sheet alone holds sufficient water volume to provide a sea level rise of about 60 m, should all of that ice be melted. Assuming that the added load of this water upon the oceanic crust caused an isostatic downwarping of 20 m, the net sea level rise would still be about 40 m. Estimates of the total sea level rise that would accompany the melting of all existing glacial ice run to 60 m, or somewhat higher. The disastrous effect of a 60-m rise of sea level upon the heavily populated lowlands of the Atlantic and Gulf coastal plains has been a favorite theme of journalists and writers, and can scarcely have escaped your attention. Of more interest, scientifically, is the very real lowering of sea level that accompanied each glaciation as enormous ice sheets spread over the continents in middle latitudes.

The records of sea level lowering in the last major advance of ice sheets in the Wisconsinan Glaciation and its subsequent rapid rise in the Holocene Epoch are quite well documented. Evidence consists of radiocarbon dates of such materials as salt-marsh peat, oyster shells, coral rock, and lithified beachrock (carbonate-cemented beach materials). At about −18,000 y., when the ice was at its greatest extent, sea level was lower than present by perhaps 60 to 80 m. On the Atlantic seaboard of the United States, this lowering of sea level exposed a broad area of the continental shelf, while the ocean shoreline receded far to the east. Peat samples dredged from the bottom sediments show that a forest of fir, spruce, pine, and oak existed at that time. Teeth of Pleistocene elephants (mastodons and mammoths) have also been dredged up from these sediments.

Carrying back the curve of sea-level changes to earlier glaciations and interglaciations has met with many difficulties for two reasons. First is that the radiocarbon method has been limited to about −40,000 y. Second, rising and sinking of the crust by epeirogenic and tectonic movements may have greatly complicated the record. Despite these difficulties, several investigators have drawn rough curves of eustatic sea level changes through the Pleistocene Epoch. A conservative opinion would perhaps be that sea level lowering during earlier glaciations was probably not much greater than that which occurred during the maximum Wisconsinan Glaciation—namely, on the order of −50 to −100 m.

If the oxygen-isotope curve gives a direct measurement of global ice volume, as we have stated, the same curve should indicate the eustatic fall and rise of sea level in terms of both timing and general magnitude of the changes.

One important effect of the lowering of sea level during glaciations was on the major rivers. Their mouths were extended out across the exposed coastal shelves. They became trenched into these shelves and carried their loads to the heads of submarine canyons, which in turn fed debris into deep-sea cones. Rivers that drained areas affected by glaciation carried great volumes of glacially eroded debris to the oceans. A particularly good example was the Hudson River. During lowered sea level, it scoured a channel 25 to 40 m below the floor of the exposed continental shelf. This channel, which lies southeast of New York harbor, fed directly into the head of the Hudson Submarine Canyon, which was eroded deeply into the outer edge of the shelf by turbidity currents.

Changes of sea level also affected the activity of major rivers in their lower courses. During high stands of sea level in interglaciations, these streams aggraded their valleys, filling them with alluvium. During lowered sea levels of glaciation, the same streams degraded their channels, trenching the alluvium previously deposited. There is evidence that the crust was gradually rising as these eustatic reversals occurred, so many river valleys contain a succession of stepped terraces and nested alluvial fills, as shown schematically in Figure 18.40.

Causes of Ice Ages

Few geological puzzles have proved as intriguing and as difficult to solve as the cause or causes of multiple glaciations and interglaciations. Despite scientific advances on many fronts in the earth sciences and an enormous gain in our knowledge of how the earth works through plate tectonics, the debate as to what causes glaciations continues unabated in its fervor and in the wide range of ideas proposed.

We must approach this puzzle on two quite different levels. First is an overall set of conditions that seems to have favored the repeated growth and disappearance of large ice sheets on the continents over a time span as long as two to three million years. This entire cold period can be called an *ice age*. In this broad view of the problem, we need to take into account ancient ice ages, such as the early and late Paleozoic glaciations so well documented on the Gondwana continent (Chapter 13). The second level is that of the immediate, or forcing causes that are responsible for actually precipitating a glaciation at a particular point in time, for causing a deglaciation to follow, and for this cycle to be repeated. The questions are these: Why are there cycles of glaciation and deglaciation? What controls the length of these cycles? What causes them to be initiated or triggered? In answering these questions, we may find ourselves considering several quite different triggering mechanisms and favorable factors acting together to cause glaciation.

A meaningful inquiry into the causes and factors involved in glaciations and interglaciations requires a good working knowledge of atmospheric and oceanic sciences, because climatic change is the key to the glacial cycles. To understand the many theories and arguments being debated you would need to know how atmospheric processes operate and how the global circulation of both the atmosphere and oceans works. We will limit our discussion to some rather general ideas, with emphasis on the geologic factors.

Fundamental Causes of an Ice Age

Under fundamental factors tending to cause an entire ice age lasting two to three million years, we may consider the following possibilities: (1) a favorable position of the continents with respect to the polar regions; (2) a withdrawal of oceans from continental cratons accompanying widespread epeirogenic uplift; (3) a sustained period of increased volcanic activity; and (4) a sustained period of diminished intensity of solar energy reaching the earth.

Of these four fundamental factors, the first three are geologic in nature. As Alfred Wegener realized, the late Paleozoic ice age saw a great sheet on the continent of Gondwana. The continent itself was located at polar latitudes and included the south pole, which paleomagnetic evidence confirms. Recall from Chapter 13 that glaciation also occurred in western Africa in late Ordovician time, when the pole was located in that area (see Figure 13.14). By late Paleozoic time, the pole path was located in what was then the single continental nucleus of southern Africa, southern South America, Antarctica, India, and Australia. This ice age ended when breakup of Gondwana was followed by the movement of most of its fragments to lower latitudes. Only Antarctica ended up in the south polar position.

Look next at the north polar region in late Paleozoic time, as shown in Figure 13.15*A*. In Permian time, only the northern tip of the Eurasian continent projected into the polar zone. As the continental plates rifted apart and the Atlantic basin opened up, North America moved westward and poleward to a position opposite Eurasia, while Greenland took up a position between North America and Europe. The effect of these plate motions was to bring an enormous landmass area to a high latitude and to surround a polar ocean with land.

The reasons why a large land area at high latitude favors the onset of an ice age are fairly simple. First, the landmasses on which ice sheets can grow become located in a cold climate zone where the rate of snowfall can greatly exceed ablation. Second and perhaps less obvious, the presence of the landmasses blocks the poleward flow of warm ocean currents that might otherwise tend to make the polar climate relatively mild. At present, the polar sea (the Arctic Ocean) is connected with the Atlantic and Pacific oceans only by narrow straits. The polar ocean, with its year-round cover of sea ice, maintains a cold climate, which extends well into the fringes of the bordering landmasses. To summarize, the most important single factor in making an ice age possible is a geologic event—the arrival of a large expanse of continental surface at

FIGURE 18.40 A highly diagrammatic representation of stepped terraces and nested alluvial fills resulting from glacial–eustatic sea-level fluctuations superimposed upon persistent crustal uplift. The actual deposits would be fragmentary.

FIGURE 18.40

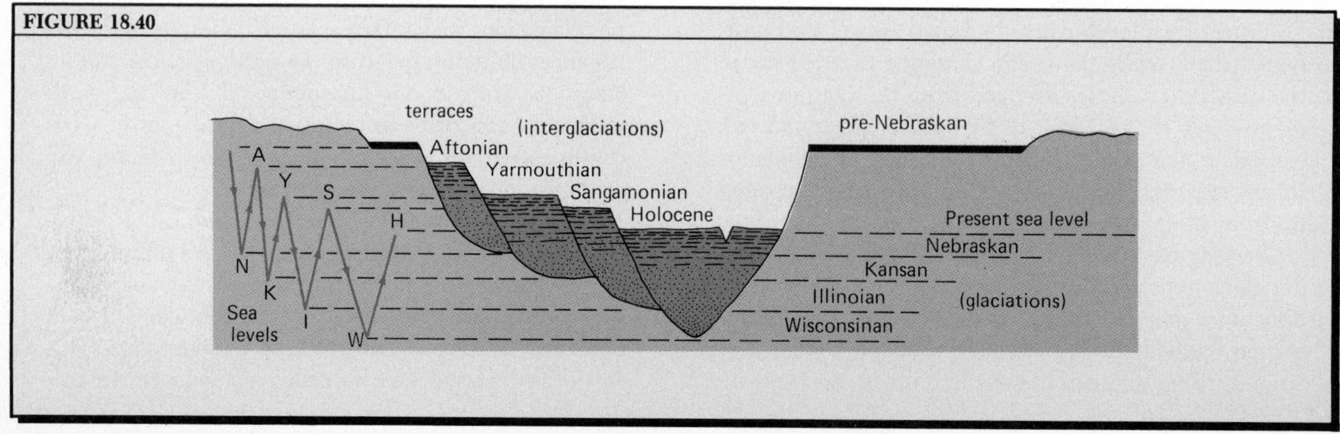

a polar position. There was, however, a long lapse of time in the late Cenozoic Era between the time the landmasses surrounded the north pole and the time the first ice sheets began to form. Clearly, other factors brought on the start of the ice sheet growth.

The second factor on our list is a general withdrawal of shallow seas from the continental margins in late Cenozoic time. In late Cretaceous time—near the close of the Mesozoic Era—shallow continental seas were extensive. By Miocene time, however, the seas had retreated to the edges of the continents. This lowering of sea level relative to the continents is usually attributed to a general positive epeirogenic movement of the continental plates. On the other hand, it may have been caused by an increase in the average depth of the ocean basins as the rate of seafloor spreading decreased following a high rate of spreading in the Cretaceous Period.

Why would a relative lowering of sea level and a reduction in the surface expanse of ocean water tend to promote an ice age? The reasons are not by any means simple, because many different effects are involved. Large landmasses tend to have colder winters (but warmer summers) than adjacent oceans in the same latitude zone. An increase in surface elevation of a continent would tend to lower the average temperature near the ground surface, but the amount of uplift over continental interiors may have been too small to be of consequence.

Some geologists have argued that an increase in tectonic and volcanic activity in the late Cenozoic caused the rise of lofty mountain ranges capable of trapping large amounts of snowfall, and thus made possible the first growth of icecaps. Others argue that with more or less continuous plate subduction going on throughout the Phanerozoic Eon, there were high mountain ranges on continental borders more or less continually at one place or another. Thus, the growth of high mountains is not in itself a cause of continental glaciation.

Our third factor, also geologic, is an increase in the intensity of volcanic activity on a global basis over a sustained time span. The link between volcanic activity and glaciation is through the emission of vast amounts of extremely fine volcanic dust during explosive volcanic eruptions, particularly the eruption of composite volcanoes and caldera explosions. The dust rapidly reaches the upper atmosphere and spreads over the entire globe within months or a year or two following an eruption. The dust particles, in turn, cause a climate change that may be global in scope.

Following the eruption of the volcano Krakatoa in 1883, a veil of volcanic dust formed in the stratosphere and spread into high latitudes. In Europe and North America, the dust veil brought on a large number of extremely brilliant sunsets and attracted a great deal of attention. Solar observatories recorded a 20% drop in the intensity of solar energy reaching the earth's surface in the first year following the explosion. For each of the next three years, the reduction was about 10%. Eventually most of the dust settled out into the lower atmosphere and the blocking effect disappeared. Some atmospheric scientists agree that the climatic effect of a stratospheric dust veil is to cause a cooling of the average atmospheric temperatures near the earth's surface. If we accept this conclusion, we can then examine the geologic record with regard to the intensity of volcanic activity before and during the Pleistocene ice age.

A record of past volcanic activity can be obtained from deep-sea cores. A major volcanic eruption is recorded as a thin layer of volcanic ash consisting of minute shards of volcanic glass. Some indication of the general intensity of volcanic activity can be had by determining the number of ash layers per thousand years. Some research scientists claim the evidence points to greatly increased volcanic activity in the Pleistocene Epoch. One study of deep-sea cores led to the conclusion that volcanic activity in the Pleistocene was at a level four times higher than the average for the 20-million-year span covered by the cores. There was also a moderately high level of volcanic activity in mid-Pliocene time and a number of minor peaks in the Miocene Epoch. These and other attempts to relate volcanic activity to the cause of an ice age have been strongly disputed by other scientists.

Our fourth possible cause of an ice age is in the realm of astronomy and we shall refer to it only briefly. The hypothesis states that our sun, as it travels through space in a rotary motion with the rest of the Milky Way galaxy, occasionally passes through a cold interstellar dust cloud. As the dust is drawn into the sun's surface, it is highly heated and increases the sun's brightness. Thus, the sun's energy output is increased for a time. The effect is to increase the input of energy into the earth's atmosphere, and this may supply the added precipitation needed for ice sheets to form and grow. Perhaps, on the other hand, the increased solar input would also raise the average air temperature—an opposite effect.

*Causes of Glaciations and Interglaciations

Our second level of approach to the ice-age problem deals with cycles of glaciation and interglaciation that are known to have been repeated at least several times, and perhaps as many as 20 to 30 times. A general lowering of air temperatures on a global scale toward the close of the Cenozoic Era is fairly well documented by geologic evidence. This overall cooling is the result of one or more of the fundamental ice-age causes we have already reviewed. Reduced temperatures generally through the Pleistocene are recorded in a lowering of the snow line, which is the elevation above which snowbanks remain throughout the year on the summits of mountains. The lowering amounted to about 600 m in equatorial latitudes and about 900 to 1200 m in middle latitudes. And, of course, we have seen that the general cooling is well documented by evidence from deep-sea cores.

Here is a short list of hypotheses of cyclic glaciations currently being debated: (1) triggering by bursts of volcanic activity; (2) control by astronomical cycles of the earth's tilt and orbital motion; (3) control by changes in arctic sea-ice cover; and (4) control by changes in the light-reflecting ability (albedo) of the land and water surfaces, largely from presence or absence of snow cover.

Because the last two hypotheses require a good working knowledge of atmospheric and oceanic science (meteorology and oceanography), we shall make no attempt to explain them here. The first hypothesis can be quickly reviewed. The principle is the same as the general effect of increased volcanic dust on global air temperature. Some investigators have tried to correlate specific events of increased volcanic activity with the onset of specific glaciations. The validity of any such correlation is much in doubt.

The control of cycles of glaciations and interglaciations by astronomical cycles is a strong contender for acceptance at the present time, despite its being soundly denounced by a number of capable scientists. The *astronomical hypothesis*, as it has often been called, has been under consideration on and off for more than 40 years. It is based on facts about the motions of the earth in its orbit around the sun. These facts are well established by astronomical observations and are not the subject of debate. Despite the need to understand solar-system astronomy to appreciate these facts, we can attempt a brief description of the concepts involved.

Two factors are involved here: (1) the changing distance between earth and sun; and (2) the changing angle of tilt of the earth's axis of rotation. As to the first factor, we can simply say that the distance that separates the earth and sun at summer solstice, June 21, undergoes a cyclic variation. During a single cycle, which lasts 21,000 years, the earth–sun distance may vary from 1 to 5% greater than the average distance to 1 to 5% less than the average distance. As to the second factor, we need to know that the earth's axis is now inclined by an angle of 23.5° from a line perpendicular to the sun's rays. This axial tilt results in the annual cycle of winter and summer seasons in middle and high latitudes. The axial tilt undergoes a 40,000-year cycle of change in which the tilt angle may be increased to as much as about 24° and decreased to as little as about 22°.

If we pick a point on the earth at, say, latitude 65° N, we can calculate the total cycle of change in the intensity of incoming solar radiation, or *insolation*, on a day in midsummer (June 21). This cycle of incoming solar radiation combines the cycle of changing earth–sun distance and the cycle of axial tilt. The intensity of insolation is measured in units known as *langleys*. (One langley is defined as one gram calorie per square centimeter.) Figure 18.41 is a graph showing the cyclic changes in summer daily insolation at latitude 65° N. This graph has been named the *Milankovitch curve*, after the astronomer who calculated the variations and published the results in 1938; he proposed that these cycles of insolation have controlled the glacial cycles. In its simplest form, the Milankovitch hypothesis tells us that a northern-hemisphere glaciation should coincide with each major low point in the insolation curve, while an interglaciation should coincide with each major high point. Actually, the glacial events might lag behind the insolation peaks and valleys. The greatest variations in one cycle represent a difference of about 5% in the total quantity of daily insolation at a given latitude. The capability of this amount of variation to produce the glacial cycle is a subject of continued debate.

The Milankovitch curve was taken up again in the mid-1960s by Professor Wallace S. Broecker, a geochemist, and his collaborators at the Lamont–Doherty Geological Observatory of Columbia University. They noted that the high insolation peaks represent a cycle averaging about 80,000 to 90,000 years. Each high insolation peak was thought to be associated with a rapid termination, i.e., a deglaciation. During intervening long periods of lower insolation, ice sheets would grow in volume. These investigators claimed to have found close correlations between the insolation cycle and cycles of both sea-level changes and oxygen-isotope ratios over the past 140,000 years. As you might expect, Broecker's interpretations had both supporters and detractors.

About ten years later, the insolation hypothesis based on the Milankovitch curve was revived by a team of scientists working on major global climate changes and their causes. Using oxygen-isotope ratios and percentages of indicator microorganisms in two deep-sea cores, these scientists found strong correlations between global climatic changes and the insolation cycle. So impressive did they find this evidence that they announced they were virtually certain the insolation cycle is responsible for glaciations and interglaciations. Of course, their data were strongly challenged, and their opponents have declared with equal assurance that no such relationships can be derived from these data.

In any case, there remains unanswered the mechanism by means of which the insolation changes actually cause ice sheets to grow or to disappear. Perhaps the Antarctic Ice Sheet acts as the intermediate control mechanism for northern-hemisphere glaciations. It has been suggested that a warm cycle in the southern hemisphere might set off a great surge of the Antarctic Ice Sheet, which would then spread over a large surrounding area of sea surface as an ice shelf. The global climate change induced by this region of floating shelf ice might trigger the growth of ice sheets in the northern hemisphere, which would be simultaneously subjected to a period of reduced insolation.

We bring to a close this overview of the causes of ice ages and glaciations leaving unmentioned a number of interesting hypotheses. The entire subject is so complex as to be difficult for even a research scientist to grasp fully. Interactions between the atmosphere, the oceans, and the continental surfaces (including the ice sheets) are numerous and closely interrelated. Many threads of cause and effect cross and recross these earth realms. Changes that occur in one realm are fed back to the other realms in a most complex manner. Modern research into the causes of ice ages and glaciations is an exciting area of the earth sciences, but it is an arena we would not want to recommend to timid souls! If you choose to enter this arena, be prepared to be bombarded from all sides by some very tough-minded critics.

FIGURE 18.41 The Milankovitch curve. It shows fluctuations in summer daily insolation at latitude 65° N over the past 500,000 years. The calculations were made by A. D. Vernekar, 1968. The zero value represents the present solstice insolation for that latitude. (Based on data of W. Broecker and J. van Donk, 1970, *Reviews of Geophysics*, vol. 8, p. 190, Figure 10.)

*Holocene Climate Cycles

The elapsed time span of about 10,000 years since the Wisconsinan Glaciation ended makes up the Holocene Epoch; it began with the rapid warming of ocean surface temperatures. Continental climate zones shifted rapidly poleward. Soil-forming processes began to act upon new parent matter of glacial deposits in midlatitudes. Plants became reestablished in glaciated areas in a succession of climate stages. The first of these is known as the Boreal stage. Boreal refers to the present subarctic region where needleleaf forests dominate the vegetation. The history of climate and vegetation throughout the Holocene has been interpreted through a study of spores and pollens found in layered order from bottom to top in postglacial bogs. (This study is called palynology.) Plants can be identified, and ages of samples can be determined. A dominant tree was spruce. Interpretation of pollens indicates that the Boreal stage in midlatitudes had a vegetation similar to that now found in the region of boreal forest climate.

There followed a general warming of climate until the Atlantic climatic stage was reached about −8000 y. Lasting for about 3000 years, the Atlantic stage had average air temperatures somewhat higher than those today—perhaps on the order of 2.5 C° higher. We call such a period a climatic optimum with reference to the midlatitude zone of North America and Europe. There followed a period of temperatures below average, the Subboreal climatic stage, in which alpine glaciers showed a period of readvance. In this stage, which spanned the age range −5000 to −2000

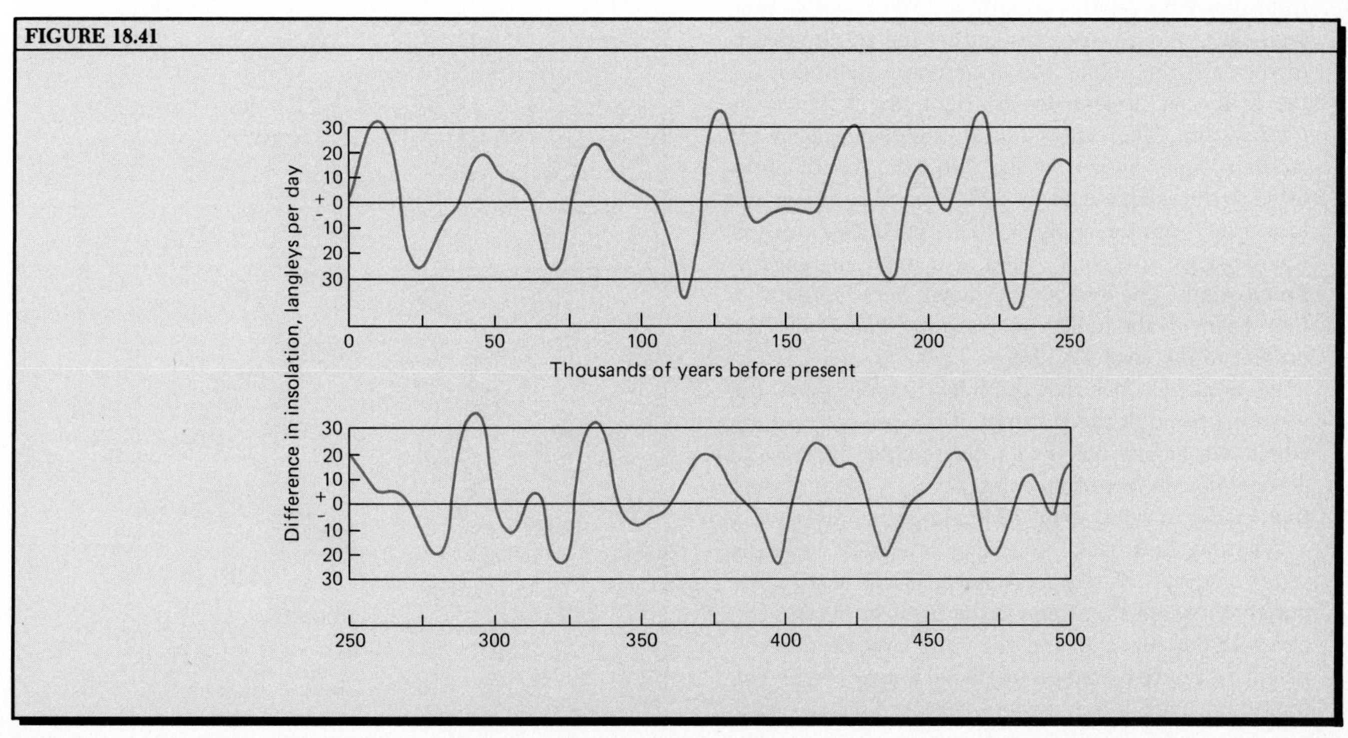

y., sea level—drawn far down from the level during glaciation—had returned to a position close to that of the present, and coastal submergence of the continents was largely completed.

The past 2000 years, from the time of Christ to the present, show climatic cycles on a finer scale than those we have described as Holocene climatic stages. This refinement in detail of climatic fluctuations is a consequence of the availability of historical records and of more detailed evidence generally. A secondary climatic optimum occurred in the period A.D. 1000 to 1200 (−1000 to −800 y.). This warm episode was followed by the Little Ice Age (A.D.1450–1850; −550 to −150 y.). During this time, valley glaciers made new advances to lower levels. In the process, the ice overrode nearby forests and thus left a mark of its maximum extent.

An independent record of climate fluctuations through the Holocene Epoch is available from oxygen-isotope ratios measured in ice layers of the Greenland and Antarctic ice sheets. Air temperature at the time of ice formation influences the O^{18}/O^{16} ratio. A lowering of air temperature results in a decreasing proportion of O^{18} in the ice molecules. Consequently, a given snow or ice layer carries with it a permanent record of average atmospheric temperature prevailing during the year in which it was formed. It is not possible to assign a specific temperature value to a particular isotope ratio, but temperature fluctuations from warmer to colder periods and vice versa can be readily recognized from a succession of samples taken at intervals along an ice core.

Figure 18.42 shows ice-core data derived from the Greenland Ice Sheet in 1966. The bar graph at the right shows the relative amount of O^{18} present in each segment of the ice core. The farther the graph projects to the left, the colder the atmospheric temperature; the farther it projects to the right, the warmer the temperature. The center line represents an arbitrary median value. Shown at the left is a smooth curve fitted to the isotope data by a mathematical procedure known as "Fourier analysis." This particular analysis was designed to reveal cycles of a medium range of time periods. The total record shown here represents a time span of about 800 years, going back from the present to the year A.D.1200.

We notice at once that the smooth curve shows the warming trend of the first half of the present century, which can be attributed to a documented increase in carbon dioxide from fuel combustion. A cooling trend that has set in since about 1940 also shows clearly.

When we look back into earlier centuries, we find a series of cycles of temperature variation of approximately the same amplitude as the latest variations. Cycles with periods of 78 and 181 years have been recognized by Fourier analysis of these isotope data. The important point is that air-temperature fluctuations of the past 800 years are of the same order of magnitude that we observe in our century. In other words, the recent trends of warming and cooling are within the normal range found throughout the record. What some scientists have interpreted as human-made changes in global atmospheric temperature may, in fact, be caused by natural variations in the earth's radiation balance.

Glaciers and Glaciations in Review

This chapter has covered three major glacier topics. First is the alpine glacier. Sculptors of high mountain scenery, alpine glaciers have given us national parks of great beauty in the Rocky Mountains and the Sierra Nevada, and they are the shapers of the spectacular fiord coasts. Living alpine glaciers are abundant in Alaska; some are very close to coastal cities.

The second of our glacial topics is the great ice sheets existing on Greenland and Antarctica. Holding

FIGURE 18.42 Variations in oxygen-isotope ratio within an ice core from the Greenland Ice Sheet. (From S. J. Johnsen et al., 1970, *Nature*, vol. 227, p. 483, Figure 1. Copyright by Macmillan (Journals) Limited.)

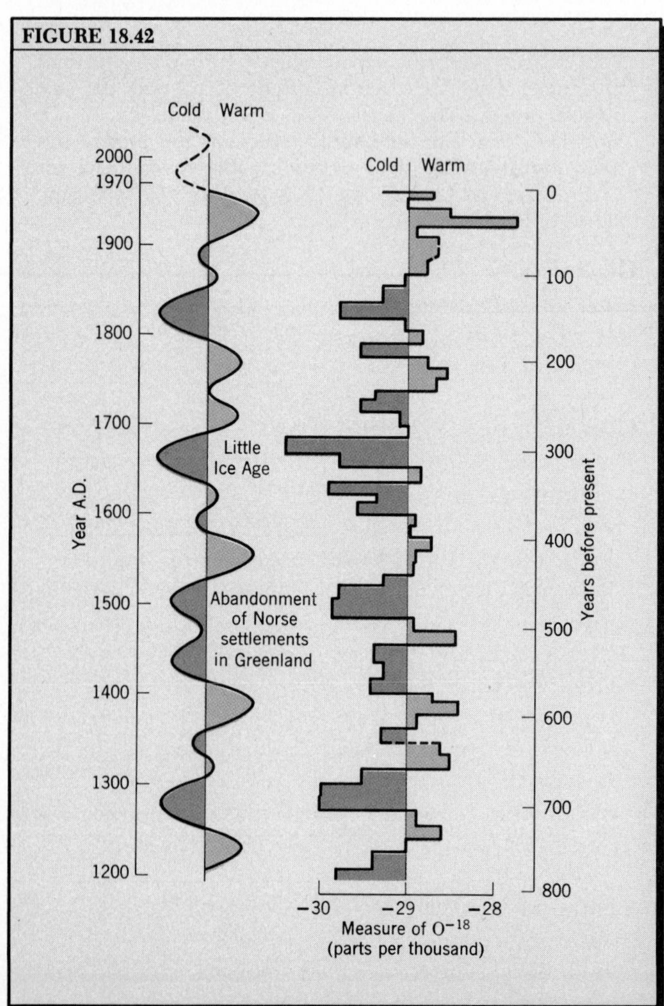

most of the world's fresh water in storage, these thick ice masses are beyond hope of yielding new living space for the growing global population. Perhaps someday icebergs from these ice sheets will be towed to our major port cities to provide fresh water.

The third major topic of our chapter is continental glaciation of the Pleistocene Epoch. For people living in the north central and northeastern United States and in southern Canada, the glacial deposits of the Pleistocene ice sheets make up the entire land surface they farm and on which they build their cities.

The human race evolved to its present state as the supreme species of mammal while the Pleistocene ice sheets were advancing and retreating. Very likely, the rigorous glacial climate and the migrations required by the ice advances stimulated this amazing evolution in a way that might not have occurred otherwise.

Key Facts and Concepts

Glaciers A *glacier* is a large natural ice accumulation capable of flowage under gravity. Glacier ice is formed by snow accumulation as *firn*. *Alpine glaciers,* long, narrow, streamlike bodies, differ from *icecaps,* which are platelike masses. Of continental proportions are *ice sheets* (*continental glaciers*), spreading far beyond the region of nourishment.

Alpine glaciers A *cirque*, occupied by a *firn field,* holds the upper end of an alpine glacier. When a glacier passes over a *rock step* in the bedrock floor, an *ice fall* is formed, with *crevasses* in the rigid upper ice zone. Near the *glacier terminus*, ice is overthrust in low-angle shear planes.

Glacier flow Rates of glacier flow, studied by *glaciologists,* show ice motion to be fastest where ice is deepest, usually in the central portion. Movement by *basal slip* augments flowage movement. *Surges* are episodes of very rapid flow.

Glaciers as open systems A glacier is a flow system of matter (snow) entering in the *zone of accumulation,* moving as ice by gravity flowage to a lower level and leaving the system by melting and evaporation in the *zone of ablation*. In *steady state,* the glacier terminus remains fixed in position, a condition of *glacier equilibrium*. Increased nourishment causes a *glacier advance;* reduced nourishment and/or increased ablation leads to *glacier recession*.

Glacier erosion and transportation Glaciers erode bedrock by *glacial plucking* and by *grinding,* leaving *glacial striations, chatter marks,* and often *glacial grooves*. *Glaciated rock knobs* show the direction of ice flow. Glacier load accumulates in *lateral, medial,* and *terminal moraines* and may be carried out as *rock flour* by meltwater streams.

Landforms carved by alpine glaciation Landforms of alpine glacier erosion include cirques, *horns, glacial troughs* (*hanging troughs*), with *rock steps* and *rock basins* with *tarns*. Drowned troughs form *fiords*.

Greenland and Antarctic Ice Sheets Great existing ice sheets are found today on Greenland and Antarctica. Ice thicknesses exceed 3 km. The ice sheets have broadly rounded summits and spread to the sea through *outlet glaciers*. Large floating *ice shelves* border Antarctica.

Pleistocene ice sheets In the Pleistocene Epoch, at least four major *glaciations* are interpreted from continental deposits. Each ended in *deglaciation* and was separated from the next by an *interglaciation*. In North America, the glaciations were: Wisconsinan (youngest), Illinoian, Kansan, Nebraskan. Pleistocene ice sheets covered large portions of North America and Europe, while highlands had expanded ice caps.

Glacial drift *Glacial drift* is of two classes: (1) *stratified drift* of sorted water-deposited sediment and (2) *till* of unsorted material, including *erratic boulders*. Drift is thick in many areas of the north central United States once covered by Pleistocene ice sheets.

Landforms of continental glaciation Landforms shaped beneath the ice sheet include *ground moraine*, terminal, *recessional,* and *interlobate moraines* (*knob and kettle*), *drumlins,* and *eskers*. Deposits built beyond the ice limit include *outwash plains,* sediments (*varves*) of *marginal lakes, kames, delta kames,* and *kame terraces*.

Origin of the Great Lakes The Great Lakes came into existence through a complex series of changes of levels and outlets during final ice retreat stages of the Wisconsinan Glaciation. Crustal warping was a major factor in determining lake outlets.

Ice sheets and isostasy Isostatic crustal depression caused by loading of an ice sheet is seen in depressed subglacial rock surfaces under existing ice sheets. Disappearance of an ice sheet is followed by isostatic uplift, documented in the Baltic Sea and Hudson Bay regions by a succession of elevated beach ridges.

Pluvial lakes of the Pleistocene In glaciations, *pluvial lakes* formed in closed basins of the Basin-and-Range region because of increased precipitation and reduced evaporation. The largest pluvial lake was Lake Bonneville, with numerous elevated shorelines.

Radiocarbon dating of Pleistocene events The *radiocarbon method* of dating deposits and events uses *carbon-14,* with a half-life of about 6000 years. Carbon-14 is formed in the upper atmosphere by cosmic ray impacts upon nitrogen. It is incorporated into organic tissues and thereafter decays steadily. Ratios of ordinary carbon to carbon-14 are used to derive the age of the sample material. Discrepancies between radiocarbon dates and direct dates from artifacts and tree-ring analysis (*dendrochronology*) have been resolved by a corrected radiocarbon time scale.

Pleistocene record on the ocean floors Deep-sea sediment cores preserve a complete record of alternate glaciations and interglaciations through climate indicators within the sediment layers. Radioisotope ratios and paleomagnetic reversals provide a fix on chronology within cores.

Glacial cycles and foraminifera Tests of foraminifera extracted from core sediments include faunas diagnostic of cold water or warm water in the surface layer. Radiolarian species are also used as indicators of sea-surface temperatures.

Glacial cycles and oxygen isotopes Ratio of oxygen-18 to oxygen-16 in carbonate matter of cores reflects to some extent surface water temperatures (*paleotemperatures*), but more particularly the global quantity of glacial ice in storage. Fluctuations in isotope ratio thus provide a *paleoglaciation curve*, consisting of a succession of *isotope-ratio stages* extending back as far as perhaps −3 m.y. A total of more than 30 Cenozoic glaciations is indicated by the complete paleoglaciation curve. Glacial-marine sediments found in cores of subantarctic latitudes show buildup of glacial ice on Antarctica as early as −11 to −14 m.y.

Pleistocene changes in sea level Eustatic changes of sea level accompanied growth and melting of ice sheets. Lowering of 50 to 100 m may have accompanied the maximum glaciations, corresponding with the oxygen-isotope paleoglaciation curve. During lowered stands of sea level, lower courses of rivers were trenched and extended across exposed shelves. Stepped terraces resulted from alternate trenching and alluvial filling.

Causes of ice ages An *ice age* is the entire period of colder global climate within which glaciations and interglaciations occurred, for example, the late Cenozoic ice age and the late Carboniferous ice age. Fundamental causes of an ice age may include: (1) favorable position of the continents with respect to the polar regions, as governed by plate tectonics; (2) withdrawal of oceans from continental cratons with epeirogenic uplift; (3) sustained increased volcanic activity; and (4) sustained diminished intensity of solar energy reaching earth. The first two causes seem well documented; the second two are debatable.

Causes of glaciations and interglaciations Some of the hypotheses currently under consideration for cyclic glaciations are: (1) triggering by bursts of volcanic activity, (2) control by astronomical cycles of earth's tilt and orbital motions (*astronomical hypothesis*), (3) changes in arctic sea-ice cover, and (4) changes in albedo of land and water surfaces. The astronomical hypothesis focuses on changes in *insolation* in summer at a high northerly latitude (65° N), represented by an insolation curve (*Milankovitch curve*). Close correlation of the insolation curve with cycles of the oxygen-isotope paleoglaciation curve are claimed, but the causative mechanism is obscure.

Questions and Problems

1. Devise a schematic flow diagram depicting an alpine glacier as an open energy-flow system. Construct a second and independent diagram to show the alpine glacier as an open material-flow system. For each system, use arrows to indicate flow paths; a circle (with labels) to represent a change of energy form or matter state, and a box to show a storage of energy or matter. Every change in energy form or state of matter must be followed in sequence by a storage. Use a rectangular frame line to show the system boundary, with inputs and outputs through the boundary. Prepare separate equations for the energy balance and the material balance, giving the conditions of steady state for each.
2. What physical or botanical evidence might allow you to recognize an alpine glacier advance that occurred during the Little Ice Age (a cold climate period that occurred between A.D. 1450 and 1850)?
3. In what ways do the landforms of alpine glaciation contribute to an ideal site for a ski resort and its ski runs? Be specific. Could you safely carry out a cross-country ski trip across an alpine glacier? If so, what part of the glacier could be crossed?
4. Suppose that you are a land developer who buys, subdivides, and builds upon large tracts of agricultural land in the northeastern United States. What classes of glacial deposits and landforms would be best suited to a large high-density development of low- or medium-cost dwellings? Give your reasons. What glaciated terrain would be best suited to a development of luxury homes on large lots? Why? Give specific examples of such developments, if possible.
5. Does postglacial isostatic uplift (rebound) of the crust following deglaciation produce any noticeable seismic phenomena? Where in North America might you expect such effects?
6. In correlating oxygen-isotope cycles of the paleoglaciation curve with cycles of insolation in the Milankovitch curve, it has been found that a glaciation lags in time behind the point when insolation reaches its minimum value. What is the possible significance of such a time lag?

particles. Loess is being deposited today in uplands adjacent to braided river channels in Alaska (Figure 19.6).

Strong support for the wind-deposited origin for much of the American loess is found in the patterns of oess distribution and thickness (Figure 19.7). Most is ound in a blanket over the north central states, with n important extension southward along the east side the Mississippi alluvial valley and with patches over e High Plains of Oklahoma and Texas. Most of this ea lies within and immediately adjacent to the glaci-d region. Another important loess area in North nerica is the Palouse region of Washington and egon.

study of loess thicknesses in the central United es shows a characteristic pattern in which the test thicknesses, 20 to 35 m, are in bluffs immedi- adjacent to the eastern edges of river floodplains. knesses diminish very rapidly eastward from the s, as shown in Figure 19.8. Close association of

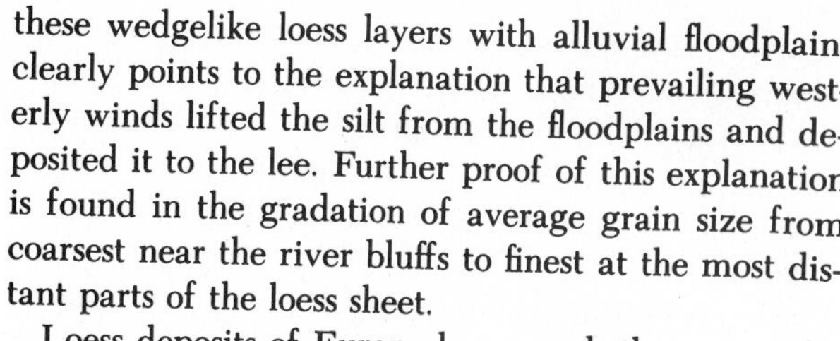

these wedgelike loess layers with alluvial floodplains clearly points to the explanation that prevailing westerly winds lifted the silt from the floodplains and deposited it to the lee. Further proof of this explanation is found in the gradation of average grain size from coarsest near the river bluffs to finest at the most distant parts of the loess sheet.

Loess deposits of Europe bear much the same relation to the glaciated region as do those in the United States. Loess forms a nearly continuous sheet in Euro-

RE 19.5 An exposure of loess of Wisconsinan age, show- rtical cleavage. At the top is a brown soil (Mollisol) as formed on the loess; it shows prismatic structure. on County, Colorado. (H. E. Malde, U.S. Geological .)

FIGURE 19.6

E 19.6 Dust clouds raised by strong winds blowing getation-free bars of the braided channel of the Delta ntral Alaska. Silt carried in this way builds modern posits on adjacent upland surfaces. (U.S. Navy photo- om T. L. Péwé, 1951, *Journal of Geology*, vol. 59,

FIGURE 19.7

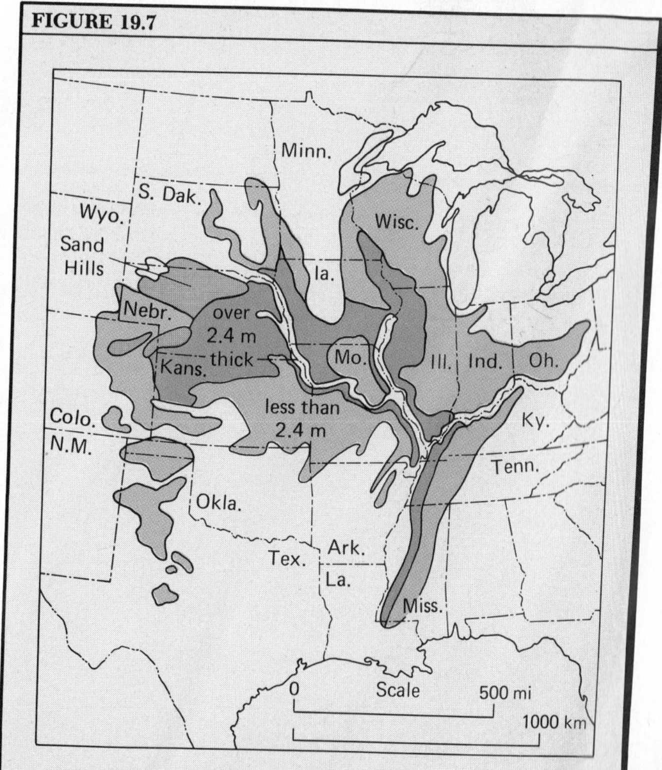

19.7 Map of loess distribution in the central United edrawn, by permission of the publisher, from A. N. *Physical Geography*, 4th ed., Figure 33.16. Copy- 975 by John Wiley & Sons, Inc.)

19

CHAPTER NINETEEN

The Geologic Work of Wind and Waves

Both wind and waves act on the continents in geologic roles of agents of denudation. Their action is, however, based on totally different types of energy systems from those of streams and glaciers, which we studied in the previous two chapters. Streams and glaciers are gravity-flow systems in which water (as a liquid or as ice) flows from higher to lower levels under the force of gravity. In contrast, in both wind and waves, kinetic energy travels more or less horizontally along the surface of the oceans or the lands. This kinetic energy is present in the motion of either a basal air layer or a surface water layer, but it works in very different physical ways in each of the two media. This difference will become obvious as we treat first the action of wind and second the action of waves. We will find that most water waves are generated by wind and represent an energy transfer from air to water. With this transfer, there occurs a change in motion, from one-way flow of the air (a gas) to a cyclic motion in water (a fluid).

Wind as a Geologic Agent

Wind action is capable of producing many distinctive landforms—some formed by erosion, others by deposition. Wind is also capable of transporting sediment, and this has also been a major geologic role of wind action. We have already seen that the finest dust particles carried by prevailing winds settle out over the oceans and eventually become part of the detrital sediment layers found in deep-sea cores. Silt particles are transported many tens of kilometers and have accumulated as layers many meters thick upon the land surface in many parts of the globe. Coarser particles—sand and gravel—do not travel far under the force of the wind, but the dunes that are built of these materials can accumulate locally into deposits hundreds of meters thick. We know this from examination of some thick formations of sedimentary strata made up of ancient dune sands.

Wind Erosion

The flow of air over a solid or liquid surface exerts a dragging force, or shearing stress, against that surface. Wind moving over a solid mineral surface, such as bedrock or hardened clay, is unable to cause an appreciable change because the cohesive strength of the material exceeds the stress exerted by the wind. Only where small mineral grains are lying loose upon an exposed surface can wind exploit its full powers of erosion and transportation.

One form of wind erosion is *deflation*, the lifting and transporting of loose particles of clay and silt sizes, collectively referred to as *dust*. The particles are suspended in turbulent eddies in the wind structure. The proccess is much like that of suspension of fine sediment in stream flow. Grains are carried up by vertical currents that exceed the settling velocities of the grains in still air. The dust is diffused upward into the atmosphere to heights ranging from a few meters to several kilometers. The height depends upon intensity of wind turbulence, duration of the wind, and fineness of the particles. The result may be a dense cloud, called a *dust storm* (Figure 19.1).

Deflation occurs where clays and silts in a thoroughly dried state are exposed on barren land surfaces. Such conditions exist in semiarid and desert regions generally, and locally in dried floodplains, tidal flats, and lake beds. Even on actively forming glacial outwash plains, deflation occurs in times of cold, dry weather.

Deflation often results in excavation of shallow depressions, called *deflation hollows*, or simply *blowouts* (Figure 19.2). Hollows develop where the natural vegetative cover of shrubs or grasses is broken down, exposing the bare soil or regolith to wind. Once formed, deflation hollows hold water after heavy rains. This attracts grazing animals whose trampling further loosens and disrupts the soil, preparing it for deflation when the soil is again dried out. Storm runoff attacks the sloping margins of the depressions, washing sediment into the bottom, and it is later removed by deflation. Thus, deflation hollows tend to deepen and enlarge, sometimes growing to widths of 1 km or more and to depths of 5 to 20 m.

Commonly, deflation produces no distinctive landform but merely removes a layer of uniform thickness from the surface of a plain. Left behind, however, are grains of gravel and pebbles too large to be moved. These remnants accumulate into a sheet that ultimately covers the finer-grained material beneath and protects it from further deflation. Such a residual gravel sheet is termed a *desert pavement*. Desert pavements develop rapidly upon alluvial-fan and terrace surfaces (Figure 19.3). Exposed surfaces of the pebbles may become coated with a nearly black iridescent substance called *desert varnish*. In some localities, the evaporation of capillary water brought surfaceward through the soil leaves behind a deposit of calcium carbonate (caliche) or gypsum which acts as a

FIGURE 19.1

FIGURE 19.1 This rapidly moving cloud is the leading edge of a dust storm. The dust is suspended within turbulent air of a cold front, Coconino Plateau, Arizona. (D. L. Babenroth.)

FIGURE 19.2

FIGURE 19.2 A large blowout on the plains of Nebraska. Deflation has lowered the surface in the foreground. The amount of removal can be judged from the height of the remaining knobs of layered sand and silt. (U.S. Geological Survey.)

cement, hardening the pavement into a conglomerate-like slab.

A second form of wind erosion is *sandblast action*, in which hard mineral grains (usually of quartz) of sand size are driven against exposed rock surfaces projecting above a plain. Because sand grains travel close to the ground, their erosive action is limited to surfaces lying with a meter or two of the flat ground over which the sand is being driven. Sandblast action probably does not erode resistant bedrock to depths of more than a few centimeters and is respo[...] for minor features such as notches and ho[...] base of a cliff or a boulder.

Sandblast erosion creates curiously sha[...] known as *ventifacts*. When developed [...] these objects take the form of elongate, [...] *dreikanter* (from the German words f[...] "edges"), with curved faces intersecting [...] edges (Figure 19.4). Dreikanter origina[...] rounded pebbles of hard fine-grain[...] quartzite, chert, or obsidian. Sandbl[...] the windward face of a partially b[...] pebble then turns to rest upon t[...] permitting a second face to be cut, [...] Dreikanter and other distinctively [...] polished stones give evidence of [...] when found buried in a sediment [...]

FIGURE 19.3 This desert pavement of quartzite fragments was formed by action of both wind and water on the surface of an alluvial fan in the desert of southeastern California. Fragments range in size from 2 to 30 cm. Silt underlies the layer of stones. (C. S. Denny, U.S. Geological Survey.)

FIGURE 19.4 A ventifact resting in place on a windswept gravel surface. A 15-cm ruler indicates size. Wright Valley, McMurdo Sound, Antarctica. (R. L. Nichols.)

FIGURE 19.3

FIGURE 19.4

Loess

Thick deposits of wind-tran[...] mulate under favorable condit[...] middle latitudes have surface [...] friable, yellowish sediment [...] fragments. (The word "loess [...] can be pronounced "less.") P[...] in the size range of silt, 4 [...] though 5 to 30% of the loe[...]

Where it forms a layer [...] areas of plains and low pl[...] layered structure or strat[...] it commonly has a natu[...] age, along which masse[...] undercut bank or cliff ([...] tion indicates uniform[...] upon grass-covered su[...] soil disturbance by pl[...] sition. The vertical [...] shrinkage cracks r[...] deposition.

Loess is commor[...] verized fragments [...] some feldspar, m[...] present. Carbona[...] in loess. Small n[...] distributed thro[...] cur in large nu[...]

The grain si[...] the middle lat[...] are best expla[...] blown dust o[...] from alluvial [...] the limits of [...] uncovered [...] Pleistocene [...] from a du[...] difference [...]

FIGURE 19.5

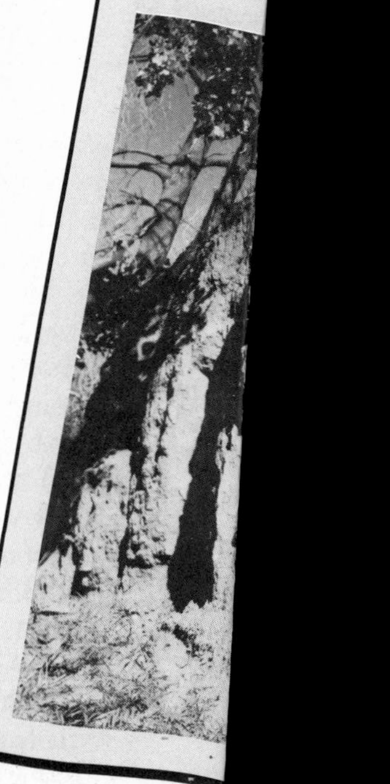

pean Russia, where the plains topography was most favorable and the region occupied a leeward position with respect to the ice sheets of northern Europe. The Russian loess is from 10 to 15 m thick over vast areas. Loess related to ice sheets is also found in central Asia.

In northern China, loess reaches thicknesses commonly over 30 m, and in some places as much as 90 m. Just how much of this loess is directly of eolian origin, and how much reworked by streams, is not agreed upon. The dust storms that brought this loess originated from nonglacial regions in the arid interior region of Asia. Loess deposits are also found in North Africa, Argentina, and New Zealand.

The great importance of loess to humankind is that it forms the parent material of some of the richest agricultural soils on earth. Soils developed on loess are largely the Mollisols (described in Chapter 14), rich in nutrients needed for grain crops. Corn and wheat thrive on soils of loess origin in the United States. Loess is easily cultivated, but is also subject to severe soil erosion and deep gullying when abused.

Transport of Sand by Wind

Removal of dust in suspension by wind from a mixture of grades, such as an alluvial deposit, leaves the sand and gravel sizes behind. The sand grains travel downwind, staying close to the surface, and are gradually separated from the gravel particles, which are too heavy to be moved very far by wind. Thus, there comes into existence a distinctive body of sediment which we call *eolian sand,* or *dune sand.* Most of the grains are from 0.1 to 1 mm in diameter. Wind is thus a very effective sediment-sorting agent.

Most dune sands are of the mineral quartz, whose hardness and resistance to chemical decay make it the most durable of the abundant rock-forming minerals. Rarely, dunes are formed of shell fragments, particles of volcanic ash, or grains of gypsum or of heavy minerals such as magnetite. Such unusual compositions are explained by a local abundance of the particular material.

Quartz grains of dune sand are beautifully rounded into spherical and egg-shaped grains, with a frosted surface texture (see Figure 5.11). Although transport by wind is not able to break down the grains into smaller sizes, countless impacts among grains increase their sphericity. These impacts cause the grain surfaces to be covered by microscopic impact fractures.

When strong winds blow over a surface of dune sand, the grains are carried along in low clouds gliding like a carpet, a phenomenon known as a *sand storm.* Most of the sand is moving in a layer only a few centimeters off the ground at most, but scattered grains rise as high as a meter or two. Although your eye cannot follow the movement of single grains, the particle paths have been photographed under controlled conditions in laboratory wind tunnels. The photographs show that the grains make long leaps downwind (Figure 19.9). After impact with grains on the surface, a single grain may rebound high into the air or merely glance off at a low angle. The process of leaping by rebound is termed *saltation.* In this way, the wind transfers kinetic energy to the grain, which, on impact with the sand surface, dislodges other grains and may project them into the air.

In addition to sand movement by saltation, grain impacts cause a slow forward *surface creep* of the sand. The energy of a striking grain is such that it can move a surface grain as large as six times its own diameter or 200 times its own weight. Surface creep causes slow downwind movement of grains too large to travel in saltation. However, the smaller grains are traveling much faster in saltation. For this reason, the smaller grains tend to become separated from the coarse. Separation results in development of sheets of coarse sand or fine gravel from which the fine sand has been largely removed.

We are all familiar with the rippled appearance of dune surfaces. The ripples are small ridges and troughs

FIGURE 19.8 Schematic cross section of loess thickness along a west-to-east line through uplands east of the lower Mississippi River alluvial plain. (Data from Map of Pleistocene Eolian Deposits of the United States, Geological Society of America, 1952.)

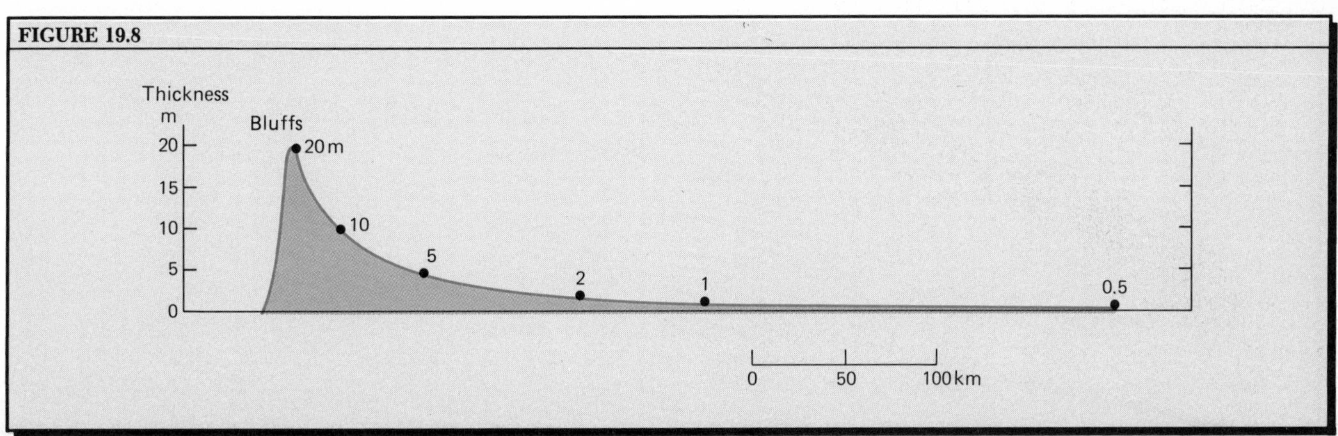

running transversely to the direction of the wind, and they result from uneven movement of grains by surface creep (Figure 19.10).

The rate of movement of sand by wind for any given type of surface and particle size grade depends on wind speed. Figure 19.11 is a graph showing the relation of rate of sand movement (metric tons/day/meter width of cross section) to wind speed. The graph uses logarithmic scales on both axes. In this case, wind speed is measured at a height of 1 m above the surface. The steepness of the line shows that rate of movement increases disproportionately with increase in wind speed. Actually, the rate of movement varies about as the cube of wind speed. Thus, a wind blowing at 15 m/sec can move in one day the same quantity of sand that a wind of half that speed will move in three weeks.

Sand Drifts and Sand Dunes

Where saltation and surface creep have had freedom to act on large supplies of loose sand, the sand is heaped into depositional landforms that may grow into dominant features of the landscape. We recognize two landforms of free sand. A *sand drift* is an accumulation of sand formed in the lee of some fixed obstruction, such as a rock or bush. Drifts do not move, but remain attached to the obstacle. A *sand dune* is an individual mound or hill of loose sand rising to a single summit and independent of any fixed surface feature. Dunes are capable of movement downwind while maintaining a characteristic shape. Dunes are sometimes completely isolated from one another upon a flat surface of gravel, hard clay, or bare rock, but they may also be joined into a continuous sand layer completely concealing whatever material lies beneath.

Suppose that we are on a flat, arid plain observing a small sand drift that has formed over an obstacle. Sand is moving rapidly in saltation across the flat ground, which is pebble strewn. The pebbles cause high rebounds of the elastic grains. As the grains fall upon the drift of fine sand, their rebounds are softened and their rate of movement sharply reduced. The grains are thus delayed as they pass over the drift, much as automobiles are delayed and tend to congregate at the toll booths of a highway. The drift grows in area and height, simultaneously becoming an increasingly efficient trap for more sand. Soon the heap has grown far larger than the original drift and no longer depends on the original obstacle. It can then begin a slow downwind migration. At this point a solitary dune has been formed and moves off, leaving behind a small drift about the obstacle to develop another dune.

Our new sand dune at first has a smoothly rounded longitudinal profile, as shown in the first line of Figure 19.12. Gradually the profile changes so that the downwind side becomes steeper (dotted line, 2). A sharp crest develops and grows higher (line 3), creating a

FIGURE 19.10 Sand ripples on dune surfaces, Qatif, Saudi Arabia. (Arabian American Oil Company.)

FIGURE 19.10

FIGURE 19.9 Sand movement by saltation and surface creep. (Based on data of R. A. Bagnold. Redrawn, by permission of the publisher, from A. N. Strahler, *Physical Geography*, 4th ed., Figure 33.4. Copyright © 1975 by John Wiley & Sons, Inc.)

FIGURE 19.9

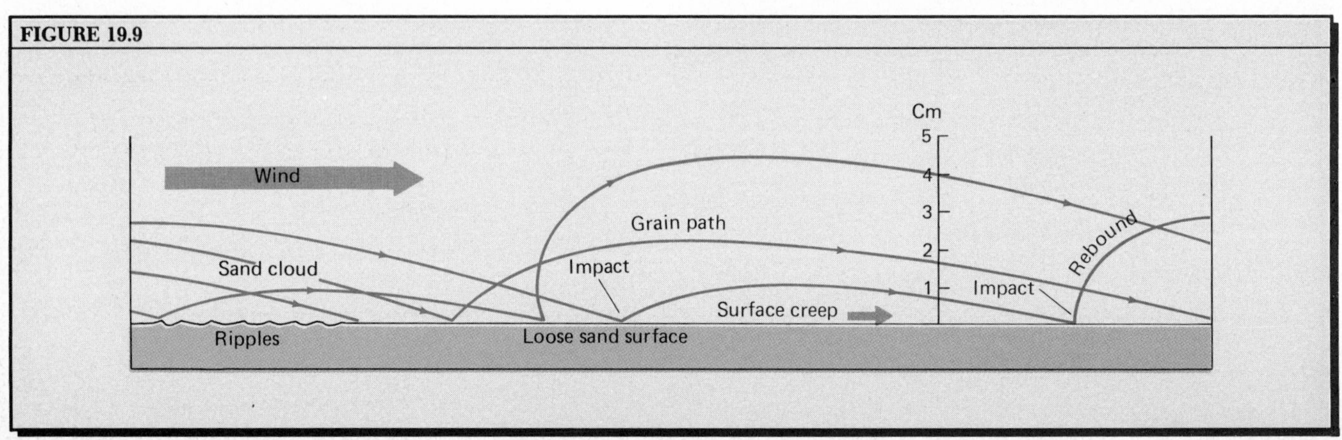

steep lee slope called the *slip face*. Next to the slip face is a shadow zone of quiet air. The slip face has an angle of slope of about 35°, representing the angle of repose for the grains that leap over the dune crest and land within the wind shadow. The slip face is gradually steepened by the rain of sand grains landing on its upper portion. When the slope angle is built up to a point of instability, a thin layer of sand slides down to the slope base, lowering the slope angle slightly to achieve a stable slope. The slip face advances by a succession of such slides.

Dune Forms

One of the simplest of dune forms is the *barchan dune*. It is an isolated dune of crescent shape (Figure 19.13). The two points, or "horns," of the barchan point downwind, and contain between them a steep, concave slip face (Figure 19.14). Barchans grow to heights as great as 30 m and widths up to 350 m. Movement of the dune often amounts to a few centimeters per day, and individual barchans have been found to travel as far as 15 m in a year.

In deserts where a great supply of sand is present, dune sand covers the entire surface. The sand surface is formed into wavelike ridges separated by troughs and hollows, much as if sea waves in a time of storm were frozen into place (Figure 19.15). Such sand waves are described as *transverse dunes*, and the entire assemblage as a *sand sea*. Both the structure and the origin of transverse dunes are similar to those of barchans, for slip faces of a sand sea are crescent-shaped and face downwind. Toward the edge of a sand sea, where the sand layer thins out, individual transverse ridges become separated by bare ground. The ridges then become segmented into individual barchans, as we see in the lower-right portion of Figure 19.15. Transverse dunes commonly occur in a

FIGURE 19.11 The rate of mass transport of sand rises very rapidly with an increase in wind speed. Both scales are logarithmic. (Based on data of R. A. Bagnold.)

FIGURE 19.12 Growth of a dune and development of the slip face. (After R. A. Bagnold, 1941, *The Physics of Blown Sand and Desert Dunes*, Methuen, London, p. 202, Figure 71.)

FIGURE 19.13 The horns of barchan dunes point downwind. (Redrawn, by permission of the publisher, from A. N. Strahler, *Physical Geography*, 4th ed., Figure 33.5. Copyright © 1975 by John Wiley & Sons, Inc.)

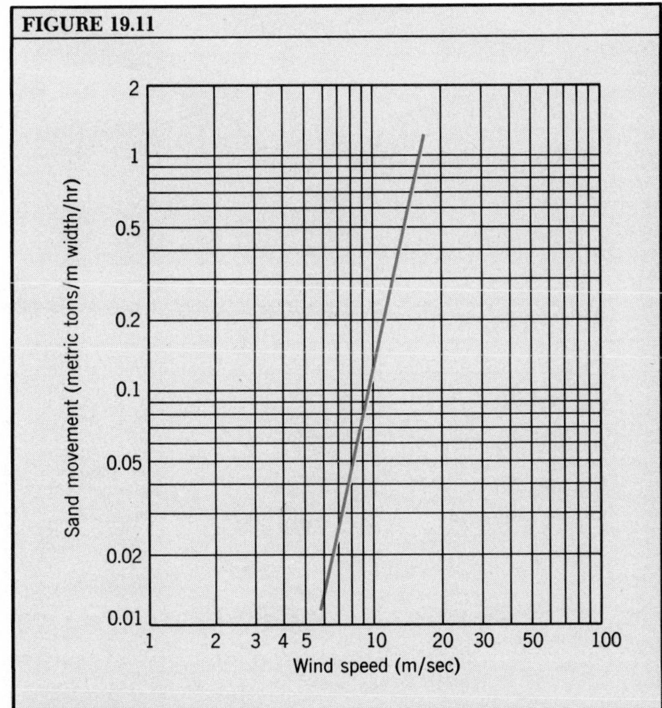

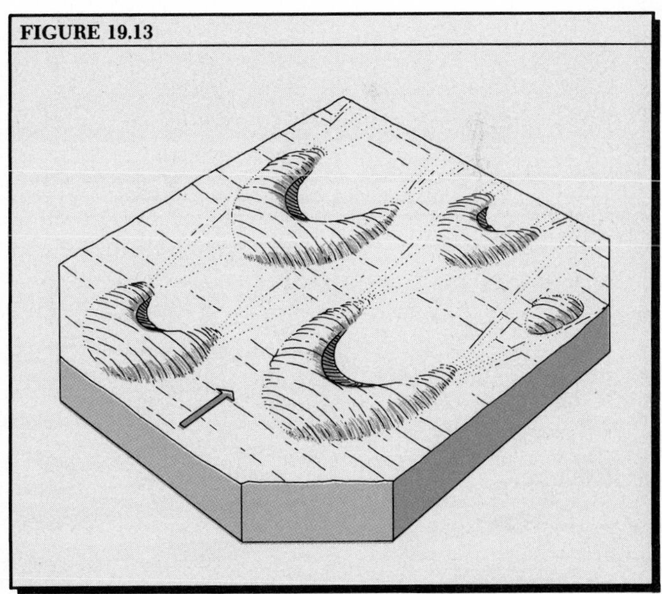

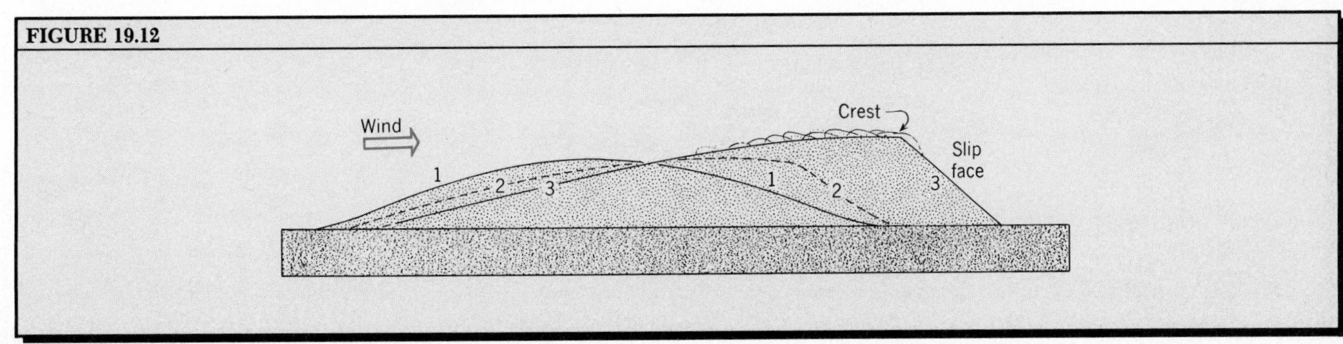

coastal belt situated inland from beaches. They are best developed where sand supply is large and vegetation is absent.

Another common dune of free sand consists of long, narrow sand ridges that trend parallel with the prevailing wind direction. These are *longitudinal dunes* (Figure 19.16 and Color Plate E.2). An interesting dune form of the Sahara and Arabian Deserts is the *star dune*, which takes the form of an isolated hill of sand. In plan, the dune base resembles a several-pointed star (Figure 19.17), with sharp-crested ridges which converge from the basal points to a central peak in some cases as high as 100 m above the surrounding plain. Star dunes can remain fixed in one position for centuries and have come to serve as desert landmarks, each bearing a local Arabic name.

Sand Dunes in the Geologic Record

Dune sands have accumulated in great thicknesses—hundreds of meters in some cases—at certain times in the geologic past in favorable localities. One of the most impressive of these is a great sandstone formation of lower Jurassic age, exposed in the Colorado Plateau.

FIGURE 19.14 These barchan dunes formed on a terrace bordering the Columbia River, near Briggs, Oregon. (U.S. Geological Survey.)

FIGURE 19.15 A great sea of transverse dunes in Imperial County between Calexico, California, and Yuma, Arizona. Some barchan dunes can be seen at the lower right. (Spence Air Photos.)

FIGURE 19.14

FIGURE 19.15

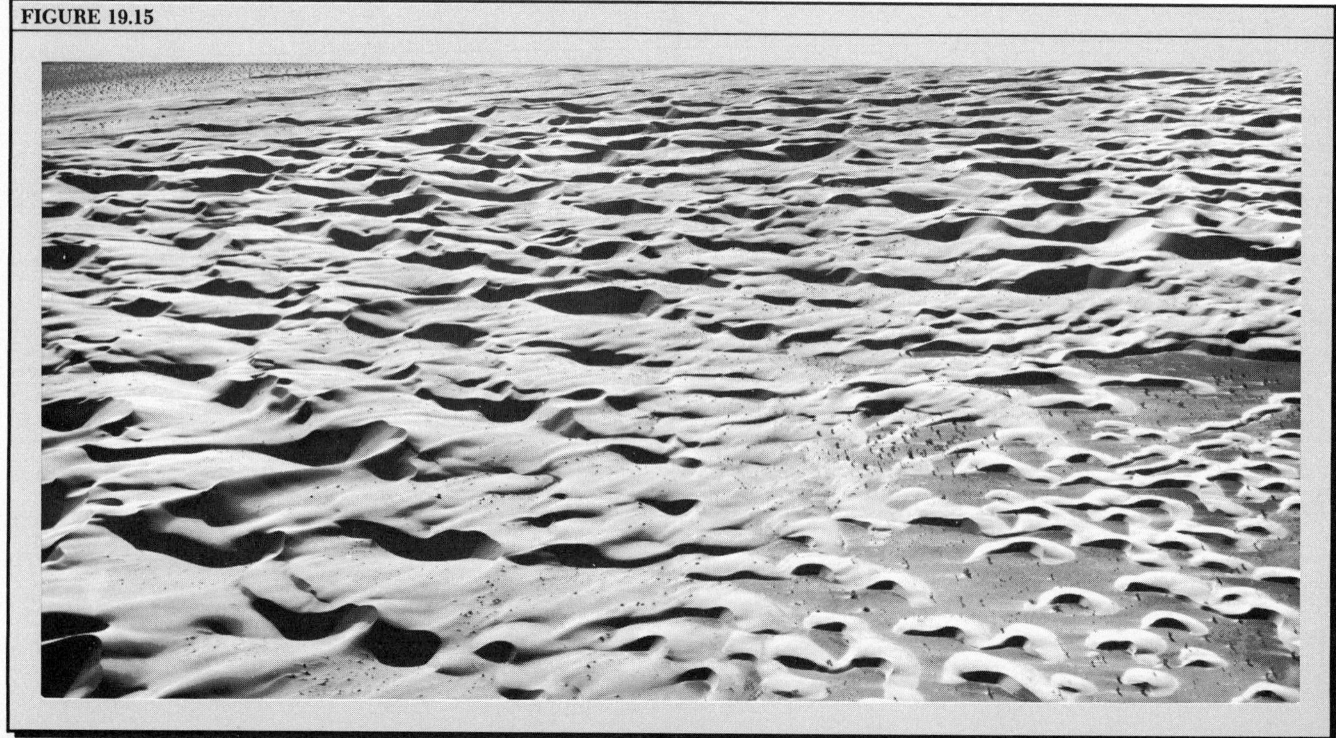

Known as the Navajo sandstone, it is exposed in sheer cliffs that form the walls of Zion Canyon and several other deep canyons in southern Utah. As the sandstone weathers, sloping laminations are etched into sharp relief in a structure known as *cross-lamination* (Figure 19.18). What we see here are cross sections through ancient dunes of quartz sand. The steeper laminations represent the former slip faces along which the dunes accumulated. The more gently dipping planes that cut across the steeper layers are surfaces of wind erosion that formed after one dune was deposited and before the next was deposited upon it. Over 500 m of dune sands accumulated here. Little is known of the circumstances surrounding this accumulation, but there must have been an extensive desert basin here to receive the sand.

Going back to the Permian Period in this same region, we find another ancient dune accumulation, the Coconino formation, about 90 m thick (see Figure 6.3). It is a nearly white quartz sandstone with well-developed cross-lamination.

How do geologists tell cross-lamination of former sand dunes from other kinds of cross-lamination? A very common form of cross-lamination is produced by an aggrading stream. The stream channel is scoured at time of high water, then filled during the ensuing period of diminishing discharge (see Figure 16.18). With each cycle of scour and fill, the stream erodes a trough, which it then fills with sand and gravel. The trough-type laminations are curved in an upward concavity, illustrated in Figure 19.19. In contrast, the laminations of dunes are nearly straight lines representing exposures of the edges of the almost planar surfaces of the slip faces. (Other forms of cross-lamination are produced by wave and current action on sand beaches.)

In the case of the Coconino sandstone, a detailed field study showed that the dominant dip of the steep laminations was southward, making it clear that the

FIGURE 19.16

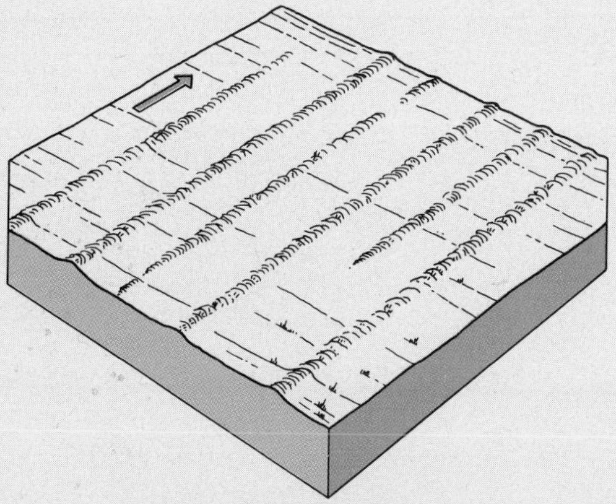

FIGURE 19.16 Longitudinal dunes trend parallel with the direction of the prevailing wind, indicated by the arrow. (Redrawn, by permission of the publisher, from A. N. Strahler, *Physical Geography*, 4th ed., Figure 33.8. Copyright © 1975 by John Wiley & Sons, Inc.)

FIGURE 19.17

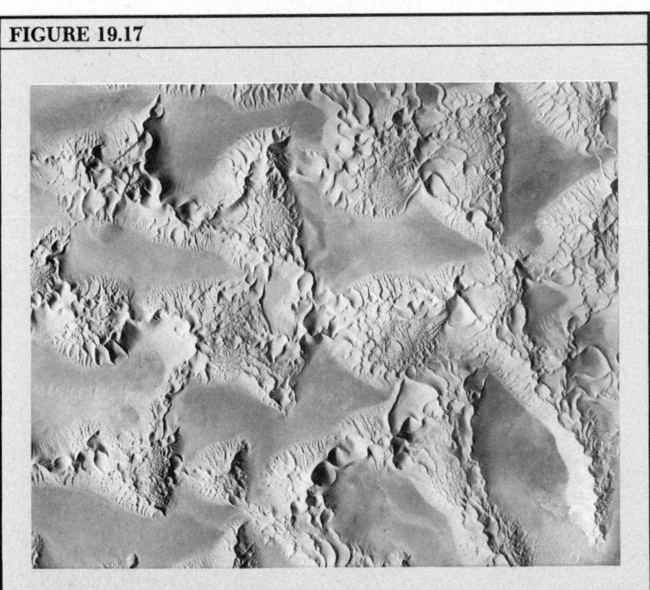

FIGURE 19.17 Star dunes of the Libyan Desert, seen from an altitude of 10 km. Dune peaks rise to heights of 100 to 200 m or more above the intervening level ground. (Aero Service Division, Western Geophysical Co. of America.)

FIGURE 19.18

FIGURE 19.18 Planar cross-lamination in ancient dune sands of the Navajo sandstone, Zion Canyon, Utah. (Douglas Johnson.)

sand had been carried from north to south from some great sand source. Looking back at the position of the Colorado Plateau region in Permian time on the map of Pangaea, Figure 13.15*A*, we see that it was located about at latitude 15° N. This position is about the same as we find the southern edge of the Sahara Desert today. In early Jurassic time, when the Navajo sandstone was deposited, the Navajo dune latitude was about 35° N, which is also well within the limits of ancient subtropical deserts. Alfred Wegener was well aware of these deserts, for he shows them on his 1929 map of the Permian, which we reproduced as Figure 13.12.

Dunes Controlled by Vegetation

Many common dune forms develop while bearing a scanty cover of grasses or small shrubs which serve as sand traps. Here and there are plant-free hollows and slopes of loose sand. Simplest of these vegetation-controlled dunes are the coastal *foredunes*. They are found adjacent to the inner edge of a sand beach (Figure 19.20). Beachgrass upon these dunes is capable of maintaining itself as the sand accumulates. Foredunes are irregular in form, with numerous blowout hollows from which sand is excavated and carried to the dune summits.

FIGURE 19.19 Two varieties of cross-lamination. The heavy line at the base of each set of laminations represents a surface of erosion truncating older sets of laminations beneath. Upper block shows planar lamination, seen in dune deposits. Lower block shows trough variety typical of streambed deposits. (Based on data of E. D. McKee and G. W. Weir, 1953, *Geological Society of America, Bull.*, vol. 64, p. 387, Figure 2.)

FIGURE 19.19

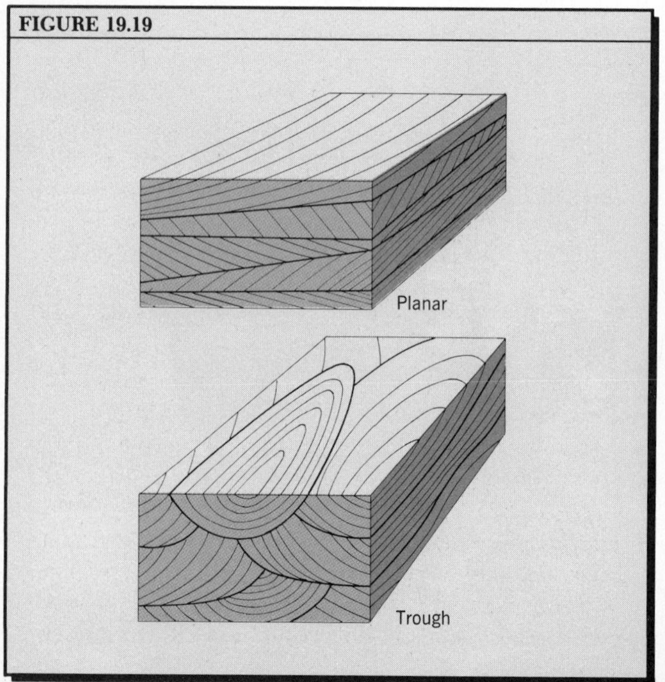

FIGURE 19.20 Beachgrass protects coastal foredunes on the Provincelands of Cape Cod, Massachusetts. (A. N. Strahler.)

FIGURE 19.21 Dunes controlled by vegetation. Wind direction is from lower right. *A*. Coastal blowout dunes. *B*. Parabolic dunes of a semiarid steppe region. (Redrawn, by permission of the publisher, from A. N. Strahler, *Physical Geography*, 4th ed., Figure 33.8. Copyright © 1975 by John Wiley & Sons, Inc.)

FIGURE 19.20

FIGURE 19.21

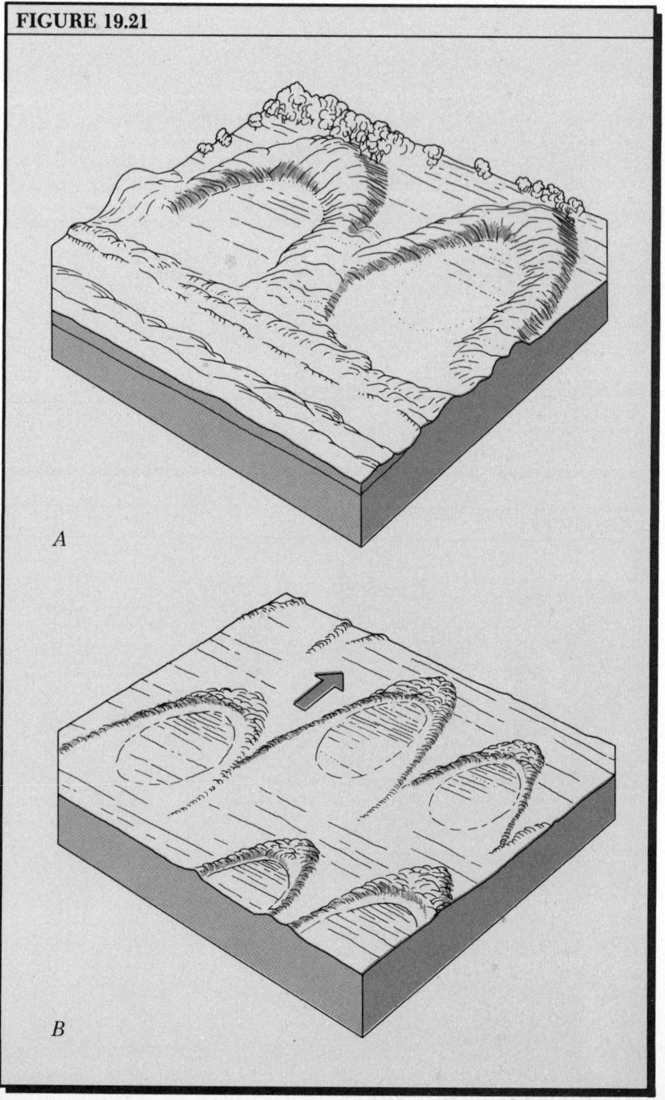

Some other vegetation-controlled dune types have distinctive shapes, easily recognized when seen from the air. The two block diagrams in Figure 19.21 illustrate a distinctive dune family characterized by parabolic outlines. These *parabolic dunes* are convexly bowed in the downwind direction (opposite to the curvature of the barchan). One variety of parabolic dune forms along coasts, landward of beaches, where strong onshore winds are supplied with abundant sand (upper block). The sparse cover of protective grasses and shrubs is locally broken, permitting a deep deflation hollow to form. Sand is carried out of the hollow, building a curved embankment, or rim, which grows higher as the blowout is enlarged. When well developed, this *coastal blowout dune* may rise to heights of 30 m. On the landward side it has a slip face that can override an inland forest, killing the trees (Figure 19.22).

Low parabolic dunes are formed in great numbers on semiarid plains where the regolith is sandy (lower block in Figure 19.21). The dune ridge, which is only 1 to 3 m high, is covered by grasses and shrubs which serve to trap sand (or coarse silt) derived by deflation from a shallow depression.

*Induced Deflation and Dune Activation

Human activities have aggravated the work of wind where modern farming practices and intensified livestock grazing have been extended into semiarid climate zones. Here the plant cover is at best sparse and is easily broken down, baring the soil to deflation. The exposed soil of a harvested wheat field has little protection except the stubble of the previous crop. On rangelands, cattle are often allowed to graze in numbers that exceed the capacity of the native grasses to sustain growth, and the trampling of the soil under the hooves of heavy animals induces easy deflation.

Now a new environmental impact has come into the picture. Use of the off-the-road vehicle (ORV) is increasing in popularity as a form of recreation. Trail bikes and four-wheel-drive vehicles churn the desert surfaces, destroying both the plant cover and the protective pebble coating of the desert pavement, and allowing clouds of dust to rise.

The coastal foredune ridge plays an important role in coastal protection. It absorbs much of the energy of storm surf, which reaches to heights several meters above the usual limit of ordinary waves. The dune ridge may be cut back many meters in a single storm. It will normally be rebuilt in longer intervening periods of normal wave levels. Dune buggies are driven heedlessly over coastal foredunes, however, destroying the beachgrass and inducing the rapid landward movement of dunes.

FIGURE 19.22 The steep landward slope of a coastal dune is advancing over a forest at Cape Henry, Virginia. (Douglas Johnson.)

FIGURE 19.22

Ocean Waves

The second part of this chapter is devoted to the action of waves and currents on the shorelines of the oceans and inland lakes. Ocean waves as well as the waves of large freshwater lakes transport enormous quantities of kinetic energy toward their shores where further energy transformations take place. The wave forms are modified in such a way as to become forward surges of water and other kinds of currents capable of eroding and transporting sediment.

By far the most important class of ocean waves—judged by their continued presence over the world ocean—are those waves generated by the action of wind. Ocean waves produced by wind belong to a class described as progressive waves, because the wave form moves rapidly through the water. The progressive waves produced by the wind are of a type called *oscillatory waves* because, as each wave passes a fixed reference point, the water particles travel through a vertical cycle of motion and return approximately to their original positions. They are similar to seismic Rayleigh waves, which are those surface waves in which particles move in vertical orbits (Chapter 8).

Figure 19.23 shows that each water particle travels in a circular vertical orbit, making one circuit with each wave period. On the wave crest particles are moving forward in the direction of wave travel, while in the trough they are moving backward, opposite to the direction of wave travel. Halfway between crest and trough, particles are moving vertically, up or down. Wave orbits die out rapidly with depth. At a depth equal to about one-half the wave length the orbit diameter is reduced to about 1/23 of its surface diameter. Actual ocean waves are much more complex in form than the diagram shows, although a long ocean swell moving through calm water is very much like the ideal waves. Large, steep waves moving under the force of strong winds are often sharp-crested and asymmetrical. The water in these steep waves creeps forward very slowly in the same direction as the wind.

Waves in Shallow Water

When a train of ocean waves arrives at the coast of a continent or island, it encounters shallow water. Now the presence of the bottom begins to interfere with the progress of the wave. New principles come into play in the shallow-water zone.

Consider first a simple series of waves entering a zone of gently shoaling water. The wave is said to feel bottom when the orbital motions of water particles are interfered with by the solid bottom. As a rule of thumb, we can assume that when water depth is one-half the wave length, the effect of the bottom will produce appreciable changes in the wave form and speed (Figure 19.24).

Wave orbits in shallow water are modified into ellipses (Figure 19.25) which become progressively flatter as the bottom is approached. Immediately adjacent to the bottom, water particles move in flat paths paralleling the bottom, alternately shifting landward (under the passing crest) and seaward (under the pass-

FIGURE 19.23 Circular orbits in simple low waves in deep water. (Redrawn, by permission of the publisher, from A. N. Strahler, *Physical Geography,* 3rd ed., Figure 10.4. Copyright © 1969 by John Wiley & Sons, Inc.)

FIGURE 19.24 A shoaling wave undergoes changes in form and finally breaks. (Redrawn, by permission of the publisher, from A. N. Strahler, *Physical Geography,* 4th ed., Figure 32.1. Copyright © 1975 by John Wiley & Sons, Inc.)

FIGURE 19.25 Elliptical wave orbits in shallow water. (Data from H. A. Panofsky, 1960, *Oceanography for the Navy Meteorologist;* Norfolk, Va., U.S. Naval Weather Research Facility.)

FIGURE 19.23

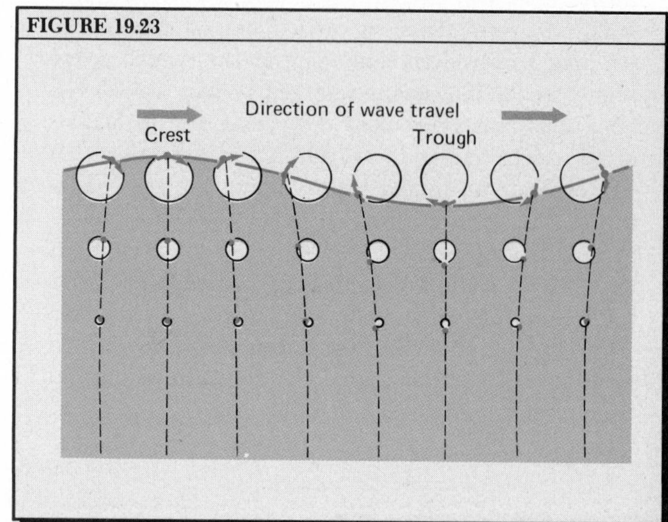

FIGURE 19.24

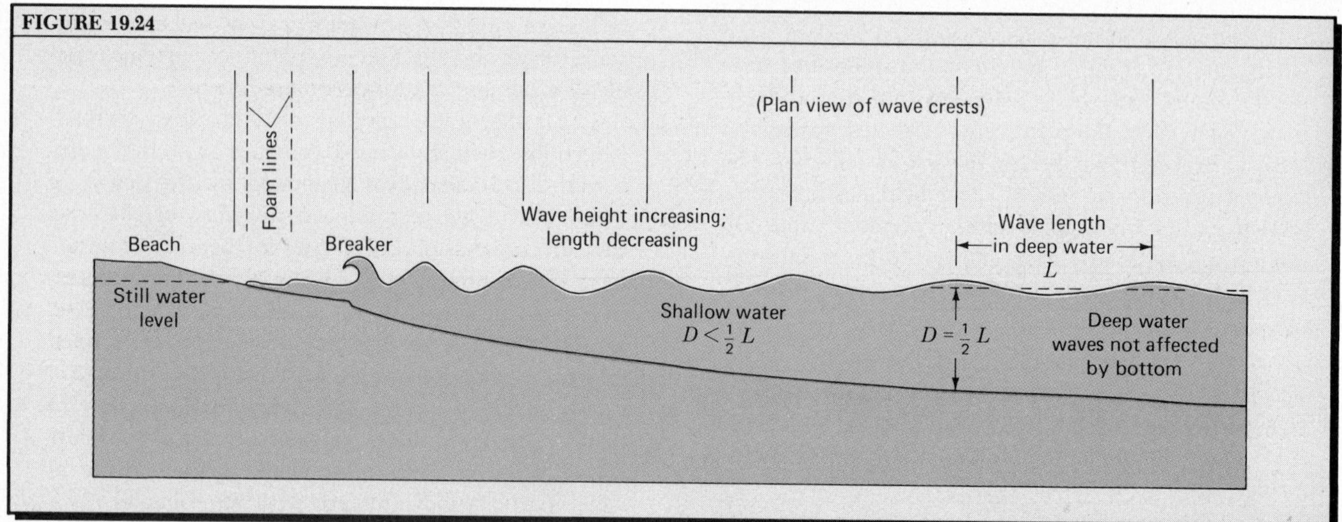

FIGURE 19.25

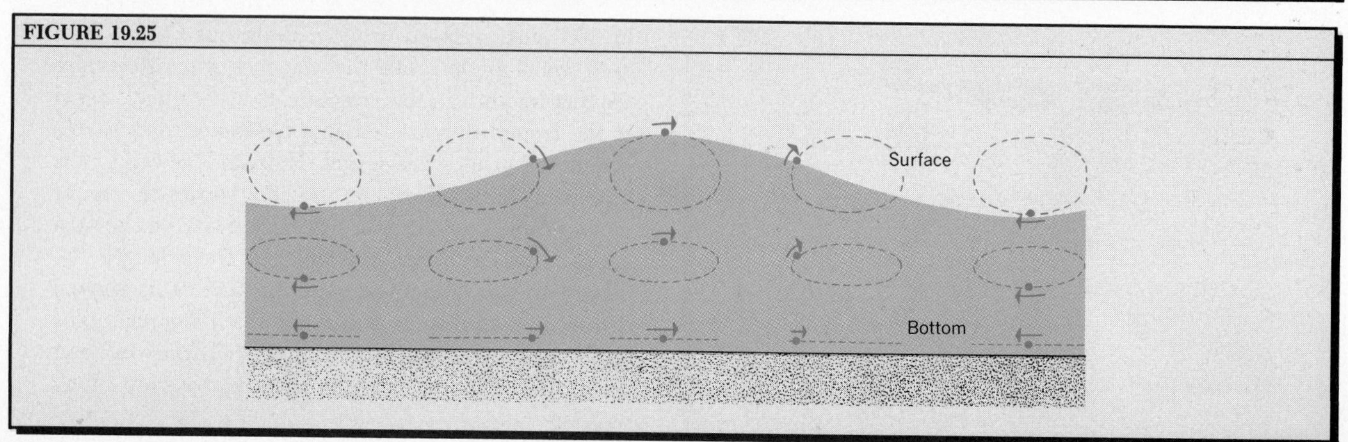

ing trough). This water motion causes a frictional drag, or shear, against the bottom and causes loss of wave energy. Wave speed is decreased, and certain changes in wave height and length occur as the wave travels shoreward into shallower water.

Although wave speed diminishes shoreward, the wave period must remain constant—that is, the same number of waves must pass a fixed point each minute. The obvious result is that the wave crests become more and more closely spaced, which is to say that the wave length diminishes (Figure 19.24). This effect is quite nicely illustrated by the flow of automobile traffic approaching a toll gate. As the cars slow, they become more closely spaced, until at the toll gate they are creeping along bumper to bumper.

Wave height also undergoes a change as the wave moves into shoaling water, first, by decreasing slightly. But when the water has shoaled to a depth about one-twentieth of the deep-water wave length, the wave height begins to increase rather sharply. This height increase continues to the point of wave breaking. A height increase of more than 50% may ultimately result. Perhaps this is why, when a low swell is approaching the shore and the sea is otherwise calm, the swell seems to rise mysteriously from the water only a few meters from the beach. Actually the swell came from deep water, but was so low there as to be scarcely noticeable.

FIGURE 19.26 On a straight shoreline, crests of waves approaching obliquely are bent so as to meet the shoreline more directly. (Redrawn, by permission of the publisher, from A. N. and A. H. Strahler, *Modern Physical Geography*, Figure 22.13. Copyright © 1978 by John Wiley & Sons, Inc.)

FIGURE 19.27 Wave refraction along an embayed coast. (Redrawn, by permission of the publisher, from A. N. Strahler, *Physical Geography*, 4th ed., Figure 32.9. Copyright © 1975 by John Wiley & Sons, Inc.)

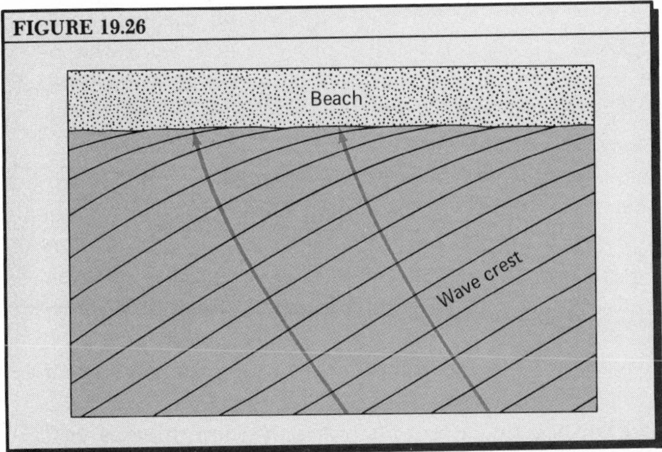

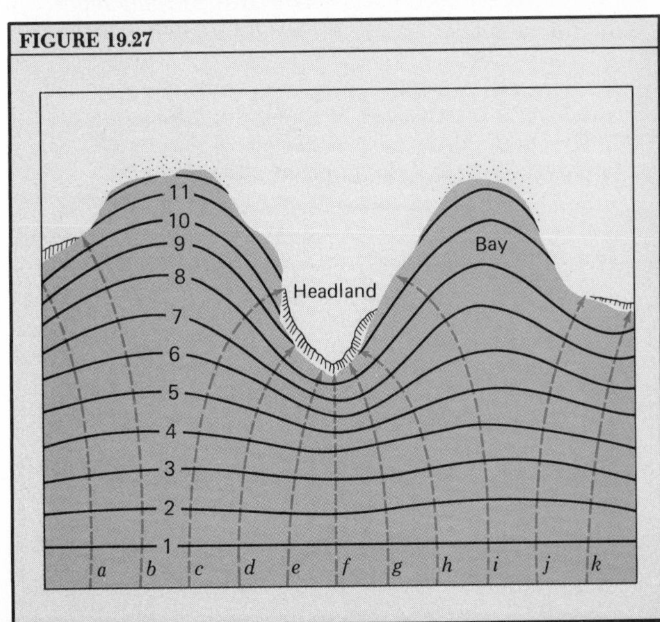

Wave Refraction

Suppose that waves were to approach a perfectly straight shoreline with crests parallel with the shore. Suppose, also, that the bottom were uniformly sloping seaward at all points. Under these conditions the sea surface as viewed from directly above would show parallel wave crests. You would also find that the crest lines become more closely spaced as they approach the shoreline. Conditions such as we have supposed are most unlikely in nature because wave crests usually approach the shore at an angle, or the shore is irregularly curved or has unequal bottom slopes and depths from place to place. Because the water depth under a single wave crest is different from point to point, the reduction of wave speed does not take place evenly. As a result the wave becomes curved by the process of *wave refraction*.

The case of a straight shoreline with uniformly sloping bottom is shown in Figure 19.26. That part of the wave which has arrived first has been slowed most, and its wave crest is bent into a curve convex toward the shore. Although the waves tend to be turned to approach the shoreline directly, the degree of turning is not great enough to achieve this limit. As a result, the waves break obliquely on the beach.

The case of an irregular shoreline, illustrated in Figure 19.27, is that of refraction about a projection of land, or *headland*, flanked by bays on either side. Notice that the wave crests are bent concavely toward the projecting headland, but convexly toward the axis of the bay. From the close spacing of the wave crests opposite the headland we infer that shallow water, encountered here first, reduces the wave speed while waves are able to travel into the deeper water of the bays with little delay. In a general way, the wave crests are being bent into rough conformity with the features of the shoreline.

Because of refraction, the distribution of energy contained in the waves is strikingly changed as the shore is approached. Assume, first, that the waves in question are large, uniform swells with even crests parallel to one another in deep water, as at the bottom of the map. We can divide the wave into equal units, shown in Figure 19.27 by the points *a*, *b*, *c*, *d*, etc. Each segment of the wave will now contain the same amount of energy, about half of which is traveling landward with the wave form. From each of the lettered points, a dashed line is drawn landward in such a way that the line always crosses the wave crests at right angles. We can call these energy-flow lines. Between any two such lines lies the zone of landward travel of the original quantity of wave energy. Where the energy-flow lines converge, as they do in front of the headland, wave energy is crowded into a narrower zone and concentrated, much as a convex lens gathers the sun's rays to a focal point. Here the waves increase greatly in height and produce large breakers upon the shore.

In contrast, where energy-flow lines diverge, as they do in the bays, the original unit of wave energy is spread more thinly along a greater length of wave crest, causing the wave to be weakened and to lose height. As a result, waves break with little force along the head of the bay. If the bay is long and narrow, it may have little or no surf on its beaches.

Currents Caused by Breaking Waves

As a steepening wave continues to travel shoreward, encountering still shallower water, the crest height increases sharply, and the forward slope of the wave becomes greatly steepened. At a critical point, the wave crest seems to leap forward and the wave form disintegrates into a mass of turbulent water, the *breaker* (Figures 19.24 and 19.28).

After a breaker has formed and collapsed, the wave is transformed into a landward moving sheet of highly turbulent water, the *swash* (or *uprush*) (Figure 19.28). Most shores have a rather gently sloping surface of sand or rock up which the swash of large waves can surge to reach a point a meter or more above the still water level. Opposing the landward flow is a component of the force of gravity acting in the downslope (seaward) direction. The swash is also acted upon by frictional resistance with the surface over which it passes, and thus it is rapidly slowed and finally stopped. The water then begins to pour seaward down the slope in a reverse flow termed the *backwash* (or *backrush*). Return flow is generally less turbulent and shallower than the swash.

Swash and backwash together constitute alternating water currents exerting a frictional drag against the surface over which they move. The currents are capable of moving rock particles of a wide range of sizes, from fine sand to cobblestones and boulders, depending upon size of the breaking waves and steepness of slope of the beach.

In the nearshore zone of steepening and breaking waves, another set of local currents is established. These currents are very important in shaping the shoreline and in transporting sand from one place to another.

The oblique approach of waves wind-driven toward a uniformly sloping straight shore sets up a current parallel with the shore and most strongly developed within the breaker zone (Figure 19.29). Called a *longshore current*, this flow owes its development to the slow shoreward movement of surface water by steep waves. Water brought to shore tends to raise the water level and to escape by flowing parallel to the shore. The shoreward drift of water by surface drag of the wind may also contribute to the rise of water level. Current speeds of 0.5 to 1 m/sec are commonly developed by a longshore current.

Another kind of current in the breaker zone flows in narrow, jetlike streams directly seaward at nearly right angles to the shoreline (Figure 19.30). Called a *rip current*, it crosses the breaker zone at irregular intervals of time and of horizontal spacing, although it sometimes occurs with a rather regular period and even spacing. Upon reaching deeper water, the rip current slackens and spreads out into a broadened *rip head*. Flow of a rip current may last only a few minutes, or it may continue for an hour or two. Rip currents represent the surgelike escape of water brought laterally

FIGURE 19.28 Development of a plunging breaker. (From W. M. Davis, 1912, *The Explanatory Description of Landforms*, Teubner, Leipzig, Figure 185.)

along the shore zone by feeder currents set up by unequal breaker heights. Where feeder currents converge from both directions at a point of low breaker height, the flow is diverted seaward in a narrow jet. A single rip current may have a length of 100 to 800 m and speeds as high as 1 m/sec; thus, they may carry unwary swimmers out to sea. The effects of erosion by rip currents are seen in shallow channels scoured through submarine sandbars.

FIGURE 19.29 Longshore currents set in motion by oblique approach of waves. (Redrawn, by permission of the publisher, from A. N. Strahler, *Physical Geography,* 4th ed., Figure 32.12. Copyright © 1975 by John Wiley & Sons, Inc.)

FIGURE 19.30 Sketch map of rip currents and longshore currents. (Based on data of U.S. Department of the Navy.)

Storm Wave Erosion

Storm waves breaking against a natural rock cliff or an artificial seawall deliver tremendous, pistonlike blows. The attack is capable of undermining the cliff and dislodging boulder-sized masses of rock or masonry (Figure 19.31). The force of a breaking storm wave can run as high as 10,000 newtons per square meter. Blocks of stone and concrete weighing from 2 to 10 metric tons, have been moved about and lifted 2 to 3 m vertically by storm waves striking a seawall.

Storm wave erosion, particularly severe in weak

FIGURE 19.31 Storm waves breaking against a seawall at Hastings, England. (Photographer not known.)

FIGURE 19.29

FIGURE 19.31

FIGURE 19.30

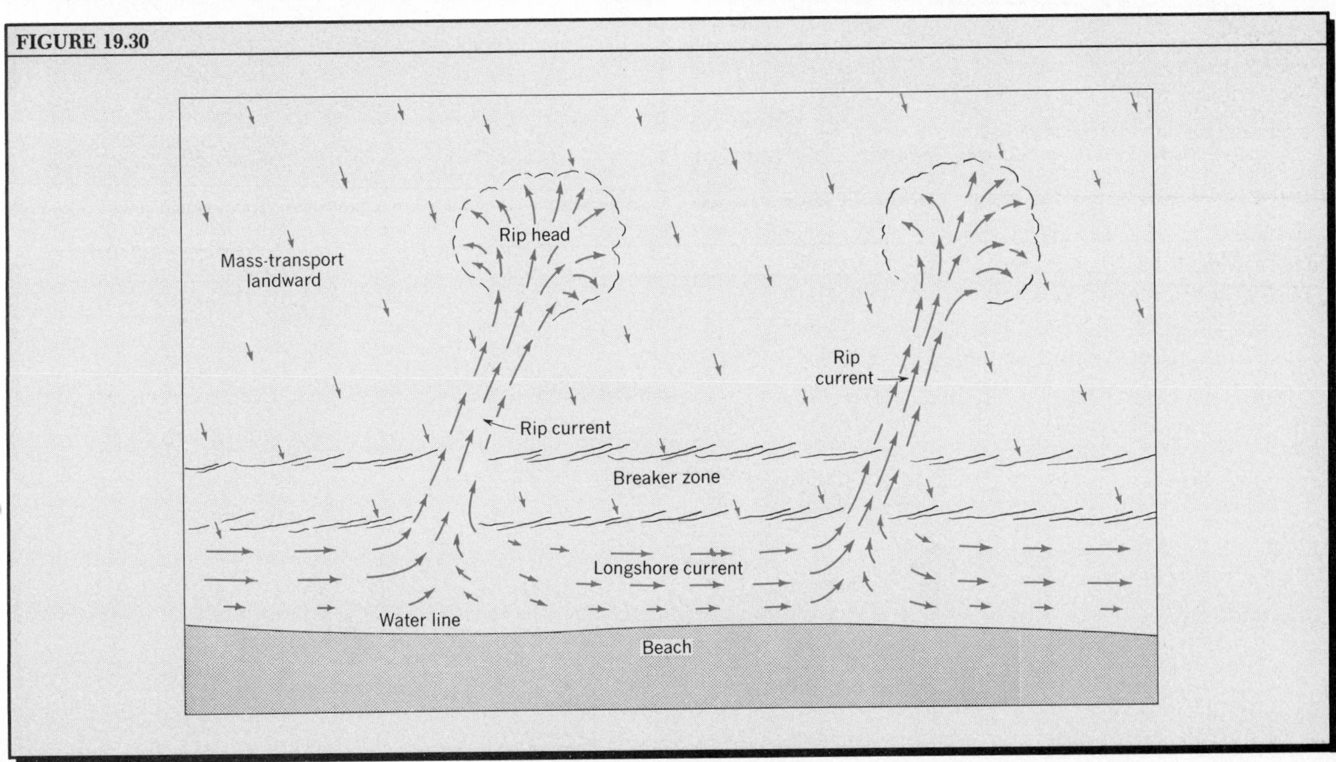

regolith and in glacial drift, affects humans adversely on exposed coasts (Figure 19.32). The outer shore of Cape Cod, facing easterly Atlantic storms, is a particularly good example. Formed of glacial sands and clays, this coastline has retreated at a rate of about 1 m/year for at least a century. The Highland Light, a Coast Guard beacon located at the brink of a 40-m marine scarp of glacial deposits, has had to be moved back repeatedly.

Throughout this chapter, we will use the term *shoreline* to mean the shifting line of contact between water and land. The broader term *coastline*, or simply *coast*, includes not only all of the beach zone, but also rock cliffs (where present), dunes, and such related features as may parallel the beach.

FIGURE 19.32 Storm waves rapidly undermined this marine cliff composed of weak glacial sands and gravels. Suffolk, England. (H. M. Geological Survey. Crown copyright reserved.)

FIGURE 19.32

Marine Cliffs

Along a coast of hard bedrock, a gently inclined rock surface, the *abrasion platform*, is gradually carved to accommodate the swash and backwash. It is usually completely covered at high tide but exposed in a broad zone, often many meters wide, at low tide (Figure 19.33). Fragments the size of pebbles and cobbles, serving as tools for abrasion, litter the surface of the wave-abraded platform, but fine particles are moved seaward into deeper water.

A shoreline being carved into bedrock develops a steep *marine cliff*, rising abruptly from the inner

FIGURE 19.33 This wide abrasion platform lies exposed at low tide. Notice the pocket beach at lower left. Pacific coast, south of Cape Flattery, Washington. (Photographer not known.)

FIGURE 19.34 Landforms of sea cliffs. A = arch; S = stacks; C = cave; N = notch; P = abrasion platform. (Drawn by E. Raisz. Redrawn, by permission of the publisher, from A. N. Strahler, *Physical Geography*, 4th ed., Figure 32.5. Copyright © 1975 by John Wiley & Sons, Inc.)

FIGURE 19.33

FIGURE 19.34

edge of the abrasion platform (Figure 19.34). The swash of storm waves thrusts rock fragments with great violence against the cliff base, eroding the cliff base and developing a *wave-cut notch.* Undermining leads to the fall of masses of bedrock from the cliff face, furnishing blocks for fragmentation into sediment. Weaker places in the bedrock are more rapidly excavated, resulting in the formation of crevices and *sea caves.* Remnants of bedrock projecting seaward are sometimes cut through to produce *sea arches;* these may collapse, leaving columnar rock *stacks* (see Figure 5.17).

FIGURE 19.35 Cobblestones form this multiple-crested beach at Smith Cove, Guysboro, Nova Scotia. The beach forms a crescent between rocky headlands. (Maurice L. Schwartz.)

FIGURE 19.36 Characteristic elements and zones of the profile of a sand beach.

FIGURE 19.35

Beaches

A *beach,* a relatively thick accumulation of sand, gravel, or cobbles in the zone of breakers and surf, is a depositional landform in contrast to the abrasion platform and marine cliff, which are erosional landforms. In certain respects, beaches are analogous to the alluvial deposits of a floodplain, particularly to the point-bar deposits of meandering rivers or to alluvial fans. In all these cases, moving water shapes excess quantities of detritus into sorted and layered deposits.

Beaches are formed of sediment produced by wave erosion or brought to the shore at various points by streams draining the land. Clay and silt, readily held in suspension, are unable to remain in the zone of breakers and surf. These fine sediments diffuse seaward to settle upon the continental shelf in deeper water.

The first beach to form on a coastline with an abrasion platform and cliffs is the *shingle beach,* composed of well-rounded fragments the size of pebbles, cobbles, or, rarely, small boulders. Shingle beaches are narrow and have very steep seaward and landward slopes (Figure 19.35). As swash rides up a shingle beach the water rapidly percolates into the coarse shingle. As a result, all the water may be absorbed and no backwash can form. Particles are thrust up the beach and accumulate in a steep slope. Shingle beaches usually form first in the most sheltered locations—in reentrants and bayheads between rocky promontories. These are called *pocket beaches* (Figure 19.33). They are commonly crescentic in plan and are concave toward the sea.

Large quantities of coastal sediment are available where large rivers are building their deltas or where the coastline is composed of sandy sedimentary strata or of glacial outwash sands. On such coasts beaches are broad and continuous. Figure 19.36 shows in profile the characteristic form elements of such sand beaches. The *foreshore,* bounded on the landward side by the limit reached by swash at high tide and on the sea-

FIGURE 19.36

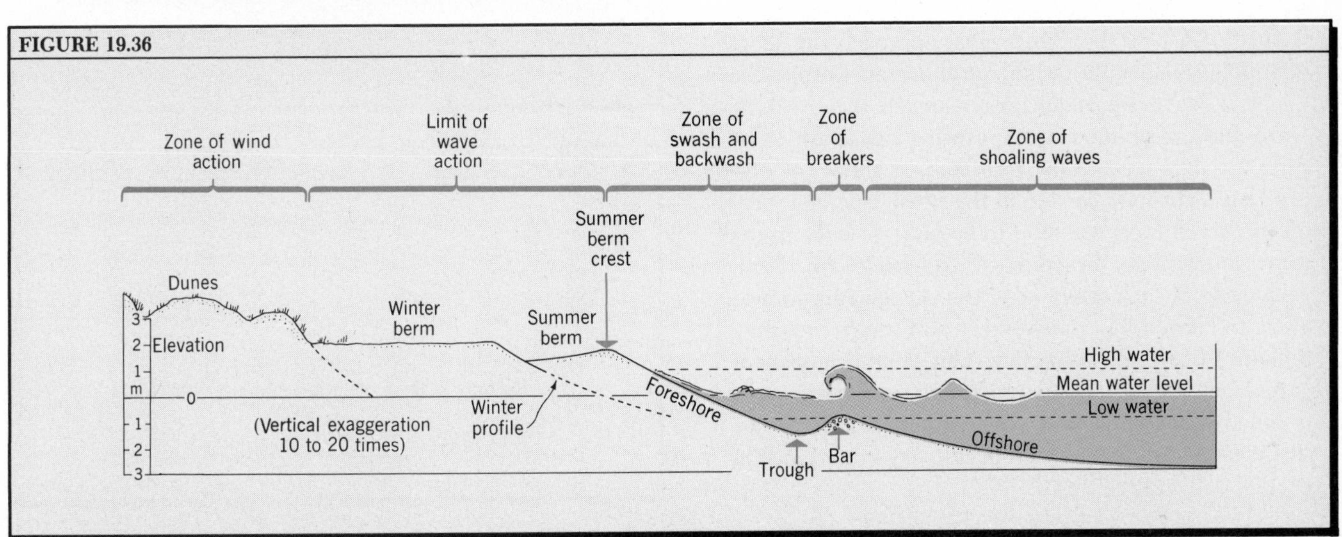

ward side by the breaker zone, is the sloping zone over which the swash and backwash act. Commonly a gravel *bar* is present beneath the point where the breakers form. Although usually covered by water, the bar crest is exposed at low tide. Between the bar and the foreshore slope is a *trough*, representing the zone of extreme turbulence at the point where the breakers collapse. Seaward of the bar lies the *offshore*, a sloping sand surface extending out into deeper water and covered at all times by water. No clearly defined outer limit can be set for the offshore. Two, three, or more bars separated by troughs exist in the offshore zone of some coasts.

During the summer, when waves are small most of the time, swash action moves sand inshore from deeper water, causing a summer accumulation, or *progradation*, of the beach. In the winter, large, steep storm waves tend to cause a sand movement in the seaward direction, depleting the beach and increasing the water depth in a process called *retrogradation*.

Progradation by summer waves builds a new sand deposit, the *summer berm*, taking the form of a bench several meters wide (Figure 19.36). The berm crest, or line of contact between berm and foreshore, is commonly slightly higher than the inner part of the berm, and swash of high tides may spill over the berm. Landward of the summer berm there is typically present a somewhat higher bench, the *winter berm*, built by the swash of large waves. In rare storms of great violence, the entire winter berm may be cut away and the waves may reach the belt of foredunes lying landward of the beach. After such extreme retrogradation, the winter berm will be built back.

Littoral Drift

So far, in describing beaches, we have assumed that the wave crests are parallel with the water line and that each wave breaks at the same instant along its entire length. If that were the case, sediment moved by swash and backwash would travel landward and seaward straight up and down the slope of the beach. Actually, on most beaches, waves approach the shore obliquely (diagonally) at almost all times. Despite the effect of wave refraction, waves reach the breaking point with an oblique trend, causing the swash to be directed obliquely up the foreshore, as shown in Figure 19.37. Particles carried in the swash ride obliquely on the beach face, but tend to be brought back in the direct downslope direction by the backwash. With each cycle of such movement, the particles are moved along the beach by an increment of distance that may amount to several decimeters. This lateral movement is multiplied by countless repetitions, and the process is called *beach drift*. It accounts for transport of vast quantities of sediment and is of primary importance in develpment of various kinds of beach deposits along a coast.

In the offshore zone, the oblique approach of waves also results in lateral movement of sediment. As we showed in Figure 19.25, wave orbits in shallow water are transformed into simple back-and-forth water movements at the bottom. Sand thus dragged back and forth forms into very small ridges and troughs known as *oscillation ripples* (Figure 19.38). Now, if the wave approach is oblique to the contour of the slope, the sand not only will move back and forth, but will also creep parallel with the shoreline. This type of movement is similar to the beach drifting but very much slower.

Sediment movement in the offshore zone is also caused by longshore currents (see Figure 19.29), which are most strongly felt in the breaker zone. Sediment movement by this process is termed *longshore drift*. In deeper water of the offshore zone, beyond the breakers, weak longshore currents are superimposed on the

FIGURE 19.37 Beach drift of sand, caused by oblique approach of swash. (Redrawn, by permission of the publisher, from A. N. Strahler, *Physical Geography*, 4th ed., Figure 32.12. Copyright © 1975 by John Wiley & Sons, Inc.)

FIGURE 19.38 Oscillation ripples in sand. (From *A Geologist's View of Cape Cod*, copyright © 1966 by Arthur N. Strahler. Reproduced by permission of Doubleday & Co., Inc.)

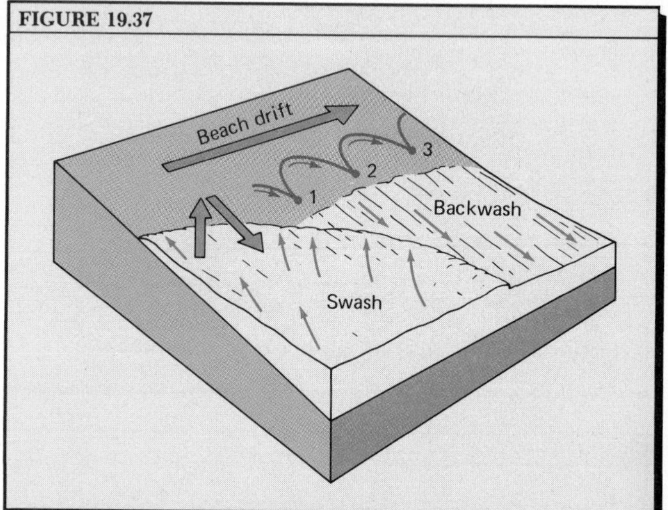

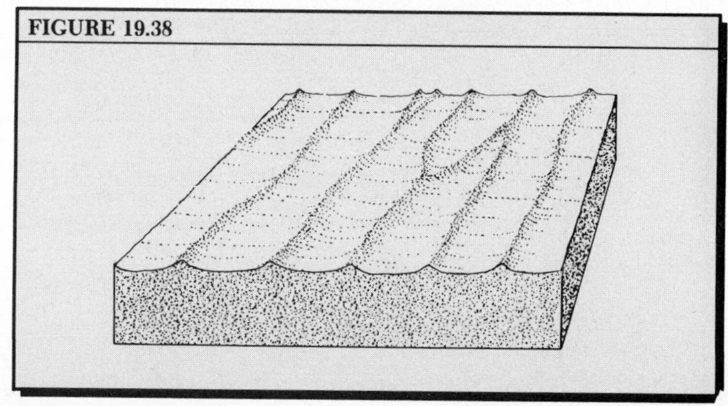

oscillations caused by shoaling waves. Here ripple marks show the current effect by developing an asymmetrical ridge form (Figure 19.39). The steep sides of the sand ridges lie on the downcurrent (lee) sides. Geologists make use of these ripple forms on bedding planes to plot the directions of currents in shallow marine waters that received sandy sediment in past geologic periods.

Both beach drift and longshore drift operate at the same time and in the same direction. Their combined effect in moving sediment is called *littoral drift;* it may operate on both straight and embayed coastlines, as illustrated in Figures 19.40 and 19.41. Along a straight coast, drift will carry sediment continuously along the beach, often for many tens of kilometers, much as bed load is carried by a river. However, if the coastline undergoes an abrupt change in direction, where a bay is encountered, sediment is carried out into open water. Here the sediment forms a *sandspit,* a fingerlike extension of the beach. Refraction of waves around the end of the spit causes the end to curve landward in a characteristic spiral of lessening radius, described as recurved. The spit may be thought of as analogous to a stream's delta, for it is a growing deposit of sediment built into open water.

The case of an embayed coast is shown in Figure 19.41. Wave refraction causes a height increase in waves and breakers against the promontories, resulting in erosion of an abrasion platform and a marine cliff. Sediment produced by abrasion moves by littoral drift along the sides of the bays, where wave approach is oblique. Sediment movement is directed along both sides of the bay toward the bayhead and accumulates there, forming the crescentic pocket beach, previously referred to in our discussion of shingle beaches.

We can readily predict the trend in evolution of an embayed coastline by imagining that the promontories are progressively eroded back while the bayheads are being filled by a widening beach. In due time, the result will be a nearly straight shoreline formed of sections of wave-cut cliff alternating with sections of broad beach. Thus, we recognize a fundamental law of shoreline evolution: Any shoreline of irregular map outline tends to be reduced in time to a simple straight (or broadly curving) shoreline along which the drift of sediment is continuously in one direction for any given direction of wave approach. Removal of irregularities in shoreline plan by wave action resembles in some ways the removal of falls and rapids by a stream to reach the condition of grade.

FIGURE 19.39 These ripple marks were formed on a nearly horizontal seafloor, but have since been tilted by mountain-making movements to an almost vertical attitude. Current direction is shown by an arrow. Precambrian quartzite strata in the Baraboo Range, Wisconsin. (A. N. Strahler.)

FIGURE 19.40 Littoral drift along a straight section of coastline, ending in a bay.

FIGURE 19.41 Littoral drift along bay sides on an irregular coastline.

FIGURE 19.39

FIGURE 19.40

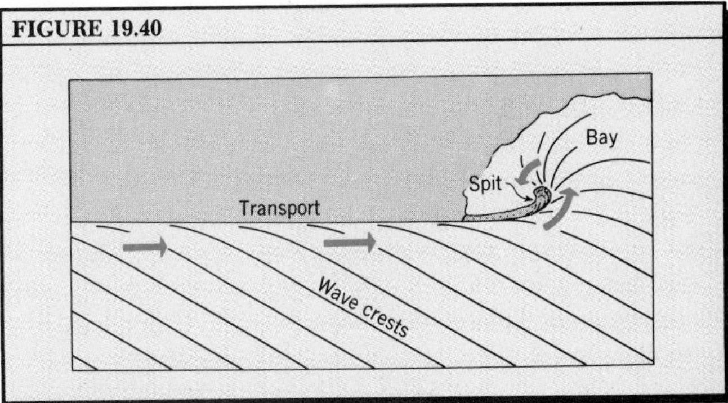

FIGURE 19.41

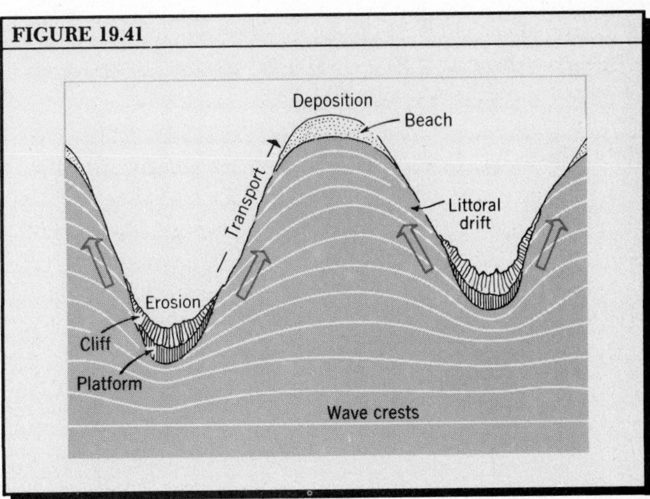

The Shore Profile of Equilibrium

Just as in the case of a graded stream, the continued action of waves produces an equilibrium beach profile where there is abundant sand and gravel with which to shape the profile and modify it rapidly in response to changing wave conditions. The continual modifications are analogous to the continual scour and fill of a stream channel in response to changes in discharge and load.

Figure 19.42 shows an idealized shore profile of equilibrium developed on a lens of sand, called a *surf lens*. This active lens, which can be reworked by wave action, extends seaward to the *surf base* at a depth of about 12 to 15 m. The profile is so adjusted that landward push of sand by forward drag of wave orbits and swash just balances the seaward movement of sand by backwash and reverse drag of wave orbits. At the same time, littoral drift of sand will be occurring in a direction parallel with the shoreline, but transverse to the profile. When equilibrium prevails, the mass flow of sand through the profile by littoral drift is so adjusted that sand neither accumulates nor is removed.

We can expect seasonal adjustments of the equilibrium profile such as we have already described, and there will also be adjustments accompanying alternate periods of storm waves and weak waves with low swells. These continual changes, normal in the regime of any shoreline, are cyclic fluctuations and cause no net long-term change in the equilibrium profile.

Such long-term changes in the equilibrium profile can be brought about by changes in rates of littoral drift of sand through the system. If more sand arrives than leaves, there is an accumulation upon the surf lens. The profile is then built out in the seaward direction. The resulting progradation leaves a series of *beach ridges*, representing older berms (Figure 19.43*A*).

Progradation is sometimes concentrated in one locality along a coast because littoral drift converges upon that locality from both directions. A *cuspate foreland* is then constructed and builds seaward as a prominent cape. One striking example is Cape Canaveral in Florida, conveniently formed and situated to serve as a launching base for spacecraft (Figure 19.44).

Where more sand leaves than enters the surf zone by littoral drift, the profile will be moved landward. This change is one of retrogradation and may result in complete removal of the surf lens (Figure 19.43*B*). Then wave attack will begin again upon the underlying coastal materials.

*Wave Erosion and Shore Protection

Postglacial rise of sea level has brought the powerful erosive action of storm waves to bear high upon continental coasts. Great lengths of coastline consist of easily erodible materials, particularly glacial drift and alluvium. But once people have occupied these coastlines with expensive buildings and highways, they feel obligated to resist the wave action that cuts away their land.

FIGURE 19.43 *A*. Progradation, producing successive beach ridges. *B*. Retrogradation, resulting in cutting back of berm and removal of beach. (Drawn by A. N. Strahler.)

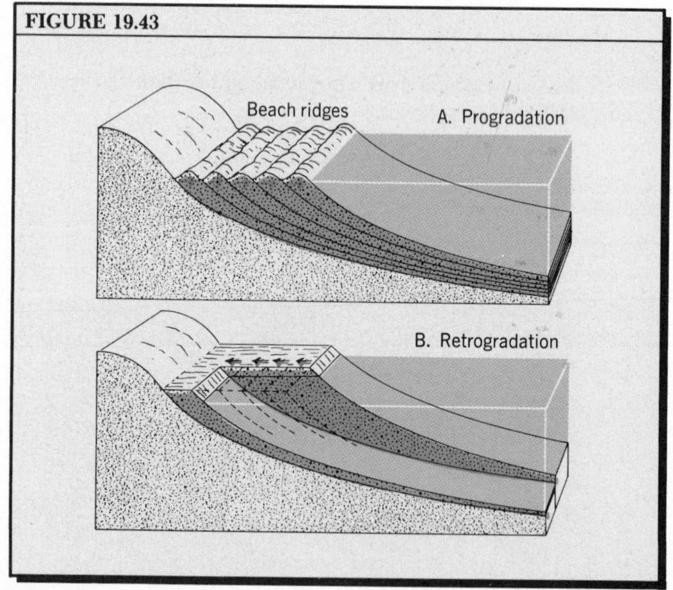

FIGURE 19.42 Idealized diagram of the equilibrium shore profile. Wave size is exaggerated. (Drawn by A. N. Strahler. Based on concepts of R. S. Dietz, 1963, *Geological Society of America, Bull.*, vol. 74, pp. 971–989.)

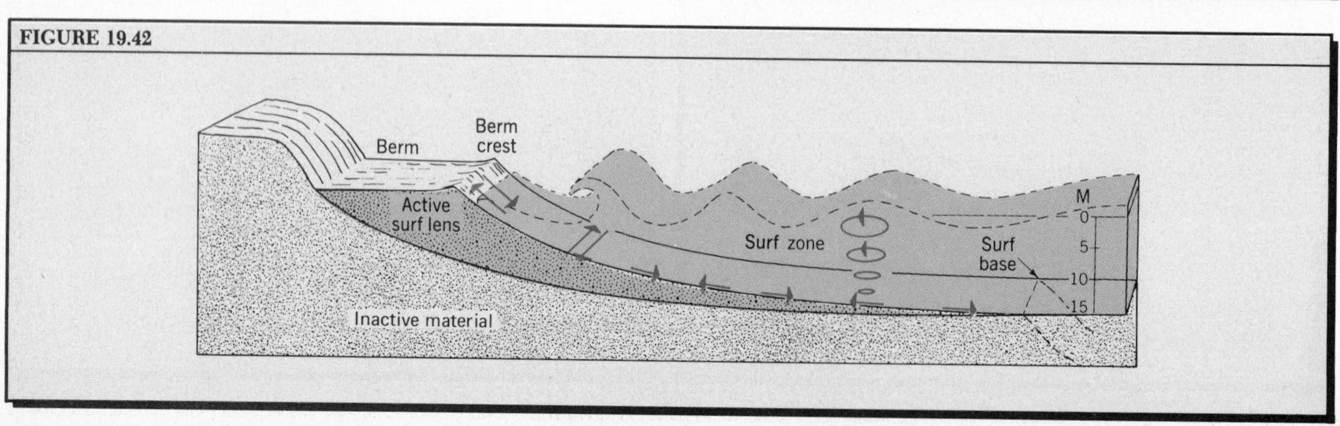

FIGURE 19.44 Map of Cape Canaveral, Florida, as it appeared about 1910, prior to human-induced modification. Ridges close to the Atlantic shore (*right*) are beach ridges, whereas those farther inland are dune ridges built upon older beach ridges. (Based on data of D. W. Johnson, 1919, *Shore Processes and Shoreline Development*, John Wiley & Sons, Inc., New York, p. 420, Figure 129.)

FIGURE 19.44

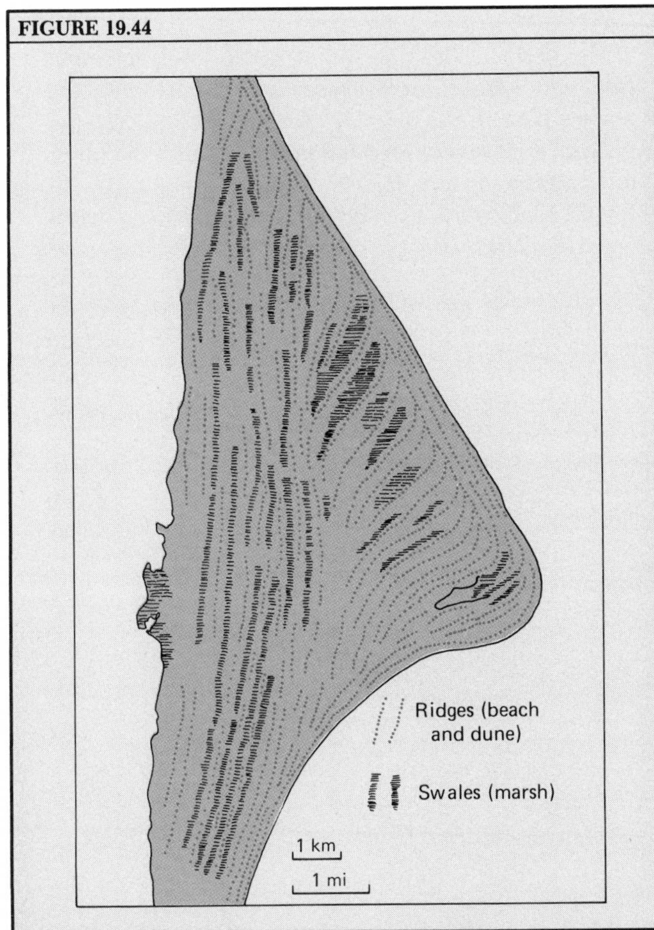

One example of human interference with the natural littoral drift is seen in the use of groins to collect beach sand. The widened beach forms a buffer zone that can absorb the short-lived but intense attacks of storm waves. A *groin* is a broad wall of huge rock masses built perpendicular to the shoreline, which interferes with the littoral drift of sand, causing a crescentic beach deposit to accumulate on the updrift side of the groin (Figure 19.45). In the case of the single groin shown in the diagram, the trapping of sand results in a deficiency of sand along the downdrift side, where the beach is depleted, and the shoreline is receding. Over a long stretch of beach, a large number of closely spaced groins is needed to retain the sand.

*Concept of the Littoral Cell

A continental coastline can be viewed as a succession of distinct compartments, called *littoral cells*. Each cell represents a complete open system of input of sediment to the shore zone from the land, transport of sediment along the shore, and output of sediment to the deep floor of the ocean.

Figure 19.46 is a schematic block diagram showing the major parts of a typical littoral cell of the Pacific coast of North America. Prevailing littoral drift is

FIGURE 19.45 Construction of a groin causes marked changes in configuration of the sand beach.

FIGURE 19.45

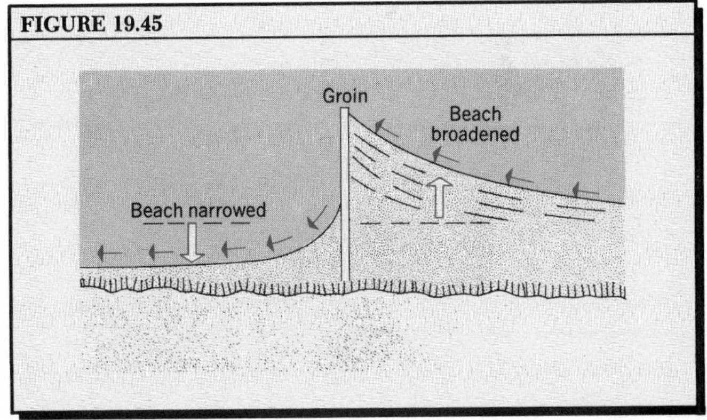

FIGURE 19.46 Schematic block diagram of a littoral cell, such as might be found along the coast of southern California. (Drawn by A. N. Strahler. Based on data of D. L. Inman and B. M. Brush, 1973, *Science*, vol. 181, p. 26.)

FIGURE 19.46

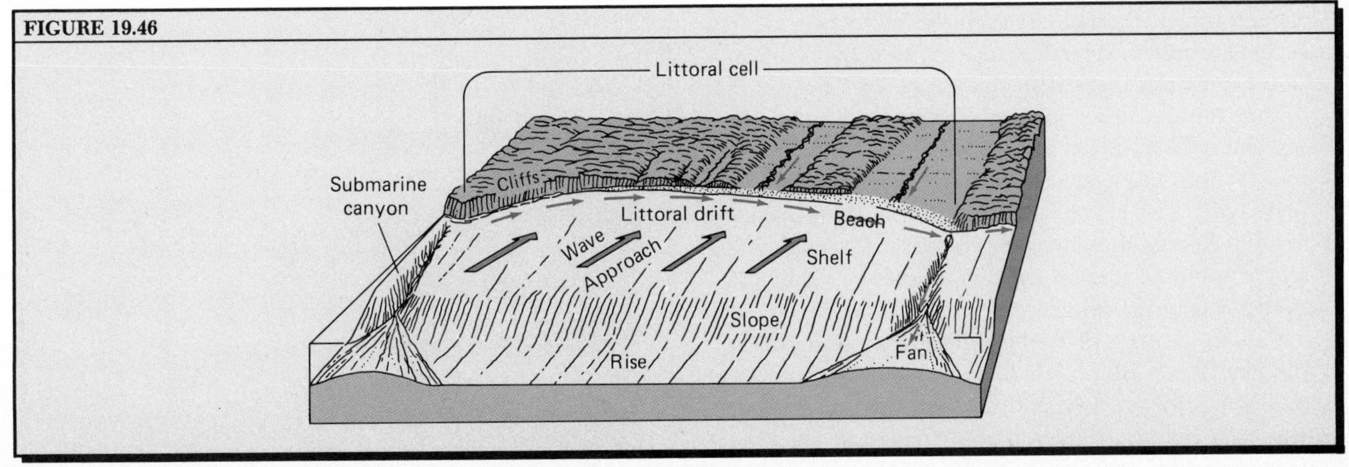

from left to right. Wave attack on rocky cliffs furnishes sediment for littoral transport. At some point along the coast, the rock abrasion platform begins to show a narrow beach, where streams contribute sediment in larger quantities, and a surf lens is formed. The beach then becomes a broad deposit. At the right, a deep submarine canyon, which trenches the narrow continental shelf, receives the sediment from the beach and conducts it to a deep-water fan built out from the continental slope. Farther along the coast are the rocky cliffs of the next littoral cell.

For most littoral cells, the major contribution of sediment (about 95% on a world basis) is from streams, rather than from cliff erosion. Many coasts lack cliffs entirely. Not shown in the diagram is a possible sediment contribution by wind because, as a rule, sand in coastal dunes moves from beaches toward the land. This movement is a local reversal of the main flow within the cell, but ultimately streams will carry the dune sand back to the shore.

Because beaches depend largely on streams for their input of sand, the construction of large dams poses a threat to beaches. Sediment that is trapped and held behind dams fails to reach the ocean and the beaches become undernourished, narrower, and may disappear entirely. With the beaches removed, wave action on marine cliffs is intensified and the cliffs retreat more rapidly. These unfavorable responses to dam-building are particularly marked along the California coast, where most of the larger streams have been dammed to provide local water storage reservoirs.

Elevated Shorelines and Marine Terraces

The development of a shoreline is sometimes interrupted by a sudden rise of the coast, forming an *elevated shoreline*. The marine cliff and abrasion platform are abruptly raised above the level of wave action, and the former abrasion platform is now a *marine terrace* (Figure 19.47). Denudation processes begin to destroy the terrace and eventually will obliterate it entirely.

Elevated shorelines are common along the continental and island coasts of the Pacific Ocean where rapid tectonic uplift is occurring adjacent to active subduction boundaries. Repeated uplifts result in a series of marine terraces in a steplike arrangement. Block faulting along the California coast is another form of tectonic activity, associated with dislocation of the Pacific plate lying west of the San Andreas transform boundary. Offshore islands along this coast show remarkable sets of multiple marine terraces, as on San Clemente Island (Figure 19.48).

Eustatic changes of sea level that took place during the Pleistocene Epoch left their marks in the form of wave-cut benches both above and below the present sea level. Because these eustatic changes of sea level were worldwide, they were superimposed on the more-or-less uniform tectonic uplift of coasts associated with active plate boundaries. Thus, the full interpretation of multiple marine terraces is an extremely complex problem.

The Ocean Tide and Tidal Currents

The periodic rise and fall of ocean level, or *ocean tide*, was known for centuries to be related to the moon's path in the sky, but it was not until Newton published his law of gravitation in 1686 that the physical explanation was understood. The broad arrow in Figure 19.49 represents the attraction the moon exerts

FIGURE 19.47 A marine terrace is the old abrasion platform of an elevated shoreline. Alluvial fan deposits are beginning to form a cover. (Drawn by A. N. Strahler.)

FIGURE 19.48 These elevated marine terraces form a staircase on the western slope of San Clemente Island, off the southern California coast. There are more than 20 terraces in this series; the highest is 400 m above sea level. (John S. Shelton.)

FIGURE 19.47

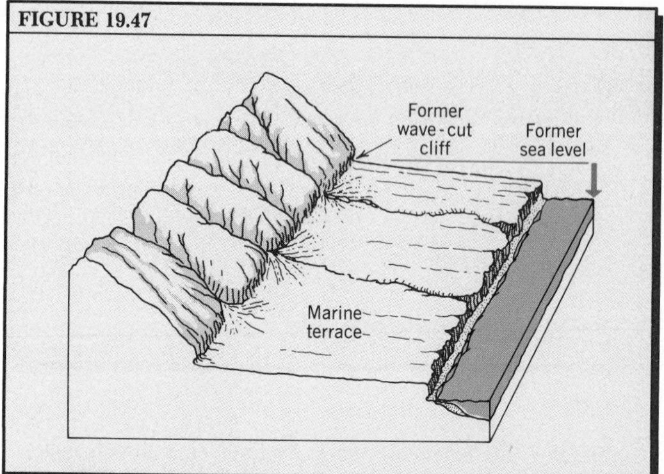

FIGURE 19.48

upon the earth. On the side nearest the moon, point *T*, the attraction is stronger than at the earth's center, point *C*, because gravitational force decreases with an increase in the distance separating two masses. For the same reason, the attractive force is even less at point *A*. These differences in force tend to distort the spherical shape of the earth into a prolate ellipsoid, shaped somewhat like an American football.

Although the solid lithosphere makes only a very small response to this earth-stretching force, the oceans respond freely. As Figure 19.50 shows, the ocean water tends to move along the lines indicated by the surface arrows. The water moves toward two centers, one at *A* and one at *T*, but moves away from a belt girdling the globe on line passing about through the poles (*N* and *S*).

Next, we realize that the earth is rotating on its axis. This means that the two centers of tidal accumulation, or "tidal bulges," will be traveling continually around the earth. They will twice daily pass by a fixed point on the globe. Each bulge produces a rise of ocean level to a maximum value, or high water, and the actual interval between high waters is close to 12.5 hours. The belt of depressed surface will also pass by twice daily, so that we will have a minimum water level, or low water, approximately 6.25 hours following a high water. This is in fact the common tidal cycle found along most coastlines of the world.

Figure 19.51 shows a typical *tide curve* recorded throughout a 24-hr period. The difference in level between high water and low water, or tide range, was about 3 m in this particular case.

The rising tide sets in motion in bays and estuaries currents of water known as *tidal currents*. The relationships between tidal currents and the tide curve are shown in Figure 19.52. When the tide begins to fall, an *ebb current* sets in. This flow ceases about the time when the tide is at its lowest point. As the tide begins to rise, a landward current, the *flood current*, begins to flow. Tidal currents, when confined in narrow passages, are often very swift. They are capable of eroding and transporting large amounts of coarse sediment. At low water, large current ripple marks are exposed, indicating the vigor of the ebb current (Figure 19.53).

FIGURE 19.49 The moon's attractive force tends to distort the earth into an ellipsoid. (Redrawn, by permission of the publisher, from A. N. Strahler, *Physical Geography*, 4th ed., Figure 6.6. Copyright © 1975 by John Wiley & Sons, Inc.)

FIGURE 19.50 Tidal forces tend to move the ocean water toward two centers, on opposite sides of the globe, along a line connecting earth and moon.

FIGURE 19.49

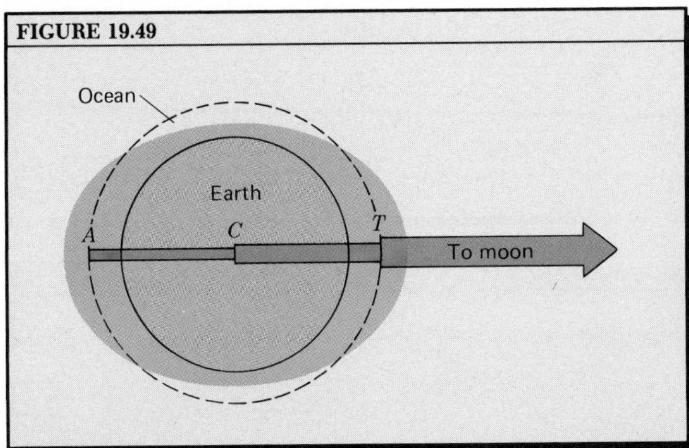

FIGURE 19.50

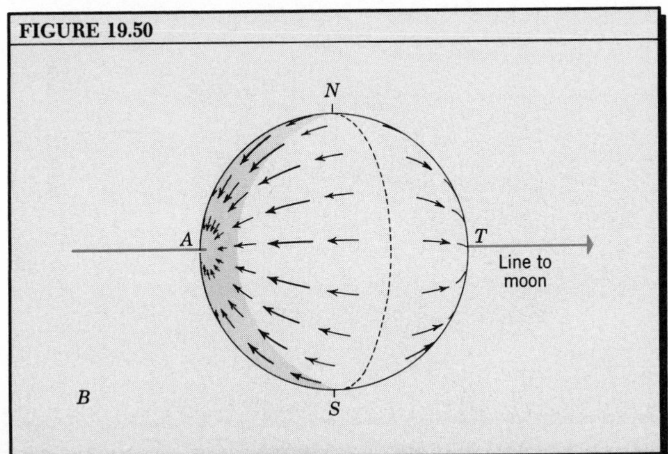

Tidal Flats and Salt Marshes

Bays and lagoons largely shut off from the open ocean are gradually filled by layers of clay and silt brought into the quiet water by streams draining the land. The sediment is distributed over the bay by tidal currents.

Organic matter, both that carried in suspension in

FIGURE 19.51 A graph of the rise and fall of tide at Boston Harbor for a 24-hr period. Dots show water level at half-hour intervals. (Redrawn, by permission of the publisher, from A. N. Strahler, *Physical Geography*, 4th ed., Figure 6.8. Copyright © 1975 by John Wiley & Sons, Inc.)

FIGURE 19.51

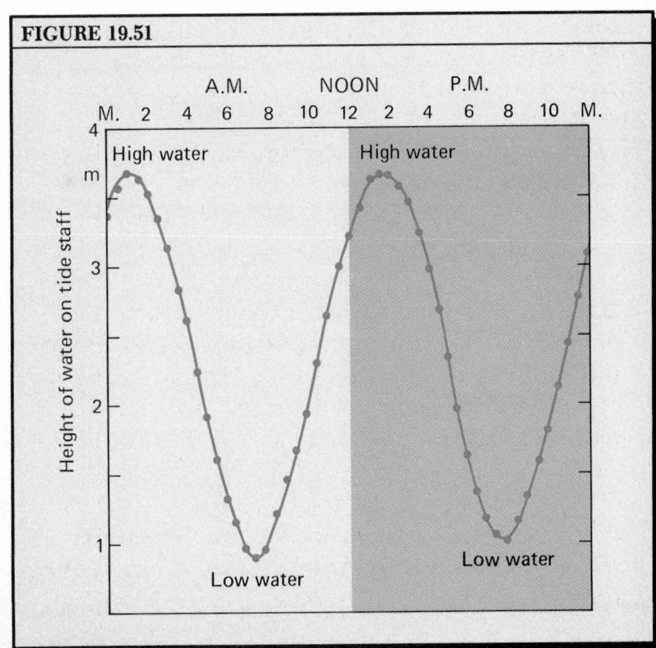

streams and that produced by growth of plants and animals on the bottom, makes up a substantial proportion of the sediment. Gradually the sediment is built upward, until the upper surface is approximately at the level of low tide. The result is a mud flat, or *tidal flat*, exposed at low tide but covered at high water.

Salt-tolerant vegetation takes hold upon the tidal mud flats, eventually forming a resistant mat of plant roots. More sediment is trapped by plant stems, and the deposit is built up to the level of high tide by a layer of peat. The completed surface is described as a *salt marsh*. Tidal flats and tidal marshes encroach in succession upon the open water until the entire bay or lagoon is filled, except for a system of meandering tidal channels (Figure 19.54).

Barrier Islands and Their Beaches

In extreme contrast to embayed coasts of strong relief with marine cliffs of bedrock are those coasts along which the water line has come to rest against a gently sloping plain of very low relief. This kind of coast might result from the epeirogenic uplift of a coastal plain, causing the inner edge of the continental shelf to emerge. Because the shelf is a site of accumulation of sedimentary strata, we should expect a very smooth surface to appear and the resulting water line to be simple in plan.

Figure 19.55 shows the development of a part of the Gulf Coast of Texas, along which an epeirogenic upwarping brought marine strata of late Cenozoic age above sea level. As we know—from the fact that wells are drilled for petroleum many kilometers out to sea along the Gulf Coast—the submarine topography is one of very gradual deepening with distance from shore. Block *A* represents conditions about 5000 years ago, when sea level had risen rapidly following the final melting of the Wisconsinan ice sheets. The shoreline then rested against a gentle slope. Breaking storm waves quickly threw up a sand bar immediately

FIGURE 19.52 The ebb current flows seaward as tide level falls. The flood current flows landward as tide level rises. (Redrawn, by permission of the publisher, from A. N. and A. H. Strahler, *Modern Physical Geography*, Figure 22.26. Copyright © 1978 by John Wiley & Sons, Inc.)

FIGURE 19.53 Large current ripples produced by the ebb tide, Avon Estuary, Nova Scotia. Current direction is indicated by the arrow. (Canada Dept. of Mines and Technical Survey, Ottawa.)

FIGURE 19.54 Tidal creek with sinuous meanders in coastal salt marsh, Rock Creek, Orleans, Massachusetts. (Harold L. R. Cooper, Cape Cod Photos, Orleans, Mass.)

FIGURE 19.52

Height of tide (m)
2
1
0
1
2
High water
Low water
No current
No current
Ebb current (seaward)
Flood current (landward)
0
3
6
9
12
15
Time (hours)

FIGURE 19.53

FIGURE 19.54

seaward of the breaker line. Scour took place over a broad zone seaward of the bar. The new bar was rapidly built above sea level by swash, creating a narrow *barrier island,* which cut off from the open sea a narrow zone of shallow water, the *lagoon,* lying between the barrier and the mainland.

Sea level continued to rise, and the barrier island continued to be built upward by the addition of sediment brought by littoral drift from mouths of larger streams. Simultaneously, the lagoon floor received sediment brought by many smaller streams that drained into it, so the lagoon remained shallow despite the rising sea level. The barrier island became wider as new storm berms were added, and these have given a corrugated form to the island. Sand dunes were formed by strong onshore winds. The sand was blown into the lagoon beyond, where it now forms a deposit overlying silts and clays previously laid down. As we find this coast today (Block *B*), large areas of the lagoon are filled to elevations of 0.5 to 1.0 m above sea level by sands, silts, and clays, forming mud flats.

Barrier islands occur along much of the Atlantic and Gulf coastal plain of the United States, along the south shore of Long Island, New York, and in other parts of the world. The explanation we have given for the Texas barrier island is only one of several possible explanations that have been advanced for barrier islands.

Barrier islands are usually broken every few kilometers by a gap called a *tidal inlet*. Strong ebb and flood currents flow through these inlets, scouring them strongly and keeping them from being closed by littoral drift of sand along the outer shoreline. Looking down on such an inlet, we will usually see a fan-shaped *tidal delta* built into the lagoon (Figure 19.56).

FIGURE 19.56

FIGURE 19.55 Development of a barrier island along the Gulf Coast of Texas. (Suggested by data of H. N. Fisk, 1959, *Proc. Second Coastal Geography Conf.*, National Academy of Science, Washington, D.C.)

FIGURE 19.56 A great storm in March 1931 cut this breach through Fire Island, a barrier island on the south shore of Long Island, New York. Tidal currents quickly carried sediment from the open Atlantic Ocean (*left*) into the lagoon (*right*) and constructed a tidal delta. The entire area shown is about 1.6 km across. (U.S. Air Force Photograph.)

FIGURE 19.55

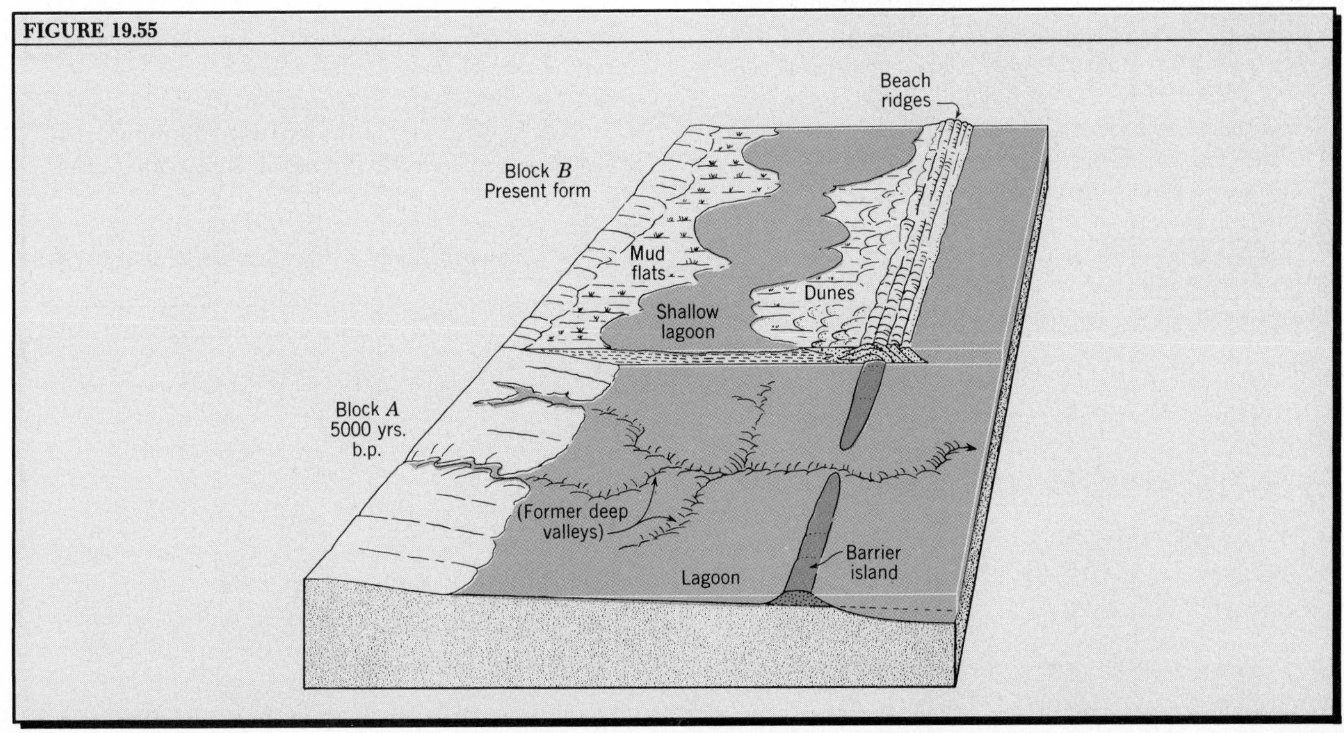

Tidal inlets sometimes form during great storms, when swash overtops the barrier and erodes a trench into the lagoon. The inlets may be closed later by rapid littoral drift. Littoral drift may also cause the barrier island to show an overlap pattern at an inlet (Figure 19.57), and the inlet gradually migrates in the direction of littoral drift.

Common Types of Coastlines

There are many kinds of coastlines. They reflect many of the geologic processes we have covered in previous chapters. Each variety of coastline is unique because of the distinctive landmass against which the shoreline has come to rest. A rising of the sea level or a sinking of the crust results in partial drowning, or *coastal submergence*, of landforms shaped by such terrestrial agents as running water or glacial ice. On the other hand, *coastal emergence*, caused by either a lowering of sea level or a crustal rise, results in exposure of submarine landforms. Emergence and submergence each produce distinct coastlines. Still another class of coastlines results when new land is built out into the ocean by volcanoes and lava flows, or by growth of deltas or coral reefs.

We will describe a few important types of coastlines as examples and illustrate them in Figure 19.58. A deeply embayed coast, resulting from submergence of a fluvially dissected landmass, is called a *ria coast* (*A*); it has many offshore islands. A *fiord coast* results from the submergence of deeply carved glacial troughs (*B*); it has long, narrow bays with steep rock walls. Emergence of a smooth, gently sloping coastal plain is often accompanied by the construction of an offshore barrier island, paralleling the inner shoreline; this is a *barrier-island coast* (*C*). Large rivers build elaborate deltas, producing *delta coasts* (*D*). Volcanoes and lava flows, built up from the ocean floor, produce a *volcano coast* (*E*), while reef-building corals create new land and make a *coral-reef coast* (*F*). Downfaulting of a landmass can bring the shoreline to rest against a fault scarp, producing a *fault coast* (*G*).

*Marine Planation of Continents

It is conceivable that, given sufficient time and stability of the crust, wave action in the surf zone might ultimately plane off an entire continent, reducing it to a shallow submarine platform, in a process called *marine planation*. The supply of wave energy is practically limitless, since it depends upon wind energy, rather than upon the gravity-flow mechanism required by streams. Moreover, the deep ocean basins provide a sink for almost limitless quantities of detrital material derived from continental denudation.

The prospect of marine planation as a real event in the geologic past becomes somewhat more likely when we consider that fluvial denudation would have been active under the same set of stable conditions, and reduction of the landmass to a peneplain would minimize the volume of rock that wave action would need to remove. Even so, the process of marine planation would be extremely slow, because the equilibrium profile of the shore is adjusted to dissipate most of the wave energy in frictional resistance.

There is, however, a set of conditions under which marine planation may have been effective. The conditions are that a continent reduced to a peneplain undergoes a slow crustal sinking (a negative epeirogenic movement), or there is a slow eustatic rise of sea level. Slow submergence of either cause assists the action of waves in removing a substantial layer of soil, regolith, and bedrock. Deepening of the water by progressive submergence allows more wave energy to act in the surf zone, while at the same time making available space in which the sediments can be deposited.

Because marine planation is an event to be anticipated during the slow submergence of a landmass, we should look for evidences of such action in the unconformities of the geologic record. Two great unconformities of the inner Grand Canyon have been described in some detail in Chapter 6. (Refer to Figure 6.3 and 6.7.) The older of the two unconformities, that which bevels the Archean rocks, is remarkably even.

FIGURE 19.57 Migration of an inlet in a barrier island.

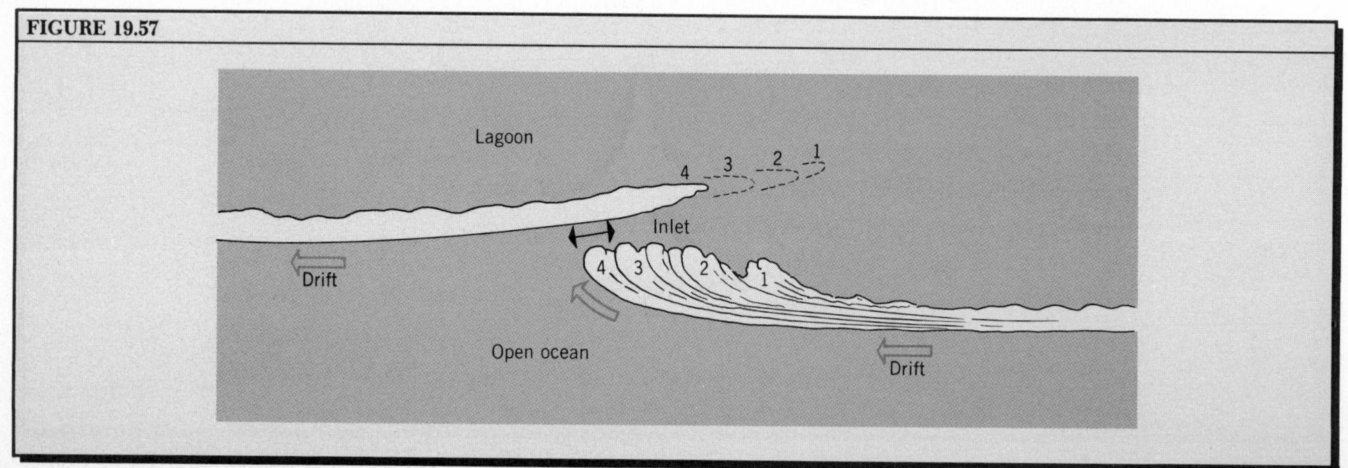

This tilted nonconformity appears to be a true peneplain of fluvial denudation, for it still retains a zone of weathered rock and regolith. Wave action of the advancing seas of Algonkian time was effective in removing only a small part of this weathered material. Thus, submergence was rapid in this case.

The second unconformity lies beneath the Cambrian Tapeats sandstone. It is an angular unconformity that cuts across both Archean and Algonkian rocks. This erosion surface shows many knobs of resistant quartzite and was evidently quite hilly when submergence began. Marine action was only moderately effective in modifying this erosion surface. In places, all weathered rock was removed, for the rock beneath the unconformity is locally quite fresh. Knobs of quartzite show definite indications of wave cutting, for they have steep slopes. They also have basal accumulations of coarse debris, now conglomerate and breccia (Figure 19.59). Possibly wave action cut off large parts of the tops of the knobs as submergence progressed.

FIGURE 19.58 Some common types of coastlines. (Redrawn, by permission of the publisher, from A. N. and A. H. Strahler, *Modern Physical Geography*, Figure 22.29. Copyright © 1978 by John Wiley & Sons, Inc.)

The evidence from the Grand Canyon does not give support to the concept of marine planation as an important process in removing a rock layer of the continental crust. Nevertheless, there remains the possibility that marine planation has been important at other times in the geologic past.

FIGURE 19.59 Diagram of unconformity between Tapeats formation of Cambrian age and Algonkian strata of Precambrian age, Grand Canyon, Arizona. A large knob of Algonkian strata (*center*) resisted attack by waves of rising sea, approaching from the right. The section is about 200 m long. (Based on data of R. P. Sharp, 1940, *Geological Society of America, Bull.*, vol. 51, p. 1263, Figure 8.)

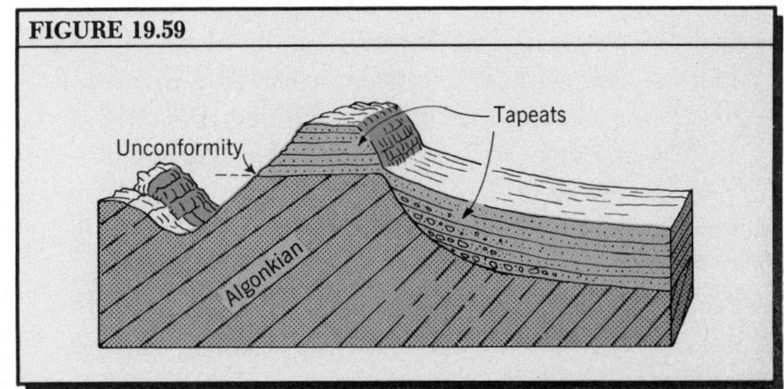

FIGURE 19.58

A. Ria coast

B. Fiord coast

C. Barrier-island coast

D. Delta coast

E. Volcano coast (*left*)
F. Coral-reef coast (*right*)

G. Fault coast

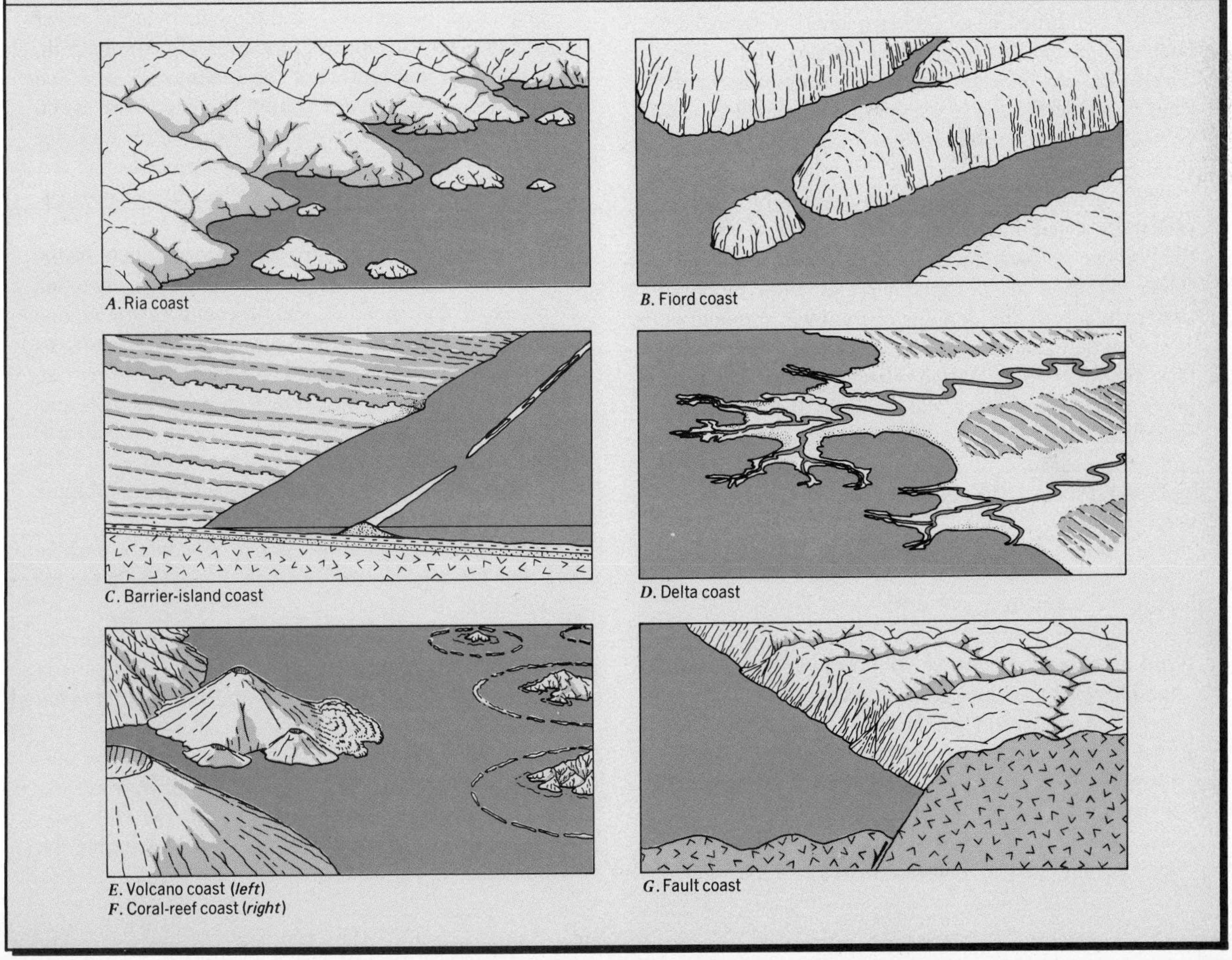

Wind and Waves as Geologic Agents

The intensive work of wind is limited to those specific places where unattached mineral particles are available in quantity, such as desert plains and coastlines. Where they take place, wind erosion and transportation are significant both in terms of environment of today and in the stratigraphic records of the geologic past.

During the Pleistocene Epoch, when ice sheets were widespread and large expanses of unprotected detrital materials were exposed to the air, the geologic role of wind was much more important than we find it today. Going far back into geologic time, the great fossil dune beds of Permian and Jurassic age in the Colorado Plateau bear witness to a scale of wind transport and deposition of sand that dwarfs anything presently occurring on earth. Perhaps our judgment that wind is a geologic agent of secondary importance needs to be tempered by the evidence of the past.

Waves and coastal currents are unquestioned major agents of landform development not only of the present but also of the past. Sedimentary deposits transported and deposited along ancient shorelines are now part of the record of stratigraphy. Many important sandstone and conglomerate formations prove to have been lithified beaches. Many ancient deltas have been recognized in the geological record, and massive coral reefs can be identified within limestone formations. Some of these ancient shoreline deposits are important to us today because they form valuable petroleum reservoirs.

Wave erosion may have been responsible for the planation of continents in the final stages of denudation. Often a peneplain, seen as an unconformity, shows evidence of abrasion by waves prior to burial under the advancing deposits of marine sediments.

The great variety in modern coastlines is due to the fact that we live at a time when the crust has been moving up in one place and down in another. Today volcanism and faulting are widespread. The growth and disappearance of Pleistocene ice sheets set off wide swings of sea level, ending in a final rise to the present level.

Key Facts and Concepts

Wind erosion One form of wind erosion is *deflation*, the lifting of *dust* into the air, creating a *dust storm*. Deflation can result in the formation of *blowouts* and *desert pavements*. *Sand-blast action*, a second form of wind erosion, acts on low rock surfaces and is capable of carving *ventifacts*.

Loess Thick deposits of wind-transported silt make up *loess*. Thick loess of Pleistocene age is found extensively over the central United States and is the parent matter of fertile, grain-producing soils.

Transport of sand by wind *Eolian sand (dune sand)* is extremely well sorted and usually consists of well-rounded quartz grains. In *saltation*, sand grains moving under the force of wind make repeated rebounds, causing slow grain movement by *surface creep*, forming *sand ripples*. Rate of transport of sand by wind increases about as the cube of wind speed.

Sand drifts and sand dunes A *sand drift* is fixed to an obstruction; a *sand dune* is capable of forward motion. By saltation, dunes of free sand develop a steep *slip face*.

Dune forms Common dune forms include the *barchan dune*, *transverse dune* of *sand seas*, and isolated *star dunes*. Sand dunes in the geologic record are recognized by *cross-lamination* with nearly planar surfaces of former slip faces.

Dunes controlled by vegetation Sparse grasses and shrubs trap sand to produce coastal *foredunes*. On arid plains, *parabolic dunes* form adjacent to deflation hollows. The *coastal blowout dune* is a related coastal type. Breakdown of protective plant cover by human activities allows soils to experience deflation and fixed sand dunes to become activated.

Ocean waves Most ocean waves are *oscillatory waves* generated by stress of wind blowing over the water surface. Wave orbits are in a vertical plane with little net forward motion of the water.

Shoaling waves Waves entering shallow water experience modification of the orbital motion and lose energy through bottom friction, decreasing the forward wave speed. Wave height increases, leading to wave cresting and breaking.

Wave refraction Waves approaching obliquely to the contours of a shallow bottom are refracted, changing direction and speed in the process of *wave refraction*. Wave energy is concentrated on *headlands* but spread more thinly along bay heads.

Currents caused by breaking waves Water produced by a *breaker* moves up a beach as *swash (uprush)* but returns as *backwash (backrush)*, carrying sand alternately up and down the beach slope. Oblique approach of waves sets up a *longshore current* parallel with the shoreline. Jetlike streams of surface water flowing seaward through the surf zone are *rip currents*.

Storm wave erosion Storm waves exert high impact pressures upon exposed rock cliffs and human-made structures, causing rapid coastal retreat in weak materials.

Marine cliffs *Marine cliffs* of resistant rock are bordered by an *abrasion platform*, carved by wave action. A *wave-cut notch* is formed at the cliff base. The *coastline*, or *coast*, is a broad zone including the *shoreline*.

Beaches *Beaches*, formed as accumulations of sand, gravel, or cobbles (*shingle beach*), absorb the energy of breaking waves. *Pocket beaches* form first in protected coves. Elements of sand beaches include the *foreshore* and *offshore* zones. *Progradation* adds to the beach and may build a *berm; retrogradation* depletes the beach and increases water depth.

Littoral drift Oblique approach of waves causes the swash to move sand parallel to the shoreline by *beach drift*, while longshore currents cause corresponding *longshore drift*. The combined process, *littoral drift*, transports sediment great distances along the shoreline and may construct a *sandspit* in open water. Littoral drift also moves sediment from headlands toward bay heads. The total effect is to reduce the irregularities in the shoreline, producing a straight or broadly curved shoreline.

The shore profile of equilibrium An *equilibrium beach profile* forms in a *surf lens* extending seaward to *surf base*. Progradation results in multiple *beach ridges*. Combined with littoral drift, progradation may produce a *cuspate foreland*.

Wave erosion and shore protection Littoral drift can be partially impeded by use of *groins*, causing beach sand to accumulate on the updrift side of the groin, but resulting in beach depletion on the downdrift side.

The littoral cell The continental coastline can be visualized as a succession of *littoral cells*, each an open system with sediment input from coastal sources and an output to the deep ocean floor.

Elevated shorelines and marine terraces Sudden tectonic rise of a coast forms an *elevated shoreline*, bordered by a *marine terrace*.

Ocean tide and tidal currents The *ocean tide*, or periodic rise and fall of ocean level, is a response to tide-raising gravitational forces exerted by moon and sun upon the earth. With earth rotation, two tidal bulges of raised ocean level travel around the earth. Alternating high waters and low waters make up the semidaily *tide curve*. *Tidal currents* consist of alternating *ebb current* and *flood current*, capable of sediment transport in narrow inlets.

Tidal flats and tidal marshes In bays and lagoons, fine sediment accumulates to form a *tidal flat*, upon which a *salt marsh* is built.

Barrier islands *Barrier islands* separated from the mainland by *lagoons* have resulted from beach upbuilding during postglacial rise of sea level. *Tidal inlets*, cutting through the barrier, are scoured by strong tidal currents, building *tidal deltas*.

Common types of coastlines Coastlines change because of *submergence* or *emergence*, resulting from either a change of sea level or crustal movement, or from both effects acting simultaneously. Important coastline types include: *ria coast*, *fiord coast*, *barrier-island coast*, *delta coast*, *volcano coast*, *coral-reef coast*, and *fault coast*.

Marine planation As a shoreline advances upon a slowly submerging continent, wave action produces a submarine platform by *marine planation*. Subsequent sediment deposition may result in an unconformity in the geologic record.

Questions and Problems

1. A sandstone butte (see Chapter 17) in an arid region shows numerous small cavities and niches high on the sheer rock wall some 100 to 200 m above the base of the cliff, which rises from a flat valley floor. A guide tells you that these cavities are the work of abrasion by sand carried in the wind. Do you accept this explanation? Give your reasons.
2. During the siege of the city of Vicksburg in the Civil War, sappers of both Confederate and Union forces were actively digging tunnels through to points beneath the enemy positions, where explosives could be planted and detonated. What geologic conditions were particularly favorable to this mode of siege warfare? Explain.
3. How is the dominant wind direction shown by the form of a barchan dune? Of transverse dunes? Of a parabolic dune?
4. Using everyday language of surf bathers, is the "undertow" the same phenomenon as the "rip current"? How should a swimmer caught in a rip current maneuver so as the escape from the current? How does wave refraction enter into the practice of surfing (surfboard riding)?
5. Make a schematic flow diagram of a littoral cell, treating it as an open material-flow system. Use arrows to show movement of sediment and boxes to indicate places of sediment storage. Show the return of sand to the mainland by wind action as a subcircuit. Include a subcircuit for the seasonal migration of sand from shallow to deep water and return from winter to summer.
6. How might it be possible to decide whether a particular marine terrace is the result of coastal uplift (tectonic) or of eustatic lowering of sea level following a high stand during an interglaciation? What evidence would you look for?

CHAPTER TWENTY

Astrogeology—the Geology of Outer Space

It may seem farfetched at first to extend the word "geology" to include a study of the Moon, planets, and other solid objects of the solar system. The prefix *geo*, after all, comes from the Greek word meaning "Earth."* Still, the Moon and closer planets are composed of minerals and rocks, and they can be expected to have some of the geological features of our Earth, such as volcanoes and lava flows, faults and earthquakes, crusts, mantles, and cores, and perhaps even tectonic belts. There may even exist sediments and sedimentary strata of some sorts on these bodies. Perhaps, then, it is not at all unreasonable to extend the science of geology to outer space.

Astrogeology is the name established by the U.S. Geological Survey for the application of principles and methods of geology to all condensed matter and gases of the solar system outside the Earth.

The Moon and Mercury, and to some extent Mars as well, have one feature in common that makes geological investigations especially fruitful: They have practically no atmosphere. None of the three has running streams or oceans of free water. On Earth, processes of rock weathering, erosion by streams, waves and wind, and sedimentation quickly remove or bury features produced by volcanism, tectonic activity, and the impact of large bodies from outer space. Where a planet has little or no atmosphere or hydrosphere, these processes do not act, or they act very, very slowly. As a result, surface forms and structures produced by volcanic and tectonic activity and by impact from outer space remain little changed through eons of geologic time on the Moon, Mars, and Mercury. Perhaps scientific study of these planetary features can shed some light on the early history of our own planet.

It is not surprising that earth scientists mounted a major effort to explore the Moon's surface and to bring back samples for analysis. Lunar geology held great promise for new knowledge, not only of the Moon itself, but of early Earth history as well. One leading scientist has likened the Moon to the Rosetta stone, which gave scholars the key to translation of Egyptian hieroglyphics and so opened up the field of Egyptian history.

In this chapter, we will apply geological principles to the understanding of the Moon, the inner planets, the asteroids, and the meteorites.

The Sun—Hub of Our Planetary System

Our solar system consists of solid objects moving in more or less circular or elliptical orbits about the Sun. To understand the concepts of origin of the planets, we need to have in mind some facts about the Sun itself, because the planets came into existence during final stages of the origin of the Sun as a star.

*In this chapter we capitalize the first letter of Earth, Moon, and Sun to make them proper names, consistent with other planets and stars.

Our Sun is an enormous sphere of incandescent gas with a diameter of about 1,400,000 km, which is about 100 times the diameter of Earth. Figure 20.1 shows the relative diameter of the Sun and major planets. We can get some idea of the enormity of the Sun when we realize that it has a mass more than 330,000 times that of Earth and a volume 1,300,000 times that of Earth. This huge solar mass produces a surface gravity 34 times as great as that felt at the Earth's surface.

The average distance from Earth to Sun is about 150 million km (see Figure 20.5). Traveling at the speed of light, which is roughly 300,000 km/sec, solar radiation takes just over eight minutes to reach the Earth.

The visible surface layer of the Sun is called the *photosphere*. The outer limit of the photosphere forms the edge of the Sun's disk as seen in white light. Gases in the photosphere are at a density less than that of Earth's atmosphere at sea level.

Temperature at the base of the photosphere is about 6000 °K but decreases to about 4300 °K at the outer photosphere boundary. Light production is extremely intense within the photosphere. Beneath the photosphere, temperatures and pressures increase to enormously high values in the Sun's interior, or nucleus. Here temperatures are between 13 and 18 million °K.

Above the photosphere lies a low solar atmosphere, the *chromosphere*. The region includes rosy, spikelike clouds of hydrogen gas called solar prominences. Still farther above the Sun's surface is the *corona*, a region of pearly-gray streamers of light, which can be thought of as the Sun's outer atmosphere (Figure 20.2). At times, the solar prominences reach far out into the corona as luminous archlike bodies (Figure 20.3) rising to heights of over one million km. Temperatures increase outward through the chromosphere and the corona until values as high as 2 million °K are reached. Surprisingly, the photosphere, or Sun's surface, is its coolest layer.

The corona extends far out through the solar system and envelops the planets. It is known as the *solar wind* in the region surrounding the inner planets, and consists of charged particles—electrons and protons—derived from the breakup of solar hydrogen atoms. The solar wind brings energetic particles to the magnetosphere, where they are entrapped by the lines of force of the Earth's magnetic field.

Although almost all the known elements can be detected by spectral analysis of the Sun's light, hydrogen is the predominant constituent of the Sun, with helium also abundant. It is estimated that hydrogen constitutes at least 90% of the Sun, and hydrogen and helium together total about 98%.

The source of the Sun's energy is the conversion of hydrogen into helium within the Sun's interior. The process of production of energy within the Sun is that of nuclear fusion, in which hydrogen is transformed into helium and mass is converted into energy. At temperatures over 4 million °K within the interior of a star, several forms of reactions occur in which helium is produced. The quantity of energy produced by conversion of matter is enormous. At its present rate of energy production, the mass of the Sun will diminish

FIGURE 20.1 Relative diameters of the Sun and planets. Figures give diameters in kilometers.

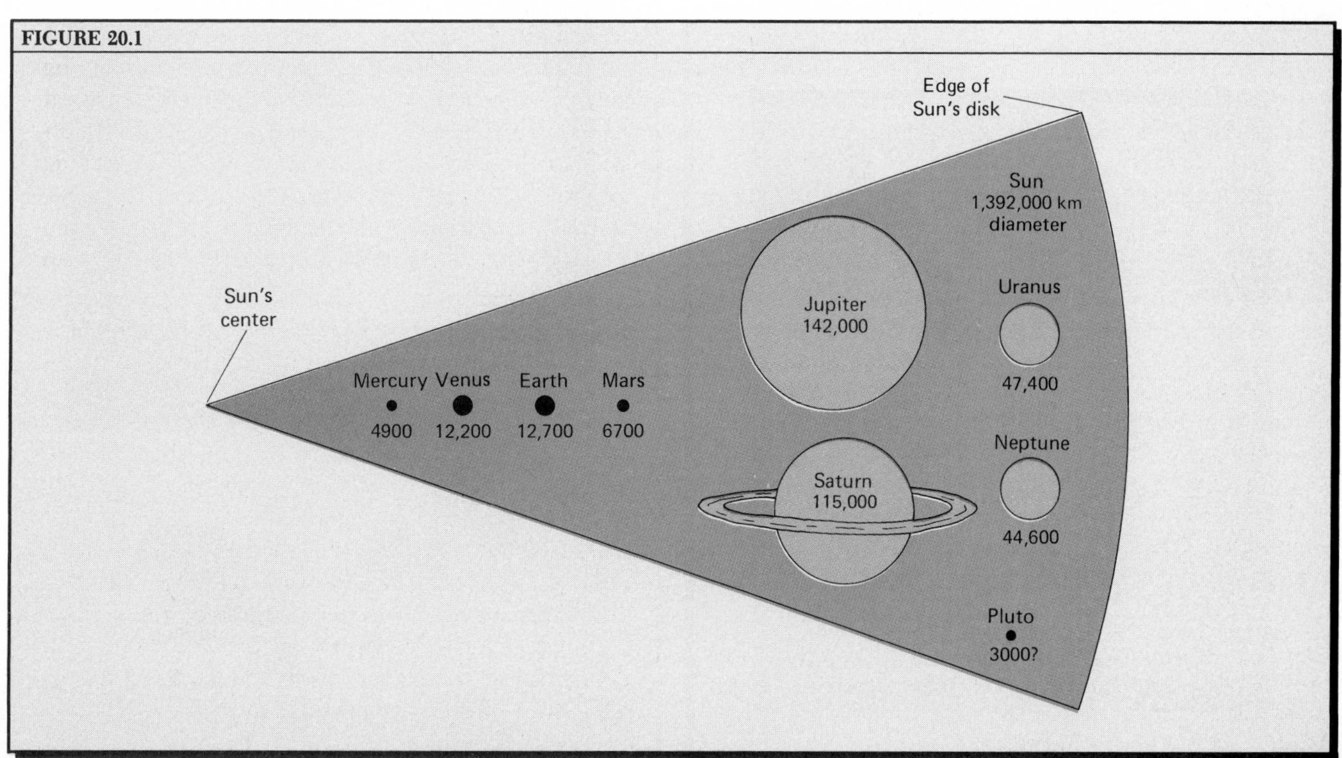

only one-millionth part in 15 million years.

Heat produced in the Sun's innermost core region moves outward by a process of radiation through the extremely dense gas of the interior. In a zone nearer the Sun's exterior, a process of convection (mixing) transports the heat to the surface.

A *sunspot* is a dark spot on the Sun's photosphere. The spot normally consists of a darker central region, the umbra, surrounded by a somewhat lighter border, the penumbra (Figure 20.4). A single sunspot may be 1000 to 100,000 km across and represents a strong disturbance extending far down into the Sun's interior. The spot has a somewhat lower temperature than the surrounding photosphere. Sunspots form and disappear over a span of several days to several weeks, during which time they can be seen to move with the sun's rotation. The frequency of sunspots follows a cycle with an average period of about 11 years.

It has been found that the sunspots have powerful magnetic fields associated with them—several thousand times as great in intensity as the magnetic field at the Earth's surface. This magnetism takes the form of strong poles, and adjacent spots of a pair in the same hemisphere have opposite polarity.

The same intense magnetic fields that are associated

FIGURE 20.2

FIGURE 20.2 This photograph of the Sun's outer corona was taken during a total eclipse. The Moon's disk completely covers the Sun, permitting this pearly white tenuous outer layer of gases to be seen. (Mount Wilson Observatory photograph. Mount Wilson and Las Campanas Observatories, Carnegie Institution of Washington.)

FIGURE 20.3

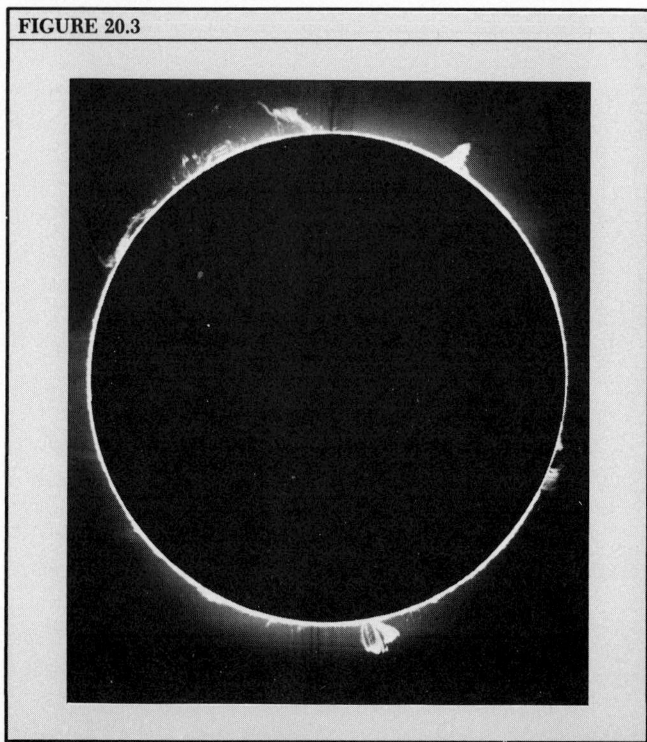

FIGURE 20.3 This photograph of the entire edge of the Sun shows several prominences. (Mount Wilson Observatory photograph. Mount Wilson and Las Campanas Observatories, Carnegie Institution of Washington.)

FIGURE 20.4

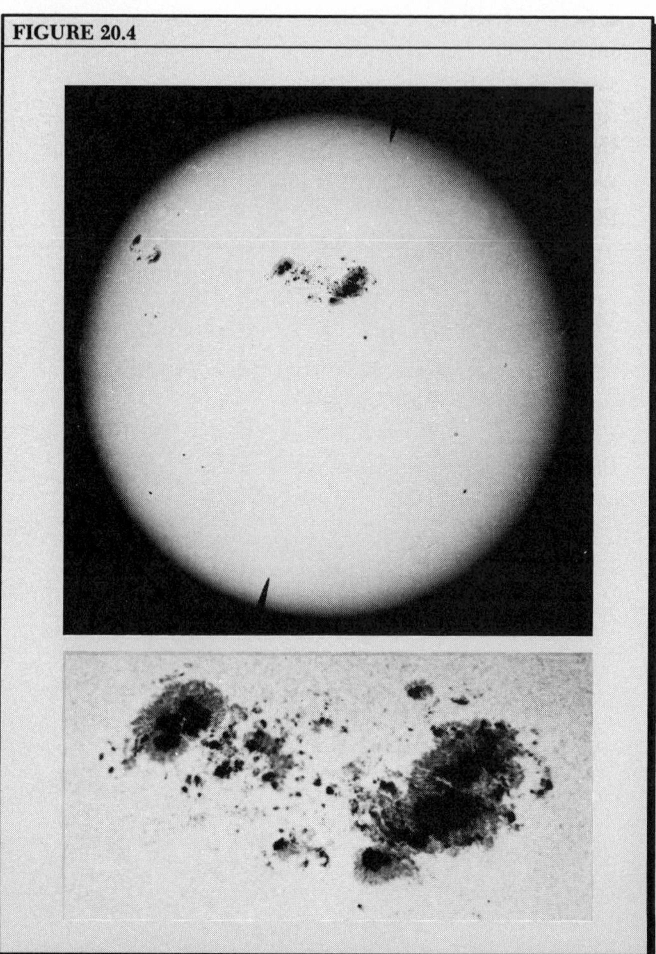

FIGURE 20.4 The whole disk of the Sun (*above*) shows a large sunspot group. Below is an enlargement of the group of spots, showing the umbra and penumbra regions. (Mount Wilson Observatory photograph. Mount Wilson and Las Campanas Observatories, Carnegie Institution of Washington.)

Table 20.1 **The Principal Planets**

Name	Distance from Sun, millions of km (10^6 km)	Period of revolution, sidereal	Diameter, thousands of km (10^3 km)	Mass, relative to Earth	Mean density g/cm^3	Period of rotation	Number of moons
Inner planets (terrestrial planets)		**Earth days**					
Mercury	58	88	4.9	0.06	5.4	58^d16^h	0
Venus	108	225	12.2	0.80	5.2	243^d	0
Earth	150	365¼	12.7	1.00	5.5	23^h56^m	1
Mars	228	687	6.8	0.11	4.0	24^h37^m	2
Outer planets (great planets)		**Earth years**					
Jupiter	779	12	142	315	1.3	9^h50^m	13
Saturn	1430	29½	115	94	0.7	10^h14^m	11+
Uranus	2870	84	47.4	15	1.6	10^h42^m	5+
Neptune	4500	165	44.6	17	1.6	15^h48^m	2
Pluto	5900	248	3(?)	0.11	1.0(?)	6^d	1(?)

with sunspots also produce *solar flares*. X rays are emitted from a flare, along with a stream of charged particles—protons and electrons. The particles reach the Earth about a day later. At such times the magnetic field at the Earth's surface is severely disrupted. This phenomenon is a magnetic storm; it interferes seriously with radio communication.

Solar flares occur in much greater numbers than sunspots. As many as 2000 to 4000 flares occur per year during times of maximum sunspot activity. Flares are about 20 times more frequent than sunspots, but their duration is correspondingly much shorter. A single sunspot group in the course of its duration will produce as many as 40 flares.

The Terrestrial Planets

Several kinds of solid objects orbit the Sun, the largest of which are the nine *major planets*. In order of distance from the Sun, these are Mercury, Venus, Earth, Mars, Jupiter, Saturn, Uranus, Neptune, and Pluto. Their relative sizes are shown in Figure 20.1 and Table 20.1.

Our major concern in this chapter is with the first four planets on this list; they are often called the inner planets, or the terrestrial planets (Figure 20.5). From the geological standpoint, the second label is the more interesting because it implies that Mercury, Venus, and Mars resemble the Earth in mass, density, and gross composition. You might say that the terrestrial planets are "rock" planets, for they consist largely of silicate minerals surrounding dense cores of iron. Average densities of the four planets in g/cc are: Mercury, 5.4; Venus, 5.2; Earth, 5.5; and Mars, 4.0. Mars' lower density implies that it has only a small core, perhaps composed of iron sulfide.

In comparing the sizes of the four terrestrial planets, we encounter major dissimilarities. Taking the Earth's mass as 100%, Venus is comparable in size, with 80% as great a mass as Earth. However, Mars is much smaller, 11%; Mercury is still smaller, 6%. Actually, Earth and Venus are closely matched in diameter, mass, and density.

The Outer Planets

The orbits of the outer planets are shown in Figure 20.6, using a scale of distance very different from that of Figure 20.5. How do the earthlike planets compare with the four great outer planets—Jupiter, Saturn, Uranus, and Neptune? Using Table 20.1, we find that even the smallest of the four, Neptune, has almost four times the diameter and 15 times the mass of Earth; whereas the giant of the group, Jupiter, has 11 times the diameter and 300 times Earth's mass. Apart from their size, a second striking difference in these two groups of planets is that of density (Table 20.1). The least dense is Saturn (0.7 g/cc), about one-eighth the density of Earth and less than three-fourths that of liquid water at the earth's surface. The other three have densities of 1.3 to 1.7 g/cc, values only one-fourth that of Earth.

A third striking difference in the two groups of planets is the extremely low prevailing temperature on the surfaces of the great planets, ranging from −138 °C on Jupiter to −201 °C on Neptune. A fourth striking difference is in composition. The four terrestrial planets are probably all composed of a rock mantle surrounding an iron core and have either no atmosphere or atmospheres of almost insignificant mass. In contrast, the four great planets have massive atmospheres of hydrogen and helium, with some ammonia, methane, and water. These substances in the solid state also make up most of the mass of each planet.

By analysis of ultraviolet light reflected from Jupiter's atmosphere, it has been determined that the outer layer consists of 84% hydrogen and 15% helium, with a remainder of largely methane and ammonia. The abundance of hydrogen in the atmosphere suggests that the planet as a whole has hydrogen as its predominant constituent. Jupiter may have a mantle of hydrogen in an extremely dense, metallic state and a rocky core about the size of the Earth.

Uranus and Neptune are nearly twins so far as diameter and mass are concerned. Under spectroscopic analysis, both planets show methane to be the dominant atmospheric constituent, whereas ammonia appears only in a trace. In addition to hydrogen compounds and helium in the solid state, both Uranus and Neptune may have rock cores, since their densities are somewhat greater than those of Jupiter and Saturn.

Because of its small size and great distance, very little is known about Pluto. Its surface temperature is judged to be not far above absolute zero, and spectroscopic data indicate that its surface may be covered with methane ice. Pluto may have a satellite (called Charon).

Besides the major planets, two classes of smaller objects orbiting the Sun are of interest to astrogeologists: asteroids and meteoroids.

The Asteroids

The *asteroids*, which have also been called the minor planets, follow the planetary laws of motion and are true planets of the sun despite their small size.

Names and diameters of the four largest asteroids are as follows:

Ceres	1018 km
Pallas	629 km
Vesta	548 km
Hygiea	450 km

Most asteroids are very much smaller—a few kilometers in diameter or less. The total number of asteroids runs into the tens of thousands, some 40,000 of which can be detected on photographs. The great majority of asteroids follow orbits between Mars and Jupiter, but some cut inside the orbit of Venus; one is known to sweep outward almost to Saturn's orbit. Their combined mass is perhaps 1/1000 to 1/500 that of Earth.

Of particular interest is Eros, an irregularly shaped asteroid about 25 km long. Its orbit is highly eccentric, at times bringing it as close as 22 million km to Earth; in 1975, it passed within 23 million km. Observation at that time proved it to have an elliptical outline and a very rough surface. The asteroid Icarus came within about 6.5 million km of Earth on June 14, 1968, when

FIGURE 20.5 Orbits of the four inner planets. A black dot represents the point at which the planet is closest to the Sun. The orbits are drawn as circles and do not show the true elliptical form of each orbit.

FIGURE 20.6 Orbits of the outer planets. The innermost circle represents Mars' orbit, and the dashed circle represents the zone of asteroid orbits. Pluto will not collide with Neptune because Pluto's orbit is inclined more than 17 degrees with respect to the plane of Neptune's orbit.

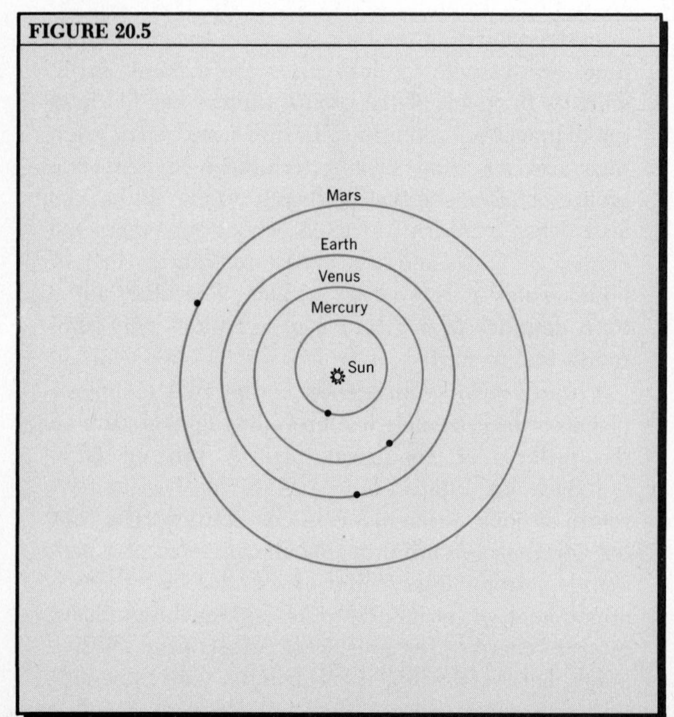

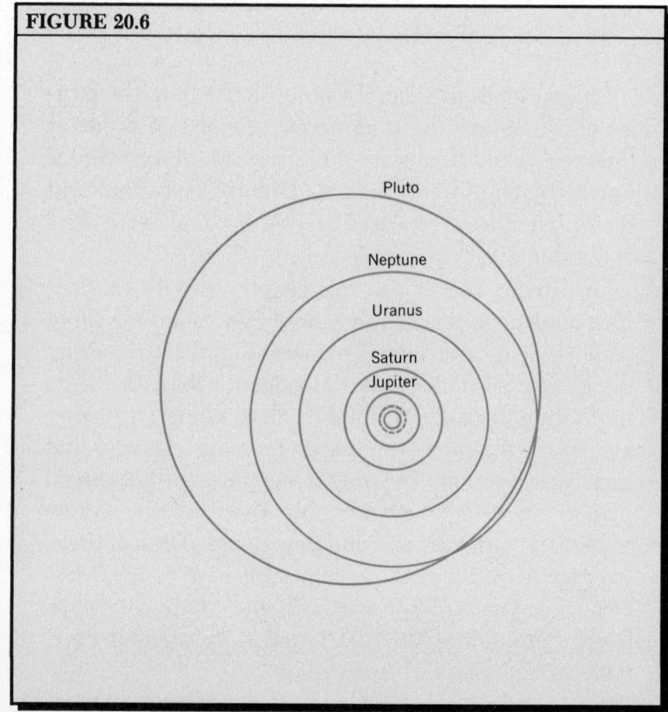

observations showed that it is irregular in shape, less than 2 km in width, and may be composed in part of iron. The smallest asteroids we can observe are about 1 km in diameter, but many are probably much smaller than this. Composition of the asteroids is thought to vary from nickel–iron to silicate rock. An early hypothesis no longer in vogue is that the asteroids represent the fragments of a single major planet which once orbited between Mars and Jupiter.

Meteorites

A fragment of solid matter entering the Earth's atmosphere from outer space is called a *meteoroid.* Most are extremely tiny particles and vaporize as they penetrate the atmosphere, leaving only a thin trail of light, called a *meteor.* Occasionally one of the meteoroids is large enough so that, though partially vaporized, it may still reach the Earth's surface. Such exotic rock objects are then called *meteorites.*

The fall of a meteorite is accompanied by a brilliant flash of light, called a *fireball,* and may also be accompanied by loud sounds. Frictional resistance with the atmosphere causes the outer surface of the object to be intensely heated. However, this heat does not penetrate to the interior of a meteorite, which reaches the Earth with its original composition and structure unchanged, even though the single mass may explode before the impact, showering fragments over a wide area. Examples of very large meteorites are shown in Figure 20.7.

FIGURE 20.7 Two kinds of meteorites. *A.* This stony meteorite, weighing 338 kg, struck the ground at Paragould, Arkansas, on February 17, 1930, forming a huge fireball visible over thousands of square kilometers. Height of the meteorite is about 0.6 m. (Yerkes Observatory.) *B.* The Willamette meteorite, an iron meteorite, weighs 14 metric tons and is over 3 m long. The huge cavities were produced by rapid melting of iron–sulfide inclusions during fall through the atmosphere. (American Museum of Natural History—Hayden Planetarium.)

FIGURE 20.7

Meteorites have been intensively studied by geologists, not only as to chemical composition and structure, but also as to age. They fall into three classes. (1) The *irons* are composed almost entirely of a nickel–iron alloy, in which the nickel content ranges from 4% to 20%. (2) At the other end of the series are the *stones,* consisting largely of silicate minerals, mostly olivine and pyroxene, with only 20% or less nickel–iron. Plagioclase feldspar may also be present. (3) An intermediate class of meteorites consists of the *stony irons.* In these, silicate minerals and nickel–iron may form a continuous medium in which spherical bodies of silicate minerals are enclosed.

The similarities to and differences from Earth rocks in the structures of the meteorites have aroused much interest. The nickel–iron of an iron meteorite typically shows crystalline structure. When a polished surface is etched, there appear distinctive line-patterns known as *Widmanstätten structure* (Figure 20.8*A*). These patterns are almost unknown in terrestrial iron, and the interpretation can be made that the alloy has cooled very slowly from a high temperature. This evidence suggests that the iron meteorites are disrupted fragments of larger original masses, such as might be found in the cores of large asteroids.

One class of stony meteorites, the *chondrites,* possesses an internal structure never observed in rocks of Earth. Olivine or pyroxene crystals occur in small rounded bodies (*chondrules*) on the order of 1 mm in diameter (Figure 20.8*B*). This structure is certainly important in the origin of the stony meteorites, but its significance is not yet known. However, another group of stony meteorites (*achondrites*) possesses a coarse-grained structure that resembles the structure of Earth's plutonic igneous rocks.

When the meteorites of observed falls are cataloged, their relative abundance turns out to be about as follows: stones, 94%; irons, 4.5%; stony irons, 1.5%. The stony meteorites are thus dominant in terms of bulk, and most of these are chondrites possessing the unique chondrule structure. These facts suggest as a working hypothesis that the chondrites represent fragments of the earliest planetary bodies to be formed in the solar system. The nickel–iron meteorites can be interpreted as cores of these planetary bodies in which the process of differentiation had taken place. Stony meteorites having textures resembling terrestrial igneous rocks point to the possibility that melting and recrystallization had taken place to some degree in these original planetary bodies.

Age determinations of meteorites have been made, using the uranium–lead, potassium–argon, and rubidium–strontium methods described in Chapter 7. There is a high degree of agreement in the results pointing to the time of formation of all types of meteorites at 4.5 billion years before present (−4.5 b.y.). This age is about 0.7 billion years greater than that of the oldest known crustal rocks on Earth.

An exciting new scientific development has been the finding of many meteorites on the surface of the Antarctic Ice Sheet. Beginning quite by chance in the early 1970s, Japanese scientists began to collect these meteorites in surprisingly large numbers. They lie exposed on the surface of blue glacial ice that has been subjected to long and severe ablation. Apparently, the meteorites landed on deep snow in the high interior region of accumulation, then moved within the ice for long distances, finally to emerge in ablation zones (see Chapter 18). A single season of searching yielded as many as 300 specimens. Similar success greeted American search parties in the late 1970s. The specimens are greatly valued because of their remarkable preservation from weathering in a deep-freeze environment of perpetual severe cold. The meteorites are handled and stored under sterile conditions, just as are the Moon rocks, and are kept frozen during transit and storage so as to minimize interaction with atmospheric moisture under warm conditions. Scientists have high hopes that intensive chemical research on these remarkable specimens will shed new light on the origin of the solar system.

Perhaps we can conclude from the study of meteorites that the first large, solid objects of the solar system were formed rapidly by aggregation of iron–magnesium silicates and nickel–iron (and of many less abundant elements) and that this event took place about −4.5 b.y. These first solid bodies were perhaps of sizes comparable to large asteroids. Within some of the asteroids, melting and recrystallization took place, accompanied by differentiation of the nickel–iron into core material. There is evidence that disruptive collisions reduced some of the larger asteroids to smaller fragments throughout most of ensuing geologic time.

The assumption that the terrestrial planets were created out of the same substance as the meteorites has strongly reinforced the conclusion that the Earth's core is composed of nickel–iron and the mantle of iron–magnesium silicates. This conclusion is now accepted in light of independent evidence of the density and related physical properties of the Earth's interior derived from study of earthquake waves (Chapter 8).

Accretion of the Planets

This is a good point at which to fill in some details of the origin of the planets, a topic that was rather broadly sketched in Chapter 1. The solar nebula which we described in Chapter 1 (Figure 1.17) can be thought of as having two basic regions or portions. The bulk of the hot gases had contracted into a dense inner mass that was to become the Sun. This solar body was intensely hot but had not yet reached the critical size

FIGURE 20.8 *A.* Widmanstätten structure etched into an iron meteorite. (Smithsonian Astrophysical Observatory photograph, courtesy of J. A. Wood.) *B.* Microscopic photograph of a thin slice of a chondrite meteorite, showing an area about 1 cm across. (From John A. Wood, 1968, *Meteorites and the Origin of Planets*, McGraw-Hill, New York, p. 18, Figure 2–3. Used with permission of the publisher.)

FIGURE 20.8

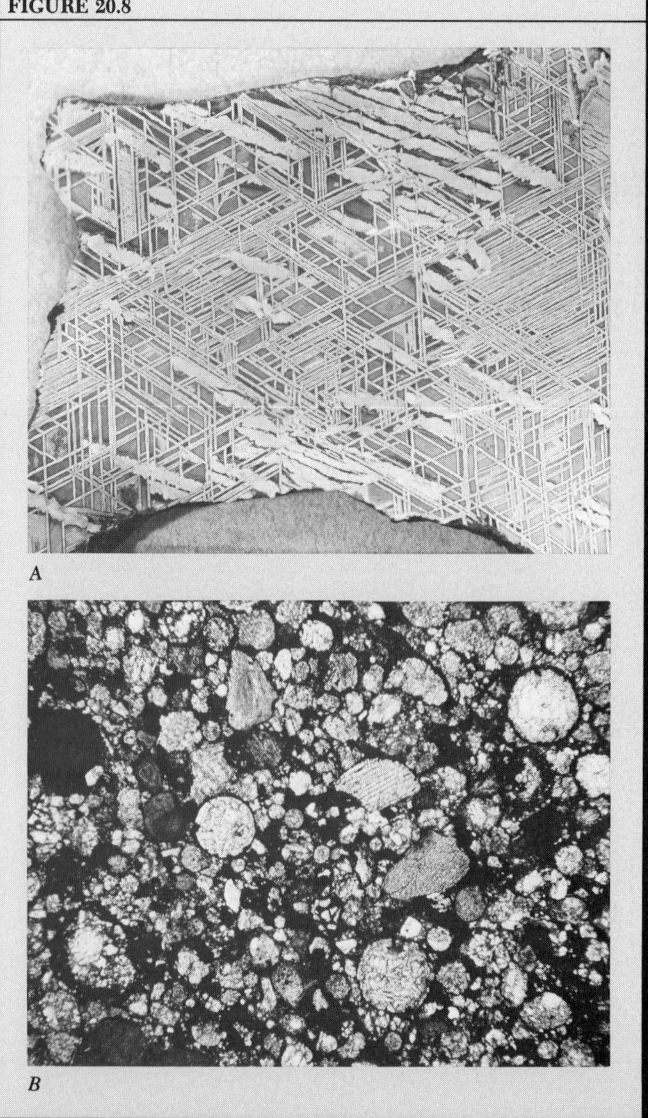

and density necessary for nuclear fusion to start. Surrounding the solar body was a thin, disklike layer of rapidly cooling gas and dust that was to evolve into the planets and their satellites. Temperature within the disk was highest near the solar body and decreased outward. Although the thin disk initially consisted largely of hydrogen and helium, it also contained a small proportion of heavier elements, including silicon, iron, magnesium, aluminum, calcium, and other elements from which the silicate mantles and nickel–iron cores of the inner planets were later to be formed.

As temperatures within the nebula fell, the gaseous phase of each of these constituents was to condense into liquid droplets or crystalline grains. However, because each group of elements (oxides of the metals, for example) has its own condensation temperature for a given pressure, there would have been a distinctive order of condensation as cooling continued. It is supposed that when the gas had cooled to a temperature of 1400 °K to 1600 °K, oxides of aluminum and calcium, along with nickel–iron would have begun to condense as mineral grains. When gas temperature dropped to about 1300 °K, the common silicate minerals would have condensed into grains. With further cooling down to between 300 °K and 100 °K, crystals of water, ammonia, and methane would have formed; we can refer to all these crystalline forms of the volatile compounds as ice.

Now a new factor may have entered the picture. When the newly formed Sun had contracted and heated to a critical point, nuclear fusion began to occur in its interior. This new heat production resulted in the Sun's surface giving off a strong solar wind of ionized particles. As the solar wind moved radially outward through the nebular disk with its mineral grains, ice particles, and gases, it selectively pushed out into distant space the remaining gas—mostly hydrogen and helium. This winnowing action left behind the solid particles. In the inner part of the disk, there remained the grains of silicate minerals and nickel–iron that were later to clot together into larger bodies that became the inner planets. Farther out in the disk, where cold conditions prevailed, the bulk of the solidified material consisted of ices of water, ammonia, and methane, which remained in sufficient quantities to form the outer planets. Although this selective condensation and winnowing process is generally agreed upon by scientists, details of the actual growth of the present planets is speculative and several alternative possibilities have been put forward.

One scenario that has been advanced for evolution of the planets calls for the accretion (coming together) of mineral and ice grains to form *planetesimals:* These were objects about the size of large asteroids, a few kilometers in diameter. Later, the planetesimals underwent accretion into larger objects, which finally grew into the major planets. The few remaining large masses rapidly swept up the smaller fragments, but a few remain today in orbit as meteoroids and asteroids. Another scenario calls for a single step of accretion of the grains into large spherical bodies by mutual gravitational attraction of the particles, a process sometimes called *gravitational collapse*. Several such scenarios can serve as reasonable models of planetary growth, and there is no agreement on the subject as yet. It is, however, agreed that the planets formed quite rapidly once the accretion process had begun. By "rapidly" we mean a time span on the order of ten million years, or even less. The place in time assigned to condensation has been determined as between −4.55 b.y. and −4.47 b.y.

Once the inner planets had reached their full size, their internal development followed the changes we outlined in Chapter 7, in which partial melting occurred and the dense iron or iron sulfide sank to the core region of each planet, while the remaining silicate minerals came to form the surrounding mantle. The time at which this differentiation into mantle and core was completed is not known with certainty, but it may have occurred between −4.3 and −4.1 b.y.

So far, we have not mentioned the possible origin of the Moon, which is an object comparable in some ways to the planet Mercury. This is a topic we will postpone until after we have investigated the geology of the Moon.

Impact Features on Earth

Have meteorites of great size struck the Earth's surface to produce recognizable impact features? The largest known single meteorite is the Hoba iron, found in Southwest Africa; it weighs about 60 metric tons and measures 3 × 3 × 1 m. The next five in order of size are irons weighing roughly half as much as the Hoba meteorite. However, no stony meteorites have been found whose weight is over one metric ton. The largest single stony meteorite observed to fall, pictured in Figure 20.7, weighs one-third of a ton.

Rarely has an observed meteorite fall produced craters of measurable dimension, but one which did was the Siberian Sikhote Alin fall of February 12, 1947, witnessed by many persons. The largest iron fragment recovered weighed 1.8 metric tons. Funnel-shaped craters as large as 28 m in diameter were produced by the larger iron fragments. Soviet scientists deduced from the observed trajectory of the meteoroid trail that it was a small asteroid traveling at a speed of 40 km/sec.

So we turn to prehistoric impacts of enormous meteorites capable of producing rimmed craters with diameters of 100 to 1000 m and larger. Several fine examples of almost perfectly circular large craters with sharply defined rims have been found and examined over the continental surfaces of Earth. In addition, there are many more circular rock structures which prove upon examination to show intense disturbance and alteration of the rock in which they occur. Altogether, perhaps fewer than one dozen large, rimmed craters and about 50 circular structures are known. The widespread availability of air photographs, combined with satellite photography, has brought the discovery of many structures of possible meteoric impact origin. We are, of course, excluding from this discussion all known volcanic craters constructed of lava and volcanic ash, as well as obvious calderas, the large craters produced by explosive demolition of a preexisting volcanic cone (Chapter 4).

At the outset, we must recognize that there exists a wide range of opinion as to the origin of circular structures, and of certain sharp-rimmed deep craters as well. Because we are treating these features under a discussion of meteorite impacts, it might seem that impact is the favored hypothesis, but a substantial number of geologists who have given intensive study to circular structures hold that some may have been formed by internal earth processes—for example, by uplift under pressure of rising magma, followed by collapse. Explosion by volcanic gases is also postulated as a cratering mechanism and has had many supporters, who refer to some circular forms as *cryptovolcanic structures*. Those who hold that the circular structures are the result of impact of large objects from space refer to the same forms as *astroblemes* (freely translated as "star-wounds"). While originally possessing sharp-rimmed craters, astroblemes seen today would represent only the deeply eroded basal parts of the original impact structures.

Perhaps the finest example of a large, circular rimmed crater of almost certain meteoritic impact origin is the Barringer Crater (formerly known as Meteor Crater) in Arizona (Figure 20.9). The diameter of this crater is 1200 m and its depth almost 180 m. The rim rises about 45 m above the surrounding plateau surface, which consists of almost horizontal limestone and sandstone strata. Rock fragments have been found scattered over a radius of 10 km from the crater center, while meteoritic iron fragments numbering in the thousands have been collected from the immediate area.

Other evidence of severe shock forces and high temperatures comes from the finding of closely fractured rock, silica glass, and unique silica minerals known as *coesite* and *stishovite*. These minerals were produced by severe shock pressures from a pure sandstone formation underlying the limestone of the plateau. Although boreholes and shafts were put down in the bottom of the crater in an attempt to locate a large iron body, none was found. It has been calculated that an impacting meteorite capable of producing such a crater would have disintegrated and partially vaporized during impact, leaving no single large mass intact. Carefully derived estimates of the impacting object specify that it was an iron meteorite or asteroid with a mass of about 56 metric tons. It was perhaps 30 m in long dimension, and was traveling at a speed of about 15 km/sec when impact occurred.

Figure 20.10 shows inferred steps in the formation of a simple meteorite crater of moderate size, such as the Barringer Crater. Kinetic energy, estimated to be on the order of 10^{21} to 10^{28} ergs, is almost instantly transferred to the ground by a shock wave, which intensely fractures and disintegrates the rock around the point of impact. A similar shock wave in the meteorite causes it to fragment into thousands of pieces and to partly vaporize as well. Large amounts of fragmental debris are then thrown out, while the solid bedrock is forced upward and outward to create the crater rim. A great deal of rock material falls back into the crater, filling in the bottom, where melted rock may be concealed.

A number of other craters of generally accepted meteoritic origin deserve mention. Comparatively small but abundantly endowed with nickel–iron fragments are the Odessa Craters in Texas, evidently formed by two large impacting fragments. A group of smaller craters in Argentina, the Campo del Cielo swarm, are associated with meteoritic iron fragments and silica glass. Within Australia, four outstanding crater localities are known, all with meteoritic iron. One of these, the Wolf Creek Crater, is remarkably perfect in form, the rim being 850 m in diameter (Figure 20.11). In Estonia, the Kaalijarv Crater is accepted as being of meteoritic origin, as is the Aouelloul Crater in Mauritania. Altogether, the list consists of only eight craters (or crater groups) with which meteoritic iron is associated.

Of those well-formed, large craters whose origin is disputed, and which have not revealed meteoritic iron, the New Quebec Crater of Canada is perhaps the most

FIGURE 20.9 An aerial view of Barringer Crater in northern Arizona. (W. B. Hamilton, U.S. Geological Survey.)

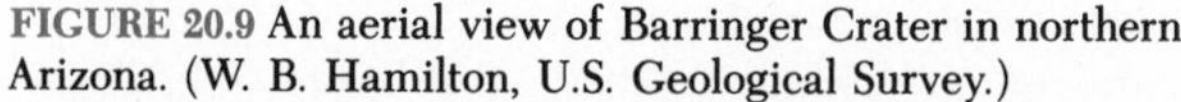

outstanding (Figure 20.12). It is exceptionally large: 3.2 km in diameter, and 400 m deep. The rim rises 100 to 150 m above the surrounding land surface. The bedrock, which consists of gneiss and granite of the Canadian Shield, is intensely fractured. Similar in some respects is the Ashanti Crater of Ghana, which contains a lake over 10 km in diameter. This crater also lies in ancient shield rocks.

Large circular structures lacking a rim or crater (presumably because of erosional removal) run to much larger diameters than the craters listed above. About a dozen such structures have been found in the

FIGURE 20.10 Hypothetical stages in the formation of a meteorite impact crater. (Data of E. M. Shoemaker and U.S. Geological Survey.)

FIGURE 20.10

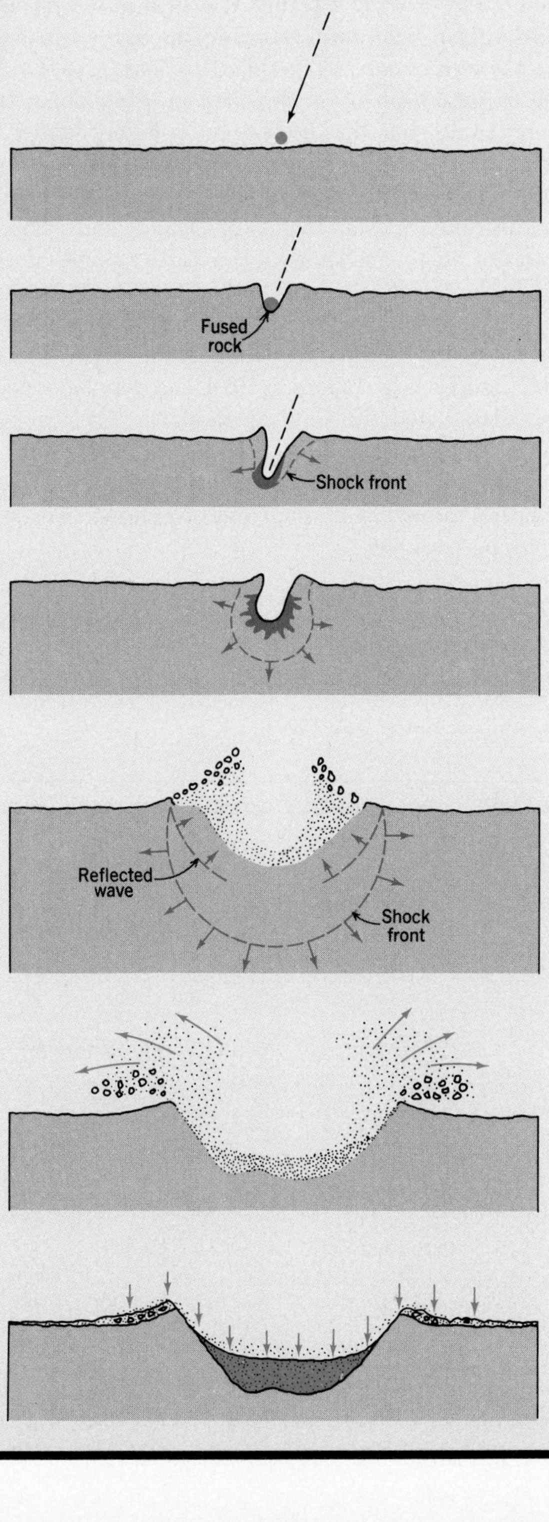

FIGURE 20.11 Oblique air view of Wolf Creek Crater, Western Australia. The crater is about 0.8 km in diameter, and the rim rises about 60 m above the flat sediment-filled floor. (Courtesy of R. M. L. Elliott, West Australian Petroleum Pty. Limited.)

FIGURE 20.12 Vertical air view of New Quebec Crater (Chubb Crater), located at about lat. 61° N, long. 73.5° W in northern Quebec. The depression, 3.2 km wide and 400 m deep, is formed in gneisses of Precambrian age. (Mosaic air photograph, K. L. Currie, Geological Survey of Canada, Dept. of Energy, Mines & Resources.)

FIGURE 20.11

FIGURE 20.12

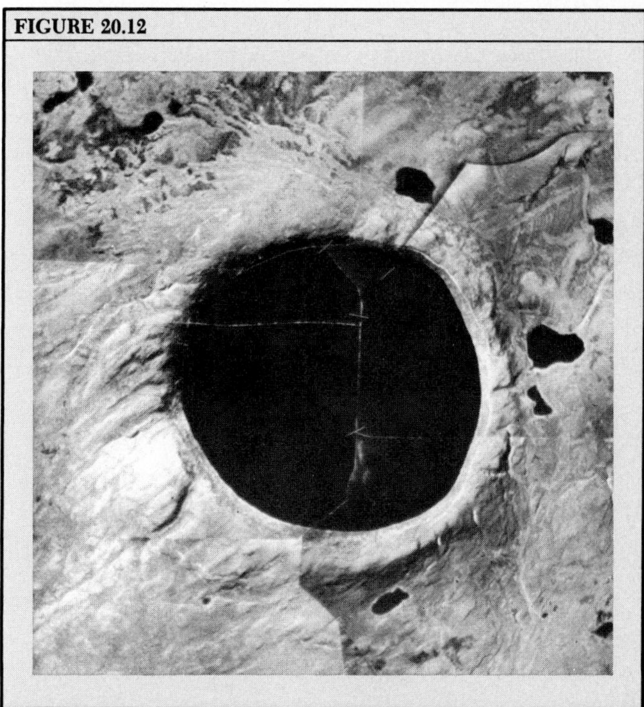

Canadian Shield. Typically, they are represented by circular depressions containing lakes (Figure 20.13). If these structures are astroblemes, as some investigators claim, they represent extremely ancient impacts, perhaps dating back as far as −200 to −300 m.y. when the Canadian Shield had been reduced to a low continental platform.

From a review of craters and circular structures of our Earth, we can draw the conclusion that impacts by large meteorites have been extremely rare events in recent geologic time. Perhaps the number has been on the order of a half-dozen occurrences in a million years. Moreover, possible meteor impact features of greater age are widely scattered and the number of impacts appears to have been small over the past half-billion years. But these frequencies are what we might expect, on the assumption that meteorites are not being produced in the modern solar system. If they represent space debris that continues to be swept up by planets and satellites, the frequency of impacts should be small in later geologic eras compared with a high frequency in the early stages of planetary formation.

Great earth-impacts by objects the size of asteroids may have played an important role in plate tectonics—or at least this possibility has been seriously raised in recent years. It has been suggested that a great impact might, by forming a huge crater, remove such a large piece of crust that it would set off the formation of a mantle plume. The reasoning behind this conclusion is that there would be a rapid isostatic rise of the lithosphere to compensate for the loss of crustal mass. The asthenosphere beneath would move rapidly upward as well, setting off a rising vertical motion that would develop into a plume.

Some geologists have gone even further by speculating that a group of asteroids or large meteorites, striking the Earth at one point in time, could have split a lithospheric plate and initiated continental rifting. The succession of mantle plumes that resulted from the impacts would have determined the zigzag line of a new rift system, such as we described in Chapter 10. The supposed impact craters that started the rifting would probably have been destroyed by downfaulting and erosion, then buried under the geoclinal sediments of the continental margins that bordered the rift.

Looking back even further into time, it seems quite likely that enormous asteroid impacts severely affected the Earth's surface during the early Precambrian time, when continents were just beginning to be formed. The huge craters that these impacts created may have set up major convection currents in the mantle and led to the first phases of lithospheric plate motion. Speculations of this kind are useful when based on established laws of physical behavior of the universe because they lead to the search for obscure new forms of scientific evidence we might not otherwise have investigated.

We must next look to a celestial body on which impact features would have been preserved with little or no erosion for the entire 4.5 b.y. since Earth and other planets formed. Three such bodies are available for study. They are our Moon and the planets Mars and Mercury. Venus is so completely concealed by a dense cloudy atmosphere as to be an unlikely prospect.

Our Moon and Other Planetary Satellites

A *planetary satellite*, or more simply, a *moon*, is a solid object orbiting a planet. It is held in the planet's gravitational field and also orbits the Sun along with the planet it circles. Altogether 34 moons (give or take one or two) have been identified in orbits about the nine planets, but the distribution is highly varied in terms of numbers of moons per planet (see Table 20.1). The two innermost planets, Mercury and Venus, have no moons. Mars has two, Deimos and Phobos. Both are much smaller than the Earth's single Moon, and both orbit much closer to the parent body. Jupiter has thirteen or more moons, of which four are large. Galileo, using the first astronomical telescope, discovered the four large moons in 1610, and they have since been designated the Galilean satellites. Their names are Io, Europa, Ganymede, and Callisto, stated in order outward (Figure 20.14). Saturn has ten or more moons; Uranus has perhaps five; Neptune, two; and Pluto, perhaps one.

FIGURE 20.13 Large circular structures of the Canadian Shield. Solid lines are lake shorelines. New Quebec Crater, shown for scale comparison, is pictured in Figure 20.12. (After M. R. Dence, 1965, *Annals, New York Academy of Science*, vol. 123, p. 943, Figure 2.)

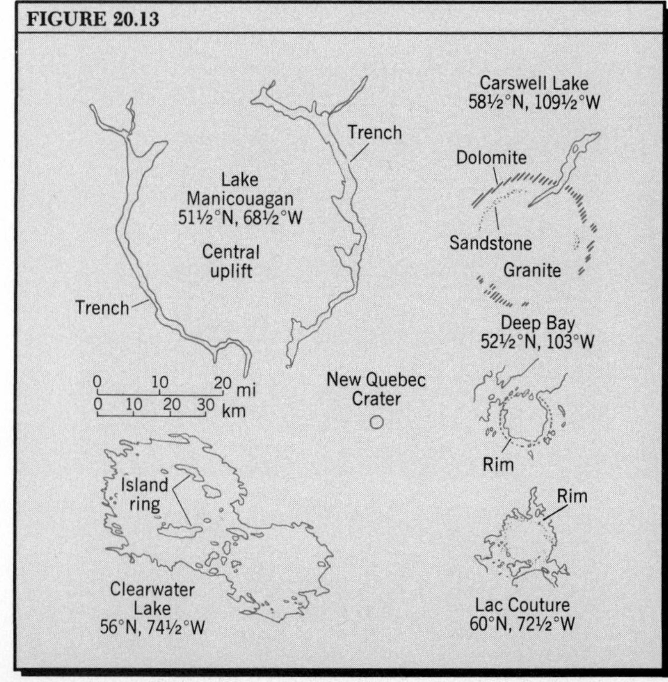

Of all the satellites, Earth's Moon is unique in that it is a very large body in ratio with the planet which it orbits. With the Moon's diameter of 3476 km, compared with an Earth diameter of about 13,000 km, the ratio of diameters is about 1:4. Our Moon has a mass of about 1/81 that of Earth. Because of the Moon's relatively large size, compared to the planet it orbits, astronomers have commented that the Earth–Moon system can be considered a "binary planet."

Figure 20.15 shows the Moon's orbit, an eccentric ellipse, and gives distances from Earth. The average radius of the Moon's orbit is about 382,000 km. When closest to Earth, the Moon is said to be in perigee; when most distant, in apogee—in fact, these terms apply to corresponding positions of any satellite, including artificial orbiting satellites. Notice that the Moon revolves about the Earth in a direction that can be described as counterclockwise, when we imagine ourselves to be viewing the system from a point above the Earth's north pole. Rotation of both Earth and Moon is also counterclockwise.

FIGURE 20.14 The four Galilean satellites of Jupiter as they might appear through binoculars or a small telescope. Distances from Jupiter are shown to correct scale, as if all four satellites were in a line at right angles to the observer. Diameters are not to scale.

FIGURE 20.15 The Moon's orbit is an ellipse. Distances shown are from center of Earth to center of the Moon.

FIGURE 20.14

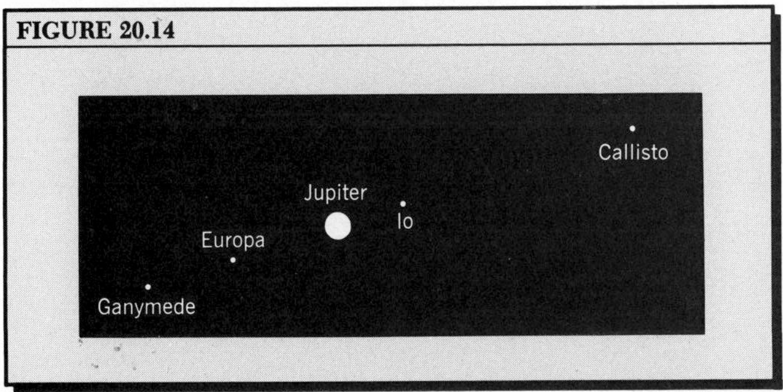

FIGURE 20.15

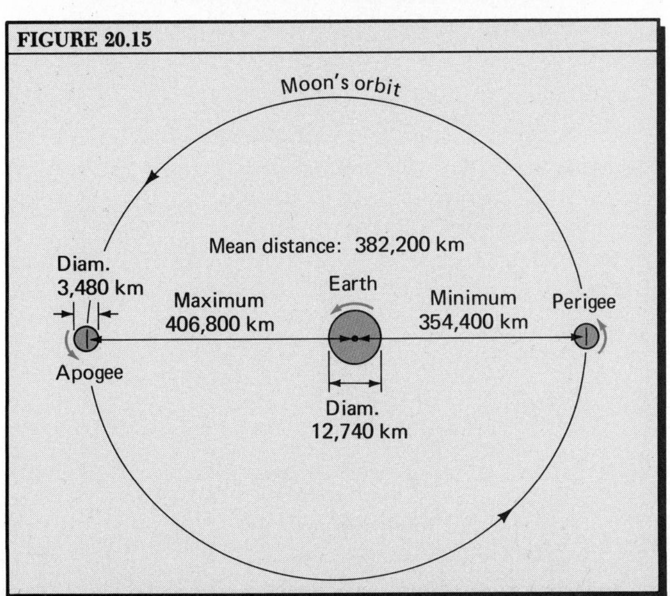

The period of the Moon's revolution, calculated in terms of 360° of angle with reference to the fixed stars, is 27.32 days and is referred to as the sidereal month. However, when we measure the revolution in terms of reference of the Sun's position in the sky (from one new moon to the next) the period averages 29.5 days and is called the synodic month. The monthly rhythm of the tides follows the synodic month (Chapter 19).

Of interest in the history of both Earth and Moon is the fact that the Moon always shows the same face to observers on Earth. Until it was seen by the astronauts in the manned Apollo spacecraft, the opposite side of the Moon had never been directly viewed by a human being, although photographs returned by unmanned satellites had previously revealed many details of that surface.

The fact that we can see only one side of the Moon requires that the Moon's period of rotation upon its axis be exactly the same as its period of revolution with respect to the stars. So we can say that the Moon rotates once in 27.32 days. Such a coincidence could scarcely be the result of mere chance. In fact, the Moon is slightly lopsided, with a concentration of denser rock bodies on the side facing the Earth. Gravitational attraction for this excess mass has locked the Moon's rotation into a period equal to its revolution.

The Lunar Surface Environment

To interpret correctly the physical features of the Moon's surface, we must first consider the surface environment of that satellite. Environmental factors include (1) the Moon's gravity field, (2) a lack of both atmosphere and hydrosphere, (3) intensity of incoming and outgoing solar radiation, and (4) surface temperatures. All these factors show striking differences when compared with the surface environment of Earth.

Recall that gravity is defined as the attraction which Earth or Moon exerts upon a very small mass located at its surface. Gravity on the Moon's surface is about one-sixth as great as that on Earth, so an object that weighs 6 newtons (N) on Earth will weight only 1 N on the Moon. This relatively weak gravity is of human interest when we watch films of our astronauts cavorting on the lunar surface. The weak gravity is of great geologic importance in interpreting the Moon's surface and history. For example, rock of the same strength as rock on Earth could stand without collapse in much higher cliffs and peaks on the Moon than on Earth. Objects thrown upward at an angle from the Moon (as when the Moon is struck by a large meteoroid) will travel much higher and farther than objects under the same impetus on Earth.

Lack of sufficient gravity has cost the Moon any atmosphere that it may once have possessed, since gas molecules of any earlier atmosphere could have escaped readily into space.

Lacking an atmosphere, the Moon's surface intercepts the Sun's radiation on a perpendicular surface with the full value of the Sun's output. Moreover, the Moon absorbs most of this incoming energy. The effect is to cause intense surface heating during the long lunar day (about two weeks' duration). When the Sun's rays have been striking at a high angle for several days continuously, surface temperatures at lunar noon reach about 100 °C. Correspondingly, conditions on the dark side of the Moon reach opposite extremes of cold during the long lunar night. With no atmosphere to block the escape of heat, surface temperatures drop to values estimated at below −173 °C.

Of particular interest is the sudden drop in lunar-surface temperature when sunlight is abruptly cut off by a lunar eclipse. In one such instance, the surface temperature fell from 71 °C to −79 °C in only one hour. This drop of 150 C° is vastly greater than any natural temperature drop on the Earth's surface in a comparable period of time.

From the geologist's standpoint, such vast temperature ranges are of interest because of the possible effect upon minerals exposed at the Moon's surface. The expansion and contraction that crystalline minerals undergo when heated and cooled can bring about the disintegration of solid rock into small particles. These volume changes can also cause loose particles to creep gradually to lower levels on a sloping ground surface. These effects may be unusually important on the Moon, because without an atmosphere and running water, ordinary terrestrial processes of weathering,

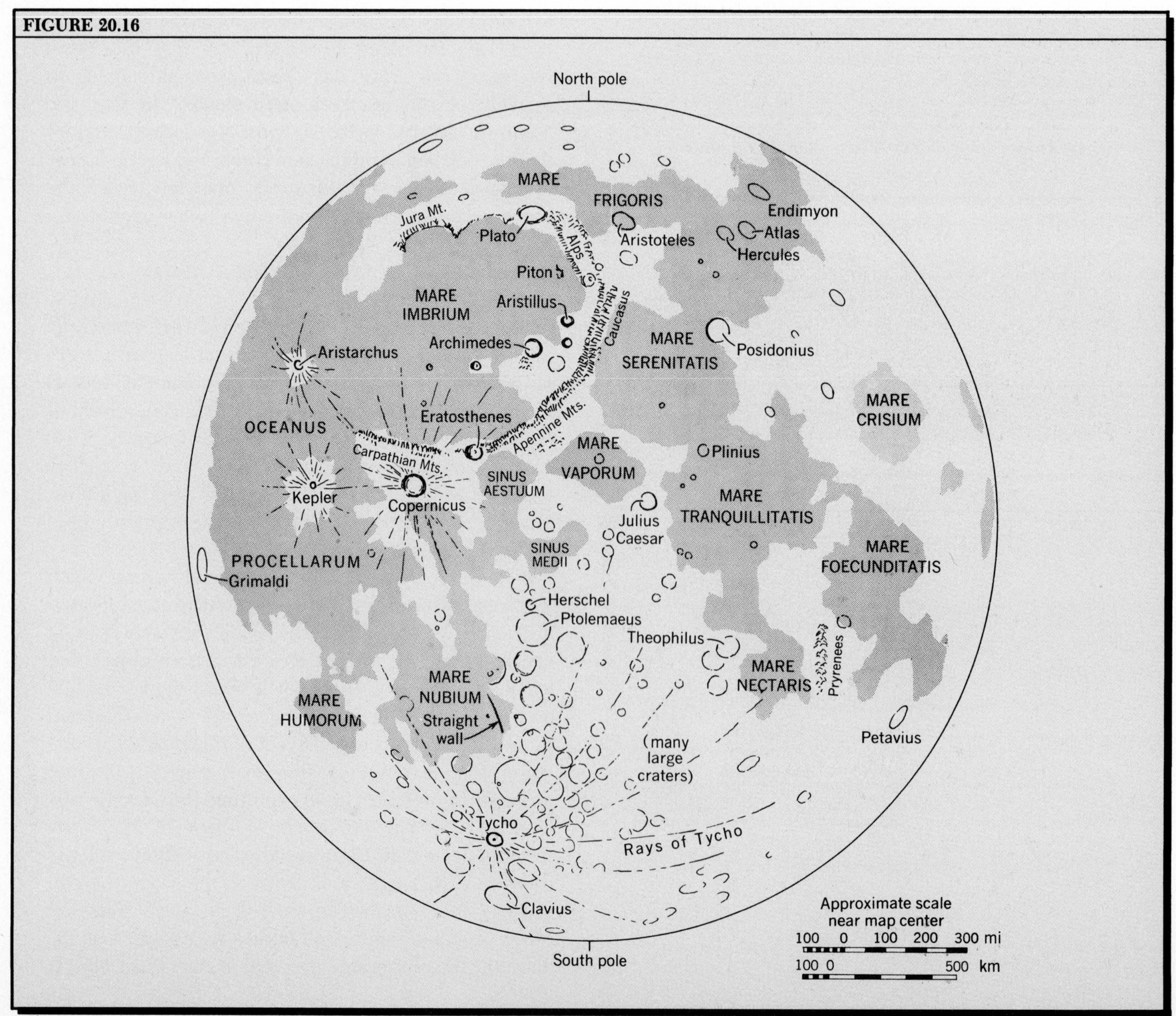

FIGURE 20.16 Sketch map of the major relief features of the Moon. Use this diagram as an aid in identifying areas and subjects in the lunar photographs of this chapter.

erosion, and transportation cannot act. Without streams and standing water bodies, the Moon has no mechanisms or receiving areas for accumulation of water-laid sedimentary strata. Obviously, in comparison with Earth, surface changes must be extremely slow on the Moon.

A Decade of Lunar Exploration

The final splashdown of *Apollo 17* in December 1972 brought to a close our first exploration of the Moon's surface. The vast store of rock and soil samples, photographs, and other observational data accumulated during the Apollo program will take many more years to analyze fully. Although certain basic geological facts about the Moon were established beyond a doubt, many more deep and complex scientific puzzles have emerged than could have been imagined.

The decade of firsthand lunar exploration began in 1963, when a U.S. Ranger spacecraft impacted the lunar surface, sending back thousands of pictures from a wide range of altitudes. There followed landings by Surveyor spacecraft, making direct physical and chemical tests of lunar surface materials as well as photographs of the ground immediately surrounding the vehicle. Then came a series of Lunar Orbiter vehicles on missions of photographing the lunar surface from distances as close as 56 km. Vertical photographs at least ten times sharper than the best taken by telescopes from Earth were obtained from both the nearside of the Moon and the previously unknown farside.

In 1969 and 1970, human-operated space vehicles of Apollo missions circled the Moon at low levels and descended to the lunar surface, permitting photographic negatives in color to be brought back, along with samples of lunar materials.

FIGURE 20.17 The Moon as it would appear if it could be uniformly illuminated. This is a composite of many photographs. (NASA.)

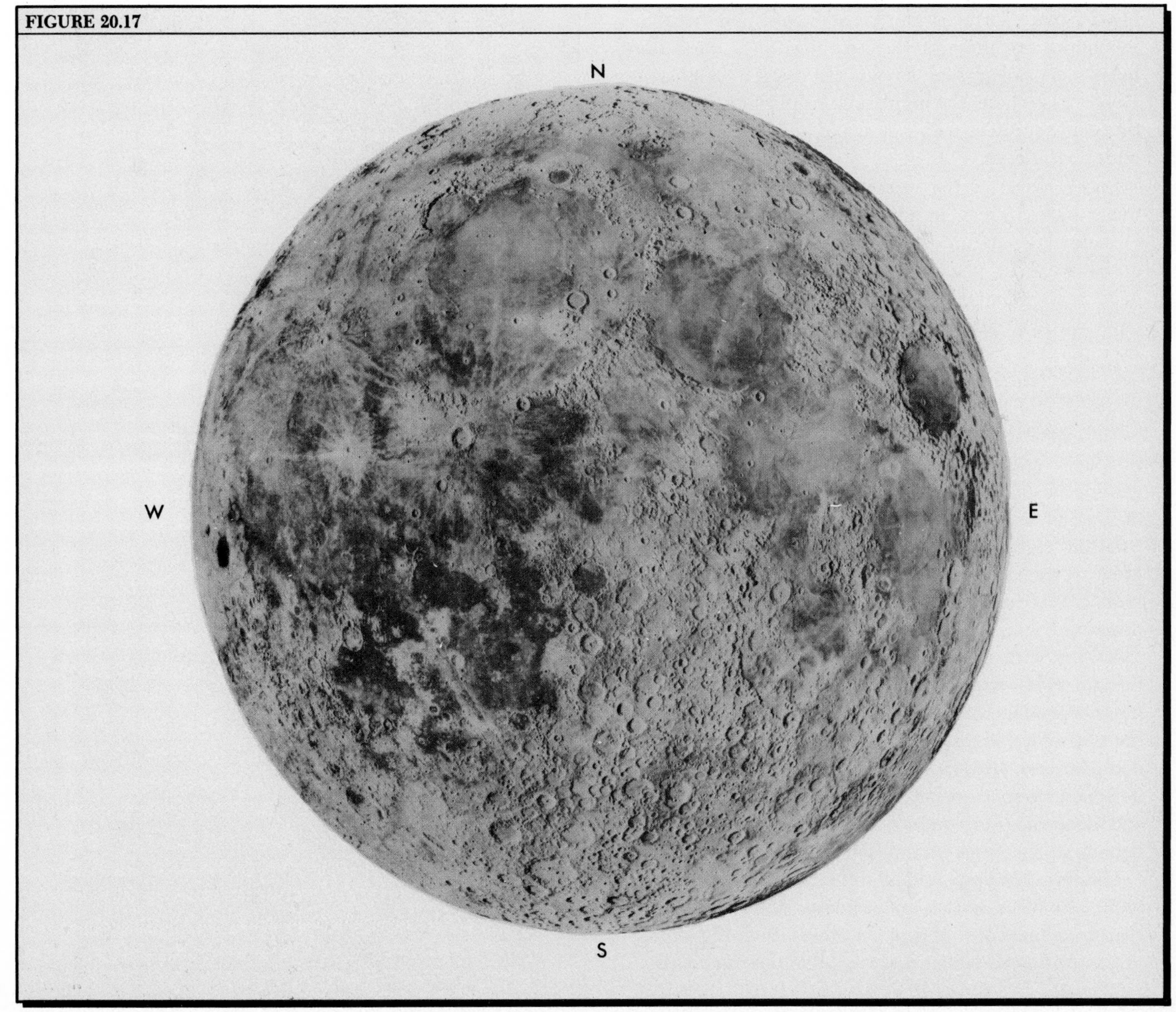

The Lunar Maria

As anyone can easily see, using a small telescope or binoculars, the first major subdivision of the Moon's surface is into light-colored areas and dark areas (Figures 20.16 and 20.17). The former constitute the relatively higher surfaces, or *lunar highlands*. For many years, the light-colored areas were called collectively the *terrae*. This Latin word reflects the earliest interpretations of Galileo—that these areas were dry lands. The dark areas—the low-lying, smooth lunar plains—are the *lunar maria*, plural of the Latin word *mare*, or "sea"; Galileo applied this term to what he believed to be true seas of liquid. Of the nearside of the Moon, about 60% is terrae and about 40% maria.

Lunar highland areas exhibit a wide range of relief features. Most outstanding are the great mountain ranges, of which there are 20 major groups on the nearside. Perhaps the most spectacular of these are the Appenines, rimming the Mare Imbrium on the southwest side (Figure 20.18). Several peaks within the Appenines rise to heights of 4 to 5 km above the nearby mare surface. The highest and most massive mountains are those of the Leibnitz range, near the lunar south pole, which have peaks rising to heights of 11 km. Elsewhere, the highland terrain consists of gently rolling surfaces with low slopes and of rough areas with steep slopes.

Lunar maria of the nearside are divided into ten major named areas, in addition to a single vast area, Oceanus Procellarum (Figure 20.16). Although maria outlines are in places highly irregular, with many bays, a circular outline is persistent for several, and particularly striking for Mare Imbrium.

The Lunar Craters

Most spectacular of the Moon's surface features are the *lunar craters*, abundant over both the highland and maria surfaces. Even under the low magnification of a small telescope, the large craters form an awe-inspiring sight when seen in a partial phase of the Moon. Using telescopes alone, it is possible to count some 30,000 craters on the nearside with diameters down to 3 km, and an estimated 200,000 have been identified with space-vehicle photography, including recognizable craters as small as 1 m across. On the lunar nearside, there are 150 craters of diameter larger than 80 km. Largest of these is Clavius, 235 km in diameter and surrounded by a rim rising 6 km above its floor. (See south polar area in Figure 20.16.) Tycho and Copernicus are among the most recent and most striking of the large craters (Figure 20.19).

The forms of large craters fall into several types. In some, the floor is flat and smooth, and the rim is sharply defined and abrupt. In others, such as Copernicus, the floor is saucer shaped, while the rim consists

FIGURE 20.18 Mare Imbrium, with its bordering mountains, the Jura, Alps, Caucasus, Apennines, and Carpathians; and the great craters Plato, Aristillus, Archimedes, Eratosthenes. Note the spinelike peak, Piton. Identify these features with the aid of Figure 20.16. (Mount Wilson Observatory photograph. Mount Wilson and Las Campanas Observatories, Carnegie Institution of Washington.)

FIGURE 20.19 Copernicus, the great lunar crater lying south of Mare Imbrium (see Figure 20.16). Note the rays—radial streaks of lighter colored material. The conspicuous crater lying east-northeast of Copernicus is Eratosthenes. (Mount Wilson Observatory photograph. Mount Wilson and Las Campanas Observatories, Carnegie Institution of Washington.)

FIGURE 20.18

FIGURE 20.19

FIGURE 20.20 Censorinus (*arrow*), one of the freshest craters on the Moon's nearside, shows a sharp rim and a surrounding zone of light-colored soil surface. Crater rim is about 7 km in diameter. (*Apollo 10* photograph by NASA.)

FIGURE 20.21 The Moon photographed by *Lunar Orbiter IV* spacecraft from a distance of 8000 km. The heavily cratered lunar farside lies to the left. (NASA.)

FIGURE 20.20

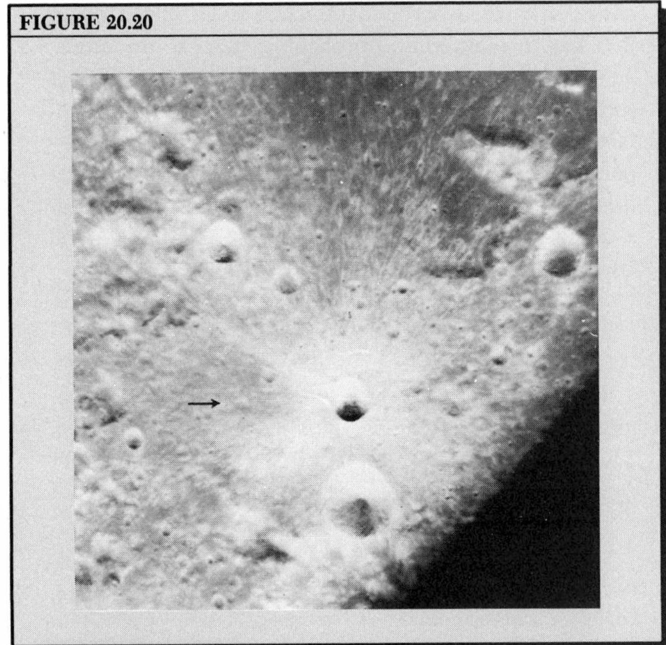

FIGURE 20.21

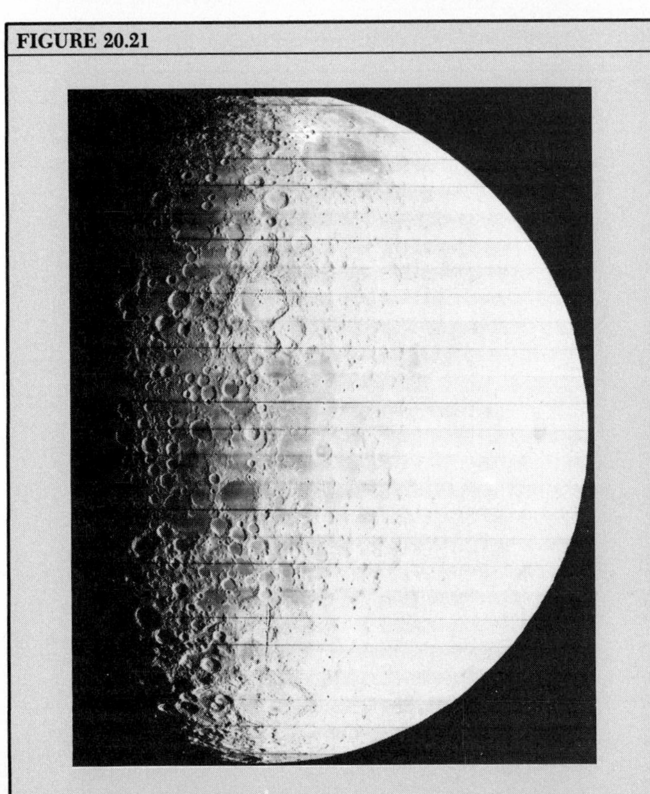

of multiple concentric ridges. Of particular interest are systems of *lunar rays* of lighter colored surface radiating from certain of the larger craters—for example, Copernicus (Figure 20.19). In several of the large craters, there is a sharply defined central peak, which must be taken into account in interpreting the origin of craters. That of Eratosthenes shows up particularly well in Figure 20.19.

The impact origin for almost all of these smaller craters is generally accepted. Many small lunar craters are almost perfectly circular and have a cup-shaped interior and a prominent rim (Figure 20.20). Others are less distinct and have low slopes, or are mere shallow depressions. A few are fresh in appearance, with a litter of boulders on the rim and within the crater itself.

The newer craters are typically lighter in color than the surrounding surface. The more subdued craters appear to be older and to have lost their sharpness through slow processes of mass wasting and meteoroid bombardment. The oldest forms have lost their rims and show only a shallow depression. A few small craters, which are elliptical in outline, may represent the secondary fall of masses dislodged by much larger impacts.

It has been shown that the number of lunar craters per unit of area is inversely proportional to the square root of the crater diameter. Consequently, the number of large craters is small and that of very small craters is legion.

The farside of the Moon, photographed in detail by Lunar Orbiter spacecraft and by astronauts of Apollo spacecraft, is heavily cratered and lacking in extensive maria (Figure 20.21).

Rilles and Walls

Yet another class of distinctive lunar-surface features is the narrow, canyonlike *rilles*. Some are remarkably straight, taking the form of a trench up to 240 km long. Others are irregular in plan, and a few are sinuous, suggestive of terrestrial meandering rivers (Figure 20.22). Astronauts of *Apollo 15* inspected the walls of Hadley's Rille at close range and discovered that layers of rock are exposed, suggesting that an erosion agent carved the rille into older deposits. Most investigators agree now that the sinuous rilles are collapsed lava tunnels, while most of the straight rilles are interpreted as fracture features in brittle rock of the lunar crust. Related features are the straight cliffs, or *lunar walls*, which may be fault escarpments. A particularly striking example is the Straight Wall in Mare Nubium (Figure 20.16), which is 240 m high and may represent a fault that broke the mare surface.

Study of Lunar Orbiter photographs has revealed many classes of minor details that are as yet little understood. Particularly prevalent in areas of rough terrain are minute systems of parallel ridges and troughs. Many of the steep slopes exhibit steplike terraces near the base.

Origin of the Great Craters and Maria

Origin of the great craters and the circular-rimmed maria brings into conflict two widely divergent hypotheses, both with supporters throughout the long history of lunar geology.

One extreme view holds that the lunar craters and maria are largely of volcanic origin. This hypothesis implies a long history of intrusion and extrusion of molten rock from the Moon's interior. Circular depressions of the great craters are explained by the collapse of volcanoes as magma was removed from below, and the maria are interpreted as vast lava fields produced by extrusion of basalt magma from deep sources.

The volcanic interpretation of craters encounters difficulty on a number of points. True volcanic cones, such as those found in abundance on Earth, are very rare, and lunar photographs rarely show features that can be interpreted as lava flows. Except for fracture lines, which are numerous, mountain-making forms typical of zones of lithospheric plate spreading and subduction seem to be totally lacking. The high lunar mountain ranges seem, instead, to constitute rims of circular maria or of large craters (see Figure 20.23). Because of these facts, the volcanic hypothesis has few supporters today.

The second major hypothesis of the Moon's geology may be described as the *meteoritic hypothesis*. It includes the assumption—in direct opposition to the first hypothesis—that the Moon is a body without internal rock melting or volcanic action, although some surface volcanism is recognized. One of the strong advocates of the meteoritic theory was the distinguished American geologist Grove Karl Gilbert, who published his explanation in 1893. Among the modern group of scientists who have contributed details to the impact hypothesis is Harold C. Urey, a Nobel Prize winner, whose modified explanation is often referred to as the Urey–Gilbert hypothesis. According to Gilbert, the great craters were produced by large meteorites in the manner which we have previously discussed in connection with terrestrial meteorite craters, even though some of the lunar craters are much larger than any recognized on Earth.

The characteristic central peak, which is also observed in several of the larger terrestrial craters, requires special explanation. One hypothesis is that it is formed of volcanic rock extruded following the impact, which melted a body of rock below the floor of the crater. Another hypothesis is that the floor of the crater was uplifted by isostatic rebound, triggered by the sudden loss of rock mass. Others have visualized a process of elastic rebound in which underlying rock, compressed elastically by the sudden impact, expanded instantly to its former bulk. The rays which emanate from several large craters are explained as debris deposits thrown out over long distances from the explosion centers (Figure 20.19).

Under the Urey–Gilbert hypothesis, the rimmed maria, of which there are at least five on the lunar nearside, are the old impact scars of enormous masses, probably asteroids (Figure 20.23). The case of Mare Imbrium is particularly striking, as Gilbert pointed out. Urey also reconstructed the Imbrium collision, considering it as involving impact of an object about 200 km in diameter, approaching from a low angle. The impact raised an enormous wave of rock that spread outward, coming to rest in a great arc of mountain ridges. Gilbert had proposed that the heat of impact melted a vast quantity of lunar rock that subsequently solidified as basalt over the mare surface, but it now seems that these flood basalts erupted much later.

Lunar Surface Materials

The first samples of lunar rock and soil were obtained in 1969 by astronauts of the *Apollo 11* mission. There, on Mare Tranquillitatis, as at subsequent Apollo landing sites, surface materials consist of

FIGURE 20.22 Hadley's Rille, a sinuous, canyonlike feature, begins in a highland area and crosses a cratered plain. This *Lunar Orbiter V* photograph spans an area about 50 km wide. (NASA.)

FIGURE 20.22

unsorted fragmental debris. The fragments range in size from dust to blocks many meters in diameter. The layer of loose, fragmental mineral debris is called *lunar regolith.* The uppermost few centimeters are a brownish to grayish, cohesive, powdery substance; it consists of grains in the size range from silt to fine sand. The material is easily penetrated. Upon compaction, it becomes stronger, easily supporting the weight of the astronauts and their equipment. Figure 20.24 shows the regolith compacted into a clear footprint.

All observed lunar material sufficiently hard to be called "rock" has proved to consist only of fragments or clusters of fragments. Nowhere was there found a massive outcropping of a large body of solid rock in place. In other words, exposed bedrock as we know it on Earth does not exist on the Moon, so far as is known.

All of the lunar rock fragments have proved to be of igneous origin, with pyroxene, plagioclase feldspar, and olivine being the most abundant minerals. There are many other accessory minerals, usually of secondary importance so far as abundance is concerned. Some examples familiar from the earlier descriptions of igneous rocks are ilmenite, quartz, potash feldspar, and amphibole. Iron and titanium are constituent elements in quite a few of the accessory minerals. Some of the minerals also contain uranium and thorium, which are radioactive.

FIGURE 20.23 The great lunar maria, dark in color and with sharp, curved rims, show particularly well in this photograph of the Moon in gibbous phase, age 11 days in the synodic month. (Mount Wilson Observatory photograph. Mount Wilson and Las Campanas Observatories, Carnegie Institution of Washington.)

FIGURE 20.23

Lunar Rocks

One way in which to classify the lunar rocks is to divide them into two groups: First are fragments of igneous rocks, perhaps broken from some unknown parent mass of solidified magma. These come in both fine-grained and coarse-grained textures (Figure 20.25). Second—and the great majority of lunar rocks collected by the astronauts—are *lunar breccias,* a breccia being a rock composed of many small, angular fragments. Individual pieces of rock making up the breccia are themselves fragments of igneous rock or other breccias. Consequently, a single chunk of lunar breccia may contain representatives of igneous rock from many different sources. It is believed that the lunar breccias originated from intense shock, which bound together particles that had previously been fragmented by shock, in a process called *impact metamorphism.* Meteoroid impacts are considered to be the shock mechanism, both for fragmenting rock into regolith and for bringing it together into breccia.

We will, for the sake of simplicity, identify only two major groups of the igneous rock fragments: basalt and a variety of anorthosite. Basalt is familiar as a common terrestrial rock composed largely of calcic plagioclase feldspar, pyroxene, and commonly olivine as well. Lunar basalt, shown in Figure 20.26, is quite similar, except that the basalts underlying the maria are richer in iron and poorer in silica than are Earth basalts. Maria basalt magma was much more fluid than Earth basalt, and probably flowed very freely as it filled the maria

FIGURE 20.24 Astronaut's footprint in the lunar soil. Notice the many miniature craters in the surrounding surface. (*Apollo 11* photograph by NASA.)

FIGURE 20.24

basins. Anorthosite occurs on Earth as a plutonic igneous rock, but true anorthosite is rare on the Moon. Rock of the lunar highlands is best described as *anorthositic gabbro*, a rock of intermediate composition between anorthosite and gabbro.

The lunar regolith proved to consist of tiny rock fragments, including both basalt and anorthositic gabbro, and numerous tiny glass beads, called *spherules* (Figure 20.27). The spherules are considered as "splash" phenomena, resulting from the sudden cooling of igneous rock melted by meteoroid impact and hurled from the impact crater as a spray of droplets, though some of the spherules may be of volcanic origin.

A particularly interesting feature of the rock samples is the presence of surface pits, averaging less than 1 mm across, lined with glass. The pits—referred to as "zap pits"—along with the rounding of gross shape of the fragments and a lighter-colored surface than the inside rock suggest that a process of erosion has been in operation and is most probably attrition by impacts of small particles.

If the Moon is smothered under a blanket of impact debris, how can we even guess at what kind of solid bedrock lies below, constituting the lunar crust? The answer lies in sampling the bouldery rim debris of craters (Figure 20.28), with the idea that a meteoroid, upon penetrating the regolith and exploding, tore loose and threw out boulders of the bedrock below. Acting on this supposition, geologists have inferred that the uppermost part of the lunar crust consists in some places of basalt and in other places of anorthositic gabbro.

As a broad generalization, we can say that basalt underlies the floors of the maria, while anorthositic gabbro underlies much of the lunar highlands. So, in a fashion somewhat like the Earth's twofold division into a felsic continental crust and a mafic oceanic crust, the Moon's crust has been differentiated into a denser basaltic crust and a less dense crust of anorthositic gabbro. At least, this arrangement applies for shallow depths, within the range of disturbance by meteoroid impacts.

A Geologic History of the Moon

Radiometric age determination has been a mainstay in attempts to work out a geologic history of the Moon. One rather surprising finding from lunar samples is that, so far, no rock has turned up with an age of crystallization from its magma younger than −3 b.y. Impact fragmentation and formation of breccia has certainly been going on since that time, but it seems to have been only a mechanical reworking of the ancient igneous rocks. Practically all samples have shown an age of crystallization in the time span from −4.2 b.y. to −3.1 b.y.

Geologists kept hoping that some specimen would prove to be older—say −4.5 b.y. or thereabouts, but they were disappointed up to the very last minute. Finally, on the last Apollo mission, astronauts Eugene Cernan and Harrison Schmitt, the latter a geologist, brought home a blue-gray breccia, labeled as coming from "Boulder No. 2." From within the breccia were

FIGURE 20.25 A lunar rock sample collected by astronauts of the *Apollo 11* mission at Tranquillity Base. It is an igneous rock of granular texture, showing glass-lined surface cavities. Specimen is about 20 cm high. (*Apollo 11* photograph by NASA.)

FIGURE 20.26 This specimen of lunar basalt was named the Goodwill Rock by Astronauts of *Apollo 17* mission. The cavities are original gas-bubble vesicles in the basalt. The specimen is about 10 cm high. (*Apollo 17* photograph by NASA.)

FIGURE 20.25

FIGURE 20.26

extracted several fragments of a greenish mineral which proved to be fragments of the ultramafic rock dunite, composed mostly of the mineral olivine. Their age was no less than −4.6 b.y. Here at last was evidence of the original igneous material with which the Moon had been put together. As yet, on Earth, no rock of this age has been discovered. In fact, no Earth rock is even as old as the most abundant Moon rocks, in the age span −4.2 to −3.8 b.y. Clearly the Moon's geologic record falls into a very different age bracket from that of Earth, and perhaps the Moon can tell us what may have happened on Earth before −3.8 b.y., where the Earth record starts. The ancient record on Earth has probably vanished for good, consumed by crustal melting.

Table 20.2 presents a broad-brush picture of lunar history as a number of investigators tentatively figure it. We refer to these happenings as "events," and the numbers assigned to them are purely arbitrary.

The Lunar Interior

The Moon's interior structure and composition are still subject to intense speculation, with some quite different interpretations being proposed simultaneously. One fact is established: The Moon as a whole is deficient in iron and almost certainly lacks the iron core present in the Earth. The Moon's low average density, 3.34 g/cc, is about the same as that of the Earth's mantle rock. If an iron core existed in the Moon, the average density would be considerably higher.

It is also agreed that the Moon has a low-density crust, averaging perhaps 2.9 g/cc. The average crustal composition is described by some researchers as "anorthositic–gabbro," by others as "gabbroic–anorthosite." Thus, the description of the crust implies a rock of average composition intermediate between basalt and anorthosite. Beneath the crust there probably lies

FIGURE 20.27

FIGURE 20.27 Glass spherules from the lunar regolith. Largest spherule is 0.4 mm in diameter. Spherules rest on an aluminum dish with a scratched surface. (*Apollo 11* photograph by NASA.)

FIGURE 20.28 Large boulder and *Apollo 17* astronaut Harrison H. Schmitt, Station 6, Taurus Littrow. (NASA.)

Table 20.2 Geologic History of the Moon

Event	Age (b.y.)	
One	−4.7 to −4.6	**Accretion of Moon.** This occurred at about the same time as for Earth, and perhaps not far away in the solar system.
Two	−4.5 to −4.3	**Big thermal event.** Moon became hot and melted down to a depth of about 300 to 400 km.
Three	−4.2 to −4.0	**Original lunar crust formed.** The crust solidified from molten magma. Differentiation of magma produced the outer crust of anorthositic gabbro, leaving mafic rock beneath. Crust is about 80 km thick.
Four	−3.9	**The big bombardment.** Moon was clobbered by many huge asteroids. Moon was then closer to Earth than now. Bombardment was uneven, mostly impacting the leading side of Moon as it moved in orbit. Maria impact basins were formed.
Five	−3.8 to −3.1	**Maria flood basalts.** The maria basins were filled by outpourings of basaltic lava. Magma originated deep within Moon.
Six	−3.0 to present	No further significant igneous activity. Countless meteoroid impacts, forming breccias and causing repeated lifting and dropping of particles.

a mantle of mafic rock composed of olivine and pyroxene.

As shown in a hypothetical lunar cross section, Figure 20.29, and as mentioned earlier, the Moon is lopsided. The crust is calculated to be over 100 km thick on the farside but only about 60 km on the nearside, beneath the maria. (This thickness is enormously exaggerated in the figure.) Evidence for crustal thickness on the nearside comes from seismic waves recorded by seismographs implanted on the Moon's surface. Minor fault movements and the impacts of meteoroids set off *moonquakes*. As these pass through the Moon, they undergo velocity changes depending upon changes in density of the rock, and thus the layered structure can be interpreted.

As Figure 20.29 shows, the center of the Moon's mass is offset by about 2 km from the geometric center of the Moon's outline and toward the lunar nearside—the side facing the earth. The explanation lies in the lenslike masses of dense maria basalt and a thinner crust on the lunar nearside. On Earth, similar masses of dense rock would have sunk down on a soft asthenosphere, coming into isostatic equilibrium. However, as the Moon's interior is strong and rigid down to a depth of several hundred kilometers, the maria basalts cannot sink. Consequently, the Moon's mass distribution is lopsided, which explains why the Moon, turning once on its axis with each revolution around Earth, keeps its nearside always toward Earth. As we noted earlier, gravitational attraction for the denser part of the crust has locked the Moon's rotation to the Earth's gravity field.

Below the lunar crust is an interior region most probably consisting of olivine and pyroxene. In other words, the interior is made up of ultramafic rock, perhaps much like that of the Earth's mantle. The density of the lunar interior probably ranges from 3.2 g/cc near the crust to over 3.5 g/cc near the center.

A major point of debate concerns the temperature of the Moon's interior. The hypothesis of a "cold" Moon, its temperature everywhere well below the melting point, had been widely held. Even at a central temperature as high as 1400 °C, the rock would remain solid, and this value was considered by some geologists to be much higher than would actually exist. However, some newer lines of evidence suggest that interior temperatures may reach a maximum value of 1600 °C at a depth of about 1100 km. Perhaps below that depth there is a layer of partial melting, similar in some respects to the soft layer of the Earth's mantle. Most moonquakes originate at the base of the outer rigid layer, and the deep moonquakes have a periodic rhythm that suggests they are caused by tidal stresses.

FIGURE 20.29 A schematic diagram of the Moon's interior. As the dimensions show, the crust is greatly exaggerated in thickness to show difference between nearside and farside. (Based on data of D. L. Anderson, 1974, *Physics Today*, vol. 27, and U. B. Marvin, 1973, *Technol. Rev.*, vol. 75.)

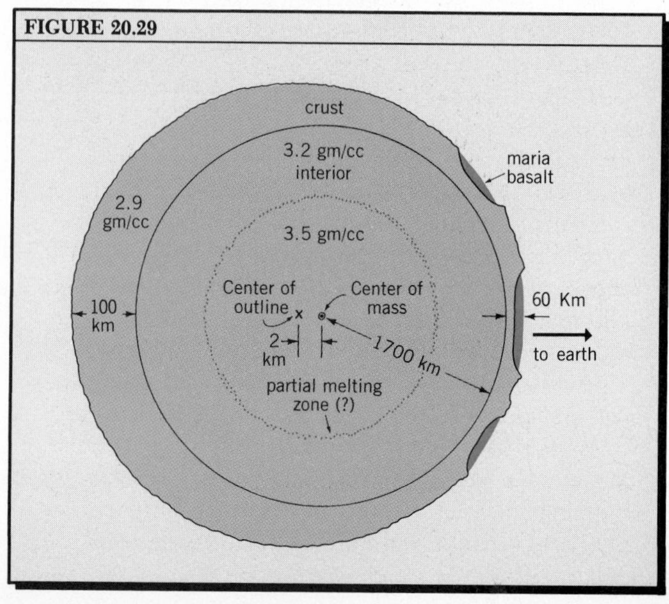

No crustal activity resembling plate tectonics on Earth is called for, and there is no surface evidence of the presence of plates or of plate motions.

Origin of the Moon

One problem that seems to have remained entirely unsettled by the Apollo program is that of the Moon's origin. One hypothesis proposes that the Moon and Earth grew by accretion of nebular material as a binary planet, linked by gravitational attraction. However, the Moon's low density, and lack of iron and similar elements, are hard to explain if its accretion occurred at the same time as the Earth's, under similar physical and chemical conditions.

A second possibility, widely debated, is known as the *capture hypothesis*. It is supposed that the Moon formed by accretion in another part of the solar nebula, where conditions were quite different from those prevailing where the inner planets were forming. Later, the Earth captured the Moon, when by chance the Moon passed close to the Earth and was trapped by Earth's gravitational field. It has been suggested that capture occurred about −3.7 to −3.6 b.y. and corresponded with the rise of the maria flood basalts. However, the mechanism of capture of so large a planetary body by Earth faces strong objections.

A third hypothesis holds that the Moon was formed of material broken away from a fast-spinning Earth. This *fission hypothesis* was put forward in the 1890s by Sir George Darwin, son of Charles Darwin and a noted authority on tides. He calculated that at some point in time the Earth was spinning much faster than today, and that the centrifugal force of rotation produced a large bulge in the Earth. The bulge then separated from the Earth and moved away to become the Moon. Darwin and others even went so far as to suggest that the Pacific Ocean basin was the source of the lunar material.

The fission hypothesis was revived in the early 1960s. Under this hypothesis, the Earth's iron core had already been formed, so the material thrown off was part of the mantle. In this way the Moon's low density and lack of a large core are explained. Although the fission hypothesis seems satisfactory on a number of basic counts, there are strong objections. Certain minor elements show very different patterns of abundance when lunar and Earth rocks are compared from a chemical standpoint, and scientists argue that these differences make it most unlikely that the lunar material was ever part of Earth.

Another still more recent revival of the fission hypothesis makes use of a tremendous impact on Earth by a planetary body about the size of the present Moon, or a bit larger. This object struck the rapidly spinning Earth at a time following the formation of the Earth's mantle and iron core, and the force of the impact tore away a large mass of the Earth's mantle. This detached mass was pulverized into dust by the impact and orbited the Earth. Accretion of the orbiting dust into a single solid sphere formed the Moon. This suggested history seems to conflict with the dates we gave earlier in the chapter for the formation of the Earth's iron core and mantle, namely, −4.3 to −4.1 b.y. If our figure of −4.6 to −4.7 b.y. holds for the accretion of the Moon, and if the impact hypothesis is valid, the Earth's core must have formed very rapidly, immediately after the Earth's accretion.

And so the speculation may revert back to accretion of the Moon as a separate planetary body, or another, new theory may be put forward. As of now, a satisfactory explanation seems almost as elusive as before the astronauts brought back their samples.

The Surface of Mars

Although *Mariner 9*, an orbiting spacecraft, circled Mars in 1971 and 1972, sending back thousands of excellent photographs, it was not until the later Viking missions that landing vehicles sent back data from the Martian surface. In 1976, two simultaneous Viking spacecraft orbited the planet continuously while their respective landing craft sent back data from the surface. The Chryse landing site of *Viking I*, located at about lat. 25° N, was the first to be reached. *Viking II* landed farther poleward, at about lat. 46° N. Thousands of high-quality photographs taken by two orbiters added to knowledge gained by the earlier *Mariner 9* mission.

Many important Martian features are strikingly different from those found on Earth and the Moon, although some others are recognizably similar. Half of Mars, roughly equivalent to its southern hemisphere, is heavily cratered, much as is the Moon's surface, whereas the other hemisphere is only lightly cratered. Although no samples have been returned from Mars, it is thought by some that the history of cratering of Mars parallels that of the Moon in age and frequency of impacts. Just why much of the northern half of Mars consists of smooth, uncratered plains is a major puzzle.

Major geological features of Mars include volcanoes and troughlike rifts of enormous size by Earth standards. One huge shield volcano, Olympus Mons, is 500 km wide at the base and has a summit, occupied by a caldera, rising 25 km above the base (Figure 20.30). Three other volcanoes are of comparable dimensions. These and lava flows indicate local accumulations of heat within the planet—possibly heat of radiogenic origin—leading to magma formation and extrusion in Mars' history.

Mars also has tectonic features in the form of numerous faults and a great rift valley system. The largest rift, Valles Marineris, appears as a straight-walled trench 100 km wide and up to 6 km deep in places (Figure 20.31). The entire rift zone is some 5000 km long. Although the Martian lithosphere seems to have been pulled apart along the rift zone, there is no sign of subduction zones and compressional mountain belts such as those on Earth.

Among the most puzzling surface features of Mars are what appear to be channels formed by fluid flow, largely concentrated in one area, north of the rift zone. Some of the channels show a braided pattern in which flow lines repeatedly divide and join, like the shallow channels of Earth streams in dry regions (Figure 20.32). Mars is now devoid of surface water in liquid form, yet the channels suggest great water floods, acting over short periods. One suggestion is that water ice, held frozen in the ground, underwent an episode of sudden melting, releasing great floods. These floods eroded the channels and left the channel floors strewn with rock debris.

Mars has an extremely thin atmosphere; the surface barometric pressure is only about one percent that on Earth. Nevertheless, winds on Mars are capable of raising great dust storms. One such storm obscured the Martian surface for weeks following the arrival of *Mariner 9*. Signs of erosion of exposed rocks by impact of windblown sand and silt were observed at both Viking landing sites, while numerous small dunes and drifts were clearly shown (see Figure 20.35). Enormous expanses of dunes, somewhat like sand seas of the Earth's deserts, also testify to the importance of the wind as a geologic agent on Mars (see Figure 20.34).

Perhaps the strangest of the many strange Martian features are the polar caps (Figure 20.33). The white polar caps consist of a thin surface coating of carbon-dioxide ice. Each cap undergoes a seasonal change in size, expanding greatly during the winter of the re-

FIGURE 20.30 Olympus Mons, an enormous Martian shield volcano with a caldera at the summit. The basal diameter is about 500 km. (NASA.)

FIGURE 20.31 Valles Marineris, a deep, steep-sided trench near the equator of Mars. North is toward the top of the photo. Width of the trough, north to south, is about 100 km; depths go to 6 km. Large landslides have occurred along the north rim of the trench, while gullylike erosion features have deeply indented the south rim. (NASA.)

FIGURE 20.32 Amazonis Channel, a braided feature resembling a dry stream channel and possibly produced by a water flood. (NASA.)

FIGURE 20.31

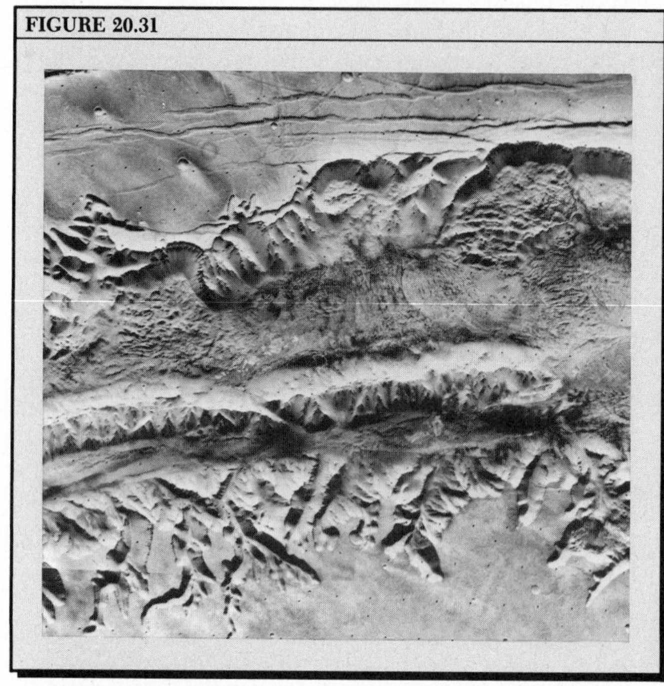

FIGURE 20.30

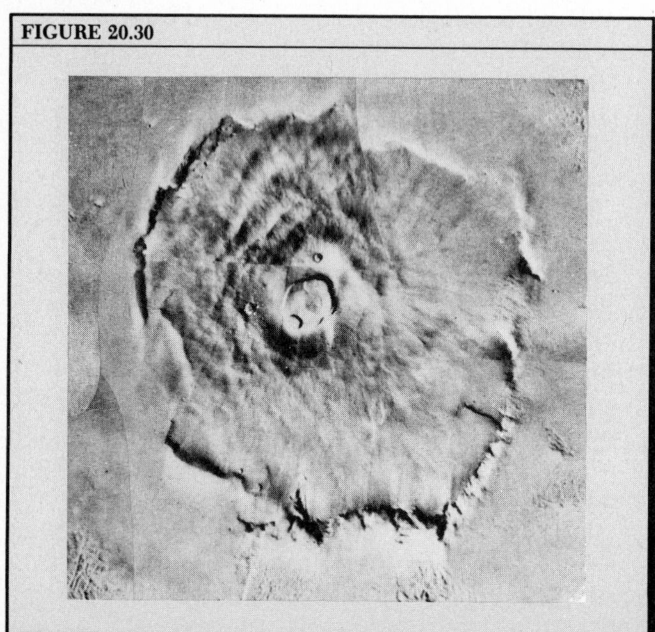

FIGURE 20.32

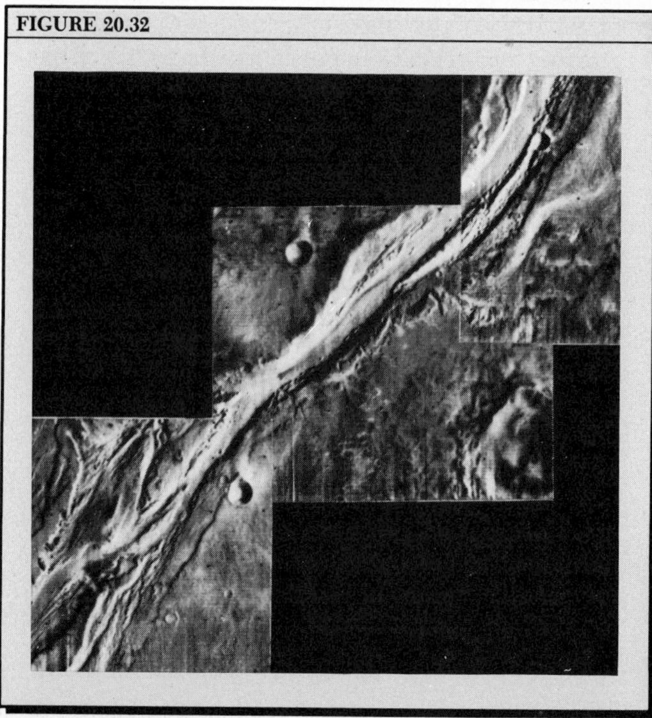

spective pole and shrinking to a small area during the summer. The small residual polar cap, left at the end of summer, is thought to consist of an accumulation of water ice (Figure 20.34). The residual cap areas have strange, platelike terrain features, looking something like a collapsed stack of poker chips arranged in a spiral form. The plates consist of light and dark laminations, which are probably alternating layers of water ice and dust. A vast belt of wavelike sand dunes surrounds the north polar cap (Figure 20.34).

At both Viking landing sites, details of the surface are quite similar (Figure 20.35). Rock fragments of a wide range of sizes from pebbles to boulders are strewn over a surface of fine-grained loose regolith. Many of the rocks are angular; many show the wind-eroded form of ventifacts common on the Earth's deserts; many of the fragments show wind-tails of fine material formed as miniature drifts to the lee of each fragment. Some of the larger rocks, particularly at the Utopia site of *Viking II*, are highly porous and have what appears to be the vesicular structure of a scoriaceous lava. Other rock surfaces appear to be fine-grained. The color of all rock surfaces and much of the regolith is a uniform brick-red, which has been attributed to thin coatings of limonite.

Although the Viking landers were not equipped to sample rock fragments, small scoopfuls of regolith close to the vehicles were subjected to analysis, using an X-ray fluorescence spectrometer. The most abundant elements are iron (13 to 14%), silicon (15 to 30%), calcium (3 to 8%), and aluminum (2 to 7%). Also present are important amounts of potassium, sulfur, chlorine, and titanium, which suggests that the rock may be of ultramafic composition, rich in olivine and pyroxene. Data for the two landing sites were almost identical. There seems to be general agreement that most Martian surface rock is probably mafic or ultra-

FIGURE 20.33

FIGURE 20.33 The Martian polar cap, with what appears to be overlapping terrain plates surrounding the north pole. (NASA.)

FIGURE 20.34

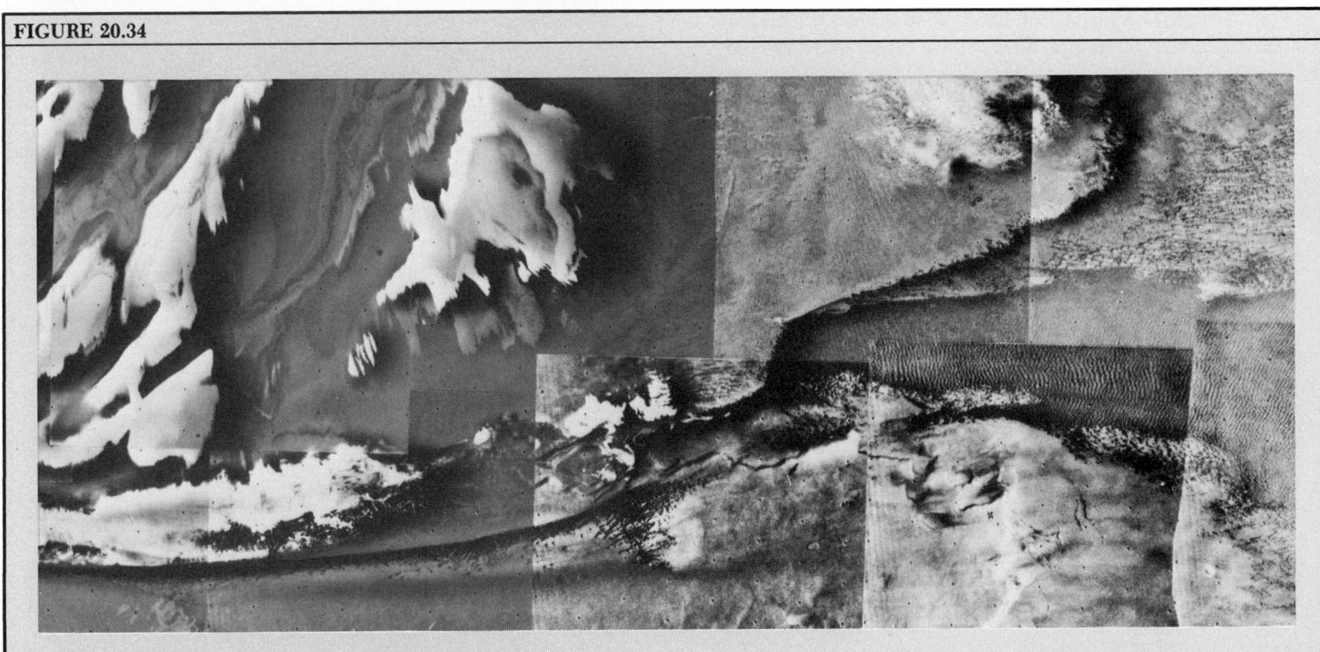

FIGURE 20.34 A closer view of the margin of the Martian north polar ice cap shows elongated patches of water ice (white areas) separated by ice-free terraced slopes. A dune field lies at the right, its rippled surface indicating wind flow from left to right (west to east). (NASA.)

mafic extrusive rock, a conclusion in agreement with the presence of volcanoes and lava flows.

The Martian surface features and materials lead to the conclusion that volcanic, tectonic, and erosional activities have all been active on Mars, but to a much lesser degree than on Earth. Thus, in many respects, Mars is geologically intermediate in geologic activity between Earth and Moon.

Viking orbiting spacecraft have passed close to the two Martian moons, Phobos and Deimos, taking many photographs. Figure 20.36 is a composite photograph of Phobos showing its irregular outline and heavily cratered surface. This moon measures about 25 km in its longest diameter, and its density has been measured as about 2 g/cc, which is about the same as for a meteorite of the chondrite class. Deimos, like Phobos, is heavily cratered and irregularly shaped, while it is only a little smaller. Closeup photos of the surface of Deimos show a rather smooth regolith of fine-textured material, littered with large blocky objects resembling boulders. The surface resembles the Moon's surface in several respects. The small size and irregular shapes of these Martian moons suggest that they are small asteroids that were captured by that planet.

The Surface of Mercury

Mariner 10 passed close to the planet Mercury in March 1974, sending back the first photographs to ever reveal the planetary surface. Strikingly like the Moon, the heavily cratered surface of Mercury shows countless lunar-type craters of many sizes and ages (Figure 20.37). Some craters are comparatively recent in age—light in color and having well-defined ray systems. Others are obviously very old, large craters with flat floors. Also seen are dark, plainlike areas suggestive of the lunar maria. Infrared radiation analyzed by instruments on the space vehicle suggest that Mercury has a regolith of low density, perhaps also much like that on the Moon. Many long, straight scarps were identified, along with long, narrow ridges suggestive of crustal compression (Figure 20.38). However, no sinuous rilles like those on the Moon were observed.

As expected in the absence of water and all but an extremely tenuous atmosphere, the surface of Mercury revealed no signs of the eroding action of fluids such as the braided channels of Mars. The magnetic field of

FIGURE 20.36 A photomosaic of Phobos, the inner of two Martian satellites, made up of photos taken by *Viking Orbiter I* from a distance of about 500 km. Besides the numerous impact craters, the surface bears sets of parallel striations or grooves, some of which are visible at the far right. (NASA.)

FIGURE 20.36

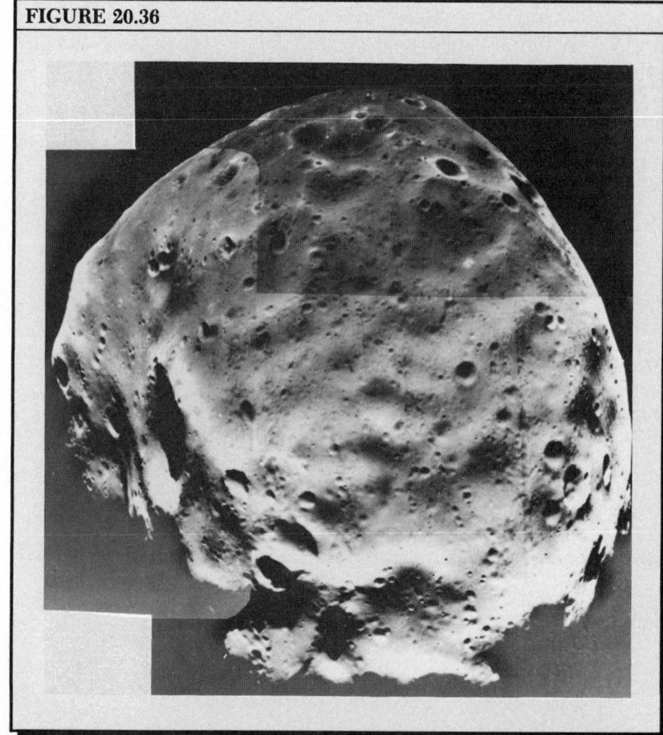

FIGURE 20.35 The surface of Mars, photographed from the *Viking I* lander in August 1976. Small dunes and drifts of loose regolith partly cover a bouldery surface. Many of the larger rock fragments show effects of wind abrasion. (NASA.)

FIGURE 20.35

FIGURE 20.37 The surface of Mercury, compiled from *Mariner 10* photographs. Notice the striking similarity between this surface and that of the farside of the Moon, shown in Figure 20.21. Rays emanate from some of the younger craters at the right side of the photograph. (NASA.)

FIGURE 20.38 Between craters, the surface of Mercury shows irregular fractures and ridges forming a blocklike pattern. The width of the area shown is about 500 km. (*Mariner 10* photograph by NASA.)

FIGURE 20.37

FIGURE 20.38

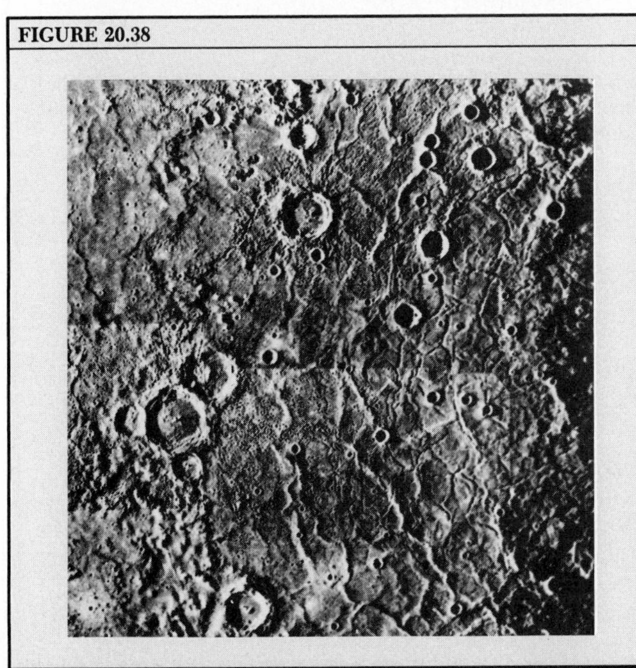

Mercury is extremely weak. Although Mercury may have an iron core, the planet's extremely slow rotation would probably not give rise to a dynamo effect such as that generated in the Earth's liquid core.

The Surface of Venus

Because of its extremely dense atmosphere and high surface temperatures (420 °C to 485 °C), Venus seemed for years as a most unlikely candidate for clear and detailed surface photography. Then, in 1975, the Soviet spacecraft *Venera 9* landed on Venus, sending back the first photograph of the solid planetary surface. What appeared on the image was a blocky ground surface that can be interpreted as composed of hard rocks fractured into angular blocks. Individual blocks seem to be on the order of 30 to 40 cm across; they cast sharp shadows (Figure 20.39).

Later in 1975, a second Soviet space vehicle, *Venera 10,* reached the surface of Venus and sent back another photographic image. The scene was described by Soviet news releases as an old mountain formation with smooth, rounded rocks. The two Soviet images offer, at best, very little information upon which to draw meaningful conclusions.

Some information on large topographic features of Venus's surface has come from radar images. Images obtained from the orbiting American Pioneer Venus spacecraft revealed a plateau, 3 to 5 km high, of pear-shaped outline and with a surface extent three times larger than the Colorado Plateau on Earth. Another feature, called Maxwell, consists of a ridge of 6-km height adjacent to a broad plateau surface rising 3 to 5 km above the surrounding terrain. The two plateau-like features cover a total area 3000 km by 1550 km, about twice the size of the Tibetan Plateau. The plateau complex is interpreted as a tectonic uplift; the high ridge may be of volcanic or tectonic origin. Other features mapped by radar include possible large craters; deep, wide channellike features; a possible shield volcano; and curved parallel ridges. The large size and variety of topographic features suggest that Venus is (or was) tectonically active.

Chemical analysis of surface material, made by Soviet landing craft (*Venera 8, 9,* and *10*), gave abundances of uranium, thorium, and potassium similar to those in Earth basalts or rocks of a composition intermediate between basalt and granite. Density of the surface material—about 2.7 g/cc—is also suggestive of the presence of igneous rocks of silicate mineral composition. There is evidence of a dust accumulation at the site of a landing probe from the Venus Pioneer spacecraft. Because of the high density of the atmosphere on Venus, it has been supposed that wind action may be important in transporting sediment and shaping surface landforms, but as yet, no positive evidence of wind action has been obtained.

Jupiter's Moons

Jupiter's four Galilean satellites—Io, Europa, Ganymede, and Callisto—have recently undergone close scrutiny from two fly-by space probes (Figure 20.40). *Voyager I* and *Voyager II* spacecraft encountered the giant planet in rapid succession in 1979, revealing startling discoveries, not only about the surface of Jupiter, but also about the character of its major moons. Because an understanding of Jupiter's dense atmosphere is primarily based in atmospheric science, we shall limit this description to the geologic aspects of the moons, all of which are in the solid state.

The following table lists the approximate diameters and densities of the four moons:

	Diameter, km	Density, g/cc
Io	3640	3.4
Europa	3100	3.1
Ganymede	5300	1.9
Callisto	4840	1.8

The range in diameters is rather small. Compare their diameters with that of Mercury, 4900 km (about the same as Ganymede and Callisto) and with the Earth's Moon, 3840 (about the same as Europa). On the other hand, their densities are all less than those of the inner planets (4.0 to 5.5 g/cc), probably in part because the Jupiter moons do not have metallic cores. The Earth's Moon, with a density of 3.34 g/cc, is in the same range as Io and Europa, which are probably largely formed of silicate rock. Ganymede and Callisto are probably more than 50% water ice, the remainder probably being largely silicate rock. Because surface temperatures are extremely low, all of the moons except Io probably have an outer ice-crust or a surface layer rich in ice.

Io, the innermost moon, is yellowish to reddish in color and completely lacking in craters (Figure 20.41). It has a smoothly contoured surface that suggests a coating of regolith of some sort. The most striking feature of Io is its volcanic activity, which was first noticed as a mushroomlike (or umbrellalike) plume of gas and dust seen on the rim of the moon's image. The plume rises to an elevation of 275 km above the rim. The erupting material may be largely sulfur, for ionized sulfur forms an atmospheric ring about the moon. Eight active volcanic vents have been counted, along with one inactive vent. The source of heat for internal igneous activity is not known, but has been attributed to tidal flexing. It seems possible that the regolith on Io's surface is a layer of evaporite salts or sulfur, or a sulfur compound.

Europa is orange-hued and seems to have an ice crust at the surface (Figure 20.42). There are few if any impact craters but, instead, the smooth surface shows many interesting linear features, some of which are 100 km wide and 1000 km long. These seem to be flat streaks or stripes, without relief, and may be merely superficial bands of coloration, but they have been interpreted as extension fractures in an ice crust.

Ganymede, the largest moon, probably has a surface crust of ice, perhaps mixed with fragmented rock debris. Craters are present, but none is of the great size of the lunar maria. The most striking feature of Ganymede's surface is a large number of linear bands that may be multiple faults (Figure 20.43). The bands appear to consist of parallel grooves and are suggestive of complex grabens or rifts, and they form an intersecting pattern. There is clear evidence of offsetting of the grooves, as if lateral movement has occurred on faults.

Callisto, the outermost moon, is most striking for its heavily cratered surface (Figure 20.44). Craters are densely and uniformly distributed over the surface and are probably formed in an ice-rich crust. At least one enormous multirimmed crater with basinlike form is conspicuous; it is about 1500 km in diameter. Perhaps Callisto retains a pristine primeval surface.

FIGURE 20.39 This Soviet photo is the first ever taken of the solid surface of Venus. The blocky objects, about 30 cm across, are interpreted as angular rocks. (SOVFOTO.)

FIGURE 20.39

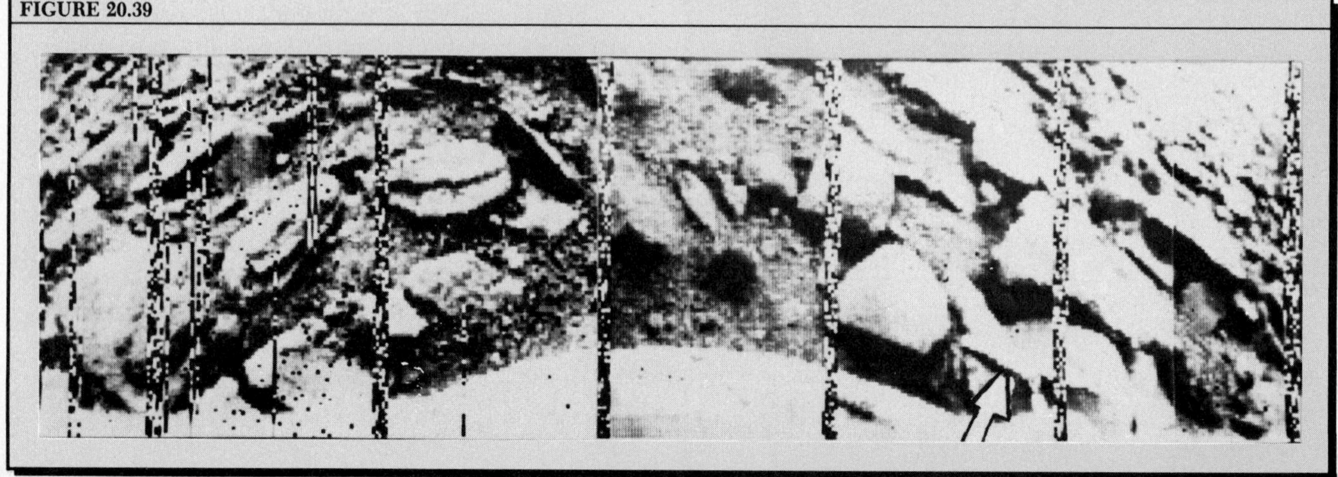

FIGURE 20.40 The four Galilean satellites of Jupiter, photographed by *Voyager I* early in 1979. Io (top left), Europa (top right), Ganymede (bottom left), and Callisto (bottom right) are shown in their correct relative sizes. (NASA.)

FIGURE 20.41 The smooth surface to Io seems to bear surface deposits that may be sulfur or mixtures of salts. The black spots may be volcanic craters, but impact craters are lacking. (NASA.)

FIGURE 20.42 Europa's surface shows a complex pattern of what seem to be intersecting cracks (extension fractures) in a brittle surface layer that may consist of water ice. Very few impact craters can be seen. (NASA.)

FIGURE 20.43 Two photos showing details of the surface of Ganymede. The intersecting bands of closely-spaced parallel ridges and grooves are of mysterious origin, but show offsetting that suggests they are intersecting multiple fault systems. (NASA.)

FIGURE 20.44 A photomosaic of Callisto, showing the uniform distribution of numerous impact craters. An enormous multirimmed impact feature can be seen near the edge of the photo at the upper right. (NASA.)

FIGURE 20.40

FIGURE 20.41

FIGURE 20.42

FIGURE 20.43

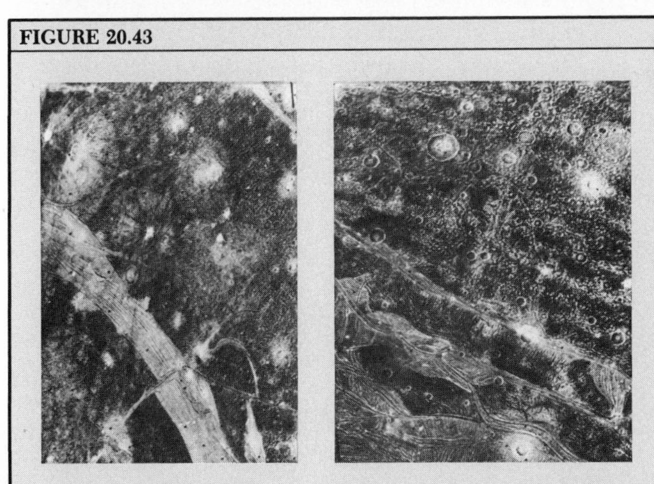

FIGURE 20.44

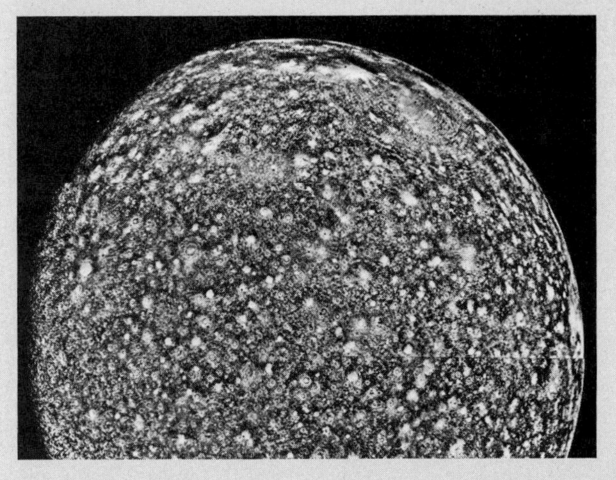

A Geology of Outer Space

Within only two decades of space exploration, the extension of geology as a science to the Moon and inner planets has passed from infancy to vigorous youth. Perhaps the greatest gain in our knowledge of the Earth's history comes from the very ancient lunar rocks, almost all of which crystallized in a billion-year period predating any known terrestrial rock. Earth's record of the first billion years of geologic time is almost totally missing, because Earth has had an active lithosphere and vigorous atmospheric processes.

While Earth rocks were continually remelted and recycled, destroying all of the ancient crust, the primitive crustal features of the Moon, Mars, and Mercury went into cold storage, so to speak. Only meteoroid impacts have disturbed the surface of the Moon and Mercury since about −3.0 b.y., and even on Mars, half the planet retains the early impacted surface. Now we have discovered that Callisto, a moon of Jupiter, is another heavily cratered object much like Mercury. We thus have not one, but four Rosetta stones from which to read the early history of our planet.

Of the inner planets, only Venus remains completely obscured from geological examination by satellite cameras. Beneath the dense, clouded atmosphere of Venus lies a varied surface possibly showing intense tectonic and volcanic activity, for Venus is a close match to Earth in size and density and may well prove to have a "plate tectonics" of its own.

Key Facts and Concepts

Geology of outer space *Astrogeology* applies principles and methods of geology to condensed matter and gases of the solar system outside the Earth.

The Sun Above the Sun's visible outer layer, the *photosphere*, lie the *chromosphere* and the *corona*, which extends out into space as the *solar wind* of charged particles. Solar energy originates in conversion of hydrogen to helium by nuclear fusion in the Sun's interior. *Sunspots* are associated with intense magnetic fields that produce *solar flares*, which send out X rays and charged particles reaching the Earth.

Terrestrial planets Of the nine *major planets*, the four inner (terrestrial) planets resemble Earth in having iron cores and silicate rock mantles. The four are Mercury, Venus, Earth, and Mars. The Moon, a small planetary body, compares with Mercury in size.

The outer planets The four great outer planets—Jupiter, Saturn, Uranus, and Neptune—are of low density and are made up largely of hydrogen and helium, with some ammonia, methane, and water.

The asteroids The *asteroids*, also called minor planets, orbit mostly between Mars and Jupiter. Some are composed of nickel–iron, others of silicate rock.

Meteorites *Meteorites*, solid fragments from space reaching the earth's surface, fall into three classes: (1) *irons* of nickel–alloy composition, (2) *stones* of silicate mineral composition, and (3) *stony irons*, mixtures of nickel–iron and silicate minerals. *Chondrites*, a class of stony meteorites, contain *chondrules* of olivine or pyroxene crystals. Meteorites may represent fragments of the earliest-formed planetary bodies in the solar system. Their age is −4.5 b.y., much older than the oldest known crustal rocks on Earth.

Accretion of the planets The solar nebula consisted of a dense, hot inner mass (solar mass) and a thin disk of cooling dust and gas. Silicates in the gases condensed first, followed by the volatiles. When nuclear fusion began in the Sun, the solar wind pushed the remaining hydrogen and helium gas out of the disk and into outer space. Accretion of mineral and ice grains produced *planetesimals*, which ultimately grew into the planets. *Gravitational collapse* may have been important in the accretion process.

Impact features Besides known meteorite impact craters, *astroblemes* include circular structures of possible impact origin found in shield areas. Barringer Crater, a known impact crater, has numerous meteoritic iron fragments. Also present are *coesite* and *stishovite*, silica minerals produced from sandstone by shock pressures. Asteroid impacts may have been severe in early stages of the Earth's history.

The Moon The Earth's Moon, a *planetary satellite*, is one of more than 34 moons of the planets. Large in relation to the Earth, the Moon and Earth form a binary planet system. Rotation of the Moon is the same as its period of revolution.

Lunar surface environment The Moon's *gravity* is weak, one-sixth that of Earth. The Moon lacks both atmosphere and hydrosphere. Surface temperatures reach 100 °C during lunar day and fall below −175 °C during lunar night. Processes of erosion and transportation by water or wind are lacking on the Moon.

Lunar exploration Lunar exploration began with photography by unmanned Ranger and Lunar Orbiter space vehicles. Manned Apollo missions allowed the Moon's surface to be directly studied and sampled.

Lunar maria The lunar surface consists of *lunar highlands* and *lunar maria* (dark plains). Maria are rimmed by high mountain ranges.

Lunar craters *Lunar craters* in vast numbers range in size from a few meters to over 100 km in diameter. Many large craters have flat floors with central peaks; others have multiple concentric ridges. Lighter colored *rays* of impact debris radiate from many large craters.

Rilles and walls Narrow, canyonlike *rilles* may be straight or sinuous. High straight *walls* are possibly fault scarps.

Origin of craters and maria The *meteoritic hypothesis* is generally accepted for most lunar craters. The maria were first formed as enormous impact craters by infall of asteroid-sized bodies; they were later flooded with basalts.

Lunar surface materials *Lunar regolith,* a fine-textured debris layer, consists of impact-fragmented mineral matter. All observed rocklike materials consist of rock fragments of igneous origin, with no massive bedrock observed in place.

Lunar rocks The majority of lunar rocks collected are *lunar breccias,* formed by *impact metamorphism.* Basalt and anorthositic gabbro are represented. Exposed rock surfaces show impact pits.

Geologic history of the Moon All lunar rock samples date older than −3.0 b.y.; most are in the age span −3.8 to −4.2 b.y., much older than Earth rocks. After accretion (−4.6 b.y.), the Moon was melted to depth of 200 km, followed by the formation of lunar crust. A big asteroid bombardment followed (−3.9 b.y.), forming maria depressions; then came maria flood basalts. Few changes occurred after −3.0 b.y.

Lunar interior The Moon lacks an iron core; it has a low-density crust and an ultramafic silicate rock interior. *Moonquakes* suggest a deep layer near the melting point.

Origin of the Moon Hypotheses of the Moon's origin include the *capture hypothesis, fission hypothesis,* and growth by accretion near Earth.

Surface of Mars Mars has craters, large volcanoes, a rift-valley system, channels shaped by fluid action, sand dunes, and white polar caps. Mars may be intermediate between Earth and Moon in terms of volcanic and tectonic activity.

Surface of Mercury Much like the Moon, Mercury's surface is heavily cratered and shows many fracture lines.

Surface of Venus Venus's dense atmosphere prevents orbiting satellite photography of surface features. A Soviet photo shows a surface littered with angular, rocklike fragments.

Jupiter's moons Jupiter's four major moons—the Galilean satellites—are comparable in diameter with the Earth's Moon and Mercury, but are less dense and do not have metallic cores. Water ice and silicate rock probably make up most of the interiors of the Jupiter moons, with surface ice crusts. Varied surface features include sulfur volcanoes, fracture systems, and impact craters.

Questions and Problems

1. The Sun consists almost entirely of hydrogen and helium, whereas the percentage of these elements in the Earth is exceedingly small. Explain how this difference came about in terms of origin of the solar system.

2. Compare the age of meteorites with the ages of lunar rocks and the age of the oldest rock so far discovered on Earth. What does this information tell us about the earliest eon of Earth history? Compare the age of the youngest known lunar rocks with ages of rocks of the Earth's Phanerozoic Eon. Does plate tectonics enter into the understanding of these rock-age differences? Explain.

3. Suppose that you pick up a strange rock—about the size of a baseball—and think perhaps you have found a meteorite. What characteristics of the rock might lead you to this conclusion? With what kinds of terrestrial rocks or minerals might a true meteorite be confused?

4. Geologists have discovered a circular structure in which strata of Paleozoic age form an upturned rim. Is it a crypto-volcanic structure or a true astrobleme? What evidence would you search for in order to answer this question?

5. Do you think that the scientific results of lunar exploration justified the enormous cost of the Apollo program? Has the geologic information obtained during this program had any practical value, as yet? Mars has been explored by means of unmanned Viking landing craft. What added information, if any, might be obtained by manned landings and field traverses on Mars?

CHAPTER TWENTY-ONE

Geologic Resources of Materials and Energy

Our modern technological society is totally dependent for its existence on a wide variety of mineral resources derived from the lithosphere. Dependence upon these resources came about very gradually at first, then with great rapidity during the Industrial Revolution, beginning about the middle of the eighteenth century. Coal, perhaps more than any other single mineral substance, symbolizes the Industrial Revolution. Coal powered the new steam engines invented in England to turn the shafts and wheels of textile mills; coal supplied the coke used to produce iron from which those machines could be made. Soon a wide variety of other metals became necessities; these included copper, lead, zinc, and tin, and the many alloy metals needed to impart strength to steel. For better or for worse, western civilization was committed to a course that required the ceaseless search for new mineral deposits.

Through the early 1900s, petroleum and natural gas came to replace coal as a major source of energy. Petroleum also supplied the hydrocarbon compounds for synthesizing a host of new chemical compounds—petrochemicals—that serve as essential materials in manufacture.

Nonrenewable Earth Resources

The mineral deposits we are about to examine in this chapter are largely *nonrenewable earth resources*—nonrenewable because the geologic processes that formed them required millions of years to act. So very slow are present-day processes of mineral accumulation that there is no possibility whatsoever of seeing useful new deposits form in spans of time as short as centuries. In contrast, the renewable earth resources involve rapid production cycles. Fresh water of the lands is an example, because water moves rapidly through the hydrologic cycle (Chapter 15); our supply is replenished in the yearly cycle of rain and snow. Organic resources—food, lumber, fiber—are rapidly cycled and replenished by organic growth processes.

Nonrenewable earth resources covered in this chapter can be grouped about as follows:

Metalliferous deposits (examples: ores of iron, copper, tin)
Nonmetallic deposits, including
  Structural materials (examples: building stone, gravel, and sand)
  Materials used chemically (examples: sulfur, salts)
Fossil fuels (coal, petroleum, oil shale, natural gas)
Nuclear fuels (uranium, thorium)

Notice that the last two groups represent sources of energy, whereas the first two groups are sources of materials.

An important geologic energy resource is *geothermal energy,* making use of heat in rock or ground water beneath the surface. Because of its vastness, the reservoir of internal heat is not subject to appreciable depletion by human use.

Nonrenewable resources that come from the earth's crust are the subjects of study of a major branch of geology, often called *economic geology*. This is the practical and applied side of geology, without which there would be no industrial civilization such as we have today.

Metals in the Earth's Crust

Metals occur in useful concentrations as ores. An *ore* is a mineral accumulation that can be extracted at a profit for refinement and industrial use. A number of important metallic elements are listed in Table 21.1 with their abundances, as percentage by weight, in the average crustal rock. Magmas are the primary source of many metals. Our concern here is with the natural geological processes of concentration of metallic elements and compounds into ores of various kinds. Whereas aluminum and iron are relatively abundant, most of the essential metals of our industrial civilization are present in extremely small proportions—witness mercury and silver, with abundances of only about 0.000008 and 0.000007% respectively.

In a classification of metals by uses, iron stands by itself in terms of total tonnage used in the production of steel. Related to iron is a group of ferro-alloy metals, which are used principally as alloys with iron to create steels with special properties. The ferro-alloys include titanium, manganese, vanadium, chromium, nickel, cobalt, molybdenum, and tungsten, listed in order of appearance in Table 21.1. Other important metals (nonferrous metals), standing apart individually with respect to industrial uses, are aluminum, magnesium, zinc, copper, lead, and tin. A minor group listed in Table 21.1 includes antimony, silver, platinum, and gold. Finally, there are metals which are radioactive, including uranium, thorium, and radium.

While a few metals, among them gold, silver, platinum, and copper, occur as elements—that is, as native metals—most occur as compounds. Oxides and sulfides are the most common forms, but more complex forms are present in many ores. Table 21.2 lists some important ore minerals with their compositions.

The crustal abundance of a metal is an abstraction of no practical value to economic geologists. They are interested in its abundance in the form of the ore of its usual occurrence, as either an element or a compound.

Obviously, most metals must occur greatly concentrated in ores to be extractable at a profit, as compared with the extremely small average crustal abundances shown in Table 21.1. For example, chromium has an average crustal abundance of only 0.01%; it must be concentrated by a factor of about 1500 times to become sufficiently rich to be extracted. For lead, the crustal abundance must be concentrated by a factor of 2500 times to become an ore. Ores of sufficient richness to be extracted have required very special geologic processes to come into existence. We shall, where possible, relate the occurrences of ores to plate tectonics.

Table 21.1 Average Metallic Abundances in Crustal Rock

Symbol	Element name	Abundance (percentage by weight)
Al	Aluminum	8.1
Fe	Iron	5.0
Mg	Magnesium	2.1
Ti	Titanium	0.44
Mn	Manganese	0.10
V	Vanadium	0.014
Cr	Chromium	0.010
Ni	Nickel	0.0075
Zn	Zinc	0.0070
Cu	Copper	0.0055
Co	Cobalt	0.0025
Pb	Lead	0.0013
Sn	Tin	0.00020
U	Uranium	0.00018
Mo	Molybdenum	0.00015
W	Tungsten	0.00015
Sb	Antimony	0.00002
Hg	Mercury	0.000008
Ag	Silver	0.000007
Pt	Platinum	0.000001
Au	Gold	0.0000004

Source: Data from B. Mason (1966), *Principles of Geochemistry,* 3d ed., New York, John Wiley, pp. 45–46.

Actually, industrial processes must carry out the final stages of mineral concentration, as in the case of nuclear fuels, copper ores, or most iron ores. In terms of percentages by weight in the earth's crust, these deposits as mined represent an extraordinary degree of concentration, but it is still not enough.

Metalliferous Ores

One major class of ore deposits is formed within magmas by *magmatic segregation,* in which mineral grains of greater density sink through the fluid magma while crystallization is still in progress. Masses or layers of a single mineral accumulate in this way. One example is chromite, the principal ore of chromium, with a density of 4.4 g/cc. Bands of chromite ore are sometimes found near the base of an igneous body.

Another example is seen in the nickel ores of Sudbury, Ontario. These sulfides of nickel apparently became segregated from a saucer-shaped magma body and were concentrated in a basal layer (Figure 21.1). Magnetite is another ore mineral that has been segregated from a magma to result in an ore body of major importance.

Table 21.2 Representative Metallic Ore Minerals

Metal	Symbol	Mineral name	Composition
Aluminum	Al	Bauxite (not a single mineral)	Hydrous aluminum oxide
Iron	Fe	Magnetite	Oxide of iron
		Hematite	Oxide of iron
		Limonite (not a single mineral)	Hydrous oxide of iron
		Pyrite	Sulfide of iron
Titanium	Ti	Ilmenite	Oxide of iron and titanium
Manganese	Mn	Pyrolusite	Oxide of manganese
		Manganite	Hydrous oxide of manganese
Chromium	Cr	Chromite	Oxide of iron and chromium
Zinc	Zn	Sphalerite	Sulfide of zinc and iron
Copper	Cu	Native copper	Metallic copper
		Chalcopyrite	Sulfide of copper and iron
		Chalcocite	Sulfide of copper
Lead	Pb	Galena	Sulfide of lead
Tin	Sn	Cassiterite	Oxide of tin
Molybdenum	Mo	Molybdenite	Sulfide of molybdenum
Tungsten	W	Wolframite	Oxide of tungsten with iron and manganese
Uranium	U	Pitchblende (uraninite)	Hydrous uranium oxide
Mercury	Hg	Cinnabar	Sulfide of mercury
Silver	Ag	Native silver	Metallic silver
		Argentite	Sulfide of silver
Gold	Au	Native gold	Alloy of gold and silver
Platinum	Pt	Platinum	Metallic platinum

The process of contact metamorphism, described in Chapter 7, is a second source of important ore deposits. Hydrothermal solutions from within the magma soak into the surrounding country rock, introducing ore minerals in exchange for components of the rock. For example, a limestone layer may have been replaced by iron ore consisting of hematite and magnetite (Figure 21.2). Ores of copper, zinc, and lead have also been produced in this manner, as have valuable deposits of nonmetallic minerals.

A third type of ore deposit is produced by direct deposition of minerals from hydrothermal solutions that leave a magma during the final stages of its crystallization. These minerals are deposited in fractures to produce mineral *veins* (Figure 21.2), some of which are sharply defined and evidently represent the filling of open cracks with layers of minerals. Where veins occur in exceptional thicknesses and numbers, they may constitute a *lode*. Pegmatite veins and dikes, described in Chapter 4, are an important class of ore-bearing deposits of hydrothermal origin (see Figures 4.15 and 4.16).

Hydrothermal solutions produce yet another important type of ore accumulation, the *disseminated de-*

FIGURE 21.1 Cross section of a sulfide nickel ore deposit at Sudbury, Ontario. The ore layer lies at the base of a body of gabbro, overlying an older granite basement.

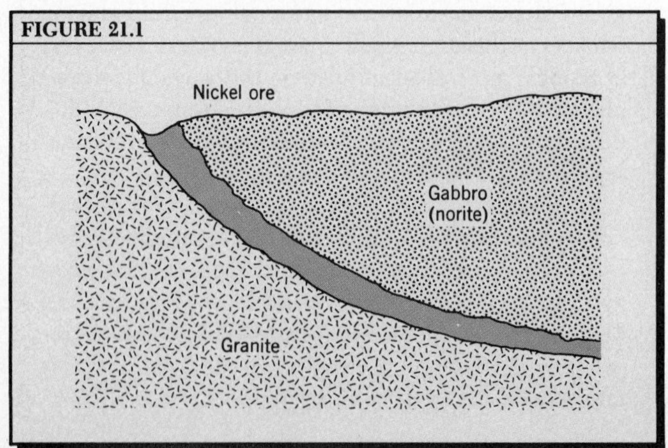

FIGURE 21.2 Schematic cross section of vein deposits and contact metamorphic deposits adjacent to an intrusive igneous body.

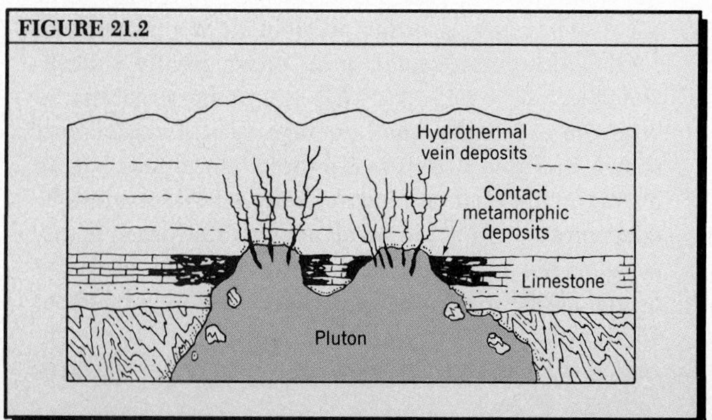

posit, in which the ore is distributed in minute masses throughout a very large mass of rock. Certain of the great copper deposits of the disseminated type are referred to as *porphyry copper* deposits. Here the ore has entered a large body of igneous rock of porphyritic texture (Chapter 4 and Figure 4.5), which had in some manner been shattered into small blocks that permitted entry of the solutions. One of the most celebrated of these deposits is at Bingham Canyon, Utah (Figure 21.3).

Hydrothermal solutions may also rise toward the surface, making vein deposits in a shallow zone and even emerging as hot springs. Many valuable ores of gold and silver are from shallow deposits. Particularly interesting is the occurence of mercury ore, in the form of the mineral *cinnabar* (HgS), as a shallow hydrothermal deposit. Most renowned are the deposits of the Almadén district in Spain, where mercury has been mined for centuries and has provided most of the world's supply of that metal.

A fourth category of ore deposits embodies the effects of downward-moving solutions in the unsaturated zone and the ground-water zone (see Chapter 15). Enrichment of mineral deposits in this manner produces the *secondary ores.*

Consider first a vein containing primary minerals of magmatic origin (Figure 21.4), mostly sulfides of copper, lead, zinc, and silver, along with native gold. They are originally disseminated through the vein rock and may not exist in concentrations sufficient to qualify as ores. Through long-continued denudation of the region, the ground surface truncates the vein, which was formerly deeply buried. Assuming a humid climate, there will exist a water table and a ground-water zone, above which is the unsaturated zone. Water, arriving as rain or melted snow, moves down through the unsaturated zone. The geologist refers to this water as *meteoric water,* since in atmospheric science, rain and snow particles are classed as "meteors." The meteoric water becomes a weak acid, because it contains dissolved carbon dioxide (carbonic acid), and will also gain sulfuric acid by reactions involving iron sulfide. Pyrite, for example, will yield sulfuric acid.

The result of downward percolation of meteoric water is to cause three forms of enrichment and thus yield ore bodies. First, in the zone closest to the surface, soluble minerals of no economic value (waste minerals) are removed. When this happens, there may accumulate certain insoluble minerals, among them gold and compounds of silver or lead. Under favorable conditions, these valuable minerals may attain sufficiently rich concentration to form a type of ore deposit known as a *gossan* (Figure 21.4). Iron oxide and quartz will also accumulate in the gossan. In American Colonial times, iron-rich gossans provided iron ore deposits that could be mined, but they have long since been exhausted.

FIGURE 21.3 Open-pit mine at Bingham Canyon, Utah. (Kennecott Copper Corporation.)

FIGURE 21.3

Leaching of other minerals carries them down into a *zone of oxidation* which lies above the water table. In this second zone, there may accumulate a number of oxides of zinc, copper, iron, and lead, along with native silver, copper, and gold. A third zone is that of *sulfide enrichment*, which lies within the upper part of the ground-water zone, just beneath the water table (Figure 21.4). Sulfides of iron, copper, lead, and zinc may be heavily concentrated in this zone. (Mineral examples are pyrite, chalcopyrite, chalcocite, galena, and sphalerite.) Sulfide enrichment may also affect large primary ore bodies of the disseminated type, such as the porphyry copper of Bingham Canyon, Utah, where the enriched layer has already been removed, and mining has progressed into low-grade primary ore beneath.

Also in the category of secondary ores is *bauxite*, a principal ore of aluminum which accumulates as a near-surface deposit in tropical regions where old oxisols accumulate (Chapter 14). Bauxite is a mixture of hydrous oxides of aluminum derived from the alteration of aluminosilicate minerals; diaspore is a principal mineral component. Bauxite is practically insoluble under the prevailing climatic conditions, so it can accumulate indefinitely as the denudation ot the land surface progresses. Produced under similar climatic conditions are residual ores of manganese (mineral, manganite) and of iron (mineral, limonite). The term *laterite* is commonly applied to these residual ore deposits.

A fifth category of ore deposit is a mineral concentration that has occurred through fluid agents of transportation: streams and waves. Certain of the insoluble heavy minerals derived from rock weathering are swept as small fragments into stream channels and carried downvalley with the sand and gravel as bed load. Because of their greater density, these minerals become concentrated in layers and lenses of gravel to form *placer deposits* (Figure 21.5). Native gold and platinum are among the minerals extensively extracted from placer deposits (Figure 21.6). A third mineral which forms important placer deposits is an oxide of tin (mineral, cassiterite). Diamonds, too, are concentrated in placer deposits, as are other gemstones. Transported by streams to the ocean, gravels bearing the heavy minerals are spread along the coast

FIGURE 21.4 A schematic cross section of secondary ore deposits formed by enrichment of minerals in a vein.

FIGURE 21.5 A gold dredge scoops up gravel from the river floodplain, washes the gravel to secure gold particles, then dumps the waste to form curious piles like fallen stacks of poker chips. Yukon River, Alaska. (Bradford Washburn.)

FIGURE 21.6 Hydraulic mining of gold-bearing gravels in the Sierra Nevada of California. The powerful water jet washes the gravel into a sluice, where the gold particles are collected on a series of riffles. The age of this photo is not known, but it probably falls in the early decades of this century. (California Division of Mines and Geology.)

FIGURE 21.4

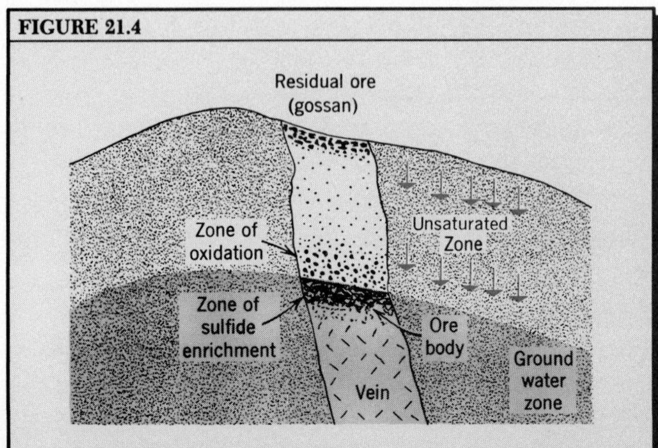

FIGURE 21.5

FIGURE 21.6

in beaches where they form a second type of placer deposit, the *marine placer.*

Finally, we can recognize a sixth group of ore deposits in the hydrogenic category of sediments, explained in Chapter 5. For the most part, sediment deposition is the principal source of nonmetallic mineral deposits (described below), but some important metalliferous deposits are of this origin. Iron, particularly, occurs as sedimentary ores in enormous quantities. Sedimentary iron ores are oxides of iron—usually hematite. A particularly striking example is iron ore of the Clinton formation of Silurian age, widespread in the Appalachian region. For reasons not well understood, unusually large quantities of iron oxides, derived by weathering of mafic minerals in rocks exposed in bordering lands, were brought to the seafloor and were precipitated as hematite. Another metal, manganese, has been concentrated by depositional processes into important sedimentary ores.

Ore Deposits and Plate Tectonics

The revolution in earth sciences—plate tectonics—has had many reverberations and repercussions throughout geology, often impacting areas of that science which were not at first involved in the concepts of plate motions. One such example is the application of plate tectonics to the occurrence of metallic ore deposits; another is the interpretation of large accumulations of petroleum.

We have learned that the primary source of metallic ore deposits is magmas, rising from great depths to penetrate the crust. Hydrothermal deposits and pegmatites illustrate mineral concentrations from igneous sources. Turning to plate tectonics, we realize that igneous activity is largely concentrated in three types of environments or locations: (1) spreading (accreting) plate boundaries, where new oceanic crust is being created; (2) subduction zones, where magma is being produced by deep melting and is giving rise to intrusion and extrusion; and (3) hot spots of rising mantle rock, which may be found at almost any location within a lithospheric plate, far from its boundaries. One of the most recent developments in plate tectonics has been the recognition of occurrences of important ore deposits associated with each of these zones of igneous activity.

FIGURE 21.7 Cyprus-type ore deposits occur within an ophiolitic suite of rocks formed at accreting plate margins. The arrangement of rocks is the same as in Figure 10.28. (Based on data of P. M. Hurley, 1975, *Technol. Rev.,* vol. 77, no. 5.)

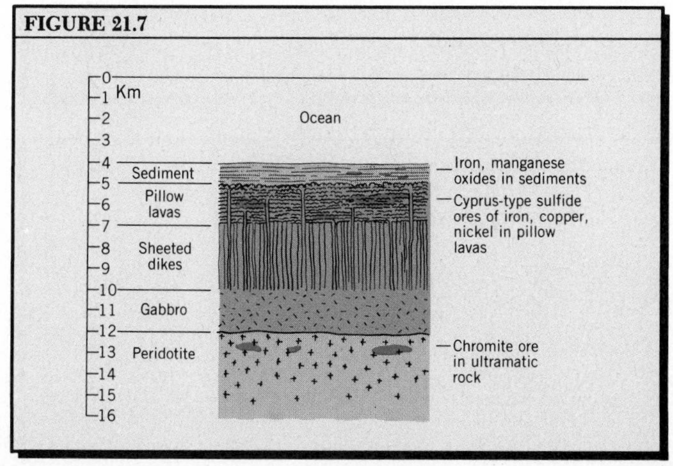

Our new classification of metallic ore deposits recognizes five basic types. We shall give each a tentative name, keeping in mind that we are dealing with an area of scientific exploration in its early, formative stages.

1. Cyprus-type Ore Deposits. On the island of Cyprus in the eastern Mediterranean Sea, a huge deposit of copper sulfide ore occurs in the Troodos Massif, a broad band of mafic igneous rock cutting across the central part of the island. The ore bodies occur within a zone of pillow lavas of basalt composition, near the northern margin of igneous rock zone. Adjacent to the pillow lava and grading into it is a zone of sheeted basaltic dikes, which in turn gives way to an adjacent zone of coarsely crystalline mafic rock of gabbro composition. Geologists now realize that what they are seeing on Cyprus is an *ophiolitic suite.*

Recall from Chapter 12 that ophiolites are masses of oceanic crust produced along accreting plate margins in the spreading zone of an ancient mid-oceanic rift. The geologic section closely resembles that which we described in Chapter 10 and in Figure 10.28. The relationship of ore deposits to the igneous rocks of the oceanic crust is shown in Figure 21.7.

Cyprus-type ore bodies, which are usually sulfides of iron, copper, or nickel, seem to have been deposited in the pillow lava layer by hydrothermal solutions rising from magma in the mantle beneath, a process explained in Chapter 13 and illustrated by an example from the Red Sea. Sediments deposited upon the pillow lavas also receive hydrothermal deposits, and these are commonly enriched with compounds of iron and manganese. Bodies of chromite (chromium ore) are sometimes found within the ultramafic rock at the base of the ophiolitic suite, as shown in Figure 21.7. These bodies may have formed by magmatic segregation in pockets of ultramafic magma at the same time that the seafloor spreading was in progress.

If the Cyprus ophiolite rocks were indeed formed at accreting plate margins in a spreading zone, how did they come to be elevated and exposed as a mountain range? We have pointed out that during subduction the oceanic crust slides beneath the continental margins and is consumed by melting at great depth. During continental collision, masses of oceanic crust are broken away from the descending plate and slide up along overthrust faults to take an elevated position among the highly deformed strata of the tectonic belt, as we showed in Figure 12.27. Later erosion exposes the fragment of oceanic crust, as in the Cyprus example.

Here we have a remarkable example of the principle of uniformity in geologic history. The hydrothermal process operating today in the Red Sea may be duplicating the process by which many ore deposits came into existence in ophiolitic suites of the geologic past, even as far back as Precambrian time, in countless spreading zones long since vanished.

2. Andean-type Ore Deposits. A second class of ore deposits related to plate tectonics has been recognized by geologists in continental crust lying above a descending plate adjacent to a subduction zone. As shown in Figure 21.8, magma bodies rising from the descending plate reach the upper crust to solidify as felsic plutons or to emerge as volcanic rocks such as andesite and rhyolite. The rising magmas bring up a supply of metals, which occur now as ores in the igneous rocks. Because this class of deposits occurs in the Andes Mountains of South America as important tin and copper ores, it is referred to as the Andean-type of ore deposit.

Magmas produced at various depths along the descending plate differ in dominant metallic element content. Consequently, the ore deposits show a distinct zoning by dominant metallic elements from continental margin to continental interior, as shown in Figure 21.8. Ores of iron are typical of a belt closest to the plate margin. Then come successive zones: copper and gold; silver, lead, and zinc; tin and molybdenum. This sequence has been recognized in the western United States and has been related to past episodes of subduction of the Farallon plate beneath the North American plate. Porphyry copper deposits are thought to be one of the forms of the Andean-type ore deposits. A possible example would be the porphyry copper at Bingham Canyon, Utah, which was described earlier in this chapter and illustrated in Figure 21.3. Porphyry copper also occurs in the Andes and beneath ancient subduction zones in Eurasia. The concept has guided recent searches for new porphyry ore deposits.

3. Island-arc-type Ore Deposits. A third group of metallic ore deposits occurs within massive accumulations of felsic lavas—andesite and rhyolite—within volcanic island arcs. An example is seen in important copper sulfide ores of Japan, one of the few rich mineral resources with which that island nation is endowed. The ore occurs in massive bodies and is usually a mixture of sulfides of copper, zinc, and lead, along with lesser amounts of silver and gold.

4. Intracontinental Ore Deposits. Some important ore deposits occur far from plate boundaries and lie in the heart of large areas of continental crust. An example is the lead and zinc ore deposits of the Mississippi Valley region. These ores are found in sedimentary strata of the continental platform. Although the explanation is highly speculative, it seems possible that ores of this type are formed by hydrothermal solutions rising from a hot spot in the mantle below.

The same concept of a hot spot beneath continental lithosphere has been called upon to explain the occurrence of diamonds within narrow, vertical, pipelike bodies of an ultramafic rock known as *kimberlite.* Fragments of garnet crystals within this rock provide evidence, through chemical analysis, of having been crystallized at a temperature of about 1360 °C. These conditions might be expected at a depth of some 150 to 200 km, well within the asthenosphere. The rock

FIGURE 21.8

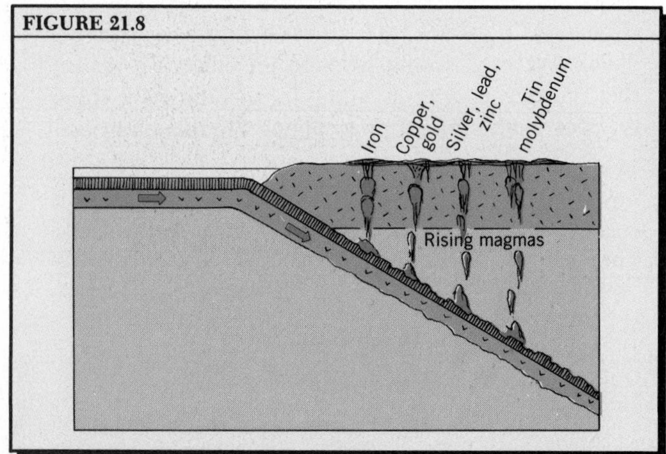

FIGURE 21.8 Andean-type metallic ore deposits are formed by magmas rising from an oceanic lithospheric plate undergoing subduction. Magma reaches the surface zone of the overlying continental lithosphere.

FIGURE 21.9

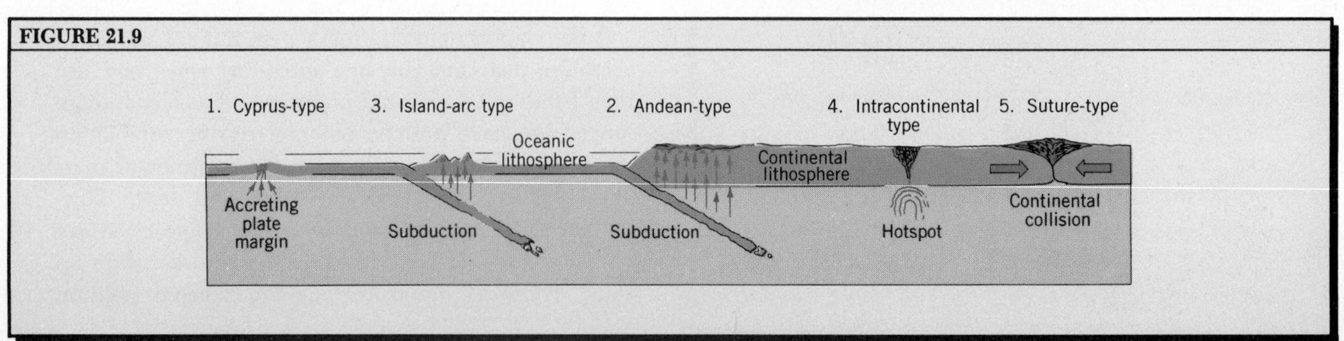

FIGURE 21.9 A schematic diagram showing five types of ore deposits in the frame of reference of plate tectonics.

was probably blown upward through the crust in a powerful jetlike blast in a very short time, propelled, it is thought, by the pressure of expanding water or carbon dioxide. Geologists call the pipelike structure a diatreme (Chapter 4), and though many diatremes are known, only a few have yielded diamonds. Diatremes at Kimberly, South Africa, are certainly the most famous diamond producers of the world. In the United States, a single diatreme located at Murfreesboro, Arkansas, has yielded many diamonds, the largest weighing just over 40 carats. But today the Arkansas diamond-bearing pipe, having exhausted its supply, is merely a tourist attraction.

5. Suture-type Ore Deposits. Any one of the four ore deposit types previously named could, in the geologic past, have become involved in continental collisions, which we described in Chapter 12. Today the rock types which bear these ores are found crumpled, overthrust, and metamorphosed in sutures within the continental shields. Take the case of the ore deposits of Newfoundland. Throughout the Paleozoic Era, the Appalachian belt, in which Newfoundland lies, suffered continental collisions interspersed with episodes of continental rifting, as the primitive Atlantic Ocean basin alternately opened and closed (Chapter 13). The final suturing occurred at the close of the Paleozoic Era. Geologists can recognize in the rocks of Newfoundland the typical ore deposits of the Cyprus type (ophiolitic suites), island-arc ore deposits in great volcanic rock masses, and even porphyry ores of the Andean type that occur above subduction zones.

FIGURE 21.10 Apportionment of world recoverable reserves of ores of iron, copper, aluminum, mercury, and chromium between the Sino–Soviet nations and the United States.

FIGURE 21.10

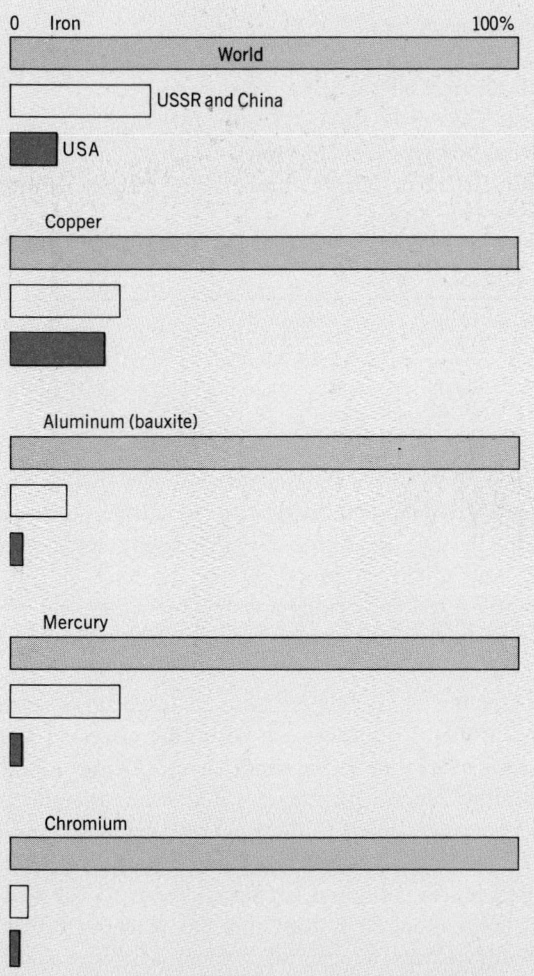

0 Iron 100%

World

USSR and China

USA

Copper

Aluminum (bauxite)

Mercury

Chromium

Figure 21.9 shows schematically the five types of ore deposits. Although still in its early stages, the tie-in of ore deposits with plate tectonics has produced a great sweep of optimism within the ranks of economic geologists engaged in exploration for new mineral deposits. Now, for the first time, they have a frame of reference in which to place the major types of ore deposits. Interpretation of a particular rock sequence in terms of plate tectonics allows the geologist to predict what type of ore deposit may occur there. As with any search program, knowing what to look for and where to look for it gives the searcher a great advantage.

U.S. Metal Demands and the World Supply

In the last two decades of the twentieth century, demands by the United States for essential industrial metals will increase greatly. A doubling of need will be felt for many metals, among them iron, manganese, zinc, cobalt, lead, antimony, mercury, and silver. For others, the factor of increased demand will be much greater—for example, our need for aluminum will increase about sixfold; for titanium, twelvefold; and for uranium, twentyfold. The answer to whether these increased demands can be met from reserves within the United States is an emphatic NO!

Even by the early 1970s, the United States was dependent upon foreign imports for a large proportion of nearly all industrial metals consumed (Table 21.3). For example, of the aluminum we consumed, we produced within the United States only about 4%; of manganese, 5%; of zinc, mercury, and silver, about 50%; of iron and lead, about 75%. We now produce practically none of the platinum, chromium, and strontium we use. We do well on uranium, mining substantially more than we need at the moment. Molybdenum is about the only essential metal we produce with an important exportable surplus—about as much extra as we consume.

Figure 21.10 shows the apportionment of known recoverable reserves of ores of five metals between the Sino–Soviet bloc and the United States in comparison with the total world resources. In the case of the essen-

Table 21.3 Proportion of U.S. Mineral Requirements Imported During 1972

Mineral	Percentage imported	Major foreign sources
Platinum group	100	U.K., U.S.S.R., South Africa, Canada, Japan, Norway
Mica (sheet)	100	India, Brazil, Malagasy
Chromium	100	U.S.S.R., South Africa, Turkey
Strontium	100	Mexico, Spain
Cobalt	98	Zaïre, Belgium, Luxembourg, Finland, Canada, Norway
Tantalum	97	Nigeria, Canada, Zaïre
Aluminum (ores and metal)	96	Jamaica, Surinam, Canada, Australia
Manganese	95	Brazil, Gabon, South Africa, Zaïre
Fluorine	87	Mexico, Spain, Italy, South Africa
Titanium (rutile)	86	Australia
Asbestos	85	Canada, South Africa
Tin	77	Malaysia, Thailand, Bolivia
Bismuth	75	Mexico, Japan, Peru, U.K., Korea
Nickel	74	Canada, Norway
Columbium	67	Brazil, Nigeria, Malagasy, Thailand
Antimony	65	South Africa, Mexico, U.K., Bolivia
Gold	61	Canada, Switzerland, U.S.S.R.
Potassium	60	Canada
Mercury	58	Canada, Mexico
Zinc	52	Canada, Mexico, Peru
Silver	44	Canada, Peru, Mexico, Honduras, Australia
Barium	43	Peru, Ireland, Mexico, Greece
Gypsum	39	Canada, Mexico, Jamaica
Selenium	37	Canada, Japan, Mexico, U.K.
Tellurium	36	Peru, Canada
Vanadium	32	South Africa, Chile, U.S.S.R.
Iron	28	Canada, Venezuela, Japan, Common Market (EEC)
Lead	26	Canada, Australia, Peru, Mexico
Cadmium	25	Mexico, Australia, Belgium, Luxembourg, Canada, Peru
Copper	18	Canada, Peru, Chile
Titanium (ilmenite)	18	Canada, Australia
Rare earths	14	Australia, Malaysia, India

Source: Data from *Mining and Minerals Policy 1973,* report of the Secretary of the Interior to the Congress.

tial metal mercury, for which there is no known substitute in many industrial uses, most of the world reserves lie in Spain, Italy, and the Sino–Soviet bloc.

The dependence of the United States upon foreign sources is heavy indeed for a long list of metals. International tensions are bound to rise where mineral resources lie in developing nations and in nations with politically unstable or unfriendly governments.

Recycling of Metals

Of increasing importance in manufacturing today is the secondary production of metals through reprocessing of durable metal goods manufactured in the past 10 to 100 years. We are not referring here to new scrap metal, derived as cuttings during initial manufacture, but to old scrap, those salvaged materials from discarded products, such as automobiles and refrigerators.

Metals can be reclaimed from old scrap by processes of distillation, electrometallurgy, mechanical separation, and chemical processes. As the total output of manufactured goods increases through time, the input of metals from secondary sources will also rise in volume.

Recycling of metals is rising in importance as national mineral resources are becoming depleted at increasing rates and as the grade of ores being mined is declining. *Metals recycling* is defined as the ratio of old scrap metal used to the total use of primary metal plus new scrap. Currently, metals recycling in the United States is highest for silver, about 65%. Lead is next, with about 35%, most of which is recovered from plates of discarded batteries. Copper recycling is about

25%; that of iron, tin, and mercury, between 15 and 20%. For aluminum, recycling is very low, about 3%.

Recovered metals are largely in the form of alloys. For example, most of the recovered copper is in brass and bronze; most of the recovered lead is antimonial lead from battery plates; most of the recovered zinc is in brass and bronze; practically all of the recovered aluminum is in alloys. These facts indicate that secondary metal sources are not, in general, capable of furnishing substantial quantities of pure metals under prevailing conditions of recovery technology.

Mineral Resources from the Seabed

If the prospect of eventually running out of various metallic mineral resources from the lands seems all too real, we may want to consider possible substitutions of mineral resources from the sea. Seawater has always been available as a resource, and it has long provided the bulk of the world's supply of magnesium and bromine, as well as much of the sodium chloride.

The list of elements present in seawater includes most of those known, and despite their small concentrations, these are potential supplies for future development. It is thought that sodium, sulfur, potassium, and iodine lie in the category of recoverable elements, but it seems beyond reason to hope for extraction of ferrous metals (principally iron) and the ferro-alloy metals in significant quantities to provide substitutes for ore deposits of the continents.

The continental margins, with their shallow continental shelves and shallow inland seas, are already being exploited for mineral production. An example is the working of placer deposits of platinum, gold, and tin in shallow waters. The petroleum resources of the North American continental shelf are already under development along the Gulf coast; zones of potential development are believed to exist on the shelf off the Atlantic coast as well. The possibility exists of finding and using mineral deposits of continental crystalline rocks submerged to shallow depths, although this has not yet happened.

Exploration of the deep ocean floor as a source of minerals is still in an early stage. Already a layer of manganese nodules found in parts of all of the oceans is regarded by some as a major future source of manganese and of a number of other metals (Chapter 9 and Figure 9.36). Nickel will doubtless prove to be the most valuable metal extracted from manganese nodules, but copper and cobalt may also be extracted profitably from nodules at some future date.

In reviewing the overall prospects of mineral resources from the oceans and ocean basins, we are only being realistic in concluding that contributions from seawater itself are limited to only a few substances. Most of the contributions of the seafloor will be from shallow continental shelves, where petroleum and natural gas are the major resources. Prospects of substantial metallic mineral contributions from the deep ocean floor are rather poor at this time. In light of these conclusions, the need for conservation and careful planning for the use of the mineral resources of the lands becomes all the more evident.

Nonmetallic Mineral Deposits

Nonmetallic mineral deposits (not including fossil and nuclear fuels) contain a large and diverse assemblage of substances and cover a wide range of uses. It would be impossible to do the subject justice in a few paragraphs. In outline form, we offer some examples of these mineral deposits classified by use categories:

Structural Materials

Clay: For use in brick, tile, pipe, chinaware, stoneware, porcelain, paper filler, and cement. Examples: kaolin (for china manufacture) from residual deposits produced by weathering of felsic rock; shales, marine and glaciolacustrine clays for brick and tile.

Portland Cement: Made by fusion of limestone with clay or blast-furnace slag. Suitable limestone formations and clay sources are widely distributed and are of many geologic ages.

Building Stone: Many rock varieties are used, including granite, marble, limestone, and sandstone. Slate is used as a roofing material.

Crushed Stone: Limestone and "trap rock" (gabbro, basalt) are crushed and graded for aggregate in concrete and in macadam pavements.

Sand and Gravel: Used in building and paving materials such as mortar and concrete, asphaltic pavements, and base courses under pavements. Sources lie in fluvial and glaciofluvial deposits and in beaches and dunes. Specialized sand uses include molding sands for metal casting, glass sand for manufacture of glass, and filter sand for filtering water supplies.

Gypsum: Major use is in calcined form for wallboard and as plaster, and as a retarder in portland cement. Source is largely in gypsum or anhydrite beds in sedimentary strata associated with red beds and evaporites.

Lime: Calcium oxide obtained by heating of limestone has uses in mortar and plaster, in smelting operations, in paper, and in many chemical processes.

Pigments: Compounds of lead, zinc, barium, titanium, and carbon, both manufactured and of natural mineral origin, are widely used in paints.

Asphalt: Asphalt occurs naturally, but most is derived from the refining of petroleum. It is used in paving, and in roofing materials.

Asbestos: Fibrous forms of four silicate minerals, used in various fireproofing materials.

Mineral Deposits Used Chemically and in Other Industrial Uses

Sulfur: Principal source is free sulfur occurring as beds in sedimentary strata in association with evaporites. Chief use is for manufacture of sulfuric acid.

Salt: Naturally occurring rock salt, or halite, is largely sodium chloride but includes small amounts of calcium, magnesium, and sulfate. It occurs in salt beds in sedimentary strata and in salt domes. Major uses include manufacture of sodium salts, chlorine, and hydrochloric acid.

Fertilizers: Some natural mineral fertilizers are phosphate rock, of sedimentary origin; potash derived from rock salt deposits and by treatment of brines; and nitrates, occurring as sodium nitrate in deserts (Atacama Desert of Chile).

Sodium Salts: Found in dry lake beds (playas) of the western United States are various salts of sodium, such as borax (sodium borate). These have a wide range of chemical uses. Also important are sodium carbonate and sodium sulfate, found in other dry lake accumulations.

Fluorite: The mineral fluorite is calcium fluoride. It is found in veins of both sedimentary and igneous rocks. Uses are metallurgical and chemical, for example, to make hydrofluoric acid.

Barite: Barite is barium sulfate and occurs as a mineral in sedimentary and other rocks. It is used as a filler in many manufactured substances and as a source of barium salts required in chemical manufacture.

Abrasives: A wide variety of minerals and rocks have been used as abrasives and polishing agents. Examples are seen in garnet, used in abrasive paper or cloth; and diamond, for facing many kinds of drilling, cutting, and grinding tools.

The above list is by no means complete. It can serve only to give you an appreciation of the strong dependence of industry and agriculture upon mineral deposits and the products manufactured from them.

Solar and Tidal Energy

Before looking into the sources of energy that are derived from the solid earth, let us review the full picture of world energy resources to gain a better perspective. Sources of energy are found in both sustained-yield and exhaustible categories. A *sustained-yield energy source* is one that undergoes no appreciable diminution of energy supply during the period of projected use.

Solar energy is a sustained-yield source of extreme constancy and reliability. Stated in terms of power, solar radiation intercepted by one hemisphere is calculated to be about 100,000 times as great as the total existing electric power-generating capacity. The problem is, of course, that solar radiation derived from a large receiving area must be concentrated into a very small distribution center. To produce power equivalent to that of a large generating plant (about 1000 megawatts capacity) would require at an average location a collecting surface of about ten square kilometers. While there seems to be no technological barrier to building such a plant, the cost at present is too high to make this energy source a practical one. At the present time, one of the most practical and important uses of direct solar energy is for space heating and hot water heating in small units individually designed for residential and commercial buildings.

Solar energy, broadly defined, includes many secondary energy sources that depend upon the sun's energy. One such source, often called solar sea power, makes use of the temperature gradient from the warm ocean surface downward to cold water in tropical latitudes. Hydropower uses the kinetic energy of flowing streams; windpower uses the kinetic energy of winds; wave power uses the kinetic energy of waves. In all three cases, the energy comes from the same basic source—solar radiation.

Another category of solar energy resources is that of biomass energy systems. Plants use and store solar energy in the form of organic molecules within the plant tissues. Animals ingest these plant materials as food and also store energy in body tissues. Uses of this stored energy as a fuel and as a source of secondary fuels (for example, methane or alcohol) are important energy technologies, already rather well developed to a practical degree.

Water power under gravity flow, or *hydropower,* is a familiar topic through our investigation of streams and their valley forms (Chapter 16). This power source has been developed to a point just over one-quarter of its estimated ultimate maximum capacity in the United States. Presently, water power supplies about 4% of the total energy production of the United States. Since we have a good knowledge of stream flow, the estimate of maximum capacity is probably not much in error. For the world as a whole, present development is estimated to be about 5% of the ultimate maximum capacity. Potential hydropower is particularly great in South America and Africa, where coal is in very short supply.

A serious defect of hydropower estimates is that the capacity of artificial reservoirs declines through sedimentation. Most large reservoirs behind big dams have an estimated useful life of a century or two at most. Perhaps, after all, water power should not be categorized as a "sustained" source of energy.

Tidal power is another sustained-yield energy source. To utilize this power, a bay is chosen along a coast subject to a large range of tide. Narrowing of the connection between bay and open ocean intensifies the differences of water level that are developing during the rise and fall of tide. A strong hydraulic current is produced and alternates in direction of flow. The flow is used to drive turbines and electrical generators, with a maximum efficiency of about 20% to 25%. Assessment of the world total of annual energy potentially available by exploitation of all suitable sites comes to only 1% of the energy potential available through hydropower development.

Geothermal Energy

A potentially great energy resource is *geothermal energy* drawn from heat sources beneath the earth's surface. For geothermal energy to be useful and practical as an energy source, it must be concentrated in high-temperature occurrences within reach of modern well-drilling techniques. We begin with those places where steam and hot water are emitted from vents at the surface, namely, hot springs and geysers. At other locations, very hot water, capable of being flashed into steam, lies within easy range of drilled wells. Still deeper, beneath some selected areas, the rock is dry but hot enough that if water is pumped down from the surface through drilled wells, it can be brought back up hot enough to flash into steam. An entirely different set of conditions occurs in some thick sequences of sedimentary strata in which hot water that is under pressure of natural gas can be tapped. We will describe briefly each of these types of occurrences and evaluate their present and future importance as sources of energy.

Steam and Hot Water Sources

As we explained in Chapter 15, at some widely separated geothermal localities over the globe, ground water reaches the surface in hot springs at temperatures not far below the boiling point of water, which is 100 °C at sea level. At some of these same places, jetlike emissions of steam and hot water occur at intervals from small vents; these are geysers (see Figure 15.24). The water that emerges from hot springs and geysers is largely ground water that has been heated in contact with hot rock and forced to the surface. In other words, this water is recycled surface water. Little, if any, is water that was originally held in rising bodies of magma.

A phenomenon related to hot springs and geysers is the fumarole, a jet of gases issuing from a small vent, and common in regions of current and recent volcanic activity. Gas temperatures in fumaroles are extremely high, up to 320 °C. Most of the gas (over 99%) is water. In other words, fumaroles emit largely superheated steam.

The natural hot water and steam localities were the first type of geothermal energy source to be developed and currently account for nearly all geothermal production of electric power. Wells are drilled to tap the

FIGURE 21.11 An electricity generating plant at The Geysers, California. Steam pipes in the foreground lead to the plant. After use in generating turbines, the steam is condensed in the large cylindrical towers. (Pacific Gas & Electric Company.)

FIGURE 21.11

hot water, which flashes into steam as it reaches the surface under reduced atmospheric pressure. The steam is separated from the water in large towers and fed into generating turbines to produce electricity (Figure 21.11). The hot water is usually released into surface stream flow, where it may create a pollution problem.

The larger steam fields have sufficient energy to generate at least 15 megawatts of electric power, and a few can generate 200 megawatts or more. Table 21.4 lists the world's existing geothermal power developments with their 1975 generating capacities. The total capacity of these fields is only about 1400 megawatts, about twice what a single nuclear power plant develops, and about the same as the hydropower output of Hoover Dam. When planned additions are completed at the localities listed in Table 21.4, and a number of new power developments are completed at other sites, the added power output will be about 640 megawatts, for a total world output of just over 2000 megawatts—abut the same as the hydropower output of the Grand Coulee Dam by itself. Obviously, energy contributions from shallow hot water and steam fields are very small, and prospects for substantial increases are indeed poor. We must turn to other occurrences of geothermal energy in search of larger energy supplies.

Deep Hot Water Sources

Much greater energy sources than those we have just described lie in deeper zones of hot ground water, but these must be tapped by deep drilling. One region of deep, hot ground water, currently under investigation, is beneath the Imperial Valley of southern California. An area of 5000 sq km is involved, and it extends over the border into Mexico in the Mexicali Valley. This region is tectonically active and has been interpreted as a zone of crustal spreading in which the lithospheric plate is being fractured. Rising basalt magma, found elsewhere in active spreading zones such as Iceland in the North Atlantic, may be responsible for the geothermal condition, but this interpretation is speculative. Test wells show that a large reservoir of extremely hot ground water—260 °C to 370 °C—is present here. This water readily flashes into steam when penetrated by a drill hole. Steam pressure forces both steam and hot water to the surface, much like the action of a coffee percolator. Near Mexicali, Mexico, this resource has already been developed, with the daily flow of steam and water amounting to 2000 metric tons. The prospect of a large power development beneath the Imperial Valley looks very good, while the salinity of the hot water is quite low. If fully developed, the Imperial Valley geothermal field could probably produce as much as 20,000 megawatts of electricity, which is an amount equal in 1975 to about 10% of the total U.S. production of electricity, or about 50% of the power output of U.S. nuclear power plants.

Table 21.4 Geothermal Generating Capacity

Country	Name of field	Capacity (megawatts)
El Salvador	Ahuachapan	30
Iceland	Namafjell	3
Italy	Larderello	400
	Monte Amiata	25
Japan	Matsukawa	20
	Otake	11
	Hachimantai	10
Mexico	Pathé	3
	Cerro Prieto	75
New Zealand	Wairakei	190
	Kawerau	10
United States	The Geysers	600
U.S.S.R.	Pauzhetsk	5
	Paratunka	1
	Total	1383

Source: A. J. Ellis (1975) *American Scientist*, vol. 63, p. 515, Table 1.

The needs of southern California could be fully met by this source alone. Heat remaining in the water after it has been used to generate electricity could be used to distill the waste water and produce a substantial yield of irrigation water, a valuable commodity in this desert agricultural region.

Hot Dry Rock Energy Sources

In certain areas, the intrusion of magma has been sufficiently recent that solid igneous rock of a batholith is still hot in a depth range of perhaps 2 to 5 km. At this depth, the rock is strongly compressed and contains little, if any, ground water. Rock in this zone may be as hot as 300 °C and could supply an enormous quantity of heat energy. The planned development of this resource includes drilling into the hot zone and then shattering the surrounding rock by hydrofracture—a method using water under pressure which is widely practiced in petroleum development. Surface water would be pumped down one well into the fracture zone and heated water pumped up another well.

Experimental holes in northern New Mexico were drilled in the middle 1970s, and the dry granite rock at the bottom fractured by water injected under pressure. This successful experiment has led to optimism that deeper zones of hot rock can be developed. Promising locations have been found in Montana and Idaho. The potential for electrical power generation from deep hot rock areas is believed to be many times greater than for hot water areas. One recent guess places the output of these dry rock wells at a possible 130,000 megawatts by 1985 and nearly 400,000 megawatts by the year 2000. The latter figure is about double the entire electricity generating capacity of the United States for 1975.

Geopressurized Energy Sources

Finally, on the list of occurrences of geothermal energy, we come to an energy source trapped in thick continental shelf sediments of the continental margin. Recall from Chapter 12 that these sediments form a wedge, thickening toward the ocean basin. Rich petroleum deposits have been developed in the continental shelf wedge of the Gulf Coast region. Here, in a region extending from Texas to Louisiana, the strata in some areas have been found to hold hot water under high pressure and to contain dissolved natural gas. Subsurface bodies of this type are described as *geopressurized zones*, and they are avoided in drilling for petroleum because the high pressure makes it difficult to control the well. These geopressurized areas should be explored, according to recent proposals, as sources of hot water for generating electricity. One estimate of the total energy resource in the Gulf area alone is a generating capability of between 30,000 and 115,000 megawatts of power. Development of this resource is probably not economically feasible at this time, but the ultimate potential is great.

FIGURE 21.12 Generalized map of coal fields of the United States and southern Canada. (Redrawn, by permission of the publisher, from A. N. and A. H. Strahler, *Modern Physical Geography*, Figure 26.10. Copyright © 1978 by John Wiley & Sons, Inc.)

Coal and Lignite

We turn next to our major source of energy for the past century, the *fossil fuels:* coal and lignite, petroleum, and natural gas. The composition, geologic occurrence, and origin of coal and lignite were discussed in Chapter 5, treating them as sediments of biologic origin.

The oldest important occurrences of coal are in strata of Carboniferous age. During this and the following Permian periods, coals of great thickness and extent were formed in many parts of Pangaea. These now have widespread distribution on the fragmented continents. Coal is also found in Triassic and Jurassic strata, but to a limited extent. Second in importance to the Carboniferous coals are those deposited in the Cretaceous Period, at which time continental separation was in progress. Strata of Cenozoic age contain most of the world's lignite, but some coals of high quality were also produced in that era.

Figure 21.12 shows the distribution of coal and lignite fields in the United States and southern Canada. The most important U.S. coal producer thus far has been the Appalachian field, with an extent of some 180,000 sq km. Here coal is in Carboniferous strata, which are folded into open synclines in the narrow Ridge-and-Valley belt (along the eastern margin of the field) but are flat lying in the Appalachian Plateaus to the west. The Anthracite field of eastern Pennsylvania,

FIGURE 21.12

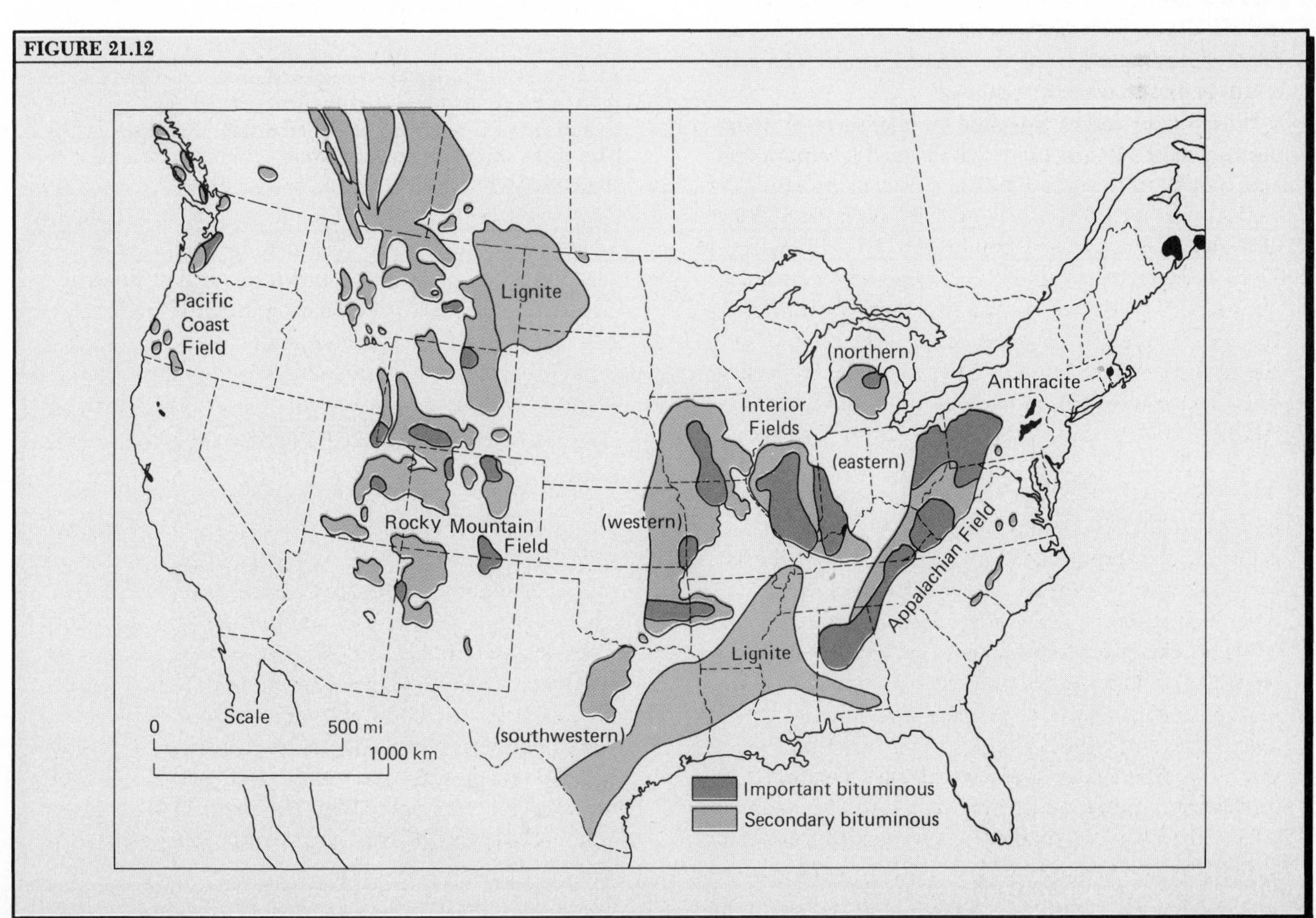

in which anthracite occurred along the axes of deeply eroded synclines, was formerly an important producer, but these deposits are almost exhausted (see Figure 17.33). The Interior Fields are now important producers of bituminous coal in parts of Indiana, Illinois, and Kentucky, where Carboniferous strata are nearly horizontal. The Rocky Mountain Field will become a great coal producer of the future, for vast reserves of low-sulfur coal lie untouched here in strata of Cretaceous age (Figure 21.13). One reserve of lignite of Cenozoic age lies beneath the Dakotas and eastern Montana; another, beneath the Gulf coastal plain in Cenozoic strata.

Coal Mining

To extract coal from horizontal strata deep beneath the surface, vertical shafts are first driven down to reach the coal. Horizontal *mine drifts* are then opened out along the coal seams. Where terrain is mountainous, coal seams outcrop along the contour of the mountainsides, and here the drifts are driven directly into the exposed seams.

Removal of coal at depth follows one of two common methods. In the *room-and-pillar method*, illustrated in Figure 21.14, about half the coal is left behind in the form of supporting pillars. A second method, which can recover a much larger proportion of the coal, is the *long-wall method*. It begins with removal along the circumference of a large circle around the entire shaft, and all of the coal is removed, proceeding inward from the starting circle. The roof collapses in the wake of removal.

Where coal seams lie close to the surface, *strip-mining* methods are used. Earth-moving equipment removes the overlying soil and rock, called *overburden*, to bare the coal seam. Power shovels then remove the coal (Figure 21.13). Two types of strip mining are used, depending upon terrain conditions. Where the land is level and strata are horizontal, overburden is removed from a long, straight trench and piled in a ridge to one side (Figure 21.15A). After the coal has been removed from the floor of the trench, another belt of adjacent overburden is removed and piled in the first trench. In this system, known as area strip mining, the overburden accumulates in a series of parallel spoil ridges, as shown in Figure 21.16. In mountainous terrain, coal is removed by contour strip mining, shown in Figure 21.15*B*. The overburden is excavated as far back into the mountainside as economically feasible, and the exposed coal is removed. Additional coal is secured by means of large augers, driven as far as possible into the mountainside. The spoil from contour strip mining is deposited in a bank on the downslope side of the excavation, where it may slide far down into the valley below. Between the spoil bank and the clifflike rock face, or *high wall*, a trench remains, and this may follow the contour of the mountainside for many kilometers, winding in and out of one valleyhead after another.

World Coal Reserves

World coal reserves are most unevenly distributed among the continents. Figure 21.17 shows a break-

FIGURE 21.13 This great coal seam, ranging from 20 to 30 m thick, is being strip-mined at Wyodak, Wyoming. The upper power shovel is removing the overburden; the shovel at the base of the seam is removing the coal. (Bureau of Mines, U.S. Dept. of the Interior.)

FIGURE 21.13

down into eight world regions; figures are in billions of metric tons. This estimate shows the U.S.S.R. and the United States in very strong positions, while Canada and Western Europe are also very favorably endowed with coal. In contrast, Africa, Australia, Japan, and the Latin American countries have poor reserves.

Note that these figures include coal lying at depths as great as 1200 m and occurring in seams as thin as 36 cm. With present mining technologies, only about one-

FIGURE 21.14 Plan (map) of a room-and-pillar mine showing rooms (white areas) and coal (black). About half of the coal remains to support the roof. (Illinois Geological Survey, Bull. 56.)

FIGURE 21.15 *A*. Area strip mining. *B*. Contour strip mining. (Redrawn, by permission of the publisher, from A. N. Strahler, *Physical Geography*, 4th ed., Figure 28.20. Copyright © 1975 by John Wiley & Sons, Inc.)

FIGURE 21.16 Area strip mining in Ohio County, Kentucky. Dragline in background is removing overburden and piling it at the right. Loader in foreground is removing the exposed coal. (TVA.)

FIGURE 21.17 Estimated world coal reserves, in billions of metric tons. (Based on data of the U.S. Geological Survey.)

FIGURE 21.14

FIGURE 21.15

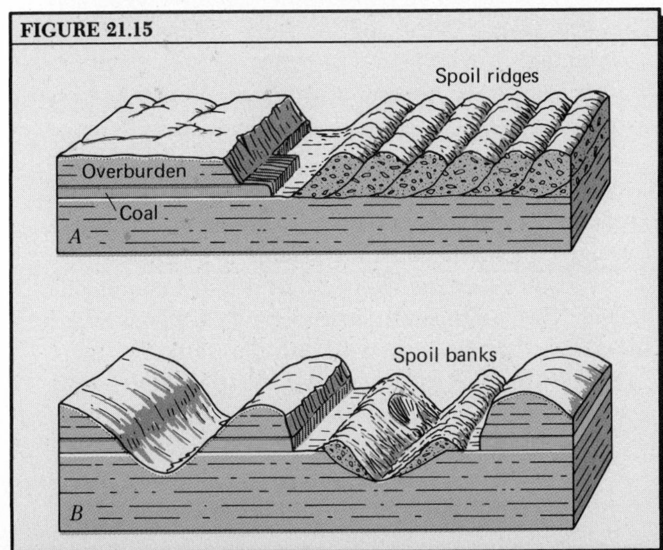

FIGURE 21.16

FIGURE 21.17

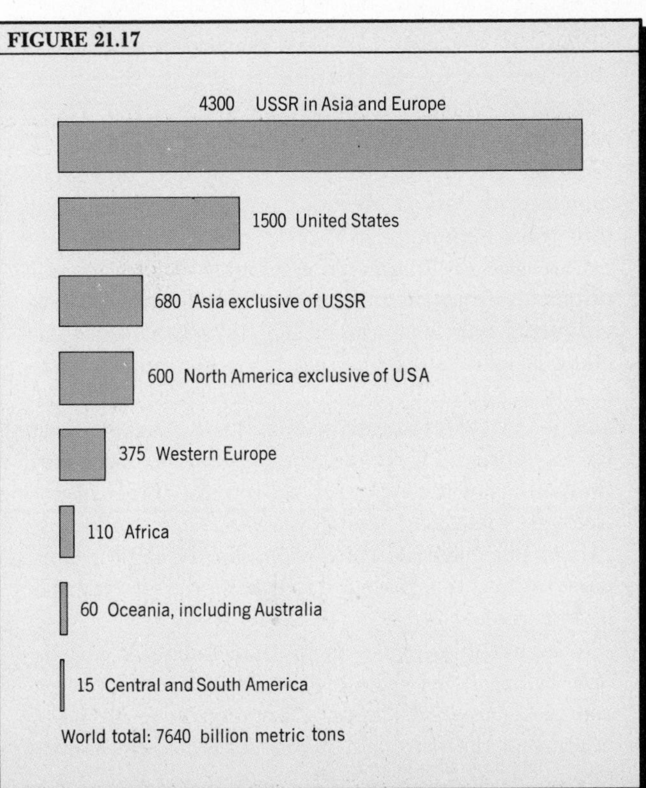

tenth of this amount of coal can be mined on an economically successful basis. So the figure of 1500 billion metric tons for the United States reduces to 150 billion metric tons, in terms of existing mining methods. Converted into crude oil energy equivalent, 150 billion metric tons represents about 640 billion barrels. This quantity is roughly 18 times larger than the proven crude oil reserves of the United States.

Petroleum—Origin and Geologic Occurrence

The general term *petroleum* spans the range from crude oil to natural gas in the one direction, and to asphalt and related semisolid hydrocarbon substances in the other. Table 21.5 gives the range of carbon, hydrogen, sulfur, nitrogen, and oxygen in typical analyses of the three forms of petroleum. *Crude oil* in the natural state is a mixture of a large number of hydrocarbon compounds. More than 200 compounds have been isolated and analyzed in crude oil. The range and abundance of hydrocarbon compounds differs greatly from one oilfield to another.

Crude oils differ in terms of the relative abundances of various hydrocarbon groups. Generally speaking, the paraffin compounds are the most abundant of hydrocarbons in both liquid petroleum and natural gas. Crude oil is described as *paraffin-base* when paraffins are dominant; it is of low density and typically yields good lubricants and a large proportion of kerosene. An example is the paraffin-base crude oil of the Pennsylvania fields. *Asphalt-base* crude oil has a high density and is referred to as a *heavy oil;* its primary yield is in the form of fuel oils.

Natural gas, found in close association with accumulations of crude oil, is a mixture of gases, the principal one of which is methane (marsh gas, CH_4). There are minor amounts of ethane, propane, and butane, all of which are also hydrocarbon compounds, and small amounts of carbon dioxide, nitrogen, oxygen, and sometimes helium.

Geologists are in general agreement that petroleum originates from organic matter buried within thick sediments that accumulate in geosynclines and geoclines of continental margins. Clays and muds deposited in a chemically reducing environment are favorable for petroleum accumulation. As the organic matter is buried, it is converted into hydrocarbon compounds of the types related to those found in petroleum. Increasing temperature plays an important part in this chemical conversion, but too high a temperature and too great a depth of burial destroy the hydrocarbons.

A second phase involves the movement, or *migration,* of petroleum from the source rock into a porous and permeable rock mass, a *reservoir rock.* It is believed that the petroleum is carried out of the source rock along with water forced out of compacting sediment. Thus, petroleum moves both upward and laterally through the sediments, and ends up in reservoir rocks, most of which are sedimentary types. Sand and sandstone formations make excellent reservoirs, and carbonate rocks may also be good, particularly where solution removal has created large, interconnected cavities. It is essential that the reservoir rock be capped or surrounded by an impermeable rock, forming a *reservoir trap,* or else the petroleum would steadily escape to the surface. Shale is a common *cap rock,* because of its low permeability.

Table 21.5 Element Composition of Typical Petroleum

Element	Percentage by weight		
	Crude oil	Asphalt	Natural gas
Carbon	82–87	80–85	65–80
Hydrogen	12–15	8.5–11	1–25
Sulfur	0.1–5.5	2–8	trace–0.2
Nitrogen	0.1–1.5	0–2	1–15
Oxygen	0.1–4.5	—	—

Source: A. I. Levorsen (1967), *Geology of Petroleum,* 2d ed., San Francisco, Freeman, p. 177, Table 5.5.
Note: Figures rounded to the nearest one-half percent.

One of the simplest sedimentary traps favorable to oil accumulation is an uparching of strata in either a dome or an anticline (Figure 21.18). Natural gas accumulates above the oil, while water saturates the zone beneath the oil. Various other favorable arrangements or rock units also form petroleum traps.

An example of an important oil-producing dome structure is in the Dominguez Hills area of southern California (Figure 21.19), where a sandstone formation of Pliocene age, lying some 1200 m beneath the surface, has trapped oil under an impermeable covering layer of shale. The strata are gently arched and give rise to a low, topographic dome at the surface. Other well-known oil pools formed in dome structures are the Teapot Dome and the Rock Springs Dome of Wyoming.

A particularly important sequence of sedimentary strata favorable to petroleum accumulation is found beneath the coastal plain of the Gulf and Atlantic coasts of the United States. Here a great thickness of sedimentary layers takes the form of an enormous wedge tapering from a thin edge on the continent to a thick body under the shallow waters of the continental shelf. This continental shelf deposit, a miogeocline, was described in Chapter 12 and illustrated in Figure 12.9. Within the wedge, individual strata usually become thinner as they are followed updip in the inland direction. Where a sandstone layer thins to the point of disappearance, it forms a type of oilbearing structure known as a *pinch-out trap* (Figure 21.20). A large number of oil pools in pinch-out traps on the coastal plain of Texas are shown in Figure 21.21, where pools occur in two broadly curved belts paralleling the coastline.

In certain areas of very thick sedimentary strata, beds of halite buried at great depth have been forced by the pressure of overlying strata to rise in stalklike columns. These salt columns, or diapirs, slowly penetrate the overlying strata and eventually come to rest with the top of the column at or close to the surface. The *salt plug*, as this odd structure is known, may have a vertical extent of several kilometers but a diameter of only a kilometer or two (Figure 21.22). Typically the salt plug has a cap rock of limestone, gypsum, and anhydrite forced up by the rising salt, and at the surface, the strata are sometimes arched into an overlying dome, called a *salt dome*. Salt domes are numerous in the Gulf coastal plain of the United States, usually with important oil pools occurring along their flanks. As shown in Figure 21.22, the oil is trapped in upturned sandstone strata adjacent to the salt plug or in the cavernous limestone of the cap rock.

Yet another form of petroleum accumulation is held in a *fault trap* (Figure 21.23). Displacement of strata along a fault brings the edge of a dipping sandstone formation against impervious shale. Petroleum migrating up the dipping bed accumulates against the fault

FIGURE 21.18 Idealized cross section of an oil pool in an anticline or a dome structure in sedimentary strata. Well *A* will draw gas; well *B* will draw oil; and well *C* will draw water. The cap rock is shale; the reservoir rock is sandstone.

FIGURE 21.19 This low dome, in Dominguez Hills, California, provided a trap for an important petroleum pool. (Based on data of H. W. Hoots and the U.S. Geological Survey.)

FIGURE 21.20 Petroleum trapped by pinch out of sand strata in the updip direction. (Redrawn, by permission of the publisher, from A. N. Strahler, *Physical Geography,* 4th ed., Figure 28.11. Copyright © 1975 by John Wiley & Sons, Inc.)

FIGURE 21.18

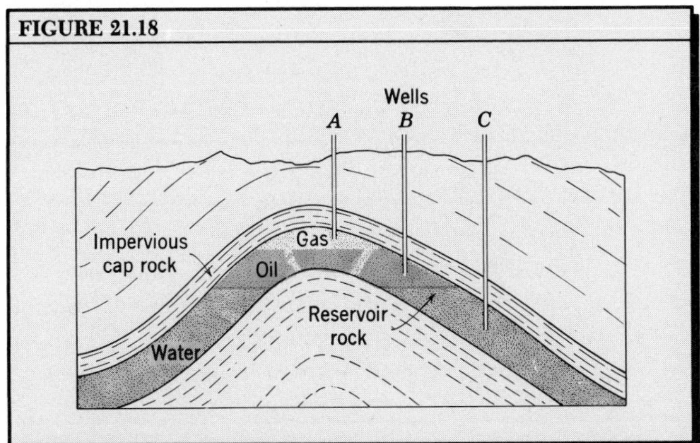

FIGURE 21.19

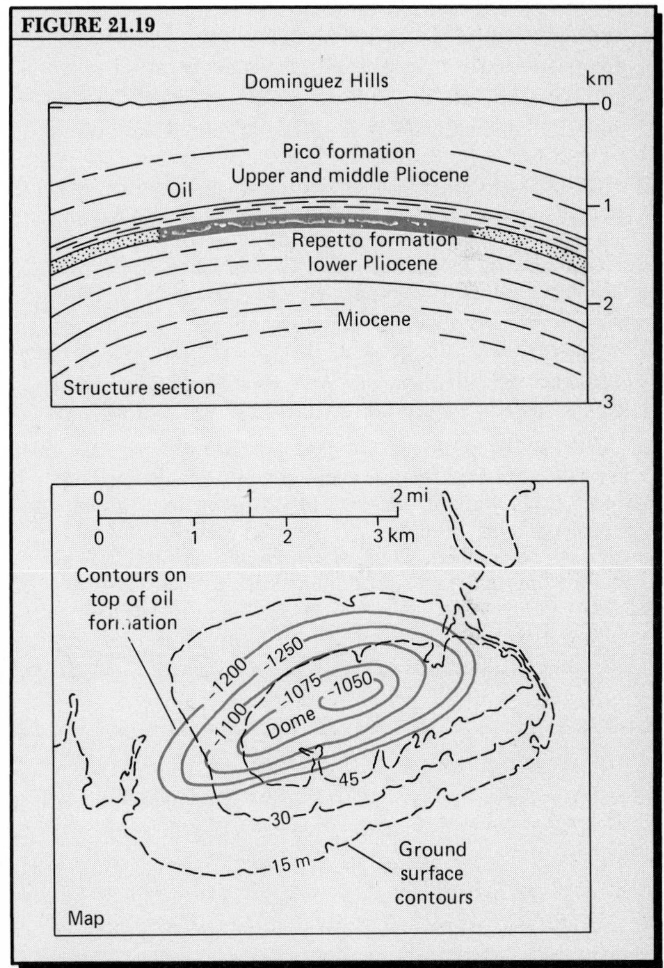

FIGURE 21.20

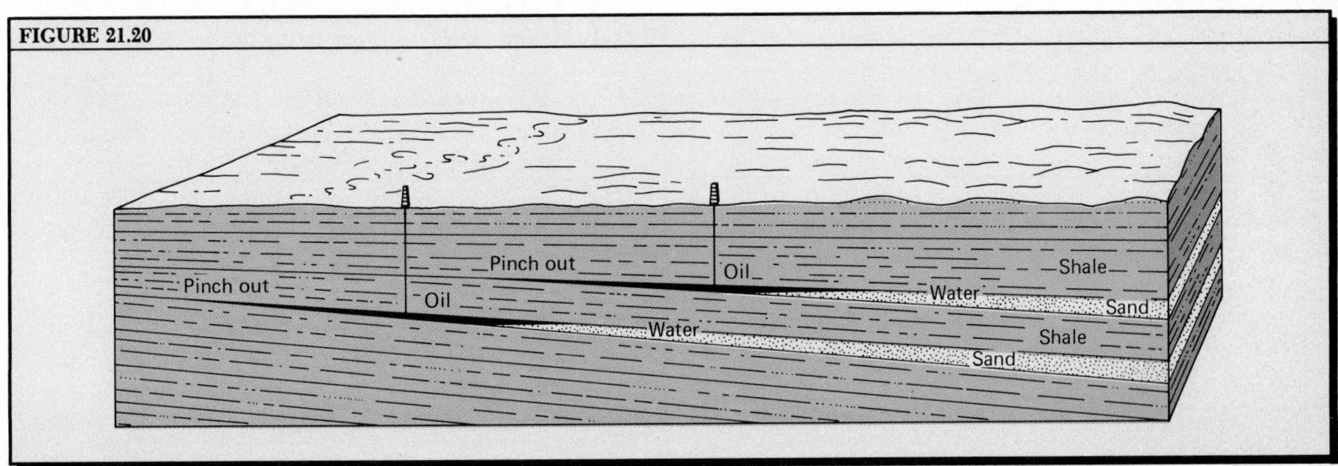

and the overlying shale. Various other types of petroleum traps exist, but the examples given here are enough to give you a general picture of the geological aspects of petroleum resources.

Petroleum Accumulations and Plate Tectonics

Plate tectonics has made its impact upon the search for new petroleum accumulations. Geologists have long been aware that the organic matter required for petroleum to be formed has accumulated in geosynclines along with thick marine sediments at the continental margins. Large delta deposits of major rivers are particularly favorable for the incorporation of organic matter into the sediment. One geologist has suggested that crude oil is derived largely from animal proteins, while natural gas is derived from altered plant matter. Both plants and animals are abundant in shallow coastal waters of continental shelves and deltas. This organic matter, along with the mineral sediment, has also been swept into deep water of the continental rise by turbidity currents, there to accumulate in thick turbidite wedges.

A second requirement for the occurrence of petroleum is that the sediment body be heated by just the right amount and for the right duration. Here is where plate tectonics comes in. One possible source of heat is the hot spot, which may be located over a plume of rising mantle rock within the asthenosphere. It is theorized that, as an organic-rich sediment wedge passes over a hot spot, heating converts the organic matter

FIGURE 21.21 Oil pools, shown in black, form distinct belts along two zones of updip pinch out. Pools of zone *AA′* are in sands of Eocene age, those of zone *BB′* in sands of Oligocene age. (Redrawn, by permission of the publisher, from A. N. Strahler, *Physical Geography*, 4th ed., Figure 28.12. Copyright © 1975 by John Wiley & Sons, Inc.)

FIGURE 21.22 Structure of a salt dome. (Redrawn, by permission of the publisher, from A. N. Strahler, *Physical Geography*, 4th ed., Figure 28.13. Copyright © 1975 by John Wiley & Sons, Inc.)

FIGURE 21.23 Petroleum accumulation in a fault trap, produced by faulting of impermeable shales against a dipping sandstone layer. (Redrawn, by permission of the publisher, from A. N. Strahler, *Physical Geography*, 4th ed., Figure 29.14. Copyright © 1975 by John Wiley & Sons, Inc.)

FIGURE 21.21

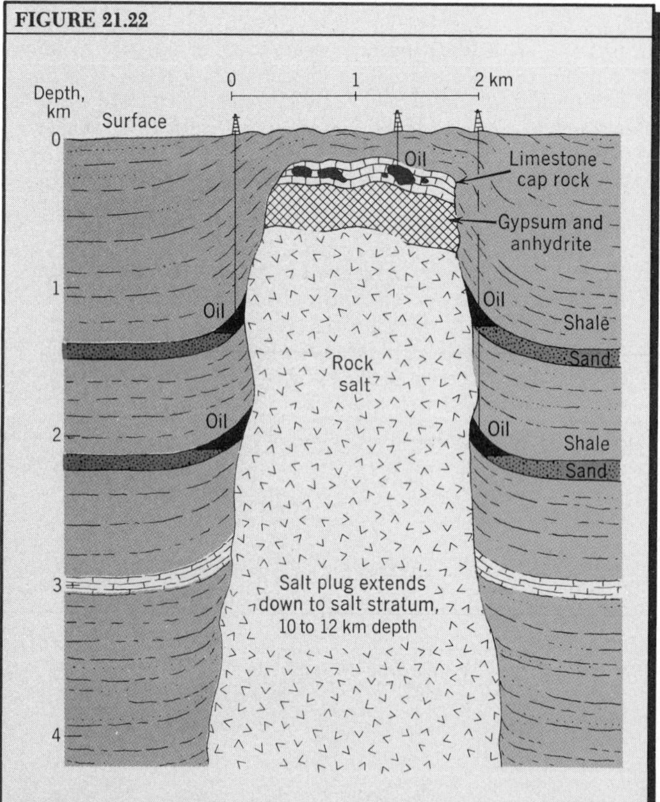

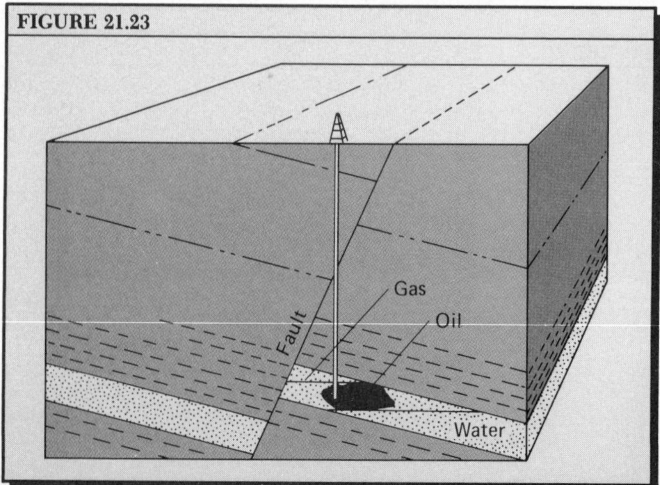

into petroleum, which later migrates through the parent mass to reach permeable reservoirs.

Another possibility for heating of the sediment body is the process of continental rifting, producing an ocean basin and a new oceanic crust. The concept here is that sediments accumulate in the new ocean basin, where heat flow is high because of the past history of rising basalt magma by which the new crust was created. Figure 21.24 shows the steps. The scenario involves first the accumulation in the narrow ocean basin of thick evaporite (salt) beds, which may reach thicknesses of several kilometers (Diagram *A*). An example is the floor of the Red Sea, a spreading basin, in which evaporite beds up to 5 km in thickness have been measured (Chapter 13).

The next step is a change of environment in the widening ocean basin from one of high salinity to one of normal salinity in a freely circulating ocean (Diagram *B*). Now the salt deposits, which lie off the continental margins in deep water, are buried under continental shelf wedges and turbidite wedges containing sediments rich in organic matter. Heat flow remains high below the wedges and the cooking process takes place, forming crude oil and natural gas within the sediments. Meanwhile, portions of the underlying salt beds begin to rise toward the surface in tall, stalklike diapirs, penetrating the soft sediments to produce salt domes (Diagram *C*). The rising salt towers deform the sediments in ways that produce traps for the petroleum. As we have already found, oil and gas are trapped between upturned strata and the solid salt column (see Figure 21.22). Oil and gas in the continental

FIGURE 21.24 Schematic diagram showing origin of thick salt beds below turbidite sediments.

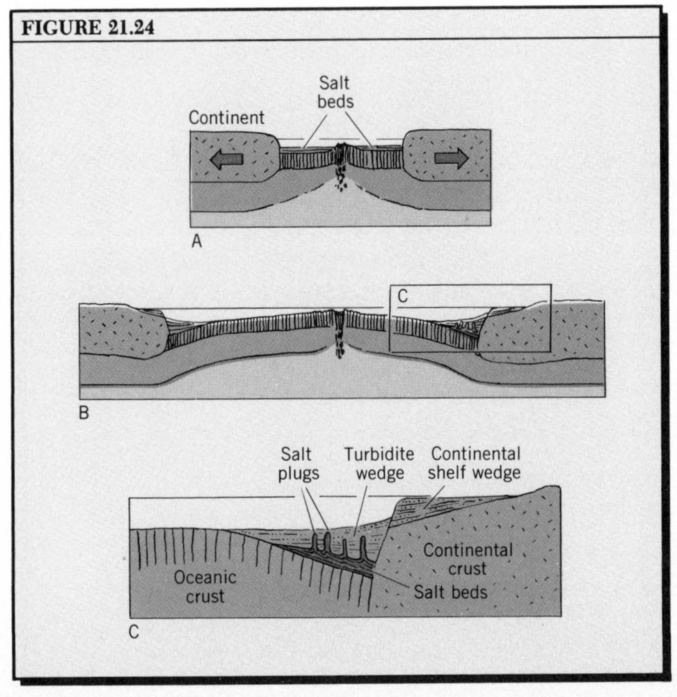

shelf strata migrate into other types of traps, such as the updip pinch-out structure shown in Figure 21.20.

As the Atlantic Ocean first opened, the narrow ocean basin was the site of thick salt deposits because ocean circulation was greatly restricted, much as in the Red Sea Trench today. Then, as the ocean widened, the salt accumulation was divided into two parts, each of which remained close to its respective continent. Accumulation of sediment wedges then followed, as shown in Figure 21.24. Today, salt dome structures are found both off the western coast of Africa and in the Gulf of Mexico. Geologists consider the prospects for discovery of new oil reserves to be extremely favorable in the African offshore zone. Petroleum accumulations of the Texas and Louisiana Gulf Coast have already been tapped in the shallow offshore zone, but there are indications of salt domes in much deeper deposits in the Gulf of Mexico, suggesting that additional petroleum accumulations remain undiscovered and untapped. In recent years, offshore exploration has turned to the Atlantic margin off the east coast of the United States. We showed the evolution of this continental margin in Figure 13.36.

Where continental collisions have occurred in the geologic past, forming sutures, petroleum-bearing strata may be expected to occur on both sides of the suture. Thus, rich Russian oilfields lie on both sides of the Ural Mountains, a suture zone. It is also suspected that new oilfields being exploited in South America on the eastern side of the Andes may represent sediment wedges caught in an ancient continental collision that predated the opening of the Atlantic Ocean.

World Oil and Gas Reserves

Proven petroleum reserves are mostly concentrated in a few world regions. As Figure 21.25 shows, the Middle East holds more than half the known world reserves of crude oil. (The standard measure of crude oil volume is the barrel, equivalent to 42 U.S. gallons or about 159 liters.) Of this enormous accumulation—some 350 billion barrels—Saudi Arabia has about 40% (140 billion barrels), which is almost four times as much as U.S. reserves (39 billion barrels), and Kuwait, Iran, and Iraq hold most of the remainder. Another important center of oil accumulation is in lands surrounding the Gulf of Mexico and Caribbean Sea, with major reserves in the U.S. Gulf Coast region and in Venezuela.

Many qualified specialists have made estimates of the total ultimately recoverable crude oil resources of the world. Several independent estimates made in the 1970s ranged from 1800 to 2000 billion barrels, which is over three times the known reserves of about 560 billion barrels shown in Figure 21.25. Figure 21.26 is a recent estimate made by Richard L. Jodry, based upon a detailed analysis of every potential petroleum-producing area in the world. Other recent estimates give a rather similar set of data. Notice that for the United States ultimately recoverable reserves are already more than half consumed, whereas the proportion consumed is much smaller for the other geographical units.

How do natural gas reserves compare with crude oil reserves? For the United States alone, known reserves that can be produced and sold at current prices amount to the energy equivalent of about 50 billion barrels of crude oil. This quantity is somewhat larger than the proven U.S. reserves of crude oil. Undiscovered U.S. natural gas reserves have been estimated by the U.S. Geological Survey to be more than double the known reserves, but some critics feel that the survey estimate is much too high. The U.S. natural gas reserves are perhaps higher than those of any nation except the U.S.S.R. (which may have more than twice as much as the United States) and Iran (which may have slightly more than the United States). Algeria and the Netherlands also rank high in gas reserves.

Bituminous Deposits

In many localities, petroleum of tarlike consistency is found occupying the pore spaces of sand or porous sandstone layers. This form of petroleum goes under various names, such as bitumen, tar, asphalt, or pitch. *Bitumen* is a highly viscous liquid which remains immobile in the enclosing sand and will flow only when heated. Outcrops of *bituminous sand* exposed to the sun will show bleeding of the bitumen. There are two geologic processes by which bitumen has come to oc-

FIGURE 21.25 The world petroleum pie. These figures by the National Petroleum Council give proven reserves of crude oil in billions of barrels.

FIGURE 21.26 Estimated world reserves of ultimately recoverable crude oil. Figures in billions of barrels. (Data of Richard L. Jodry. Diagram after M. King Hubbert, 1974, U.S. energy resources, a review as of 1972, *Part 1; U.S. Congress, Senate Committee on Interior and Insular Affairs, A National Fuels and Energy Policy,* U.S. Government Printing Office, Washington, D.C.)

FIGURE 21.25

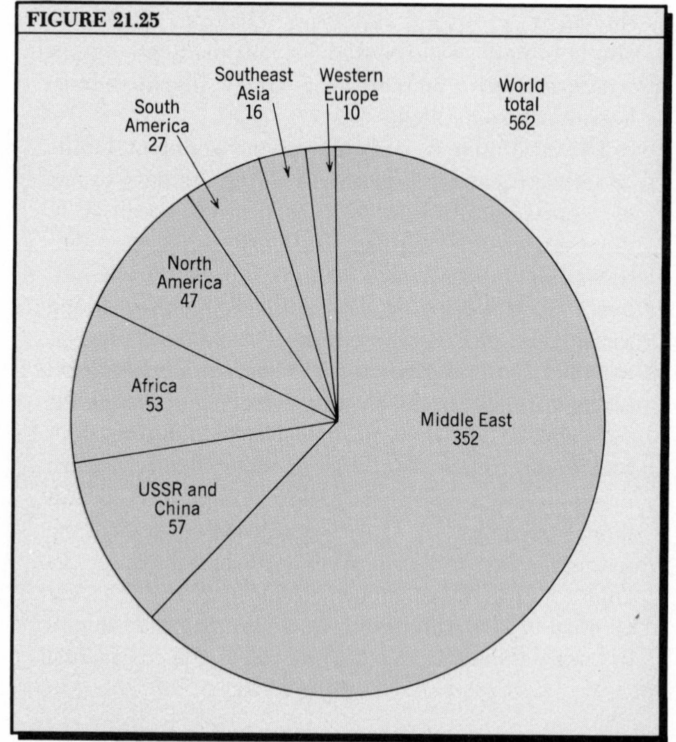

FIGURE 21.26

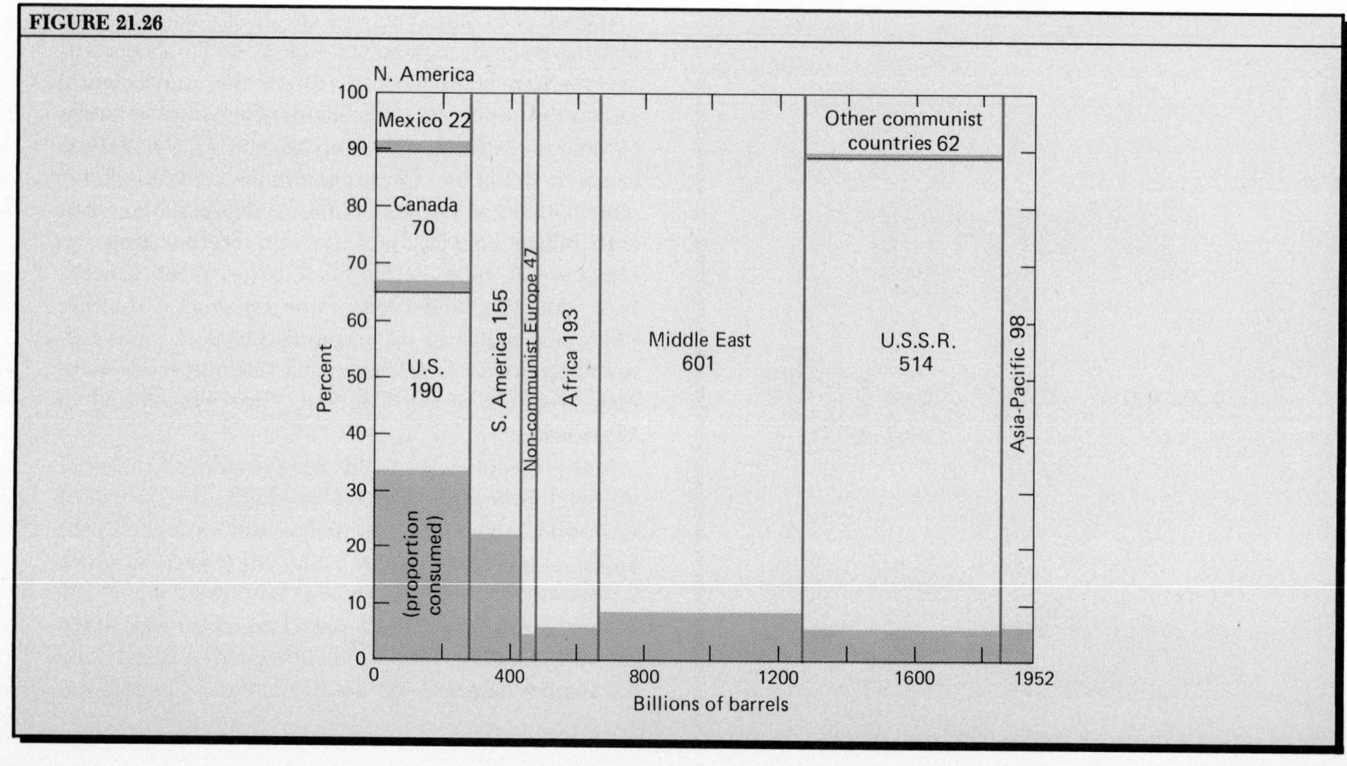

cupy the pore spaces in a sand formation. Some occurrences are explained through the drying out of a liquid petroleum accumulation in its reservoir rock. Other deposits are interpreted as sedimentary accumulations of a mixture of sand grains and bitumen particles deposited simultaneously.

Perhaps best known of the great bituminous deposits are those occurring in Alberta, Canada. The largest is known as the Athabasca Deposit. Locally, the deposits are known as "oil sands" and occur in strata of Cretaceous age. The four major oil sand deposits of Alberta contain an estimated total of 1350 billion barrels of bitumen, more than twice the total world petroleum reserves shown in Figure 21.25. The Athabasca Deposit is exposed along the banks of the Athabasca River, where the oil sand is extracted at two commercial surface mines. When agitated in hot water, the sand grains fall to the bottom while the oil comes to the surface as a froth. Only a small fraction of the total Athabasca Deposit lies close enough to the surface (50 m) to be profitably mined. Extraction of oil from wells will require that the oil sand be heated by steam or other heat sources, such as combustion in place. Many difficult technological problems must be solved before this vast energy source can become a major contributor to the world supply.

Oil Shale

Almost everyone who reads the papers has heard of *oil shale* and of the tremendous reserve of hydrocarbon fuel it holds. The truth is that this sedimentary rock in the Rocky Mountain region is not really shale at all, and the hydrocarbon it holds is not really petroleum. Srata of the Rocky Mountains called oil shales are of calcium carbonate and magnesium carbonate composition and were formed as lake deposits of lime mud (marl) in a Cenozoic lake. These soft, laminated deposits belong to the Green River formation, which is of Eocene age (Figure 21.27).

The hydrocarbon material of the Green River formation occurs in a particular bed, the Mahogany Zone, about 20 m thick. It is a waxy substance, called *kerogen*, which adheres to the tiny grains of carbonate material. When oil shale is crushed and heated to a temperature of 480 °C, the kerogen is altered to petroleum and driven off as a liquid. The shale may be mined and processed in surface plants or burned in underground mines, from which the oil is pumped to the surface.

A pilot plant operated by the Union Oil Company along Parachute Creek in Colorado successfully extracted from 300 to 1000 metric tons of petroleum per day from shale. Spurred by the energy crisis, plans are being made for extensive development of this resource. It is estimated that the equivalent of some 1800 billion barrels of petroleum lie in reserve in the entire shale formation, and that about 120 billion barrels lie in the prime beds, 10 m thick and capable of yielding 115 liters of oil per metric ton of shale.

FIGURE 21.27 These cliffs of oil shale, near Rifle, Colorado, are the site of test mining operations. (U.S. Bureau of Mines.)

FIGURE 21.27

The Future of Fossil Fuel Resources

The quantity of stored hydrocarbons in the earth's crust is finite, while the rate of geologic production and accumulation of new hydrocarbons is immeasurably small in comparison with the rate of their consumption. The fact is that the ultimate exhaustion of this energy source is inescapable. When this event will happen is, however, a very difficult thing to predict, since we have to project into the future two independent curves. First is the rate of production, which has been increasing by about 6%/year for crude oil. Until recently, discoveries of new oil reserves more than kept pace with production, so there was a moderate increase of known reserves. By 1960, however, the rate of proved discoveries of U.S. reserves began to decline, and once a decline has set in, reserves will dwindle and eventually be entirely used, after which point production itself must begin to decline and will ultimately approach zero.

The future of U.S. crude oil production has been projected by a leading geophysicist, M. King Hubbert, in the form of a bell-shaped curve, shown in Figure 21.28. The dashed line smooths out production through the year 1970, then follows a predicted rapid decline to 20% in the year 2000, and a near-zero value is reached by the year 2060. The peak of production has already passed. A similar projection for natural gas gives much the same schedule, with a production peak about in 1980 and a decline to 20% by the year 2015.

The future of world crude oil production has also been predicted by Hubbert, as shown in Figure 21.29. The production peak is shown to be reached in about 1995, which is 25 years later than for the United States, while decline to the 20% level is delayed by about the same time span.

What of the future of U.S. coal reserves, which we have set at the equivalent of 640 billion barrels of crude oil? These reserves are fully ten times larger than the proven U.S. crude oil reserves. Using the Hubbert method, the bell-shaped production curve of U.S. coal production peaks about in the year 2200, which is two centuries after the peak of crude oil production. A similar curve for world coal production peaks about in the year 2150.

The conclusion we must draw from known reserves and any reasonable schedule of their depletion is that coal will become our main fossil-fuel energy resource by the year 2000 or thereabouts, and will continue in that role for perhaps another century or two, until it, too, is exhausted.

Nuclear Energy as a Resource

The controlled release of energy from concentrated radioactive isotopes can be achieved through one of two processes: fission and fusion. *Nuclear fission* makes use of uranium-235, a rare isotope of a very rare element. The fission of 1 g of this substance yields an amount of heat equivalent to the combustion of about 3 metric tons of coal, or about 14 barrels of crude oil.

The U.S. reserve of uranium ore profitably recoverable with existing technology is estimated to have the

FIGURE 21.28 The complete cycle of crude oil production for the United States, exclusive of Alaska. (After M. K. Hubbert, 1974, U.S. energy resources, a review as of 1972, *Part 1; U.S. Congress, Senate Committee on Interior and Insular Affairs, A National Fuels and Energy Policy,* U.S. Government Printing Office, Washington, D.C.)

FIGURE 21.29 The estimated cycle of world crude oil production. (After M. K. Hubbert, 1974, *Part 1; U.S. Congress, Senate Committee on Interior and Insular Affairs, A National Fuels and Energy Policy,* U.S. Government Printing Office, Washington, D.C.)

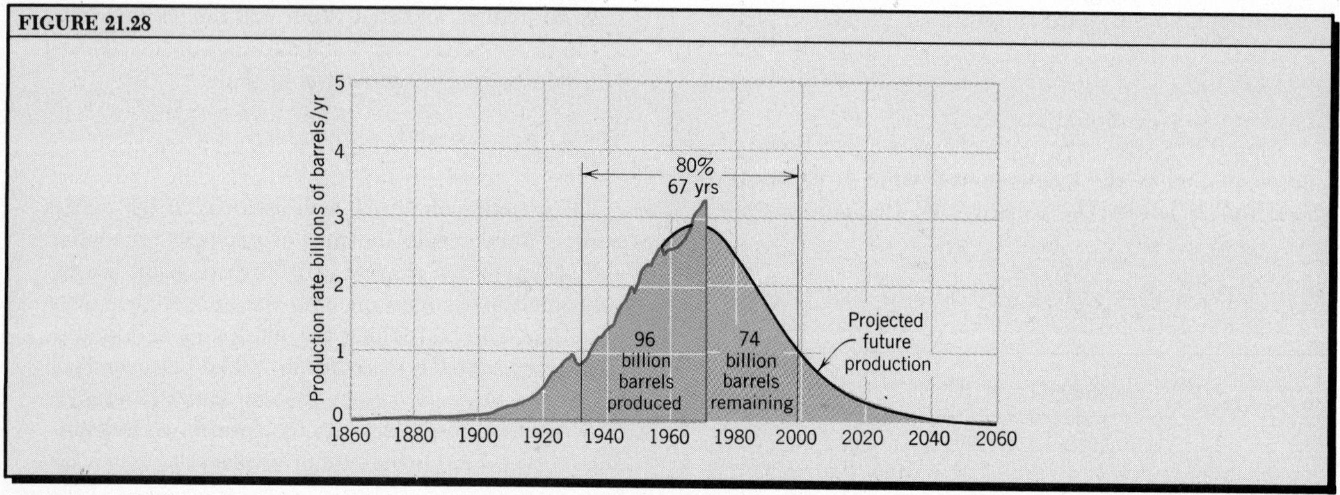

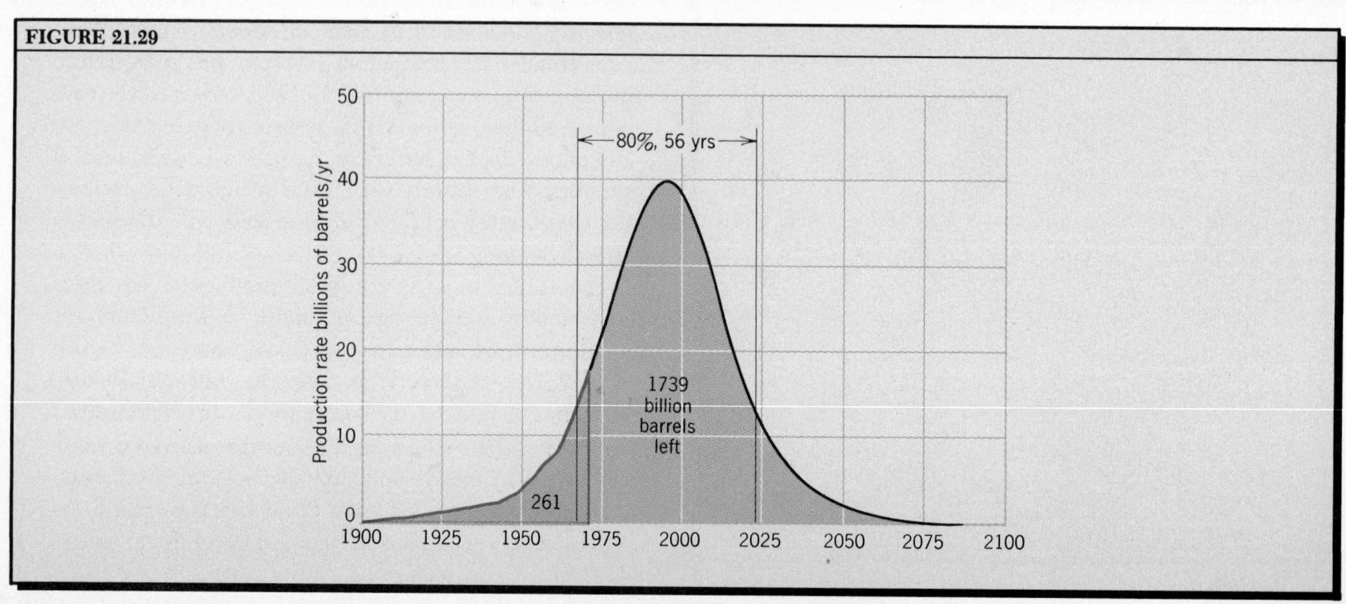

equivalent energy of 85 billion barrels of crude oil. Compare this figure with a U.S. proven crude oil reserve of 35 billion barrels. Compare this figure also with our estimate of recoverable U.S. coal reserves, equivalent to about 640 billion barrels of oil. Coal appears to be much the greatest U.S. energy resource.

The world supply of uranium ore would be rapidly exhausted if nothing else were used. It is, however, possible to induce fission in other isotopes, notably other isotopes of uranium and of plutonium and thorium. This induced fission, known as breeding, can greatly reduce the expenditure of uranium-235. At one time, geologists stressed the necessity of developing breeder reactors to conserve uranium. If such development were carried out, low-grade deposits of uranium could be exploited, making available a source of energy judged to range from hundreds to thousands of times greater than all reserves of fossil fuels. But because the breeder produces plutonium, a deadly radioactive substance, waste disposal is hazardous, and the future development of breeder reactors is in jeopardy.

Energy from *nuclear fusion* depends upon fusing isotopes of hydrogen into helium, with a consequent large release of energy. As everyone knows, the explosive release of enormous quantities of energy has been achieved through the hydrogen (thermonuclear) bomb. As yet, controlled release of energy through hydrogen fusion has not been achieved, although research is in progress. In theory, the quantities of energy available through fusion could exceed that of all fossil fuels by a factor ranging into the hundreds of thousands.

Nuclear-energy development is accompanied by a host of difficulties and attendant environmental problems. One is the generation of unwanted heat that raises water temperatures of rivers or lakes into which it is discharged, an activity known as thermal pollution, whose importance is the subject of debate. Of more far-reaching importance is the environmental problem of disposal of radioactive wastes from reactors and chemical plants that process nuclear fuels. There is also the possibility of accidental release of radioactive substances from reactors in nuclear power plants. Consideration of these problems brings into play all of the concepts of systems of water circulation within the hydrosphere, including runoff, groundwater movement, and oceanic circulation. In addition, atmospheric circulation systems may become involved, as in the case of fallout of radioactive particles released into the atmosphere by nuclear explosions.

FIGURE 21.30 Schematic diagram of petroleum seepage along a fault zone penetrated by an offshore well. (Redrawn, by permission of the publisher, from A. N. and A. H. Strahler, *Geography and Man's Environment*, Figure 61.6. Copyright © 1977 by John Wiley & Sons, Inc.)

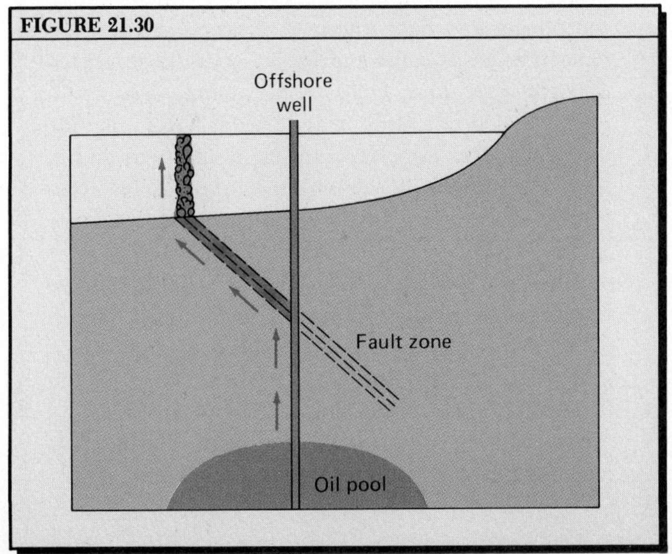

The Environmental Impact of Mineral Extraction

Extraction of mineral resources from the earth has many serious environmental impacts. We shall mention only a few of its effects here.

Deep scarring of the land and the accumulation of great heaps of waste rock (spoil) go hand in hand with open-pit mining of ores, quarrying of structural materials, and the strip mining of coal, clay, and phosphate rock. The strip mining of coal, in particular, poses one of the most formidable environmental problems, since it seems likely that much of our energy supply in the next few decades will come from coal. While the greatest part of this resource lies too deep for strip mining, there is strong pressure to strip mine all shallow coal as soon as possible. Besides deeply scarring the natural land surface and burying it under great masses of rock spoil, strip mining adversely affects the quality of water of streams and lakes deriving flow from strip-mining areas. Deep mining for coal also produces large spoil heaps, and there may be subsidence of the land above mine workings.

The extraction of petroleum may seem to have little impact on the land surface above an oil reservoir, but there are important undesirable side effects in some instances. As petroleum is withdrawn, compaction of the strata may set in, causing the land surface to be lowered. *Land subsidence,* as this phenomenon is known, amounted to more than 8 m in the harbor area of Long Beach, California, as a result of oil production over a quarter of a century. In a low-lying coastal area such as this, subsidence of the land may lead to flooding by ocean water, and expensive dikes and levees must be constructed to keep the salt water out.

Another side effect is the *oil spill,* occurring from offshore wells drilled in search of petroleum beneath continental shelves. Oil under pressure may force its way through faults in weak rock surrounding the well, to emerge uncontrolled from the ocean floor in a type of break often called a *blowout*. Figure 21.30 is a schematic drawing of such oil seepage. Many persons are familiar with the Santa Barbara, California, accident of 1969, in which a blowout occurred during drilling operations from an offshore platform. A large quantity of crude oil was released and severely polluted the beaches and harbor of that city. Renewal of drilling operations from offshore platforms threatens to bring more oil spills to this coastline.

Subsurface mining for any mineral resource, whether it be metallic ores or coal, has always been a hazardous business. Serious health hazards to miners exist through the inhalation of dusts consisting of pulverized coal or silicate minerals. In the case of uranium mining, there is the added hazard of exposure to ionizing radiation from the radioactive radon gas released in the natural decay of uranium. Another dangerous daughter product of radioactive decay is radium, which may be leached from the rock waste of uranium mining and processing operations to reach streams and render the water toxic. Added to these mining hazards are the possibilities of pollution of streams and lakes by leakage of radioactive isotopes generated in the processing of nuclear fuels and in the operation of nuclear power plants.

The combustion of fossil fuels, as we all know, is our major source of air pollution generally. Besides the emission of pollutant substances such as hydrocarbon compounds and oxides of nitrogen and sulfur, fuel combustion releases large amounts of heat into the atmosphere. Urban climates are known to be substantially changed as a result, and there may be important global effects upon climate in the future, as the quantities of fuel burned annually increase greatly. In addition, smelting of ores, particularly the sulfide ores, has been an important source of air pollution in areas surrounding smelters, and such pollution can have severe effects on vegetation and surface water.

Yet another environmental impact of the use of fossil fuels and nuclear fuels arises from the disposal of heat from large power-generating plants. Where water is used in large quantities as a coolant, it is released into streams, lakes, or estuaries at a much higher temperature than when it is withdrawn. The release of heated water gives rise to thermal pollution, often with serious consequences to aquatic life.

Earth Resources in Review

This chapter carries a message of warning to humankind. Industrial nations are ever more rapidly consuming natural resources that required millions of years of geologic time to bring into existence. As demands for materials and energy rise steeply in decades to come, rates of extraction of ore deposits and mineral fuels will be intensified. The richer, more easily extractable deposits will become exhausted, sending us in search of new deposits. We have stressed the dependence of the United States upon foreign nations for most of our industrial metals and for much of our petroleum.

Plate tectonics offers hope for the discovery of new, rich ore deposits. But unless this promise materializes, ores of poorer grade, and those occurring at greater depths, will be put into use, but with higher production costs and greater environmental damage. New oil reserves will be sought by geologists in localities much more difficult to develop, as in the case of offshore drilling in deeper waters, and in the case of the North Slope of North America, under a forbidding arctic climate.

Opinions also differ as to the quantities of mineral resources as yet undiscovered. In following the debate, your knowledge of the nature and occurrence of mineral resources should give you increased ability to evaluate the arguments accurately. You will also be more keenly aware of some of the environmental impacts to follow from increased intensity of mineral extraction.

Key Facts and Concepts

Nonrenewable earth resources *Nonrenewable earth resources* include *metalliferous deposits* and *nonmetallic deposits, fossil fuels,* and *nuclear fuels.* These deposits required millions of years to accumulate by slow geologic processes. *Economic geology* is the scientific study of mineral resources and their origin.

Metals in the earth's crust Metals in useful quantities occur as *ores,* sufficiently concentrated to be extracted at a profit. Magmas are the primary source of metallic ores. Whereas aluminum, iron, magnesium, and titanium are relatively abundant in crustal rocks, many essential industrial metals occur in extremely minute percentages, except where greatly concentrated in ores.

Metalliferous ores Major classes of ore deposits are produced by *magmatic segregation,* contact metamorphism, and *hydrothermal solutions.* Ores occur concentrated as *veins* and *lodes,* or as *disseminated deposits (porphyry copper). Secondary ores* result from action of subsurface water and include *laterite. Placer deposits* form in stream beds and in beaches *(marine placer). Sedimentary ores,* commonly of iron or manganese, occur in sedimentary rocks.

Ore deposits and plate tectonics Ore deposits have been related to plate tectonics in five basic types: (1) Cyprus-type deposits are formed by hydrothermal solutions in *ophiolitic suites* at spreading (accreting) plate boundaries. (2) Andean-type and (3) island-arc-type deposits form by rising magmas above subduction zones. (4) Intracontinental deposits may result from hot spots. (5) Suture-type deposits occur in zones of continental collisions. Diamond-bearing *kimberlite* occurs in *diatremes*.

U.S. metal demands and the world supply The United States is heavily dependent upon foreign supplies for many of the industrial metals it uses (examples: platinum, chromium, aluminum, manganese, tin, nickel). *Metals recycling* can substantially reduce the U.S. need to use certain imported primary metals.

Mineral resources from the seabed Several elements can be profitably extracted from seawater. Shallow continental shelves will become important sources of petroleum. Manganese nodules on the deep ocean floor offer promise of future exploitation. The prospect for major contributions of industrial minerals from the ocean floors in the near future is rather poor.

Nonmetallic mineral deposits Nonmetallic deposits include structural materials and minerals used chemically and in other industrial uses.

Sources of energy *Sustained-yield energy sources* include *solar energy, hydropower,* and *tidal power.* Constancy and reliability characterize solar energy, with excellent prospects for development. Hydropower and tidal power will be limited to minor roles.

Geothermal energy *Geothermal energy* presently exploited for electrical power is from steam and hot water sources in localities where hot springs, geysers, and fumaroles occur naturally. Deep hot water sources are found in zones of crustal spreading (rifting), where magma is rising at shallow depths. Hot, dry energy sources make use of deep holes to reach heated dry rock, such as granite, retaining magmatic heat. *Geopressurized zones* of hot water deep within the continental shelf wedge are a large potential energy source.

Coal and lignite The *fossil fuels*—coal, lignite, petroleum, and natural gas—are the present major energy resources. Major coal resources are in strata of Carboniferous and Cretaceous age; lignite, in Cenozoic strata.

Coal mining Subsurface mining follows *drifts;* coal is removed by the *room-and-pillar method* or the *long-wall method. Strip-mining* removes *overburden* to reveal the coal seam.

World coal reserves World coal reserves greatly exceed petroleum reserves, but are very unevenly distributed among the continents and political units. The U.S.S.R. and the United States are generously endowed with coal; Africa and Latin America have very little.

Petroleum and natural gas *Petroleum* and *natural gas* originate from organic matter accumulating in thick sediments, then migrate to reservoir rocks, where petroleum forms *oil pools,* usually overlain by an impervious *cap rock. Petroleum traps* include anticlines, sedimentary domes, *pinch-out trap* structures, *salt domes* above *salt plugs,* and *fault traps.*

Petroleum and plate tectonics Petroleum deposits can be interpreted through plate tectonics. In one case, continental rifting leads to salt deposition in a narrow ocean, followed by thick, organic-rich sediment accumulations. Rise of salt in diapirs then forms salt-dome traps. Continental collisions may incorporate petroleum accumulations into suture zones.

World petroleum reserves Proven world reserves of crude oil are largely concentrated in the Middle East, with important but much smaller reserves in the U.S.S.R. and China, Africa, and North America. Estimated world reserves are roughly three times greater than proven reserves.

Bituminous deposits *Bitumen,* petroleum of tarlike consistency, is held in *bituminous sand* deposits. These oil sands hold enormous reserves of energy, but require special technologies to extract petroleum in liquid form.

Oil shale *Oil shale* of the Rocky Mountain regions consists of *kerogen* disseminated through carbonate strata. When highly heated, kerogen becomes petroleum. Within the U.S. Rocky Mountain region alone, oil shale reservoirs in prime beds contain approximately four times as much petroleum as the proven U.S. petroleum (crude oil) reserves.

Future of fossil fuel resources According to one theory, production (and consumption) of petroleum, natural gas, and coal will follow bell-shaped curves, each rising to a peak and declining to near zero. Petroleum will peak first (about the year 2000), followed a century or two later by coal. Thus, coal promises to become our major hydrocarbon resource by the year 2050.

Nuclear energy as a resource *Nuclear fission,* our present method of using nuclear energy as a fuel, will rapidly consume our uranium resources unless *breeding* (induced fission) is soon developed. *Nuclear fusion,* while promising ultimately to provide unlimited energy, is far from achieving technological feasibility. Many environmental problems attend the development of nuclear energy.

Environmental impact of mineral extraction Mining creates environmental impact through deep scarring and large waste accumulations. Petroleum withdrawal may result in *land subsidence,* or in marine oil spills through *blowouts.* Air and water pollution, including *thermal pollution,* often accompany mining and smelting operations and the operation of power plants using fossil fuels and nuclear fuels.

Questions and Problems

1. Is it likely that iron (as steel) and the ferro-alloy metals will be largely replaced in future decades in

industrial use by synthetic substances, such as plastics, made from organic compounds and fossil fuels? Explain.

2. What major environmental impact can be expected from greatly increased combustion of coal to generate electricity as a replacement for dwindling petroleum resources and to replace hazardous nuclear power plants? Has solar power any unique long-term environmental advantages over other major energy sources? If so, what are they?
3. Of the geothermal energy sources, which offer the greatest potential in terms of quantity of available energy, assuming technological problems of recovery are solved? Would the withdrawal and use of dry, hot rock energy involve any major long-term environmental problems?
4. What are the major environmental problems connected with the large-scale development of coal resources in the Rocky Mountain region? Does oil shale development present similar problems?
5. Environmentalists have strongly opposed the expansion of oil extraction in the Santa Barbara Channel, where a disastrous oil spill occurred. Others contend that if the petroleum reservoirs of this field are developed and exploited, subsurface fluid pressure will be relieved and the danger of future oil spills will be eliminated. Where do you stand on this issue?

Bibliography

General References

Glossary of Geology. 2nd ed., R. L. Bates and J. A. Jackson, Eds., Amer. Geological Institute, Falls Church, Va., 749 pp., 1980.

Geology Illustrated. John S. Shelton, Freeman, San Francisco, 434 pp., 1966.

McGraw-Hill Encyclopedia of the Geological Sciences. D. N. Lapedes, Ed., McGraw-Hill, New York, 915 pp., 1978.

Chapter 1 Introduction to Geology

Adventures in Earth History. Preston Cloud, Ed., Freeman, San Francisco, 1015 pp., 1970. See Section II. Origin of the universe, solar system, and planets.

Continental Drift: A Study of the Earth's Moving Surface. Revised ed., Don and Maureen Tarling, Anchor Press/ Doubleday, Garden City, N.Y., 142 pp., 1975.

Continents in Motion: The New Earth Debate. Walter Sullivan, McGraw-Hill, New York, 399 pp., 1974.

Continents Adrift (Readings from *Scientific American*). Freeman, San Francisco, 172 pp., 1970.

Debate About the Earth. Revised ed., H. Takeuchi, S. Uyeda, and H. Kanamori, Freeman, San Francisco, 281 pp., 1970.

The New View of the Earth: Moving Continents and Moving Oceans. Seiya Uyeda, Freeman, San Francisco, 217 pp., 1978.

Chapter 2 Physical Science

College Physical Science. 3rd ed., V. W. Miles, H. O. Hooper, E. J. Kaczor, and W. H. Parsons, Harper & Row, New York, 655 pp., 1974.

Essentials of Physical Science. K. R. Atkins, J. R. Holum, and A. N. Strahler, Wiley, New York, 546 pp., 1978.

General College Chemistry. 6th ed., C. W. Keenan, D. C. Kleinfelter, and J. H. Wood, Harper & Row, New York, 880 pp., 1980.

Chapter 3 Minerals

Earth Materials. W. G. Ernst, Prentice-Hall, Englewood Cliffs, N.J., 149 pp., 1969.

An Introduction to the Rock-Forming Minerals. W. A. Deer, R. A. Howie, and J. Zussman, Wiley, New York, 528 pp., 1966.

Manual of Mineralogy (after James D. Dana). 19th ed., C. S. Hurlbut and C. Klein, Wiley, New York, 532 pp., 1977.

Chapter 4 Igneous Rocks, Geochemistry, General Petrology

Chemistry of the Earth. Karl K. Turekian, Holt, Rinehart and Winston, New York, 131 pp., 1972.

Encyclopedia of Geochemistry and Environmental Sciences. R. W. Fairbridge, Ed., Van Nostrand Reinhold, 1321 pp., 1972.

Guide to the Study of Rocks. 2nd ed., L. E. Spock, Harper & Row, New York, 298 pp., 1962.

Introduction to Petrology. Brian Bayly, Prentice-Hall, Englewood Cliffs, N.J., 371 pp., 1968.

Principles of Geochemistry. 3rd ed., Brian Mason, Wiley, New York, 329 pp., 1966.

Principles of Petrology. 12th ed., G. W. Tyrell, Chapman & Hall, London, 364 pp., 1972.

Rocks and Minerals. Brian Simpson, Pergamon Press, Oxford, 302 pp., 1966.

Textbook of Lithology. Kern C. Jackson, McGraw-Hill, New York, 552 pp., 1970.

Volcanoes. Cliff Ollier, M.I.T. Press, Cambridge, Mass., 177 pp., 1972.

Volcanoes of the Earth. Revised ed., Fred M. Bullard, Univ. of Texas Press, Austin, 579 pp., 1976.

Chapter 5 Sedimentology, Sedimentary Rocks

The Encyclopedia of Sedimentology. R. W. Fairbridge and J. Bourgeois, Eds., Dowden, Hutchinson & Ross, Stroudsburg, Pa., 901 pp., 1978.

Principles of Sedimentology. G. M. Friedman and J. E. Sanders, Wiley, New York, 972 pp., 1978.

Sedimentary Rocks. 3rd ed., F. J. Pettijohn, Harper & Row, New York, 628 pp., 1975.

Chapter 6 Stratigraphy, Structural Geology

Ancient Environments. Léo F. Laporte, Prentice-Hall, Englewood Cliffs, N.J., 115 pp., 1968.

Earth History and Plate Tectonics. 2nd ed., C. K. Seyfert and L. A. Sirkin, Harper & Row, New York, 600 pp., 1979.

Elements of Structural Geology. 2nd ed., E. Sherbon Hills, Chapman & Hall, London, 512 pp., 1972.

An Outline of Structural Geology. B. E. Hobbs, W. D. Means, and P. F. Williams, Wiley, New York, 571 pp., 1976.

Principles of Stratigraphy. C. O. Dunbar and J. Rodgers, Wiley, New York, 356 pp., 1957.

Stratigraphic Principles and Practice. J. Marvin Weller, Harper & Row, New York, 725 pp., 1960.

Chapter 7 Metamorphic Rocks, Geochronometry, Continental Crust

Evolution of the Earth's Crust. D. H. Tarling, Academic Press, London, 443 pp., 1978.

Geologic Time. 2nd ed., Don L. Eicher, Prentice-Hall, Englewood Cliffs, N.J., 150 pp., 1976.

Metamorphic Processes. R. H. Vernon, Wiley, New York, 247 pp., 1975.

Metamorphism: A Study of Transformations of Rock Masses. 4th ed., Alfred Harker, Chapman & Hall, London, 376 pp., 1974.

Chapter 8 Earthquakes and Seismology

The Alaska Earthquake March 27, 1964: Lessons and Conclusions. Edwin B. Eckel, U.S. Geological Survey Professional Paper 546, U.S. Government Printing Office, Washington, 564 pp., 1970.

Earthquake Country. Robert Iacopi, Lane Books, Menlo Park, Calif., 160 pp., 1971.

Earthquakes: A Primer. Bruce A. Bolt, Freeman, San Francisco, 241 pp., 1978.

Earthquakes and Earth Structure. John H. Hodgson, Prentice-Hall, Englewood Cliffs, N.J., 166 pp., 1964.

Elementary Seismology. Charles F. Richter, Freeman, San Francisco, 768 pp., 1958.

Predicting Earthquakes: A Scientific and Technical Evaluation—with Implications. Panel on Earthquake Prediction of the Committee on Seismology, National Research Council, Nat. Academy of Sciences, Washington, 62 pp., 1976.

The San Fernando, California, Earthquake of February 9, 1971. U.S. Geological Survey Professional Paper 733, U.S. Government Printing Office, Washington, 254 pp., 1971.

Chapter 9 Marine Geology, Oceanography

The Face of the Deep. B. C. Heezen and C. D. Hollister, Oxford University Press, New York, 659 pp., 1971.

Introduction to Marine Geology and Geomorphology. C. A. M. King, Edward Arnold, London, 309 pp., 1975.

Oceanography: An Introduction to the Marine Environment. Peter K. Weyl, Wiley, New York, 535 pp., 1970.

Oceans. Karl T. Turekian, Prentice-Hall, Englewood Cliffs, N.J., 120 pp., 1968.

Ocean Science (Readings from *Scientific American*). Freeman, San Francisco, 307 pp., 1977.

Submarine Geology. 3rd ed., Francis P. Shepard, Harper & Row, New York, 517 pp., 1973.

Chapters 10, 11, 12, and 13 Plate Tectonics

The Dynamic Earth: Textbook in Geosciences. Peter J. Wyllie, Wiley, New York, 416 pp., 1971.

Geological and Geophysical Investigations of Continental Margins. J. S. Watkins, L. Montadert, and P. W. Dickerson, Eds., A.A.P.G. Memoir 29, Amer. Assoc. of Petroleum Geologists, Tulsa, Okla., 472 pp., 1979.

The History of the Earth's Crust: A Symposium. Robert A. Phinney, Ed., Princeton Univ. Press, Princeton, N.J., 244 pp., 1968.

Island Arcs, Deep Sea Trenches, and Back-arc Basins. M. Talwani and W. C. Pitman III, Eds., Maurice Ewing Series 1, Amer. Geophysical Union, Washington, D.C., 470 pp., 1977.

Our Wandering Continents: An Hypothesis of Continental Drifting. Alexander L. DuToit, Hafner, New York, 266 pp., 1937.

Physics and Geology. 2nd ed., J. A. Jacobs, R. D. Russell, and J. T. Wilson, McGraw-Hill, New York, 622 pp., 1974.

Plate Tectonics (Selected papers from *The Journal of Geophysical Research*). J. M. Bird and B. Isacks, Eds., Amer. Geophysical Union, Washington, D.C., 951 pp., 1972.

Plate Tectonics (Selected papers from publications of the American Geophysical Union). 2nd ed., enlarged, John M. Bird, Ed., Amer. Geophysical Union, Washington, D.C., 986 pp., 1980.

Plate Tectonics and Geomagnetic Reversals. Allan Cox, Ed., W. H. Freeman, San Francisco, 702 pp., 1973.

A Revolution in the Earth Sciences: From Continental Drift to Plate Tectonics. A. Hallam, Clarendon Press, Oxford, 127 pp., 1973.

Tectonics of the Indonesian Region. Warren Hamilton, U.S. Geological Survey Professional Paper 1078, U.S. Government Printing Office, Washington, D.C., 345 pp., 1979.

Theory of Continental Drift: A Symposium on the Origin and Movement of Land Masses both Inter-continental and Intra-continental, as Proposed by Alfred Wegener. Amer. Assoc. of Petroleum Geologists, Tulsa, Okla., 240 pp., 1928.

Chapter 14 Geomorphology, Weathering, Mass Wasting, Soils

Encyclopedia of Geomorphology. R. W. Fairbridge, Ed., Reinhold, New York, 1295 pp., 1968.

Geomorphology from the Earth. Karl W. Butzer, Harper & Row, New York, 463 pp., 1976.

Landslides and Related Phenomena. C. F. S. Sharpe, Columbia Univ. Press, New York, 137 pp., 1938.

The Periglacial Environment. Troy L. Péwé, Ed., McGill-Queen's Univ. Press, Montreal, 487 pp., 1969.

Principles of Geomorphology. 2nd ed., William D. Thornbury, Wiley, New York, 594 pp., 1969.

Soil Genesis and Classification. S. W. Buol, F. D. Hole, and R. J. McCracken, Iowa State Univ. Press, Ames, 360 pp., 1973.

The Surface of the Earth. Arthur L. Bloom, Prentice-Hall, Englewood Cliffs, N.J., 152 pp., 1969.

Chapter 15 Ground Water, Karst

Ground Water Hydrology. D. K. Todd, Wiley, New York, 336 pp., 1959.

Karst. J. N. Jennings, The M.I.T. Press, Cambridge, Mass., 252 pp., 1971.

Chapters 16, 17 Fluvial Processes and Landforms

Desert Landforms. J. A. Mabbutt, The M.I.T. Press, Cambridge, Mass., 340 pp., 1977.

Fluvial Processes in Geomorphology. L. B. Leopold, M. G. Wolman, and J. P. Miller, Freeman, San Francisco, 522 pp., 1964.

The Fluvial System. Stanley A. Schumm, Wiley, New York, 338 pp., 1977.

Humid Landforms. Ian Douglas, The M.I.T. Press, Cambridge, Mass., 288 pp., 1977.

River Morphology. Stanley A. Schumm, Ed., Dowden, Hutchinson & Ross, Stroudsburg, Pa., 429 pp., 1972.

Slope Morphology. S. A. Schumm and M. P. Mosley, Eds., Dowden, Hutchinson & Ross, Stroudsburg, Pa., 454 pp., 1973.

Slopes. Anthony Young, Oliver & Boyd, Edinburgh, 288 pp., 1972.

Streams: Their Dynamics and Morphology. Marie Morisawa, McGraw-Hill, New York, 175 pp., 1968.

Chapter 18 Glaciers and Glaciation

Geology of the Great Lakes. Jack L. Hough, Univ. of Illinois Press, Urbana, 313 pp., 1958.

Glacial Geomorphology. C. Embleton and C. A. M. King, Edward Arnold, London, 573 pp., 1975.

Glacial and Quaternary Geology. R. F. Flint, Wiley, New York, 892 pp., 1971.

Glaciers and Landscape: A Geomorphological Approach. D. E. Sugden and B. S. John, Wiley, New York, 376 pp., 1976.

Ice Ages: Solving the Mystery. J. Imbrie and K. P. Imbrie, Enslow Publishers, Short Hills, N.J., 224 pp., 1979.

The Physics of Glaciers. W. S. B. Paterson, Pergamon Press, Oxford, 250 pp., 1969.

The Winters of the World: Earth Under the Ice Ages. Brian S. John, Ed., Wiley, New York, 256 pp., 1979.

The World of Ice. James L. Dyson, A. Knopf, New York, 243 pp., 1979.

Chapter 19 Wave Action, Wind Action

Beaches and Coasts. C. A. M. King, Edward Arnold, London, 403 pp., 1959.

Beach Processes and Coastal Hydrodynamics. J. S. Fisher and R. Dolan, Eds., Dowden, Hutchinson & Ross, Stroudsburg, Pa., 382 pp., 1977.

Coasts. E. C. F. Bird, The M.I.T. Press, Cambridge, Mass., 246 pp., 1969.

The Physics of Blown Sand and Desert Dunes. R. A. Bagnold, Methuen, London, 265 pp., 1941.

A Study of Global Sand Seas. Edwin D. McKee, Ed., U.S. Geological Survey Professional Paper 1052, U.S. Government Printing Office, Washington, 429 pp., 1979.

Waves and Beaches: The Dynamics of the Ocean Surface. Willard Bascom, Doubleday, Garden City, N.Y., 260 pp., 1964.

Chapter 20 Astrogeology

Geology of the Moon: A Stratigraphic View. Thomas A. Mutch, Princeton Univ. Press, Princeton, N.J., 324 pp., 1970.

An Introduction to Planetary Physics. William M. Kaula, Wiley, New York, 490 pp., 1968.

The Lunar Rocks. B. Mason and W. G. Melson, Wiley, New York, 179 pp., 1970.

Meteorites and the Origin of Planets. John A. Wood, McGraw-Hill, New York, 117 pp., 1968.

Moons and Planets: An Introduction to Planetary Science. William K. Hartmann, Wadsworth, Belmont, Calif., 404 pp., 1972.

Chapter 21 Mineral and Energy Resources

Earth Resources. Brian J. Skinner, Prentice-Hall, Englewood Cliffs, N.J., 149 pp., 1969.

Geology of Petroleum. 2nd ed., A. I. Levorsen, Freeman, San Francisco, 724 pp., 1967.

Geothermal Energy. H. C. H. Armstead, Chapman & Hall, London, 256 pp., 1978.

Geothermal Resources. Robert Bowen, Wiley, New York, 243 pp., 1979.

Mineral Resources. Kenneth Warren, Wiley, New York, 272 pp., 1973.

The Mineral Resources of the Sea. John L. Mero, Elsevier, Amsterdam, 312 pp., 1964.

Geologic Hazards, Environmental Geology

Cities and Geology. Robert F. Legget, McGraw-Hill, New York, 624 pp., 1973.

Environmental Geology: Conservation, Land-use Planning, and Resource Management. Peter T. Flawn, Harper & Row, New York, 313 pp., 1970.

Environmental Geomorphology. Donald R. Coates, Ed., Publ. in Geomorphology, State Univ. of New York, Binghamton, 262 pp., 1971.

Focus on Environmental Geology. 2nd ed., Ronald Tank, Ed., Oxford Univ. Press, New York, 538 pp., 1976.

Geological Hazards: Earthquakes, Tsunamis, Volcanoes, Avalanches, Landslides, Floods. 2nd ed., B. A. Bolt, W. L. Horn, G. A. Macdonald, R. F. Scott, Springer-Verlag, New York, 330 pp., 1977.

Geology in Environmental Planning. A. D. Howard and I. Remson, McGraw-Hill, New York, 478 pp., 1978.

Glossary of Geologic Terms

Glossary terms have been selected to emphasize principles, processes, and forms important in modern geology, with special attention to plate tectonics. To avoid excessive length, terms in the following categories do not appear: geologic time units and events; terms from basic physics, chemistry, and biology; mineral species, some minor rock types; minor textures, structures, and landforms; many terms in optional chapters 20 and 21.

A term not found in the glossary can be found in the text through the Index, in which an italicized page number locates the definition of the term.

ablation Wastage of glacial ice by both melting and evaporation.

abrasion Erosion of bedrock of a stream channel by impact of particles carried in a stream and by rolling of larger rock fragments over the stream bed. Abrasion is also an activity of glacial ice, waves, and wind.

absolute age Age of a mineral, rock, fossil, or geologic event in terms of actual number of years-before-present. (See also *relative age.*)

absolute plate motion Direction and rate of motion of a lithospheric plate relative to the entire lower mantle and core. (See also *relative plate motion.*)

abyssal plain Large expanse of very smooth, flat ocean floor found at depths of 4600 to 5500 m.

accreting margin Lithospheric plate margin along which new lithosphere is being formed by rise of basaltic magma from the underlying mantle in a zone of seafloor spreading.

accretion (planetary) Process of coming together under gravitation and adhering of particles of the solar nebula to form planetesimals.

accretion of lithosphere Production of new oceanic lithosphere at an active spreading plate boundary by the rise and solidification of magma of basaltic composition.

accretionary prism Mass of deformed trench sediments accumulated in imbricate wedgelike slices on the underside of the overlying plate by understuffing accompanying offscraping of the subducting plate. (See also *mêlange.*)

achondrites A class of stony meteorites possessing coarse-grained structure resembling terrestrial plutonic rocks; to be distinguished from the *chondrites,* another class of stony meteorites.

acidic lava Lava of felsic mineral composition, such as rhyolite.

active agents Fluid geomorphic agents, including running water, glacial ice, waves and currents, and winds; they perform erosion, transportation, and deposition.

active continental margins Continental margins that coincide with tectonically active plate boundaries. (See also *continental margins, passive continental margins.*)

aeon A time span of one billion years. This spelling used in Chapter 1 to designate the five billion-year units of elapsed geologic time. Can also be spelled *eon.* (See also *eon.*)

aftershocks Earthquakes that occur in a time period of hours to weeks or months immediately following the main earthquake and originating at or near the same focus. (See also *foreshocks.*)

aggradation Upbuilding of the floor of a stream channel by continued deposition of bed load.

alkalic feldspars Plagioclase feldspars (albite and oligoclase) with a high proportion of sodium as compared to the calcic feldspars with a high proportion of calcium.

alkali feldspars Feldspars in a solid solution group of aluminum silicates with sodium or potassium or some proportion of both. Examples: orthoclase, microcline.

alluvial fan Low, gently sloping, conical accumulation of coarse alluvium deposited by a braided stream undergoing aggradation below the point of emergence of the channel from a narrow canyon.

alluvial meanders Sinuous bends of a graded stream flowing in the alluvial deposit of a floodplain.

alluvial river River (stream) of low gradient flowing upon thick deposits of alluvium and experiencing approximately annual overbank flooding of the adjacent floodplain.

alluvial terrace Terrace carved in alluvium by a stream during degradation.

alluvium Any stream-laid sediment deposit found in a stream channel and in low parts of a stream valley subject to flooding.

alpine glacier Long narrow mountain glacier on a steep downgrade, occupying the floor of a troughlike valley.

alpine structure Intensely deformed strata characterized by the presence of recumbent folds, thrust sheets, and nappes, as in the European Alps and other similar orogens produced by continental collisions.

aluminosilicates Silicate minerals containing aluminum as an essential element. Example: feldspars.

amorphous mineral Mineral in the noncrystalline state, which may be a glass or gel. Example: opal.

amphibole group (amphiboles) Group of complex aluminosilicate minerals rich in calcium, magnesium, and iron, dark in color and classed as mafic minerals. Example: hornblende.

amphibolite Variety of metamorphic rock characterized by the presence of hornblende (an amphibole) and representing an intermediate grade of dynamothermal metamorphism on a regional scale.

andesite Extrusive igneous rock of diorite composition dominated by plagioclase feldspar of intermediate composition and with amphibole and pyroxene as important constituents; the extrusive equivalent of diorite.

angular unconformity Variety of unconformity in which younger strata above the unconformity lie upon the truncated edges of older strata with an angular discordance. (See also *unconformity, disconformity, nonconformity.*)

anorthosite Variety of igneous rock formed largely of plagioclase feldspar; a coarse-textured felsic igneous rock, usually occurring as a pluton.

anthracite Grade of coal very high in fixed carbon content, with little volatile matter, and of metamorphic development; also called hard coal.

anticline Upfold of strata or other layered rocks in an archlike structure; a class of folds. (See also *syncline.*)

aphanitic texture Igneous rock texture class in which crystals are too small to be distinguished as individual particles without the aid of a microscope; the crystals are smaller than about 0.05 mm in diameter. (See also *phaneritic texture.*)

aquiclude Rock mass or layer of low permeability that impedes or largely prevents the gravitational movement of ground water.

aquifer Rock mass or layer of both high porosity and high permeability that readily transmits and holds ground water.

arkose Variety of sandstone, also called *feldspathic sandstone,* characterized by having 30 percent or more of feldspar grains; it is derived by disintegration and partial chemical weathering of coarse-textured igneous or metamorphic rocks.

artesian flow Spontaneous rise of water in a well or fracture zone above the level of the surrounding water table.

ash flow A highly heated mixture of volcanic gases and frothed lava particles that moves rapidly downslope as a *glowing avalanche* (*nuée ardente*); upon coming to rest it becomes a layer of tuff or welded tuff.

asthenosphere Soft layer of the upper mantle, beneath the rigid lithosphere. Rock of the asthenosphere is close to its melting point and has low strength.

astrobleme An ancient, deeply eroded impact feature on the earth's solid surface produced by a large meteorite or an asteroid; literally translated as a "star wound."

astrogeology Branch of geology applying principles and methods of geology to all condensed matter and gases of the solar system outside the earth.

astronomical hypothesis Explanation for glaciations and interglaciations making use of cyclic variations in the form of the earth's orbit and the angle of inclination of the earth's axis as controls of the cyclic variations in the intensity of solar energy received at the earth's surface.

atoll Circular or closed-loop coral reef enclosing a lagoon of open water with no central island.

aulacogen Variety of geosyncline formed in a downfaulted continental crustal basin that may have originated as the failed arm of a triple junction produced during continental rupture. (See also *taphrogen.*)

available landmass Landmass above base level that can be consumed by fluvial denudation.

axial planar cleavage Cleavage in strongly folded shale or slate (slaty cleavage) paralleling the axial surfaces of the folds.

axial rift Narrow, trenchlike depression situated along the center line of the mid-oceanic ridge and identified with active seafloor spreading.

axial surface Imaginary surface that passes through the hinge lines of all bedding planes of an anticline or syncline.

axis of rotation of lithospheric plate Imaginary earth axis passing through the pole of rotation of a lithospheric plate.

backarc basin Comparatively small ocean basin, underlain by oceanic crust and lithosphere, lying between an island arc and the continental mainland or between two island arcs. Example: Sea of Japan.

backarc trough Long, narrow submarine trough lying between a volcanic arc and the continental craton or foreland; it contains a contemporary geosyncline associated with an active subduction boundary. Example: Banda Sea. (See also *forearc trough.*)

backarc trough deposit Thick deposit of terrigenous sediments accumulating in a backarc trough; a variety of geosyncline.

bank caving Incorporation of masses of alluvium or other weak bank materials into a stream channel because of undermining, usually in high flow stages.

barrier island Long narrow island, built largely of beach sand and dune sand, parallel with the mainland and separated from it by a lagoon.

basal slip Blocklike downvalley motion of an entire glacier, or large segment of a glacier, slipping over the bed.

basalt Extrusive igneous rock of gabbro composition; occurs as lava flows, sheeted dikes, shield volcanoes, or cinder cones.

basaltic cinder cone Small conical volcano built of basaltic tephra, usually without lava flows.

basaltic shield volcano (See *shield volcano.*)

base level Lower limiting surface or level that can ultimately be attained by a stream under conditions of stability of the earth's crust and sea level; an imaginary surface equivalent to sea level projected inland. (See also *local base level.*)

basic lava Lava of mafic mineral composition, usually basalt.

batholith Large discordant pluton (body of intrusive igneous rock), with an area of surface exposure greater than 100 sq km, and usually consisting of coarse-grained rock, such as granite.

beach Wedge-shaped accumulation of sand, gravel, or cobbles in the foreshore and offshore zones.

beach drift Transport of sand in the foreshore zone of a beach parallel with the shoreline by an alternating succession of landward and seaward water movements at times when the swash approaches obliquely.

bedding planes (See *stratification planes.*)

bed load That fraction of the total load of a stream being moved in traction, close to the bed of the channel.

bedrock Solid rock in place with respect to the surrounding and underlying rock and relatively unaltered or softened by weathering processes.

beds (See *strata.*)

Benioff zone. (See *Wadati-Benioff zone.*)

biofacies That aspect of the facies of a sedimentary rock unit relating to the character of the fossil fauna it contains. (See also *lithofacies.*)

biogenesis The event of origin of life on earth in Precambrian time. (Alternatively, the concept that all life forms originate from previously existing life forms.)

biogenic-pelagic sediments Class of deep-sea sediments consisting principally of calcareous or siliceous mineral matter secreted by organisms.

biogenic sediments (See *organically derived sediments.*)

bituminous coal Grade of coal with substantial content of volatiles; called soft coal.

block faulting Faulting accompanying crustal extension within the continents, leading to the dislocation of crustal blocks along normal faults and the occurrence of massive grabens, horsts, and fault block mountains.

block mountains (See *fault block mountains.*)

blueschist Variety of regional metamorphic rock formed under high confining pressure but comparatively low temperature; named for the presence of glaucophane, a bluish variety of amphibole.

body waves Seismic waves that travel within the solid earth, as a class different from surface waves. Body waves include *primary waves* and *secondary waves.*

bottom currents Currents sweeping over the deep floor of the ocean and capable of transporting coarse sediment, often along the contour of the bottom. (See also *contour currents.*)

bottom-transported detrital sediments Class of deep-sea sediments that includes *turbidites, contourites,* and *terrigenous muds.*

Bouguer anomaly Gravity anomaly remaining after the free-air, Bouguer, and topographic corrections have been made upon an observed gravity reading. (See also *gravity anomaly.*)

Bouguer correction A correction applied to a gravity reading to compensate for the effects of crustal masses lying above the ideal ellipsoidal surface or for deficiencies of mass below that surface.

Bowen reaction series A reaction series within a cooling silicate magma involving two converging branches: one of discontinuous reaction, the other of continuous reaction.

breccia General term for a rock consisting of angular rock fragments in a matrix of finer particles. (See *sedimentary breccia, volcanic breccia.*)

brittle solid Solid that fails by fracture (rupture) under applied shear stress when its elastic limit is exceeded. (See also *ductile solid.*)

brown clay Ocean-floor inorganic pelagic sediment; a detrital sediment consisting mostly of clay minerals.

Bunsen theory of geyser action Geyser action explained by the spontaneous phase change of hot water to steam as a water column is lifted in a subterranean tube, suddenly reducing the confining pressure and boiling point.

calcic feldspar Plagioclase feldspars rich in calcium, at the calcic end of the solid solution series.

caldera Large, steep-sided circular depression resulting from the terminal explosion and subsidence of a large composite volcano.

Caledonides Orogens associated with the Caledonian Orogeny, occurring near the close of the Silurian Period; they form ancient mountain roots in western Europe and northeastern North America.

capillary fringe Saturated layer lying at the top of the ground water zone and consisting of water held by capillary tension above the level that would represent a surface of hydrostatic equilibrium under gravity alone.

capillary water Water in the unsaturated zone and soil-water zone held in the form of capillary films attached to mineral particles.

carbonate rocks Rocks consisting of a high proportion of carbonate minerals, principally limestone and dolomite (dolostone).

carbonates (carbonate minerals) Minerals that are carbonate compounds of calcium or magnesium or both, i.e., calcite (calcium carbonate) or dolomite (magnesium carbonate).

carbonate platform High-standing, flat-topped submarine plateau consisting of reef carbonates and related carbonate materials formed by upbuilding during prolonged crustal subsidence of a passive continental margin. Example: Bahamas Platform.

carbonation Chemical reaction of carbonic acid in rainwater, soil water, and ground water with minerals; most strongly affects carbonate minerals and rocks, such as limestone and marble; an activity of chemical weathering.

carbon-14 Radioactive isotope of carbon, mass number 14, formed by cosmic ray bombardment of nitrogen in the upper atmosphere. (Synonymous with *radiocarbon.*)

cataclastic rocks Metamorphic rocks resulting from mechanical deformation without appreciable chemical change and recrystallization. (See also *friction breccia, mylonite.*)

caverns (See *limestone caverns.*)

cementation Lithification of sediment by mineral deposition in the interstices between mineral and rock grains.

chains of tetrahedra In the silicate minerals, the linkage of silicon-oxygen tetrahedrons by oxygen sharing to form chains that may be *single chains* or *double chains.*

chalk Variety of limestone that is soft, earthy, and white, formed of hard parts of various planktonic marine organisms (foraminifera, algae).

channel (See *stream channel.*)

channel flow (stream flow) Flowage of runoff to a lower level in a stream channel.

channel network (stream network) The total branching system of stream channels contributing discharge to a single trunk stream at its mouth or junction with a larger stream.

channel segments Within a channel network, the individual segments, or links, each of which lies between two consecutive channel junctions.

chemical weathering Chemical change in rock-forming minerals through exposure to atmospheric conditions in the presence of water; mainly involving oxidation, hydrolysis, carbonic acid reaction, and direct solution.

chert Variety of sedimentary rock composed largely of chalcedony and various impurities, in form of nodules and layers.

Chilean type subduction mode Pattern of plate subduction characterized by a low dip angle for the subducting slab, which presses strongly against the opposing, overlying plate and generates great earthquakes. (See also *Mariana type subduction mode.*)

chondrites A class of stony meteorites characterized by the presence of *chondrules,* rounded bodies consisting of olivine or pyroxene crystals.

cinder cone (See *basaltic cinder cone.*)

circum-Pacific belt Chains of andesitic volcanoes making up mountain arcs and island arcs surrounding the Pacific Ocean basin.

clastic sediment Sediment consisting of particles broken away physically from a parent rock source.

clastic sedimentary rocks Class of sedimentary rocks formed from clastic sediments through compaction and/or cementation.

clay minerals Class of minerals, produced by alteration of silicate minerals, having layer lattice atomic structure and plastic properties when moist.

claystone Sedimentary rock formed by lithification of clay and lacking fissile structure.

cleavage of mineral Property of a mineral to split readily along a set of parallel planes or along two or three intersecting sets of parallel planes. (See also *slaty cleavage.*)

cleavage planes Planar mineral surfaces along which parting occurs because of weaker atomic bonding in the direction of those surfaces.

coal Solid hydrocarbon compounds formed of compacted, lithified, and altered accumulations of terrestrial plant remains (peat), and interbedded with sedimentary rocks. (See also *anthracite, bituminous coal, lignite.*)

coal measures Stratigraphic sequences consisting of coal seams interbedded with shale, sandstone, and limestone.

coast (coastline) Zone in which coastal processes operate or have a strong influence.

coastal emergence The exposure of a belt of seafloor as a new land surface through crustal uplift (epeirogenic uplift, tectonic uplift) or a eustatic lowering of sea level. (See also *coastal submergence.*)

coastal plain Coastal belt, emerged from beneath the sea as a former continental shelf, underlain by strata with gentle dip seaward.

coastal submergence The drowning of a coastal belt of former land surface through crustal subsidence (epeirogenic downwarping, tectonic downfolding or downfaulting) or by eustatic rise of sea level. (See also *coastal emergence.*)

coastline (See *coast.*)

colliding resistance force Opposing (resisting) force exerted by a continental lithospheric plate against a subducting plate driven by a plate-moving force, such as a *ridge-push force.*

colloids (mineral) Mineral particles of extremely small size, capable of remaining indefinitely in suspension in water; typically in the form of thin plates or scales.

colluvium Deposit of sediment particles accumulating from overland flow at the base of a slope and originating from higher slopes where sheet erosion is in progress. (See also *alluvium.*)

columnar jointing Joint system of flat-faced columnlike masses of fine-textured igneous rock of a lava flow, sill, or dike, typically with 5 or 6 sides per column and transverse to the cooling surfaces, produced by volume shrinkage accompanying cooling.

compensation depth Oceanic depth, in the range of 3500 to 4000 m, at which sinking calcium carbonate tests begin to dissolve in seawater.

competent beds Beds (strata) that behave as brittle solids and fracture when subjected to folding, in contrast to the ductile behavior of *incompetent beds* in the same sequence.

composite volcano Volcano constructed of alternate layers of lava and tephra (volcanic ash); also called a *stratovolcano.*

compressional stress Stress tending to bring two reference points closer together along the line (stress axis) on which they lie. (See also *shear stress, tensional stress.*)

compressional waves (seismic) Seismic body waves in which the particles transmitting the wave move only forward and backward in the direction of wave travel, causing alternate compression and expansion of the medium; also known as *primary waves* or *P-waves.*

concentric folds Folds of strata in which the bedding surfaces maintain their original distances of separation as measured at right angles to those reference surfaces. (See also *similar folds.*)

concordant pluton Pluton bounded by natural surfaces of separation (such as bedding planes) in the enclosing country rock; it is thus a *tabular intrusion.* Example: sill. (See also *discordant pluton.*)

cone of depression Conical configuration of the lowered water table around a well from which water is being rapidly withdrawn.

confined aquifer Aquifer above which is an impermeable rock layer, an aquiclude, commonly of clay or shale in a sedimentary sequence.

conformable strata An unbroken succession of strata deposited in more or less continuous sequence under the same general environmental conditions.

conglomerate Clastic rock consisting of pebbles, cobbles, or boulders usually well rounded, in a matrix of sand or silt.

consequent stream Stream that takes its course down the slope of an initial landform, such as a newly emerged coastal plain or a volcano.

consumption of plate Destruction or disappearance of a subducting lithospheric plate in the asthenosphere, in part by melting of the upper surface, but largely by softening because of heating to the temperature of the surrounding mantle rock.

contact metamorphic rocks Class of metamorphic rocks formed by contact metamorphism.

contact metamorphism Rock metamorphism localized in country rock adjacent to intruding magma, resulting from intense heating or infusion of chemical solutions.

continental collision Event in plate tectonics in which subduction brings two segments of continental lithosphere into head-on contact, closing the intervening ocean basin and causing suturing with the formation of a continental suture.

continental crust Crust of the continents, of felsic composition in the upper part, thicker and less dense than oceanic crust.

continental denudation (See *denudation.*)

continental drift Hypothesis introduced by Alfred Wegener and others early in the 1900s of the breakup of a parent continent, Pangaea, starting near the close of the Mesozoic Era, and by drifting apart of the fragments resulting in the present arrangement of continents and intervening ocean basins.

continental foreland General term for that part of the stable continental crust immediately adjacent to an active tectonic belt, such as a subduction zone or a collision zone.

continental glacier (See *ice sheet.*)

continental lithosphere Lithosphere that bears continental crust. (See also *oceanic lithosphere.*)

continental margins (1) Topographic: one of three major divisions of the ocean basins, being the zones directly adjacent to the continent and including the continental shelf, continental slope, and continental rise. (2) Tectonic: marginal belt of continental crust and lithosphere that is in contact with oceanic crust and lithosphere, with or without an active plate boundary being present at the contact. (See also *active continental margin, passive continental margin.*)

continental nuclei Oldest crustal masses or patches of the continental shields, generally older than about 2½ billion years.

continental platform Sedimentary cover of Paleozoic and younger strata lying upon ancient continental shield rocks.

continental rise Gently sloping deep-seafloor lying at the foot of the continental slope and leading gradually into the abyssal plain.

continental rupture Extension affecting the continental lithosphere and its crust, so as to cause a rift-valley system to appear and to widen, eventually creating a new belt of oceanic lithosphere and a new ocean basin.

continental shelf Shallow, gently sloping belt of seafloor adjacent to the continental shoreline and terminating at its outer edge in the continental slope.

continental shields Deformed crustal rock masses of the continents, largely of felsic igneous and metamorphic rocks, and mostly of Precambrian age.

continental slope Steeply descending belt of seafloor between the continental shelf and the continental rise.

continental suture Long narrow zone of crustal deformation, including underthrusts and alpine structure, produced by continental collision. Example: Himalayas.

continental suturing Process of formation of a suture during continental collision.

continental underthrusting Forced descent of continental crust and lithosphere along a downsloping fault surface so as to pass beneath the edge of an adjacent continental lithospheric plate or its crust during a continental collision.

continents High-standing areas of continental lithosphere capped by continental crust. (See also *ocean basins.*)

continuity (See *principle of continuity.*)

continuous reaction Reaction within a cooling magma involving gradual composition change in a crystalline mineral by ion substitutions. (See also *discontinuous reaction.*)

contour currents Deep ocean currents flowing over the continental rise in a direction paralleling the contour of the slope, i.e., along the base of the continental slope.

contourites Class of deep-sea detrital sediments formed by the winnowing action of contour currents; the sediment is typically a well-sorted sand or coarse silt.

convection, convection currents General term for any overturning motions within fluids. Referring to the solid earth, convection currents are slow rising and sinking motions within the asthenosphere and deeper mantle.

convection system Total arrangement of a convection pattern into a system of closed loops of fluid motion.

converging plate boundary Boundary along which two lithospheric plates are coming together, requiring one plate to pass beneath the other by subduction; same as *subduction boundary.*

coral reef Rocklike accumulation of carbonate mineral matter secreted by corals and algae in shallow water along a marine shoreline.

core of earth Spherical central mass of the earth composed largely of iron and consisting of an outer liquid zone and an interior solid zone.

core of ocean floor Long, narrow cylindrical sample of ocean bottom sediment or rock, obtained by penetrating the seafloor with a length of open pipe, or by rotary drilling methods.

corner-flow model Ideal system of mantle convection in which rock of the asthenosphere is forced to flow in a tight bend between overlying and underlying plates of a subduction zone, causing heating and melting of the upper surface of the lower plate and producing magma.

country rock Rock that encloses a body of intrusive igneous rock (a pluton) and predates the intrusion.

crater (See *volcanic crater, lunar craters.*)

craton Stable continental crust, including both shield and continental platform areas, most of which have not been affected by significant tectonic activity since the close of the Paleozoic Era.

cross-bedding (See *cross-lamination.*)

cross-cutting relationships General class of structural and igneous features in which one rock body or tectonic structure intrudes or cuts across an older body or structure.

cross-lamination System of sloping or curved layers (laminae) within a massive sandstone unit, indicating deposition in a sand dune, a delta, or a sand bar in a stream bed or beach; also known as *cross-bedding.*

crustal roots (mountain roots) Those portions

of the continental crust extending deeply into the mantle beneath high-standing mountain and plateau areas.

crust of earth Outermost solid sheet or layer of the earth, composed largely of silicate minerals and ranging in rock composition from felsic rocks in the upper continental crust to mafic rocks in the oceanic crust. (See also *continental crust, oceanic crust.*)

cryptocrystalline Crystalline rock texture in which mineral crystals are too small to be distinguished except with the aid of a high-power microscope.

cryptovolcanic structures Small circular areas of highly disturbed and brecciated rock possibly produced by volcanic gas explosions.

crystal Ideal external form of a crystalline solid as seen in pure mass of the mineral compound allowed to grow freely in a surrounding gaseous or liquid medium.

crystal faces Planar surfaces forming the exterior of a crystal.

crystal habit The tendency of a crystalline mineral to repeatedly show preference for a particular growth form or combination of forms.

crystal lattice The three-dimensional internal structural arrangement of atoms or ions that forms a crystalline mineral.

crystalline limestone Granular limestone consisting of interlocking crystals of calcite, often formed by recrystallization of preexisting carbonate minerals during diagenesis under high confining pressure.

crystalline solid Matter in the solid state consisting of atoms locked into a regularly repeating three-dimensional space-lattice pattern.

crystallographic axes Imaginary axial straight lines that determine the position and orientation of crystal faces and cleavage planes for each of the several crystal systems.

crystallography Science of crystals and crystal structure.

crystal symmetry The similarity or balance of crystal faces on opposite sides of a plane of symmetry.

crystal system One of six classes of mineral crystals determined by a unique set and orientation of crystallographic axes.

cumulo-dome (See *viscous lava dome.*)

Curie point Critical temperature during cooling of magma, below which crystallized iron and titanium minerals become magnetized by lines of force of the earth's magnetic field.

dacite Extrusive igneous rock consisting of intermediate plagioclase feldspar (about 50 percent), amphibole, pyroxene, and minor quartz.

Darcy's law Average velocity of motion of ground water in a given flow path between two points is directly proportional to the hydraulic head, inversely proportional to the length of the flow path, and proportional to a coefficient of permeability.

daughter product (isotope) Isotope produced by radioactive decay of another element.

debris flood Streamlike flow of muddy water heavily charged with sediment of a wide range of size grades, including boulders, generated by sporadic torrential rains upon steep mountain watersheds.

decay series Regular succession of steps in radioactive decay of a given radioactive isotope, leading to a final stable isotope.

decomposition Alteration of rock-forming minerals in which those minerals are changed in chemical composition during exposure to the atmospheric environment. (See *chemical weathering.*)

deep-focus earthquakes Earthquakes with foci at depths between 300 and 650 km, largely confined to the Wadati-Benioff zone beneath converging plate boundaries.

deep-sea ooze (See *ooze.*)

deflation Lifting and transport in turbulent suspension by wind of loose particles of soil or regolith from dry ground surfaces.

deglaciation Widespread recession of ice sheets during a period of warming global climate, leading to an interglaciation. (See also *glaciation, interglaciation.*)

degradation Downcutting by a stream, causing the stream channel to be lowered in altitude, resulting in trenching of alluvial deposits or cutting of a bedrock gorge.

delta Sediment deposit built by a stream entering a body of standing water and formed of the stream's load.

dendritic pattern Drainage pattern of treelike branched form, in which the smaller streams take a wide variety of directions and show no parallelism or dominant trend.

dendrochronology Science of study of tree rings as a means to interpret past climate changes and establish the absolute ages of events in Holocene time.

density current Fluid current resulting from the tendency of a more dense fluid (or fluid suspension) to flow to lower levels under the influence of gravity, moving on the solid floor beneath the fluid body. (See also *turbidity current.*)

denudation Total action of all external processes whereby the exposed rocks of the continents are worn down and the resulting sediments are transported to the sea by the fluid agents; it includes also weathering and mass wasting.

depleted mantle Rock of the upper mantle that has been depleted of certain of its original chemical components by the rise of magma formed by partial melting to become new basaltic crust along spreading plate boundaries.

depositional landforms Sequential landforms created by the deposition of sediment by a fluid agent of denudation.

depth of compensation (See *compensation depth.*)

depth recorder (See *precision depth recorder.*)

detrital remnant magnetism Paleomagnetism in detrital sediments deposited in quiet water of lakes or the oceans, occurring when particles of magnetic minerals have oriented themselves with their axes parallel with the earth's magnetic force lines. (See also *thermal remnant magnetism.*)

detrital sediment Sediment consisting of mineral fragments derived by weathering of preexisting rock and transported to places of accumulation by currents of air, water, or ice.

detrital sedimentary rocks Sedimentary rocks formed from detrital sediment.

diagenesis Overall process of change in sediment in the process of transformation to sedimentary rock, including both physical and chemical changes. (See also *lithification.*)

diamond pipe Pipelike body of ultrabasic igneous rock (kimberlite) enclosing diamond crystals; the igneous body is also called a *diatreme.*

diatreme Narrow, pipelike body of ultramafic intrusive igneous rock formed by gaseous explosion emanating from great depth in the crust or upper mantle. (See also *diamond pipe.*)

dike Discordant pluton, often near-vertical or with a steep dip, occupying a widened fracture in the country rock, and typically cutting across older rock planes.

dike swarm Occurrence of a large number of igneous dikes in one locality, with parallel or radial strike orientation.

dilatancy An increase in the volume of pore space within a body of rock, resulting from strain applied to rock close to an active fault.

dilatation Form of rock deformation involving only a change in volume (expansion, contraction) but without change in shape.

diorite Intrusive igneous rock consisting dominantly of intermediate plagioclase feldspar and pyroxene, with some amphibole and biotite; a felsic igneous rock, it occurs as a pluton.

dip Acute angle between an inclined natural rock plane or surface and an imaginary horizontal plane of reference; always measured perpendicular to the strike. (Also a verb, meaning "to incline toward.")

dip-slip fault Geometrical class of fault in which all relative motion is in the direction of dip of the fault plane. (See also *oblique-slip fault, strike-slip fault.*)

discharge Volume of flow moving through a given cross section of a stream in a given unit of time; commonly given in cubic meters per second.

disconformity Surface of separation between

two formations of parallel strata, representing a large time gap, usually including an episode of erosional removal of part of the lower formation; a variety of *unconformity*. (See also *angular unconformity, nonconformity.*)

discontinuity As applied to the earth's interior structure, a horizontal surface or layer in which a rather abrupt change takes place in the speed of seismic waves; it represents an abrupt change in rigidity of the rock, associated with a change in density or in chemical composition. (See also *M-discontinuity, Moho.*)

discontinuous reaction Form of reaction in a cooling magma in which an early-crystallized mineral reacts with the remaining liquid to form a mineral of a different composition. (See also *continuous reaction, reaction pair.*)

discordant pluton Pluton with outer boundary cutting across structures of the country rock. Examples: batholith, stock, dike. (See also *concordant pluton.*)

disintegration Breakup of rock by mechanical weathering processes.

dolomite Carbonate mineral or sedimentary rock having the composition calcium-magnesium carbonate.

dolomitization Formation of dolomite from calcium carbonate seafloor sediments by substitution or addition of magnesium ions from seawater.

dolostone Alternative term for dolomite as a kind of rock.

dome (See *salt dome, stratigraphic dome.*)

double couple Two force couples oriented perpendicular to one another.

drainage divide Crest line surrounding a drainage basin and defining the limit of a watershed.

drainage pattern Geometrical pattern formed by a total stream channel network as depicted on a map. (See also *dendritic drainage pattern, rectangular drainage pattern, trellis drainage pattern.*)

drainage system A branched network of stream channels and adjacent ground slopes, bounded by a drainage divide and converging to a single channel at the outlet.

drawdown Difference in height between base of cone of depression and original water table surface.

drift (See *glacial drift, stratified drift.*)

drumlin Hill of glacial till, oval or elliptical in basal outline and with smoothly rounded summit, formed by plastering of till beneath moving, debris-ladened glacial ice.

ductile solid Solid substance that yields by plastic flowage without rupture under applied shear stress. Examples: most native metals. (See also *brittle solid.*)

ductility Property of ductile solids allowing them to be reshaped or stretched without rupture.

dunes (See *sand dunes.*)

dunite Ultramafic igneous rock consisting almost entirely of olivine.

dynamic metamorphism Form of rock metamorphism involving mechanical disruption without appreciable chemical change and giving rise to *cataclastic rocks*.

dynamo theory Earth magnetism explained by dynamo action within the earth's rotating liquid core.

dynamothermal metamorphism Regional rock metamorphism taking place at great depth through large-scale shearing under high pressure and high temperature during tectonic activity, with the gain or loss of mineral components.

earthflow Moderately rapid downhill flowage of masses of water-saturated regolith, clay, or weak shale, typically forming a steplike terrace at the top and a bulging toe at the base.

earthquake A trembling or shaking of the ground produced by the passage of *seismic waves*. (See *deep-focus earthquake, intermediate-focus earthquake, shallow-focus earthquake.*)

earthquake focus (pl. foci) Point within the earth at which the energy of an earthquake is first released by rupture and from which seismic waves emanate.

earthquake intensity scale Scale of numbers indicating the intensity of earth shaking felt at a particular place during an earthquake. (See *Mercalli scale, modified.*)

earthquake magnitude Strain energy released as kinetic energy by a single earthquake at its focus, in units of ergs or joules. (See also *Richter scale.*)

earthquake swarm Occurrence of many small earthquakes in a short time span from a small area; in some cases immediately preceding a large earthquake occurring in the same area.

eclogite A regional metamorphic rock containing an unusual green pyroxene rich in sodium and calcium and a red-brown variety of garnet (almandite) rich in magnesium; it forms under conditions of extremely high pressure and moderate temperature.

elastic limit The limit of strength of an elastic solid when strained, beyond which bending gives way to sudden rupture on shear fractures or undergoes some other permanent change in shape.

elastico-viscous Dual property of deep-seated rock to behave either as an elastic solid or a viscous fluid, depending upon the nature and duration of the applied stresses; applies to mantle rock in the asthenosphere.

elastic-rebound effect Behavior of crustal rock on either side of an active fault in which the rock is gradually elastically deformed prior to the rupture that produces an earthquake, but rebounds instantly to its original configuration when rupture occurs. (Part of the total rebound may occur gradually by creep over a long period of time.)

elastic solid Solid that exhibits strain in direct proportion to applied stress, but returns to its original shape when the stress is removed, provided that the elastic limit has not been exceeded.

elastic strain Strain characteristic of an *elastic solid*.

elevated shoreline Inactive (defunct) shoreline following its uplift above the level of wave action by a sudden crustal rise.

entrapment hypothesis Backarc basin explained by the formation of a new subduction boundary on the deep ocean floor, cutting off an area of oceanic lithosphere between the subduction boundary and the old continental margin. (See also *lateral drift.*)

entrenched meanders Winding, sinuous valley bends produced by degradation of a stream with trenching of the bedrock by downcutting.

eon Largest subdivision of geologic time, the two eons being the Cryptozoic and the Phanerozoic. *Eon* is also used to denote a time span of one billion years. (See also *aeon.*)

epeirogenic movement (epeirogeny) Slow rising or sinking of the crust over a large area, without appreciable faulting or folding.

epicenter Ground surface point directly above the earthquake focus.

epicontinental marginal sea Body of ocean water lying well within the continental limits and deeper than the continental shelves. Example: Gulf of Maine.

epoch (1) Unit of geologic time into which a period is subdivided. (2) Paleomagnetic time unit. (See *normal epoch, reversed epoch.*)

equigranular texture Igneous rock texture in which crystals are all within a comparatively small range of diameters.

era Subdivision of the eon of geologic time; the Phanerozoic Eon is subdivided into Paleozoic, Mesozoic, and Cenozoic eras.

erosion General term for the removal of mineral particles from exposed surfaces of bedrock or regolith by impact of a fluid—water, air, or ice—and by impact of solid particles carried by the fluid (abrasion). (See also *stream erosion, wind abrasion.*)

erosional landform Landform shaped by erosion.

erosion velocity Threshold of increasing velocity at which particles of a given grade size are first set in motion from traction on the bed of a stream.

eugeocline Geocline (geosyncline) of a passive continental margin consisting of a wedge-shaped detrital sediment deposit of the continental rise, resting on oceanic crust and thinning outward to the abyssal ocean floor. (See also *geocline, miogeocline.*)

eustatic sea-level change Worldwide change in sea level relative to the continental and insular shoreline by a rise or fall of the ocean surface, either because the volume of ocean water is changed or the total capacity of the ocean basins is changed.

evaporites Chemically precipitated sediments and sedimentary rocks composed of soluble salts deposited from evaporating salt water bodies.

event of magnetic polarity (See *magnetic event.*)

evolution (See *organic evolution.*)

exfoliation Development of sheets, shells, or scales of rock upon exposed bedrock outcrops or boulders, usually in concentric spheroidal arrangements.

exotic blocks Blocks of diverse kinds of solid rock enclosed in the native matrix of a mélange; for example, blocks of oceanic basalt in a matrix of deformed trench sediments. (See also *matrix, mêlange, native blocks.*)

expanding clays Clay minerals with a strong tendency to expand in volume when allowed to adsorb water. Example: montmorillonite.

explosion crater Volcanic crater formed by an outward explosion of volcanic gas rising from beneath. The term may also be applied to craters not associated with igneous rock but believed to have been formed by gas explosions. (See also *cryptovolcanic structure.*)

extension fracture Fracture formed by the yielding of a brittle solid to tensional stress.

external earth processes Processes powered by solar energy and acting upon rocks and minerals exposed to the atmosphere and hydrosphere; the processes involved in denudation. (See also *internal earth processes.*)

extinct transform (See *healed transform fault.*)

extrusive igneous rock Rock produced by the solidification of lava or ejected igneous fragments (tephra). (See also *intrusive igneous rock.*)

facies General term referring to the total or overall aspect of a complex structural arrangement of properties or substances, such as a series or sequence of strata. (See *biofacies, lithofacies.*)

failed arm Arm of a triple rift junction that was abandoned as a widening continental rift followed the other two arms to produce a new ocean basin. (See also *aulacogen.*)

fault Shear fracture surface in rock or regolith with displacement (shear) of block on one side with respect to the adjacent block. (See *normal fault, overthrust fault, transcurrent fault, transform fault.*)

fault block Blocklike crustal mass lying between two parallel normal faults. (See also *graben, horst.*)

fault creep More or less continuous slippage on a fault plane, relieving some of the accumulated strain.

faulting Process of formation of a fault or of continued movement on a fault; a form of tectonic activity. (See also *block faulting.*)

fault-line scarp Erosion scarp developed upon an inactive fault line.

fault plane Surface of shear fracture or slippage between two earth blocks moving relative to each other during faulting.

fault scarp Clifflike surface feature produced by faulting and exposing the fault plane; commonly associated with a normal fault. (See also *fault-line scarp.*)

fauna A natural assemblage of several animal species, used as a group to identify the age of a unit of strata. (See also *faunal zone, succession of faunas.*)

faunal zone Occurrence of a particular fauna at a certain level in a sequence of strata.

feldspar group Aluminosilicate mineral group containing one or two of the metals potassium, sodium, or calcium. (See *plagioclase feldspar, potash feldspar.*)

feldspathic sandstone (See *arkose.*)

feldspathoids Aluminosilicate minerals with sodium and potassium, but with less quartz than in the alkali feldspars, to which they are closely related. Example: *nepheline.*

felsic igneous rock Igneous rock dominantly composed of felsic minerals.

felsic minerals, felsic mineral group Quartz and feldspars treated as a silicate mineral group showing pale colors and relatively low density. (See also *mafic minerals.*)

fiord Narrow, deep ocean embayment partially filling a glacial trough.

firn Granular old snow forming a surface layer in the zone of accumulation of a glacier.

first motion The initial impulse that generates a seismic wave of the compressional type (P-wave); it reveals the relative direction of shearing motion on a fault plane.

fission hypothesis Hypothesis that the moon was formed out of the earth's mantle, torn from the earth by the centrifugal force of rapid earth rotation, by gravitational attraction exerted by a passing planetary object, or by massive impact of a planetary object.

fissure flows Basaltic lava flows that issue from fissures.

flocculation The clotting together of colloidal mineral particles to form larger particles; it occurs when a colloidal suspension in fresh water mixes with seawater.

flood basalts Large-scale outpourings of basalt lava to produce thick accumulations of basalt layers over a large area.

floodplain Belt of low, flat ground, present on one or both sides of a graded stream channel, subject to inundation by a flood about once annually and underlain by alluvium.

fluid agents Fluids that erode, transport, and deposit mineral and organic matter; they are running water, waves and coastal currents, glacial ice, and wind.

fluvial denudation Denudation occurring largely through the action of running water as runoff in overland flow and stream flow, together with weathering and mass wasting.

flysch Fine-grained detrital sediment that accumulates rapidly in great thickness in deep, narrow ocean troughs in early stages of an orogeny, when high adjacent mountain ranges have been elevated by the building of nappes. Turbidites are usually an important constituent of flysch.

focus (See *earthquake focus.*)

folding Process of formation of folds in layered rocks; a form of tectonic activity.

folds Wavelike corrugations of strata (or other layered rock masses) as a result of crustal compression, a form of tectonic activity. (See also *anticline, syncline.*)

foliation Crude layering structure or parting imposed on rock by dynamothermal metamorphism; seen commonly in schist.

foot wall (See *hanging wall.*)

force couple Pair of forces acting in parallel but opposite directions on the two sides of a shear plane or fault. (See also *double couple.*)

forearc trough Long, narrow submarine trough lying between a tectonic arc and a volcanic arc; it contains a contemporary geosyncline associated with an active subduction boundary along which an accretionary prism is being formed. Example: Savu Sea. (See also *backarc trough.*)

forearc trough deposit Thick accumulation of terrigenous sediments in a forearc trough; a variety of geosyncline.

foredeep Any deep narrow trough on the ocean floor, such as a trench formed along a subduction boundary, lying adjacent to a tectonic arc or volcanic arc.

foreland basement deformation Formation of isolated structural domes, anticlinal uplifts, and structural basins in platform strata of a craton by tectonic processes in a zone that is adjacent to an orogen affected by severe folding and thrusting.

foreland trough deposit Thick accumulation of coarse clastic sediments (molasse) in a subsiding trough over continental crust adjacent to rapidly elevated tectonic structures (thrust sheets, nappes) of an orogen resulting from continental collision; can be recognized as a variety of geosyncline. (See also *molasse.*)

foreshocks Earthquakes, usually of faint to low magnitude, that may occur in advance of a major earthquake. (See also *earthquake swarm.*)

foreshore Sloping face of a beach that is

within the zone of swash and backwash. (See also *offshore*.)

formation Distinctive layer or sequence of physically similar sedimentary strata deposited in a comparatively short span of geologic time and identified by a proper name.

fossil Ancient plant and animal remains or their traces or impressions preserved in sedimentary rocks. (See also *index fossil*.)

fossil fuels Collective term for coal, petroleum, and natural gas capable of being utilized by combustion as energy sources.

fractionation of magma Change in magma composition during cooling as crystallized minerals are physically separated from the melt.

fracture zone, oceanic Name first applied to linear fracturelike ocean floor features (scarps, ridges) offsetting the mid-oceanic ridge and its axial rift. Most of these are now recognized as active transform faults or healed transform faults. (See also *healed transform fault, transform fault*.)

frameworks Continuous networks of silicon-oxygen tetrahedra in certain silicate minerals, for example, quartz and the feldspars.

free-air correction Correction of a gravity reading for the elevation of the observing station above or below the reference ellipsoid; it amounts to a decrease of about 0.3 milligal per meter of ascent.

friction breccia Variety of cataclastic metamorphic rock produced by crushing and grinding of rock along a fault plane.

frost action Disintegration of rock or regolith by forces accompanying the freezing of water in pore spaces and fractures.

frost heaving Lifting of regolith or individual pebbles and cobbles in regolith by growth of vertical needlelike ice crystals formed during freezing of soil water.

fumarole Vent in earth's surface emitting volcanic gases, mostly steam, at high temperature.

gabbro Intrusive igneous rock consisting largely of pyroxene and calcic plagioclase feldspar, with variable amounts of olivine; a mafic igneous rock, occurring as a pluton.

garnet group Silicate minerals with aluminum, calcium, magnesium, iron or manganese, crystallizing in the isometric system, and commonly produced in dynamothermal metamorphism.

geanticline Long crustal arch, a product of tectonic upwarping, thought to have existed within a continental margin adjacent to a geosyncline. The term is used in the older or classical context of geosynclines.

gel mineral Noncrystalline form of mineral matter representing a gel that has become solidified by loss of water.

geochronometry Measurement of absolute ages of minerals and rocks, usually by radiometric age determinations.

geocline Class or type of geosyncline formed on a subsiding passive continental margin and the adjacent oceanic crust; it includes the *miogeocline* and the *eugeocline*.

geode Spherical mass of rock containing a central cavity lined with mineral crystals having freely developed faces and ends.

geologic map Map showing the surface extent and boundaries of rock units of different ages or lithologic types.

geology Science of the solid earth, including the earth's origin and history, materials comprising the earth, and processes acting within the earth and upon its surface. (See also *historical geology, physical geology*.)

geomagnetic axis Central axis of the earth's magnetic field, defining the magnetic poles.

geomagnetic polarity reversals Reversing of the north and south geomagnetic poles to define the beginning or end of a particular geomagnetic polarity epoch or event. (See also *normal epoch, reversed epoch*.)

geomagnetism Magnetic phenomenon of the earth as a planet, including the internal and external magnetic fields believed to be generated by motions within the liquid core.

geomorphology Science of landforms, including their history and processes of origin.

geophysics Branch of geoscience that applies principles and methods of physics to the study of the earth; includes specialties of seismology, earth magnetism, and gravity.

geopressurized zone Stratigraphic zone deep within the strata of a geocline beneath a coastal plain in which abnormally high fluid pressure exists in heated ground water.

geoscience Common synonym for geology, but sometimes used to include atmospheric, oceanic, and space sciences along with traditional geology.

geosyncline Thick accumulation of sediments, lens-shaped in cross section, deposited in a long, narrow trough paralleling the continental margin to the accompaniment of slow crustal subsidence.

geothermal energy Heat energy of igneous origin drawn from steam, hot water, or dry hot rock beneath the earth's surface.

geothermal gradient The rate at which temperature increases with increasing depth beneath the earth's solid surface.

geothermal locality Place where geothermal heat reaches the earth's surface, emanating from a deep-seated magma body in the crust.

geyser Periodic jetlike emission of hot water and steam from a narrow vent at a geothermal locality.

glacial-control theory Hypothesis of origin of barrier reefs and atolls stating that wave planation at time of lowered sea level during glaciations was followed by upbuilding of reefs during post-glacial rise of sea level.

glacial drift General term for all varieties and forms of rock debris deposited in close association with Pleistocene ice sheets.

glacial-marine sediments Detrital sediments enclosed in icebergs which drift into the open ocean and melt, allowing the solid particles to fall to the deep ocean floor and become a sediment layer.

glacial plucking Removal of masses of bedrock from the floor beneath an alpine glacier or ice sheet as ice moves forward suddenly by basal slip.

glacial trough Deep, steep-sided rock trench of U-shaped cross section formed by alpine glacier erosion.

glaciation (1) General term for the total process of growth of glaciers and the landform modifications they make. (2) Single episode or time period in which ice sheets form and spread widely, as contrasted with an interglaciation.

glacier Large natural accumulation of land ice affected by present or past flowage. (See also *alpine glacier, outlet glacier*.)

glacier equilibrium State of balance in the activity of a glacier in which the rate of nourishment equals the rate of ablation and the terminus holds a fixed position.

glowing avalanche Rapidly moving tongue of highly heated gases and ash produced by explosive eruption of a composite volcano; also called a *nuée ardente*.

gneiss Textural class of regional metamorphic rock showing banding and commonly rich in quartz and feldspar. (See also *granite gneiss*.)

Gondwana (Gondwanaland) Hypothetical southern hemisphere continent existing throughout Paleozoic and early Mesozoic eras, consisting of continental shields of South America, Africa, India, Madagascar, Australia, and Antarctica; part of Pangaea. (See also *Laurasia*.)

graben Trenchlike depression representing the surface of a fault block dropped down between two opposed, infacing normal faults. (See also *rift valley*.)

graded bedding Sedimentary structure within a turbidite in which particle size grades from coarse at the base to fine at the top, representing the continuous deposition from a single turbidity current.

graded channel (graded stream) Stream with its gradient adjusted to achieve a balanced state in which the average bed load transport rate is matched to the average bed load input rate.

granite Intrusive igneous rock consisting largely of quartz, potash feldspar, and sodic plagioclase feldspar, with minor amounts of biotite and hornblende; a felsic igneous rock, occurring as a pluton.

granite-gabbro series Igneous rock series ranging in composition from felsic igneous rock (granite) to mafic igneous rock (gabbro).

granite gneiss Variety of gneiss having the composition of granite.

granite pegmatite Pegmatite of the approximate composition of granite.

granitic rock General term for rock of the upper layer of the continental crust, composed largely of felsic igneous and metamorphic rock; rock of composition similar to that of granite.

granitization Transformation of preexisting crustal rock into granite by metasomatism (infusion of chemical solutions) but without melting; an alternative to the magmatic origin of granite.

granodiorite Felsic igneous rock of composition intermediate between granite and diorite; it contains about equal proportions of quartz, plagioclase feldspar, and potash feldspar.

granular disintegration Grain-by-grain breakup of the outer surface of coarse-grained rock, yielding gravel and leaving behind rounded boulders.

granulite Textural term for metamorphic rock consisting of small mineral grains of more or less equal size, developed through shearing action in regional metamorphism. (See also *pyroxene granulite*.)

gravitational collapse In a condensing solar nebula, the coming together of particles of matter through their mutual gravitational attraction; the process by which the sun became the central mass of the solar system.

gravity Gravitational attraction of the earth upon any small mass near the earth's surface; also applies to the moon. In practical usage or geophysics, it is the acceleration of gravity, g, at the earth's surface.

gravity anomaly Difference between observed and predicted values of gravity at a given station, after certain corrections have been applied. (See also *Bouguer anomaly, isostatic gravity anomaly*.)

gravity anomaly profile Graph showing the variations in a particular gravity anomaly observed along a line of traverse.

gravity gliding The forward movement of a thrust sheet or nappe on a low downgrade under the influence of gravity during orogeny.

gravity well Drilled or dug well in which the standing water level coincides with the level of the surrounding water table.

graywacke Traditional name for lithic sandstone, containing fine-grained rock particles and gray to black color. Graywacke is now widely recognized as consisting of turbidites. (See *lithic sandstone, turbidites*.)

greenschist Regional metamorphic rock of low grade with schist texture in which dominant minerals are chlorite, muscovite mica, biotite, sodic plagioclase, and quartz.

ground ice General term for ice crystals and ice bands formed within the soil or regolith.

groundmass In a porphyry, the fine-textured or aphanitic rock material that surrounds phenocrysts.

ground water Subsurface water occupying the saturated zone and moving under the force of gravity.

ground water recharge Replenishment of ground water by downward movement of water through the unsaturated zone, or from stream channels, or through recharge wells.

guyot Flat-topped seamount; a former volcano, presumed to have been beveled by wave action and later submerged by crustal subsidence.

hanging wall In the case of an inclined fault, the rock surface that would overhang the vertical if it were to be exposed by mining excavation; it opposes the *foot wall*.

healed transform fault Trail or scar left on the surface of the oceanic crust as movement continues on a transform fault between the offset ends of a segment of an oceanic spreading boundary; also called a *transform scar* or an *extinct transform*. When first discovered, healed transform faults were identified as submarine *fracture zones*.

heavy minerals Group of minerals having exceptionally high density, usually 4 g/cm^3 and greater, typically occurring in detrital sediments. Example: magnetite.

hinge line In folds of strata, the line connecting all points of maximum curvature on a single bedding surface at the crest or trough of the fold.

historical geology Major division of the science of geology dealing with the events of geological history from earliest Precambrian time to the present, with particular emphasis on paleogeography as interpreted from strata (stratigraphy) and the progress of organic evolution revealed through the study of fossils (paleontology). (See also *physical geology*.)

hornfels Contact metamorphic rock, fine-grained, dark in color, and with hornlike texture, commonly derived from shale in direct contact with an invading pluton.

horst Fault block uplifted between two normal faults.

hot spot Center of persistent volcanic activity usually located within a lithospheric plate and not related to volcanism of either a subduction boundary (volcanic arc) or an oceanic spreading boundary; postulated to be formed over a rising mantle plume. Examples: Hawaii, Yellowstone Park.

hot springs Springs discharging heated ground water at a temperature approaching the boiling point; usually related to igneous rock intrusion at depth. (See also *warm springs*.)

hydraulic action Stream erosion by impact force of the flowing water upon alluvium or regolith exposed in the bed and banks of the stream channel.

hydraulic gradient In ground-water flow, the ratio between the hydraulic head and the length of flow path. (See also *Darcy's law, hydraulic head*.)

hydraulic head Difference in level of the water table between one point and another, or of height of one point on the ground-water flow path with respect to a lower point on that path, setting up a pressure difference and causing the flow of ground water.

hydrogenic sediments Sediments that are chemical precipitates from an aqueous solution.

hydrologic cycle Total plan of movement, exchange, and storage of the earth's free water in gaseous, liquid, and solid states.

hydrolysis Chemical union of water molecules with minerals to form different, usually more stable mineral compounds, classed as hydroxides.

hydropower Renewable energy source using the kinetic energy of water moving to lower levels in streams under the influence of gravity.

hydrosphere Total water realm of the earth's surface zone, including the oceans, surface waters of the lands, ground water, and water held in the atmosphere.

hydrostatic pressure surface Imaginary surface that represents the level to which ground water will rise if free to do so in a well or tube penetrating the ground water body.

hydrothermal alteration Chemical change in rocks and minerals caused by the action of hydrothermal (hot-water) solutions rich in volatiles that rise from a cooling magma body.

hydrothermal stage Final stage in magma crystallization when hydrothermal solutions carrying volatiles rise to become deposited as veins in the solidified magma or in the adjacent country rock. (See also *granite pegmatite*.)

hydrous silicates Class of silicate minerals containing the hydroxide ion or water molecule in their chemical composition, derived by hydrolysis of other silicate minerals. (See also *hydrolysis*.)

hydroxides Class of minerals that are hydrous oxides of the metals, most commonly hydroxides of aluminum, iron, and manganese. Example: limonite.

ice age A span of geologic time, usually on the order of one to three million years, or longer, in which glaciations alternate with interglaciations repeatedly in rhythm with cyclic global climate changes. (See also *interglaciation, glaciation*.)

icecap Platelike mass of glacial ice limited to

the high summit region of a mountain range or plateau; a variety of glacier.

ice sheet Large thick plate of glacial ice moving outward in all directions from a central region of accumulation; also called a *continental glacier.* Example: Greenland Ice Sheet.

ice shelf Thick plate of partially floating glacial ice attached to an ice sheet and fed by the ice sheet and by snow accumulation.

igneous rock Rock solidified from a high-temperature molten state; rock formed by cooling of magma. (See *extrusive igneous rock, felsic igneous rocks, intrusive igneous rocks, mafic igneous rocks, ultramafic igneous rock.*)

illite group Clay mineral group derived by chemical weathering from such silicate minerals as feldspar and muscovite mica.

imbricate thrusts Succession of thrust sheets stacked one against the next in an overlapping arrangement.

impact metamorphism Formation of lunar breccia by fragmentation and lithification of lunar rock during impacts that formed lunar craters.

impermeable rock Rock of very low permeability. Example: *shale.* (See *aquiclude, permeability.*)

inactive continental margins (See *passive continental margins.*)

inclined folds Folds in which the axial surfaces are dipping. (See also *overturned folds, recumbent folds, vertical folds.*)

incompetent beds Beds (strata), usually of shale, that show plastic behavior when deformed between enclosing *competent beds* during folding.

independent tetrahedra (isolated tetrahedra) In the silicate minerals, crystal lattice structure in which the silicon-oxygen tetrahedra are isolated from one another and do not share oxygen ions.

index fossil Fossil species particularly well suited, because of its limited range in time, to establish the geologic age of a stratigraphic unit.

index mineral In regional metamorphic rocks, a distinctive mineral whose first appearance defines a *metamorphic zone.*

initial landforms Landforms produced directly by internal earth processes of volcanism and tectonic activity. Examples: volcano, fault scarp. (See also *sequential landform.*)

injection gneiss (See *migmatite.*)

intensity scale (See *earthquake intensity scale.*)

interglaciation Within an ice age, a time interval of mild global climate in which continental ice sheets were largely absent or were limited in extent to the Greenland and Antarctic ice sheets; the interval between two glaciations. (See also *deglaciation, glaciation.*)

intermediate-focus earthquake Earthquake with focus at a depth between 55 and 240 km.

intermediate lavas Lavas of composition intermediate between felsic and mafic igneous rocks; the common variety is rhyolite.

internal earth processes Geologic processes acting within the earth's interior, powered by internal earth forces and deriving energy from internal heat sources, such as radioactivity (radiogenic heat). Volcanism and tectonic activity are the manifestations of the internal processes. (See also *external earth processes.*)

intrusive igneous rocks Igneous rocks formed by the solidification of magma beneath the surface in contact with older rock (country rock). (See also *extrusive igneous rocks.*)

irons Class of meteorites composed almost entirely of nickel-iron alloy.

island arc Chain of islands paralleling a subduction boundary and formed of volcanic rocks or rocks of accretionary prisms. (See also *tectonic arc, volcanic arc.*)

isochrone On a stratigraphic cross section, a line drawn through all points where sediment deposition was occurring simultaneously.

isoclinal folds Tight folds in which the opposing limbs are more or less parallel.

isograd Line on a geologic map showing the appearance of each index mineral in a sequence of metamorphic grades representing increasing temperature of recrystallization. (See also *index mineral, metamorphic grades.*)

isolated tetrahedra (See *independent tetrahedra.*)

isoseismal Line on a map drawn through all surface points having equal value of earthquake intensity, usually given in numbers on the modified Mercalli scale.

isostasy Equilibrium state, resembling hydrostatic flotation, in which crustal and lithospheric masses stand at relative levels determined by their thickness and density, equilibrium being achieved by plastic flowage of denser mantle rock of the underlying asthenosphere.

isostatic compensation Crustal rise or sinking in response to unloading by denudation or loading by sediment deposition, following the principle of isostasy.

isostatic correction Gravity correction taking into account the presence of crustal and mantle layers of differing thicknesses and densities. (See also *isostatic gravity anomaly.*)

isostatic equilibrium (See *isostasy.*)

isostatic gravity anomaly Gravity anomaly remaining after an isostatic correction has been added to (or subtracted from) all other corrections through the Bouguer correction.

isotope-ratio stages Time units of the geologic past determined by cyclic fluctuations in the *oxygen-isotope ratio;* they have been numbered from 1 through 41 to cover the past two million years of record based on plankton from deep-sea cores.

jointing Presence of one or more systems of joints in bedrock.

joints Internal bedrock fracture surfaces of hairline thickness along which no shearing has occurred.

kame Hill composed of sorted coarse water-laid glacial drift, largely sand and gravel, built into an impounded water body within stagnant ice or against the margin of an ice sheet.

kaolinite group Mineral group of which kaolinite is a principal member.

karst Landscape or class of topography dominated by surface features of limestone solution and underlain by limestone cavern systems.

kimberlite Intrusive igneous rock of peridotite composition found in a diatreme, or volcanic pipe, and which may contain diamond crystals.

laccolith Concordant pluton, lenslike in cross section and circular in plan, with a flat floor and domed upper surface, intruded into sedimentary rocks so as to lift the overlying beds.

laccolithic dome Structure of uparched sedimentary rocks caused by intrusion of a laccolith. (See also *stratigraphic dome.*)

laminar flow (See *streamline flow.*)

landforms Configurations of the land surface taking distinctive forms and produced by natural processes. Examples: hill, valley, plateau. (See *depositional landforms, erosional landforms, initial landforms, sequential landforms.*)

landmass Large area of continental crust lying above sea level (base level) and thus available for removal by denudation. (See also *available landmass.*)

landslide Rapid sliding of large masses of bedrock on steep mountain slopes or from high cliffs. (See also *rockslide, slump, slump blocks.*)

lateral drift of island arc Slow relative horizontal motion of an island arc and its subduction zone away from a continental margin, so as to open up a new backarc basin. (See also *entrapment hypothesis.*)

lateral planation Sidewise cutting of a shifting stream channel, usually occurring on the outside of a bend.

laterite Rocklike layer rich in sesquioxide of aluminum and iron, including bauxite and limonite, found in low latitudes.

Laue photograph X-ray photograph of a mineral crystal in which the diffracted rays form a regular geometric pattern of spots or dots, thereby revealing the crystal lattice structure.

Laurasia Northern hemisphere continent existing throughout the late Paleozoic and early Mesozoic eras, consisting of what are now continental shields of North America and Eurasia; part of Pangaea. (See also *Gondwana.*)

lava Magma emerging on the earth's solid surface, exposed to air or water. (See also *acidic lavas, basic lava, intermediate lavas.*)

lava flow Outpouring of molten lava upon the earth's surface or seafloor, congealing to form extrusive igneous rock.

lattice structure (See *sheet structure.*)

lignite Low grade of coal, intermediate in development between peat and coal; also called brown coal.

limb of fold Flank or side of a fold in strata; that part of the bedding midway between the adjacent regions of the hinge lines.

limestone Nonclastic carbonate sedimentary rock in which calcite is the predominant mineral, and with varying minor amounts of magnesium carbonate, silica, and clay minerals. (See also *cherty limestone, crystalline limestone, fragmental limestone, oolitic limestone, reef limestone.*)

limestone caverns Subterranean systems of interconnected large cavities formed in limestone by the carbonic acid action (carbonation) of circulating ground water.

lineation Structure in gneiss in which individual mineral crystals have been drawn out into parallel pencillike bodies.

liquefaction of clays (See *spontaneous liquefaction.*)

lithic sandstone Type of sandstone containing 15 percent or more of fine-grained rock particles; it is dark in color and speckled in appearance; structure is typically that of turbidites. (See also *graywacke.*)

lithification Process of hardening of sediment to produce sedimentary rock. (See also *diagenesis.*)

lithofacies The sum of the physical qualities of a particular unit of strata, including texture, mineral composition, type of bedding, and typical structures within beds. (See also *biofacies, facies.*)

lithofacies-time diagram Graphical representation of changes in sedimentary lithofacies with the passage of time along a fixed line of cross section.

lithosphere (1) General term for the entire solid earth realm. (2) In plate tectonics, the strong brittle outermost earth shell, lying above the asthenosphere.

lithospheric plate Large segment of the lithosphere moving as a unit in contact with adjacent lithospheric plates along plate boundaries.

littoral cell Coastal sediment flow system in which fluviatile sediment provides the basic input and discharge of beach sediment through a submarine canyon to the deep ocean floor provides the output.

littoral drift Transport of sediment parallel with the shoreline by the combined action of beach drift and longshore current transport.

littoral environment Environment of sediment deposition occupying the intertidal zone of the marine coastline. (See also *marine environment, terrestrial environment.*)

load (See *stream load.*)

local baselevel Threshold of resistant bedrock at a point on the channel of a stream serving as a control on gradation of the upstream portion of the stream profile.

locked fault Segment of a transcurrent fault on which elastic strain is accumulating without being relieved by slippage; term is applied to sections of the San Andreas Fault.

loess Yellowish to buff-colored, fine-grained sediment, largely of silt grade, deposited upon upland surfaces after transport by wind in a dust storm.

longitudinal stream profile Graphic representation of the descending course of a stream or stream channel; altitude is plotted on the vertical scale, downstream distance on the horizontal scale.

longitudinal wave Form of seismic wave motion found in the primary wave (P-wave).

longshore current Current in the breaker zone, set up by the oblique approach of waves and running parallel with the shoreline.

longshore drift Littoral drift caused by action of a longshore current.

lopolith Concordant pluton, more or less circular in outline and saucer-shaped, conforming with the downward curvature of enclosing rock layers above and below.

Love wave Type of surface seismic wave in which motion is entirely horizontal and at right angles to the direction of wave motion. (See also *Rayleigh wave.*)

lower mantle That part of the mantle lying below the P-wave discontinuity, which occurs at a depth of 650 km and may signify a phase transition. (See also *mantle, upper mantle.*)

low velocity zone Zone in the upper mantle, starting at a depth of about 60 km and centered at about 150 km, through which there is a downward decrease in seismic wave velocity, suggesting a layer of diminished rock strength identified with the asthenosphere.

lunar breccia Lunar rock consisting of angular rock fragments enclosed in smaller rock fragments, formed by impact metamorphism during lunar cratering.

lunar craters Circular, rimmed depressions on the moon's surface, largely formed by the impact of large meteoroids and asteroids.

lunar highlands (terrae) Light-colored, upland regions of the moon's surface, as distinguished from the dark areas of maria.

lunar maria Dark-colored plains of the moon's surface underlain by basalt. (Latin *mare:* sea.)

lunar regolith Particles of finely divided rock in a loose state forming a surface layer over most of the lunar surface.

mafic igneous rocks Igneous rocks composed dominantly of mafic minerals.

mafic minerals, mafic mineral group Rock-forming minerals, largely silicates, rich in magnesium and iron, dark in color, and of relatively high density. (See also *felsic minerals.*)

magma Mobile, high-temperature molten state of rock, usually of silicate mineral composition and with dissolved gases and other volatiles.

magma reservoir Large pocket of magma situated just below the feeder pipe of a volcano.

magmatic differentiation Concept of splitting of a single parent magma into two or more magmas of different composition.

magmatic segregation Segregation of certain components of a magma by crystallization of minerals, which then sink because of greater density to form a discrete rock or mineral layer. Partial melting of mantle rock, with rise of the melt fraction is an alternative mode of segregation.

magnetic anomaly Any departure of magnetic intensity or inclination from an average or normal value, as observed by use of the magnetometer.

magnetic event Relatively brief period within either a normal epoch or a reversed magnetic polarity epoch during which magnetic polarity was opposite to that of the epoch.

magnetic inclination Vertical angle between geomagnetic lines of force and the horizontal, as measured by a dip needle.

magnetic minerals Minerals susceptible to becoming magnetized; most are oxides of iron and/or titanium. Example: magnetite.

magnetic polarity reversals (See *geomagnetic polarity reversals.*)

magnetic quiet interval Long time span in which no polarity reversal occurred, as inferred from the lack of magnetic anomalies over parts of the ocean floor.

magnetosphere External portion of the earth's magnetic field, shaped by pressure of the solar wind.

manganese nodules Mineral masses resembling pebbles or cobbles, found on the deep ocean floor and consisting of manganese oxides and iron oxides, or of coatings of those minerals around nuclei of volcanic rock.

mantle Rock layer or shell of the earth beneath the crust and surrounding the core, composed of ultramafic rock of silicate mineral composition. (See *lower mantle, upper mantle.*)

mantle convection (See *convection, convection system.*)

mantle-drag force Force exerted by a current in the asthenosphere and tending to move a horizontal lithospheric plate in a horizontal direction.

mantle plume A hypothetical columnlike or stalklike rising of heated mantle rock, thought to be the cause of a hot spot in the overlying lithospheric plate.

marble Variety of metamorphic rock derived from limestone or dolomite by recrystallization under pressure of regional metamorphism.

maria (See *lunar maria.*)

Mariana type subduction mode Pattern of plate subduction characterized by a steeply dipping lithospheric slab and the presence of an actively spreading backarc basin, relieving the plate contact of compression and possibly leading to tectonic erosion. (See also *Chilean type subduction mode.*)

marine environment Environment of sediment deposition consisting of the ocean bottom lying below the level of low tide. (See also *littoral environment, terrestrial environment.*)

marine geology Branch of geology dealing with the ocean basins and ocean floors; also called *submarine geology.*

marine planation Reduction of a low-lying continental surface to a broad abrasion plane by wave action during gradual submergence.

marine terrace Former abrasion platform elevated to become a benchlike coastal landform.

mass wasting Spontaneous downward movement of soil, regolith, and bedrock under the influence of gravity; does not include the action of fluid agents.

matrix In a breccia, conglomerate, mudflow, or mélange, the finer-textured sediment or rock that encloses larger particles or masses.

M-discontinuity (See *Moho.*)

meanders (See *alluvial meanders.*)

meander sweep Slow downvalley migration of a series of alluvial meanders because of the persistent influence of the floodplain gradient.

mechanical (physical) weathering Breakup of massive rock into blocks and smaller particles through the action of physical stresses at or near the earth's surface. (See also *weathering.*)

mélange Metamorphic rock, originating as a form of breccia consisting of blocks of various sizes and of varied lithologic types enclosed in a strongly sheared matrix of deformed plastic (ductile) sediments that may consist of turbidites or clays of terrigenous origin; it is interpreted as the material of which an accretionary prism is constructed above a subducting plate. (See also *accretionary prism, offscraping, understuffing.*)

Mercalli scale, modified Earthquake intensity scale, modified in 1956 by C. Richter, using 12 intensity levels and relating each to phenomena observed during an earthquake. (See also *earthquake intensity scale.*)

metamorphic grades Levels or stages in a sequence of regional metamorphic rocks showing the effects of increasing temperature during recrystallization; each grade being represented by an index mineral and shown on a map by isograds. (See *index mineral, isograd.*)

metamorphic rocks Rocks altered in physical structure and/or chemical (mineral) composition while in the solid state (without melting) by action of heat, pressure, shearing stress, or infusion or loss of elements (metasomatism), all taking place at substantial depth beneath the earth's surface.

metamorphic zones Belts of different kinds of regional metamorphic rocks reflecting in order the increased temperature that accompanied their recrystallization. (See also *index mineral.*)

metamorphism Process of formation of metamorphic rocks. (See also *contact metamorphism, dynamic metamorphism, dynamothermal metamorphism, impact metamorphism.*)

metasomatism Import or export of mineral substances (elements, ions) into or out of a solid rock mass from or to the surrounding rock by gaseous diffusion or movement of solutions; essentially a replacement process that does not necessarily change the total rock volume and usually leaves structural features more or less intact.

meteoric water Ground water that originated as precipitation from the atmosphere, as distinguished from water that has its origin in a rising magma (juvenile water).

meteorite Meteoroid that has landed on earth, becoming a variety of rock.

meteoritic hypothesis Interpretation of the lunar craters as impact features produced by the infall of large meteoroids or asteroids.

mica group Aluminosilicate mineral group of complex chemical formula having perfect cleavage into thin sheets.

microcracks Minute, closely spaced irregular cracks in rock produced by rock strain in the zone near an active fault.

microcrystalline Crystalline rock texture in which individual crystals are too small to be seen except through a microscope. (See also *cryptocrystalline.*)

microseisms Trains of extremely small seismic waves, sometimes produced by atmospheric disturbances; they are often present on a seismogram and may be unrelated to seismic activity.

mid-oceanic ridge One of three major topographic divisions of the ocean basins, the central belt of submarine mountain topography with a characteristic axial rift which marks a spreading plate boundary.

migmatite A metamorphic rock consisting of bands that appear to be metamorphic rock alternating with bands having the texture and composition of granite.

Milankovitch curve Curve showing the fluctuations in insolation at a given latitude resulting from the combined effects of variations in the form and precession of the earth's orbit and tilt of the earth's axis, for a period extending back to a half million years or more before present.

mineral (mineral species) Naturally occurring homogeneous inorganic solid substance, having either a definite chemical composition and an orderly atomic structure or one that is variable between stated limits. (See also *felsic minerals, mafic minerals, silicate minerals.*)

mineral hardness The degree to which a mineral resists being scratched or pulverized by other minerals or solid substances. (See also *Mohs scale.*)

mineral luster Characteristic surface appearance of a mineral under reflected light. Examples: metallic, vitreous, pearly, silky.

mineralogy Science of minerals, including their formation, composition, structure, and classification.

mineraloids Amorphous minerals that may be gels. (See *gel mineral.*)

mineral replacement Action of slowly moving ground water in which molecule-by-molecule replacement of one mineral by another takes place.

mineral streak Track left when a mineral specimen is drawn across an unglazed white ceramic plate.

mineral vein (See *vein.*)

miogeocline Geocline (geosyncline) of a passive continental margin consisting of a wedge-shaped shallow-water sedimentary sequence, thickening seaward, and resting on continental crust beneath the continental shelf and the continental slope. (See also *eugeocline, geocline.*)

mobile core In a hypothetical model of the Cordilleran-type orogeny, a body of highly heated mantle rock that rises above the subducting plate, producing migmatites and numerous plutons and causing an updoming of the crust.

Moho (Mohorovičić discontinuity, M-discontinuity) Contact surface between the earth's crust and mantle; named for A. Mohorovičić, the seismologist who discovered the discontinuity.

Mohs scale Scale of mineral hardness using a set of ten minerals to establish ten levels of hardness.

molasse Thick accumulation of clastic sediments, which may be continental (nonmarine), in a foreland crustal trough adjacent to a recently formed orogen. (See also *foreland trough deposit.*)

monolith Large mass of fresh bedrock free of joints or other fractures or flaws.

montmorillonite group Clay mineral group, hydrous aluminosilicates derived by the chemical alteration of silicate minerals in various igneous rocks; they expand greatly when adsorbing water. (Also called *smectite group.*)

moonquake Seismic disturbance set off within the moon, analogous to an earthquake.

moraine Accumulation of rock debris previously carried by an alpine glacier or an ice sheet and deposited by the ice to become a depositional landform.

mountain arc Curved (arcuate) segment of an alpine mountain chain, usually of complex geologic structure and associated with either a subduction boundary or a continental suture.

mudflow Rapid flowage of a mud stream down a canyon floor and spreading out upon a plain at the foot of a mountain range; often contributing to the building of an alluvial fan.

mudstone Sedimentary rock formed by the lithification of mud.

mylonite Variety of cataclastic metamorphic rock that is fine-grained and dense with a streaked or banded appearance and is produced by movement on a fault plane.

nappe Large, sheetlike recumbent fold or thrust sheet that has moved horizontally or on a low downgrade many kilometers or tens of kilometers; an orogenic structure typical of continental collision tectonics.

native blocks In a mélange, blocks of more brittle materials from the same sources as the sedimentary matrix; typically they are turbidite lithic sandstones.

nebula (See *solar nebula.*)

nonclastic sedimentary rocks Class of sedimentary rocks consisting of chemical precipitates (hydrogenic sediments) and organically derived sediments (biogenic sediments). (See also *clastic sedimentary rocks.*)

nonconformity Variety of unconformity in which younger strata have been deposited upon the erosionally truncated upper surface of older igneous or metamorphic rocks. (See also *angular unconformity, disconformity, unconformity.*)

normal epoch Time span in geologic record in which the polarity of the earth's geomagnetic field was like that of the present time. (See also *reversed epoch.*)

normal fault Variety of fault in the dip-slip class in which the side having the foot wall moves up, while the side with the hanging wall moves down. (See also *reverse fault.*)

normal sea-level gravity Value of gravity adjusted for latitude on the ellipsoid of reference.

nuée ardente (See *glowing avalanche.*)

oblique-slip fault Geometrical class of fault in which relative motion combines that of dip slip and strike slip. (See also *dip-slip fault, strike-slip fault.*)

ocean-basin floors One of the major topographic divisions of the ocean basins, consisting of the deep portions, including abyssal plains, seamounts, and low hills.

ocean basins Deep, roughly flat-floored depressions of subglobal dimension, underlain largely by oceanic lithosphere and crust and holding the water of the oceans. (See also *continents.*)

oceanic crust Crust of basaltic composition beneath the ocean floors, capping oceanic lithosphere. (See also *continental crust.*)

oceanic lithosphere Lithosphere bearing oceanic crust. (See also *continental lithosphere.*)

oceanic rise Topographic unit of the ocean floor that is higher than the surrounding abyssal plains and may have relief (hills) ranging from subdued to rugged. Example: Bermuda Rise.

oceanic trench Long, narrow, deep, troughlike depression in the ocean floor representing the line of subduction of oceanic lithosphere beneath the margin of continental lithosphere.

offscraping Forcible removal of pelagic sediments, trench sediments, and masses of oceanic bedrock from the upper surface of a subducting plate through the dragging action exerted by the stationary overlying plate of the continental margin and leading to the accumulation of an accretionary prism by understuffing.

offshore That part of the beach profile lying in the zone of shoaling waves and below the level of low tide. (See also *foreshore.*)

olivine basalt (olivine gabbro) Variety of basalt or gabbro containing olivine.

olivine group Mineral group consisting of silicates of magnesium, iron, manganese, and calcium; it includes *olivine, forsterite,* and *fayalite.*

oolite (oolitic limestone) Sediment or sedimentary rock consisting of small round calcareous particles (oolites or ooliths) resembling fish roe.

ooze, deep-sea Ocean-floor pelagic sediment rich in tests of plankton, which may be calcareous or siliceous.

open folds Folds in which the opposed limbs form a large angle. (See also *isoclinal folds, tight folds.*)

ophiolite, ophiolitic suite Sequence of igneous rocks with ultramafic rocks at the base, followed in order upward by coarse-grained gabbro, sheeted dikes, pillow lavas, and associated deep-sea sediments.

organically derived sediment Class of sediments consisting of the remains of nonliving plants or animals, or of mineral matter produced by the activities of plants or animals. (Also known as *biogenic sediment.*)

organic evolution Continual change in the form and structure of living organisms with the passage of time, as a result of changes within genetic materials.

orogen The mass of tectonically deformed rocks and related igneous rocks produced during an *orogeny.*

orogeny Major episode of severe tectonic deformation with attendant igneous intrusion occurring in a long, relatively narrow belt of the continental margin as a result of formation of a new subduction boundary on the adjacent ocean basin or by continental collision.

outcrop Surface exposure of bedrock.

outgassing Process of exudation of volatiles as gases from the earth's crust, largely through volcanic activity, to become a part of the earth's hydrosphere and atmosphere.

outlet glacier Tonguelike ice stream, resembling an alpine glacier, fed by ice from the margin of an ice sheet.

outlier Isolated erosional remnant of a thrust sheet found beyond the limits of the main body of the thrust sheet.

overland flow Gravity flow of a surface layer of water over a sloping ground surface at times when the infiltration rate is exceeded by the precipitation rate; a form of runoff.

overthrust fault Fault characterized by the overriding of one fault block or thrust sheet over the adjacent crustal mass along a gently inclined fault plane.

overturned folds Folds in which the axial surface has assumed a low dip and the steeper limb of the fold is inclined beyond the vertical. (See also *inclined folds, recumbent folds.*)

oxygen-isotope ratio Ratio of oxygen-18 to oxygen-16 as determined from foraminifera tests of deep-ocean cores and from ice layers of the Greenland Ice Sheet. (See also *isotope-ratio stages.*)

paleoclimate Climate as it existed at a given point in geologic time.

paleogeographic map Map compiled to show various geographic and geologic features for a given region and time in the geologic past; it may show such features as shorelines, cratons, geosynclines, tectonic and volcanic arcs, and sedimentary facies.

paleoglaciation curve Curve of oxygen-isotope ratios plotted back into time so as to reveal fluctuations interpreted as synchronous with fluctuations in the total volume of glacier ice present on the globe at any given time. (See also *isotope-ratio stages.*)

paleomagnetism Relict magnetism within magnetic minerals in rocks, representing the state of the earth's magnetic field at the time the rock was formed.

paleontology Study of ancient life based on fossil remains of plants and animals.

paleosols Soils and their profiles produced under past conditions of climate and now preserved by burial beneath younger sediments or colluvial materials.

paleotemperatures Determinations of the temperature of the atmosphere or surface ocean water for a point in time in the geologic past; can be determined by analysis of oxygen-isotope ratios and ratios of plankton species in deep-sea cores, or by other methods.

Pangaea Hypothetical parent continent, enduring from late Paleozoic time to late in the Mesozoic Era, consisting of the continental shields of Laurasia and Gondwana joined into a single unit. (See also *continental drift.*)

Panthalassa The great world ocean that surrounded Pangaea.

parasitic folds Small secondary folds formed in an incompetent bed (shale) between two competent beds on the limb of a larger fold.

partial melting Melting affecting certain rock-forming minerals selectively as a slow rise of temperature occurs in deep-seated rock of the lower crust or upper mantle; thought to be a possible source of basaltic (gabbroic) magma rising to form new oceanic crust beneath spreading plate boundaries.

passive continental margins Continental margins lacking active plate boundaries at the contact of continental crust with oceanic crust. A passive margin thus lies within a single lithospheric plate. Example: Atlantic continental margin of North America. (See also *continental margins, active continental margins.*)

pediment Gently sloping, rock-floored surface found at the base of a mountain mass or cliff in an arid region.

pediplain Desert land surface of low relief composed in part of pediment surfaces and in part of alluvial fan and playa surfaces.

pegmatite General textural term for igneous rock consisting of very large mineral crystals occupying veins, dikes, or irregular pockets. (See also *pegmatite dike.*)

pegmatite dike Pegmatite occurring in the form of a dike.

pelagic-detrital sediment Particles of nonbiogenic matter that have settled to the deep ocean floor from the near-surface layer of the ocean; they include volcanic and terrestrial dusts, particles transported from stream mouths by ocean currents and debris carried in icebergs. (See also *detrital sediment, pelagic sediment.*)

pelagic sediment Sediment of the deep ocean floor settling from suspension in the near-surface water layer; it may be biogenic sediment (e.g., tests of plankton) or detrital sediment. (See also *ooze, brown clay.*)

peneplain Land surface of low elevation and slight relief produced in the late stage of landmass denudation.

perched water table Water table of a layer or lens of ground water formed over an aquiclude and occupying a position above the main water table.

peridotite Igneous rock consisting largely of olivine and pyroxene; an ultramafic igneous rock occurring as a pluton, also thought to comprise much of the upper mantle.

period of geologic time Time subdivision of the era, each ranging in duration between about 35 and 70 million years.

permafrost Condition of permanently frozen water in the regolith and bedrock in cold climates of subarctic and arctic regions.

permeability Property of relative ease of movement of ground water through rock when a hydraulic head exists.

petroleum Natural mixture of many complex hydrocarbon compounds occurring in rock; it includes natural gas, crude oil, and bitumens, but in common usage the term is synonymous with crude oil.

phaneritic texture Igneous rock texture class in which crystals are large enough (more than 0.05 mm diameter) to be distinguished with the unaided eye or with the help of a hand lens. (See also *aphanitic texture.*)

phase transition As applied to the lower crust and mantle, a physical change in the atomic structure of mineral matter affecting the density of packing of the atoms and thus explaining an abrupt change in rock density without requiring a change in the element composition of the rock.

phenocrysts Large, whole mineral crystals enclosed in a fine-textured groundmass in a porphyry. (See also *groundmass, porphyry.*)

physical geology Major division of geology dealing with physical processes and forces acting within the earth and upon its surface; it includes geophysics and geochemistry, but not historical geology, stratigraphy, or paleontology.

physical weathering (See *mechanical weathering.*)

pillow lava Basalt lava that has congealed in bulbous, pillowlike forms upon contact with seawater.

plagioclase feldspar Feldspar subgroup that is a solid solution series between pure albite (sodium aluminum silicate) and pure anorthite (calcium aluminum silicate); those two minerals represent respectively the alkalic and calcic end members of the series.

planetesimals Aggregations of matter formed into discrete solid bodies during accretion of the solar nebula.

plastic flow Mode of deformation of a soft or ductile solid in which shearing takes place along surfaces of microscopic or molecular thickness; it is the kind of rock flowage believed to occur in the asthenosphere.

plastic layer of mantle (See *asthenosphere.*)

plastic solid Ductile solid that deforms by *plastic flow* under suitable conditions of temperature and confining pressure.

plate boundary segment The continuous line separating two lithospheric plates and terminating at either end in a triple junction.

plate tectonics Theory of tectonic activity of the earth's lithospheric plates, their present and past interactions, and the influence of their activity upon all aspects of geology.

platform (See *continental platform.*)

playa Flat-floored central area of a desert basin, usually the site of an ephemeral lake and underlain by fine-textured detrital sediments and evaporites.

plucking (See *glacial plucking.*)

plume (See *mantle plume.*)

plunge of fold Descent in altitude of the hinge line of a fold.

plunging fold Fold with a descending or rising hinge line.

pluton Body of intrusive igneous rock. (See also *concordant pluton, discordant pluton.*)

pluvial lake Lake that reached full development during past climatic periods of relatively high ratio of precipitation to evaporation and is presently extinct or a small remnant of its former extent. The term applies specifically to Pleistocene lakes that formerly occupied closed depressions in the floors of basins in the Basin-and-Range Province. Example: Lake Bonneville.

polar wandering path Line on a map or globe connecting former positions of the pole of rotation of an individual fragment of continental lithosphere. (See also *pole of rotation.*)

pole of rotation Point on the globe that is the center of the circular path describing the relative motion of a point on a lithospheric plate; the pole lies at one end of the *axis of rotation.*

polymorphism The ability of a given chemical element or compound to assume either of two different crystalline atomic structures and thus to crystallize in either of two crystal systems. Example: carbon as diamond (isometric system) or as graphite (hexagonal system).

ponding of sediment Accumulation of sediment in closed depressions on the ocean floor.

porosity Total volume of pore space within a given volume of rock; a ratio, expressed as a percentage.

porphyritic texture Igneous rock having the texture of a porphyry.

porphyroblast Relatively large mineral crystal

that has grown by crystallization within schist during metamorphism.

porphyry Texture class of igneous rock consisting of widely separated large crystals (phenocrysts) enclosed in a fine-grained groundmass.

porphyry copper Copper ore in the form of a disseminated deposit occurring within a body of porphyry.

potash feldspar A class of feldspars in which potassium is a dominant metallic ion along with sodium in a solid solution series. Example: orthoclase.

precision depth recorder Instrument that measures water depth below a vessel, using reflected sound waves.

primary magma An imagined single deep-seated magma body capable of yielding by magmatic differentiaton the separate magma types needed to produce the full range of igneous rock varieties. (The concept now appears inadequate or invalid in the light of plate tectonics.)

primary waves (P-waves) Variety of seismic body waves in which motion is forward and backward (longitudinal) in the path of wave travel, alternately causing compression and expansion of the rock; also called longitudinal waves. (See also *secondary waves*.)

principle of continuity Principle of stratigraphy that strata at one locality can be matched in age with strata at a distant locality by tracing the unbroken layer sequence across the intervening country.

principle of superposition Principle of stratigraphy that strata are arranged in order of decreasing age upward within any sequence of strata that has not been subsequently overturned.

principle of uniformity Governing principle that all phenomena of geology must be explained through laws of science we accept as valid and ruling out any supernatural causes or forces; a modern interpretation of the concept of *uniformitarianism*.

progradation Shoreward building of a beach, bar, or sandspit by addition of coarse detrital sediment carried by littoral drift or brought from deeper water offshore. (See also *retrogradation*.)

pumice Glassy, light-colored scoria of very low density, resembling solidified foam; it commonly has the composition of rhyolite.

P-waves (See *primary waves*.)

pyroclastic sediment Sediment consisting of particles thrown into the air by volcanic explosion in the form of tephra or volcanic ash.

pyroclasts (pyroclastic materials) Rock and mineral fragments blown out of a volcanic vent under pressure of rapidly expanding gases during a volcanic eruption. (See also *tephra*.)

pyroxene granulite High-grade regional metamorphic rock of granulite texture in which pyroxene is an important constituent and sillimanite is an index mineral. (See also *granulite*.)

pyroxenes, pyroxene group Group of complex aluminosilicate minerals rich in calcium, magnesium, and iron, dark in color, relatively high in density, classed as mafic minerals. Example: augite.

quartzite Metamorphic rock consisting largely of quartz. (Some forms of quartzite are recognized as sedimentary rocks.)

quick clays Clay layers that spontaneously change from a solid condition to a near-liquid condition when disturbed, a process called spontaneous liquefaction.

radiocarbon (See *carbon-14*.)

radiocarbon method Method of absolute age determination by analysis of the ratio of carbon-14 to ordinary carbon in organic materials.

radiogenic heat Heat energy generated by radioactive decay.

radiometric age Age of a mineral or rock determined by the measurement of ratios of radioactive isotopes to their daughter products.

Rayleigh wave Type of surface seismic wave in which all particle motion is in the vertical plane and consists of a retrograde elliptical orbit lying in the plane of the wave path. (See also *Love wave*.)

reaction in magma Changes in composition of crystallized silicate minerals in a magma as cooling proceeds. (See also *continuous reaction, discontinuous reaction*.)

reaction pair Two silicate minerals that are formed in sequence by discontinuous reaction during cooling of a magma. (See also *discontinuous reaction*.)

reaction series An orderly sequence of chemical reactions within a cooling silicate magma. (See also *Bowen reaction series*.)

recharge (See *ground water recharge*.)

recrystallized rocks Major group of metamorphic rocks that have undergone recrystallization, a process considered to be a form of chemical change; the group includes contact metamorphic rocks and regional metamorphic rocks.

recumbent folds Folds in which the axial surfaces are in a nearly horizontal attitude. (See also *overturned folds*.)

red beds Accumulations of reddish sedimentary rock, largely shales with interbedded sandstones, colored by reddish ferric oxides of iron (hematite) and indicating a terrestrial source environment favoring extreme oxidation.

reef limestone Limestone consisting of skeletal structures secreted by corals and algae in coral reefs, and accumulations of detrital fragments of reefs.

regional metamorphic rocks Recrystallized metamorphic rocks that are the product of dynamothermal metamorphic metamorphism; they occur over a large area and affect a substantial thickness of crust.

regolith Layer of mineral particles overlying the bedrock; it may be derived by weathering of underlying bedrock or transported from other locations by fluid agents. (See also *lunar regolith, residual regolith, transported regolith*.)

regression (marine) In stratigraphy, the withdrawal of ocean water from the continental margin or craton as a result of either epeirogenic uplift or negative eustatic change of sea level.

rejuvenation of landmass Episode of rapid fluvial denudation set off by a rapid crustal rise or fall of sea level, increasing the available landmass. (See also *rejuvenation of stream*.)

rejuvenation of stream Episode of rapid degradation (downcutting) by a stream in an attempt to reestablish grade at a lower level following crustal uplift or a fall of sea level. (See also *rejuvenation of landmass*.)

relative age Age of a rock unit, such as a sedimentary formation or an igneous intrusion, compared with the age of another rock unit, as to whether one is older or younger than the other. (See also *absolute age*.)

relative plate motion Motion of a lithospheric plate with respect to an adjoining plate, where the two plates share a common active boundary and a common pole of rotation.

relief Measure of the average elevation difference between adjacent high and low points in a land surface, as for example hilltops and valley bottoms.

residence time As applied to the oceans, the average time that a single ion of a particular element remains free in the ocean before being precipitated as sediment.

residual regolith Regolith formed in place by weathering of bedrock. (See also *transported regolith*.)

restored stratigraphic section Stratigraphic section that has been replotted and redrawn, using as a horizontal upper line of reference the upper surface of the youngest stratigraphic unit of the entire series shown.

retrogradation Cutting back (retreat) of a shoreline, beach, marine cliff, or marine scarp by wave action. (See also *progradation*.)

reversed epoch Time span in geologic record in which the polarity of the earth's magnetic field was reversed from that of the present time. (See also *normal epoch*.)

reverse fault Variety of fault in the dip-slip class in which the side having the hanging wall moves up while the side having the foot wall moves down. (See also *normal fault*.)

rhyolite Extrusive igneous rock of granite composition, classed as an acidic lava.

ria coast Deeply embayed coast formed by partial submergence of a landmass previously shaped by fluvial denudation.

Richter scale Logarithmic scale of magnitude numbers stating the relative quantity of energy released by an earthquake.

ridge-push force A horizontal component of the gravitational force acting upon the uplifted edge of an oceanic lithospheric plate, causing the entire plate to move horizontally away from the spreading boundary.

rift Long, narrow, trenchlike surface feature produced during the pulling apart (extension) of the crust, as in the case of seafloor spreading. (See also *rift valley*.)

rift valley Trenchlike valley with steep, parallel sides; essentially a graben between two normal faults; associated with crustal spreading. (See also *axial rift*.)

ring dike Igneous dike exposed at the surface as a circular structure and extending down into the country rock in the form of a cone.

rip current Strong, narrow seaward current flowing in the breaker zone at right angles to the shoreline.

rock Natural aggregate of minerals in the solid state; often hard and consisting of one or more mineral varieties.

rock deformation Any change in shape or volume of a rock mass in response to tectonic or gravitational stresses.

rockfall Free fall of particles or masses of bedrock from a steep cliff face, typically accumulating at the cliff base in the form of talus.

rock metamorphism (See *metamorphism*.)

rockslide Form of landslide consisting of the slippage of a bedrock mass on a dipping rock plane.

rock terrace Terrace carved in bedrock during the degradation of a stream channel, induced by the crustal rise or a fall of the sea level. (See also *alluvial terrace, marine terrace*.)

rock transformation cycle Total cycle of changes in which rock of any one of the three major rock classes—igneous, sedimentary, metamorphic—is transformed into rock of one of the other classes.

roof pendant Remnant of country rock extending down into a batholith.

runoff Flow of water from continents to oceans by way of stream flow and ground water flow; a term in the water balance of the hydrologic cycle. In a more restricted sense, runoff refers to surface flow by overland flow and channel flow.

rupture Breakage (fracture) of a solid substance along a discrete fracture that may be a *shear fracture* or an *extension fracture*.

sag pond Small closed depression of tectonic origin in the floor of a valley developed on an active transcurrent fault; a typical feature of some reaches of the San Andreas Fault.

saltation Leaping, impacting, and rebounding of spherical sand grains transported over a sand or pebble surface by wind.

salt dome Dome structure in sedimentary rock, produced by forcible rise of a salt plug.

salt plug Stalklike or columnar mass of rock salt, a *diapir*, that has forced its way up through a thick sequence of sedimentary rocks, forming a salt dome structure at the top.

sand drift Accumulation of wind-transported sand in the lee of an obstacle.

sand dune Hill or ridge of loose, well-sorted sand shaped by wind and usually capable of slow downwind motion.

sandstone Textural class of detrital sedimentary rock consisting dominantly of mineral particles of sand grade size. (See also *feldspathic sandstone; lithic sandstone, quartz sandstone*.)

sand storm Dense, low cloud of sand grains traveling in saltation over the surface of a sand dune or beach.

saprolite Surface layer of deeply decayed igneous rock or metamorphic rock, rich in clay minerals, typically formed in warm, humid climates; a form of residual regolith.

saturated zone Zone beneath the land surface in which all pores of the bedrock or regolith are filled with ground water; synonymous with *ground water zone*.

schist Textural class of metamorphic rock characterized by *foliation* in which mica flakes are typically found oriented parallel with foliation surfaces.

schistosity Foliated texture characteristic of a schist.

scoria Lava or tephra containing numerous cavities produced by expanding gases during cooling. (See also *pumice*.)

scoriaceous texture Texture associated with scoria.

scour marks Troughlike or cuplike depressions on the bedding surface of a layer of fine-textured sediment, such as the top layer of a turbidite, eroded by swift bottom currents. (Scour marks are usually seen in lithified sediments as *sole marks*, the molds or counterparts preserved on the underside of the overlying sandy bed.)

seafloor spreading Pulling apart of the oceanic crust along the axial rift of the mid-oceanic ridge, and representing the continual separation of two lithospheric plates made up of oceanic lithosphere.

seamount Prominent, isolated conical or peaklike submarine landform rising above an abyssal plain; usually identified as an extinct submarine volcano or *guyot*.

secondary waves (S-waves) Variety of seismic body waves in which motion is back and forth at right angles (sidewise) to the direction of wave travel; also called *shear waves*. (See also *primary waves*.)

sediment Finely divided mineral matter derived directly or indirectly from preexisting rock and organic matter produced by life processes. (See *detrital sediment, organically derived sediment, pelagic sediment*.)

sedimentary basin Large circular or elliptical crustal depression within a craton in which a thick series of sedimentary rocks accumulates as the basin floor subsides. Example: Michigan Basin.

sedimentary breccia Class of detrital sediment consisting of angular fragments in a matrix of finer fragments.

sedimentary rock Rock formed by the accumulation and diagenesis of sediment.

seismic gap Segment of an active plate boundary, particularly a subduction boundary, along which no major earthquake has occurred for at least several decades and which is thus a likely site for a large earthquake in the near future.

seismicity Level of seismic activity (earthquake activity).

seismic profiler traverse Continuous profile of the ocean floor and its sediment layers drawn from a series of reflected seismic-wave impulses.

seismic reflection Turning back of a seismic wave impulse at a rock or sediment interface so as to return the impulse to a recording instrument.

seismic sea wave (tsunami) Train of sea waves set off by an earthquake or other seafloor disturbance and traveling over the ocean surface with a wave velocity proportional to the square root of the ocean depth.

seismic waves Waves sent out during an earthquake by faulting or other form of crustal rupture from an earthquake focus and propagated through the solid earth.

seismogram Record of seismic waves as produced by a seismograph.

seismograph Instrument for detecting and recording seismic waves.

seismology Study of earthquakes and the interpretation of seismic waves in terms of the physical properties of earth layers through which the waves pass; a branch of geophysics.

sequential landforms Landforms produced by external earth processes in the total activity of denudation. Examples: canyon, alluvial fan, floodplain.

serpentine group Mineral group, hydrous silicates of magnesium, derived by hydrolysis of mafic minerals such as pyroxene and amphibole; it includes antigorite and chrysotile, the latter commonly known as asbestos.

serpentinite Metamorphic rock, rich in serpentine, formed by hydrothermal alteration of mafic rocks, such as peridotite.

serpentinization Process of formation of serpentinite by hydrothermal alteration.

sesquioxides Oxides of aluminum or iron with a ratio of two atoms of aluminum or iron to three atoms of oxygen. (See also *ferric iron oxide*.)

settling velocity Rate of free fall of a sediment particle in quiet air or water.

shadow zone Zone of earth's surface, lying between 103 and 143 degrees of arc distant from an earthquake epicenter, within which no direct P-waves or S-waves are received.

shale Fissile sedimentary rock of mud or clay composition, showing lamination.

shallow-focus earthquake Earthquake with focus less than 55 km beneath the earth's surface.

shear fracture Form of shearing affecting a brittle solid stressed beyond its elastic limit in which a shear fracture (a fault) is formed and motion takes place in the shear plane.

shearing Form of strain in which relative motion takes place between parallel layers, each of which slides or glides past the adjacent layer, whether the substance be a fluid, plastic solid, or brittle solid. (See also *shear fracture*.)

shear plane Plane on which shear fracture takes place.

shear stress Stress that tends to deform a substance by shearing; it can be represented by parallel arrows pointing in opposite directions. (See also *compressional stress, tensional stress*.)

shear waves (See *secondary waves*.)

sheeted dikes Multiple basalt dikes parallel with one another in vertical orientation, formed in a seafloor spreading zone, at the accreting margins of the lithospheric plates.

sheeting structure Thick, subparallel layers of massive bedrock formed by spontaneous expansion (dilatation) accompanying unloading.

sheet structure Crystal lattice structure of silicate minerals in which silicon-oxygen tetrahedra are linked into continuous sheets, as in the micas.

shields (See *continental shields*.)

shield volcano Domelike accumulation of basalt lava flows emerging from radial fissures.

shoreline Shifting line of contact between land and the water of a lake or ocean.

sial Obsolete term referring to crustal rock that is felsic in composition, i.e., granitic. (See also *sima*.)

silicate magma Magma from which silicate minerals are formed.

silicates, silicate minerals Minerals containing silicon and oxygen atoms, occurring in the crystal space lattice as tetrahedra consisting of four oxygen atoms bonded to each silicon atom.

siliceous sinter Mineral matter, largely silica, forming encrustations and terraces around and near hot springs and geysers.

silicic lava Lava of felsic composition, also called *acidic lava*.

silicon-oxygen tetrahedron Crystal lattice unit of silicate minerals, consisting of four oxygen ions surrounding a central silicon ion to form a tetrahedron.

sill Tabular concordant pluton in the form of a plate, representing magma forced into a natural parting in the country rock, such as a bedding surface in a sequence of sedimentary rocks.

sima Obsolete term referring to crustal rock of mafic composition, i.e., basaltic. (See also *sial*.)

similar folds Folds of strata in which the wave form as seen in cross section is repeated throughout the sequence of layers; thus the beds appear to be thicker on the hinges and thinner on the limbs. (See also *concentric folds*.)

slab-pull force Component of the gravitational force acting upon a subducting lithospheric plate in the direction of motion of that plate.

slaking Spontaneous disintegration of certain types of shale, clay-rich sandstone, and coal (especially lignite) upon exposure to the atmosphere.

slate Compact, fine-grained textural class of metamorphic rock, derived from shale and showing well-developed slaty cleavage.

slaty cleavage Type of natural rock parting characteristic of slate and resulting from shearing during folding.

slip face Steep face of an active sand dune, receiving sand by saltation over the dune crest and repeatedly sliding because of oversteepening; it assumes the angle of repose of noncohesive mineral particles on a free slope.

slump Form of landslide in which a single large block of bedrock moves downward with backward rotation upon an upwardly concave fracture surface.

soft layer of mantle (See *asthenosphere*.)

soil Surface layer over the lands, formed of inorganic (mineral) matter, organic matter (humus), and living organisms possessing a set of physical, chemical, and organic properties favorable to the support of plants.

soil creep Extremely slow downhill movement of regolith as a result of continued agitation and disturbance of the particles by such activities as frost action, temperature changes, or wetting and drying of the soil.

soil erosion Removal of particles of soil or other regolith by the eroding action of water moving down a hillslope as overland flow.

solar nebula Primordial body of diffuse gas and dust that through condensation gave rise to the solar system.

sole (of thrust) Fault plane beneath a low-angle overthrust fault or thrust sheet.

sole marks The molds or counterparts of scour marks. (See *scour marks*.)

solid solution series In certain mineral groups, a continuously variable ratio in the proportion of two ions or elements, either of which can occupy the same space position in the crystal lattice. Example: plagioclase feldspars.

solifluction Arctic tundra variety of earthflow in which the saturated thawed layer over permafrost flows slowly downhill to produce multiple terraces and lobes.

solution of limestone (See *carbonation*.)

sorting Separation of one grade size of sediment particles from another by the action of currents of air or water.

space lattice (See *crystal lattice*.)

specific gravity Ratio of weight of 1 cc of a mineral to 1 cc of pure water at a temperature of 4°C; may also be called *relative density* since it has the same numerical value as density (g/cc), but is a dimensionless quantity.

spontaneous expansion Volume expansion (dilatation) of bedrock as denudation uncovers the rock, relieving the overlying confining pressure.

spontaneous liquefaction In quick clays, the sudden loss of internal strength when shearing occurs during initial stages of flowage or shaking by an earthquake.

spreading plate boundary Lithospheric plate boundary along which two plates of oceanic lithosphere are undergoing separation, while at the same time new lithosphere is being formed by accretion. (See also *converging plate boundary, transform plate boundary*.)

spring Discharge of ground water from a point on the land surface, on the floor of a stream or lake, or at a shoreline.

step faulting Arrangement of multiple parallel normal faults so as to form stepped fault blocks.

stick-slip friction Mechanism of repeated fault movements in which slow buildup of elastic rock strain is punctuated by sudden rupture that generates an earthquake.

stock Small discordant pluton, area of surface exposure less than 100 sq km, possibly representing an upward projection on a batholith.

stone polygons (rings) Linked ringlike ridges of cobbles or boulders lying on the ground surface in arctic and alpine tundra regions.

stones Meteorites composed largely of silicate minerals, usually olivine and pyroxene, with only 20 percent or less nickel-iron.

stony irons Meteorites of composition intermediate between *irons* and *stones*.

stoping Upward igneous intrusion, wrenching loose masses of country rock and incorporating them into the rising magma.

strain Yielding of a solid to applied stress, whether by change in volume or shape, or by fracturing. (See also *elastic strain.*)

strata Layers of sediment or sedimentary rock separated from one another along *stratification planes* (bedding planes).

stratification planes (bedding planes) Planes of separation between individual strata in a sequence of sedimentary rocks.

stratified drift Glacial drift made up of sorted and layered clay, silt, sand, or gravel deposited from meltwater in stream channels or in marginal lakes close to the ice front.

stratigraphic correlation Establishment of the relative ages and depositional sequences of strata exposed at various widely separated localities.

stratigraphic cross section Graphical presentation of the sequence of strata underlying a line of traverse, showing the thicknesses of the units on the vertical scale and indicating lithologies present. (See also *restored stratigraphic section.*)

stratigraphic dome Uparched strata forming a circular structure with domed summit and flanks with moderate to steep outward dip.

stratigraphy Branch of historical geology dealing with the sequence of events in the earth's history as interpreted from the evidence found in sedimentary rocks.

stratovolcano (See *composite volcano.*)

stream Long, narrow body of flowing surface water moving to lower elevations in an open, troughlike *stream channel.* (Stream and stream channel are often used interchangeably.) (See also *consequent stream, graded stream, subsequent stream.*)

stream capacity Maximum load of solid matter that can be carried by a stream for a given discharge.

stream channel Long, narrow, troughlike depression occupied and shaped by a stream moving to progressively lower levels.

stream competence Ability of a stream to transport load, measured in terms of the rock particle of largest diameter that can be set in motion on the stream bed. (Competence in the same sense may also apply to transport by wind in saltation.)

stream erosion Progressive removal of mineral particles from the floor or sides of a stream channel by drag force of the moving water, or by abrasion, or by chemical reaction with ions in stream water.

stream flow (See *channel flow.*)

stream gradient Rate of descent to lower elevations along the profile of a stream or stream channel, stated in m/km, degrees, or percent.

streamline (laminar) flow Fluid flow in which all particles of the fluid move in parallel paths conforming with the solid boundaries confining the flow, and without turbulence.

stream load Solid matter carried by a stream in dissolved form, in turbulent suspension, and as bed load. (See also *bed load, suspended load.*)

strike Compass direction of the line of intersection of an inclined rock plane and a horizontal plane of reference. (See also *dip.*)

strike-slip fault Geometrical class of fault in which all relative motion is in the direction of the strike of the fault plane. (See also *dip-slip fault, oblique-slip fault, transcurrent fault.*)

strong motion seismograph Seismograph, often portable and self-activating, that records the maximum ground acceleration caused by an earthquake.

structural geology Subject area of geology dealing with geologic structures (particularly those of tectonic origin) and the processes and forces that produce them.

structural high Crest or summit zone of an accretionary prism, often separating a forearc basin from the inner trench slope; it may be represented by a chain of islands or a narrow landmass.

structural uplift An updoming or arching of the crust affecting platform strata as well as the shield crust beneath. Example: Black Hills uplift.

structure section Cross section in the vertical plane depicting rock masses and varieties and their structures from the surface down to a given depth.

subduction Descent of the downbent edge of a lithospheric plate into the asthenosphere so as to pass beneath the edge of the adjoining plate along an active plate boundary.

subduction boundary (See *converging plate boundary.*)

subduction complex The structurally and lithologically complex rock mass that constitutes an *accretionary prism.* (See also *mélange.*)

submarine canyon Narrow, V-shaped submarine valley cut into the continental slope, usually attributed to erosion by turbidity currents.

submarine fan (cone) An accumulation of coarse-textured turbidites in a large fan-shaped deposit on the deep ocean floor, usually situated at the lower end of a submarine canyon system leading down the outer slopes of a major river delta on the continental shelf.

submarine geology (See *marine geology.*)

submergence, marine Inundation or partial drowning of a former land surface by a rise of sea level or a sinking of the crust or both.

subplate Portion of a lithospheric plate bounded by active tectonic structures (transcurrent faults, rift valleys) that suggest it may be capable of independent relative motion and may eventually become an independent plate. Example: Somalian subplate.

subsequent stream Stream that develops its course by erosion along a zone or belt of weaker rock, such as the crushed zone of a fault line.

subsidence theory Hypothesis advanced by Charles Darwin to explain an atoll by subsidence of the oceanic crust and continued upbuilding of coral reefs upon the summit of a seamount or guyot.

subsurface water Water of the lands held below the surface in regolith or bedrock.

succession of faunas Faunas differing one from the next within a succession of strata.

superposition (See *principle of superposition.*)

surface water Water of the lands flowing exposed as streams or impounded as ponds, lakes, or marshes.

surface waves Class of seismic waves that travel at the earth's solid surface. (See *Love wave, Rayleigh wave.*)

surf base Lower limiting water depth in which loose sediment on the ocean floor can be moved about by shoaling waves and wave-induced currents.

surf lens Body or prism of beach sand that can be reworked by shoaling storm waves.

surge of glacier Episode of very rapid down-valley movement within an alpine glacier.

suspended load That part of the stream load carried in turbulent suspension.

suspension Form of stream or wind transport in which mineral particles are sustained in upward currents in the eddies of turbulent flow.

suture (See *continental suture.*)

suturing Process of formation of a *continental suture* by collision of two plates consisting of continental lithosphere.

syenite (syenite group) Group of igneous rocks deficient in silica and composed dominantly of potash feldspar, with only a small amount of quartz or no quartz, and with minor amounts of plagioclase feldspar, biotite, and hornblende.

syncline Downfold of strata or other layered rock in a troughlike structure; a type of fold. (See also *anticline.*)

tabular intrusion Pluton that is platelike in form, with flat, parallel sides. Example: sill.

talus Accumulation of loose rock fragments (slide rock) derived by rockfall from a cliff.

taphrogen Fault block in the form of a large graben, subsiding between normal fault boundaries and receiving a great thickness of detrital sediments; it is associated with crustal thinning by extension in early stages of continental rifting; a type of geosyncline. (See also *aulacogen.*)

tectonic Relating to tectonic activity and tectonics.

tectonic activity Crustal processes of bending (folding) and breaking (faulting), usually concentrated on or near active lithospheric plate boundaries. (See also *plate tectonics, tectonic, tectonics.*)

tectonic arc Long, narrow chain of islands or mountains or a narrow submarine ridge adjacent to a subduction boundary and its trench, formed by tectonic processes, such as the construction and rise of an accretionary prisms.

tectonic erosion Removal of masses of rock from the lower edge of a lithospheric plate by downdrag exerted by a subducting plate passing beneath it.

tectonics Branch of geology relating to tectonic activity and the features it produces. (See also *plate tectonics, tectonic activity.*)

tensional stress Stress that tends to increase the distance separating two reference points on the line (stress axis) along which they lie. (See also *compressional stress, shear stress.*)

tephra Collective term for all size grades of particles of solidified magma blown out under gas pressure from a volcanic vent.

terrace Any steplike landform having a flattened tread bounded by a rising steeper slope on one side and a descending steeper slope on the other. (See *alluvial terrace, rock terrace, marine terrace.*)

terrae (See *lunar highlands.*)

terrestrial environment Environment of sediment deposition that includes all land surfaces lying above high-tide level. (See also *littoral environment, marine environment.*)

terrigenous muds Silty muds found on the deep-ocean floor close to sources of terrigenous sediment and commonly accumulating in the trench of a subduction zone or on the upper surface of an accretionary prism.

texture (of rock) Physical property of rock pertaining to the sizes, shapes, and arrangements of mineral grains it contains, e.g., coarse-grained or fine-grained.

thermal remnant magnetism Paleomagnetism permanently locked within magma as it solidifies and cools below the Curie point. (See also *detrital remnant magnetism.*)

thermal springs Natural springs of warm water rising to the surface through favorable rock structures and heated by the normal warm temperature of deep rock, but not related to volcanic activity. (See also *hot springs.*)

thermal uparching Crustal uplift postulated to occur over a descending lithospheric plate because of expansion following heating of the plate by currents in the asthenosphere and the production of rising magma.

thermister probe Temperature-sensing device attached to a coring device and capable of revealing the temperature within deep-sea sediment.

throw of fault Vertical displacement of one side of a fault plane with respect to the other.

thrust sheet Sheetlike mass of rock moving forward over a low-angle overthrust fault.

tight folds Folds in which the angle between opposed limbs is small and may approach zero. (See also *isoclinal folds, open folds.*)

till Heterogeneous mixture of rock fragments ranging in size from clay to boulders, deposited beneath moving glacier ice or directly from the melting in place of stagnant glacier ice.

tillite Lithified glacial till, a type of sedimentary rock.

topographic correction Correction of a gravity reading taking into account the presence of local topographic features that will either add to or subtract from the observed reading.

traction Form of transport in which particles of sand, gravel, or cobbles are moved along the floor of a stream channel by drag force of the flowing water; the mechanism by which bed load is moved.

transcurrent fault Variety of fault on which the relative motion is dominantly horizontal, in the direction of the strike of the fault plane; a *strike-slip fault.* (See also *left-lateral transcurrent fault, right-lateral transcurrent fault.*)

transform fault Special case of a transcurrent fault making up the boundary of two moving lithospheric plates, most commonly found transverse to the mid-oceanic ridge where it connects the offset segments of the spreading plate boundary.

transform plate boundary Lithospheric plate boundary along which two plates are in contact on a *transform fault*, the relative motion being that of a strike-slip fault.

transform resistance force A form of frictional resistance to the movement of one lithospheric plate past another along a transform boundary.

transform scar (See *healed transform fault.*)

transgression (marine) The inundation of a continental margin or craton by shallow ocean accompanying eustatic rise of sea level or negative epeirogenic movement of the continental crust. (See also *regression.*)

transition zone of mantle Basal region of the upper mantle below a depth of 400 km through which P-wave velocity increases steadily until the 650-km discontinuity is reached. (See also *mantle, upper mantle.*)

transported regolith Regolith formed of mineral matter carried by fluid agents from a distant source and deposited upon the bedrock or upon older regolith. Examples: floodplain silt, lake clay, beach sand. (See also *residual regolith.*)

travel-time curve Curve on a graph showing the relationship between travel time of a particular kind of seismic wave and the distance of travel in degrees of arc or kilometers, measured over the earth's surface from focus to observing station.

travertine Carbonate mineral matter, usually calcite, accumulating upon limestone cavern surfaces situated in the unsaturated zone.

trench (See *oceanic trench.*)

trench wedge Thick sediment accumulation on the floor of an oceanic trench along an active subduction boundary.

triple junction Common meeting point of three lithospheric plate boundaries.

trough (See *glacial trough.*)

tsunami (See *seismic sea wave.*)

tuff Consolidated or lithified deposit of volcanic ash; a form of pyroclastic sediment.

turbidite The sediment deposit of a single turbidity-current flow, characterized in the lower part by a graded bed of clastic sediment and overlying layers of sand, silty clay, and clay. (See also *turbidites.*)

turbidites Deep-sea terrigenous sediments formed by repeated turbidity-current flows. (See also *turbidite.*)

turbidity current Rapid downslope streamlike flow of turbid (muddy) seawater close to the seabed, and often confined within a submarine canyon on the continental shelf, or flowing down the wall of an oceanic trench; a form of density current.

turbulent flow Mode of fluid flow in which individual fluid particles (molecules) move in complex eddies, superimposed on the average downstream flow path. (See also *streamline flow.*)

ultramafic igneous rocks Igneous rocks composed almost entirely of mafic minerals, usually olivine or pyroxene.

unconfined aquifer Aquifer in which the water table is free to receive recharge from above and to fluctuate up or down.

unconformity Discordant relationship between strata of one geologic age and older rocks lying beneath, the surface of separation representing denudation through a large gap in time, often following orogeny that affected the older rock mass. (See *angular unconformity, disconformity, nonconformity.*)

understuffing During formation of an accretionary prism, the process of adding in succession new imbricate wedges of deformed sediment obtained by *offscraping* of the upper surface of the subducting plate.

underthrusting Descent of a crustal mass beneath an overlying rock mass along a downdipping fault plane; essentially the same tectonic process as overthrusting, but with emphasis on the relative downward motion of the mass below the thrust plane.

uniformitarianism Concept introduced by James Hutton in the late 1700s that geologic processes acting in the past are essentially the same as those seen in action today; opposed to the view of catastrophists of that day. (See also *principle of uniformity.*)

unit cell (1) In the atomic structure of crystalline minerals the simplest geometrical grouping of bonded ions that is repeated to form the space lattice. (2) In a drainage basin, the watershed that contributes runoff to a single stream segment of the first order.

unloading Process of removal of overlying rock load from bedrock by denudation, accompanied by spontaneous expansion and often leading to the development of sheeting structure.

unsaturated zone Subsurface water zone in which pores are not fully saturated, except at times when infiltration is very rapid; it lies above the saturated zone.

upper mantle Upper division of the mantle, extending from the Moho (M-discontinuity) at the base of the crust down to a depth of 650 km, where a phase discontinuity is encountered; it includes the asthenosphere and a transition zone between 400 and 650 km. (See also *lower mantle, mantle, transition zone of mantle.*)

varves Annual bands in the form of alternate light and dark layers found in fine-textured glaciolacustrine sediments.

vein Thin, sheetlike layer of minerals or igneous rock, intruded or precipitated in fractures in the country rock; often comprising complex intersecting systems; a common form of occurrence of ores.

vent, volcanic (See *volcanic vent.*)

ventifact Pebble or cobble shaped by sandblast action, often with three curved faces joining in three sharp edges and two points, i.e., a *dreikanter.*

vertical folds Folds in which the axial surfaces are in the vertical plane, a condition rare in nature. (See also *inclined folds.*)

vesicles Cavities of spherical outline formed in lava by the expansion of gas contained in the magma.

vesicular texture Extrusive igneous rock texture characterized by abundant vesicles throughout the mass. (See also *scoria.*)

viscosity The relative ease and rapidity with which a fluid or plastic solid will flow in response to unequally applied stresses.

viscous lava dome Dome-shaped mass of highly viscous lava of the acidic type that rises within the vent of a volcano, where it may solidify with a rough, blocky exterior.

volatiles Elements and compounds normally existing in the gaseous state under atmospheric conditions, dissolved in magma. (See also *outgassing.*)

volcanic arc Long, narrow chain of composite volcanoes on an active continental margin or island arc paralleling an active subduction boundary and formed by the rise of magma originally generated on or above the upper surface of the descending lithospheric plate. (See also *tectonic arc.*)

volcanic ash Finely divided extrusive igneous rock blown under gas pressure from a volcanic vent; a form of *tephra.*

volcanic bomb Lava fragment of cobble or boulder size, plastic when thrown from a volcanic vent; often formed into a spheroidal or spindle shape.

volcanic breccia Breccia consisting of pyroclastic sediment.

volcanic crater Central summit depression associated with the principal vent of a volcano. (See also *lunar craters.*)

volcanic dust Extremely fine mineral particles emitted during a volcanic eruption and capable of rising into the upper atmosphere to remain suspended for months or years, until washed to the earth's surface by precipitation.

volcanic eruption Explosive activity of a volcano in which lava flows or tephra emerge under the pressure of expanding gases in the rising magma.

volcanic glass Lava of glassy (noncrystalline) texture, resulting from rapid cooling of magma.

volcanic island chain (See *volcanic arc.*)

volcanic pipe Feeder tube through which magma rises in a volcano to reach the crater; may also refer to the volcanic rock that solidifies within the feeder tube. (See also *volcanic plug.*)

volcanic plug Mass of solidified magma occupying the feeder pipe of a volcano and resisting the further rise of magma from the reservoir below. (See also *volcanic pipe.*)

volcanic vent Place of emergence of lava, tephra, and gases from a volcano.

volcanism General term for volcano building and related forms of extrusive igneous activity.

volcano Conical, peaklike hill or mountain built by accumulations of lava flows and tephra, including volcanic ash. (See *composite volcano, shield volcano.*)

Wadati-Benioff zone A dipping zone of earthquake foci within the upper mantle associated with subduction of a lithospheric plate.

warm springs (See *thermal springs.*)

water-laid tuff Tuff that has settled out from suspension in water.

watershed In a drainage system, the total surface area enclosed by a continuous drainage divide.

water table Upper boundary surface of the saturated zone; the upper limit of the ground water body. (See also *perched water table.*)

weathering Total of all processes acting at or near the earth's surface to cause physical disruption and chemical decomposition of rock. (See *chemical weathering, mechanical weathering.*)

well (See *gravity well.*)

welded tuff Rock formed of indurated volcanic ash and pumice carried as a glowing avalanche (nuée ardente), literally welded into solid rock by heat and pressure.

Wentworth scale System of particle size grades applied to detrital sediments, in which grades are named and defined according to a logarithmic scale.

Widmanstätten structure Distinctive pattern of intersecting lines revealed by etching the polished cut surface of an iron meteorite.

Wilson cycle Series of stages in ocean basin evolution beginning with the opening of a new ocean basin by continental rifting and seafloor spreading, followed by the closing of the basin through subduction, and ending in continental collision and suturing.

window Closed-in valley eroded through a thrust sheet to expose rock lying below the fault plane.

xenolith Fragment of country rock enclosed by invading magma of a pluton and retaining its identity after solidification of the magma.

yield stress Threshold of shear stress that must be applied to a plastic solid to cause a transition from elastic strain to plastic flowage.

zeolites Mineral group, hydrous aluminosilicates of calcium and sodium, typical of an early stage in regional metamorphism when temperatures and pressures are comparatively low.

zone of ablation Lower portion of glacier in which ablation exceeds gain of mass by snowfall; the zone of wastage of a glacier.

zone of accumulation Upper portion of a glacier in which the firn becomes transformed into glacial ice; the zone of nourishment of a glacier.

zone of oxidation Near surface zone in which oxidation affects metallic minerals, forming one type of secondary ore.

Index

Number in italic type gives page on which term is defined or explained.

82 83 9 8 7 6 5 4 3

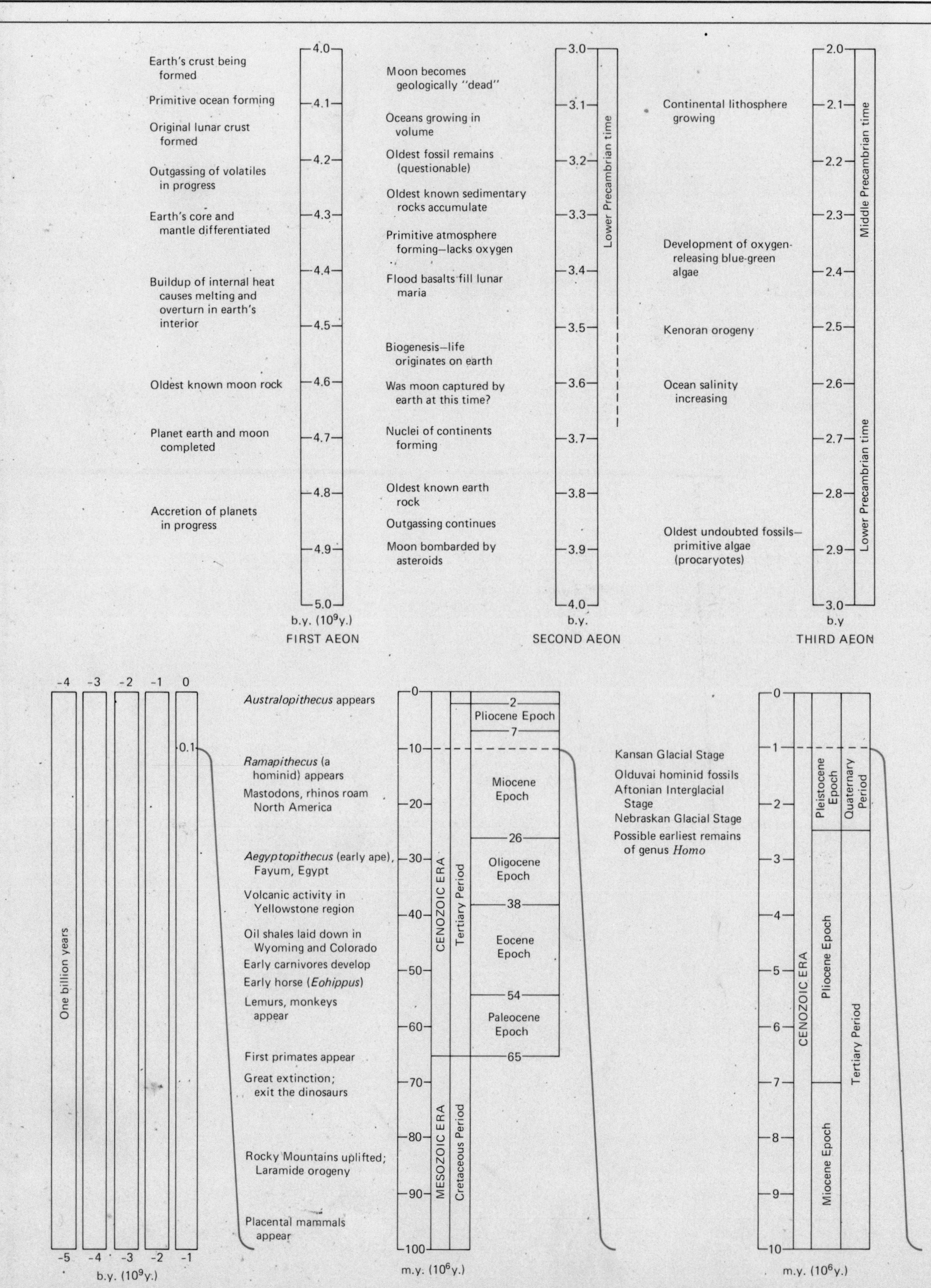

An annotated table of geologic time.